Molecular Biology of Photosynthesis

Molecular Biology of Photosynthesis

Edited by

GOVINDJEE, Editor-in-Chief
University of Illinois at Urbana, U.S.A.

Reprinted from *Photosynthesis Research*,
Vols. 16 – 19, 1988 – 1989.

Kluwer Academic Publishers
DORDRECHT / BOSTON / LONDON

Library of Congress Cataloging in Publication Data

```
Molecular biology of photosynthesis / edited by Govindjee.
     p.   cm.
   Reprinted from Photosynthesis research.
   Includes index.
   ISBN 0-7923-0097-1
   1. Photosynthesis.  2. Plant molecular genetics.   I. Govindjee,
 1933-   .
 QK882.M833   1989
 581.1'3342--dc19
```
                                                         88-34046
                                                            CIP

ISBN 0–7923–0097–1

Published by Kluwer Academic Publishers,
P.O. Box 17, 3300 AA Dordrecht, The Netherlands.

Kluwer Academic Publishers incorporates
the publishing programmes of
D. Reidel, Martinus Nijhoff, Dr W. Junk and MTP Press.

Sold and distributed in the U.S.A. and Canada
by Kluwer Academic Publishers,
101 Philip Drive, Norwell, MA 02061, U.S.A.

In all other countries, sold and distributed
by Kluwer Academic Publishers Group,
P.O. Box 322, 3300 AH Dordrecht, The Netherlands.

Printed in The Netherlands

Preface

Molecular biology, particularly molecular genetics, is among the newest and most powerful approach in modern photosynthesis research. Development of molecular biology techniques has provided new methods to solve old problems in many biological disciplines. Molecular biology has its greatest potential for contribution when applied in combination with other disciplines, to focus not just on genes and molecules, but on the complex interaction between them and the biochemical pathways in the whole organism. Photosynthesis is surely the best studied research area in plant biology, making this field the foremost candidate for successfully employing molecular genetic techniques. Already, the success of molecular biology in photosynthesis has been nothing short of spectacular. Work performed over the last few years, much of which is summarized in this volume, stands in evidence. Techniques such as site-specific mutagenesis have helped us in examining the roles of individual protein domains in the function of multiunit complexes such as the enzyme ribulose-1,5-bisphosphate carboxylase/oxygenase (RUBISCO) and the oxygen evolving photosystem (the photosystem II). The techniques of molecular biology have been very important in advancing the state of knowledge of the reaction center from the photosynthetic bacteria whose structure has been elegantly deduced by H. Michel and J. Deisenhofer from the X-ray studies of its crystals.

Molecular biology can significantly aid in the design of new experiments in photosynthesis research, experiments that can lead to the verification of mechanisms, to the design of new hypotheses which subsequently may be tested by genetic/protein engineering and studies of engineered organisms. Considering the wealth of our knowledge on how chemical energy is generated from light energy (see books edited by J. Amesz, J. Barber and Govindjee during the last several years), it seems that molecular biology has come of age at the most appropriate time for its full exploitation in understanding the molecular mechanisms of photosynthesis.

Problems that have plagued our thoughts for decades are beginning to yield to these powerful techniques. One specific example of the successful application of molecular genetics to photosynthesis has been the recent chemical identification of an intermediate labeled "D" (or "F"), that acts as a slow electron donor to the reaction center chlorophyll a P680, as a tyrosine on the D_2 protein rather than the previously presumed plastoquinol (Debus et al.; Vermaas et al.). This has allowed the suggestion that the normal electron donor "Z" may also be a tyrosine molecule. Furthermore, the amino acid similarity between the "L" and "M" subunits of the bacterial reaction centers and the D_1 and D_2 subunits of plant photosystem II has led to suggestions as to the binding sites of the various chromophores and of the herbicides in the plant photosystem II. In contrast, differences have led to suggestions as to the binding sites of the components of the oxygen evolving complex on the lumenal side of D_1 and D_2 since only plants, not photosynthetic bacteria, evolve oxygen. The cloning of genes and examination of control of gene expression has led to information on the translational control of the synthesis of chlorophyll-binding proteins when chlorophyll is not synthesized. Additionally, use of gene-fusion techniques has enabled us to raise specific antibodies against individual protein components and to use them to advance functional studies. We suspect that each of you will have a personal list of important problems to which molecular genetics has contributed long-awaited insights.

There is a much wider and continuing technological revolution in the molecular biology of photosynthesis. The entire nucleotide sequence of the chloroplast genome of two plants (tobacco and the liverwort [*Marchantia polymorpha*]) is now available. Several other complete sequences will inevitably be generated. The publication of this book, a compilation of four special issues of the journal **Photosynthesis Research**, "Molecular Biology of Photosynthesis", marks the achievements of this field. The book is unique in capturing this current excitement in its entirety. Twenty-eight authoritative minireviews and ten original regular papers on this and related topics are included in this book. Topics range from the nature and the composition of the genes, molecular genetic methods, use of molecular biology in understanding the antenna system (the light harvesting system), the reaction centers, herbicide resistance, photosystem I, photosystem II, bacterial system, electron transport complexes and components, ATP synthase, carbon fixation, RUBISCO, Crassulacean Acid Metabolism (CAM), photorespiration, and transport of proteins.

We hope that a collection of these "burning" topics in photosynthesis research, as assembled in this volume, will be of great use to students of agriculture, biotechnology, biochemistry, biophysics, plant biology, plant physiology, cell and molecular biology. Several articles contain discussions of newer techniques. Technical terms, which are frequently used in this field, as well as a subject index, have been included to increase the usefulness of this book.

We feel that a rapid transfer of technology from the area of basic molecular genetics to the functional studies on photosynthesis is currently taking place. We are fortunate to be part of this exciting revolution in the field of photosynthesis. This is, however, only a beginning and much more remains to be done. We are convinced that for years to come, the continued development of molecular biology as a research tool will be an important contributor in our efforts to thoroughly understand how plants convert the solar energy into chemical energy.

The editors of this book are grateful to the chief editors Drs. R. Marcelle and R. Blankenship and all the associate editors of **Photosynthesis Research**, Mr. Ir. A.C.Plaizier, Ms. Josje Dominicus and Ms. Maria Vermeulen-James of Kluwer Publishers for their cooperation and help during the publication of this book. We also thank Dr. John C. Cushman (University of Arizona) for the glossary of terms and Dr. Rajni Govindjee (University of Illinois) for the subject index.

EDITORS

Contents

III. Reaction centers; herbicide resistance

Contributors

Abresch, Edward C., *Department of Physics B-019, University of California at San Diego, LaJolla CA 92093, USA* (pp 321–342)

Anderson, Lamont K., *Carnegie Institution of Washington, 290 Panama Street, Stanford CA 94305, USA* (pp 161–194)

Blackwell, Ray D., *Department of Biological Sciences, University of Lancaster, Lancaster LA1 4YQ, UK* (pp 677–698)

Blubaugh, Danny J., *Department of Agronomy, N-212 Ag. Sci. N., University of Kentucky, Lexington KY 40546, USA* (pp 441–484)

Bohnert, Hans J., *Biochemistry Department, University of Arizona, Tucson AZ 85721, USA* (pp 699–711)

Borrias, Mies, *Department of Molecular Cell Biology and Institute of Molecular Biology, University of Utrecht, Padualaan 8, 3584 CH Utrecht, The Netherlands* (pp 517–542)

Boyer, Scott K., *Biological Sciences Department, Purdue University, Lilly Hall of Life Sciences, West Lafayette IN 47907, USA* (pp 43–59)

Bradley, Douglas, *Department of Molecular Genetics, Plant Breeding Institute, Trumpington, Cambridge CB2 2LQ, UK* (pp 105–120)

Bruns, Brigitte U., *Carnegie Institution of Washington, 290 Panama Street, Stanford CA 94305, USA* (pp 161–194)

Brusslan, Judy, *Department of Molecular Genetics and Cell Biology, University of Chicago, Chicago IL 60637, USA* (pp 343–352)

Bryant, Donald A., *Department of Molecular and Cell Biology, S-101 Frear Building, Penn State University, University Park PA 16802, USA* (pp 353–369 & 371–387)

Buetow, Dennis E., *Department of Physiology and Biophysics, University of Illinois at Urbana-Champaign, 407 South Goodwin Avenue – 524 Burrill Hall, Urbana IL 61801, USA* (pp 283–319)

Bullerjahn George, *Division of Biological Sciences, University of Missouri, Tucker Hall, Columbia MO 65211, USA* (pp 229–257)

Buzby, Jeffrey S., *Department of Biology, University of California, Los Angeles CA 90024, USA* (pp 353–369)

Cantrell, Amanda, *Department of Molecular and Cell Biology, S-101 Frear Building, Penn State University, University Park PA 16802, USA* (pp 371–387)

Capuano, Véronique, *Department of Molecular and Cellular Biology, Penn State University, University Park PA 16802, USA* (pp 195–228)

Chen, Houqi, *Department of Physiology and Biophysics, University of Illinois at Urbana-Champaign, 407 South Goodwin Avenue – 524 Burrill Hall, Urbana IL 61801, USA* (pp 283–319)

Chitnis, Parag R., *Biology Department and Molecular Biology Institute, University of California, Los Angeles CA 90024, USA* (pp 259–281)

Conley, Pamela B., *Carnegie Institution of Washington, 290 Panama Street, Stanford CA 94305, USA* (pp 161–194)

Cook, William B., *Department of Biological Sciences, 108 Tucker Hall, University of Missouri, Columbia MO 65211, USA* (pp 77–103)

Cremers, Fons, *Department of Molecular Cell Biology and Institute of Molecular Biology, University of Utrecht, Padualaan 8, 3584 CH Utrecht, The Netherlands* (pp 517–542)

Curtis, Stephanie E., *Department of Genetics, Box 7614, North Carolina State University, Raleigh NC 27695-7614, USA* (pp 543–564)

Cushman John C., *Biochemistry Department, University of Arizona, Tucson AZ 85721, USA*

Damerval, Thierry, *Département de Biochimie et Génétique Moléculaire, Unité de Physiologie Microbiènne (CNRS UA 1129), Institut Pasteur, 28, rue du Docteur Roux, 75724 Paris Cédex 15, France* (pp 195–228)

Daniell, Henry, *Biochemistry/Biophysics Program, Washington State University, Pullman WA 99164-4660, USA* (pp 121–135)

De Groot, Rolf, *Department of Molecular Cell Biology and Insitute of Molecular Biology, University of Utrecht, Padualaan 8, 3584 CH Utrecht, The Netherlands* (pp 517–542)

Depka, Brigitte, *Department of Biology, Ruhr-University of Bochum, P.O. Box 10 21 48, D-4630 Bochum 1, FRG* (pp 407–421)

Donohue, Timothy J., *Bacteriology Department, University of Wisconsin, 1550 Linden Drive, Madison WA 53706, USA* (pp 137–159)

Ellis, R. John, *Department of Biological Sciences, University of Warwick, Coventry CV4 7AL, UK* (pp 661–675)

Erdös, Géza, *Department of Physiology and Biophysics, University of Illinois at Urbana-Champaign, 407 South Goodwin Avenue – 524 Burrill Hall, Urbana IL 61801, USA* (pp 283–319)

Feher, George, *Department of Physics B-019, University of California at San Diego, LaJolla CA 92093, USA* (pp 321–342)

Fukuzawa, Hideya, *Institute of Applied Microbiology, University of Tokyo, Tokyo 113, Japan* (pp 27–42)

Gatenby, Anthony A., *Central Research and Development Department, Experimental Station, E.I. DuPont de Nemours & Co., Wilmington DE 19898, USA* (pp 105–120 & 607–619)

Gingrich, Jeffrey C., *Chemical Biodynamics Division, Lawrence Berkely Laboratory, University of California, Berkeley CA 94720, USA* (pp 353–369)

Gnanam, A., *Department of Plant Sciences, School of Biological Sciences, Madurai Kamaraj University, Madurai 625021, India* (pp 777–800)

Govindjee, *Departments of Plant Biology, Physiology and Biophysics,*

*University of Illinois at Urbana-Champaign, 505 South Goodwin Avenue –
289 Morrill Hall, Urbana IL 61801, USA* (pp 441–484)

Gray, John C., *Botany School, University of Cambridge, Downing Street,
Cambridge CB2 3EA, UK* (pp 43–59 & 497–516)

Grossman, Arthur G., *Carnegie Institution of Washington, 290 Panama
Street, Stanford CA 94305, USA* (pp 161–194)

Gruissem, Wilhelm, *Department of Botany, University of California,
Berkeley CA 94720, USA* (pp 621–644)

Guglielmi, Gérard, *Département de Biochimie et Génétique Moléculaire,
Unité de Physiologie Microbiènne (CNRS UA 1129), Institut Pasteur, 28,
rue du Docteur Roux, 75724 Paris Cédex 15, France* (pp 195–228)

Hall, Nigel P., *Biochemistry Department, Rothamsted Experimental Station.
Harpenden Herts AL5 2JQ, UK* (pp 677–698)

Hallenbeck, Paul L., *Department of Microbiology, University of Illinois, 407
South Goodwin Avenue – 136 Burrill Hall, Urbana IL 61801, USA* (pp
583–591)

Haselkorn, Robert, *Department of Molecular Genetics and Cell Biology,
University of Chicago, Chicago IL 60637, USA* (pp 343–352)

Hayashida, Nobuaki, *Center for Gene Research, Nagoya University,
Chikusa, Nagoya 464, Japan* (pp 1–25)

Houmard, Jean, *Département de Biochimie et Génétique Moléculaire, Unité
de Physiologie Microbiènne (CNRS UA 1129), Institut Pasteur, 28, rue du
Docteur Roux, 75724 Paris Cédex 15, France* (pp 195–228)

Hudson Graham S., *Division of Plant Industry, CSIRO, GPO Box 1600,
Canberra, A.C.T. 2601, Australia* (pp 565–582)

Hughes, J.E., *Department of Botany, University of Georgia, Athens GA
30602, USA* (pp 423–439)

Ikeuchi, M., *Solar Energy Research Group, RIKEN, Hirosawa 2-1, Wako-
shi, Saitama 351-01, Japan* (pp 389–405)

Inoue, Y., *Solar Energy Research Group, RIKEN, Hirosawa 2-1, Wako-shi,
Saitama 351-01, Japan* (pp 389–405)

Johanningmeier, Udo, *Department of Biology, Ruhr-University of Bochum,
P.O. Box 10 21 48, D-4630 Bochum 1, FRG* (pp 407–421)

Kaplan, Samuel, *Department of Microbiology, University of Illinois at Urbana-Champaign, 407 South Goodwin Avenue – 136 Burrill Hall, Urbana
IL 61801, USA* (pp 137–159 & 583–591)

Keegstra, Kenneth, *Botany Department, University of Wisconsin, Madison
WI 53706, USA* (pp 713–734)

Kendall, Alan C., *Biochemistry Department, Rothamsted Experimental
Station, Harpenden, Herts AL5 2JQ, UK* (pp 677–698)

Kiley, Patricia J., *Microbiology Department, University of Illinois, 407 South Goodwin Avenue - 136 Burrill Hall, Urbana IL 61801, USA* (pp 137–159)

Kohchi, Takayuki, *Research Center for Cell and Tissue Culture, Faculty of Agriculture, Kyoto University, Kyoto 606, Japan* (pp 27–42)

Kraft, Bernd, *Department of Biology, Ruhr-University of Bochum, P.O. Box 10 21 48, D-4630 Bochum 1, FRG* (pp 407–421)

Lea, Peter J., *Department of Biological Sciences, University of Lancaster, Lancaster LA1 4YQ, UK* (pp 677–698)

Lemaux, Peggy G., *Carnegie Institution of Washington, 290 Panama Street, Stanford CA 94305, USA* (pp 161–194)

Link, G., *Pflanzliche Zellphysiologie, Ruhr-University of Bochum, P.O. Box 10 21 48, D-4630 Bochum 1, FRG* (pp 423–439)

Lubben, Thomas H., *Botany Department, University of Wisconsin, Madison WI 53706, USA* (pp 713–734)

Mannan, R. Mannar, *Department of Plant Sciences, School of Biological Sciences, Madurai Kamaraj University, Madurai 625021, India* (pp 777–800)

Manzara, Thianda, *Department of Botany, University of California, Berkeley CA 94720, USA* (pp 621–644)

Mason, John G., *Division of Plant Industry, CSIRO, GPO Box 1600, Canberra, A.C.T. 2601, Australia* (pp 565–582)

Mazel, Didier *Département de Biochimie et Génétique Moléculaire, Unité de Physiologie Microbiènne (CNRS UA 1129), Institut Pasteur, 28, rue du Docteur Roux, 75724 Paris Cédex 15, France* (pp 195–228)

McFadden, Bruce A., *Biochemistry/Biophysics Program, Washington State University, Pullman WA 99164-4660, USA* (pp 121–135 & 645–660)

Michalowski, Christine, *Biochemistry Department, University of Arizona, Tucson AZ 85721, USA* (pp 699–711)

Miles, Donald, *Department of Biological Sciences, 108 Tucker Hall, University of Missouri, Columbia MO 65211, USA* (pp 77–103)

Mishkind, Michel L., *Department of Biochemistry and Microbiology, Lipman Hall, Cook College, Rutgers University, New Brunswick NJ 08903, USA* (pp 745–776)

Mullet, John E., *Department of Biochemistry & Biophysics, Texas A & M University, College Station TX 77843-2128, USA* (pp 43–59)

Murray, Alan J.S., *Department of Biological Sciences, University of Lancaster, Lancaster LA1 4YQ, UK* (pp 677–698)

Nelson, Nathan, *Roche Institute of Molecular Biology, Roche Research Center, Nutley NJ 07110, USA* (pp 485–496)

Ohyama, Kanji, *Research Center for Cell and Tissue Culture, Faculty of Agriculture, Kyoto University, Kyoto 606, Japan* (pp 27–42)

Okamura, Melvin Y., *Department of Physics B-019, University of California at San Diego, LaJolla CA 92093, USA* (pp 321–342)

Omata, Tatsuo, *Solar Energy Research Group, The Institute of Physical and Chemical Research (RIKEN), Wako-shi, Saitama, 351-01, Japan* (pp 593–606)

Ozeki, Haruo, *Department of Biophysics, Faculty of Science, Kyoto University, Kyoto 606, Japan* (pp 27–42)

Paddock, Mark L., *Department of Physics B-019, University of California at San Diego, LaJolla CA 92093, USA* (pp 321–342)

Pierce, John, *Central Research and Development Department, E.I. DuPont de Nemours and Co., Experimental Station, Building 402, Room 2230, Wilmington DE 19898, USA* (pp 593–606)

Plant, Aine L., *Botany School, University of Cambridge, Downing Street, Cambridge CB2 3EA, UK* (pp 43–59)

Reddy, K.J., *Division of Biological Sciences, University of Missouri, Tucker Hall, Columbia MO 65211, USA* (pp 229–257)

Reilly, Patricia, *Roche Institute of Molecular Biology, Roche Research Center, Nutley NJ 07110, USA* (pp 485–496)

Riethman, Harold, *Division of Biological Sciences, University of Missouri, Tucker Hall, Columbia MO 65211, USA* (pp 229–257)

Rongey, Scott H., *Department of Physics B-019, University of California at San Diego, LaJolla CA 92093, USA* (pp 321–342)

Rothstein, Steven J., *Department of Molecular Biology & Genetics, University of Guelph, Ontario N1G 2W1, Canada* (pp 105–120)

Sano, Tohru, *Research Center for Cell and Tissue Culture, Faculty of Agriculture, Kyoto University, Kyoto 606, Japan* (pp 27–42)

Schmitt, Jürgen M., *Botanisches Institüt der Universität Würzburg, 8700 Würzburg, FRG* (pp 699–711)

Scioli, Scott E., *Department of Biochemistry and Microbiology, Lipman Hall, Cook College, Rutgers University, New Brunswick NJ 08903, USA* (pp 745–776)

Sherman, Louis A., *Division of Biological Sciences, University of Missouri, Tucker Hall, Columbia MO 65211, USA* (pp 229–257)

Shinozaki, Kazuo, *Center for Gene Research, Nagoya University, Chikusa, Nagoya 464, Japan* (pp 1–25)

Small, Christopher L., *Biochemistry/Biophysics Program, Washington State University, Pullman WA 99164-4660, USA* (pp 645–660)

Smeekens, Sjef, *Department of Molecular Cell Biology and Institute of Molecular Biology, University of Utrecht, Padualaan 8, 3584 CH Utrecht, The Netherlands* (pp 735–744)

Stirewalt, Veronica L., *Department of Molecular and Cell Biology, S-101*

Frear Building, Penn State University, University Park PA 16802, USA (pp 353–369)

Subbaiah, C.C., *Department of Plant Sciences, School of Biological Sciences, Madurai Kamaraj University, Madurai 625021, India* (pp 777–800)

Sugiura, Masahiro, *Center for Gene Research, Nagoya University, Chikusa, Nagoya 464, Japan* (pp 1–25)

Tandeau De Marsac, Nicole, *Département de Biochimie et Génétique Moléculaire, Unité de Physiologie Microbiènne (CNRS UA 1129), Institut Pasteur, 28, rue du Docteur Roux, 75724 Paris Cédex Roux, France* (pp 195–228)

Theg, Steven M., *Department of Botany, University of California at Davis, Davis CA 95616, USA* (pp 713–734)

Thornber, J. Philip, *Biology Department and Molecular Biology Institute, University of Calfornia, Los Angeles CA 90024, USA* (pp 259–281)

Trebst, Achim, *Department of Biology, Ruhr-University of Bochum, P.O. Box 10 21 48, D-4630 Bochum 1, FRG* (pp 407–421)

Turner, Janice C., *Biochemistry Department, Rothamsted Experimental Station, Harpenden, Herts AL5 2JQ, UK* (pp 677–698)

Umesono, Kazuhiko, *Department of Biophysics, Faculty of Science, Kyoto University, Kyoto 606, Japan* (pp 27–42)

Van Arkel, Gerard, *Department of Molecular Cell Biology and Institute of Molecular Biology, University of Utrecht, Padualaan 8, 3584 CH Utrecht, The Netherlands* (pp 517–542)

Van Der Plas, Jan, *Department of Molecular Cell Biology and Institute of Molecular Biology, University of Utrecht, Padualaan 8, 3584 CH Utrecht, The Netherlands* (pp 517–542)

Van der Vies, Saskia M., *Department of Biological Sciences, University of Warwick, Coventry CV4 7AL, UK* (pp 661–675)

Vermaas, W.F.J., *Department of Botany, Arizona State University, Tempe AZ 85787-1601, USA* (pp 389–405)

Wallsgrove, Roger M., *Biochemistry Department, Rothamsted Experimental Station, Harpenden, Herts AL5 2JQ, UK* (pp 677–698)

Weisbeek, Peter, *Department of Molecular Cell Biology and Institute of Molecular Biology, University of Utrecht, Padualaan 8, 3584 CH Utrecht, The Netherlands* (pp 517–542 & 735–744)

Willey, David L., *Botany School, University of Cambridge, Downing Street, Cambridge CB2 3EA, UK* (pp 497–516)

Woortman, Martin, *Department of Molecular Cell Biology and Institute of Molecular Biology, University of Utrecht, Padualaan 8, 3584 CH Utrecht, The Netherlands* (pp 517–542)

Yi, Lee S.H., *Department of Physiology and Biophysics, University of Illinois at Urbana-Champaign, 407 South Goodwin Avenue – 524 Burrill Hall, Urbana IL 61801, USA* (pp 283–319)

Govindjee et al. (eds), Molecular Biology of Photosynthesis: xvii
© 1988 Kluwer Academic Publishers

Glossary of terms

Compiled by: John C. Cushman
Department of Biochemistry, University of Arizona, Tucson, Arizona 85721, USA

Adapter
Small synthetic oligonucleotides which when annealed and ligated to blunt ended DNA molecules from "Preformed" protruding termini which do not require cleavage with a restriction enzyme to create a cohesive end (see linker).

Agarose gel electrophoresis
A technique used to separate according to size, to identify and purify DNA fragments. Used in conjunction with Southern blotting and Northern blotting (for RNA).

Alkaline phosphatase (bacterial or calf intestinal)
An enzyme that catalyses the removal of 5′ phosphate residues from DNA or RNA. Used in constructing recombinant molecules to prevent circularization of vectors without inserts.

Amplification
The process of propagating a primary or secondary library created in plasmids or bacteriophage (lambda) vectors which greatly increases the number of clones. Amplification may also refer to the production of additional copies of a chromosomal sequence.

Anticodon
Trinucleotide sequence present on tRNA molecules that is complementary to the codon representing a specific amino acid.

ATA
Aurintricarboxylic Acid. Used as a potent inhibitor of RNAses during preparation of RNA.

Bacteriophage (phage)
A virus that infects and may kill bacterial cells. See lambda phage and M13.

Bal31 nuclease
An enzyme that catalyses the removal of nucleotides from both 5′- and 3′-termini of double-stranded and single-stranded DNA creating shorter molecules. Used in restriction mapping and in vitro mutagenesis.

Blunt end ligation
   See ligation.
Blunt (flush) ends
   Completely base paired termini of double-stranded DNA molecules.
CAAT Box
   A conserved sequence located about 75 bp upstream from the start point
   of transcription in eukaryotes.
cDNA (complementary DNA)
   A DNA molecule that is complementary to an RNA molecule. Usually
   synthesized by the enzyme reverse transcriptase.
cDNA cloning
   A technique which copies mRNA into DNA using reverse transcriptase
   followed by the synthesis of a second strand of DNA and insertion of
   double-strand DNA into a cloning vector.
Chloroplast gene nomenclature
   Chloroplast genes are named according to the recommendations set forth
   by: Hallick R.B. and Bottomley, W. 1983. Proposals for the Naming of
   Chloroplast Genes. Plant Molecular Reporter 1: 38–43.
Chromatin
   The complex of DNA and protein in the nucleus of a eukaryote cell or
   nucleoid of a bacterial cell.
Chromosome
   A discrete unit of the genome consisting of proteins and a very long DNA
   molecule carrying many genes.
Clone
   A member of a population of identical DNA molecules or of genetically
   identical cells or organisms derived from a single individual by asexual
   processes. Cloning is the term used to designate the generation and
   propagation of discrete DNA molecules.
Codon
   A triplet of nucleotides that specify a particular amino acid or a termina-
   tion signal (stop codon). As initiation codon, usually ATG (AUG in
   RNA) is used. Termination codons are TAA, TAG and TGA. While
   codon assignment is, in principle, universal, some mitochondria have
   specific different codon assignments.
Cohesive (sticky) ends
   Single-stranded termini of double stranded DNA molecules that are
   complementary to one another. Some restriction endonucleases cut DNA
   so that cohesive ends are formed, others result in blunt ends.
Colony hybridization
   A technique used to detect the presence of a cloned DNA segment in a
   bacterial colony (see screening).

Concatemer

Two or more identical sequences joined tandemly head to tail.

Cosmids

Plasmid vectors which contain phage lambda DNA sequences (COS sites) allowing the DNA to be packaged in vitro into lambda phage particles. The size of DNA inserts that can be cloned is greater than that of lambda vectors.

DNA Ligase (T_4)

An enzyme that catalyses the fomation of phosphodiester bonds between adjacent 3'-OH and 5' phosphate termini in double-stranded DNA. Used to join double-stranded DNA molecules with compatible cohesive termini (sticky ends) or to join blunt-ended double-stranded DNA molecules or linkers. (See RNA Ligase (T_4).)

DNA polymerase I (*E. coli*)

Polymerizes DNA (needs primer, template and nucleotides) in a 5'- to 3'-direction. Also contains a 5'- to 3'-exonuclease, and a 3'- to 5'-exonuclease for proofreading. Used for labeling via nick-translation.

DNA polymerase (*T4*)

See T_4 DNA polymerase.

DNase I

An endonuclease isolated from bovine pancreas that breaks double or single stranded DNA. In the presence of Mg^{++}, DNase I makes single strand nicks in double-stranded DNA; in the presence of Mn^{++} DNAse cuts double-strand DNA resulting in fragments of approximately the same size. Used in characterizing nucleosomes, and for making single-strand nicks in DNA prior to nick-translation.

End Labelling

A technique that introduces radioactively labeled nucleotides (dNTPs) which are usually radioactively labeled using Klenow fragment or T_4 DNA polymerase or T_4 polynucleotide kinase to transfer phosphate groups to the termini of DNA molecules.

Enhancer

A specific DNA sequence element that increases the utilization of a promoter in cis. Enhancers can function independently of location and orientation relative to the promoter.

Exon

˙Any segment of an interrupted gene represented in fully processed, mature mRNA (see intron) that is translated into amino acids.

Exonuclease III

An enzyme that catalyses the stepwise removal of 5' mononucleotides in a 3' to 5' direction from 3'-OH ends of double-stranded DNA. Used for preparing single-stranded DNA probes, and for deleting sequences from the ends of DNA fragments in conjunction with a S_1 nuclease. Commonly used for the production of overlapping deletion subclones used in sequencing large DNA fragments.

Exonuclease VII

A processive exonuclease that removes small oligonucleotides from the 3′ and 5′ ends of single-stranded DNA.

Expression vector

A vector that allows the synthesis of RNA or protein from a coding region after engineering to contain promoters or other control elements for efficient synthesis in the particular host organism.

Footprinting

A technique used to identify regions of DNA bound by specific domains of some proteins by virtue of protection against nuclease attack.

Gene

A segment of DNA involved in producing a polypeptide chain including a promoter and other regulatory elements. A gene includes regions preceding (5′-leader sequence) and following (3′-trailing sequence) individual coding segments (exons).

Gene Cassette

A gene or portions of a gene bordered by suitable restriction sites that can be interchanged to produce a desired vector or gene fusion.

Gene Fusion (Chimeric genes)

Gene constructs in which different gene cassettes have been combined to produce composite transcriptional units or translational reading frames.

Gene nomenclature

Bacterial gene nomenclature should follow the conventions set forth by: Demerec, M., E.A. Adelberg, A.J. Clark, and Philip E. Hartman. 1966. A proposal for a Uniform Nomenclature in Bacterial Genetics. Genetics 54: 61–76. Increasingly, this nomenclature is also being used for eukaryotic genes.

Genome

The entire complement of chromosomal DNA of a given organism. Genomic clones are derived directly from genomic DNA, not from a cDNA copy.

Genomic Southern

A technique in which total genomic DNA is digested with a restriction endonuclease, separated by agarose gel electrophoresis and transferred to filters. See Southern blotting.

Guanidinium chloride or isothiocyanate

Potent chaotropic agents used to inactivate nucleases and disintegrate cell structures and nucleoproteins from nucleic acids. Used to isolate intact RNA from tissues that are typically rich in RNase.

Hybridization

Association by base pairing that occurs between two complementary strands of either DNA and/or RNA. Often used to identify a gene from an organism by using a probe from an identical, analogous gene of another organism. The strength of hybridization depends on nucleotide sequence similarity between the DNAs.

Hybridoma

A cell produced as a result of fusing a myeloma with a lymphocyte which will indefinitely express the immunoglobulins of both parent cells.

Inducer

A small molecule that triggers transcription of a gene by binding to a regulator protein.

Insert

Any piece of DNA that has been introduced into a specific site (restriction site) of a vector.

Introns (intervening sequences)

Segments of DNA that are transcribed, but that are removed from the mature mRNA. The remaining pieces of RNA (exons) are joined together by splicing.

In vitro mutagenesis

Modifications made to isolated and purified genes by chemical or physical methods. Some modifications made include deletions, insertions, inversions and transpositions.

Isoschizomer

Different restriction endonuclease (type II) which recognize(s) and cleave(s) the same target sequences.

Kb (kilobase)

An abbreviation for 1000 base pairs of DNA or 1000 bases of RNA.

Klenow fragment

Derived from DNA polymerase I of E. coli by treatment with subtilisin (a proteolytic enzyme). Some firms now offer a product that is made by recombinant DNA techniques. Lacks 5′ to 3′ exonuclease activity of the holoenzyme. Used for Sanger DNA sequencing (see sequencing), filling in the 3′ recessed termini (5′ overhangs) created by some restriction enzymes, and for second-strand cDNA synthesis.

Lambda exonuclease

Digests DNA in a 5′ to 3′ direction.

Lambda phage

A tempelate phage containing about 50 kb of DNA. Derivatives of this phage provide a useful vector for cloning large segments of DNA (up to about 20 kb). Genes, or parts of genes, are used in plasmids (e.g.) for promoter constructions or as repressors and terminators of transcription.

Leader

The nontranslated sequence at the 5′ end of a transcribed mRNA that precedes the initiator codon.

Library

A collection of cloned random fragments of DNA representing the entire genome. Sometimes called a gene bank. Both cDNA libraries (a collection of the DNA versions of a group of RNAs) and genomic libraries are commonly encountered.

Ligation

The formation of a phosphodiester bond covalently linking two adjacent bases separated by a nick in one strand of a DNA duplex. The term blunt-end ligation refers to the reaction wherein two blunt-ended DNA duplex molecules are joined covalently at their ends.

Linker

Self complementary synthetic oligonucleotides of defined sequence containing the cleavage site of one or more restriction endonucleases. After these fragments are blunt-end ligated to DNA molecules, the molecule contains the recognition sequence of a restriction endonuclease and can be subsequently cleaved.

Lysogen

A bacterium that possesses a repressed prophage integrated into its genome.

M13

A filamentous bacteriophage that carries a single strand of DNA whose life cycle includes a double-stranded DNA stage. Used in DNA sequencing and for in vitro mutagenesis.

Methylases

A number of enzymes that introduce methyl groups into DNA. These enzymes may be used to modify specific sites in DNA and prevent their being cut with certain restriction enzymes.

Micrococcal nuclease

An endonuclease that cleaves DNA and RNA. In chromatin, micrococcal nuclease preferentially cleaves between nucleosomes.

mRNA

RNA that serves as a template for protein synthesis. Called 'messenger' RNA. Its sequence may differ from that of the gene from which it was transcribed because of the removal of introns and other processing events.

Mung-bean nuclease

Enzyme with properties similar to S1 nuclease, but gentler in its action. Used for converting protruding termini to blunt ends.

Nick

The absence of a phosphodiester bond between two adjacent nucleotides on one strand of duplex DNA.

Nick translation

Nick translation makes use of the 5′ and 3′ exonuclease activity of DNA polymerase I to move ('translate') a nick in DNA from one position to another. The polymerase starts synthesis of new strand at the nick, degrading the DNA ahead of it. Nick translation is used to label DNA so that it can be used as a probe.

Northern blots

A technique for analysis of RNA. The RNA is run out on an agarose or acrylamide gel and then transferred to a suitable substrate filter (nitro-cellulose or nylon membrane). The RNA of interest is then detected by hybridization with an appropriately labeled probe. See "Southern blots".

Nucleosome

The basic structural subunit of chromatin consisting of about 200 bp of DNA and an octamer of histone proteins.

Nucleotide (base)

The basic structural unit of nucleic acids. A nucleotide consists of a heterocyclic ring of carbon and nitrogen atoms (nitrogenous base), a five carbon sugar ring (pentose) and a phosphate group. The pentose ring distinguishes DNA and RNA. The pentose ring in DNA is 2-deoxyribose, whereas in RNA it is ribose. The nitrogenous bases are of two types: the pyrimidines (cytosine, uracil and thymine) have six-membraned rings and the purines (adenine and guanine) have fused five and six-membered rings.

Oligonucleotide

DNA or RNA molecules of variable (6 to, usually, less than 100 bases) length that may be synthesized in vitro to serve as linkers, adaptors or probes, or templates for site-directed mutagenesis.

Papovaviruses

Animal viruses containing small circular genomes. Examples are SV40 and polyoma.

Phagemid

A plasmid vector which contains regions of DNA required for the pack-aging of the plasmid as single-stranded DNA by an M13 bacteriophage derivative called a 'helper phage'.

Plasmid

An autonomously replicating, circular, extrachromosomal genetic element.

Plasmid Nomenclature

Cloned DNA fragments contained in a vector can be named in a variety of ways, always preceeded by 'p', denoting either a person (pBR), a laboratory (pEMBL), a function of the cloned DNA (ptac), or combinations of those terms usually followed by numbers.

Polyacrylamide gel electrophoresis

A simple technique to analyze and purify DNA fragments less than 1 kb length. Polyacrylamide gels are also used for analyzing proteins and for sequencing.

Polyadenylation

Addition of poly A to the 3′-end of an RNA molecule by a mechanism that does not involve transcription.

Polynucleotide kinase (T_4)

An enzyme that catalyses transfer of the gamma-phosphate of ATP to the 5′-hydroxyl at 5′-termini of RNA or DNA for sequencing by the Maxam-Gilbert technique. Used for labeling DNA and RNA (with ^{32}P-phosphate labeled rATP), and for restoring the phosphate removed by alkaline phosphatase.

Polysome

A string of ribosomes attached to a single mRNA molecule.

Primer

In DNA synthesis, a single stranded DNA, often an oligonucleotide (oligomer), that serves as a starting point for polymerization of a second chain. The primer base pairs with the template (the other strand) and is extended by a DNA polymerase to form a complementary strand.

Probe

A gene specific DNA fragment from coding, leader or trailer regions which is usually radioactively labeled and used via hybridization to locate and define another gene or mRNA of unknown character. The term is also used for the specific antibodies used to screen gene libraries (see Screening).

Promoter

In prokaryotes, a region of DNA that is involved in DNA-dependent RNA polymerase recognition and binding. The promoter regulates the site of initiation of transcription as well as its efficiency. In eukaryotes, promoter sequences do not directly bind RNA polymerase. Instead, they are the site of binding of a complex of proteins that are recognized by the polymerase.

Prophage

A phage genome covalently integrated as a linear part of the bacterial chromosome.

Pseudogene

Inactive, but stable version(s) of a gene derived by mutation from an ancestral active gene.

Replisome

A multiprotein complex containing DNA polymerase and other proteins that assembles at the bacterial replication fork to conduct DNA synthesis.

Restriction Endonucleases (Type II)

Enzymes that recognize a particular sequence (usually 4–8 nucleotides with a twofold axis of symmetry) in double-stranded DNA and cut, in or near that sequence, leaving 5′ phosphate and 3′ hydroxyl protruding cohesive termini or blunt ends. Used to cut DNA molecules into pieces of defined sizes with specific kinds of ends.

Restriction Endonuclease Nomenclature

Restriction endonucleases are named according to the proposal by: Smith, O. and Nathans. 1973. A Suggested Nomenclature for Bacterial Host Modification and Restriction Systems and Their Enzymes. J. Mol. Current Biol. 81: 419–423. A listing of all known restriction endonucleases can be found in: Roberts, R.J. 1988. Restriction Enzymes and Their Isoschizomers. Nucleic Acids Research. Sequences Supplements 16: r271–r313.

Reverse Transcriptase (RNA-Dependent DNA polymerase)

An enzyme that catalyzes the polymerization of DNA 5′ to 3′ using an oligonucleotide primer or single-stranded RNA as a template (DNA may also be used). Used in the first synthesis of cDNA (first strand). The enzymes also contain a processive 5′ and 3′ ribonuclease specific for RNA: DNA hybrids.

RFLP – Restriction fragment length polymorphism

Term used to describe differences in the DNA restriction pattern that distinguishes several genes in an organism or in individuals of a population. RFLP is used to monitor the number of genes for one character, the evolution of genes and the linkage between DNA rearrangements and mutant characters.

Ribonuclease A

This enzyme isolated from bovine pancreas is an endoribonuclease which specifically attacks the 3′-phosphate groups of pyrimidine residues cleaving the 5′-phosphate linkage of adjacent nucleotides. Used to remove RNA from DNA preparations.

Ribonuclease T_1

This enzyme isolated from *Aspergillus oryzae* is an endoribonuclease which specifically attacks the 3′-phosphate groups of guanosine residues cleaving the 5'-phosphate linkage of adjacent nucleotides. (Used to remove large pieces of RNA from DNA preparations).

Ribosome

Ribonucleoprotein particles consisting of two subunits that catalyze the assembly of amino acids into polypeptides using mRNA as a template.

RNA ligase (T4)

An enzyme that catalyzes the covalent linkage of the 5'-phosphate of one molecule (single-stranded DNA or RNA) with the 3' OH of another. Reported to increase the efficiency of blunt-end ligation of DNA molecules using T_4 ligase.

RNase

A group of enzymes with different specificities for the degradation of RNA molecules.

RNasin

A commercially available enzyme that is a potent inhibitor of RNase. It can be included in enzymatic reactions and easily extracted with phenol.

S_1 mapping

A technique used to locate the ends of RNA molecules in relation to specific sites (e.g. restriction sites) within the template DNA.

S_1 nuclease

Single-strand specific endonuclease isolated from *Aspergillus oryzae*. DNA/DNA, DNA/RNA, and RNA/RNA hybrids are resistant to its action at low concentrations of the enzyme. Used in "S_1 mapping", RNA protection experiments, and cDNA cloning to remove non-complementary areas of DNA/DNA, DNA/RNA, and RNA/RNA hybrids.

Screening

The act of searching for and isolating a particular cell phenotype under particular conditions confered by a selectable marker or searching for a particular gene using a nucleic acid or antibody probe.

Selectable Marker

A gene or genes that encodes a protein that confers upon a cell the ability to survive under conditions of selection (e.g. antibiotic resistance genes).

Sequenase

A modified form of T_7 DNA polymerase, which possesses greater processivity than Klenow fragment, used in Sanger-type sequencing.

Sequencing

A technique for determining the nucleotide sequence of a specific fragment of DNA or RNA employing Klenow, sequenase, or reverse transcriptase and chain terminating dideoxynucleotides (Sanger method) or chemical modification base specific chemical cleavage reactions (Maxam-Gilbert method).

Shine-Dalgarno Sequence (Ribosome binding site)

A poly purine stretch with the consensus prokaryotic feature AGGAGG located just upstream (usually 7–10 bases) of the AUG initiation codon. This sequence is complementary to the sequence at the 3′-end of 16S rRNA and is involved in the binding of ribosomes to mRNA.

Shuttle vector

A plasmid that is capable of growing in two or more hosts.

Site-directed mutagenesis

Construction of altered genes or gene products using methods that allow defined changes at a predetermined position in a gene. Examples are nucleotide exchanges which may lead to amino acid exchanges.

SnRNA (small nuclear RNA)

Small RNA molecules confined to the nucleus involved in splicing.

SnuRPS

Small nuclear ribonucleoproteins associated with SnRNA involved in splicing and splicosome formation.

Southern blotting

A technique originated by E. Southern. DNA is transferred from gels after electrophoresis to filters where specific sequences may be detected using radioactively labeled nucleic acid probes.

Splicing

The removal of introns and joining of exons in RNA molecules.

Splicosome

A ribonucleoprotein particle of about 40S consisting of about 40S consisting of SnRNAs and 7–10 small basic proteins ("SnuRPS") that catalyzes the splicing out of introns and splicing together of exon.

TATA Box (Hogness box)

A conserved A–T rich region located about 25 bp upstream from the transcriptional start site thought to position RNA polymerase II for transcriptional initiation.

T₄ DNA polymerase

An enzyme used to label recessed 3′-termini created by some restriction endonucleases. It possesses a more powerful 3′- to 5′-exonuclease activity than Klenow making. Also possesses the 5′- to 3′-polymerase activity.

Telomere

The natural end of a chromosome. Plasmids containing telomeres are maintained as one copy per cell.

Terminal transferase (Terminal Deoxynucleotidyl Transferase)

An enzyme that adds deoxynucleotides to the 3′-OH end of single-stranded DNA molecules or double-stranded DNA with protruding 3′-OH termini. Used to add homopolymer tails onto DNA before ligation, and for end-labeling the 3′-ends of DNA fragments.

Trailer
The nontranslated sequence located at the 3′-end of a mRNA following the termination codon.

Transacting factor
A protein that mediates transcription by binding to promoters, enhancers or other cis-acting regulatory elements.

Transcription
Synthesis of RNA directed by a DNA template using the enzyme RNA-polymerase.

Translation
Synthesis of protein by ribosomes directed by mRNA.

Transfer RNA (tRNA)
A small RNA molecule of 75–85 nucleotides that recognizes and becomes covalently attached to only one amino-acid to form an amino acyl-tRNA and contains the anticodon for recognizing the codon present in mRNA that represents the specific amino acid.

Transfection
Uptake of purified viral DNA by (most often) bacterial cells.

Transformation
Induced uptake of DNA of any kind by cells.

Transposase
The enzyme activity involved in the insertion of a transposon at a new site.

Transposon (Transposable elements)
A DNA sequence usually carrying a gene or genes that is able to replicate and insert a copy of itself at a new location in the genome of prokaryotes or eukaryotes.

Vector
Any plasmid, cosmid or phage into which a foreign piece of DNA (insert) may be ligated to be cloned.

Western blotting (Immunoblotting)
A technique in which proteins are resolved by polyacrylamide gel electrophoresis and transferred to a filter and probed with antibodies [identified by binding to a secondary labeled antibody] to identify a specific protein of interest.

Govindjee et al. (eds), Molecular Biology of Photosynthesis: 1–25
© 1988 Kluwer Academic Publishers

Minireview

Nicotiana chloroplast genes for components of the photosynthetic apparatus

KAZUO SHINOZAKI, NOBUAKI HAYASHIDA & MASAHIRO SUGIURA[1]

Center for Gene Research, Nagoya University, Chikusa, Nagoya 464, Japan; [1] *author for correspondence*

Received 23 September 1987; accepted 31 December 1987

Key words: *Nicotiana*, chloroplast genome, photosynthesis, NADH dehydrogenase, protein structure, gene expression

Abstract. In order to understand more fully chloroplast genetic systems, we have determined the complete nucleotide sequence (155, 844 bp) of tobacco (*Nicotiana tabacum* var. Bright Yellow 4) chloroplast DNA. It contains two copies of an identical 25,339 bp inverted repeat, which are separated by 86, 684 bp and 18,482 bp single-copy regions. The genes for 4 different rRNAs, 30 different tRNAs, 44 different proteins and 9 other predicted protein-coding genes have been located. Fifteen different genes contain introns.

Twenty-two genes for components of the photosynthetic apparatus have so far been identified. Most of the genes (except the gene for the large subunit of ribulose-1,5-bisphosphate carboxylase/oxygenase) code for thylakoid membrane proteins. Twenty of them are located in the large single-copy region and one gene for a 9-kd polypeptide of photosystem I is located in the small single-copy region. The gene for the 32-kd protein of photosystem II as well as the gene for the large subunit of ribulose-1,5-bisphosphate carboxylase/oxygenase have strong promoters and are transcribed monocistronically while the other genes are transcribed polycistronically. We have found that the predicted amino acid sequences of six DNA sequences resemble those of components of the respiratory-chain NADH dehydrogenase from human mitochondria. As these six sequences are highly transcribed in tobacco chloroplasts, they are probably genes for components of a chloroplast NADH dehydrogenase. These observations suggest the existence of a respiratory-chain in the chloroplast of higher plants.

Introduction

In eukaryotes photosynthesis takes place in organelles called chloroplasts. The discovery of non-Mendelian mutants of the chloroplast phenotype at the beginning of this century suggested the existence of an extranuclear genetic system in chloroplasts. Since the demonstration of a unique DNA species in *Chlamydomonas* chloroplasts (Sager and Ishida 1963), intensive studies of structure and expression of chloroplast DNA have been made.

The availability of recombinant DNA techniques and nucleotide sequence analyses have simplified the anslysis of gene structures, and have led to an explosion of information on the structure and expression of chloroplast genomes at the molecular level.

To understand more fully the organization and expression of chloroplast genes, we have determined the entire DNA sequence of tobacco (*Nicotiana tabacum*) chloroplast genome (Shinozaki et al. 1986c). We have chosen tobacco for our study (Sugiura and Kusuda 1979) because this plant has been favored for studies of inheritance and evolution (Smith 1974). There are many interspecific hybrids, chloroplast mutants and cell lines with altered chloroplast ribosomes (Hughes 1982). Moreover, because of the recent technical advances in cell and protoplast cultures and protoplast fusion (Takebe and Nagata 1984, Medgyesy et al. 1985), tobacco cells provide a model system for studying somatic cell genetics.

In this article we will discuss the most recent studies on tobacco chloroplast genes for components of the photosynthetic apparatus. Structures of the other genes and gene expression will be also discussed. The nucleotide sequence of the liverwort chloroplast genome has also been determined (Ohyama et al. 1986). This chloroplast genome has essentially the same organization as that of the tobacco chloroplast.

Readers interested in chloroplast genomes are referred to several recent reviews (Dyer 1984, Crouse et al. 1984, 1985, Palmer, 1985, Shinozaki and Sugiura 1986a, Sugiura 1987a,b)

Nicotiana chloroplast DNA

Chloroplast DNAs of higher plants are circular molecules with a size of 120 to 217 kbp. One of the outstanding features of most chloroplast DNAs is the presence of large inverted repeats. These sequences are separated by a large and a small single-copy region. Chloroplast DNAs are known to contain all the chloroplast rRNA genes and tRNA genes and all the genes for proteins which are synthesized in the chloroplast (Dyer 1984, Gray et al 1984).

Molecular cloning techniques have led to well-characterized clone banks of *Nicotiana* chloroplast DNA (Zhu et al. 1982, Fluhr et al. 1983, Hildebrand et al. 1985). The entire *N. tabacum* (var Bright Yellow 4) chloroplast DNA was cloned as a set of overlapping restriction endonuclease fragments in order to sequence the entire molecule (Sugiura et al. 1986). Overlapping DNA fragments which cover the entire genome were essential for this task to ensure, that short restriction fragments were not overlooked. This clone

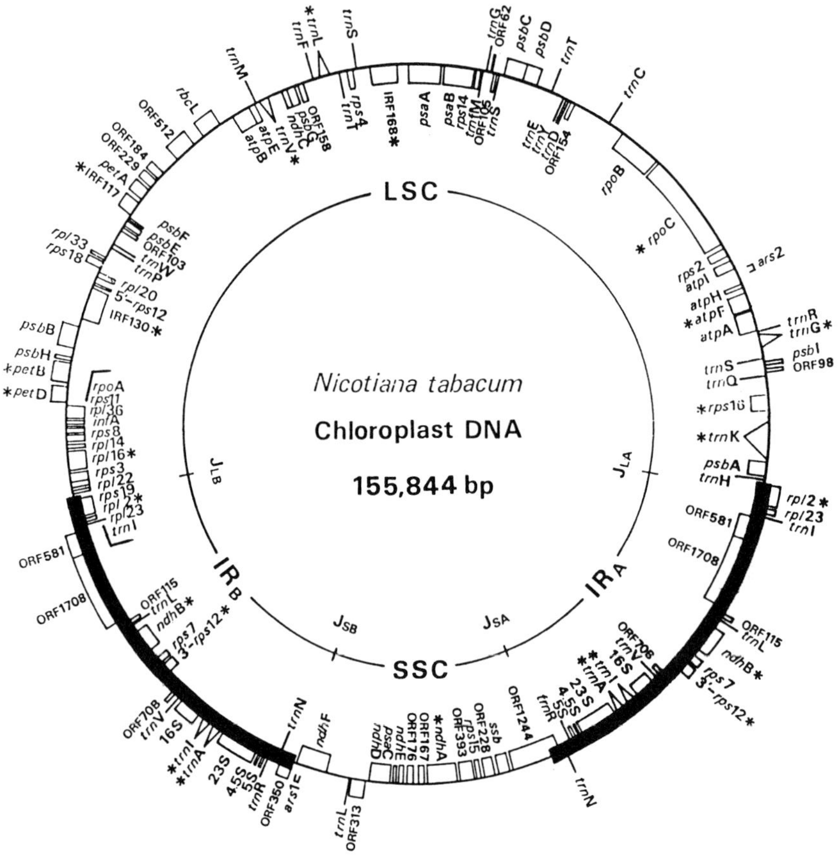

Fig. 1. Circular gene map of the *N. tabacum* chloroplast genome. Genes shown outside the circle are on the A strand and transcribed counterclockwise. Genes shown inside the circle are on the B strand and transcribed clockwise. Asterisks indicate split genes. Major ORFs are included. From Shinozaki et al. (1986d) with minor modifications.

bank was used to determine the complete nucleotide sequence of *N. tabacum* chloroplast DNA (Shinozaki et al. 1986c and d).

The *N. tabacum* chloroplast DNA is divided into four regions (LSC, SSC, IR$_A$ and IR$_B$) (Sugita et al. 1984). The large and small single-copy regions (LSC and SSC, respectively) are 86,684 bp and 18,482 bp long, respectively. The inverted repeats (IR$_A$ and IR$_B$) have been sequenced separately and found to be completely identical (25,339 bp). The entire genome is thus 155,844 bp. Figure 1 shows a circular gene map of the *N. tabacum* chloroplast genome. The DNA strand which codes for the large subunit of ribulose-1,5-bisphosphate carboxylase/oxygenase (RuBisCO) has been desig-

4

Table 1. N. tabacum chloroplast genes.

Genes	Gene products
RNA genes	
rDNA	rRNA (16S, 23S, 4.5S, 5S)
trn	tRNA 30 species
Photosynthesis	
rbcL	Rubisco large subunit
atpA, B, E,	ATP synthetase CF_1 α, β, ε subunits
F, H, I	CF_0 I, III, IV subunits
psaA, B, C,	Photosystem I A1, A2, 9kd proteins
psbA, B, C, D, E,	Photosystem II D1, 51kd, 44kd, D2, Cytb559–9kd,
F, G, H, I	Cytb559–4kd, G, 10kd Pi, I proteins
petA, B, D	Electron transport Cytf, Cytb6, IV subunits
Respiration	
ndhA, B, C, D, E, F	NADH dehydrogenase (ND) subunits 1, 2, 3, 4, 4L, 5
Gene expression	
rps2, 3, 4, 7, 8, 11	30S ribosomal proteins (CS) 2, 3, 4, 7, 8, 11,
12, 14, 15, 16,	12, 14, 15, 16,
18, 19	18, 19
rpl2, 14, 16, 20, 22,	50S ribosomal proteins (CL) 2, 14, 16, 20, 22,
23, 33, 36	23, 33, 36
rpoA, B, C	RNA polymerase α, β, β' subunits
*inf*A	Initiation factor 1
ssb	ssDNA binding protein

nated as A and the complementary strand as B (Deno et al. 1983). The nomenclature for genes follows the proposal of Hallick and Bottomley (1983). The genes for 4 different rRNAs, 30 different tRNAs, 44 different proteins and 9 other predicted protein-coding genes have been localized. Fifteen of these genes contain introns. Blot hybridization analyses revealed that all rRNA, tRNA and protein genes are transcribed in the chloroplast. Table 1 summarizes those chloroplast genes of *N. tabacum* which have been identified.

Genes for components of photosynthesis

Photosynthesis consists of both dark and light reactions. In the dark reaction, a variety of enzymes in the Calvin-Benson cycle function in CO_2 fixation. Among them the large subunit of RuBisCO is the only identified protein encoded in the chloroplast DNA. Thylakoid membranes of higher

Table 2. N. tabacum chloroplast genes for components of the photosynthetic apparatus.

Genes	Gene products	Intron	Transcription	Total amino acids[a]	N-terminal processing[b]	Molecular weight (kd)[c]
	RuBisCO					
*rbc*L	large subunit	–	monocistronic	477	14[1]	51.402 (52.895)
	H+ ATPase					
*atp*A	α subunit	–	polycistronic	507	n.d.	(55.435)
B	β subunit	–	polycistronic	498	n.d.	(53.551)
E	ε subunit	–	polycistronic	133	n.d.	(14.606)
F	I subunit	695bp	polycistronic	184	17[2]	19.084 (20.941)
H	III subunit	–	polycistronic	81	n.d.	(7.990)
I	IV subunit	–	polycistronic	247	18[3]	25.048 (27.001)
	Photosystem I					
*psa*A	A1 subunit	–	polycistronic	750	n.d.	(82.985)
B	A2 subunit	–	polycistronic	734	0 or 1[4]	(82.403)
C	9kd protein	–	polycistronic	81	1[5]	8.907 (9.038)
	Photosystem II					
*psb*A	32kd protein	–	monocistronic	353	C terminus[6,7]	(39.970)
B	51kd protein	–	polycistronic	508	n.d.	(56.011)
C	44kd protein	–	polycistronic	473	n.d.	(51.906)
D	D2 protein	–	polycistronic	353	n.d.	(39.532)
E	cytochrome b559	–	polycistronic	83	1[8]	9.265 (93.906)
F	cytochrome b559	–	polycistronic	39	1[9]	4.353 (4.484)
G	G protein	–	polycistronic	284	n.d.	(32.323)
H	10kd phosphoprotein	–	polycistronic	73	1[10]	7.627 (7.759)
I	I protein	–	polycistronic	52	n.d.	(6.065)
	Electron transfer					
*pet*A	cytochrome f	–	polycistronic	320	35[11]	31.229 (35.244)
B	cytochrome b6	753bp	polycistronic	215	n.d.	(24.135)
D	subunit 4	742bp	polycistronic	160	n.d.	(17.458)

[a] Number of amino acid residues deduced from DNA sequences.
[b] N-terminal amino acid residues removed.
[c] Molecular weights of mature proteins and those of primary products in parentheses.
[1] Amiri et al. 1984; [2] Bird et al. 1985; [3] Fromme et al. 1987; [4] Fish and Bogorad 1986; [5] Hayashida et al. 1987; [6] Marder et al. 1984; [7] Eyal et al. 1987; [8] Widger et al. 1984; [9] Widger et al. 1985; [10] Farchaus and Dilley 1986; [11] Ho and Krogmann 1980.
n.d. = not determined.

plants have five functionally distinct complexes involved in the light reaction (Herrmann et al. 1985). These are the proton-translocating ATPase (ATP synthase), the photosystem I complex (PSI), the photosystem II complex (PSII), the cytochrome b/f complex and the light-harvesting chlorophyll protein complex. The proteins of the latter complex are all nuclear-coded.

Table 2 summarizes chloroplast-encoded proteins involved in photosynthesis. so far 22 genes for proteins involved directly in photosynthesis have been found in the *N. tabacum* chloroplast genome. These genes have been identified by using the corresponding spinach genes as probes (gifts from Dr R.G. Herrmann) and/or through their homology with published nucleotide and amino acid sequences of genes from other chloroplasts and from *Escherichia coli*. Figure 2 shows a map of genes for proteins involved in

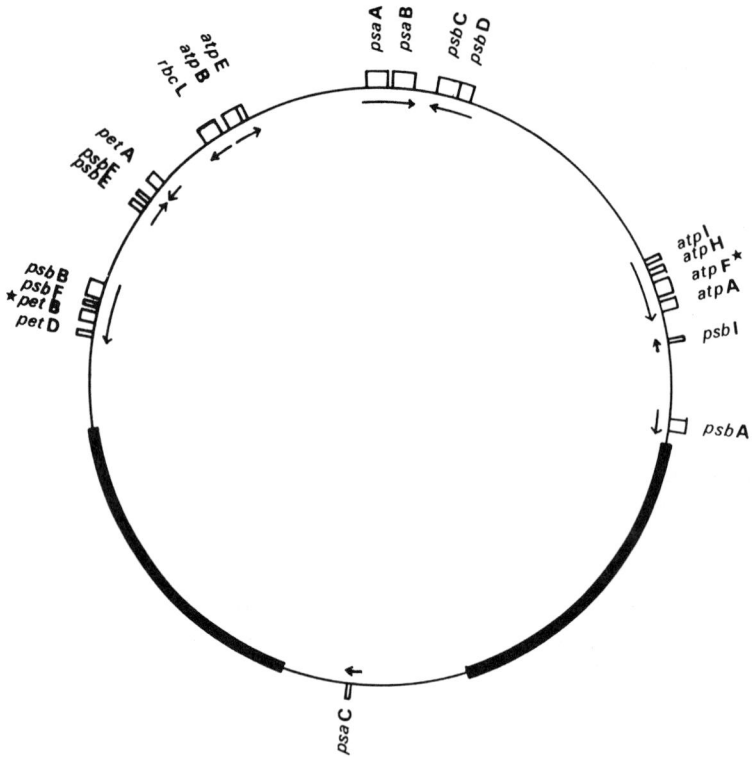

Fig. 2. A map of *N. tabacum* chloroplast genes for components of the photosynthetic apparatus. Arrows indicate directions of transcription. Asterisks indicate genes containing introns. From Shinozaki and Sugiura (1986b) with minor modifications.

photosynthesis. Most of the genes for components of the photosynthetic apparatus are located in LSC except a gene for a 9-kd polypeptide of PSI.

Ribulose-1,5-bisphosphate carboxylase/oxygenase
RuBisCO, or Fraction I protein, is the major soluble protein of chloroplasts. This enzyme is composed of large and small subunits. The large subunit (LS) is encoded in the chloroplast DNA while the small subunit (SS) is encoded in the nuclear DNA. The first *rbc*L gene to be cloned and sequenced was that of maize (McIntosh et al. 1980). In *Nicotiana*, *rbc*Ls for *N. tabacum* (Shinozaki and Sugiura 1982), *N. otophora* and *N. acuminata* (Lin et al. 1986) have been analyzed. *N. tabacum* is the only plant for which the complete amino acid sequence of LS has been determined chemically (Amiri et al. 1984). this sequence is (except for position 284, glycine instead of cysteine) in agreement with that deduced from the DNA sequence. Polymorphism of the *N. tabacum* LS has been found from the amino acid sequence

```
N. tabacum -   MSPQTETKASVGFKAGVKEYKLTYYTPEYQTKDTDILAAFRVTPQPGVPPEEAGAAVAAESSTGTWTTVWTDGLTSLDRYKGRCYRIERVVGEKDQYIAY  100
A. nidulans -  . KQSA..Y      D        DTP  L     FS   AD     I              L  DM    K H  P Q ENS F F              97


              VAYPLDLFEEGSVTNMFTSIVGNVFGFKALRALRLEDLRIPPAYVKTFQGPPHGIQVERDKLNKYGRPLLGCTIKPKLGLSAKNYGRAVYECLRGGLDFT  200
              I            IL      IS      IFVL           L          M                                             197


              KDDENVNSQPFMRWRDRFLFCAEALYKAQAETGEIKGHYLNATAGTCEEMIKRAVFARELGVPIVMHDYLTGGFTANTSLAHYCRDNGLLLHIHRAMHAV  300
                 I    Q        V D IH S        V  P        M E K      I F A        T KW      V                      297


              IDRQKNHGIHFRVLAKALRMSGGDHIHSGTVVGKLEGERDITLGFVDLLRDDFVEQDRSRGIYFTQDWVSLPGVLPVASGGIHVWHMPALTEIFGDDSVL  400
              R       C     L   L            DKAS      M E HI A    VF   A M                 V                      397


              QFGGGTLGHPWGNAPGAVANRVALEACVKARNEGRDLAQEGNEIIREACKWSPELAAACEVWKEIVFNFAAVDVLDK  477
                    T     Q           YR  GD L   G         LDL     K E ETM K ..           472
```

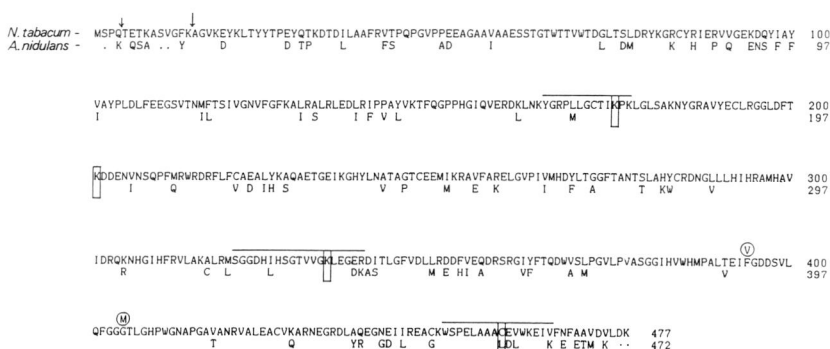

Fig. 3. Comparison of the deduced amino acid sequences of the RuBisCO large subunit of N. tabacum and A. nidulans. Arrows indicate tobacco processing sites (Amiri et al. 1984). Circled amino acid residues show multimeric forms of the N. tabacum LS derived from amino acid sequencing. Overlined and boxed amino acid residues show active site residues. From Sugiura (1987c) with minor modifications.

analysis (Amiri et al. 1984). Figure 3 shows the LS amino acid sequences of N. tabacum and the cyanobacterium Anacystis nidulans 6301 (Shinozaki et al. 1983b). The amino acid sequence of the A. nidulans LS shows 80% homology with that of N. tabacum, while the sequence of the A. nidulans SS shows 40% homology with that of N. tabacum (Shinozaki and Sugiura 1983). The rbcL and rbcS (a gene for SS) are cotranscribed in cyanobacteria (Shinozaki and Sugiura 1983, 1985, Nierzwicki-Bauer et al. 1984). In higher plants rbcL and rbcS are transcribed monocistronically in chloroplasts and nuclei, respectively. Both the rbcL and psbA genes have strong light-inducible promoters (Sugita and Sugiura 1984).

Proton-translocating ATPase complex
ATP synthase or H⁺-ATPase of chloroplasts is an essential component of light-driven ATP synthesis. It consists of two parts, CF_1 and CF_0. CF_1, located on the outer surface of the thylakoid membrane, is composed of five different subunits (α, β, γ, δ and ε). CF_0 is located in the membrane and is composed of four different subunits (I, II, III and IV). The genes for 6 ATP synthase subunits (α, β, ε, I, III and IV) are encoded in the tobacco chloroplast genome (Table 2). Genes for ATP synthase are located separately in two regions of LSC of tobacco chloroplast DNA (Fig. 2). The genes for the β and ε subunits (atpB and atpE) are located upstream from rbcL on the opposite DNA strand (Krebbers et al. 1982, Zurawski et al. 1982b, Shinozaki et al. 1983a). Genes for subunits IV, III, I and α (atpI, atpH, atpF and atpA in this order, are located 40 kbp away from atpB and atpE (Deno et al.

a. atpA

```
        10         20         30         40         50         60         70         80         90        100
MVTIRADEIS NIIRERIEQY NREVKIVNTG TVLQVGDGIA RIHGLDEVMA GELVEFEEGT IGIALNLESN NVGVVLMGDG LLIQEGSSVK ATGRIAQIPV

       110        120        130        140        150        160        170        180        190        200
SEAYLGRVIN ALAKPIDGRG EISASEFRLI ESAAPGIISR RSVYEPLQTG LIAIDSMIPI GRGQRELIIG DRQTGKTAVA TDTILNQQGQ NVICVYVAIG

       210        220        230        240        250        260        270        280        290        300
QKASSVAQVV TTLQERGAME YTIVVAETAD SPATLQYLAP YTGAALAEYF MYRERHTLII YDDPSKQAQA YRQMSLLLRR PPGREAYLGD VFYLHSRLLE

       310        320        330        340        350        360        370        380        390        400
RAAKLSSSLG EGSMTALPIV ETQSGDVSAY IPTNVISITD GQIFLSADLF NSGIRPAINV GISVSRVGSA AQIKAMKQVA GKLKLELAQF AELEAFAQFA

       410        420        430        440        450        460        470        480        490        500
SDLDKATQNQ LARGQRLREL LKQSQSAPLT VEEQIMTIYT GTNGYLDSLE VGQVRKFLVE LRTYLKTNKP QFQEIISSTK TFTEEAEALL KEAIQEQMDR

FILQEQA   507
```

b. atpB

```
        10         20         30         40         50         60         70         80         90        100
MRINPTTSGS GVSTLEKKNP GRVVQIIGPV LDVAFPPGKM PNIYNALVVQ GRDSVGQPIN VACEVQQLLG NNRVRAIAMS ATEGLTRGME VIDTGAPISV

       110        120        130        140        150        160        170        180        190        200
PVGGATLGRI FNVLGEPVDN LGPVDTSTTS PIHRSAPAFI QLDTKLSIFE TGIEVVDLLA PYRRGGKIGL FGGAGVGKTV LIMELINNIA KAHGGVSVFG

       210        220        230        240        250        260        270        280        290        300
GVGERTREGN DLYMEMKESG VINEENIAES KVALVYGQMN EPPGARMRVG LTALTMAEYF RDVNEQDVLL FIDNIFRFVQ AGSEVSALLG RMPSAVGYQP

       310        320        330        340        350        360        370        380        390        400
TLSTEMGSLQ ERITSTKEGS ITSIQAVYVP ADDLTDPAPA TTFAHLDATT VLSRGLAAKG IYPAVDPLDS TSTMLQPRIV GEEHYETAQR VKQTLQRYKE

       410        420        430        440        450        460        470        480        490
LQDIIAILGL DELSEEDRLL VARARKIERF LSQPFFVAEV FTGSPGKYVG LAETIRGFQL ILSGELDGLP EQAFYLVGNI DEATAKAMNL EMESNLKK   498
```

c. atpE

```
        10         20         30         40         50         60         70         80         90        100
MTLNLSVLTP NRIVWDSEVE EIVLSTNSGQ IGILPNHAPI ATAVDIGILR IRLNDQWLTM ALMGGFARIG NNEITVLVND AEKGSDIDPQ EAQQTLELAE

       110        120        130
ANVKKAEGRR QKIEANLALR RARTRVEAIN PIS   133
```

d. atpF

```
        10       | 20         30         40         50         60         70         80         90        100
MKNVTDSFVS LGHWPSAGSF GFNTDILATN PINLSVVLGV LIFFGKGVLS DLLDNRKQRI LNTIRNSEEL RGGAIEQLEK ARSRLRKVES EAEQFRVNGY

       110        120        130        140        150        160        170        180
SEIEREKLNL INSTYKTLEQ LENYKNETIQ FEQQRAINQV RQRVFQQALR GALGTLNSCL NNELHLRTIR SNIGMLGTMK EITD   184
```

e. atpH

```
        10         20         30         40         50         60         70         80
MNPLISAASV IAAGLAVGLA SIGPGVGQGT AAGQAVEGIA RQPEAEGKIR GTLLLSLAFM EALTIYGLVV ALALLFANPF V   81
```

f. atpI

```
        10       |20         30         40         50         60         70         80         90        100
MNVLSCSINT LKGLYDISGV EVGQHFYWQI GGFQVHGQVL ITSWVVIAIL LGSATIAVRN PQTIPTGGQN FFEYVLEFIR DVSKTQIGEE YGPWVPFIGT

       110        120        130        140        150        160        170        180        190        200
MFLFIFVSNW SGALLPWKII QLPHGELAAP TNDINTTVAL ALLTSVAYFY AGLTKKGLGY FGKYIQPTPI LLPINILEDF TKPLSLSFRL FGNILADELV

       210        220        230        240
VVVLVSLVPL VVPIPVMLLG LFTSGIQALI FATLAAAYIG ESMEGHH   247
```

g. psaA

```
        10         20         30         40         50         60         70         80         90        100
MIIRSPEPEV KILVDRDPVK TSFEEWARPG HFSRTIAKGP DTTTWIWNLH ADAHDFDSHT SDLEEISRKV FSAHFGQLSI IFLWLSGMYF HGARFSNYEA

       110        120        130        140        150        160        170        180        190        200
WLSDPTHIGP SAQVVWPIVG QEILNGDVGG GFRGIQITSG FFQIWRASGI TSELQLYCTA IGALVFAALM LFAGWFHYHK AAPKLAWFQD VESMLNHHLA

       210        220        230        240        250        260        270        280        290        300
GLLGLGLSLW AGHQVHVSLP INQFLNAGVD PKEIPLPHEF ILNRDLLAQL YPSFAEGATP FFTLNWSKYA DFLTFRGGLD PVTGGLWLTD IAHHHLAIAI

       310        320        330        340        350        360        370        380        390        400
LFLIAGHMYR TNWGIGHGLK DILEAHKGPF TGQGHKGLYE ILTTSWHAQL SLNLAMLGSL TIVVAHHMYS MPPYPYLATD YGTQLSLFTH HMWIGGFLIV

       410        420        430        440        450        460        470        480        490        500
GAAAHAAIFM VRDYDPTTRY NDLLDRVLRH RDAIISHLNW ACIFLGFHSF GLYIHNDTMS ALGRPQDMFS DTAIQLQPVF AQWIQNTHAL APGATAPGAT

       510        520        530        540        550        560        570        580        590        600
ASTSLTWGGG DLVAVGGKVA LLPIPLGTAD FLVHHIHAFT IHVTALILLK GVLFARSSRL TPDKANLGFR FPCDGPGRGG TCQVSAWDHV FLGLFWMYNA

       610        620        630        640        650        660        670        680        690        700
ISVVIFHFSW KMQSDVWGSV SDQGVVTHIT GGNFAQSSIT INGWLRDFLW AQASQVIQSY GSSLSAYGLF FLGAHFVWAF SLMFLFSGRG YWQELIESIV

       710        720        730        740        750
WAHNKLKVAP ATQPRALSII QGRAVGVTHY LLGGIATTWA FFLARIIAVG   750
```

h. psaB

```
        10         20         30         40         50         60         70         80         90        100
MALRFPRFSQ GLAQDPTTRR IWFGIATAHD FESHDDITEE RLYQNIFASH FGQLAIIFLW TSGNLFHVAW QGNFESWVQD PLHVRPIAHA IWDPHFGQPA

       110        120        130        140        150        160        170        180        190        200
VEAFTRGGAL GPVNIAYSGV YQWWYTIGLR TNEDLYTGAL FLLFLSAISL IAGWLHLQPK WKPSVSWFKN AESRLNHHLS GLFGVSSLAW TGHLVHVAIP

       210        220        230        240        250        260        270        280        290        300
ASRGEYVRWN NFLDVLPHPQ GLGPLFTGQW NLYAQNPDSS SHLFGTAQGA GTAILTLLGG FHPQTQSLWL TDIAHHHLAI AFIFLVAGHM YRTNFGIGHS

       310        320        330        340        350        360        370        380        390        400
MKDLLDAHIP PGGRLGRGHK GLYDTINNSL HFQLGLALAS LGVITSLVAQ HMYSLPAYAF IAQDFTTQAA LYTHHQYIAG FIMTGAFAHG AIFFIRDYNP

       410        420        430        440        450        460        470        480        490        500
EQNEDNVLAR MLEHKEAIIS HLSWASLFLG FHTLGLYVHN DVMLAFGTPE KQILIEPIFA QWIQSAHGKT SYGFDVLLSS TSGPAFNAGR SIWLPGWLNA

       510        520        530        540        550        560        570        580        590        600
VNENSNSLFL TIGPGDFLVH HAIALGLHTT TLILVKGALD ARGSKLMPDK KDFGYSFPCD GPGRGGTCDI SAWDAFYLAV FWMLNTIGWV TFYWHWKHIT

       610        620        630        640        650        660        670        680        690        700
LWQGNVSQFN ESSTYLMGWL RDYLWLNSSQ LINGYNPFGM NSLSVWAWMF LFGHLVWATG FMFLISWRGY WQELIETLAW AHERTPLANL IRWRDKPVAL

       710        720        730
SIVQARLVGL AHFSVGYIFT YAAFLIASTS GKFG   734
```

i. psaC

```
     ↓   10         20         30         40         50         60         70         80
MSHSVKIYDT CIGCTQCVRA CPTDVLEMIP WDGCKAKQIA SAPRTEDCVG CKRCESACPT DFLSVRVYLW HETTRSMGLA Y    81
```

j. psbA

```
        10         20         30         40         50         60         70         80         90        100
MTAILERRES ESLWGRFCNW ITSTENRLYI GWFGVLMIPT LLTATSVFII AFIAAPPVDI DGIREPVSGS LLYGNNIISG AIIPTSAAIG LHFYPIWEAA

       110        120        130        140        150        160        170        180        190        200
SVHEWLYNGG PYELIVLHFL LGVACYMGRE WELSFRLGMR PWIAVAYSAP VAAATAVFLI YPIGQGSFSD GMPLGISGTF NFMIVFQAEH NILMHPFHML

       210        220        230        240        250        260        270        280        290        300
GVAGVFGGSL FSAMHGSLVT SSLIRETTEN ESANEGYRFG QEEETYNIVA AHGYFGRLIF QYASFNNSRS LHFFLAAWPV VGIWFTALGI STMAFNLNGF

       310        320        330        340        350
NFNQSVVDSQ GRVINTWADI INRANLGMEV MHERNAHNFP LDLAAIEAPS TNG   353
```

k. psbB

```
        10         20         30         40         50         60         70         80         90        100
MGLPWYRVHT VVLNDPGRLL SVHIMHTALV AGWAGSMALY ELAVFDPSDP VLDPMWRQGM FVIPFMTHLG ITNSWGGWSI TGGTVTNPGI WSYEGVAGAH

       110        120        130        140        150        160        170        180        190        200
IVFSGLCFLA AIWHWVYWDL EIFCDERTGK PSLDLPKIFG IHLFLSGVAC FGFGAFHVTG LYGPGIWVSD PYGLTGKVQP VNPAWGVEGF DPFVPGGIAS

       210        220        230        240        250        260        270        280        290        300
HHIAAGTLGI LAGLFHLSVR PPQRLYKGLR MGNIETVLSS SIAAVFFAAF VVAGTMWYGS ATTPIELFGP TRYQWDQGYF QQEIYRRVSA GLAENQSLSE

       310        320        330        340        350        360        370        380        390        400
AWSKIPEKLA FYDYIGNNPA KGGLFRAGSM DNGDGIAVGW LGHPIFRDKE GRELFVRRMP TFFETFPVVL VDGDGIVRAD VPFRRAESKY SVEQVGVTVE

       410        420        430        440        450        460        470        480        490        500
FYGGELNGVS YSDPATVKKY ARRAQLGEIF ELDRATLKSD GVFRSSPRGW FTFGHASFAL LFFFGHIWHG ARTLFRDVFA GIDPDLDAQV EFGAFQKLGD

PTTKRQAA   508
```

l. psbC

```
        10         20         30         40         50         60         70         80         90        100
MKTLYSLRRF YHVETLFNGT LALAGRDQET TGFAWWAGNA RLINLSGKLL GAHVAHAGLI VFWAGAMNLF EVAHFVPEKP MYEQGLILLP HLATLGWGVG

       110        120        130        140        150        160        170        180        190        200
PGGEVIDTFP YFVSGVLHLI SSAVLGFGGI YHALLGPETL EESFPFFGYV WKDRNKMTTI LGIHLILLGL GAFLLVFKAL YFGGVYDTWA PGGGDVRKIT

       210        220        230        240        250        260        270        280        290        300
NLTLSPSIIF GYLLKSPFGG EGWIVSVDDL EDIIGGHVWL GSICILGGIW HILTKPFAWA RRALVWSGEA YLSYSLGALS VFGFIACCFV WFNNTAYPSE

       310        320        330        340        350        360        370        380        390        400
FYGPTGPEAS QAQAFTFLVR DQRLGANVGS AQGPTGLGKY LMRSPTGEVI FGGETMRFWD LRAPWLEPLR GPNGLDLSRL KKDIQPWQER RSAEYMTHAP

       410        420        430        440        450        460        470
LGSLNSVGGV ATEINAVNYV SPRSWLATSH FVLGFFFFVG HLWHAGRARA AAAGFEKGID RDFEPVLSMT PLN   473
```

m. psbD

```
        10         20         30         40         50         60         70         80         90        100
MTIALGKFTK DENDLFDIMD DWLRRDRFVF VGWSGLLLFP CAYFAVGGWF TGTTFVTSWY THGLASSYLE GCNFLTAAVS TPANSLAHSL LLLWGPEAQG

       110        120        130        140        150        160        170        180        190        200
DFTRWCQLGG LWTFVALHGA FGLIGFMLRQ FELARSVQLR PYNAIAFSGP IAVFVSVFLI YPLGQSGWFF APSFGVAAIF RFILFFQGFH NWTLNPFHMM

       210        220        230        240        250        260        270        280        290        300
GVAGVLGAAL LCAIHGATVE NTLFEDGDGA NTFRAFNPTQ AEETYSMVTA NRFWSQIFGV AFSNKRWLHF FMLFVPVTGL WMSALGVVGL ALNLRAYDFV

       310        320        330        340        350
SQEIRAAEDP EFETFYTKNI LLNEGIRAWM AAQDQPHENL IFPEEVLPRG NAL   353
```

n. psbE

↓   10   20   30   40   50   60   70   80
MSGSTGERSF ADIITSIRYW VIHSITIPSL FIAGWLFVST GLAYDVFGSP RPNEYFTESR QGIPLITGRF DPLEQLDEFS RSF   83

o. psbF

↓   10   20   30
MTIDRTYPIF TVRWLAVHGL AVPTVFFLGS ISAMQFIQR   39

p. psbG

     10   20   30   40   50   60   70   80   90   100
MGNEFRRIGC ICIYRSFHFR AYLNYWFSLC MAKGGIGMVL APEYSDNKKK NGKNKIETVM NSIQFPLLDR TTQNSVISTT LNDLSNWSRL SSLWPLLYGT

     110   120   130   140   150   160   170   180   190   200
SCCFIEFASL IGSRFDFDRY GLVPRSSPRQ ADLILTAGTV TMKMAPSLVR LYEQMPEPKY VIAMGACTIT GGMFSTDSYS TVRGVDKLIP VDVYLPGCPP

     210   220   230   240   250   260   270   280
KPEAVIDAIT KLRKKISREL YEDRIRSQRA NRCFTTNHKF HVQHSIHTGN YDQRVLYQPP STSEIPTEIF FKYKNSVSSP ELVN   284

q. psbH

↓   10   20   30   40   50   60   70
MATQTVENSS RSGPRRTAVG DLLKPLNSEY GKVAPGWGTT PLMGVAMALF AVFLSIILEI YNSSVLLDGI SMN   73

r. psbI

     10   20   30   40   50
MIYSLFFKKN HLGDCVMLTL KLFVYTVVIF FVSLFIFGFL SNDPGRNPGR ΞE   52

s. petA

     10   20   30   ↓ 40   50   60   70   80   90   100
MQTRNAFSWL KKQITRSISV SLMIYILTRT SISSAYPIFA QQGYENPREA TGRIVCANCH LANKPVEIEV PQAVLPDTVF EAVVRIPYDM QLKQVLANGK

     110   120   130   140   150   160   170   180   190   200
RGGLNVGAVL ILPEGFELAP PDRISPEMKE KIGNLSFQSY RPNKKNILVI GPVPGQKYSE ITFPILSPDP ATKKDVHFLK YPIYVGGNRG RGQIYPDGSK

     210   220   230   240   250   260   270   280   290   300
SNNTVYNATA AGIVSKIIRK EKGGYEITIT DASDGRQVVD IIPPGPELLV SEGESIKFDQ PLTSNPNVGG FGQGDAEIVL QDPLRVQGLL FFLASVILAQ

     310   320
IFLVLKKKQF EKVQLAEMNF   320

t. petB

     10   20   30   40   50   60   70   80   90   100
MSKVYDWFEE RLEIQAIADD ITSKYVPPHV NIFYCLGGIT LTCFLVQVAT GFAMTFYYRP TVTEAFASVQ YIMTEANFGW LIRSVHRWSA SMMVLMMILH

     110   120   130   140   150   160   170   180   190   200
VFRVYLTGGF KKPRELTWVT GVVLAVLTAS FGVTGYSLPW DQVGYWAVKI VTGVPDAIPV IGSPLVELLR GSASVGQSTL TRFYSLHTFV LPLLTAVFML

     210
MHFPMIRKQG ISGPL   215

u. petD

     10   20   30   40   50   60   70   80   90   100
MGVTKKPDLN DPVLRAKLAK GMGHNYYGEP AWPNDLLYIF PVVILGTIAC NVGLAVLEPS MIGEPADPFA TPLEILPEWY FFPVFQILRT VPNKLLGVLL

     110   120   130   140   150   160
MVSVPAGLLT VPFLENVNKF QNPFRRPVAT TVFLIGTAVA LWLGIGATLP IDKSLTLGLF   160

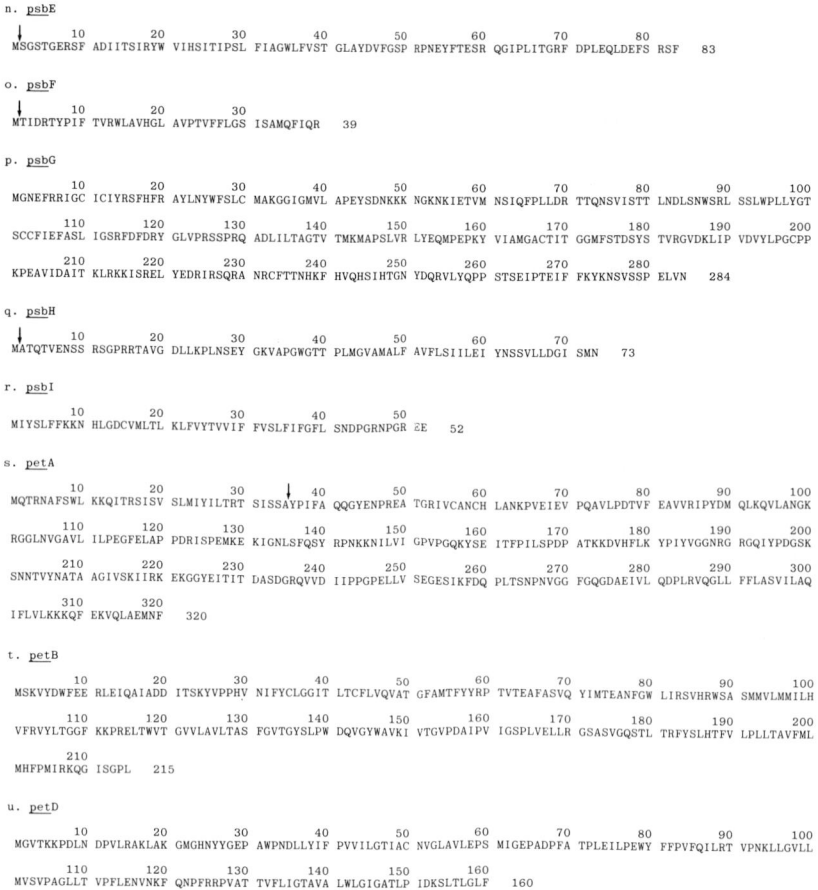

Fig. 4. Amino acid sequences of components of the thylakoid membrane involved in the light-reaction. The amino acid sequences are derived from DNA sequences. Arrows indicate processing sites of primary products.

1983, 1984, Shinozaki et al. 1986b). The *atp*B and *atp*E genes overlap by 4 bp, so that the first two bases of the TGA stop codon of *atp*B and the adenosine preceding it form the ATG initiation codon of *atp*E. These two genes constitute a single operon. The *atp*I, *atp*H, *atp*F and *atp*A genes are cotranscribed.

*Atp*F contains a single 695-bp intron. Amino acid sequences of the α, β, ε, I, III and IV subunits of ATP synthase as deduced from the DNA sequences are shown in Fig. 4. Amino acid sequences of tobacco chloroplast α, β, ε and III subunits have 54%, 62%, 27% and 28% homology, respectively, with the sequences of their corresponding subunits of *E. coli*. Chloroplast subunits I and IV (formally designated as a) have a local sequence

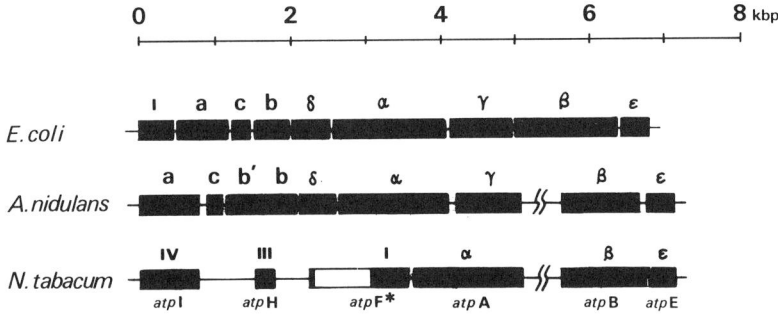

Fig. 5. Organization of the genes for subunits of ATP synthase in *E. coli*, *A. nidulans* and *N. tabacum* chloroplasts.

hmology with the b and a subunits of *E. coli* (Bird et al. 1985, Shinozaki et al. 1986b, Cozens et al. 1986). Recently nucleotide sequences of the genes for the subunits of a cyanobacterial ATP synthase were determined. The gene arrangement resembles that of chloroplasts, i.e., *atp*B-*atp*E and *atp*I-*atp*III-*atp*II-*atp*II′-*atp*D-*atp*A-*atp*C are located far from each other and constitute different operons (Fig. 5; Cozens et al. 1986, Cozens and Walker 1987). In contrast, genes for subunits of ATP synthase of *E. coli* constitute a single operon. These results support the idea of endosymbiotic origin of chloroplast genomes, i.e. chloroplasts were derived through endosymbiosis of ancient cyanobacteria.

Photosystem I complex

At least three components of the PSI complex are encoded in *N. tabacum* chloroplast DNA. Genes for two subunits (A1 and A2) of the P700 apoprotein of PSI (*psa*A and *psa*B) were first found to be located in the middle of LSC of maize chloroplast DNA (Fish et al. 1985). Tobacco *psa*A and *psa*B are also located in the middle of LSC, are separated by a 25-bp spacer sequence, and are cotranscribed (Shinozaki et al. 1986c, Meng et al. unpublished). Amino acid sequences of A1 and A2 have 45% homology (Fig. 4). Recently a gene for a 9-kd polypeptide (an apoprotein for the iron-sulfur centers A and B) of PSI (*psa*C) has been located in the SSC of tobacco chloroplast DNA (Fig. 2) (Hayashida et al. 1987). This gene was identified by comparing the N-terminal amino acid sequence of the spinach 9-kd polypeptide with the entire sequence of the tobacco chloroplast genome. The *psa*C is the first gene for a component of the photosynthetic apparatus to be found in the small single copy region of a chloroplast DNA. The *psa*C is cotranscribed with a gene for ND4 subunit of a putative NADH dehydrogenase (*ndh*D). The predicted polypeptide of *psa*C is rich in cysteine residues and contains a unique repeated sequence.

a

```
            1      2      3      4     5 kbp

         psb B         psb H  pet B      pet D
```

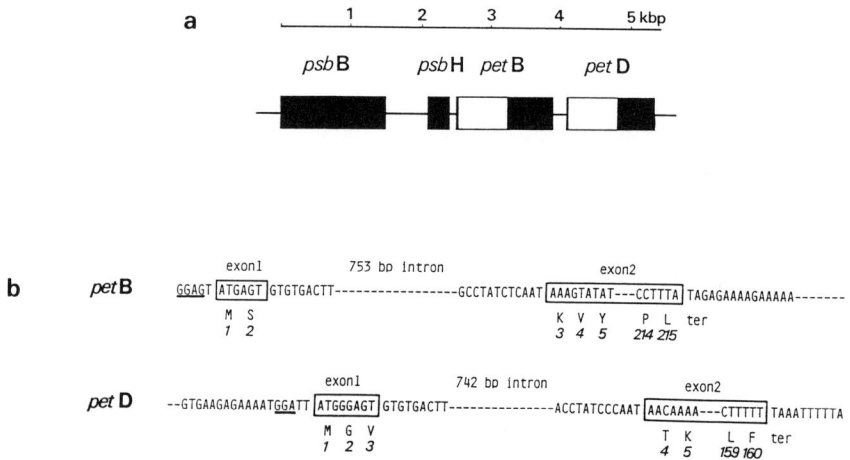

b   petB

```
              exon1         753 bp intron                    exon2
      GGAGT ATGAGT GTGTGACTT----------------GCCTATCTCAAT AAAGTATAT---CCTTTA TAGAGAAAAGAAAAA-------
            M S                                           K V Y      P L  ter
            1 2                                           3 4 5    214 215
```

petD

```
              exon1              742 bp intron                   exon2
   --GTGAAGAGAAAATGGATT ATGGGAGT GTGTGACTT--------------ACCTATCCCAAT AACAAAA---CTTTTT TAAATTTTTA
                        M G V                                        T K     L F  ter
                        1 2 3                                        4 5   159 160
```

Fig. 6. Organization of the N. tabacum gene cluster for psbB, psbH, petB and petD (psbB operon). a. Gene organization; b. the intron–exon boundary sequences of petB and petD. From Tanaka et al. (1987) with minor modifications.

Photosystem II complex

At least 9 components of the PSII complex are encoded in N. tabacum chloroplast DNA (Table 2). The 33-kd, 24-kd and 18-kd polypeptides which function as the water-splitting apparatus are encoded in the nuclear DNA and their cDNAs have been cloned (Tittgen et al. 1986). Tobacco chloroplast genes for PSII components are located at six distinct sites in LSC (Fig. 2).

A gene for the tobacco 32-kd thylakoid membrane protein (psbA) was first sequenced in N. debneyi (Zurawski et al. 1982) and then in N. tabacum (Sugita and Sugiura 1984). The psbA gene is located near the IR$_A$ and transcribed monocistronically. The psbA mRNA is more stable in the dark than that of rbcL although the 32-kd protein is quite unstable and is degraded rapidly in the dark (Edelman et al. 1984). The 32-kd protein is turned over rapidly and present only in very small amounts so that it has been difficult to study its structure. However, nucleotide sequence analyses of psbA enable us to deduce its protein structure. The coding region of psbA is now thought to contain 317 codons (Fig. 4) (Marder et al. 1984, Eyal et al. 1987). The 34-kd precursor polypeptide is synthesized from the psbA mRNA and seems to be processed at its C-terminal end to produce the mature 32-kd protein. The 32-kd protein is now called D1 protein and with D2 and cytochrome b559 proteins is thought to constitute the PSII reaction center (Nanba and Satoh 1987). The 32-kd protein has been found to bind herbicides such as atrazine (Steinback et al. 1981). PsbA genes have been isolated from herbicide-resistant mutants of higher plants (Hirschber and

McIntosh 1983) and *Chlamydomonas* (Erickson et al. 1984). The mutant genes have point mutations resulting in substitution of a specific amino acid residue of the 32-kd protein. Therefore, the *psb*A gene is one of the more important genes in plant breeding.

Genes for a 51-kd protein (*psb*B) and a 10-kd phosphoprotein (*psb*H, formerly designated as *psb*F, Shinozaki et al. 1986c) are located upstream from genes for cytochrome b6 (*pet*B) followed by subunit IV (*pet*D). The four genes, *psb*B, *psb*H, *pet*B and *pet*D, constitute a single operon (*psb*B operon) (Fig. 6) (Westhoff et al. 1983, Shinozaki et al. 1986c). The *pet*B and *pet*D genes contain single introns (Tanaka et al. 1987). Transcription and processing of the *psb*B operon have been investigated in detail in *N. tabacum* (Tanaka et al. 1987) and spinach (Morris and Herrmann 1984, Heinemyer et al. 1984). The intervening sequences of *pet*B and *pet*D are removed very rapidly from the primary transcript. Northern and S1 mapping analyses have shown that the spliced RNA molecules are subjected to a stepwise processing to produce several discrete smaller RNA species. Processing of the primary transcripts to produce multiple mRNAs has been observed in other chloroplast gene clusters e.g., *atp* clusters (Deno et al. 1984) and ribosomal protein gene clusters (Tanaka et al. 1986).

Genes for the 44-kd (*psb*C) and D2 (*psb*D) proteins overlap by 53 bp and are located close to the *psa*A–*psa*B operon (Holschuh et al. 1984, Alt et al. 1984, Rasmussen et al. 1984). Cytochrome b559 has two subunits of 9 kd and 4 kd (Widger et al. 1985). Genes for subunits of cytochrome b559 (*psb*E and *psb*F) are located downstream and on the opposite strand from the gene for cytochrome f and are cotranscribed (Herrmann et al. 1984, Hird et al. 1986, Hayashida et al. unpublished). Genes for G and I proteins were first mapped in maize chloroplast DNA (*psb*G and *psb*I, respectively) (Steinmetz et al. 1986, Kato et al. 1987). These two genes were identified as genes for PSII components by Western blotting analyses using antibodies against synthetic oligopeptides. The *psb*G gene is cotranscribed with a gene for ND3, a subunit of a putative NADH dehydrogenase (*ndh*C) (Matsubayashi et al. 1987). Gene *psb*I corresponds to the ORF2 of 52 codons located between *trn*S and *trn*Q (Deno and Sugiura 1983).

The 51-kd protein has several hydrophobic regions which are probably located in the membrane (Morris and Herrmann 1984). There are several homologous regions between the 51-kd protein and 44-kd protein. The D2 protein also has several homologous regions with the 32-kd protein (Holschuh et al. 1984. Alt et al. 1984). Functions of the G and I proteins have not been determined. Amino acid sequences of PSII components deduced from DNA sequences are summarized in Fig. 4.

Cytochrome b/f complex

The cytochrome b/f complex functions in electron transport between PSII and PSI in higher plants. Six components have been shown to function in the electron transport pathway. Three genes, cytochrome f (*pet*A), cytrochrome b6 (*pet*B) and subunit IV (*pet*D), are encoded in chloroplast DNA (Alt et al. 1983, Alt and Herrmann 1984, Willey et al. 1984a,b, Heinemeyer et al. 1984, Phillips and Gray 1984, Shinozaki et al. 1986c). The remaining three genes, Rieske Fe-S protein (*pet*C), plastocyanin (*pet*E) and ferredoxin-NADP oxidoreductase (*pet*F), are nuclear-coded (Dyer 1984). Plastocyanin and ferredoxin-NADP oxidoreductase function in electron transport but they are not components of the cytochrome b/f complex. The *pet*A gene is located about 4 kbp downstream from *rbc*L on the same strand. Tobacco chloroplast *pet*A is cotranscribed with neighboring ORFs (Hayashida et al. unpublished). After translation, the N-terminal 35 amino acids of cytochrome f are removed to yield the mature cytochrome f protein (Willey et al. 1984a). Cytochrome f has a hydrophobic region at the C-terminus where it binds to the thylakoid membrane (Willey et al. 1984a). Genes *pet*B and *pet*D are cotranscribed with *psb*B and *psb*H (Tanaka et al. 1987). Both *pet*B and *pet*D have single introns of 753 bp and 742 bp, respectively (Fig. 6). The first exons of *pet*B and *pet*D are only 6 bp and 8 bp long, respectively. It is interesting that these mini-exons have been conserved in the chloroplast genome during evolution. Cytochrome b6 and subunit IV of cytochrome b6/f have homologous regions with the mitochondrial cytochrome b (Widger et al. 1984).

Gene for NADH dehydrogenase subunits

Mitochondria and chloroplasts are the major sites for energy conversion in the plant cell. The respiratory-chain of mitochondria contains a large number of electron-carrying proteins that act in sequence to transfer electrons from a reduced substrate to oxygen. In the mitochondrial respiratory chain NADH dehydrogenase (complex I) accepts electrons from NADH and transfers them to ubiquinone. Evidence for an analogous NADH dehydrogenase and a respiratory chain in the chloroplast of *Chlamydomonas reinhardtii* has been presented (Godde 1982, Bennoun 1982). However, the existence of a NADH dehydrogenase has not been reported in higher plant chloroplasts.

Six *N. tabacum* chloroplast DNA sequences whose predicted amino acid sequences resemble those of components (ND1, 2, 3, 4, 4L and 5) of the respiratory-chain NADH dehydrogenase from human mitochondria

Table 3. Amino acid sequence homologies between *N. tabacum* chloroplast NDs and human and *Chlamydomonas* mitochondrial NDs. From Matsubayashi et al. (1987) with minor modifications.

N. tabacum			Human mitochondria		Chlamydomonas mitochondria	
Genes	Subunits	Amino acids	Amino acids	Homology (%) with tobacco NDs	Amino acids	Homology (%) with tobacco NDs
*ndh*A	ND1	364	318	31(42)		
*ndh*B	ND2	387	347	19(30)	301	15(27)
*ndh*C	ND3	120	115	24(48)		
*ndh*D	ND4	509	459	21(34)		
*ndh*E	ND4L	101	98	27(42)		
*ndh*F	ND5	710	603	22(31)	546	22(30)

(Chomyn et al. 1985) have been found (Shinozaki et al. 1986c, Meng et al. 1986) (Table 3). As these sequences are highly transcribed in tobacco chloroplasts, they are likely to be the genes for components of a chloroplast NADH dehydrogenase (*ndh*A, B, C, D, E and F) (Matsubayashi et al. 1987). The *ndh*A, *ndh*D, *ndh*E and *ndh*F genes are located in SSC, *ndh*B is in IRs and *ndh*C is in the middle of LSC (Fig. 7). The *ndh*A and *ndh*B genes contain single introns and their transcripts are spliced very quickly in tobacco leaves (Matsubayashi et al. 1987).

We have shown that sequences homologous to the six tobacco *ndh* genes are located essentially in the same position in rice, sugar beet and broad bean chloroplast genomes (Meng et al. 1986). Similar sequences have also been

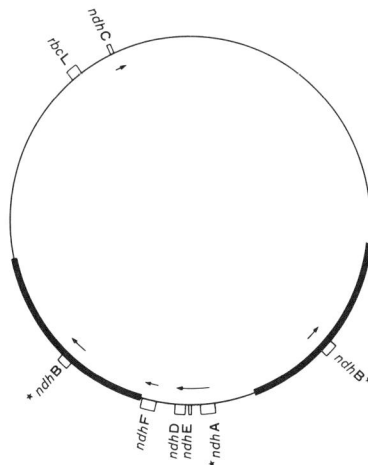

Fig. 7. A map of *N. tabacum* chloroplast genes for subunits of a putative NADH dehydrogenase. Asterisks indicate genes containing introns. From Shinozaki and Sugiura (1986b).

found in the liverwort chloroplast genome (Ohyama et al. 1986). These observations suggest that a respiratory-chain exists in the chloroplast of higher plants. One possibility is that the chloroplast *ndh* sequences are an example of the transposition of genes from the mitochondria genome into the chloroplast genome. However this seems not to be the case as the *N. tabacum* chloroplast ND1 gene (*ndh*A) contains a single intron while the water melon mitochondrial ND1 (URF1) has multiple introns. Moreover, the homology between these two ND1 genes is low (approximately 20%) (Matsubayashi et al. 1987).

Genes for rRNAs, tRNAs and proteins involved in the gene expression

In this section we will summarize the organization and expression of genes for rRNAs, tRNAs and proteins involved in the gene expression.

Four rRNA genes are arranged in the order of 16S, 23S, 4.5S and 5S in each IR. There are consequently two copies of each gene (Kössel et al. 1983). The nucleotide sequences of the rRNA genes have the highest homology with those of *Anacystis nidulans* (Tomioka and Sugiura 1983, Kumano et al. 1983, 1986).

Thirty different tRNA genes have been identified in the *N. tabacum* chloroplast DNA sequence. Seven of them are in the IR and, therefore, the total number of tRNA genes in the chloroplast genome is 37. These 30 differnt tRNAs are likely to be sufficient to read all codons in the tobacco system using 'two out of three' and 'U:N wobble' mechanisms (Shinozaki et al. 1986c, Wakasugi et al. 1986). No tRNA genes code for the 3'-CCA end. Six different tRNA genes contain long single introns (503–2526 bp). All the tRNA sequences show higher homology with the corresponding bacterial tRNAs than with the corresponding eukaryotic cytoplasmic and mitochondrial tRNAs.

Chloroplast ribosomes are of the 70S type and contain 58 to 62 ribosomal proteins (Capel and Bourque 1982). We have found 20 different genes for ribosomal proteins (Table 1) (Shinozaki et al. 1986c, Sugiura et al. 1987). three of these genes are located in IR and, therefore, the total number of ribosomal protein genes is 23. Four different genes contain introns. The deduced amino acid sequences of the chloroplast ribosomal proteins show high sequence homology with those of *E. coli*. Several ribosomal protein gene clusters have gene orders similar to those of *E. coli*. The chloroplast *rpl*23-*rps*11 genes (*rpl*2 operon) are clustered in the same order as the homologous genes in the *E. coli* S10 and *spc* operons (Tanaka et al. 1986).

The 3'-rps12 and rps7 genes are arranged as in the *E. coli str* operon (Fromm et al. 1986, Shinozaki et al. 1986c).

It has been suggested that the RNA polymerase in higher plant chloroplasts is nuclearly encoded (Lerbs et al. 1985). ORF1070 was first designated as a gene for an RNA polymerase subunit, the β subunit (*rpo*B) (Ohme et al. 1986). ORF337 located at the end of the *rpl*2 cluster has sequence homology with the α subunit of *E. coli* RNA polymerase (*rpo*A) (Sijben-Muller et al. 1986). A series of ORFs located downstream from *rpo*B show homology with the gene for the β' subunit of *E. coli* RNA polymerase. These ORFs may constitute a split gene for the β' subunit (*rpo*C) (Shinozaki et al. 1986c). These findings raise the possibility that the chloroplast is an additional site of synthesis of its RNA polymerase subunits (Ohme et al. 1986. Shinozaki et al. 1986c).

Sequence analyses of chloroplast genes for rRNAs, tRNAs and the proteins involved in gene expression have shown that the structures of these genes are homologous with the structure of the analogous prokaryotic genes.

Gene expression

Chloroplast genes are transcribed by the chloroplast RNA polymerase. Some of the genes have been shown to be co-transcribed (e.g. *atp*I-*atp*A, *trn*E-*trn*Y-*trn*D, *atp*B-*atp*E, *rpl*2 cluster, 3'-*rps*12-*rps*7, *psb*B cluster and *rrn*) while others are transcribed monocistronically (e.g. *psb*A, *trn*K, *rps*16, *trn*G-UCC, *trn*V-UAC and *rbc*L).

Transcriptional initiation sites of the *N. tabacum psb*A, *trn*G-UCC, *trn*EYD, *atp*BE, *psb*B and *rbc*L genes have been identified by S1 mapping, Upstream of these sites there are sequences highly homologous to bacterial '-10' and '-35' regions. Essential regions in the spinach *trn*M2, *rbc*L, *atp*BE and *psb*A promoters have been identified experimentally as similar to the prokaryotic '-35' and '-10' regions (Gruissem and Zurawski 1985a,b). Many other genes also contain sequences preceeding coding regions that are similar to prokaryotic promoters. It is likely that these sequences are promoters for these genes, however, they have not yet been defined functionally (Crouse et al. 1984, Kung and Lin 1985). Some genes (e.g. *trn*K, *rps*16, *trn*V-UAC and *rrn*) seem to have multiple promoters. Figure 8 lists promoter sequences of several chloroplast genes. The '-10 region'-like sequences of chloroplast promoters occur 3 to 9 bp upstream from the transcriptional initiationsites, and the distance between the '-10 region' and '-35 region'-like sequences of chloroplast promoters is 17 to 21 bp (Shinozaki and Sugiura 1986a).

```
                              "-35"                                      "-10"
rbcL                                                                                    +1
    Tobacco           ATTGGG TTGCGC TATATATATGAAAGAGTA      TACAAT AATG      ATTAT
    Spinach           GTTGGG TTGCGC CATATATATGAAAGAGTA      TACAAT AATG      ATGTAT
    Maize             TTTGGG TTGCGC TATATCTATCAAAGAGTA      TACAAT AATG      ATGGAT
    Chlamydomonas     TGCTAG TTTACA TTATTTTTTATTTCTAAATA    TATAAT ATATTT    AAATGT
psbA                                                                                    +1
    Tobacco           CCTTGG TTGACA CGAGTATATAAGTCATGT      TATACT GTTG      AATAAA
    Spinach           TATTGG TTGACA CGGGCATATAAGGCATGT      TATACT GTTG      AATAAC
    Soy bean          TATTGG TTGACA CTGGTATATAAGTCATGT      TATACT GTTG      AATAAC
    Mustard           CATTGG TTGACA TGGCTATATAAGTCATGT      TATACT GTTC      AATAAC
atpB                                                                                    +1
    Tobacco           ATAGAA TAGATA ATATGGATGGGATTGTC       TATAAT GATA      GACAAA
    Spinach           TTAGTC TTGACA GTGGTATATGTTGTATATGTA   TATCCT AGA       ꞬGTGAA
    Maize             TCTCTG TTGACA GCAATCTATGCTTCACAG      TAGTAT ATA       TTTTGT
psbB                                                                                    +1
    Spinach           CTCCGA TTGCGT ATTGCTACTTATCGAGTA      TAGAAT AGATTT    GTTTCT
rDNA                                                                                    +1
    Maize             GTGGGA TTGACG TGATAGGGTAGGGTTGGC      TATACT GCTGGT    GGCG
    Spinach           GTGGGA TTGACG TGAGGGGGCAGGGATGGC      TATATT TCTG      GGAGCG
    Duckweed          GTGGGA TTGACG TGATAGGGTAGGGATGGC      TATATT GCTGGGAGC C
trn                                                                                     +1
    Tobacco trnV      GTGGCG TTGAGT TTTCTCGACCCTTTGACT      TAGGAT TAGTC     AGTTCT
    Tobacco trnG      CCCTGT TCGACA AAAGTTGCATTTGTA         TACAAT AATCGGG   ATTGTA
    Tobacco trnE      ATTATA TTGACA ATTTCAAAAAACTGATCA      TACTAT GATCATA   GTATGA


Chloroplast                                                                             +1
conserved sequence        ---ꞡ TTGaca --------------------      TAtAaT a-----      a
                                              17 - 21bp                      3 - 9bp


E. coli                                                                                 +1
conserved sequence        ---- TTGaca --------------------      TAtAaT ------      a
                                              14 - 21bp                      4 - 10bp
```

Fig. 8. Promoter sequences of chloroplast genes. Transcription initiation sites are positioned as + 1. In conserved sequences capital letters show nucleotides which appear at a frequency of over 80% and small letters those appering at a frequency of 50%. From Shinozaki and Sugiura (1986b).

Palindromic sequences which can form stem and loop structures in the RNA transcripts are required for rho-factor independent transcription termination of prokaryotic genes. Transcriptional termination sites of the *psb*A, *trn*EYD, *atp*BE and *rbc*L genes have been identified by S1 mapping. Short inverted repeat sequences have been found preceeding transcription stop sites. Such sites are known to be important for transcription termination in prokaryotic organisms.

Introns in chloroplast genes were first reported for the 23S rDNA of *Chlamydomonas* (Rochaix and Malnoe 1978). Long introns (0.5–2.5 kbp) have also been found in six different tRNA genes and nine different protein genes in *N. tabacum* plants (Shinozaki et al. 1986c, Sugiura et al. 1987). Most of the chloroplast introns have conserved sequences at their boundaries. The conserved boundary sequences are GTGCGNY at the 5′ ends and ATCNRYY(N)YYAY at the 3′ ends (Sugita et al. 1985, Shinozaki et al. 1986a). These conserved boundary sequences resemble those of *Euglena gracilis* chloroplast introns (Hallick et al. 1985). They also resemble the conserved boundary sequences of nuclear gene introns and mitochondrial class II introns.

Tobacco *rps*12 represents an unusual type of gene organization. It is divided into one copy of 5′-*rps*12 and two copies of 3′-*rps*12 (Torazawa et al. 1986, Fromm et al. 1986). The 5′-*rps*12 contains exon 1 of 38 codons. The 3′-*rps*12 consists of an exon 2 of 78 codons, a 536-bp intron and an exon 3 of 7 codons. Northern blot hybridization revealed that both 5′-*rps*12 and 3′-*rps*12 are transcribed in the chloroplasts. Reverse transcription analysis indicated that trans splicing between the 5′-*rps*12 and 3′-*rps*12 transcripts occurs in vivo (Zaita et al. 1987, Sugiura et al. 1987). Trans-spliced CS12 mRNA has also been observed by electron microscopy using an artificial construct consisting of DNAs of exons 1, 2 and 3 (Koller et al. 1987). The 3′-flanking sequence of exon 1 has been designated as 'transon' 1 and the 5′-flanking sequence of exon 2 as 'transon' 2 (Zaita et al. 1987). The sequence GTGCGAC at the 5′-end of transon 1 and the sequence GTCAACTTTTCC at the 3′-end of transon 2 fit the conserved boundary sequences of most of the chloroplast introns. A long complementary structure (about 70 bp) can be constructed between transon 1 and transon 2. This structure may enhance or may be necessary for trans splicing according to results obtained in in vitro experiments (Solnick 1985, Konarska et al. 1985).

References

Alt J and Herrmann RG (1984) Nucleotide sequence of the gene for preapocytochrome *f* in the spinach plastid chromosome. Curr Genet 8: 551–557

Alt J, Morris J, Westhoff P and Herrmann RG (1984) Nucleotide sequence of the clustered genes for the 44 kd chlorophyll *a* apoprotein and the '32 kd'-like protein of the photosystem II reaction center in the spinach plastid chromosome. Curr Genet 8: 597–606

Alt J, Westhoff P, Sears BB, Nelson N, Hurt E, Hauska G and Herrmann RG (1983) Genes and transcripts for the polypeptides of the cytochrome *b6/f* complex from spinach thylakoid membranes. EMBO J 2: 979–986

Amiri I, Salnikow J and Vater J (1984) Amino-acid sequence of the large subunit of D-ribulosebisphosphate carboxylase/oxygenase from *Nicotiana tabacum*. Biochim Biophys Acta 784: 116–123

Bennoun P (1982) Evidence for a respiratory chain in the chloroplast. Proc Natl Acad Sci USA 79: 4352–4356

Bird CR, Koller B, Auffret AD, Huttly AK, Howe CJ, Dyer TA and Gray JC (1985) The wheat chloroplast gene for CF_0 subunit of ATP synthase contains a long intron. EMBO J 4: 1381–1388

Capel MS and Bourque DP (1982) Characterization of *Nicotiana tabacum* chloroplast and cytoplasmic ribosomal proteins. J Biol Chem 257: 7746–7755

Chomyn A, Mariottini, P, Cleeter MWJ, Ragan CI, Matsuno-Yagi A, Hatefi Y, Doolittle RF and Attardi G (1985) Six unidentified reading frames of human mitochondrial DNA encode components of the respiratory-chain NADH dehydrogenase. Nature 314: 592–597

Cozens AL and Walker JE (1987) The organization and sequence of the genes for ATP synthase subunits in the cyanobacterium *Synechococcus* 6301 support for an endosymbiotic origin of chloroplasts. J Mol Biol 194: 359–383

Cozens AL, Walker JE, Phillips AL, Huttly AK and Gray JC (1986) A sixth subunit of ATP synthase, an F_0 component, is encoded in the pea chloroplast genome. EMBO J 5: 217–222

Crouse EJ, Bohnert HJ and Schmitt JM (1984) Chloroplast RNA synthesis. In: Ellis RJ (ed.) Chloroplast Biogenesis, pp 83–136. Cambridge: Cambridge University Press

Crouse EJ, Schmitt JM and Bohnert HJ (1985) Chloroplast and cyanobacterial genomes, genes and RNAs: a compilation. Plant Mol Biol Rep 3: 43–89

Deno H, Shinozaki K and Sugiura M (1983) Nucleotide sequence of tobacco chloroplast gene for the α subunit of proton-translocating ATPase. Nucl Acids Res 11: 2185–2191

Deno H, Shinozaki K and Sugiura M (1984) Structure and transcription pattern of a tobacco chloroplast gene coding for subunit III of proton-translocating ATPase. Gene 32: 195–201

Deno H and Sugiura M (1983) The nucleotide sequences of tRNA^{Ser}(GCU) and tRNA^{Gln}(UUG) genes from tobacco chloroplasts. Nucl Acids Res 11: 8407–8414

Dyer TA (1984) The chloroplast genome: Its nature and role in development. In: Baker NR and Barber J (eds) Chloroplast Biogenesis, pp 23–69. Amsterdam: Elsevier

Edelman M, Mattoo K and Marder JB (1984) Three hats of the rapidly metabolized 32 kd protein of thylakoids. In: Ellis RJ (ed.) Chloroplast Biogenesis, pp 283–302. Cambridge: Cambridge University Press

Erickson JM, Rahire M, Bennoun P, Delepelaire P, Diner B and Rochaix JD (1984) Herbicide resistance in *Chlamydomonas reinhardii* results from a mutation in the chloroplast gene for the 32-kilodalton protein of photosystem II. Proc Natl Acad Sci USA 81: 3617–3621

Eyal Y, Goloubinoff P and Edelman M (1987) The amino terminal region delimited by Met, and Met37 is an integral part of the 32 kDa herbicide binding protein. Plant Mol Biol 8: 337–343

Farchaus J and Dilley RA (1986) Purification and partial sequence of the Mr 10,000 phosphoprotein from spinach thylakoids. Arch Biochem Biophys 244: 94–101

Fish LE and Bogorad L (1986) Identification and analysis of the maize P700 chlorophyll *a* apoproteins PSI-A1 and PSI-A2 by high pressure liquid chromatography analysis and partial sequence determination. J Biol Chem 261: 8134–8139

Fish LE, Kuck U and Bogorad L (1985) Two partially homologous adjacent light-inducible maize chloroplast genes encoding polypeptides of the P700 chlorophyll *a*-protein complex of photosystem I. J Biol Chem 260: 1413–1421

Fluhr R, Fromm H and Edelman M (1983) Clone bank of *Nicotiana tabacum* chloroplast DNA: mapping of the alpha, beta and epsilon subunits of the ATPase coupling factor, the large subunit of ribulosebisphosphate carboxylase, and the 32-kDal membrane protein. Gene 25: 271–280

Fromm H, Edelman M, Koller B, Goloubinoff P and Galun E (1986) The enigma of the gene coding for ribosomal protein S12 in the chloroplast of *Nicotiana*. Nucl Acids Res 14: 883–898

Fromme P, Graber P and Salnikow J (1987) Isolation and identification of a fourth subunit in the membrane part of the chloroplast ATP-synthase. FEBS 218: 27–30

Godde D (1982) Evidence for a membrane bound NADH-plastoquinone-oxidoreductase in *Chlamydomonas reinhardii* CW-15. Arch Microbiol 131: 197–202

Gray JC, Phillips AL and Smith AG (1984) Protein synthesis by chloroplasts. In: Ellis RJ (ed.) Chloroplast Biogenesis, pp 137–163. Cambridge: Cambridge University Press

Gruissem W and Zurawski G (1985a) Identification and mutational analysis of the promoter for a spinach chloroplast transfer RNA gene. EMBO J 4: 1637–1644

Gruissem W and Zurawski G (1985b) Analysis of promoter regions for the spinach chloroplast *rbc*L, and *atp*B and *psb*A genes. EMBO J 4: 3375–3383

Hallick RB and Bottomley W (1983) Proposals for the naming of chloroplast genes. Plant Mol Biol Rep 1: 38–43

Hallick RB, Gingrich JC, Johanningmeier U and Passavant CW (1985) Introns in *Euglena* and *Nicotiana* chloroplast protein genes. In: Van Vloten-Doting L, Groot GSO and Hall TC (eds) Molecular Form and Function of the Plant Genome, pp 211–231. New York: Plenum Press

Hayashida N, Matsubayashi T, Shinozaki K, Sugiura M, Inoue K and Hiyama T 91987) The gene for the 9 kd polypeptide, a possible apoprotein for the iron-sulfur centers A and B of the photosystem I complex, in tobacco chloroplast DNA. Curr Genet 12: 247–250

Heinemeyer W, Alt J and Herrmann RG (1984) Nucleotide sequence of the clustered genes for apocytochrome b6 and subunit 4 of the cytochrome b/f complex in the spinach plastid chromosome. Curr Genet 8: 543–549

Herrmann RG, Alt J, Schiller B, Widger WR and Cramer WA (1984) Nucleotide sequence of the gene for apocytochrome b-559 on the spinach plastid chromosome: implications for the structure of the membrane protein. FEBS Lett 176: 239–244

Herrmann RG, Westhoff P, Alt J, Tittgen J and Nelson N (1985) Thylakoid membrane protein and their genes. In: Van Vloten-Doting L, Groot GSP and Hall TC (eds) Molecular Form and Function of the Plant Genome, pp 233–256. New York: Plenum Press

Hildebrand M, Jurgenson JE, Ramage RT and Bourque DP (1985) Derivation of a physical map of chloroplast DNA from *Nicotiana tabacum* by two-dimensional gel and computer-aided restriction analysis. Plasmid 14: 64–79

Hird SM, Willey DL, Dyer TA and Gray JC (1986) Location and nucleotide sequence of the gene for cytochrome b559 in wheat chloroplast DNA. Mol Gen Genet 203: 95–100

Hirschberg J and McIntosh L (1983) Molecular basis of herbicide resistance in *Amaranthus hybridus*. Science 222: 1346–1349

Ho KK and Krogmann DW (1980) Cytochrome *f* from spinach and cyanobacteria: Purification and characterization. J Biol Chem 255: 3855–361

Holschuh K, bottomley W and Whitfeld PR (1984) Structure of the spinach chloroplast genes for the D2 and 44 kd reaction-centr proteins of photosystem II and for tRNA^{Ser}(UGA). Nucl Acids Res 12: 8819–8834

Hughes KW (ed) (1982) Plant Mol Biol Newsletter 3: No 2, a special issue for *Nicotiana*

Kato K, Sayer RT and Bogorad L (1987) Expression of genes for PSII and PSII omponents in maize chloroplasts. Proc Ann Meeting Jpn Soc Plant Physiol, p 208. Urawa: (in Japanese)

Koller B, Fromm H, Galun E and Edelman M (1987) Evidence for in vivo trans splicing of pre-mRNAs in tobacco chloroplasts. Cell 48: 111–119

Konarska MM, Padgett RA and Sharp PA (1985) Trans splicing of mRNA precursors in vitro. Cell 42: 165–171

Kössell H, Edwards K, Fritzsche E, Koch W and Schwarz Zs (1983) Phylogenetic significance of nucleotide sequence analysis. In: Jensen U and Fairbrothers DE (eds) Proteins and Nucleic Aids in Plant Systematics, pp 36–57. Berlin: Springer-Verlg

Krebbers ET, Larrinua IM, McIntosh L and Bogorad L (1982) The maize chloroplast genes for the β and ε subunits of the photosynthetic coupling factor CF_1 are fused. Nucl Acids Res 10: 4985–5002

Kumano M, Tomioka N, Shinozaki K and Sugiura M (1986) Analysis of the promoter regions in the *rrn*A operon from a blue-green alga, *Anacystis nidulans* 6301. Mol Gen Genet 202: 173–178

Kumano M, tomioka N and Sugiura M (1983) the complete nucleotide sequence of a 23S rRNA gene from a glue-green alga, *Anacystis nidulans*. Gene 24: 219–225

Kung SD and Lin CM (1985) Chloroplast promoters from higher plants. Nucl Acids Res 13: 7543–7549

Lerbs S, Brautigam E and Parthier B (1985) Polypeptides of DNA-dependent RNA polymerase of spinach chloroplast: characterization by antibody-linked polymerase assay and determination of sites synthesis. EMBO J 444: 1661–1666

Lin CM, Liu ZQ and Kung SD (1986) *Nicotiana* chloroplast genome: X. Correlation between the DNA sequences and the isoelectric focusing patterns of the LS of Rubisco. Plant Mol Biol 6: 81–87

Marder JB, Goloubinoff P and Edelman M(1984) Molecular architeture of the rapidly metabolized 32-kilodalton protein of photosystem II. J Biol Chem 259: 3900–3908

Matsubayashi T, Wakasugi T, Shinozaki K, Yamaguchi-Shinozaki K, Zaita N, Hidaka T, Meng BY, Ohto C, Tanaka M, Kato A, Maruyama T and Sugiura M (1987) Six chloroplast gene (ndhA to F) homologous to human mitochondrial genes encoding components of the respiratory-chain NADH dehydrogenase are actively expressed: Determination of the splice-sites in ndhA and ndhB pre-mRNAs. Mol Gen Genet 210: 385–393

McIntosh L, Poulsen C and Bogorad L (1980) Chloroplast gene sequence for the large subunit of ribulose bisphosphate carboxylase of maize. Nature 288: 556–560

Medgyesy P, Fejes E and Maliga P (1985) Interspecific chloroplast recombination in a Nicotiana somatic hybrid. Proc Natl Acad Sci USA 82: 6960–6964

Meng BY, Matsubayashi T, Wakasugi T, Shinozaki K, Sugiura M, Hirai A, Mikami T, Kishima Y and Kinoshita T (1986) Ubiquity of the genes for components of a NADH dehydrogenase in higher plant chloroplast genomes. Plant Sci 47: 181–184

Morris J and Herrmann RG (1984) Nucleotide sequence of the gene for the P680 chlorophyll a apoprotein of the photosystem II reaction center from spinach. Nucl Acids Res 12: 2837–2850

Nanba O and Satoh K (1987) Isolationof a photosystem II reaction center consisting of D-1 and D-2 polypeptides and cytochrome b-559. Proc Natl Acad Sci USA 84: 109–112

Nierzwicki-Bauer SA, Curtis SE and Haselkorn R (1984) Cotranscription of gene encoding the small and large subunits of ribulose-1,5-bisphosphate carboxylase in cyanobacterium Anabaena 7120. Proc Natl Acad Sci USA 81: 5961–5965

Ohme M, Tanaka M, Chunwongse J, Shinozaki K and Sugiura M (1986) A tobacco chloroplast DNA sequence possibly coding for a polypeptide similar to E. coli RNA polymerase β subunit. FEBS Lett 200: 87–90

Ohyama K, Fukuzawa H, Kohchi T, Shirai H, Sano T, Sano S, Umesono K, Shiki Y, Takeuchi M, Chang Z, Aota S, Inokuchi H and Ozeki H (1986) Chloroplast gene organization deduced from complete sequence of liverwort Marchantia polymorpha chloroplast DNA. Nature 322: 572–574

Palmer JD (1985) Comparative organization of chloroplast genomes. Annu Rev genet 19: 325–354

Phillips AL and Gray JC 91984) Location and nucleotide sequence of the gene for the 15.2 kDa polypeptide of the cytochrome b-f complex from pea chloroplast. Mol Gen Genet 194: 477–484

Rasmussen OF, Bookjans G, Strummann M and Henningsen KH (1984) Localization and nucleotide sequence of the gene for the membrane polypeptide D2 from pea chloroplast DNA. Plant Mol Biol 3: 191–199

Rochaix JD and Malnoe P (1978) Anatomy of the chloroplast ribosomal DNA of Chlamydomonas reinhardtii. Cell 15: 661–670

Sager R and Ishida MR (1963) Chloroplast DNA in Chlamydomonas. Proc Natl Acad Sci USA 50: 725–730

Shinozaki K, Deno H, Kato A and Sugiura M (1983a) Overlap and cotranscription of the genes for the beta and epsilon subunits of tobacco chloroplast ATPase. Gene 24: 147–155.

Shinozaki K, Deno H, Sugita M, Kuramitsu S and Sugiura M (1986a) Intron in the gene for the ribosomal proteins S16 of tobacco chloroplast and its conserved boundary sequences. Mol Gen Genet 202: 1–5

Shinozaki K, Deno H, Wakasugi T and Sugiura M (1986b) Tobacco chloroplast gene coding for subunit I of proton-translocating ATPase: comparison with the wheat subunit I and E. coli subunit b. Curr Genet 10: 421–423

Shinozaki K, Ohme M, Tanaka M, Wakasugi T, Hayashida N, Matsubayashi T, Zaita N, Chunwongse J, Obokata J, Yamaguchi-Shinozaki K, Ohto C, Torazawa K, Meng BY,

Sugita M, Deno H, Kamogashira T, Yamada K, Kusuda J, Takaiwa F, Kato A, Tohdoh N, Shimada H and Sugiura M (1986c) The complete nucleotide sequence of the tobacco chloroplast genome: its gene organization and expression. EMBO J 5: 2043–2049

Shinozaki K, Ohme M, Tanaka M, Wakasugi T, Hayashida N, Matsubayashi T, Zaita N, Chunwongse J, Obokata J, Yamaguchi-Shinozaki K, Ohto C, Torazawa K, Meng BY, Sugita M, Deno H, Kamogashira T, Yamada K, Kusuda J, Takaiwa F, Kato A, Tohdoh N, Shimada H and Sugiura M (1986d) The complete nucleotide sequence of tobacco chloroplast genome. Plant Mol Biol Rep 4: 111–147

Shinozaki K and Sugiura M (1982a) The nucleotide sequence of the tobacco chloroplast gene for the large subunit of ribulose-1,5-bisphosphate carboxylase/oxygenase. Gene 20: 91–102

Shinozaki K and Sugiura M (1982b) Sequence of the intercistronic region between the ribulose-1,5-bisphosphate carboxylase/oxygenase large subunit and the coupling factor β subunit genes. Nucl Acids Res 10: 4923–4934

Shinozaki K and Sugiura M (1983) The gene for the small subunit of ribulose-1,5-bisphosphate carboxylase/oxygenase is located close to the gene for the large subunit in the cyanobacterium *Anacystis nidulans* 6301. Nucl Acids Res 11: 6957–6964

Shinozaki K and Sugiura M (1985) Genes for the large and small subunits of ribulose-1,5-bisphosphate carboxylase/oxygenase construct a single operon in a cyanobaterium *Anacystis nidulans* 6301. Mol Gen Genet 200: 27–32

Shinozaki K and Sugiura M (1986a) Organization of chloroplast genomes. Adv Biophys 21: 57–78

Shinozaki K and Sugiura M (1986b) Structure and expression of chloroplast genomes (in Japanese). Jpn J Genet 61: 371–409

Shinozaki K, Yamada C, Takahata N and Sugiura M (1983b) Molecular cloning and sequence analysis of the cyanobacterial gene for the large subunit of ribulose-1,5-bisphosphate carboxylase/oxygenase. Proc Natl Acad Sci USA 80: 4050–4054

Sijben-Muller G, Hallick RB, Alt J, Westhoff P and Herrmann RG (1986) Spinach plastid genes coding for initiation factor IF-1, ribosomal protein S11 and RNA polymerase α-subunit. Nucl Acids Res 14: 1029–1044

Smith HH (1974) *Nicotiana*. In: King RC (ed.) Handbook of Genetics 2, pp 281–314. New York: Plenum Press

Solnick D (1985) Trans splicing of mRNA precursors. Cell 42: 157–164

Steinback KE, McIntosh L, Bogorad L and Arntzen CJ (1981) Identification of the triazine receptor protein as a chloroplast gene product. Proc Natl Acad Sci USA 78: 7463–7467

Steinmetz AA, Castroviejo M, Sayre RT and Bogorad L (1986) Protein PSII-G: An additional component of photosystem II identified through its plastid gene in maize. J Biol Chem 26-: 2485–2488

Sugita M, Kato A, Shimada H and Sugiura M (1984) Sequence analysis of the junctions between a large inverted repeat and single-copy regions in tobacco chloroplast DNA. Mol Gen Genet 194: 200–205

Sugita M, Shinozaki K and Sugiura M (1985) Tobacco chloroplast tRNA^{Lys}(UUU) gene contains a 2.5-kilobase-pair intron: An open reading frame and a conserved boundary sequence in the intron. Proc Natl Acad Sci USA 82: 3557–3561

Sugita M and Sugiura M (1984) Nucleotide sequence and transcription of the gene for the 32,000 dalton thylakoid membrane protein from *Nicotiana tabacum*. Mol Gen Genet 195: 308–313

Sugiura M (1987a) Organization and expression of chloroplast genome. In: Kung SD and Arntzen CJ (eds) Plant Biotechnology. Stoneham: Butterworth Publishers, in press

Sugiura M (1987b) Structure and function of the tobacco chloroplast genome. Tokyo: Bot Mag Tokyo 100: 407–436

Sugiura M (1987c) Chloroplast DNA. In: Akazawa T, Sugiura M and Nishimura M (eds) Methods of Plant Cell Biology (in Japanese), pp 288–307. Tokyo: Kyoritsu Shuppan Co

Sugiura M and Kusuda J (1979) Molecular cloning of tobacco chloroplast ribosomal RNA genes. Mol Gen Genet 172: 137–141

Sugiura M, Shinozaki K, Tanaka M, Hayashida N, Wakasugi T, Matsubayashi T, Ohto C, torazawa K, Meng BY, Hidaka T and Zaita N (1987) split genes and cis/trans splicing in tobacco chloroplasts. In: Von Wettstein D and Chua NH, (eds) Plant Molecular Biology, pp 65–76. New York: Plenum Press

Sugiura M, Shinozaki K, Zaita N, Kusuda M and Kumano M (1986) Clone bank of the tobacco (*Nicotiana tabacum*) chloroplast genome as a set of overlapping restriction endonuclease fragments: mapping of eleven ribosomal protein genes. Plant Sci 44: 211–216

Takebe I and Nagata T (1984) Isolation and culture of protoplasts: tobacco. In: Vasil IK (ed) Cell Culture and Somatic Cell Genetics of Plants, pp 328–339. New York: Academic Press

Tanaka M, Obokata J, Chunwongse J, Shinozaki K and Sugiura M (1987) Rapid splicing and stepwise processing of a transcript from the *psb*B operon in tobacco chloroplasts: Determination of the intron sites in *pet*B and *pet*D. Mol Gen Genet 209: 427–431

Tanaka M, Wakasugi T, Sugita M, Shinozaki K and Sugiura M (1986) Genes for the eight ribosomal proteins are clustered on the chloroplast genome of tobacco (*Nicotiana tabacum*): Similarity to the S10 and *spc* operons of *Escherichia coli*. Proc Natl Acad Sci USA 83: 6030–6034

Tittgen J, Hermans J, Steppuhn J, Jansen T, Jansson C, Andersson B, Nuchushtai R, Nelson N and Herrmann RG (1986) Isolation of cDNA clones for fourteen nuclear-encoded thylakoid membrane proteins. Mol Gen Genet 204: 258–265

Tomioka N and Sugiura M (1983) The complete nucleotide sequence of a 16S ribosomal RNA gene from a blue-green alga, *Anacystis nidulans*. Mol Gen Genet 191: 46–50

Torazawa K, Hayashida N, Obokata J, Shinozaki K and Sugiura M (1986) The 5′ part of the gene for ribosomal protein S12 is located 30 kbp downstream from its 3′ part in tobacco chloroplast genome. Nucl Acid Res 14: 3143

Wakasugi T, Ohme M, Shinozaki K and Sugiura M (1986) Structures of tobacco chloroplast genes for tRNA^{Ile}(CAU), tRNA^{Leu}(CAA), tRNA^{Cys}(GCA), tRNA^{Ser}(UGA) and tRNA^{Thr}(GGU): a compilation of tRNA genes from tobacco chloroplasts. Plant Mol Biol 7: 385–392

Westhoff P, Alt J and Herrmann RG (1983) Localization of the genes for the two chrorophyll *a*-conjugated polypeptides (mol. wt. 51 and 44 kd) of the photosystem II reaction center on the spinach plastid chromosome. EMBO J 2: 2229–2237

Widger WR, Cramer WA, Hermodson M and Herrmann RG (1985) Evidence for a hetero-oligomeric structure of the chloroplast cytochrome *b*-559. FEBS Lett 191: 186–190

Widger WR, Cramer WA, Hermodson M, Meyer D and Gullifor M (1984) Purification and partial amino acid sequence of the chloroplast cytochrome *b*-559. J Biol Chem 259: 3870–3876

Willey DL, Auffret AD and Gray JC (1984a) Structure and topology of cytochrome f in pea chloroplast membranes. Cell 36: 555–562

Willey DL, Howe CJ, Auffret AD, Bowman CM, Dyer TA and Gray JC (1984b) Location and nucleotide sequence of the gene for cytochrome *f* in wheat chloroplast DNA. Mol Gen Genet 194: 416–422

Zaita N, Torazawa K, Shinozaki K and Sugiura M (1987) Trans splicing in vivo: joining of transcripts from the 'divided' gene for ribosomal protein S12 in the chloroplasts of tobacco. FEBS Lett 210: 153–156

Zhu YS, Duvall EJ, Lovett PS and Kung SD (1982) *Nicotiana* chloroplast genome: V. Consturction, mapping and expression of clone library of *N. otophora* chloroplast DNA. Mol Gen Genet 187: 61–66

Zuraswski G, Bohnert HJ, Whitfeld PR and Bottomley W (1982a) Nucleotide sequence of the gene for the M_R 32,000 thylakoid membrane protein from *Spinacia oleracea* and *Nicotiana debneyi* predicts a totally concerved primary translation product of M_R 38,950. Proc Natl Acad Sci USA 79: 7699–7703

Zuraswski G, Bottomley W and Whitfeld PR (1982b) Structures of the gene for the β and e subunits of spinach chloroplast ATPase indicate a dicistronic mRNA and overlapping translation stop/start signal. Proc Natl Acad Sci USA 79: 6260–6264

Govindjee et al. (eds), Molecular Biology of Photosynthesis: 27–42
© 1988 Kluwer Academic Publishers

Minireview

Gene organization and newly identified groups of genes of the chloroplast genome from a liverwort, *Marchantia polymorpha*

KANJI OHYAMA,[1]* TAKAYUKI KOHCHI,[1] HIDEYA FUKUZAWA,[1]** TOHRU SANO,[1] KAZUHIKO UMESONO[2] & HARUO OZEKI[2]

[1]Research Center for Cell and Tissue Culture, Faculty of Agriculture, Kyoto University, Kyoto 606, Japan; [2]Department of Biophysics, Faculty of Science, Kyoto University, Kyoto 606, Japan; To whom all correspondence should be sent; **Present address: Institute of Applied Microbiology, University of Tokyo, Tokyo 113, Japan*

Received 28 September 1987; accepted 1 December 1987

Key words: chloroplasts, iron-sulfur protein, membrane transport protein, *Marchantia polymorpha*, NADH(PQ) oxidoreductase

Abstract. The complete nucleotide sequence of chloroplast DNA from a liverwort, *Marchantia polymorpha* has made clear the entire gene organization of the chloroplast genome. Quite a few genes encoding components of photosynthesis and protein synthesis machinery have been identified by comparative computer analysis. Other genes involved in photosynthesis, respiratory electron transport, and membrane-associated transport in chloroplasts were predicted by the amino acid sequence homology and secondary structure of gene products. Thirty-three open reading frames in the liverwort chloroplast genome remain unidentified. However, most of these open reading frames are also conserved in the chloroplast genomes of two species, a liverwort, *Marchantia polymorpha*, and tobacco, *Nicotiana tabacum*, indicating their active functions in chloroplasts.

Abbreviations: bp – base pair, kDa – kilodalton, IR – inverted repeat, ORF – open reading frame, DALA – δ-aminolevulinate

Introduction

The complete nucleotide sequences of chloroplast DNA from two green plants, a liverwort, *Marchantia polymorpha* (Ohyama et al. 1986), and tobacco, *Nicotiana tabacum* (Shinozaki et al. 1986), are known. The overall gene composition of both organisms is similar despite the large difference in genome size. Now we are able to integrate information on chloroplast photosynthetic mechanisms with the structure and function of chloroplast genes (Ellis 1981, Steinback et al. 1985). In this article, we will describe the overall gene organization and newly identified groups of genes, of which some are unique to liverwort, for proteins in photosynthesis, electron transport, and membrane-associated transport systems in chloroplasts.

Overall gene organization

The complete nucleotide sequence of the liverwort chloroplast DNA consists of 121,024 base pairs (bp) in a double-stranded circular form with a set of large inverted repeats (IR_A and IR_B, each of 10,058 bp), a large single copy region (LSC, 81,095 bp), and a small single copy region (SSC, 19,813 bp). The total G + C content of the liverwort chloroplast genome is 28.8% (G + C, 69,726; A + T, 172,322). The coding sequences for stable RNA genes have a higher G + C content, 52.6% for tRNAs and 52.1% for rRNAs. On the other hand, the coding sequences for proteins have 28.5% G + C, and the spacer regions between the coding sequences have much less G + C (19.5%). We have detected 136 possible genes of which 103 gene products are related to known stable RNAs or proteins. Stable RNA genes for four species of ribosomal RNAs (rRNAs) and 32 species of transfer RNAs (tRNAs) (one of them, proline-like tRNA, is a pseudogene) have been located on the liverwort chloroplast genome. Twenty-five genes had structural similarity to three subunits (α, β, and β' subunits) of *Escherichia coli* RNA polymerase, of which β' subunit gene in *E. coli* may have been split into two liverwort chloroplast genes (*rpoC1* and *rpoC2*), 19 ribosomal proteins (*rpl* and *rps* genes), and two related proteins (*infA* and *secX* genes). Twenty genes encoding polypeptides involved in photosynthesis, electron transport, and membrane-associated transport systems were deduced by comparison with known chloroplast or *E. coli* genes. Interestingly, seven open reading frames (ORFs) had a high degree of homology to human mitochondrial NADH dehydrogenase genes. We could further deduce two ORFs for bacterial 4Fe-4S type ferredoxin, one with homology to a component of nitrogenase, two for the components of inner membrane permease in *E. coli*, and one for an antenna protein of a light harvesting complex in cyanobacteria. The other 33 ORFs (29 to 2136 codons) remain unidentified despite an extensive computer-aided search for homology.

Genes for tRNAs and rRNAs in liverwort chloroplast genome

The coding sequences for four kinds of rRNA genes (16S, 23S, 4.5S, and 5S) and 32 species of tRNA genes (proline $tRNA_{GGG}$ is a pseudogene with incomplete amino-acyl stem structure) have been deduced from the entire liverwort chloroplast nucleotide sequence. Duplicated rRNA genes are located in the IR regions (IR$_A$ and IR$_B$). Five duplicate tRNA genes, trnV(GAC), trnI(GAU), trnA(UGC), trnR(ACG), and trnN(GUU), are also present in the inverted repeat regions (Fig. 1). None of the tRNA genes has a CCA sequence at the 3′ end of their coding sequences. The tRNA genes were scattered throughout the liverwort genome. Six of the tRNA genes, trnA(UGC), trnI(GAU), trnG(UCC), trnK(UUU), trnL(UAA), and-trnV(UAC), have introns in their coding sequences (Fig. 1).

All rRNA and tRNA molecules seem to be generated within the chloroplasts because it is generally believed that no RNA molecule can enter the

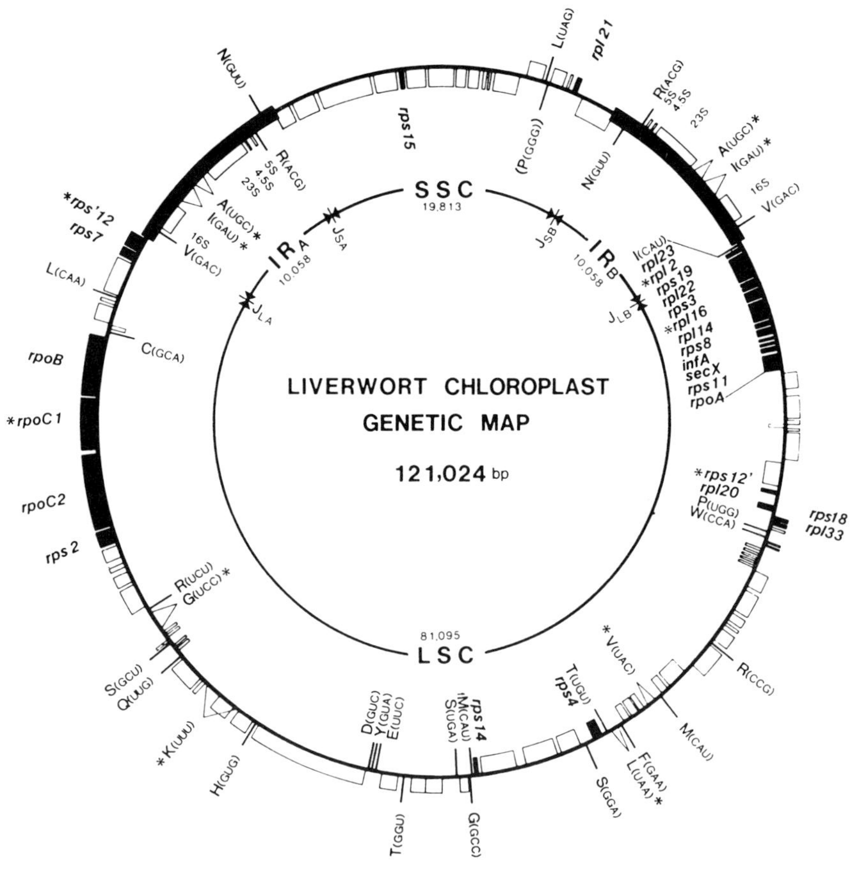

Fig. 1. Genetic map of genes for protein synthesis and related genes in the liverwort chloroplast genome. Nomenclature of chloroplast genes follows that of Crouse et al. (1985). Asterisks indicate genes with introns in their coding sequences. Genes rpo, infA, and secX are for subunits of RNA polymerase, initiation factor, and ribosomal protein of large subunits (L36; Wada and Sako 1987), respectively.

chloroplasts from the cytoplasm. Therefore, the 31 species of tRNAs deduced from DNA sequence analysis must satisfy codon-anticodon recognition in the chloroplasts (Table 1). The number of tRNA species in liverwort chloroplasts is smaller than the estimated number (about 50) of *E. coli* or *Bacillus subtilis* (Fournier and Ozeki 1985), but higher than the 24 species in yeast mitochondria (Bonitz et al. 1980) and the 22 species in human mitochondria (Anderson et al. 1981). In mitochondria, however, the codon table is not universal. Therefore, this is the first time that a complete set of tRNAs in a genetic system has been elucidated to clarify universal codon table (Crick 1966). In tobacco chloroplasts, 30 species of tRNA genes have been described with the difference that arginine tRNA$_{CCG}$ is present in liverwort but not in tobacco (Shinozaki et al. 1986). The liverwort proline tRNA$_{GGG}$ pseudogene was not detected in tobacco either. This implies that the number of tRNA genes in chloroplasts is probably being reduced in the course of evolution as seen with mitochondria.

Table 1.   Codon table and tRNAs encoded by liverwort chloroplast genome

Codon	tRNA	Codon	tRNA	Codon	tRNA	Codon	tRNA
Phe UUU		Ser UCU		Tyr UAU		Cys UGU	
UUC	GAA	UCC	GGA	UAC	GUA	UGC	GCA
Leu UUA	UAA*	UCA	UGA	Ter UAA		Ter UGA	
UUG	CAA	UCG		UAG		Trp UGG	CCA
Leu CUU		Pro CCU		His CAU		Arg CGU	ACG
CUC		CCC	(GGG)	CAC	GUG	CGC	
CUA	UAG	CCA	UGG	Gln CAA	UUG	CGA	
CUG		CCG		CAG		CGG	CCG
Ile AUU		Thr ACU		Asn AAU		Ser AGU	
AUC	GAU*	ACC	GGU	AAC	GUU	AGC	GCU
AUA	CAU	ACA	UGU	Lys AAA	UUU*	Arg AGA	UCU
Met fMet AUG	CAU CAU	ACG		AAG		AGG	
Val GUU		Ala GCU		Asp GAU		Gly GGU	
GUC	GAC	GCC		GAC	GUC	GGC	GCC
GUA	UAC*	GCA	UGC*	Glu GAA	UUC	GGA	UCC*
GUG		GCG		GAG		GGG	

Anticodons are expressed with unmodified bases. The AUG codon is an initiation codon. Termination codons (UAA, UAG, and UGA) are indicated by Ter. Asterisks indicate tRNA genes with introns in their coding sequences. Amino acids are shown by three-letter symbols. Proline tRNA$_{GGG}$ is a pseudogene.

A unique glutamate tRNA with an unusual modified anticodon is involved in chlorophyll synthesis in chloroplasts (Schon et al. 1986). A molecule of chlorophyll is synthesized from eight molecules of δ-aminolevulinate (DALA), the universal precursor of porphyrins. The components involved in chlorophyll synthesis contain an RNA as the essential component, identified as a chloroplast glutamate tRNA. The nucleotide sequence of the *trnQ*(UUG) gene in the liverwort chloroplast genome had a high degree of homology with that reported for RNA[DALA] in higher plants. This *trnQ*(UUG) must have dual functions for both protein and chlorophyll synthesis in the chloroplasts, because it is the only gene for glutamate tRNA in the liverwort chloroplast genome.

We have analyzed codon usages of all the genes including the unidentified ORFs. A distinctive feature of the codon usage is that 88.1% of the third letters of codons are either A or U. This A or U preference in the codon usage coincides with the overall high A + T content (72.2%) in the liverwort chloroplast genome, and it facilitated the precise identification of ORFs throughout the liverwort chloroplast genome. We observed a peculiar codon usage in the *psbA* gene (Table 2). As described above, either A or U is preferred at the third letter of codons in liverwort protein genes, but the codon usage of *psbA* is different in the choice of the third letter, especially for pyrimidine two-codon boxes such as asparagine, aspartic acid, cysteine, histidine, phenylalanine, serine, and tyrosine, where the *psbA* gene used cytosine more often than uracil in the third letter of the codons. This peculiar codon usage in the *psbA* gene may be correlated with the translational efficiency (stronger codon-anticodon association by G-C pairs than A-U pairs) of the mRNA molecules.

Genes for protein synthesis and related genes

We have identified coding sequences for 19 ribosomal proteins (large subunit proteins L33, L20, L14, L16, L22, L2, L23, and L21; small subunit proteins S7, S2, S14, S4, S18, S12, S11, S8, S3, S19, and S15). In addition, the genes *infA* (*E. coli* initiation factor 1), *secX* (recently identified as the ribosomal large subunit protein L36; Wada and Sako 1987) and *rpo* (*rpoA*, *rpoB*, and *rpoC1-rpoC2* for α, β, and β' subunits of RNA polymerase, respectively) were on the liverwort chloroplast genome (Fig. 1). It is of interest that ribosomal protein and related genes from a large cluster in the chloroplast genome (*rpl23-rpl2-rps19-rpl22-rps3-rpl16-rpl14-rps8-infA-secX-rps11-rpoA* genes), although others are scattered throughout the genome. The genes in the cluster have a similar order to the clusters reported

Table 2. Peculiar codon usage of psbA gene in liverwort chloroplast genome

	Codon	psbA	Total		Codon	psbA	Total		Codon	psbA	Total		Codon	psbA	Total
Phe	UUU	8	1,539	Ser	UCU	12	616	Tyr	UAU	2	824	Cys	UGU	0	219
	UUC	17	80		UCC	0	71		UAC	11	84		UGC	2	39
Leu	UUA	15	1,852		UCA	4	355	Ter	UAA	1	83	Ter	UGA	0	2
	UUG	5	198		UCG	0	48		UAG	0	5	Trp	UGG	10	431
Leu	CUU	8	516	Pro	CCU	12	465	His	CAU	5	383	Arg	CGU	10	347
	CUC	0	25		CCC	0	39		CAC	5	57		CGC	2	45
	CUA	2	141		CCA	3	364	Gln	CAA	6	881		CGA	0	258
	CUG	0	25		CCG	0	49		CAG	0	53		CGG	0	22
Ile	AUU	17	1,502	Thr	ACU	14	602	Asn	AAU	7	1,249	Ser	AGU	3	411
	AUC	12	99		ACC	2	55		AAC	14	161		AGC	7	39
	AUA	0	708		ACA	1	498	Lys	AAA	0	1,764	Arg	AGA	2	380
Met	AUG	12	509		ACG	0	42		AAG	1	77		AGG	0	24
Val	GUU	11	637	Ala	GCU	32	747	Asp	GAU	4	731	Gly	GGU	29	598
	GUC	0	47		GCC	0	66		GAC	4	68		GGC	1	81
	GUA	13	437		GCA	6	446	Glu	GAA	17	1,116		GGA	3	675
	GUG	0	48		GCG	0	50		GAG	2	83		GGG	0	88

Numbers indicate frequency of codons used. Total codon usage indicates all codons in all ORFs predicted minus psbA codons. Amino acids are expressed by three-letter symbols.

for the *E. coli* ribosomal protein operons such as S10 (S10-L3-L4-*rp123*-*rp12*-*rps19*-*rp122*-*rps3*-*rp116*-L29-S17; Zurawski and Zurawski 1985), *spc* (*rp114*-L24-L5-S14-*rps8*-L6-L18-S5-L30-*secY*-*secX*; Cerretti et al. 1983) and α (S13-*rps11*-S4-*rpoA*-L17; Bedwell et al. 1985). Three additional clusters of ribosomal protein genes are seen with the *rps12'*(exon 1)-*rp120*, *rp133*-*rps18*, and *rps'12*(exons 2 and 3)-*rps7* genes. The ribosomal protein S12 gene (*rps12*) has been reported to be split between different DNA strands indicating in vivo *trans*-splicing (Fukuzawa et al. 1986). It is well known that the nuclear genome encodes genes for chloroplast proteins, and that proteins synthesized in the cytoplasm are then transported into chloroplasts and assembled with the organelle-synthesized proteins to form functional complexes such as ribosome particles. It is interesting that the ribosomal proteins (S15, S4, S8, and S7) in the 30S subunits encoded on the liverwort chloroplast genome are essential for the initial assembly of ribosomal protein-rRNA molecules (Dorne et al. 1984). Large ribosomal subunit components (L2 and L20) also form the major center of the assembly (Wittmann 1983).

The gene encoding ribosomal protein L21 (*rp121*) was not seen in tobacco, and the gene for ribosomal protein S16 (*rps16*) was not deduced from the liverwort chloroplast genome. This indicates that the chloroplast genomes of these two species may have undergone different gene arrangements in the course of evolution.

Genes for photosynthetic and electron transport systems

Genes for photosynthetic and electron transport polypeptides of the liverwort chloroplast genome were identified by comparison of the amino acid sequences of the ORFs with those of photosynthetic and electron transport proteins with structures deposited in the NBRF-PIR data base (Fig. 2). Genes (*psaA* and *psaB*) for the P700 chlorophyll *a* apoproteins of photosystem I (Fish et al. 1985) are closely linked with a very short spacer region (26 bp). The nucleotide sequence in this intergenic spacer is almost identical to the sequences of different species of plants, but the 5′ and 3′ flanking regions of the genes do not show any simlarities. Ths spacer region may have a control signal for transcription or translation of the *psaB* gene downstream. Eight genes (*psbA*, *psbB*, *psbC*, *psbD*, *psbE*, *psbF*, *psbG*, and *psbH*) for photosystem II polypeptides were in the large single-copy region. Two of them (*psbD-psbC*) (Alt et al. 1984, Holschuh et al. 1984) are in tandem with overlapping of the 3′-*psbD* and 5′-*psbC* ends. A 32-kDa protein gene, *psbA* (Zurawski et al. 1982) retains a highly conservative structure among

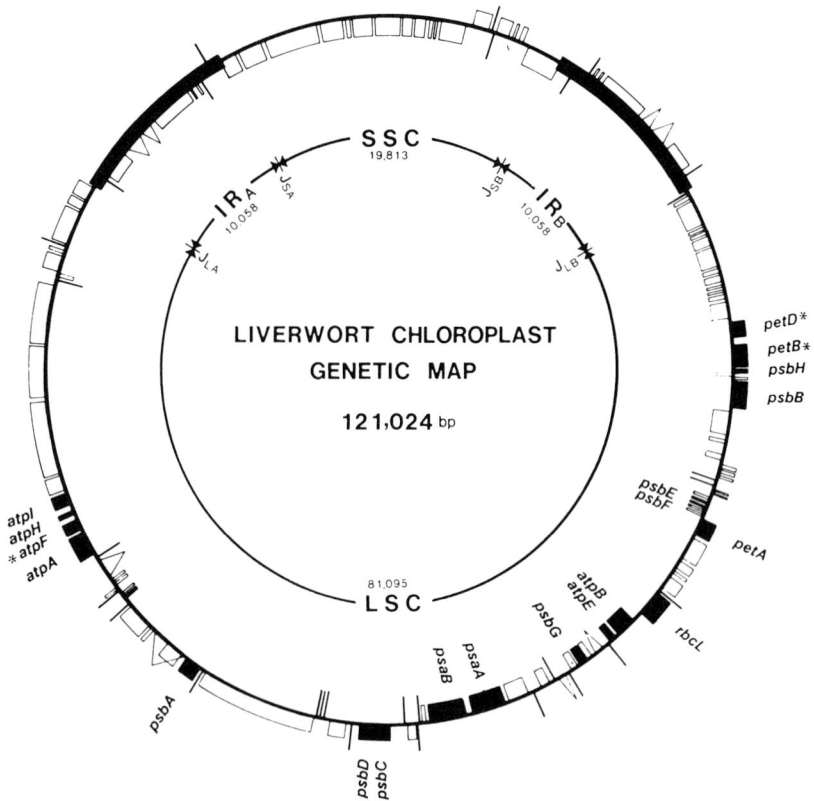

Fig. 2. Genetic map of genes for photosynthesis in chloroplasts. Nomenclature of photosynthetic genes follows that of Crouse et al. (1985). Asterisks indicate genes with introns in their coding sequences.

higher plants; it is believed to be responsible for resistance of atrazine herbicide (Hirschberg and McIntosh 1983). The liverwort *psbG* gene encoding 243 amino acids was identified by the amino acid sequence of its maize counterpart, which is located in the photosystem II particles (Steinmetz et al. 1986). However, our cluster analysis of genes by computer showed that the *psbG* gene behaved differently from other components of the photosystem II particles (Ohyama et al. unpublished results). The coding regions for the *psbE* and *psbF* genes (Herrmann et al. 1984) are clustered on the opposite DNA strand next to the *petA* gene (Alt and Herrmann 1984). The *psbB* operon consisted of *psbB*, ORF35, *psbH*, *petB*, and *petD* genes, which are cotranscribed as a single precursor RNA (Kohchi et al. in preparation). The amino acid sequence of liverwort *psbB* protein shows a high degree of homology to that of the 51-kDa chlorophyll *a* apoprotein in spinach (Morris and Herrmann 1984). The *psbH* gene product is a 10-kDa phosphoprotein associated with photosystem II (Westhoff et al. 1986). The ORF35 remains unidentified. We deduced the ORF43 gene divergently on the opposite DNA strand of the spacer region between ORF35 and the *psbH* genes. This ORF gene is actively transcribed during illumination (Kohchi et al. in preparation).

Genes for cytochrome f protein (*petA*) (Alt and Herrmann 1984), cytochrome *b*6 (*petB*) (Heinemeyer et al. 1984), and cytochrome b_6/f complex subunit IV (*petD*) (Phillips and Gray 1984) are found in the liverwort chloroplast genome. The *petA* protein deduced from the DNA sequence may have an N-terminal sequence of 35 amino acids as a signal peptide, because the N-terminal portion of the protein is species-specific, as in other species of plants (Alt and Herrmann 1984). DNA sequence analysis also showed that the *petB* and *petD* genes have group II introns in their coding sequences (Fukuzawa et al. 1987). The *petB* and *petD* genes are cotranscribed with genes in the *psbB* operon and the primary transcript is processed down to a bicistronic mRNA containing the *petB-petD* genes (Kohchi et al. in preparation).

Six genes (*atpI-atpH-atpF-atpA*, and *atpB-atpE*) for the chloroplast H^+-ATPase (Henning and Herrmann 1986, Cozen et al. 1986), which consists of nine non-identical subunits, were identified by comparison of their amino acid sequences with those of genes in higher-plant chloroplast genomes. The *atp* gene organization in the liverwort chloroplast genome was the same as those in higher plants and similar to those of cyanobacteria and *E. coli*.

Newly identified groups of genes in liverwort chloroplast genome

There are a few genes common to the liverwort and tobacco chloroplast genomes deduced from entire DNA sequences the gene products of which have not been described in chloroplasts. On the other hand, there are several genes that are detected in the liverwort chloroplast genome, but not in tobacco, and vice versa.

Seven newly identified genes (*ndh1*, *ndh2*, *ndh3*, *ndh4*, *ndh4L*, *ndh5*, and *ndh6*) in the liverwort chloroplast genome correspond to components of human mitochondria NADH dehydrogenase (ND1, ND2, ND3, ND4, ND4L, ND5, and ND6, respectively) (Chomyn et al. 1985, Chomyn et al. 1986). In tobacco, the six genes *ndhA*, *ndhB*, *ndhC*, *ndhD*, *ndhE*, and *ndhF* correspond to ND1, ND2, ND3, ND4, ND4L, and ND5, respectively. The two continuous tobacco ORF138 and ORF99B may be made to correspond

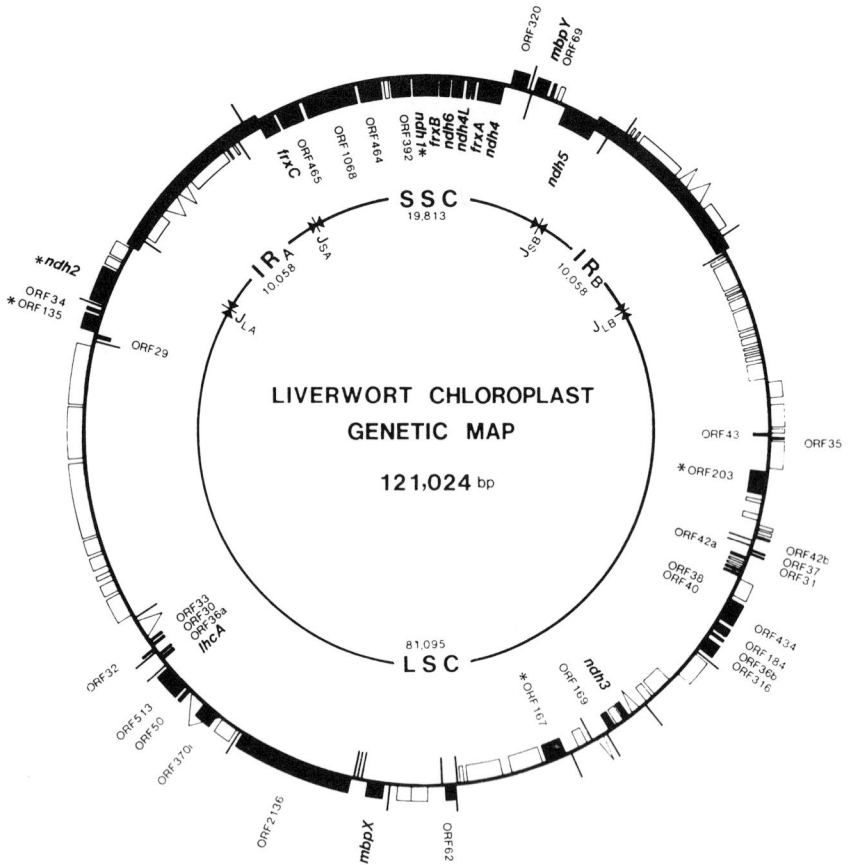

Fig. 3. Newly identified groups of genes and unidentified ORFs in the liverwort chloroplast genome. Genes *ndh*, *frx*, *mbp*, and *lhcA* encode subunits of NADH(PQ) oxidoreductase, iron-sulfur binding proteins, subunits of membrane-associated transport system, and light harvesting protein, respectively. ORFn indicates an open reading frame with a certain number (n) of amino acids.

to the human mitochondrial Nd6 by shifting the frame in the tobacco sequence to obtain maximum matching (Shinozaki et al. 1986). Northern hybridization analysis showed that there is active transcription of *ndh* genes in liverwort chloroplasts (Kohchi et al., unpublished results). DNA sequence analysis of watermelon mitochondria DNA fragments has found a portion of a watermelon mitochondrial gene corresponding to the ND1 gene in human mitochondria (Stern et al. 1986). Thus, the amino acid sequence of three species (liverwort chloroplast *ndh1*, human mitochondria ND1, and watermelon mitochondria URF1) had more than 35% homology to each other. This means that the liverwort chloroplast *ndh1* gene is probably not a pseudogene of mitochondria and that it has diverged independently from a common origin. Reports dealing with the hydrogenase-dependent processes, photoreduction and the dark oxyhydrogen reaction, in the chloroplasts of *Chlamydomonas reinhardtii* suggest that NADH(PQ) oxidoreductase is involved in the electron transport system for chloroplast respiration (Bennoun 1982, Maione and Gibbs 1986). This indicates that the products of chloroplast *ndh* genes may be components of the enzyme.

The *frx* genes have been elucidated from the DNA sequence as iron-sulfur proteins based on the characteristic distribution of cysteine residues in the polypeptides deduced from the DNA sequence. In particular, *frxA* and *frxB* gene products have repeated units of -C-X-X-C-X-X-C-X-X-X-C-P-, a typical amino acid sequence in bacterial ferredoxin (Howard et al. 1983, Minami et al. 1985). The gene product corresponding to *frxA* (*psaC*) has been identified in the Fe–S center of the photosystem I complex (Oh-oka et al. 1987, Høj et al. 1987). The *frxA* gene originally was not described in the tobacco genome (Shinozaki et al. 1986). The *frxA* (*psaC*) gene in the tobacco genome has been deduced by the insertion of one nucleotide and shifting a frame in the tobacco sequence (Hayashida et al. 1987). Tobacco ORF167 in the single-copy region corresponded to the liverwort *frxB* gene. This observation indicates that the *frxA* and *frxB* genes are common to chloroplast genomes. The *frxC* gene product, which contains nine cysteine residues, has a high degree of homology to the Fe-binding protein in nitrogen-fixing bacteria, the *nifH* gene products in *Azotobacter vinelandii* (Hausinger and Howard 1982) and the F202 gene in *Rhodopseudomonas capsulata* (Youvan et al. 1984, Hearst et al. 1985). The amino acid sequence (-G-X-X-X-G-K-S-) in the N-terminal portion is a conserved nucleotide binding site seen in the ATP-binding proteins (Higgins et al. 1986). Curiously, this gene was not detected in the tobacco chloroplast genome, indicating the possibility of a pseudogene for the nitrogenase component. However, immunological analysis gave a positive reaction for the liverwort chloroplast extract with antibody against *frxC* gene product made in *E. coli* (Takahashi, personal communication).

We have deduced *mbpX* and *mbpY* genes from the liverwort chloroplast genome. The amino acid sequence of *mbpX* shows similarity to the sequences of inner membrane components of bacterial permeases such as the products of the genes *hisP* (Higgins et al. 1982), *malK* (Gilson et al. 1982), and *oppD* (Higgins et al. 1985) for the histodine, maltose, and oligopeptide transport systems, respectively. The *mbpX* gene product near the N-terminal portion has a typical amino acid sequence which is a consensus nucleotide binding site, indicating a membrane transport system by nucleotide-driven energy in the liverwort chloroplasts. The hydrophobicity of the *mbpY* gene product is also similar to that of another component of the inner membrane permease encoded by *hisQ* in *Salmonella typhimurium* (Higgins et al. 1982) and *malF* of *E. coli* (Froshauer and Beckwith 1984). These gene products may associate with components encoded on the nuclear genome and form a transport complex in chloroplast membranes. These genes could not be deduced from the tobacco chloroplast genome (Shinozaki et al. 1986).

Unidentified open reading frames

Thirty-three open reading frames ranging from 29 (ORF29) to 2136 amino acid residues (ORF2136) remain unidentified in the liverwort chloroplast genome, although we attempted to identify their genes by extensive computer analysis. Most of the ORFs seem to be conserved in the liverwort and tobacco chloroplast genomes, indicating their active functions in chloroplasts (Table 3). Membrane spanning analysis with an algorithm of Klein et

Table 3. Predicted location of ORF gene products in chloroplasts and liverwort ORFs conserved in the chloroplast genomes of tobacco and elsewhere.

Liverwort ORF	Predicted location	Remarks	Reference
ORF34	I	+	
ORF135*	I	–	
ORF29	I	+	
ORF33	I	–	
ORF30	I	–	
ORF32	I	–	
ORF36a	I	ORF2	Deno and Sugiura, 1983
ORF513	P	–	
ORF50	I	–	
ORF370i	P	ORF509	Sugita *et al.*, 1985
ORF2136	I	ORF1708	Shinozaki *et al.*, 1986
ORF62	I	URF62 in wheat	Quigley and Weil, 1985
ORF167*	P	ORF82	Shinozaki *et al.*, 1986
ORF169	P	ORF158	Shinozaki *et al.*, 1986
ORF316	P	ORF512	Shinozaki *et al.*, 1986
ORF36b	I	+	
ORF184	I	ORF184	Shinozaki *et al.*, 1986
		ORF149 in *E. gracilis*	Montandon and Stutz, 1983
ORF434	I	ORF229	Shinozaki *et al.*, 1986
ORF40	I	ORF40	D. A. Bryant, personal
		in *Cyanophora paradoxa*	communication
ORF38	I	+	
ORF42a	P	–	
ORF31	I	+	
ORF37	I	+	
ORF42b	I	+	
ORF203*	I	X gene in spinach Westhoff, 1985	
		ORF73 + ORF74B	Shinozaki *et al.*, 1986
ORF35	I	+, in *psbB* operon Kohchi *et al.*, unpublished	
ORF43	I	+, divergent to *psbB* operon, light-inducible	
			Kohchi *et al.*, unpublished
ORF69	P	+, ribosomal protein	
			Kohchi *et al.*, unpublished
ORF320	I	ORF313	Shinozaki *et al.*, 1986
ORF392	P	ORF393	Shinozaki *et al.*, 1986
ORF464	I	ORF228	Shinozaki *et al.*, 1986
ORF1068	I	ORF1244	Shinozaki *et al.*, 1986
ORF465	I	–	

Letters I and P show integral and peripheral locations of the ORFs in chloroplasts, respectively, predicted by computer-aided membrane spanning analysis (Klein *et al.*, 1985). Symbols + or – indicate the presence or absence of the corresponding ORFs in the tobacco DNA sequence (EMBL data library, release 12.0). Asterisks indicate ORF with introns.

al. (1985) showed the predicted location of ORF products in the chloroplasts (Table 3). Functionally related genes are likely to form a cluster on the chloroplast genome (see Fig. 2). These findings suggest that an ORF in a gene cluster is related functionally to nearby genes. Further computer-aided cluster anlaysis of the amino acid composition divided the liverwort chloroplast genes into four groups (two for photosynthetic and two for housekeeping genes) (Ohyama et al., unpublished results).

The complete nucleotide sequences of chloroplast genomes from two species of plants gave us a tremendous amount of information on photosynthetic and chloroplast genetic systems. Identification of new genes (ORFs) and the elucidation of their function has been made possible by the information on DNA sequences. However, we were not able to gather any information on the chloroplast DNA replication system (DNA replication origin and DNA polymerase) from the complete DNA sequences, among other topics. For this investigation, we need a DNA transformation system in chloroplasts.

Acknowledgements

This research was supported in part by a Grant-in-Aid for Special Research Projects from the Ministry of Education, Science, and Culture, Japan. We appreciate the continued encouragement of Prof. Dr Y. Yamada, Director of the Research Center for Cell and Tissue Culture, Faculty of Agriculture, Kyoto University. We also thank Prof. Dr M. Kanehisa of Institute of Chemical Research, Kyoto University, for his help with computer analysis.

References

Alt J and Herrmann RG (1984) Nucleotide sequence of the gene for preapocytochrome f in the spinach plastid chromosome. Curr Genet 8: 551–557

Alt J, Morris J, Westhoff P and Herrmann RG (1984) Nucleotide sequence of the clustered genes for the 44 kd chlorophyll a apoprotein and the '32 kd'-like protein of the photosystem II reaction center in the spinach plastid chromosome. Curr Genet 8: 597–606

Anderson S, Bankier AT, Barrell BG, de Bruijn MHL, Coulson AR, Drouin J, Eperon IC, Nierlich DP, Roe BA, Sanger F, Schreier PH, Smith AJH, Staden R and Young IG (1981) Sequence and organization of the human mitochondrial genome. Nature 290: 457–465

Bedwell D, Davis G, Gosink M, Post L, Nomura M, Kestler H, Zengel JM and Lindahl L (1985) Nucleotide sequence of the alpha ribosomal protein operon of *Escherichia coli*. Nucl Acids Res 13: 3891–3903

Bennoun P (1982) Evidence for a respiratory chain in the chloroplast. Proc Natl Acad Sci USA 79: 4352–4356

40

Bonitz SG, Berlani R, Coruzzi G, Li M, Macino G, Nobrega FG, Nobrega MP, Thalenfeld BE and Tzagoloff A (1980) Codon recognition rules in yeast mitochondria. Proc Natl Acad Sci USA 77: 3167–3170

Cerretti DP, Dean D, Davis GR, Bedwell DM and Nomura M (1983) The *spc* ribosomal protein operon of *Escherichia coli*: sequence and cotranscription of the ribosomal protein genes and a protein export gene. Nucl Acids Res 11: 2599–2616

Chomyn A, Mariottini P, Cleeter MWJ, Ragan CI, Matsuno-Yagi A, Hatefi Y, Doolittle RF and Attardi G (1985) Six unidentified reading frames of human mitochondrial DNA encode components of the respiratory-chain NADH dehydrogenase. Nature 314: 592–597

Chomyn A, Cleeter MWJ, Ragan CI, Riley M, Doolittle RF and Attardi G (1986) URF6, last unidentified reading frame of human mtDNA, codes for an NADH dehydrogenase subunit. Science 234: 614–618

Cozens AL, Walker JE, Phillips AL, Huttly AK and Gray JC (1986) A sixth subunit of ATP synthase, an F_0 component, is encoded in the pea chloroplast genome. EMBO J 5: 217–222

Crick FHC (1966) The genetic code: III. Sci Amer 254(4): 55–62.

Crouse EJ, Schmitt JM and Bohnert HJ (1985) Chloroplast and cyanobacterial genomes, genes and RNAs: a compilation. Plant Mol Biol Reporter 3: 43–89

Deno H and Sugiura M (1983) The nucleotide sequences of tRNA^{Ser}(GCU) and tRNA^{Gln}(UUG) genes from tobacco chloroplasts. Nucl Acids Res 11: 2185–2192

Dorne AM, Lescure AM and Mache R (1984) Site of synthesis of spinach chloroplast ribosomal proteins and formation of incomplete ribosomal particles in isolated chloroplasts. Plant Mol Biol 3: 83–90

Ellis RJ (1981) Chloroplast proteins: synthesis, transport and assembly. Ann Rev Plant Physiol 32: 111–137

Fish LE, Kuck U and Bogorad L (1985) Two partially homologous adjacent light-inducible maize chloroplast genes encoding polypeptides of the p700 chlorophyll *a*-protein complex of photosystem I. J Biol Chem 260: 1413–1421

Fournier MJ and Ozeki H (1985) Structure and organization of the transfer ribonucleic acid genes of *Escherichia coli* K-12. Microbiol Rev 49: 379–397

Froshauer S and Beckwith J (1984) The nucleotide sequence of the gene for *malF* protein, an inner membrane component of the maltose transport system of *Escherichia coli*. J Biol Chem 259: 10896–10903

Fukuzawa H, Kohchi T, Shirai H, Ohyama K, Umesono K, Inokuchi H and Ozeki H (1986) Coding sequences for chloroplast ribosomal protein S12 from the liverwort, *Marchantia polymorpha*, are separated far apart on the different DNA strands. FEBS Lett 198: 11–15

Fukuzawa H, Yoshida T, Kohchi T, Okumura T, Sawano Y and Ohyama K (1987) Splicing of group II introns in mRNAs coding cytochrome b_6 and subunit IV in the liverwort *Marchantia polymorpha* chloroplast genome: Exon specifying a region coding for two genes with the spacer region. FEBS Lett 220: 61–66

Gilson E, Nikaido H and Hofnung M (1982) Sequence of the *malK* gene in *E. coli* K12. Nucl Acids Res 10: 7449–7458

Hausinger RP and Howard JB (1982) The amino acid sequence of the nitrogenase iron protein from *Azotobacter vinelandii*. J Biol Chem 257: 2483–2487

Hayashida N, Matsubayashi T, Shinozaki K, Sugiura M, Inoue K and Hiyama T (1987) The gene for the 9 kd polypeptide, a possible apoprotein for the iron-sulfur centers A and B of the photosystem I complex, in tobacco chloroplast DNA. Curr Genet 12: 247–250

Hearst JE, Alberti M and Doolittle RF (1985) A putative nitrogenase reductase gene found in the nucleotide sequences from the photosynthetic gene cluster of *R. capsulata*. Cell 40: 219–220

Heinemeyer W, Alt J and Herrmann RG (1984) Nucleotide sequence of the clustered genes for apocytochrome *b6* and subunit 4 of the cytochrome *b/f* complex in the spinach plastid chromosome. Curr Genet 8: 543–549

Henning J and Herrmann RG (1986) Chloroplast ATP synthase of spinach contains nine nonidentical subunit species, six of which are encoded by plastid chromosomes in two operons in a phylogenetically conserved arrangement. Mol Gen Genet 203: 117–128

Herrmann RG, Alt J, Schiller B, Widger WR and Cramer WA (1984) Nucleotide sequence of the gene for apocytochrome *b*-559 on the spinach plastid chromosome: implications for the structure of the membrane protein. FEBS Lett 176: 239–244

Higgins CF, Haag PD, Nikaido K, Ardeshir F, Garcia G and Ames GFL (1982) Complete nucleotide sequence and identification of membrane components of the histidine transport operon of *S. typhimurium*. Nature 298: 723–727

Higgins CF, Hiles ID, Whalley K and Jamieson DJ (1985) Nucleotide binding by membrane components of bacterial periplasmic binding protein-dependent transport systems. EMBO J 4: 1033–1040

Higgins CF, Hiles ID, Salmond GPC, Gill DR, Downie JA, Evans IJ, Holland IB, Gray L, Buckel SD, Bell AW and Hermodson MA (1986) A family of related ATP-binding subunits coupled to many distinct biological processes in bacteria. Nature 323: 448–450

Hirschberg J and McIntosh L (1983) Molecular basis of herbicide resistance in *Amaranthus hybridus*. Science 222: 1346–1349

Høj PB, Svendsen I, Scheller HV and Moller BL (1987) Identification of a chloroplast-encoded 9-kDa polypeptide as a 2[4Fe-4S] protein carrying centers A and B of photosystem I. J Biol Chem 262: 12676–12684

Holschuh K, Bottomley W and Whitfeld PR (1984) Structure of the spinach chloroplast genes for the D2 and 44 kd reaction-center proteins of photosystem II and for tRNA^{Ser}(UGA). Nucl Acids Res 12: 8819–8834

Howard JB, Lorsbach TW, Ghosh D, Melis K and Stout CD (1983) Structure of *Azotobacter vinelandii* 7Fe ferredoxin. J Biol Chem 258: 508–522

Klein P, Kanehisa M and DeLisi C (1985) The detection and classification of membrane-spanning proteins. Biochim Biophys Acta 815: 468–476

Maione TE and Gibbs M (1986) Association of the chloroplastic respiratory and photosynthetic electron transport chains of *Chlamydomonas reinhardtii* with photoreduction and the oxyhydrogen reaction. Plant Physiol 80: 364–368

Minami Y, Wakabayashi S, Wada K, Matsubara H, Kerscher L and Oesterhelt D (1985) Amino acid sequence of a ferredoxin from thermoacidophilic Archaebacterium, *Sulfolobus acidocaldarius*. Presence of an N^6-monomethyllysine and phyletic consideration of Archaebacteria. J Biochem 97: 745–753

Montandon PE and Stutz E (1983) Nucleotide sequence of a *Euglena gracilis* chloroplast genome region coding for the elongation factor Tu; evidence for a spliced mRNA. Nucl Acids Res 11: 5877–5891

Morris J and Herrmann RG (1984) Nucleotide sequence of the gene for the P_{680} chlorophyll *a* apoprotein of the photosystem II reaction center from spinach. Nucl Acids Res 12: 2837–2850

Oh-oka H, Takahashi Y, Wada K, Matsubara H, Ohyama K and Ozeki H (1987) The 8 kDa polypeptide in photosystem I is a probable candidate of an iron-sulfur center protein coded by the chloroplast gene *frxA*. FEBS Lett 218: 52–54

Ohyama K, Fukuzawa H, Kohchi T, Shirai H, Sano T, Sano S, Umesono K, Shiki Y, Takeuchi M, Chang Z, Aota S, Inokuchi H and Ozeki H (1986) Chloroplast gene organization deduced from complete sequence of liverwort *Marchantia polymorpha* chloroplast DNA. Nature 322: 572–574

Phillips AL and Gray JC (1984) Location and nucleotide sequence of the gene for the 15.2 kDa polypeptide of the cytochrome *b-f* complex from pea chloroplasts. Mol Gen Genet 194: 477–484

Quigley F and Weil JH (1985) Organization and sequence of five tRNA genes and of an unidentified reading frame in the wheat chloroplast genome: evidence for gene rearrangements during the evolution of chloroplast genomes. Curr Genet 9: 495–503

Schon A, Krupp G, Gough S, Berry-Lowe S, Kannangara CG and Soll D (1986) The RNA required in the first step of chlorophyll biosynthesis is a chloroplast glutamate tRNA. Nature 322: 281–284

Shinozaki K, Ohme M, Tanaka M, Wakasugi T, Hayashida N, Matsubayashi T, Zaita N, Chunwongse J, Obokata J, Yamaguchi-Shinozaki K, Ohto C, Torazawa K, Meng BY, Sugita M, Deno H, Kamogashira T, Yamada K, Kusuda J, Takaiwa F, Kato A, Tohdoh N, Shimada H and Sugiura M (1986) The Complete nucleotide sequence of the tobacco chloroplast genomes: its gene organization and expression. EMBO J 5: 2043–2049

Steinback KE, Bonitz S, Arntzen CJ and Bogorad L (1985) (eds) Molecular Biology of the Photosynthetic Apparatus. Cold Spring Harbor Laboratory, Cold Spring Harbor, NY

Steinmetz AA, Castroviejo M, Sayre RT and Bogorad L (1986) Protein PSII-G: An additional component of photosystem II identified through its plasmid gene in maize. J Biol Chem 261: 2485–2488

Stern DB, Bang AG and Thompson WF (1986) The watermelon mitochondrial URF-1 gene: evidence for a complex structure. Curr Genet 10: 857–869

Sugita M, Shinozaki K and Sugiura M (1985) Tobacco chloroplast tRNA^{Lys}(UUU) conserved boundary sequence in the intron. Proc Natl Acad Sci USA 82: 3557–3561

Wada A and Sako T (1987) Primary structures of and genes for new ribosomal proteins A and B in *Escherichia coli*. J Biochem 101: 817–820

Westhoff P (1985) Transcription of the gene encoding the 51 kd chlorophyll *a*-apoprotein of the photosystem II reaction center from spinach. Mol Gen Genet 201: 115–123

Westhoff P, Farchaus JW and Herrmann RG (1986) The gene for the M_r 10,000 phosphoprotein associated with photosystem II is part of the *psbB* operon of the spinach plastid chromosome. Curr Genet 11: 165–169

Wittmann HG (1983) Architecture of prokaryotic ribosomes. Ann Rev Biochem 52: 35–65

Youvan DC, Bylina EJ, Alberti M, Begusch H and Hearst JE (1984) Nucleotide and deduced polypeptide sequences of the photosynthetic reaction-center, B8970 antenna, and flanking polypeptides from *R. capsulata*. Cell 37: 949–957

Zurawski G, Bohnert HJ, Whitfeld PR and Bottomley W (1982) Nucleotide sequence of the gene for the M_r 32,000 thylakoid membrane protein from *Spinacia oleracea* and *Nicotiana debneyi* predicts a totally conserved primary translation product of M_r 38,950. Proc Natl Acad Sci USA 79: 7699–7703

Zurawski G and Zurawski SM (1985) Structure of the *Escherichia coli* S10 ribosomal protein operon. Nucl Acids Res 13: 4521–4526

Govindjee et al. (eds), Molecular Biology of Photosynthesis: 43–59
© 1988 Kluwer Academic Publishers

Minireview

Introns in chloroplast protein-coding genes of land plants

AINE L. PLANT & JOHN C. GRAY

Botany School, University of Cambridge, Downing Street, Cambridge CB2 3EA, UK

Received 14 September 1987; accepted 1 December 1987

Key words: intron, splice, exon, lariat

Abstract. Several protein-coding genes from land plant chloroplasts have been shown to contain introns. The majority of these introns resemble the fungal mitochondrial group II introns due to considerable nucleotide sequence homology at their 5′ and 3′ ends and they can readily be folded to form six hairpins characteristic of the predicted secondary structure of the mitochondrial group II introns. Recently it has been demonstrated that some mitochondrial group II introns are capable of self-splicing in vitro in the absence of protein co-factors. However evidence presented in this overview suggests that this is probably not the case for chloroplast introns and that *trans*-acting factors are almost certainly involved in their processing reactions.

Abbreviations: kbp — kilobase pairs, ORF — Open Reading Frame, pre-RNA — precursor ribonucleic acid

Introduction

Introns were first discovered in the ovalbumin and β-globin genes of chickens and mammals respectively (Breathnach et al. 1977, Jeffreys and Flavell 1977). They are DNA sequences within genes, that interrupt and thereby split the protein-coding regions (exons). Introns are transcribed together with the exons into precursor RNA molecules and are subsequently removed in a series of reactions leading to splicing of the exons to produce the mature transcript. Based on their nucleotide sequence, introns can be divided into 4 different classes. One class comprises the introns of nuclear mRNA precursors. These include the ovalbumin and β-globin gene introns mentioned above and are characterized by invariant GU and AG dinucleotides at their boundaries (Breathnach et al. 1978). Splicing of these introns requires assembly of pre-RNA into a complex structure involving ribonucleoprotein particles (Brody and Abelson 1985, Frendewey and Keller 1985, Grabowski et al. 1985). Self-splicing introns typified by the *Tetrahymena* rRNA gene (Kruger et al. 1982) form another class. Similar introns are also found in fungal mitochondrial genes; these introns have been termed group I introns on the basis of conserved sequence elements (Davies et al. 1982, Michel and Dujon 1983). Another group of self-splicing introns occurring in yeast mitochondrial genes form another class. These introns were termed group II introns and possess a set of conserved sequences distinct from those of group I introns (Michel and Dujon 1983). Finally there are the introns characteristic of nuclear tRNA genes (Abelson 1979).

Introns in chloroplast protein-coding genes

The first protein-coding gene from higher plant chloroplasts reported to contain an intron was the *rpl2* gene for the ribosomal protein L2 from *Nicotiana debneyi* (Zurawski et al. 1984). This intron was discovered as a result of a direct comparison between the nucleotide sequences of the *rpl2* genes from *Spinacea oleracea* and *Nicotiana debneyi* and showed that the *Nicotiana debneyi* gene contained a 666 bp insertion which was absent from the *Spinacea oleracea* gene. Prior to this, the discovery of introns in chloroplast genes was confined to tRNA genes (Koch et al. 1981, Deno et al. 1982, Steinmetz et al. 1982, Bonnard et al. 1984, Deno and Sugiura 1984) and those present in the rRNA and protein-coding genes of *Chlamydomonas reinhardii* and *Euglena gracilis* (Rochaix and Malnoe 1978, Erickson et al. 1984, Karabin et al. 1984, Koller et al. 1984). It has been estimated that the *Euglena gracilis* chloroplast genome contains a minimum of 50 introns accounting for approximately 32 kbp of the 145 kbp genome (Koller and Delius 1984); the *rbcL* gene, coding for the large subunit of ribulose 1,5-bisphosphate carboxylase, alone has a total of 9 introns (Koller et al. 1984, Gingrich and Hallick 1985). The abundance of introns in protein-coding genes of *Euglena gracilis* chloroplasts contrasted sharply with their rare occurrence in genes from land plant chloroplasts. However, since the discovery of the intron in *rpl2*, a number of introns have been reported in chloroplast genes (see Table 1). Of these, however, only a few have been shown to have the intron correctly removed with ligation of the exons in vivo. For the *Triticum aestivum atpF* gene it was shown by Sl nuclease analysis that the mature RNA has the intron removed at the predicted exon-intron boundaries (Bird et al. 1985). Sequencing of *Spinacea oleracea* chloroplast cDNA clones also revealed that the exons of *atp* F were correctly ligated (Hudson et al. 1987). Similarly, primer extension of *Marchantia polymorpha* chloroplast RNA using exon-specific primers has shown that the *pet*B, *pet*D and *trans*-spliced *rps*12 introns are precisely removed with correct ligation of the exons (Fukuzawa et al. 1987, Zaita et al. 1987). The

Table 1. Introns present in protein-coding genes of land plant chloroplasts.

Gene	Species	Intron size (bp)	Reference
*atp*F	*Triticum aestivum*	823	Bird et al. (1985)
*atp*F	*Spinacea oleracea*	764	Hennig and Herrmann (1986)
*atp*F	*Pisum sativum*	690	Hudson et al. (1987)
*atp*F	*Nicotiana tabacum*	695	Shinozaki et al. (1986a)
*atp*F	*Marchantia polymorpha*	588	Ohyama et al. (1986)
*ndh*A	*Nicotiana tabacum*	1,242	Shinozaki et al. (1986c)
*ndh*A	*Marchantia polymorpha*	713	Ohyama et al. (1986)
*ndh*B	*Nicotiana tabacum*	757	Shinozaki et al. (1986c)
*ndh*B	*Marchantia polymorpha*	537	Ohyama et al. (1986)
*pet*B	*Triticum aestivum*	744	S. Hird, unpublished
*pet*B	*Pisum sativum*	810	C. Eccles, unpublished
*pet*B	*Nicotiana tabacum*	759	Shinozaki et al. (1986c)
*pet*B	*Spinacea oleracea*	777	Westhoff et al. (1986)
*pet*B	*Marchantia polymorpha*	495	Fukuzawa et al. (1987)
*pet*D	*Triticum aestivum*	746	S. Hird, unpublished
*pet*D	*Pisum sativum*	715	C. Eccles, unpublished
*pet*D	*Nicotiana tabacum*	742	Shinozaki et al. (1986c)
*pet*D	*Spinacea oleracea*	743	Heinemeyer et al. (1984)
*pet*D	*Marchantia polymorpha*	493	Fukuzawa et al. (1987)
*rpl*16	*Nicotiana tabacum*	1,020	Shinozaki et al. (1986c)
*rpl*16	*Marchantia polymorpha*	536	Ohyama et al. (1986)
*rpl*2	*Nicotiana tabacum*	666	Shinozaki et al. (1986c)
*rpl*2	*Nicotiana debneyi*	666	Zurawski et al. (1984)
*rpl*2	*Marchantia polymorpha*	545	Ohyama et al. (1986)
*rps*12	*Nicotiana tabacum*		Fromm et al. (1986)
exons I–II		trans-split	Torazawa et al. (1986)
exons II–III		540	
*rps*12	*Marchantia polymorpha*		Fukuzawa et al. (1986)
exons I–II		trans-split	
exons II–III		499	
*rps*16	*Nicotiana tabacum*	860	Shinozaki et al. (1986b)
*rpo*C	*Marchantia polymorpha*	597	Ohyama et al. (1986)

presence of introns in the other genes listed in Table 1 was predicted by sequence homology within the introns, comparison of their amino acid sequences with those of the corresponding *Escherichia coli* genes or the detection of precursor and mature RNA molecules in vivo.

Trans-splicing of the chloroplast *rps*12 gene

The chloroplast *rps*12 gene is split by two introns into 3 exons. In the *Nicotiana tabacum* genome exons II and III are separated by a 540 bp intron and are present as two copies in the inverted repeat region, whereas exon 1 is located in the large single copy region, 90 kbp and 126 kbp from the two copies of exons II and III (Shinozaki et al. 1986c). In the *Marchantia polymorpha* chloroplast genome, exons II and III are located in the large single copy region and exon I is found 60 kbp away on the opposite strand (Ohyama et al. 1986). Thus, expression of *rps*12 may require *trans*-splicing of its exons. Koller et al. (1987) have shown, by electron microscope analysis of RNA:DNA hybrids, that in *Nicotiana tabacum* exon I is transcribed as part of a polycistronic RNA separately from exons II and III which also form part of a polycistronic RNA. Furthermore, the most abundant class of hybridized RNA molecules contained exon I covalently linked to exons II and III in the absence of sequences transcribed downstream of exon I and upstream of exons II and III, thus providing definitive evidence for a *trans*-splicing reaction between exons I and II. Further evidence for *trans*-splicing of exons I and II was provided by Zaita et al. (1987) who demonstrated, by reverse transcription of total chloroplast RNA from *Nicotiana tabacum* using a primer specific to exon II, that exons I and II were correctly ligated. Such a *trans*-splicing reaction absolutely requires a specific interaction between pre-RNAs, and as such Zaita et al. (1987) have shown that a base-pairing interaction is possible between sequences downstream of exon I and sequences upstream of exon II. *Trans*-splicing has also been shown to occur for the nuclear adenovirus and β-globin pre-RNAs (Konarska et al. 1985; Solnick, 1985).

Chloroplast group II introns

The majority of the introns from land plant chloroplasts resemble the group II introns from yeast mitochondrial genes (Ohyama et al. 1986). The only exceptions to this to date are the *trn* L (UAA) genes for leucine tRNA from *Zea mays* and *Vicia faba* (Steinmetz et al. 1982, Bonnard et al. 1984) which appear to contain group I type introns. The introns from the *Chlamydomonas reinhardii* rRNA and *psb*A genes also resemble the group I introns (Erickson et al. 1984, Rochaix et al. 1985), while those from *Euglena gracilis* share some features of the group II introns (Keller and Michel 1985).

Group II introns are characterised by considerable sequence homology at their 3′ ends and a common $G_U^A G_C^U G$ sequence motif at their 5′ ends (Michel and Dujon 1983, Keller and Michel 1985). They have the potential to form a core made up of six helical structures, the last two of which involve well-conserved sequences which fall within the last 100 or so 3′ residues and invariably form a 14 bp hairpin with a GAAA terminal loop and ·G bulge

Table 2. Alignment of the 5′ and 3′ conserved sequences of the Yeast mitochondrial a/5c group II intron with the 5′ and 3′ sequences of the introns present in protein coding genes from higher plant chloroplasts. Vertical arrows show the 5′ and 3′ splice junctions. Underlined nucleotides form the typical 3′ consensus sequence motif of group II introns.

5′ conserved sequence	Gap	3′ conserved sequence	Gene
GUGCG....CGUGAGCCGUAUGCGAUGAAAAGCGCACGUACGGUCCUUACCGGUCGAAAAACUU←	4nts	→AGGUCUACCUAUCGGGAU	Yeast a/5c (1)
GUGCG....UGAGAGCCAAAUGAAUCGAAAGAGAAUCAUGUUUGUUCCGGAAGGGAUCAUAAAA←	13nts	→AAAAUAAUCUACUUUCAU	T. aestivum atpF (2)
GUGCG....UGAGAGCCAAAUGAAUGAAAUUCACGUUUGUUCCGGAAGGGAUCAUGAAU←	17nts	→GAGAUAAUCUACUUUCAU	S. oleracea atpF (3)
GUGCG....UGAGAGCCAAAUGAAAUGAAAUUGAAAGAUUCCGGAAGAGAUCAUAGAC←	17nts	→AAGAGAAUCUACUUUCAU	P. sativum atpF (4)
GUGCG....UGAGAGCCAAAUGAAAUGAAACUGAAAGAGAAUGUUUGUUCCGGAUAUUGGGAU←	17nts	→AAAAUAAUCUACUUUCAU	N. tabacum atpF (5)
GUGCG....AGAAAGCCGAUGAAAUUGAAAGUGGCAUGUUCGGUUUUGGGGGAGAGAUUAUAAAA←	4nts	→UAUAUAAUCUACUUUCAU	M. polymorpha atpF (6)
GUGCG....GAGGAGCCGUAUGAGGUGAAAAAGCGGUCGUUACGGUCGUUGGGGAGAGUGCAGUAAG←	6nts	→UUAUCUGUCAACUUUCC	N. tabacum rps12 exons 1-2 (7,8)
GUGCG....GAGAGCGAUGUAGGAUAAUUAAUCAGUAAGAUUGUUAAAGUGCAAUUUA←	6nts	→UUAUUUGUCAACUUUCC	M. polymorpha rps12 exons 1-2 (9)
GUGCG....GGAAAGCCGUAUUCGAUGAAAAGCGUUGCGUUACGGCUGGAGGAGAUCUUUCAU←	3nts	→UUUCGAGAUCCACCCUAC	N. tabacum rps12 exons 2-3 (7,8)
GUGCG....AAAAAGCCGUAUUCGUUGAAAAACGGAGGUACGGUUGGAGGAGAGAUAAAAAAAA←	No Gap	→UCCACCCUAC	M. polymorpha rps12 exons 2-3 (9)
GUGUG....CUCGAGCCGUACGAGAGAAACUGCCAUAUACGGUUGGGGGCCCCUUGA←	No Gap	→AUCUACACCAUCCCAU	N. tabacum rps16 (10)
GUGUG....CUUGAGCCUUNCGAGNUGAANNUUUCAUAUACGGGGGGGACUCGGGA←	No Gap	→CUAUCUCAAU	N. tabacum petB (11)
GUGUG....UUUGAGCCGUACGAGACAGAUUUUCUAUACGGUUCUAUACGGGUUCGGCCCCUUG←	No Gap	→GUUACCUAUCUCAAU	T. aestivum petB (12)
GUGUG....CUUGAGCCGUAUGAAUGAAAGCUCCAUAUACGGGUUUAGGGGGUUUAAUAG←	No Gap	→UUUACCUAUCUCAAU	P. sativum petB (13)
GUGUG....UUUAAGCCGUAAGAGAUAUAAUAAUCAUUGAUUUCAGUUUCAGGGGGAACUUUA←	No Gap	→GUAACCUAUCUCAAU	S. oleracea petB (14)
GUGUG....CUCGAGCCGAUGAUGAUGAUUUACAGUUCCAGUUUCCGGGUUUUGAGGGGGUUUGGGA←	4nts	→GAUUCACCUAUCCCAAU	M. polymorpha petB (15)
GUGUG....CUCCAGCCGAUGAUGAUAAUAAAAUUCCAUGUCCGGUUCUCCGGGUUUGGGGGA←	9nts	→GAGUUCACCUAUCCCAAU	N. tabacum petD (11)
GUGUG....CUCGAGCCGAUGAUGAUGAGAAAUAUUAUUCAUGUUCCGGUUCGUCCCAUUUGGGGGGCACUCU←	No Gap	→GUUCCACCUACCCAAU	T. aestivum petD (12)
GUGUG....CUGGAGCCGAUGAUGAGAAGAAACUCUGCAUGUGCUGGUCUUCGGAGAUGCAUCU←	4nts	→GAUUCACCUAUCCCAAU	P. sativum petD (13)
GUGUG....CUUGAGCCGAUGAUUCAUGUCUCGGUUCUGGGAGGUCUUUGGGGGGCACUCUU←	No Gap	→GAUUCACCUAUCCCAAU	S. oleracea petD (16)
GUGUG....UUGGAGCCGAUGAUAUUUAAAUUUAUACAUGUCCCGAUCUUUGGGGGCACUUUUU←	No Gap	→UAAUCUACCUAUAU	M. polymorpha petD (15)
GUGUG....GAGGAGCCGAUGAAUUUAAAA-UUUCACGUUUGUUUGGGAUAGUGAUGGAAU←	3nts	→GAAACACCUACACUAU	N. tabacum rpl16 (11)
GUGUG....GAGGAGCCGAUGAAUUAAAA-UUUCACGUUUUGAAAGGCGACAUAAUAAC←	No Gap	→CCACUAUAAC	M. polymorpha rpl16 (6)
GUGCG....AGAAAGAAGUAUCCUUUGGAAGAGCUUGUACAGUUUGGGAAGGGUUUUGGAUU←	19nts	→CAAAAGAAGAAUCUACUU	N. tabacum rpl2 (11)
GUGCG....AGAAAGCCGUAUGCUUGGAGUUUACAGUUUGGGAAGGGUUUUGAUU←	2nts	→CAAAAGAGAAUCUACUU	N. debneyi rpl2 (17)
GUGCG....GAAAAGCCGUAUGCUUGAAAA-AAGCUGUACUGUUUACUGUUUACUCUU←	3nts	→AAAAUUUAAAAUCUACUU	M. polymorpha rpl2 (6)

(1) Bonitz et al., 1980. (2) Bird et al., 1985. (3) Hennig and Herrmann, 1986. (4) Hudson et al., 1987. (5) Shinozaki et al., 1986a. (6) Ohyama et al., 1986. (7) Fromm et al., 1986. (8) Torazawa et al., 1986. (9) Fukuzawa et al., 1986. (10) Shinozaki et al., 1986b. (11) Shinozaki et al., 1986c. (12) S. Hird, unpublished. (13) C. Eccles unpublished. (14) Westhoff et al., 1986. (15) Fukuzawa et al., 1987. (16) Heinemeyer et al., 1984. (17) Zurawski et al., 1984.

on its 3′ side (Michel and Dujon 1983). The consensus sequence of this 14 bp hairpin is GAGC...RUR..R.gaaa.U..YAygY...GUUY (base-pairing nucleotides are in capitals, R = purine, Y = pyrimidine). Table 2, line 1 shows the 5′ and 3′ sequences of the group II a/5c intron from the yeast mitochondrial cytochrome *c* oxidase subunit 1 gene (Bonitz et al. 1980). Underlined nucleotides are common to other yeast mitochondrial group II introns and include the 14 bp hairpin consensus and a YUAYY.Y...AY consensus sequence positioned at the extreme 3′ end (Michel and Dujon 1983, Keller and Michel 1985). Aligned beneath this sequence are the 5′ and 3′ nucleotide sequences from a number of introns found in the chloroplast protein-coding genes of some land plants. The 5′ sequences of these chloroplast introns compare very well with the consensus for the yeast mitochondrial group II introns. Direct comparison of the chloroplast intron 3′ sequences with that of the a/5c intron shows clearly that the consensus sequence of the 14 bp hairpin is present. Base changes, where they occur, are nearly all compensatory, thereby maintaining CG, AU or GU base-pairing. Where the base changes are not compensatory it is still possible to form the characteristic 14 bp hairpin; however, the bulge on its 3′ side does not always

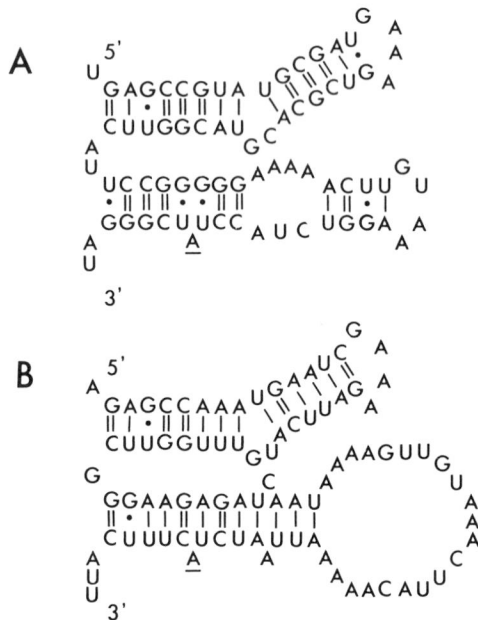

Fig. 1. Predicted secondary structure of the yeast mitochondrial a/5c (A) and *Triticum aestivum atp*F (B) introns obtained by folding the 3′ conserved sequences presented in Table 2. Note the 14 bp hairpin with the GAAA terminal loop, the underlined adenosine residue is the nucleotide thought to form the stem of the branch of the lariat.

contain a G residue. Furthermore, for some of the chloroplast introns the terminal loop does not contain the nucleotides GAAA but may consist of UAAA, AAAA or GGAA. The sequence YUAYY.Y...AY at the 3′ boundary of the intron is also apparent for the majority of the chloroplast introns. It is however, only poorly represented for the introns in the *Nicotiana tabacum* and *Marchantia polymorpha rp*12 genes, although if the 3′ exon boundary was moved 3 nucleotides downstream these sequences would then align very well and the reading frame would be maintained. Figure 1 shows the predicted secondary structure of the 3′ ends of the yeast mitochondrial a/5c and *Triticum aestivum atp*F introns. These structures were obtained by folding the 3′ sequences shown in Table 2 and are typical of group II introns (Michel and Dujon 1983). From the sequences presented in Table 2 it is clear that the introns so far detected in protein-coding genes of land plant chloroplasts can be classified as group II. This grouping however differs from that proposed by Shinozaki et al. (1986b), who classified the introns from *rp*12, *rps*12, *atp*F and *rps*16 into a group distinct from the yeast mitochondrial group II introns and called these group III introns. These introns were distinguished by the conserved boundary sequences GUGC-GNY at the 5′ end and AUCNRYY(N)YYAY at the 3′ end as well as their resemblance to the introns in *Euglena gracilis* chloroplast genes (Hallick et al. 1985). However, these consensus sequences also resemble those for group II introns and furthermore, introns from *Euglena gracilis* chloroplast genes have been shown to share some structural homologies with the yeast mitochondrial group II introns (Keller and Michel 1985).

Self-splicing of group II introns in vitro

Two mitochondrial group II introns have recently been shown to be capable of self-splicing in vitro: the a/5c intron of the yeast gene encoding subunit 1 of cytochrome *c* oxidase (Peebles et al. 1986; Van Der Veen et al. 1986) and the first intron (b/1) of the yeast gene encoding apocytochrome *b* (Schmelzer and Schweyen 1986). Self-splicing had previously been demonstrated only for the large rRNA gene of *Tetrahymena* (Kruger et al. 1982), a gene containing an intron with close structural homologies to the group I introns of fungal mitochondria, some of which have since been shown to self-splice in vitro (Garriga and Lambowitz 1984, Van Der Horst and Tabak 1985). Self-splicing of group II introns occurs at an optimum temperature of 45 °C, requires only low concentrations of Mg^{2+} (5–10 mM) and is enhanced by spermidine (2 mM) (Peebles et al. 1986, Van Der Veen et al. 1986). The most important difference from the self-splicing of group I introns is a complete lack of requirement for an added guanosine nucleotide (Peebles et al. 1986, Van Der Veen et al. 1986). The self-splicing reaction occurs with excision of the intron as a stable circular structure together with ligation of the exons. The circular introns were in fact shown to be branched circles or lariats, held together by a 2′–5′ phosphodiester bond (Wallace and Edmonds 1983) and are characteristic of those formed during the splicing of nuclear pre-RNA's (Grabowski et al. 1984, Padgett et al. 1984, Ruskin et al. 1984). The branch point of the lariat formed during the self-splicing reaction of intron a/5c has been mapped 8 nucleotides upstream from the 3′ splice junction, to a site which contains the sequence AUC and falls within a region that forms a highly conserved hairpin in group II introns (Van Der Veen et al. 1986). The adenine residue, thought to form the stem of the branch is present in the bulge in the hairpin and is conserved in group II introns (see Fig. 1). Indeed, this adenine is similarly positioned in all the chloroplast introns listed in Table 2 with the exception of the second intron of the *trans*-split *rps*12 gene, although moving the 3′ splice junction 3 bases downstream would correct this.

Self-splicing of group II introns, like group I introns, requires no external source of energy and the reaction is therefore thought to occur via a series of transesterification events (Peebles et al. 1986, Van Der Veen et al. 1986). The 3′-OH of the nucleotide cofactor GTP, necessary for self-splicing of group I introns in vitro, serves as an attacking group for the first phospho-transfer (Cech 1986). For group II introns this is provided by the 2′-OH group at the branch site which attacks the 5′ splice-site junction, resulting in a transesterification reaction, cleavage of the 5′ exon-intron boundary and lariat formation (see Fig. 2). The specificity of this attack may be provided by the conserved GUGCG sequence motif at the 5′ end of the intron (Table 1). Cleavage at the 3′ exon-intron boundary results from nucleophilic attack by the 3′-OH group of the 5′ exon, thereby releasing the lariat and spliced exons (Fig. 2).

The splicing pathway outlined in Fig. 2 requires that the 5′ exon is associated with the intron-3′ exon intermediate in order for exon ligation to occur. For nuclear pre-RNA splicing this is achieved by the ribonucleoprotein particles within the spliceosome structure, while for group I introns there is a 5′-exon binding site within the intron which serves this purpose (Been and Cech 1986, Waring et al. 1986). Recent *trans*-splicing experiments with intron-3′ exon transcripts and separate 5′ exon molecules have indicated that an interaction occurs between the a/5c intron and its 5′ exon (Jacquier and Robash 1986). Further evidence for such an interaction was provided by a series of experiments which demonstrated that > 18 nucleotides of the 5′ exon were necessary for efficient splicing of the a/5c intron

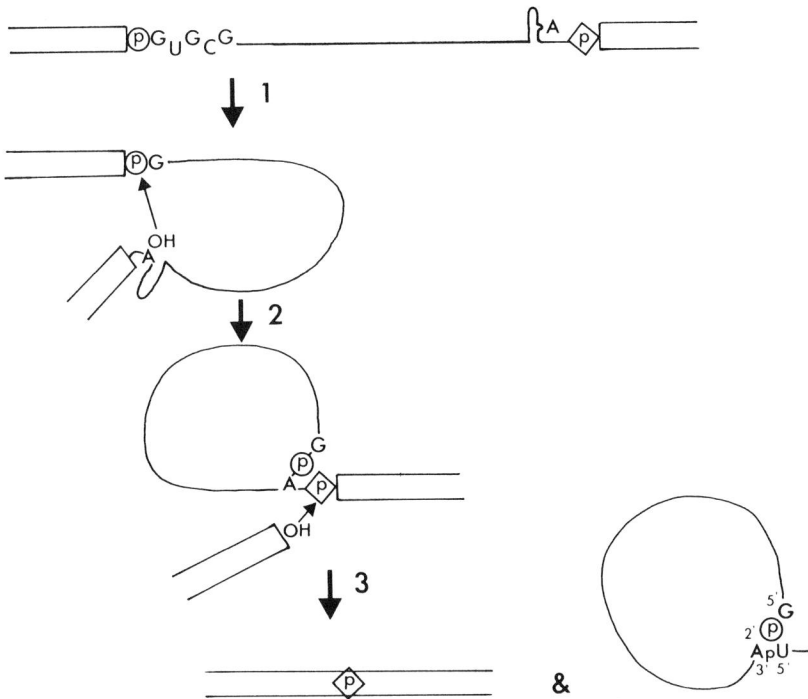

Fig. 2. Putative pathway for the self-splicing of group II introns in vitro (adapted from Van Der Veen et al. 1986). (1) shows the 2′ OH group of the branch point adenosine nucleotide attacking the 5′ splice site junction. (2) shows 3′-OH group of the cleaved 5′ exon attacking the 3′ splice-site junction resulting in (3) religated exons and the excised intron lariat. It is thought that the phosphate group in the 5′–2′ bond of the lariat branch is donated by the 5′ exon.

(Van Der Veen et al. 1987). Finally, Jacquier and Michel (1987) have defined, by sequence and mutational analysis of the a/5c and b/1 introns, two sites within the 3′ end of the 5′ exon capable of base-pairing with two sites within the 5′ end of the intron. The first pairing involves the last 5–8 bases of the 5′ exon (IBS1 = Intron Binding site 1) and sequences within the peripheral domains of the first intron hairpin EBS1 (Exon Binding Site 1). The second pairing involves an average of 6 nucleotides of the 5′ exon, located just upstream of IBS1, and another site within the first intron hairpin. These binding sites have been detected in the introns within the *Spinacea oleracea petD* and *Marchantia polymorpha* chloroplast genes (Jacquier and Michel 1987). Binding sites are also present within the first conserved hairpin of the *Triticum aestivum atpF* intron capable of base-pairing with the 3′ end of the 5′ exon, and these sequences are present too in the *Nicotiana tabacum, Spinacea oleracea* and *Marchantia polymorpha atpF* introns. Indeed such an interaction between the intron and upstream exon might explain how *trans*-splicing could occur in plant chloroplasts. Interestingly, these sites are absent from the group II-like introns of *Euglena gracilis* chloroplast genes (Jacquier and Michel 1987).

Due to the similarity between the introns of chloroplast protein-coding genes and yeast mitochondrial group II introns it might be expected that chloroplast introns self-splice in vitro. To examine this possibility the *Triticum aestivum atp*F gene intron, together with 103 bp of the 5′ exon and 41 bp of the 3′ exon, were cloned behind the SP6 promotor of pSP65 (Melton, 1984) to produce pSP65/*atp*F. Linearization of pSP65/*atp*F with *Pvu* II followed by transcription with the SP6 polymerase in the presence of α-^{32}P-UTP produced a 1.2 kb precursor RNA, which after purification from a 4% polyacrylamide/8M urea gel was used as a substrate for self-splicing in vitro. Reactions were carried out in a buffer containing 4mM Tris-HCl (pH 8.3), 5mM $MgCl_2$ and 2mM spermidine for 45 minutes at 45 °C and the products were analysed on 4% polyacrylamide/8M urea gels. Under these conditions there was no evidence for the production of a lariat intron structure to indicate splicing of the *atp*F pre-RNA (Fig. 3, lanes 1,2). Furthermore, self-splicing of the yeast intron a/5c pre-RNA (transcribed in vitro from the clone pSP64/6 kindly provided by L. Grivell) under identical conditions yielded the excised intron lariat (Fig. 3, lanes 3,4). Further attempts to self-splice the *atp*F pre-RNA were made by varying the conditions of the reaction. Incubations were performed at a number of physiologically-significant temperatures ranging from 15 to 45 °C, the $MgCl_2$ concentration in the buffer was varied from 5 to 100mM and the nucleotide GTP was added in amounts varying from 0.1 to 1mM in a group I-type self-splicing reaction (performed at 30 °C for 2h in 30 mM Tris-HCl pH 7.5, 100mM $(NH_4)_2SO_4$, 10mM $MgCl_2$; Burke et al. 1986). In all cases there was no observed lariat production from the *atp*F intron. Changing the cross-linking as well as the percentage composition of polyacrylamide in the gels (conditions under which a lariat exhibits differences in mobility; Padgett et al. 1984, Ruskin et al. 1984) did not reveal a lariat product which might have been obscured by the full-length 1.2 kb pre-RNA or its degradation products. Therefore, it was concluded that the *atp*F pre-RNA probably does not self-splice to produce a lariat intron in vitro.

To investigate further the mode of splicing of the *atp*F exons, purified pre-RNA was added to various crude chloroplast extracts. Transcriptionally-active extracts were prepared from the chloroplasts of *Pisum sativum* or *Euglena gracilis* following the methods of Gruissem et al. (1983), Greenberg et al. (1984) and Orozco et al. (1985). The purified *atp*F pre-RNA was added to these extracts under the conditions specified for transcription

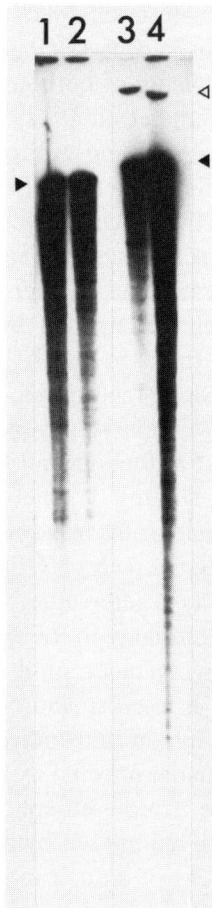

Fig. 3. In vitro self-splicing reactions of the pSP65/*atp*F (lanes 1, 2) and pSP64/6 (lanes 3, 4) pre-RNAs. The solid arrow head indicates the full-length pre-RNA (~ 1.2 kb for pSP65/*atp*F and ~ 1.6 kb for pSP64/6). The open arrow head indicates the lariat intron of pSP64/6.

in vitro, except the nucleotides CTP and UTP were omitted, while ATP was added to 5mM, GTP to 2.5 mM and creatine phosphate to 50 mM. Incubation was for 1h at 25 °C following which the products were analysed on 4% polyacrylamide/8M urea gels as before. There was no observed lariat production from the *atp*F intron. However, a major problem with these extracts was endogenous ribonuclease activity which often resulted in some degradation of the pre-RNA, thereby possibly obscuring any spliced products. Incubation of the *atp*F pre-RNA in crude chloroplast lysates prepared from *Pisum sativum* or *Euglena gracilis* produced similar negative results. As such therefore, it has not yet been possible to analyse the splicing of the *atp*F intron in vitro.

The inability of the *atp*F intron to self-splice in vitro suggests the involvement of *trans*-acting factor(s) in its processing pathway. *Trans*-acting factors have been implicated in the splicing of both group I (Lazowski et al. 1980, Anziano et al. 1982, De La Salle et al. 1982, Weiss-Brummer et al. 1982, Guiso et al. 1984, Jacq et al. 1984) and group II introns (Carignani et al. 1983, Schmelzer et al. 1983, Schmelzer and Schweyen 1986). *Trans*-acting factors or proteins involved in the splicing of yeast and fungal mitochondrial genes may be encoded by Open Reading Frames (ORFs) present in the introns themselves or in nuclear genes (Carignani et al. 1983, Schmelzer et al. 1983). Those proteins that are encoded by introns have been termed RNA maturases. Splicing of the a/5c and b/1 group II introns is almost certainly facilitated by *trans*-acting factors since the reaction in vitro is slow and incomplete while that in vivo is very rapid (Schmelzer and Schweyen 1986). Furthermore, two yeast mitochondrial group II introns a/1 and a/2 of the cytochrome *c* oxidase subunit 1 gene contain long ORFs, the expression of that for at least a/1 being necessary for excision of its intron (Carignani et al. 1983). Indeed, the *trn*K (UUU) gene for lysine tRNA from *Nicotiana tabacum* chloroplasts is interrupted by a group II intron which contains an ORF with local homology to the ORF present in the a/2 intron (Shinozaki et al. 1986c). Involvement of a nuclear-gene product with splicing of the mitochondrial b/1 self-splicing intron was demonstrated by a mutational analysis of the intron and indicated that the nuclear factor interacted with the intron sequence affected by the mutation (Schmelzer and Schweyen 1986). Similarly, the group I self-splicing intron of the *Neurospora* apocytochrome *b* gene is inhibited in vivo by a recessive nuclear mutation, thereby implicating involvement of a nuclear-encoded factor (Garriga and Lambowitz 1984, Collins and Lambowitz 1985). If it is presumed that all group I and group II introns use the same respective splicing pathways, a plausible role for the *trans*-acting factors is that they facilitate folding of precursor RNAs and stabilize the resulting secondary structure (Cech 1986, Kreike et al. 1987). As such therefore, absence of these factors could account for the lack of self-splicing of some group I introns and the inefficient self-splicing of some group II introns (Schmelzer and Schweyen 1986). Furthermore, involvement of *trans*-acting factors in the splicing of chloroplast introns would explain the inability of the *atp*F intron to self-splice in vitro.

Due to the close homology between the conserved sequences of their introns, the splicing of chloroplast introns can be assumed to follow a similar pathway to that outlined for the yeast mitochondrial group II introns (Fig. 2). Although *trans*-acting factors are almost certainly involved, no ORFs present in the chloroplast introns of protein-coding genes have been shown to encode maturase-like molecules. However, it is possible that the *trans*-acting factors involved resemble the ribonucleoprotein particles of nuclear pre-RNA splicing. This raises the possibility that proteins and small RNA molecules may be involved in the splicing reactions in chloroplasts. The stage at which these factors might interact with the intron and relevant exon sequences is not known and it is possible that the splicing events may be coupled to transcription in vivo. Further work is necessary to investigate these possibilities.

Acknowledgements

We thank Professor L. Grivell for the gift of plasmids containing the mitochondrial group II intron, Sean Hird and Chris Eccles for access to their unpublished results, Hugh Salter and Dave Willey for help with computer analysis, and Chris Chalk and David Last for the photography. This work was supported by an SERC grant.

References

Abelson J (1979) RNA processing and the intervening sequence problem. Ann Rev Biochem 48: 1035–1069

Anziano PQ, Hanson DL, Mahler HR and Perlman PS (1982) Functional domains in introns: *trans*-acting and *cis*-acting regions of intron 4 of the *cob* gene. Cell 30: 925–932

Been MD and Cech TR (1986) One binding site determines sequence specificity of Tetrahymena pre-rRNA self-splicing, *trans*-splicing, and RNA enzyme activity. Cell 47: 207–216

Bird CR, Koller B, Auffret AD, Huttly AK, Howe CJ, Dyer TA and Gray JC (1985) The wheat chloroplast gene for CF_o subunit 1 of the ATP synthase contains a large intron. EMBO J 4: 1381–1388

Bonitz SG, Coruzzi G, Thalenfeld BE, Tzagoloff A and Macino G (1980) Assembly of the mitochondrial membrane system. Structure and nucleotide sequence of the gene coding for subunit 1 of yeast cytochrome oxidase. J Biol Chem 255: 11925–11941

Bonnard G, Michel F, Weil JH and Steinmetz A (1984) Nucleotide sequence of the split tRNA $^{Leu}_{UAA}$ gene from *Vicia faba* chloroplasts: evidence for structural homologies of the chloroplast tRNA^{Leu} intron with the intron from the autosplicable *Tetrahymena* ribosomal RNA precursor. Mol Gen Genet 194: 330–336

Breathnach R, Mandel JL and Chambon P (1977) Ovalbumin gene is split in chicken DNA. Nature 270: 314–319

Breathnach R, Benoist C, O'Hare K, Gannon F and Chambon P (1978) Ovalbumin gene: evidence for a leader sequence in mRNA and DNA sequences at the exon-intron boundaries. Proc Natl Acad Sci USA 75: 4853–4857

Brody E and Abelson J (1985) The "spliceosome": yeast pre-messenger RNA associates with a 40S complex in a splicing-dependent reaction. Science 228; 963–967

Burke JM, Irvine KD, Kaneko KJ, Kerker BJ, Oettgen AB, Tierney WM, Williamson CL, Zaug AJ and Cech TR (1986) Role of conserved sequence elements 9L and 2 in self-splicing of the Tetrahymena ribosomal RNA precursor. Cell 45: 167–176

56

Carignani G, Groudinsky O, Frezza D, Schiaron E, Bergantino E and Slonimski PP (1983) An mRNA maturase is encoded by the first intron of the mitochondrial gene for the subunit 1 of cytochrome oxidase in S. cerevisiae. Cell 35: 733–742

Cech TR (1986) The generality of self-splicing RNA: relationship to nuclear mRNA splicing. Cell 44: 207–210

Collins RA and Lambowitz AM (1985) RNA splicing in *Neurospora* mitochondria. Defective splicing of mitochondrial mRNA precursors in the nuclear mutant *Cyt* 18-1. J Mol Biol 184: 413–428

Davies RW, Waring RB, Ray JA, Brown TA and Scazzocchio C. (1982) Making ends meet: a model for RNA splicing in fungal mitochondria. Nature 300: 719–724

De La Salle H, Jacq C and Slonimski PP (1982) Critical sequences within mitochondrial introns: pleiotropic mRNA maturase and *cis*-dominant signals of the *box* intron controlling reductase and oxidase. Cell 28: 721–732

Deno H, Kato A, Shinozaki K and Sugiura M (1982) Nucleotide sequences of tobacco chloroplast genes for elongator $tRNA^{Met}$ and $tRNA^{Val}$ (UAC): the $tRNA^{Val}$ (UAC) gene contains a long intron. Nucl Acids Res 10: 7511–7520

Deno H and Sugiura M (1984) Chloroplast $tRNA^{Gly}$ gene contains a long intron in the D stem: nucleotide sequences of tobacco chloroplast genes for $tRNA^{Gly}$ (UCC) and $tRNA^{Arg}$ (UCU). Proc Natl Acad Sci USA 81: 405–408

Erickson JM, Rahire M and Rochaix JD (1984) *Chlamydomonas reinhardii* gene for the 32000 mol. wt. protein of photosystem II contains four large introns and is located entirely within the chloroplast inverted repeat. EMBO J 3: 2753–2762

Frendewey D and Keller W (1985) Stepwise assembly of a pre-mRNA splicing complex requires U-snRNPs and specific intron sequences. Cell 42: 355–367

Fromm H, Edelman M, Koller B, Goloubinoff P and Galun E (1986) The enigma of the gene coding for ribosomal protein S12 in the chloroplasts of *Nicotiana*. Nucl Acids Res 14: 883–898

Fukuzawa H, Kohchi T, Shirai H, Ohyama K, Umesono K, Inokuchi H and Ozeki H (1986) Coding sequences for chloroplast ribosomal protein S12 from the liverwort, *Marchantia polymorpha*, are separated far apart on the different DNA strands. FEBS Lett 198: 11–15

Fukuzawa H, Yoshida T, Kohchi T, Okumura T, Sawano Y and Ohyama K (1987) Splicing of group II introns in mRNAs coding for cytochrome *b*6 and subunit IV in liverwort *Marchantia polymorpha* chloroplast genome: exon specifying a region coding for two genes with the spacer region. FEBS Lett 220: 61–66

Garriga G and Lambowitz AM (1984) RNA splicing in Neurospora mitochondria: self-splicing of a mitochondrial intron in vitro. Cell 39: 631–641

Gingrich JC and Hallick RB (1985) The *Euglena gracilis* chloroplast ribulose-1,5-bisphosphate carboxylase gene 1. Complete DNA sequence and analysis of the nine intervening sequences. J Biol Chem 260: 16156–16161

Grabowski PJ, Padgett RA and Sharp PA (1984) Messenger RNA splicing in vitro: an excised intervening sequence and a potential intermediate. Cell 37: 415–427

Grabowski PJ, Seiler SR and Sharp PA (1985) A multi-component complex is involved in the splicing of messenger RNA precursors. Cell 42: 345–353

Greenberg BM, Narita JO, De Luca-Flaherty C, Gruissem W, Rushlow KA and Hallick RB (1984) Evidence for two RNA polymerase activities in *Englena gracilis* chloroplasts. J Biol Chem 259: 14880–14887

Gruissem W, Greenberg BM, Zurawski G, Prescott DM and Hallick RB (1983) Biosynthesis of chloroplast transfer RNA in a spinach chloroplast transcription system. Cell 35: 815–828

Guiso N, Dreyfus M, Siffert O, Danchin A, Spyridakis A, Gargouri A, Claise M and Slonimski PP (1984) Antibodies against synthetic oligopeptides allow identification of the mRNA-maturase encoded by the second intron of the yeast *cob-box* gene. EMBO J 3: 1769–1772

Hallick RB, Gingrich JC, Johanningmeier U and Passavant CW (1985) Introns in *Euglena* and *Nicotiana* chloroplast protein genes. In: Vloten-Doting L, Groot GSP and Hall TC, eds. Molecular Form and Function of the Plant Genome, pp 211–220 NATO ASI series, Plenum

Heinemeyer W, Alt J and Herrmann RG (1984) Nucleotide sequence of the clustered genes for apocytochrome b6 and subunit 4 of the cytochrome b/f complex in the spinach plastid genome. Curr Genet 8: 543–549

Hennig J and Herrmann RG (1986) Chloroplast ATP synthase of spinach contains nine non-identical subunit species, six of which are encoded by plastid chromosomes in two operons in a phylogenetically conserved arrangement. Mol Gen Genet 203: 117–128

Hudson GS, Mason JG, Holton TA, Koller B, Cox GR, Whitfeld PR and Bottomley W (1987) A gene cluster in the spinach and pea chloroplast genomes encoding one CF_1 and three CF_0 subunits of the H^+-ATP synthase complex and the ribosomal protein S2. J Mol Biol 196: 283–298

Jacq C, Banroques J, Becam AM, Slonimski PP, Guiso N and Danchin A (1984) Antibodies against a fused '*lacZ*-yeast mitochondrial intron' gene product allow identification of the mRNA maturase encoded by the fourth intron of the yeast *cob-box* gene. EMBO J 3: 1567–1572

Jacquier A and Rosbash M (1986) Efficient trans-splicing of a yeast mitochondrial RNA group II intron implicates a strong 5′exon-intron interaction. Science 234: 1099–1104.

Jacquier A and Michel F (1987) Multiple exon-binding sites in class II self-splicing introns. Cell 50: 17–29

Jeffreys AJ and Flavell RA (1977) The rabbit β-globin gene contains a large insert in the coding sequence. Cell 12: 1097–1108

Karabin GD, Farley M and Hallick RB (1984) Chloroplast gene for M_r 32000 polypeptide of photosystem II in *Euglena gracilis* is interrupted by four introns with conserved boundary sequences. Nucl Acids Res 12: 5801–5812

Keller M and Michel F (1985) The introns of the *Euglena gracilis* chloroplast gene which codes for the 32-kDa protein of photosystem II. Evidence for structural homologies with Class II introns. FEBS Lett 179: 69–73

Koch W, Edwards K and Kossel H (1981) Sequencing of the 16S–23S spacer in a ribosomal RNA operon of Zea mays chloroplast DNA reveals two split tRNA genes. Cell 25: 203–213

Koller B and Delius H (1984) Intervening sequences in chloroplast genomes. Cell 36: 613–622

Koller B, Gingrich JC, Stiegler GL, Farley MA, Delius H and Hallick RB (1984) Nine introns with conserved boundary sequences in the Euglena gracilis chloroplast ribulose-1,5-bisphosphate carboxylase gene. Cell 36: 545–553

Koller B, Fromm H, Galun E and Edelman M (1987) Evidence for in vivo *trans* splicing of pre-mRNAs in tobacco chloroplasts. Cell 48: 111–119

Konarska MM, Padgett RA and Sharp PA (1985) *Trans* splicing of mRNA precursors in vitro. Cell 42: 165–171

Kreike J, Schulze M, Ahne F and Lang BF (1987) A yeast nuclear gene, MRS1, involved in mitochondrial RNA splicing: nucleotide sequence and mutational analysis of two overlapping open reading frames on opposite strands. EMBO J 6: 2123–2129

Kruger K, Grabowski PJ, Zaug AJ, Sands J, Gottschling DE and Cech TR (1982) Self splicing RNA: autoexcision and autocyclization of the ribosomal RNA intervening sequences of *Tetrahymena*. Cell 31: 147–157

Lazowski J, Jacq C and Slonimski PP (1980) Sequence of introns and flanking exons in wild-type and *box 3* mutants of cytochrome b reveals an interlaced splicing protein coded by an intron. Cell 22: 333–348

Melton DA, Krieg PA, Rebagliatti MR, Maniatis T, Zinn K and Green ME (1984) Efficient in vitro synthesis of biologically active RNA and RNA hybridization probes from plasmids containing a bacteriophage SP6 promoter. Nucl Acids Res 12: 7035–7056

Michel F and Dujon B (1983) Conservation of RNA secondary structures in two intron families including mitochondrial-, chloroplast- and nuclear-encoded members. EMBO J 2: 33–38

Ohyama K, Fukuzawa H, Kohchi T, Shirai H, Sano T, Sano S, Umesono K, Shiki Y, Takeuchi M, Chang Z, Aota S, Inokuchi H and Ozeki H (1986) Chloroplast gene organization deduced from complete sequence of liverwort *Marchantia polymorpha* chloroplast DNA. Nature 322: 572–574

Orozco EM, Mullet JE and Chua N-H (1985) An in vitro system for accurate transcription initiation of chloroplast protein genes. Nucl Acids Res 13: 1283–1302

Padgett RA, Konarska MM, Grabowski PJ, Hardy SF and Sharp PA (1984) Lariat RNA's as intermediates and products in the splicing of messenger RNA precursors. Science 225: 898–903

Peebles CL, Perlman PS, Mecklenburg KL, Petrillo ML, Tabor JH, Jarrell KA and Cheng H-L (1986) A self-splicing RNA excises an intron lariat. Cell 44: 213–223

Rochaix JD and Malnoe P (1978) Anatomy of the chloroplast ribosomal DNA of Chlamydomonas reinhardii. Cell 15: 661–670

Rochaix JD, Rahire M and Michel F (1985) The chloroplast ribosomal intron of *Chlamydomonas reinhardii* codes for a polypeptide related to mitochondrial maturases. Nucl Acids Res 13: 975–984

Ruskin B, Krainer AR, Maniatis T and Green MR (1984) Excision of an intact intron as a novel lariat structure during pre-mRNA splicing in vitro. Cell 38: 317–331

Schmelzer C, Haid A, Grosch G, Schweyen RJ and Kaudewitz F (1981) Pathways of transcript splicing in yeast mitochondria. Mutations in intervening sequences of the split gene cob reveal a requirement for intervening sequence-encoded products. J Biol Chem 256: 7610–7619

Schmelzer C and Schweyen RJ (1986) Self-splicing of group II introns in vitro: mapping of the branch point and mutational inhibition of lariat formation. Cell 46: 557–565

Schmelzer C, Schmidt C, May K and Schweyen RJ (1983) Determination of functional domains in intron b/1 of yeast mitochondrial RNA by studies of mitochondrial mutations and a nuclear suppressor. EMBO J 2: 2047–2052

Shinozaki K, Deno H, Wakasugi T and Sugiura M (1986a) Tobacco chloroplast gene coding for subunit 1 of proton-translocating ATPase: comparison with the wheat subunit 1 and *E. coli* subunit b. Curr Genet 10: 421–423

Shinozaki K, Deno H, Sugita M, Kuramitsu S and Sugiura M (1986b) Intron in the gene for the ribosomal protein S16 of tobacco chloroplast and its conserved boundary sequences. Mol Gen Genet 202: 1–5

Shinozaki K, Ohme M, Tanaka M, Wakasugi T, Hayashida N, Matsubayashi T, Zaita N, Chunwongse J, Obokata J, Yamaguchi-Shinozaki K, Ohto C, Torazawa K, Meng BY, Sugita M, Deno H, Kamogashira T, Yamada K, Kusuda J, Takaiwa F, Kato A, Tohdoh N, Shimada H and Suguira M (1986c) The complete nucleotide sequence of the tobacco chloroplast genome: its gene organization and expression. EMBO J 5: 2043–2049

Solnick D (1985) *Trans* splicing of mRNA precursors. Cell 42: 157–164

Steinmetz A, Gubbins EJ and Bogorad L (1982) The anticodon of the maize chloroplast gene for tRNA$_{UAA}^{leu}$ is split by a large intron. Nucl Acid Res 10: 3027–3037

Torazawa K, Hayashida N, Obokata J, Shinozaki K and Sugiura M (1986) The 5′ part of the gene for ribosomal protein S12 is located 30 kbp down-stream from its 3′ part in tobacco chloroplast genome. Nucl Acids Res 14: 3143

Van Der Horst G and Tabak HF (1985) Self-splicing of yeast mitochondrial ribosomal and messenger RNA precursors. Cell 40: 759–766

Van Der Veen R, Arnberg AC, Van Der Horst G, Bonen L, Tabak HF and Grivell LA (1986) Excised group II introns in yeast mitochondria are lariats and can be formed by self-splicing in vitro. Cell 44: 225–234

Van Der Veen R, Arnberg AC and Grivell LA (1987) Self-splicing of a group II intron in yeast mitochondria: dependence on 5′ exon sequences. EMBO J 6: 1079–1084

Wallace JC and Edmonds M (1983) Polyadenylated nuclear RNA contains branches. Proc Natl Acad Sci USA 80: 950–954

Waring RB, Towner P, Minter SJ and Davies RW (1986) Splice-site selection by a self-splicing RNA of *Tetrahymena*. Nature 321: 133–139

Weiss-Brummer B, Rodel G, Schweyen RJ and Kaudewitz F (1982) Expression of the split gene *cob* in yeast: evidence for a precursor of a "maturase" protein translated from intron 4 and preceding exon. Cell 29: 527–536

Westhoff P, Farchaus JW and Herrmann RG (1986) The gene for the M_r 10,000 phosphoprotein associated with photosystem II is part of the psb B operon of the spinach plastid chromosome. Curr Genet 11: 165–169

Zaita N, Torazawa K, Shinozaki K and Suguira M (1987) *Trans*-splicing in vivo: joining of transcripts from the 'divided' gene for ribosomal protein S12 in the chloroplasts of tobacco. FEBS Lett 210: 153–156

Zurawski G, Bottomley W and Whitfeld PR (1984) Junctions of the large single copy region and the inverted repeats in *Spinacia oleracea* and *Nicotiana debneyi* chloroplast DNA: sequence of the genes for tRNA^{His} and the ribosomal proteins S19 and L2. Nucl Acids Res 12: 6547–6558

Govindjee et al. (eds), Molecular Biology of Photosynthesis: 61–76
© 1988 Kluwer Academic Publishers

Regular Paper

Pea chloroplast tRNA^{Lys}(UUU) gene: transcription and analysis of an intron-containing gene

SCOTT K. BOYER[1] & JOHN E. MULLET[2]

[1]*Purdue University, Biological Sciences Department, Lilly hall of Life Sciences, West Lafayette, Indiana 47907, USA;* [2]*Author for correspondence; Department of Biochemistry and Biophysics, Texas A&M University, College Station, Texas 77843-2128, USA*

Received October 1, 1987 accepted 15 December, 1987

Key words: Chloroplast, intron, *Pisum sativum*, transcription, tRNA

Abstract. The pea chloroplast *trnK* gene which encodes tRNA^{Lys}(UUU) was sequenced. *TrnK* is located 210 bp upstream from the promoter of *psbA* and immediately downstream from the 3′-end of *rbcL*. The gene is transcribed from the same DNA strand as *psbA* and *rbcL*. A 2447 bp intron with class II features is located in the *trnK* anticodon loop. The intron contains a 506 amino acid open reading frame which could encode an RNA maturase. The primary transcript of *trnK* is 2.9 kb long; its 5′-end was identified as a site of transcription initiation by *in vitro* transcription experiments. The 5′-terminus is adjacent to DNA sequences previously identified as transcription promoter elements. The most abundant *trnK* transcript is 2.5 kb long with termini corresponding to the 5′ and 3′ ends of the *trnK* exons. Intron specific RNAs were not detected. This suggests that RNA processing which produces tRNA^{Lys} leads to rapid degradation of intron sequences.

Introduction

Chloroplasts of most vascular plants contain a circular genome of 120 to 180 kbp (Whitfield and Bottomley 1983). Chloroplast DNA encodes approximately 120 genes including genes for proteins involved in photosynthesis, transcription and translation (Shinozaki et al. 1986). All of the chloroplast tRNAs are thought to be encoded on chloroplast DNA (Chu et al. 1985, Crouse et al. 1986). In tobacco, 30 genes which encode different tRNAs have been identified (Wakasugi et al. 1986). These genes are scattered throughout the genome and unlike Euglena (Hallick et al. 1984), there are few clusters of tRNA genes.

The sequences of chloroplast tRNAs show high homology to tRNAs described in prokaryotes (Gauss and Sprinzel 1984). Some of the chloroplast tRNAs have promoters comprised of DNA sequences 10 and 35 bp upstream of their transcription initiation site which are similar to promoter elements of chloroplast protein genes (Gruissem and Zurawski 1985, Hanley-Bowdoin and Chua 1987). Two chloroplast tRNA genes (*trnS1, trnR1*) have been reported to lack these sequences (Gruissem et al. 1987). Chloroplast tRNA genes have also been found to contain introns (Koch et al. 1981, Deno et al. 1982, Steinmetz et al. 1982, Bonnard et al. 1984, Deno and Sugiura. 1984, Krebbers et al. 1984, Bonnard et al. 1985, Sugita et al. 1985, Yamada et al 1986, Neuhaus and Link 1987). These introns are large (451 to 2574 bp) compared to introns in nucleur tRNA genes (13–60 bp) (Ogden et al. 1980).

Only one gene which encodes tRNA for lysine has been found in chloroplast DNA, the *trnK* (UUU) gene (Fig. 1A). The *trnK* gene is sufficient for the incorporation of all lysines into proteins because the "wobble" position of the tRNA anticodon allows it to recognize both lysine codons AAA and AAG. This gene was first characterized by Sugita et al. (1985) in tobacco. It was determined that tobacco *trnK* was interrupted by a 2.5 kb intron and within this intron was an open reading frame of 509 codons. Recently, Neuhaus and Link (1987) have found a similar situation in mustard and have shown a low amount of sequence homology between the intron open reading frame and maturases found in the introns of some yeast mitochondrial genes (Bonitz et al. 1980); (Osiewacz and Esser 1984).

We have previously reported the presence of the *trnK* gene upstream from *psbA* in pea chloroplast DNA (Boyer and Mullet 1986). In this paper a detailed analysis of the pea *trnK* transcription unit is described.

Materials and methods

DNA sequencing and preparation of ^{32}P labeled nucleic acid
DNA sequencing was performed using the M13/dideoxy sequencing method (Messing 1980). DNA, RNA, and protein sequences were analyzed on a VAX 11/750 (Digital Equipment Corp.) using *Intelligenetics*™ (Intelligenetics, Inc.) computer programs. Linear DNA molecules or oligonucleotides were 5'-end labelled with γ-^{32}P-ATP and T_4 polynucleotide kinase or 3'-end labeled with the appropriate α-^{32}P-dNTP and Klenow fragment of DNA polymerase (Maniatis et al. 1982). Radiolabelled RNA and non-radioactive RNA transcripts were produced in vitro by incubation of T_7 RNA polymerase and linearized pTZ-derived plasmids under conditions specified by Promega Technical Bulletin No. 001. When necessary, ^{32}P-labeled nucleic acid was gel purified (Maniatis et al. 1982).

Plasmid construction

A 307 bp HaeIII-DraI fragment from the 5′-end of pea *trnK* was inserted into pUC18 and pUC19 to generate pPHD118 and pPHD119, respectively (Fig. 6). These two plasmids were used in in vitro transcription experiments. The plasmid pPURF319 was constructed by limited exonuclease III digestion of a DNA fragment which contained the entire pea *trnK* gene. T_7 polymerase produced transcripts in vitro from pPURF319 which were similar in sequence to the in vivo *trnK* primary transcript. Transcripts produced by pPURF319 were used to test for in vitro self-splicing. A derivative of pPURF319 was constructed which contained approximately 600 bp of the 5′-end of the *trnK* gene. This plasmid, pPMB319 was used to generate RNAs in vitro which were tested for secondary structure and the ability to generate S_1 nuclease artifacts.

RNA analysis

Northern blot analysis was performed using denaturing formaldehyde gels (Maniatis et al. 1982) and subsequent transfer of the denatured RNA molecules to nylon membranes by blotting using 25 mM sodium phosphate buffer (pH 6.5). Prehybridization and hybridization were performed in a shaking incubator at 50° to 55°C. The prehybridization solution consisted of 0.27 M NaCl, 1.5 mM EDTA, 15 mM $NaPO_4$ (pH 7), 50% formamide, 2% SDS, 0.5% Blotto, and 0.5 mg/ml of denatured salmon sperm DNA. Hybridization solutions contained the same ingredients as the prehybridization solutions with the addition of 1 to 100 × 10^6 cpm, of ^{32}P-labeled riboprobes (specific activity of approximately 10^8 to 10^9 per μg RNA). Solutions used with RNA samples were treated with diethylpyrocarbonate to inhibit RNAses. S_1 nuclease protection experiments, primer extension assays, in vitro transcription and transcription capping experiments were performed as previously described (Boyer and Mullet 1986).

Results

DNA sequence analysis of the trnK region

The 3′-end of *rbcL* and the 5′-end of *psbA* were previously localized to the ends of a 5.2 kbp PstI fragment of pea ctDNA (Fig. 1B). Sequence analysis upstream from the 5′-end of *psbA* revealed the presence of a 3′-exon of *trnK* (Boyer and Mullet, 1986). In order to localize the 5′-exon of *trnK* the DNA region between the 3′-end of *rbcL* and the 3′-exon of *trnK* was mapped and sequenced (Figs. 1B, 2). The sequence analysis showed the presence of the

A

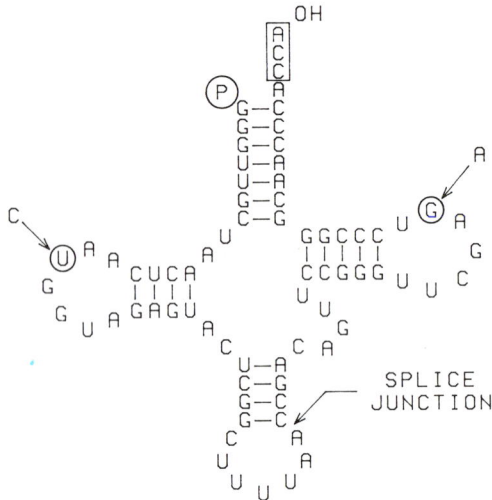

Fig. 1A. The cloverleaf structure of tRNA^Lys(UUU) without modified bases. Also indicated are sequence differences between *N. tabacum* (Sugita et al. 1985) and *P. sativum* with the position of the nucleotide substitution in the pea *trnK* circled. The site of the putative intron splice-junction is indicated is indicated by the arrow pointing to the anti-codon loop.

5′-exon of the *trnK* gene and confirmed that pea *trnK* contains a large intron similar to *trnK* genes of tobacco (Sugita et al. 1985) and mustard (Neuhaus) and Link 1987). Sequences near the putative 3′-splice junction could be folded into a secondary structure which has been observed in class II introns (Fig. 3). Other features which are consistent with this assignment include a conserved sequence motif "UGCGAC" at the 5′ splice junction and the sequence "GAAA" in a stem-loop near the 3′-splice junction (Michel and Dujon 1983). Furthermore, an open reading frame of 506 deduced amino acids was observed within the intron sequences (Fig. 2). The open reading frame was similar in sequence to open reading frames located in the *trnK* introns of tobacco and mustard (Sugita et al. 1985), Neuhaus and Link 1987).

Northern blot analysis of the trnK region
Numerous transcripts hybridize to the 5.2 kb PstI fragment (Fig. 4). As a first step in the analysis of these RNAs, Northern blots were done using single stranded probes specific for *rbcL, trnK* and *psbA* (Fig. 4). These blots

B

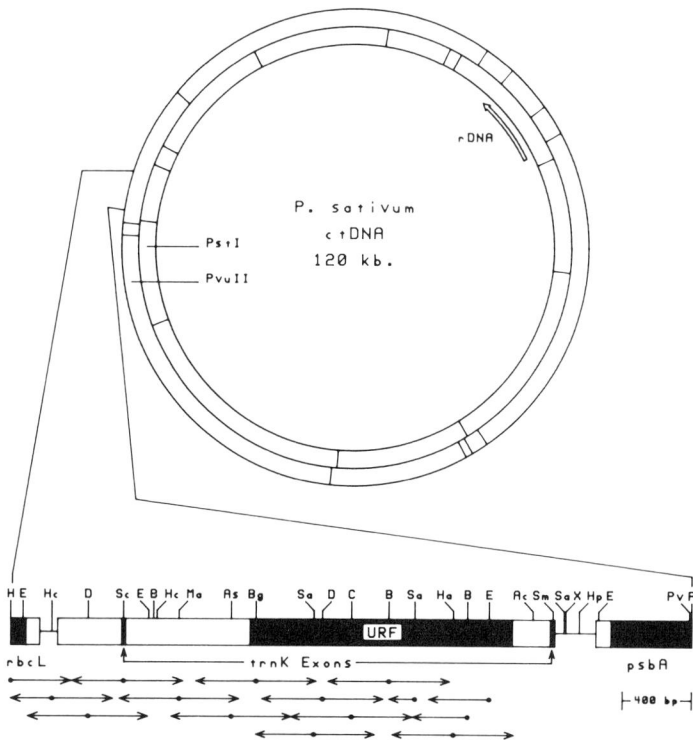

Fig. 1B. Restriction map, gene map, and sequencing strategy for the 4.0 kpb HindIII-PstI fragment of the *P. sativum* genome which encodes *trnK*. The boxed regions indicate mRNA coding regions with shaded regions as open reading frames. The sequencing strategy is shown below the map. Note that only portions of *psbA* and *rbcL* are shown. The restriction sites used for sequencing are abbreviated as follows: Ac-AccI, As- AsuII, B-BamHI, C-ClaI. D-DraI, E-EcoRI, H-HindIII, Ha-HaeIII, Hp-HphI, Ma-MaeII, P-PstI, Pv-PvuII, Sa-Sau3AI, Sc-ScaI, Sm-SmaI, and X-XbaI.

showed that a 1.8 kb RNA hybridized to *rbcL*, a 1.3 kb RNA hybridized to psbA and RNAs varying from 2.9 to 0.07 kb hybridize to *trnK*. It should be noted that the time of autoradiography of individual blots was varied in order to optimize visualization of the major transcripts. Identification of minor RNA species which hybridize to *trnK* was only possible using high specific activity RNA probes and long exposure times. For example, lanes labeled *psbA* and *rbcL* required only 15 and 30 min exposures at room temperature, whereas the lane labeled *trnK* required a 36 hour exposure at −80°C with an intensifying screen. Since gels were loaded on an equal RNA basis and probes were of similar specific activity, this indicates that tran-

```
   1 AAGCTTGCAAATGGAGTCCTGAATTAGCTGCTGCTTGTGAAGTCTGGAAGGAAATCAAATTTGAATTCC   69
  70 CAGCAATGGATACTTTGTAATCCAGTAATAATCATTCGTTCTATTAATTTCCATTAAACTCGGCCCAAT   138
                           ⊽
 139 CTTTTACTAAAAGGATTGAGCCGAATACTGTACACAATACTTGTTGTTTGTATCTATAATAAATCCTAT   207
 208 CGATCCGAAGGATCGATATATATATATTTTGTCAACACTAATATTGACAACAGTGTATCAAACAAATAT   276
                                             ‾‾‾‾‾‾              -10
                                               -35
 277 AATCTGATCGTGAATAAATGGATCGATTGACTCATGTCAGATAGGTTTTTTACTTAATCTTAATGGTGG   345
     ‾‾‾                                               ‾‾‾‾‾‾‾‾‾‾‾‾‾‾‾‾‾‾
 346 ATTCGTTGTATTTTGTTTGATACAAACAAGAAGTTTTTTTTTTATTCAAAGTGAAATAGAAATATGAAA   414
 415 AATATGTAAAAAATAATAATAATTTAAAACTATAGTAATACACAAATAACTACTGTCCTAAATAAGTGG   483
     ‾‾‾
 484 TCTATCATTTTTGATTCTAATTTTTTTTCTGTTTTTTTTTCATTGTAATTTAATGTAATATGTAACATTT   552
 553 GATTCTATTATTTATAGAATTGCATCTTTTTTTTTTTATTGCGATTGGCTATACACATCCTTGTTTGTAA   621
 622 TTTGATTAGGGTTGCTAACTCAATGGTAGAGTACTCGGCTTTTAAGTGCGACTCCATTTTTTACACATT   690
            ‾‾‾‾‾‾‾‾‾‾‾‾‾‾‾‾‾‾‾‾‾‾‾‾‾‾‾‾‾‾
            5'-Exon
 691 TCTATGAACTAATTGGTTCATCCCTACCATCGGTAGGGTTTGTAAGACCACGACTGATCCAGAAAGGAA   759
 760 TGAATGGAAAAAGCAGCATGTCGTATCAACGGTGAATTCTAAAAATTTTTCATTCATATTGGATCCGAC   828
                   ‾‾‾‾‾‾‾‾‾‾‾‾‾‾‾‾‾‾‾‾‾‾‾
 829 CAATATTTTTCTTTGAATTGTTGACTCGGAACAAATGAATTACGTTGGGTTGAATTAATAAAGGGATAG   897
 898 AGCTTAGCTGCTCCAATTATAAGGAAACAAAAAGCAACGAGCTTTCATTTTTTATTTGAATGATTACCA   966
 967 CATCTAATTATACGTTAAAAAAGGATTAGTGCTTGCTGTGGAAAAACTTTTTCCCACGAATGGATTAAG  1035
1036 GATTTTTGTTATGAGTCCTAATTATTAGCTATACCCCATTATGTTATGGGGTAGCAAGAAATATGTAGA  1104
1105 AGAAAGAGTATATTGATAAAGATATTTTTTTTCCAAAATCAAAAGAGCGATTGGTCCAAAAAATAAAGGA  1173
1174 TTTCTAACTAGCTTGTTAAAATGAGAAGGAACAAGCGAGGGGAGATAGTCCATTAATCGGTTTTAACTC  1242
1243 GTTTTCTAGGTATCTATTCATATTTGTTTTTGATTTCAATTCGAATATAGATATAGAATACCTTGTTTT  1311
1312 GACTGTATCGCACTATGTATCATTTGAGAATCCAATGAATCTCTGATCCTTTGACCAAATCGAATTTAG  1380
1381 AAAATGAAGGAATATCAAGTATATTTAGAACGATCTAGATCTCGCCAACAGGACTTCCTATACCCACTT  1449
         ‾‾‾
1450 CTTTTTAGGGAGTATATTTATGGACTTGCTTATAGTCATCATTTTCATAGATCCATTTTTTTGGAAAAT  1518
1519 GTAGGTTATGACAATAAATATAGTTTACTAATTGTAAAACGGTTAATTACTCGAATGTATCAACAGAAT  1587
1588 CATTTAATCATTTCGGCTAATGATTCTCCCAAAAATCCATTTTGGGGGTATAATAAGAATTTGGATTGT  1656
1657 CAAATAATATCAGAGGGTTTGCCCATCGTCGTGGAAATTCCATTTTTCCGACAATTAAGCTCTTCCTTC  1725
1726 GAGGAGGCAGAAATCCTCAAATCTTTTAAAAATTTGAGATCAATTCATTCCATTTTCCCCTTTTTGGAA  1794
1795 GATAAATTTACATATTTAAATTATGTGTCAGATATACGAATACCCTATCCTATCCATCTGGAAATCTTA  1863
1864 GTTCAAATCCTTCGATACTGGGTGAAAGATGCCCCTTTTTTTCATTTATTACGGCTGTTTCTTTATCAT  1932
1933 TTTTGTAATTGGAATAGTTTTTATTACTACCAAAAAATCGATTTCGACTTTTTCAAAAAGTAATCCAAGA  2001
2002 TTATTTTTGTTCCTCCATAATTTTTATGTATGTGAATATGAATATATCCTCGTTTTTCTACGTAATAAA  2070
2071 TCCTCTCATTTAGGATTCAAATCTTTTAGCATTTTTTTTGAGCGAATTTTTTTTTTATGGAAAAAGAGAG  2139
2140 CATCTTGGAAAAGTTTTTGCTAAAGATTTTTCGTATCCTTTAACATTCTTCAAGGATCCTAACATTCAT  2208
2209 TATGTTCGATATCAAGGAAAATGCATTCTGGCTTCAAAGAATGCGCCTTTTTTGATGAATAAATGGAAA  2277
2278 CACTATTTTATCCATTTATGGCAATGTTTTTTTGATGTTTGGTCTCAACCAAGAATGATCAATATAAAC  2346
2347 CCATTATCCGAACATTCATTTCAGCTTTTAGGCTATTTTTCAAATGTGCGACTAAATCGTTCAGTGGTA  2415
2416 CGGAGTCAAATGCTGCAAAATACATTTCTAATCGAAATTGTTATCAAAAAATTGGATATAATAGTTCCA  2484
2485 ATTATTCCTCTAATTAGATCCTTGGCTAAAGCGAAATTTTGTAATGTATTAGGGGAGCCCATTAGTAAG  2553
2554 CCGGTCTGGGCCGATTCATCCGATTTTGATATTATTGATCGATTTTTGCGAATATGCAGAAATCTTTCT  2622
2623 CATTATTATAATGGATCCTCAAAAAAAAAAGTTTGTATCGAATAAAATATATACTTCGGCTTTCTTGT  2691
2692 ATTAAAACTTTGGCTTGTAAACACAAAGTACTGTACGCGCTTTTTTGAAAAGATCAGGTTCAGAAGAA  2760
2761 TTATTGCAAGAATTCTTTACAGAGGAAGAAGAAATTATTTCTTTGATTTTTCCAAGAGATTCCTCTACT  2829
2830 TTGCAGAGGTTACATAGAAATCGTATTTGGTATTTGGATATTCTTTTCAGTAACGATCTGGTCCATGAT  2898
2899 GAATGATTGGTTATGATACTTAAATATTATATATATCTATATAGAAAGGATTATGAAATGTTAAGAGTT  2967
          ‾‾‾‾‾‾
2968 GAGTTTCAAGATTCCTCTTCTGGAAAAGTAACTCAGGTTTAGATGTATACATAGGGAAAGCCGTGTGCA  3036
3037 ATGAAAAATGCATGCACGGCTTGGGGAGGGGTTTTTTTATTGTTTTATTTGATTAAACATGGAATTTTCT  3105
                                                 ⊽
3106 ACTCCATCCGACTAGTTCCGGGTTCGAGTCCCGGGCAACCCATTCTAATTAATAGATAAATTATATATT  3174
          ‾‾‾‾‾‾‾‾‾‾‾‾‾‾‾‾‾‾‾‾‾‾‾‾‾‾‾‾‾‾‾‾‾‾‾‾
          3'-Exon
3175 ATAATTAATATAGCGAAAGAATGAATAGATCACTATTACATATCATAGCGAAGTCATATCTAGAGAATA  3243
3244 TAGAAAACCTTTTTTCTTTTTTTTTTGAATGGATGGTGAAATGAGGTAAAAAAATAAAATATGTCTGAAT  3312
3313 CTAGATCAATAACAGGATACGGTGGATATTGGTATTGGGTGACACCCGTATATAAGTCATGTTATACTG  3381
                                      ‾‾‾‾‾‾                       ‾‾‾‾‾‾
                                       -35                          -10
3382 TTTTATAACAAGCCCTTAATTCTATAGTTATAGAGAATTC  3421
```

Fig. 2. The DNA sequence of the 3.3 kbp HindIII-HphI fragment where *trnK*, a 506 amino acid open reading frame located in the *trnK* intron, and promoter sequences (-35 to -10 regions) are boxed. Solid arrows indicate the 5'-termini of the *trnK* primary transcript and the open arrows with an inset "t" (*trnK*) or inset "r" (*rbcL*) are transcript 3'-ends. Primers used for primer extension assays are underlined.

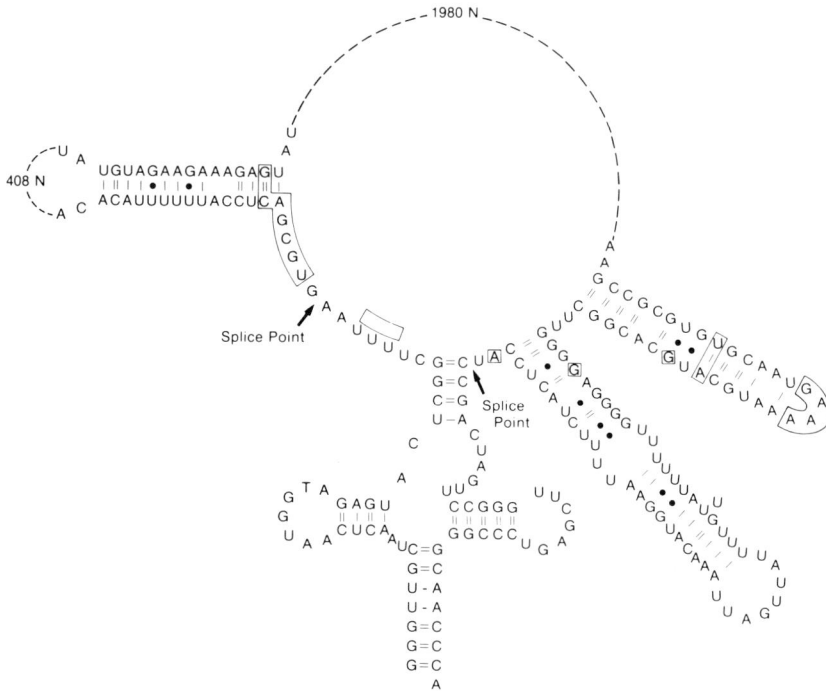

Fig. 3. Secondary structure model of the pea *trnK* 2.5 kb transcript. Putative splice points are indicated by arrows. The anticodon loop (UUU) has a box over it. The boxed "A" near the 3'-splice junction is conserved in class II introns and has been implicated in lariat formation in yeast mitochondria Van der Veen et al. 1986). The other boxed sequences are characteristic of class II introns (Michel and Dujon 1983).

scripts hybridizing to *trnK* are much less abundant than transcripts which hybridize to *psbA* or *rbcL*.

The transcripts which hybridize to *trnK* were further localized using Northern probes to specific regions of the gene (Fig. 4), top; probes 1–8, bottom; lanes 1–8. Probe 1 was derived from sequences between the *trnK* 5'-exon and the 3'-end of *rbcL*. This probe hybridized to RNAs 2.9, 2.4, and 1.9 kb in size plus several lower abundance RNAs (Fig. 4, lane 1). Several of the low abundance RNAs, 0.5 to 1.9 kb in size, were also detected with probes 2 to 4 (Fig. 4, lanes 2–4, marked with dots). It should be noted that the Northern blot in lane 1 was exposed to X-ray film 10 times longer than lanes 2 to 8. Probes 2 to 7 hybridized to a 2.5 kb transcript of relatively high abundance and to transcripts which were 2.9 and 2.4 kb in length (Fig. 4). Probe 7 also hybridized to an RNA of tRNA size (lane 7). Probe 8, derived from sequences between the 3'-exon of *trnK* and *psbA* did not show detectable hybridization to any RNAs.

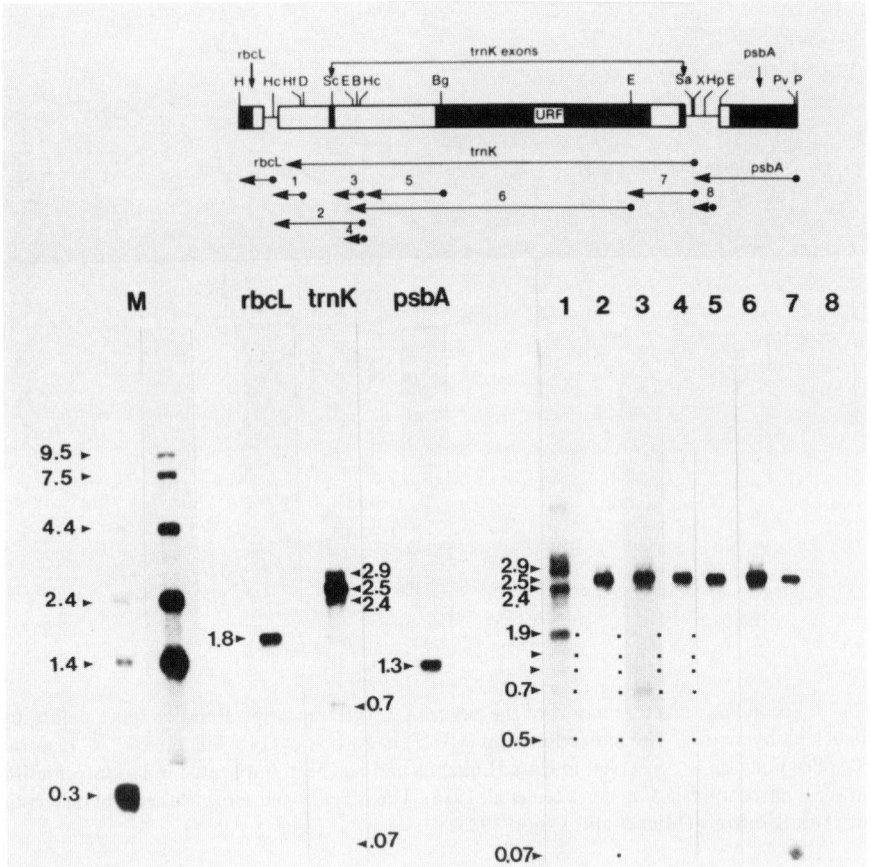

Fig. 4. Northern analysis of *trnK* and surrounding DNA region. The two lanes marked M show RNA markers (kb) using two different exposures. Lanes *rbcL, trnK* and *psbA* show Northern blots using riboprobes from the respective open reading frames. Lanes 1–8 are Northern blots using single stranded probes from the *trnK* transcription unit. The DNA regions from which single stranded probes were derived are shown at the top of the figure. The probes are as follows; lane 1, HinfI-HincII; lane 2, HincII-HincII; lane 3, HincII-ScaI; lane 4, HincII-FokI (FokI is at position 696 in Fig.2); lane 5 BglII-HincII; lane 6, EcoRI-EcoRI; lane 7, XbaI-EcoRI; lane 8, HphI-Sau3AI. *trnK* is transcribed from left to right in the figure.

Mapping of transcript termini with S_1 nuclease and reverse transcriptase
The location of the 3'-end of *rbcL* transcripts was determined by S_1 nuclease assays (Fig. 5A). This experiment showed that *rbcL* transcript 3'-termini map to position 162 in the sequence shown in Fig. 2 (marked with an r enclosed by an arrow). Sequences near the *rbcL* transcript 3'-end could be folded into a stem loop structure which may function in transcript termination or as a barrier to exonucleases (data not shown).

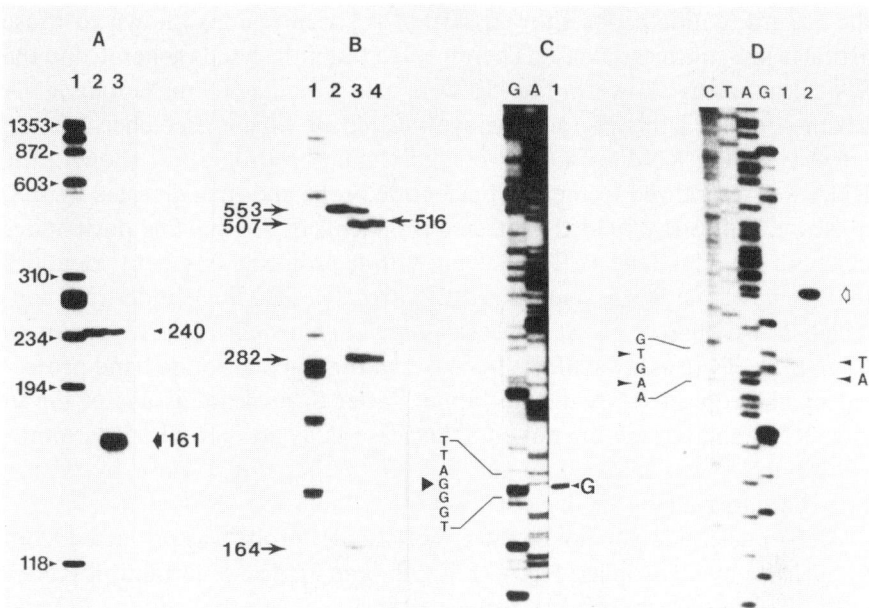

Fig. 5A. Autoradiagram of an S_1 nuclease assay which identifies the 3′-end of *rbcL* transcripts. Lane 1, end-labeled ØX174 HaeIII molecular weight markers; lane 2, a 240 nucleotide HincII-HindIII DNA fragment, 3′-end labeled at the HindIII site; lane 3, S_1 nuclease assay using the HincII-HindIII probe and pea chloroplast RNA. The approximate sizes of the S_1 nuclease protected fragments are indicated next to lane 3; Fig. 5B. Analysis of *trnK* transcript 5′-ends. Lane 1, ØX174 HaeIII molecular weight markers; lane 2, a 553 bp HincII-EcoRI DNA fragment from the 5′-end of *trnK* which was 5′-end labeled at the EcoRI site; lane 3, S_1 nuclease assay where the HincII-EcoRI DNA probe was hybridized to chloroplast RNA; lane 4, S_1 nuclease assay where the HincII-EcoRI DNA probe was hybridized to RNA synthesized in vitro from a plasmid which contains 516 nucleotides of the HincII-EcoRI DNA sequence. The approximate size (in nucleotides) of S_1-protected DNAs are indicated at the sides of the figure; Fig. 5c. Determination of the 5′-terminus of the *trnK* transcript localized near the *trnK* 5′-exon. Lanes G and A show dideoxy sequencing reactions using a 22 bp primer (see Fig. 2 for location); lane 1, primer extension assay using the 22 bp primer described above hybridized to chloroplast RNA; Fig. 5D. Determination of the 5′-terminus of the 2.9 kb *trnK* transcript. Lanes C, T, A, G show dideoxy sequencing reactions using a 19 bp primer from a sequence upstream from the 5′-exon of *trnK* (see Fig. 2). Lanes 1 and 2 are primer extension assays where the primer was hybridized to chloroplast RNA (lane 1) or an RNA synthesized from *trnK* sequences in vitro (lane 2). Open arrow marks the 5′-end of the RNA synthesized in vitro from plasmid pPMB319.

S_1 nuclease analysis of *trnK* transcript 5′-ends was done using a 553 nucleotide probe (HincII-EcoRI) which extends from within the *trnK* intron to 389 nucleotides upstream of the *trnK* 5′-exon (Fig. 5B, lane 2). Assays using the 553 nucleotide probe and pea chloroplast RNA revealed the presence of three putative *trnK* transcripts 5′-termini which protect 507, 282, or 164 nucleotides of the probe (Fig. 5B, lane 3). The sequences covered by the 553 nucleotide probe showed several AT rich regions known to cause artifacts in S_1 nuclease assays. Therefore, to test if the bands generated in the S_1 nuclease assay shown in Fig. 5B were artifactual, a 600 nucleotide RNA which contained this region was synthesized in vitro. After checking the integrity of the RNA on a urea-polyacrylamide gel (data not shown), the RNA was hybridized to the 553 nucleotide probe and then digested with S_1 nuclease. This assay yielded protected fragments of 516 and 282 nucleotides in length (Fig. 5B, lane 4). The 516 nucleotide protected fragment, identifies the 5′-end of the RNA synthesized in vitro. The 282 nucleotide fragment, which is also observed in S_1 nuclease assays of chloroplast RNA, appears to be an S_1 nuclease assay artifact. In contrast, the 164 nucleotide band protected by chloroplast RNA did not appear after S_1 nuclease assays of the in vitro RNA and is therefore not an S_1 nuclease assay artifact. The 5′-terminus of this RNA was mapped to the first G of the *trnK* 5′-exon by primer extension assays (Fig. 5C).

A 507 nucleotide portion of the HincII-EcoRI probe was protected from S_1 nuclease by chloroplast RNA (Fig. 5B, lane 3). The 5′-termini of RNAs in this region were mapped by primer extension assays to positions 287 and 289 in the sequence presented in Fig. 2 (T and A respectively, Fig 5D, lane 1) To test if the 5′-termini mapped at positions 287 and 289 were due to RNA secondary structure, primer extension assays were done using the 600 nucleotide RNA synthesized in vitro from the 5′-end of *trnK*. Primer extension assays mapped the 5′-terminus of the 600 nucleotide transcript to a position 9 nucleotides, upstream from the *trnK* in vivo transcript termini (Fig. 5D lane 2). Furthermore,this assay showed that the *trnK* transcript 5′-ends at positions 287 and 289 were not due to RNA secondary structure.

Site of trnK transcription initiation
Sequences 10 and 35 bp upstream from the transcript 5′-ends at positions 287 to 289 show homology to elements which promote transcription in procaryotes (Bujard 1980) and in chloroplast extracts (Hanley-Bowdoin and Chua 1987). These sequences are boxed in Fig. 2. A first indication that the RNA termini at positions 287 to 289 were generated by transcription initiation, was obtained by capping chloroplast RNA using guanylyltransferase and a α-^{32}P-GTP (Mullet et al 1985) and subjecting the capped RNA to S_1 nuclease analysis. This experiment showed that cappable transcripts have their 5′-ends located at positions 287 to 289 (data not shown). However, due to the low abundance of these transcripts the capping results required verification. Therefore, two plasmids (pPHD118 and pPHD119) which contained the putative *trnK* promoter region were constructed (Fig. 6) and used in in vitro transcription assays (Fig. 7). Both templates were able

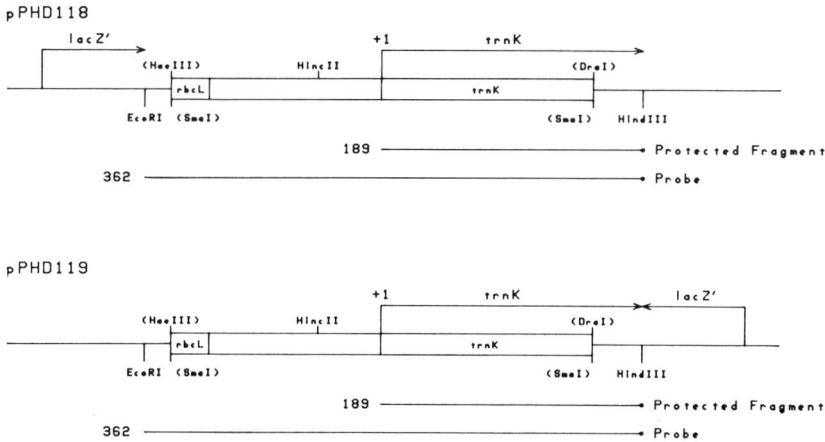

Fig. 6. Plasmid constructions used as templates for in vitro transcription in pea chloroplast extracts. Single lines indicate vector sequences and boxed regions indicate sequences of chloroplast origin. The restriction sites are indicated above the figure for chloroplast-encoded sites and below the figure for vector-encoded sites. Sites listed in parenthesis are no longer recognized due to the construction of the plasmids. Arrows above the figure indicate the start site and direction of transcription of the *trnK* and *lacZ* genes. The probes and S_1 protected fragments are shown for each of the plasmids with the sizes given in nucleotides (see Fig. 7).

to promote transcription in pea chloroplast extracts and produce transcripts which had 5'-termini similar to in vivo *trnK* transcript termini mapped to position 287 to 289 (Fig. 7).

Discussion

TrnK in pea, tobacco (Sigita et al. 1985), and mustard (Neuhaus and Link 1987) ctDNA is located adjacent to and at the 5'-side of *psbA*. In these plants, *trnK* is transcribed on the same DNA strand as *psbA* and in each case *trnK* contains a large (2.5 kb) intron in the anticodon loop. The *trnK* transcription unit in pea has a site of transcription initiation approximately 342 bp upstream from the 5' exon of *trnK*. The transcript which has its 5'-end at this site was identified as a primary transcript by three criteria. First, sequences 10 and 35 bp 5' to the RNA terminus are homologous to chloroplast promoter sequence elements (Hanley-Bowdoin and Chua 1987). Second, the DNA region in question was found to promote transcription initiation in vitro at a site approximately 342 bp upstream of the *trnK* 5'-exon. Third, the transcript could be capped using guanylyltransferase (data not shown). In mustard, a site of transcription initiation was identified 120 bp upstream of the 5'-exon of *trnK* based on S_1 nuclease assays and

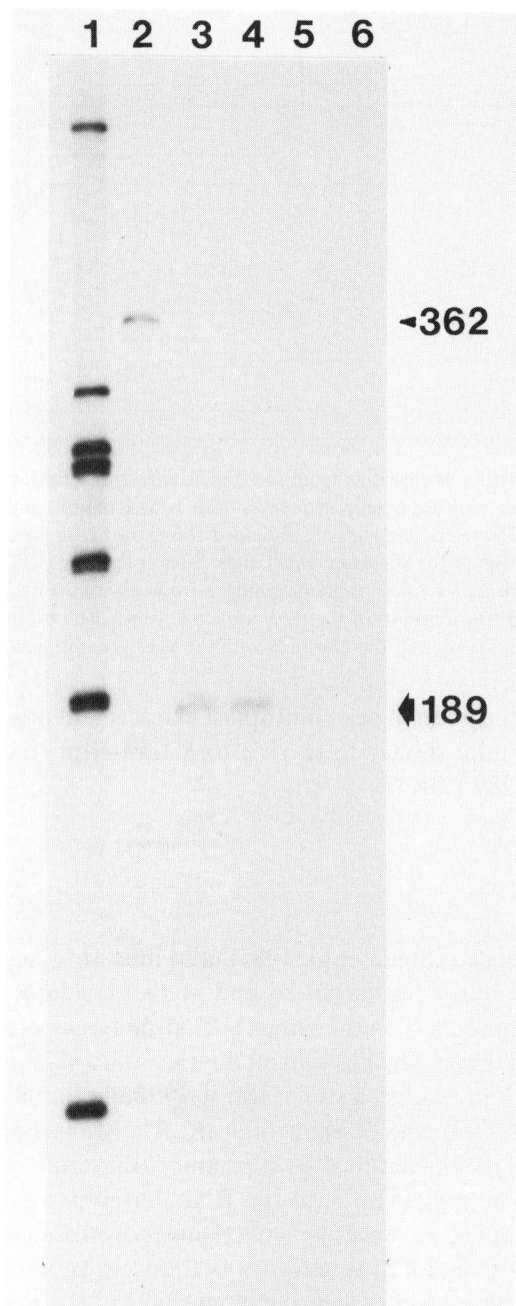

Fig. 7. In vitro transcription of the *trnK* constructs shown in Fig. 6 using a pea chloroplast in vitro transcription extract. Lane 1 shows DNA molecular weight markers and lane 2 the 362 nucleotide, HindIII-EcoRI S$_1$ probe. S$_1$ nuclease assays of transcripts produced in vitro are shown in lanes 3 and 4 for the templates pPHD118 and pPHD119, respectively. The 189 nucleotide band (lanes 3 and 4) was not detected in transcription reactions with no template (lane 6) and reactions with template but no chloroplast extract (lane 5).

sequence data (Neuhaus and Link 1987). The 3′-end of the *trnK* transcripts was located near the 3′-end of the *trnK* 3′-exon. It is not known if the transcript 3′-end is generated by transcription termination or by RNA processing.

Northern analysis of *trnK* in pea showed that three large transcripts (2.9, 2.5 and 2.4 kb) hybridized to this gene. Northern analysis and S_1-nuclease assays indicate that the 5′-end of the 2.9 kb RNA is approximately 342 nucleotides upstream of the 5′-exon of *trnK* and that this RNA extends to the 3′-side of the *trnK* 3′-exon (see Fig. 8 for a summary). The 2.5 kb *trnK* transcript has the *trnK* 5′-exon at its 5′-end and the 3′-exon at its 3′-end. This transcript is more abundant than other large transcripts which hybridize to *trnK*. This may be due to the presence of *trnK* exons at its termini which could stabilize this transcript. It is likely that the 2.5 kb RNA is derived from the 2.9 kb primary transcript by RNA processing as shown in Fig. 8. An RNAseP-like activity, recently found in chloroplasts (Yamaguchi-Shinozaki et al 1987), could carry out the proposed RNA processing.

The 2.5 kb *trnK* transcript contains an intron with features which suggest that it is a class II intron. These include an "A" adjacent to the 3′-splice

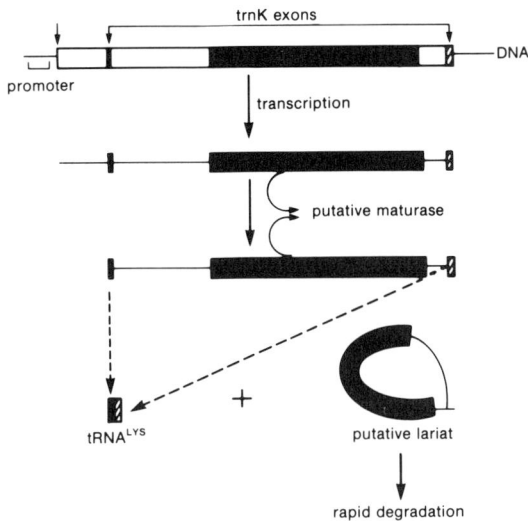

Fig. 8. Proposed transcription and RNA processing pathway for pea *trnK*. Top line; *trnK* transcription unit with promoter and *trnK* exons indicated. The boxed region corresponds to transcribed DNA. The large black box denotes the location of the open reading frame located in the *trnK* intron. Line 2; 2.9 kb primary transcript which could be translated to produce a putative maturase. Line 3; 2.5 kb transcript produced from the primary transcript by RNA processing. Line 4; splicing reaction to produce tRNA^{Lys} and a putative lariat RNA which is rapidly degraded.

74

junction, a stem loop structure near the 3′-end of the intron which contains a terminal "GAAA" sequence, and a conserved sequence near the 5′-splice junction (Fig. 3). These assignments are based on consensus sequences of class II introns generated by Michel and Dujon (1983). The 2.5 kb *trnK* transcript is likely to be spliced to form a small RNA which will become tRNA$_{Lys}$ (after modification of bases and addition of a terminal CCA) and a lariat RNA (Fig. 8). No intron specific RNAs were observed in Northern blots. This suggests that intron sequences are rapidly degraded. It should be noted that the proposed RNA processing pathway does not address the origin of low abundance RNAs 2.4 to 0.5 kb in size which hybridize to *trnK* probes. An in vitro RNA processing extract capable of splicing *trnK* transcripts may be required to address the origin of these RNAs.

Seventeen different chloroplast genes of higher plants have been found to contain introns (Shinozaki et al 1986). The introns in most of these genes are class II introns. Two exceptions are *trnL* (UUA) in *Z. mays* (Steinmetz et al. 1982) and *V. faba* (Bonnard et al. 1984) which contain class I introns. Class II introns of yeast mitochondrial genes can self-splice under appropriate conditions (i.e. 45°C, 5 mM MgCl$_2$, 2 mM spermidine). However, there is also evidence that other proteins facilitate splicing reactions. These include nuclear-encoded proteins (Schmelzer et al. 1983) and maturases which are encoded within class II introns (Caragnani et al. 1983).

Chloroplast RNAs have not been shown to self-splice. Unsuccessful splicing reactions were attempted in this study using *trnK* transcripts synthesized in vitro (data not shown). Our lack of success in this regard may reflect unusual salt or pH requirements for self-splicing of chloroplast *trnK* RNA. Alternatively, it may indicate that a protein(s) is required for splicing. One protein which may facilitate *trnK* transcript splicing could be the polypeptide encoded by the *trnK* intron. The protein encoded within the mustard *trnK* intron has some homology to yeast mitochondrial RNA maturases which facilitate intron processing (Nauhaus and Link 1987). Therefore, the *trnK* open reading frame could encode a protein which is required to process not only *trnK* but also other chloroplast RNAs which contain introns. In addition, the level of putative maturase might well limit RNA processing because removal of *trnK* intron sequences leads to rapid degradation of the RNA which encodes the putative maturase.

Acknowledgements

This work was supported by NSF Grant No. DCB 86-16156.

References

Bonitz S, Coruzzi G, Thalenfeld B, Tzagoloff A and Macino G (1980) Assembly of the mitochondrial system: structure and nucleotide sequence of the gene coding for subunit I of yeast cytochrome oxidase. J Biol Chem 225: 11927–11941

Bonnard G, Michel F, Weil JH and Steinmetz A (1984) Nucleotide sequence of the split tRNA^{Leu}(AAA) gene from *Vicia faba* chloroplasts: evidence for structural homologies of the chloroplast tRNA^{Leu} intron with the intron from the autosplicable *Tetrahymena* ribosomal RNA precursor. Mol Gen Genet 194: 330–336

Bonnard G, Weil JH and Steinmetz A (1985) The intragenic region between *Vicia faba* chloroplast tRNA^{Leu}(CAA) and tRNA^{Leu}(UAA) genes contains a partial copy of the split tRNA^{Leu}(UAA) gene. Curr Genet 9: 417–422

Boyer SK and Mullet JE (1986) Characterization of *P. sativum* chloroplast psbA transcripts produced *in vivo, in vitro* and in *E. coli*. Plant Mol Biol 6: 229–43

Bujard H (1980) The interactions of *E. coli* RNA polymerase with promoters. Trends Biochem Sci 5: 274–278

Carignani G, Groudinsky O, Frezza D, Schiaron E, Bergantino E and Slonimski PP (1983) An RNA maturase is encoded by the first intron of the mitochondrial gene for the subunit I of cytochrome oxidase in S. cerevisiae. Cell 35: 733–742

Chu NM, Shapiro DR, Oishi KK and Tewari KK (1985) Distribution of transfer RNA genes in the *Psium sativum* chloroplast DNA. Plant Mol Biol 4: 65–79

Crouse, EJ, Mubumbila M, Stummann BM, Bookjans G, Mickalowsi C, Bohnert HJ, Weil J–H and Henningsen KW (1986) Divergence of chloroplast gene organization in three legumes. Plant Mol Biol 7: 143-147

Deno H, Kato A, Shinozaki K and Sigiura M (1982) Nucleotide sequences of tobacco chloroplast genes tRNA^{Met} and tRNA^{Val}(UAC): the tRNA^{Val}(UAC) gene contains a long intron. Nuc Acids Res 10: 7511–7521

Deno H and Sugiura M (1984) Chloroplast tRNA^{Gly} gene contains a long intron in the D stem: nucleotide sequences of tobacco chloroplast genes for tRNA^{Gly}(UCC) and tRNA^{Arg}(UCU). Proc Natl Acad Sci USA 81: 405–408

Gauss DH and Sprinzel M (1984) Compilation of tRNA sequences. Nuc Acids Res 12: rl–r131

Gruissem W, Elsner-Menzelc, Latshaw S, Narita JO, Shaffer MA and Zurawski G (1987) A subpopulation of spinach chloroplast tRNA genes does not require upstream promoter elements for transcription. Nuc Acids Res 14: 7541–7556

Gruissem W and Zurawski G (1985) Identification and mutational analysis of the promoter for a spinach chloroplast transfer RNA gene. EMBO J 4: 1637–1644

Hallick RB, Hollingsworth MJ and Nickoloff JA (1984) Transfer RNA genes of *Euglena gracilis* chloroplast DNA. Plant Mol Biol 3: 169–175

Hanley-Bowdoin L and Chua N-H (1987) Chloroplast promoters. Trends Biochem Sci 12: 67–70

Koch W, Edwards K and Kössel H (1981) Sequencing of the 16S–23S spacer in a ribosomal RNA operon of *Zea mays* chloroplast DNA reveals two split tRNA genes. Cell 25: 203–213

Krebbers E, Steinmetz A and Bogorad L (1984) DNA sequences for the *Zea mays* tRNA genes tV-UAC and tS-UGA: tV-UAC contains a large intron. Plant Mol Biol 3: 13–20

Maniatis T, Fritsch EF, Sambrook J (1982) Molecular Cloning: a laboratory manual. Cold Spring Harbor Laboratory, Cold Spring Harbor, New York

Messing J (1980) New M13 vectors for cloning. In: Colowich SP, and Kaplan NO (eds) Methods in Enzymology, Vol 101C. Academic Press, New York, pp 20–77

Michel F and Dujon B (1983) Conservation of RNA secondary structures in two intron families including mitochondrial, chloroplast and nuclear-encoded members EMBO J 2: 33–38

Mullet JE, Orozco EM, Chua N-H (1985) Multiple transcripts for higher plant *rbcL* and *atpB* genes and localization of the transcription initiation site of rbcL gene. Plant Mol Biol 4: 39–54

Neuhaus H and Link G (1987) The chloroplast tRNA^{Lys}(UUU) gene from mustard *(Sinapsis alba)* contains a class II intron potentially coding for a maturase polypeptide. Curr Genet 11: 251-257

Ogden RC, Knapp G, Peebles CL, Kang HS, Beckmann JS, Johnson PF, Fuhrman SA and Abelson JN (1980) In Söll D, Abelson JN and Schimmel PR (eds) Transfer RNA: Biological Aspects, pp 173-190, Cold Spring Harbor, NY, Cold Spring Harbor Laboratory

Osiewacz HD and Esser K (1984) The mitochondrial plasmid of *Podospora anserina:* A mobile intron of a mitochondrial gene. Curr Genet 8: 299–305

Schmelzer C, Schmidt C, May K and Schweyen RJ (1983) Determination of functional domains in intron b/1 of yeast mitochondrial RNA by studies of mitochondrial mutations and a nuclear suppressor. EMBO J 2: 2047–2052

Shinozaki K, Ohme M, Tanaka M, Wakasugi T, Hayashida N, Matsubayash T, Zaita N, Chunwongse J, Obokato J, Yamaguchi-Shinozaki K, Ohto C, Torozawa K, Meng BY, Sugita M, Deno H, Kamogashira T, Yamada K, Kusada J, Takaiwa F, Kato A, Tohdoh N, Shimada H and Sugiura M (1986) The complete nucleotide sequence of the tobacco chloroplast genome, its gene organization and expression EMBO J 5: 2043–2049.

Steinmetz A, Gubbins EJ and Bogorad L (1982) The anticodon of the maize chloroplast gene for tRNA^{Leu}(UAA) is split by a large intron. Nuc Acids Res 10: 3027–3037

Sugita M, Shinozaki K and Sugiura M (1985) Tobacco chloroplast tRNA^{Lys}(UUU) gene contains a 2.5-kilobase pair intron: An open reading frame and a conserved boundary sequence in the intron. Proc Natl Acad Sci 82: 3557–3561

Van der Veen R, Arnberg AC, Van der Horst G, Bonen L, Tabak HF and Grivell LA (1986) Excised group II introns in yeast mitochondria are lariats and can be formed by self splicing in vitro. Cell 44: 225-234

Wakasugi T, Ohme M, Shinozaki K and Sugiura M (1986) Structures of tobacco chloroplast genes for tRNA^{Ile}(CAU), tRNA^{Leu}(CAA), tRNA^{Cys}(GCA), tRNA^{Ser}(UGA), and tRNA^{Thr}(CGU): A compilation of tRNA genes from tobacco chloroplasts. Plant Mol Biol 7: 385–392

Whitfield PR and Bottomley W (1983) Organization and structure of chloroplast genes. Ann Rev Plant Physiol 34: 279–310

Yamaguchi-Shinozaki K, Shinozaki K and Sugiura M (1987) Processing of precursor tRNAs in a chloroplast lysate. FEBS Lett 215: 132–136

Yamada K, Shinozaki K and Sugiura M (1986) DNA sequences of tobacco chloroplast genes for tRNA^{Ser}(GAA), tRNA^{Thr}(UGU), tRNA^{Leu}(UAA): the tRNA^{Leu} gene contains a 503 bp intron. Plant Mol Biol 6: 193–199

Govindjee et al. (eds), Molecular Biology of Photosynthesis: 77–103
© 1988 Kluwer Academic Publishers

Minireview

Transposon mutagenesis of nuclear photosynthetic genes in *Zea mays*

WILLIAM B. COOK & DONALD MILES
Department of Biological Sciences, 108 Tucker Hall, University of Missouri, Columbia, MO 65211, USA

Received 29 September 1987; accepted 25 January 1988

Key words: cloning, mutation, photosynthesis mutant, transposon

Abstract. The discovery of a new maize (*Zea mays* L.) transposon system, *Mutator*, and the cloning of the 1.4 kilobase transposon, *Mul*, have made feasible the isolation of nuclear photosynthetic genes which are recognized only by their mutant phenotype. Mutant maize plants which express a *high chlorophyll fluorescent* (*hcf*) phenotype due to a defect in the electron transport or photophosphorylation apparatus have been isolated following mutagenesis with an active *Mutator* stock. The affected genes and their products in these mutants are inaccessible to classical methods of analysis. However, mutagenesis with the *Mutator* transposon makes it possible to isolate these genes.

Although the PSII-deficient mutant *hcf3* has been thoroughly studied by classical photobiological methods, the nature of the lesion which results in the observed phenotype has not been established. A *Mutator*-induced allele of *hcf3* has been isolated. A fragment of genomic DNA has been identified which is homologous to *Mul* and co-segregates with the mutant phenotype. This fragment is expected to contain a portion of the *hcf3* locus which will be used to clone the normal gene. Direct study of the gene can provide insight into the nature and function of its polypeptide product.

This approach can be used to study any photosynthetic gene which has been interrupted by a transposon. The isolation of more than 100 different chemically-induced *hcf* mutants, most of which can not be fully characterized using classical means, indicates the wealth of information which can be obtained using a transposon tagging technique.

Introduction

Transposable DNA elements, or transposons, have become standard tools of the molecular biologist for gene analysis. The technique of transposon mutagenesis (transposon tagging) has made possible the isolation of genes identified by nothing more than a mutant phenotype. Many genes which are involved in the development, regulation and functioning of the chloroplast, and the photosynthetic apparatus in particular, fall into this category. A large number of chemically-induced mutants now exist which affect the proper expression of the photosynthetic mechanism but for which the primary lesions are unknown. Transposon mutagenesis provides a means by which a mutant gene which causes a defect in the photosynthetic apparatus can be isolated and studied directly.

This paper presents a strategy by which nuclear photosynthetic genes, which are otherwise inaccessible, may be isolated and studied. The first data demonstrating the utility of this approach will be presented. Transposons will be considered as practical tools in this report. Their fascinating biological characteristics and evolutionary implications will not be covered in detail here. For those who wish to explore the topic of transposons in greater depth, a number of recent reviews are available (Calos and Miller 1980, Campbell 1981, Kleckner 1981, Federoff 1983, Heffron 1983, Iida et al. 1983, Freeling 1984, Lillis and Freeling 1986, Nevers et al. 1986, Walbot et al. 1987).

hcf *mutants in maize*

Mutant plants which are unable to utilize the energy of singlet state excited chlorophyll are characterized by a *high chlorophyll fluorescent* (*hcf*) phenotype. More than 100 different nuclear *hcf* mutants have been induced by chemical (EMS) mutagenesis. Among these are many mutants with lesions which specifically affect one or more of the thylakoid membrane polypeptide complexes; PSI, PSII or the Cyt b_6/f or ATPase complexes. Mutants affecting soluble proteins, such as ribulose-1,5-bisphosphate carboxylase/oxygenase have also been isolated on the basis of the *hcf* phenotype (Miles 1980, Miles 1982, Miles and Randall 1983).

At least three different nuclear maize mutants have been isolated which specifically affect the Cyt b_6/f complex (Metz et al 1985, Barkan et al. 1986, Cook et al. 1987, Taylor et al. 1987). However, only one structural component of the Cyt b_6/f complex is encoded in the nucleus. Therefore, there must be several nuclear-encoded activities involved in the proper synthesis and assembly of this complex. The same argument may be made for the specifically affect the Cyt b_6/f complex (Metz et al. 1985, Barkan et al. 1986, ponents of those complexes have not been absolutely determined.

The chemically-induced *hcf* mutants have shown that nuclear-encoded functions other than subunit synthesis are required for assembly of a functional complex. However, they have not proven to be useful in deciphering the nature of these functions. The molecular phenotype of the Cyt b_6/f mutants generally includes the loss of the entire complex as well as the electron transport activity associated with it.

The process of assembling a polypeptide complex within the thylakoid membrane may be thought of as analogous to a biochemical pathway. Several sequential steps contribute to the development of a normal complex so that the loss of each one of these steps results in the same phenotype; failure of the complex to be assembled. Therefore, it is not possible to work backward from the phenotype (loss of the entire complex) to the particular activity which has been lost (any of several conceivable activities) in a particular mutant. The loss of an entire complex, as occurs in many *hcf* mutants, clearly demonstrates that the product of the affected gene is essential to the proper assembly or stability of that complex. However, the similarities among the phenotypes caused by several different nuclear mutations prevent the identification of the primary lesions resulting from those mutations.

The harnessing of transposons in higher plants has provided a route by which the processes involved in the assembly and functioning of the photosynthetic apparatus may be studied. When a gene which is necessary to normal photosynthetic function is interrupted by a transposon, a mutant phenotype results. Unlike a gene which has been mutated by a chemical mutagen, a gene which is interrupted by a transposon bears a molecular tag which permits it to be identified and isolated. Once the gene is isolated, its nucleotide sequence and transcriptional regulation signals can be studied. These data can, in turn, be used to attempt to classify the gene product either directly or by analogy with other well characterized polypeptides.

Transposons

DNA sequences which are able to move from place to place within the genome have been referred to by several different names. The maize *controlling elements* of McClintock (see Federoff 1983), named for their ability to control the expression of the genes which they affect, are now known to resemble the *transposable elements* or *transposons* of prokaryotic (Calos and Miller 1980, Kleckner 1981, Heffron 1983, Iida et al. 1983) and of lower eucaryotic systems (Roeder and Fink 1983, Rubin 1983) in some respects. This paper will consider primarily the controlling elements of maize and another more recently discovered maize system. In line with current convention and for the sake of clarity the term *transposon* will be used to refer to discrete DNA sequences which are able to move from one site to another within the genome and to alter the expression of genes with which they become associated.

Discovery and incidence

The first transposon was discovered in maize (*Zea mays* L.) (see Federoff 1983). Its presence was deduced when a segment of chromosome 9 was lost due to the activity of the element. In a series of remarkable classical genetic studies that followed, McClintock demonstrated that there were indeed

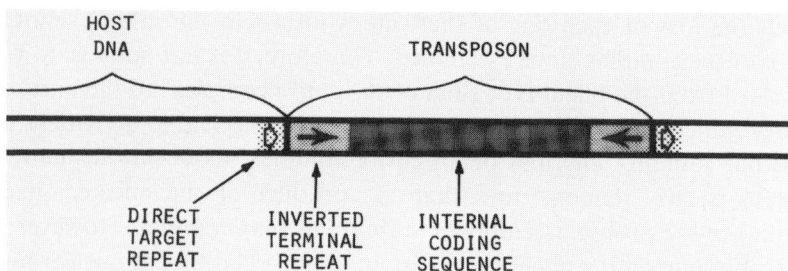

Fig. 1. Transposon model showing features common to most prokaryotic and eucaryotic elements. The direct target repeats are due to a duplication of host genomic DNA at the site of insertion. They vary from 3 to 12 base pairs in length. Terminal inverted repeats may be a few base pairs or several dozen base pairs in length. They form the boundaries of the transposon. DNA sequences between the terminal inverted repeats may encode the enzymes which catalyze transposition of the element, products which are unrelated to the transposition of the element, or, as in the case of non-autonomous elements, no functional product.

genetic elements capable of transposition and able to induce unstable mutations, chromosomal breaks, deletions, rearrangements and other phenomena (McClintock 1984).

It was not until similar elements were discovered in prokaryotic organisms that they were widely accepted as other than genetic oddities peculiar to maize. Bacterial Insertion Sequences (IS element) and Transposons (Tn elements) provided the basis for a molecular characterization of transposons and their activities (Calos and Miller 1980).

Since their 'validation' in prokaryotes, transposons have been discovered in many additional organisms, both prokaryotic and eucaryotic. Some of the most thoroughly characterized eucaryotic systems are the *Ty* elements of yeast (Roeder and Fink 1983) and the *copia* and *P* elements of *Drosophila* (Rubin 1983). Unstable mutations resulting in variegated pigmentation and attributable to transposon activity have been reported in more than 30 plant species (Nevers et al. 1986). Aside from those in maize, the most thoroughly characterized transposons in plants are the *Tam* elements of *Antirrhinum majus* (Carpenter et al. 1987, Hehl et al. 1987, Hudson et al. 1987).

Molecular description

Transposons, whether prokaryotic or eucaryotic, share two features at a molecular level. Each is bounded by terminal inverted repeat (TIR) sequences and each is flanked by Direct Target Repeat (DTR) sequences (Fig. 1). TIRs may be a few base pairs or several dozen base pairs in length. The TIRs apparently are, or contain, the recognition sites for the enzymes which catalyze transposition. DTRs occur at the site of insertion as products of the insertion mechanism. They are usually three to eleven base pairs in length.

The sequences of DNA located between the TIRs of transposons often encode the genes required for transposition of the elements. In the case of bacterial transposons, other, unrelated genes which encode enzymes for antibiotic resistance are also often found in the internal DNA sequences. Some elements do not encode any functional product in the sequences between the TIRs.

Transposon families
Six or more distinct transposon systems have been reported in maize (Nevers et al. 1986). Typically, these transposon systems occur as families of related elements. A family of transposons usually includes two types of elements; those which are autonomous and those which are not. Autonomous elements are capable of catalyzing their own transposition activities and the transposition of other related elements present in the same genome. Non-autonomous elements are capable of the same transposition activities as autonomous elements *only* when a related autonomous element is present in the same cell. Otherwise they remain stably integrated into the genome. A transposon family has come to be defined as all of those elements which are activated by a particular type of autonomous element.

Non-autonomous elements often arise as deletion derivatives of autonomous elements (Federoff et al. 1983, Pohlman et al. 1984). Presumably the loss of a segment of DNA from within a coding region in the internal portion of the element inactivates one or more of the enzymes required for transposition. This eliminates the element's ability to sponsor transposition activity but not its ability to respond to a related autonomous element which is present in the same genome.

Not all non-autonomous elements are related to the corresponding autonomous elements by their internal DNA sequences. Some non-autonomous elements are known to be entirely unrelated to the autonomous elements in their family *except* for their terminal inverted repeat sequences (Freeling 1984). Homology among the terminal sequences of transposons provides a molecular explanation of the phenomenological definition of transposon families referred to above.

Inactivation
While only autonomous transposons are able to sponsor the transposition of related elements, they are not always capable of doing so. Autonomous elements can be inactivated and remain in a quiescent state within the genome for many generations (Federoff 1983). The cause of inactivation is not known, but the mechanism appears to involve methylation of some or all of the cytosine residues in an element (Chandler and Walbot 1986,

Fig. 2. Maize seedling leaf showing the reversion of a mutant phenotype to normal. Each dark sector represents a clone of cells arising from a single progenitor in which a transposon(*Mu*)-induced mutant allele reverts to normal by excision of the transposon. The dark sectors were a normal green color while the lighter background was a pale yellow-green.

Schwartz and Dennis 1986, Bennetzen 1987, Bennetzen et al. 1987). While inactive, the autonomous element can catalyze neither its own transposition nor that of related non-autonomous elements. Reactivation apparently requires an event which shocks the genome such as chromosome breakage (McClintock 1984).

Transposition mechanism

The mechanisms by which transposons move about the genome are not well understood. There is evidence that different prokaryotic systems employ different transposition mechanisms (Grindley and Reed 1985). One mechanism is replicative in nature, that is, the donor element remains in place, is copied and the copy inserts into a new location. The second mechanism involves conservative transposition. The donor element is excised from its location in the genome and inserted into a new location. The best evidence for a maize system suggests that its elements move predominantly by conservative transposition (see Federoff 1983). Neither of these mechanisms is universal for prokaryotes or eucaryotes. As more is learned about the mechanisms involved in transposition, variations on these two general themes are certain to be found.

Regardless of the mechanism by which transposons move about the genome, they all share the ability to excise from their genomic locations. In higher plants this excision may occur in either somatic tissue (that giving rise to vegetative structures) or germinal tissue (that giving rise to reproductive structures). Excision of a transposon from a locus in germinal tissue results in a heritable change. If excision is from a gene characterized by a visible phenotype, the progeny of the plant in which it occurs will be recognized as having reverted from a mutant to a normal phenotype (in the simplest case). If excision from such a gene occurs in somatic tissue the result will be a clone of normal cells (the cell in which the event occurs plus its progeny) on a background of mutant cells. For example, excision of a transposon from a gene required for chlorophyll synthesis will result in green sectors (the cells in which the excision occurs and their progeny will appear normal) on a white (or yellow) leaf (most of the transposons in the cells of a given leaf will not excise) (Fig. 2).

Transposon tagging strategy

The first genes to be cloned by modern molecular techniques were isolated by a strategy which has become the classical method (e.g. see Young and Davis 1983). First, the product of the gene of interest is isolated and purified and antibodies are raised against it. Then, messenger RNA (representing the coding sequences of genes that are being actively transcribed) is isolated and used to produce a population (library) of DNA copies (copy DNAs or cDNAs) using the enzyme *reverse transcriptase*. The cDNAs are inserted into a bacteriophage vector which is capable of expressing the encoded message following infection of a suitable host bacterium. The cDNA library in the form of multiple populations (plaques) of bacteriophage, each containing and expressing a different cDNA clone, is then screened with the antibody that was raised against the product of the gene of interest. cDNA clones which produce the protein of interest are propagated and used as probes to isolate the native gene from a library consisting of fragments of normal genomic DNA inserted into a vector.

This classical method has been used to clone a number of nuclear photosynthetic genes which encode structural components of the electron transport chain; genes for which the protein products are known and can be isolated (Tittgen et al. 1986). This method has also been used to clone several transposable elements which were inserted into genes with known products (Federoff et al. 1983, Shure et al. 1983, Bennetzen et al. 1984). The drawback of the classical method from the point of view of cloning nuclear-encoded photosynthetic genes is that, aside from the several already cloned, most of these genes are known only by complex phenotypes which mask the identities of the polypeptide products and their activities. Some of the most thoroughly characterized photosynthetic mutants fall into this category (Leto and Miles 1980, Metz and Miles 1982, Metz et al. 1983).

Transposon mutagenesis provides a method by which the classical dependence on knowledge of a gene product can be by-passed and a gene can be cloned on the basis of a complex mutant phenotype alone. To accomplish this, mutations are induced by crossing a transposon-bearing plant to a normal plant and screening progeny for the presence of a new mutant. DNAs from several individuals of a population segregating the mutation are isolated. The DNAs are hydrolyzed with a restriction endonuclease which produces a unique set of fragments by cleaving at a specific sequence of from four to six nucleotides wherever it occurs. The restriction fragments are size fractionated by agarose gel electrophoresis, and immobilized on a membrane. The immobilized DNA fragments are then hybridized with a DNA probe consisting of a single stranded sequence derived from the transposon which was used to induce the mutation. Following hybridization a single DNA fragment is identified which is present in each individual which is carrying the mutant allele and is absent from all individuals which are homozygous for the normal gene. This DNA fragment includes the transposon which caused the mutation. The transposon should be flanked on one or both sides by host genomic DNA including at least a part of the gene of interest. The segregating, transposon-homologous fragment can then be cloned. The genomic portion of the clone can be subcloned and used to screen a library of normal genomic DNA fragments to isolate the uninterrupted gene.

Generation of mutants

Many mutant alleles of genes have been isolated genetically since the time when work began on maize transposons. Transposon-induced alleles were commonly called *mutable* alleles, referring to their tendency to revert to normal during development of the plant. As a result, many of these alleles bear the designation *m*; for example, *wx-m9* is an allele of the *Wx(Waxy)* locus caused by a *Ds* element of the *Ac–Ds* transposon family. Many transposon-induced alleles of genes that are important to classical genetics are available, in particular, those producing non-lethal, easily scored kernal mutations. Therefore, these genes are accessible to cloning without the necessity of generating mutant alleles de novo. Seedling-lethal photosynthetic mutants were not of particular interest to classical geneticists and so, for the most part, the cloning of photosynthetic genes requires the production of new mutations.

The need to induce new mutations requires that a choice be made from among the several transposon systems which are available. Elements from three different maize transposon families have been cloned and are available as hybridization probes: *Activator-Dissociation* (*Ac–Ds*) (Federoff et al. 1983), *Suppressor-Mutator* (*Spm*) (Schwarz-Sommer et al. 1984) and Robertson's *Mutator* (*Mu*) (Barker et al. 1984, Bennetzen et al. 1984). Two

Table 1. Three maize transposon families and some of their component elements.

Family	Element	Length (bp)	Inverted terminal repeat	Direct target repeat	Reference
Ac–Ds	*Ac9*	4563	11	8	Pohlman et al. (1984)
	Ac	4500[a]	11	8	see Nevers et al. (1986)
	Ds9	4369	11	8	Pohlman et al. (1984)
	Dsl	405	11	8	Sachs et al. (1983)
Spm (En/I)	*Enl*	8287	13	3	Pereira et al. (1985)
	Spm-I8	2242	13	3	Schwarz-Sommer et al. (1984)
Mutator	*Mul*	1367	213	9	Bennetzen et al. (1984)
	Mul.7	1700[a]	213	9	Taylor et al. (1987)

[a] The lengths of these elements are estimates; all other lengths reported have been determined by nucleotide sequencing.

unrelated *Tam* elements have also been cloned from *Antirrhinum* (Bonas et al. 1984, Nevers et al. 1986) as has the *Tgml* sequence from soybean (*Glycine max*) (Goldberg et al. 1983).

In choosing an element for use as a transposon tag several factors must be considered. The ease with which the organism is manipulated genetically and the genetic database available are of importance during the production of new mutants as well as in the future when it may become necessary to characterize the cloned gene as a part of the genome from which it was taken. The rate at which new mutations are produced is another crucial factor in choosing a transposon system. Since mutant induction is essentially random (in the absence of any information about the location of interesting genes) a higher rate of production of new mutants will yield more new mutants per round of mutagenesis and will be more likely to yield mutants of a particular type quickly.

The complexity of the transposon family also affects the ease with which genes can be cloned. A line of maize containing a transposon family which consists of many diverse elements may yield a large number of new mutations per generation, but the probe(s) available to screen for elements in that family may not hybridize to all of the different elements belonging to the family. Consequently, while interesting new mutants might be produced, there would be no guarantee that the genes could be isolated using the available transposon probe(s).

The number of copies of the transposon in the genome is of interest as well. If many copies are present the identification of a single copy which is inserted into the gene of interest will require a complex sorting process. On the other hand, a DNA sequence which is known to be interrupted by a transposon that is present in the genome in only a few copies can be isolated easily.

Maize transposon families

The transposon families which are available as transposon tagging systems each have advantages and disadvantages in terms of the considerations discussed above. The three most thoroughly characterized maize transposon families (Table 1) will be considered in terms of their suitability for the job of cloning photosynthetic genes. Each has been used to clone one or more genes from maize.

Because of the wealth of genetic information available about maize and because of the ease with which maize can be manipulated genetically, we believe that it is presently the best available system for this type of gene tagging program. However, the strategy that has been presented for the isolation of photosynthetic genes can be applied to any plant species in which an active transposon system exists.

The Ac–Ds *family*

The *Activator–Dissociation* family (*Ac–Ds*, which includes the *Mp* system) was the first two element system to be be studied in depth by McClintock (see Federoff 1983). It is made up of autonomous (*Ac*) elements and related non-autonomous (*Ds*) elements. Many mutations induced by members of this family have been isolated genetically and are available to be cloned without the time required for selecting new mutants (Federoff et al. 1983, Nevers 1986).

Many of the mutations which have been propagated affect genes necessary for the production of anthocyanin pigments. One of these, the *Acbz-m2* allele of the *bronze* (*bz*) locus was the first gene to be cloned by transposon tagging in maize (Federoff et al. 1984). A portion of a cloned *Ac* element was used to probe DNA fragments from a plant carrying the *Acbz-m2* allele for *Ac*-homologous sequences. The portion of the *Ac* element that was used as a probe was known to hybridize only to active *Ac* elements. Active *Ac* elements are present in only one or a few copies per genome so only a small number of sequences were identified by the probe. This simplified the identification of the DNA fragment containing a part of the *bz* locus. The genomic portion of this fragment was subcloned and used to isolate a normal copy of the *bz* gene.

This procedure was simplified by the fact that an *Ac*-induced allele of the *bz* locus did not need to be generated de novo. It was also important that the mutation was known to be induced by an *Ac* element and not by some unspecified member of the *Ac–Ds* family which was present in the genome. A drawback to the use of *Ac* for our purposes is that the rate at which new mutations are produced is quite low. Since maize is a diploid organism, new recessive mutations can not be observed in the M_1 (F_1) generation. Each mutagenized M_1 kernal must be grown, the resulting plant self-pollinated, and the M_2 progeny grown to the seedling stage to identify individual plants which are homozygous for a new mutation. The investment of so much time in the isolation of new mutants requires that the return be significant. The rate at which new mutations can be retrieved using *Ac* may not be high enough to justify its use for the induction of mutants whose isolation will involve a substantial screening program.

A second possible drawback to the use of *Ac* as a transposon tag is that it tends to transpose from one location to another on the same chomosome (Federoff 1983). Since there are not generally enough *Ac* elements available in a genome to occupy each of the ten maize chromosomes, all of the possible photosynthetic gene loci, which are presumed to be randomly distributed about the genome, can not be effectively targeted by this system. However, if a particular photosynthetic gene, whose chromosome location is known, is the target of mutagenesis, then a maize stock carrying an active *Ac* element on the same chromosome might be used to tag the gene.

The Spm (*En/I*) family

The *Suppressor–Mutator* family (*Spm*, also called *Enhancer–Inhibitor En-I*) is another two element system. As is the case with *Ac–Ds*, many mutations have been isolated using the *Spm/En-I* family due to the fact that it has been studied intensively over many years (see Federoff 1983). An *En* element was used to clone the *al* locus, another gene essential to the production of anthocyanin pigments (O'Reilly et al. 1984). One reason that an *En* element was used was that no *Ac*-induced *al* allele existed. The *En* probe that was used did not share with *Ac* the convenience of a low number of sequences in the genome to which it would hybridize. Instead many *En*-homologous sequences were present and had to be sorted through to identify the DNA fragment containing a part of the *al-m(papu)* allele. The sorting was simplified by simultaneously screening DNA fragments isolated from the *al-Mum2* allele which was induced by yet another transposon system, Robertson's *Mutator*. Clones which were found to hybridize to one or the other of the two transposon probes were cross hybridized to identify a genomic sequence which had been tagged independently by the two transposons. This dual-tagging strategy greatly simplified the cloning of *al*.

A second locus, *Cl*, which is involved in the regulation of anthocyanin biosynthesis, was cloned by two different groups using *Spm* elements (Cone et al. 1986, Paz-Ares et al. 1986). Cone et al. used a novel approach to simplify the cloning. It has been observed that many transposon-homologous sequences in the maize genome represent inactive elements which do not participate in the production of new mutations. These elements tend to be methylated at a level similar to that of the surrounding genomic DNA. Active elements, as well as the genomic DNA which flanks them, tend to be under-methylated relative to the inactive elements. Methylation inhibits the ability of some restriction enzymes to hydrolyze DNA. Genomic DNA from the *c1-m5* allele was hydrolyzed by one of these methylation-sensitive enzymes. It failed to excise the inactive, methylated elements, but did cleave the DNA in the vicinity of the active, under-methylated elements normally. When the restricted DNA was size fractionated, immobilized on a membrane and hybridized with an *Spm* probe, only a few fragments were detected as distinct hybridization bands. The fragment of interest was selected from among these on the basis of segregation with the mutant phenotype.

Unfortunately, this transposon family also produces new mutations at a relatively low rate. Its elements also tend to transpose preferentially along the donor chromosome although less so than do *Ac* elements (Nevers et al. 1986)

The Robertson's mutator *family*

In 1978, Robertson reported a line of maize that produced new mutations at a rate up to 50 fold higher than normal maize lines (Robertson 1978). He called the line *Mutator*. An *alcohol dehydrogenase 1 (Adh1)* allele isolated from this line was cloned by classical methods and a 1.4 kilobase (kb) insert was found within the gene (Barker et al. 1984). The insert possessed the terminal inverted repeats and direct target repeats typical of transposons and contained four open reading frames within its internal sequence. The insert was called *Mu1*. *Mu1* and *Mu1.7*, a closely related sequence containing an additional 300 base pairs of DNA, are apparently responsible for most or all of the mutable alleles so far isolated from *Mutator* lines which have been examined at the molecular level.

The *Mutator* system differs from *Ac–Ds* and *Spm* in at least two ways. No autonomous element has been found associated with *Mutator* lines. *Mu1* and *Mu1.7* apparently do not encoded the enzymes necessary for their own transposition. While it is possible that an autonomous element may be identified, it has been suggested that *Mu* elements may transpose by a novel mechanism (Lillis and Freeling 1986).

Mu elements are present in the genome of *Mutator* lines in high numbers (10–60 copies) (Lillis and Freeling 1986, Barker et al. 1984). This may partly explain the high rate at which new mutations are generated by these lines. The high mutation rate is advantageous for the purpose of producing a variety of new mutations quickly. However a high number of *Mu* copies can complicate the isolation of the one copy which has caused the mutation under investigation.

Methods have been deduced by which the number of copies of *Mu* elements can be reduced over a period of several generations. When a *Mutator* line is outcrossed to a non-*Mutator* line, 90% of the progeny retain *Mutator* activity (Robertson 1978). The remaining 10% lose *Mutator* activity until they are again crossed to an active *Mutator* line. By repeatedly backcrossing the inactivated *Mutator* stock to non-*Mutator* lines the number of copies of *Mu* elements in the progeny is cut approximately in half with each generation (Lillis and Freeling 1986). This process can be continued until the number of copies is low enough to be practically workable.

Another way to reduce copy number is to initially increase copy number (Alleman and Freeling 1986, Lillis and Freeling 1986). When two *Mutator* lines are intercrossed the F_1 progeny are found to contain twice the number of *Mu* elements of either parent. Within three generations the progeny lose all *Mutator* activity permanently. These inactive individuals can then be outcrossed as described above, to reduce copy number by segregation. The process of reducing copy number is time consuming but it is not necessary with all *Mutator* lines since not all lines contain unworkably high numbers of *Mu* elements. When it is necessary, the time required to reduce the copy number may be offset by the high rate at which new mutants are produced by the *Mutator* system.

Materials and methods

Induction of mutants and stock maintenance

The F_1 progeny of a cross between the standard maize line Q60 and an active *Mutator* line (DR8091-10 × 7092-8) were supplied as seed by Don Robertson (Genetics Dept, Iowa State University) as a source of *Mutator* activity. In the summer of 1983 new mutations were induced when pollen from the *Mutator* plants was crossed onto the silks of standard inbred lines Mo17 and W23. During the winter of 1983–84 and in subsequent crop seasons, M_1 (F_1) progeny of the mutagenic crosses were self-pollinated to recover recessive nuclear mutations. The resulting M_2 (F_2) progeny were grown and screened as seedlings to identify photosynthetic mutants (Fig. 3). After growth producing three leaves, plants which were homozygous for a mutation

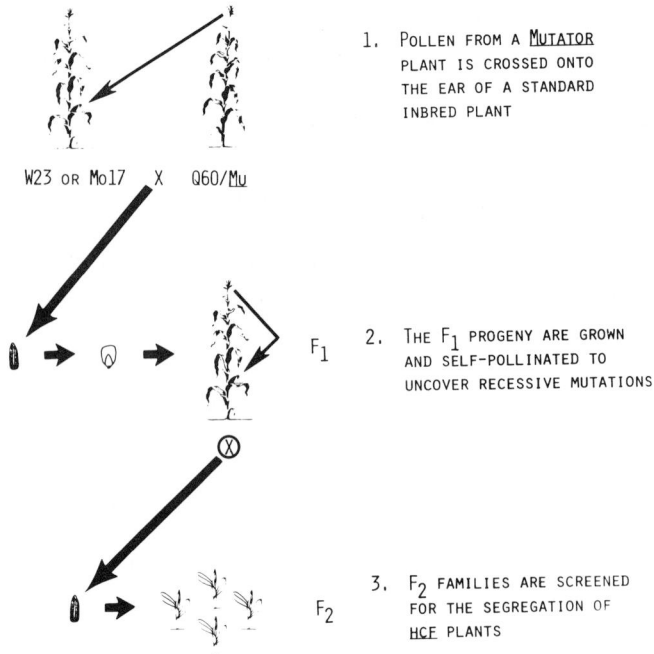

1. POLLEN FROM A **MUTATOR** PLANT IS CROSSED ONTO THE EAR OF A STANDARD INBRED PLANT

W23 OR Mo17 X Q60/**Mu**

F_1

2. THE F_1 PROGENY ARE GROWN AND SELF-POLLINATED TO UNCOVER RECESSIVE MUTATIONS

F_2

3. F_2 FAMILIES ARE SCREENED FOR THE SEGREGATION OF HCF PLANTS

Fig. 3. The mutagenesis of maize with the *Mutator* transposon system. The procedure requires two growing seasons plus two weeks for the initial screening step.

disrupting photosynthesis had exhausted the kernal endosperm supply on which they subsisted and subsequently died. All biochemical, photochemical and molecular analyses of mutant plants were carried out using pre-lethal seedlings.

When a desirable mutation was isolated it was subsequently propagated by self-pollinating the normal sibling plants from the family in which mutant plants segregated and concurrently outcrossing the same plants to standard inbred lines. Two thirds of the normal sibling plants which were self-pollinated were heterozygous for the mutant allele; the rest were homozygous for the normal allele. The progeny of the self-pollinations were then screened to identify the parents which carried the mutation. This informtion was then used to identify which of the outcross progeny carried the mutation. These progeny were then grown, self-pollinated and outcrossed as described.

Mutant isolation and characterization

Screening
Thirty progeny from each self-pollination were germinated and grown to the three leaf stage before screening for the presence of *high chlorophyll fluores-*

cent (*hcf*) individuals (Miles 1980, 1982). In a darkened room, the seedlings were illuminated with long-wave UV light. Plants with lesions in the photosynthetic apparatus affecting the normal use of singlet state excited chlorophyll energy fluoresced with a bright red visible color which was easily distinguishable from the dull reddish-purple color of normal siblings.

When a new *hcf* mutant was found to segregate within an M_2 family it was characterized according to three criteria: whole leaf fluorescence induction kinetics, the banding patterns produced by PAGE of thylakoid proteins and the rates of electron transport through the reaction center complexes individually and through the electron transport chain as a whole.

Fluorescence induction kinetics
Fluorescence induction kinetics were evaluated by illuminating whole leaf segments (Miles 1980, 1982). Actinic light was filtered through a broad bandpass blue filter with peak illumination of 450 nm and a half-bandwidth of 110 nm. Fluorescent light was measured at 45° to the leaf surface by a photodiode preceded by a red interference filter passing 683–695 nm light. The signal was amplified and recorded on a fast response X–Y chart recorder.

LDS-PAGE
Thylakoid membrane polypeptides were isolated and prepared as described (Metz and Miles 1982). Samples were solubilized with lithium dodecyl sulfate, loaded on a chlorophyll basis and separated on a 12–18% gradient polyacrylamide gel at a constant power of 4.5 W at 4 °C. Polypeptides were stained with Commassie Brilliant Blue R-250 or with 3,3′,5,5′ tetramethyl benzidine to visualize cytochromes (Thomas et al. 1976).

Electron transport
Electron transport rates were measured as O_2 uptake or evolution by thylakoids in the presence of artifical electron donors and acceptors using a Clark-type O_2 electrode (Yellow Springs Instruments) as previously described (Krueger and Miles 1981).

DNA hybridization
Genomic DNA was probed for the presence of sequences homologous to the *Mu1* element using the protocol of Chandler et al. (1986). DNA was isolated from the leaf tissue of individual plants by a modification (Chen 1986) of the method of Shure et al. (1983). Restriction enzymes were purchased from Promega (St. Louis, MO). DNA fragments were size fractionated by electrophoresis through 0.8% agarose gels. DNA was transferred to Genatran

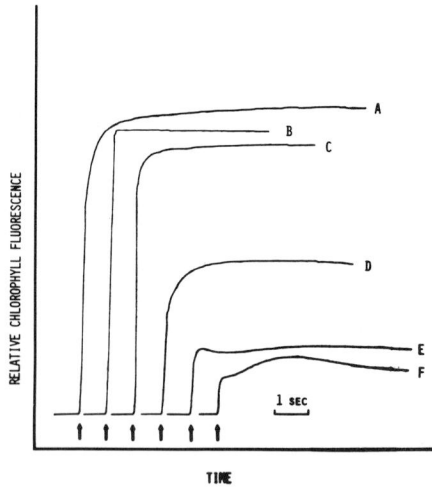

Fig. 4. A composite of fluorescence induction kinetics from five *hcf* mutants. A. *hcf101*: typical PSI-deficient mutant kinetics. B. *hcf3-Mu*: square kinetics indicative of a block in electron transport on the reducing side of PSII prior to the reduction of the plastoquinone pool. C. *hcf113*: the kinetics of a mutant which is defective in electron transport through PSI and Cyt b_6/f. D. *hcf102*: typical Cyt b_6/f complex-deficient mutant kinetics. E. *hcf108*: the kinetics of a mutant lacking the ATPase complex. F. Normal kinetics. The up arrows indicate the points at which the actinic light was turned on.

(Plasco; Woburn, MA) nylon membrane by the method of Southern (1975). Immobilized DNA fragments were probed for *Mu*-homologous sequences with a 729 base pair fragment of the pMJ9 plasmid (Barker et al. 1984) which is released by hydrolysis with the restriction enzymes Taql and Aval. The probe was labeled with ^{32}P-dCTP using the oligolabeling reaction of Feinberg and Vogelstein (1983).

Results

Mutagenesis

Seventeen *Mutator* plants with the genetic constitution Q60/*Mu* were crossed as male parents onto the silks of 37 standard inbred plants. 1087 M_1 progeny of these crosses were grown and self-pollinated during the spring of 1984 and the summers of 1984 and 1986. A variety of new visually detectable recessive mutations were observed among the M_2 progeny, including several albino, virescent and yellow-green mutants, and a number of defective kernal mutants as well as one blue fluorescent, one dwarf and two glossy

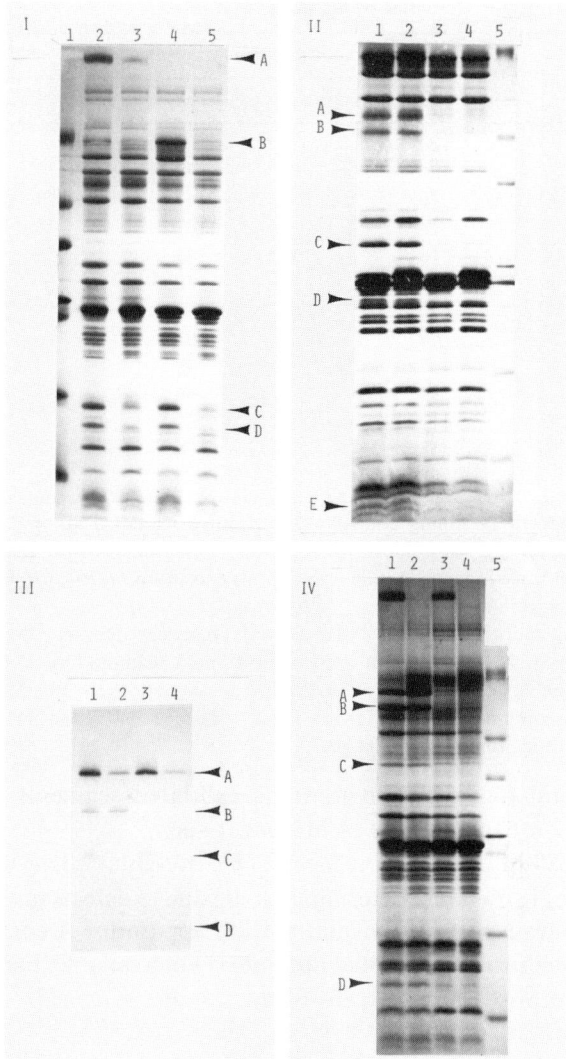

Fig. 5. LDS-PAGE profiles of thylakoid polypeptides from four *hcf* mutants. Arrows indicate bands which are reduced or missing in the mutant lanes. Mol wt standards are 66, 45, 36, 29, 24, 20 and 14 × 10^3 Da. I. *hcf101*: lane 1 contains mol wt standards (mws), lanes 2 and 4 contain normal thylakoids, 3 and 5 contain mutant thylakoids. Lanes 4 and 5 were heated at 70 °C for 4 min before electrophoresis. A = CP1, B = 68 kDa CP1-apoprotein, C and D = 18 and 16 kDa polypeptides associated with PSI. II. *hcf3-Mu*: lanes 1 and 2 contain normal thylakoids, 3 and 4 contain mutant thylakoids, 5 contains mws. Lanes 2 and 4 were heated as in I. A and B = 49 and 45 kDa chl binding proteins, C = 33 kDa, D = 22 kDa and D = 10 kDa polypeptides, all associated with PSII. III. *hcf102*: lanes are as in I with normal thylakoids in 1 and 3 and mutant thylakoids in 2 and 4. TMBZ-stained gel shows A = Cyt f and C = Cyt b_6 associated with the Cyt b_6/f complex as well as B = LHCII and D = Cyt b_{559} associated with PSII. IV. *hcf107*: lanes are as in II. A, B and C = α, β and γ subunits of the CF$_1$ ATPase and D is probably the epsilon subunit of the same complex.

Table 2. Electron transport activities of *hcf* maize mutants.[a]

Mutant	Photosynthetic electron transport[b]		
	PSII	PSI	PSII + PSI
Control	100	100	100
hcf101	126	72	84
hcf104	76	41	40
hcf3-Mu	18	107	5
hcf103	18	143	0
hcf102	81	128	16
hcf109	100	130	4
hcf110	97	101	11
hcf105	57	37	0
hcf111	91	66	12
hcf112	112	31	24
hcf113	58	76	38
hcf108	73	75	61

[a] Expressed as percent of sibling control value. Control values: PSII = 259, PSI = 849, PSII + PSI = 136μmol O_2 (mg chlorophyll)$^{-1}$ hour^{-1}, evolved or consumed. Values are the averages of at least three independent determinations except for *hcf105*, for which the results of two experiments were averaged.

[b] Electron donors and acceptors for PSII: 0.4 mM phenylene diamine, 2.0 mM potassium ferricyanide; for PSI: 2.0 mM sodium ascorbate, 0.5 mM diaminodurene, 0.1 mM methyl viologen, 8.0μM DCMU; for PSII + PSI: 0.1 mM methyl viologen, 10 mM methyl amine. The reaction mix contained 40 mM Tricine pH 8.0, 60 mM NaCl and 4 mM $MgCl_2$. See Materials and methods for further details.

mutants. Several pigmentation mutants exhibited segments of normally colored tissue on backgrounds of mutant tissue.

From the 1087 M_2 families screened, 23 *hcf* mutants (2.1%) were isolated. Eleven of these (1.0%) were determined to be due to defects in the photosynthetic electron transport or phosphorylation apparatus. A complete record of families which segregated visual mutants is not available, but those which were noted comprised nine percent of the families screened.

Fluorescence induction kinetics

Fluorescence induction kinetics traces typical of mutants defective in PSI (*hcf101*), PSII (*hcf3-Mu*), the Cyt b_6/f complex (*hcf102*) and the ATPase complex (*hcf108*) as well as one affected at several points in the electron transport apparatus are shown (Fig. 4). *hcf3-Mu* and *hcf103* both exhibit traces typical of PSII mutants while *hcf101*, *hcf102*, *hcf104*, *hcf105*, *hcf109*, *hcf110*, *hcf111*, *hcf112* and *hcf113* produce traces indicating that they suffer blocks in electron transport beyond the plastoquinone (PQ) pool. The trace of *hcf108* does not correspond to the pattern of any mutant characterized previously.

LDS-PAGE

Representative thylakoid protein profiles produced by LDS-PAGE are shown in Fig. 5. The predominant characteristic of the mutant profiles is the loss of one or more sets of bands associated with particular membrane complexes. *hcf101* and *hcf104* have lost some or all of the chlorophyll protein band associated with PSI (CP1) as well as the 68 kilodalton band representing the reaction center apoprotein in denatured samples. *hcf3-Mu* and *hcf103* each lack several bands associated with PSII. *hcf102*, *hcf109* and *hcf110* each lack cytochromes b_6 and f, indicating the loss of the Cyt b_6/f complex. *hcf107* and *hcf108* are each missing the α andβ subunits of the coupling factor complex. *hcf105*, *hcf111* and *hcf112* lack bands associated with PSI as well as bands associated with the Cyt b_6/f complex.

Electron transport

Electron transport rates through the reaction center complexes and through the whole electron transport chain are presented in Table 2. Mutants blocked at a particular complex are not necessarily completely blocked. Most mutants exhibit a low rate of activity through the affected complex. Electron transport data are in general agreement with the diagnoses based on PAGE results.

hcf104 possesses a substantial rate of PSI activity which agrees with the amounts of CP1 and 68 kDa reaction center apoprotein detectable following electrophoresis. *hcf102*, *hcf109* and *hcf110* all possess near-normal rates of electron transport through the reaction center complexes but support only low rates of whole-chain electron transport. *hcf3-Mu* and *hcf103* each lack PSII activity. In contrast to these examples, *hcf101* exhibits a very high rate of PSI electron transport in the absence of CP1 or the 68 kDa bands following PAGE. These data on *hcf101* are being presented in detail elsewhere (Cook and Miles, in preparation).

hcf105, *hcf111*, *hcf112* and *hcf113* each reflect the loss of PSI reaction center but because an independent evaluation of the Cyt b_6/f complex requires functional PSI and PSII complexes this defect is not revealed by the data.

hcf108 electron transport data are equivocal. It is not clear whether the lesion in the ATPase complex is capable of interfering with electron transport as observed or whether the mutation has a more pleiotropic effect, disrupting electron transport as well as the ATPase complex. Preliminary measurements of photophosphorylation and ATP hydrolysis by *hcf107* and *hcf108* indicate that each mutant is substantially reduced in both activities (data not shown).

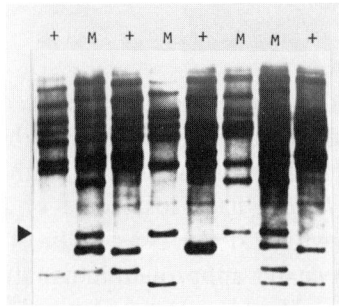

Fig. 6. A *Mu1*-homologous hybridization band which segregates with the *hcf3-Mu* phenotype. DNA isolated from plants carrying the *hcf3-Mu* allele and plants homozygous for the normal gene was hydrolyzed with HindIII and hybridized with a 729 nucleotide TaqI-AvaI internal *Mu1* probe. + indicates DNA from a homozygous-normal plant. *M* indicates DNA from a plant carrying the mutant allele. The arrow to the left of the lanes indicates a 4.4 kilobase HindIII fragment which is present in plants bearing the mutant allele and absent from normal samples. No other co-segregating band has been identified. Segregation of the 4.4 kb fragment has been observed in 39 individuals.

Genetic analysis

hcf3-Mu was tested for allelism to the chemically (EMS)-induced PSII-defective *hcf3*. Pollen from normal M_2 individuals from the family which segregated *hcf3-Mu* was crossed onto the ears of progeny of the original *hcf3* isolate. Eight of 50 progeny of this cross exhibited the *hcf* trait and fluorescence induction kinetics characteristic of the two parents.

Southern hybridization

DNA was isolated from 39 normal plants from families segregating the *hcf3-Mu* phenotype, hydrolyzed with HindIII and probed with an internal fragment of *Mu1*. A representative autoradiograph is presented (Fig. 6). A single hybridization band of 4.4 kilobases is found to co-segregate with the mutant phenotype, suggesting that this DNA fragment contains at least a portion of the gene which is interrupted in the *hcf3-Mu* mutants.

Discussion

Two primary obstacles have stood in the way of the isolation of photosynthetic genes by means of transposon tagging. The first was the lack of a transposon system which produced new mutants at a sufficiently high rate. The second was the demonstration that photosynthetic genes so tagged could be isolated and identified.

The first obstacle was overcome when Robertson's *Mutator* system was applied to the problem. The variety of visible mutations isolated with this system is comparable to that obtained using chemical mutagens. Furthermore, the rate at which new mutations arise is very high and a significant proportion of these are *hcf* mutants (2% in these experiments). At this rate of production, a steady supply of new mutants can be obtained with a moderate investment of time, energy and field space. The primary drawback to the use of *Mutator* was expected to be the high number of copies of the element in *Mu* lines of maize. The *Mu* lines which have been examined in conjunction with this work have contained only 10–20 copies of *Mu1*-homologous elements. This number of sequences has not presented a serious obstacle to the identification of a band co-segregating with a mutant phenotype. Apparently, careful selection of a *Mutator* source line with a low number of copies of *Mu1* prior to mutagenesis can eliminate the need to reduce copy number.

More than 100 *hcf* mutants have been isolated using chemical mutagens. There is no evidence that this number approaches the total number of nuclear mutations resulting in the *hcf* phenotype. When a new transposon-induced allele of a well-characterized chemically-induced mutation is isolated, the time required for its characterization is shortened. In the case of *hcf3*, enough photobiological characterization has been done that the cloning of the gene is a next reasonable step. New transposon-induced mutants can be characterized in the same way as chemically-induced mutants with the added incentive that the gene responsible for the phenotype is accessible to cloning.

The standard photochemical characterization presented in the Results section can also be supplemented. *hcf-Mu106, hcf120, hcf121* and *hcf122* were evaluated by immunoblot analysis for the presence of components of the thylakoid complexes. Each mutant was deficient in at least one complex but none was completely lacking the components of any complex. (Taylor et al. 1988, Barkan et al. 1986). *hcf-Mu106* lacks the components of PSI, PSII and the Cyt b_6/f complex, but the trace amounts of the polypeptides which are present occur in stoichiometric proportions. *hcf120* lacks the components of the Cyt b_6/f complex but, in contrast to *hcf-Mu106*, one component (Cyt *f*) is completely absent, while two other components are present in trace amounts. When chloroplast mRNAs which encode the thylakoid polypeptides were analyzed in *hcf-Mu106, hcf120, hcf121* and *hcf122*, they were present in normal amounts and sizes. However, when a chemically induced mutant, *hcf38* was evaluated similarly, many of the chloroplast mRNAs were reduced or missing.

The variety of phenotypic expression among these mutants reflects the many possible sites at which lesions might result in defective photosynthetic electron transport, including gene transcription or translation, post-translational protein modification, transport or membrane insertion and stabilization of assembled membrane complexes.

The development of a transposon mutagenesis system capable of producing new mutants at a high rate has been accomplished. The advantage of this mutagenesis system is that it renders the newly mutated genes accessible to cloning. The theoretical potential of the system has not yet been realized in the field of photosynthesis, but many genes have been cloned in maize and in other organisms using transposon tagging strategies. The first step in the process of cloning a photosynthetic gene, the identification a DNA fragment which putatively contains a portion of the gene of interest, has been taken. *Mu*-homologous fragments which co-segregte with the mutant phenotype have been identified for *hcf3-Mu* and for *hcf-Mu106* (Martienssen et al. 1987). The identification of a *Mu*-homologous fragment which co-segregates with a mutant phenotype provides preliminary evidence that a *Mutator* element has caused the mutation. However, such evidence can only demonstrate linkage between the affected gene and the location of the element. Increasing the number of individuals which are tested for co-segregation can demonstrate tighter linkage, but this approach can not prove that the mutation is caused by insertion of the linked element. Linkage of a map unit or less may represent separation by many kilobases.

Insertion of a transposon can often be demonstrated by the instability of the resulting mutation. The excision of the element from cells in somatic tissue may result in a characteristic pattern of variegation. To date, however, no *hcf* mutant among our stocks has exhibited somatic instability. The high mutation rate and the presence of somatic instability among many of the mutants arising in this line along with the co-segregation of a *Mu*-homologous fragment with the *hcf3-Mu* phenotype do not constitute conclusive evidence that the *hcf3-Mu* allele has been tagged by a *Mutator* element. However, the evidence is sufficiently strong to begin cloning the locus.

The actual molecular cloning of the DNA fragment is a straightforward, though not trivial, process. DNA from an individual carrying the *hcf3-Mu* mutation which does not carry any other *Mu*-homologous HindIII restriction fragment near 4.4 kb in size will be hydrolyzed with HindIII and size fractionated on an agarose gel. Following electrophoresis, a portion of the gel containing restriction fragments of approximately 4.4 kb will be excised. The DNA fragments will be removed from the gel and ligated into a vector. The desired clone will be selected from among the library of recombinants on the basis of its homology to *Mu1* since only one *Mu*-homologous fragment will be cloned. Once the fragment has been cloned, a number of approaches can be followed to verify that it does indeed contain a fragment of the gene of interest. The versatility of maize genetics can be most clearly appreciated at this point.

hcf3 has been mapped to the short arm of chromosome 1. As a result, a simple test of allelism mapped the newly isolated *Mu*-induced *hcf3* allele to the same arm. However, many new mutations which are not allelic to previously-mapped mutations can themselves be mapped to a chromosome arm in a single generation using B–A translocation stocks (Beckett 1978). Each of these stocks lacks a portion of one of the 20 maize chromosome arms. When one of these stocks is crossed to a plant bearing a recessive mutation, the mutant phenotype can be detected in the F_1 progeny if the B–A parent carried a deficiency in the chromosome arm on which the mutation resides. B–A translocation stocks lacking portions of 19 out of the 20 maize chromosome arms are available.

When the chromosome location of a mutation has been determined genetically, the cloned DNA fragment can be mapped to ensure that it is located on the same chromosome. Two different methods of mapping a cloned DNA fragment exist in maize. One method involves the use of DNA isolated from monosomic maize stocks (Webber 1986). These stocks each lack one copy of one of the ten maize chromosomes. DNA isolated from these stocks is probed with the isolated DNA fragment and the source chromosome is identified by the intensity of the hybridization bands produced. The second mapping method makes use of collections of restriction fragment length polymorphism (RFLP) probes. These probes identify undefined fragments of DNA which are hydrolyzed differently in different maize lines. The probes are mapped to the chromosome arms on which the RFLPs are found. The probes are then used to map other cloned fragments within the maize genome by linkage analysis (Helentjaris et al. 1985). RFLP mapping may allow a very precise location of a clone, depending on the relative positions of the available probes.

Mapping the cloned fragment provides evidence that the correct fragment – the fragment which co-segregates with the phenotype — has been isolated. A further experiment can provide functional evidence that the proper fragment has been cloned. A gene defined by an *hcf* phenotype should normally produce an mRNA molecule. The mRNA is likely to be missing or modified in a mutant which is caused by insertion of a transposon. To test whether this is true for a particular mutant, a portion of the cloned DNA fragment which flanks the transposon sequence is used to probe RNA isolated from several normal and several mutant plants. If the probe identifies an mRNA species which is missing or of an altered size in mutant plants, it is quite likely that the gene of interest has been cloned.

One drawback to the use of maize as the organism in which to carry out these experiments is that maize has not, as yet, been successfully transformed and regenerated. These techniques are necessary for the reintroduction of altered forms of the cloned gene or other manipulations related to its further characterization. This has provided much of the motivation for several groups to attempt the transfer of a transposon from maize or another organism into a transformable plant species. The attempt has been successful in at least one case. The *Ti* plasmid of *Agrobacterium tumefaciens* was used as an insertion cassette to introduce the *Ac* element into tobacco (*Nicotiana tobaccum*) (Baker et al. 1986). Evidence of active transposition of the *Ac* element indicates that this approach may provide a productive alternative to the isolation of genes in maize and their subsequent transfer to a different species for transformation studies.

Conclusion

hcf3 has been referred to as the most thoroughly characterized PSII mutant from among higher plants (Somerville 1986). Nevertheless, nothing is known about the exact lesion which causes the well characterized phenotype. The value of the *hcf3-Mu* mutation is that it provides access to the affected gene and its product, allowing the characterization of the specific process which is aborted in the mutant. The same potential exists for each of the other mutants isolated from a *Mutator* background or from any line containing a transposon system.

Over 100 *hcf* mutants have been isolated following EMS mutagenesis. This demonstrates the complexity involved in the development of a functional photosynthetic apparatus. Many of the essential processes, such as that mediated by the *hcf3* locus, can not be studied by classical biochemical or photochemical approaches. Transposon mutagenesis offers a means by which these processes can be approached and eventually understood.

Acknowledgements

This work has been supported by USDA grants SE84 CRCR 1-1480 and SE86 CRCR 1-2028. W.B. Cook is supported by USDA grant 84-GRAD-9-0033.

References

Alleman M and Freeling M (1986) The *Mu* transposable elements of maize: Evidence for transposition and copy number regulation during development. Genetics 112: 107–119

Baker B, Schell J, Lorz H and Federoff NV (1986) Transposition of the maize controlling element 'Activator' in tobacco. Proc Natl Acad Sci USA 83: 4844–4848

Barkan A, Miles D and Taylor WC (1986) Chloroplast gene expression in nuclear, photosynthetic mutants in maize. EMBO J 5: 1421–1427

Barker RF, Thompson DV, Talbot DR, Swanson J and Bennetzen JL (1984) Nucleotide sequence of the maize transposable element *Mu1*. Nucleic Acids Res 12: 5955–5967

Beckett JB (1978) B–A translocation in maize. I. Use in locating genes by chromosome arms. J Heredity 69: 27–36.

Bennetzen JL, Swanson J, Taylor WC and Freeling WC (1984) DNA insertion in the first intron of maize *Adh1* affects message levels: Cloning of progenitor and mutant alleles. Proc Natl Acad Sci USA 81: 4125–4128

Bennetzen JL (1987) Covalent DNA modification and the regulation of *Mutator* element transposition in maize. Mol Gen Genet 208: 45–51

Bennetzen JL, Fracasso RP, Morris DW, Robertson DS and Skogen-Hagenson MJ (1987) Concomitant regulation of *Mu1* transposition and *Mutator* activity in maize. Mol Gen Genet 208: 57–62

Bonas U, Sommer H, Harrison BJ and Saedler H (1984) The transposable element Tam1 of *Antirrhinum majus* is 17 kb long. Mol Gen Genet 194: 138–143

Calos MP and Miller JH (1980) Transposable elements. Cell 20: 579–595

Campbell A (1981) Evolutionary significance of accessory DNA elements in bacteria. Ann Rev Microbiol 35: 55–83

Carpenter R, Martin C and Coen ES (1987) Comparison of genetic behavior of the transposable element Tam3 at two unliked pigment loci in *Antirrhinum majus*. Mol Gen Genet 207: 82–89

Chandler VL and Walbot V (1986) DNA modification of a maize transposable element correlates with loss of activity. Proc Natl Acad Sci USA 83: 1767–1771

Chandler VL, Rivin C and Walbot V (1986) Stable non-*Mutator* stocks of maize have sequences homologous to the *Mu1* transposable element. Genetics 114: 1007–1021

Chen J (1986) Plant genomic DNA preparation. In: Plant Gene Cloning Manual. Cold Spring Harbor: Maize molecular genetics group, Cold Spring Harbor Laboratory

Cone KC, Burr FA and Burr B (1986) Molecular analysis of the maize anthocyanin regulatory locus *C1*. Proc Natl Acad Sci USA 83: 9631–9635

Cook B and Miles D (1987) Mutator-induced PSII photosynthesis mutant is allelic to *hcf3*. Maize Genet Coop Newslett 61: 44

Cook B, Hunt M and Miles D (1987) *Mutator*-induced mutation on 8L affects the chloroplast cytochrome b_6/f complex. Maize Genet Coop Newslett 61: 44

Federoff NV (1983) Controlling elements in maize. In: Shapiro JA (ed) Mobile Genetic Elements, pp 1–65. New York: Academic Press/Harcourt Brace Jovanovich

Federoff NV, Wessler S and Shure M (1983) Isolation of the transposable maize controlling elements *Ac* and *Ds*, Cell 35: 235–242

Federoff NV, Furtek DB and Nelson OE (1984) Cloning of the *bronze* locus in maize by a simple and generalizable procedure using the transposable controlling element *Activator* (*Ac*). Proc Natl Acad Sci USA 81: 3825–3829

Feinberg AP and Vogelstein B (1983) A technique for radiolabeling DNA restriction endonuclease fragments to high specific activity. Anal Bioch 132: 6–13 and Addendum Anal Bioch 137: 266–267

Freeling M (1984) Plant transposable elements and insertion sequences. Ann Rev Plant Physiol 35: 277–298

Goldberg RB, Hoschek G and Vodkin LO (1983) An insertion sequence blocks the expression of a soybean lectin gene. Cell 33: 465–475

Grindley NDF and Reed RR (1985) Transpositional recombination in prokaryotes. Ann Rev Biochem 54: 863–896

Guikema JA and Sherman LA (1980) Electrophoretic profiles of cyanobacterial membrane polypeptides showing heme-dependent peroxidase activity. Biochem Biophys Acta 637: 189–201

Heffron F (1983) Tn3 and its relatives. In: Shapiro JA (ed) Mobile Genetic Elements, pp 223–260. New York: Academic Press/Harcourt Brace Jovanovich

Hehl R, Sommer H and Saedler H (1987) Interaction between Tam1 and Tam2 transposable elements of *Antirrhinum majus*. Mol Gen Genet 207: 47–53

Helentjaris T, King G, Slocum M, Siedenstrang C and Wegman S (1985) Restriction fragment polymorphisms as probes for plant diversity and their development as tools for applied plant breeding. Plant Mol Biol 5: 109–118

Hudson A, Carpenter R and Coen ES (1987) De novo activation of the transposable element Tam2 of *Antirrhinum majus*. Mol Gen Genet 207: 54–59

Iida S, Meyer J and Arber W (1983) Prokaryotic IS elements. In: Shapiro JA (ed) Mobile Genetic Elements, pp 159–222. New York: Academic Press/Harcourt Brace Jovanovich

Kleckner N (1981) Transposable elements in prokaryotes. Ann Rev Genet 15: 341–404

Krueger RW and Miles D (1981) Photosynthesis in tall fescue. I. High rates of electron transport and photophosphorylation in chloroplasts of hexaploid plants. Plant Physiol 67: 763–767

Leto KJ and Miles D (1980) Characterization of three Photosystem II mutants in *Zea mays* L. lacking a 32,000 Dalton lamellar polypeptide. Plant Physiol 66: 18–24

Lillis M and Freeling M (1986) *Mu* transposons in maize. Trends Genet 2: 183–188

Martienssen RA, Barkan A, Scriven A and Taylor WC (1987) Identification of a nuclear gene involved in thylakoid structure. In: Leaver C and Sze H (Eds) Plant Membranes: Structures, Function and biogenesis (UCLA Symposia on Molecular and Cellular Biology 63: 118–194). New York: Alan R Liss

McClintock B (1984) the significance of responses of the genome to challenge. Science 226: 792–801

Metz JG and Miles D (1982) Use of a nuclear mutant of maize to identify components of photosystem II. Biochem Biophys Acta 681: 95–102

Metz JG, Miles D and Rutherford AW (1983) Characterization of nuclear mutants of maize which lack the cytochrome f/b-563 complex. Plant Physiol 73: 452–459

Miles D (1980) Mutants of higher plants: Maize. In: San Pietro A (ed) Photosynthesis, Part C, Methods in Enzymology 69: 3–22. New York: Academic Press

Miles D (1982) the use of mutations to probe photosynthesis in higher plants. In: Edelman M, Hallick RB and Chua N-H (eds) Methods in Chloroplast Molecular Biology, pp 75–107. Amsterdam: Elsevier Biomedical Press

Miles D and Randall D (1983) Nuclear mutants of maize altering the large subunit of ribulose-1,5-bisphosphate carboxylase. In: Randall DD, Blevins DG and Larson R (eds) Current Topics in Plant Biochemistry and Physiology 1: 231. Columbia: Interdisciplinary Plant Biochemistry and Physiology Program

Miles D, Leto KJ, Neuffer MG, Polacco M, Hanks JF and Hunt MA (1985) Chromosome arm location of photosynthesis mutants in *Zea mays* L. using B-A translocations. In: Steinbeck KE, Arntzen CJ and Bogorad J (Eds) Molecular biology of the photosynthetic apparatus, pp 361–365. Cold spring Harbor: Cold Spring Harbor Laboratory

Nevers, P, Shepherd NS and Saedler H (1986) Plant transposable elements. Adv Bot Res 12: 103–203

O'Reilly C, Shepherd NS, Pereira A, Schwarz-Sommer Zs, Bertram I, Robertson DS, Peterson PA and Saedler H. (1985) Molecular cloning of the *a1* locus of *Zea mays* using the transposable elements *En* and *Mu1*. EMBO J 4: 877–882

Paz-Ares J, Wienand U, Peterson PA and Saedler H (1986) Molecular cloning of the *c* locus of *Zea mays*: a locus regulating the anthocyanin pathway. EMBO J 5: 829–833

Pereira A, Schwarz-Sommer Zs, Gierl A, Bertram I, Peterson PA and Saedler H (1985) Genetic and molecular analysis of the Enhancer (En) transposable element system of *Zea mays*. EMBO J 4: 17–23

Pohlman RF, Federoff NV and Mesing J (1984) the nucleotide sequence of the maize controlling element *Activator*. Cell 37: 635–643

Robertson DS (1978) Characterization of a mutator system in maize. Mutation Res 51: 21–28

Roeder GS and Fink GR (1983) Transposable elements in yeast. In: Shapiro JA (ed) Mobile Genetic Elements, pp 300–328. New York: Academic Press/Harcourt Brace Jovanovich

Rubin GM (1983) Dispersed repetitive DNAs in *Drosophila*. In: Shapiro JA (ed) Mobile Genetic Elements, pp 329–362. New York: Academic Press/Harcourt Brace Jovanovich

Sachs MM, Peacock WJ, Dennis ES and Gerlack WL (1983) Maize *Ac/Ds* controlling elements. A molecular viewpoint. Maydica 28: 289–303

Schwartz D and Dennis E (1986) Transposase activity of the *Ac* controlling element in maize is regulated by its degree of methylation. Mol Gen Genet 206: 476–482

Schwarz-Sommer Zs, Gierl A, Klosgen RB, Wienand U, Peterson PA and Saedler H (1984) The Spm (En) transposable element controls the excision of a 2-kb DNA insert at the wx^{m-8} allele of *Zea mays*. EMBO J 3: 1021–1028

Shapiro JA (1983) Genetic reorganization in cell lineages. In: Shapiro JA, (ed) Mobile Genetic Elements, pp xi–xvi. New York: Academic Press/Harcourt Brace Jovanovich

Shure M, Wessler S and Federoff N (1983) Molecular identification and isolation of the *Waxy* locus in maize. Cell 35: 225–233

Somerville CR (1986) Analysis of photosynthesis with mutants of higher plants and algae. Ann Rev Plant Physiol 37: 467–507

Southern EM (1975) Detection of specific sequences among DNA fragments separated by gel electrophoresis. J Mol Biol 98: 503–517

Taylor LP, Chandler VL and Walbot V (1986) Insertion of 1.4 kb and 1.7 kb *Mu* elements into the *Bronze-1* gene of *Zea mays* L. Maydica 31: 31–45

Taylor WC, Barkan A and Martienssen RA (1988) The use of nuclear mutants in the analysis of chloroplast development. Developmental Genet 8: 305–320

Thomas PE, Rayn D and Levin W (1976) An improved staining procedure for the detection of the peroxidase activity of cytochrome P-450 on sodium dodecyl sulfate polyacrylamide gels. Anal Biochem 75: 168–176

Tittgen J, Hermans J, Steppuhn J, Jansen T, Jannson C, Anderson B, Nechushtai R, Nelson N and Herrmann RG (1986) Isolation of cDNA clones for fourteen nuclear-encoded thylakoid membrane proteins. Mol Gen Genet 204: 258–265

Walbot V, Chandler VL, Taylor LP and McLaughlin P (1987) Regulation of transposable element activities during the development and evolution of *Zea mays* L. In: Development as an Evolutionary Process, pp 265–284. New York: Alan R Liss

Webber DF (1986) The production and utilisation of monosomic *Zea mays* in cytogenetic studies. In: Reddy GM and Co EH (eds) Gene Structure and Function in Higher Plants, pp 191–204. Dehli: Oxford and IBH Publishing

Young RA and Davis RW (1983) Efficient isolation of genes by using antibody probes. Proc Natl Acad Sci USA 80: 1194–1198

Govindjee et al. (eds), Molecular Biology of Photosynthesis: 105–120
© 1988 Kluwer Academic Publishers

Minireview

Using bacteria to analyze sequences involved in chloroplast gene expression

ANTHONY A. GATENBY,[1] STEVEN J. ROTHSTEIN[2] & DOUGLAS BRADLEY[3]

[1] *Central Research and Development Department, Experimental Station, E.I. du Pont de Nemours & Co., Wilmington, DE 19898 USA;* [2] *Department of Molecular Biology and Genetics, University of Guelph, Guelph, Ontario, Canada, N1G 2W1;* [3] *Department of Molecular Genetics, Plant Breeding Institute, Trumpington, Cambridge CB2 2LQ, UK*

Received 21 September 1987; accepted 24 March 1988
mutagenesis, promoters, secretion, transcription

Abstract. The expression of higher plant chloroplast genes in prokaryotic cells has been used to examine organelle sequences involved in promoter recognition by RNA polymerase, and protein translocation through membranes. The similarity in sequence structure between *Escherichia coli* promoters and the maize chloroplast *atpB* promoter has been investigated using deletion and single base pair substitution mutants. The *atpB* mutants were mainly isolated by a selection system in *E. coli*, and then used as templates for the analysis of transcription using chloroplast RNA polymerase. It was found that both the bacterial and chloroplast RNA polymerases behaved in a similar fashion with the wild-type and mutant promoters, indicating that the sequences involved in promoter recognition share a considerable degree of homology. Signal peptide recognition of pea cytochrome *f* has also been examined in *E. coli*. This signal peptide, which is probably responsible for insertion of the protein into the thylakoid membrane, is efficiently recognized in *E. coli* leading to the inner membrane insertion of *petA::lacZ* fusion proteins. This process requires the bacterial SecA protein and points to a general similarity in the mechanisms of protein translocation within chloroplasts and bacteria.

Gene symbols: *atpB* – ATPase F_1 β subunit, *lacZ* – β-galactosidase, *petA* – cytochrome *f*, *psbA* – photosystem II 32K protein; *rbcL* – large subunit of ribulose bisphosphate carboxylase, *trn*M2 – $tRNA_2$ Met

Introduction

Chloroplast genes exhibit considerable homology with *Escherichia coli* genes for sequences involved in recognition by RNA polymerase, and subsequent interaction with ribosomes (Whitfeld and Bottomley 1983; Rochaix 1985). This has provided a special opportunity to use bacterial systems to identify features that are involved in chloroplast gene expression. Clearly, examining the regulatory features of chloroplast gene expression in bacteria is not a completely faithful elucidation of the events occurring in the plant organelle, but it can provide valuable information on the interaction of bacterial RNA polymerase, 70S ribosomes and membrane targeting mechanisms with the chloroplast sequences. It should therefore be considered as an additional tool to aid in the dissection of the molecular events occurring within the chloroplast. A particularly promising approach using this methodology is the ability to select and screen mutations of chloroplast gene function in bacteria, and then to test the mutations in chloroplast derived transcription systems (Bradley and Gatenby 1985). If faithful chloroplast translation and membrane insertion reactions can also be developed, so that mutants generated and selected in bacteria can be tested, then a wide range of events concerning chloroplast biogenesis can be studied. In this review we shall describe two areas in which chloroplast gene expression has been studied using bacterial systems to gain insight into the events occurring in the organelle viz. promoter recognition, and protein translocation.

Characterization of promoters

In early experiments it was demonstrated that isolated chloroplast DNA from spinach could direct the expression of a chloroplast protein in vitro using an *E. coli* cell-free coupled transcription-translation system (Bottomley and Whitfield 1979), suggesting that the bacterial RNA polymerase could bind and initiate transcription at sites on the chloroplast genome. These findings were supported by in vivo expression studies in which synthesis of chloroplast proteins could be detected in *E. coli* using cloned chloroplast genes, in a manner that was consistent with trancription initiating at sites on the chloroplast DNA, rather than occurring from readthrough transcription from vector sequences (Gatenby et al. 1981; Gatenby and Cuellar 1985). Using these in vitro or in vivo methods numerous chloroplast genes from several different plant species have been successfully transcribed by *E. coli* RNA polymerase (Howe et al. 1982; Willey et al. 1983; Fluhr et al. 1983; Bovenberg et al. 1984; Woessner et al. 1984; Zhu et al. 1984; Gatenby and Rothstein 1986). Chloroplast DNA restriction fragments have also been used to provide a sequence for *E. coli* RNA polymerase to recognize, resulting in the in vivo expression of a downstream promoter deficient gene such as β-galactosidase (Gatenby et al. 1981; Fukuzawa et al. 1985) or galactokinase (Kong et al. 1984). Although such an approach may identify a promoter that is active in both the chloroplast and *E. coli*, a problem arises when sequences fortuitously function as promoters in bacteria, but which are not authentic chloroplast gene promoters. For example, it has been observed that a maize *rbcL* restriction fragment that is known not to contain the authentic *rbcL* transcription initiation site, nevertheless has a measurable level of promoter activity in *E. coli* (A.A. Gatenby and R.E. Cuellar unpublished).

The clearest evidence for authentic chloroplast promoter recognition by *E. coli* RNA polymerase has been obtained from experiments using high resolution transcript mapping, sometimes in combination with mutagenesis. Thus, Shinozaki and Sugiura (1982) using tobacco *rbcL* and *atpB* genes, Tohdoh et al. (1981) using tobacco rRNA genes, Erion et al. (1983) using spinach *rbcL* and Boyer and Mullet (1986) using pea *psbA*, were able to demonstrate that purified *E. coli* RNA polymerase would initiate transcription in the same position on chloroplast DNA as would the corresponding chloroplast RNA polymerase. Fukuzawa et al. (1985) also observed recognition of a chloroplast promoter from a liverwort with the bacterial RNA polymerase. These studies have shown that the transcription initiation site is located downstream from regions of -10 and -35 base pairs that have a high degree of homology to the consensus sequence found for *E. coli* promoters (Rosenberg and Court 1979).

Several authors have combined both mutagenesis and transcript mapping to characterize chloroplast promoter sequences. Hanley-Bowdoin et al. (1985) using the *rbcL* gene and chloroplast RNA polymerase suggested that chloroplast and prokaryotic promoters share sequence homology. This observation was supported by introducing a mutation to the maize *rbcL* promoter that increased the spacing between the putative -10 and -35 promoter elements from 18 to 20 base pairs, but without changing the structure of the elements themselves. Transcription of the insertion mutant by homologous RNA polymerase was depressed in vitro relative to wild-type levels, indicating that the spacing of the -10 and -35 region is important for effficient transcription by RNA polymerase. Gruissem and Zurawski (1985b) also obtained data that are consistent with a prokaryotic model for chloroplast promoter function. Using a chloroplast transcription extract and synthetic DNA fragments containing the defined transcriptional start sites of *rbcL*, *atpB* and *psbA* they were able to use an in vitro assay to examine the relative strengths of wild-type and hybrid -10 and -35 canonical sequences. The introduction of single base pair changes into the -10 region of the *psbA* promoter reduced transcription levels in the chloroplast extract and were analogous to similar base pair changes which lower promoter efficiency in *E. coli*. A chloroplast hybrid promoter that had absolute homology to the canonical -10 and -35 region, was the most efficient in the in vitro transcription system. A mutational analysis of the

Bal 31 Deletion Mutants

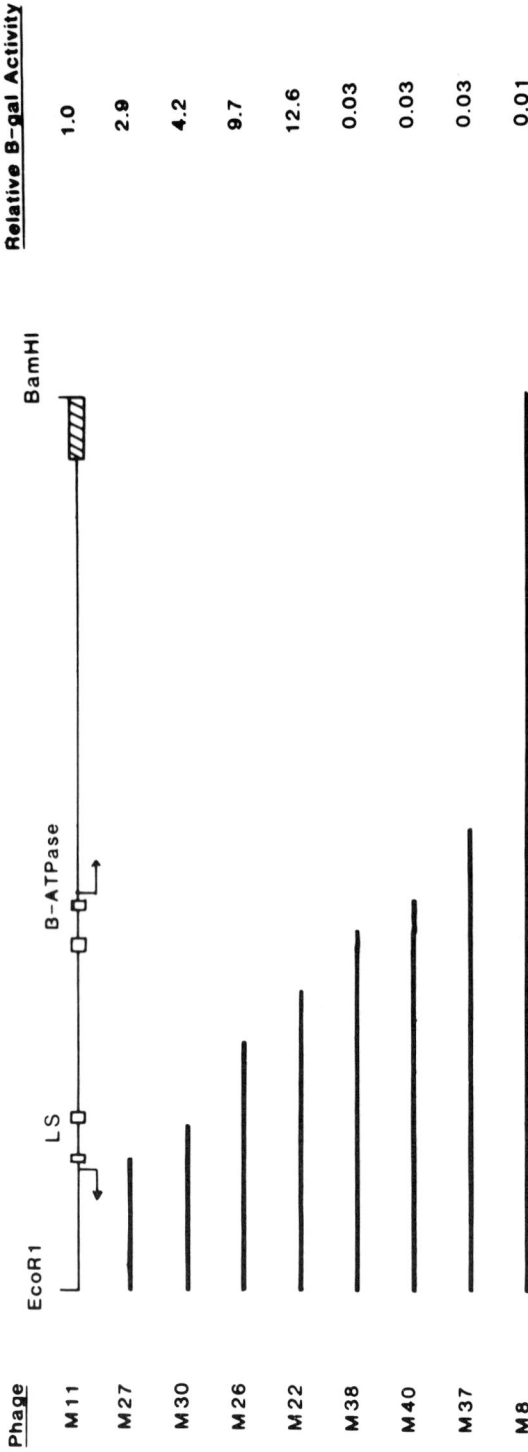

Phage	Relative B-gal Activity
M11	1.0
M27	2.9
M30	4.2
M26	9.7
M22	12.6
M38	0.03
M40	0.03
M37	0.03
M8	0.01

EcoR1 LS B-ATPase BamHI

Fig. 1. Construction and analysis of *Bal*31 deletion mutants on the expression of *atpB::lacZ* gene fusions transcribed from the chloroplast *atpB* promoter in *E. coli*. An *Eco*RI/*Bam*HI fragment of maize chloroplast DNA is shown for phage M11 that contains the promoter for the large subunit gene of ribulose bisphosphate carboxylase (LS) transcribing towards the left, and the β-ATPase gene promoter transcribing towards the right. The transcription initiation sites are marked by arrows, the -10 regions by small open boxes and the -35 regions by large open boxes. The striped box represents the sequence encoding the amino terminus of the β-ATPase protein and the heavy lines represent DNA sequences deleted from each phage. The levels of phage directed β-galactosidase were measured and normalized to the average value obtained for phage M11. Reprinted from Bradley and Gatenby (1985).

spinach *trn*M2 gene also supported the view that the arrangement of DNA sequences recognized by the chloroplast RNA polymerase resembles the prokaryotic promoter organization (Gruissem and Zurawski 1985a).

Given the close similarity in structure between the chloroplast and *E. coli* promoters, as outlined above, an attempt was made to see if a selection scheme for the isolation of chloroplast promoter mutants in *E. coli* could be devised (Bradley and Gatenby 1985). To do this the 5′ end of the maize *atpB* gene containing the promoter, ribosome binding site and the first ten codons of the ATPase β subunit were fused in-frame to *lacZ* in an M13 phage vector to give M11 (Fig. 1). A series of *Bal*31 deletion mutants were also made by treating the DNA with this nuclease prior to cloning into the M13 vector. Transfectants were examined on plates containing 5-bromo-4-chloro-3-indolyl-β-D-galactopyranoside for differences in phage β-galactosidase levels. A wide range of blue phenotypes were recovered, indicating that the chloroplast *atpB* DNA sequences that were operating as a promoter in *E. coli* were doing so in a complex fashion. The mutant deletion end points and the relative levels of β-galactosidase produced by each phage were measured (Fig. 1). The deletion of sequences > 120 base pairs upstream of the putative *atpB* promoter (phage M27) caused the production of three times more β-galactosidase activity than the parent phage (M11). The M27 deletion removed two nucleotides from the *rbcL* -10 promoter region that was adjacent to the *atpB* gene, but which is transcribed divergently. Further deletion of this *rbcL* promoter region resulted in even higher levels of β-galactosidase activity. These results suggest that, at least in *E. coli*, loss of the *rbcL* promoter improves the transcriptional efficiency of the adjacent *atpB* promoter. Possibly the close juxtaposition of the *rbcL* and *atpB* genes leads to mutual interference of RNA polymerase binding and that once the *rbcL* promoter is deleted this occlusion effect is abolished. Deletions can extend into the chloroplast insert as far as the *atpB* -35 region (M27, M30, M26, M22) without a negative effect on *atpB* expression. However, when sequences inclusive of the -35 region are deleted, expression of the *atpB::lacZ* gene product is reduced to almost background levels (M38, M40, M37), suggesting that a critical promoter region must reside between the M22 and M38 deletions.

To assay directly and compare the *atpB* promoter activity of each *Bal*31 mutant, S1 nuclease protection experiments were carried out using RNA from both phage infected *E. coli* cells and from maize chloroplast in vitro transcription reactions. The size of the S1 protected RNA indicated that transcription is initiated in *E. coli* at the same site as that used in maize chloroplasts. The level of *atpB* transcription measured in *E. coli* correlated well with the β-galactosidase activities recorded in Fig. 1. When the in vitro transcription reactions were carried out in a maize chloroplast extract, the levels of β-ATPase specific transcription obtained from each of the deletion templates exhibit variation that is quantitatively similar to that seen in *E. coli* cells. Again, deletions that remove the *rbcL* promoter region seem to have an enhancing effect on the transcriptional activity of the *atpB* promoter. As with transcription in *E. coli*, when a deletion extends into the -35 region of the *atpB* promoter (phage M38), *atpB* transcription was not detected using maize chloroplast RNA polymerase, showing that the region is important in the identification of a chloroplast promoter by its RNA polymerase.

To define more accurately those sequences that comprise a chloroplast promoter, single base substitution mutants were sought that would alter *atpB* promoter function in *E. coli*. The effects of the mutations on chloroplast RNA polymerase promoter recognition were then assayed using the chloroplast in vitro transcription system. Several different methods were used to isolate point mutations in the maize *atpB* promoter (Bradley and Gatenby 1985), but with all of these the mutant phenotype was selected in vivo using altered levels of *atpB* promoter transcribed β-galactosidase activity following phage infection as a screen. The DNA sequences of the single base change mutants recovered using these techniques is shown in Table 1, together with the relative levels of β-galactosidase produced. The mutations M50, M51, M54 and M75 were all isolated independently at least twice, suggesting that target site saturation had been achieved with this

Table 1. The DNA sequence and relative levels of *lacZ* activity of single base pair substitution mutants of the maize *atpB* promoter. The LacZ activity of promoter mutant M113 is presented as a percentage of the value obtained for the wild-type promoter phage M114. The LacZ activities of the remaining promoter mutants are all expressed as a percentage of phage M22. Underlined nucleotides designate the sites of mutations. The *E. coli* consensus sequence is that of Hawley and McClure (1983). Reprinted from Bradley and Gatenby (1985).

Phage	-35	-10	Relative LacZ activity
M114	TTGACA	TAGTAT	100%
M113	TCGACA	TAGTAT	0.4%
M22	TTGACA	TAGTAT	100%
M50	TTGGCA	TAGTAT	0.9%
M51	TTGAAA	TAGTAT	0.9%
M54	TTAACA	TAGTAT	0.3%
M75	CTGACA	TAGTAT	0.5%
M76	TTTACA	TAGTAT	7.3%
M38	Gene fusion, no *atpB* promoter		0.2%
E. coli consensus sequence			
	TTGACA	TATAAT	

mutant selection. The levels of *atpB* specific transcription from genes with single base changes in their promoters were determined in S1 nuclease protection experiments. RNA purified from *E. coli* cells infected with each of the mutant genes gave levels of S1-protected RNA that mirrored the β-galactosidase activities presented in Table 1. The *atpB* point mutants M50, M51, M54, M75 and M113 produced no detectable *atpB* specific RNA. The results of chloroplast in vitro transcription reactions on each of the *atpB* promoter mutants showed that transcription is reduced. These data indicated that there are no sequences essential for *atpB* transcription upstream from the − 35 region. They also demonstrate that not only is the − 35 region important for chloroplast promoter function, but that the sequence is similar in detail to the − 35 region of *E. coli* promoters. Deletions can extend to within one nucleotide of the *atpB* − 35 region without loss of promoter activity in either *E. coli* or in a chloroplast in vitro transcription system. The *atpB* − 10 sequence by itself is not sufficient for promoter recognition in the chloroplast transcription system or in *E. coli*. Similar conclusions have been reached by Link (1984) and Gruissem and Zurawski (1985a). The six different point mutations isolated within the − 35 region of *atpB* (Table 1) affect five of the six positions that make up the *E. coli* − 35 consensus sequence. The 3′-most nucleotide of the six-base − 35 consensus sequence, TTGACA, is the least conserved position (Hawley and McClure 1983), thus it is not surprising that chloroplast − 35 down mutations were not found at this site. No mutations were recovered within the − 10 region of the *atpB* promoter using a strong selection for promoter down mutants. This could be due to the fact that the *atpB* − 35 region is an optimal *E. coli* − 35 sequence and mutations in the − 10 region do not diminish promoter function sufficiently to be picked up by the promoter mutant selection method. Alternatively, the − 10 region may be repeated within the *atpB* promoter, and a second, less optimal − 10 region can be found. The apparent similarity between chloroplast and *E. coli* gene control signals can be successfully exploited in order to isolate in *E. coli* chloroplast mutations that are of functional importance. It should be possible to carry out similar genetic studies in *E. coli* to investigate chloroplast transcription termination sites and chloroplast ribosome binding sequences, thus allowing us to gain important information about chloroplast gene expression.

Although the resemblance between chloroplast and bacterial promoters has enabled a selection technique to be devised to obtain mutant chloroplast promoters in *E. coli* it is important to realise that there are a number of differences between chloroplast and bacterial promoters. These differences have been reviewed by Hanley-Bowdoin and Chua (1987) and they include sensitivity to spacing mutations, level of specificity, and promoter preference by the RNA polymerases. The chloroplast and *E. coli* RNA polymerases also have a number of different physical properties. It has also been found that not all chloroplast genes possess the -10 and -35 consensus sequences. Gruissem et al. (1986) have described a subpopulation of tRNA genes that do not require these upstream promoter sequences for their transcription.

Protein translocation

The proteins of the chloroplast are, for the most part, encoded and synthesized outside the organelle, and are transported through the chloroplast envelope after translation is completed (reviewed by Schmidt and Mishkind 1986). A few proteins are chloroplast encoded and are synthesized in the stroma. Whether imported, or synthesized within the chloroplast, there are several organelle compartments in which the various proteins will finally reside. These are the outer and inner membranes of the envelope, the envelope intermembrane space, the stroma, thylakoid membrane, or thylakoid lumen. The major multi-subunit complexes of the thylakoid membranes are the cytochrome f/b_6 complex, ATP synthase and photosystems I and II, and each complex contains both nuclear and chloroplast encoded polypeptides.

It has been demonstrated that during import of the precursor to the light-harvesting chlorophyll *a/b* protein (pre-LHCP) into isolated chloroplasts the precursors were converted to their final size and became integrated into the thylakoid membrane in a chlorophyll-protein complex (Schmidt et al. 1981). The existence of a pathway within chloroplasts for incorporating soluble precursor proteins into thylakoid membranes has been suggested by Cline (1986) who identified an activity in chloroplast lysates that incorporates pre-LHCP into thylakoid membranes. Hageman et al. (1986) described a novel protease located in the thylakoids that processes an intermediate form of preplastocyanin to the mature protein. It was suggested that a thylakoid-transfer domain exists in preplastocyanin and that this shared a resemblance to a prokaryotic signal peptide, with the inference that the thylakoid protease would function analogously to the prokaryotic signal peptidase. Kirwin et al. (1987) have partially purified the thylakoid protease involved in plastocyanin biogenesis and have considered the possibility that the protease may process some chloroplast-synthesized proteins, in addition to the imported preplastocyanin. We are also interested in the mechanism which translocates proteins through thylakoid membranes, and in the remainder of this paper will be described experiments in which the expression of pea cytochrome *f* in *E. coli* has been used to examine the properties of the signal peptide, which is thought to initiate membrane insertion.

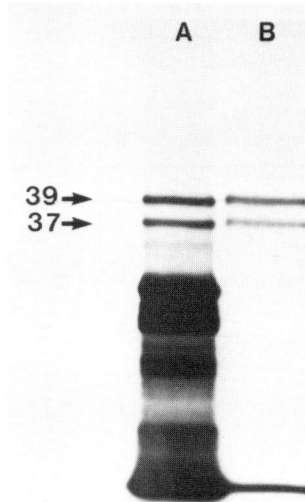

Fig. 2. Identification of pea cytochrome *f* synthesized in *E. coli* minicells containing the chloroplast *petA* gene. Minicells were labelled with [³⁵S]methionine and the cytochrome *f* polypeptides were identified by immunoprecipitation. Lanes: A, total plasmid-encoded products from minicells; B, immunoprecipitate obtained with antibodies to charlock (*Sinapsis arvensis*) cytochrome *f* and protein A-Sepharose. The 37 kDa polypeptide is the same molecular weight as mature cytochrome *f*. Reprinted from Rothstein et al. (1985).

Cytochrome *f* is a chloroplast-encoded membrane protein that is synthesized within the organelle on membrane-bound ribosomes. It is a component of the membrane cytochrome *b-f* complex involved in photosynthetic electron transfer between photosystem II and photosystem I. The polypeptide has a transmembrane arrangement in the thylakoid membrane, with the *N*-terminal region in the intrathylakoid space, and a 15 amino acid *C*-terminal sequence in the stroma (Willey et al. 1984). A single membrane-spanning region near the *C*-terminus holds the polypeptide in the membrane. Cytochrome *f* is synthesized as a higher molecular weight form in an *E. coli* transcription-translation system, and it was suggested that a signal peptide may be present to enable the *N*-terminal part of the protein to be secreted across the thylakoid membrane (Willey et al. 1983). This suggestion was supported by DNA sequence analysis, which demonstrated that the amino acid sequence immediately preceding the mature *N*-terminus had properties similar to signal peptides used to direct polypeptides to membrane locations (Willey et al. 1984).

To determine whether the higher molecular weight form of pea cytochrome *f* observed in an *E. coli* cell free transcription-translation system could also

Cytochrome f – lac Z gene fusions

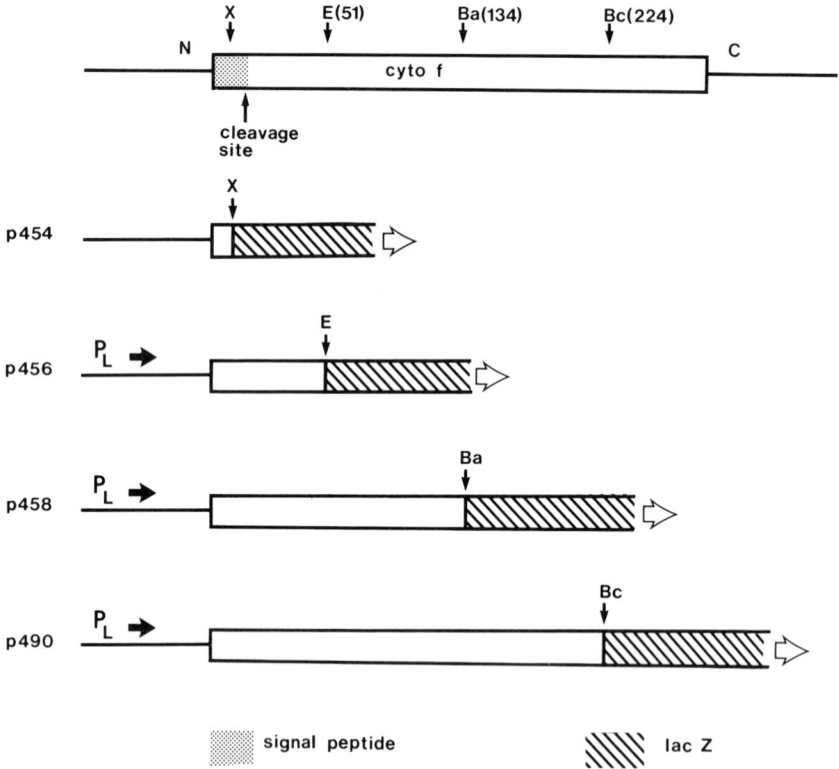

Fig. 3. Construction of *petA::lacZ* gene fusions. Various parts of the 5′ end of the cytochrome *f* gene were fused in the correct reading frame with *lacZ* (striped line) at the *Xho*I (X), *Eco*RI (E), *Bam*HI (Ba), and *Bcl*I (Bc) sites. The region encoding the signal peptide (stippled) and the cleavage site giving mature cytochrome *f* in the chloroplast are shown. All the plasmids except p454 encode the complete signal peptide and cleavage site. Numbers in parentheses indicate the number of amino acid residues of cytochrome *f* encoded from the cleavage site to the restriction enzyme site. Reprinted from Rothstein et al. (1985).

be synthesized in vivo in bacteria, an *E. coli* minicell-producing strain was transformed with a plasmid encoding the chloroplast gene (*petA*). Two polypeptides were immunoprecipitated from ^{35}S-labelled minicells using an antiserum raised against cytochrome *f* (Rothstein et al. 1985). The larger polypeptide (39 kDa) was similar in size to the form synthesized in an in vitro transcription-translation reaction, and the smaller polypeptide (37 kDa) is similar in size to mature cytochrome *f* found in chloroplast thylakoid membranes (Fig. 2). The demonstration of precursor and mature

Table 2. Cell fractionation of pea cytochrome *f-β*-galactosidase fusion proteins. *E. coli* cells containing *petA::lacZ* gene fusions, or induced wild-type *lacZ*, were disrupted by sonication. The cell-free extract was separated into a washed membrane fraction and a cytoplasmic fraction by ultracentrifugation and then asasayed for *β*-galactosidase activity. The figures show the percentage distribution of total *β*-galactosidase activity in the original cell extract. Reprinted from Rothstein et al. (1985).

	Membrane	Cytoplasm
w.t. control	2	82
p454	43	36
p456	80	9
p458	85	10
p490	78	20

Figures represent percentage of total *β*-galactosidase activity.

sized cytochrome *f* in *E. coli* raised the intriguing possibility that processing was occurring. Minicells have been shown to process secretory proteins poorly, which might account for the presence of the precursor-size cytochrome *f*. However, caution is required in this interpretation because of difficulties encountered in identifying the correct initiation codon for the pea *petA* gene. The DNA sequence reveals three in-frame AUG codons upstream from the mature *N*-terminus, although only one is preceded by a recognizable ribosome binding site (Willey et al. 1984). The two forms of cytochrome *f* synthesized in *E. coli* could arise by the use of two different translational start sites by the heterologous bacterial system.

Due to the very low levels of expression of cytochrome *f* in *E. coli*, it was difficult to carry out cell fractionation or pulse-labeling experiments to determine whether signal peptide recognition was occurring. To circumvent this problem, in-frame gene fusions were constructed with *lacZ* such that four different hybrid *β*-galactosidase molecules could be synthesized (Fig. 3). This enabled *β*-galactosidase assays to be used in cell fractionation and mutant analysis experiments (Rothstein et al. 1985). The shortest fusion (p454) was within the signal sequence, and the other three fusions all contained an intact signal peptide and contained 51 (p456), 134 (p458) or 224 (p490) *N*-terminal amino acids of the mature cytochrome *f*.

If the cytochrome *f* signal peptide is recognized by the *E. coli.* secretory pathway, it would be expected that the *β*-galactosidase fusion proteins would behave in a manner similar to proteins containing a bacterial signal peptide fused to *β*-galactosidase. A characteristic feature of these fusions is that they usually become trapped in the cytoplasmic membrane, probably because membrane incompatible amino acid sequences in *β*-galactosidase jam export sites (Silhavy et al. 1983). Table 2 shows the result of assaying a washed membrane fraction and a cytoplasmic fraction from cells contain-

sucrose step gradients

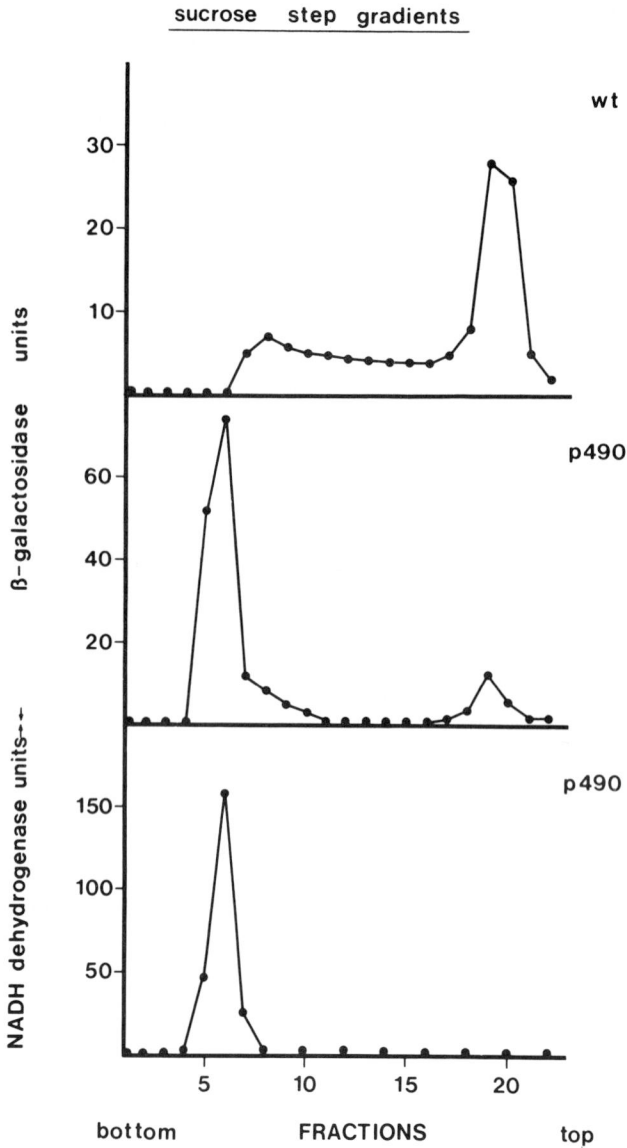

Fig. 4. Analysis of cytochrome *f*-β-galactosidase binding to membranes by use of sucrose gradient centrifugation. A crude sonicated cell extract was layered on the gradients and centrifuged. The gradients were fractionated and assayed for enzyme activity. Top, wild-type β-galactosidase; middle, p490-encoded fusion protein; bottom, the gradient containing the p490 transformant extract assayed for the cytoplasmic membrane marker NADH dehydrogenase. β-galactosidase units are nmol of *o*-nitrophenylgalactoside hydrolyzed per min per fraction, and NADH dehydrogenase units are nmol of NADH oxidized per min per fraction. Reprinted from Rothstein et al. (1985).

ing *petA::lacZ* fusions, or wild-type *lacZ*. Induced wild-type β-galactosidase activity is located primarily in the cytoplasm. Cytochrome *f* fusion proteins that contain the signal peptide (encoded by p456, p458 and p490), however, are located mainly in the membrane fraction. The p454 encoded fusion protein, which has a partially disrupted signal peptide, is distributed between the cytoplasm and membrane fractions. To demonstrate that co-sedimentation of the signal peptide fusion proteins with the membrane fraction during differential centrifugation was not due to general insolublity, cell extracts were centrifuged on sucrose gradients (Fig. 4). The wild-type enzyme is located mainly in the soluble fraction at the top of the gradient, but the enzyme fused to cytochrome *f* is located in the lower part of the gradient. The gradients were also assayed for the inner membrane marker NADH dehydrogenase and the peak fraction coincided with the p490 encoded β-galactosidase activity. The location of the fusion proteins in the inner membrane was demonstrated by using the differential solubility properties of the inner and outer membranes in the presence of sarkosyl and EDTA. Most of the membrane bound *petA::lacZ* fusion proteins are released under conditions that solubilize the inner membranes of *E. coli*.

The results obtained from cell-fractionation experiments indicate that export of cytochrome *f* through the bacterial cytoplasmic membrane was initiated but then blocked by the structure of β-galactosidase. This suggests that the chloroplast signal peptide is recognized as such in *E. coli*, particularly since disruption of the signal peptide (as encoded in p454) leads to inefficient membrane localization (Table 2). If this interpretation is correct, then *E. coli* mutants known to disrupt secretion of bacterial proteins would be expected to have an effect on *petA* signal peptide recognition and initiation of export. The *secA* gene of *E. coli* is thought to encode a component of the bacterial secretion machinery, and mutations in this gene selectively interfere with the synthesis or export of secreted proteins. The strain MM52 is a temperature sensitive conditional-lethal *secA* mutant (Oliver and Beckwith 1981). At the non-permissive temperature, the precursors of a number of secreted proteins accumulate in the cytoplasm, although some periplasmic proteins are correctly localized.

MM52(*secA*) or MC4100(wt) cells containing the *petA::lacZ* gene fusions were grown to early log phase at 30 °C and then were either shifted to 41 °C, or maintained at 30 °C, and were assayed for the cellular location of the β-galactosidase activity at various times. At 30 °C the bulk of the enzyme activity ws found in the washed membrane fraction in both MM52 and MC4100. However, at 41 °C over a period of 3–4 hr in strain MM52, most of the enzyme was found in the soluble fraction and was not associated with the membranes. In the control strain MC4100 at 41 °C, most of the enzyme activity remained in the enzyme fraction. At this higher temperature the temperature-sensitive SecA lesion is known to lead to a defect in export. It would appear that membrane insertion of the *petA::lacZ* fusion protein is also affected by this defect.

These results show that the signal peptide of pea cytochrome *f* is recognized by the *E. coli* secretory pathway, leading to initiation of export through the bacterial cytoplasmic membrane. This view is reinforced by the observation that deletion of the processing site and part of the signal peptide (p454) leads to inefficient membrane localization. There is, therefore, a considerable degree of functional homology between a chloroplast signal peptide and the secretory mechanisms of bacteria. Possibly a protein with a function analogous to the SecA protein may be involved in insertion of cytochrome *f* into the chloroplast thylakoid membrane.

References

Bottomley W and Whitfeld PR (1979) Cell-free transcription and translation of total spinach chloroplast DNA. Eur J Biochem 93: 31–39

Bovenberg WA, Howe CJ, Kool AJ and Nijkamp HJJ (1984) Physical mapping of genes for chloroplast DNA encoded subunit polypeptides of the ATP synthase complex from *Petunia hybrida*. Curr Gen 8: 283–290

Boyer SK and Mullet JE (1986) Characterization of *P. sativum* chloroplast psbA transcripts produced in vivo, in vitro and in *E. coli*. Plant Mol Biol 6: 229–243

Bradley D and Gatenby AA (1985) Mutational analysis of the maize chloroplast ATPase-β subunit gene promoter: the isolation of promoter mutants in *E. coli* and their characterization in a chloroplast in vitro transcription system. EMBO J 4: 3641–3648

Cline K (1986) Import of proteins into chloroplasts. Membrane integration of a thylakoid precursor protein reconstituted in chloroplast lysates. J Biol Chem 261: 14804–14810

Erion JL, Tarnowski J, Peacock P, Caldwell P, Redfield B, Brot N and Weissbach H (1983) Synthesis of the large subunit of ribulose-1,5-bisphosphate carboxylase in an in vitro partially defined *E. coli* system. Plant Mol Biol 2: 279–290

Fluhr R, Fromm H and Edelman M (1983) Clone bank of *Nicotiana tabacum* chloroplast DNA: mapping of the alpha, beta and epsilon subunits of the ATPase coupling factor, the large subunit of ribulose bisphosphate carboxylase, and the 32-kDal membrane protein. Gene 25: 271–280

Fukuzawa H, Uchida Y, Yamano Y, Ohyama K and Komano T (1985) Molecular cloning of promoters functional in *Escherichia coli* from chloroplast DNA of a liverwort, *Marchantia polymorpha*. Agric Biol Chem 49: 2725–2731

Gatenby AA, Castleton JA and Saul MW (1981) Expression in *E. coli* of maize and wheat chloroplast genes for large subunit of ribulose bisphosphate carboxylase. Nature 291: 117–121

Gatenby AA and Cuellar RE (1985) Antitermination is required for readthrough transcription of the maize *rbcL* gene by a bacteriophage promoter in *Escherichia coli*. Eur J Biochem 153: 355–359

Gatenby AA and Rothstein SJ (1986) Synthesis of maize chloroplast ATP-synthase β-subunit fusion proteins in *Escherichia coli* and binding to the inner membrane. Gene 41: 241–247

Gruissem W, Elsner-Menzel C, Latshaw S, Narita JO, Schaffer MA and Zurawski G (1986) A subpopulation of spinach tRNA genes does not require upstream promoter elements for transcription. Nucl Acids Res 14: 7541–7556

Gruissem W and Zurawski G (1985a) Identification and mutational analysis of the promoter for a spinach chloroplast transfer RNA gene. EMBO J 4: 1637–1644

Gruissem W and Zurawski G (1985b) Analysis of promoter regions for the spinach chloroplast *rbcL*, *atpB* and *psbA* genes. EMBO J 4: 3375–3383

Hageman J, Robinson C, Smeekens S and Weisbeek P (1986) A thylakoid processing protease is required for complete maturation of the lumen protein plastocyanin. Nature 324: 567–569

Hanley-Bowdoin L and Chua N-H (1987) Chloroplast promoters. Trends Biochem Sci 12: 67–70

Hanley-Bowdoin L, Orozco EM and Chua N-H (1985) In vitro synthesis and processing of a maize chloroplast transcript encoded by the ribulose 1,5-bisphosphate carboxylase large subunit gene. Mol Cell Biol 5: 2733–2745

Hawley DK and McClure WR (1983) Compilation and analysis of *Escherichia coli* promoter DNA sequences. Nucl Acids Res 11: 2237–2255

Howe CJ, Bowman CM, Dyer TA and Gray JC (1982) Localization of wheat chloroplast genes for the beta and epsilon subunits of ATP synthase. Mol Gen Genet 186: 525–530

Kirwin PM, Elderfield PD and Robinson C (1987) Transport of proteins into chloroplasts. Partial purification of a thylakoidal processing peptidase involved in plastocyanin biogenesis. J Biol Chem 262: 16386–16390

Kong XF, Lovett PS and Kung SD (1984) The *Nicotiana* chloroplast genome IX. Identification of regions active as prokaryotic promoters in *Escherichia coli*. Gene 31: 23–30

Link G (1984) DNA sequence requirements for the accurate transcription of a protein-coding plastid gene in a plastid in vitro system from mustard (*Sinapis alba* L.). EMBO J 3: 1697–1704

Oliver DB and Beckwith J (1981) *E. coli* mutant pleiotropically defective in the export of secreted proteins. Cell 25: 765–772

Rochaix JD (1985) Genetic organization of the chloroplast. Int Rev Cytol 93: 57–91

Rosenberg M and Court D (1979) Regulatory sequences involved in the promotion and termination of RNA transcription. Ann Rev Gen 13: 319–353

Rothstein SJ, Gatenby AA, Willey DL and Gray JC (1985) Binding of pea cytochrome *f* to the inner membrane of *Escherichia coli* requires the bacterial *secA* gene product. Proc Natl Acad Sci USA 82: 7955–7959

Schmidt GW, Bartlett SG, Grossman AR, Cashmore AR and Chua N-H (1981) Biosynthetic pathways of two polypeptide subunits of the light-harvesting chlorophyll *a/b* protein complex. J Cell Biol 91: 468–478

Schmidt GW and Mishkind ML (1986) The transport of proteins into chloroplasts. Ann Rev Biochem 55: 879–912

Shinozaki K and Sugiura M (1982) Sequence of the intercistronic region between the ribulose-1,5-bisphosphate carboxylase/oxygenase large subunit and the coupling factor β subunit gene. Nucl Acid Res 10: 4923–4934

Silhavy TJ, Benson SA and Emr SD (1983) Mechanisms of protein localization. Micro Biol Rev 47: 313–344

Tohdoh N, Shinozaki K and Sugiura M (1981) Sequence of a putative promoter region for the rRNA genes of tobacco chloroplast DNA. Nucl Acids Res 9: 5399–5406

Whitfeld PR and Bottomley W (1983) Organization and structure of chloroplast genes. Ann Rev Plant Physiol 34: 279–310

Willey DL, Huttly AK, Phillips AL and Gray JC (1983) Localization of the gene for cytochrome *f* in pea chloroplast DNA. Mol Gen Genet 189: 85–89

Willey DL, Auffret AD and Gray JC (1984) Structure and topology of cytochrome f in pea chloroplast membranes. Cell 36: 555–562

Woessner JP, Masson A, Harris EH, Bennoun P, Gillham NW and Boynton JE (1984) Molecular and genetic analysis of the chloroplast ATPase of *Chlamydomonas*. Plant Mol Biol 3: 177–190

Zhu YS, Lovett PS, Williams DM and Kung SD (1984) *Nicotiana* chloroplast genome 7. Expression in *E. coli* and *B. subtilis* of tobacco and *Chlamydomonas* chloroplast DNA sequences coding for the large subunit of RuBP carboxylase. Theor Appl Genet 67: 333–336

Govindjee et al. (eds), Molecular Biology of Photosynthesis: 121–135
© 1988 Kluwer Academic Publishers

Minireview

Binding, uptake and expression of foreign DNA by cyanobacteria and isolated etioplasts

BRUCE A. McFADDEN & HENRY DANIELL

Biochemistry/Biophysics Program, Washington State University, Pullman, WA 99164-4660, USA

Received 24 August 1987; accepted 11 March 1988

Key words: binding, uptake, transformation, DNA, cyanobacteria, etioplasts

Abstract. Discoveries of the uptake and expression of various *Escherichia coli* plasmids by the cyanobacterium *Anacystis nidulans* and isolated cumber etioplasts are reviewed. In particular, the binding and uptake of nick-translated ^{32}P-labeled plasmids and the expression of genes in the native plasmids are considered.

Permeaplasts of *A. nidulans* 6301 and isolated EDTA-washed cucumber etioplasts exhibit binding and uptake of DNA that is unaffected by uncouplers of photophosphorylation or by dissipators of transmembrane proton graident. ATP inhibits both binding and udptake by permeaplasts or EDTA-washed etioplasts but the analog AMP-PNP (non-hydrolzable) is noninhibitory. With permeaplasts there is no effect of 20 mM Mg^{2+} (in the light) upon intake, whereas with EDTA-washed etioplasts, Mg^{2+} at the same concentration inhibits uptake as does 20 mM Ca^{2+}.

The transformation of *A. nidulans* 6301 to ampicillin-resistance by the plasmid pBR322 is much enhanced in permeaplasts. Indeed extracts of transformed cells catalyze the hydrolosis of the β-lactam nitrocefin. Transfromation of *A. nidulans* to antibiotic resistance may also be achieved with the plasmids pHUB4 and pCH1. The effect of light on transformation of *A. nidulans* 6301 differs with different plasmids. In pBR322 transformants the expression of ribulose bisphosphate carboxylase-oxygenase (RuBisCO) is markedly elevated. In these transformants, the foreign plasmid replicates by a pathway involving chromosomal integration and dissociation.

The plasmid pCS75, a derivative of pUC9 (and therefore of pBR322) containing a *Pst*1 insert carrying genes for the large and small (S) subunits of RuBisCO from *A. nidulans*, is taken up and expressed in EDTA-washed cucumber cotyledon etioplasts. Expression is evidenced by the hydrolysis of nitrocefin and immunoprecipitation of labeled S subunits of RuBisCO (utilizing etioplasts which have been labeled with ^{35}S−methionine after incubation with pCS75). The plasmid pUC9-CM carrying a *cat* gene is also expressed in cucumber etioplasts in a manner that demonstrates dependence both on the duration of etioplast washing by EDTA and plasmid concentration. Translation (as measured by ^{35}S-methionine incorporation) by EDTA-washed etioplasts increases with cotyledon greening. However the enhancement of translation by prior incubation of EDTA-washed plastids with pCS75 decreases to zero during 24 hr of cotyledon greening. Results suggest that the expression of foreign DNA in plastids may depend critically upon their developmental state.

Abbreviations: AMP-PNP – adenyl-5-yl imidodiphosphate, AP^r – amplicillin resistance, *cat*– chloramphenicol acetyltransferase, RuBisCO – ribulose bisphosphate carboxylase/oxygenase.

Introduction and historical perspective

The evolution of cyanobacteria to chloroplasts was suggested in 1905 (Mereshkowsky 1905). It was postulated that the chloroplast originated from a blue-green alga which had become endosymbiotic with a non-photosynthetic organism. Fossil records and other considerations suggest that cyanobacteria (blue-green algae) arose about 2.5×10^9 years ago (Schopf and Walter 1982) with photosynthetic eukaryotes following $1 - 1.5 \times 10^9$ years later (Echlin 1970). Intracellular organelles termed cyanelles, found in the photosynthetic biflagellated protist *Cyanophora paradoxa,* may actually reflect endosymbiotic cyanobacteria that lost the outer lipopolysaccharide cell wall layer characteristic of all free-living cyanobacteria (Herdman and Stanier 1977).

Recognizing the structural differences between cyanobacteria and chloroplasts, we have been impressed by their relatedness (Schwartz and Dayhoff 1981, McFadden et al. 1986) and the hypothesized evolutionary origin of chloroplasts. In 1984, we therefore set out to examine the expression of *Escherichia coli* plasmids in cyanobacteria as a model system for analogous expression in chloroplasts.

The present *overview* traces a progression of experiments demonstrating, sequentially, the transformation of cyanobacteria and isolated etioplasts. The use of techniques of molecular biology will be evident throughout and prospects for genetic engineering, especially of higher plants, will be explored. *For the sake of brevity, citations of the literature will focus upon our contributions in which primary references can be found.*

Cyanobacteria

Transformation

Anacystis nidulans 6301 is a unicellular cyanobacterium containing two endogenous plasmids pUH24 [8.0 kilobase pairs (kbp) and pUH25 (48 kbp)]. Prior to 1986, attempts to transfer plaslmids of *E. coli* into cyanobacteria had been unsuccessful. In order to circumvent this, Van den Hondel and coworkers (Daniell et al. 1986) constructed a plasmid, pCH1 (12.9 kbp), that contains an Ap[r] gene insert in the native *A. nidulans* plasmid pUH24. However, neither the resultant pCH1 nor its deletion derivative pUC1 could be used to transform *E. coli* because of the absence of a functional replicon. In order to achieve successful transformation by a plasmid in either host, a number of different *E. coli-A. nidulans* shuttle cloning vectors have been developed in recent years. However, most of these hybrid plasmids (Daniell et al. 1986) suffered from one or more of the following disadvantages: (i) they were relatively large (10 – 14 kbp), thereby reducing the size of potential inserts; (ii) they carried few unique restriction sites for cloning; (iii) they lacked an easy screening system for detecting the presence of DNA inserts; and (iv) they had to be recombined with the endogenous plasmid pUH24 to potentiate transformation. More recently, Lau and Straus (Daniell et al. 1986) reported construction of smaller shuttle vectors (7.3–7.8 kbp) by utilizing the cyanobacterial origin of replication of the endogenous plasmid pUH24 from *A. nidulans*.

In yet another approach *Synechococcus* R_2 was transformed to antibiotic resistance by chimeric DNA molecules consisting of *Synechococcus* chromosomal DNA linked to antibiotic resistance genes from *E. coli* (Daniell et al. 1986). Recently, random fragments of chromosomal DNA from *Agmenellum quadruplicatum* PR-6 ligated with an Ap^r gene were used to generate ectopic mutants; these ectopic mutants were used as recipients in transformation to kanamycin resistance by biphasic vectors derived from the endogenous cryptic plasmid pAQ1 (Buzby et al. 1985). In a different approach, irradiation of the cyanobacterium *Synechocystis* in low UV light resulted in stable integration of foreign DNA, under conditions in which there was neither autonomous plasmid replication or homologous recombination (Dzelzkalns and Bogorad 1986).

Although these results were collectively promising, the transformation of a cyanobacterium by the extremely well characterized *E. coli* plasmids belonging to the ColE1 group had not been achieved prior to 1986. We now summarize our research which demonstrated the incorporation of the parental plasmid of this group, pBR322, into *Anacystics nidulans* 6301 (Daniell et al. 1986, McFadden et al. 1986).

Transformation of A. nidulans 6301 by ColE1 plasmids of pCH1

Use of permeaplasts. Reasoning that prior failures to achieve transformation of cyanobacteria with *E. coli* plasmids may have been due to a permeability barrier to foreign DNA, the production of permeaplasts of *A. nidulans* 6301 was explored. It was established that cells which had been incubated for 1 to 2 hr with 1 mM EDTA plus lysozyme (2 mg/ml) at 36°C repaired and grew normally. In contrast, cells incubated for 3–8 hr grew poorly or not at all after removal of EDTA/lysozyme. By analogy to the findings of Ward and Myers, cells of *A. nidulans* 6301 which had been treated for 2 hr with EDTA/lysozyme were referred to as permeaplasts (Daniell et al 1986). Using pBR322 or the plasmid pCH1 permeaplasts were transformed to Ap^r with high efficiency (expressed as transformants/10^3 cells) which peaked for pBR322 at 16% after 30 hr contact (with shaking) at 32° C between the plasmid and permeaplasts in the light (Daniell et al. 1986). No transformation of permeaplasts was observed in the dark by pBR322 (Daniell and McFadden 1986). Of special importance was the observation that the *transformation efficiency for cells* in the light by pBR322 or pCH1 was reduced 54- and 10-fold, respectively, in comparison to that for permeaplasts (Daniell and McFadden 1986). Remarkably, the efficiency of transformation of permeaplasts by the plasmid pCH1 was elevated from 8% in the light to 59% in the dark.

To recapitulate, these results established conclusively that permeaplasts are more readily transformed than intact cells of *A. nidulans* 6301. Indeed, our permeaplast method of transformation has been very recently used to transfer herbicide resistance genes to the filamentous cyanobacterium *Nostoc muscorum* (Singh et al. 1987). Permeaplasts are potentially viable structures of unusually high permeability and capacity for cell-wall regeneration and may therefore take up DNA at very high rates and subsequently repair and divide.

The effect of light upon transformation of permeaplasts of *A. nidulans* remains poorly understood. In related studies, transformation of *A. nidulans* cells by pHUB4, which is a derivative of pBR322, was more than 4-fold more efficient in the dark than in the light although permeaplasts were not examined (Daniell et al. 1987a). Uptake of this 6.5 kbp-ColE1 plasmid, which carries a thermally inducible phage lambda promoter, P_L, conferred kanamycin-resistant growth upon transformants that was considerably elevated after thermal induction at 45°C (Daniell et al. 1987a).

Characterization of transformation of whole cells. In studies of transformation of intact cells of *A. nidulans* 6301 at 32°C acquisition of Ap^r reflected plasmid-dependent synthesis of β-lactamase. This was measured spectrophotometrically by using the β-lactam substrate nitrocefin which undergoes a shift in absorption maximum from 390 nm to 490 nm upon hydrolysis (Daniell et al. 1986). Expression of the β-lactamase gene of pBR322 or pCH1 in transformants was measured both by this method and in the pBR322-transformants by immunoprecipitation using anti-β lactamase after labeling cells with ^{35}S-methionine. Curiously, this antiserum also precipitated labeled large (L) 55-kDa and small (S) 12-kDa peptides (Daniell et al. 1986) that were subunits of ribulose bisphosphate carboxylase/oxygenase (RuBisCO). Immunological probes of ^{35}S-labeled RuBisCO from pBR322-transformants using rabbit anti-serum against the spinach enzyme also resulted in the precipitation of an ^{35}S-labelled 30-kDa protein (presumably β-lactamase) or the detection of this 30-kDa protein by Western blotting (Daniell and McFadden unpublished observation). Whether RuBisCO and β-lactamase have common epitopes remains to be established. In pBR322-transformants there was considerable elevation of L and S subunits of RuBisCO, an enzyme encoded by chromosomal DNA, but this was not the case in pCH1-transformants. Instead, in the latter the elevation of several soluble proteins of molecular masses of 17, 22, 25, 63 and 66 kDa was evident. In cells which had been transformed either by pBR322 or pCH1, there was a substantial increase in DNA per cell of approximately 5-fold (Daniell et al. 1986).

Although plasmids were not isolated from pCH1-transformants of *A. nidulans* 6301, 4.3 and 6.5 kbp plasmids were isolated in good yield from pBR322- (Daniell et al. 1986) and pHUB4-transformants (Daniell and McFadden unpublished observation). The size of each these plasmids was identical to that of the input plasmid (either pBR322 or pHUB4). The yield and size of both recovered plasmids suggested that plasmid replication had occurred during transformation to give back closely similar (or identical) plasmids, presumably via chromosomal integration.

The Mechanism of transformation of whole cells. Amplification of L and S subunits of RuBisCO and of the chromosomal content/cell in pBR322-transformants suggested that transformation had enhanced (and perhaps altered) chromosome expression. In recent research (Daniell et al. 1987b), we have established that the activity of RuBisCO is indeed elevated from 6- to 12-fold in pBR322-transformants grown in the presence of ampicillin. These transformants, grown in the absence of ampicillin for 122 generations, resumed their original growth rate in the presence of this antibiotic (at 1 μg/ml). The specific activity of RuBisCO in transformants was essentially unaltered by the absence of antibiotic. Hence the pBR322-transformant population was stable. Of special significance was our finding that nick-translated [^{32}P]-labeled pBR322 hybridized at moderatley high stringency with chromosomal DNA from either untransformed *A. nidulans* 6301 or pBR322-transformants [Daniell et al. 1987b]. Moreover, [^{32}P]-pCS75, which is a pUC9 derivative containing an insert of L and S subunit genes (for RuBisCO) from *A. nidulans,* hybridized at very high stringency with restriction fragments of chromosomal DNA from both untransformed and transformed cells. The hybridization patterns suggested the dispersion of L and S subunit genes, of which there is normally one copy, in chromosomal DNA from pBR322-transformants.

Taken together, our data suggest that at least one segment of the plasmid pBR322 is integrated within chromosomal DNA by homologous recombination during transformation of *Anacystis nidulans*. This transformation results in both RuBisCO amplification and gene dispersion and an increase in the cellular content of DNA. Because plasmids of the same size as those used in transformation can be isolated from transformants, replication of integrated plasmids must occur and a dissociative pathway yielding free plasmids must contribute. Whether these plasmids isolated from transformant populations will prove to be subtly different from pBR322 and pHUB4 remains to be established. Our findings parallel those reported from several laboratories for the transformation of *Bacillus subtilis* and *Streptococcus pneumoniae* by plasmids that contain regions which are homologous to chromosomal segments (Daniell et al. 1986, 1987b).

Our demonstration of the transformation of *A. nidulans* 6301 by ColE1 plasmids may open a new way to genetic engineering of the cyanobacteria. This plasmid family includes many ideal cloning vectors of known sequence and multiple cloning sites such as the pUC series. By the insertion of genetically engineered genes such as those for L and S subunits of RuBisCO, it may be possible to replace the resident counterparts by homologous recombination. Genetically engineered stable transformants may result. Such an approach may prove to be useful in altering genes for photosynthesis and nitrogen fixation in the cyanobacteria.

DNA binding and uptake by A. nidulans

In parallel with the studies of transformation described, an examination of the binding and uptake of nick-translated [^{32}P]-pBR322 by *A. nidulans* 6301 was conducted (Daniell and McFadden 1986). Total ^{32}P-labeled DNA associated with permeaplasts (or cells) was measured by separation of labeled permeaplasts (or cells) after sedimentation through a discontinuous sucrose gradient. In previous research, association of DNA with *B. subtilis* had been stopped by the addition of 4 mM EDTA plus calf thymus DNA at 1 mg/ml (Joenje and Venema 1975). As became evident, however, in our studies the pretreatment of cells of *A. nidulans* with EDTA and lysozyme to form permeaplasts greatly enhanced the association of nick translated ^{32}P-labeled pBR322 with this organism. To quench association of labeled DNA with permeaplasts, a large excess of nonradioactive calf thymus DNA was therefore added. Because 20% of the label associated with permeaplasts was liberated by the quench with nonradioactive DNA, the experimental values for labeled DNA bound to permeaplasts after treatment with non-labeled DNA were multiplied by 1.25 to compute the total labeled DNA originally associated with permeaplasts. DNA binding was computed from this latter value minus DNA uptake measured as described below.

Cells, untreated or treated with EDTA or lysozyme plus EDTA, were exposed to ^{32}P-DNA at 32°C for various durations. Further association was stopped by transferring the incubation mixture to an equal volume of buffer containing 4 mM EDTA and calf thymus DNA. Each sample was layered on an ice-cold discontinuous sucrose gradient and centrifuged. After decanting the supernatants, the pellets were extensively washed to release weakly bound donor DNA. After drying the tubes, the pellets were resuspended in buffer. Radioactivity was determined in suitable aliquots and total DNA originally associated with permeaplsts or cells was calculated as described.

In replicate portions of the incubation mixture, DNA uptake was measured after DNAse treatment of permeaplasts or cells. These and subsequent steps to isolate labeled cells including sucrose gradient centrifugation were exactly as described in detail for the procedure to measure DNA-association (Daniell and McFadden 1986).

Finally, the extent of breakdown of intracellular and extracellular DNA during incubations was determined by mixing still another replicate portion of the incubation mixture with an equal volume of ice-cold 6% perchloric acid and incubating for 30 min on ice. Subsequently, the samples were centrifuged and suitable aliquots were used to determine radioactivity in the acid-soluble supernatant fraction (Daniell and McFadden 1986).

Both binding and uptake of DNA were considerably enhanced in permeaplasts compared to cells confirming parallel studies of transformation. The breakdown of labeled DNA, as reflected in the perchloric acid-soluble fraction, was not correlated with binding or uptake by permeaplasts or cells. Uptake of DNA by permeaplasts was unaffected by: Mg^{2+} or Ca^{2+}, light, or inhibitors of photophosphorylation such as valinomycin or gramicidin D in the presence or absence of NH_4Cl. ATP at 2.5–10 mM inhibited both binding and uptake of labeled DNA by permeaplasts of *A. nidulans* whereas the ATP analog adenyl-5-yl imido-diphosphate was non-inhibitory in the same concentration range (Daniell and McFadden 1986).

On the basis of previous studies of transformation of *B. subtilis*, it had been suggested that breakdown of donor DNA, reflected in the perchloric acid-soluble fraction, may be an essential requirement for transformation (see, for example, Gunge and Sakgachi 1979). However, in studies of yeast protoplasts binding of plasmid DNA was not accompanied by its degradation as no trichloroacetic acid-soluble radioactivity was found by Brzobohaty and Kovac in the solution after the incubation of radioactive plasmid with protoplasts (Daniell and McFadden 1986). With *Chlamydomonas* cells no breakdown of DNA was observed by Lurquin and Behki but a CW15 cell-wall lacking mutant extensively degraded the donor DNA (Daniell and McFadden 1986). In this context, it is important to reiterate that in our work, the capacity of *A. nidulans* cells to breakdown donor DNA was unaffected by prior treatment with EDTA, lysozme, or lysozyme-EDTA

(Daniel and McFadden 1986). In contrast, *in B. subtilis* this process was abolished in the presence of EDTA (Joenje and Venema 1975).

Our achievement of high rates of uptake and binding of DNA by *A. nidulans* enabled a probe of the factors that might influence these processes. In *B. subtilis* cells, it had been suggested that DNA uptake or binding was an active process requiring the driving force of proton motive force and especially the ΔpH component (Daniell and McFadden 1986). On the other hand, it had also been suggested that DNA uptake by yeast protoplasts was a passive process because it was independent of temperature (in the interval of 0–30°C), cell metabolism and pH and was nonspecific (Daniell and McFadden 1986). In *A. nidulans* DNA uptake and binding by permeaplasts also appeared to be passive as each was unaffect by light, Mg^{2+}, or uncouplers of photophosphorylation. In this regard, the striking inhibition of DNA binding (and binding-dependent uptake) by ATP is an enigma. Because the ATP analog AMP-PNP was non-inhibitory, externally added ATP may have phosphorylated one or more membrane components which are required for DNA binding to permeaplasts (Daniell and McFadden 1986).

Etioplasts

As stressed in the Introduction, research was undertaken on transformation of cyanobacteria as models for chloroplasts. Results were promising and studies of chloroplasts were therefore undertaken. In this connection, technology had already been described to incorporate chloroplats into protoplasts and to generate plants from the protoplasts (Wallin 1984). It remained to be established, however, that chloroplasts could take up, transcribe and translate foreign genes as a first step in the genetic engineering of higher plants by this new route.

In our efforts to achieve gene transfer into chloroplasts, the isolation of inact organelles capable of efficient uptake, transcription and translation of foreign DNA was essential. Etioplasts from hormone-conditioned cucumber cotyledons seemed promising. For example, Daniell and Rebeiz had already isolated cucumber etioplasts capable of synthesis of protochlorophyllide and chlorophyll at extremely high rates. Also etioplasts, which had been preloaded with prothylakoid proteins by prior treatment of cucumber cotyledons with hormones, converted prothylakoids into macrograna when illuminated in a cofactor-enriched medium. Daniell and colleagues had also demonstrated the development of electron transport coupled to photo-phosphorylation in concordance with the synthesis of required polypeptides in isolated etioplasts. Finally, linear pigment biosynthesis and translation of endogenous mRNA had also observed for 8 hours (Daniell and McFadden 1987). These observations collectively established that etioplasts of cucumber cotyledons were both metabolically very active and unusually stable in their capacity for protein synthesis marking them as exceptional targets for gene incorporation and expression. The following sections summarize the binding and uptake of DNA and expression in isolated etioplasts of pUC derivatives carrying genes for the large L and small S subunits of RuBisCO of *Anacystis nidulans* or chloramphenicol acetyltransferase (cat) (Daniell and McFadden 1987).

DNA binding and uptake
For the isolation of etioplasts, cucumber seeds which had been germinated at 32°C were used. After 3 days of dark germination, cotyledons were excised with hypocotyl hooks and incubated for 20 hr in the dark with 0.5 mM kinetin and 2 mM gibberellic acid. After gentle homogenization, etioplasts were isolated by centrifugation, purified by layering on sucrose, and incubated in the presence or absence of 10 mM EDTA for 10 minutes in the dark at 0–4°C. Etioplasts were then incubated with nick-translated ^{32}P-labeled pCS75, a pUC9 derivative with an insert containing genes for the L and S subunits of RuBisCO from *A. nidulans* 6301 (Daniell and McFadden 1987). DNA binding and uptake as well as parallel breakdown, were measured after incubation at 27°C as described for *A. nidulans* (Daniell and McFadden 1986) except that calf thymus DNA was not used to quench association.

Preliminary experiments on the incubation of etioplasts with nick-translated ^{32}P-labeled pCS75 showed a linear increase in uptake or binding of DNA at 27°C for at least 120 min. Therefore, all incubations of DNA with etioplasts were carried out for a duration of 2 hr. Subsequent DNase treatment of etioplasts ensured that the high and continuing uptake of DNA observed had been due to intact organelles. The rate of uptake of ^{32}P-pCS75 by EDTA-washed etioplasts in the light, 0.86×10^4 cpm/μg protein (Daniell and McFadden 1987), was comparable to a value of 1.35×10^4 cpm/μg protein calculated for permeaplasts of *A. nidulans* using ^{32}P-pBR322 of approximatley identical specific radioactivity (Daniell and McFadden 1986). *Etioplasts which had been subjected to 10-min incubation at 0–4°C in the dark in buffer lacking EDTA showed only 3% as much uptake of DNA as was observed with EDTA-washed etioplasts* after 2 hr of incubation with nick-translated ^{32}P-pCS75. On the other hand, 42% as much binding of DNA as had been observed with EDTA-washed etioplasts was observed with etioplasts washed in the absence of EDTA under identical conditions of incubation. The presence or absence of light did not affect

DNA uptake, binding or breakdown in etioplasts washed in the presence or absence of EDTA. Cations such as those of calcuim or magnesium significantly inhibited DNA uptake (86%) in EDTA-washed etioplasts but enhanced binding (23–200%) and breakdown (163–235%) of DNA (Daniell and McFadden 1987).

As mentioned above (Section 2.3), it had been previously suggested that the calcium-dependent breakdown of donor DNA, reflected in the perchloric acid-soluble fraction, may have been an essential requirement for transformation of *B. subtilis*. However, in studies of yeast protoplasts, binding of plasmid DNA had not been accompanied by its degradation. We have demonstrated that DNA uptake in permeaplasts of *A. nidulans* is unrelated to the breakdown of donor DNA (Section 2.3 and Daniell and McFadden 1986). In our work on etioplasts, cations inhibited DNA uptake but enhanced breakdown (Daniell and McFadden 1987) suggesting that the two processes may be unrelated in this case also.

As with permeaplasts of *A. nidulans*, uncouplers that abolish either membrane potential ($\Delta\Psi$) and/or transmembrane proton gradient (ΔpH) did not affect DNA binding, uptake or breakdown by etioplasts. Also as with permeaplasts of *A. nidulans*, both DNA binding and uptake were severely inhibited by ATP. Presumably, this resulted from the hydrolysis of ATP as the poorly hydrolyzable analog adenyl-5-yl imidodiphosphate did not inhibit the uptake or binding of DNA by etioplasts (Daniell and McFadden 1987). This latter result was also reminiscent of a similar finding with permeaplasts of *A. nidulans* 6301 (Daniell and McFadden 1986). In summary, then, the data establish that comparable rates of DNA uptake to those measured for permeaplasts of *A. nidulans* can be achieved with EDTA-washed etioplasts. Moreover, binding and uptake phenomena are similar for permeaplasts and etioplasts in that neither is affected by light or uncouplers that abolish membrane potential or transmembrane proton gradient. Also, both are inhibited by ATP in a process that is presumably dependent upon hydrolysis of this triphosphate. Only in the effect of Mg^{2+} and Ca^{2+} upon DNA uptake are permeaplasts and EDTA-washed etioplasts different. Whereas with permeapasts, there is no effect of 20 mM Mg^{2+} (in the light) upon uptake (Daniell and McFadden 1986), with EDTA-washed etioplasts Mg^{2+} at the same concentration inhibits uptake by 86% as does 20 mM Ca^{2+} (Daniell and McFadden 1987). The effect of Mg^{2+} upon DNA uptake by etioplasts which had not been treated with EDTA was not investigated.

The comparisons cited established the validity of using cyanobacteria as models for chloroplasts, at least under certain experimental conditions. Similarities and differences have been noted in the preceding sections. More significantly, our studies of DNA binding and uptake by etioplasts were extremely useful in designing experiments to test the expression of plasmid DNA in these organelles to be considered in the next section.

Gene expression

In studies of the expression of foreign genes by cucumber etioplasts (Daniell and McFadden 1987), the plasmids pCS75 (containing an insert for L and S subunit genes of *A. nidulans*) and pUC9-CM carrying the gene for *cat* were used. The latter plasmid is a construction that includes the bacterial promoter for the *cat* gene. Both plasmids encode β-lactamase because they are pUC9 derivatives with an Ap^r gene.

In all cases, plasmids (with or without prior washing with 10 mM EDTA) were incubated aseptically with etioplasts at 27°C in the light. After centrifugation and washing, etioplasts were incubated for 2 hr at 27°C in a buffer (pH 8.2) containing the following sterile components: ATP, CTP, GTP, UTP, an ATP-regenerating system, tRNA (*E. coli*), pyridoxine, FAD, NADP, *p*-aminobenzoate and 19 amino acids. When labeling of proteins was desired, ^{35}S-methionine was added to this solution (termed the transcription-translation mixture) during the 2-hr incubation. In studies of *cat* expression, the inclusion of ribonuclease during the 2-hr incubation or of thermolysin after the incubation was examined to probe for organelle intactness. After incubation, etioplasts were isolated, and subjected to immunoprecipitation of ^{35}S-labeled S subunits (using rabbit antiserum to the 12 kDa-S subunits of RuBisCO from *A. nidulans*). Alternatively, they were assayed for β-lactamase using nitrocefin hydrolysis (Section 2.2.2) or for *cat* in which acetyl CoA-dependent acetylation of ^{14}C-chloramphenicol was measured by thin-layer chromatography and radioautography.

Bacterial contamination was assessed by streaking from the following fractions: (a) the original etioplast pellet, (b) the etioplasts which had been purified by layering on a sucrose gradient and (c) purified EDTA-washed etioplasts which had been incubated with plasmid and subsequently incubated in the transcription-translation mixture for 2 hr. Bacteria were enumerated at 30° and 37°C on agar plates solidified with a rich medium in the presence and absence of ampicillin (or the ampicillin analog, carbanicillin). Enumerations using LB plates incubated for 5 days at 30°C resulted in the following ratios of etioplast protein to bacterial protein for fraction: (a) 2.2×10^5; (b) 2.4×10^5; and (c) 2.0×10^5. Analogous ratios obtained after incubation of LB plates for 5 days at 37°C were 2.4×10^5, 2.0×10^5 and 2.3×10^5, respectively. These ratios corresponded to actual colony counts (per mg plastid protein) of 32000, 29000 and 34000 for fractions a, b and c, respectively, after plating out and growth at 30°C. Moreover, ampicillin-resistant bacteria could not be detected after incubation of fractions a, b or c with amipicillin (50 μg/ml) plus carbanicillin (200 μg/ml) for 5 days at 30° or 37°C.

Whereas, sonic extracts of EDTA-washed etioplasts incubated for 2 hr in the presence of pCS75 hydrolyzed nitrocefin, control extracts (from etioplasts incubated in the absence of DNA) did not. The initial rate of nitrocefin hydrolysis by transformed etioplasts was 0.12 OD_{490} units/mg protein (20°C) in comparison with a value of 0.4 (22°C) obtained for pBR322-transformants of permeaplasts of *A. nidulans*; the transformations had been conducted at 27°C and 32°C, respectively (Daniell and McFadden 1986, 1987). The level of bacterial contamination was too low to contribute to any results reflecting expression of foreign DNA in etioplasts. Moreover, no Ap[r] (or carbanicillin-resistant) bacteria were detectable as stressed previously. Thus contamination of etioplasts by ampicillin-resistant bacteria did not contribute to our detection of β-lactamase.

Labeled S subunit of RuBisCO was detected by immunoprecipitation in etioplasts which had been washed with EDTA prior to incubation with pCS75; labeled S subunits were absent in extracts of etioplasts which had been incubated without pCS75 or in another control in which pre-immune serum was used for precipitation. The 12 kDa-S subunit synthesized was readily differentiated from endogenous (unlabeled S subunits) which had a molecular weight of 14,000 (Daniell and McFadden 1987). An effect of bacterial contamination is also contraindicated by these results demonstrating expression of the S subunit gene for RuBisCO, an enzyme which is not found in heterotrophic bacteria.

Expression of the *cat* gene in etioplasts which had been incubated with pUC9-CM was enhanced by washing with 10 mM EDTA in a time-dependent manner. Moreover, this expression was dependent upon the plasmid concentration ('gene dosage') and parallel incubation with ribonuclease or subsequent incubation with thermolysin had no effect.

In organello translation of endogenous mRNA in etiochloroplasts isolated from greening cucumber cotyledons showed a continuous increase as greening progressed. In contrast, pCS-75-dependent incorporation of ^{35}S-methionine decreased dramatically as the tissue matured and only 5% of the original transcription/translation activity towards pCS75 remained after 24 hr of greening (Daniell and McFadden 1987). It is important to stress that all of the gene expression studies described were conducted on non-green cotyledons corresponding to zero-time of greening. In this context, etioplasts have been shown quite recently to be 10-fold more transcriptionally efficient than mature chloroplasts (Mullet and Klein 1987).

The results presently summarized establish that three different bacterial genes on two plasmids can be expressed in EDTA-washed intact cucumber etioplasts. Bacterial contamination does not contribute to the gene expression observed. The developmental state of chloroplasts may be crucial in obtaining optimal expression of foreign DNA because etioplasts are much more effective in DNA-dependent translation than etiochloroplasts.

In very recent experiments (unpublished), we have observed that electroporation at different voltages and capacitances does not improve expression of plasmids in (control or EDTA-washed) etioplasts.

Conclusion

Our results suggest that permeaplasts of cyanobacteria are good models for plastids from higher plants in terms of DNA binding and uptake. In addition, they demonstrate that DNA binding and uptake studies may be an invaluable prelude to investigations of the expression of foreign genes in isolated organelles. The results also establish that genes of bacterial contaminants do not add significantly to expression of plasmid genes in etioplasts under the conditions employed in our research. Finally, our results suggest that etioplasts from non-green cucumber cotyledons may exhibit transient expression of foreign plasmid genes much more readily than more mature etiochloroplasts. Whether a parallel will exist for plastids from other higher plants remains to be seen.

Our findings raise anticipation that gene uptake and expression in mitochrondria may also be achieved if systematic studies of DNA binding and uptake (and breakdown) are done. Additionally, cognizance of the developmental state may prove to be important in studies of mitochondria. The uptake and stable maintenance of mitochondria in yeast protoplasts has already been reported (Gunge and Sakguchi 1979).

Elsewhere we have proposed that our results may open a new avenue to the genetic engineering of higher plants (Daniell and McFadden 1987). For example, it seems likely that modifications of the L subunit gene of RuBisCO, which specify enhanced catalytic activity (in the gene product), will become available. Analogous modifications of the S subunit gene may also become available by directed mutagenesis (McFadden and Small 1988). Like the S subunit of RuBisCO, sensitivity to the herbicide glyphosphate is encoded by the nucleus in higher plants. On the basis of our results, it may be feasible to express glyphosate-resistant genes or modifications of S subunit genes that lead to enhanced RuBisCO in isolated etioplasts. If, after reintroduction of transformed etioplasts into protoplasts from a homologous plant, stable gene expression can be achieved, the potential will be opened to regenerate altered plants. The effect of RuBisCO upon photosyn-

134

thesis and growth and of glyphosate upon growth could be examined in such plants. Prospects for the introduction and expression of altered genes for the chloroplast-encoded L subunit of RuBisCO are less encouraging. This is because the number of genomes/chloroplast varies between 180 and 10 in mature and young spinach leaves, respectively (Scott and Possingham 1983). Each genome probably contains one L subunit gene for RuBisCO (Coen et al. 1977). Thus interesting modifications of L subunit genes would be highly diluted in a given transformed chloroplast. The magnitude of the problem is further calibrated by data which suggest that there are 1500–9500 chloroplast genomes/cell (Lamppa and Bendich 1979; Scott and Possingham 1980, 1983) and one L subunit gene per cell (Corruzzi et al. 1984). Nevertheless it may be possible to introduce functional modifications of the L subunit gene via chloroplast engineering if the expression of resident genes can be reduced while enhancing that of the introduced foreign L subunit gene(s). We anticipate that this new approach to the remodeling of plant plastid genomes will become a fertile area of research.

Acknowledgment

Much of the research summarized in this minireview was supported in part by NIH grant GM-19,972.

References

Buzby JS, Porter RD and Stevens SE, Jr (1985) Expression of the *Escherichia coli lac z* gene on a plasmid vector in a cyanobacterium. Science 230: 805–807

Coen DM, Bedbrook JR, Bogorad L and Rich A (1977) Maize chloroplast DNA fragment encoding the large subunit of ribulosebisphosphate carboxylase. Proc Natl Acad Sci USA 74: 5487–5491

Coruzzi G, Broglie R, Edwards C and Chua N-H (1984) Tissue-specific and light-regulated expression of a pea nuclear gene encoding the small subunit of ribulose-1,5-bisphosphate carboxylase. EMBO J 3: 1671–1679

Daniell H and McFadden BA (1986) Characterization of DNA uptake by the cyanobacterium *Anacystis nidulans* Mol Gen Genet 204: 243–248

Daniell H and McFadden BA (1987) Uptake and expression of bacteroa and cyanobacterial genes by isolated cucumber etioplasts. Proc Natl Acad Sci USA, 84: 6349–6353

Daniell H, Sarojini G and McFadden BA (1986) Transformation of the cyanobacterium *Anacystis nidulans* 6301 with the *Escherichia coli* plasmid pBR322. Proc Natl Acad Sci USa 83: 2546–2550

Daniell H, Sarojini G and McFadden BA (1987a) Cyanobacterial transformation: Expression of ColE1 Plasmids in *Anacystis nidulans* 6301. In: Biggins J (ed.) Progress in Photosynthesis Research, Vol. IV, pp 12.837–12.840. Dordrecht: Martinus Nijhoff

Daniell H, Torres-Ruiz JA and McFadden BA (1987b) Amplified expression of ribulose bisphosphate carboxylase/oxygenase in pBR322-transformants of *Anacystis nidulans*. Submitted

Dzelzkalns VA and Bogorad L (1986) Stable transformation of the cyanobacterium Synochocystis sp. PCC 6803 induced by UV irradiation. J. Bacteriol 165: 964–971

Echlin P (1970) The Origins of Plants. In: Harborne JB, ed. Phytochemical Phylogeny, Proceedings of the Phytochemical Society Symposium, Bristol, pp 1–19. New York: Academic Press Inc

Gunge N and Sakgachi K (1979) Fusion of mitochondria with protoplasts in *Saccharomyces cerevisiae*. Mol Gen Genet 170: 243–247

Herdman M and Stanier RY (1977) The cyanelle: chloroplast or endosymbiotic prokaryote? FEMS Microbiol Lett 1: 7–12

Joenje H and Venema G (1975) Different nuclease activities in competent and non-competent *Bacillus subtilis*. J. Bacteriol 122: 25–33

Lamppa GK and Bendich AJ (1979) Changes in chloroplast DNA levels during development of pea (*Pisum sativum*) Plant Physiol 64: 126–130

McFadden BA and Small CL (1988) Cloning, expression and mutagenesis of the genes for ribulose bisphosphate carboxylase/oxygenase. Photosynthesis Res, 18: 245–260.

McFadden BA, Torres-Ruiz J, Daniell H and Sarojini G (1986) Interaction, functional relations and evolution of large and small subunits in Rubisco from prokaryota and eukaryota. Phil Trans R Soc Lond B313: 347–358

Mereschkowsky C (1905) Uber Natur and Ursprung Chromatophoren in Pflanzenreiche. Biol Zentralbl 25: 593–604

Mullet JE and Klein RR (1987) Transcription and RNA stability are important determinants of higher plant chloroplast RNA levels. EMBO J 6: 1571–1579

Schopf JW and Walter MR (1982) Origin and Early Evolution of Cyanobacteria: the Geological Evidence. In: Carr NG and Whitton BA (eds) The Biology of Cyanobacteria, pp 543–564. Oxford: Blackwell Scientific Publications

Schwartz RM and Dayhoff MO (1981) Chloroplast Origins: Inferences from Protein and Nucleic Acid Sequences. In: Frederick JF. (ed) Origins and Evolution of Eukaryotic Intracellular Organelles, Ann NY Acad Sci, Vol. 361, pp. 260–272. New York: New York Academy of Sciences
Author note: numerous other articles establishing similarities between cyanobacteria and chloroplasts may be found in this symposium volume

Scott NS and Possingham JV (1980) Chloroplast DNA in expaning spinach leaves. J Exp Botany 31: 1081–1092

Scott NS and Possingham JV (1983) Changes in chloroplast DNA levels during growth of spinach leaves. J Exp Botany 34: 1756–1767

Singh DT, Nirmala K, Modi DR, Katiyar S and Singh NH (1987) Genetic transfer of herbicide resistance gene(s) from *Gleocapsa* spp. to *Nostoc muscorum*. Mol Gen Genet 208: 436–438

Wallin A and other contributors (1984) Uptake of Organelles. In: Vasil IK (ed.) Cell Culture and Somatic Cell Genetics of Plants, Vol. 1. Laboratory Procedures and Their Applications, pp 503–513. New York: Academic Press, Inc

Govindjee et al. (eds), Molecular Biology of Photosynthesis: 137–159
© 1988 Kluwer Academic Publishers

Minireview

The *puf* operon region of *Rhodobacter sphaeroides*

TIMOTHY J. DONOHUE,[1] PATRICIA J. KILEY[2,3] & SAMUEL
KAPLAN[2]
[1]*Bacteriology Department, University of Wisconsin, 1550 Linden Drive, Madison WI 53706,
USA; [2]Microbiology Department, University of Illinois, Urbana, IL 61801, USA.; [3]Current
address: Biochemistry Department, University of Wisconsin, Henry Mall, Madison WI
53706, USA*

Received 22 September 1987; accepted 11 March 1988

Key words: Bacteriochlorophyll-protein complex, gene expression, light-harvesting complex,
reaction center, *Rhodospirillaceae*, transcriptional control, translational control

Abstract. The *puf* operon of the purple nonsulfur photosynthetic bacterium, *Rhodobacter
sphaeroides*, contains structural gene information for at least two functionally distinct bacteriochlorophyll-protein complexes (light harvesting and reaction center) which are present in
a fixed ratio within the photosynthetic intracytoplasmic membrane. Two proximal genes
(*pufBA*) specify subunits of a long wavelength absorbing (i.e., 875 nm) light harvesting
complex which are present in the photosynthetic membrane in ≃ 15 fold excess relative to the
reaction center subunits which are encoded by the *pufLM* genes. This review summarizes
recent studies aimed at determining how expression of the *R. sphaeroides puf* operon region
relates to the ratio of individual bacteriochlorophyll-protein complexes found within the
photosynthetic membrane. These experiments indicate that *puf* operon expression may be
regulated at the transcriptional, post-transcriptional, translation and post-translational levels.
In addition, this review discusses the possible role(s) of newly identified loci upstream of *pufB*
which may be involved in regulating either synthesis or assembly of individual bacterio-
chrlorophyll-protein complexes as well as the *pufX* gene, the most distal genetic element within
the *puf* operon whose function is still unknown.

Introduction

Purple nonsulfur photosynthetic bacteria of the family *Rhodospirillaceae*
have provided a wealth of information regarding photosynthetic membrane
structure and assembly, as well as furthering our understanding of the
primary photochemical events involved in converting light to cellular energy. The physiological control of photosynthetic intracytoplasmic membrane (ICM) synthesis by oxygen and light (Kaplan 1981, Donohue and
Kaplan 1986, Kiley and Kaplan 1988) make facultative *Rhodospirillaceae*
excellent models with which to study synthesis of ICM pigment-protein
complexes. The light-harvesting (LH) antenna and reaction center (RC)

bacteriochlorophyll (Bchl)-protein complexes are the major ICM pigment-protein complexes. Although individual members of the *Rhodospirillaceae* can be distinguished either by the type and number of LH complexes or the structure of the photosynthetic membrane in which these pigment-protein complexes are housed (Imhoff et al. 1984), they all contain a fixed ratio of a specific LH complex in close association with the RC. Therefore, these bacteria also represent excellent model systems in which to study the physiological and genetic factors regulating assembly of these supramolecular membrane complexes.

Within the RC, light energy initially absorbed by the LH complexes is ultimately converted to cellular energy by light-induced oxidation-reduction reactions (Feher and Okamura 1978). The *Rhodobacter sphaeroides* and *Rhodobacter capsulatus* B875 complex (LHI) transfers exciton energy from a pool of peripherally-arranged B800–850 (LHII) complexes to a centrally-located RC complex (Monger and Parsons 1977, Meinhardt et al. 1985). *Rhodospirillum rubrum* (Berard et al. 1986) and *Rhodopseudomonas viridis* (Stark et al. 1986) contain a single antenna complex (B890 and B1015, respectively), while *Rhodopseudomonas acidophila* contains an LHI-type antenna complex in addition to two B800–850-type LH complexes and a B800–820 LH complex which allows this bacterium to efficiently harvest low intensity light (Angerhofer et al. 1986).

Aggregates of B875 and RC complexes in the *R. sphaeroides* ICM have been termed the fixed photosynthetic unit (FPU) since the $\simeq 15:1$ ratio of B875 to RC complexes is independent of light intensity (Kaplan 1981, Donohue and Kaplan 1986, Kiley and Kaplan 1988). In contrast, the cellular abundance of the B800–850 complex varies inversely with incident light intensity and the term variable photosynthetic unit (VPU) denotes aggregates of B800–850 complexes surrounding the FPU.

The VPU contains a minimum of seven unique polypeptides and the corresponding *R. sphaeroides* and *R. capsulatus* structural genes have been sequenced and placed within three transcriptional units (Kiley and Kaplan 1988, Scolnick and Marrs 1987). The minimal functional unit required for B800–850 spectral activity in *R. sphaeroides* (Cogdell and Crofts 1978) consists of six molecules of Bchl, three molecules of carotenoid and two each of two small polypeptides designated B800–850-α (5599 dalton) and B800–850-β (5448 dalton) which are encoded by the *puc* operon (Theiler et al. 1984, Zuber 1985, Kiley and Kaplan 1987). However, the absence of a third polypeptide is correlated with a B800–850-minus phenotype in *R. sphaeroides* (Kiley et al. 1988) and some purified B800–850 complexes from *R. capsulatus* contain an additional polypeptide (Drews and Reiner 1978), although the structural and/or functional role(s) of these additional polypeptides are not understood. The *R. sphaeroides* B875 complex contains the

B875-β (5457 dalton) and B875-α (6809 dalton) polypeptides in a 1:1 stoichiometry, two molecules of Bchl and two carotenoids (Broglie et al. 1980, Theiler et al. 1985, Zuber 1985). The *R. sphaeroides* RC subunits (which were initially classified by their apparent molecular weights on SDS-PAGE), RC-H, RC-M, and RC-L, are present in equimolar amounts in purified RC complexes along with four molecules of Bchl, two molecules of bacteriopheophytin, two ubiquinone molecules, a non-heme iron and a single carotenoid (Feher and Okamura 1978). The organization of chromophores and polypeptides within the RC complex from *R. viridis* (which also contains one molecule of a membrane bound c-type cytochrome, [Deisenhofer et al. 1985, Michel et al. 1986a]) and *R. sphaeroides* (Allen et al. 1986, Chang et al. 1986, Allen et al. 1987) has recently been determined by X-ray crystallography.

In spite of the fact that the RC-H, RC-M and RC-L subunits are found in equimolar amounts within purified RC complexes, the RC-H gene (*puhA*) is not linked to those genes encoding the RC-L (*pufL*) and RC-M (*pufM*) subunits (Marrs 1981, Taylor et al. 1983, Michel et al. 1985, Donohue et al. 1986a). The *pufL* and *pufM* genes are distal to the structural genes for the β (*pufB*) and α (*pufA*) subunits of the B875 (or corresponding LHI-type) complex (Williams et al. 1983, Williams et al. 1984, Youvan et al. 1984, Kiley et al. 1987). Because *puf* operon gene products (LHI and RC polypeptides) are present in defined stoichiometries within individual ICM pigment-protein complexes, and because the LHI and RC comprise the basic photosynthetic unit, much effort has been invested in determining how *puf* operon expression relates to assembly of the FPU. In addition, the crystallization of the *R. viridis* (Deisenhofer et al. 1985, Michel et al. 1986a) and *R. sphaeroides* (Allen et al. 1986, Chang et al. 1986, Allen et al. 1987) RC complexes and modern molecular genetic techniques (Bylina et al. 1986, Kiley and Kaplan 1988, Scolnick and Marrs 1987) have made possible a detailed structure-function analysis of this complex. In this regard, it is particularly important to understand how the *puf* operon and interacting genetic regions function in FPU assembly when designing experiments to probe association of the RC-L and RC-M polypeptides with each other, with chromophores in the RC complex, and with other polypeptides such as cyt c_2 (Prince et al. 1975) or antenna light-harvesting complexes (Stark et al. 1986, Monger and Parsons 1977). Our intention is to summarize recent advances on the structure and function of the *puf* operon as well as those adjacent DNA regions involved in either regulating *puf* operon expression or FPU assembly. We will focus our discussion on *R. sphaeroides*, but we will also discuss the organization and regulation of the *puf* operon in other photosynthetic bacteria where the appropriate information is available.

Structure of the *puf* operon region

The *R. sphaeroides puf* operon (Fig. 1) consists of the *pufBALMX* genes which code for, respectively, the B875-β and B875-α polypeptides, the RC-L and RC-M proteins, and an open reading frame (ORF) which could code for an approximately 8000 dalton polypeptide (*pufX*) whose physiological function is currently under investigation. This region of the genome was initially identified as *rxcA* in *R. capsulatus* (Marrs 1974, Marrs 1981, Taylor et al. 1983) and it operationally defines one presumed end of the 'photosynthetic gene cluster' in both this bacterium (Yen and Marrs 1976, Marrs 1981) and *R. sphaeroides* (Sistrom et al. 1984, Pemberton and Harding 1986). The other presently known boundary of the photosynthetic gene cluster (which also contains some genes for carotenoid and Bchl biosynthetic enzymes) is the RC-H structural gene which is some 45 kb away and transcribed in the opposite direction relative to the *puf* operon (Marrs 1981, Youvan et al. 1984). Preliminary evidence from genomic mapping by pulse field electrophoresis (Wu and Kaplan unpublished) suggest that other genetic loci

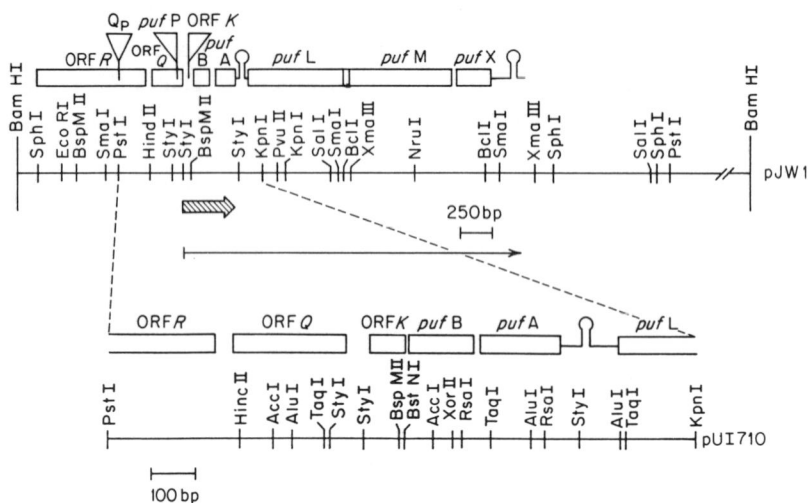

Fig. 1. Physical-genetic map of the *R. sphaeroides puf* operon region (see text for more details) originally described within an $\simeq 12$ kb *Bam*HI restriction fragment on plasmid pJW1 (Williams et al. 1983). Shown are the known (*pufBALM*) or hypothetical coding regions (ORF K and *pufX*) and the two polycistronic transcripts encoded by the *puf* operon. The *puf*0.5 (hatched arrow) and *puf*2.6 (line arrow) transcripts are depicted relative to a putative *puf* operon promoter(s)(*pufP*). Stem loop structures shown between *pufA* and *pufL* or downstream of *pufX* are upstream of the 3' ends of the *puf*0.5 and *puf*2.6 transcripts, respectively. The coding sequences for the *R* and *Q* ORF are also indicated as well as a region within *R* which has tentatively been identified to contain a promoter for the *Q* gene (*QP*).

involved in photosynthetic membrane function may be close to the *R. sphaeroides* photosynthetic gene cluster and complementation and hybridization analysis demonstrate that the cyt c_2 structural gene is downstream of *puhA* (MacGregor et al. unpublished).

Youvan et al. (1984) initially suggested that the *R. capsulatus* B875 and RC-L and RC-M structural genes constituted a polycistronic operon and they identified the *pufX* gene (open reading frame C2397 by their nomenclature) immediately downstream of *pufM*. The arrangement of LHI and RC structural genes within the *puf* operon has been conserved in *R. rubrum* (Berard et al. 1986) and *R. viridis* (Michel et al. 1986b). The *R. viridis puf* operon most likely encodes the membrane bound c-type cytochrome of the RC complex (*pufC*) since the initiator methionine for the putative *pufC* signal sequence overlaps the *pufM* gene by one nucleotide (Michel et al. 1986b, Weyer et al. 1987). However there is currently no information on whether the *R. viridis puf* operon contains the structural gene for the gamma polypeptide of the B1015 LH complex (Stark et al. 1986) or an ORF analogous to the putative *pufX* gene.

Two genes upstream of *pufB* have recently been identified in both *R. capsulatus* and *R. sphaeroides* (Fig. 1). Bauer et al. (1988) initially described the *R. capsulatus* Q gene (78 amino acid ORF) upstream of *pufB*. Similarly, a 77 amino acid gene product which terminates 131 bp upstream of *pufB* is predicted from the analogous *R. sphaeroides* Q gene DNA sequence and a protein of the expected molecular weight is synthesized in vitro when an *R. sphaeroides* coupled transcription-translation system is used with *Q*-containing DNA templates (Havelka and Kaplan unpublished). The precise physiological function, subcellular location, and relationship of the *Q* gene to the *puf* operon remains to be determined (see below). *R. sphaeroides* (Davis et al. 1988) and *R. capsulatus* (Youvan et al. 1985, Bauer et al. 1988) strains with drug resistance cartidges interrupting expression of the *Q* gene synthesize reduced amounts of Bchl-protein complexes. Bauer et al. (1987, 1988) have proposed that the *R. capsulatus Q* gene product is the putative carrier of Bchl biosynthetic intermediates initially proposed in *R. sphaeroides* (Rebeiz and Lascelles 1982) and that the *Q* gene is actually part of the *puf* operon (see below).

A second ORF designated *R* terminates 59 bp upstream of the start of the *R. sphaeroides Q* gene. The size of the *R* gene product inferred from the DNA sequence is 38 315 daltons and an approximately 39.5 kD polypeptide is synthesized from this region in vitro (Havelka and Kaplan, unpublished). The *R* gene region complements the B875⁻ phenotype of several independent *R. sphaeroides* strains (Davis and Kaplan, Sockett and Kaplan unpublished), suggesting a role for the *R* gene product in some aspect of FPU synthesis. There is no precise information available on the physical relationship between the *R* gene and the *bchA* locus of the Bchl biosynthetic pathway, since the DNA sequence for these regions is not yet complete. Published genetic and physical maps of the photosynthetic gene cluster predict that *bchA* lies ≃ 1.7 to 2.8 kb upstream of the *R. capsulatus puf* operon. These distances must be considered to be tentative since the precise site of the mutations in the original *R. capsulatus rxcA* and *bchA* alleles used for mapping are unknown (Taylor et al. 1983). In addition, there might be differences in gene organization in this region between *R. sphaeroides* and *R. capsulatus* since genetic evidence suggests that the *bchA* locus is the proximal gene on an operon consisting of *bchC* and *bchA* in *R. capsulatus* (Yen and Marrs 1976, Zsebo and Hearst 1984, Biel and Marrs 1983), while Pemberton and Harding (1986) suggest close linkage between *bchA* and *bchB* in *R. sphaeroides*. It is perhaps significant to note that the available information suggests that the region upstream of *pufB* is all transcribed in the same direction (see below).

In this regard, the working model outlined in this minireview is that expression of the region extending at least from *R* through *pufX* in *R. sphaeroides* is highly regulated at many levels to insure the stoichiometric synthesis of Bchl, Bchl-binding apoproteins within individual ICM pigment-protein complexes, and a fixed ratio of Bchl-protein complexes within the FPU. Further, we propose that transcription of this region is highly polarized such that the promoter or other *cis*-acting regulatory elements for each 'downstream' transcriptional unit are contained within the preceding upstream structural gene (**transcriptional control**). A region of secondary structure between *pufA* and *pufL* may affect stoichiometry and stability of individual *puf* operon transcripts as well as the cellular level of the cognate ICM Bchl-protein complexes (**post-transcriptional control**). Although *puf* operon-specific transcripts are present in chemoheterotrophically grown cells such cells lack detectable Bchl or pigment-binding proteins, suggesting **translational** or **post-translational** control over Bchl-binding protein synthesis. Finally, studies indicate **post-translational** control over stability of specific Bchl-binding apoproteins when they are synthesized in the absence of free Bchl.

Transcriptional control

The location of the *pufBA* genes proximal to *pufLM* originally led to the proposal these genes constituted a polycistronic operon and that some form of differential expression of the proximal genes relative to the distal genes resulted in the $\simeq 15:1$ ratio of *pufBA* gene products relative to RC complexes in the ICM (Youvan et al. 1984, Belasco et al. 1985). Subsequently, using gene specific probes Zhu and Kaplan (1985) demonstrated that the *R. sphaeroides* B875 structural genes were encoded by two stable mRNA species; a $\simeq 2600$ nucleotide (nt) low abundance transcript which also hybridized to *pufL* and *pufM*-specific probes, and a $\simeq 500$ nt high abundance transcript homologous only to *pufBA* (Fig. 1). This analysis also demonstrated that the $\simeq 500$ nt *puf* operon mRNA (hereafter referred to as *puf*0.5) was, on the average, in 12-fold excess over the 2600 nt *puf* operon transcript (*puf*2.6) under photosynthetic growth conditions (Zhu and Kaplan 1985, Zhu et al. 1986c) similar to what was observed in *R. capsulatus* (Belasco et al. 1985). The *puf*0.5:*puf*2.6 ratio is independent of light intensity and, since this approximates the stoichiometry of ICM B875 and RC complexes, it has been suggested that the size of the FPU reflects the relative abundance of *puf* operon mRNA species under photosynthetic conditions (Zhu et al. 1986c). Although this is a conceptually attractive hypothesis, it should be noted that no experiments have been performed to address the relative translational fidelity of individual *puf* operon transcripts in vivo (see below).

Different mechanisms have been proposed to account for the molar excess of the two stable *puf* operon transcripts. Belasco et al. (1985) concluded that there was a precursor-product relationship between the *R. capsulatus puf*2.6 and *puf*0.5 transcripts and that selective degradation at the 3′ end of the large transcript produced the *puf*0.5 mRNA which was stabilized due to a region of RNA secondary structure between *pufA* and *pufL*. An *R. capsulatus* strain containing an extrachromosomal *puf* operon with the inverted repeat in the *pufAL* intercistronic region deleted has recently been constructed (Klug et al. 1987, see below) and the lack of detectable *puf*0.5 mRNA in this mutant would be consistent with the model of Belasco et al. (1985). However, a rigorous demonstration of the existence or lack of a precursor-product relationship between the *puf*2.6 and *puf*0.5 transcripts is lacking in both *R. sphaeroides* and *R. capsulatus*. In contrast we have proposed (Zhu et al. 1986c) that the relative abundance of the stable *R. sphaeroides puf* operon transcripts is due, at least in part, to selective transcription termination within a region of secondary structure between *pufA* and *pufL*.

The functional and physical half lives of the stable *puf* operon transcripts in both *R. sphaeroides* (Zhu et al. 1986c) and *R. capsulatus* (Dierstein 1984, Belasco et al. 1985, Zhu et al. 1986a) indicate that *puf*0.5 is more stable ($t_{1/2} \simeq 20\text{-}25$ min) than the *puf*2.6 transcript ($t_{1/2} \simeq 9\text{-}10$ min) when transcription initiation is inhibited by the addition of rifampicin to photosynthetic cultures. It has been previously shown that exposure of photosynthetic cultures to oxygen results in the immediate cessation of Bchl-protein complex synthesis (Shepherd and Kaplan 1983) and this also decreases the $t_{1/2}$ of the $puf0.5$ and $puf2.6$ transcripts to approximately 10 and 5 min, respectively (Zhu et al. 1986c). This suggests that oxygen may directly (for example, via an oxygen activated ribonuclease) or indirectly (by blocking translation of these transcripts, see below) decrease stability of these mRNA species. However, more recent experiments have demonstrated that the $t_{1/2}$ of $puf0.5$ and $puf2.6$ mRNA in steady state chemoheterotrophic cells is indistinguishable from that of photosynthetic cells (Lee and Kaplan, unpublished), suggesting that the accelerated turnover of these transcripts observed when oxygen is introduced into a photosynthetic culture is transient. These results highlight the fact that ICM synthesis and presumably puf operon expression are extremely sensitive to small changes in oxygen tension. Experiments monitoring puf operon transcription have often been performed with cells grown in 'low oxygen' conditions. By definition these cultures are not in a steady-state environment and it is difficult to quantitatively monitor puf operon expression and ICM synthesis in such oxygen-limited cultures. This may be particularly true when one considers the apparent effects of oxygen on expression of the R and Q genes, or how transcription of R and Q relate to expression of the puf operon (see below).

Two 5' ends for the stable *R. sphaeroides* puf operon mRNA species have been identified 104 ($puf0.5$) and 75 ($puf2.6$) nt upstream of $pufB$ in both chemoheterotrophically and photosynthetically grown cells (Zhu et al. 1986c). A major stable 5' end was mapped $\simeq 110$ bp upstream of $pufB$ in *R. capsulatus*; a second 5' end (which represented 1–2% of the total in low oxygen cells) was identified $\simeq 30$ bp downstream, but this second 5' end was considered to be inconsequential to puf operon transcription (Belasco et al. 1985). The 5' end of the stable *R. sphaeroides* $puf0.5$ mRNA is $\simeq 26$ bp downstream of the end of the Q gene and $\simeq 10$ bp downstream of a region of secondary structure which could serve as a Q gene transcription terminator. Analysis of puf operon transcription is by no means complete, but our working model proposes that differential transcription initiation using unique regulatory region(s) located within the Q gene is partly involved in determining the ratio of $puf0.5$ to $puf2.6$ transcripts (see below).

Bauer et al. (1987) have used a $pufM:lacZ$ translational fusion to monitor puf operon expression in *R. capsulatus* and they have suggested that Q is the first gene in the puf operon, that the Q promoter(s) which produced mRNA with 5' ends 511 and 699 bp upstream of $pufB$ initiates transcription for the entire puf operon, and that the previously mapped 5' ends of puf operon mRNA were degradation products of these primary transcripts (Bauer et al. 1988). Transcription initiation sites for mRNA species homologous to Q or *pufB* have yet to be determined by in vitro capping of cellular mRNA in either organism. In agreement with the experiments of Bauer et al. (1988), insertion of a drug resistance cartridge in *Q* (Youvan et al. 1985, Davis et al. 1988, Havelka and Kaplan unpublished) results in *R. sphaeroides* strains which do not accumulate wild-type amounts of *puf* operon mRNA in vivo. However, sequences with promoter activity exist immediately upstream of the 5′ end of the stable *R. sphaeroides puf* operon transcripts, suggesting that *Q* and *pufB* are not on the same operon. Using an *R. sphaeroides* coupled transcription-translation system (Chory and Kaplan 1982) we demonstrated that ≃ 230 bp of DNA upstream of *pufB* (which includes the 3′ end of the *Q* gene coding sequence) is sufficient to direct synthesis of the *pufBA* gene products (Kiley et al. 1987). Deletion of ≃ 70 bp, including the 5′ end of the stable *puf*0.5 mRNA, reduced synthesis of the *pufBA* polypeptides in vitro; this would be expected if this 70 bp region contained a promoter. The quantitative effect of this deletion on synthesis of the *pufBA* proteins was dependent on the amount of additional upstream DNA on the plasmid template (Kiley et al. 1987), suggesting that sequences upstream of the deletion endpoint (which extend further into the *Q* gene) can affect *puf* operon expression in an as yet undetermined fashion. In addition we have been able to separate synthesis of the putative *Q* gene product and the *pufBA* proteins in vitro by using specific plasmid templates (Kiley et al. 1987, Havelka and Kaplan unpublished). A similar conclusion is derived from in vivo experiments with a *pufB:lacZ* fusion containing the distal 47 bp of the *Q* gene which produces β-galactosidase activity in *R. sphaeroides* when present on a low copy number plasmid (Tai et al. 1988). Synthesis of β-galactosidase activity is dependent on *R. sphaeroides* sequences since a transcriptional terminator from the omega cartridge (Prentki and Krisch 1984), which we know functions in vivo (Tai et al. 1988) is immediately upstream of the *R. sphaeroides* DNA. Finally, a run-off transcript which initiates close to the previously mapped 5′ end of one stable *puf* operon mRNA (≃ 136–141 nt upstream of *pufB*) is synthesized using purified *R. sphaeroides* RNA polymerase and DNA templates containing the distal portion of the *Q* gene into *pufB* (Kansy and Kaplan 1988). Similar analysis indicates that a promoter for *Q* maps ≃ 183–193 nt upstream of the *Q* gene (i.e., just upstream of the *Pst*I site in Fig. 1 and within the *R* gene coding sequence). By analogy, the 5′ end of the mRNA homologous to the *Q* gene is within the *R. capsulatus bchA* gene (Bauer et al. 1988).

Thus, in spite of the detailed analysis of *puf* operon expression performed to date by several laboratories, there is still an apparent paradox as to where the *puf* operon promoter(s) resides or precisely how much upstream DNA is require in *cis* for regulated in vivo expression of this operon (Beatty et al.

1986). It is perhaps important to note that many *Escherichia coli rif*[R] mutants are effected in transcription termination (Jin et al. 1988) and it has not been described whether the *rif*[R] allele in *R. capsulatus* SB1003 used by several labs (Youvan et al. 1985, Zhu et al. 1986a, b, Bauer et al. 1987, 1988) has any contribution on transcription termination in *Q* or the *puf* operon region. In the absence of any evidence showing physical linkage between the *Q* and *pufBA* transcripts, our working hypothesis is that *Q* and *pufBALMX* are not on the same operon, but that the *puf* operon promoter (and perhaps other *cis*-acting regulatory sequences) is buried within *Q*. A similar relationship may exist between the *R* and *Q* genes since sequences required for either *Q* protein synthesis (Havelka and Kaplan unpublished) or *Q*-specific run-off transcripts in vitro contain *R* coding sequences (Kansy and Kaplan 1988) and the *Q* promoter is buried within the *R. capsulatus bchA* gene (Bauer et al. 1988).

There is no detectable mRNA homologous to *R* or *Q*-specific probes when cells are grown chemoheterotropically in a 30% oxygen atmosphere (Lee and Kaplan unpublished). In contrast, there is only 3-fold less *pufBA*-specific mRNA under chemoheterotrophic growth conditions than in steady state photosynthetic cells grown in the presence of saturating light (Zhu et al. 1986b, c). Therefore, if the *puf* operon transcript initiates upstream of *Q* (or even *R*), then there must be both specific 5′ mRNA processing to produce the stable 5′ ends mapped in both *R. capsulatus* (Belasco et al. 1985) and *R. sphaeroides* (Zhu et al. 1986c) and differential oxygen-dependent regulation of this processing activity to allow detectable *Q*-specific transcripts only in photosynthetically grown cells. In addition, the model of Belasco et al. (1985) proposes specific 3′ processing of the *puf*2.6 mRNA to produce the *puf*0.5 transcript (see above). At this time we cannot unambiguously rule out the combination of 5′ and 3′ processing of a large primary transcript ($\simeq 3500$ nt) to generate the stable *puf* operon mRNA species; nor do we have any direct evidence for such a phenomenon in *R. sphaeroides*. However if these stable *puf* operon transcripts are processed species, the 5′ and 3′ processing events are very precise and subject to strong physiological control since *R* and *Q*-specific mRNA is both undetectable in chemoheterotrophically grown cells and in very low amounts relative to *pufBA*-specific mRNA under photosynthetic conditions (see above).

Our model attempts to reconcile these seemingly disparate observations by suggesting that *cis*-acting regulatory sequences (i.e. promoters and others) for the downstream operon exist within the upstream structural gene. Alternatively, transcription of an upstream gene (i.e. *R* or *Q*) may affect transcription initiation at the immediate downstream promoter (*Q* or *puf*, respectively) by influencing the structure of the DNA in the putative promoter/regulatory region. In this regard there have recently been several reports that structural constraints within DNA mediated by gyrase/topoisomerase activities can effect procaryotic gene expression, particularly under anaerobic growth conditions (Kranz and Haselkorn 1986, Rudd and Menzel 1987). In addition, *cis*-acting regulatory (McFall 1986), or transcriptional regulatory sequences well upstream of initiation sites or buried within upstream structural genes have been demonstrated in several highly regulated procaryotic and eucaryotic systems (Ptashne 1986).

Post-transcriptional control

The different half lives of the *puf*0.5 and *puf*2.6 transcripts (see above) undoubtedly contribute to the cellular ratio of mRNA specific for the B875 and RC-L and RC-M structural genes. However the approximately 15 fold excess of the *puf*0.5 transcript relative to *puf*2.6 under photosynthetic conditions, or the $\simeq$ 20–30 fold excess of this mRNA under chemoheterotrophic conditions cannot be explained solely by the $\simeq$ 2 to 4 fold measured differences in stability of the two stable *puf* operon transcripts under steady-state physiological conditions (Belasco et al. 1985, Zhu et al. 1986b, c). We have proposed that transcription termination by a subset of RNA polymerase molecules transcribing the *puf* operon contributes to establishing the *puf*0.5:*puf*2.6 ratio (Zhu et al. 1986c). The intercistronic region between *puf*A and *puf*L in *R. sphaeroides* contains two inverted repeats which could form mutually exclusive stable stem-loop structures typical of procaryotic transcription terminators (Kiley et al. 1987) and the 3' end of the *puf*0.5 mRNA has recently been mapped immediately downstream of the first of these two inverted repeats. The analogous region in *R. capsulatus* contains regions of possible secondary structure (Youvan et al. 1984) and Belasco et al. (1985) have shown that the 3' end of the *puf*0.5 transcript is downstream of an inverted repeat. However their model states that the stem loop structure stabilizes the *puf*0.5 mRNA against 3' processing rather than signalling transcription termination. It is reasonable to suggest that, in addition to acting as a transcription-terminator, the proximal stem-loop between *puf*A and *puf*L also serves to stabilize the *puf*0.5 transcript as has been observed elsewhere (Bechofer and Dubnau 1987, Newbury et al. 1987). It is important to note, however, that there is an additional inverted repeat between *puf*A and *puf*L in *R. sphaeroides* which, if the 3'-processing model were correct, does not appear to be involved in stabilizing the 0.5 kb transcript since no detectable 3' ends are found immediately downstream of this sequence.

We have proposed that the *puf*0.5 mRNA (which has a stable 5' end 104 nt upstream of *pufB*) results from transcription termination prior to *pufL* (Zhu et al. 1986c). Conversely the *puf*2.6 transcript (which has a stable 5' end 75 nt upstream of *pufB*) results from transcription through any potential regulatory signals in the *pufAL* intercistronic region. It has recently been determined that the *R. sphaeroides puf*2.6 transcript extends $\simeq$ 400 bp 3' of *pufM* which would include coding sequences for the *pufX* gene. The 3' end of the *R. capsulatus puf*2.6 transcript has not been determined but Northern blot analysis using a *pufX*-specific probe has determined that this mRNA includes the proximal portion of the *pufX* gene (Zhu et al. 1986b).

The *R. sphaeroides puf*2.6 mRNA has a 3' end near a region of secondary structure which is followed by 4 T residues in the DNA sequence. Although very little is known about transcription termination in *Rhodospirillaceae*, the structure preceding the 3' end of the *puf*2.6 transcript is typical of an *E. coli* factor-independent transcription terminator (von Hippel et al. 1987). The DNA sequence downstream of the 3' end of the *puf*0.5 mRNA is relatively A/T rich, however it only contains 2 consecutive T residues. If this region is a transcription termination site, future studies need to determine whether specific *pufBA* sequences or protein factors are involved in controlling termination (Richardson et al. 1987, Chamberlain et al. 1987) or antitermination (Greenblatt et al. 1987) of RNA synthesis at this site (see below), or how such processes relate to transcription initiation at potential *puf* operon promoters.

Inherent in all existing models is the assumption that alterations in the *pufAL* intercistronic region should affect the ratio of *puf* operon transcripts and, thereby, the size of the FPU. Klug et al. (1987) have shown that the size of the FPU decreases in *R. capsulatus* strains containing a deletion of the genomic *puf* operon and a plasmid-borne *puf* operon where the inverted repeat in the *pufAL* intercistronic region is removed. Such strains grow normally under chemoheterotrophic conditions but growth under photoheterotrophic conditions is slower than wild-type at the single light intensity tested. An *R. sphaeroides* strain was recently engineered that contains a genomic deletion of the *pufAL* intercistronic region which removes the stem loop region upstream of the 3' end of the *puf*0.5 transcript (DeHoff and Kaplan, unpublished). Growth of the resulting strain under photoheterotrophic conditions at very high light intensities or chemoheterotrophically is similar to wild type strains, but, similar to other strains which have a smaller FPU than wild-type (Meinhardt et al. 1985, Kiley et al. 1988) this strain grows slower under photosynthetic conditions at low light intensities. Preliminary experiments indicate an approximately 4 fold reduction in the level of flash-oxidizable RC complexes ($\simeq$ 2 fold reduction in RC polypeptide levels measured with specific antibodies) and an $\simeq 12$ fold reduction in B875 complexes, so that the FPU size has been changed from $\simeq 15:1$ in the wild type (Kiley et al. 1987) to approximately 1–2:1 in this mutant. At the mRNA level there is approximately 50% of the *puf*2.6 transcript relative to the wild type ($t_{1/2}$ of $\simeq 2.5$ min relative to $\simeq 8$ min in the wild type), no detectable *puf*0.5 transcript and an excess of a $\simeq 700$ nt *pufBA*-specific mRNA which has a 3′ end downstream of a region of secondary structure in *pufL* and a $t_{1/2}$ of 14 min (DeHoff et al. unpublished). In contrast, *R. capsulatus* strains with the above mentioned deletion in the *pufAL* intercistronic region as an extrachromosomal element contain approximately 50% of the wild type level of B875 β and α polypeptides and approximately 2–3 fold more flash oxidizable RC per mg of membrane protein. In addition, these *R. capsulatus* mutants accumulate wild-type levels of the *puf*2.6 mRNA but no detectable *puf*0.5 transcript (Klug et al. 1987). It will be of interest in future studies to determine the contribution of *puf* operon copy number to the apparent differences between these *R. sphaeroides* and *R. capsulatus* mutants, but strains such as these which have alterations in the size of the FPU should be very useful in spectroscopic studies on the efficiency of light energy transfer to the RC (Meinhardt et al. 1985), in genetic screens to isolate *cis* or *trans*-acting mutations which restore wild-type photosynthetic growth rates and in studies to determine the physiological control of FPU size.

Translational control

Puf operon-specific transcripts are detectable in cells grown chemoheterotrophically in a 30% oxygen atmosphere which lack detectable Bchl, visible absorbing Bchl intermediates, Bchl-binding polypeptides and ICM (Kiley and Kaplan 1988). This observation reveals that FPU synthesis is not solely mediated by oxygen-dependent transcriptional control of the *puf* operon. A similar situation exists in developing chloroplast protoplastids where mRNA for chlorophyll-binding proteins is present but translation of these polypeptides is apparently controlled by the phytochrome-mediated conversion of protochlorophyllide to chlorophyllide (Klein and Mullett 1986, Deng and Gruissen 1987).

Analysis of *R. sphaeroides* (Takemoto and Lascelles 1973) and *R. capsulatus* (Klug et al. 1985, 1986) mutants has demonstrated that Bchl synthesis is required for assembly of Bchl-protein complexes. Therefore it is not surprising to find that chemoheterotrophically grown cells lacking detectable pools of Bchl are also devoid of intact Bchl-protein complexes. The heme-dependent formation of an active translation initiation complex by globin mRNA in erythrocytes is well documented (Cox et al. 1985); by analogy, it is possible that Bchl (or some visible absorbing intermediate after protoporphyrin IX specific to the Mg-branch of tetrapyrrole synthesis) is required for translation of *puf* operon mRNA.

Translation of *puf* operon mRNA in chemoheterotrophically grown cells may also be controlled by sequences within these transcripts. For example, both the *puf*0.5 and *puf*2.6 transcripts contain a ribosome binding site and a 20 amino acid putative leader peptide (ORF K) with its stop codon 1 nt proximal to the *puf*B ATG (see Kiley et al. 1987 for DNA sequence). The putative ORF K ribosome binding site is 4 and 33 nt downstream of the 5′ end of the *puf*2.6 and *puf*0.5 transcripts, respectively. Nine of the 20 ORF K codons are rarely used in *R. sphaeroides* structural genes; in fact, several (CCA-proline [0/60], TCA-serine [0/40], TTA-leucine [0/120], and UAG-STOP [0/9]) have not been observed to date (Williams et al. 1983, 1984, 1986, Donohue et al. 1986a b, Kiley et al. 1987, Kiley and Kaplan 1987). By analogy to regulatory leader peptides in procaryotic (Bauer et al. 1983) and eucaryotic (Thireos et al. 1984, Werner et al. 1987) systems, ribosomes which become stalled in ORF K (assuming it is translated in vivo) may influence translation of the *puf*B gene since the ribosome binding site for *puf*M in *R. viridis* (Michel et al. 1986b), and the *puf*BA genes do not overlap reported changes in isoaccepting tRNA$_{phe}$ species in chemoheterotrophically grown cells adapting to photoheterotrophic conditions and recent analysis confirms that the predominance of phenylalanine codons in *R. sphaeroides* structural genes is in accord with the distribution of tRNA$_{phe}$ isoaccepting species. Attenuation control of the *Salmonella leu* (Carter et al. 1986) and *Serratia ilvGEDA* (Hsu et al. 1985) operons is controlled by rare codons in the leader peptide. Therefore, it is possible that ribosomes which may become stalled while translating ORF K under chemoheterotrophic conditions might allow translation of the *puf* operon structural genes if specific isoaccepting tRNA species for some of the rare ORF K codons are synthesized under photosynthetic growth conditions. Alternatively, by analogy to well studied systems in other organisms, the rate of translation through ORF K might serve to influence interaction of protein factors with RNA polymerase and thereby control the relative frequencies of transcription termination or anti-termination in the *puf*AL intercistronic region (see above).

The 1:1 stoichiometry of the individual gene products within the RC complex has been proposed to be mediated by the overlap in coding sequences for *puf*L and *puf*M in *R. sphaeroides* (Williams et al. 1983, 1984) and *R. capsulatus* (Youvan et al. 1984) or *puf*M and *puf*C in *R. viridis* (Michel et al. 1986b). In contrast, there is no overlap in coding sequences for *puf*L and *pufM* in *R. viridis* (Michel et al. 1986b), and the *pufBA* genes do not overlap in either *R. sphaeroides* (Kiley et al. 1987), *R. capsulatus* (Youvan et al. 1984) or *R. rubrum* (Berard et al. 1986), while the respective gene products are found in equimolar amounts in the appropriate ICM pigment-protein complexes. Therefore, until further experiments are performed, the physiological significance of overlapping structural genes in dictating stoichiometry of individual subunits of pigment-protein complexes in the ICM remains to be determined.

Post-translational control

Previous studies have shown that previously assembled Bchl-protein complexes (i.e., LH and RC) are stable in vivo even when photosynthetic cultures are exposed to high levels of oxygen (Shepherd and Kaplan 1983). However several recent studies point to post-translational control over the stability of unliganded Bchl-binding proteins. Drews and coworkers (Klug et al. 1986) have used mutants to demonstrate that B800–850 apoproteins are rapidly degraded when they are synthesized in the absence of Bchl and it has been shown that the *R. sphaeroides* B800–850-β (Kiley and Kaplan 1987) and B875-β (Kiley, 1987) polypeptides are labile when synthesized in vitro in the absence of Bchl.

Comparing the structural gene and amino acid sequences for *R. sphaeroides* and *R. capsulatus* B875 and RC subunits (Williams et al. 1983, 1984, Youvan et al. 1984, Kiley et al. 1987) did not reveal amino terminal extensions typical of signal sequences, suggesting that these polypeptides associate with the membrane and become assembled into functional pigment-protein complexes without proteolytic processing. In contrast, the deduced amino acid sequences for the *R. rubrum* B890 (Berard et al. 1986) and *R. viridis* B1015 (Michel personal communication) reveals carboxy terminal residues not present when the polypeptide is purified from intact pigment-protein complexes. Future experiments need to determine if these or other regions within individual LH polypeptides are involved in assembly of functional pigment-protein complexes. In this regard, recent experiments in *R. capsulatus* suggest that regions within individual B875 polypeptides may be involved in proper assembly of light-harvesting or RC complexes in the ICM (Jackson et al. 1986).

R. sphaeroides strain RS103 has previously been shown to lack detectable B875 spectral complexes (Meinhardt et al. 1985, Jackson et al. 1987) and a more recent analysis demonstrates that *puf* operon transcript levels in RS103 are indistinguishable from the wild type (Davis and Kaplan, unpublished).

RS103 contains $\simeq 25\%$ of the wild type level of the B875-α polypeptide under photosynthetic conditions and all of the detectable B875-α antigen in wild type cells is in the membrane fraction (Kiley et al. 1988). Therefore, functional association of the B875-α polypeptide with BchI is not required for the interaction of this protein with the membrane fraction in RS103. In addition, if one assumes that the mutation in RS103 does not effect translation of the *pufA* gene, this result suggests that turnover of B875-α protein is more rapid in RS103 than in a wild type genetic background. A more thorough analysis of the rates of synthesis and the stability of newly synthesized BchI-binding proteins in wild type and mutant strains is required to unequivocally address post-translational control of *puf* operon expression under different physiological conditions. This is particularly relevant to the previously mentioned fate of *puf* operon mRNA in chemoheterotrophically grown cells.

Summary

Given the central role of B875 and RC complexes in photosynthesis, it is perhaps, not surprising that *Rhodospirillaceae* have evolved a complex regulatory system to insure synthesis of *puf* operon gene products in the precise stoichiometries required for assembly of a functional FPU. It will be interesting to determine if synthesis and assembly of individual subunits and chromophores of the FPU are coordinated by a similar network of regulatory signals to those which control expression of the genes required for proteins and cofactors involved in other supramolecular complexes such as nitrogenase (Roberts and Brill 1981). In this regard it is of interest to note that Bauer et al. (1988) have recently reported that a potential oxygen-regulated promoter for the Q gene contains homology to the consensus sequence for *ntrA*-regulated promoters. In addition, there is recent genetic evidence that the *ntrA* sigma factor (which was initially discovered based on its ability to control transcription of nitrogen assimilatory enzymes) may also be involved in regulating transcription of *E. coli* formate hydrogenlyase under anaerobic conditions (Birkmann et al. 1987). In this review we have outlined several levels at which *puf* operon expression is regulated in vivo. The tight regulation of *puf* expression is also highlighted by the fact that derepression of FPU synthesis is not observed at the RNA, protein, or pigment-protein complex level in *R. sphaeroides* containing multiple copies of the operon on a plasmid (Davis et al. 1988). It is hopefully obvious to the reader that, although the study of the molecular mechanisms regulating *puf* operon expression has been a fertile area of research in recent years, many questions still need to be resolved.

One major question relates to the physiological function of the *pufX*, *Q* and *R* gene products and their relationship either to *puf* operon expression or FPU assembly. The DNA and inferred protein sequences for the *R. sphaeroides* (DeHoff and Kaplan unpublished) and *R. capsulatus* (Youvan et al. 1984) *pufX* gene products are highly homologous, suggesting a conserved function for either this region of the DNA or the putative gene product in *puf* operon expression or FPU assembly, respectively. Both the *pufX* and *Q* gene products contain a central hydrophobic domain (perhaps membrane associated) bounded by hydrophilic, positively charged regions at the amino and carboxy termini. It remains to be determined precisely what role the *pufX* or *Q* gene products play in synthesis of Bchl-protein complexes. Future research will undoubtedly center on resolving the location of the *puf* operon promoter(s), what *cis*-acting regulatory sequences are required for *puf* operon expression, what physiological and genetic factors set the ratio of *puf*0.5 and *puf*2.6 transcripts, how this ratio relates to FPU size, and how changes in the size of the FPU affect either photosynthetic growth or the efficiency of light energy transfer to the RC.

In spite of the new advances brought about by applying molecular biological techniques to study of photosynthetic bacteria (Scolnick and Marrs 1987, Kiley and Kaplan 1988), we should not lose sight of the ability to acquire new important information on how synthesis of Bchl-protein complexes is regulated by 'reopening' old areas of research. For example, the proposal of Bauer et al. (1988) that the *Q* gene product is the putative carrier of Bchl intermediates originally proposed by Lascelles and coworkers (Rebeiz and Lascelles 1982) highlights the significance of reinitiating the technically difficult studies on the enzymology and regulation of individual branches of the tetrapyrrole biosynthetic pathway. In addition, experiments to correlate the genotype of known *R. sphaeroides* and *R. capsulatus puf* operon mutants with their phenotype at the level of RNA and protein synthesis will yield valuable information regarding the function of specific *puf* operon sequences or gene products in synthesis or assembly of the individual Bchl-protein complexes.

The observation that mutations upstream of *pufB* affect synthesis of both the FPU and B800–850 complexes (Youvan et al. 1985, Bauer et al. 1988, Davis et al. 1988) underscores the fact that expression of the *puf* operon region is somehow connected to overall synthesis of Bchl-protein complexes. For example, although *Q*-deficient mutants synthesize reduced amounts of B800–850 complexes under low-oxygen or dark anaerobic inducing conditions using DMSO as an external electron acceptor, such strains contain more *puc*-specific mRNA than wild type strains (Davis et al. 1988). This observation is consistent with the proposal that the *Q* gene product is not required for transcription of the *puc* operon.

The ability to complement the B875⁻ phenotype of mutants such as

RS103 (Davis and Kaplan unpublished) and others (see above) with sequences upstream of *pufB* also suggests that loci exist (putative *R* gene product) which function somehow in regulating assembly of the B875 complex. Several of these B875 deficient strains are also derepressed for B800–850 complex synthesis at both the RNA and protein level, suggesting that cells respond either to the lack of B875 complexes or the resulting inefficiency of light energy transfer (Meinhardt et al. 1985) by increasing the size of the VPU (Kiley et al. 1988).

Finally, experiments with a *puhA* deletion strain have shown that these cells possess low levels of functional RC's, derepressed levels of B800–850 complexes, but no detectable B875 complex (Sockett and Kaplan unpublished). In addition, complementation analysis of this strain suggests that sequences closely linked to the *puh* operon may be indirectly involved in assembly of the FPU. These observations serve as the first indication that the *puh* operon genetic region may somehow affect expression of the *puf* operon or assembly of individual Bchl-protein complexes within the FPU. Given the information summarized in this review, it will be interesting to determine how *puf* operon expression and Bchl-protein complex synthesis is regulated in 'simple' organisms such as *R. rubrum* and *R. viridis* which contain a single LH complex; or in what promises to be a 'very complex' system such as *R. acidophila* which contains B875 and B800–850 complexes and a B800–820 complex which are synthesized specifically at low light intensities.

Acknowledgements

The authors wish to thank the members of our laboratories and others who have shared unpublished information with us to include in this review. Recent studies have been supported by Public Health Service grants GM15590 and GM31667 to S. Kaplan; USDA 85-CRCR-1-1809 to S. Kaplan, T.J. Donohue, and R. Gumport; and USDA Hatch Project WIS3028 to T.J. Donohue.

References

Allen JP, Feher G, Yeates TO, Rees DC, Deisenhofer J, Michel H and Huber R (1986) Structural homology of reaction centers from *Rhodopseudomonas sphaeroides* and *Rhodopseudomonas viridis* as determined by X-ray diffraction. Proc Natl Acad Sci USA 83: 8589–8593

Allen JP, Feher, G, Yeates TO, Komiya H and Rees DC (1987) Structure of the reaction center from *Rhodobacter sphaeroides* R-26: The protein subunits. Proc Natl Acad Sci USA 84: 6162–6166

Angerhofer A, Cogdell RJ and Hipkins MF (1986) A spectral characterisation of the light-harvesting pigment-protein complexes from *Rhodopseudomonas acidophila*. Biochim Biophys Acta 848: 333–341

Bauer CE, Carey J, Kasper LM, Lynn SP, Waechter DA and Gardner JF (1983) In: Beckwith J, Davies J and Gallant JA (eds.), Gene function in procaryotes, pp 65–89. Cold Spring Harbor Laboratory, Cold Spring Harbor, New York

Bauer CE, Eleuterio M, Young DA and Marrs BL (1987) Analysis of transcription through the *Rhodobacter capsulatus puf* operon using a translational fusion of *pufM* to the *E. coli lacZ* gene. In: Biggins J (ed.), Progress in photosynthesis research, vol. IV, pp 699–705. Martinus Nijhoff Publishers, Dordrecht, Netherlands

Bauer CE, Young DA and Marrs BL (1988) Analysis of the *Rhodobacter capsulatus puf* operon. Location of the oxygen-related promoter region and the identification of an additional puf-encoded gene. J Biol Chem 263: 4820–4827

Beatty JT, Adams CW and Cohen SN (1986) Regulation of expression of the *rxcA* operon of *Rhodopseudomonas capsulata*. In: Youvan DC and Daldal F (eds), Microbial Energy Transduction, pp 27–29. Cold Spring Harbor Press, Cold Spring Harbor, N.Y.

Bechhofer DH and Dubnau D (1987) Induced mRNA stability in *Bacillus subtilis*. Proc Natl Acad Sci USA 84: 498–502

Belasco JG, Beatty JT, Adams CW, Gabain AV and Cohen SN (1985) Differential expression of photosynthesis genes in *R. capsulata* results from segmental differences in stability within the polycistronic *rxcA* transcript. Cell 40: 171–181

Berard J, Bélanger G, Corriveau P and Gingras G (1986) Molecular cloning and sequence of the B880 holochrome genes from *Rhodospirillum rubrum*. J Biol Chem 261: 82–87

Biel AJ and Marrs BL (1983) Transcriptional regulation of several genes for bacteriochlorophyll biosynthesis in *Rhodopseudomonas capsulata* in response to oxygen. J Bacteriol 156: 686–694

Birkmann A, Sawers RG and Bock A (1987) Involvement of the *ntrA* gene product in the anaerobic metabolism of *Escherichia coli*. Mol Gen Genetics 210: 535–542

Broglie RM, Hunter CN, Delepelaire P, Niederman RA, Chua N-H and Clayton RK (1980) Isolation and characterization of the pigment-protein complexes of *Rhodopseudomas sphaeroides* by lithium dodecylsulfate polyacrylamide gel electrophoresis. Proc Natl Acad Sci USA 77: 87–91

Bylina EJ, Ismail S and Youvan DC (1986) Plasmid pU29, a vehicle for mutagenesis of the photosynthetic *puf* operon in *Rhodopseudomonas capsulata*. Plasmid 16: 175–181

Carter PW, Bartkus JM and Calvo JM (1986) Transcription attenuation in *Salmonella typhimurium*: The significance of rare leucine codons in the *leu* leader. Proc Natl Acad Sci USA 83: 8127–8131

Chamberlin MJ, Arndt KM, Briat JF, Reynolds RL and Schmidt MC (1987) Prokaryotic factors involved in termination at rho-independent termination sites, pp 347–356. In: Reznikoff WS, Burgess RR, Dahlberg JE, Gross CA, Record MT and Wickens M (eds). RNA Polymerase and the Regulation of Transcription. Elsevier Science Publishing Co., New York, N.Y.

Chang CH, Tiede D, Tang J, Smith U, Norris J and Schiffer M (1986) Structure of the *Rhodopseudomonas sphaeroides* R-26 reaction center. FEBS Lett 205: 82–86

Chory J and Kaplan S (1982) The *in vitro* transcription-translation of DNA and RNA templates by extracts of *Rhodopseudomonas sphaeroides*. J Biol Chem 257: 15110–15121

Cogdell RJ and Crofts AR (1978) Analysis of the pigment protein content of an antenna pigment-protein complex from three strains of *Rhodopseudomonas sphaeroides.* Biochim Biophys Acta 502: 409–416

Cox TM, O'Donnell MW, Aisen P and London IM (1985) Hemin inhibits internalization of transferrin by reticulocytes and promotes phosphorylation of the membrane transferrin receptor. Proc Natl Acad Sci USA 82: 5170–5174

Davis J, Donohue TJ and Kaplan S (1988) Construction, characterization and complementation of a puf⁻ mutant of *Rhodobacter sphaeroides.* J Bacteriol 170: 320–329

Deisenhofer J, Epp O, Miki K, Huber R and Michel H (1985) Structure of the protein subunits in the photosynthetic reaction centre of *Rhodopseudomonas viridis* at 3 Å resolution. Nature 318: 619–624

Deng X-W and Gruissem W (1987) Control of plastid gene expression during development: The limited role of transcriptional regulation. Cell 49: 379–387

Dierstein R (1984) Synthesis of pigment-binding protein in toluene-treated *Rhodopseudomonas capsulata* and in cell-free systems. Eur J Biochem 138: 509–518

Donohue TJ and Kaplan S (1986) Synthesis and assembly of bacterial photosynthetic membranes. In: Staehelin LA and Arnzten CJ (eds), Photosynthesis III: Photosynthetic Membranes. Encyclopedia of plant physiology, new series, vol. 19, pp 632–639, Springer-Verlag, N.Y.

Donohue, TJ, Hoger JH and Kaplan S (1986a) Cloning and expression of the *Rhodobacter sphaeroides* reaction center H gene. J Bacteriol 168: 953–961

Donohue TJ, McEwan AG and Kaplan S (1986b) Cloning, DNA sequence, and expression of the *Rhodobacter sphaeroides* cytochrome c_2 gene. J Bacteriol 168: 962–972

Drews G and Feick R (1978) Isolation and characterization of light-harvesting bacteriochlorophyll protein complexes from *Rhodopseudomonas capsulata.* Biochim Biophys Acta 501: 499–513

Feher G and Okamura MY (1978) Chemical composition and properties of reaction centers. In: Clayton RK and Sistrom WR (eds), The Photosynthetic Bacteria, pp 349–396. Plenum Publishing Corp., N.Y.

Greenblatt J, Horwitz RJ and Li J (1987) Genetic and structural analysis of an elongation control particle containing the N protein of bacteriophage lambda. In: Reznikoff WS, Burgess RR, Dahlberg JE, Gross CA, Record MT and Wickens M (eds), RNA Polymerase and Regulation of Transcription, Elsevier Science Publishing Co., New York, N.Y.

Hsu J-H, Harms E and Umbarger HE (1985) Leucine regulation of the *ilvGEDA* operon of *Serratia marcescens* by attenuation is modulated by a single leucine codon. J Bacteriol 164: 217–222

Imhoff JF, Truper HG and Pfennig N (1984) Rearrangement of the species and genera of the phototrophic 'purple nonsulphur bacteria.' Inter J System Bacteriol 34: 340–343

Jackson WJ, Prince RC, Stewart GJ and Marrs BL (1986) Energetic and topographic properties of a *Rhodopseudomonas capsulata* mutant deficient in the B870 complex. Biochemistry 25: 8440–8446

Jackson WJ, Kiley PJ, Haith CE, Kaplan S and Prince RC (1987) On the role of the light-harvesting B880 in the correct insertion of the reaction center of *Rhodobacter capsulatus* and *Rhodobacter sphaeroides.* FEBS Lett 215: 171–174

Jin DJ, Walter WA and Gross C (1988) Characterization of the termination phenotypes of rifampicin-resistant mutants of *Escherichia coli.* J Mol Biol (in press)

Kansy J and Kaplan S (1988) Purification, characterization and transcriptional analyses of RNA polymerase from *Rhodobacter sphaeroides* cells grown chemoheterotrophically and photoheterotrophically. J Biol Chem (in press)

Kaplan S (1981) Development of the membranes of photosynthetic bacteria. Photochem Photobiol 34: 769–774

Kiley PJ (1987) Ph.D. Thesis, University of Illinois

Kiley PJ, Donohue TJ, Havelka WA and Kaplan S (1987a) DNA sequence and in vitro expression of the B875 light-harvesting polypeptides of *Rhodobacter sphaeroides*. J Bacteriol 169: 742–750

Kiley PJ and Kaplan S (1987b) Cloning, DNA sequence and expression of the *Rhodobacter sphaeroides* light-harvesting B800–850-α and B800–850-β genes. J Bacteriol 169: 3268–3275

Kiley PJ and Kaplan S (1988) Molecular genetics of photosynthetic membrane biosynthesis in *Rhodobacter sphaeroides*. Microbiol Rev 52: 50–69

Kiley PJ, Varga A and Kaplan S (1988) Physiological and structural analysis of light harvesting mutants of *Rhodobacter sphaeroides*. J Bacteriol 170: 1103–1115

Klein RR and Mullet JE (1986) Regulation of chloroplast-encoded chlorophyll-binding protein translation during higher plant chloroplast biogenesis. J Biol Chem 261: 11138–11145

Klug G, Kaufmann N and Drews G (1985) Gene expression of pigment binding proteins of the bacterial photosynthetic apparatus: Transcription and assembly in the membrane of *Rhodopseudomas capsulata*. Proc Natl Acad Sci USA 82: 6485–6489

Klug G, Leibetanz R and Drews G (1986) The influence of bacteriochrlorophyll biosynthesis on formation of pigment-binding proteins and assembly of pigment protein complexes in *Rhodopseudomonas capsulata*. Arch Microbiol 146: 284–291

Klug G, Adams CW, Belasco J, Doerge B and Cohen SN (1987) Biological consequences of segmental alterations in mRNA stability: effects of deletion of the intercistronic hairpin loop region of the *R. capsulatus puf* operon. EMBO J 6: 3515–3520

Kranz RG and Haselkorn R (1986) Anaerobic regulation of nitrogen-fixation genes in *Rhodopseudomonas capsulata*. Proc Natl Acad Sci USA 83: 6805–6809

Lascelles J (1978) Regulation of pyrrole synthesis. In: Clayton RK and Sistrom WR (eds), The Photosynthetic Bacteria, pp 795–808. Plenum Publishing Corp., N.Y.

Marrs B (1974) Genetic recombination in *Rhodopseudomonas capsulata*. Proc Natl Acad Sci USA 71: 971–973

Marrs B (1981) Mobilization of the genes for photosynthesis from *Rhodopseudomonas capsulata* by a promiscuous plasmid. J Bacteriol 146: 1003–1012

McFall E (1986) *cis* Acting proteins. J Bacteriol 167: 429–432

Meinhardt SW, Kiley PJ, Kaplan S, Crofts AR and Harayama S (1985) Characterization of light-harvesting mutants of *Rhodopseudomonas sphaeroides* 1. Measurement of the efficiency of energy transfer from light-harvesting complexes to the reaction center. Arch Biochem Biophys 236: 130–139

Michel H, Epp O and Deisenhofer J (1986a) Pigment-protein interactions in the photosynthetic reaction centre from *Rhodopseudomonas viridis*. EMBO J 5: 2445–2451

Michel H, Weye KA, Gruenberg H, Donger I, Oseterhelt D and Lottspeich F (1986b) The 'light' and 'medium' subunits of the photosynthetic reaction centre from *Rhodopseudomonas viridis*: Isolation of the genes, nucleotide and amino acid sequence. EMBO J 5: 1149–1158

Michel H, Weyer KA, Gruenberg H and Lottspeich F (1985) The 'heavy' subunit of the photosynthetic reaction centre from *Rhodopseudomonas viridis*: isolation of the gene, nucleotide and amino acid sequence. EMBO Journal 4: 1667–1672

Monger TG and Parsons WW (1977) Singlet-triplet fusion in *Rhodopseudomonas sphaeroides* chromatophores: a probe of the organization of the photosynthetic apparatus. Biochim Biophys Acta 460: 393–407

Newbury SF, Smith NH, Robinson EC, Hiles ID and Higgins CF (1987) Stabilization of translationally active mRNA by prokaryotic REP sequences. Cell 48: 297–310

Pemberton JM and Harding CM (1986) Cloning of carotenoid biosynthesis genes from *Rhodopseudomonas sphaeroides*. Curr Microbiol 14: 25–29

Prentki P and Krisch HM (1984) In vitro insertional mutagenesis with a selectable DNA fragment. Gene 29: 303–313

Prince RC, Baccarini-Melandri A, Hauska GA, Melandri BA and Crofts AR (1975) Asymmetry of an energy transducing membrane: the location of cytochrome c_2 in *Rhodopseudomonas sphaeroides* and *Rhodopseudomonas capsulata*. Biochim Biophys Acta 387: 212–227

Ptashne M (1986) Gene regulation by proteins acting nearby and at a distance. Nature 322: 697–701

Rebeiz CA and Lascelles J (1982) Biosynthesis of pigments in plants and bacteria. In: Govindjee (ed.), Photosynthesis: energy conversion by plants and bacteria, vol. 1, pp 699–780. Academic Press Inc., N.Y.

Richardson JP, Ruteshouser EC and Chen C-YA (1987) Identification of upstream sequence components of rho-dependent transcription terminators. In: Reznikoff WS, Burgess RR, Dahlberg JE, Gross CA, Record MT and Wickens M (eds), RNA Polymerase and the Regulation of Transcription, pp 335–345, Elsevier Science Publishing Co., Inc., New York, N.Y.

Roberts GP and Brill WJ (1981) Genetics and regulation of nitrogen fixation. Ann Rev Microbiol 35: 207–235

Rudd KE and Menzel RE (1987) *his* operons of *Escherichia coli* and *Salmonella typhimurium* are regulated by DNA supercoiling. Proc Natl Acad Sci USA 84: 517–521

Scolnik PA and Marrs BL (1987) Genetic research with photosynthetic bacteria. Ann Rev Microbiol 41: 703–726

Shepherd WD and Kaplan S (1978) Changes in *Rhodopseudomonas sphaeroides* isoaccepting phenylalanyl-tRNA species during transitions from chemoheterotrophic to photoheterotrophic growth. Arch Microbiol 116: 161–167

Shepherd WD and Kaplan S (1983) Effect of cerulenin on macromolecule synthesis in chemoheterotrophically and photoheterotrophically grown *Rhodopseudomonas sphaeroides*. J Bacteriol 156: 1322–1331

Sistrom WR, Macalusa A and Pledger R (1984) Mutants of *Rhodopseudomonas sphaeroides* useful in genetic analysis. Arch Microbiol 138: 161–165

Stark W, Jay F and Muehlethaler K (1986) Localisation of reaction centre and light harvesting complexes in the photosynthetic unit of *Rhodopseudomonas viridis*. Arch Microbiol 146: 130–133

Tai TN, Havelka WA and Kaplan S (1988) A broad host range system for cloning and translational *lacZ* fusion analysis in *Rhodobacter sphaeroides*. Plasmid (in press)

Takemoto J and Lascelles J (1973) Coupling between bacteriochlorophyll and membrane protein synthesis in *Rhodopseudomonas sphaeroides*. Proc Natl Acad Sci USA 70: 799–803

Taylor D, Cohen SN, Clark WG and Marrs BL (1983) Alignment of genetic and restriction maps of the photosynthesis region of the *Rhodopseudomonas capsulata* chromosome by a conjugation-mediated marker rescue technique. J Bacteriol 154: 580–590

Theiler R, Suter F, Pennoyer JD, Zuber H and Niederman RA (1985) Complete amino acid sequence of the B875 light-harvesting protein of *Rhodopseudomonas sphaeroides* strain 2.4.1. FEBS Lett 184: 231–236

Theiler R, Suter F, Zuber H and Cogdell RJ (1984) A comparison of the primary structures of the two B800–850-apoproteins from wild-type *Rhodopseudomonas sphaeroides* strain 2.4.1 and a carotenoidless mutant strain R26.1. FEBS Lett 175: 231–237

Thireos G, Driscoll Penn M and Greer H (1984) 5′ untranslated sequences are required for translational control of a yeast regulatory gene. Proc Natl Acad Sci USA 81: 5096–5100

von Hippel PH, Yager TD, Bear DG, McSwiggen JA, Geiselmann J, Gill SC, Linn JD and Morgan WD (1987) Mechanistic aspects of transcript elongation and rho-dependent termination in *Escherichia coli*. In: Reznikoff WS, Burgess RR, Dahlberg JE, Gross CA, Record MT and Wickens M (eds), RNA Polymerase and the Regulation of Transcription, pp 325–334. Elsevier Science Publishing Co., New York, N.Y.

Werner M, Feller A, Messenguy F and Pierard A (1987) The leader peptide of yeast gene CPAI is essential for the translational repression of its expression. Cell 49: 805–813

Weyer KA, Lottspeich F, Gruenberg H, Lang F, Oesterhelt D and Michel H (1987) Amino acid sequence of the cytochrome subunit of the photosynthetic reaction centre from the purple bacterium *Rhodopseudomonas viridis*. EMBO J 6: 2197–2202

Williams JC, Steiner LA and Feher G (1986) Primary structure of the reaction center from *Rhodopseudomonas sphaeroides*. Proteins 1: 312–325

Williams JC, Steiner LA, Feher G and Simon MI (1984) Primary structure of the L subunit of the reaction center of *Rhodopseudomonas sphaeroides*. Proc Natl Acad Sci USA 81: 7303–7308

Williams JC, Steiner LA, Odgen RC, Simon MI and Feher G (1983) Primary structure of the M subunit of the reaction center from *Rhodopseudomonas sphaeroides*. Proc Natl Acad Sci USA 80: 6505–6509

Yen H-C and Marrs B (1976) Maps of genes for carotenoid and bacteriochlorophyll biosynthesis in *Rhodopseudomonas capsulata*. J Bacteriol 126: 619–629

Youvan DC, Ismail S and Bylina EJ (1985) Chromosomal deletion and plasmid complementation of the photosynthetic reaction center and light-harvesting genes from *Rhodopseudomas capsulata*. Gene 38: 19–30

Youvan DC, Bylina EJ, Alberti AM, Begusch H and Hearst JE (1984) Nucleotide and deduced polypeptide sequences of the photosynthetic reaction center, B870 antenna and flanking sequences from *R. capsulata*. Cell 37: 949–957

Zhu YS and Kaplan S (1985) Effects of light, oxygen and substrates on steady-state levels of mRNA coding for ribulose 1,5-bisphosphate carboxylase and light-harvesting and reaction center polypeptides in *Rhodopseudomonas sphaeroides*. J Bacteriol 162: 925–932

Zhu YS, Cook DW, Leach F, Armstrong GA, Alberti M and Hearst JE (1986a) Oxygen-regulated mRNAs for light-harvesting and reaction center complexes and for bacteriochlorophyll and carotenoid biosynthesis in *Rhodobacter capsulatus* during the shift from anaerobic to aerobic growth. J Bacteriol 168: 1180–1188

Zhu YS and Hearst J (1986b) Regulation of expression of genes for light-harvesting antenna proteins LHI and LHII; reaction center polypeptides RC-L, RC-M, and RC-H; and enzymes of bacteriochlorophyll and carotenoid biosynthesis in *Rhodobacter capsulatus* by light and oxygen. Proc Natl Acad Sci USA 83: 7613–7616

Zhu YS, Kiley PJ, Donohue TJ and Kaplan S (1986c) Origin of the mRNA stoichiometry of the *puf* operon in *Rhodobacter sphaeroides*. J Biol Chem 261: 10366–10374

Zsebo KM and Hearst JE (1984) Genetic-physical mapping of a photosynthetic gene cluster from *R. capsulata*. Cell 37: 937–947

Zuber H (1985) Structure and function of light-harvesting complexes and their polypeptides. Photochem Photobiol 42: 821–844

Govindjee et al. (eds), Molecular Biology of Photosynthesis: 161–194
© 1988 Kluwer Academic Publishers

Minireview

Characterization of phycobiliprotein and linker polypeptide genes in *Fremyella diplosiphon* and their regulated expression during complementary chromatic adaptation*

ARTHUR R. GROSSMAN,[1] PEGGY G. LEMAUX, PAMELA B.
CONLEY, BRIGITTE U. BRUNS & LAMONT K. ANDERSON
Carnegie Institution of Washington, 290 Panama Street, Stanford, California 94305, USA;
[1] *author for correspondence*

Received 20 August 1987; accepted 14 December 1987

Key words: phycobilisomes, phycobiliprotein, linker polypeptide genes, *Fremyella diplosiphon*, complementary chromatic adaptation

Abstract. Phycobilisomes, comprised of both chromophoric (phycobiliproteins) and non-chromophoric (linker polypeptides) proteins, are light-harvesting complexes present in the prokaryotic cyanobacteria and the eukaryotic red algae. Many cyanobacteria exhibit complementary chromatic adaptation, a process which enables these organisms to optimize absorption of prevalent wavelengths of light by altering the composition of the phycobilisome. To examine the mechanisms involved in adjusting the levels of phycobilisome components during complementary chromatic adaptation, we have isolated and sequenced genes encoding phycobiliprotein and linker polypeptides in the cyanobacterium *Fremyella diplosiphon*, analyzed their transcriptional characteristics (transcript sizes and abundance when *F. diplosiphon* is grown in different light qualities) and mapped transcript initiation and termination sites. Our results demonstrate that genes encoding phycobilisome components are often cotranscribed as polycistronic messenger RNAs. Light quality regulates the composition of the phycobilisome by causing changes in the abundance of transcripts encoding specific components, suggesting that regulation is at the level of transcription (although not eliminating the possibility of changes in mRNA stability). The work presented here sets the foundation for analyzing the evolution of the different phycobilisome components and exploring signal transduction from photoperception to activation of specific genes using in vivo and in vitro genetic technology.

Abbreviations: kDa – kilodalton, kb – kilobase pair, AP – allophycocyanin, PC – phycocyanin, PE – phycoerythrin, PEC – phycoerythrocyanin; an α or β given as a superscript to AP, PC or PE indicates specifically the α or β subunit of that biliprotein; $L_C^{7.9}$, $L_R^{9.7}$, L_{RC}^{31}, L_R^{35}, $L_R^{35.5}$, $L_R^{37.5}$ and L_R^{39} are the linker (L) polypeptides of the phycobilisome either located in the core (subscript C), the core-rod interface (subscript RC) or the rods (subscript R). Apparent molecular masses, in kDa, are indicated as superscripts to L.

Introduction

Phycobilisomes are accessory light-harvesting complexes found on thylakoid membranes of the prokaryotic cyanobacteria and the eukaryotic red

* C.I.W.–D.P.B. Publication No. 992

algae (Gantt 1981, Glazer 1982, 1984, 1985, Glazer et al. 1983, Zuber 1983, 1986). They are large structures (in cyanobacteria they are larger than ribosomes) that can contain 10–20 different proteins, depending on the organisms and culture conditions. Most of the phycobilisome mass consists of phycobiliproteins which are proteins that are covalently bonded with bilin chromophores. The types of bilins bound and the interactions between these bilins and the apoproteins produce four spectrally distinct classes of biliproteins, allophycocyanin (AP), phycocyanin (PC), phycoerythrin (PE) and phycoerythrocyanin (PEC). About 15% of the phycobilisome mass consists of non-pigmented linker proteins (Tandeau de Marsac and Cohen-Bazire 1977) which interact with the biliproteins forming discrete complexes that are the building blocks of the phycobilisome (Lundell et al. 1981). Phycobilisome function is a consequence of the spectral properties of different biliprotein-linker complexes and the spatial organization of the chromophores within the phycobilisome. As accessory antennae, the chromophore array in phycobilisomes extends the photosynthetic capacity of cyanobacteria and red algae by harvesting 550–650 nm light and funneling this energy into the reaction centers of photosynthesis, a process that is better than 90% efficient (Porter et al. 1978).

As an experimental system, phycobilisomes offer many biochemical advantages. They are abundant (phycobilisomes can be 40% of the cellular dry weight), easy to isolate, and their integrity can be assayed by fluorescence emission spectroscopy. Phycobilisome substructures can be generated by limited proteolysis or altering the ionic strength of the buffers used during the isolation of this light-harvesting complex. Individual phycobilisome proteins can be separated and reconstituted in vitro to form stable complexes. Reconstitution and dissociation products have been analyzed for structure, composition, spectral properties and assembly potential, providing the experimental foundation for models describing phycobilisome architecture and energy transfer through the complex (Yamanaka et al. 1980, 1982, Lundell et al. 1981a, b, Lundell and Glazer 1983a, b and c, Glazer et al. 1985, Anderson and Eiserling 1986). Since phycobilisomes are accessory pigments that are not essential for photosynthesis, phycobilisome mutants are easy to generate and maintain. Analyses of these mutants have played a major role in defining the structure, function and assembly of this light-harvesting complex (Williams et al. 1980, Gingrich et al. 1982, Cobley and Miranda 1983, Glazer et al. 1983, Anderson et al. 1984, Anderson et al. 1987).

Over the last decade detailed examination of phycobilisome structure has yielded much information concerning their architecture and the interactions among the different polypeptide and bilin constituents. All classes of biliproteins share the same subunit structure; dissimilar α and β subunits interact to form a protomer and subsequent assembly steps result in trimeric and hexameric complexes of the protomer. The hexamers have a disc-like structure (110–120 Å diameter and 60 Å height) and, with the linker proteins, assemble into stacks of discs (making cylinders) that are organized into the two structural domains of the phycobilisome, the core and the rods (Glazer et al. 1983). The core is comprised of AP biliproteins, is the site of thylakoid attachment, and contains the phycobilisome terminal energy acceptors (specialized AP biliproteins with 675–680 nm emission maxima) that must be energetically linked to membrane-bound chlorophyll complexes. In *F. diplosiphon* and many other cyanobacteria, the core is a triangular array of three cylinders, two of which are edge-on associated with the membrane; the third cylinder forms the apex of the core (Bryant et al. 1979, Glazer et al. 1983, Anderson and Eiserling 1986). The rods always contain PC biliproteins at the core-proximal positions and, depending on the organism and culture conditions, can have PE or PEC biliproteins in the core-distal regions (Gingrich et al. 1982, Glazer et al. 1983). Linker proteins are present in both the core and rod substructures. In the rods each biliprotein hexamer-disk is associated with one of the linker proteins that range in molecular masses from 27 to 39 kDa. These linkers have two functions: they mediate assembly of hexamers into rods and they modulate the spectral properties of their affiliated biliproptein hexamers, contributing to the highly efficient energy transfer that exists within the phycobilisome (Lundell et al. 1981b, Glazer et al. 1985). It is the rod biliproteins and their affiliated linkers that are regulated by light quality during chromatic adaptation (see below). The arrangement of polypeptides in the phycobilisome core is too complex to describe here, however, current models for this substructure are available in the literature (Lundell and Glazer 1983a, b and c, Anderson and Eiserling 1986).

Biliproteins have also been analyzed by X-ray crystallography. A preliminary characterization of PEC crystals from *Mastigocladus laminosus* has been reported (Rümbeli et al. 1985) and high resolution crystal structures have been solved for PC trimers from *M. laminosus* (Schirmer et al. 1985) and PC hexamers from *Agmenellum quadruplicatum* (Schirmer et al. 1986). PC trimers have three identical units ($\alpha\beta$) arranged around a three-fold symmetry axis to form a disc of 110 Å by 30 Å with a central channel of 35 Å in diameter. The individual subunits (both α and β) each have eight α-helices and irregular loops. Although analyses by Schirmer et al. (1985) demonstrated that the tertiary structure of six helices in the globular region of the PC subunits resemble the globin fold, a more recent examination of the amino acid sequences which compromise these similar structures suggest that the globin and PC gene families are not phylogenetically related (Bashford et al. 1987). The associated bilin chromophores are linear tetra-pyrroles that are in an extended conformation and appear to interact via their propionic acid side chains with arginine residues on the protein moieties. Interactions of the pyrrole nitrogen atoms with aspartate residues may also contribute to the stabilization of the chromophore orientation. An understanding of these interactions helps explain the absorptions properties of the phycobiliproteins, both in monomeric and aggregate states (Schirmer et al. 1986). Conserved residues involved in the interactions between α and β subunits and that may be important for trimer and hexamer formation have also been defined (Schirmer et al. 1986).

Recently, phycobiliprotein genes have been isolated from several organisms including the eukaryotic alga *Cyanophora paradoxa* (Bryant et al. 1985b, c, Lemaux and Grossman 1984, 1985) and the prokaryotic cyanobacteria *Synechococcus sp.* PCC 6301 (Houmard et al. 1986, Lind et al. 1985), *Synechococcus sp.* PCC 7002 (Bryant et al. 1987, 1985a, c, de Lorimier et al. 1984, Pilot and Fox 1984), *Synechocystis sp.* PCC 6701 (Anderson and Grossman 1987), *Anabaena sp.* PCC 7120 (Belknap and Haselkorn 1987), *Pseudanabaena sp.* PCC 7409 (Bryant et al. 1987, 1985a) and *Fremyella diplosiphon* (Conley et al. 1985, 1986, 1988, Mazel et al. 1986). Genes encoding linker polypeptides have also been isolated and sequenced (Belknap and Haselkorn 1987, Lomax et al. 1987). The phycobiliprotein gene set which has been most extensively characterized among different organisms encodes the PC subunits (gene designations *cpcA* and *cpcB* are for PC$^\alpha$ and PC$^\beta$ respectively). In some organisms such as *Anabaena sp.* PCC 7120 (Belknap and Haselkorn 1987) and *Synechococcus sp.* PCC 7002 (de Lorimier et al. 1984) there is a single set of genes encoding PC, while in others such as *Anacystis nidulans* (*Synechococcus sp.* PCC 6301/7942) and *Fremyella diplosiphon* (*Calothrix sp.* PCC 7601) there are at least two gene sets (Gustafsson, personal communication; Conley et al. 1986, 1987). The genes encoding phycobilisome components provide a system for the study of regulated gene expression in cyanobacteria since the biosynthesis of the different polypeptides of this complex are coordinately regulated and modulated by environmental cues. Phycobilisome biosynthesis may be modulated by nutrient status (Allen and Smith 1969, Ihlenfeldt and Gibson 1975, Lau et al. 1977, Lawry and Jensen 1979, 1986, Peterson et al. 1981, Stevens and Paone 1981), light quantity (Öquist 1974a, b, Lönneborg et al. 1985) and light quality (Bogorad 1975, Tandeau de Marsac 1977, Gendel and Bogorad 1979, Tandeau de Marsac 1983, Grossman et al. 1986). We are especially interested in the way that some cyanobacteria dramatically alter the polypeptide composition of the phycobilisome following changes in the wavelengths of light to which the organism is exposed, a process termed complementary chromatic adaptation (Bogorad 1975, Tandeau de Marsac 1983, Grossman et al. 1986). This process allows the cyanobacterium to more effectively absorb prevalent wavelengths of light in its environment. We are studying this phenomenon in the filamentous cyanobacterium *Fremyella diplosiphon* and our work on the sequence, organization and transcription of genes encoding phycobilisome constituents from this cyanobacterium is the main focus of this review. These analyses of chromatic adaptation at a molecular level are a prerequisite for work that we anticipate will define the molecular events involved in the light-wavelength-regulated expression of phycobilisome genes.

Complementary chromatic adaptation

The perception of light by photosynthetic organisms and processes involved in the transduction of radiant energy into the biochemical signals necessary for altering patterns of gene expression are crucial for developmental processes in both algae (Bennett and Bogorad 1973, Tandeau de Marsac 1983) and higher plants (Schäfer and Briggs 1986). The phenomenon of complementary chromatic adaptation offers the potential for analyzing photoperception in a relatively simple photosynthetic organism which can be genetically manipulated. Furthermore, the changes which accompany complementary chromatic adaptation are striking and involve dramatic shifts in the synthesis of the very abundant pigmented and nonpigmented components of the phycobilisome. Altered levels of phycobilisome components following a shift in light wavelength can be monitored by spectroscopic methods, at the level of protein composition of the isolated light-harvesting complex and at the level of transcript abundance. The latter may reflect both changes in the rate of transcription and mRNA degradation (these possibilities are discussed below).

The curves presented in Fig. 1A demonstrate the dramatic differences in the absorption characteristics of phycobilisomes from *F. diplosiphon* grown in red and green light. When grown in green light the cells have very high levels of PE with an absorption maximum at approximately 560 nm and relatively low levels of PC. Growth of *F. diplosiphon* in red light results in the massive accumulation of PC and very low levels (verging on the non-detectable) of PE. Under both growth conditions the level of AP remains relatively constant. Since limited action spectra reveal that the most effective wavelengths for causing the shift in synthesis from PC to PE and PE to PC are in green (540–550 nm) and red (640–650 nm) regions of the spectrum, respectively (Diakoff and Scheibe 1973, Haury and Bogorad 1977, Vogelmann and Scheibe 1978), the photoregulatory protein(s) is probably associated with bilin type chromophores, and may be encoded by a member of the phycobiliprotein gene family. Although certain AP species which exhibit photoreversible bleaching at the appropriate wavelengths were hypothesized to be the pigmented species responsible for complementary chromatic adaptation (Scheibe 1972, Björn and Björn 1976, Ohad et al. 1980), more recent results demonstrate that the photoreversible nature of the isolated AP is observed only after partial denaturation (Ohki and Fujita 1979a, b and 1981) and is not characteristic of the native molecules.

The polypeptide composition of *F. diplosiphon* phycobilisomes isolated from organisms grown in red and green light is presented in Fig. 1B. Differences in polypeptide profiles reflect changes in both the phycobiliprotein and linker polypeptide content of the organelle. For example, PE^{β} is present at high levels in cells grown in green light but is barely detectable in red light-grown cells. Furthermore, a distinct set of PC subunits, the red light-inducible PC subunits (not well resolved from other biliprotein subunits in Fig. 1B), is not detected after growth of the cyanobacterium in green light but is present after growth in red light. Two different sets of PC subunits, one red light-inducible (PC_i) and the other constitutive (PC_c), have

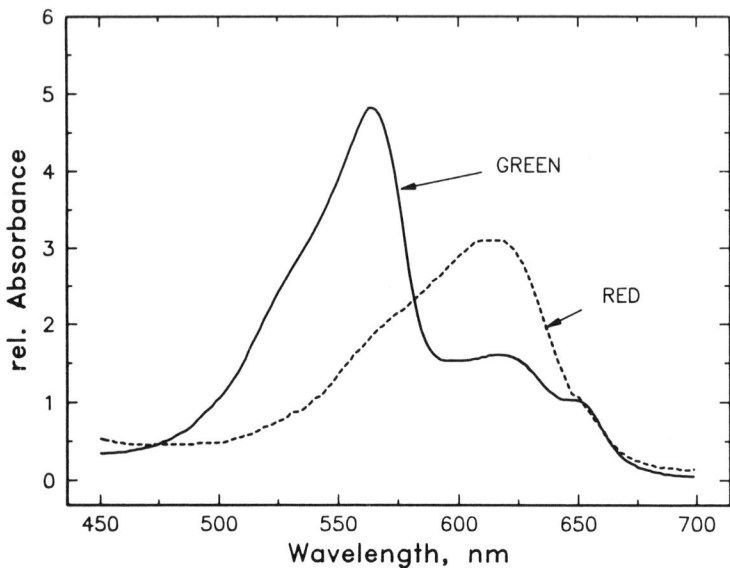

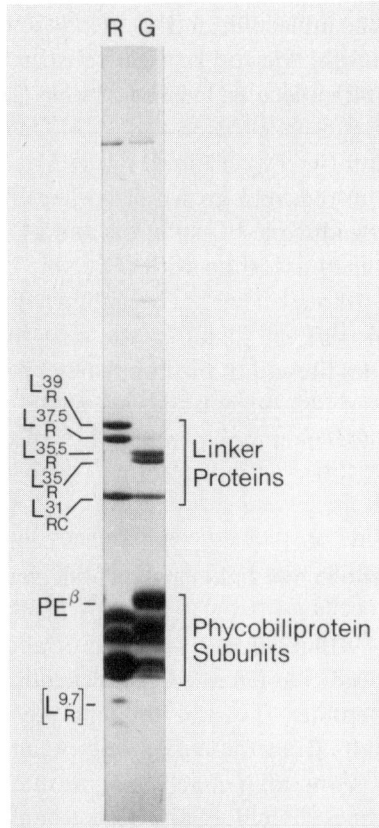

Fig. 1A. Absorption spectra of cell extracts from *Fremyella diplosiphon* grown in red and green light. *Fremyella diplosiphon* was grown at 32°C in continuous red or green light to mid-exponential phase. The cells were frozen in liquid nitrogen and treated with lysozyme (5 μg/ml final concentration, 30 min) at room temperature. Cell membranes and debris were removed by centrifugation at 14 000 × g and absorbance of the supernatant, containing the phycobili-proteins, was measured by scanning from 450 nm to 700 nm using a Cary 1756 spectro-photometer. The two curves were normalized to the absorbance of allophycocyanin.

Fig. 1B. SDS-PAGE of *Fremyella diplosiphon* phycobilisomes isolated from cells grown in red (R) or green (G) light. The name of each linker polypeptide, L (Glazer, 1985), is indicated on the left of the figure. The subscripts R and RC denote the positions of the linkers in either the rod substructure or at the rod-core interface, respectively. The superscripts denote their apparent molecular masses in kDa. The β subunit of PE (PE$^\beta$) is also indicated. L$_R^{9.7}$ was placed in brackets since we are uncertain of the position of this polypeptide in the polyacrylamide gel.

previously been observed in the complementary chromatically adapting organisms *Pseudanabaena sp.* PCC 7409 and *F. diplosiphon* (Bryant 1981, Bryant and Cohen-Bazire 1981). Five rod linker polypeptides over 30 kDa in molecular weight are resolved on the polyacrylamide gel presented in Fig. 1B. The linker polypeptides are named according to Glazer (1985) where the superscripts designate the molecular masses in kDa and the subscripts, R or C, designate the location in the rod or core substructures, respectively. A linker polypeptide with a molecular mass of 31 kDa (L_{RC}^{31}) is present in cells grown in red or green light while the 35 and 35.5 kDa linkers (L_R^{35} and $L_R^{35.5}$) are abundant in green light-grown cells and the 37.5 and 39 kDa linkers ($L_R^{37.5}$ and L_R^{39}) are abundant in red light-grown cells. The coordinate appearance of $L_R^{37.5}$ and L_R^{39} with the inducible PC subunits and L_R^{35} and $L_R^{35.5}$ with the PE subunits suggests that they associate with inducible PC hexamers and PE hexamers, respectively (although it has not been shown that hexamers containing solely PC_i or PC_c exist in red light grown cells in vivo). Lower molecular weight polypeptides present in phycobilisomes isolated from red light-grown cells but not green light-grown cells are also evident in Fig. 1B. One of these is probably the linker polypeptide ($L_R^{9.7}$) involved in placing the terminal PC trimer onto the rod substructure (see below).

Isolation of phycobiliprotein and linker polypeptide genes

To study the process of complementary chromatic adaptation, at the molecular level, we isolated and characterized genes encoding both phycobiliproteins and linker polypeptides. Toward this end we isolated the gene set encoding the PC subunits from the eukaryotic organism *Cyanophora paradoxa*. In *C. paradoxa* (Grossman et al. 1983), as in other eukaryotic algae containing phycobilisomes (Egelhoff and Grossman 1983), phycobiliproteins are translated in the plastid and therefore probably encoded on the plastid DNA. Since the plastid genome is only approximately 130 kb (Löffelhardt et al. 1983, Lambert et al. 1985) and readily separated from nuclear DNA, the isolation of phycobiliprotein gene sequences from *C. paradoxa* is much easier than the isolation of analogous sequences from cyanobacteria (where genomes are 5000 kb or more (Herdman 1982)). Since the individual phycobiliprotein classes are highly conserved between eukaryotic and prokaryotic species, it was likely that a biliprotein gene from one organism could be used to clone analogous sequences from other organisms. To identify clones encoding phycobiliprotein polypeptides we screened 300 *E. coli* transformants carrying *Sau*3A fragments of *C. paradoxa* plastid DNA in pUC8 with antibodies raised against a mixture of PC and AP subunits from the red alga *Porphyridium aeruginosum*. Four of the colonies exhibited a positive immunological reaction. The insert DNA present in these four clones overlapped and DNA sequence analysis revealed the presence of the genes encoding PC^α and PC^β (gene designations *cpcA* and *cpcB*, respectively) contiguous on the genome (Lemaux and Grossman 1984, 1985). All biliproteins are related and the PC subunits exhibit approximately 30% sequence

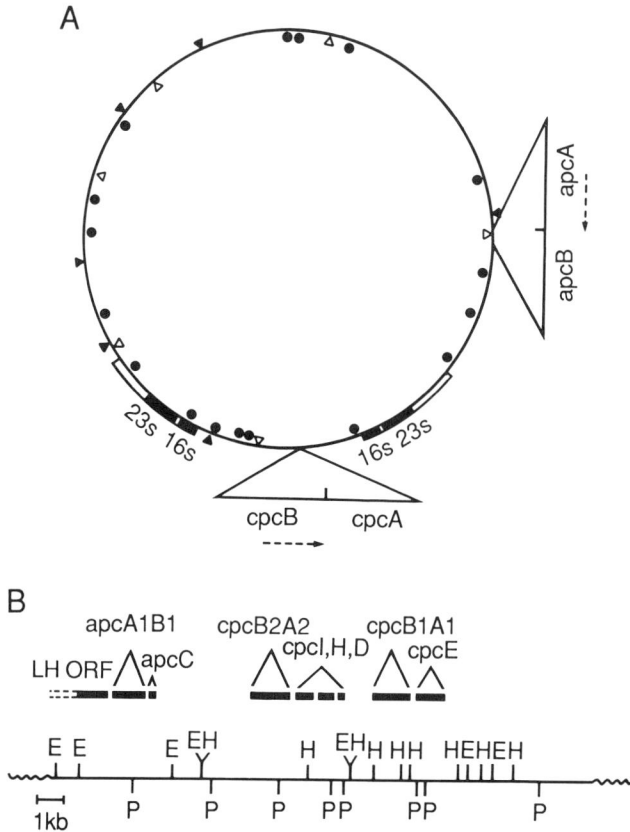

Fig. 2A. Location and direction of transcription of the phycobiliprotein genes on the plastid genome of *Cyanophora paradoxa*. The plastid DNA map has been redrawn from Kuntz et al. (1984) and Bohnert et al. (1983). The bracketed areas at the bottom of the map represent the inverted repeat regions containing the 16S and 23S rDNA cistrons. Symbols are (△), inner circle, *Sal*I; (▲), outer circle, *Bam*HI; (●), *Xho*I (Lemaux and Grossman 1985). Arrows indicate the direction of transcription from the *cpc* and *apc* gene sets.

Fig. 2B. Restriction endonuclease map of *F. diplosiphon* chromosomal DNA containing a cluster of genes encoding phycobilisome components (Conley et al. 1986). The filled boxes delineate the coding regions of the individual genes. Not all of the *Pst*I sites are indicated on the map. Abbreviations: H – *Hind*III, P – *Pst*I, E – *Eco*RI, ORF – open reading frame, LH – linker homology. Gene designations are described in the text.

homology with the AP subunits, with greater homologies in specific regions of the polypeptides (compare sequences in Fig. 3A). This allowed us to use the *cpc* gene set as a hybridization probe under conditions of moderate to low stringencies (allowing 35–45% mismatch) to locate the genes encoding AP^α and AP^β (gene designations *apcA* and *apcB*, respectively), also contiguous on the *C. paradoxa* plastid genome. Mapping of the phycobiliprotein genes to the plastid genome placed *cpcBA* in the small single copy region and *apcAB* approximately 90° away in the large single copy region (Lemaux

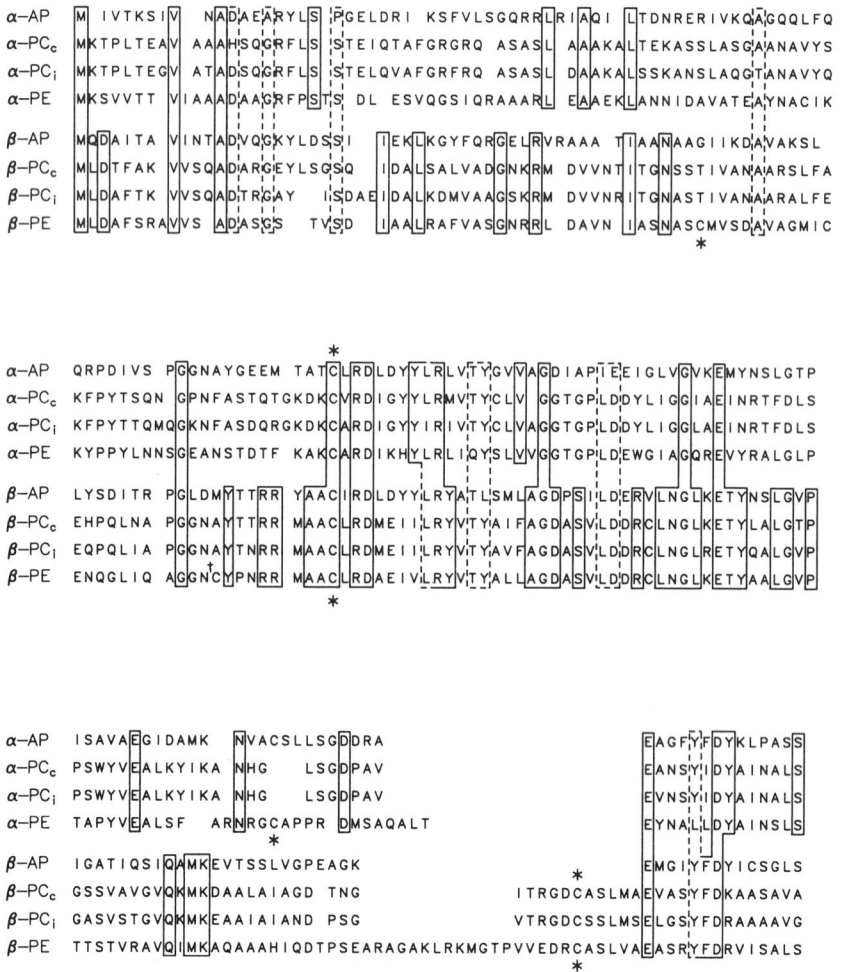

```
α–AP    M IVTKSIV  NADAEARYLS PGELDRI KSFVLSGQRRLRIAQI LTDNRERIVKQAGQQLFQ
α–PCc   MKTPLTEAV AAAHSQGRFLS STEIQTAFGRGRQ ASASL AAAKALTEKASSLASGANAVYS
α–PCi   MKTPLTEGV ATADSQGRFLS STELQVAFGRFRQ ASASL DAAKALSSKANSLAQGTANAVYQ
α–PE    MKSVVTT VIAAADAAGRFPSTS DL ESVQGSIQRAAARL EAAEKLANNIDAVATEAYNACIK

β–AP    MQDAITA VINTADVQGKYLDSSI  IEKLKGYFQRGELRVRAAA TIAANAAGIIKDAVAKSL
β–PCc   MLDTFAK VVSQADARGEYLSGSQ  IDALSALVADGNKRM DVVNTITGNSSTIVANAARSLFA
β–PCi   MLDAFTK VVSQADTRGAY  ISDAEIDALKDMVAAGSKRM DVVNRITGNASTIVANAARALFE
β–PE    MLDAFSRAVVS ADASGS  TVSD LIAALRAFVASGNRRL DAVN IIASNASCMVSDAVAGMIC
                                                                    *

                                      *
α–AP    QRPDIVS PGGNAYGEEM TATCLRDLDYYLRLVTYGVVAGDIAPIEEIGLVGVKEMYNSLGTP
α–PCc   KFPYTSQN GPNFASTQTGKDKCVRDIGYYLRMVTYCLV GGTGPLDDYLIGGIAEINRTFDLS
α–PCi   KFPYTTQMQGKNFASDQRGKDKCARDIGYYIRIVTYCLVAGGTGPLDDYLIGGLAEINRTFDLS
α–PE    KYPPYLNNSGEANSTDTF KAKCARDIKHYLRLIQYSLVVGGTGPLDEWGIAGQREVYRALGLP

β–AP    LYSDITR PGLDMYTTRR YAACIRDLDYYLRYATLSMLAGDPSIILDERVLNGLKETYNSLGVP
β–PCc   EHPQLNA PGGNAYTTRR MAACLRDMEII LRYVTYAIFAGDASVLDDRCLNGLKETYLALGTP
β–PCi   EQPQLIA PGGNAYTNRR MAACLRDMEII LRYVTYAVFAGDASVLDDRCLNGLRETYQALGVP
β–PE    ENQGLIQ AGGNCYPNRR MAACLRDAEIVLRYVTYALLAGDASVLDDRCLNGLKETYAALGVP
                            *

α–AP    ISAVAEGIDAMK NVACSLLSGDDRA              EAGFYFDYKLPASS
α–PCc   PSWYVEALKYIKA NHG  LSGDPAV              EANSYIDYAINALS
α–PCi   PSWYVEALKYIKA NHG  LSGDPAV              EVNSYIDYAINALS
α–PE    TAPYVEALSF  ARNRGCAPPR DMSAQALT        EYNALLDYAINSLS
                                    *
β–AP    IGATIQSIQAMKEVTSSLVGPEAGK              EMGIYFDYICSGLS
β–PCc   GSSVAVGVQKMKDAALAIAGD TNG       ITRGDCASLMAEVASYFDKAASAVA
β–PCi   GASVSTGVQKMKEAAIAIAND PSG       VTRGDCSSLMSELGSYFDRAAAAVG
β–PE    TTSTVRAVQIMKAQAAAHIQDTPSEARAGAKLRKMGTPVVEDRCASLVAEASRYFDRVISALS
                                          *                  *
```

Fig. 3A. Comparisons of amino acid sequences, as deduced from nucleotide sequences, of *Fremyella diplosiphon* phycobiliproteins. The α subunits are displayed on the top portion of each section, β subunits on the bottom. Optimal alignments of α and β phycobiliprotein subunits were done using the FastP program (Lipman and Pearson 1985) and manual analysis. The solid boxes indicate residues conserved among all eight phycobiliprotein subunits or residues conserved among either all α or all β subunits. The dashed boxes indicate positions where seven of eight amino acid residues are identical. An asterisk (*) above or below a cysteine (C) residue indicates a bilin chromophore attachment site. Amino acids are designated with the standard single letter code. The amino acid at position 70 of PE[β], N[†], is γ-N-methylasparagine (Rümbeli et al. 1987).

B.

$L_R^{37.5}$ MTSSTAARQLGFEPFASTAPTELRAS   SDVPAVIHAAYRQVFGNDHVMQSERLTSAESLLQQGNISVRDFVRLQ
$L_R^{39.0}$ PITSAASRLGTTAYQ TNPIELRPNWTAEDAKIVIQAVYRQVLGNDYLMQSERLTSLESLLTNGKLSVRDFVRAK

$L_R^{37.5}$ SELYRQKFFYSTPQVRFIELNYKHLLGRAPYDESEISYHVNLYTEKGYEAEINSYIDSAEYQESFGERIVPHYRGF
$L_R^{39.0}$ SELYKRKFLYPHFQTRVIELNFKHLLGRAPYDESEVIEHLDRYQNQGFDADIDSYIDSAEYDTYFGDSIVPYYRDL

$L_R^{37.5}$ ETQPGQKTVGFNRMFQIYRGYANSDRSQGKNKSAWLTQDLALNLASNIQTPNFG   KGLTG
$L_R^{39.0}$ VTTGVGQRTVGFTRMFRLYRGYANSDRSQLAGSSSRLASDLATNSATAIIAPSGGTQGWSYLPSKQGTAPSRTFGR
$L_R^{9.7}$ MLGSVLTRR

$L_R^{37.5}$ VVAGDRGQLYRVRVIQA   DRGRTTQIRRSIQEYLV   SYDQLSPTLQRLNQRGSRVVNISPA
$L_R^{39.0}$ SSQGSTPRLYRIEVTGI   SLPRYPKVRRSNKEFIV   PYEQLSSTLQQINKLGGKVASITFAQ
$L_R^{9.7}$ SSSGSDNRVFVYEVEGLRQNEQTDNNRYQIRNSSTIEIQV   PYSRMNEEDRRITRLGGRIVNIRPAGENPTED
$L_C^{7.9}$ MGRLFKVTACVP   SQTRIRTQRELQNTYFTKLVPFENWFREQQRIMKMGGKIVKVELATGKQGTNTGL

Fig. 3B. Comparisons of amino acid sequences, as deduced from the nucleotide sequences, for three PC-associated rod linkers ($L_R^{37.5}$, L_R^{39} and $L_R^{9.7}$) and the small core linker ($L_C^{7.9}$). Identical amino acids are indicated by a colon (:) and conserved substitutions by a period (.). Amino acid designations are as in (A). Boxed residues indicate homologous sequences among all four proteins, deduced from analyses using the FastP program.

and Grossman 1984, 1985, Lambert et al. 1985). A map showing the positions of the phycobiliprotein genes on the plastid genome of *C. paradoxa* and their direction of transcription is presented in Fig. 2A.

Analyses of transcripts encoding the PC and AP subunits of *C. paradoxa* demonstrate that the genes encoding the α and β subunits for each of these biliproteins are transcribed together as a dicistronic mRNA (Lemaux and Grossman 1985). A similar situation has been observed for transcription of biliprotein gene sets from several different organisms (Belknap and Haselkorn 1986, Conley et al. 1985, 1986, Mazel et al. 1986, Pilot and Fox 1984). Cotranscription of genes encoding dissimilar subunits of a specific phycobiliprotein class will result in equal levels of RNA sequences for these subunits, and if efficiencies of translation of these RNA sequences and post-translational modifications (e.g., chromophore attachment) of the resulting apoproteins are similar, a 1:1 ratio of the polypeptide subunits (this is the ratio observed in the phycobilisome) will be ensured.

Since we were interested in studying complementary chromatic adaptation, a process not observed in *C. paradoxa*, we isolated phycobiliprotein genes from the cyanobacterium *Fremyella diplosiphon*, an organism in which complementary chromatic adaptation has been partially characterized (Bennett and Bogorad 1973, Gendel et al. 1978, Vogelmann and Scheibe 1978, Haury and Bogorad 1977). We used the *C. paradoxa* gene encoding PC$^\beta$ as a heterologous probe to screen a library of *F. diplosiphon* DNA in λ EMBL3.

Two of the isolated clones (clone 37 and 4–10) were overlapping and extensive characterization of these clones indicated that they encoded two different sets of PC subunits, a single set of AP subunits and at least four linker polypeptides. A map of this region of the *F. diplosiphon* genome showing the relative positions of the different genes is presented in Fig. 2B. The two *cpc* gene sets are contiguous on the cyanobacterial genome. One set of *cpc* genes is expressed during growth of *F. diplosiphon* in both red and green light (*cpcB1A1* encoding PC_c^β and PC_c^α, constitutive PC subunits) and the other set is expressed at high levels only after growth of *F. diplosiphon* in red light (*cpcB2A2* encoding PC_i^β and PC_i^α, inducible PC subunits). As indicated on the map, the *apc* gene set is located approximately 3 kb upstream from *cpcB2A2*. Genes encoding linker polypeptides are located downstream from both *apcB1* and *cpcA2*. The gene (*apcC*) encoding the small linker polypeptide $L_C^{7.9}$, which is localized to the core substructure of the phycobilisome, is immediately downstream of *apcB1* and cotranscribed with the *apcA1B1* gene set (see below), while genes (*cpcI, cpcH* and *cpcD*, respectively) encoding the three linker polypeptides $L_R^{37.5}$, L_R^{39}, and $L_R^{9.7}$, which probably associate with PC subunits, are downstream from, and cotranscribed with the inducible *cpcB2A2* gene set as a polycistronic mRNA only detected in cells maintained in red light (Lomax et al. 1987) (see below). The clustering of phycobiliprotein and linker polypeptide genes in *F. diplosiphon* and other cyanobacteria contrasts with the situation observed in eukaryotic algae. In *Cyanophora paradoxa* (Grossman et al. 1983) and certain red algae (Egelhoff and Grossman 1983) the linker polypeptides (except for the anchor protein) appear to be synthesized on 80S cytoplasmic ribosomes suggesting that they are encoded by the nuclear genome rather than the plastid genome. This is further supported by experiments in which no homology was obtained (using Southern analyses) between linker genes from *F. diplosiphon* and the plastid DNA of *C. paradoxa*. Therefore, in the context of the endosymbiont theory of evolution (Gray and Doolittle 1982), the genes encoding the linker polypeptides had to be moved from the genome of the invading endosymbiont to the nuclear genome of the host organism during the course of evolution. Furthermore, some communication between the two genomes may exist to ensure coordinate synthesis of the different phycobilisome constituents in *C. paradoxa*.

Several other λ clones containing *F. diplosiphon* DNA which encode phycobilisome components have been isolated and analyses of the phycobiliprotein and linker polypeptide gene sequences have begun. A list of some of these clones and the characteristics (genes encoded on the clones, complete or partial sequence, expression in red and green light, mapping of 3' and 5' ends of the transcripts) of the genes on them is given in Table 1.

Table 1. Isolated genes encoding phycobilisome polypeptides and their characteristics from *F. diplosiphon*

Protein	Gene name	Clone designations	Sequence status	Transcriptional properties
$PC_i^{\beta,\alpha}$	*cpcB2, cpcA2*	Clone 37	Complete sequence	Polycistronic mRNA, inducible, same operon as linker genes 5′ and 3′ ends mapped
$PC_c^{\beta,\alpha}$	*cpcB1, cpcA1*	Clone 37	Complete sequence	Dicistronic mRNA, constitutive, 5′ and 3′ ends mapped
$PC^{\beta,\alpha}$	*cpcB3, cpcA3*	Clone 11	Partial sequence	Uncharacterized
Linkers contiguous to *cpcB3A3*	*cpcL, cpcM*	Clone 11	Complete sequence of one	Uncharacterized
$PE^{\beta,\alpha}$	*cpeB, cpeA*	Clone 3	Complete sequence	Dicistronic mRNA, inducible, 5′ end mapped
$AP^{\alpha,\beta}$	*apcA1, apcB1*	Clone 37	Complete sequence	Polycistronic mRNA, constitutive same operon as core linker gene, 5′ end mapped
7.9 kDa core linker	*apc*	Clone 37	Complete sequence	Polycistronic mRNA, constitutive, same operon as AP
Rod linkers associated with PC	*cpcH, cpcI, cpcD*	Clone 37	Complete sequences	On polycistronic mRNA with sequences encoding $PC_i^{\beta,\alpha}$
30 kDa ORF	*cpcE*	Clone 4–10	Complete sequence	Immediately downstream of *cpcB1A1.* No transcript detected.

PC_i = inducible PC;
PC_c = constitutive PC;
ORF = open reading frame;
Several other λ clones overlapping with those noted in the table have been partially characterized.

In addition to the phycobiliprotein-linker gene cluster shown in Fig. 2B and discussed above, clone 3 encodes the *cpe* gene set and clone 11 encodes a 'cpc-like' gene set (*cpcB3A3*) plus genes encoding linker polypeptides. The *cpe* gene set has also been completely sequenced (by us and others, Mazel et al. 1986) and the 5′ end of the transcript mapped (see below). No genes encoding linker sequences (based on low stringency hybridizations with probes constructed from the linker genes located downstream from *cpcB2A2*) are adjacent to the *cpe* gene set. Therefore, unlike the situation established for genes encoding the inducible PC subunits and associated linkers, the linkers which interact with PE hexamers are probably encoded by genes having their own promoter(s). Recently several λ clones containing sequences which share homology with known linker polypeptide genes have been isolated from *F. diplosiphon.* Differential hybridization of RNA and cDNA prepared from red and green light-grown *F. diplosiphon* is being used to identify the linker genes whose products accumulate in coordination with the PE subunits. (This approach assumes that the levels of transcripts from these genes correlate with the levels of the gene products, as observed for the red light-inducible PC linkers and the inducible PE subunits.) No linkage

A.

```
                    10        20        30        40        50        60
Agmenellum      F I R     A  EF  SDKLE   KV  E T  S A S M N   S T   Q
Anabaena        MT  V     S  EFL NEQL  ANV KE N   L       S   A T
Mastigocladus   AY V      S  EFL NEQL  ANV KE N   L       S       T
Cyanidium          A  AA  A  EFL NTQLN  SK  SE N   L       S   A T

Inducible       MLDAFTKVVSQADTRGAYISDAEIDALKDMVAAGSKRMDVVNRITGNASTIVANAARALF
                ::::.:.:::::::.::.:.:..:::...::.:.::::::::::::::.:::::::::.::
Constitutive    MLDTFAKVVSQADARGEYLSGSQIDALSALVADGNKRMDVVNRITGNSSTIVANAARSLF
                    10        20        30        40        50        60


                    70        80        90       100       110       120
Agmenellum      AD                             T T     N           V
Anabaena                                       IL      I           Y
Mastigocladus              S TR GT             I  IL    I
Cyanidium       S     Q    T  D                 S  II    S I

Inducible       EEQPQLIAPGGNAYTNRRMAACLRDMEIILRYVTYAVFAGDASVLDDRCLNGLRETYQAL
                .:.::: :::::::.::::::::::::.::.::::.:.:::::::::::::::::.:::.::
Constitutive    AEHPQLNAPGGNAYTTRRMAACLRDMEIILRYVTYAIFAGDASVLDDRCLNGLKETYLAL
                    70        80        90       100       110       120


                   130       140       150       160       170
Agmenellum      AA   RA GK  V  VM PA  S       QQ IEL  FT  K  E
Anabaena        T S  AV     D  VG     N I K   Q I  VA       -
Mastigocladus   T S  AV I      N      N I K   A I  VA           A
Cyanidium            AV IE   DS        I T     A  AV T      T  Q

Inducible       GVPGASVSTGVQKMKEAAIAIANDPSGVTRGDCSSLMSELGSYFDRAAAAVG
                :.::.::..:::::::.::.:::::...:.:::::.:::.:..:::::.::.::.
Constitutive    GTPGSSVAVGVQKMKDAALAIAGDTNGITRGDCASLMAEVASYFDKAASAVA
                   130       140       150       160       170
```

B.

```
                    10        20        30        40        50        60
Agmenellum        A  L       N   YLY  L  GAFA E  QT TA  DT VN A Q
Anabaena        MV  I  AI A  T   GN   S R  YER A  E  RG T N QR ID ATQ
Mastigocladus   V   I DAI A  T   N   AVN  YQR A  E  R  TAN QR ID A Q
Cyanidium           I AI A  N    N   AVN  YQR A  E  RS T N ER IN A Q

Inducible       MKTPLTEGVATADSQGRFLSSTELQVAFGRFRQASASLDAAKALSSKANSLAQGTANAVY
                :::::::.:.:::.:::::::::.::.:::::::::::::.:::::...:: :.:::::
Constitutive    MKTPLTEAVAAAHSQGRFLSSTEIQTAFGRFRQASASLAAAKALTEKASSLASGAANAVY
                    10        20        30        40        50        60


                    70        80        90       100       110       120
Agmenellum      S    STP N   A              L M  C       M E     A VD
Anabaena             TP PQ  A S    S     V H L  I   S       E     A      S
Mastigocladus        LI TS P YA A   S       H L  I   S       E     A N   D
Cyanidium       S    LI TS PQY  SAV A       L M     V       M E   A E

Inducible       QKFPYTTQMQGKNFASDQRGKDKCARDIGYYIRIVTYCLVAGGTGPLDDYLIGGLAEINR
                ::::::.: .: :::::.: :::::.::::::.:.::::: :::::::::::::::.::::
Constitutive    SKFPYTSQ-NGPNFASTQTGKDKCVRDIGYYLRMVTYCLV-GGTGPLDDYLIGGIAEINR
                    70        80        90       100       110


                   130       140       150       160
Agmenellum                   -H      T  A T T N
Anabaena                     H       QA N A T
Mastigocladus   A E     I            QA N A T      V
Cyanidium                    N       QA N A T

Inducible       TFDLSPSWYVEALKYIKANHGLSGDPAVEVNSYIDYAINALS
                :::::::::::::::::::::::::::::.:::::::::::::
Constitutive    TFDLSPSWYVEALKYIKANHGLSGDPAVEANSYIDYAINALS
                120       130       140       150       160
```

has been established between the *cpe* gene set and the *cpc-apc* gene cluster, based on the characterization of several λ clones encoding phycobiliprotein subunits. However, no systematic attempt has been made to use 3' and 5' regions of characterized λ clones toward this end. Like the *cpe* gene set, *cpcB3A3* on clone 11 has not been linked to genes encoding other phycobiliproteins. The polypeptide sequences of the 'PC-like' subunits encoded on clone 11 (sequences of the genes are nearly complete), as derived from nucleotide sequences, also share homology with PEC. Genes encoding linker polypeptides are also present on clone 11, although their positions relative to *cpcB3A3* have not been established. One of these linker genes has been completely sequenced and the derived amino acid sequence is similar to that of $L_R^{37.5}$ and L_R^{39} (greater than 50% homology) with relatively strong homology over the entire sequence. Generally, homologies amongst all the linkers characterized in *F. diplosiphon* are strongest at the carboxyl termini (Fig. 4B), although the amino acid sequenced of $L_R^{37.5}$ and L_R^{39} are most conserved in the central regions of the polypeptides.

Characterization of the arrangement of phycobiliprotein and linker polypeptide genes on the genome of *F. diplosiphon* have raised several interesting points and suggest additional experiments. Examination of the region of genomic DNA encoding the phycobiliprotein-linker polypeptide gene cluster could be extended by using 5' and 3' specific probes to isolate segments of genomic DNA linked to this cluster. The fact that the gene cluster extends 5' from the cloned sequences is indicated with the identification of a gene upstream from the *apcA1B1* genes whose translation product (based on nucleotide sequence) shares homologies with linker polypeptides. This gene, which is interrupted at the cloning site, is transcribed into a major 3100 base RNA. Close to the 3' end of the cluster and immediately downstream from *cpcB1A1* is a gene for a 30 kDa ORF, similarly located in both *Synechococcus sp.* PCC 7002 (Bryant et al. 1987, gene designation *cpcE*) and *Anabaena sp.* PCC 7120 (Belknap and Haselkorn 1987). The function of the 30 kDa polypeptide in phycobilisome biosynthesis is still unclear although based on transcript analyses Belknap and Haselkorn have suggested that this polypeptide is a linker. However, we find no homology between this sequence and that of other linker polypeptides of *F. diplosiphon*, and unlike all of the

Fig. 4. Comparisons of the amino acid sequences of two PC$^\alpha$ and PC$^\beta$ subunits of *Fremyella diplosiphon* with each other, and with amino acid sequences of PC subunits from other species. Sequences of PC$^\beta$ (A) and PC$^\alpha$ (B) are compared to analogous sequences from *Agmenellum quadruplicatum* (de Lorimier et al. 1984, Pilot and Fox 1984), *Anabaena sp.* PCC 7120 (Belknap and Haselkorn 1987), *Mastigocladus laminosus* (Sidler et al. 1986) and *Cyanidium caldarium* (Offner et al. 1981, Troxler et al. 1981). Amino acid abbreviations and homology designations are as in Fig. 3. For the other cyanobacteria, only those residues which differ from the residues at that position of the *Fremyella diplosiphon* inducible PC subunit are shown.

rod linker polypeptides characterized thusfar, it has an acidic isoelectric point (Lomax et al. 1987). Furthermore, in *F. diplosiphon* transcripts from this ORF have not been detected using probes labelled to high specific activities, although the coding region of the transcript is preceded by a strong ribosome binding site. These findings suggest that the 30 kDa ORF is not a linker. However, its proximity to the constitutive *cpcB1A1* gene set in the three organisms for which it has been analyzed thusfar, and its cotranscription with the *cpc* gene set in *Anabaena sp.* PCC 7120, suggest that it may play a role in the biosynthesis of the phycobilisome. Recently, inactivation of the analogous gene in *Synechococcus sp.* PCC 7002 via interruption of the coding sequence with an antibiotic resistance gene has resulted in a lack of chromophorylation of the α subunit of PC (Bryant, personal communication). Therefore, this gene product may be either directly or indirectly involved in the chromophorylation of PC^{α}. It is uncertain if other sequences involved in phycobilisome biosynthesis are linked to the 30 kDa ORF since only a limited region downstream from it has been characterized in *F. diplosiphon*.

While the levels of transcripts encoding the PE subunits are strongly modulated by wavelength, little is known about the transcription of *cpcB3A3* and the associated linker polypeptide genes on clone 11. Although no major transcripts from either have been observed in RNA from *F. diplosiphon* maintained in green or red light, other environmental cues may be important for expression of these genes. For example, in *Anabaena sp.* PCC 7120 there are changes in the levels of *cpc* transcripts with changing levels of illumination (Belknap and Haselkorn 1987). In the eukaryotic alga *Callothamnion roseum*, the ratio of phycoerythrobilin to phycourobilin associated with PE changes under different intensities of illumination; phycourobilin is favoured by low light intensities (Yu et al. 1981). Since phycourobilin has a higher extinction coefficient than phycoerythrobilin, this alteration enables the phycobilisome to be more effective in absorbing light quanta. These results suggest that either the conversion of phycoerythrobilin to phycourobilin (isomerization) is sensitive to light intensity or that different PE apoproteins which bond with different chromophores are synthesized. Changes in phycobiliprotein gene expression also occur in response to nutrient deprivation. Depriving cyanobacteria of nitrogen (Allen and Smith 1969, Yamanada and Glazer 1980), sulfur (Lawry and Jensen 1979) or phosphorus (Ihlenfeldt and Gibson 1975) leads to the rapid degradation of phycobiliproteins while continued deprivation results in reduced chlorophyll levels and elevated carotenoid levels (Lau et al. 1977). The molecular events involved in altering phycobiliprotein levels during changes in the nutrient status of the cells are not well defined, although they probably involve specific proteases (Wood and Haselkorn 1979, 1980) and reduced levels of phycobiliprotein transcripts (de Lorimier et al. 1984). With more extensive characterization of the transcription of genes encoding the different phycobiliprotein and linker polypeptides during changing environmental conditions, a role for each of these genes (e.g. the *cpcB3A3* gene set on clone 11) in the biosynthesis of the phycobilisome may become apparent.

Amino acid sequences of *F. diplosiphon* phycobiliproteins

Comparisons of amino acid sequences among the different phycobiliproteins in a single organism, and among analogous phycobiliprotein subunits from different organisms, have helped establish residues that are highly conserved and that may have roles in determining the architectural and biophysical characteristics of a phycobilisome. Protein sequences (derived from the gene sequences) of the different phycobiliproteins and linker polypeptides from *F. diplosiphon* which have been characterized, are presented in Fig. 3A and B respectively. The amino acids identical among the different phycobiliprotein subunits (Fig. 3A) are boxed and the chromophore binding sites are indicated with stars. The greatest degree of homology among all of the biliproteins is in the central region of the proteins (from approximately amino acid 65 to 120) and there is generally more conservation among the β subunits. Homologies among the phycobiliprotein subunits from the different biliprotein classes range from 22 to 57%. The functions of many of these conserved residues are discussed by Schirmer et al. (1986). Insertions or deletions have occurred throughout the biliproteins and are probably responsible for differences surrounding the last chromophore binding sites (closest to the carboxy' terminus) in both PC^β and PE^β.

As shown in Fig. 4, the two different sets of PC subunits (inducible and constitutive) are very similar, with the α and β sequences being 85% and 77% conserved, respectively, at the amino acid level (Conley et al. 1988). Exactly the same degree of conservation is observed at the nucleotide level. Conservation at the nucleotide level may be important in creating a biased codon usage to ensure efficient translation of the subunits or for maintaining features of the RNA secondary structure which impart stability to the transcript. The inducible and constitutive PC sequences share a similar degree of conservation with each other (PC_c^α and PC_c^β compared to PC_i^α and PC_i^β, respectively) as when compared to PC subunits from other prokaryotic and eukaryotic organisms (ranging from 67–78% as deduced from the data shown in Fig. 4). Therefore, it is likely that the *cpc* gene duplication in *Fremyella diplosiphon* occurred early in evolution. This is in contrast to the *cpc* gene duplication observed in *Synechococcus sp.* PCC 6301 in which no differences in the partial nucleotide sequences of the coding regions of the two gene sets have been observed (Gustafsson, personal communication).

The sequences of the PE subunits, derived from the nucleotide sequence of the *cpe* gene set (Fig. 3A), matches the protein sequence reported by Sidler et al. (1986) as modified by Rümbeli et al. (1987). The gene set has also been sequenced by Mazel et al. (1986). Interestingly, the asparagine at position 70 of PE^β is methylated (Rümbeli et al. 1987).

Amino acid sequences of the linker polypeptides are shown in Fig. 3B. The molecular weights assigned to $L_R^{37.5}$ and L_R^{39} based on migration on denaturing polyacrylamide gels are considerably different from those deduced from the amino acid sequences (Lomax et al. 1987). The discrepancy in molecular weights may be the result of anomalous migration in denaturing polyacrylamide gels or post-translational modifications of the polypeptides (e.g. glycosylation, Riethman et al. 1987). In contrast to the acidic phycobiliproteins (Bryant et al. 1976), the linker polypeptides $L_R^{37.5}$, L_R^{39} and $L_C^{7.9}$ have basic isoelectric points. The isoelectric point of $L_R^{9.7}$ is slightly acidic. For the linkers associated with the rod substructure in red light ($L_R^{37.5}$ and L_R^{39}) the carboxyl termini of the proteins (amino acids 166 to 269 and 168 to 288 for $L_R^{37.5}$ and L_R^{39}, respectively) are more basic than the remainder of the protein (isoelectric points greater than 11). This region might be important in the interaction with the acidic residues of phycobiliprotein hexamers. A comparison of the sequences encoding $L_R^{37.5}$ and L_R^{39} reveals very strong homology (53%) with the greatest divergence being at the amino terminus. Certain regions within these proteins are nearly identical; amino acids 29 to 72 of L_R^{39} are 70% homologous to amino acids 28 to 71 of $L_R^{37.5}$ and amino acids 92 to 111 of L_R^{39} are 90% homologous to amino acids 91 to 110 of $L_R^{37.5}$. The difference in size between the two polypeptides is almost totally accounted for by an insertion of sixteen amino acids following amino acid 205 of $L_R^{37.5}$. The small linker polypeptides $L_C^{7.9}$ and $L_R^{9.7}$ are homologous to the carboxyl termini of the larger linker proteins. $L_R^{9.7}$ is associated with the rod substructure and is probably involved in adding the final PC trimer to the rods (Bryant et al. 1987 and personal communication) while $L_C^{7.9}$ is associated with the AP trimers of the core (Lundell and Glazer 1983b, Anderson and Eiserling 1986). Because of limited homology between the linker polypeptides and the phycobiliprotein subunits it was proposed that genes encoding linkers arose via fusions of genes encoding α and β phycobiliprotein subunits (Füglistaller et al. 1985b). Comparisons of the inducible PC subunits (or any of the other phycobiliprotein subunits) of *F. diplosiphon* with $L_R^{37.5}$ and L_R^{39} reveal no significant homology, suggesting that the phycobiliprotein and linker polypeptides may each form a gene family having arisen from separate progenitor genes.

Table 2. Combined codon usage for phycobiliproteins and linkers.

Codon		P	%	L	%
UUU	F	16	53	19	68
UUC	F	14	47	9	32
UUA	L	35	31	12	21
UUG	L	38	33	16	27
CUU	L	4	3	8	14
CUC	L	9	8	3	5
CUA	L	20	17	8	14
CUG	L	8	7	11	19
AUU	I	25	36	8	42
AUC	I	44	63	3	16
AUA	I	1	1	8	42
AUG	M	31	100	10	100
GUU	V	58	62	12	27
GUC	V	4	4	13	29
GUA	V	29	31	15	34
GUG	V	3	3	4	9
UCU	S	38	32	17	26
UCC	S	31	26	7	11
UCA	S	3	2	8	12
UCG	S	0	0	4	6
AGU	S	10	8	12	18
AGC	S	36	30	17	26
CCU	P	19	53	10	36
CCC	P	15	42	7	25
CCA	P	2	5	9	32
CCG	P	0	0	2	7
ACU	T	14	18	20	39
ACC	T	51	67	15	29
ACA	T	10	13	15	29
ACG	T	1	1	1	2
GCU	A	138	68	15	34
GCC	A	10	5	4	9
GCA	A	50	25	21	48
GCG	A	4	2	4	9

Codon		P	%	L	%
UAU	Y	11	17	17	45
UAC	Y	54	83	21	55
CAU	H	2	33	6	75
CAC	H	4	67	2	25
CAA	Q	38	54	3	16
CAG	Q	32	46	16	84
AAU	N	12	23	18	51
AAC	N	41	77	17	49
GAU	D	42	51	19	68
GAC	D	41	49	9	32
GAA	E	53	91	32	76
GAG	E	5	9	10	24
UGU	C	8	38	1	100
UGC	C	13	62	0	0
UGG	W	3	100	4	100
CGU	R	37	46	17	26
CGC	R	31	39	20	31
CGA	R	3	4	5	8
CGG	R	3	7	9	14
AGA	R	6	7	14	21
AGG	R	0	0	0	0
GGU	G	70	65	25	42
GGC	G	28	26	10	17
GGA	G	9	8	12	20
GGG	G	0	0	3	5
UAA	–	2	25	2	50
UAG	–	6	75	1	25
UGA	–	0	0	1	25

A compilation of codons used in genes encoding phycobiliprotein sub-units (Conley et al. 1988, unpublished) and linker polypeptides (Lomax et al. 1987, unpublished) in F diplosiphon is shown in Table 2. Phycobiliprotein codon usage is biased toward the use of a limited number of codons. A few codons such as AUA (isoleucine), GUG and GUC (valine), UGA (serine), CCA (proline), ACG (threonine), and CGA and CGG (arginine) are rarely used while the codons UCG (serine), CCG (proline), AGG (arginine) and GGG (glycine) are never used. Codon usage for the linker polypeptides, which are also relatively abundant polypeptides (about 15% as abundant as the biliprotein subunits), is not quite as biased. All the codons with the exception of AGG (arginine) and UGC (cysteine) are used at least once. (However, only a single cysteine residue is present in all of the linker polypeptides sequenced.) While much of the regulated synthesis of the phycobiliprotein subunit and linker and polypeptide genes is reflected in the relative levels of the transcripts encoded by these genes (Conley et al. 1985, Mazel et al. 1986, Lomax et al. 1987), some control may be exerted at the post-transcriptional level (de Boer and Kastelein 1986, Stormo 1986, Lomax et al. 1987), and a greater range of codons used in the translation of the linker polypeptides may reduce their rate of synthesis (see de Boer and Kastelein 1986). A correlation between codon bias and protein abundance has been made in *E. coli* (de Boer and Kastelein 1986).

Transcription of phycobiliprotein and linker polypeptide genes

Examination of phycobilisome and linker polypeptide transcripts suggests potential mechanisms of the controlled synthesis of phycobilisome constituents that occur during complementary chromatic adaptation. The results of hybridizations of specific phycobiliprotein and linker polypeptide genes to RNA isolated from cells grown in either red or green light are presented in Fig. 5. A DNA fragment internal to the *cpcB1A1* gene set hybridizes to a single species of 1500 bases (Panel A) whose 5′ and 3′ ends have been mapped using S1 nuclease. No larger transcript has been observed. Although the transcript encoding the constitutive PC subunits is present in both red and green light-grown cells, we frequently although not consistently observe two times the level of transcript in red compared to green light-grown cells (intensities were equalized for the two light qualities and cultures used for RNA isolations were of approximately the same cell density). In contrast, a DNA fragment encoding the inducible PC subunits hybridizes to two transcripts detected in cells maintained in red light but not in green light-grown cells (Panel B). The smaller transcript is 1600 bases and encodes solely the inducible PC$^\alpha$ and PC$^\beta$. The larger transcript, approximately 15% as abundant, also encodes the three linker polypeptides L$_R^{39}$, L$_R^{37.5}$ and L$_R^{9.7}$. DNA fragments internal to the linker genes hybridize only to the larger transcript. Therefore, all of the polypeptides required to complete the rod substructure in red light are probably encoded on the 3800-base transcript. The *cpe* gene set (Panel C) hybridizes to a 1500-base transcript which is very abundant in cells maintained in green light but barely detectable in cells maintained in red light. In contrast to the *cpcB2A2* gene set, no linker polypeptide genes are observed downstream from the *cpe* gene set (based on hybridizations at low stringencies to linker sequences downstream from *cpcB2A2*) and no larger transcript hybridizes to a *cpe*-specific probe. These results indicate that the genes encoding linker polypeptides which associate

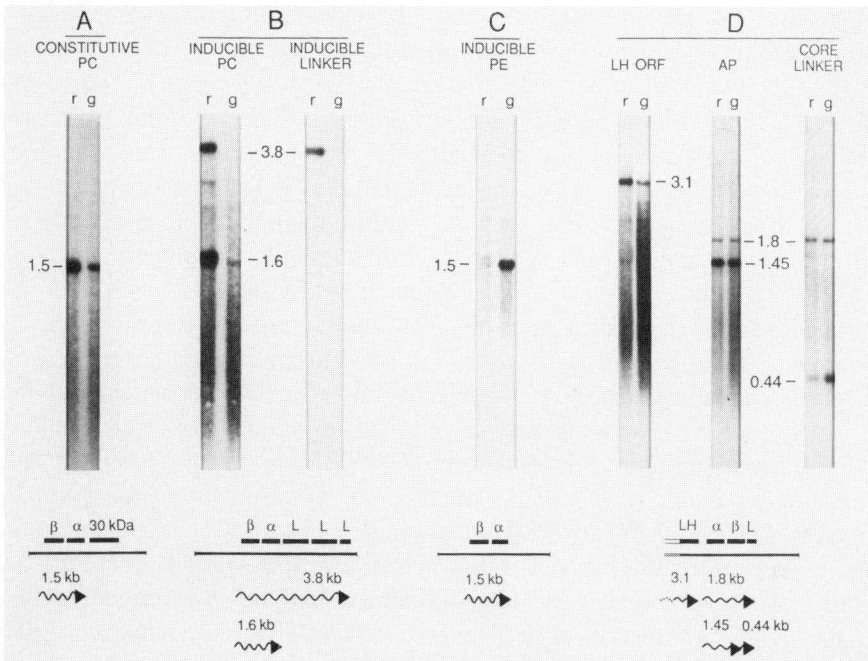

Fig. 5. Transcript analyses of *Fremyella diplosiphon* phycobiliprotein and linker polypeptide genes. The upper portion of each panel (A–D) displays hybridizations of gene-specific probes to RNA isolated from cells maintained in either red (r) or green (g) light. The positions and sizes of the predominant hybridizing species are indicated. The DNA fragments used for hybridizations encoded (A) constitutive PC^β and PC^α, (B) red light-induced PC^β and PC^α, and three associated linker polypeptides, (C) green light-induced PE^β and PE^α, (D) a portion of the 30 kDa ORF, upstream of AP, which has linker homology (LH) and AP^α and AP^β and the small core linker polypeptide. The lower portion of each panel depicts the arrangement of the genes from which the transcripts are derived, and the proteins encoded on the transcripts. Wavy arrows define the boundaries of the transcripts.

with PE are not cotranscribed with the PE subunit genes and are under the control of their own promoter(s).

The α and β subunits of AP, like the inducible PC subunits, are encoded on two transcripts, equally abundant in red and green light. The larger transcript is 1800 bases while the smaller is 1450 bases (Panel D). In addition to encoding the AP subunits, the larger transcript encodes the core linker protein, $L_C^{7.9}$, which is also encoded on a separate 440 base transcript (Panel D). Finally, an open reading frame which precedes the *apcA1B1* gene set and shares homology with the PC linkers is encoded on a 3100 base transcript. The nature of the protein defined by this open reading frame and its role in the biosynthesis of the phycobilisome is still unclear although recent results indicate that the gene for the anchor protein (terminal energy acceptor in the core of the phycobilisome) is upstream from the *apcAB* genes in *Nostoc sp.* strain MAC (Zilinska, personal communication). The size of the 3100 base transcript would be sufficient to encode the high molecular weight anchor protein. Recently, we have demonstrated that messages approximately 5200 and 5900 bases, which are not very abundant, include the sequence of the 3100-base transcript as well as those encoding the AP subunits and the core linker polypeptide (Lemaux and Grossman, unpublished).

The above analyses of transcripts raise some interesting questions about control of phycobilisome biosynthesis in red and green light. Since there is a correlation between the level of protein for a particular component and the quantity of RNA encoding that component, it is likely that much of the control during complementary chromatic adaptation is at the level of transcription while fine tuning of the process may involve post-transcriptional processes like RNA turnover. RNA processing events may be important in the generation of the different biliprotein transcripts as well. There are at least two ways in which the two differently-sized mRNAs encoding the inducible PC subunits could be generated. RNA polymerase may transcribe the *cpcB2A2* gene set and terminate transcription, approximately 85% of the time, at the 3′ end of the gene encoding PC$^\alpha$. The remainder of the time the RNA polymerase would read through the termination signal (hairpin loop structure, see below) and only be released after completing transcription of the genes encoding the three linker polypeptides (resulting in the generation of a 3800-base transcript). Alternatively, the larger species, which may be the primary transcript, could be processed via exonucleolytic and/or endonucleolytic enzymes to the more stable 1600-base transcript. This latter mechanism would imply segmental stability of the RNA with the region encoding the biliprotein subunits having a longer half-life than the linker polypeptide region. Segmental stability of a 2700-base transcript encoding the light-harvesting and reaction center proteins of *Rhodobacter capsulatus* has been invoked as the cause of the differential accumulation of light-harvesting and reaction center polypeptides (Belasco et al. 1985). Characterization of sequences which terminate the 1600-base transcript provides a clue to the events that generate the two transcripts encoding the inducible PC subunits. The 3′ end of the 1600-base transcript contains a stem-loop structure followed by a string of U residues (Conley et al. 1987). This structure is characteristic of a prokaryotic termination signal (Platt 1986) and suggests that the 1600-base transcript is generated by termination of RNA synthesis, and the 3800-base transcript results from readthrough a small and characteristic percentage of the time. While the ratio of the 1600-

to the 3800-base transcript correlates approximately with the ratio of the inducible PC subunits to the associated linker polypeptides, post-transcriptional processes, such as the rates of translation initiation and elongation of the nascent polypeptide, may also be important (Lomax et al. 1987).

Analyses of transcripts encoding the AP subunits and the core linker polypeptide also suggest that RNA processing is important in the biosynthesis of the phycobilisome. The 1800-base transcript encoding both the core linker and the AP subunits may be a primary transcript which is subsequently cleaved into two species, a 1400-base species encoding only AP^α and AP^β, and a 440-base species encoding the core linker polypeptide. Alternatively, RNA polymerase might terminate transcription most of the time following the coding region for AP^β with readthrough occurring occasionally. In this case, to generate the 440-base species, RNA polymerase would have to reinitiate transcription at a separate promoter preceding the core linker coding region. Higher molecular weight transcripts (5200 and 5900 bases), mentioned above, may also be the primary products of transcription. Capping of mRNA (Furuichi and Shatkin 1979) may be one approach to determine the mechanism by which the different biliprotein transcripts are generated. Such experiments rely on the fact that only the primary transcripts, which contain a terminal di- or triphosphate at their 5' ends, can be capped. If the 5' end is derived from processing of a larger transcript, no di- or triphosphate will be present and no capping could occur. Pulse-chase experiments using inhibitors of transcription and determinations of the half-lives of each of the mRNA species (Belasco et al. 1985) may also help decipher precursor/product relationships. At this stage, some results from analyses of the *apcA1B1C* transcripts are puzzling since the small transcript encoding the core linker protein is about as abundant as the *apcA1B1* transcript, under certain conditions, in contrast to the ratio of AP to the core linker polypeptides (approximately 5:1) in the phycobilisome (Glazer et al. 1983). The unexpectedly high levels of the core linker transcript may occur because of decreased translational efficiency relative to the *apcA1B1* transcript or the need to generate more of the core linker polypeptide because this protein may have more than one role in the biosynthesis of the phycobilisome under specific conditions. This protein is peripheral in the core substructure (Anderson and Eiserling 1986) which may make it susceptible to turnover under certain circumstances. Further experiments will be required to determine the rate of transcription of the core linker gene, turnover of the core linker polypeptide and the function that it serves in the light-harvesting complex.

The 5' ends of the transcripts from the *apc*, *cpe* and two *cpc* gene sets have now been determined by high resolution S1 mapping. Figure 6 shows the

```
         -150                                            -101
cpcB1A1   AGCGATCGAGTATTAATTAATAAGTGTTAAGCATTAAAAATTTTGTTAAT

cpcB2A2   TGTAACTCTAATTTAAGTCTAAAGACAAAACGAAAATATAAACAATCCTT

apcA1B1   TTAGGTAACTAATACTCTCGTTTGTCACTAATTAACACTCACAGGTCAGG

cpeBA     CCCCAATCCCCATTACCCCTTATCCCCAGAGGTTCCCCGAGTTCCCCAGT
          - - - - - - - - ▶            ─────────▶      ─────▶
```

```
         -100                                             -51
cpcB1A1   AAAGGCAAAAAGCCTTACTGGTTAAGCAATTCAGCCTCTTAGTATGACTA

cpcB2A2   GCTATCAAAGGGTTTCAGCATTTGCTTAATATTTTTTAACTAATTATCTG

apcA1B1   AGTTCATGCCTTGGACTCTTGACCAAATTCTGTTTTGAGTACTGAGTGCT

cpeBA     CCCCAGTCCCCAATCCTGACTGGGGATTTTTTGTTAAGGATTGTTACTTA
          ─────────▶ - - - - - - - ▶
```

```
         -50                                               -1
                                                            ↓
cpcB1A1   ACTTGACAATTCGTAATAAACAAACGATCCAACGATATAGTATAAACAAG

cpcB2A2   AAAACACATTTAAATTTGCACAAAATTTAACACAAACGAATTGCTTTAAT

apcA1B1   GATTTCTGGGCGTGAACCAGCACTGTTGACTTTTGACAAGAGCTTGGCAA

cpeBA     GTTTCTCATAACTGAGACTGAGATAGCTTTCATCTTTTATGTTCTATATT
          · · · · · ▶
                · · · · · · ▶
```

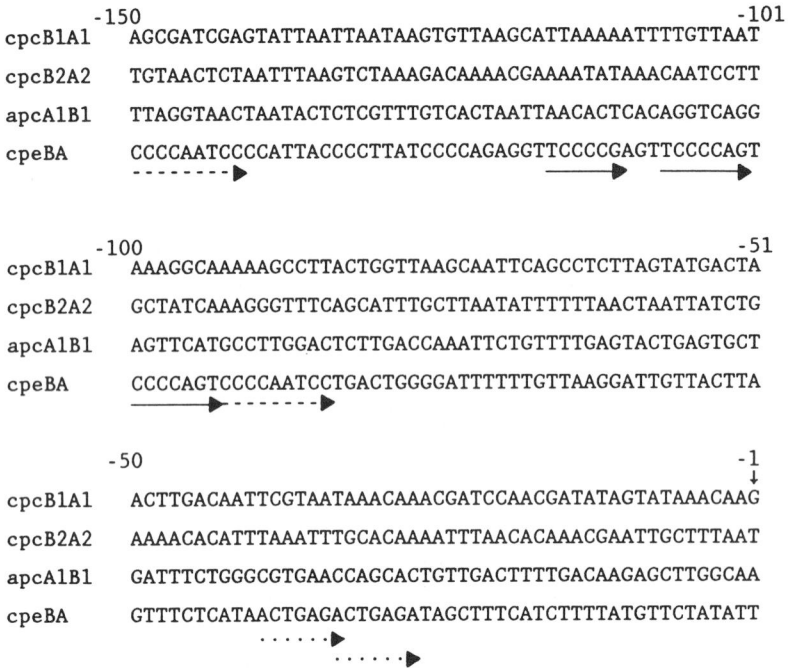

Fig. 6. Comparisons of promoter regions of *F. diplosiphon* phycobiliprotein genes. The 5′ ends of the transcripts are indicated by arrows at − 1. The genes from which the promoter regions were derived are given at the left-hand side of each sequence. Three types of direct repeats are indicated with different arrows (solid, dashed and dotted) below the nucleotide sequence from *cpeBA*. The sites of transcription initiation are 251, 359, 305 and 62 nucleotides from the initiation codons of *cpcB1A1*, *cpcB2A2*, *apcA1B1* and *cpeBA* gene sets, respectively.

nucleotide sequences (150 bp) upstream from the sites where the transcripts for the four biliprotein gene sets begin (indicated by an arrow). Although all of the mRNAs are abundant transcripts in the cell, some are expressed in both red and green light (mRNA encoding AP and PC$_c$) while others are expressed only in red or green light (mRNA encoding PC$_i$ and PE, respectively). We anticipated that the promoter regions of the genes expressed under similar light regimes might share some common sequences. While there is little homology immediately preceding the putative sites of transcription initiation of the *cpcB1A1* and *apcA1B1* gene sets, a comparison of this region of *cpcB1A1* from *Fremyella* and *cpcBA* of *Anabaena sp.* PCC 7120 exhibits 73% homology (Conley et al. 1987). Limited sequence homologies between regions upstream of the transcription initiation sites *cpcB2A2* and *cpe* gene sets are also observed (these genes are expressed under very different light conditions). Some of the sequence similarities may be misleading since the regions which precede transcription initiation are generally AT-rich. After extensive analyses of the 5′ sequences, we can find no sequences that might be responsible for the high levels of transcription of all four of these gene sets. Interestingly, in only some of the promoter regions can sequences similar to the *E. coli* − 10 sequence (TATaat) or − 35 sequence (TTGaca) be found. In the *cpe* gene set there is a TATGTT at − 13 to − 8, in *cpcB1A1* there is a TATAGT at − 15 to − 10 and TATAAA at − 10 to − 5, and in *cpcB2A2* there is a TTGCAC at − 35 to − 29. At − 48 to − 43 of *cpcB1A1* there is a sequence (TTGACA) which perfectly matches the *E. coli* − 35 consensus sequence. One interesting feature of the *cpe* promoter region is the presence of several unusual repeat units just upstream from the transcription initiation site (indicated by arrows in Fig. 6). Hybridization of this repeated DNA to digests of *F. diplosiphon* genomic DNA indicates that there are many copies of these repeats in other regions of the cyanobacterial genome. Identical repeat units (different from those shown in Fig. 6) are also found 3′ to the *cpe* and *apc* genes. It is unclear whether these sequences play a role in regulating expression from the *cpe* gene set or whether they are solely part of a family of repetitive genetic elements (which may or may not be mobile). Recently, mobile genetic elements have been observed in *F. diplosiphon*. In a spontaneous pigment mutant of *F. diplosiphon*, a 1.4 kb insertion element was found downstream of the *cpcA1B1* gene set (Tandeau de Marsac, Cyanobacterial Genetics Workshop, St. Louis, MO). This element has both direct and inverted repeats similar to other prokaryotic transposons and may ultimately serve as a useful tool in the analysis of genes important for phycobilisome biosynthesis.

Future directions

The two most immediate areas where we can improve our understanding of phycobilisome biosynthesis concern the precise sequence of events that result in the construction of a phycobilisome and the mechanistic details involved in regulating expression from phycobiliprotein and linker polypeptide genes. The former provokes several questions: are the individual subunits chromophorylated post- or cotranslationally? Does chromophorylation precede or follow assembly? Does the biosynthesis of a phycobilisome require a series of ordered assembly events beginning with the construction of the core complex followed by the sequential positioning of PC and PE trimers (with their associated linker polypeptides) onto the core to form the rod substructures? The generation of assembly mutants and measurements of the levels of the different phycobilisome components in them may help us piece together the events required for assembly, and identify the different genes and proteins involved. Such mutants may also help identify genes encoding chromophorylating enzymes, enzymes which modify or glycosylate phycobilisome constituents and proteins involved in regulating the synthesis of phycobilisome components. Our laboratory is most interested in delineating the molecular mechanisms responsible for regulating phycobilisome biosynthesis during complementary chromatic adaptation and stress responses. To understand these events, it is necessary to analyze them both in vivo and in vitro. In vivo analysis requires using methods for gene transfer into cyanobacteria. Transformation has been achieved in a number of unicellular cyanobacteria. The prevalent species used for transformation are *Anacystis nidulans*, R-2 (Kuhlemeier et al. 1983, Golden and Sherman 1984, Sherman and de Putte 1982), *Synechococcus sp.* PCC 7002 (Stevens and Porter 1980) and *Synechocystis sp.* PCC 6803 (Chauvat et al. 1986). Furthermore, conjugation has been successfully used to introduce DNA into a number of filamentous cyanobacteria (Wolk et al. 1984). In such organisms genes encoding the different phycobiliprotein components can be interrupted in vitro with a drug resistance marker, the interrupted sequence used to replace the wild type sequence and the phenotype of the mutant analyzed. The precise targeting of the mutation followed by phenotypic analysis may be useful in defining steps in the assembly process but also may suggest if assembly inhibition alters expression from phycobiliprotein subunit and/or linker polypeptide genes. For example, if assembly of PC into the rod substructure is blocked, are levels of PC reduced, and if so, at what point in the synthesis of the protein (transcriptional, post-transcriptional) is control exercised? Control at the transcriptional level may be inferred from measurements of transcript levels for specific phycobilisome components coupled with measurements of the rates of RNA turnover. Translational regulation may also be important. Autogenous control of the translation of ribosomal proteins offers an example of this (Nomura et al. 1984, Cole and Nomura 1986). Certain ribosomal proteins, when unassembled, can inhibit protein synthesis from their own messenger RNAs.

Gene transfer may also be useful for defining sequences in the promoter regions of the biliprotein genes which are important for their regulated transcription. The *cpe* or *cpcB2A2* promoter could be fused to a reporter gene such as that encoding chloramphenicol acetyltransferase (*Cat* for gene and CAT for protein), placed in to a replicating vector and transferred into a filamentous cyanobacterium exhibiting complementary chromatic adaptation, such as *F. diplosiphon*. Recently, transfer of an autonomously replicating plasmid into *F. diplosiphon* by conjugation has been reported (Cobley 1985). The effect of light on CAT levels can be evaluated using a relatively easy assay procedure. Deletions or site-directed mutations could be introduced into the regions preceding the transcription initiation site and then expression from the *Cat* gene assayed. Demonstrations of alterations in the levels of CAT activity would be a good indication that the regulation of the gene is at least partly at the level of transcription, although, post-transcriptional events (e.g. RNA turnover, translation and protein turnover) would also have to be evaluated.

Gene transfer techniques could also be used to isolate both the gene encoding the photoreceptor and any *trans*-acting proteins which might be involved in altering transcription from biliprotein promoters. This approach would entail fusing a light-regulated promoter (e.g. from *cpeBA* or *cpcB2A2* gene sets) to a gene which, when expressed, would be lethal to the host cyanobacterium. For example, the *cpe* promoter could be fused to a promoterless gene encoding *Eco*R1 (or perhaps genes encoding bacteriocins), ligated into a replicating plasmid and conjugated into *F. diplosiphon*. The lethal gene would only be expressed in green light and therefore killing would only occur in green light and not in red light. However, if the host organism were mutated prior to gene transfer, exconjugants surviving green illumination would be incapable of activating the *cpe* promoter. This may either be a result of a lesion in the photoreceptor or another component of the sensory transduction chain. Characterization of the mutant phenotype and isolation and characterization of the altered gene would strengthen our understanding of the regulatory processes involved in both photoperception and signal transduction.

To determine biochemically the events required for regulated transcription, it is important to develop and exploit an in vitro transcription system that reflects in vivo regulation. Characterization of the effects of defined lesions in the promoter regions of the light-regulated genes on their transcription in vitro would lend support to results from in vivo analyses. Such a system would be important for isolating proteins or specific RNA polymerases required for the regulated expression of phycobiliprotein genes. In vitro transcription has been used to define *trans*-acting proteins involved in activating *ntr* promoters (Reitzer and Magasanak 1986, Hirschman et al. 1985) and for the isolation of specific RNA polymerases necessary for transcription from heat shock (Grossman et al. 1984, Taylor et al. 1984), *ntr* (Hirschman et al. 1985) and sporulation-specific promoters (Losick et al. 1986), as well as for identifying and isolating *trans*-acting elements that activate eukaryotic genes (Dynan and Tjian 1985, Heberlein et al. 1985, Wu 1985, Sen and Baltimore 1986, Sen et al. 1986). Proteins that activate transcription in vitro could be used in binding studies to determine the specificity of the interaction and the segment of DNA required for binding (footprint analyses; Galas and Schmitz 1978, Van Dyke and Dervan 1983).

188

Many sophisticated molecular techniques are becoming available which will help in deciphering how light is converted into biochemical signals that either activate or suppress the expression of specific genes. While light regulation of gene expression has been studied extensively in higher plants, primarily with respect to phytochrome (Schäfer and Briggs 1986, Thompson et al. 1986), several difficulties in these analyses have been encountered because of the complexity of the responses, the differential activation of genes which may comprise multigene families, and the modulation of the transcription of numerous, coordinately-regulated genes. Furthermore, it is difficult to transform certain plants, and if transformed, the introduced DNA might enter the genome in a number of different places yielding variable copy number and variable rates of transcription (depending upon positions in the genome). Both the presence of multigene families and the inability to obtain homologous recombination into the plant genome makes it impossible to target specific lesions and analyze phenotypes. Finally, it is very time consuming to grow up transformed tissue, analyze its genome for the presence of introduced sequences and characterize the phenotypic affects. Our hope is that chromatic adaptation may provide a system analogous to the phytochrome system of higher plants (it is a photoreversible system as is the phytochrome system) but with the ease of analysis of systems generally associated with prokaryotic organisms.

Acknowledgements

We are grateful to Glenn Ford, Loretta Tayabas and Sabrina Robbins for excellent technical assistance. The work presented was supported by a Public Health Service Grant GM-334336-01 from the National Institute of Health, a National Science Foundation Grant DCB-86 15606 and the Carnegie Institution of Washington. L.K.A. was supported by a plant molecular biology postdoctoral fellowship from the National Science Foundation DMB-8508808.

References

Allen MM and Smith AJ (1969) Nitrogen chlorosis in blue-green algae. Arch Microbiol 69: 111–120

Anderson LK and Eiserling FA (1986) Asymmetrical core structure in phycobilisomes of the cyanobacterium *Synechocystis* 6701. J Mol Biol 191: 441–451

Anderson LK and Grossman AR (1987) Phycocyanin genes in the cyanobacterium *Synechocystis* 6701 and a potential gene rearrangement in a pigment variant. In: Biggins J (ed.)

Progress in photosynthesis (Proceedings of the VIIth International Congress of Photosynthesis) Vol 4, pp 817–820. The Hague: Martinus Nijhoff Publishers

Anderson LK, Rayner MC and Eiserling FA (1987) Mutations that affect structure and assembly of light-harvesting proteins in the cyanobacterium *Synechocystis* sp. strain 6701. J Bacteriol 169: 102–109

Anderson LK, Rayner MC and Eiserling FA (1984) Ultraviolet mutagenesis of *Synechocystis* sp. 6701: Mutations in chromatic adaptation and phycobilisome assembly. Arch Microbiol 138: 237–243

Bashford D, Chothea C and Lesk ML (1987) Determinants of a protein fold. Unique features of the globin amino acid sequences. J Mol Biol 196: 199–216

Belasco JG, Beatty T, Adams CW, von Gabain A and Cohen SN (1985) Differential expression of photosynthesis genes in *Rhodopseudomonas capsulata* results from segmental differences in stability within the polycistronic rxc transcript. Cell 40: 171–181

Belknap WR and Haselkorn R (1987) Cloning and light regulation of expression of the phycocyanin operon of the cyanobacterium *Anabaena*. EMBO J 6: 871–884

Bennett A and Bogorad L (1973) Complementary chromatic adaptation in a filamentous blue-green alga. J Cell Biol 58: 419–435

Björn GS and Björn LO (1976) Photochromic pigments from blue-green algae: phycochromes a, b and c. Physiol Plant 36: 297–304

Bogorad L (1975) Phycobiliproteins and complementary chromatic adaptation. Annu Rev Plant Physiol 26: 369–401

Bohnert HJ, Michalowski C, Koller B, Delius H, Mucke H and Löffelhardt W (1983) The cyanelle genome from *Cyanophora paradoxa*. In: Schenk HEA (ed.) Endocytobiology II, pp 433–448. Berlin: de Gruyter

Bryant DA (1981) The photoregulated expression of multiple phycocyanin species. General mechanism for control of phycocyanin synthesis in chromatically adapting cyanobacteria. Eur J Biochem 119: 425–429

Bryant DA and Cohen-Bazire G (1981) Effects of chromatic illumination on cyanobacterial phycobilisomes. Evidence for the specific induction of a second pair of phycocyanin subunits in *Pseudanabaena* 7409 grown in red light. Eur J Biochem 119: 415–424

Bryant DA, de Lorimier R, Guglielmi G, Stirewalt VL, Cantrell A and Stevens SE Jr (1987) The cyanobacterial photosynthetic apparatus. In: Biggins J (ed.) Progress in Photosynthesis (Proceedings of the VIIth International Congress on Photosynthesis), Vol 4, pp 749–755. The Hague: Martinus Nijhoff Publishers

Bryant DA, de Lorimier R, Guglielmi G, Stirewalt VL, Dubbs JM, Illman B, Porter RD and Stevens SE Jr (1985a) Genes for phycobilisome components in *Synechococcus* 7002, *Pseudanabaena* 7409 and *Mastigocladus laminosus*. In: V International Symposium on Photosynthetic Prokaryotes (Abstract) 103

Bryant DA, de Lorimier R, Lambert DH, Dubbs JM, Stirewalt VL, Stevens SE Jr, Porter RD, Tam J and Jay E (1985b) Molecular cloning and nucleotide sequence of the α and β subunits of allophycocyanin from the cyanelle genome of *Cyanophora paradoxa*. Proc Natl Acad Sci USA 82: 3242–3246

Bryant DA, de Lorimier R, Porter RD, Lambert DH, Dubbs JM, Stirewalt VL, Field PI, Stevens SE Jr, Liu W-Y, Tam J and Jay EWK (1985c) Phycobiliprotein genes in cyanobacteria and cyanelles. In: Arntzen C, Bogorad L, Bonitz S and Steinback K (eds) Molecular Biology of the Photosynthetic Apparatus, pp 249–258. Cold Spring Harbor Laboratory Publication, Cold Spring Harbor, NY

Bryant DA, Glazer AN and Eiserling FA (1976) Characterization and structural properties of the major biliproteins of *Anabaena* sp. Arch Microbiol 110: 61–75

Bryant DA, Guglielmi G, Tandeau de Marsac N, Castets A-M and Cohen-Bazire G (1979) The structure of cyanobacterial phycobilisomes: A model. Arch Microbiol 123: 113–127

Chauvat F, De Vries L, Van der Ende A, and Van Arkel GA (1986) A host vector system for gene cloning in the cyanobacterium *Synechocystis* PCC 6803. Mol Gen Genet 204: 185–191

Cobley JG (1985) Chromatic adaptation in *Fremyella diplosiphon*. I. Mutants hypersensitive to green light. II. Construction of mobilizable vectors lacking sites for FDI I and II. In V International Symposium on Photosynthetic Prokaryotes (Abstract) 105

Cobley JG and Miranda RD (1983) Mutations affecting chromatic adaptation in the cyanobacterium *Fremyella diplosiphon*. J Bact 153: 1486–1492

Cole JR and Nomura M (1986) Translational regulation is responsible for growth-rate-dependent and stringent control of the synthesis of ribosomal proteins L11 and L1 in *Escherichia coli*. Proc Natl Acad Sci USA 83: 4129–4133

Conley PB, Lemaux PG and Grossman AG (1988) Molecular characterization and evolution of sequences encoding light harvesting components in the chromatically adapting cyanobacterium *Fremyella diplosiphon*. J Mol Biol 199: 447–465

Conley PB, Lemaux PG and Grossman AR (1985) Cyanobacterial light-harvesting complex subunits encoded in two red light-induced transcripts. Science 230: 550–553

Conley PB, Lemaux PG, Lomax TL and Grossman AR (1986) Genes encoding major light-harvesting polypeptides are clustered on the genome of the cyanobacterium *Fremyella diplosiphon*. Proc Natl Acad Sci USA 83: 3924–3928

De Boer HA and Kastelein RA (1986) Biased codon usage: An explanation of its role in optimizing translation. In: Reznikoff W and Gold L (eds) Maximizing gene expression. pp 225–285. Boston: Butterworth Press

De Lorimier R, Bryant DA, Porter RD, Liu W-Y, Jay E and Stevens SE Jr (1984) Genes for α and β phycocyanin. Proc Natl Acad Sci USA 81: 7946–7950

Diakoff S and Schiebe J (1973) Action spectrum for chromatic adaptation in *Tolypothrix tenuis*. Plant Physiol 51: 382–385

Dynan WS and Tjian R (1985) Control of eukaryotic messenger RNA synthesis by sequence-specific DNA-binding proteins. Nature 316: 774–778

Egelhoff T and Grossman AR (1983) Cytoplasmic and chloroplast synthesis of phycobilisome polypeptides. Proc Natl Acad Sci USA 80: 3339–3343

Füglistaller P, Rümbeli R, Suter F and Zuber H (1985) Minor polypeptides from the phycobilisome of the cyanobacterium *Mastigocladus laminosus*. Hoppe-Seyler's Z Physiol Chem 365: 1085–1096

Füglistaller P, Suter F and Zuber H (1985) Linker polypeptides of the phycobilisome from the cyanobacterium *Mastigocladus laminosys*: amino acid sequence and relationships. Biol Chem Hoppe-Seyler 366: 993–1001

Furuichi Y and Shatkin AJ (1976) Differential synthesis of blocked and unblocked 5'-termini in reovirus mRNA: Effect of pyrophosphate and pyrophosphatase. Proc Natl Acad Sci USA 73: 3448–3452

Galas DJ and Schmitz A (1978) DNAse footprinting: a simple method for the detection of protein-DNA binding specificity. Nucl Acids Res 5: 3157–3170

Gantt E (1981) Phycobilisomes. Annu Rev Plant Physiol 32: 327–347

Gendel S, Ohad I and Bogorad L (1979) Control of phycoerythrin synthesis during chromatic adaptation. Plant Physiol 64: 786–790

Gray MW and Doolittle WF (1982) Has the endosymbiont hypothesis been proven? Microbiol Rev 46: 1–42

Gingrick JC, Williams RC and Glazer AN (1982) Rod substructures in cyanobacterial phycobilisomes: phycoerythrin assembly in *Synechocystis* 6701 phycobilisomes. J Cell Biol 95: 170–178

Glazer AN (1985) Light harvesting by phycobilisomes. Annu Rev Biochem 14: 47–77

Glazer AN (1984) A macromolecular complex optimized for light energy transfer. Biochim Biophys Acta 768: 29–51

Glazer AN (1982) Phycobilisomes: structure and dynamics. Annu Rev Microbiol 36: 173–198

Glazer AN, Lundell DJ, Yamanaka G and Williams RC (1983) The structure of a simple phycobilisome. Ann Microbiol (Inst Pasteur) 134B: 159–180

Glazer AN, Yeh SW, Webb SP and Clark JH (1985) Disk-to-disk transfer as the rate limiting step for energy flow in phycobilisomes. Science 227: 419–423

Golden SS and Sherman LA (1984) Optimal conditions for genetic transformation of the cyanobacterium *Anacystis nidulans* R2. J Bact 158: 36–42

Grossman AD, Erickson JW and Gross CA (1984) The *htpR* gene product of *E. coli* is a sigma factor for heat shock promoters. Cell 38: 383–390

Grossman AR, Talbot L and Egelhoff T (1983) Biosynthesis of phycobilisome polypeptides of *Porphyridium aerugineum* and *Cyanophora paradoxa*. In: Carnegie Institution of Washington Year Book 82: 112–116

Grossman AR, Lemaux PG and Conley PB (1986) Regulated synthesis of phycobilisome components. Photochem Photobiol 44: 827–837

Haury JF and Bogorad L (1977) Action spectra for phycobiliprotein synthesis in a chromatically adapting cyanophyte. Plant Physiol 60: 835–839

Heberlein U, England B and Tjian R (1985) Characterization of Drosophila transcription factors that activate the tandem promoters of alcohol dehydrogenase gene. Cell 41: 965–977

Herdman M (1982) Evolution and genetic properties of cyanobacterial genomes. In: Carr NG and Whitton BA (eds) The Biology of Cyanobacteria, pp 263–306. Berkeley and Los Angeles: University of California Press

Hirschman J, Wong P-K, Sei K, Keene J and Kustu S (1985) Products of nitrogen regulatory genes *ntrA* and *ntrC* of enteric bacteria activate *glnA* transcription in vitro: evidence that the *ntrA* product is a sigma factor. Proc Natl Acad Sci USA 82: 7525–7529

Houmard J, Mazel D, Moguet C, Bryant DA and Tandeau de Marsac N (1986) Organization and nucleotide sequence of genes encoding core components of the phycobilisomes from *Synechococcus* 6301. Mol Gen Genet 205: 404–410

Ihlenfeldt MJA and Gibson J (1975) Phosphate utilization and alkaline phosphatase activity in *Anacystis nidulans* (*Synechococcus*). Arch Microbiol 102: 23–28

Kuhlemeier CJ, Thomas AAM, Van Der Ende A, Van Leen RW, Borrias WE, Van Den Hondel CAMJJ and Van Arkel GA (1983) A host-vector system for gene cloning in the cyanobacterium *Anacystis nidulans* R2. Plasmid 10: 156–163

Kuntz M, Crouse EJ, Mubumbila M, Burkard G, Weil J-H, Bohnert HJ, Mucke H and Löffelhardt W (1984) Transfer RNA gene mapping studies on cyanelle DNA from *Cyanophora paradoxa*. Mol Gen Genet 194: 508–512

Lambert DH, Bryant DA, Stirewalt VL, Dubbs JM, Stevens SE Jr and Porter RD (1985) Gene map for the *Cyanophora paradoxa* cyanelle genome. J Bact 164: 659–664

Lau RJ, Mackenzie MM and Doolittle WF (1977) Phycocyanin synthesis and degradation in the blue-green bacterium *Anacystis nidulans*. J Bacteriol 132: 771–778

Lawry NH and Jensen TE (1979) Deposition of condensed phosphate as an effect of varying sulfur deficiency in the cyanobacterium *Synechococcus* sp (*Anacystis nidulans*). Arch Microbiol 120: 1–7

Lawry NH and Jensen TE (1986) Condensed phosphate deposition, sulfur amino acid use, and unidirectional transsulfuration in *Synechococcus leopoliensis*. Arch Microbiol 144: 317–323

Lemaux PG and Grossman AR (1985) Major light-harvesting polypeptides encoded in polycistronic transcript in eukaryotic algae. EMBO J 4: 1911–1919

Lemaux PG and Grossman AR (1984) Isolation and characterization of a gene for a major light-harvesting polypeptide from *Cyanophora paradoxa*. Proc Natl Acad Sci USA 81: 4100–4104

Lipman DJ and Pearson WR (1985) Rapid and sensitive protein similarity searches. Science 227: 1435–141

Lind LK, Kalla SR, Lönneborg A, Öquist G and Gustafsson P (1985) Cloning of the β phycocyanin gene from *Anacystis nidulans*. FEBS Lett 188: 27–32

Löffelhardt W, Mucke H, Crouse EJ and Bohnert HJ (1983) Comparison of the cyanelle DNA from two different strains of *Cyanophora paradoxa*. Curr Genetics 7: 139–144

Lomax TL, Conley PB, Schilling J and Grossman AR (1987) Isolation and characterization of light-regulated phycobilisome linker polypeptide genes and their transcription as a polycistronic mRNA. J Bacteriol 169: 2675–2684

Lönneborg A, Lind LK, Kalla SR, Gustafsson P and Öquist G (1985) Acclimation processes in the light-harvesting system of the cyanobacterium *Anacystis nidulans* following a light shift from white to red. Plant Physiol 78: 110–114

Losick R, Youngman P and Piggot P (1986) Genetics of endospore formation in *Bacillus subtilis*. Annu Rev Genet 20: 625–669

Lundell DJ and Glazer AN (1983a) Molecular architecture of a light-harvesting antenna. Quarternary interactions in the *Synechococcus* 6301 phycobilisome core as revealed by partial tryptic digestion and circular dichroism studies. J Biol Chem 258: 8708–8713

Lundell DJ and Glazer AN (1983b) Molecular architecture of light-harvesting antenna. Core substructure in *Synechococcus* 6301 phycobilisomes: Two new allophycocyanin and allophycocyanin B complexes. J Biol Chem 258: 902–908

Lundell DJ and Glazer AN (1983c) Molecular architecture of a light-harvesting antenna. Structure of the 18S core-rod subassembly of *Synechococcus* 6301 phycobilisomes. J Biol Chem 258: 894–901

Lundell DJ, Yamanaka G and Glazer AN (1981a) A terminal energy acceptor of the phycobilisome; a new biliprotein. J Cell Biol 91: 315–319

Lundell DJ, Williams RC and Glazer AN (1981b) Molecular architecture of light harvesting antenna. In vitro assembly of the rod substructures of *Synechococcus* 6301 phycobilisomes. J Biol Chem 256: 3580–3592

Mazel D, Guglielmi G, Houmard J, Sidler W, Bryant DA and Tandeau de Marsac N (1986) Green light induces transcription of phycoerythrin operon in the cyanobacterium *Calothrix* 7601. Nucl Acids Res 14: 8279–8290

Nomura M, Gourse R and Baughman G (1984) Regulation of synthesis of ribosomes and ribosome components. Annu Rev Biochem 53: 75–117

Offner GD, Brown-Mason AS, Ehrhardt MM and Troxler RF (1981) Primary structure of phycocyanin from the unicellular rhodophyte *Cyanidium caldarium*. Complete amino acid sequence of the α subunit. J Biol Chem 256: 12167–12175

Offner GD and Troxler RF (1983) Primary structure of allophycocyanin from the unicellular rhodophyte, *Cyanidium caldarium*. The complete amino acid sequences of the α and β subunits. J Biol Chem 258: 9931–9940

Ohad I, Schneider HAW, Gendel S and Bogorad L (1980) Light-induced changes in allophycocyanin. Plant Physiol 65: 6–12

Ohki K and Fujita Y (1979a) In vivo transformation of phycobiliproteins during photobleaching of *Tolypothrix tenuis* to forms active in photoreversible absorption changes. Plant Cell Physiol 20: 1341–1347

Ohki K and Fujita Y (1979b) Photoreversible absorption changes of guanidine-HCl treated phycocyanin and allophycocyanin isolated from the blue-green alga *Tolypothrix tenuis*. Plant Cell Physiol 20: 483–490

Ohki K and Fujita Y (1981) On the relationship between photocontrol of phycoerythrin formation and photoreversible pigments of Scheibe. Plant Cell Physiol 22: 347–357

Öquist G (1974) Light induced changes in pigment composition of photosynthetic lamellae and cell-free extracts obtained from the blue-green alga *Anacystis nidulans*. Physiol Plant 30: 45–48

Peterson RB, Dolan E, Calvert HE and Ke B (1981) Energy transfer from phycobiliproteins to photosystem I in vegetative cells and heterocysts of *Anabaena variabilis*. Biochim Biophys Acta 634: 237–248

Pilot TJ and Fox JL (1984) Cloning and sequencing of the genes encoding α and β subunits from the cyanobacterium *Agmenellum quadruplicatum*. Proc Natl Acad Sci USA 81: 6983–6987

Platt T (1986) Transcription termination and the regulation of gene expression. Annu Rev Biochem 55: 339–372

Porter G, Tredwell CJ, Searle GFW and Barber J (1978) Picosecond time-resolved energy transfer in *Porphyridium cruentum*. Part I. In the intact chloroplast. Biochim Biophys 501: 532–545

Reithman HC, Mawhinnet TP and Sherman LA (1987) Phycobilisome associated glycoproteins in the cyanobacterium *Anacystis nidulans* R2. FEBS Lett 215: 209–214

Reitzer LJ and Magasanak B (1986) Transcription of *glnA* in *E. coli* is stimulated by activators bound to sites far from the promoters. Cell 45: 785–792

Rümbeli R, Schirmer T, Bode W, Sidler W and Zuber H (1985) Crystallization of phycoerythrocyanin from the cyanobacterium *Mastigocladus laminosus* and a preliminary characterization of two crystal forms. J Mol Biol 186: 197–200

Rümbeli R, Suter S, Wirth M, Sidler W and Zuber H (1987) γ-N-Methylasparagine in phycobilisomes from the cyanobacteria *Mastigocladus laminosus* and *Calothrix*. FEBS Lett 221: 1–2

Schäfer E and Briggs WR (1986) Photomorphogenesis from signal perception to gene expression. Photobiochem Photobiophys 12: 305–320

Scheibe J (1972) Photoreversible pigment occurrence in blue green alga. Science 176: 1037–1039

Schirmer T, Bode W, Huber R, Sidler W and Zuber H (1985) X-ray crystallographic structure of the light-harvesting biliprotein C-phycocyanin from thermophilic cyanobacterium *Mastigocladus laminosus* and its resemblance to globin structure. J Mol Biol 184: 257–277

Schirmer T, Huber R, Schneider M, Bode W, Miller M and Hackert M (1986) Crystal structure analysis and refinement at 2.5 Å of hexameric C-phycocyanin from the cyanobacterium *Agmenellum quadruplicatum*. The molecular model and its implication for light-harvesting. J Mol Biol 188: 651–676

Sen R and Baltimore D (1986) Multiple nuclear factors interact with immunoglobulin enhancer sequences. Science 46: 705–716

Sen R, Baltimore D and Sharp PA (1986) A nuclear factor that binds a conserved sequence motif in transcriptional control elements of immunoglobulin genes. Nature 319: 154–158

Sherman LA and Van den Putte P (1982) Construction of a hybrid plasmid capable of replication in the bacterium *E. coli* and the cyanobacterium *Anacystis nidulans*. J Bact 150: 410–413

Sidler W, Kumpf B, Rudiger W and Zuber H (1986) The complete amino acid sequence of C-phycoerythrin from the cyanobacterium *Fremyella diplosiphon*. Biol Chem Hoppe-Seyler 367: 627–647

Stevens SE Jr, and Paone DAM (1981) Accumulation of cyanophycin granules as a result of phosphate limitation in *Agmenellum quadruplicatum*. Plant Physiol 67: 716–719

Stevens SE Jr and Porter RD (1980) Transformation in *Agmenellum quadruplicatum*. Proc Natl Acad Sci USA 77: 6052–6056

Stormo GD (1986) Translation initiation. In: Reznikoff W and Gold L (eds) Maximizing gene expression. London: Butterworth Publishers

Tandeau de Marsac N (1977) Occurrence and nature of chromatic adaptation in cyanobacteria. J Bact 130: 82–91

Tandeau de Marsac N (1983) Phycobilisomes and complementary chromatic adaptation in cyanobacteria. Bull Inst Pasteur 81: 201–254

Tandeau de Marsac N and Cohen-Bazire G (1977) Molecular composition of cyanobacterial phycobilisomes. Proc Natl Acad Sci USA 74: 1635–1639

Taylor WE, Strauss DB, Grossman AD, Buston ZF, Gross CA and Burgess RR (1984) Transcription from a heat-inducible promoter causes heat shock regulation of the sigma subunit of RNA polymerase. Cell 38: 371–381

Thompson WF, Kaufman LS and Watson JC (1986) Induction of plant gene expression by light. BioEssays 3: 153–159

Troxler RF, Ehrhardt M, Brown-Mason AS and Offner GD (1981) Primary structure of phycocyanin from the unicellular rhodophyte *Cyanidium caldarium*. Complete amino acid sequence of the β subunit. J Biol Chem 256: 12176–12184

Van Dyke MW and Dervan PB (1983) Methidiumpropyl-EDTA-Fe(II) and DNase I footprinting report different small molecule binding site sizes on DNA. Nucl Acids Res 11: 5555–5567

Vogelmann TC and Scheibe J (1978) Action spectrum for chromatic adaptation in the blue-green alga *Fremyella diplosiphon*. Planta 143: 233–239

Williams RC, Gingrich JC and Glazer AN (1980) Cynobacterial phycobilisomes. Particles of *Synechocystis* 6701 and two pigment mutants. J Cell Biol 85: 558–566

Wolk FC, Vonshak A, Kehoe P and Elhai J (1984) Construction of shuttle vectors capable of conjugative transfer from *Escherichia coli* to nitrogen-fixing filamentous cyanobacteria. Proc Natl Acad Sci USA 81: 1561–1565

Wood NB and Haselkorn R (1979) Proteinase activity during heterocyst differentiation in nitrogen-fixing cyanobacteria. In: Cohen GN and Holzer H (eds) Limited Proteolysis in Microorganisms, pp 159–166. US DHEW Publication No. (NIH) 79—1591, Bethesda, MD

Wood NB and Haselkorn R (1980) Control of phycobiliprotein proteolysis and heterocyst differentiation in *Anabaena*. J Bacteriol 141: 1375–1385

Wu C (1985) An exonuclease protection assay reveals heat shock element and TATA box DNA-binding proteins in crude nuclear extracts. Nature 317: 84–87

Yamanaka G and Glazer AN (1980) Dynamic aspect of phycobilisome structure. Phycobilisome turnover during nitrogen starvation in *Synechococcus* sp. Arch Microbiol 124: 39–47

Yamanaka G and Glazer AN (1981) Dynamic aspects of phycobilisome structure: Modulation of phycocyanin content of *Synechococcus* phycobilisomes. Arch Microbiol 130: 23–30

Yamanaka G, Lundell DJ and Glazer AN (1982) Molecular architecture of a light-harvesting antenna. Isolation and characterization of phycobilisome subassembly particles. J Biol Chem 257: 4077–4086

Yamanaka G, Glazer AN and Williams RC (1980) Molecular and architecture of a light-harvesting antenna. Comparison of wild type and mutant *Synechococcus* 6301 phycobilisomes. J Biol Chem 255: 11004–11010

Yu M-H, Glazer AN, Spencer KG and West JA (1981) Phycoerythrins of the red algae *Callithamnion*. Variation in phycoerythrobilin and phycourobilin content. Plant Physiol 68: 482–488

Zuber H (1983) Structure and function of the light-harvesting phycobiliproteins from the cyanobacterium *Mastigocladus laminosus*. In: Papageorgiou G and Packer L (eds) Photosynthetic Prokaryotes: Cell Differentiation and Function, pp 23–41. Amsterdam: Elsevier Science Publishing Co

Zuber H (1986) Structure of light harvesting antenna complexes of photosynthetic bacteria, cyanobacteria and red algae. Trends Biochem Sci 11: 414–419

Govindjee et al. (eds), Molecular Biology of Photosynthesis: 195–228
© 1988 Kluwer Academic Publishers

Minireview

Photoregulation of gene expression in the filamentous cyanobacterium *Calothrix* sp. PCC 7601: light-harvesting complexes and cell differentiation

NICOLE TANDEAU DE MARSAC, DIDIER MAZEL, THIERRY
DAMERVAL, GÉRARD GUGLIELMI, VÉRONIQUE CAPUANO &
JEAN HOUMARD
*Unité de Physiologie Microbienne (C.N.R.S., U.A. 1129), Département de Biochimie et
Génétique Moléculaire, Institut Pasteur, 28 rue du Docteur Roux, 75724 Paris Cedex 15,
France*

Received 6 October 1987; Accepted 23 December 1987

Key words: *Calothrix* sp. PCC 7601, gas vesicles, hormogonia, photoregulation, phycobiliproteins, phycobilisomes

Abstract. Light plays a major role in many physiological processes in cyanobacteria. In *Calothrix* sp. PCC 7601, these include the biosynthesis of the components of the light-harvesting antenna (phycobilisomes) and the differentiation of the vegetative trichomes into hormogonia (short chains of smaller cells). In order to study the molecular basis for the photoregulation of gene expression, physiological studies have been coupled with the characterization of genes involved either in the formation of phycobilisomes or in the synthesis of gas vesicles, which are only present at the hormogonial stage.

In each system, a number of genes have been isolated and sequenced. This demonstrated the existence of multigene families, as well as of gene products which have not yet been identified biochemically. Further studies have also established the occurrence of both transcriptional and post-transcriptional regulation. The transcription of genes encoding components of the phycobilisome rods is light-wavelength dependent, while translation of the phycocyanin genes may require the synthesis of another gene product irrespective of the light regime. In this report, we propose two hypothetical models which might be part of the complex regulatory mechanisms involved in the formation of functional phycobilisomes. On the other hand, transcription of genes involved in the gas vesicles formation (*gvp* genes) is turned on during hormogonia differentiation, while that of phycobiliprotein genes is simultaneously turned off. In addition, an antisense RNA which might modulate the translation of the *gvp* mRNAs is synthezised.

Abbreviations: AP – allophycocyanin, APB – allophycocyanin B, bp – base pair, GVP – gas vesicle protein, kb – kilobase, kDa – kilodalton, L – linker polypeptide, PC – phycocyanin, PCC – Pasteur Culture Collection, PE – phycoerythrin, UTEX – University of Texas algal collection

1. Introduction

In cyanobacteria, like in higher plants, light is not only used to perform photosynthesis but also to trigger and modulate various developmental and regulatory processes (Stanier and Cohen-Bazire 1977). The best known examples of photoregulation in cyanobacteria are those which affect pigment content and cell differentiation. No photoreceptor involved in such processes has yet been isolated in these prokaryotic organisms, although physiological evidence strongly suggests that such molecules exist to mediate responses to light (for reviews, see Tandeau de Marsac 1983 and Grossman et al. 1986). The purpose of this review is to present our recent progress in this research area. No attempt has been made to extensively cover the literature but we have chosen to discuss the molecular aspects since most of the physiological and biochemical studies performed in this field have often been reviewed during the last decade. When appropriate, the reader will be directed to specialized recent reviews in which more detailed information and original references are available.

In contrast to higher plants, cyanobacteria contain chlorophyll *a* but not chlorophyll *b*, and, like in rhodophyta, phycobiliproteins constitute the major light-harvesting antenna. These chromoproteins are organized into multimolecular structures called phycobillisomes which, although functionally analogous to the light-harvesting chlorophyll-protein complexes of higher plants, are attached to instead of being embedded in the photosynthetic membranes. Phycobilisomes are composed of two domains: the central core proximal to the photosynthetic membrane, and six rods which radiate from the core. This generally results in a hemidiscoidal structure. In addition, linker polypeptides, which are specifically associated with each phycobiliprotein complex of the phycobilisome, maintain its physical integrity and slightly modify the intrinsic spectral characteristics of the constituent phycobiliproteins, resulting in an optimization of the energy transfer to the terminal acceptor(s) of the phycobilisome. This structure is thus particularly well adapted to the trapping and funnelling of the photon energy to the photosystem II reaction centers (for reviews on phycobiliproteins and phycobilisomes, see for example, Glazer 1984 and 1985).

It is known that in general, in photosynthetic organisms, antenna size is inversely proportional to the light intensity received by the cultures during growth. Similarly, cyanobacteria respond to light intensity by increasing (under low light intensity) or decreasing (under high light intensity) their chlorophyll *a* and phycobiliprotein content, but the mechanisms by which light regulates these processes remain unknown. In contrast, the phenomenon of complementary chromatic adaptation, in which the synthesis of some components of the phycobilisome rods can be regulated by the incident light wavelength, has been more extensively studied and is undoubtedly governed by at least one photoreversible pigment. This photoreceptor presents some analogy with the phytochrome of higher plants, although the most efficient wavelengths for its interconversion are different (green versus red radiation, instead of red versus far-red radiation for phytochrome; for a review on complementary chromatic adaptation, see Tandeau de Marsac 1983).

In some filamentous cyanobacterial strains, light also controls differentiation processes such as hormogonia differentiation. This process, which is part of a developmental cycle, is a prerequisite for the dispersal of these species in their natural habitats. Indeed, in contrast to vegetative trichomes from which they differentiate, hormogonia are more resistant to environmental stresses; they are also motile, the motility being correlated with the appearance of pili at their cell surface (G. Guglielmi unpublished data) and they are filled with gas vesicles which provide them with buoyancy (Walsby 1981). Interestingly, this differentiation process has been shown to depend upon several environmental factors, among which are red radiations (Armstrong et al. 1983, Tandeau de Marsac 1983, Rippka and Herdman 1985). Whether such a phenomenon is controlled by the same photoreceptor as the one involved in complementary chromatic adaptation is an open question.

With the aim of understanding the molecular basis of these physiological adaptations to the environment, and to determine if one or several photoreceptors are involved in these regulatory processes, we have undertaken the characterization of the structure, organization, and expression of the genes encoding phycobilisome components, as well as those involved in hormogonia differentiation in *Calothrix* 7601 (also called *Fremyella diplosiphon* UTEX 481)*. This filamentous cyanobacterium belongs to chromatic adapters of group III (Tandeau de Marsac 1983).

2. The phycobilisome components: phycobiliproteins and linker polypeptides

2.1 Features emerging from biochemical and physiological data

Biochemical and physiological studies have shown that phycobiliproteins are stable chromoproteins, which can represent up to 50% of the total cell protein content and are organized in phycobilisomes. In response to changes

*Cyanobacterial strains are designated throughout this review by their genus name followed by their number in the Pasteur Culture Collection when available (for example, *Calothrix* 7601 = *Calothrix* sp. PCC 7601).

of the incident light, the number of phycobilisomes per cell, as well as the molecular composition and the size of the rods of the phycobilisomes, may vary (Raps et al. 1985; for reviews, see Cohen-Bazire and Bryant 1982, Glazer 1982, Tandeau de Marsac 1983).

Analysis of *Calothrix* 7601 phycobilisomes, purified from fully red- or green-light-adapted cells, has allowed a rather precise determination of their structure and composition under these two different light regimes. The core of both red- and green-light phycobilisomes seems invariant and is composed of at least five chromophoric polypeptides and one non-chromophoric linker polypeptide: the allophycocyanin subunits (αAP and βAP), allophycocyanin B (αAPB), the 18.3 kDa polypeptide (β18.3), the 92 kDa anchor polypeptide (L_{CM}^{92}) and the AP-associated linker ($L_C^{7.8}$). By analogy with *Synechocystis* 6701 phycobilisomes (Gingrich et al. 1983, Glazer and Clark 1986), it is thought that these polypeptides are organized into 12 trimers within the core of *Calothrix* 7601, the total number of molecules being 32, 34, 2, 2, 2, 6 for αAP, βAP, αAPB, β18.3 L_{CM}^{92} and $L_C^{7.8}$, respectively. In contrast to the core, the rods of red- and green-light phycobilisomes of *Calothrix* 7601 differ in composition (Tandeau de Marsac 1983, G. Guglielmi unpublished data). Rods of red-light phycobilisomes are composed of at least four chromophoric polypeptides: the phycocyanin-1 subunits (αPC1 and β PC1) and the phycocyanin-2 subunits (αPC2 and βPC2), the absorption maxima of which are situated at approximately 620 nm. These polypeptides are organized into hexamers ($\alpha\beta$)6. In addition, there are three PC-associated rod linker polypeptides (L_R^{38}, L_R^{39} and $L_R^{9.7}$) and one PC-associated core-rod linker polypeptide (L_{CR}^{30}). In rods of green-light phycobilisomes, the basal disc is a PC hexamer-L_{CR}^{30} complex like in red-light phycobilisome rods, while the more distal discs are PE hexamers whose absorption maxima are situated at approximately 565 nm. Two specific rod linker polypeptides (L_R^{35} and L_R^{36}) are associated with these hexamers of PE. A schematic representation of red- and green-light phycobilisomes from *Calothrix* 7601 is presented in Fig. 1.

No precise analysis has been performed on the composition of phycobilisomes from *Calothrix* 7601 cells grown under different light intensities. However, from information obtained with other cyanobacteria (Yamanaka and Glazer 1981, Raps et al. 1985), we anticipate that the number of phycobilisomes will be greater, and/or the size of their rods longer, in cells grown under low light intensity than in cells grown under high light intensity. In addition, a change in the PC/PE ratio could occur, if the level of synthesis of these two phycobiliproteins is not modified to the same extent in response to a given light intensity. A specific degradation of phycobilisome components, mainly PC, has been shown to occur in several *Anabaena* species (Foulds and Carr 1977, Wood and Haselkorn 1980), in

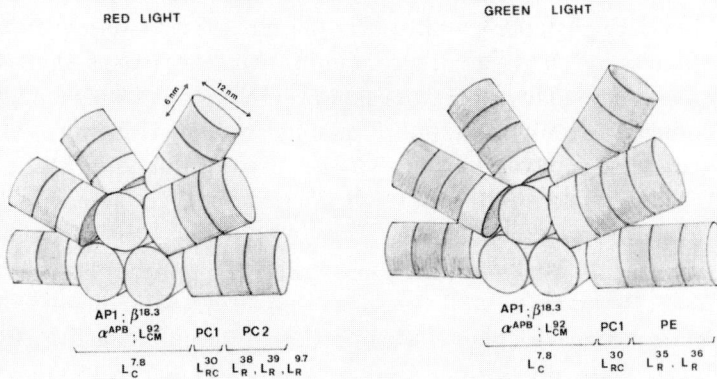

Fig. 1. Schematic representation of the *Calothrix* 7601 phycobilisomes purified from cells grown under red- or green-light conditions. For phycobiliproteins, the abbreviations AP1, PC1, PC2 and PE refer to the α and β subunits of allophycocyanin-1, phycocyanin-1, phycocyanin-2 and phycoerythrin, respectively. αAPB and β18.3 denote the α-type allophycocyanin B and the β-type phycobiliprotein of MM 18.3 kDa, respectively. Linker polypeptides are abbreviated, with a superscript denoting their apparent molecular mass in kDa and a subscript that specifies their location in the phycobilisome: R, rod substructure; RC, rod-core junction; C, core. The anchor polypeptide (MM 92 kDa) is denoted L_{CM}^{92}, with CM for core-membrane junction. The size of a disk (6 $\times$ 12 nm) corresponds to an hexamer ($\alpha\beta$) 6 of PC or PE in the rod. The location of the linker polypeptides within the rods has not yet been precisely established.

Synechococcus 6301 (Lau et al. 1977, Yamanaka and Glazer 1980), *Synechococcus* 7002 (Stevens et al. 1981) and *Synechocystis* 6803 (Elmorjani and Herdman 1987) in response to nitrogen starvation. In contrast, adaptation of cyanobacterial cells to changes in light wavelength, and most probably in light intensity, does not result in the induction of specific degradative processes (Bennett and Bogorad 1973). This suggests that most of the events which lead to changes in the phycobilisome content of the cells, or in their composition, are very likely to result from precise transcriptional (and/or translational) controls of the expression of the genes encoding phycobilisome components.

2.2 Functional organization of the genes involved in the formation of phycobilisomes

2.2.1 Gene isolation and characterization

Comparisons of partial or complete amino acid sequences of phycobiliproteins have shown that individual subunits are highly conserved among different cyanobacterial species (about 80% identity) and that each phycobiliprotein subunit shares homology with all the others, especially in the region of the conserved chromophore attachment site (Cohen-Bazire and Bryant 1982, Füglistaller et al. 1983, Glazer 1984). Consequently, our strategy for cloning the different phycobiliprotein genes from *Calothrix* 7601 was mostly based on the use of heterologous DNA probes which include the region coding for the chromophore attachment site. However, when the amino acid sequence was known, we also used synthetic oligonucleotide probes. Hybridization analyses of restriction enzyme digests were performed by using an *apcAB* probe (αAP and βAP) from *Synechococcus* 6301 (Houmard et al. 1986), *cpcA* (αPC) and *cpcB* (βPC) probes from *Synechococcus* 7002 (De Lorimier et al. 1984), as well as synthetic oligonucleotides (17 nucleotides long) designed from a portion of the amino acid sequence of the *Calothrix* 7601 αPE subunit (Sidler et al. 1986). Seven *Calothrix* 7601 *Eco*RI fragments of 9.5, 7.0, 6.5, 6.3, 4.5, 3.7 and 3.5 kb, cloned either from a λ EMBL3 genomic library or from partial libraries constructed in *E. coli* plasmid vectors, were totally or partially sequenced (for references, see below). The predicted amino acid sequence of each open reading frame has been compared with the sequences previously determined for proteins purified from *Calothrix* 7601 or from other cyanobacteria. Their transcription has been examined by hybridization experiments of total RNA extracted from *Calothrix* 7601 cells grown under green- or red-light conditions. Most of these open reading frames have thus been identified. As shown in Fig. 2, each of the cloned *Eco*RI fragments carries one or two genes encoding phycobiliprotein subunits and, for three of them, genes encoding linker polypeptides have been found. Some additional open reading frames, which do not correspond to any known phycobilisome components, were found to be involved in the formation of functional phycobilisome rods.

Phycobiliprotein genes. As expected from hybridization experiments, the 7.0 kb *Eco*RI fragment carries two genes, *cpeB* (552 bp) and *cpeA* (492 bp), which encode βPE and αPE, respectively (Mazel et al. 1986). These two genes are 79 nucleotides apart. It is surprising that, although only two different PCs (the 'constitutive' PC1 and 'inducible' PC2) have been characterized biochemically, three *Eco*RI fragments were found to harbor complete sets of PC genes. The first PC gene cluster, *cpcB1* (516 bp) and *cpcA1* (486 bp), is carried by the 4.5 kb *Eco*RI fragment and encodes βPC1 and αPC1, respectively (Conley et al. 1986, Mazel et al. 1988). The second PC gene cluster, *cpcB2* (516 bp) and *cpcA2* (486 bp), carried by the 6.5 kb *Eco*RI fragment, encodes βPC2 and αPC2, respectively (Conley et al. 1985, V. Capuano et al. 1988). The third PC gene cluster, *cpcB3* (519 bp) and *cpcA3* (486 bp) carried by the 9.5 kb *Eco*RI fragment encodes the β and α subunits

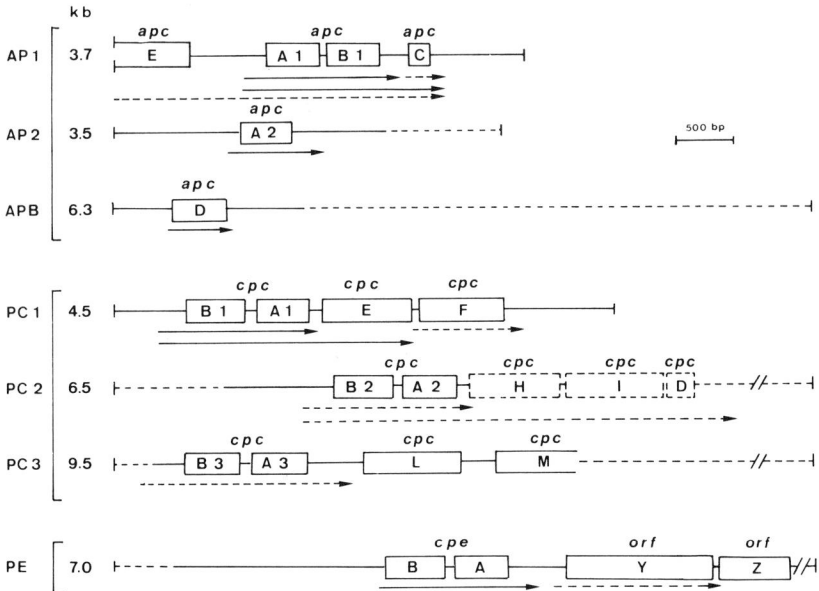

Fig. 2. Physical organization and transcription of genes involved in the biosynthesis of *Calothrix* 7601 phycobilisomes. Each *Eco*R1 fragment is denoted by its size in kilobases (kb). Solid or dotted lines represent totally or partially sequenced DNA, respectively. Solid or dotted arrows represent mRNA mapped or only identified in Northern hybridization experiments, respectively. The different genes are designated as follows: *apcA* and *apcB*, for the α and β subunits of the allophycocyanins (AP1 and AP2); *apcD* for the allophycocyanin B (APB); *cpcA* and *cpcB*, for the α and β subunits of the phycocyanins (PC1, PC2 and PC3); *cpeA* and *cpeB*, for the α and β subunits of the phycoerythrin (PE); *apcE*, for the anchor polypeptide L_{CM}^{92}; *apcC*, for the linker polypeptide $L_C^{7.8}$; *cpcL* and *cpcM*, for rod linker polypeptides; *cpcE, cpcF, orfY* and *orfZ*, for the open reading frames which correspond to unidentified gene products. The genes designated *cpcH, cpcI* and *cpcD* which encode the linker polypeptides L_R^{38}, L_R^{39} and $L_R^{9.7}$, respectively, have been sequenced by Lomax et al. (1987). Some of the previously published gene designations have been changed according to the nomenclature rules recently proposed for cyanobacterial genes (Houmard and Tandeau de Marsac 1988).

of a second 'constitutive' PC species, provisionally called PC3 (Mazel et al.1988). In each PC gene cluster, the *cpcB* gene is located upstream from the *cpcA* gene. The intergenic regions of the *cpc1, cpc2* and *cpc3* gene clusters are 118, 69 and 116 nucleotides long, respectively, and exhibit little or no sequence homology. The deduced amino acid sequences of the *cpcB3* and *cpcA3* genes are 75–78 and 84–85% identical to those deduced from the *cpcB1* or *cpcB2* and from the *cpcA1* or *cpcA2* genes, respectively. Possible roles and location in the phycobilisome structure of the new type of PC (PC3) will be discussed in section 2.3.

Three genes encoding an α-type subunit of AP have been isolated in *Calothrix* 7601, while only two different α-type subunits of APs (αAP1 and αAPB) have been biochemically characterized in cyanobacterial phycobilisomes. The first gene, apcA1 (483 bp), is located 70 nucleotides upstream from the gene encoding βAP1, apcB1 (486 bp). The two other genes, apcA2 and apcD, which are each 483 nucleotides long, are carried by the 3.5 and 6.3 kb EcoRI fragments, respectively. No open reading frame which could encode phycobiliprotein subunits has been found in the vicinity of these two genes (Houmard et al. 1988). The deduced amino acid sequence of the predicted apcD product is about 70% homologous to the sequence of the APB purified from the cyanobacterium Synechococcus 6301 (Suter et al. 1987). It contains the two tryptophan residues specifically found in this phycobiliprotein. The predicted amino acid sequence of the apcA2 gene shares 59 and 43% identity with those deduced from apcA1 and apcD, respectively (J. Houmard and T. Coursin unpublished data). Additionally, an open reading frame (apcE), located upstream from apcA1, has been partially sequenced. By analogy to Nostoc sp. (Zilinskas et al. 1987), our preliminary data indicate that the putative truncated gene product corresponds to the high molecular weight protein L_{CM}^{92} (J. Houmard and V. Capuano unpublished data).

With the exception of the 3.7, 6.5 and 4.5 kb EcoRI fragments which are adjacent on the chromosome (Conley et al. 1986), the other EcoRI fragments are not in close proximity to each other (J. Houmard and T. Coursin unpublished data). An alignment of the predicted amino acid sequences of the phycobiliprotein genes is presented in the Appendix (Fig. 11A).

Linker polypeptide genes. Some linker polypeptides have been sequenced from purified proteins (Füglistaller et al. 1984, 1985, 1986). Based on their homology with the deduced amino acid sequences of different open reading frames found in Calothrix 7601, three linker polypeptide genes have been identified. The apcC gene (204 bp) located 240 nucleotides downstream from apcB1 encodes the small linker polypeptide $L_C^{7.8}$ associated with AP in the core of the phycobilisome. Its deduced amino acid sequence is 73 and 91% homologous to those from Synechococcus 6301 (Houmard et al. 1986) and Mastigocladus laminosus (Füglistaller et al. 1984), respectively. The cpcL gene (978 bp) and the partially sequenced cpcM gene are located downstream from the cpcA3 gene. These genes encode two polypeptides which are highly homologous to the linker polypeptides L_R^{33} from Synechococcus 7002 (D.A. Bryant, cited in Glazer 1987) and $L_R^{34.5}$ from Mastigocladus laminosus (Füglistaller et al. 1985, 1986), respectively. Moreover, these genes share approximately 55% identity with the two genes which are located downstream from cpcA2 and encode the linker polypeptides L_R^{38} and L_R^{39} associated with PC2 in red-light phycobilisomes (Lomax et al. 1987). The cpcL and *cpcM* genes are thus likely to encode additional linker polypeptides associated with PCs in the phycobilisome rods. A compilation of the predicted amino acid sequences of the linker polypeptides, including those published by Lomax et al. (1987), is presented in the Appendix (Fig. 11B).

Unassigned open reading frames. Four other open reading frames were found by sequence analysis which are likely involved in phycobilisome formation. The first two are *cpcE* (882 bp), located 73 nucleotides downstream from *cpcA1* (Mazel et al. 1988), and *cpcF* (783 bp) which is located 51 nucleotides downstream from *cpcE*. As will be discussed later (see *Regulation of gene expression*), *cpcE* is cotranscribed with the *cpcB1* and *cpcA1* genes in contrast to *cpcF* which is independently transcribed. The two other open reading frames, *orfY* (1290 bp) and *orfZ* (618 bp), have been also completely sequenced. The first one, *orfY,* is located 522 nucleotides downstream from *cpeA*, while *orfZ* is 51 nucleotides downstream from *orfY*. The putative gene products of these four open reading frames do not share significant homology either with known linker polypeptides or with identified phycobiliproteins. Moreover, they could not be assigned to any known proteins by comparison with sequences available in DNA (Los Alamos and EMBL) and protein (NBRF and PseqIP) data banks (D. Mazel unpublished data). Their deduced amino acid sequences are presented in the Appendix (Fig. 11C).

2.2.2 Transcription analysis

The characterization of the mRNA species corresponding to most of the *Calothrix* 7601 genes that we have identified has been carried out by Northern hybridization, S1 nuclease mapping and primer extension experiments using total RNA extracted from cells grown under red- or green-light conditions. The identified mRNA species are presented in Fig. 2.

From this analysis, it appears that only *apcA2, apcD, cpcF* and *orfY* are transcribed as monocistronic units, the sizes of the transcripts being approximately 0.65, 0.55, 1.1 and 1.5 kb, respectively. The 5′ extremities of the *apcA2* and *apcD* transcripts are located 97 and 24 nucleotides respectively upstream from the initiation codon of the corresponding genes. The *apcA1B1C, cpcB1A1E* and *cpcB3A3* gene clusters are transcribed as polycistronic units. The *apcA1B1C* operon is transcribed into three major mRNA species of approximately 1.7 (*apcA1B1C*), 1.4 (*apcA1B1*) and 0.25 kb (*apcC*). The 1.7 and 1.4 kb transcripts have the same 5′ extremity, 193 nucleotides upstream from the initiation codon of the *apcA1* gene. The 5′ end of the shortest transcript has not been mapped, but is likely to be located between the *apcB1* and *apcC* genes. In addition, two transcripts of 5.7 and 5.5 kb, which are about one hundred times less abundant than the 1.4 kb mRNA species, can also be detected. These correspond to the cotranscription of both the *apcE* gene and the *apcA1B1C* operon. Thus, these large transcripts include most of the genes which encode components of the phycobilisome core. The *cpc1* operon is transcribed as two mRNA species of 1.45 and 2.4 kb corresponding to *cpcB1A1* and *cpcB1A1E*, respectively. These transcripts have the same 5′ extremity which occurs 255 nucleotides upstream from the initiation codon of the *cpcB1* gene. The *cpcB3A3* operon is transcribed as a unique mRNA species of approximately 2 kb. All the genes mentioned above are transcribed at roughly similar levels in cells grown under red or green light conditions and, consequently, are probably independent of the light wavelength control exerted during complementary chromatic adaptation.

The remaining two gene clusters, *cpc2* and *cpe*, which respectively encode the 'inducible' PC2 and PE subunits, are also transcribed as polycistronic units, but only under specific chromatic illumination (Conley et al. 1985, Mazel et al. 1986, our laboratory unpublished data). Two red-light specific transcripts of the *cpcB2A2* operon can be detected. One (1.5 kb) is produced by the cotranscription of *cpcB2* and *cpcA2*. The second transcript (3.8 kb) extends downstream from *cpcA2* and includes the genes encoding the PC2-associated linker polypeptides: L_R^{38}, L_R^{39} and $L_R^{9.7}$ (Lomax et al. 1987, D. Mazel unpublished data). The *cpeBA* operon is transcribed as a unique green-light specific 1.5 kb mRNA species, whose 5′ end occurs 66 nucleotides upstream from the initiation codon of *cpeB* (Mazel et al. 1986, D. Mazel unpublished data).

With the exception of the *cpe* operon in which *E. coli*-like promoter sequences are found, analysis of the nucleotide sequences upstream from the potential start sites of transcription of the different mRNA species identified in *Calothrix* 7601, did not reveal sequences similar to the *E. coli* consensus promoter sequences (Fig. 3A). Moreover, no consensus sequence which might correspond to the binding site for the *Calothrix* 7601 RNA polymerase could be reasonably deduced from the comparison of all the sequences which are available today (Fig. 3A). In contrast, a striking homology (56%) is found between a 43 bp long sequence (located 41 nucleotides upstream of and including the first two nucleotides of the *Calothrix* 7601 *cpc1* transcripts) and a sequence of *Synechococcus* 7002 (332 to 374 nucleotides upstream from the AUG of the *cpcB* gene), which, according to Pilot and Fox (1984), probably corresponds to the promoter region (Fig. 3B). Similarly, in *Anabaena* 7120, the nucleotide sequence (nucleotides −287 to −243 in Belknap and Haselkorn 1987) which precedes and includes the transcription start site (nucleotide −243) is 73% homologous to that of *Calothrix* 7601 (Fig. 3B). In contrast, no other homology has been noticed

A

```
            -50       -40       -30       -20       -10
apcA1B1C    TAATATTACAAAATATTAAGAGCAGTCATAAATGCTCAACAGAATGCCGGAGAATGTTTT...193bp...ATG

apcA2       CAGATTATTTTTCATATCGTAGTTAACCAAATCAACCAAGAAATGCAACAGCGCCAAGTC....97bp...ATG

apcD        AATACATCGCTGCGTATAACCCTTCACTTGTGTACCTCCAACTCGTTAAACTAGCAAATA....24bp...ATG

cpcB1A1E    CCTCTTAGTATGACTAACTTGACAATTCGTAATAAACAAACGATCCAACGATATAGTATA...255bp...ATG

cpeBA       AAGGATTGTTACTTAGTTTCTCATAACTGAGACTGAGATAGCTTTCATCTTTTATGTTCT....66bp...ATG

gvpA1A2C    GTTTACAGTATTTTGGGTGTGATTCTATTTACATTTCAATCGAGTGTTAATAGTATTGTT....56bp...ATG

gvp(antiRNA) ATTGCTAACTCAGCAATTCTGATATTATGGACTCTGCTTAATGATTAAGCAGGTACTGCA
```

B

```
                         -40       -30       -20       -10      ↓
Synechococcus 7002   AGATCTTTTTACAAGATGTAATGTTTAAATGC-CGGCAGACGTTGTATAACATTTACCTA
                       *  ** ****  *****    *** *  *    ** * ******        *
Calothrix     7601   GACTAACTTGACAATTCGTAATAAACAAACGATCCAACGATATAGTATAaACAAGTAATG
                       * ** ****** **** ***  ** ****  *    ** ********    **
Anabaena      7120   TTATTTATTCACAATTTGTAACAAAATAAGGATC-TATAGCATTGTATAAACaTAAGCTG
```

Fig. 3. Nucleotide sequences of the promotor regions. A. Comparison of the sequences from the *Calothrix* 7601 genes. Distances from the first ATG of the coding sequences are indicated in base pairs (bp). Numbering of nucleotides is based on the assignment of the first base of the transcripts as nucleotide + 1. B. Comparison of the sequence of the promoter regions of the *Calothrix* 7601 *cpcB1A1E* operon with those of the *cpc* operon from *Synechococcus* 7002 (nucleotides 1 to 59 in De Lorimier et al. 1984) and from *Anabaena* 7120 (nucleotides –294 to –236 in Belknap and Haselkorn 1987). Lower cases correspond to the start sites of transcription determined for *Calothrix* 7601 (also indicated by an arrow) and for *Anabaena* 7120. Stars indicate nucleotide identities. Dashes indicate gaps inserted to maximize homology. Numbering of nucleotides is based on the assignment of the first base of the *Calothrix* 7601 *cpc* transcripts as nucleotide + 1.

within the 5′ untranslated regions of the transcripts. It is thus tempting to hypothesize that the homologous part of these sequences plays a role in initiating transcription of the PC genes in these cyanobacterial strains (Mazel et al. 1988).

The 3′ end of the *cpcB1A1* transcript has also been mapped. Transcripts terminate a few nucleotides downstream from a thermodynamically stable stem-and-loop structure which is not followed by an AU-rich sequence (Mazel et al. 1988). This result is similar to those obtained for the 3′ ends of the transcripts of the *gvpA1A2C* operon (see *Transcription of the gvp genes*).

2.2.3 Regulation of gene expression

Phycobilisomes are supramolecular structures in which the stoichiometry of the different components and their positioning within the structure are essential for optimizing their light-harvesting capacity and their ability to transfer energy to the reaction centers of photosystem II. Moreover, the energy trapped by the phycobilisomal units generally does not exceed the functional capacity of the photosynthetic reaction centers. To do this, some cyanobacteria are not only able to alter the number and the size of their phycobilisomes in response to the incident light energy, but also to modify their rod composition in response to light wavelength. These properties rely on an accurate photoperception system coupled to precise coordination of the expression of the different genes involved in these light-dependent processes. At the molecular level, this can be performed by varying the copy number and/or by regulatory mechanisms which can operate at different levels: transcriptional (level of transcription and/or mRNA turnover), translational (occurrence or modulation of the protein synthesis) and post-translational (maturation and/or degradation of proteins).

Based on physiological studies performed on exponentially growing cells, it has been established that phycobiliproteins are stable proteins which do not turnover (Bennett and Bogorad 1973, Tandeau de Marsac 1977). After transfer from red to green illumination, or *vice versa*, changes in the phycobiliprotein content of the cells are due to *de novo* synthesis, the preexisting phycobiliproteins disappearing only by dilution during growth. Experiments performed using transcription inhibitors indicated that adaptation to new illumination conditions were probably due to transcriptional regulation of the expression of the genes encoding PE and most probably PC2 (Gendel et al. 1979, Tandeau de Marsac 1983).

With the recent characterization of the genes encoding components of the phycobilisomes in *Calothrix* 7601 and the analysis of their transcription using total RNA from cells grown under red- or green-light conditions, it has been definitely established that the transcription of the *cpcB2A2HID* and *cpeBA* operons is red- and green-light specific, respectively (Fig. 2; Conley et al. 1985, Mazel et al. 1986, our laboratory unpublished data). However, this regulatory mode constitutes only part of the complex transcriptional regulatory system which occurs in *Calothrix* 7601. Indeed, at least three operons, *cpcB1A1E*, *cpcB2A2HID* and *apcEA1B1C*, give rise to segmented transcripts which are present at different levels. It is also worth noting that in each of these three operons, and also in the *cpeBA* and *cpcB3A3* operons, the genes coding for the α and β subunits of a specific phycobiliprotein are always cotranscribed, ensuring that both subunits can be made in an equimolar ratio. Besides the *cpcB3A3* operon and the *apcA2* gene, which are transcribed about 50 times less efficiently than the *cpcB1A1* and the *apcA1B1* genes, respectively, and to which no function has yet been assigned, we generally observe a good correlation between the levels of transcription of the different operons and the proportions of the corresponding phycobiliproteins in the phycobilisome. This suggests that the promoters of these operons may have different efficiencies and/or require specific effectors. The occurrence of segmented transcripts for the *apcEA1B1C* and *cpcB2A2H1D* operons, and the higher level of the mRNA species which do not contain the linker sequences, constitute another means of coordinating the expression of genes whose products are required in different molar ratios (Grossman et al. 1986, J. Houmard and V. Capuano unpublished data). The occurrence of a small mRNA species (0.25 kb) corresponding to the *apcC* gene which encodes the AP-specific linker $L_C^{7.8}$, is somewhat more surprising. As also noticed by Grossman et al. (1986), we observe that this small transcript is more abundant in cells grown under red light than under green light, which is unexpected given the invariance of the core composition during complementary chromatic adaptation. Conversely, the 1.7 kb mRNA (*apcA1B1C*) is less abundant in cells grown under red light than under green light, while the amount of the 1.4 kb species (*apcA1B1*) is almost unchanged. Since the sum of the 0.25 kb and 1.7 kb transcripts is about constant under both light regimes, this suggests that the small transcript might result from the processing of the longer one.

In *Synechococcus* 7002 (Bryant et al. 1986), the *cpc* gene cluster consists of six adjacent genes, namely *cpcB* (βPC), *cpcA* (αPC), *cpcC* (L_R^{33}) and *cpcD* ($L_R^{9.7}$), followed by two open reading frames, *cpcE* and *cpcF*. In *Anabaena* 7120 (Belknap and Haselkorn 1987), at least five genes have been found which are organized similarly and are highly homologous to the *cpcBACDE* genes from *Synechococcus* 7002. Surprisingly, the *Calothrix* 7601 *cpcB1* and *cpcA1* genes are followed by two open reading frames whose deduced amino acid sequences bear no significant homology with any of the known linker polypeptide sequences. However, the first open reading frame, *cpcE*, which is cotranscribed with the *cpcB1* and *cpcA1* genes, shares approximately 65% identity with the *cpcE* gene located immediately downstream from *cpcD* in *Synechococcus* 7002 and *Anabaena* 7120 (Belknap and Haselkorn 1987, D.A. Bryant personal communication). Finally, unlike *Synechococcus* 7002 in which there is a short open reading frame (114 bp long) within the transcript upstream from the *cpcB* gene (De Lorimier et al. 1984, Pilot and Fox 1984), no open reading frame has been found upstream from *cpcB1* in *Calothrix* 7601, although this untranslated region is also rather long (255 bp).

As shown in Fig. 4, an interesting feature of the *cpcB1A1E* operon is that 22 nucleotides (sequence 1), located at the 5′ extremity of the transcript, are able to form a stem-and-loop structure ($\Delta G = -14.6$ kcal/mol) with the 22 nucleotides (sequence 3) located 35 nucleotides downstream. In addition, 11 nucleotides (sequence 2), located inside the loop of this secondary structure, are able to form a stem-and-loop structure ($\Delta G = -9.8$ kcal/mol) with part of sequence 1 (sequence 1′) which is involved in the formation of the stem of the structure 1–3. In contrast to structure 1′–2, the larger and

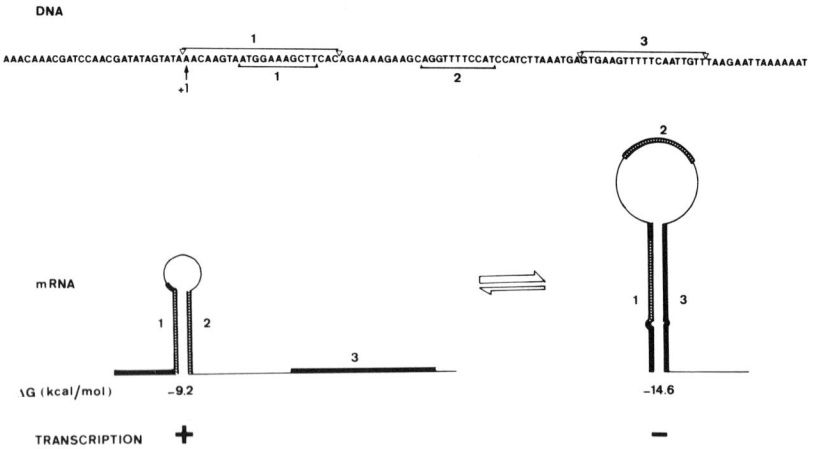

Fig. 4. Hypothetical attenuation-like regulatory mechanism for the transcription of the *Calothrix* 7601 *cpcB1A1E* operon. The arrow denotes the start site of transcription (+1). Numbers indicate the DNA sequences which are involved in the formation of secondary structures. The upper line coresponds to the DNA sequence and the bottom of the figure presents the equilibrium between the two potential secondary structures which can be formed in the 5′ untranslated region of the mRNAs. The free energy values (ΔG) were calculated according to Cech et al. (1983).

more stable one (structure 1–3) is followed by a very AT rich region. These observations, together with the length of the 5′ untranslated region of the mRNA, lead us to propose an 'attenuation' model, in which structure 1–3 could act as an early transcription terminator, while structure 1′–2 would allow transcription to proceed. A positive effector could stabilize structure 1′–2, as soon as the corresponding sequences are transcribed, in order to prevent the formation of the potential 'terminator' (structure 1–3). In this hypothetical model, the equilibrium between these two potential structures would modulate the transcription rate of the *cpcB1A1E* operon in response to environmental factors, such as light intensity or nutrient availability (Mazel et al. 1988).

On the other hand, while the *cpcB1A1* and *cpcB1A1E* transcripts have the same 5′ extremity, they are present in different amounts. Thus, there must be a difference either in the termination efficiency or in the stability of these two mRNAs. Potential secondary structures in the mRNAs have been postulated to play an important role in transcription termination in prokaryotes (Platt 1986). However, recent results suggest that stem-and-loop structures might be important in protecting against 3′ exoribonucleases, rather than in terminating transcription (Brawerman 1987). Indeed, it has been clearly demonstrated that these structures can influence the rate of mRNA decay and increase their stability by preventing degradation (Belasco et al. 1985, Wong and Chang 1986, Newbury et al. 1987). In the case of the *cpcB1A1* transcript, an extremely stable stem-and-loop structure ($\Delta G = -30\,\text{kcal/mol}$) could be formed ending two nucleotides upstream from its 3′ end. In contrast, no such stable secondary structure exists downstream from the *cpcE* stop codon and, on Northern blots, the band corresponding to the *cpcB1A1E* mRNA is more diffuse than that of *cpcB1A1* mRNA, suggesting a lower stability of the *cpcB1A1E* mRNA. Although we cannot rule out that these hairpin structures play a role in the release of the RNA polymerase, it is tempting to correlate their presence with the stability and abundance of the two mRNA species corresponding to *cpcB1A1* and *cpcB1A1E*.

Finally, the fact that the *cpcE* gene is cotranscribed with the *cpcB1* and *cpcA1* genes but does not encode a phycobilisomal protein-like product, raises the questions concerning the function of the *cpcE* gene product. Among other possibilities, it might encode a factor involved in a post-translational process, such as the linkage of the chromophore to the apoproteins or the maturation of the polypeptides. Examples of post-translational modifications of the phycobilisome components have recently been described, including methylation of the phycobiliprotein β subunits (Minami et al. 1985, Klotz et al. 1986, Rümbeli et al. 1987), reversible phosphorylation (Allen et al. 1985) and glycosylation of some linker polypeptides (Riethman et al. 1987). Some of these modifications might be necessary to prevent proteolytic degradation of the phycobilisome components and/or to confer a functional configuration. Analysis of a mutant obtained by interposon mutagenesis (Bryant 1988) indicated that the *cpcE* gene is probably implicated in chromophore attachment to the PC α subunit in *Synechococcus* 7002. This would be in agreement with our first hypothesis. However, confirmation of this result awaits further transcription analysis in order to rule out a possible polar effect of the mutation studied on the expression of the *cpcF* gene located downstream from the inactivated *cpcE* gene in *Synechococcus* 7002.

In *Calothrix* 7601, the *cpcF* gene found downstream from *cpcE*, is transcribed as a unique mRNA species and could have thus been considered as a gene unrelated to the formation of phycobilisomes. However, analysis of spontaneous pigmentation mutants from *Calothrix* 7601 strongly suggests that the *cpcF* gene product is involved in some regulatory processes in the synthesis of phycobilisomes. Indeed, the characterization of the pigmentation mutant, GY3, revealed that a typical bacterial insertion sequence, IS*701*, had spontaneously inserted in the 5′ part of *cpcF*, leading to transcription termination. Phycobilisomes purified from this mutant are smaller than those from the wild-type strain. They are practically devoid of PCs and

PE and, consequently, do not contain rods, while linker polypeptides appear to be present. Although red-light phycobilisomes from this mutant contain less than 10% of PC2 when compared with the wild-type phycobilisomes and green-light phycobilisomes contain only traces of PE, the synthesis of these phycobiliproteins remains under the chromatic light control (D. Mazel unpublished data).

Even more interesting are the results of the anlaysis of the transcription of the phycobiliprotein genes in the mutant GY3. They revealed that the *apcA1B1, cpcB1A1* and *cpcB2A2* operons are transcribed to the same extent as in the wild-type strain, while the transcription of the *cpeBA* operon is dramatically decreased. However, no free PCs (or apoproteins) have been detected by biochemical and immunological techniques although the AP1 concentration is unmodified. This suggests that the *cpc* genes are transcribed but are not translated or are only poorly translated. One cannot totally exclude, however, that the apoproteins are very rapidly degraded. Since a single gene product is unlikely to be directly involved in the translational regulation of one operon and in the transcriptional regulation of another one, we propose, as a working hypothesis, that the *cpcF* gene product is primarily acting on the translation of the *cpc* genes, the effect on the transcription of the *cpe* genes being a secondary consequence of this translational regulation. Indeed, since PE hexamers cannot attach directly to the phycobilisome core, an autoregulation of the transcription of the *cpe* genes by free PE could occur. (D. Mazel unpublished data).

If our hypothesis is correct, how could the *cpcF* gene product act on the translation of the *cpc* genes? Potential secondary structures in the long 5′ untranslated region of the *cpcB1A1* operon and in the region between *cpcB1* and *cpcA1* might mask the ribosome binding site sequences. We thus propose a model of regulation (Fig. 5) in which the *cpcF* gene product acts as a positive effector destabilizing the hairpin structures upstream from *cpcB1* and allowing the ribosome to bind. Subsequently, the migration of the ribosomes along the mRNA would liberate the ribosome binding site upstream from the *cpcA1* gene allowing translation of this gene to proceed. Such a translational regulatory mode, which is independent of the light wavelength, would permit rapid changes in the synthesis of PC1 in response to environmental factors. Since hairpin structures are also present in the same region of the transcripts for the *cpcB2A2* and *cpcB3A3* operons, this suggests that these operons could also be regulated by a similar mechanism.

2.3 Conclusions

With the exception of the gene encoding one of the phycobiliproteins of the phycobilisome core, β18.3, all of the *Calothrix* 7601 phycobiliprotein genes

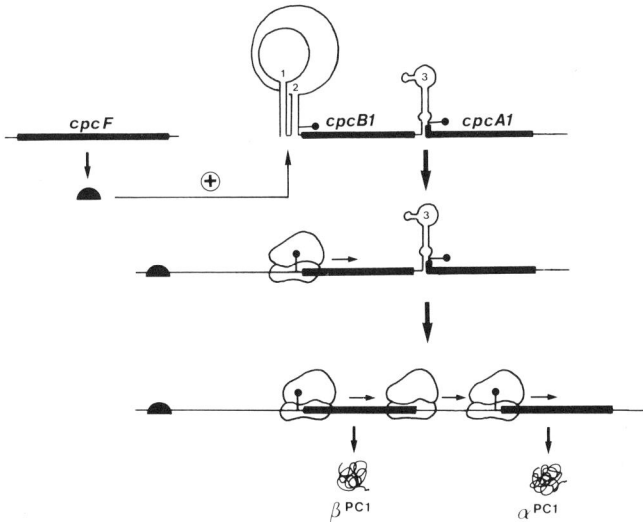

Fig. 5. Hypothetical model for post-transcriptional regulation of the *cpcB1A1E* operon mediated by the *cpcF* gene product. Boxed regions denote the coding regions of the genes. Hairpins 1, 2 and 3 are the potential secondary structures which might form within the *cpcB1A1* and *cpcB1A1E* transcripts (free energy values $\Delta G = -14.6$, -7.5 and -22.3 kcal/mol, respectively, calculated according to Cech et al. 1983). ↑ indicates the mRNA sequences to which ribosomes bind to initiate translation. The filled half-circle represents the *cpcF* gene product.

have now been identified, as have most of the genes encoding linker polypeptides. At least two levels of control are involved in the regulation of the expression of these genes: transcriptional and translational. The first mode of control occurs in response to changes in the chromatic light, but, according to our proposed 'attenuation' model, it might also occur in response to other environmental changes. The translational mode appears to be independent of the chromatic illumination, and is likely to be a more general process involved in modulating the synthesis of phycobilisome components.

Three sets of genes encoding both the α and β subunits of PCs and three genes encoding α-type subunits of AP have been found in *Calothrix* 7601. The functions of *cpcB3A3* and of *apcA2* remain to be determined. If their gene products are accommodated within the phycobilisome structure, the divergence of their amino acid sequences, as compared with those of the other *cpc* and *apc* gene products, suggests that they may have acquired different roles and/or lead to heterogeneity in the phycobilisome building blocks. Alternatively, these phycobiliprotein gene products might be specific for different cellular types. This second hypothesis can probably be ruled out, since results from experiments performed with *Calothrix* 7601 and 7504 (which, in contrast to *Calothrix* 7601, is still able to differentiate fully functional heterocysts) have shown that the level of transcription of the *cpcB3A3* genes is similar both in heterocysts and in vegetative cells (D. Mazel unpublished data). In contrast, detailed analyses of the structure and composition of the phycobilisome core from *Synechococcus* 6301 and *Synechocystis* 6701 have shown that the constituent trimers are heterogeneous. This heterogeneity is correlated with the function of the different phycobiliproteins in the transfer of energy to the photosynthetic reaction centers (Glazer 1982, Anderson and Eiserling 1986). It might well be that some of the AP α-type subunits found in the core correspond to the *Calothrix* 7601 *apcA2* gene product. Similarly, although the only example of rod heterogeneity reported so far is that of the PE- or PC-rich rods present in phycobilisomes from *Nostoc* sp. MAC (Glick and Zilinskas 1982, Anderson et al. 1983), heterogeneity of the 'constitutive' PC molecules could exist either within different hexamers or within one hexamer of the peripheral rods. In fact, this hypothesis would be in full agreement with the results obtained by G. Guglielmi and D.A. Bryant (unpublished data). These authors observed that PE-rich rods from *Nostoc* 8009 phycobilisomes purified from white-light grown cells only contain one PC hexamer which is composed of three types of PCs (two 'constitutive' PCs and one 'inducible' PC).

Several other genes, besides the genes encoding phycobilisome components, are required for the synthesis of the chromophores and their linkage to the apoproteins. Regulatory genes might also be necessary to ensure coordinated synthesis of the chromophore(s) and of the apoprotein. Furthermore, in the case of chromatically adapting strains, several other genes are expected to be involved in the synthesis of the photoreceptor. Particular efforts should now be directed towards their identification. The study of the function of the *apcA2* gene product is of particular interest with regards to that problem. Indeed, we cannot exclude that the *apcA2* gene product corresponds to the photoresponsive AP found in *Calothrix* 7601 by Ohad et al. (1979, 1980). Nevertheless, the question of the relationship between such a photoresponsive AP and the photoreversible pigment involved in complementary chromatic adaptation remains to be solved.

3. Hormogonia differentiation

As shown in Fig. 6 (A and B), vegetative filaments of *Calothrix* 7601 are morphologically different from differentiated hormogonia. Hormogonia cells are smaller and filled with refringent granules which are aggregates of gas vesicles providing cells with buoyancy. Pili are produced at the

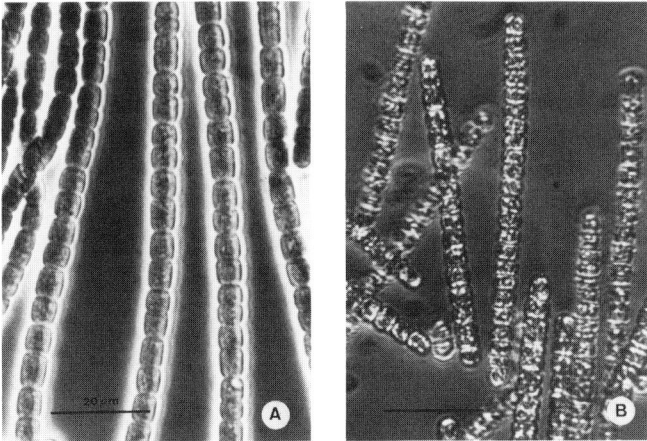

Fig. 6. Light micrographs of *Calothrix* 7601 vegetative trichomes (A) and differentiated hormogonia (B). Bar marker indicates 20 μm.

cell surface. Their presence is correlated with hormogonia motility (G. Guglielmi unpublished data).

3.1 Physiological characteristics of hormogonia differentiation

Physiological studies of hormogonia differentiation in *Calothrix* 7601 revealed that it may begin in any cell within the filament and then progresses to neighboring cells along the filament. Moreover, the percentage of differentiated cells has been shown to be highly dependent on i) the spectral quality of the light, ii) the growth phase of the preculture and iii) the incubation temperature. Indeed, 90 to 100% differentiation is obtained if cells in early exponential phase are transferred to red illumination and incubated at 25–30 °C. If, under similar conditions, cells are incubated either under green light or under 50% red light plus 50% green light, instead of red light alone, the percentage of differentiation decreases to less than 30% or to approximately 50%, respectively. Interestingly the efficiency of differentiation is lower for cells in mid- or late-exponential phase. Indeed, 90–100% differentiation is only obtained if such cells are transferred into fresh culture medium and then incubated under red light (T. Damerval and G. Guglielmi unpublished data).

Thin sections of cells collected at different times after the induction of the differentiation (Fig. 7) show that cells collected 6 h after the induction are dividing with a reduced elongation of the longitudinal cell wall. They produce gas vesicles which appear as electron-transparent cylindrical structures. After 12 h, cell divisions (usually two) are completed and aggregates

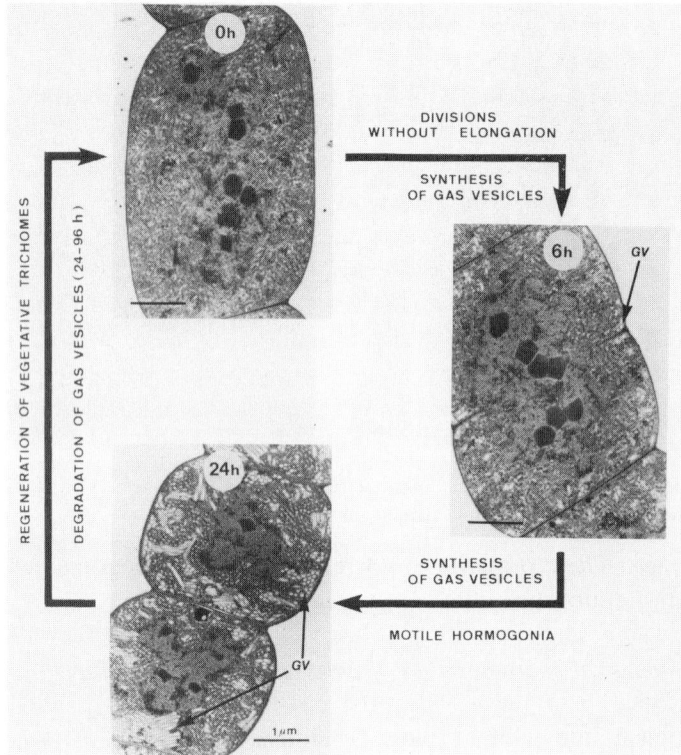

Fig. 7. Developmental cycle of hormogonia differentiation in *Calothrix* 7601. Electron micrographs of thin section of cells collected at various times (h) after the induction of the differentiation. Gas vesicles (GV) are visible as electron transparent structures. Bar marker indicates 1 μm.

of gas vesicles increase in size until 24 h. This hormogonia state is transient; as growth proceeds, regeneration of vegetative filaments occurs with a concomittant loss of the gas vesicles (T. Damerval and G. Guglielmi unpublished data). Figure 8 presents an electron micrograph of a gas vesicle isolated from *Calothrix* 7601.

3.2 Functional organization of the genes involved in gas vesicle formation

In order to elucidate the molecular basis of hormogonia differentiation, we looked for an appropriate marker. Since the formation of gas vesicles appears to be correlated with the other morphological changes which occur during hormogonia differentiation, we chose to isolate the gene encoding the major structural protein (GVP) of the gas vesicle.

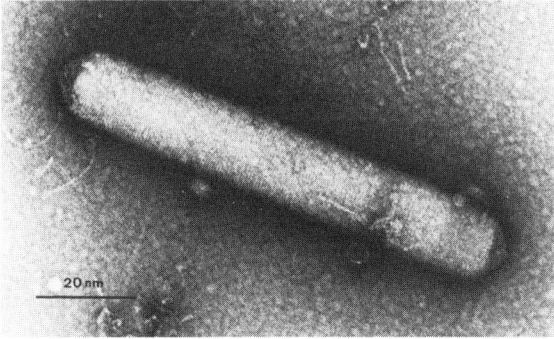

Fig. 8. Electron micrograph of a negatively stained gas vesicle isolated from *Calothrix* 7601.

3.2.1 Characterization of the gvp genes

Gas vesicle structure and function have been analyzed in detail (for reviews see Walsby 1978, 1981, 1987). According to studies by Walsby and coworkers, these vesicles, which are impermeable to water and permeable to gases, are made up of only one protein species whose amino acid sequence is highly conserved among cyanobacteria and other gas-vacuolate bacteria (Walker et al. 1984). It was thus possible to design a synthetic 29 nucleotide-long oligodeoxynucleotide sequence in order to probe the *Calothrix* 7601 genome. Two *Hin*dIII fragments of 2.6 and 2.3 kb hybridized with this probe. Both fragments were cloned from a genomic library constructed in λEMBL3 and the genes carried by these two DNA fragments were further analyzed. The 2.6 kb fragment bears two genes encoding the structural gas-vesicle protein. These genes, *gvpA1* and *gvpA2* are identical in length (216 bp) and are separated by 105 nucleotides. Their nucleotide sequences are 91.5% homologous. The 18 nucleotide differences are scattered along the length of the genes, but generally affect only the third base of a codon (Tandeau de Marsac et al. 1985, Damerval et al. 1987). These genes most likely arose by duplication of an ancestral gene followed by silent mutations. In contrast, the upstream and downstream regions of each of these two copies share no homology. Each of these two genes encode a protein of 7375 daltons (see Appendix, Fig. 12). The first methionine residue is apparently proteolytically processed (Tandeau de Marsac et al. 1985), since it has not been found in the amino acid sequence performed on the purified protein (Hayes et al. 1986). As expected from biochemical studies (Walsby 1978, Walker et al. 1984), the *gvpA* gene product is highly hydrophobic and shares a very high degree of homology with the total or partial amino acid sequences determined for the gas vesicles protein from other cyanobacteria and from halobacteria (Tandeau de Marsac et al. 1985, Damerval et al.1987, Das Sarma et al. 1987).

A third open reading frame has been found downstream from the *gvpA2* gene. It could not be correlated to any known gene or product, but has been called *gvpC* since it is cotranscribed with *gvpA1* and *gvpA2* (see Section 3.2.2). A potential ribosome binding site, GGAG, located 6 nucleotides upstream from an ATG sequence suggests that the *gvpC* gene starts 215 bp downstream from the end of *gvpA2* and would be 489 bp long. Unlike the gas vesicle protein, the *gvpC* gene product is predominantly hydrophilic. Interestingly, the internal part of the *gvpC* gene is composed of four contiguous repeats, each 99 bp long, which form highly homologous repeats in the deduced amino acid sequence (see Appendix, Fig. 12). Another kind of periodicity has been detected inside the 99 bp repeats, suggesting that the *gvpC* gene might have evolved by amplification of a 33 bp long primordial building block (Damerval et al. 1987).

A fourth gene involved in gas vesicle formation is present in the *Calothrix* 7601 genome. Indeed, the 2.3 kb fragment which hybridized with the 29-mer oligonucleotide also hybridized with a probe corresponding to an internal region of the *gvpA1* gene. This gene, designated *gvpD*, is located approximately 5 kb upstream from the other *gvp* genes. Preliminary sequence data indicate that *gvpD* is 252 nucleotides long and could encode a protein of approximately 8600 daltons. Consequently, the *gvpD* gene product is, like the *Halobacterium halobium* GVP (Das Sarma et al. 1987), a few amino acids longer than that of the *gvpA* genes. The sequence of the N-terminal 70 amino acid residues is about 80% identical to that of the *gvpA* gene products (T. Damerval unpublished data).

3.2.2 Transcription of the gvp genes

As expected for a differentiation process, the *gvp* genes are transcribed specifically in hormogonia and not in vegetative cells. For the purpose of a detailed transcriptional analysis of the different mRNA species corresponding to the *gvp* genes, total RNA was extracted from hormogonia 6 h after the induction of differentiation (see section 3.3). Hybridization experiments, coupled with mapping of the 5' and 3' ends of the transcripts, demonstrated that the 0.3, 0.8 and 1.4 kb long mRNA species correspond to the transcription of *gvpA1*, *gvpA1A2* and *gvpA1A2C*, respectively. The fourth mRNA species, 0.6 kb long, corresponds to the transcription of *gvpD* (Csiszàr et al. 1987). Thus, only the first three genes are organized in an operon which is segmentally transcribed. As shown in Fig. 9, these transcripts have the same 5' extremity, but different 3' ends located, in each case, a few nucleotides downstream from stem-and-loop structures, the thermodynamic stability of

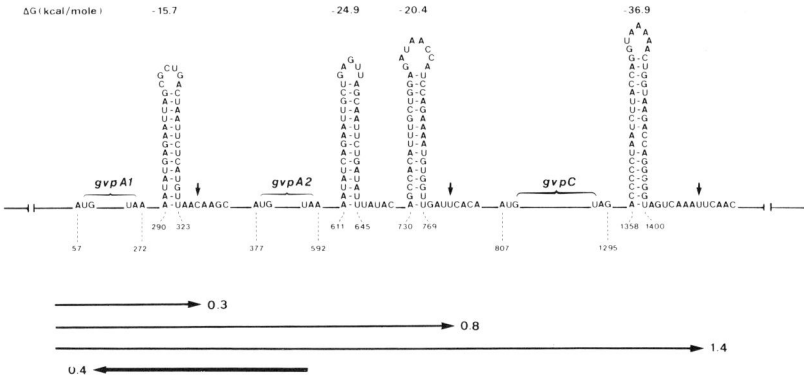

Fig. 9. Potential secondary structures in relation to the transcription of the *gvpA1A2C* operon. Vertical arrows indicate the 3′ ends of the transcripts. Numbering of nucleotides is based on the assignment of the first base of the *gvpA1A2C* operon transcript as nucleotide +1. The horizontal arrows depict the three mRNA species which start at nucleotide +1, as well as the antisense RNA which starts at nucleotide 576. The sizes of the transcripts are indicated in kilobases. The free energy values (ΔG) of the potential secondary structures were calculated according to Cech et al. (1983). The *gvpA2* gene was previously designated *gvpB* (Damerval et al. 1987).

which increases gradually towards the end of the operon. The start site of transcription of the *gvpA1A2C* operon is located 56 nucleotides upstream from the initiation codon of the *gvpA1* gene. *E. coli*-like promotor sequences are present upstream from this start site of transcription (TAGTAT as '–10' and TTTACA as '–35'; Fig. 3A). Interestingly, we also found an antisense RNA, approximately 0.4 kb long, which starts in *gvpA2* and ends in *gvpA1*. This antisense RNA can form a perfectly matched duplex with the three mRNA species corresponding to the *gvpA1A2C* operon. No sequences sharing homology with either the *gvpA1A2C* or the *E. coli* promotor sequences were found upstream from its mapped transcription start site (Fig. 3A; Csiszàr et al. 1987).

3.3 Regulation of gene expression during hormogonia differentiation

Hormogonia differentiation is a complex developmental process which leads to rapid changes in the metabolic state of the vegetative cells and probably involves coordinated regulatory mechanisms. In a first attempt to elucidate these mechanisms, it was of interest to examine the kinetics of transcription of the genes involved in the gas vesicle formation, as well as those involved in the synthesis of the phycobiliproteins after the induction of and during this differentiation process. For this purpose, total RNA extracted from cells, collected at different times after the induction of the differentiation, was hybridized with probes specific for either the gas vesicle (*gvpA1*) or the phycobiliprotein (*cpcB1A1* and *apcA1B1*) genes. The latter two probes were chosen because the corresponding genes are expressed in vegetative cells irrespective of the light conditions (see section 2.2.2). Similar results were obtained with these two probes. Figure 10 presents the results of this kinetic study. At the time of the transfer (0 h), no transcript is revealed with the *gvpA1* probe. Transcription of the *gvp* genes and of the antisense RNA becomes maximal between 3 and 6 h later. The preexisting transcripts are then rapidly degraded (Fig. 10). After 9 to 12 h, probes internal to either the *gvpA1* or the *gvpC* genes revealed the same smear of short mRNA species (data not shown). This indicates that mRNA degradation probably results from both endonucleolytic cleavages and 3' exoribonucleases activities. In contrast, the amount of the antisense RNA apepars more constant over the same time period. This suggests that there are differences in the kinetics of appearance and/or stability of these two types of RNAs. Conversely, the transcription of genes encoding phycobiliproteins is totally and specifically arrested only when the genes involved in the formation of the gas vesicles are transcribed (Fig. 10). These results clearly demonstrate that the expression of both the *gvp* and the phycobiliprotein genes is regulated at the level of transcription during hormogonia differentiation (T. Damerval and G. Guglielmi unpublished data). Some of the genes are turned on while others are turned off. Our results suggest that different RNA polymerase sigma

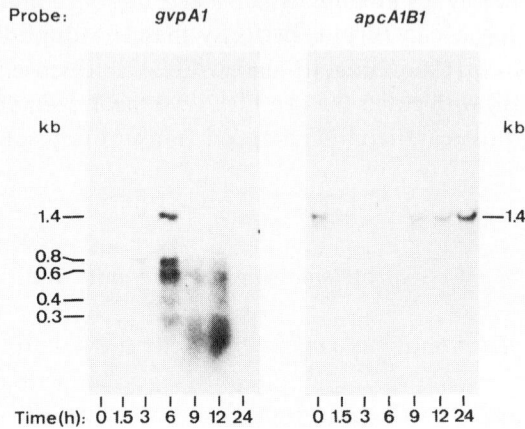

Fig. 10. Kinetics of the transcription of the gas vesicle protein (*gvp*) and of the allophycocyanin (*apc*) genes in *Calothrix* 7601. Total RNA was extracted at various times (h) after the induction of hormogonia differentiation and hybridized with probes internal to the *gvpA1* and *apcA1B1* genes. The size of the transcripts is indicated in kilobases (kb).

factors may exist allowing transcription to occur at the different developmental stages. Moreover, the presence of an antisense RNA, together wtih the rapid degradation of the transcripts corresponding to the *gvp* genes, indicate that the expression of the *gvpA1A2C* operon is also regulated at a post-transcriptional level (Csiszàr et al. 1987).

2.4 Conclusions

Until recently, it was thought that gas vesicle formation results from an autocatalytic phenomenon involving a single protein species (Jost and Jones 1970, Walsby 1981). We have shown that four genes are involved in gas vesicle formation in the cyanobacterium *Calothrix* 7601. The two structural genes, *gvpA1* and *gvpA2*, specify identical products despite the mutations which have occurred within the coding sequences. The very small size and very particular function of this protein may allow only few modifications to occur in its primary structure without altering the function of the molecule. In addition, the fact that this gene has been duplicated might be related to the very high level of expression required during the differentiation process. Indeed, large amounts of protein must be synthesized within a very short time. Multigene families are usually found whenever expression must be faster than that which can be achieved by enhanced expression from a single copy (Stark and Wahl 1984). On the other hand, the existence of the *gvpD* gene, which is very homologous to *gvpA* indicates that two protein species are most probably required to make the two different parts of the gas vesicle: the cylindrical body and the conical ends (Fig. 8).

Another unexpected result revealed by our studies concerns the discovery of the *gvpC* gene. Although its product has not yet been characterized biochemically, there are indications that this gene is translated in *Calothrix* 7601. Essentially two roles can be anticipated for this putative gene product: binding of the gas vesicle protein so as to maintain it in a soluble form suitable for assembly, or regulation of the gas vesicle formation. The primary structure of the *gvpC* gene product and the results of a systematic search for the presence of the *gvpC* gene in different cyanobacterial strains which produce gas vesicles either 'constitutively' or after induction, lead us to favor the first hypothesis (T. Damerval and A.-M. Castets unpublished data). Indeed, the role of this protein could be to increase the efficiency of the gas vesicle assembly, the combination of the gas vesicle proteins with this binding protein being a dispensable step in the assembly process.

Finally, we have shown that the *gvp* genes are rapidly turned on during hormogonia differentiation, while those encoding phycobiliproteins are simultaneously and temporarily turned off. The antisense RNA could form double-stranded duplexes with part of the *gvpA1A2C* transcripts, thereby favoring their rapid degradation by double-stranded RNA-specific endoribonucleases. The same antisense RNA could also block the translation of the *gvpA1* and *gvpA2* genes by impeding the migration of the ribosomes. These results constitute the first evidence that this complex differentiation process is regulated at both transcriptional and post-transcriptional levels.

4. Concluding remarks

Molecular studies of *Calothrix* 7601 genes involved in phycobilisome synthesis and hormogonia differentiation have yielded important new information concerning their physical organization and the regulation of their expression. However, major questions related to these photoregulated processes remain unanswered. The most intriguing ones are: how does light act at the molecular level; does the same photoreceptor trigger the modulation of gene expression or do two or more systems coexist in this cyanobacterium; are different effectors (sigma factors?) involved in these two processes? At present speculations, but no definite answers, can be formulated. Undoubtedly, light wavelength plays an important role in both processes. However, it is worth noting that there are differences in the way cells respond to the chromatic light stimuli. In the case of complementary chromatic adaptation, PC2 is synthesized under red illumination but also in darkness, while PE synthesis can only occur under green illumination (Tandeau de Marsac 1983). Furthermore, this antagonistic effect of red versus green illumination is efficient and fully reversible at any stage of exponential cell growth. Red light also stimulates hormogonia differentiation, but, in contrast to chromatic adaptation, hormogonia differentiation does not occur in the dark under our experimental conditions, i.e. when the preculture is grown under white light. Moreover, as growth of differentiated hormogonia proceeds in red light, regeneration of vegetative filaments occurs without any green light requirement. Hormogonia differentiation thus appears to be rather similar to the way in which red light triggers a cascade of events, which lead to the final biological response in higher plants, such as the stimulation of seed germination (Cone and Kendrick 1986).

In fact the differences between the two systems, phycobilisome biosynthesis and hormogonia differentiation, reinforce the interest in their study in the same organism. Recent and extremely important progress has been made by Wolk and coworkers (Wolk et al. 1984, Flores and Wolk 1985) in establishing a conjugation system for filamentous cyanobacteria. This genetic transfer system has been adapted to *Calothrix* 7601 by J.G. Cobley (personal communication). In addition, numerous mutants are now available. With these genetic tools, we expect to obtain deeper insights into these complex but fascinating photoregulatory processes in the near future.

Acknowledgements

We are very grateful to Dr G. Cohen-Bazire for valuable comments and support, to Dr D.A. Bryant for his very fruitful collaboration in this research project, to K. Csiszàr for her valuable contribution to this work and to T. Coursin and A.-M. Castets for their skillful technical assistance. We would like also to thank Dr A. Pugsley for a critical reading of the manuscript. The work presented in this review was supported by ATP-CNRS grant (Microbiologie 955353, 1984) and AI-CNRS 990019.

Appendix

Access to the sequences in the EMBL sequence data library

Several of the nucleotide sequences of the *Calothrix* 7601 genes which are mentioned in this paper are or will soon be available in the EMBL/GenBank under the following mnemonics:

FDAPCD	– Y00539	for *apcD*,
FDCPCABE	– X06084	for *cpcB1A1E,*
FDCPCBA2	– X06451	for *cpcB2A2*
FDCPCAB3	– X06083	for *cpcB3A3*
FDGVPA	– X03101; X06085	for *gvpA1A2C*.

A

```
                              .        20         .         40          .
apcA1.AA  --MSIVTKSI VNADAEARYL SPGELDRIKS FVSGGERRLR IAQILTENRE
apcA2.AA  --MSIITKMI LNADAEVRYL TPGELDQINI FVKSSQRRLQ LVEALTQSRA
apcD.AA   --MTVISQVI LQADDELRYP SSGELKSIQA FMQTGVKRTR IASTLAENEK
cpcA1.AA  -MKTPLTEAV AAADSQGRFL SSTEIQTAFG RFRQASASLA AAKALTEKAS
cpcA2.AA  -MKTPLTEAV ATADSQGRFL SSTELQVAFG RFRQASASLD AAKALSSKAN
cpcA3.AA  MTKTPLTEAV VSADSQGRFL S-TELQVAFG RFRQAGSSLE AAKALSKKAS
cpeA.AA   -MKSVVTTVI AAADAAGRFP STSDLESVQG SIQRAAARLE AAEKLANNID

apcB1.AA  MAQDAITSVI NSADVQGKYL DSAALDKLKG YFGTGELRVR AASTISANAA
cpcB1.AA  -MLDAFAKVV SQADARGEYL SGSQIDALSA LVADGNKRMD VVNRITGNSS
cpcB2.AA  -MLDAFTKVV SQADTRGAYI SDAEIDALKT MVAAGSKRMD VVNRITGNAS
cpcB3.AA  MVQDAFSKVV SQADARGEYL SDGQLDALIN LVKEGNKRVD VVNRISSNAS
cpeB.AA   -MLDAFSRAV VSADASTSTV SD--IAALRA FVASGNRRLD AVNAIASNAS

                              .        70         .         90          .
apcA1.AA  RLVKQAGEQV FQKRPDVVSP GGNAYGQEL- -TATCLRDLD YYLRLVTYGI
apcA2.AA  TIVKQAGKDI FQRFPRLVAP GGNAYGENM- -TATCLRDMD YYLRLITYSV
apcD.AA   KIVQEATKQL WQKRPDFISP GGNAYGERQ- -RSLCIRDFG WYLRLITYGV
cpcA1.AA  SLASGAANAV YSKFPYTTSQ NGPNFASTQT GKDKCVRDIG YYLRMVTYCL
cpcA2.AA  SLAQGAVNAV YQKFPYTTQM QGKNFASDQR GKDKCARDIG YYIRIVTYCL
cpcA3.AA  SLAEAAANAV YQKFPYTTTT SGPNYASTQT GKDKCVRDIG YYLRIVTYGL
cpeA.AA   AVATEAYNAC IKKYPYLNNS GEANSTDTF- -KAKCARDIK HYLRLIQYSL

apcB1.AA  AIVKEAVAKS LL-YSDVTRP GGNMYTTRR- -YAACIRDLD YYFRYATYAM
cpcB1.AA  TIVANAARSL FAEQPQLIAP GGNAYTSRR- -MAACLRDME IILRYVTYAI
cpcB2.AA  TIVANAARAL FEEQPQLIAP GGNAYTNRR- -MAACLRDME IILRYVTYAV
cpcB3.AA  SIVANAARSL FAEQPQLIAP GGNAYTSRR- -AAACVRDLE IILRYVTYAI
cpeB.AA   CMVSDAVAGM ICENQGLIQA GGSCYPNRR- -MAACLRDAE IVLRYVTYAL

                              .        120        .         140         .
apcA1.AA  VSGDVTPIEE IGVIGAREMY KSLGTPIEGI TEGIRALKSG ASSLLSGED-
apcA2.AA  AAGDTTPIQE IGIVGVRQMY RSLGTPIDAV AESVRAMKNI TTSMLSGED-
apcD.AA   LAGDKEPIEK IGLIGVREMY NSLGVPVPGM VEAIASLKKA ALDLLSAED-
cpcA1.AA  VVGGTGPLDD YLIGGIAEIN RTFDLSPSWY VEALKYIKAN HG--LSGDP-
cpcA2.AA  VAGGTGPLDD YLIGGLAEIN RTFDLSPSWY VEALKYIKAN HG--LSGDP-
cpcA3.AA  VVGGTGPIDD YLIGGLAEIN RTFELSPSWY IEALKYIKAN HG--LSGDP-
cpeA.AA   VVGGTGPLDE WGIAGQREVY RALGLPTAPY VEALSFARNR GCA-PRDMSA

apcB1.AA  LAGDPSILDE RVLNGLKETY NSLGVPVSST VQAIQAIKEV TASLVGSDA-
cpcB1.AA  FAGDASVLDD RCLNGLKETY LALGTPGSSV AVGVQKMKDA ALAIAGDTNG
cpcB2.AA  FAGDASVLDD RCLNGLRETY QALGVPGASV STGVQKMKEA AIAIANDPSG
cpcB3.AA  FAGDASVLDD RALNGLRETY LALGTPGASV AVGIQKLKES SIAIANDPNG
cpeB.AA   LAGDASVLDD RCLNGLKETY AALGVPTTST VRAVQIMKAQ AAAHIQDTPS

                              .        170        .
apcA1.AA  ---------- ---------- ---AAEAGSY FDYVVGALS
apcA2.AA  ---------- ---------- ---ASEVGTY FDYLITNLQ
apcD.AA   ---------- ---------- ---AAEASPY FDYIIQAMS
cpcA1.AA  ---------- ---------- ---AVEANSY IDYAINALS
cpcA2.AA  ---------- ---------- ---AVEANSY IDYAINALS
cpcA3.AA  ---------- ---------- ---AVEANSY IDYIINALS
cpeA.AA   QA-------- ---------- ---LTEYNAL LDYAINSLS

apcB1.AA  ---------- ---------- ---GKEMGVY LDYISSGLS
cpcB1.AA  ITRGD----- ---------C ASLMAEVASY FDKAASAVA
cpcB2.AA  VTRGD----- ---------C SSLMSELGSY FDRAAAAVG
cpcB3.AA  ITRGD----- ---------C SSLIAEVSGY FDRAAAAVA
cpeB.AA   EARAGAKLRK MGTPVVEDRC ASLVAEASSY FDRVISALS
```

B

```
                     .        20         .        40         .
apcC.AA   MARLFKVTAC VPSQTRIRTQ RELQNTYFTK LVPFENWFRE QQRIMKMGGK

          IVKVELATGK QGTNTGLL


                     .        20         .        40         .
cpcL.AA   MAPLTEASRL GVRPFADSDK VELRFVKTAE EVRSVIWSAY RQVLGNEHLF

cpcM.AA   MPITTAASRL GTSAFSNAAP IELRSNTNKA EIAQVIAAIY RQVLGNVIVL

cpcH.AA   MTSSTAARQL GFEPFASTAP TELRASS--- DVPAVIHAAY RQVFGNDHVM

cpcI.AA   MPITSAASRL GTTAYQ-TNP IELRPNWTAE DAKIVIQAVY RQVLGNDYLM


                     .        70         .        90         .
cpcL.AA   ESERLSSAES LLQQAQISVR DFVRAIAQSE LYRQKFFYSN SQVRFIELNY

cpcM.AA   QSERLKGLES LLTNGNITVQ EFVRQLAKSI YTSSFL

cpcH.AA   QSERLTSAES LLQQGNISVR DFVRLLAQSE LYRQKFFYST PQVRFIELNY

cpcI.AA   QSERLTSLES LLTNGKLSVR DFVRAVAKSE LYKTKFLYPH FQTRVIELNF


                     .        120        .        140        .
cpcL.AA   KHLLGRAPYD ESEIAYHVDI YTSQGYEAEI NSYIDSVEYQ QNFGDSIVPY

cpcM.AA

cpcH.AA   KHLNGRAPYD ESEISYHVNL YTEKGYEAEI NSYIDSAEYQ ESFGERIVPH

cpcI.AA   KHLLGRAPYD ESEVIEHLDR YQNQGFDADI DSYIDSAEYD TYFGDSIVPY


                     .        170        .        190        .
cpcL.AA   YRGYQTT-VG QKTAGFPRFF QLYRGYANRD R-QNKSKGQ- LTWDLAKNLV

cpcM.AA

cpcH.AA   YRGFETQ-PG QKTVGFNRMF QIYRGYANSD RSQGKNKSAW LTQDLALNLA

cpcI.AA   YRDLVTTGVG QRTVGFTRMF RLYRGYANSD RSQLAGSSSR LASDLATNSA


                     .        220        .        240        .
cpcL.AA   SPIY-PADA- ---------- -----GSLTG VSTGNRGNTY RIRTTQAASP

cpcM.AA

cpcH.AA   SNIQTPNFG- ---------- -----KGLTG VVAGDRGQLY RVRVIQADRG

cpcI.AA   TAIIAPSGGT QGWSYLPSKQ GTAPSRTFGR SSQGSTPRLY RIEVTGISLP


                     .        270        .
cpcL.AA   NSPRIRQSIS EVVVPFDQLS NLLQQLNRQG GKVISIALS

cpcM.AA

cpcH.AA   RTTQIRRSIQ EYLVSYDQLS PTLQRLNQRG SRVVNISPA

cpcI.AA   RYPKVRRSNK EFIVPYEQLS STLQQINKLG GKVASITFAQ


                     .        20         .        40         .
cpcD.AA   MLGSVLTRRS SSGSDNRVFV YEVEGLRQNE QTDNNRYQIR NSSTIEIQVP

          YSRMNEEDRR ITRLGGRIVN IRPAGENPTE DASEN
```

C

```
                 .        20         .        40         .
cpcE.AA   MYRHLSEGIE DHREQEQKVE NAANIQDDNQ LTVEQAIANL QGEDLGLRVY

          AAWWLGRFRV DAPEAIDVLI QALEDEDDRT NVGGYPLRRN AARALGKLGE

          KRAVPALIKA LECSDFYVRE AAAQSLEMLG DSSSIPRLIE LLNDQVPGTL

          PAPEPPQLTQ PFDAIIEALG TLGASDAIPI IQEFLEHTVP RIQYAAARAM

          YQLTSESTAG YNQYGDRLVQ ALAQDDLQLR RAVLSDLGAI GYLPAAEAIA

          DTLAENSLKL ISLKGLLEKQ FQPTKPEDLS PGAIKVMQLM DALL


                 .        20         .        40         .
cpcF.AA   MPDSLNSLIR AVEDANSSIL LQEAVKNLAA ARLEGAIPTL IAALSYNNPG

          AAVAAVDGLI QIGEPAVPSL LELLDMHNYT ARSWAIRTLA GIGDPRGLVT

          LLGAATADFA LSVRRAAAKG LGMMKWHWFP EELLEIAQAE AMEALLFVAQ

          EDEEWVVRYS AIVGLQFLAN AIAVSHPDWR SQILSNFEQI AAKEESWPVR

          ARVLLAQQEL QQITATIPTQ DIENRPSPLS SMDWQKIMED LYGRKGQERL

          VFAEGDPRRI


                 .        20         .        40         .
orfY.AA   MDKRFFNFFN LTEDQAIALL DTPQDQLSEN DSRYIAASHL VNFPTERSIN

          ALIRAVQQTD PSLDNRIVRR KSVETLGRLK ATTALPFIRI CLFDEDCYTV

          ENAAWAIGEI GTQDTDILED VAQLLEKPGQ TYRVIIHTLT KFNYQPALER

          IRKFVNDSDP PTASAAIAAV CRLTGDYSQM AKVVQILQHP NVLGRRLSIQ

          DLMDARYYDA IPDIAKCPVS LVFRLRGLRT LAEAGISEGA ITFAKIQPYL

          EQTLYDHPQD LNLVHSYDRL PTLEILIRGL YETDFGRCYL ATKTILEHYA

          DAAAEALFAT YAAEANNDYG AHFHVIKLFG WLKHAPAYDL IVEGLHNKQP

          QFQKSRAAAA IALAELGDPK AIPELKACLE TKIWDLKYAT LMALEKLGDI

          SEHKQAAQDS DWLIARKASS TLKNQEITA


                 .        20         .        40         .
orfZ.AA   MPTTEELFQQ LKHPNPHLRD QAMWELAENP DETTIPRLMS ILDEEDTTYR

          RAAVKALGAI GPDAITPLVQ ALLNSDNVTV RGSAAKALAQ VAINHPDVPF

          AAEGVQGLKT ALDDPNPVVH IAAVMALGEI GSPVVDVLIE ALQTTDNPAL

          GISIVNALGS IGDSRGVEVL QSLIENESTD SYVRESATSA LSRLEMTTKF

          QRGEK
```

Fig. 11. Predicted amino acid sequences of the polypeptides involved in the formation of phycobilisomes in *Calothrix* 7601. The single-letter code is that recommended by IUPAC/IUB. Some of the previously published gene designations have been changed according to the nomenclature rules proposed for cyanobacterial genes (Houmard and Tandeau de Marsac 1988).
A. Phycobiliprotein subunits: αAP1 (*apcA1*.AA), αAP2 (*apcA2*.AA), αAPB (*apcD*.AA), αPC1 (*cpcA1*.AA), αPC2 (*cpcA2*.AA), αPC3 (*cpcA3*.AA), αPE (*cpeA*), βAP1 (*apcB1*.AA), βPC1 (*cpcB1*.AA), βPC2 (*cpcB2*.AA), βPC3 (*cpcB3*.AA) and βPE (*cpeB*).
B. Linker polypeptides: $L_C^{7.8}$ (*apcC*.AA), L_R^{38}-like polypeptide (*cpcL*.AA), L_R-like polypeptide (*cpcM*.AA), L_R^{38} (*cpcH*.AA), L_R^{39} (*cpcI*.AA) and $L_R^{9.7}$ (*cpcD*.AA). The last three sequences are taken from Lomax et al. 1987.
C. Unassigned open reading frames.

```
                 .        20         .        40         .
gvpA1.AA  MAVEKTNSSS SLAEVIDRIL DKGIVVDAWV RVSLVGIELL AIEARIVIAS

          VETYLKYAEA VGLTQSAAVP A


                 .        20         .        40         .
gvpC.AA   MTPLMIRIRQ EHRGIAEEVT QLFKDTQEFL SVTTAQRQAQ AKEQAENLHQ

          FHKDLEKDTE EFLTDTAKER MAKAKQQAED LFQFHKEMAE NTQEFLSETA

          KERMAQAQEQ ARQLREFHQN LEQTTNEFLA DTAKERMAQA QEQKQQLHQF

          RQDLFAEIFG TF
```

Fig. 12. Predicted amino acid sequences of the polypeptides involved in gas vesicle formation in *Calothrix* 7601. The amino acid sequence of the *gvpA2* gene is strictly identical to that of *gvpA1*. The single letter code is that recommended by IUPAC/IUB.

References

Allen JF, Sanders CE and Holmes NG (1985) Correlation of membrane protein phosphorylation with excitation energy distribution in the cyanobacterium *Synechococcus* 6301. FEBS Lett 193: 271–275

Anderson LK, Rayner MC, Sweet RM and Eiserling FA (1983) Regulation of *Nostoc* sp. phycobilisome structure by light and temperature. J Bacteriol 155: 1407–1416

Anderson LK and Eiserling FA (1986) Asymmetrical core structure in phycobilisomes of the cyanobacterium *Synechocystis* 6701. J Mol Biol 191: 441–451

Armstrong RE, Hayes PK and Walsby AE (1983) Gas vacuole formation in hormogonia of *Nostoc muscorum*. J Gen Microbiol 128: 263–270

Belasco JG, Beatty JT, Adams CW, von Gabain A and Cohen SN (1985) Differential expression of photosynthesis genes in *R. capsulata* results from segmental differences in stability within the polycistronic *rxcA* transcript. Cell 40: 171–181

Belknap WR and Haselkorn R (1987) Cloning and light regulation of expression of the phycocyanin operon of the cyanobacterium *Anabaena*. EMBO J 6: 871–884

Bennett A and Bogorad L (1973) Complementary chromatic adaptation in a filamentous blue-green alga. J Cell Biol. 58: 419–435

Brawerman G (1987) Determinants of messenger RNA stability. Cell 48: 5–6

Bryant DA, de Lorimier R, Guglielmi G, Stirewalt VL, Dubbs JM, Illman B, Gasparich G, Buzby JS, Cantrell A, Murphy RC, Gingrich J, Porter RD and Stevens SE Jr (1986) The cyanobacterial photosynthetic apparatus: a molecular genetic analysis. In: Youvan DC and Daldal F (eds) Microbial Energy Transduction. Genetics, Structure, and Function of Membrane Proteins. pp 39–46. Cold Spring Harbor: Cold Spring Harbor Laboratory

Bryant DA (1988) Phycobilisomes of *Synechococcus* sp. PCC 7002, *Pseudanabaena* sp. PCC 7409, and *Cyanophora paradoxa*: an analysis by molecular genetics. In: Scheer H and Schneider S (eds) Photosynthetic Light-harvesting Systems. Structure and function. Berlin: de Gruyter & Co, in press

Capuano V, Mazel D, Tandeau de Marsac N and Houmard J (1988) Complete nucleotide sequence of the red-light specific set of phycocyanin genes from the cyanobacterium PCC 7601. Nucleic Acids Res 16: 1626

Cech TR, Tanner NK, Tinoco I Jr, Weir BR, Zuker M and Perlman PS (1983) Secondary structure of the *Tetrahymena* ribosomal RNA intervening sequence: Structural homology with fungal mitochondiral intervening sequences. Proc Natl Acad Sci USA 80: 3903–3907

Cohen-Bazire G and Bryant DA (1982) Phycobilisomes: composition and structure. In: Carr NG and Whitton BA (eds) The Biology of Cyanobacteria, pp 143–190. Oxford: Blackwell Scientific Publications

Cone JW and Kendrick RE (1986) Photocontrol of seed germination. In: Kendrick RE and Kronenberg GHM (eds) Photomorphogenesis in plants, pp 443–465. Dordrecht: Martinus Nijhoff Publishers

Conley PB, Lemaux PG and Grossman AR (1985)· Cyanobacterial light-harvesting complex subunits encoded in two red light-induced transcripts. Science 230: 550–553

Conley PB, Lemaux PG, Lomax TL and Grossman AR (1986) Genes encoding major light-harvesting polypeptides are clustered on the genome of the cyanobacterium *Fremyella diplosiphon*. Proc Natl Acad Sci USA 83: 3924–3928

Csiszàr K, Houmard J, Damerval T and Tandeau de Marsac N (1987) Transcriptional analysis of the cyanobacterial *gvpABC* operon in differentiated cells: occurrence of an antisense RNA complementary to three overlapping transcripts. Gene 60: 29–38

Damerval T, Houmard J, Guglielmi G, Csiszàr K and Tandeau de Marsac N (1987) A developmentally regulated *gvpABC* operon is involved in the formation of gas vesicles in the cyanobacterium *Calothrix* 7601. Gene 54: 83–92

DasSarma S, Damerval T, Jones JG and Tandeau de Marsac N (1987) A plasmid-encoded gas vesicle protein gene in a halophilic archaebacterium. Mol Microbiol 1: 365–370

De Lorimier R, Bryant DA, Porter RD, Liu W-Y, Jay E and Stevens SE Jr (1984) Genes for the α and β subunits of phycocyanin. Proc Natl Acad Sci USA 81: 7946–7950

Elmorjani K and Herdman M (1987) Metabolic control of phycocyanin degradation in the cyanobacterium *Synechocystis* PCC 6803: a glucose effect. J Gen Microbiol 133: 1685–1694

Flores E and Wolk CP (1985) Identification of facultatively heterotrophic, N_2-fixing cyanobacteria able to receive plasmid vectors from *Escherichia coli* by conjugation. J Bacteriol 162: 1339–1341

Foulds IJ and Carr NG (1977) A proteolytic enzyme degrading phycocyanin in the cyanobacterium *Anabaena cylindrica*. FEMS Microbiol Lett 2: 117–119

Füglistaller P, Suter F and Zuber H (1983) The complete amino-acid sequence of both subunits of phycoerythrocyanin from the thermophilic cyanobacterium *Mastigocladus laminosus*. Hoppe-Seyler's Z Physiol Chem 364: 691–712

Füglistaller P, Rümbeli R, Suter F and Zuber H (1984) Minor polypeptides from the phycobilisome of the cyanobacterium *Mastigocladus laminosus*. Isolation, characterization and amino-acid sequences of a colourless 8.9 kDa polypeptide and of a 16.2 kDa phycobiliprotein. Hoppe-Seyler's Z Physiol Chem 365: 1085–1096

Füglistaller P, Suter F and Zuber H (1985) Linker polypeptides of the phycobilisome from the cyanobacterium *Mastigocladus laminosus*: amino-acid sequences and relationships. Biol Chem Hope-Seyler 366: 993–1001

Füglistaller P, Suter F and Zuber H (1986) Linker polypeptides of the phycobilisome from the cyanobacterium *Mastigocladus laminosus*. II Amino-acid sequences and functions. Biol Chem Hoppe-Seyler 367: 615–626

Gendel S, Ohad I and Bogorad L (1979) Control of phycoerythrin synthesis during chromatic adaptation. Plant Physiol 64: 786–790

Gingrich JC, Lundell DJ and Glazer AN (1983) Core substructure in cyanobacterial phycobilisomes. J Cell Biochem 22; 1–14

Glazer AN (1982) Phycobilisomes: structure and dynamics. Ann Rev Microbiol 36: 173–198

Glazer AN (1984) Phycobilisome. A macromolecular complex optimized for light energy transfer. Biochim Biophys Acta 768: 29–51

Glazer AN (1985) Light harvesting by phycobiliscmes. Ann Rev Biophys Biophys Chem 14: 4–77

Glazer AN and Clark JH (1986) Phycobilisomes. Macromolecular structure and energy flow dynamics. Biophys J 49: 115–116

Glazer AN (1987) Phycobilisomes: assembly and attachment. In: Fay P and Van Baalen C (eds) The Cyanobacteria, pp 69–94. Amsterdam: Elsevier Science Publishers

Glick RE and Zilinskas BA (1982) Role of the colorless polypeptides in phycobilisome reconstitution from separated phycobiliproteins. Plant Physiol 69: 991–997

Grossman AR, Lemaux PG and Conley PB (1986) Regulated synthesis of phycobilisome components. Photochem Photobiol 44: 827–837

Hayes PK, Walsby AE and Walker JE (1986) Complete amino acid sequence of cyanobacterial gas vesicle protein indicates a 70-residue molecule that corresponds in size to the crystallographic unit cell. Biochem J 236: 31--36

Houmard J, Mazel D, Moguet C, Bryant DA and Tandeau de Marsac N (1986) Organization and nucleotide sequence of genes encoding core components of the phycobilisomes from *Synechococcus* 6301. Mol Gen Genet 205: 404–410

Houmard J, Capuano V, Coursin T and Tandeau de Marsac N (1988) Isolation and molecular characterization of the gene encoding allophycocyanin B, a terminal energy acceptor in cyanobacterial phycobilisomes. Mol Microbiol 2: 101–107

Houmard J and Tandeau de Marsac N (1988) Cyanobacterial Genetic Tools: Current Status. In: Packer L and Glazer AN (eds) Methods in Enzymology: Cyanobacteria. Academic Press, (in press)

Jost M and Jones DD (1970) Morphological parameters and macromolecular organization of gas vacuole membranes of *Microcystis aeruginosa* Kuetz. emend. Elenkin. Can J Microbiol 16: 159–164

Klotz AV, Leary JA and Glazer AN (1986) Post-translational methylation of asparaginyl residues. Identification of β-71γ-N-methylasparagine in allophycocyanin. J Biol Chem 261: 15891–15894

Lau RH, MacKenzie MM and Doolittle WF (1977) Phycocyanin synthesis and degradation in the blue-green bacterium *Anacystis nidulans*. J Bacteriol 132: 771–778

Lomax TL, Conley PB, Schilling J and Grossman AR (1987) Isolation and characterization of light-regulated phycobilisome linker polypeptide genes and their transcription as a polycistronic mRNA. J Bacteriol 169: 2675–2684

Mazel D, Guglielmi G, Houmard J, Sidler W, Bryant DA and Tandeau de Marsac N (1986) Green light induces transcription of the phycoerythrin operon in the cyanobacterium *Calothrix* 7601. Nucleic Acids Res 14: 8279–8290

Mazel D, Houmard J and Tandeau de Marsac N (1988) A multigene family in *Calothrix* sp. PCC 7601 encodes phycocyanin, the major component of the cyanobacterial light harvesting antenna. Mol Gen Genet 211: 296–304

Minami Y, Yamada F, Hase T, Matsubara H, Murakami A, Fujita Y, Takao T and Shimonishi Y (1985) Amino acid sequences of allophycocyanin α- and β-subunits isolated from *Anabaena cylindrica* FEBS Lett 191: 216–220

Newbury SF, Smith NH, Robinson EC, Hiles ID and Higgins CF (1987) Stabilization of translationally active mRNA by prokaryotic REP sequences. Cell 48: 297–310

Ohad I, Clayton RK and Bogorad L (1979) Photoreversible absorbance changes in solutions of allophycocyanin purified from *Fremyella diplosiphon*: temperature dependence and quantum efficiency. Proc Natl Acad Sci USA 76: 5655–5659

Ohad I, Schneider HAW, Gendel S and Bogorad L (1980) Light-induced changes in allophycocyanin. Plant Physiol 65: 6–12

Pilot TJ and Fox JL (1984) Cloning and sequencing of the genes encoding the α- and β-subunits of C-phycocyanin from the cyanobacterium *Agmenellum quadruplicatum*. Proc Natl Acad Sci USA 81: 6983–6987

Platt T (1986) Transcription termination and the regulation of gene expression. Ann Rev Biochem 55: 339–372

Raps S, Kycia JH, Ledbetter MC and Siegelman HW (1985) Light intensity adaptation and phycobilisome composition of *Microcystis aeruginosa*. Plant Physiol 79: 983–987

Riethman HC, Mawhinney TP and Sherman LA (1987) Phycobilisome-associated glycoproteins in the cyanobacterium *Anacystis nidulans* R2. FEBS Lett 215: 209–214

Rippka R and Herdman M (1985) Division patterns and cellular differentiation in cyanobacteria. Ann Microbiol (Inst Pasteur) 136A: 33–39

Rümbeli R, Suter F, Wirth M, Sidler W and Zuber H (1987) γ-N-Methylasparagine in phycobiliproteins from the cyanobacteria *Mastigocladus laminosus* and *Calothrix*. FEBS Lett 221: 1–2

Sidler W, Kumpf B, Rüdiger W and Zuber H (1986) The complete amino-acid sequence of C-phycoerythrin from the cyanobacterium *Fremyella diplosiphon*. Biol Chem Hoppe-Seyler 367: 627–642

Stanier RY and Cohen-Bazire G (1977) Phototrophic prokaryotes: The Cyanobacteria. Ann Rev Microbiol 31: 225–274

Stark GR and Wahl GM (1984) Gene amplification. Ann Rev Biochem 53: 447–491

Stevens SE Jr, Balkwill DL and Paone DAM (1981) The effects of nitrogen limitation on the ultrastructure of the cyanobacterium *Agmenellum quadruplicatum*. Arch Microbiol 130: 204–212

Suter F, Füglistaller P, Lundell DJ, Glazer AN and Zuber H (1987) Amino acid sequences of α-allophycocyanin B from *Synechococcus* 6301 and *Mastigocladus laminosus*. FEBS Lett 217: 279–282

Tandeau de Marsac N (1977) Occurrence and nature of chromatic adaptation in cyanobacteria. J Bacteriol 130: 82–91

Tandeau de Marsac N (1983) Phycobilisomes and complementary chromatic adaptation in cyanobacteria. Bull Inst Pasteur 81: 201–254

Tandeau de Marsac N, Mazel D, Bryant DA and Houmard J (1985) Molecular cloning and nucleotide sequence of a developmentally regulated gene from the cyanobacterium *Calothrix* PCC 7601: a gas vesicle protein gene. Nucleic Acids Res 13: 7223–7236

Walker JE, Hayes PK and Walsby AE (1984) Homology of gas vesicle proteins in cyanobacteria and halobacteria. J Gen Microbiol 130: 2709–2715

Walsby AE (1978) The gas vesicles of aquatic prokaryotes. Symp Soc Gen Microbiol 28: 327–358

Walsby AE (1981) Cyanobacteria: planktonic gas-vacuolate forms. In: Starr M, Stolp H, Trüper H, Balows A and Schlegel HG (eds) The Prokaryotes, pp 224–235. New York: Springer-Verlag

Walsby AE (1987) Mechanisms of buoyancy regulation by planktonic cyanobacteria with gas vesicles. In: Fay P and Van Baalen C (eds) The Cyanobacteria, pp 377–392. Amsterdam: Elsevier Science Publishers

Wolk CP, Vonshak A, Kehoe P and Elhai J (1984) Construction of shuttle vectors capable of conjugative transfer from *Escherichia coli* to nitrogen-fixing filamentous cyanobacteria. Proc Natl Acad Sci USA 81: 1561–1565

Wong HC and Chang S (1986) Identification of a positive retroregulator that stabilizes mRNAs in bacteria. Proc Natl Acad Sci USA 83: 3233–3237

Wood NB and Haselkorn R (1980) Control of phycobiliprotein proteolysis and heterocyst differentiation in *Anabaena*. J Bacteriol 141: 1375–1385

Yamanaka G and Glazer AN (1980) Dynamic aspects of phycobilisome structure. Phycobilisome turnover during nitrogen starvation in *Synechococcus* sp. Arch Microbiol 124: 39–47

Yamanaka G and Glazer AN (1981) Dynamic aspects of phycobilisome structure: modulation of phycocyanin content of *Synechococcus* phycobilisomes. Arch Microbiol 130: 23–30

Zilinskas BA, Chen KH and Howell DA (1987) Cloning genes for phycobilisome core components. Plant Physiol (Suppl) 83: 60

Govindjee et al. (eds), Molecular Biology of Photosynthesis: 229–257
© 1988 Kluwer Academic Publishers

Minireview

Regulation of cyanobacterial pigment-protein composition and organization by environmental factors

HAROLD RIETHMAN, GEORGE BULLERJAHN, K.J. REDDY & LOUIS A. SHERMAN
University of Missouri, Division of Biological Sciences, Tucker Hall, Columbia, MO 65211, USA

Received 30 September 1987; accepted 24 December 1987

Key words: chlorophyll-protein, cyanobacteria, iron stress, phycobilisome

Abstract. The coordinate expression of stress-specific genes is a common response of all organisms to altered environmental conditions. In cyanobacteria, the physiological consequences of stress are often reflected in both the ultrastructure of the cell and in photosynthesis-related properties. This review will focus on the alterations in cyanobacterial pigment-protein organization which occur under different growth conditions, and how several molecular genetic aproaches are being used in this laboratory to investigate the regulatory mechanisms underlying these alterations. We will discuss in detail the response to iron starvation, and present a testable hypothesis for the mechanism of thylakoid reorganization mediated by this response.

Abbreviations: ConA – concanavalin A, CP – chlorophyll-protein complex, LHCP – light harvesting chlorophyll protein, MSP – manganese-stabilizing protein

I Introduction

Environmental factors affecting cyanobacterial pigment-proteins
The pigment systems of photosynthetic organisms have evolved to fit the local environments of the particular species. Land plants have evolved specialized chloroplasts containing copious amounts of LHCPII to harvest sunlight, whereas many algal species have evolved unusual carotenoid and Chl-like pigments to allow efficient light-harvesting at particular depths in the water or in special terrestrial environments. Cyanobacteria have evolved to fill nearly every imaginable ecological niche from symbiotic associations with fungi in antarctic sandstone to geothermal hot springs. Not surprisingly, cyanobacterial pigments reflect this ecological diversity. Phycobiliproteins and carotenoids are present in the various cyanobacterial light-harvesting systems which can absorb light from most regions of the visible spec-

trum. These pigments are organized into phycobilisomes, intrinsic membrane Chl-carotenoid-protein complexes, and carotenoprotein complexes (reviewed by Thornber 1986, Bryant 1987, Wyman and Fay 1987).

In addition to the genotypically evolved pigment systems, most cyanobacteria can physiologically acclimate to rapid changes in their environments by altering their cellular pigment-protein composition. These altered 'stress' conditions are probably a routine occurrence for cyanobacteria in their natural environments (reviewed in Grossman et al. 1986, Bryant 1987, Sherman et al. 1987a).

A. Adaptation to changes in light. A well-studied acclimation of cyanobacteria to altered light conditions is that of complementary chromatic adaptation; certain species of cyanobacteria change their phycobilisome composition to maximize their absorption of the most prevalent spectral region of incident light (Grossman et al. 1986, Wyman and Fay 1987). Specifically, the pigment-protein phycoerythrin is synthesized in green light, whereas phycocyanin is made in red light. These specific phycobiliproteins are synthesized, and separate linker proteins are necessary to assemble each type of pigment-protein into phycobilisomes. The study of the regulatory mechanisms involved in chromatic adaptation is a very active topic. However, the strains studied in our laboratory, *Anacystis nidulans* R2, *Synechocystis* sp. PCC6714 and *Synechocystis* sp. PCC6803, cannot make phycoerythrin, and thus do not undergo complementary chromatic adaptation.

A. nidulans does, however, alter the ratio of its photosystems in response to changes in light wavelength (Jones and Meyers 1965, Fujita et al. 1985, Manodori and Melis 1986). The ratio of PSII to PSI reaction centers in *A. nidulans* was shown to vary from about 0.30 in cells grown in yellow light (absorbed primarily by phycobilisomes) to 0.45 in cells grown in white light (harvested by both photosystems) to nearly 1.0 (cells grown in red light, preferentially absorbed by PSI). These changes can be rationalized in terms of the cell adjusting its PSII and PSI levels in order to compensate for the changed spectral quality, since the two photosystems have distinct action spectra. Whether only PSII, only PSI, or both photosystems change in abundance in response to light quality is a matter of debate (Fujita et al. 1985, Manodori and Melis 1986). Nevertheless, the cell must have mechanisms for sensing these spectral changes and signalling the biosynthetic machinery to adjust the levels of these multisubunit complexes. It is generally assumed that only the quantity of each photosystem changes, and not the composition of the photosystem itself. Although there is biophysical evidence supporting this hypothesis (Myers et al. 1980, Fujita et al. 1985, Manodori and Melis 1986) it has not been demonstrated biochemically in many circumstances. It is possible that the absorption cross-section of a photosystem might change by an alteration in the size of the phycobilisome or the number of Chl molecules feeding into a photosystem. Most studies have been primarily spectroscopic analyses of photosynthetic parameters, and the conclusions from these studies still need to be validated by biochemical means.

Light intensity can also change the pigment-protein organization of cyanobacterial thylakoids. In many ways, these results are very similar to those found with altered light wavelength; in this case, low light favors high PSI/PSII ratios, whereas high light favors low PSI/PSII ratios. As is the case for light wavelength, some researchers argue that PSI levels stay constant and PSII levels change (Kuwamura et al. 1979), whereas others suggest that the opposite is true (Khanna et al. 1983, Barlow and Alberte 1985). An additional effect of light intensity was suggested by Lönneborg et al. (1985). This group transferred cells from high intensity white light to low intensity red light, and their spectral and biochemical data indicated both an increase in size of the phycobilisome and an increase in the quantity of phycobilisomes upon shifting to low intensity red light. Light wavelength and light quantity effects are coordinately regulated by additional growth conditions such as temperature and nutrient status (Grossman et al. 1986). Temperature has been shown to affect phycobilisome composition during chromatic adaptation in *Synechocystis* sp. PCC6701 (Anderson et al. 1983) and other altered growth conditions (Bryant 1987). The nutrient status of the cell can severely affect the phycobilisome pigment organization and quantity, and, at least in the case of iron deficiency, the Chl-protein composition as well (Pakrasi et al. 1985a, b). All of these factors need to be considered when speaking of 'pigment changes due to light'.

The carotenoid content and composition of cyanobacteria also change with environmental conditions (Bryant 1987); this particular aspect of acclimation to stress has not been examined thoroughly. The effects of oxidative damage to cell components are expected to be unusually high in oxygen-producing organisms such as cyanobacteria, and the anti-oxidant properties of carotenoids may be particularly important at high light intensities. A detailed discussion of the proteins which bind carotenoid pigments in cyanobacteria will be presented in section II D of this review.

B. Adaptation to nutrient stress. Nutrient deprivation often has a more dramatic effect on the pigment organization and composition than light effects in cyanobacteria. The rapid degradation of phycobilisomes in cells deprived of either sulfate, nitrate, carbon, phosphate, or iron (Allen and Smith 1969, Öquist 1971, Lawry and Jensen 1979, Miller and Holt 1977,

Stevens et al. 1981a, b) suggests a common mechanism for the effects these stresses have on the major cyanobacterial pigment-protein complex (up to 45% of the soluble cellular protein is associated with the phycobilisome). The degradation of phycobilisomes may in fact be one aspect of a more general cyanobacterial stress response, although some have argued that it is exclusively an effect of nitrate deficiency (Boyer et al. 1987). If this were the case, the other stress conditions would presumably limit nitrate assimilation enzymes, causing this phenotype indirectly (Boyer et al. 1987).

Water stress, salt stress, and other environmental changes have not been characterized in sufficient detail to assess their full effects on pigment-proteins. Water stress does cause a depletion of phycobilisomes and an accumulation of intracellular polysaccharides in *Nostoc commune* (Potts 1985), possibly following the pattern seen with nutrient stresses. Additionally, since most cyanobacteria are photoautotrophic, they must be able to survive the night portion of the diurnal cycle. Indeed, mutational analysis suggested that specific proteins are necessary for the survival of *A. nidulans* in prolonged darkness (Doolittle and Singer 1974). Whether these proteins overlap with those effecting light intensity or nutrient deficiency alterations in pigment-proteins remains to be seen. However, the bulk of this review will concentrate on the mechanisms involved in the adaptation of the cyanobacterial photosynthetic mechanism to nutrient starvation. As iron stress has been studied most extensively in our laboratory, much of the work described will focus on the iron stress response as a model system for understanding the regulation of synthesis and biogenesis of the thylakoid membrane.

II Cyanobacterial pigment-proteins

A. Strategies employed to detect and analyze pigment-proteins
Polyacrylamide gel electrophoresis under conditions of limited denaturation has proven to be a powerful tool in the identification and characterization of Chl-binding proteins in cyanobacteria and chloroplasts. Detergent solubilization of thylakoid samples at low (4 °C) temperatures has enabled the electrophoretic separation of intrinsic thylakoid complexes which contain functionally bound Chl. Following electrophoresis, the Chl-protein complexes can be excised from the gel and examined by fluorescence spectroscopy, denaturing electrophoresis and by immunoblotting procedures; this approach is reviewed in detail by Thornber (1986). It should be stressed that the type of solubilizing detergent plays an extremely important role in yielding cyanobacterial green complexes which can be resolved in gels. Whereas PSI Chl is very stably liganded to protein in the presence of ionic

detergents such as sodium and lithium dodecyl sulfates (Takahashi et al. 1982, Guikema and Sherman 1983a, Takahashi et al. 1985), cyanobacterial PSII Chl-protein complexes are not easily obtained following thylakoid extraction with such detergents. This is likely due to two major factors: firstly, the PSII Chl may be more labile during electrophoresis after SDS extraction; and secondly, PSII Chl (under most growth conditions) represents a small fraction of the total cellular Chl. Any dissociation of PSII Chl may lower the yield of PSII complexes to undetectable levels. Use of other detergents has allowed the isolation of large amounts of PSII Chl-protein complexes following thylakoid solubilization with dodecyl-β-D-maltoside (Pakrasi et al. 1985b, Bullerjahn et al. 1985) and lauryldimethyla-mine-N-oxide (Stewart 1980). Indeed, only after dodecyl maltoside extraction were the PSII Chl-proteins of A. nidulans R2 and Synechocystis sp. identified and characterized on nondenaturing gels (Pakrasi et al. 1985b, Bullerjahn and Sherman 1986a). The major PSI and PSII Chl-proteins of cyanobacteria are summarized in a recent review (Sherman et al. 1987a). As will be apparent later, the electrophoretic analysis of the Chl-protein complexes has enabled the identification of specific pigment components which either accumulate or are depleted under conditions of environmental stress.

B. Major chlorophyll-proteins – PSI and PSII core
The Chl-binding proteins of the PSI core are the large subunits having apparent molecular masses of approximately 60–65 kDa (Muster et al. 1984). Analysis of PSI preparations from chloroplasts and cyanobacteria has demonstrated that the large subunits comprise the PSI reaction center as well as bind the acceptors A_0, A_1 and the iron sulfur center 'X' (Muster et al. 1984, Golbeck and Warden 1986, Warden and Golbeck 1986, Høj and Moller 1986). Katoh and coworkers have resolved 5 PSI complexes from a thermophilic strain of *Synechococcus* sp. and analyzed each band with respect to their carotenoid, phylloquinone and iron contents (Takahashi et al. 1982, Takahashi et al. 1985). Similarly, Guikema and Sherman (1983a) and Pakrasi et al. (1985b) also demonstrated that PSI in *A. nidulans* yields at least four electrophoretic forms, CPI–CPIV (Fig. 1A, B; lanes 1). CPI and CPII have the same polypeptide composition as a highly purified PSI reaction center, whereas CPIII and CPIV contain an additional polypeptide of 36 kDa (see Table 1). It is thought that these higher molecular mass complexes correspond to higher order oligomers of PSI; this is based on the observation that PSI preparations can self-aggregate in solution (Williams et al. 1983).

The major cyanobacterial Chl-binding proteins associated with the PSII core are two polypeptides which have been shown by immunological and

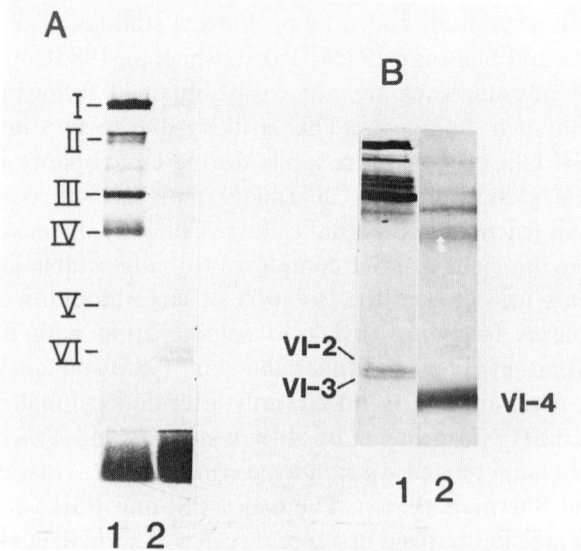

Fig. 1. Electrophoretic pattern of *A. nidulans* R2 Chl-protein complexes on nondenaturing gels. A) Dodecyl-maltoside-solubilized thylakoids from normally-grown (lane 1) and iron-limited (lane 2) cultures. The green complexes were resolved on a 5–15% acrylamide gradient gel. B) same as A) except that the samples were loaded onto a 10–15% acrylamide gradient gel. Note the accumulation of complex CPVI-4 under iron-limited growth.

spectroscopic studies to be homologs to the chloroplast proteins CP47 and CP43. In *A. nidulans* R2 these complexes are termed CPVI-2 and CPVI-3, respectively (Pakrasi et al. 1985b). Recent studies have shown that these proteins are not involved in primary photochemistry but instead likely comprise a proximal antenna to the reaction center of PSII (Yamagishi and Katoh 1984, Nanba and Satoh 1987). The 77 K fluorescence emission spectra obtained from bands excised from Chl-protein gels have shown that the 696 nm fluorescence associated with PSII arises from CP47 and its homologs (Nakatani et al. 1984, Pakrasi et al. 1985b), whereas CP43 fluoresces at 686 nm. These two Chl-proteins have been characterized as green complexes in a variety of mesophilic and thermophilic cyanobacteria (Sherman et al. 1987a). It should be mentioned that the dimer composed of the D1 and D2 polypeptides, which has been demonstrated to carry the P680 reaction center Chl (Nanba and Satoh 1987, Satoh et al. 1987) has yet to be identified as a green complex on non-denaturing gels. However, Machold (1986) has reported a physical association between CP43 and the D1 polypeptide on green gels of *Vicia* sp. thylakoid extracts.

C. Accessory Chl-proteins of cyanobacteria

Recent work on PSII-enriched dodecyl-maltoside extracts of cyanobacterial thylakoids has identified a novel Chl-protein complex in strains of both *A.*

Table 1. Pigment-proteins discussed in this review.

Pigment-protein	Organism	Protein compositon	Function	Reference
CPI–CPII	*A. nidulans*	65, 10, 14.5, 16, 16.5 kDa	PSI reaction center	Guikema and Sherman 1983a
CPIII–CPIV	*A. nidulans*	65 kDa	PSI reaction center	Guikema and Sherman 1983a
		36 kDa, four small proteins (10, 14.5, 16, 16.5 kDa)		
CPVI-1	*A. nidulans*	71, 42 kDa	PSII antenna associated	Pakrasi et al. 1985b
CPVI-2	*A. nidulans*	48 kDa	PSII core antenna (CP47 homolog)	Pakrasi et al. 1985b
CPVI-3	*A. nidulans*	45 kDa	PSII core antenna (CP43 homolog)	
CPVI-4	*A. nidulans*	36, 34 kDa	iron stress specific CP	Pakrasi et al. 1985b
CPIIIb	*Synechococcus* sp.	36, 34 kDa	accumulates under iron stress	Bullerjahn et al. 1985
CPIII'	*Synechococcus* sp.	65, 54 kDa	accumulates under iron stress	Bullerjahn et al. 1985
45 kDa carotenoprotein	*Synechococcus* sp.	45 kDa	accumulates under oxygenic conditions	Bullerjahn and Sherman 1986b
42 kDa carotenoprotein	*A. nidulans*	42 kDa	accumulates in response to high light levels	Masamoto et al. 1987
16 kDa carotenoprotein	*Aphanizomenon* sp. *Microcystis* sp.	16 kDa	?	Holt and Krogmann 1981

nidulans and *Synechocystis* sp. (Pakrasi et al. 1985b, Bullerjahn et al. 1985). This complex, termed CPVI-4 in *A. nidulans* R2 and CPIIIb in *Synechocystis* sp., absorbs maximally in the red at 671 nm and yields 685 nm fluorescence at 77 K. Denaturing electrophoresis has determined that the complex contains polypeptides of 36 and 34 kDa; it is thought that these two proteins both participate in the binding of Chl. Immunological studies have confirmed that CPVI-4 and CPIIIb are homologous and that the 36 and 34 kDa proteins are not breakdown products of the CPVI-2 or CPVI-3 complexes (Bullerjahn and Sherman 1986a). The function of this complex has remained elusive, although the observation that it copurifies with PSII particles has led us to suggest that this component can serve as an antenna to PSII. Perhaps the most interesting feature of CPVI-4 was demonstrated by Pakrasi et al. (1985b). They showed that CPVI-4 accumulates to high levels in iron stressed *A. nidulans* R2 to become the major Chl-binding species during severe iron deprivation (Fig. 1A, B; lanes 2). Since it has been previously shown that iron-starved cells exhibit spectral alterations consistent with the reorganization of Chl in the membrane (Öquist 1974b, Guikema and Sherman 1983b), the accumulation of CPVI-4 under iron stress is likely responsible for these changes. In agreement with these observations, iron-starved *Synechocystis* sp. PCC6803 also accumulate CPIIIb to high levels demonstrating that CPVI-4 and CPIIIb share similar properties (Bullerjahn and Sherman, unpublished).

CPVI-4 has also been biochemically purified by chromatography on DEAE-Sephacel. The spectroscopic characteristics of this preparation were identical to the material obtained from Chl-protein gels (Riethman 1987). Subsequent electrophoresis of the biochemical preparation on such gels yielded several electrophoretic forms of CPVI-4. Analysis of these distinct bands revealed that the 36 kDa polypeptide was enriched in the bands exhibiting lower electrophoretic mobility. From these data it was concluded that the 36 kDa polypeptide is important in forming aggregates of CPVI-4. This point will become important later, when we discuss possible roles for this complex during iron starvation and recovery from iron stress.

In addition to CPVI-4, several minor Chl-proteins have been observed in cyanobacteria. Pakrasi et al. (1985b) have identified a Chl-containing complex from *A. nidulans* R2, CPVI-1, which is composed of the 71 kDa phycobilisome anchor protein and a 42 kDa polypeptide. Whether the *A. nidulans* phycobilisome anchor alone is capable of binding Chl is still unclear, but Redlinger and Gantt (1982) have described a green preparation from the red alga, *Porphyridium cruentum*, which contains only the 95 kDa phycobilisome anchor polypeptide. Perhaps the interaction between the 42 kDa and 71 kDa proteins is necessary for Chl binding in *A. nidulans,* whereas the 95 kDa anchor is sufficient for pigment attachment in *P. cruentum.*

Recently, another complex, CPIII′ has been identified in *Synechocystis* sp. PCC 6714 and PCC 6803 (Bullerjahn et al. 1985, Bullerjahn and Sherman 1986a). The major apoproteins of this complex are polypeptides of 64 kD and 54 kDa. CPIII′ copurifies with PSII following dodecyl-maltoside extraction, but can be removed from PSII by chromatography without loss of photochemical activity; these results suggest that CPIII′ may serve as a PSII antenna. Interestingly CPIII′ accumulates together with CPIIIb under conditions of iron starvation in *Synechocystis* sp. (Bullerjahn and Sherman, unpublished).

To date, immunological studies have not identified any homologs to complexes CPVI-1, CPVI-4, CPIII′ or CPIIIb in higher plant chloroplasts. However, the recent discovery of prokaryotes which contain the higher plant pigment Chl *b,* has revealed Chl-protein complexes which may prove to be distantly related homologs to the cyanobacterial complexes. One such organism, *Prochlorothrix hollandica* (Burger-Wiersma et al. 1986), has been shown to have thylakoid organization very similar to that in *Prochloron* sp. (Schuster et al. 1984, Hiller and Larkum 1985, Miller et al. 1988). The major Chl *a/b* antenna protein of *P. hollandica* is a 33 kDa polypeptide which cross-reacts weakly to an antibody prepared against the CPVI-4 complex of *A. nidulans* R2 (Bullerjahn et al. 1987a). The fact that *Prochloron* sp. also bears a major Chl *a/b* antenna apoprotein of similar molecular mass raises the possibility that these accessory Chl-proteins found in *A. nidulans* R2, *Synechocystis* sp., *Prochloron* sp. and *P. hollandica* are homologous, thus representing a family of Chl-proteins restricted to oxygenic, photosynthetic prokaryotes. Alternatively, these proteins could be homologous to Chl-proteins in higher plants; it remains possible that the homologs are too highly divergent to yield cross-reactivity with our antisera. At any rate, these speculations point to further work aimed at sequencing the genes encoding cyanobacterial Chl-proteins.

D. Carotenoid-binding proteins of cyanobacteria

It has been known for many years that carotenoids are important pigments both in photosynthetic bacteria and chloroplasts. Not only do carotenoids serve as accessory pigments in harvesting light energy, they are essential in quenching triplet Chl and excited singlet O_2 that arise from endogenous photosensitization (reviewed by Siefermann-Harms 1987). Many reports have documented the presence of carotenoids in the cyanobacterial cell wall (Jürgens and Weckesser 1985), the cytoplasmic membrane (Omata and Murata 1983) and the thylakoid membrane (Omata and Murata 1984); recently, several papers have demonstrated the presence of specific carotenoid-binding proteins in these membrane systems. Bullerjahn and Sherman (1986b) purified a 45 kDa carotenoprotein from the cytoplasmic membrane of *Synechocystis* sp. PCC6714; immunological studies confirmed the presence of this polypeptide in carotenoid-containing extracts of *A. nidulans* R2 (Bullerjahn et al. 1987b) and in the prochlorophyte *P. hollandica* (unpublished observations). The protein accumulates only under autotrophic growth conditions in the heterotrophic strain *Synechocystis* sp. PCC6714, suggesting that the polypeptide is present only when the cell is evolving oxygen at high rates (Bullerjahn and Sherman 1986b). This fact had led us to speculate that this pigment-protein acts to quench diffusible singlet oxygen produced as a toxic byproduct of PSII activity. Furthermore, the observation that the carotenoprotein is present in *P. hollandica*, an organism likely phylogenetically distinct from cyanobacteria, indicates that the protein may play an important role in photoprotection under oxygenic conditions.

A thylakoid-associated carotenoprotein has recently been described in *A. nidulans* R2 (Masamoto et al. 1987). This polypeptide has a molecular mass of 42 kDa and accumulates under high irradiance levels. The pattern by which the protein appears in the thylakoid membrane is inconsistent with this protein serving an antenna function. It is more likely that the 42 kDa carotenoprotein acts to deactivate products of photosensitization under high light stress. Moreover, preliminary DNA sequence data of the gene encoding this protein suggest that this polypeptide may be the same as the 42 kDa protein which is synthesized in *A. nidulans* grown under CO_2 limitation (Omata and Ogawa 1986). If so, this protein is a candidate for a general stress-induced polypeptide in cyanobacteria. Certainly, a detailed analysis of the genes encoding regulating the synthesis of the 45 kDa and 42 kDa carotenoproteins will help us understand the mechanisms by which cyanobacteria sense environmental oxygen and irradiance levels.

Holt and Krogmann (1981) have also documented the presence of a carotenoprotein in *Aphanizomenon flos-aquae*, *Spirulina maxima* and *Microcystis aeruginosa*. The apoprotein has a molecular mass of 16 kDa as judged by denaturing electrophoresis and it is found in the soluble fraction of cell lysates. More recently, a carotenoprotein sharing similar features has been purified from *A. nidulans* (Diversé-Pierluissi and Krogmann 1988). It is not known whether these proteins are synthesized in response to environmental changes, but the fact that the protein is present in aqueous extracts is interesting due to the hydrophobic nature of the bound pigment.

E. Phycobilisomes

Under optimal growth conditions, the major light harvesting structure is the phycobilisome; this complex can represent up to 45% of the cell mass. Many recent reviews have described the composition, assembly and function of phycobilisomes with respect to growth under different light regimes (Glazer 1985, Gantt 1986, Zilinskas and Greenwald 1986); this topic will not be discussed in detail here. In contrast, we are interested in the fate of the phycobilisome when cells are grown under nutrient-limitation. One common feature of cyanobacteria grown under iron, carbon, nitrogen, phosphate or sulphate stress is that the phycobilisome is degraded (Miller and Holt 1977, Stevens et al. 1981a, b, Sherman and Sherman 1983, Wanner et al. 1986). This degradation is thought to be due, in part, to the fact that the phycobilisome represents a large pool of carbon skeletons and amino acids which can be mobilized under suboptimal conditions (Lau et al. 1977, Stevens et al. 1981a). An event often noted during the disappearance of the phycobilisomes is the accumulation of glycogen granules replacing the phycobilisomes at the thylakoid membrane surface (Stevens et al. 1981a, Sherman and Sherman 1983). Riethman et al. (1987) have demonstrated that the major unpigmented polypeptides of the phycobilisome are heavily glycosylated, and have discussed the possibility that the glycogen granules may represent a pool of carbohydrate at least partially derived from the degraded phycobilisome.

We have previously demonstrated that in iron-limited *A. nidulans*, a novel Chl-protein complex, CPVI-4, accumulates to high levels. We are currently investigating the possibility that one of the functions of CPVI-4 is to act as a compensatory antenna in stressed cells to allow energy capture in the absence of the phycobilisome. This topic, along with the discussion on the glycosylation of the phycobilisome, will be presented in more detail in sections III C and D of this review.

The pigment-protein complexes described in this paper are listed in Table 1. The recent identification of the Chl-protein complexes has only been achieved as methods for the electrophoretic separation of submembrane particles have improved. As a result, we have yet to determine the precise role of many of the Chl-protein complexes. The analysis of mutants defective in the synthesis of such pigment-proteins will likely reveal their functions in cyanobacteria grown under different environmental conditions.

III Effects of iron stress on cyanobacteria

The results obtained in this laboratory bear on a number of important questions regarding the cyanobacterial photosynthetic apparatus and the structural changes which occur in it upon acclimation to stress conditions. The possible role(s) which the intrinsic Chl-protein CPVI-4 plays in the stress response is a key question, as are the possible reasons for the specific glycosylation of the phycobilisome anchor and linker proteins. This section will discuss these two topics, relate our results to what is known of these topics, and conclude possible rationales for what we have seen.

A. Structural and physiological consequences of iron stress
The coordinate expression of sets of genes is a common response of all organisms to various types of stress. In cyanobacteria, the physiological consequences of stress are often directly reflected in both the ultrastructure of the cell and in altered photosynthetic-related functions (Allen and Smith 1969, Öquist 1971, Prakash and Kumar 1971, Miller and Holt 1977, Stevens et al. 1981a, b, Samuelsson et al. 1985, Sandmann 1985, Suranyi et al. 1987). Global regulatory mechanisms are likely invoked to coordinate essential cellular functions such as Chl synthesis, energy production, DNA replication, lipid and sugar biosynthesis, etc. under stress conditions. Especially interesting are the regulatory mechanisms and biochemical consequences of stress impinging on the structure and function of the photosynthetic apparatus. *A. nidulans* R2 is a transformable, unicellular obligate photoautotroph; its membrane composition has been studied extensively, and the genetic systems recently worked out for this organism (Golden and Sherman 1983, Kuhlemeier et al. 1983, Williams and Szalay 1983) make it an attractive one in which to investigate these questions.

Öquist (1971) first noted that *A. nidulans* cells grown in iron-deficient media displayed dramatically different photosynthetic properties as compared to normal cells, yet were capable of growth at nearly the same rate as normal cells. Subsequent work by Öquist (1974a, b), Sherman and collaborators (Guikema and Sherman 1983b, 1984, Pakrasi et al. 1985a, Sherman and Sherman 1983), and others (summarized in Sherman et al. 1987a) has revealed remarkable changes upon iron stress in the architecture of the photosynthetic membrane and in other essential cellular components. These changes are summarized below:

1. Iron-stressed cells have one to two thylakoid membranes per cell, whereas normally grown cells have 3–4 concentric rings of thylakoids.
2. Carboxysomes are depleted in iron-stressed cells.
3. Phycobilisomes are depleted in iron-stressed cells.
4. 'Low iron cells are filled with glycogen granules that replace phycobilisomes in the intermembrane spaces and sometimes fill the nucleoplasm' (Sherman and Sherman 1983).
5. Phycocyanin and Chl content of iron-stressed cells are depressed.
6. The absorption maximum of membrane-bound Chl is blue-shifted from 680 nm to 672 nm.

7. Iron-stressed cells show a large increase in original and in maximal Chl fluorescence when monitored at room temperature.

8. Low temperature (77 K) Chl fluorescence peaks at 696 nm and 716 nm decrease dramatically relative to the 685 nm peak.

9. The amount of Chl bound to the PSI reaction center decreases relative to that bound to other thylakoid Chl-proteins; this is reflected in a depletion of bands CPI-IV (Fig. 1).

10. Specific polypeptide patterns are present in membranes from iron-deficient cells which are very different from those found in membranes from normal cells.

Remarkably, these pronounced effects do not influence the viability of the cells. When iron is restored to the growth media, the cells return to their normal phenotype within one to two generations in a series of discrete steps, during which the different alterations induced by iron stress have a differing time course during return to normalcy. Clearly, this controlled system offers an opportunity to investigate the regulated biosynthesis and organization of multisubunit structures present in the thylakoid membrane. Much of the work in our laboratory was done with the goal of biochemically and immunologically defining some of the key proteins regulated by iron stress which are constituent polypeptides of three interacting multisubunit thylakoid complexes: PSII, the phycobilisome and CPVI-4. An equally important topic which is examined is the possible relation of these changes in pigment protein structures with other stress-induced processes. The conditional expression of CPVI-4 and phycobilisome components in *A. nidulans* R2 makes the mechanisms involved in both the assembly of these complexes and their interactions with PSII uniquely accessible.

B. Molecular genetic analysis and mutants

An understanding of iron stress and the genetic regulation of this phenomenon can best be addressed using molecular genetics. We have employed a λgt11 library to clone genes which play important roles both in photosynthesis and in adaptation to low-iron growth. The major advantage afforded by the λgt11 system is that antisera specific for the protein of interest can be used to clone the gene directly (Young and Davis 1983, Young et al. 1985). Whereas many cyanobacterial genes have been cloned by using heterologous DNA probes from plant and algal sources, this approach has allowed the cloning of iron-regulated genes which may be unique to cyanobacteria. Furthermore, the library has been useful in cloning cyanobacterial genes for which heterologous antibodies show high specificity for the protein; recently Kuwabara et al. (1987) have isolated the gene encoding the extrinsic PSII manganese-stabilizing protein (MSP) of *A. nidulans* R2 by probing with antibody to the spinach homolog. The gene, named *woxA* (for *w*ater *ox*idation), encodes a polypeptide of 29.3 kDa which is synthesized as a protein bearing a putative signal sequence of 28 amino acids (Fig. 2A). By comparing hydropathy profiles of other processed higher plant nuclear gene products, the signal sequence bears structural similarity to the thylakoid transfer domains of the transit peptides of plastocyanin and spinach MSP (Smeekens et al. 1986, Tyagi et al. 1987; Fig. 2B). These data reveal the first evidence for a signal sequence in cyanobacteria, and also

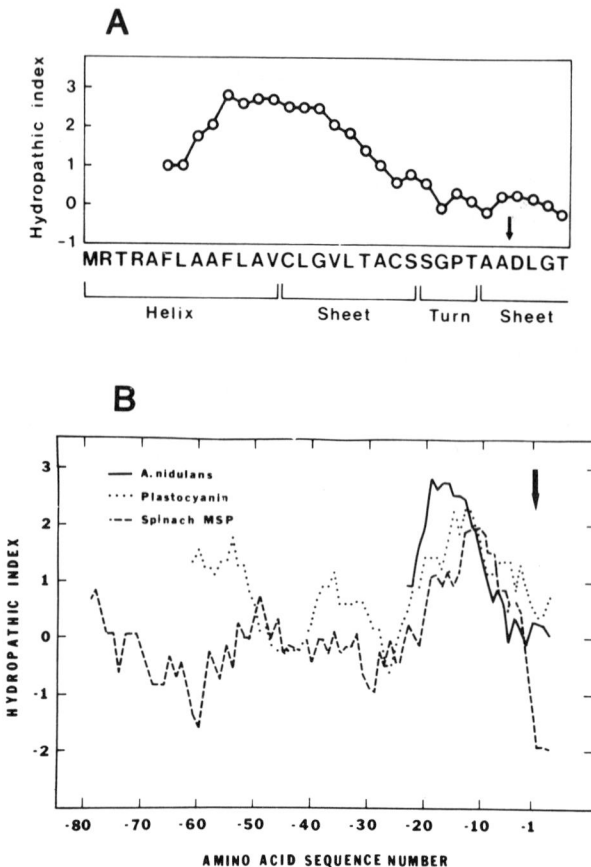

Fig. 2. A) Structural profile of the N-terminal region of *A. nidulans* R2 precursor MSP; the putative processing site is shown by an arrow. Hydropathic indices were calculated according to Kyte and Doolittle (1982) with a window of 11 amino acid residues. B) Comparison of the hydropathy profiles of the N-terminal regions of *A. nidulans* MSP, spinach MSP (Tyagi et al. 1987) and plastocyanin (Smeekens et al. 1986). The arrow indicates the cleavage site for spinach and plastocyanin and the proposed cleavage site for *A. nidulans* MSP.

suggest a conserved domain involved in directing precursor proteins to the thylakoid lumen of chloroplasts and cyanobacteria.

We have recently begun to employ antibodies to retrieve λgt11 clones containing genes which encode iron-stress-associated proteins. The first such gene to be isolated and sequenced, *irpA* (for *i*ron *r*egulated *p*rotein), encodes a 36 kDa intrinsic membrane protein which copurifies with cytoplasmic membranes (Sherman et al. 1987b). The hydropathy profile of the protein suggests that the polypeptide has only limited hydrophobic stretches, although the hydrophobic N-terminus may be responsible for anchoring the protein to the membrane; alternatively, it could represent another signal sequence (Fig. 3). An interesting feature of this gene was revealed by sequencing the region immediately 5′ to the coding sequence (Fig. 4); a putative operator was discovered which is structurally very similar to the iron-regulated *fur* operator in *Escherichia coli* (de Lorenzo et al. 1987). To our knowledge, this is the first candidate for an operator sequence in cyanobacteria. Analysis of other genes which are regulated by levels of iron (such as the CPVI-4 apoproteins) may share the same putative regulatory motif.

In order to assess whether *irpA* is an essential iron-regulated function, we constructed a strain lacking a functional *irpA* gene by insertion of Tn5 into the *irpA* coding sequence. The transposon mutagenesis of the *irpA* gene was as follows: the 0.7 kb *Eco*RI/*Ava*I fragment carrying a portion of the coding sequence was sub-cloned into pUC 8. The resulting plasmid, pRB96, expressed a 29 kDa protein in *E. coli* which was recognized by the CPVI-4

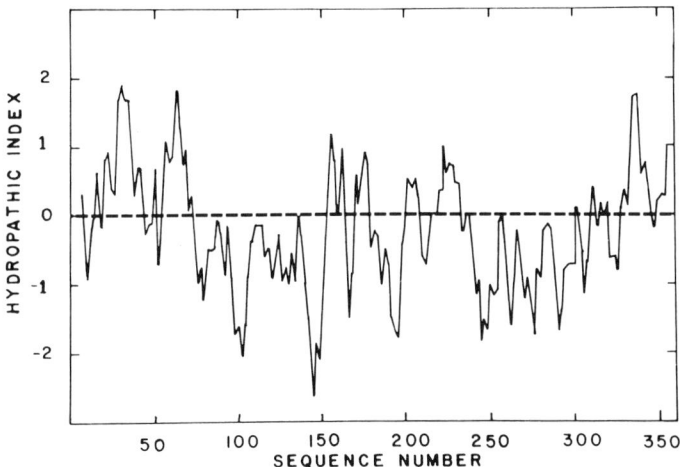

Fig. 3. Hydropathy profile of the *irpA* gene product. Hydropathic indices were calculated as in Fig. 2.

5′ GAT AATGA TAA TCATT ATC 3′ E . COLI CONSENSUS FOR

CTA TTACT ATT AGTAA TAG FUR OPERATOR

−185

5′ TAAA AATGA TTATTATTC TCATT TTTA 3′

ATTT TTACT AATAATAAG AGTAA AAAT A . NIDULANS R2

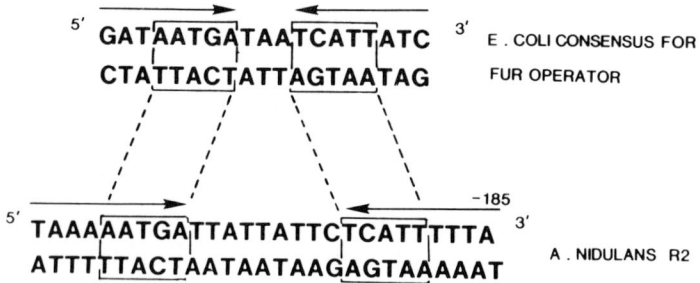

Fig. 4. Sequence comparison between the iron-regulated *fur* operator of *E. coli* and the 5′ upstream region of the *irpA* gene of *A. nidulans* R2.

antibody. pRB96 was subjected to transposon mutagenesis using λ::Tn5 (de Bruijn and Lupski 1984, Sherman et al. 1987b), and the Tn5-interrupted gene was inserted into the *A. nidulans* R2 chromosome by genetic transformation and homologous recombination. This resulted in the construction of a site-directed *irpA*::Tn5 mutant which is impaired in low-iron growth and showed increased sensitivity to iron chelating agents, such as 2,2′-bipyridyl (Fig. 5).

Due to the data suggesting that the *irpA* gene product is localized in the cytoplasmic membrane, we propose that the irpA protein is not directly involved in thylakoid organization under iron stress, but instead plays an essential role in high-affinity iron transport. If so, the *irpA*::Tn5 mutant will be a useful strain for genetic examination of the events which control thylakoid development during iron limitation, because pathways controlling iron uptake from those mediating pigment-protein accumulation or degradation are uncoupled. A detailed physiological study of this mutant will help determine the function of the 36 kDa protein in establishing or maintaining the iron-stress response in *A. nidulans*.

To date, the λgt11 expression library has also been successfully used for isolating genes for the 42 kDa carotenoid-binding protein (Masamoto et al. 1987; see section II). Currently, we are expanding this approach to enable the isolation, characterization and mutagenesis of the genes encoding the 34 kDa and 36 kDa Chl-binding components of the CPVI-4 complex. Additionally, we plan to clone and analyze the gene encoding GP35, another major stress-related *A. nidulans* protein (Riethman 1987; see section IIID). This polypeptide is an extrinsic, basic glycoprotein which accumulates to high levels during iron starvation. As demonstrated by the isolation of the *irpA* gene, the expression vector system presents the most direct route for cloning and mutagenizing the genes encoding stress-induced proteins unique to cyanobacteria.

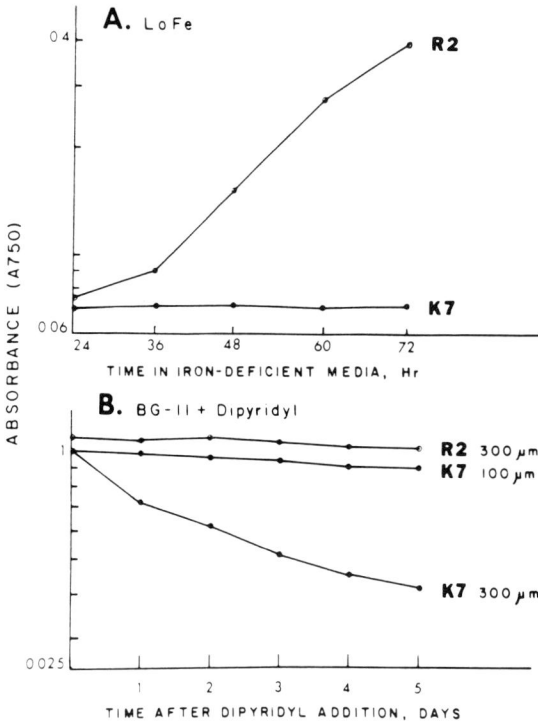

Fig. 5. A) Growth characteristics of *A. nidulans* R2 and the *irpA*::Tn5 derivative, K7. *A. nidulans* R2 and K7 cells grown in iron-containing medium were diluted 1:1000 into low-iron BG-11 (Pakrasi et al. 1985a) and subcultured once more at a 1:1000 dilution and monitored for growth by light scattering at 750nm. B) Sensitivity of *A. nidulans* R2 and K7 to 2,2'-bipyridyl. Exponential-phase cells grown in iron-containing medium were treated with bipyridyl at the indicated concentrations. Note that bipyridyl inhibits growth of the wild-type cells; however, only the K7 culture is killed at the 300 μM concentration.

C. Accumulation of complex CPVI-4

The major Chl-protein in iron-stressed cells is complex CPVI-4, which has a 77 K Chl fluorescence emission peak at 685 nm when assayed immediately after electrophoresis on green gels (Pakrasi et al. 1985b). CPVI-4 was barely detectable in normally-grown *A. nidulans* cells, and from a re-analysis of these experimental results it appears that the CPVI-4 present in normal cells is different from that found in iron-stressed cells. The only protein consistently found in CPVI-4 from normal cells is one which has the electrophoretic properties of the 36 kDa CPVI-4 apoprotein, whereas in iron-stressed cells both the 36 kDa and the 34 kDa apoproteins of CPVI-4 are present.

The fluorescence properties of PSII-enriched preparations from normal and iron-stressed cells were consistent with those found for isolated PSII

Chl-proteins. PSII preparations were enriched in CPVI-2 and CPVI-3 in normal cells, and displayed a 685 nm 77 K Chl emission peak with a 695 nm shoulder. PSII particles from iron-stressed cells contained CPVI-4 as well as CPVI-2 and CPVI-3, and contained a much greater 685 nm 77 K peak relative to the 695 nm shoulder. The combination of this 685 nm peak and a physical association with PSII suggest that the fraction of CPVI-4 associated with these preparations is serving as an antenna to PSII.

The most highly purified PSI from normally grown cells (Riethman 1987) exhibited mainly a 716 nm emission peak at 77 K, with a smaller peak sometimes present at 676 nm (Nechushtai et al. 1986). Less pure PSI from normal cells contained a protein which had properties identical to the 36 kDa protein of CPVI-4. This 36 kDa protein had previously been found associated with the *A. nidulans* PSI-containing Chl-protein bands CPIII and CPIV (Guikema and Sherman 1983).

PSI from iron-stressed cells was obtained from dodecyl-maltoside-solubilized membranes after DEAE cellulose chromatography (Riethman 1987). In addition to the ca. 65 kDa apoproteins, this sample also contained substantial amounts of CPVI-4, indicating that a fraction of CPVI-4 may be part of a PSI submembrane particle in iron-stressed cells. This fraction emitted 77 K Chl fluorescence mainly at 685 nm, but did contain some 714 nm fluorescence as well. Since most of the Chl of this fraction was associated with the PSI subunit, we would have expected a much greater 715 nm peak than that found. This is probably due to the fact that the small amount of CPVI-4 associated with PSI is extremely fluorescent, and overwhelms the fluorescence from the PSI reaction center Chl.

The presence of small amounts of the CPVI-4 36 kDa protein in CPIII, CPIV, and PSI particles of normal cells may indicate an important function of this polypeptide in the organization of thylakoid structure. The CPVI-4 aggregates present in green gels of biochemically purified CPVI-4 were enriched in this 36 kDa protein relative to the 34 kDa CPVI-4 apoprotein; possibly the 36 kDa CPVI-4 apoprotein also causes aggregation of Chl-proteins in normal cells. A testable hypothesis is that this 36 kDa protein functions in the formation of CPIII and CPIV in normal cells, whereas in iron-stressed cells it is associated primarily with the 34 kDa protein in CPVI-4. Thus, the 36 kDa polypeptide might interact specifically with each of the Chl-proteins of the thylakoid. Analysis of the membrane development of *A. nidulans* R2 upon recovery from iron stress could provide a critical test of this hypothesis. A corollary of this hypothesis is that the 36 kDa protein could mediate the insertion of nascent apoproteins into the thylakoid during recovery from iron stress, serving a dual function in both providing Chl and carotenoid from CPVI-4 to stabilize newly inserted apoproteins and serving as a translocator protein (Singer et al. 1987) for the polypeptides of the thylakoid membrane.

Singer et al. (1987) proposed a model for the transfer of integral membrane proteins into membranes. In this model, an integral membrane 'translocator protein' was proposed to mediate insertion of new peptides (either cotranslationally or post-translationally) into the membrane in a step-wise, energy-dependent mechanism. The translocation protein would presumably be multisubunit and capable of forming a channel for transloca- tion of peptide segments during insertion. An unstated requirement for this model is that, in rapidly developing membranes, one can expect relatively large quantities of such translocator proteins. The massive accumulation of newly synthesized photosynthetic membrane proteins in the thylakoid upon recovery from iron stress requires the insertion of each of these proteins into multisubunit complexes containing noncovalently bound Chl, carotenoid, and other cofactors (both covalently and noncovalently bound). A translo- cator protein coupled to a Chl and carotenoid reservoir (CPVI-4) would fulfill many of the theoretical requirements expected of such a mechanism.

An assessment of the functional status of CPVI-4 in iron-stressed cells and in cells recovering from iron stress can illuminate this hypothesis. From the 685 nm 77 K fluorescence peak of CPVI-4 and its enrichment in PSII parti- cles, we had concluded that CPVI-4 is probably an antenna for PSII in iron-stressed cells (Pakrasi et al. 1985b). However, PSI antenna Chl-proteins can also fluoresce at 685 nm and CPVI-4 is present in a PSI particle from iron-stressed cells. Thus, CPVI-4 may be capable of transferring its captured light energy to both photosystems.

An unresolved problem in apparent conflict with the proposed light- harvesting function of CPVI-4 is the data obtained on the room temperature Chl fluorescence of iron-stressed cells (Guikema and Sherman 1984, Guikema 1985). In normally grown *A. nidulans* R2, the Chl fluorescence induction curve is similar to that found in plant chloroplasts. There is a moderate F_o and a typical rise to F_{max} following blockage of electron flow to the PQ pool with DCMU. In iron stressed cells, however, there is a very large F_o and little or no detectable F_v; the Chl fluorescence reaches its maximum very rapidly, and is not affected appreciably by the blockage of the PQ pool with DCMU (Guikema and Sherman 1983b, 1984, Guikema 1985). Upon recovery from iron stress, the fluorescence induction properties begin returning to normal within 1–2 h and have returned to normal by about 10 to 12 h (Guikema 1985).

These fluorescence induction characteristics can be explained, in part, by a decrease in the number of trapping centers relative to the Chl pool feeding into them in iron-stressed cells. A second possibility is that there is a PSII electron transport rate limitation other than the PQ pool in iron stressed cells (possibly, the absence of the non-heme iron between Q_A and Q_B).

The explanation that there is a large excess of Chl in membranes from iron-stressed cells relative to the number of trapping centers seems the most reasonable one, based on a comparison of levels of PSII Chl-proteins, PSI Chl-proteins, and CPVI-4 in iron-stressed cells (Riethman 1987). The decreased number of P680 centers cannot handle the flux of photons from CPVI-4, and these photons are therefore immediately re-emitted as fluorescence, resulting in a very high F_o. The pool of excess Chl in these membranes might be dedicated to providing Chl for the stabilization of nascent apoproteins synthesized during the early stages of recovery. As the initial recovery events proceed, CPVI-4-associated Chl would be transferred to new Chl-proteins. If this is true, then inhibition of Chl biosynthesis should not prevent the initial spectral changes seen upon recovery from iron stress.

Gabaculine specifically inhibits Chl biosynthesis in *A. nidulans* R2 (Guikema et al. 1986), and when it was added along with iron to iron-stressed cultures, the normal decrease in the fluorescence yield (both F_o and F_{max}) during recovery was not affected (Guikema 1985). These experiments demonstrated that, as long as protein synthesis was allowed to occur, the spectral changes in room temperature Chl fluorescence upon recovery from iron stress (membrane development) occurred normally, even in the absence of new Chl. Guikema (1985) concluded that recovery of this parameter was independent of Chl biosynthesis and likely dependent on repair or turnover of existing pigment-protein structures. Since the levels of 695 nm and 715 nm fluorescence at 77 K were not monitored in this experiment, it is difficult to correlate the decrease in room temperature fluorescence with the presumably related rise in these signals. It is possible that during intermediate stages of recovery, CPVI-4 goes from an inefficient antenna to an efficient antenna, then is finally phased out as phycobilisomes assemble and take over the main light-harvesting function. The Chl-protein patterns were not monitored during recovery in these presence of gabaculine, so we cannot assess how far the reorganization of the thylakoid can proceed in the absence of new Chl. This is a key experiment for testing whether a major role of CPVI-4 is to provide Chl for newly synthesized apoproteins, or if their accumulation requires new Chl. Moreover, site-directed mutants which either lack or produce altered CPVI-4 complexes will be useful in the future to test the above hypothesis. If CPVI-4 is playing a role in membrane biogenesis, mutants which cannot assemble a functional CPVI-4 would likely be developmentally blocked at some stage of the membrane reorganization pathway initiated after iron restoration. Such mutants would yield altered physiological responses to iron stress and recovery; these responses could be monitored by examining fluorescence data and nondenaturing green gels.

Akoyunoglou and co-workers (Argyroudi-Akoyunoglou et al. 1982, Akoyunoglou and Akoyunoglou 1985) noted that, if chloroplast thylakoids still in the process of development were transferred to the dark, a reorganization of the Chl already in the thylakoid occurred. This reorganization shares some properties with the changes in Chl organization postulated to occur during recovery from iron stress in *A. nidulans* R2. Specifically, Chl associated with integral membrane Chl-protein complexes (in this case, LHCPI or LHCPII) is transferred to nascent PSII and PSI core Chl-proteins which continue to be synthesized in the dark (where no new Chl synthesis occurs). This process occurs in spite of the presence of newly synthesized LHCP apoproteins, which are apparently degraded in the absence of new Chl. Akoyunoglou and Akoyunoglou (1985) postulated that two steps were required for these events in the dark; the disorganization of LHCP complexes, then the utilization of the Chl *a* from these complexes by new synthesized apoproteins of PSI and PSII. If our hypothesis is correct, these events would occur simultaneously as the apoproteins are inserted into the membrane. Thus, LHCPI and LHCPII might both be associated with major translocator proteins (Singer et al. 1987) for thylakoid membrane systems. The multisubunit nature of LHCPI and LHCPII are in agreement with the postulated mechanism for translocation described by Singer et al. (1987), and the available pool of Chl and carotenoid associated with these complexes suggest a similar function in assembly as proposed for CPVI-4. Minor Chl-proteins found in association with LHCPII or with the PSII core (Dunahay et al. 1987) may be important in the formation of a multisubunit translocation channel in close proximity or within the LCHPII complex.

Based on the association of CPVI-4 with PSI, it is possible that CPVI-4 and its analogs in the facultatively photoheterotrophic cyanobacteria *Synechocystis* sp. PCC6714 and PCC6803 may function as a PSI antenna as well as a PSII antenna. *Synechocystis* sp. PCC6714 and PCC6803 make large amounts of the CPVI-4 analog (CPIIIb; see Table 1) constitutively, although its levels are enhanced by iron stress (Bullerjahn et al. 1985). If the *Synechocystis* sp. PCC6714 CPVI-4 analog serves both PSI and PSII, then the constitutive accumulation of this Chl-protein in *Synechocystis* sp. might be due to the importance of PSI in photoheterotrophic growth, where electrons cycle through PSI to provide energy for ATP synthesis. It is possible that the constitutive presence of a CPVI-4-like Chl-protein is a universal property of cyanobacteria capable of photoheterotrophic growth.

D. Regulation of phycobilisome and glycogen metabolism

The most immediately noted effect of many stress conditions in cyanobacteria is the degradation of phycobilisomes and the concurrent accumulation of glycogen (Wanner et al. 1987). Only in the case of nitrogen-deficiency in heterocyst-forming cyanobacteria has a rationale for this phenomenon been supported by experimental data. Glycogen breakdown can supply electrons in support of nitrogenase activity (Böhme and Schrautemeier 1987), and presumbaly the nitrogen present in phycobilisomes is utilized by heterocysts to make other cellular components (heterocysts lack oxygen-evolving PSII activity, and phycobilisomes would presumably be expendable). Results obtained in our laboratory suggest that phycobilisome breakdown and glycogen synthesis might in fact be related events. We have shown conclusively that the anchor and linker subunits of the phycobilisome are glycosylated (Fig. 6; Riethman et al. 1987), and two of the phycobilisome rod linker proteins are preferentially glycosylated near the ends of the molecules (Riethman 1987). These are regions which have been shown to be important for the assembly of the phycobilisome rods and the attachment of the phycobilisome rod to the core (Zilinskas and Greenwald 1986). The phycobilisome anchor protein, which is believed to attach the phycobilisome to PSII, is extremely susceptible to protease activity; cleavage of the anchor releases an intact phycobilisome from the membrane (Hiller et al. 1983). It is the anchor protein which is the most heavily glycosylated subunit of the phycobilisome (Riethman 1987). Thus, the proteins most important for phycobilisome assembly and membrane attachment are glycosylated, and apparently glycosylated in key regions most exposed to cellular enzymes in vivo.

An extremely abundant extrinsic membrane glycoprotein was found in iron-stressed *A. nidulans* R2. Designated GP35 (glycoprotein of 35 kDa), this protein is very basic and is ConA-reactive, properties shared with the phycobilisome anchor and linker subunit (Riethman 1987). Its massive accumulation upon iron stress coincides with the accumulation of glycogen granules of the outer surface of the thylakoid membrane, a region formerly occupied by the phycobilisomes in normal cells. Taken together, these results suggest a direct relation between GP35, glycogen granule formation, and phycobilisome degradation. GP35 has many of the properties of a phycobilisome linker protein, and in fact might be a modified linker.

As mentioned above, the anchor polypeptide is particularly susceptible to proteolysis. Limited trypsin proteolysis of intact phycobilisome or isolated cores led to the formation of 40 kDa and 11 kDa fragments of *A. nidulans* anchor (Lundell and Glazer 1983a, b). Both of these anchor fragments remained associated with the phycobilisome, and the 40 kDa fragment was shown to contain the covalently attached bilin. The remaining 24 kDa of the

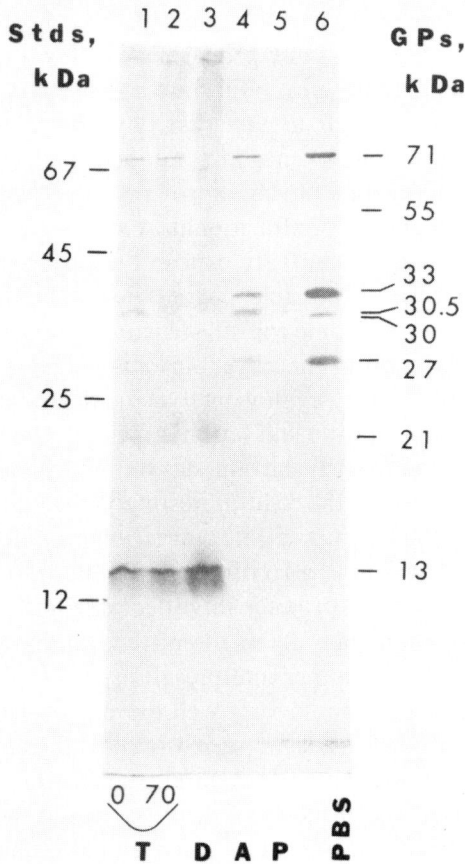

Fig. 6. ConA-reactive membrane components from *A. nidulans* R2. The material in each lane was separated by denaturing electrophoresis, transferred to nitrocellulose and stained for ConA-reactive components (Clegg 1982). In addition, membranes were fractionated into intrinsic membrane, extrinsic membrane and detergent-insoluble fractions by using Triton X-114 phase partitioning (Bordier 1981). Lanes 1 and 2: total thylakoid material (T) solubilized at 0 and 70 °C, respectively; lanes 3–5: detergent (intrinsic membrane), aqueous (extrinsic membrane) and Triton X-114 insoluble fractions after phase partitioning, respectively; lane 6: purified phycobilisomes.

anchor was postulated to function in membrane attachment (Lundell and Glazer 1983a). Proteolytic mapping of ConA-reactive regions of the *A. nidulans* anchor protein demonstrated that a 40 KDa region of the anchor protein lacked ConA reactivity (Riethman 1987). It is tempting to speculate that this non-ConA-reactive region corresponds to the 40 kDa, protease-resistant region which contains the attached bilin, and that the heavily glycosylated region of the anchor includes the 24 kDa fragment. Thus, this heavily glycosylated portion might function in membrane attachment and be exposed to cytosolic enzymes in vivo.

Heterocyst formation has some properties which would tend to support the above hypothesis. Intact phycobilisome rod substructures are found in some heterocysts, in spite of the fact that phycobilisomes are absent and allophycocyanin core components are not detectable (reviewed by Van Baalen 1987). The remaining phycobiliproteins can still feed light into PSI, although PSII is apparently nonfunctional. A stress-specific cleavage of the anchor protein would preferentially expose the phycobilisome allophyco-cyanin-containing core to possible degradation. Thus, in heterocysts of some species, the phycobilisome rod substructures might reorganize around PSI after the degradation of the phycobilisome core.

In non-heterocyst-forming cyanobacteria, the degradation of the phyco-bilisome in response to nitrate starvation appears to be more complete. A phycocyaninase is induced by nitrate starvation which rapidly degrades most of the phycocyanin in the cell. Interestingly, the addition of glucose to *Synechocystis* PCC6803 during nitrate starvation inhibited the degradation of phycocyanin (Elmorjani and Herdman 1987). Because a non-metaboliz-able glucose analog did not produce this effect, the authors concluded that metabolite would irreversibly inhibit the phycocyaninase.

In light of our results, another interpretation of this glucose effect seems likely. The added glucose could very well drive a reversible glycosylation reaction such that a glycosidase activated by nitrate starvation to remove glucose from the anchor (or linkers) would fail to do so. The end result would be a failure of the proteolysis of phycocyanin, as it would remain assembled in the phycobilisome. The glucose analog might not be recog-nized by the glycosidase or might fail to block hydrolysis of glucose from the phycobilisome, and so it might not mimic the glucose effect.

The results of other stress conditions on the phycobilisome-glycogen co-regulation are likely to be similar to those postulated for nitrate starva-tion. Whether these stresses each produce a common signal, or if they are mediated by an indirect nitrate starvation signal, the end result would be the same.

Since so little is known about the regulation of phycobilisome linker and anchor glycosylation and its relation to glycogen metabolism, much of the above discussion is highly speculative. Nevertheless, such an analysis clearly suggests some fairly simple experiments which can test many of these speculations. It is unlikely that such an unusual (for a prokaryote) post-translational modification would be unimportant; the above arguments seem reasonable ones, and will hopefully provide a framework for further experimentation.

References

Akoyunoglou A and Akoyunoglou G (1985) Reorganization of thylakoid components during chloroplast development in higher plants after transfer to darkness. Plant Physiol 79: 425–431

Allen MM and Smith AJ (1969) Nitrogen chlorosis in blue-green algae. Arch Mikrobiol 69: 114–120

Anderson LK, Rayner MC, Sweet RM and Eiserling FA (1983) Regulation of *Nostoc* sp. phycobilisome structure by light and temperature. J Bacteriol 155: 1407-14-6

Argyroudi-Akoyunoglou JH, Akoyunoglou A, Kalosakas K and Akoyunoglou G (1982) Reorganization of the photosystem II unit in developing thylakoids of higher plants after transfer to darkness. Changes in chlorophyll β, content, and grana stacking. Plant Physiol 70: 1242–1248

Barlow RG and Alberte RS (1985) Photosynthetic characteristics of phycoerythrin-containing *Synechococcus* spp. I. Responses to growth photon flux density. Mar Biol 86: 63–74

Böhme H and Schrautemeier B (1987) Electron donation to nitrogenase in a cell-free system from heterocysts of *Anabaena variabilis*. Biochim Biophys Acta 891: 115–120

Bordier C (1981) Phase separation of integral membrane proteins in Triton X-114 solution. J Biol Chem 256: 1604–1607

Boyer GL, Gillam AH and Trick C (1987) Iron chelation and uptake. In: Fay P and Van Baalen C (eds) The Cyanobacteria, pp 415–436. Amsterdam: Elsevier Publihsers

Bryant DA (1987) The cyanobacterial photosynthetic apparatus: comparison of those of higher plants and photosynthetic bacteria. Can Bull Fish Aquat Sci 214: 423–500

Bullerjahn GS, Riethman HC and Sherman LA (1985) Organization of the thylakoid membrane from the heterotrophic cyanobacterium, *Aphanocapsa* 6714. Biochim Biophys Acta 810: 148–157

Bullerjahn GS and Sherman LA (1986a) Immunological characterization of photosystem II chlorophyll-binding proteins from the cyanobacterium, *Aphanocapsa* 6714. J Bioenerg Biomembr 18: 285–293

Bullerjahn GS and Sherman LA (1986b) Identification of a carotenoid-binding protein in the cytoplasm membrane from the heterotrophic cyanobacterium *Synechocystis* sp. strain PCC6714. J Bacteriol 167: 396–399

Bullerjahn GS, Matthijs HCP, Mur LR and Sherman LA (1987a) Chlorophyll-protein composition of the thylakoid membrane from *Prochlorothrix hollandica*, a prokaryote containing chlorophyll *b*. Eur J Biochem 168: 295–300

Bullerjahn GS, Riethman HC and Sherman LA (1987b) Isolation and immunological characterization of a carotenoid-binding protein from the cyanobacteria *Synechocystis* sp. PCC6714 and *A. nidulans* R2. In: Biggins J (ed) Progress in Photosynthesis Research, Vol II, pp 145–148. Dordrecht: Martinus Nijhoff Publishers

Burger-Wiersma T, Veenhuis M, Korthals HJ, Van de Wiel CCM and Mur LR (1986) A new prokaryote containing chlorophylls *a* and *b*. Nature 320: 262–264

Clegg JCS (1982) Glycoprotein detection in nitrocellulose transfers of electrophoretically separated protein mixtures using concanavalin A and peroxidase: application to arenavirus and flavivirus proteins. Anal Biochem 127: 389–394

De Bruijn FJ and Lupski JR (1984) The use of transposon Tn5 mutagenesis in the rapid generation of correlated physical and genetic maps of DNA segments cloned into multicopy plasmids – a review. Gene 27: 131–149

DeLorenzo V, Wee S, Herrero M and Neilands JB (1987) Operator sequences of the aerobactin operon of plasmid ColV-K30 binding the ferric uptake regulation (*fur*) repressor. J Bacteriol 169: 2624–2630

254

Diversé-Pierluissi M and Krogmann DW (1988) A zeaxanthin-protein from *Anacystis nidulans*. Biochim Biophys Acta, in press

Doolittle WF and Singer RA (1974) Mutational analysis of dark endogenous metabolism in the blue-green bacterium *Anacystis nidulans*. J Bacteriol 119: 677–683

Dunahay TG, Schuster G and Staehelin LA (1987) Phosphorylation of spinach chlorophyll-protein complexes. CPII* but not CP29, CP27 or CP24 is phosphorylated in vitro. FEBS Let 215: 25–30

Elmorjani K and Herdman M (1987) Metabolic control of phycocyanin degradation in the cyanobacterium *Synechocystis* PCC6803: a glucose effect. J Gen Microbiol 133: 1685–1694

Foulds IJ and Carr NG (1977) A proteolytic enzyme degrading phycocyanin in the cyanobacterium *Anabaena cylindrica*. FEMS Microbiol Lett 2: 117–119

Fujita Y, Ohki K and Murakami A (1985) Chromatic regulation of photosystem composition in the photosynthetic system of red and blue-green algae. Plant Cell Physiol 26: 1541–1548

Gantt E (1986) Phycobilisomes. In: Staehelin LA and Arntzen CJ (eds) Encyclopedia of Plant Physiology: Photosynthesis III, pp 260–268. Berlin: Springer-Verlag Publishers

Glazer AN (1985) Light harvesting by phycobilisomes. Ann Rev Biophys Chem 14: 47–77

Golbeck JH and Warden JT (1986) Photosystem I charge separation in the absence of centers A and B. I Optical characterization of center A_2 and evidence for its association with a 64 kDa peptide. Biochim Biophys Acta 849: 16–24

Golden SS and Sherman LA (1983) A hybrid plasmid is a stable cloning vector for the cyanobacterium *Anacystis nidulans* R2. J Bacteriol 155: 966—972

Grossman AR, Lemaux PG and Conley PB (1986) Regulated synthesis of phycobilisome components. Photochem Photobiol 44: 827–837

Guikema JA and Sherman LA (1983a) Chlorophyll-protein organization of membranes from the cyanobacterium *Anacystis nidulans*. Arch Biochem Biophys 220: 155–166

Guikema JA and Sherman LA (1983b) Organization and function of chlorophyll in membranes of cyanobacteria during iron starvation. Plant Physiol 73: 250–256

Guikema JA and Sherman LA (1984) Influence of iron deprivation on the membrane composition of *Anacystis nidulans*. Plant Physiol 74: 90–95

Guikema JA (1985) Fluorescence induction characteristics of *Anacystis nidulans* during recovery from iron deprivation. J Plant Nutr 8: 891–908

Guikema JA, Freeman L and Fleming EH (1986) Effects of gabaculin on pigment biosynthesis in normal and nutrient deficient cells of *Anacystis nidulans*. Plant Physiol 82: 280–284

Hiller RG, Post A and Stewart AC (1983) Isolation of intact detergent-free phycobilisomes by trypsin. FEBS Lett 156: 180–184

Hiller RG and Larkum AWD (1985) The chlorophyll-protein complexes of *Prochloron* sp. (Prochlorophyta). Biochim Biophys Acta 806: 107–115

Høj PB and Moller BL (1986) The 110 kDa reaction center protein of photosystem I, P700-chlorophyll*a* protein 1, is an iron-sulfur protein. J Biol Chem 261: 14292–14300

Holt TK and Krogmann DW (1981) A carotenoid-protein from cyanobacteria. Biochim. Biophys Acta 637: 408–414

Jones LW and Myers J (1965) Pigment variation in *Anacystis nidulans* induced by light of selected wavelengths. J Phycol 1: 7–13

Jürgens UJ and Weckesser J (1985) Characterization of the cell wall of the unicellular cyanobacterium *Synechocystis* PCC6714. Arch Microbiol 142: 168–174

Kawamura M, Mimuro M and Fujita Y (1979) Quantitative relationship between two reaction centers in the photosynthetic system of blue-green algae. Plant Cell Physiol. 20: 697–705

Khanna R, Graham JR, Myers J and Gantt E (1983) Phycobilisome composition and possible relationship to reaction centers. Arch Biochem Biophys 224: 534–542

Kuhlemeier CJ, Thomas AAM, Van der Ende A, Van Leen RW, Borrias WE, Van den Hondel CAMJJ and Van Arkel GA (1983) A host-vector system for gene cloning in the cyanobacterium *Anacystis nidulans* R2. Plasmid 10: 156–163

Kuwabara T, Reddy KJ and Sherman LA (1987) Nucleotide sequence of the gene encoding the Mn-stabilizing protein involved in photosystem II water oxidation in the cyanobacterium *Anacystis nidulans* R2. Proc Natl Acad Sci USA 84: 8230–8234

Kyte J and Doolittle RF (1982) A simple method for displaying the hydropathic character of a protein. J Mol Biol 157: 105–132

Lau RH, Mackenzie MM and Doolittle WF (1977) Phycocyanin synthesis and degradation in the blue-green bacterium *Anacystis nidulans*. J Bacteriol 132: 771–778

Lawry NH and Jensen TE (1979) Deposition of condensed phosphate as an effect of varying sulfur deficiency in the cyanobacterium *Synechococcus* sp. (*A. nidulans*). Arch Microbiol 120: 1–7

Lönneborg A, Lind LK, Kalla SR, Gustafsson P and Öquist G (1985) Acclimation processes in the light-harvesting system of the cyanobacterium *Anacystis nidulans* following a light shift from white to red light. Plant Physiol 78: 110–114

Lundell DJ and Glazer AN (1983a) Molecular architecture of a light-harvesting antenna. Structure of the 18S core-rod subassembly of the *Synechococcus* 6301 phycobilisome. J Biol Chem 258: 894–901

Lundell DJ and Glazer AN (1983b) Molecular architecture of a light-harvesting antenna. Quarternary interactions in the *Synechococcus* 6301 phycobilisome core as revealed by partial tryptic digestion and circular dichroism studies. J Biol Chem 258: 8708–8713

Machold O (1986) Relationship between the 43 kDa chlorophyll-protein of PSII and the rapidly metabolized 32 kDa Qb protein. FEBS Lett. 204: 363–367

Manodori A and Melis A (1986) Cyanobacterial acclimation to photosystem I or photosystem II light. Plant Physiol 82: 185–189

Masamoto K, Riethman HC and Sherman LA (1987) Isolation and characterization of a carotenoid-associated thylakoid protein from the cyanobacterium *Anacystis nidulans* R2. Plant Physiol 84: 633–639

Miller KR, Jacob JR and Matthijs HCP (1988) Photosynthetic membrane structure in *Prochlorothrix holandica*: a chlorophyll *b*-containing prokaryote. Eur J Cell Biol., in press

Miller LS and Holt SC (1977) Effect of carbon dioxide on pigment and membrane content in *Synechococcus lividus*. Arch Microbiol 115: 185–198

Muster P, Binder A and Bachofen R (1984) A single subunit P-700 reaction center of the thermophilic cyanobacterium *Mastigocladus laminosus*. FEBS Lett 166: 160–164

Myers J, Graham JR and Wang RT (1980) Light harvesting in *Anacystis nidulans* studied in pigment mutants. Plant Physiol 66: 1144–1149

Nakatani HY, Ke B, Dolan E and Arntzen CJ (1984) Identity of the photosystem II reaction center polypeptide. Biochim Biophys Acta 765: 347–352

Nanba O and Satoh K (1987) Isolation of a photosystem II reaction center consisting of D1 and D2 polypeptides and cytochrome b-559. Proc Nat Acad Sci USA 84: 109–112

Nechushtai R, Nourizadeh SD and Thornber JP (1986) A reevaluation of the fluorescence of the core chlorophylls of photosystem I. Biochim Biophys Acta 848: 193–200

Omata T and Murata N (1983) Isolation and characterization of the cytoplasmic membranes from the blue-green agla (cyanobacterium) *Anacystis nidulans*. Plant Cell Physiol 24: 1101–1112

Omata T and Murata N (1984) Isolation and characterization of three types of membranes from the cyanobacterium (blue-green algae) *Synechocystis* PCC6714. Arch Microbiol 139: 113–116

Omata T and Ogawa T (1986) Biosynthesis of a 42 kDa polypeptide in the cytoplasmic

256

membrane of the cyanobacterium *Anacystis nidulans* strain R2 during adaptation to low CO_2 concentration. Plant Physiol 80: 525–530

Öquist G (1971) Changes in pigment composition and photosynthesis induced by iron-deficiency in the blue-green alga *Anacystis nidulans*. Physiol Plant 25: 188–191

Öquist G (1974a) Iron deficiency in the blue-green alga, *Anacystis nidulans*. Changes in pigmentation and photosynthesis. Physiol Plant 30: 30–37

Öquist G (1974b) Iron deficiency in the blue-green alga, *Anacystis nidulans*. Fluorescence and absorption spectra recorded at 77 K. Physiol Plant 31: 55–58

Pakrasi HB, Goldenberg A and Sherman LA (1985a) Membrane development in the cyanobacterium *Anacystis nidulans* during recovery from iron starvation. Plant Physiol 79: 290–295

Pakrasi HB, Riethman HC and Sherman LA (1985b) Organization of pigment-proteins in the photosystem II complex of the cyanobacterium *Anacystis nidulans* R2. Proc Natl Acad Sci USA 82: 6903–6907

Potts M (1985) Protein synthesis and proteolysis in immobilized cells of the cyanobacterium *Nostoc commune* UTEX 584 exposed to matric water stress. J Bacteriol 164: 1025–1031

Prakash G and Kumar HD (1971) Studies on sulphur-selenium antagonism in blue-green algae. Arch Mikrobiol 77: 196–202

Reddy KJ, Vann C and Sherman LA (1987) Cloning and characterization of a gene encoding an iron-regulated membrane protein in *Anacystis nidulans* R2. In: Biggins J (ed) Progress in Photosynthesis Research, Vol IV, pp 777–780. Dordrecht: Martinus Nijhoff Publishers

Redlinger T and Gantt E (1982) A M 95,000 polypeptide in *Porphyridium cruentum* phycobilisomes and thylakoids: possible function in linkage of phycobilisomes to thylakoids and in energy transfer. Proc Natl Acad Sci USA 79: 5542–5546

Riethman HC (1987) Characterization of pigment-protein complexes in the cyanobacterium *Anacystis nidulans* R2, 286 pp. PhD dissertion, University of Missouri, Columbia

Riethman HC, Mawhinney TP and Sherman LA (1987) Phycobilisome-associated glycoproteins in the cyanobacterium *Anacystis nidulans* R2. FEBS Lett 215: 209–214

Samuelsson G, Lönneborg A, Rosenqvist E, Gustafsson P and Öquist G (1985) Photoinhibition and reactivation of photosynthesis in the cyanobacterium *Anacystis nidulans*. Plant Physiol 79: 992–995

Sandmann G (1985) Consequences of iron deficiency on photosynthetic and respiratory electron transport in blue-green algae. Photosynth Res 6: 261–271

Satoh K, Fujii Y, Aoshima T and Tado T (1987) Immunological identification of the polypeptide bands in the SDS-polyacrylamide gel electrophoresis of photosystem II preparations. FEBS Lett 216: 7–10

Schuster G, Owens GC, Cohen Y and Ohad I (1984) Thylakoid polypeptide compoisiton and light-independent phosphorylation of the chlorophyll *a/b* protein of *Prochloron* sp., a prokaryote exhibiting oxygenic photosynthesis. Biochem Biophys Acta 767: 596–605

Sherman DA and Sherman LA (1983) The effects of iron deficiency and iron restoration on the ultrastructure of the cyanobacterium *Anacystis nidulans*. J Bacteriol 156: 393–401

Sherman LA, Bricker T, Guikema J and Pakrasi H (1987a) The protein composition of the photosynthetic complexes from the cyanobacterial thylakoid membrane. In: Fay P and Van Baalen C (eds) The Cyanobacteria, pp 1–34. Amsterdam: Elsevier Publishers

Sherman LA, Reddy KJ, Riethman HC and Bullerjahn GS (1987b) Genetic analysis of a cyanobacterial gene encoding a membrane protein which accumulates under iron stress. In Biggins J (ed) Progress in Photosynthesis Research, Vol IV, pp 773–776. Dordrecht: Martinus Nijhoff Publishers

Sicfermann-Harms D (1987) The light-harvesting and protective functions of carotenoids in photosynthetic membranes. Physiol Plant 69: 561–568

Singer SJ, Maher PA and Yaffe MP (1987) On the translocation of proteins across membranes. Proc Natl Acad Sci USA 84: 1015–1019

Smeekens S, Bauerle C, Hagemann J, Keegstra K and Weisbeek P (1986) The role of the transit peptide in the routing of precursors toward different chloroplast compartments. Cell 46: 365–375

Stevens SE, Balkwill DL and Paone DAM (1981a) The effects of nitrogen limitation on the ultrastructure of the cyanobacterium *Agmenellum quadruplicatum*. Arch Microbiol 130: 204–212

Stevens SE, Paone DAM and Balkwill DL (1981b) Accumulation of cyanophycin granules as a result of phosphate limitation in *Agmenellum quadruplicatum*. Plant Physiol 67: 716–719

Stewart AC (1980) The chlorophyll-proteins of a thermophilic blue-green alga. FEBS Lett 114: 67–72

Suranyi G, Korcz A, Palfi Z and Borbély G (1987) Effect of light deprivation on RNA synthesis, accumulation of guanosine 3'(2')-diphosphate 5' diphosphate, and protein synthesis in heat-shocked *Synechococcus* sp. strain PCC 6301, a cyanobacterium. J Bacteriol 169: 632–639

Takahashi Y, Koike H and Katoh S (1982) Multiple forms of chlorophyll-protein complexes from a thermophilic cyanobacterium *Synechococcus* sp. Arch Biochem Biophys 219: 219–227

Takahashi Y, Hirota Y and Katoh S (1985) Multiple forms of P700-chlorophyll *a* protein complexes from *Synechococcus* sp. – the iron, quinone and carotenoid contents. Photosynth Res 6: 183–192

Thornber JP (1986) Biochemical characterization and structure of pigment-proteins of photosynthetic organisms. In: Staehelin LA and Arntzen CJ (eds) Encyclopedia of Plant Physiology: Photosynthesis III, pp 98–142. Berlin: Springer-Verlag Publishers

Tyagi A, Hermans J, Stepphun J, Jansson C, Vater F and Herrmann RG (1987) Nucleotide sequence of cDNA clones encoding the complete '33 kDa' precursor protein associated with the photosynthetic oxygen-evolving complex from spinach. Mol Gen Genet 207: 288–293

Van Baalen C (1987) Nitrogen fixation. In: Fay P and Van Baalen C (eds) The Cyanobacteria. pp 187–198. Amsterdam: Elsevier Publishers

Wanner G, Henkelmann G, Schmidt A and Köst HP (1986) Nitrogen and sulfur starvation of the cyanobacterium *Synechococcus* 6301 – an ultrastructural, morphometrical and biochemical comparison. Z Naturforsch 41c: 741–750

Warden JT and Golbeck JH (1986) Photosystem I charge separation in the absence of centers A and B. II. ESR spectral characterization of center 'X' and correlation with optical signal 'A₂'. Biochim Biophys Acta 849: 25–31

Williams RC, Glazer AN and Lundell DJ (1983) Cyanobacterial photosystem I: morphology and aggregation behavior. Proc Natl Acad Sci USA 80: 5923–5926

Williams JGK and Szalay AA (1983) Stable integration of foreign DNA into the chromosome of the cyanobacterium *Synechococcus* R2. Gene 24: 37–51

Wyman M and Fay P (1987) Acclimation to the natural light climate. In: Fay P and Van Baalen C (eds) The Cyanobacteria, pp 347–376. Amsterdam: Elsevier Publishers

Yamagishi A and Katoh S (1984) A photoactive photosystem II reaction center complex lacking a chlorophyll-binding 40 kDa subunit from the thermophilic cyanobacterium *Synechococcus* sp. Biochim Biophys Acta 765: 118–124

Young RA and Davis RW (1983) Efficient isolation of genes by using antibody probes. Proc Natl Acad Sci USA 80: 1194–1198

Young RA, Bloom BR, Grosskinski CM, Ivanyi J, Thomas D and Davis RW (1985) Dissection of *Mycobacterium tuberculosis* antigens using recombinant DNA. Proc Natl Acad Sci USA 82: 2583–2587

Zilinskas BA and Greenwald LS (1986) Phycobilisome structure and function. Photosynth Res 10: 7–36

Govindjee et al. (eds), Molecular Biology of Photosynthesis: 259–281
© 1988 Kluwer Academic Publishers

Minireview

The major light-harvesting complex of Photosystem II: aspects of its molecular and cell biology

PARAG R. CHITNIS & J. PHILIP THORNBER

Biology Department and Molecular Biology Institute, University of California, Los Angeles, CA 90024, USA

Received 27 August 1987; Accepted 30 November 1987

Key words: photosynthesis, chlorophyll-proteins, chloroplasts, assembly of membrane proteins, *cab* genes, gene regulation, protein sorting, protein import by organelles

Abstract. The light-harvesting complex of photosystem II (LHC II) contains one major (LHC IIb) and at least three minor chlorophyll-protein components. The apoproteins of LHC IIb (LHCP) are encoded by nuclear genes and synthesized in the cytoplasm as a higher molecular weight precursor(s) (pLHCP). Several genes coding for pLHCP have been cloned from various higher plant species. The expression of these genes is dependent upon a variety of factors such as light, the developmental stage of the plastids and the plant. After its synthesis in the cytoplasm, pLHCP is imported into plastids, inserted into thylakoids, processed to its mature form, and assembled into LHC IIb. The pathway of assembly of LHC IIb in the thylakoid membranes is currently being investigated in several laboratories. We present a model that gives some details of the steps in the assembly process. Many of the steps involved in the synthesis and assembly are dependent on light and the stage of plastid development.

Abbreviations: PS — Photosystem, LHC II — Light-harvesting complex of PS II, LHCP — Apoproteins of LHC IIb, pLHCP — Precursor of LHCP, PAGE — Polyacrylamide gel electrophoresis

Introduction

Higher plant chloroplasts have distinct subcompartments which include two envelope membranes, the space between them, the stroma, a network of thylakoids and the lumen within them. Most thylakoid membrane polypeptides are organized into discrete multicomponent units that are embedded in the lipid bilayer (Steinbeck et al. 1985). These units consist of the photosystem (PS) I and II complexes which harvest light and perform the primary charge separations, and other complexes that lack photosynthetic pigments (cytochrome b/f and ATPase complexes). The latter process the products of the primary events to yield the ATP and NADPH necessary for biosynthesis. Both photosystems contain chlorophyll *a* and *b* molecules that are coordinated with, but not covalently linked to, several specific membrane-associated proteins (Markwell et al. 1979, Thornber 1986). The proteins function to orient and space precisely their associated chlorophyll and carotenoid molecules so that energy absorbed by any one of these often abundant pigment-proteins is transferred efficiently to the PS I or PS II reaction centers. Each photosystem can be thought of as being composed of two distinct parts: 1) a core complex that contains those polypeptides and cofactors required for a plastid to have a stable and functional primary photochemical event; and 2) a light-harvesting complex (LHC) that houses most of the antenna pigments in the photosystem, but is not essential for the structure and function of the core complex (Thornber 1986). Despite this simple conceptual organization, at least eight rather than the anticipated four (i.e. core complex I, LHC I, core complex II and LHC II), different complexes containing chlorophyll have been isolated from higher plants (Peter and Thornber 1988).

LHC II is the most abundant of these pigmented complexes in the thylakoids. LHC II can be resolved into several pigment-protein components (Table 1) of which LHC IIb represents almost half of the chlorophyll and one-third of the protein in green plant thylakoids (Bellemare et al. 1982, Bennett et al. 1981, Camm and Green 1980, Machold and Meister 1979). It functions not only as an antenna but it and/or some other LHC II component also contributes to the stacking of the thylakoids to form grana (Mullet 1983). LHC IIb has also been implicated in regulating, via its phosphorylation, the proportion of absorbed excitation energy directed to each of the two photosystems; i.e., the state I–state II transition (see Bennett 1983, for review). The apoproteins of LHC IIb are often termed the light-harvesting chlorophyll a/b-binding proteins (LHCP); such a name is, however, not strictly correct since there are other chlorophyll-proteins than LHC IIb in higher plants that contain chlorophylls a and b.

LHC IIb was one of the first higher plant pigment-protein complexes to be described (see Thornber 1986). Since it is the most abundant protein in the thylakoids and has been considered to be a typical thylakoid protein, it has been extensively studied over the last twenty years. Only recently has its molecular biology been thoroughly investigated, however. Several reviews on its biochemical characteristics are available (Bennett 1983, Machold 1984, Thornber 1986). LHCP is now widely considered as a good representative of a thylakoid protein for studying (a) gene regulation in plants; (b) coordination between chloroplast and nuclear genomes during plastid development; and (c) import of proteins by chloroplasts and their assembly into the correct suborganellar compartment. These aspects of the molecular and cell biology of LHC IIb are considered in the present review.

Composition and structure of LHC IIb

A chlorophyll *a/b* ratio between 1.0 and 1.4 has been generally reported for preparations of LHC IIb (Thornber et al. 1979). A consensus composition of an LHC IIb monomer unit is 5–7 chlorophyll *a*, 4–6 chlorophyll *b* and 2–3 xanthophyll molecules per polypeptide of 25–28 kDa (Peter and Thornber 1988); however, ratios of 6–13 chlorophyll molecules per polypeptide have been reported (e.g., Kuhlbrandt 1984, Ryrie et al. 1980, Thornber et al. 1979). There are multiple, typically three, apoproteins of 25–28 kDa in LHC IIb that apparently do not occur in a simple stoichiometry (cf. Peter and Thornber 1988). The number of different LHC IIb apoproteins is apparently not constant among different plant species. Some confusion about their exact number might have arisen, however, because other LHC II components have apoproteins that also migrate in the 20–30 kDa region during PAGE (Table 1) and can be mistaken for LHC IIb apoproteins (cf. Peter and Thornber 1988). The in situ form of this caroteno-chlorophyll-protein is not known. It almost certainly occurs in an oligomeric form, and although some evidence exists that it occurs as a trimeric structure (see below), it is likely that the 8 nm particle putatively identified as LHC II in freeze-fracture micrographs of thylakoid membranes (Staehelin 1986), is composed of an even higher oligomer. Physical biochemical data have indicated that three of the chlorophyll *b* molecules in an LHC IIb monomer are situated close enough to interact excitonically, and are surrounded by chlorophyll *a* molecules (Knox and Van Metter 1979, Li 1985). The first model of LHC IIb (Knox and Van Metter 1979), showing the relative arrangement of its chromophores, was based on this evidence.

Crystals of higher plant chlorophyll-proteins, suitable for high resolution X-ray analysis of the structure, have not yet been reported (cf. Kuhlbrandt 1984, 1987; Witt et al. 1987). In their absence, computer analyses of the predicted amino acid sequences of their polypeptides have allowed hypotheses of their folding with respect to the lipid bilayer (Karlin-Neumann et al. 1985, Kirsch et al. 1986, Trebst 1986); in particular, the parts of the polypeptide chains that could form membrane-spanning alpha-helices have been identified. For LHCP, we proposed (Karlin-Neumann et al. 1985) that the polypeptide chain has three hydrophobic, alpha-helical segments, each of which traverses the lipid bilayer, anchoring the complex in the membrane. This simple model has been embellished with predictions of other secondary structural features of the polypeptide as well as with the sites of cofactor attachment (Karlin-Neumann et al. 1985, Kohorn et al. 1986, Kohorn and Tobin 1987). An outcome of such predictions is that a large part (at least half) of the protein chain is apparently located outside the thylakoid lipid bilayer. In addition to computer predictions, two-dimensional arrays of LHC IIb have been reconstituted into artificial membranes and examined in the electron microscope (Kuhlbrandt 1984, Li 1985). This work together with immunological studies provide some evidence that LHC IIb exists as a trimer and that its polypeptide chain in the membrane protrudes some 2 nm on the stromal side and 0.7 nm on the lumen side of the thylakoid membrane bilayer (cf. Kuhlbrandt 1984, Li 1985).

The chlorophyll molecules are thought to be closely associated with the membrane-spanning alpha-helices, not only in LHC IIb but also in the other hydrophobic pigment-protein complexes of photosynthetic membranes. Histidine residues coordinate many of the chlorophyll molecules to their apoproteins in bacterial pigment-proteins (e.g. Deisenhofer et al. 1985, Matthews et al. 1979, Zuber 1985). While this may also be the major mode of binding chlorophyll molecules to protein in the core components of plant photosystems, which are relatively rich in histidine residues (Kirsch et al. 1986), it cannot be so for the light-harvesting complexes which have a paucity of histidine residues. The involvement of glutamine and asparagine residues in addition to histidine, in coordinating the chlorophyll molecules to the polypeptide (Wechsler et al. 1985), seems a good possibility to us.

Our predicted structure of LHC IIb has been tested by us and others: The recent advances in molecular biology have enabled us to test these predictions using in vitro mutagenesis. Deletion mutations and site-specific mutations in a gene from *Lemna gibba* were used to synthesize mutated pLHCP which was used for in vitro import by chloroplasts, and the fate of the imported protein was analyzed (Kohorn et al. 1986, Kohorn and Tobin 1987). The results from these experiments are consistent with the model of folding of LHC IIb proposed by us (Karlin-Neumann et al. 1985). Burgi et al. (1987) have examined the labelling of tyrosine residues in LHC IIb in the right-side-out and inside-out vesicles of thylakoid membranes. Their results support a folded protein that spans the membrane three times, whereas Anderson and Goodchild (1987), probing right-side-out and inside-out thylakoids with an antibody to the C-terminal sequence, conclude that four

Table 1. Some characteristics of the light-harvesting complex of photosystem II.[1]

	Apparent size (kDa) of holocomplex	Apoprotein(s)	Chl a/b	% Total Chl in *L. gibba*
LHC IIa	35	29	2.5	4
LHC IIb	72 (oligomer)	3 at 25–28	1.33	42
LHC IIc	27–32	27 or 31	$a + b$	3
LHC IId	24	21	0.93	3

[1] From Peter and Thornber (1988).

membrane-spanning segments occur. The four proposed helices require, however, the inclusion within them, and hence in the lipid bilayer, of more than ten charged residues, which seems an unlikely situation. Furthermore, the proposed helices are not compatible with the in vitro mutagenesis studies (Kohorn et al. 1986; Kohorn and Tobin 1987).

Genes encoding LHC IIb apoproteins

LHCPs are encoded by nuclear genes. A classical Mendelian pattern of inheritance was demonstrated for LHCPs in tobacco (Kung et al. 1972). Molecular genetics approaches have been applied to isolate and sequence many of the genes encoding LHC IIb polypeptides. These genes, often termed *cab* genes, have typical eukaryotic 5' and 3' flanking regions (TATA and CAAT boxes and polyadenylation sites). They are members of a multigene family containing between 3 (Leutwiler et al. 1986) and 16 (Dunsmuir and Bedbrook 1983, Dunsmuir et al. 1983) genes depending on the plant species. Some information about *cab* gene families from different plants is

Table 2. Some characteristics of *cab* genes encoding LHCPs.

Species	Type	# of genes		# of amino acids in		Untranslated region (bp)		References
		in genome	sequenced	transit	mature	5'	3'	
Triticum sp. (?)	I		1	34	232	70	240	a
Pisum sativum	I	8	1	39	230			b, c, d
Petunia hybrida	I	16	10	34–36	232–233			e–h
	II	1	1	36–37	228–229	80		i
Arabidopsis thaliana	I	3–4	3	35–35	232–233	52		j
Lemna gibba	I	12	1	33	233	67	134	k,l
	II		1	35	229	63	130	m
Silene pratensis	II		1	36	211			n
Lycopersicon esculentum	I	9–10	7	34	233			o–q
	II	2	2	36	229			o–q
Cucumis sativus	I	2	2		231			r
Hordeum vulgare	I		1	34	230			s
Zea mays	I		1	31	234	64	105	t

[a] Lamppa et al. (1985); [b]Broglie et al. (1981); [c]Cashmore (1984); [d]Coruzzi et al. (1983); [e]Dunsmuir (1985); [f]Dunsmuir and Bedbrook (1982); [g]Dunsmuir and Bedbrook (1983); [h]Dunsmuir et al. (1983); [i]Stayton et al. (1986); [j]Leutwiler et al. (1986); [k]Kohorn et al. (1986); [l]Tobin et al. (1984); [m]Karlin-Neumann et al. (1985); [n]Weisbeek et al. (1986); [o]Pichersky et al. (1985); [p]Pichersky et al. (1987); [q]Piechulla et al. (1986); [r]Greenland et al. (1987); [s]Chitnis, unpublished results; [t]Matsuoka et al. (1987).

264

```
                          1                                                 50
                          |                                                  |
LEMNA ab-30       MAA----SMALSSPSLVGK-AVKLAPAA--SE-VF-GEGRVSMRKTAGKPKPV-SSGSPWYGPDRVKYLGPFSGEAPSYLTGEFAGDYGWDTAGLSADPETFAKNRELEVIHARWAMLGALGCVFPELLARNG
BARLEY            ....AT.....STFA.....NLSSS....Q.DA..........ATK.VG............S.....P..............G.....
WHEAT ab-1.6      ...-TT.S...S.FA.....NL.SS--AL-I--DA..N......A.A.Q..S.....S...L..L..P....P..............C.....
MAIZE             .S--ST....TAFA....-NVPSSS--------A..T.......A.A.AAA........L...L..P....P.....................
ARABIDOPSIS 3 Genes ....T.....AFA.....*.S..........L-.S...T....V-A.PKG-P......S..............P......R...........
PEA ab-80         ..SSSS......T.A..-QL.N.SS-Q.-L..AA.PT...S.TTK.VA.............S.........P..............S....S...
PEA pab-96        ...............TK.VA-..S..H.................S.........P..............S...
PETUNIA Cab-91R   ...-AT......FA......FS.SS----.IT.-.N.KAT....VT.A.............P.....E.................C.....
PETUNIA Cab-13    ...-AT...S.FA...-.NV-.SS----.IT.RN.K.T....VT.A.............P.....A....K...C..T.....
PETUNIA Cab-2     ...-AT..I..S.FA...-NV-.SS--Q-IT-.N.KAT....VT.A.............P..............C.....
PETUNIA Cab-22L   ...-AT.....STFA.-V..S.SS----.IT-.N.KAT....T.A...............P..............C.....
PETUNIA Cab-22R   ...-TT.....S.FA.....SSSS-----.IT-.N.K.T....VT.A...S........PS.............
TOMATO Cab-1B     ...-AT.....FA.Q......S.SS----.IS-.N.TT....AV-.A.S-.P.S.........S.........P..............C.....
TOMATO Cab-3C     .-T---.T....STFA.....S.SS----.II-.N...T....TTA...............S.........P..............C.....
CUCUMBER          .-..T.A.N....S.N.-.P.-IQ-.NAKFT..........S.S................P.....P..............S....
LEMNA ab-19       ..........IQ.SAFA.QT.L.QRDELVRKV-GV-SD..F...R-V-.AV.---Q.I..A.P.F....EQT.......P.....S.........I...SK..
SILENE            .T-----.TIQQSAFA.QTLL.PQNEL.KV-GG-NG....-.R.I-.SA.---E.II....P.F....EQT........P......C.......K.
PETUNIA Cab-1B    .T-----.IQQSAFA.QT.L.SQNEL.KI-GSF.G..AT...R.I-.SA.---Q.I..E.P.F....EQT.......P.....R...........I.K..
TOMATO Cab-4      .T--C--.IQQSAF..QAVG.SQNEFI.KVGN......IT...R.V-.VS.---Q.I..E.P.F....EQT.......P......R...........C...SK..
TOMATO Cab-5      .............S.E-G..TT..R.V-.VS.---Q.I..E.P.F....EQT.......P......R...........C...I.SK..
```

REF

```
                 100                      150                      200
                  |                        |                        |
VKFGEAVWFKAGSQIFSEGGLDYLGNPSLVHAQSILAIWATQVVLMGAVEGYRVAG-GPLGEVVDPLYPGGSFDPLGLADDPEAFAELKVKEIKNGRLAMFSMFGFFVQAIVTGKGPLENLADHLADPVNNNAWAFATNFVPGK-COOH   (a)
........K........Q...........C......................................................A............................                                      (b)
........G........D........L..C......................I.......................ER.Q...............................LI....                                  (c)
.................................C..................I.................GD....I.............Y....                                                        (d)
.................I................................................G....L.K...L.....I.....                                                             (e)
...........D.....................I.....N..AE.L.......T.......................I...................                                                      (f)
...........................Q..I...Q.......................................SY....                                                                      (g)
.L...............I.............I..................EV........L....................SY...                                                                 (h)
...............K..................C............E.........................SY...                                                                        (h)
.................K...............C................I.......................                                                                             (h)
.................................C................I.......................                                                                             (h)
A..............A.................C..............V.....E......T..........V......SY....R.                                                                (h)
I...........................C..................E.......................SY...                                                                           (i)
...........Q.....................C................I.......E....................                                                                        (i)
...........Q.....................C................I.......................                                                                             (i)
.................................C...............T.I.......L.......................Y....                                                               (j)
.................................N..............LI....G....A....................I....A....                                                             (k)
.................Q...............C...........G....GL.Q...A....E......                                                                                  (l)
.T...............N.I.............A....F....G....GL.KI...A....E....................I..Y..V....A....                                                     (m)
.................N................F....G....GL.KI...A................................S..IN...A....Y....                                                (n)
.T...............Q........N.I....S....F....G....GL.KI...A..........................I..S..I...A....Y....                                                (n)
```

* N in two genes and K in one

[a]Kohorn et al.,1986; [b]Chitnis,unpublished data; [c]Lampa et al.,1985; [d]Matsuoka et al.,1987; [e]Leutwiler et al.,1986; [f]Timko et al.,1985; [g]Coruzzi et al.,1983; [h]Dunsmuir,1985; [i]Pichersky et al.,1987; [j]Greenland et al.,1987; [k]Karlin-Neumann et al.,1985; [l]Smeekens et al.,1986; [m]Stayton et al.,1986; [n]Pichersky et al.,1985.

Fig. 1. A comparison of the available amino acid sequences of the precursor form of the apoprotein of LHC IIb as deduced from gene sequences. The last five sequences are those deduced from type II genes (see text). A vertical line in the upper half of the figure separates the transit peptide sequence from that of the mature protein. The mature sequence is numbered; however, it should be noted that several gaps (marked by —) are placed in some sequences to assist in aligning them. Undetermined portions of the sequences are represented by blank spaces. All three genes from *Arabidopsis thaliana* give the same sequence except for one residue in the transit peptide (*) which is N in two genes and K in one.

summarized in Table 2. Each of the genes characterized so far codes for a precursor polypeptide (pLHCP) which has a transit peptide of 33–35 amino acids; the mature polypeptide consists of approximately 233 amino acids. The transit peptide sequences deduced from different genes show a prevalence of basic amino acid residues, and they differ more from each other than do the sequences of the mature protein which are highly conserved among different species (Fig. 1) (Karlin-Neumann and Tobin 1986).

The 3′ and 5′ untranslated sequences of these genes diverge from each other. Depending on this divergence, members of a gene family can be grouped into subfamilies; for example, tomato (Pichersky et al. 1985, 1987; Piechulla et al. 1986) as well as petunia (Dunsmuir 1985) genes have been divided into five subfamilies. The genes encoding LHC IIb can also be classified on the basis of the presence or absence of introns (Stayton et al. 1986). Type I genes contain no intron, are more numerous in most plant species (Table 2), code for a slightly longer polypeptide and have significantly different transit peptides as compared to type II genes that contain one intron (Karlin-Neumann et al. 1985, Stayton et al 1986). The transit peptides of both types of genes are functional in import of pLHCP by plastids (Kohorn and Tobin 1986). The divergence found in the flanking sequences of members of *cab* gene families could be responsible for the differential regulation of their expression.

Regulation of expression of cab genes

Since LHC IIb polypeptides are among the most abundant proteins in plants, the genes encoding them were among the first plant genes to be cloned and the mechanism of their regulation characterized. Many techniques have been used to probe different steps in the expression of these genes: One of the earliest ways was to immunoprecipitate translation products of polyadenylated mRNAs (e.g. Tobin 1978, 1981b). This procedure gives levels of translatable mRNAs for this protein in a tissue. Northern blots are used to measure the level of a particular message in the total RNA population (e.g. Stiekema et al. 1983). To determine the rate of transcription, in vitro nuclear run-off transcription has been used by many laboratories (e.g. Silverthorne and Tobin 1984). Using these and other biochemical techniques, the influence of several intrinsic (e.g. developmental cues) and extrinsic (e.g. light) factors on the expression of *cab* genes has been studied (see summary in Table 3).

The regulation of *cab* genes has been reported to occur at many different levels of gene expression. For example, phytochrome affects transcription of

Table 3. Regulation of expression of *cab* genes (expanded and modified from Tobin and Silverthorne 1985).

Species	Level	Reference
White light		
Hordeum vulgare	translatable mRNA	Apel and Kloppstech (1978)
Lemna gibba	translatable mRNA	Tobin (1978, 1981)
Pisum sativum	translatable mRNA	Cuming and Bennett (1981)
	transcription	Gallagher and Ellis (1982)
	protein stability	Bennett (1981)
Zea mays	hybridizable RNA	Nelson et al. (1984)
Arabidopsis thaliana	hybridizable RNA	Leutwiler et al. (1986)
Chlamydomonas reinhardii	hybridizable RNA	Shepherd et al. (1983)
Cucumis sativus	hybridizable RNA	Greenland et al. (1987)
		Walden and Leaver (1981)
Triticum sp. (?)	hybridizable RNA	Lamppa et al. (1985)
Phytochrome		
Hordeum vulgare	translatable mRNA	Apel (1979)
	hybridizable mRNA	Batschauer and Apel (1984)
		Gollmer and Apel (1983)
	transcription	Mosinger et al. (1985)
Lemna gibba	translatable mRNA	Tobin (1981)
	hybridizable mRNA	Stiekema et al. (1983)
	transcription	Silverthorne and Tobin (1984)
Pisum sativum	hybridizable mRNA	Bennett et al. (1984)
Vigna radiata	hybridizable mRNA	Thompson et al (1983)
Arabidopsis thaliana	hybridizable mRNA	Tobin et al. (1987)
Nicotiana tabacum	hybridizable mRNA	Simpson et al. (1985)
Triticum sp. (?)	hybridizable mRNA	Nagy et al. (1986a)
Blue light receptor		
Chlamydomonas reinhardii	hybridizable RNA	Johanningmeier and Howell (1984)
Plant development		
Glycine max	hybridizable mRNA	Walling L, personal communication
	transcription	Walling L, personal communication
Tissue specificity		
Zea mays	hybridizable RNA	Broglie et al. (1984)
		Schuster et al. (1985)
		Sheen and Bogorad (1986)
	protein	Bennett (1983); Broglie et al. (1984)
		Schuster et al. (1985)
Nicotiana tabacum	hybridizable RNA	Simpson et al. (1986b)
Organ specificity		
Nicotiana tabacum	hybridizable RNA	Lamppa et al. (1985)
		Simpson et al. (1986b)

Table 3.

Species	Level	Reference
Plastid development		
Triticum sp. (?)	hybridizable RNA	Lamppa et al. (1985)
	protein	Baker et al. (1984); Leech (1984)
Zea mays	hybridizable RNA	Nelson et al. (1984)
Nicotiana tabacum	hybridizable RNA	Simpson et al. (1986b)
Light intensity		
Pisum sativum	phosphorylation	Bennett (1981), Bennett et al. (1981)
Low temperature		
Zea mays	protein	Hayden et al. (1986)
Carotenoids		
Zea mays	hybridizable RNA	Harpster et al. (1984)
		Mayfield and Taylor (1984)
Chlorophyll precursors		
Chlamydomonas reinhardii	hybridizable RNA	Johanningmeier and Howell (1984)
Chlorophyll		
Hordeum vulgare	protein stability	Apel and Kloppstech (1978)
		Bellemare et al. (1982)
Cytokinins		
Lemna gibba	RNA stability	Flores and Tobin (1986)
		Flores and Tobin (1987)

these genes (Silverthorne and Tobin 1984); intermittent red light has an influence on the translation of LHCP mRNA (Slovin and Tobin 1982); changes in light intensity affect post-translational modification (phosphorylation) (Bennett 1977, Bennett et al. 1981). The expression of *cab* genes is regulated in both a quantitative as well as a qualitative manner. For example, the *cab* genes are expressed in an organ-specific manner in tobacco; however, individual members of the gene family have different patterns of expression in the various organs (Simpson et al. 1985, 1986b). Similarly, different sets of *cab* genes are expressed during embryogenesis and during maturity in soybean (Walling, personal communication). The 5′ region of *cab* genes has been investigated in detail for its role in their differential expression. Chimeric genes under the control of the 5′-flanking sequences of a *cab* gene from pea were used to study tissue-specific and light-inducible expression of *cab* genes in transgenic tobacco plants (Simpson et al. 1986b). The results showed that about 400-base pairs of the 5′-flanking region is sufficient to give the tissue-specific pattern of expression observed for *cab* genes. The levels of expression are, however, low indicating that some elements that affect LHCP transcription quantitatively are located still further upstream. A 247–base pair element from this 400–base pair region acts as a light-inducible enhancer and also a tissue-specific silencer (Simpson et al. 1986a). Similarly, a region in the 5′-flanking sequence of a wheat *cab* gene is found to confer phytochrome responsiveness on the gene (Nagy et al. 1986a, 1986b). The tissue specificity patterns of LHCP are probably correlated to the developmental and metabolic stages of the plastids (Simpson et al. 1986b). Thus, a variety of factors affect the expression of *cab* genes at different levels in both a quantitative and a qualitative fashion, and this complexity makes a *cab* gene family an excellent system in which to investigate coordinate expression of genes.

Assembly of LHC IIb

The assembly of LHC IIb and other similar thylakoid protein complexes involves many intricate steps as well as an interplay between cytoplasmic and chloroplastic products: Translocation of pLHCP across chloroplast envelopes; its processing to its mature form; its insertion into thylakoid membranes; the binding of chlorophyll and carotenoid molecules to it; and, its association with other LHCP molecules to form an oligomeric and functional LHC IIb (Schmidt and Mishkind 1986). The assembly process is further complicated because pLHCP is water-soluble whereas the processed product is water-insoluble. Furthermore, the site of membrane translocation (envelope) and membrane integration (thylakoids) are separated by an aqueous stroma. Two plausible pathways can be envisioned to accomplish this complicated process. In the first, the imported thylakoid protein could insert into the inner envelope membrane, then travel to the thylakoid in vesicles that bud from the inner envelope membrane and subsequently fuse with the thylakoids (see Cline 1986). In the second, the precursor protein could pass through the envelope and subsequently travel to the thylakoids in a soluble form, possibly brought about by a folding of the precursor that is quite different from that of the mature polypeptide. After passage through the stroma it is integrated into or transported across the thylakoid bilayer after further modification of its folding. An exact description of each step involved in the import, processing and assembly of thylakoid proteins is still unknown. Only recently has some preliminary information become available about pLHCP that supports the second pathway for the assembly of thylakoid proteins (Chitnis et al. 1986, 1987a, 1987b; Cline 1986, Cline et al. 1985).

Several approaches have been used to dissect the different steps in the assembly of LHC IIb. In vitro uptake of labelled polypeptides obtained by in vitro translation of total poly A-RNA by isolated intact plastids is the most popular one (Bartlett et al. 1986, Chua and Schmidt 1979, Cline et al. 1985, Grossman et al. 1980, Mullet and Chua 1983, Schmidt et al. 1981). Less equivocal data are obtained if only one particular precursor polypeptide is added to the plastids. Such a polypeptide can be made by using SP6 or T7 promoters in front of the gene of interest to synthesize the specific mRNA in vitro (Chitnis et al. 1986, Kohorn et al. 1986), which is then translated in a protein-synthesizing system. The import of pLHCP into the plastids is post-translational (Schmidt et al. 1981), energy-dependent (Grossman et al. 1980) and probably mediated through specific receptors in chloroplast envelopes (Cline et al. 1985). After the uptake of the in vitro synthesized pLHCP(s) by the isolated plastids, labelled mature polypeptide(s) can be detected in the pigmented LHC IIb in the thylakoids. In vitro-import experiments using in vitro-mutated pLHCP of *L. gibba* have revealed the importance of the amino acid charge distribution in the membrane-spanning helices of the polypeptide for the stability of the newly imported LHCP in the thylakoids and for its assembly into LHC IIb (Kohorn and Tobin 1987). The precursor form of LHCP can be seen in the thylakoids and also in LHC IIb under certain conditions: for example, when either barley or maize plastids (Chitnis et al. 1986), but not when *L. gibba* etiochloroplasts (Kohorn et al. 1986), import in vitro synthesized *L. gibba* pLHCP, the mature *as well as* the precursor polypeptides are observed as integral thylakoid proteins. Both of these polypeptides migrate specifically with the LHC IIb band in partially denaturing PAGE. The occurrence of both forms is also observed when in vitro synthesized barley pLHCP is imported into barley plastids isolated from etiolated plants which have been illuminated for less than 24 h; when plastids from plants greened for 24 h are used, only LHCP is seen in the thylakoids (Chitnis, unpublished data). Thus the presence of pLHCP in thylakoids is not due solely to the heterologous nature of the system used but is largely due to the influence of the developmental stage of plastids. Pulse-chase experiments showed that the pLHCP integrated into thylakoids of intact plastids can be processed to LHCP (Chitnis, unpublished data).

Another way to study assembly is to separate and analyze the different steps in vitro and later reconstitute the entire assembly process. Some of the many steps in the assembly of LHC IIb have been successfully repeated in vitro. For example, pLHCP can be inserted into isolated thylakoids (Chitnis et al. 1987a, Cline 1986). Both Mg-ATP and a stromal factor are absolutely required for the insertion (Chitnis et al 1987b, Cline 1986). The stromal factor is a protein (Chitnis et al. 1987b). Although the exact way this factor helps in the insertion is not known, one possibility is that it is involved in attaching some hydrophobic moiety to the water-soluble pLHCP so that the modified pLHCP is then compatible with the hydrophobic environment of thylakoid membranes. Such a moiety may be palmitic acid which is known to be attached to the mature LHCP (Mattoo and Edelman 1987). Alternatively, the stromal protein may change the conformation of the pLHCP in a similar way as to make its exterior surface hydrophobic. Evidence that membrane translation is achieved by using some factor that apparently denatures/unfolds the polypeptide and permits its translation has been found in *E. coli* (Crooke and Wickner 1987) and yeast (G. Schatz, personal communication).

The processing enzyme(s) for pLHCP has not yet been identified. The site and nature of processing of pLHCP in vivo is also not unequivocally defined, although pLHCP's presence in the thylakoids and its processing to LHCP in our import system strongly suggest that processing takes place on the thylakoid membranes. Mature LHCP is not a water-soluble protein and so it is reasonable to expect it to be processed at or near its final destination, if a function of its transit peptide is to make the precursor water-soluble. There is evidence that another thylakoid membrane protein, albeit one synthesized within the chloroplast, is processed after its insertion into the membrane (Grebanier et al. 1978). The presence of pLHCP in LHC IIb indicates that processing is not required for its inclusion in the pigmented complex (Chitnis et al. 1986).

Reconstitution experiments using purified LHCP and chlorophyll molecules showed that chlorophyll molecules can bind to the apoprotein(s) in vitro and that carotenoids play an important role in permitting this binding and in the formation of LHC IIb (Plumley and Schmidt 1987). Lastly, the assembly of purified LHC IIb with the PS II core complex in membranes of intermittent light-grown plants that lack LHC IIb, has been demonstrated; however, it occurred with low efficiency (Darr and Arntzen 1986, Day et al. 1984).

Biogenesis of LHC IIb during plastid development

Chloroplast development involves intricate biochemical and morphological changes (Briggs et al. 1987, Klein et al. 1986, Leech 1984). The major problem in the study of plastid development at a biochemical level is how to obtain homogenous preparations of plastids at identical developmental stages. One method is to use the gradient of plastid development found along the leaf length in cereals. The plastids at the base of the leaf are immature while those at the tip of the leaves are fully developed (Leech 1984). Defined sections along the length of the leaf can give relatively homogenous populations of plastids of the same developmental stage (Baker et al. 1984). This method is, however, impractical for large scale isolation of plastids required for some experiments (e.g. import of polypeptides, in vitro labelling, etc.).

Genetic mutations that arrest plastid development at early stages can be used to study the effects of development on different steps in LHC IIb synthesis and assembly. In chlorophyll-deficient mutants of maize, plastids are arrested prior to mature chloroplast formation (Mascia and Robertson 1978), while carotenoid-deficient mutants contain plastids that are arrested at a rudimentary stage of development (Bachmann et al. 1973). Studies with such mutants reveal that events at early stages of plastid development, such as synthesis of pigments, influence accumulation of *cab* mRNA (Harpster et al. 1984, Taylor et al. 1986).

The most common approach used to obtain plastids of the same developmental stage is to isolate plastids from etiolated plants illuminated for a particular period (Apel et al. 1984, Apel and Kloppstech 1978, Bennett et al. 1984, Briggs et al. 1987, Chitnis et al. 1986, 1987b, Hiller et al. 1978, Hoyer-Hansen and Simpson 1977, Klein et al. 1986, Tanaka and Tsuji 1985). It is, however, often difficult to separate the effects of light and those of plastid development from each other when using this method. Different plant species show different responses during light-triggered plastid development. When barley plants are grown in the dark, neither pLHCP nor the mRNA coding for it can be detected (Apel and Kloppstech 1978, Hiller et al. 1978, Hoyer-Hansen and Simpson 1977, Tanaka and Tsuji 1985), but when such etiolated plants are transferred to light, mRNA's for pLHCPs appear in the cytoplasm (Apel and Kloppstech 1978, Briggs et al. 1987) and LHCPs start accumulating in the thylakoids (Hiller et al. 1978, Hoyer-Hansen and Simpson 1977, Tanaka and Tsuji 1985). In contrast, etiolated pea plants do not contain LHCP but do have detectable levels of mRNA coding for it (Bennett et al. 1984). The accumulation of LHCP in greening pea leaves is not primarily governed by the levels of *cab* mRNA but by post-translational stabilization, in which chlorophyll synthesis is thought to play a necessary role (Bennett et al. 1984).

During chloroplast development not only is the synthesis of pLHCP and chlorophyll triggered but so also is the machinery for the import and assembly of pLHCP. Immature plastids from interior leaves of lettuce have been found to be more efficient at importing pLHCP in vitro than mature chloroplasts from pea (Schmidt et al. 1981). Similarly when barley plastids of different developmental stages are used to import *L. gibba* pLHCP in vitro, the relative amount of precursor and processed forms observed in the thylakoids changes significantly (Chitnis et al. 1986, 1987b). So at least one of the steps involved in the assembly of LHC IIb is dependent on plastid development. The insertion of pLHCP into thylakoid membranes also depends on the stage of plastid development for both the appearance of the stromal factor and the thylakoid membrane's receptivity for insertion (Chitnis et al. 1987b). The synthesis or activity of the processing enzyme for pLHCP could also be under the control of light.

Conclusion

The structure and composition of LHC IIb has been studied for more than two decades; however, several of the major questions are still unanswered. How many different polypeptides make a functional unit of LHC IIb? How are these polypeptides folded in the membrane? How are the pigment molecules coordinated by these polypeptides? The recent advances in molecular biology and protein crystallography of membrane proteins should ultimately answer these questions. The genes coding for LHCP have been cloned from several plant species and most of them have been sequenced.

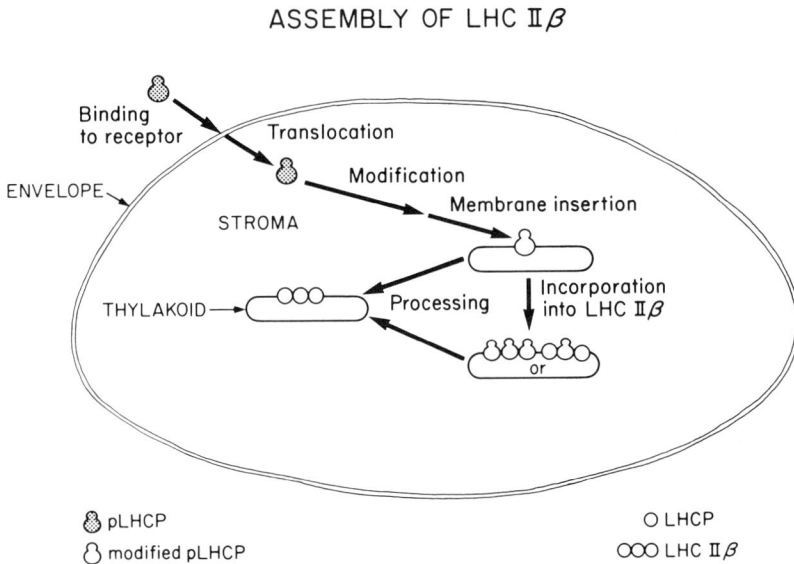

Fig. 2. A model showing steps in the assembly of LHC IIb.

The sequence data available from these genes have been used to predict some of the structural and biochemical features of LHC IIb.

It is well known that LHCPs are synthesized on the cytoplasmic ribosomes as a higher molecular weight pLHCP(s). The pLHCP is imported into plastids and then inserted into the thylakoid membrane, assembled into LHC IIb, and proteolytically processed. It also binds pigments. The exact sequence of these events is not known, but based on the recent findings, we present a model that attempts to explain the pathway leading to LHC IIb assembly (Fig. 2). We suggest the following steps occur during the assembly of LHC IIb.

a) The water-soluble precursor binds to the envelope and is then translocated, probably with the help of receptors (Cline et al. 1985). This process is energy dependent (Cline et al. 1985).

b) Sometime during its passage through the stroma, the conformation of pLHCP is changed to make it sufficiently hydrophobic to permit its insertion into the thylakoid membrane. Some stromal factor is involved in this process (Fig. 2). The change brought about in pLHCP could be attachment of a fatty acid to it by a stromal enzyme or binding of a stromal protein to pLHCP such that the resulting complex is made more hydrophobic. Modification of pLHCP and/or its insertion into thylakoids are dependent on the magnesium salt of ATP.

c) Once inserted in the membrane pLHCP can be incorporated into LHC IIb (Chitnis et al. 1986). Processing is not a prerequisite for incorporation of pLHCP into LHC IIb (Chitnis et al. 1986). Processing appears to take place in the thylakoids and can occur independently from pLHCP insertion into thylakoids and incorporation into LHC IIb.

The regulation of different steps involved in the synthesis and assembly of LHC IIb is being vigorously studied in many laboratories. Although many intrinsic and extrinsic factors have been clearly shown to influence synthesis and assembly of LHC IIb, the physiological significance and correlations of these effects are poorly understood. Plastid development seems to be the basic factor affecting expression of *cab* genes, since many factors exert their influence through their effect on plastid development (Simpson et al. 1986b). The stage of development of plastids affects not only the events occurring inside the plastids (e.g. insertion of pLHCP into the thylakoids) but also the events occurring in the cytoplasm (e.g. steady-state RNA levels for pLHCP) (Batschauer et al. 1984). The feedback mechanisms of such regulation remain to be elucidated.

Acknowledgements

We thank Daryl Morishige, Shivanthi Anandan, Gary Peter, Dr Elaine Tobin and Dr Alexander Vainstein for critically reviewing the manuscript, and Vaishali P. Chitnis for her help in preparing tables and references. We acknowledge Dr Rachel Nechushtai for her help and encouragement during our experimental work described in this review. Grants from NSF and USDA supported our research and P.R.C. was supported by ARCO and McKnight Fellowships.

References

Anderson JM and Goodchild DJ (1987) Transbilayer organization of the main chlorophyll *a/b*-protein of photosystem II of thylakoid membranes. FEBS Lett 213: 29–33

Apel K (1979) Phytochrome-induced appearance of mRNA activity for the apoprotein of the light-harvesting chlorophyll *a/b*-protein of barley (*Hordeum vulgare*). Eur J Biochem 97: 183–188

Apel K, Gollmer I and Batschauer A (1984) The light-dependent control of chloroplast development in barley (*Hordeum vulgare* L.). J Cell Biochem 23: 181–189

Apel K and Kloppstech K (1978) The plastid membranes of barley (*Hordeum vulgare*): Light-induced appearance of mRNA coding for the apoprotein of the light-harvesting chlorophyll *a/b*-protein. Eur J Biochem 85: 581–588

Apel K and Kloppstech K (1980) The effect of light on the biosynthesis of the light-harvesting chlorophyll *a/b*-protein. Evidence for the requirement of chlorophyll *a* for the stabilization of the apoprotein. Planta 150: 426–430

Bachmann MD, Robertson DS, Bowen CC and Anderson IC (1973) Chloroplast ultrastructure in pigment-deficient mutants of *Zea mays* under reduced light. J Ultrastruc Res 45: 384–406

Baker NR, Webber AN, Bradbury M, Markwell JP, Baker MG and Thornber JP (1984) Development of photochemical competence during growth of the wheat leaf. UCLA Symp Mol Cell Biol (New Ser.) 14: 237–255

Bartlett SG, Landry SJ and Pomarico SM (1986) Transport of proteins into chloroplasts. Current Topics in Plant Biochem Physiol 5: 105–115

Bassi R, Machold O and Simpson DJ (1985) Chlorophyll-proteins of two photosystem I preparations from maize. Carlsberg Res Commun 50: 145–162

Batschauer A and Apel K (1984) An inverse control by phytochrome of the expression of two nuclear genes in barley. Eur J Biochem 143: 593–597

Batschauer A, Moesinger E, Kreuz K, Doerr I and Apel K (1986) The implication of a plastid-derived factor in the transcriptional control of nuclear genes encoding the light-harvesting chlorophyll *a/b* protein. Eur J Biochem 154: 625–634

Bellemare G, Bartlett S and Chua N-H (1982) Biosynthesis of chlorophyll *a/b*-binding polypeptides in wild type and the chlorina f2 mutant of barley. J Biol Chem 257: 7762–7767

Bennett J (1977) Phosphorylation of chloroplast membrane polypeptides. Nature 269: 344–346

Bennett J (1981) Biosynthesis of the light-harvesting chlorophyll *a/b*-protein. Polypeptide turnover in darkness. Eur J Biochem 118: 61–70

Bennett J (1983) Regulation of photosynthesis by reversible phosphorylation of the light-harvesting chlorophyll *a/b* protein. Biochem J 212: 1–13

Bennett J, Jenkins GI and Hartley MR (1984) Differential regulation of the accumulation of the light-harvesting chlorophyll *a/b*-complex and ribulose bisphosphate carboxylase/oxygenase in greening pea leaves. J Cell Biochem 25: 1–13

Bennett J, Markwell JP, Skrdla MP and Thornber JP (1981) Higher plant chlorophyll *a/b*-protein complexes: studies on the phosphorylated apoproteins. FEBS Lett 131: 325–330

Briggs WR, Mosinger E, Batschauer A, Apel K and Schafer E (1987) Molecular events in photoregulated greening in barley leaves. UCLA Symp Mol Cell Biol (New Ser.) 44: 413–423

Broglie R, Bellemare G, Bartlett SG, Chua NH and Cashmore AR (1981) Cloned DNA sequences complementary to mRNAs encoding precursors to the small subunit of ribulose-1,5-bisphosphate carboxylase and a chlorophyll *a/b*-binding polypeptide. Proc Natl Acad Sci USA 78: 7304–7308

Broglie R, Coruzzi G, Keith B and Chua NH (1984) Molecular biology of C4 photosynthesis in *Zea mays*: differential localization of proteins and mRNAs in the two leaf cell types. Plant Mol Biol 3: 421–444

Burgi R, Suter F and Zuber H (1987) Arrangement of the light-harvesting chlorophyll-*a* chlorophyll-*b* protein complex in the thylakoid membrane. Biochim Biophys Acta 890: 346–351

Camm EL and Green BR (1980) Fractionation of thylakoid membranes with the nonionic detergent octyl-beta-D-glucopyranoside. Plant Physiol 66: 428–432

Cashmore AR (1984) Structure and expression of a pea nuclear gene encoding a chlorophyll *a/b*-binding polypeptide. Proc Natl Acad Sci USA 81: 2960–2964

Chitnis PR, Harel E, Kohorn BD, Tobin EM and Thornber JP (1986) Assembly of the precursor and processed light-harvesting chlorophyll *a/b* protein of *Lemna* into the light-harvesting complex II of barley etiochloroplasts. J Cell Biol 102: 982–988

Chitnis PR, Nechushtai R, Harel E and Thornber JP (1987a) Some requirements for the insertion of the precursor of apoproteins of *Lemna* light-harvesting complex II into barley thylakoids. In: Biggins J (ed.) Progress in Photosynthesis Research Vol. 4, pp. 573–576. The Hague: Martinus Nijhoff/Junk

Chitnis PR, Nechushtai R and Thornber JP (1987b) Insertion of the precursor of the light-harvesting chlorophyll *a/b*-protein into the thylakoids requires the presence of a developmentally regulated stromal factor. Plant Mol Biol 10: 3–11

Chua N-H and Schmidt GW (1979) Transport of proteins into mitochondria and chloroplasts. J Cell Biol 81: 461–483

Cline K (1986) Import of proteins into chloroplasts. Membrane integration of a thylakoid precursor protein reconstituted in chloroplast lysates. J Biol Chem 261: 14804–14810

Cline K, Werner-Washbourne M, Lubben TH and Keegstra K (1985) Precursors to two nuclear-encoded chloroplast proteins bind to the outer envelope membrane before being imported into chloroplasts. J Biol Chem 260: 3691–3696

Coruzzi G, Broglie R, Cashmore A and Chua NH (1983) Nucleotide sequences of two pea cDNA clones encoding the small subunit of ribulose 1,5-bisphosphate carboxylase and the major chlorophyll *a/b*-binding thylakoid polypeptide. J Biol chem 258: 13399–13402

Crooke E and Wickner W (1987) Trigger factor: A soluble protein that folds pro-OmpA into a membrane-assembly-competent form. Proc Natl Acad Sci USA 84: 5216–5220

Cuming AC and Bennett J (1981) Biosynthesis of the light-harvesting chlorophyll *a/b*-protein. Control of messenger RNA activity by light. Eur J Biochem 118: 71–80

Darr S and Arntzen CJ (1986) Reconstitution of the light harvesting chlorophyll *a/b* pigment-protein complex into developing chloroplast membranes using a dialyzable detergent. Plant Physiol 80: 931–937

Day DA, Ryrie I and Fuad N (1984) Investigations of the role of the main light-harvesting chlorophyll-protein complex in thylakoid membranes — reconstitution of depleted membranes from intermittent-light-grown plants with the isolated complex. J Cell Biol 97: 163–172

Diesenhofer J, Epp O, Miki O, Huber R and Michel H (1985) Structure of the protein subunits in the photosynthetic reaction center of *Rhodopseudomonas viridis* at 3A° resolution. Nature 318: 618–624

Dunahay TG and Staehelin LA (1986) Isolation and characterization of a new minor chlorophyll *a/b*-protein complex (CP24) from spinach. Plant Physiol 80: 429–434

Dunsmuir P (1985) The petunia chlorophyll *a/b* binding protein genes: a comparison of Cab genes from different gene families. Nucleic Acids Res 13: 2503–2518

Dunsmuir P and Bedbrook J (1982) Chlorophyll *a/b* binding proteins and the small subunit of ribulose bisphosphate carboxylase are encoded by multiple genes in petunia. Proc Int Congr Biochem 12: 302–306

Dunsmuir P and Bedbrook J (1983) Chlorophyll *a/b* binding proteins and the small subunit of ribulose bisphosphate carboxylase are encoded by multiple genes in petunia. NATO Adv Sci Inst Ser. A 63: 221–230

Dunsmuir P, Smith SM and Bedbrook J (1983) The major chlorophyll *a/b* binding protein of petunia is composed of several polypeptides encoded by a number of distinct nuclear genes. J Mole Appl Genet 2: 285–300

Flores S and Tobin EM (1986) Benzyladenine modulation of the expression of two genes for nuclear-encoded chloroplast proteins in *Lemna gibba*: Apparent post-transcriptional regulation. Planta 168: 340–349

Flores S and Tobin EM (1987) Benzyladenine regulation of the expression of two nuclear genes for chloroplast proteins. UCLA Symp Mol Cell Biol (New Ser.) 44: 123–132

Gallagher TF and Ellis RJ (1982) Light-stimulated transcription of genes for two chloroplast polypeptides in isolated pea leaf nuclei. EMBO J 1: 1493–1498

Ghanotakis DF, Demetriou DM and Yocum CF (1987) Isolation and characterization of an oxygen-evolving photosystem II reaction center core preparation and a 28 kDa chl-*a*-binding protein. Biochim Biophys Acta 891: 15–21

Gollmer I and Apel K (1983) The phytochrome-controlled accumulation of mRNA sequences encoding the light-harvesting chlorophyll *a/b* protein of barley (*Hordeum vulgare* L.). Eur J Biochem 133: 309–313

Grebanier AE, Coen DM, Rich A and Bogorad L (1978) Membrane protein synthesized but not processed by isolated maize chloroplasts. J Cell Biol 78: 734–746

Greenland AJ, Thomas MV and Walden RM (1987) Expression of two nuclear genes encoding chloroplast proteins during early development of cucumber seedlings. Planta 170: 99–110

Grossman A, Bartlett S and Chua, N-H (1980) Energy-dependent uptake of cytoplasmically synthesized polypeptides by chloroplasts. Nature 285: 625–628

Harpster MH, Mayfield SP and Taylor WC (1984) The effect of pigment-deficient mutants on the accumulation of photosynthetic proteins in maize. Plant Mol Biol 3: 59–71

Hayden DB, Baker NR, Percival MP and Beckwith PB (1986) Modification of the photosystem II light-harvesting chlorophyll *a/b* protein complex in maize during chill-induced photoinhibition. Biochim Biophys Acta 851: 86–92

Hiller RG, Pilger TBG and Genge S (1978) Formation of chlorophyll protein complexes during greening of etiolated barley leaves. In: G. Akoyunoglou (ed.) Chloroplast Development, pp 215–220. Amsterdam: Elsevier/North-Holland Biomedical

Hoyer-Hansen G and Simpson DJ (1977) Changes in the polypeptide composition of internal membranes of barley plastids during greening. Carlsberg Res Commun 42: 441–458

278

Johanningmeier U and Howell SH (1984) Regulation of light-harvesting chlorophyll-binding protein mRNA accumulation of *Chlamydomonas reinhardii*. J Biol Chem 259: 3541–3549

Karlin-Neumann GA and Tobin EM (1986) Transit peptides of nuclear-encoded chloroplast proteins share a common amino acid framework. EMBO J 5: 9–13

Karlin-Neumann GA, Kohorn BD, Thornber JP and Tobin EM (1985) A chlorophyll *a/b*-protein encoded by a gene containing an intron with characteristics of a transposable element. J Mol Appl Genet 3: 45–61

Kirsch W, Seyer P and Herrmann RG (1986) Nucleotide sequence of the clustered genes for two P700 chlorophyll *a* apoproteins of the photosystem I reaction center and the ribosomal protein S14 of the spinach plastid chromosome. Current Genet 10: 843–855

Klein RR, Gamble PE and Mullet JE (1986) Regulation of transcription and translation during chloroplast biogenesis. Current Topics in Plant Biochem Physiol 5: 74–87

Knox RS and Van Metter RL (1979) Fluorescence of light-harvesting chlorophyll *a/b* protein complexes: implications for the photosynthetic unit. CIBA Foundation Symp 61: 177–190

Kohorn BD, Harel E, Chitnis PR, Thornber JP and Tobin EM (1986) Functional and mutational analysis of the light-harvesting chlorophyll *a/b* protein of thylakoid membranes. J Cell Biol 102: 972–981

Kohorn BD and Tobin EM (1986) Chloroplast import of light-harvesting chlorophyll *a/b*-proteins with different amino termini and transit peptides. Plant Physiol 82: 1172–1174

Kohorn BD and Tobin EM (1987) Amino acid charge distribution influences the assembly of apoprotein into light-harvesting complex II. J Biol Chem 262: 12897–12899

Kuhlbrandt W (1984) Three dimensional structure of the light-harvesting chlorophyll *a/b*-protein complex. Nature 307: 478–480

Kuhlbrandt W (1987) Three dimensional crystals of the light-harvesting chlorophyll *a/b*-protein complex from pea chloroplasts. J Mol Biol 194: 757–762

Kung SD, Thornber JP and Wildman SG (1972) DNA codes for the photosystem II chlorophyll-protein of chloroplast membranes. FEBS Lett 24: 185–188

Lamppa G, Nagy F and Chua N-H (1985) Light-regulation and organ-specific expression of a wheat Cab gene in transgenic tobacco. Nature 316: 750–752

Lamppa GK, Morelli G and Chua N-H (1985) Structure and developmental regulation of a wheat gene encoding the major chlorophyll *a/b*-binding polypeptide. Mol Cell Biol 5: 1370–1378

Leech RM (1984) Chloroplast development in angiosperms: Current knowledge and future prospects. In: Baker NR and Barber J (eds) Chloroplast Biogenesis, pp 1–21. Amsterdam: Elsevier Science

Leutwiler LS, Meyerowitz EM and Tobin E (1986) Structure and expression of three light-harvesting chlorophyll *a/b*-binding protein genes in *Arabidopsis thaliana*. Nucleic Acids Res 14: 4051–4064

Li J (1985) Light-harvesting chlorophyll *a/b*—protein: three dimensional structure of a reconstituted membrane lattice in negative strain. Proc Natl Acad Sci USA 82: 386–390

Machold O (1984) Chlorophyll *a/b*-proteins in their relation to the light-harvesting complex. In: Sybesma C (ed.) Advances in Photosynthesis Research Vol 2, pp. 107–114 The Hague: Martinus Nijhoff/Junk

Machold O and Meister A (1979) Resolution of the light-harvesting chlorophyll *a/b*-protein of *Vicia faba* chloroplasts into two different chlorophyll-protein complexes. Biochim Biophys Acta 546: 472–480

Markwell JP, Thornber JP and Boggs RT (1979) Higher plant chloroplasts: Evidence that all the chlorophyll exists as chlorophyll-protein complexes. Proc Natl Acad Sci USA 76: 1233–1235

Mascia PN and Robertson DS (1978) Studies in chloroplast development in four mutants defective in chlorophyll biosynthesis. Planta 143: 207–211

Matsuoka M, Kano-Murakami Y and Yamamoto N (1987) Nucleotide sequence of cDNA encoding the light-harvesting chlorophyll *a/b* binding protein from maize. Nucleic Acids Res 15: 6302–6306

Matthews BW, Fenna RE, Bolognesi MC, Schmid MF and Olson JM (1979) Structure of a bacteriochlorophyll *a*-protein from the green photosynthetic bacterium *Prosthecochloris aestuarii*. J Mol biol 131: 259–285

Mattoo AK and Edelman M (1987) Intramembrane translocation and posttranslational palmitoylation of the chloroplast 32-kDa herbicide-binding protein. Proc Natl Acad Sci USA 84: 1497–1501

Mayfield SP and Taylor WC (1984) Carotenoid-deficient maize seedlings fail to accumulate light-harvesting chlorophyll *a/b* binding protein (LHCP) mRNA. Eur J Biochem 144: 79–84

Mosinger E, Batschauer A, Schafer E and Apel K (1985) Phytochrome control of in vitro transcription of specific genes in isolated nuclei in barley (*Hordeum vulgare*). Eur J Biochem 147: 137–142

Mullet JE (1983) The amino acid sequence of the polypeptide with regulates membrane adhesion (grana stacking) in chloroplasts. J Biol Chem 258: 9941–9948

Mullet JE and Chua N-H (1983) In vitro reconstitution of synthesis, uptake, and assembly of cytoplasmically synthesized chloroplast proteins. Methods Enzymol 97: 502–509

Nagy F, Fluhr R, Kuhlemeir C, Kay S, Boutry M, Green P, Poulsen C and Chua N-H (1986a) Cis-acting elements for selective expression of two photosynthetic genes in transgenic plants. Phil Trans R Soc Lond B314: 493–500

Nagy F, Kay SA, Boutry M, Hsu MY and Chua NH (1986b) Phytochrome-controlled expression of a wheat *Cab* gene in transgenic tobacco seedlings. EMBO J 5: 1119–1124.

Nelson T, Harpster M, H Mayfield, SP and Taylor W (1984) Light-regulated gene expression during maize leaf development. J Cell Biol 98: 558–564

Peter GF and Thornber JP (1988) Antenna components of photosystem II with emphasis on the major pigment-protein, LHC IIb. In: Scheer H and Schneider S (eds) Photosynthetic light-harvesting systems — Structure and Function. Berlin: W. de Gruyter and Co. Berlin In press.

Pichersky E, Bernatsky R, Tanksley SD, Breidenbach RB, Kausch AP and Cashmore AR (1985) Molecular characterization and genetic mapping of two clusters of genes encoding chlorophyll *a/b*-binding proteins in *Lycopersicon esculentum* (tomato). Gene 40: 247–258

Pichersky E, Bernatsky R, Tanksley SD, Malik VS and Cashmore AR (1987) Genomic organization and evolution of the *rbc* and *cab* gene families in tomato and other higher plants. Proc Tomato Biotechnology Symp (in press)

Piechulla B, Pichersky E, Cashmore AR and Gruissem W (1986) Expression of nuclear and plastid genes for photosynthesis-specific proteins during tomato fruit development and ripening. Plant Mol Biol 7: 367–376

Plumley FG and Schmidt GW (1987) Reconstitution of chlorophyll *a/b* light-harvesting complexes: Xanthophyll-dependent assembly and energy transfer. Proc Natl Acad Sci USA 84: 146–150

Ryrie IJ, Anderson JM and Goodchild DJ (1980) The role of light harvesting chlorophyll *a/b* protein complex in chloroplast membrane stacking. Cation-induced aggregation of recon-stituted proteoliposomes. Eur J Biochem 107: 345–354

Schmidt GW and Mishkind ML (1986) The transport of proteins into chloroplasts. Annu Rev Biochem 55: 879–912

Schmidt GW, Bartlett SG, Grossman AR, Cashmore AR and Chua N-H (1981) Biosynthetic pathways of two polypeptide subunits of the light-harvesting chlorophyll *a/b* protein complex. J Cell Biol 91: 468–478

Schuster G, Ohad I, Martineau B and Taylor WC (1985) Differentiation and development of bundle sheath and mesophyll thylakoids im maize. Thylakoid polypeptide composition, phosphorylation and organization of photosystem II. J Biol Chem 260: 11866–11873

Sheen JY and Bogorad L (1986) Differential expression of 6 light-harvesting chlorophyll a/b binding-protein genes in maize leaf cell-types. Proc Natl Acad Sci USA 83: 7811–7815

Shepherd HS, Ledoigt G and Howell SH (1983) Regulation of light-harvesting chlorophyll-binding protein (LHCP) mRNA accumulation during the cell cycle in Chlamydomonas reinhardii. Cell 32: 99–107

Silverthorne J and Tobin E (1984) Demonstration of transcriptional regulation of specific genes by phytochrome action. Proc Natl Acad Sci USA 81: 1112–1116

Simpson J, Michael TP, Cashmore AR, Schell J, Van Montagu M and Herrera-Estrella L (1985) Light-inducible and tissue-specific expression of a chimeric gene under control of the 5'-flanking sequence of a pea chlorophyll a/b binding protein gene. EMBO J 4: 2723–2729

Simpson J, Schell J, Van Montagu M and Herrera-Estrella L (1986a) Light-inducible and tissue-specific pea lhcp gene expression involves an upstream element combining enhancer- and silencer-like properties. Nature 323: 55–554

Simpson J, Van Montagu M and Herrera-Estrella L (1986b) Photosynthesis-associated gene families: Differences in response to tissue-specific and environmental factors. Science 233: 34–38

Slovin JP and Tobin EM (1982) Synthesis and turnover of the light-harvesting chlorophyll a/b-protein in Lemna gibba grown with intermittent red light: possible translational control. Planta 154: 465–472

Smeekens S, Van Ooster J, De Groot M and Weisbeek P (1986) Silene cDNA clone for a divergent chlorophyll-a/b-binding protein and a small subunit of ribulosebisphosphate carboxylase. Plant Mol Biol 7: 433–440

Staehelin LA (1986) Chloroplast structure and supramolecular organization of photosynthetic membranes. Encyclopedia of Plant Phyiol (New Ser.) 19: 1–84

Stayton MM, Black M, Bedbrook J and Dunsmuir P (1986) A novel chlorophyll a/b binding (Cab) protein gene from petunia which encodes the lower molecular weight Cab precursor protein. Nucl Acids Res 14: 9781–9796

Steinback KE, Arntzen CJ and Bogorad L (1985) The physical organization of genetic determinants of the photosynthetic apparatus of chloroplasts. In: Steinback KE, Bonitz S, Arntzen CJ and Bogorad L (eds) Molecular Biology of Photosynthetic Apparatus, pp 1–19. Cold Spring Harbor: Cold Spring Harbor Laboratories

Stiekema WJ, Wimpee CF, Silverthorne J and Tobin EM (1983) Phytochrome control of the expression of two nuclear genes encoding chloroplast proteins in Lemna gibba L. Plant Physiol 72: 717–724

Tanaka A and Tsuji H (1985) Appearance of chlorophyll-protein complexes in greening barley seedlings. Plant Cell Physiol 26: 893–902

Taylor WC, Burgess DG and Mayfield SP (1986) The use of carotenoid deficiencies to study nuclear-chloroplast regulatory interactions. Current Topics in Plant Biochem Physiol 5: 117–127

Thompson WF, Everett M, Polans NO, Jorgensen RA and Palmer JD (1983) Phytochrome control of RNA levels in developing pea and mung-bean leaves. Planta 158: 487–500

Thornber JP (1986) Biochemical characterization and structure of pigment-proteins of photosynthetic organisms. Encyclopedia of Plant Physiol (New Ser.) 19: 98–142

Thornber JP, Markwell JP and Reinman S (1979) Plant chlorophyll-protein complexes: recent advances. Photochem Photobiol 29: 1205–1216

Thornber JP, Peter GF, Nechushtai R, Chitnis PR, Hunter FA and Tobin EM (1986) Electrophoretic separation of chlorophyll-protein complexes and their apoproteins. Plant Biol 2: 249–258

Timko MP and Cashmore AR (1983) Nuclear genes encoding the constituent polypeptides of the light-harvesting chlorophyll *a/b*-protein complex from pea. UCLA Symp Mol Cell Biol (New Ser.) 12: 403–407

Timko MP, Kausch AP, Hand JM, Cashmore AR, Herrera-Estrella L, Vanden Broeck G and Van Montagu M (1985) Structure and expression of nuclear genes encoding polypeptides of the photosynthetic apparatus. In: Steinback KE, Bonitz S, Arntzen SJ and Bogorad L (eds) Molecular Biology of Photosynthetic Apparatus, pp 381–396. Cold Spring Harbor: Cold Spring Harbor Laboratories.

Tobin EM (1978) Light regulation of specific mRNA speies in *Lemna gibba* L. G-3. Proc Natl Acad Sci USA 75: 4749–4753

Tobin EM (1981a) White light effects on the mRNA for the light-harvesting chlorophyll *a/b*-protein in *Lemna gibba* L. G-3. Plant Physiol 67: 1078–1083

Tobin EM (1981b) Phytochrome-mediated regulation of messenger RNAs for the small subunit of ribulose 1,5-bisphosphate carboxylase and the light-harvesting chlorophyll *a/b*-protein in *Lemna gibba*. Plant Mol Biol 1: 35–51

Tobin EM and Silverthorne J (1985) Light regulation of gene expression of in higher plants. Annu Rev Plant Physiol 36: 569–593

Tobin EM, Silverthorne J, Flores S, Leutwiler L and Karlin-Neumann GA (1987) Regulation of the synthesis of two chloroplast proteins encoded by nuclear genes. UCLA Symp Mol Cell Biol (New Ser.) 44: 401–411

Tobin EM, Wimpee CF, Silverthorne J, Stiekema WJ, Neumann GA and Thornber JP (1984) Phytochrome regulation of the expression of two nuclear-coded chloroplast proteins. UCLA Symp Mol Cell Biol (New Ser.) 14: 325–334

Trebst A (1986) The topology of the plastoquinone and herbicide binding peptides of photosystem II in the thylakoid membrane. Z Naturforsch 41c: 240–245

Vallejos CE, Tanksley SD and Bernatsky R (1986) Localization in the tomato genome of DNA restriction fragments containing sequences homologous to the ribosomal-RNA (45S), the major chlorophyll *a/b* binding polypeptide and the ribulose bisphophate carboxylase genes. Genetics 112: 93–105

Walden R and Leaver CJ (1981) Synthesis of chloroplast proteins during germination and early development of cucumber. Plant Physiol 67: 1090–1096

Wechsler T, Suter F, Fuller RC and Zuber H (1985) The complete amino acid sequence of the bacteriochlorophyll *c* binding polypeptide from chlorosomes of the green photosynthetic bacterium *Chloroflexus aurantiacus*. FEBS Lett 181: 173–178

Weisbeek P, Hageman J, Cremers F, Keegstra K, Bauerle C and Smeekens S (1986) Nuclear-encoded chloroplast proteins: Genes, transport and localization. Current Topics in Plant Biochem Physiol 5: 88–104

Witt I, Witt HT, Gerken S, Asaenger W, Dekker JP and Rogner M (1987) Crystallization of reaction center I of photosynthesis: Low-concentration crystallization of photoactive protein complexes from the cyanobacterium *Synechococcus* sp. FEBS Lett 221: 260–264

Zuber H (1985) Structure and function of light-harvesting complexes and their polypeptides. Photochem Photobiol 42: 821–844

Govindjee et al. (eds), Molecular Biology of Photosynthesis: 283–319
© 1988 Kluwer Academic Publishers

Minireview

Regulation and expression of the multigene family coding light-harvesting chlorophyll *a/b*-binding proteins of photosystem II

DENNIS E. BUETOW,[1,2] HOUQI CHEN,[1] GÉZA ERDŐS[1] & LEE S.H. YI[1]

Departments of [1] Physiology and Biophysics and of [2] Plant Biology, University of Illinois, 524 Burrill Hall, 407 S. Goodwin Avenue, Urbana, IL 61801, USA

Received 27 October 1987; accepted 24 February 1988

Key words: genes, light-harvesting Chl *a/b* binding proteins of PSII, regulation of gene expression

Abstract. The current state of knowledge concerning the expression of the nuclear genes that code the light-harvesting chlorophyll *a/b*-binding polypeptides of photosystem II is presented. This review covers the structure of these genes, the complex multistep pathway involved in their expression, and the environmental and other factors which regulate their expression. Some of the effects of these factors are mediated, at least in part, at the level of transcription, but other effects can be explained only by the existence of multiple posttranscriptional regulatory steps.

Abbreviations: bp – base (nucleotide) pairs, BSC – bundle sheath cells of maize leaves, CAT – chloramphenicol acetyl transferase, cDNA – complementary DNA, kb – kilobases, LF – low fluence illumination, LHC – mature light-harvesting complex, LHC-II – mature light-harvesting complex of Photosystem II, LHCP-II – light-harvesting Chl a/b-binding protein of Photosystem II, MC – mesophyll cells of maize leaves, mRNA – messenger RNA, NPT II – neomycin phosphotransferase II, PS I – photosystem I, PS II – photosystem II, Rubisco – ribulose-1,5-bisphosphate carboxylase/oxygenase, VLF – very low fluence illumination

I. Introduction

The expression of plant genes is highly regulated. Products of some genes can be detected only following distinct changes in the environment while products of other genes are present only in certain tissues or only at specific stages of development. Products of still other genes always may be detected in plants or at least in certain tissues, but their levels may be modulated during the plant's life span or in response to specific environmental changes. Thus, regulation of the expression of plant genes is proving to be a complex process, and much remains to be answered. This review considers the current state of knowledge concerning the expression of the multiple nuclear genes that code the light-harvesting Chl a/b-binding polypeptides of PS II, i.e., the LHCP-IIs.

The first part of this review covers the structure of the LHCP-II genes and discusses their similarities and differences within a plant and between different plants. Potentially, the expression of these genes can be regulated at many steps from their transcription to the final assembly and integration of their mature functional protein products into the thylakoid membranes of the chloroplast. Emphasizing the possibility of multiple regulatory steps, the complex pathway of expression of LHCP-II genes is considered in the second part. Finally, environmental and other factors that are involved in the regulation of LHCP-II gene expression are considered. LHCP-II genes were last reviewed in a similar fashion in 1984 (Freyssinet and Buetow 1984), but the expression of these genes has continued to be a very active area of research. The present review emphasizes the large literature which has been published since 1983–1984. A number of reviews on related topics have appeared recently (e.g., Tobin and Silverthorne 1985, Akoyunoglou and Argyroudi-Akoyunoglou 1986, Anderson 1986, Schell and Van Montagu 1986, Schmidt and Mishkind 1986, Kuhlemeier et al. 1987, Silverthorne and Tobin 1987), and these should be referred to for additional details.

II. LHCP-II: background

Plants produce photosynthetic energy in their chloroplasts by means of an electron transport process coupled with photophosphorylation. The electron transport process, which occurs in the photosynthetic thylakoid membranes, is realized through a consumption of light energy at PS II and PS I (e.g., Givindjee and Whitmarsh 1982). Each photosystem appears to be a discrete thylakoid membrane-complex containing a reaction center with specialized chlorophyll molecules and electron acceptor molecules. Also present is the accessory Chl a/b-containing light-harvesting complex, i.e., LHC (e.g., Freyssinet and Buetow 1984, Thornber 1986).

The LHC of PS II is variously designated in the literature as the LHC, LHCP, LHC-II and CP-II (Thornber 1986). The term LHC-II will be used in this review. The LHC-II has been the subject of extensive study and many reviews (e.g., Thornber et al. 1979, Hiller and Goodschild 1981, Glazer 1983, Zuber 1985, Brecht 1986, Thornber 1986, Anderson 1986, Glazer and Melis 1987, Williams and Allen 1987). It functions as an antenna for the capture of primary light energy but also is involved in the stacking of thylakoids to form grana (Arntzen 1978) and in the control of the distribution of excitation energy between PS II and PS I (Kyle et al. 1983). The LHC-IIs contain about half of the total Chl (including most of the Chl *b*) and half of the protein of the thylakoids of green plants (Thornber 1986). Carotenoids also are present (e.g., Schmidt and Mishkind 1986). The "apoprotein" of any one LHC-II is composed of at least two, and likely three LHCP-IIs (Glazer and Melis 1987), with mol wts of 25 and/or 27–29 kDa; however, the exact sizes of the polypeptides vary to a small extent among different higher plants (Thornber 1986). In the discussion to follow, the term LHCP-II will be used to refer to any one or all of the Chl *a/b*-binding light-harvesting polypeptides of LHC-II. Most plants produce multiple nonidentical but antigenically-related LHCP-IIs (Plumley and Schmidt 1983, Sheen and Bogorad 1986). The various LHCP-IIs comprising the LHC-II "apoprotein" also have been shown to be very similar to one another in primary structure by amino acid analysis, fragmentation by cyanogen bromide or proteases, and sequencing of the genes that code them (Thornber 1986). The LHCP-IIs are known to traverse the thylakoid membrane and to protrude from the lipid bilayer on both sides (Andersson et al. 1982, Li 1985, Bürgi et al. 1987, Hinz and Welinder 1987).

III. LHCP-II genes

A. Multigene family

LHCP-IIs are encoded by nuclear DNA (Kung et al. 1972). LHCP-II genes comprise a multigene family in various plant species. Restriction mapping, S1-endonuclease mapping, heteroduplex analysis, nucleotide sequencing, and Southern-blotting all have been used to characterize these genes and their transcripts. As a result it is now known that the LHCP-II multigene family in petunia, for example, contains at least 16 genes classified into five small multigene subfamilies (Dunsmuir et al. 1983, Dunsmuir 1985, Stayton et al. 1986). Within a subfamily, the individual genes are closely related and have similar 3'- and 5'-flanking regions. However, the genes of the different subfamilies can be distinguished by sequence divergence in both flanking regions, and they code slightly different polypeptides (Dunsmuir 1985). A multigene family coding LHCP-IIs also has been demonstrated in pea (Coruzzi et al. 1983, Timko and Cashmore 1983, Polans et al. 1985), tomato (Pichersky et al. 1985, 1987), *Lemna gibba* (Karlin-Neumann et al. 1985), *Arabidopsis thaliana* (Leutwiler et al. 1986), wheat (Lamppa et al. 1985a) and maize (Sheen et al. 1986).

LHCP-II precursor polypeptides can be separated into two major size classes by one-dimensional SDS-PAGE (Schmidt et al. 1981, Dunsmuir et al. 1983). As judged from the result of hybrid selection experiments, most of the LHCP-II genes characterized so far are 'Type I' genes which encode the larger precursors (e.g., Broglie et al. 1981, Dunsmuir et al. 1983). Recently, 'Type II' LHCP-II genes encoding the smaller precursors have been demonstrated in petunia (Stayton et al. 1986), tomato (Pichersky et al. 1985), *Lemna* (Karlin-Neumann et al. 1985) and *Silene* (Smeekens et al. 1986a).

In both monocotyledonous and dicotyledonous plants, Type 1 LHCP-IIs are coded by a set of 3-20 nuclear genes per haploid genome (Broglie et al. 1981, Coruzzi et al. 1983, Dunsmuir et al. 1983, Cashmore 1984, Dunsmuir 1985, Lamppa et al. 1985, Timko et al. 1985, Kohorn et al. 1986, Leutwiler et al. 1986). Genes for Type 1 LHCP-IIs are clustered, at least in some cases. In petunia, two subfamilies contain only two genes each and in both cases the genes are closely linked in the genome (Dunsmuir et al. 1983). In pea, a cluster of LHCP-II genes is located on chromosome 2 (Polans et al. 1985). Leutwiler et al. (1986) isolated three LHCP-II genes from *A. thaliana*, all of which are clustered on an 11 kb genomic clone while Southern blot-analysis indicated that a fourth related gene also is present in *Arabidopsis*. In tomato, five independent chromosomal loci contain LHCP-II genes (Vallejos et al. 1986). Two loci (Cab-3, Cab-1) contain clusters of three and four LHCP-II genes, respectively (Pichersky et al. 1985). The Cab-3 locus also contains an additional truncated LHCP-II gene.

Type II genes are not clustered, at least in tomato (Vallejos et al. 1986). The Type II genes substantially diverge in nucleotide sequence from Type I genes, but gene structure is strongly conserved within the two types (Pichersky et al. 1987a, b). A Type II gene similar in sequence to tomato Type II genes has been demonstrated in *L. gibba* (Karlin-Neumann et al. 1985) and in *Silene pratensis* (Smeekens et al. 1986a).

B. Structure of Type I LHCP-II genes

1. Pea

A pea cDNA clone (AB96) was the first LHCP-II DNA sequenced (Coruzzi et al. 1983). Two more pea LHCP-II genes now have been sequenced, i.e., AB80 by Timko et al. (1983) and Cashmore (1984) and AB66 by Timko et al. (1985). The latter two genes are almost identical, differing in their coding regions by only two silent nucleotide substitutions. Both genes contain a large open reading frame, starting with methionine, of 269 amino acids and code identical LHCP-II precursors with a transit peptide of about 36 or 37 amino acids. However, they differ significantly in the nucleotide sequences of their 5'-flanking regions including an insertion/deletion of 281 bp. The coding regions terminate with a TAA codon located 100 nucleotides upstream of the presumed polyadenylation signal, AATACAA. A putative promoter sequence, TATAATA, is located 34 nucleotides upstream from the cap site. The mature polypeptides encoded by AB66 and AB80 differ from that encoded by AB96 by five amino acid residues.

2. Petunia

Five distinct LHCP-II cDNA clones from petunia have been characterized (Dunsmuir et al. 1983). Their maximal nucleotide divergence is 10%, occurring in the regions corresponding to the C-terminal 60 amino acids of the coded proteins. However, the 3′-untranslated regions in these clones are extremely divergent.

The structures of five genes from different LHCP-II subfamilies of petunia have been reported (Dunsmuir 1985). Each has an uninterrupted open reading frame of 266 or 267 amino acids which corresponds to a precursor polypeptide including an apparent transit peptide of 34 to 36 amino acids at the NH_2-terminus. The predicted transit peptide sequences are more divergent than the mature polypeptide sequences coded by the genes from the different subfamilies. Two regions within the mature polypeptides are conserved between all the genes: a sequence of 28 amino acids located near the NH_2-terminus and another sequence of 26 amino acids in the middle of the polypeptide. Nucleotide sequences in the 3′-untranslated regions coded by the different subfamily genes are highly divergent (Dunsmuir 1985), a finding consistent with that observed previously with LHCP-II cDNA clones (Dunsmuir et al. 1983). All the petunia LHCP-II genes have a TATA-like promoter sequence about 25 to 30 nucleotides upstream from the transcription start site as well as a CCAAT-like sequence at position -70 to -80 from the transcription start. The CCAAT sequence is thought to be involved in regulating the level of transcription of many eukaryotic genes (e.g., Nevins 1983). In three of the petunia LHCP-II genes there is a 48 nucleotide-long region of similarity centered at around -130 from the proposed transcription initiation site.

3. Wheat

Lamppa et al. (1985a) characterized a wheat LHCP-II genomic clone (AB 1.6). This gene encodes an mRNA containing nucleotide sequences for a mature polypeptide of 232 amino acids, a 70-nucleotide-long 5′-untranslated region, and a 34 amino acid-long transit peptide which is basic and rich in serines. There are no introns in AB 1.6. Putative promoter sequences are located upstream of the transcription start site at -25 (TTTAAATA) and -72 (CCAACCA).

4. *Arabidopsis*

the LHCP-II-multigene family of *A. thaliana* consists of 3 closely-related genes and, possibly, a fourth less-related gene (Leutwiler et al. 1986). The three closely-related genes (AB140, AB165 and AB180) have been isolated and sequenced. All three contain open reading frames of 267 amino acids. At the amino acid level they show essentially 100% similarity in the mature polypeptides. The only difference is the substitution of an asparagine for lysine in the transit peptide of AB140. At the nucleotide level, the similarity is 96% among the three genes within the translated regions. Also, AB165 and AB180 are highly similar in the 5′-untranslated region and through the presumptive 5′-upstream promoter region as well as in the 50 bp immediately downstream of the stop codon TGA. However, the 5′- and 3′-regions of AB165 and AB180 diverge from those of AB140. A TATA-like sequence, TATATAAT, is located upstream at -79 in both AB165 and AB180, and the sequence TATTATATATA lies at -92 in AB140. The sequence CCAAT appears 51, 55 and 51 nucleotides 5′ of the putative TATA-like sequences in AB165, AB180 and AB140, respectively. A feature of AB165 and AB180 is the presence of an adenine-rich region of 23 and 22 nucleotides (interrupted by only 2 cytosine residues) located 7 nucleotides upstream of the translation start site. However, this adenine-rich region is not found in AB140.

5. *Lemna*

Kohorn et al. (1986) sequenced an LHCP-II gene (AB30) from *L. gibba* which has an open reading frame of 266 amino acids including an apparent transit peptide of about 33 amino acids. Putative promoter sequences TATTA and CCAAT are located upstream of the transcription initiation site at -30 and -100, respectively.

6. *Tomato*

Three genes in the tomato Cab-3 locus and four genes in the Cab-1 locus have been sequenced completely or partially (50%) by Pichersky et al. (1985). Genes within a locus are similar to each other. However, the polypeptides encoded by Cab-1 genes diverge significantly from those encoded by Cab-3 genes in two regions, i.e., the 34 or 35 amino acid-long transit peptides and the first 11 to 14 amino acids at the putative NH_2-termini of the mature polypeptides. For example, the nucleotide sequences of one Cab-1 (Cab-1B) and one Cab-3 (Cab-3C) gene are 12.5% divergent from each other in the coding region. Also, they show little or no similarity at their 5′- and 3′-untranslated regions. Cab-1 and Cab-3 genes code open reading frames of 265 and 267 amino acids, respectively.

7. Maize

The nucleotide sequence of a cloned full-length maize LHCP-II cDNA has been reported by Matsuoka et al. (1987). The cDNA contains 798 bp of coding region as well as 64 and 105 bp of the 5- and 3′-untranslated regions, respectively. The coded transit peptide is 31 amino acids long and the mature polypeptide is 234 amino acids long.

8. Soybean

A nuclear DNA fragment (AB2.3) from the soybean (*Glycine max*) has been cloned and shown to contain a region coding a mature LHCP-II as well as the 3′-untranslated region and a region coding a portion of the transit peptide (Hsiao et al. 1987). The 5′-end of AB2.3 codes 245 amino acids, the first 14 of which represent a partial segment of the transit peptide and the latter 231 a mature LHCP-II.

9. Cucumber

The haploid genome of cucumber appears to contain only two LHCP-II genes (Greenland et al. 1987), a number lower than found in other higher plants so far investigated. Nucleotide sequences of two LHCP-II cDNA clones (Cu5/989 and MVTG-9) representing these two genes show that they are Type I genes. Cu5/989 is probably truncated, encoding 206 amino acids of a mature LHCP-II plus 160 nucleotides of a 3′-untranslated region. MVTG-9 codes a complete, 231 amino-acid long LHCP-II, 24 amino acids of a 5′-transit peptide and 114 nucleotides of a 3′-untranslated region. The nucleotide sequence coding the mature LHCP-II in Cu5/989 is 94% similar to the coding region in MVTG-9. Moreover, all the nucleotide substitutions are silent yielding 100% identity in deduced amino-acid sequences of the LHCP-IIs coded by these two cDNAs. In contrast, the 3′-untranslated regions diverge greatly.

10. Polyadenylation signals

A polyadenylation-signal sequence, AATAAA, is a common feature of the 3′-untranslated region of animal viral genes (Fitzgerald and Shenk 1981), animal nuclear genes themselves (Brawerman 1981) and some, but not all, plant nuclear genes (Messing et al. 1983, Hunt et al. 1987). In a telling set of experiments, Hunt et al. (1987) introduced chimeric plant genes containing polyadenylation signals from a human gene or animal virus genes into tobacco cells. In all cases, inefficient and incorrect utilization of the heterologous polyadenylation signal occurred. It appears then that plant cells do not properly recognize, at least not always, animal-type polyadenylation signals. Yet plant nuclear-coded mRNAs are polyadenylated. A hexanucleotide sequence, TTGTT, is located downstream of the termination codon in four

LHCP-II cDNA clones of petunia (Dunsmuir et al. 1983) and in an LHCP-II gene of wheat (Lamppa et al. 1985), *Arabidopsis* (Leutwiler et al. 1986) and soybean (Hsiao et al. 1987). The TTGTTT sequence is postulated to be a polyadenylation signal at least for some plant nuclear genes. The two LHCP-II genes in cucumber do not have the AATAAA sequence but their 3′-untranslated regions are rich in T and A residues (Greenland et al. 1987).

11. Transit-peptide consensus sequences
The transit-peptide sequences of the LHCP-IIs examined so far show a high degree of similarity (Schmidt and Mishkind 1986). Karlin-Neumann and Tobin (1986) have identified three major blocks of amino acid similarity shared by the transit peptides of three nuclear-coded chloroplast proteins, the LHCP-IIs, the small subunit of Rubisco and ferredoxin. As discussed by Karlin-Neumann and Tobin (1986) and by Hsiao et al. (1987), these blocks are shared by the transit peptides coded by Type I LHCP-II genes from wheat (Lamppa et al. 1985), *Lemna* (Kohorn et al. 1986), pea (Cashmore 1984), *Arabidopsis* (Leutwiler et al. 1986), petunia (Dunsmuir 1985), tomato (Pichersky et al. 1985) and soybean (Hsiao et al. 1987) as well as by the transit peptides coded by Type II LHCP-II genes (see section III C below). The recently published amino acid sequence of the transit peptide coded by a maize LHCP-II gene (Matsuoka et al. 1987) also shares these three major blocks of similarity. The partial transit peptide coded by a cucumber LHCP-II cDNA clone (MVTG-9) has 2 regions at amino acids 2 through 6 and 19 through 24 which are reminiscent of blocks II and III in other LHCP-II transit peptides (Karlin-Neumann and Tobin 1986) but appear to be more divergent (Greenland et al. 1987). In any case, the three major blocks of amino acid similarity shared by many nuclear-coded plant genes appear to form a rather common framework in transit peptide-bearing chloroplast precursors. This common framework in turn is suggested to equip transit peptides with the means to carry out common roles, i.e., the transport of nuclear-coded precursors to and into the chloroplasts and the processing of the precursors to their mature forms (Karlin-Neumann and Tobin 1986). Interestingly, this common framework does appear to be unique to transit peptides of nuclear-coded chloroplast precursors. A similar common framework has not been observed in mitochondrial leader sequences nor in the signal sequences of secretory or integral membrane proteins (Karlin-Neumann and Tobin 1986).

12. Summary
Type I LHCP-II genes contain open reading frames of 265 to 269 amino acids divided between a transit peptide of 31 to 37 amino acids and a mature

Table 1. Percentage similarities of the coding nucleotide and amino acid sequences[a] of mature LHCP-IIs (from Hsiao et al. 1988)

Plant	LHCP-II gene[b,c]	AB2.3	AB1.6	AB30	AB80	AB140	Cab91R	Cab-1B	Cab-4	Cab37	AB19
1. soybean	AB2.3	–	76.0	76.8	83.7	76.6	82.8	83.1	74.4	75.3	70.3
2. wheat	AB1.6	88.0	–	80.4	75.3	73.3	78.9	79.3	71.3	71.8	76.9
3. Lemna	AB30	91.7	90.1	–	76.2	78.1	77.6	76.9	69.6	71.0	85.8
4. pea	AB80	93.0	87.2	92.3	–	78.2	81.5	83.4	74.9	76.0	71.7
5. Arabidopsis	AB140	89.7	87.2	91.5	88.8	–	79.2	79.3	72.4	72.5	72.9
6. petunia	Cab91R	92.1	91.0	96.1	92.3	90.2	–	88.2	76.0	73.9	73.4
7. tomato	Cab-1B	91.3	91.0	94.0	91.9	91.4	94.4	–	75.8	76.4	73.4
8. tomato	Cab-4	82.7	81.5	86.5	84.4	83.7	84.7	86.6	–	87.7	72.9
9. petunia	Cab37	82.7	80.2	86.5	83.5	84.9	83.4	86.1	96.9	–	73.9
10. Lemna	AB19	86.2	81.1	89.4	85.7	85.3	83.5	85.7	92.1	92.1	–

[a] Percentage similarities of nucleotide sequences are shown in the upper right matrix and of amino acid sequences in the lower left matrix.

[b] Genes from plants 1–7 are Type I genes and those from plants 8–10 are Type II genes.

[c] References are: AB2.3 (Hsiao et al. 1988), AB1.6 (Lamppa et al. 1985), AB30 (Kohorn et al. 1986), AB80 (Cashmore 1984), AB140 (Leutwiler et al. 1986), CAB91R (Dunsmuir 1985), Cab-1B (Pichersky et al. 1985), Cab-4 (Pichersky et al. 1987), Cab37 (Stayton et al. 1986), AB19 (Karlin-Neumann et al. 1985).

LHCP-II of about 231–235 amino acids. Nucleotide sequences coding mature LHCP-IIs in a variety of plants are 70–88% similar to one another while their deduced amino acid sequences are 82–96% similar (Table 1). No introns are present. Polyadenylation signals have not been proven, but the downstream sequence TTGTTT may be involved. Putative 5'-promoters include TATA-like and CCAAT-like sequences. Transit peptides contain three major blocks of homology. Amino termini of Type I LHCP-IIs diverge somewhat from one other (Table 2) and 3'-termini diverge greatly from one another (Table 3). Certain regions within NH$_2$-termini of mature polypeptides are conserved between LHCP-II gene subfamilies in a given plant species, though subfamily-specific amino acid differences are present. Divergence at 3'- and/or 5'-untranslated regions is thought to be involved in differential regulation of the expression of related genes (e.g., Jones et al. 1985).

C. Structure of Type II LHCP-II genes

1. Lemna

Karlin-Neumann et al. (1985) sequenced a Type II LHCP-II gene (AB19) from *L. gibba* which contains an open-reading frame of 264 amino acids, including a transit peptide of 35 amino acids. Unlike Type I genes, the AB19

Table 2. Comparison of amino termini of LHCP-IIs (from Hsiao et al. 1988).

LHCP-II Plant	Gene	Amino terminus of LHCP-II[a]																
Type I gene																		
soybean	AB2.3	M	R	K	T	A	S	–	–	K	T	V	S	S	G	S	P	W
wheat	AB1.6		R	K	T	A	A	K	A	K	Q	V	S	S	G	S	P	W
Lemna	AB30	M	R	K	T	A	G	K	P	K	P	V	S	S	G	S	P	W
pea	AB80	M	R	K	S	A	T	T	K	K	V	A	S	S	G	S	P	W
pea	AB96						T	T	K	K	V	A	S	S	S	S	P	W
Arabidopsis	AB140	M	R	K	T	V	A	K	P	K	G	P	S	–	G	S	P	W
petunia	Cab91R	M	R	K	T	V	T	K	A	K	P	V	S	S	G	S	P	W
tomato	Cab-1B	M	R	K	A	V	A	–	–	K	S	A	P	S	S	S	P	W
Type II gene																		
Silene	AB1		R	R	T	I	–	–	–	K	S	A	P	E	–	S	I	W
tomato	Cab-4	M	R	R	T	V	–	–	–	K	S	A	P	Q	–	S	I	W
petunia	Cab37	M	R	R	T	V	–	–	–	K	S	A	P	Q	–	S	I	W△
Lemna	AB19	M	R	R	T	V	–	–	–	K	A	V	P	Q	–	S	I	W△

[a] A dash indicates that no residue is present at the position. A triangle denotes the start of an intron.
[b] References as in Table 1. The reference for AB96 is Coruzzi et al. (1983) and for AB1 is Smeekens et al. (1986a).

Table 3. Percentage similarities between the nucleotide sequences of the 3'-untranslated regions of LHCP-II genes (from Hsiao et al. 1988).

Plant	LHCP-II gene[a]	AB1.6	AB30	AB80	AB96	AB140	Cab91R	Cab-1B	Cab-4	Cab37	AB19
Type I gene											
soybean	AB2.3	48.6	41.5	53.2	59.1	*[b]	48.8	42.9	44.2	47.1	50.8
wheat	AB1.6	–	52.3	46.3	*	*	63.0	42.2	*	57.0	47.5
Lemna	AB30		–	53.6	41.4	*	*	55.2	55.8	58.6	57.6
pea	AB80			–	76.3	46.6	*	58.9	*	64.9	46.5
pea	AB96				–	37.5	43.4	*	*	51.0	40.1
Arabidopsis	AB140					–	48.0	50.0	56.9	56.0	51.1
petunia	Cab91R						–	53.0	45.1	51.4	51.3
tomato	Cab-1B							–	60.4	60.4	42.1
Type II gene											
tomato	Cab-4								–	*	44.9
petunia	Cab37									–	41.1
Lemna	AB19										–

[a] References as in Tables 1 and 2.
[b] Asterisk (*) indicates insufficient homology for alignment.

coding region contains an intron of 84 nucleotides. Interestingly, this intron has characteristics of a transposable element. AB19 also contains a 63 bp 5'-untranslated sequence. Eukaryotic-type promoter sequences, TATTAA and CCAAT, are located upstream from the transcription initiation site at -34 and -91. The 3'-untranslated region does not have a readily identifiable polyadenylation-signal sequence (see section III, B, 10 above). When the mature LHCP-IIs coded by AB19 and the Type I gene AB30 of *Lemna* are compared, an 89% amino-acid sequence similarity is found (Table 1). However, little similarity is seen in the NH$_2$-terminal sequences of the two mature polypeptides (Table 2). Also, the 3'-untranslated nucleotide sequences of AB19 and AB30 greatly differ (Table 3). In contrast, the transit peptides coded by these two genes share the same three major blocks of similarity (Karlin-Neumann and Tobin 1986).

2. Petunia

A petunia Type II LHCP-II cDNA (Cab37) was described by Stayton et al. (1986). Cab37 codes a LHCP-II precursor which differs from precursors encoded by petunia Type I LHCP-II genes (section III, B, 2 above). The Cab37 precursor is shorter by 2–3 amino acids. Further, the NH$_2$-terminus of the mature polypeptide coded by Cab37 is significantly diverged (Table 2). Also, Cab37 contains a 106 bp intron which is located 146 bp downstream from the translation start site. The 5'-untranslated sequence is 80 bp in length and is unrelated to the Type I LHCP-II genes in petunia. Upstream from the transcription start site, putative promoter sequences are found, i.e., TATATATA at -29 and CCAAAT at -100. Four of the five petunia LHCP-

II gene subfamilies (Dunsmuir 1985) possess a similar region centered upstream at about -130, but this region is not conserved in Cab37. The 3'-untranslated region of Cab37 also greatly differs from those of Type I LHCP-II genes (Table 3). However, Cab37 does carry the sequence TTGTAT at a position 25 bp down-stream from its TGA stop codon. The latter hexanucleotide sequence is similar to the TTGTTT sequence which may be a polyadenylation signal (see section III, B, 10 above). The transit peptide coded by Cab37 at first appears significantly diverged in amino acid sequence compared to the transit peptide coded by petunia Type I LHCP-II genes (Stayton et al. 1986). However, a careful inspection of the amino acid sequence of the Cab37 transit peptide shows that it shares the three major blocks of similarity (section III, B, 11 above) as defined by Karlin-Neumann and Tobin (1986) and that it is the interblock regions that are highly divergent. Cab37 is expressed and its mRNA is present at high steady-state levels in light-grown leaf tissue. The mature LHCP-II coded by Cab37 is about 229 amino acids long.

3. Silene

Smeekens et al. (1986a) sequenced a *S. pratensis* (white campion) cDNA clone (AB1). AB1 codes a 36 amino-acid-long transit peptide and the first 169 amino acids of a mature LHCP-II. The amino-acid sequence of the NH_2-termini of Type I LHCP-IIs but is similar to those of other Type II genes (Table 2). Though the overall transit peptide sequence of AB1 appears to differ from those found with precursors coded by Type I genes (Smeekens et al. 1986a), its amino-acid sequence does possess the three major blocks of similarity defined by Karlin-Neumann and Tobin (1986).

4. Tomato

Recently, Pichersky et al. (1987) investigated two tomato Type II cDNA clones (Cab-4 and Cab-5). The Cab-4 gene contains at least one intron. The nucleotide and amino-acid sequences of the coding region of the Cab-4 gene are about 76% and 87% similar, respectively, with the tomato Type I Cab-1B gene (Table 1). In contrast, the NH_2-termini of the LHCP-IIs coded by Cab-4 and Cab-1B are quite different from each other (Table 2) as are the nucleotide sequences of their 3'-untranslated regions (Table 3).

The tomato Type II Cab-5 cDNA codes an entire mature LHCP-II and eight amino acids of a transit peptide. The coding sequences of Cab-4 and Cab-5 are 88% homologous at the nucleotide level. The mature LHCP-IIs (each 229 amino acids) coded by Cab-4 and Cab-5 differ by only 4 amino acids (1.7% divergence). It is not known whether the Cab-5 gene contains an intron.

IV. The pathway of LHCP-II gene expression

A. Transcription

Transcription of a gene is the first step in its expression. The gene is transcribed into a precursor mRNA which is subsequently processed to the mature mRNA. Neither the transcription nor the processing of LHCP-II mRNA has been studied in any detail.

B. Translation of LHCP-II mRNA

Following its porocessing in the nucleus, mature LHCP-II mRNA is transported to the cytoplasm where it is integrated into free polysomal complexes (e.g., Freyssinet and Buetow 1984). As discussed above, the mRNA codes for a precursor LHCP-II. The precursors do not accumulate in the cytoplasm following their translation, but, rather, are imported rapidly into chloroplasts. The translated precursors are released from free cytoplasmic polysomes prior to being imported into the organelles (e.g., Schmidt and Mishkind 1986). Thus, import of chloroplast precursors into chloroplasts occurs posttranslationally.

C. Binding to chloroplast envelopes

Following their translation and release from polysomes into the aqueous phase of the cytoplasm, the precursors specifically bind to receptors located on the cytoplasmic side of the outer of the two membranes of the chloroplast envelope. Quantitative assays indicate that a chloroplast can bind up to 3000 to 5000 molecules (Pfisterer et al. 1982, Cline et al. 1985). This binding capacity can accomodate a rapid import of the large amounts of nuclear-coded chloroplast proteins synthesized, for example, during light-induced chloroplast differentiation (Pfisterer et al. 1982). Binding does not require energy in the form of ATP. Precursors are bound in preference to mature proteins. Originally it was thought that all cytoplasmically-synthesized chloroplast precursor proteins were bound by the same type of receptor. However, as discussed by Schmidt and Mishkind (1986), there appears to be more than one type because, when compared to the binding of Rubisco small-subunit precursors, the binding of LHCP-II precursors shows a different pH-dependency (Bitsch and Kloppstech 1986) and a greater sensitivity to proteolysis (Cline et al. 1985). Also, nigericin and EDTA are more inhibitory to the import of LHCP-II precursors into chloroplasts than to the import of Rubisco small-subunit precursors (Cline et al. 1985). The binding reaction is completed within 10 minutes, shows a high affinity of the reactants, and is a nonionic interaction (Pfisterer et al. 1982).

Functional domains of transit peptides, including those of LHCP-II precursors, have not been specifically identified (Schmidt and Mishkind 1986). Of the three blocks of amino acid similarity found in the transit peptides of chloroplast precursor proteins (section III, B, 11 above), Karlin-Neuman and Tobin (1986) speculate that blocks I and/or II mediate recognition, binding, and uptake of precursors into the chloroplast and that blocks II and III mediate processing of the precursor to the mature form. However, the role(s) of these blocks of similarity remain unclear. For example, Reiss et al. (1987), using deletion mutants, showed that block II is not involved in transport into the chloroplast of the precursor to the small subunit of Rubisco. Also, Smeekens et al. (1986b) studied the role of transit peptides with the aid of fusion proteins containing the transit peptide of one chloroplast precursor protein and the mature protein of another. Their results suggest that the transit peptide has two functional domains, one involved in targeting proteins to the chloroplast and the other involved in determining their subsequent localization within the organelle. In any case, it cannot yet be ruled out that targeting, maturation, localization, etc. depend on sequences in both the transit peptide and the mature protein portions of the precursors. Further, a transit sequence may not even be essential for transport. For example, polyphenoloxidase appears to be translated in the mature form without a transit sequence and still is transported into the chloroplast (Flurkey 1985).

D. Transport into the chloroplast

Unlike the binding of the precursor proteins to envelope membranes, transport of the precursors into the chloroplast requires energy. The energy required is supplied by ATP (Cline et al. 1985, Schmidt and Mishkind 1986) and not by an electrochemical gradient across the envelope membranes (Schmidt and Mishkind 1986). With the possible exception of polyphenoloxidase (section IV, C above), cytoplasmically-translated chloroplast precursor proteins appear to be proteolytically-matured during or shortly after transport into the organelle. So far, of the proteases involved in removing transit-peptide sequences from precursor proteins, only the 'transit peptidases' from pea (Robinson and Ellis 1984) and *Chlamydomonas* (e.g., Schmidt and Mishkind 1986) have been partially purified and characterized. The activity of the *Chlamydomonas* peptidase appears specific for algal proteins (Schmidt and Mishkind 1986). The pea peptidase is a soluble metallo-protease (Smith and Ellis 1979, Robinson and Ellis 1984) but has not been tested for its ability to process LHCP-II precursors.

E. Translocation through the chloroplast stroma

The LHCP-II or, for that matter, any other nuclear-coded protein destined to locate in thylakoids must be translocated through the stroma. How this translocation occurs remains obscure (e.g., Schmidt and Mishkind 1986), in part because the specific site(s) where maturation of the precursors of thylakoid proteins occurs remains unknown. How many processing steps are involved in the maturation of an LHCP-II also is not known. In any case, there appear to be two general models for the translocation of a protein through the stroma to the thylakoids. One model suggests that an LHCP-II precuror moves through the stroma as the precursor itself (Chitnis et al. 1987). It is possible also that a molecule could be moved through the stroma as some intermediate processed form (Smeekens et al. 1986b), but this latter possibility has not been studied with LHCP-IIs. It has been suggested that the LHCP-II precursor undergoes, possible with the aid of a stromal molecule, a conformational change which in turn makes it hydrophobic enough to get inserted into the thylakoid membrane (Chitnis et al. 1987). In any case, this model would require that an immature LHCP-II be processed to a mature form in the thylakoid. Recent in vitro evidence indicates that processing of an LHCP-II precursor is not a prerequisite for its incorporation into the thylakoid membranes of *Lemna* or for its subsequent insertion into an LHC-II complex (Chitnis et al. 1986). The second model (Douce et al. 1984, Schmidt and Mishkind 1986) proposes that transit peptides may be removed as soon as the LHCP-II precursors are exposed in the stroma and, further, that mature LHCP-IIs stimulate the formation of vesicles from the inner membrane of the envelope. These vesicles then would function as carriers for translocation of the mature polypeptides. This latter model further envisages that, upon being matured, the LHCP-IIs may be capable of interacting with each other to form higher order complexes and also could associate with components such as xanthophylls and other lipids that are synthesized in the inner membrane (Douce et al. 1984) and that are known to be specific components of the LHC-II (Trémolièrs et al. 1981, Braumann et al. 1982, Siefermann-Harms et al. 1982, Suss 1983). Therefore, when the membrane evaginates, multiple constituents of the thylakoid membranes would be translocated together in bulk through the stroma.

Chl *b* does not participate in translocating LHCP-IIs through the stroma but it is necessary to stabilize these polypeptides and prevent their degradation within the chloroplast (e.g., Freyssinet and Buetow 1984, Schmidt and Mishkind 1986). What role Chl *a* and/or its precursors may play in the import and maturation of LHCP-IIs is not known. However, relevant studies have been done on *Chlamydomonas* with inhibitors of Chl synthesis and with mutants that have defects in Chl synthesis (Johanningmeier and Howell 1984, Schmidt and Mishkind 1986). Results indicate that Chl precursors in some way are involved in the regulation of nuclear genes and that the complexing of Chl with precursor or mature LHCP-IIs may be an early event in the import process.

F. Insertion into thylakoid membranes

Light is clearly needed for LHCP-IIs to accumulate in thylakoid membranes (e.g., Freyssinet and Buetow 1984). The light requirement is likely explained by the fact that, in most cases at least, light is needed for the synthesis of Chl *a* and *b* with which LHCP-IIs form stable complexes. LHCP-IIs are usually degraded when not complexed to Chl. However, cartenoids complex with LHCP-IIs before Chl does (Gounaris and Wellburn 1983). Plumley and Schmidt (1987) have devised a method for the in vitro reconstitution of spinach thylakoid LHC-IIs. They show that the stable association of Chl *a* and *b* with LHCP-IIs requires xanthophyll carotenoids. For maximum yields of reconstituted LHC-IIs, three xanthophylls (lutein, violaxanthin and neoxanthin) are required, but any two support in vitro reassembly to varying degrees. Also, all three of the above xanthophylls are necessary for reconstituted LHC-IIs to show normal spectral and excitation-energy transfer characteristics (Plumley and Schmidt 1987). No other lipid appears necessary. However, additional lipid may be needed for the oligomerization of LHC-II monomers in thylakoid membranes because exogenous lipase dissociates the oligomers into monomers (Trémolièrs et al. 1981, Rémy et al. 1982).

As noted above, the processing of LHCP-II precursors to mature form is not needed for their *in vitro* insertion into thylakoid membranes or assembly within the membrane into mature LHC-IIs (Chitnis et al. 1986, 1987). Also, mutated LHCP-IIs which have various portions of the mature protein deleted are both processed and inserted into *Lemna* thylakoid membranes (Kohorn et al. 1986). However, the mature protein portion apparently must be intact for assembly to occur because these mutated LHCP-IIs are not incorporated into mature LHC-IIs.

G. Modification by phosphorylation

Phosphorylation of the mature LHCP-IIs affects the function of LHCP-II gene products. In the thylakoid membrane, the LHCP-IIs are assembled into the LHC-IIs with 1 to 2 kD of their NH_2-termini oriented toward the stroma. Phosphorylation occurs on the exposed NH_2-termini (Mullet 1983) by means of a membrane-bound protein kinase (Bennett 1979, 1981, Allen 1984; Bennett et al. 1987) whose activity is regulated by light *via* the redox state of the plastoquinone pool between the PS II and cytochrome b_6/f complexes. The protein kinase is activated when the pool of plastoquinone becomes reduced (Horton et al. 1981). Phosphorylated LHCP-IIs are reversibly hydrolysed by a light-insensitive phosphoprotein phosphatase which is associated with thylakoid membranes (Bennett 1980). This reversible phosphorylation cycle of LHCP-II induces changes in energy transfer between PS I and PS II. When phosphorylation occurs, a mobile part of the LHCP-II pool is thought to migrate (Anderson and Goodschild 1987, Williams and Allen 1987) from appressed thylakoid regions (rich in PS II) to non-appressed regions (rich in PS I) with the result that light energy is transferred to PS I. The opposite movements would occur after LHCP-IIs are dephosphorylated, and the distribution of excitation energy then would favor PS II. Such a system is thought to correct the imbalance of energy transfer between the two photosystems and to increase the efficiency of noncyclic electron transport (Kyle et al. 1983, Bennett 1984, Anderson 1986, Briantais et al. 1986, Thornber 1986).

V. Regulation of LHCP-II genes

Plants are capable of responding to and surviving changes in the external environment. A basic question to be answered is how a plant cell can modulate its concentrations of enzymes and other critical proteins in response to a change in environmental factors such as light, darkness, temperature and the availability of nutritional substrates for growth. Many environmental changes lead to alterations in cellular contents of LHCP-IIs indicating that the expression of LHCP-II genes can be regulated by external stimuli. However, attempts to determine the molecular mechanisms underlying such a regulation is complicated by the facts that the expression of LHCP-II genes is cell-specific, organ-specific, and cell-cycle specific and, also, is influenced by the developmental state of the chloroplasts which house the mature LHCP-IIs.

A. Light-regulated LHCP-II gene expression

1. Photoreceptors
The absorption of light by a photoreceptor is the first step in all photoresponses. The most studied photoreceptor is phytochrome, a red-light absorbing pigment (Jabben and Holmes 1983, Quail 1983, Sharma 1984). In

addition to phytochrome, one or more blue/UV photoreceptors play an important role in gene expression in higher plants (Senger 1982, Briggs and Iino 1983, Schäfer and Haupt 1983, Kaufman et al. 1985b, Eskins and McCarthy 1986). Algae which have been studied do not have phytochrome but do have blue and red/blue photoreceptors (e.g., Schwartzbach and Schiff 1982, Senger 1982, Buetow 1986). In tobacco suspension-culture cells, red light is inhibitory, but long-term illumination with blue light induces LHCP-II gene expression (Richter and Wessel 1985, Richter et al. 1986).

After a photoreceptor absorbs light, little is known about how the signal is transduced to bring about changes in gene expression (e.g., Thompson et al. 1985, Schäfer and Briggs 1986, Kuhlmeier et al. 1987). It is clear that the response to light can be complex at the molecular level. For example, fluence is a parameter of illumination that can differentially affect the expression of LHCP-II genes. In the case of red light, responses fit two categories: those triggered by LF and those triggered by VLF. The LF response has a threshold of about 1 μmole/m^2 of red light and is a far-red reversible phytochrome response (Kaufman et al. 1984, 1985b). The VLF response has a threshold of about 10^{-4} μmole/m^2 of red light and is not reversible by far-red light. Some plant genes respond only to light in the LF range while others respond in the VLF range (e.g., Kaufman et al. 1985b, Schäfer and Briggs 1986, Kuhlmeier et al. 1987). However, in pea (Kaufman et al. 1984, 1985a, 1986), maize (Eskins and McCarthy 1986) and barley (Schäfer and Briggs 1986), LHCP-II genes respond to both LF and VLF. In *Phaseolus* leaves, red light leads to an age-independent increase in translatable LHCP-II mRNA (Tavladoraki et al. 1986).

2. Photodamage

High intensity light can negatively affect the expression of LHCP-II genes. In pea, for example, thylakoid membranes "adapt" over a wide range of light intensities, i.e., as light intensity during growth of the plant is increased, the amount of Chl associated with LHCP-II is decreased (Leong and Anderson 1984). However, when the light level is raised enough, photodamage to the chloroplast photosynthetic system results. In turn, LHCP-II gene expression is somehow affected. This is seen in maize plants which have been subjected to a photoinhibitory light intensity (Hayden et al. 1987). Such plants are unable to process the 31 kDa LHCP-II precursor to the mature protein, and the precursor accumulates in the thylakoid membranes of mesophyll chloroplasts. The appearance of the precursor in the membranes correlates with malfunctioning LHC-IIs.

3. Light versus darkness

Apel and Kloppstech (1978) were the first to show, with an in vitro translation system, that the level of translatable LHCP-II mRNA is much higher in light-grown than in dark-grown etiolated plant seedlings (barley). Phytochrome is involved in this light-mediated increase in translatable mRNA (e.g., Tobin and Silverthorne 1985). Other studies using hybridization probes have shown also that the content of LHCP-II mRNA increases under normal light conditions (e.g., Tobin and Silverthorne 1985).

At least a low level of LHCP-II mRNA is detected in dark-grown plants (e.g., Freyssinet and Buetow 1984, Harpster and Apel 1985). Therefore, light is not needed for the transcription of LHCP-II genes. *Phaseolus* grown in the dark for 5 days after sowing contain a small amount of translatable LHCP-II mRNA, and this amount about doubles if growth in the dark is prolonged to 9 to 10 days (Tavladoraki et al. 1986). However, in *Phaseolus* as well as in most plants, light is needed for LHCP-II mRNA to accumulate to maximal levels. This light-stimulated accumulation of LHCP-II mRNA could result either from an increased rate of transcription of LHCP-II genes or from a decreased rate of breakdown of nuclear transcripts. Transcription studies with isolated plant leaf nuclei have been done to distinguish between these two possibilities (Gallagher and Ellis 1982, Silverthorne and Tobin 1984, Mösinger et al. 1985, Steinmüller et al. 1985). Results show an increased incorporation of radioactive nucleotides into LHCP-II mRNAs in nuclei isolated from light-grown leaves compared to those isolated fro dark-grown leaves. Such results do not eliminate the possibility that the genes are transcribed at equal rates in both dark- and light-grown leaves, but that the transcripts are rapidly degraded in the dark nuclei. However, pulse-labelling experiments show that LHCP-II mRNA transcripts are stable in both dark and light nuclei (e.g., Gallagher and Ellis 1982). Therefore, as a whole, the results with isolated nuclei indicate that the increased accumulation of LHCP-II mRNA in the light is mediated by an increased transcription rather than by an increased stability of the mRNA.

Light also may have an effect on the translation of LHCP-II mRNA. Though the transcription of LHCP-II genes is stimulated by white or red light, efficient translation of LHCP-II mRNA occurs in vivo in the presence of white light or in vitro in a wheat-germ cell-free system (Slovin and Tobin 1982). Therefore, a light-limited factor may be needed for translation of this mRNA. If so, this factor apparently is present in plants treated with white light (but not red light) and also apparently is present in wheat germ extracts. Alternatively, a translation inhibitor may be present in red light and absent or overcome in white light and in wheat germ extracts.

Further, and as mentioned above (sections IV, F and G), the accumulation of the translated LHCP-IIs depends on light because a) the stable assembly of LHCP-IIs in thylakoid membranes requires the concomitant light-driven synthesis of Chl, and b) the activity of the assembled LHCP-IIs is modulated by light-dependent phosphorylation.

In sum, light can regulate the expression of LHCP-II genes at multiple steps, i.e., transcription, translation, and the posttranslational stabilization and phosphorylation of the translated LHCP-IIs.

4. Cis-acting elements for light-regulated LHCP-II gene expression
Identification of any DNA sequences that mediate the effect(s) of light on genes is essential for an understanding of the molecular basis of light-induced transcription. The study of the regulation of gene expression in plants has benefitted from advances in techniques to transfect plant cells with specific and defined DNA segments and to regenerate transgenic plants from transfected cells. Transgenic plants are very useful for determining the function of cis-acting elements because they can be used for phytochrome experiments and they provide a normal ebvironment within which normal chloroplast differentiation occurs (e.g., Kuhlemeier et al. 1987). A 1.8 kb 5′-flanking sequence from a wheat LHCP-II gene (AB1.6) was fused to the coding sequence of a bacterial CAT gene followed by a 3′-untranslated region of a pea nuclear gene (Nagy et al. 1986a). The wheat flanking sequence (positions − 1816 to + 31) confers phytochrome responsiveness on the chimeric gene in transgenic tobacco seedlings. In further studies, the cis-acting element responsible for the phytochrome response was located to a 268-bp enhancer-like 5′-upstream element between − 89 and − 357 (Nagy et al. 1986a, b, 1987). Similarly in pea, Simpson et al. (1985) localized phytochrome regulation to a 2.5-kb upstream sequence of an LHCP-II gene. About 0.4 kb of this upstream element is sufficient for light-regulated expression of a chimeric gene in transgenic tobacco plants (Simpson et al. 1986a). In addition, a 247-bp sequence (− 347 to − 100) within the 0.4 kb fragment acts as a light-responsive enhancer-like sequence (Simpson et al. 1986b).

No trans-acting elements controlling the transcription of LHCP-II genes have been identified.

5. Posttranscriptional regulation of LHCP-II gene expression is important during light-induced chloroplast differentiation
a. Chloroplasts in leaves grown under normal light-dark cycles. Leaves of monocotyledons grown under normal light-dark cycles show a gradient of increasing chloroplast differentiation from the basal region where the plastids are essentially undifferentiated and exist as proplastids to the apical region where the plastids are fully differentiated into chloroplasts (Baker and Leech 1977, Wellburn 1983). Therefore, the monocotyledon leaf

provides a unique system with which the regulation of LHCP-II gene expression during chloroplast differentiation can be studied under natural developmental conditions. The LHCP-II content of wild-type barley leaves developing under a normal dark-light cycle increases from the basal segment (segment 1) to segment 4 where it reaches a steady-state level which is maintained through apical segment 6 (Viro and Kloppstech 1980). The level of translatable LHCP-II mRNA is highest in segments 2 and 3 and decreases in subsequent segments. Thus, the increase in LHCP-II level follows the increase in translatable mRNA level through segment 3. However, LHCP-IIs continue to increase to segment 4 while the mRNA decreases, a result indicating the occurrence of posttranscriptional regulation of LHCP-II gene expression during chloroplast differentiation.

b. Chloroplasts in leaves, algae and tissue culture cells first grown in the dark and then greened in continuous light. When seedlings, etiolated by growth for 4 to 10 days in the dark, are placed in continuous light, their chloroplasts differentiate and LHCP-IIs accumulate in the organelles. It is clear that light is not always needed for the transcription of LHCP-II genes because at least a low level of translatable LHCP-II mRNA is detectable in many dark-grown plants (see section V,A,3, above). Therefore, the accumulation of LHCP-II in the light has been interpreted as being mediated by phytochrome via an increased transcription of LHCP-II genes (e.g., Tobin and Silverthorne 1985). However, the situation is not always so straightforward. For example, etiolated pea seedlings already contain a substantial amount of LHCP-II mRNA prior to any illumination (Cuming and Bennett 1981, Jenkins et al. 1983, Thompson et al. 1983, Bennett et al. 1984), and nuclei from dark-grown pea seedlings transcribe LHCP-II mRNA at a substantial rate (Gallagher and Ellis 1982). Further, when etiolated pea plants are illuminated, the LHCP-II mRNA starts to accumulate only after a relatively long lag (Thompson et al. 1983, Bennett et al. 1984). Therefore, any phytochrome-mediated effect leading to an increased transcription of LHCP-II genes, at the least, need not be a direct one. Mung-bean shows even higher levels of LHCP-II mRNA in the dark than does pea, and the mung-bean mRNA shows little change in response to light (Thompson et al. 1983). Similarly, exposure of dark-grown *Euglena gracilis* to white light has no effect on the relative amount of LHCP-II mRNA even though accumulation of LHCP-IIs is induced (Devic and Schantz 1984). Therefore, in mung-bean and *Euglena* there may be no light-mediated effect at all on LHCP-II gene transcription. Any increase, then, in the level of LHCP-II in the light would be mediated posttranscriptionally. Further, during the first six hours of greening of cultured dark-grown soybean cells exposed to the light, the amount of LHCP-II increases about 70-fold whereas the level of translatable LHCP-II mRNA increases only 7-fold (Erdös et al. 1986, 1987). Again, accumulation of the LHCP-IIs in cultured soybean cells greening in the light must be regulated strongly at some posttranscriptional level(s).

Mustard seedlings, grown in the dark for 60 hours and then continuously illuminated with white, red or far-red light, show only a slight accumulation of LHCP-II mRNA (Oelmüller and Schuster 1987). Indeed, the highest amount of translatable LHCP-II mRNA appears when mustard seedlings are given a 5-minute pulse of red light and then kept in the dark for 12 hours more. Therefore, in the case of mustard seedlings at least, continuous light following growth in the dark actually is inhibitory to the accumulation of LHCP-mRNA.

C. Chloroplasts in leaves grown from germination in cotinuous light. Second leaves from maize plants grown from germination under continuous light were analysed for contents of LHCP-II mRNA and LHCP-IIs by Nelson et al. (1984). Under these constant growth conditions, the second leaf emerges from the coleoptile at 4 or 5 days after germination. As the second leaves develop over an 8-day period, their differentiating chloroplasts show a 200-fold accumulation of LHCP-II mRNA. However, the rate of accumulation of LHCP-IIs from day 4 to day 7 is much higher than the rate of increase in their mRNAs. Similar results are found with mustard cotyledons grown under continuous red light. The level of translatable LHCP-II mRNA increases and peaks at 3 to 4 days and then declines while the level of LHCP-IIs continues to increase and reaches a steady-state level at 6 days (Schmidt et al. 1987). These results with maize and mustard grown under continuous light again suggest some posttranscriptional regulation of LHCP-II gene expression during chloroplast differentiation.

6. Circadian rhythmicity in LHCP-II gene expression
Maize and pea seedlings have been grown under alternating light and dark conditions to mimic the normal day-night cycle, i.e., 12 hours of light followed by 12 hours of darkness. Under this condition, the levels of LHCP-II mRNAs in both types of seedlings varies considerably during a 24-hour period (Kloppstech 1985). Moreover, when the plants are transferred from the alternating light-dark conditions to constant light, the variation in mRNA level persists. Thus, there appears to be an endogenous circadian rhythmicity in the expression of LHCP-II genes. Also, in the light-dark cycle, the mRNA levels begin to increase two hours *before* the beginning of the light period, i.e., at a time when the residual level of active phytochrome is low at best. Therefore, any effect that phytochrome might

have on increasing the level of LHCP-II mRNA is not a direct one, at least under alternating light and dark conditions, but must rely on additional steps (Kloppstech 1985).

7. Differential expression in the light of individual genes in the LHCP-II multigene family

Most studies done with LHCP-IIs have not distinguished between the transcripts of the various genes comprising the LHCP-II multigene family. Relatively little is known, for example, whether the levels of the different LHCP-IIs reflects quantitative differences in expression of their respective genes. However, two studies have addressed the differential expression of individual LHCP-II genes during leaf development and chloroplast differentiation. Lamppa et al. (1985a) recognized at least seven different transcripts of LHCP-II genes in wheat leaf. Not all of the wheat LHCP-II genes follow the same pattern of expression along the length of the leaf. For example, one transcript is found in substantial amount in sections 1 (basal segment of leaf), 2 and 3 and only in trace amount in sections 4 and 5 (apical segment). Another transcript is lowest at the leaf base and increases toward the leaf apex. In a related study, Sheen and Bogorad (1986) examined six distinct LHCP-II cDNA clones from maize. The light-inducibility of the genes corresponding to these cDNAs differ. Two are dramatically stimulated by light, one is induced at a somewhat lower level, and three show only a moderate response to light. Clearly, there is no common pattern of regulation by light for all members of the LHCP-II multigene family in either maize or wheat.

B. Effects of other environmental factors on LHCP-II gene expression

1. Heat shock

Heat shock results in the accumulation of LHCP-II mRNA but not chlorophyll in the y-1 mutant of *Chlamydomonas reinhardii* kept in the dark (Hoober and Stegeman 1976, Hoober et al. 1982). In contrast, no induction of LHCP-II gene transcription is detected in heat-shocked etiolated or green pea plants (Meyer and Kloppstech 1986). Apparently, different species react differently to heat stress.

2. Phytohormones

Phytohormones play a central role in the induction of greening and the initiation of chloroplast development (e.g., Parthier 1979, Sharma 1984). Cytokinins stimulate the synthesis of LHCP-IIs in tobacco cell-suspension cultures (Axelos et al. 1984). Further, the levels of detectable and translatable LHCP-II mRNA in the cultured cells also increase 10-fold with cytokinin (kinetin) treatment (Teyssendier de la Serve et al. 1985, 1986). The accumulation of LHCP-II mRNA also is stimulated by cytokinin in petunia cell-suspension cultures (Funckes-Shippy and Levine 1985).

In *Lemna*, the action of cytokinin (benzyladenine) somehow interacts with the phytochrome response to enhance the induction of LHCP-II mRNA (Flores and Tobin 1986). By itself, cytokinin has little inductive effect on transcription, but specifically enhances the phytochrome response for LHCP-II mRNA apparently by modulating LHCP-II gene expression post-transcriptionally (Flores and Tobin 1986, Silverthorne and Tobin 1987, Tobin et al. 1987).

3. Nutrition

The alga, *E. gracilis*, can grow photoautotrophically or photoheterotrophically. Organic carbon-sources added to its growth medium inhibit the formation and assembly of LHCP-IIs (Koll et al. 1980, Yi et al. 1985, Brandt and Winter 1987). How a nutritional factor represses the expression of LHCP-II genes is not known.

C. Cell-specific expression of LHCP-II genes

Cell-specific expression of LHCP-II genes has been studied mainly in maize, a C4 plant. Maize leaves contain two distinct photosynthetic cell types, MC and BSC, which compartmentalize the components and metabolic pathways of C4-photosynthesis. Briefly, primary carbon-fixation by phosphoenolpyruvate carboxylase occurs in MC and the fixed CO_2 is transported as malate to BSC where CO_2 is released by decarboxylation and then refixed via the photosynthetic carbon-reduction cycle. Therefore, MC and BSC provide a model system for studying cell-specific expression of photosynthesis-related genes. The agranal chloroplasts in BSC are poor in PSII in comparison to their activity in PSI (e.g., Golbeck et al. 1981) and contain less LHCP-II and other PSII polypeptides (e.g., Broglie et al. 1984, Bassi et al. 1985, Schuster et al. 1985) than are present in the granal chloroplasts of MC.

Broglie et al. (1984) reported that the steady-state level of LHCP-II mRNA is higher in MC than in BSC. They suggested that the differential expression of LHCP-II genes in the two cell-types is controlled at the level of steady-state mRNA. Also, Schuster et al. (1985) at best detected only trace amounts of LHCP-IIs and LHCP-II mRNA in BSC. They suggested that differentiation of BSC thylakoids involves a repression of nuclear and chloroplast genes coding PSII components. However, Sheen and Bogorad (1986, 1987) determined by immunoprecipitation and Northern blot analyses that, though the levels of LHCP-II proteins and mRNA are higher in MC, significant levels of these components are present in BSC. Thus, these latter results support the interpretation of Broglie et al. (1984) that the cell-specific expression of LHCP-II genes is regulated at the level of transcript abundance. However, whether the control of the LHCP-II transcript pool size occurs at the initiation of transcription or at the level of mRNA turnover rates or both is not known (Sheen and Bogorad 1987). Also, LHCP-II levels could be affected by further translational or posttranslational controls.

The organization of LHCP-II genes into a multigene family raises the question of whether the expression of individual members of a gene family is cell-type-specific (or tissue- or organ-specific). Sheen and Bogorad (1986) investigated this question as regards the expression of different LHCP-IIs in MC and BSC of maize leaves. Six members of the LHCP-II gene family in maize are expressed in leaves. Three of these genes are preferentially expressed in MC cells and account for over 50% of the total LHCP-II mRNA detected in greening leaves. Two genes are expressed equally in MC and BSC, and one is expressed at a higher level in BSC than in MC. The accumulation of each of the five differently-sized LHCP-IIs detected by immunoblot analysis was induced by light in both cell-types. Two were about equal in MC and BSC, two were the more abundant in MC, and one was the more abundant in BSC. Therefore, the six members of the LHCP-II gene family apparently are not regulated through a common mechanism (Sheen and Bogorad 1986). Instead, the genes comprise three differentially-regulated types in terms of their cell-specificity, light-inducibility, and levels of mRNA and protein accumulation.

D. Organ-specific expression of LHCP-II genes

Thompson et al. (1981) compared the level of LHCP-II mRNA in dark-grown and illuminated pea plants. The dark level of this mRNA is high in buds but not in hooks, stems or roots. In the light, LHCP-II mRNA is present at high levels in buds, hooks and stems but not in roots. Similarly, in tomato the level of LHCP-II mRNA is highest in leaf, next-highest in stem and undetectable in root or etiolated seedlings (Piechulla et al. 1986). In barley, there is an increased DNase I sensitivity of LHCP-II genes in leaf chromatin where their transcription is high compared to endosperm where their transcription is low (Steinmüller et al. 1986).

Transgenic plants have been used to study the role of 5′-flanking sequences of LHCP-II genes in the organ-specific expression of these genes. Simpson et al. (1986a) constructed chimeric genes consisting of the 5′-flanking sequences (2.5 kb) of a pea LHCP-II gene and the coding sequences of a reporter gene, NPT II. The 5′-flanking sequence confers phytochrome-responsiveness on the chimeric gene in transgenic tobacco plants as already discussed (section V, A, 4 above). Also, NPT II activity is detected only in leaf and stem but not in root, sepal, stigma or petal. Further a 0.4 kb portion of the 5′-flanking sequence confers not only organ-specific but also cell-specific expression on the chimeric gene in the leaf. The chimeric gene is expressed in mesophyll cells but not in epidermis or midrib. In a subsequent study, the light-inducibility and the organ-specificity of the chimeric gene was localized to a 247-bp segment (-347 to -100) of the LHCP-II 5′-flanking sequence (Simpson et al. 1986b). Most interestingly, the same 247-bp segment also acts to silence the activity of the chimeric gene in roots of transgenic tobacco plants. Both the enhancer effect in the light and the silencing effect in roots are independent of the orientation of the 247-bp upstream segment. How the same sequence both enhances expression of the gene in leaves in the light and silences its expression in roots is not known. In any case, the interaction of both positive and negative trans-acting regulatory factors with the same DNA segment is suggested. Related studies have shown that 5′-flanking sequences of a wheat LHCP-II gene confer both phytochrome-responsiveness (section V, A, 4 above) and organ-specificity (Lamppa et al. 1985b) on a chimeric gene in transgenic tobacco plants.

The expression of LHCP-II genes has been studied in pericarp tissue during tomato fruit formation (Piechulla et al. 1986). The level of LHCP-II mRNA peaks early in chromoplasts during fruit development (day 7) and declines to an undetectable level in premature fruit (day 30–48). However, assays with LCHP-II-specific antibodies shows that LHCP-IIs are still present in photosynthetically-active green fruit. The observed activation and subsequent inactivation of LHCP-II genes during tomato fruit development and ripening in the presence of light may indicate that these genes are controlled at least in part by a developmental program in tomato pericarp tissue (Piechulla et al. 1986).

E. Cell-cycle specific expression of LHCP-II genes

The accumulation of LHCP-II mRNA in *Chlamydomonas reinhardi* appears to be under both light and cell-cycle control. This mRNA accumulates in a wave-like manner and appears abruptly and peaks during the mid-light period in cells synchronized on a 12 hour light-12 hour dark cycle (Shepherd et al. 1983). This accumulation is dependent on the action of light on an unidentified blue-light receptor (Johanningmeier and Howell 1984). Also, from studies with inhibitors and mutants, LHCP-II mRNA accumulation in the light appears dependent on Chl-synthesis. Indeed, normal decay of the mRNA begins late in the light period at the time accumulation of chlorophyll ceases. It was suggested that a destabilization of the LHCP-II mRNA is brought about by a buildup of Chl-synthesis intermediates late in the light period (Johanningmeier and Howell 1984).

In synchronized *E. gracilis*, synthesis of LHCP-IIs also is light-dependent and occurs and peaks in a wave-like manner in the latter half of the light period (Brandt 1984).

F. The developmental state of chloroplasts

The developmental state of chloroplasts appears to affect the expression of LHCP-II genes (Harpster et al. 1984, Mayfield and Taylor 1984b, Taylor and Mayfield 1985, Batschauer et al. 1986, Oelmüller and Mohr 1986, Oemüller et al. 1986, Piechulla et al. 1986, Simpson et al. 1986a, b, Hilditch 1987, Oelmüller and Schuster 1987). For example, plastids are deformed or missing in the leaves of plants treated with herbicides that inhibit carotenoid and abscissic acid synthesis (Batschauer et al. 1986, Oelmüller et al. 1986, Simpson et al. 1986a), carotenoid-deficient maize seedlings (Mayfield and Taylor 1984b) and in sensescing meadow fescus (Hilditch 1987). In all these cases, LHCP-II mRNA ceases or fails to accumulate during illumination. Although a postulated plastid-dependent 'signal' has not been identified, elimination of functional plastids seems to cause a specific suppression of the expression of LHCP-II genes.

G. Multiple LHCP-IIs

Multiple nonidentical but antigenically-related LHCP-IIs are found in mature LHC-IIs (see section II, above). Multiple molecular forms do result from the expression of multiple genes in the LHCP-II multigene family (e.g., Sheen and Bogorad 1986, 1987) with each gene presumably specifying a distinct LHCP-II. However, multiple forms of in vitro synthesized LHCP-IIs also are observed after the uptake into isolated *Lemna* chloroplasts of the translation product of a single LHCP-II gene (Kohorn et al. 1986, Tobin et al. 1987). It is not known how these multiple forms arise in the latter case. There appear to be two possibilities: (a) an LHCP-II-specific protease and/or modification enzyme is active during the uptake of the presumed single LHCP-II precursor and generate(s) the multiple LHCP-II forms observed, or (b) the primary transcript of a single LHCP-II gene is differentially processed into multiple mature LHCP-II mRNAs which then are translated into multiple precursors that are taken up by and matured in the chloroplasts.

VI. Concluding remarks

The LHCP-II genes of plants are like the RNA polymerase II transcription units found in other eukaryotes. Similarities include TATA- and CAAT-like sequences as well as enhancer- and silencer-like sequences in the 5'-flanking regions. The coding portions of the multiple LHCP-II genes within a single plant species and between different plant species are clearly related in nucleotide sequence. However, the 5'- and 3'-flanking and untranslated regions diverge considerably, and these regions undoubtedly confer specificity on the individual LHCP-II genes. For example, specific 5'-flanking sequences (*cis*-acting elements) are involved in phytochrome-responsiveness and organ-specificity.

LHCP-II genes respond to changes in the environment and also are regulated by cell-, organ-, and cell-cycle-specific "factors". Further, the expression of LHCP-II genes also is influenced by the developmental state of the chloroplast wherein the functional products of these genes finally locate. The molecular basis for these environmental, developmental and cell-specific effects on LHCP-II gene expression has not been delineated in any case. However, the importance of posttranscriptional regulatory steps for mediating cell- and cell-cycle-specificity as well as for mediating the effects of phytohormones and other environmental factors has become clear as the result of many studies. In addition, it should be noted that the regulation of LHCP-II genes by any one factor can be complex. Light for example regulates multiple steps in LHCP-II gene expression and affects transcription, stabilization of the LHCP-IIs and possibly also their translation. Further, the effect of light or phytochrome activation on the transcription of LHCP-II genes frequently has been interpreted as an induction of transcription even though transcription also occurs in the dark. Therefore, light appears to be a modulatory factor rather than an inducing factor, i.e., light appears to modulate LHCP-II gene transcription which is turned on by a light-independent, and apparently endogenous, process. Also, since there are multiple genes for LHCP-IIs, light could induce the expression of some and modulate the expression of others. Further, some genes could be expressed only in the dark and others only in the light.

References

Akoyunoglou G and Argyroudi-Akoyunoglou J (1986) Post-translational Regulation of Chloroplast Differentiation. In: Akoyunoglou G and Senger H (eds) Regulation of Chloroplast Differentiation, pp 571–582. New York: Alan R Liss.

Allen JF (1984) Photosynthesis and phosphorylation of light-harvesting chlorophyll a/b-protein in intact chloroplasts. FEBS Lett 166: 237–244.

Anderson JM (1986) Photoregulation of the composition, function and structure of thylakoid membranes. Annual Rev Plant Physiol 37: 93–135.

Anderson JM and Goodschild DJ (1987) Lateral distribution of the photosystem I complex between the appressed and non-appressed regions of spinach thylakoid membranes: an immunocytochemical study. In: Biggins J (ed) Progress in Photosynthesis Research, Vol II, pp 301–304. Dordrecht: Martinus Nijhoff Publishers.

Andersson B, Anderson JM and Ryrie IJ (1982) Transbilayer organization of the chlorophyll proteins of spinach thylakoids. Eur J Biochem 123: 465–472.

Apel K and Kloppstech K (1978) The plastid membranes of barley (*Hordeum vulgare*). Light induced appearance of the mRNA coding for the apoprotein of the light-harvesting chlorophyll a/b protein. Eur J Biochem 85: 581–588.

Arntzen CJ (1978) Dynamic structural features of chloroplast lamellae. Curr Top Bioenerg 8: 111–160.

Axelos M, Barbet J and Péaud-Lenoël C (1984) Influence of cytokinins on the biosynthesis of light-harvesting chlorophyll a/b proteins in tobacco cell suspensions:detection by radioimmunological assay. Plant Sci Lett 33: 201–212.

Baker NR and Leech RM (1977) Development of photosystem I and photosystem II activities in leaves of light-grown maize (*Zea mays*). Plant Physiol 60: 640–644.

Bassi R, Peruffo AdB, Barrato R and Ghisi R (1985) Differences in chlorophyll-protein complexes and composition of polypeptides between thylakoids from bundle sheath and mesophyll cells in maize. Eur J Biochem 146: 589–595.

Batschauer A, Mösinger E, Kreuz K, Dörr I and Apel K (1986) The implication of a plastid-derived factor in the transcriptional control of nuclear genes encoding the light-harvesting chlorophyll a/b protein. Eur J Biochem 154: 625–634.

Bennett J (1979) Chloroplast phosphoproteins. The protein kinase of thylakoid membranes is light-dependent. FEBS Lett 103: 342–344.

Bennett J (1980) Chloroplast phosphoproteins. Evidence for a thylakoid-bound phosphoprotein phosphatase. Eur J Biochem 104: 85–89.

Bennett J (1981) Biosynthesis of the light-harvesting chlorophyll a/b protein: polypeptide turnover in darkness. Eur J Biochem 118: 61–70.

Bennett J (1984) Chloroplast protein phosphoprylation and the regulation of photosynthesis. Physiol Plant 60: 583–590.

Bennett J (1987) Phosphorylation of thylakoid proteins and synthetic peptide analogs. Differential sensitivity to inhibition by a plastoquinone antagonist. FEBS Lett 210: 22–26.

Bennett J, Jenkins GI and Hartley MR (1984) Differential regulation of the accumulation of the light-harvesting chlorophyll a/b complex and ribulose bisphosphate carboxylase/oxygenase in greening pea leaves. J Cell Biochem 25: 1–13.

Bitsch, A and Kloppstech K (1986) Transport of protein into chloroplasts. Reconstitution of the binding capacity for nuclear-coded precursor proteins after solubilization of envelopes with detergents. Eur J Cell Biol 40: 160–166.

Brandt P (1984) Regulation of the cooperation of plastidial and cytoplasmic translation for the formation of the photosynthetic apparatus. In: Sybesma C (ed) Adv Photosyn Res, Vol IV, pp. 5.517–5.520. The Hague: Martinus Nijhoff/Dr W Junk Publishers.

Brandt P and Winter J (1987) The influence of permanent light and of intermittant light on the reconstitution of the light-harvesting system in regreening *Euglena gracilis*. Protoplasma 136: 56–62.

Braumann J, Weber G and Grimme LH (1982) Carotenoid and chlorophyll composition of light-harvesting and reaction centre proteins of the thylakoid membrane. Photobiochem Photobiophys 4: 1–8.

Brawerman G (1981) The role of the poly(A) sequence in mammalian messenger RNA. CRC Rev Biochem 10: 1–38.

Brecht E (1986) The light-harvesting chlorophyll a/b-protein complex II of higher plants: results from a twenty-year research period. Photobiochem Photobiophys 12: 37–50.

Briantais J-M, Vernotte C, Krause GH and Weis E (1986) Chlorophyll a fluorescence of higher plants: chloroplasts and leaves. In: Givindjee, Amesz J and Fork DC (eds) Light Emission by Plants and Bacteria, pp 539–583. Orlando: Academic Press.

Briggs, WR and Iino M (1983) Blue-light absorbing photoreceptors in plants. Philos Trans Roy Soc London Ser B 303: 347–354.

Broglie R, Bellemare G, Bartlett SG, Chua N-H and Cashmore AR (1981) Cloned DNA sequences complementary to mRNAs encoding precursors to the small subunit of ribulose-1,5-bisphosphate carboxylase and a chlorophyll a/b binding polypeptide. Proc Natl Acad Sci USA 78: 7304–7308.

Broglie R, Coruzzi G, Keith B and Chua N-H (1984) Molecular biology of C_4 photosynthesis in Zea mays: Differential localization of proteins and mRNAs in the two leaf cell types. Plant Mol Biol 3: 431–444.

Buetow DE (1986) Chloroplast development. In: Akoyunoglou G and Senger H (eds) Regulation of Chloroplast Differentiation, pp 427–432. New York, Alan R Liss Inc.

Bürgi R, Suter F, and Zuber H (1987) Arrangement of the light-harvesting chlorophyll a/b protein complex in the thylakoid membrane. Biochim Biophys Acta 890: 346–351.

Cashmore AR (1984) Structure and expression of a pea nuclear gene encoding a chlorophyll a/b binding polypeptide. Proc Natl Acad Sci USA 81: 2960–2964.

Chitnis PR, Harel E, Kohorn BD, Tobin EM and Thornber JP (1986) Assembly of the precursor and processed light-harvesting chlorophyll a/b-protein of Lemna into the light-harvesting complex II of barley etiochloroplasts. J Cell Biol 102: 982–988.

Chitnis PR, Nechushtai R, Harel E and Thornber JP (1987) Some requirements for the insertion of the precursor of apoproteins of Lemna light-harvesting complex II into barley thylakoids. In: Biggins J (ed) Progress in Photosynthesis Research, Vol IV, pp 573–576. Dordrecht: Martinus Nijhoff Publishers.

Cline K, Werner-Washburne OM, Lubben TH and Keegstra K (1985) Precursors to two nuclear-encoded proteins bind to the outer envelope membrane before being imported into chloroplasts. J Biol Chem 260: 3691–3696.

Coruzzi G, Broglie R, Cashmore A and Chua N-H (1983) Nucleotide sequences of two pea cDNA clones encoding the small subunit of ribulose-1,5-bisphosphate carboxylase and the major chlorophyll a/b binding thylakoid polypeptide. J Biol Chem 258: 1399–1402.

Cuming AC and Bennett J (1981) Biosynthesis of the light-harvesting chlorophyll a/b protein. Control of messenger RNA activity by light. Eur J Biochem 118: 71–80.

Devic M and Schantz R (1984) Light-induced biosynthesis of chlorophyll-binding proteins in Euglena. In: Sybesma C (ed) Adv Photosyn Res, Vol IV, pp 5.575–5.578. The Hague: Martinus Nijhoff/Dr W Junk Publishers.

Douce R, Block MA, Dorne A-J and Joyard J (1984) The plastid envelope membranes: Their structure, composition, and role in chloroplast biogenesis. Subcell Biochem 10: 1–84.

Dunsmuir P (1985) The petunia chlorophyll a/b binding protein genes: a comparison of Cab genes from different gene families. Nucleic Acids Res 13: 2503–2518.

Dunsmuir P, Smith SM and Bedbrook J (1983) The major chlorophyll a/b binding protein of petunia is composed of several polypeptides encoded by a number of distinct nuclear genes. J Mol Appl Genet 2: 285–300.

Erdös G, Shinohara K and Buetow DE (1986) Regulation of expression of nuclear genes coding plastid proteins in cultured soybean cells. In: Akoyunoglou G and Senger H (eds) Regulation of Chloroplast Differentiation, pp 505–510. New York: Alan R Liss.

Erdös G, Shinohara K, Chen H-Q, Lee S, Gillott M and Buetow DE (1987) Chloroplast development and regulation of LHCP-gene expression in greening cultured soybean cells. In: Biggins J (ed) Progress in Photosynthesis Research, Vol IV, pp 9.539–9.542. Dordrecht: Martinus Nijhoff Publishers.

Eskins K, Duysen ME and Olson L (1983) Pigment analysis of chloroplast pigment-protein complexes in wheat. Plant Physiol 71: 777–779.

Eskins K and McCarthy S (1986) Efficiency of low-irradiance red and blue light in development of corn mesophyll and bundle sheath chloroplasts. In: Akoyunoglou G and Senger H (eds) Regulation of Chloroplast Differentiation, pp 663–669. New York: Alan R Liss.

Fitzgerald M and Shenk T (1981) The sequence 5′-AAUAAA-3′ forms part of the recognition site for polyadenylation of late SV40 mRNAs. Cell 24: 251–260.

Flores S and Tobin EM (1986) Benzyladenine modulation of the expression of two genes for nuclear-coded chloroplast proteins in *Lemna gibba*: apparent post-transcriptional regulation. Planta 168: 340–349.

Flurkey WH (1985) *In vitro* biosynthesis of *Vicia faba* polyphenoloxidase. Plant Physiol 79: 564–567.

Freyssinet G and Buetow DE (1984) Regulation and expression of genes for chloroplast proteins. Israel J Bot 33: 107–131.

Funckes-Shippy CL and Levine AD (1985) Cytokinin Regulates the Expression of Nuclear Genes Required for Photosynthesis. In: Arntzen C, Bogorad L, Bonitz S and Steinback K (eds) Molecular Biology of the Photosynthetic Apparatus, pp 407–411. New York: Cold Spring Harbor Laboratory.

Gallagher TF and Ellis RJ (1982) Light-stimulated transcription of genes for two chloroplast polypeptides in isolated pea leaf nuclei. EMBO J 1: 1493–1498.

Glazer AN (1983) Comparative biochemistry of photosynthetic light-harvesting systems. Annu Rev Biochem 52: 125–157.

Glazer AN and Melis A (1987) Photochemical reaction centers: structure, organization, and function. Annu Rev Plant Physiol 38: 11–45.

Goldbeck JH, Martin IP, Velthuys BR and Radner R (1981) A critical reassessment of the photosystem II content in bundle sheath chloroplasts of young leaves of *Zea mays*. In: Akoyunoglou G (ed) Photosynthesis, Vol 5: Chloroplast Development, pp 533–546. Philadelphia: Balaban International Science Services.

Gounaris I and Wellburn AR (1983) Formation of chlorophyll-carotenoid-protein complexes in cereal plastids during greening and normal light-grown development. Biochem Physiol Pflanzen 178: 433–442.

Govindjee and Whitmarsh J (1982) Introduction to photosynthesis: energy conversion by plants and bacteria. In: Govindjee (ed.) Photosynthesis, Vol. 1, pp 1–16. New York: Academic Press.

Greenland AJ, Thomas MV and Walden RM (1987) Expression of two nuclear genes encoding chloroplast proteins during early development of cucumber seedlings. Planta 170: 99–110.

Harpster M and Apel K (1985) The light-dependent regulation of gene expression during plastid development in higher plants. Physiol Plant 64: 147–152.

Harpster MH, Mayfield SP and Taylor WC (1984) Effects of pigment-deficient mutants on the accumulation of photosynthetic proteins in maize. Plant Mol Biol 3: 59–71.

Hayden DB, Baker NR and Michael PP (1987) Photoinhibitory stress causes accumulation of a 31 kilodalton protein in the chloroplast light-harvesting apparatus. In: Biggins J (ed.) Progress in Photosynthesis Research, Vol 4, pp 51–54. Dordrecht: Martinus Nijhoff Publishers.

Hilditch P (1987) Immunological quantification of the light-harvesting chlorophyll A/B protein during leaf senescence. Biochem Soc Trans 15: 136.

314

Hiller RG and Goodschild DJ (1981) Thylakoid Membrane and Pigment Organization. In: Hatch MD and Boardman NK (eds) The Biochemistry of Plants, Vol 8, pp 1–49. New York: Academic Press.

Hinz UG, and Welinder KG (1987) The light-harvesting complex of photosystem II in barley. Structure and chlorophyll organization. Calesberg Res Commun 52; 39–54.

Hoober JK, Bednarik D, Keller BJ and Marks DB (1982) Regulatory Aspects of Thylakoid Membrane Formation in *Chlamydomonas reinhardtii* y-1. In: Akoyunoglou G, Evangelopoulos AE, Georgatsos J, Palaiologos G, Trakatellis A and Tsiganos CP (eds) Cell Function and Differentiation, Part B: Biogenesis of energy transducing membranes and membrane and protein energetics, pp 101—110. New York: Alan R Liss, Inc.

Hoober JK and Stegeman WJ (1976) Kinetics and regulation of synthesis of major polypeptides of thylakoid membranes in *Chlamydomonas reinhardtii* y-1 at elevated temperatures. J Cell Biol 70: 326–337

Horton P, Allen JF, Black MT and Bennett J (1981) Regulation of phosphorylation of chloroplast membrane polypeptides by the redox state of plastoquinone. FEBS Lett 125: 193–196.

Hsiao K-C, Erdös G and Buetow DE (1988) Cloning and nucleotide sequence of a soybean gene encoding a light-harvesting chlorophyll-a/b-binding protein of Photosystem II. Plant Mol Biol (in press).

Hunt AG, Chu NM, Odell JT, Nagy F and Chua N-H (1987) Plant cells do not properly recognize animal gene polyadenylation signals. Plant Mol Biol 8: 23–35.

Jabben M and Holmes MG (1983) Phytochrome in light-grown plants. In: Shropshire W and Mohr H (eds) Encyclopedia of Plant Physiology: Photomorphogenesis, New Series, Vol 27, pp 704–719. Berlin: Springer-Verlag.

Jenkins GI, Hartley MR and Bennett J (1983) Photoregulation of chloroplast development: transcriptional and post-transcriptional controls. Phil Trans Roy Soc London Ser B 303: 419–431.

Johanningmeier U and Howell SH (1984) Regulation of light-harvesting chlorophyll-binding protein mRNA accumulation in *Chlamydomonas reinhardi*. Possible involvement of chlorophyll synthesis precursors. J Biol Chem 259: 13541–13549.

Jones JDG, Dunsmuir P and Bedbrook J (1985) High level expression of introduced chimaeric genes in regenerated transformed plants. EMBO J 4: 2411–2418.

Karlin-Neumann GA, Kohorn BD, Thornber JP and Tobin EM (1985) A chlorophyll a/b binding protein encoded by a gene containing an intron with characteristics of a transposable element. J Mol Appl Genet 3: 45–61.

Karlin-Neumann GA and Tobin EM (1986) Transit peptides of nuclear-encoded chloroplast proteins share a common amino acid framework. EMBO J 5: 9–13.

Kaufman LS, Thompson WF and Briggs WR (1984) Different red light requirements for phytochrome-induced accumulation of *cab* RNA and *rbcS* RNA. Science 226: 1447–1449.

Kaufman LS, Briggs WR and Thompson WF (1985a) Phytochrome control of specific mRNA levels in developing pea buds. Plant Physiol 78: 388–393.

Kaufman LS, Watson JC, Briggs WR and Thompson WF (1985b) Photoregulation of nucleus-coded transcripts: blue-light regulation of specific transcript abundance. In: Arntzen C, Bogorad L, Bonitz S and Steinback K (eds) Molecular Biology of the Photosynthetic Apparatus, pp 367–372. New York: Cold Spring Harbor Laboratory.

Kaufman LS, Roberts LL, Briggs WR and Thompson WF (1986) Phytochrome control of specific mRNA levels in developing pea buds. Kinetics of accumulation, reciprocity, and escape kinetics of the low fluence reponse. Plant Physiol 81: 1033–1038.

Kloppstech K (1985) Diurnal and circadian rhythmicity in the expression of light-induced plant nuclear messenger RNAs. Planta 165: 502–506.

Kohorn BD, Harel E, Chitnis PR, Thornber JP and Tobin EM (1986) Functional and mutational analysis of the light-harvesting chlorophyll a/b protein of thylakoid membranes. J Cell Biol 102: 972–981.

Koll M, Brandt P and Wiessner W (1980) Hemmung der lichtabhängigen chloroplastentwicklung etiolierter *Euglena gracilis* durch Glucose. Protoplasma 105: 121–128.

Kuhlemeier C, Green PJ and Chua N-H (1987) Regulation of gene expression in higher plants. Annual Rev Plant Physiol 38: 221–257.

Kung SD, Thornber JP and Wildman SD (1972) Nuclear DNA codes for the photosystem II chlorophyll-protein of chloroplast membranes. FEBS Lett 24: 185–188.

Kyle DJ, Staehelin LA and Arntzen CJ (1983) Lateral mobility of the light-harvesting complex in chloroplast membranes controls excitation energy distribution in higher plants. Arch Biochem Biophys 222: 527–541.

Lamppa G, Morelli G and Chua N-H (1985a) Structure and developmental regulation of a wheat gene encoding the major chlorophyll a/b binding polypeptide. Mol Cell Biol 5: 1370–1378.

Lamppa GK, Nagy F and Chua N-H (1985b) Light-regulated and organ-specific expression of a wheat Cab gene in transgenic tobacco. Nature 316: 750–752.

Leong T-Y and Anderson JM (1984) Adaptation of the thylakoid membranes of pea chloroplasts to light intensities. I. Study on the distribution of chlorophyll-protein complexes. Photosyn Res 5: 117–128.

Leutwiler LS, Meyerowitz EM and Tobin EM (1986) Structure and expression of three light-harvesting chlorophyll a/b binding protein genes in *Arabidopsis thaliana*. Nucleic Acids Res 14: 4051–4064.

Li J (1985) Light-harvesting chlorophyll a/b proteins: Three-dimensional structure of a reconstituted membrane lattice in negative stain. Proc Natl Acad Sci USA 82: 386–390.

Matsuoka M, Kano-Murakami Y and Yamamoto N (1987) Nucleotide sequence of cDNA encoding the light-harvesting chlorophyll a/b binding protein from maize. Nucleic Acids Res 15: 6302.

Mayfield SP and Taylor WC (1984a) The appearance of photosynthetic proteins in developing maize leaves. Planta 161: 481–486.

Mayfield SP and Taylor WC (1984b) Carotenoid-deficient maize seedlings fail to accumulate light-harvesting chlorophyll a/b binding protein (LHCP) mRNA. Eur J Biochem 144: 79–84.

Messing J, Geraghty D, Heidecker G, Hu N-T, Kridl J and Rubenstein I (1983) Plant gene structure. In: Kosuge T, Meredith CP and Hollaender A (eds) Genetic Engineering of Plants, pp 211–227. New York: Plenum Press

Meyer G and Kloppstech K (1986) Nuclear-coded chloroplast heat-shock proteins in pea. In: Akoyunoglou G and Senger H (eds) Regulation of Chloroplast Differentiation, pp 731–736. New York: Alan R Liss

Mösinger E, Batschauer A, Schäfer E and Apel K (1985) Phytochrome-control of in vitro transcription of specific genes in isolated nuclei from barley (*Hordeum vulgare*). Eur J Biochem 147: 137–142.

Mullet JN (1983) The amino acid sequence of the polypeptide segment which regulates membrane adhesion (grana stacking) in chloroplasts. J Biol Chem 258: 9941–9948.

Nagy F, Kay SA, Boutry M, Hsu MY and Chua N-H (1986a) Phytochrome-controlled expression of a wheat Cab gene in transgenic tobacco seedlings. EMBO J 5: 1119–1124.

Nagy F, Fluhr R, Kuhlemeier C, Kay S, Boutry M, Green P, Poulsen C and Chua N-H (1986b) Cis-acting elements for selective expression of two photosynthetic genes in transgenic plants. Phil Trans Roy Soc London Series B 314: 493–500.

Nagy F, Boutry M, Hsu M-Y, Wong M and Chua N-H (1987) The 5'-proximal region of the wheat Cab-1 gene contains a 268-bp enhancer-like sequence for phytochrome response. EMBO J 6: 2537–2542.

316

Nelson T, Harpster MH, Mayfield SP and Taylor WC (1984) Light-regulated gene expression during maize leaf development. J Cell Biol 98: 558–564.

Nevins JR (1983) The pathway of eukaryotic mRNA formation. Annual Rev Biochem 52: 441–466.

Oelmüller R and Mohr H (1986) Photooxidative destruction of chloroplasts and its consequences for expression of nuclear genes. Planta 167: 106–113.

Oelmüller R and Schuster C (1987) Inhibition and promotion by light of the accumulation of translatable mRNA of the light-harvesting chlorophyll a/b-binding protein of photosystem II. Planta 172: 60–70.

Oelmüller R, Levitan I, Bergfeld R, Rajasekhar VK and Mohr H (1986) Expression of nuclear genes as affected by treatments acting on plastids. Planta 168: 482–492.

Parthier B (1979) The role of phytohormones (cytokinines) in chloroplast development. Biochem Physiol Pflanzen 174: 173–214.

Pfisterer J, Lachmann P and Kloppstech K (1982) Transport of proteins into chloroplasts. Binding of nuclear-coded chloroplast proteins to the chloroplast envelope. Eur J Biochem 126: 143–148.

Pichersky R, Tanksley SD, Breidenbach RB, Kausch AP and Cashmore AR (1985) Molecular characterization and genetic mapping of two clusters of genes encoding chlorophyll a/b binding proteins in *Lycopersicon esculentum* (tomato). Gene 40: 247–258.

Pichersky E, Hoffman NE, Malik VS, Bernatzky R, Tanksley SD, Szabo L and Cashmore AR (1987) The tomato Cab-4 and Cab-5 genes encode a second type of CAB polypeptides localized in photosystem II. Plant Mol Biol 9: 109–120.

Piechulla B, Pichersky E, Cashmore AR and Gruissem W (1986) Expression of nuclear and plastid genes for photosynthesis-specific proteins during tomato fruit development and ripening. Plant Mol Biol 7: 367–376.

Plumley FG and Schmidt GW (1983) Rocket and crossed immunoelectrophoresis of proteins solubilized with sodium dodecyl sulfate. Anal Biochem 134: 86–95.

Plumley FG and Schmidt GW (1987) Reconstitution of chlorophyll a/b light-harvesting complexes: Xanthophyll-dependent assembly and energy transfer. Proc Natl Acad Sci USA 84: 146–150.

Polans NO, Weeden NF and Thompson WF (1985) Inheritance, organization and mapping of rbcS and Cab multigene families in pea. Proc Natl Acad Sci USA 82: 5083–5087.

Quail PH (1983) Rapid action of phytochrome in photomorphogenesis. In: Shropshire W and Mohr H (eds) Encyclopedia of Plant Physiology: Photomorphogenesis, New Series, Vol 27, pp 178–212. Berlin: Springer-Verlag.

Reiss B, Wasmann CC and Bohnert HJ (1987) Regions in the transit peptide of SSU essential for transport into chloroplasts. Mol Gen Genet 209: 116–121.

Rémy R, Trémolières A, Duval JC, Ambard-Bretteville F and Dubacq JP (1982) Study of the supramolecular organization of light-harvesting chlorophyll-protein (LHCP). Conversion of the oligomeric form into the monomeric one by phospholipase A_2 and reconstitution with liposomes. FEBS Lett 137: 271–275.

Richter G and Wessel K (1985) Red light inhibits light-induced chloroplast development in cultured plant cells at the mRNA level. Plant Mol Biol 5: 175–182.

Richter G, Einspanier R, Hüsemann W, Dudel A and Wessel K (1986) Gene expression in blue light-dependent chloroplast differentiation of cultured plant cells. In: Akoyunoglou G and Senger H (eds) Regulation of Chloroplast Differentiation, pp 549–558. New York: Alan R Liss.

Robinson C and Ellis RJ (1984) Transport of proteins into chloroplasts: the precursor of the small subunit of ribulose bisphosphate carboxylase is processed to the mature size in two steps. Eur J Biochem 142: 337–342.

Süss K-H (1983) A new isolation method and properties of the light-harvesting chlorophyll a/b-protein complex of higher plants. Photobiochem Photobiophys 5: 317–324.

Schäfer E and Briggs WR (1986) Photomorphogenesis from signal perception to gene expression. Photobiochem Photobiophys 12: 305–320.

Schäfer E and Haupt W (1983) Blue-light effects in phytochrome-mediated responses. In: Shropshire W and Mohr H (eds) Encyclopedia of Plant Physiology: Photomorphogenesis, New Series, Vol 27, pp 723–744. Berlin: Springer-Verlag.

Schell J and Van Montagu M (1986) Regulated expression of foreign genes of plants. In: Bogorad L (ed) Molecular Developmental Biology, pp 3–13. New York, Alan R Liss.

Schmidt GW and Mishkind ML (1986) The transport of proteins into chloroplasts. Annual Rev Biochem 55: 879–912.

Schmidt GW, Bartlett SG, Grossman AR, Cashmore AR and Chua N-H (1981) Biosynthetic pathways of two polypeptide subunits of the light-harvesting chlorophyll a/b protein complex. J Cell Biol 91: 468–478.

Schmidt S, Drumm-Herrel H, Oelmüller R and Mohr H (1987) Time course of competence in phytochrome-controlled appearance of nuclear-encoded plastidic proteins and messenger RNAs. Planta 170: 400–407.

Schuster G, Ohad I, Martineau B and Taylor WC (1985) Differentiation and development of bundle sheath and mesophyll thylakoids in maize. J Biol Chem 260: 11866–11873.

Schwartzbach S and Schiff JA (1982) Photocontrol of chloroplast development in *Euglena*. In: Buetow DE (ed) The Biology of *Euglena*, Vol III, pp 313–352. New York: Academic Press.

Senger H (1982) The effect of blue light on plants and microorganisms. Photochem Photobiol 35: 911–920.

Sharma R (1984) Phytochrome – a plant photochromic sensor. J Sci Indus Res 43: 615–633.

Sheen J-Y and Bogorad L (1986) Differential expression of six light-harvesting chlorophyll a/b binding protein genes in maize leaf cell types. Proc Natl Acad Sci USA 83: 7811–7815.

Sheen J-Y and Bogorad L (1987) Regulation of levels of nuclear transcripts for C_4 photosynthesis in bundle sheath and mesophyll cells of maize leaves. Plant Mol Biol 8: 227–238.

Shepherd HS, Ledoigt G and Howell SH (1983) Regulation of light-harvesting chlorophyll-binding protein (LHCP) mRNA accumulation during the cell cycle in *Chlamydomonas reinhardi*. Cell 32: 99–107.

Siefermann-Harms D, Ross JW, Kaneshiro KH and Yamamoto HY (1982) Reconstitution by monogalactosyldiacylglycerol of energy transfer from light-harvesting chlorophyll *a/b*-protein complex to the photosystems in Triton X-100-solubilized thylakoids. FEBS Lett 149: 191–196.

Silverthorne J and Tobin EM (1984) Demonstration of transcriptional regulation of specific genes by phytochrome action. Proc Natl Acad Sci USA 81: 1112–1116.

Silverthorne J and Tobin EM (1987) Phytochrome regulation of nuclear gene expression. BioEssays 7: 18–23.

Simpson F, Van Montagu M and Herrera-Estrella L (1986a) Photosynthesis-associated gene families: differences in response to tissue-specific and environmental factors. Science 233: 34–38.

Simpson J, Schell J, Van Montagu M and Herrera-Estrella L (1986b) Light-inducible and tissue-specific pea *lhcp* gene expression involves an upstream element combining enhancer- and silencer-like properties. Nature 323: 551–554.

Simpson J, Timko MP, Cashmore AR, Schell J, Van Montagu M and Herrera-Estrella L (1985) Light-inducible and tissue-specific expression of a chimaeric gene under control of the 5′ flanking sequence of a pea chlorophyll a/b binding protein gene. EMBO J 4: 2723–2729.

Slovin JP and Tobin EM (1982) Synthesis and turnover of the light-harvesting chlorophyll a/b-protein in *Lemna gibba* grown with intermittent red light: possible translational control. Planta 154: 465–472.

Smeekens S, Van Oosten J, de Groot M and Weisbeek P (1986a) *Silene* cDNA clones for a divergent chlorophyll-a/b-binding protein and a small subunit of ribulose bisphosphate carboxylase. Plant Mol Biol 7: 433–440.

Smeekens S, Bauerle C, Hageman J, Keegstra K and Weisbeek P (1986b) The role of the transit peptide in the routing of precursors toward different chloroplast compartments. Cell 46: 365–375.

Smith SM and Ellis RJ (1979) Processing of small subunit precursor of ribulose bisphosphate carboxylase and its assembly into whole enzyme are stromal events. Nature 278: 662–664.

Stayton MM, Black M, Bedbrook J and Dunsmuir P (1986) A novel chlorophyll a/b binding (Cab) protein gene from petunia which encodes the lower molecular weight Cab precursor protein. Nucleic Acids Res 14: 9781–9796.

Steinmüller K, Batschauer A, Mösinger E, Schäfer E, Rasmussen SK and Apel K (1985) The light-induced greening of barley. In: Van Vloten-Doting L, Groot GSP and Hall TC (eds) Molecular Form and Function of the Plant Genome, pp 277–290. New York: Plenum Press.

Steinmüller K, Batschauer A and Apel K (1986) Tissue-specific and light-dependent changes of chromatin in barley (*Hordeum vulgare*). Eur J Biochem 6: 514–525.

Tavladoraki P, Akoyunoglou G, Bitsch A, Meyer G and Kloppstech K (1986) Age and phytochrome-induced changes at the level of the translatable mRNA coding for the LHC-II apoprotein of *Phaseolus vulgaris* leaves. In: Akoyunoglou G and Senger H (eds) Regulation of Chloroplast Differentiation, pp 559–564. New York: Alan R Liss.

Taylor WC and Mayfield SP (1985) Ontogenetically regulated photosynthetic genes. In: Arntzen C, Bogorad L, Bonitz S and Steinback K (eds) Molecular Biology of the Photosynthetic Apparatus, pp 413–416. New York: Cold Spring Harbor Laboratory.

Teysendier de la Serve B, Axelos M and Péaud-Lenoël C (1985) Cytokinins modulate the expression of genes encoding the protein of the light-harvesting chlorophyll a/b complex. Plant Mol Biol 5: 155–163.

Teyssendier de la Serve B, Axelos M and Péaud-Lenoël C (1986) Kinetin-induced accumulation of mRNA encoding the apoprotein of the light-harvesting chlorophyll a/b-protein complex in tobacco cell suspensions. In: Akoyunoglou G and Senger H (eds) Regulation of Chloroplast Differentiation, pp 565–570. New York: Alan R Liss.

Thompson WF, Everett M, Polans NO, Jorgensen RA and Palmer JD (1983) Phytochrome control of RNA levels in developing pea and mung-bean leaves. Planta 158: 487–500.

Thompson WF, Kaufman P and Watson JC (1985) Induction of plant gene expression by light. BioEssays 3: 153–159.

Thornber JP (1986) Biochemical characterization and structure of pigment-proteins of photosynthetic organisms. In: Staehelin LA and Arntzen CJ (eds) Encyclopedia of Plant Physiology, New Series, Vol 19, pp 98–142. Berlin: Springer-Verlag.

Thornber JP, Markwell JP and Reinman S (1979) Plant chlorophyll-protein complexes: recent advances. Photochem Photobiol 29: 1205–1216.

Timko MP and Cashmore AR (1983) Nuclear genes encoding the constituent polypeptides of the light-harvesting chlorophyll a/b protein complex from pea. In: Goldberg RB (ed) Plant Molecular Biology, pp. 403–412. New York: Alan R Liss.

Timko MP, Kausch AP, Hand JM, Cashmore AR, Herrera-Estrella L, Van den Broeck G and Montagu MV (1985) Structure and expression of nuclear genes encoding polypeptides of the photosynthetic apparatus. In: Arntzen C, Bogorad L, Bonitz S and Steinback K (eds) Molecular Biology of the Photosynthetic Apparatus, pp. 381–395. New York: Cold Spring Harbor Laboratory.

Tobin EM and Silverthorne J (1985) Light regulation of gene expression in higher plants. Annual Rev Plant Physiol 36: 569–593.

Tobin EM, Silverthorne J, Flores S, Leutwiler LS and Karlin-Neumann GA (1987) Regulation of the synthesis of two chloroplast proteins encoded by nuclear genes. In: Fox JE and Jacobs M (eds) Molecular Biology of Plant Growth Control. UCLA Symp, A Symp Mol Cell Biol, New Series, Vol 44, pp 401–411. New York: Alan R Liss.

Trémolières A, Dubacq J-P, Ambard-Bretteville F and Rémy R (1981) Lipid composition of chlorophyll-protein complexes. Specific enrichment in *trans*-hexadecenoic acid of an oligomeric form of light-harvesting chlorophyll *a/b* protein. FEBS Lett 130: 27–31.

Vallejos CE, Tanksley SD and Bernatzky R (1986) Localization in the tomato genome of DNA restriction fragments containing sequences homologous to the rRNA (45S), the major chlorophyll a/b-binding polypeptide and the ribulose bisphosphate carboxylase genes. Genetics 112: 93–105.

Wellburn AR (1982) Bioenergetic and ultrastructural changes associated with chloroplast development. Intl Rev Cytol 80: 133–191.

Williams WP and Allen JF (1987) State 1/State 2 changes in higher plants and algae. Photosyn Res 13: 19–45.

Yi LSH, Gilbert CW and Buetow DE (1985) Temporal appearance of chlorophyll-protein complexes and the N,N′-dicyclohexylcarbodiimide-binding coupling factor$_o$-subunit III in forming thylakoid membranes of *Euglena gracilis*. J Plant Physiol 118: 7–21.

Zuber H (1985) Structure and function of light-harvesting complexes and their polypeptides. Photochem Photobiol 42: 821–844.

Govindjee et al. (eds), Molecular Biology of Photosynthesis: 321–342
© 1988 Kluwer Academic Publishers

Regular paper

Reaction centers from three herbicide-resistant mutants of *Rhodobacter sphaeroides* 2.4.1: sequence analysis and preliminary characterization

MARK L. PADDOCK, SCOTT H. RONGEY, EDWARD C.
ABRESCH, GEORGE FEHER & MELVIN Y. OKAMURA
*Department of Physics – B-019, University of California, San Diego at La Jolla, CA 92093,
USA*

Received 17 November 1987; accepted 30 December 1987

Key words: bacterial photosynthesis, electron transfer, herbicide resistance, reaction center,
recombinant DNA, terbutryne

Abstract. Many herbicides that inhibit photosynthesis in plants also inhibit photosynthesis in
bacteria. We have isolated three mutants of the photosynthetic bacterium *Rhodobacter sphaeroides* that were selected for increased resistance to the herbicide terbutryne. All three mutants
also showed increased resistance to the known electron transfer inhibitor o-phenanthroline.
The primary structures of the mutants were determined by recombinant DNA techniques. All
mutations were located on the gene coding for the L-subunit resulting in these changes
$Ile^{229} \rightarrow Met$, $Ser^{223} \rightarrow Pro$ and $Tyr^{222} \rightarrow Gly$. The mutations of Ser^{223} is analogous to the
mutation of Ser^{264} in the D1 subunit of photosystem II in green plants, strengthening the
functional analogy between D1 and the bacterial L-subunit. The changed amino acids of the
mutant strains form part of the binding pocket for the secondary quinone, Q_B. This is
consistent with the idea that the herbicides are competitive inhibitors for the Q_B binding site.
The reaction centers of the mutants were characterized with respect to electron transfer rates,
inhibition constants of terbutryne and o-phenanthroline, and binding constants of the
quinone UQ_0 and the inhibitors. By correlating these results with the three-dimensional
structure obtained from x-ray analysis by Allen et al. (1987a, 1987b), the likely positions of
o-phenanthroline and terbutryne were deduced. These correspond to the positions deduced by
Michel et al. (1986a) for *Rhodopseudomonas viridis*.

Abbreviations: ATP – adenosine 5′-triphosphate, Bchl – bacteriochlorophyll, Bphe – bacteriopheophytin, bp – basepair, cyt c^{2+} – reduced form of cytochrome c, DEAE – diethylaminoethyl, EDTA – ethylenediamine tetraacetic acid, Fe^{2+} – non-heme iron atom, LDAO –
lauryl dimethylamine oxide, Pipes – piperazine-N,N′-bis-2-ethane-sulfonic acid, PSII –
photosystem II, RC – reaction center, SDS – sodium dodecylsulfate, Tris – tris(hydroxy-
methyl)aminomethane, UQ_0 – 2,3-dimethoxy-5-methyl benzoquinone, UQ_{10} – ubiquinone 50

Introduction

The reaction center (RC) is a membrane bound pigment-protein complex
that mediates the primary photochemistry in photosynthesis. RCs from the

photosynthetic bacterium *Rhodobacter* (*Rb.*) *sphaeroides* have been extensively studied by biochemical and biophysical techniques (for reviews see Feher and Okamura 1978, Michel-Beyerle 1985) including X-ray crystallography (Allen et al. 1987a, Allen et al. 1987b). RCs consist of three protein subunits L, M and H and the following cofactors: four bacteriochlorophylls (Bchl), two bacteriopheophytins (Bphe), one non-heme iron (Fe^{2+}), and two ubiquinones (UQ_{10}). A specialized bacteriochlorophyll dimer, $(Bchl)_2$, serves as the primary electron donor, a bacteriopheophytin as an intermediate electron acceptor and two ubiquinones, Q_A and Q_B, as the primary and secondary electron acceptors, respectively. Absorption of a photon initiates electron transfer from the donor to one of the Bphe molecules. Subsequent electron transfer occurs to Q_A followed by electron transfer to the secondary quinone acceptor, Q_B. Q_B is believed to act as a mobile electron carrier; after receiving two electrons and two protons (Wraight 1981) it dissociates from the RC and transports the protons across the bacterial membrane to create a pH gradient, which drives ATP synthesis (for reviews see Ort and Melandri 1982 and Ort 1986).

The electron transfer from Q_A^- to Q_B in bacterial RCs is inhibited by triazine herbicides (Wraight 1981) of which terbutryne, studied in this work, is an example. These herbicides, which also inhibit electron transfer in PSII in green plants, are believed to act by competing with Q_B for its binding site (Tischer and Strotman 1977, Pfister and Arntzen 1979, Velthuys 1981, Wraight 1981, Vermass et al. 1983, Brown et al. 1984, Diner et al. 1984, Vermaas et al. 1984, Kyle 1985). The binding site for Q_B in *Rb. sphaeroides* has been shown by X-ray crystallography to be in a region between the D and E helices of the L subunit (Allen et al. 1987b). The inhibitors o-phenanthroline and terbutryne were found by X-ray studies to bind near the homologous region in *Rhodopseudomonas* (*R.*) *viridis* RC (Michel et al. 1986a). These structural studies verified the previous proposals based on photoaffinity labeling with azido-atrazine (Brown et al. 1984, deVitry and diner 1984) that Q_B binds to the L subunit.

Herbicide resistant mutants of *Rb. sphaeroides* have been isolated by several groups (Brown et al. 1984, Okamura 1984, Stein et al. 1984, Gilbert et al. 1985, Schenck et al. 1986) and of *R. viridis* by Sinning and Michel (1987). RCs from these mutants show decreased sensitivity to inhibition of electron transfer to Q_B by triazine herbicides (like terbutryne) and altered electron transfer properties. In this work we report on the sequence analysis and preliminary characterization of RCs from three terbutryne resistant mutants of *Rb. sphaeroides* 2.4.1. The changes in the primary structure obtained from the nucleotide sequence are correlated with changes in binding properties for quinone and inhibitor taking the known three-dimensional structure into account. A preliminary account of this work has been presented (Paddock et al. 1987). A more detailed report on the kinetics and thermodynamic properties of Q_B in these mutants will be published elsewhere.

Materials and methods

Materials

T4 DNA ligase, T4 DNA polynucleotide kinase (*E. coli B*), restriction enzymes BamHI, PvuII, SalI, and XhoI and radioactive nucleotides [γ-^{32}P]ATP and [α-^{35}S]dCTP were obtained from Amersham; restriction enzymes HindIII, PstI, and NruI were from New England Biolabs; large fragment *Escherichia* (*E.*) *coli* DNA polymerase I (Klenow subfragment) was from Bethesda Research Laboratories; pUC8 (Messing and Vieira 1982), M13mp18 and M13mp19 (Yanisch-Perron et al. 1985) were from p-L Biochemicals; deoxynucleotides and dideoxynucleotides were from Pharmacia; and alkaline phosphatase from calf intestine was from Boehringer Mannheim Biochemicals. The plasmid pUC119, which contains the M13 origin for replication, and the helper phage M13K07 were kindly supplied by Jeff Vieira. Centricon 30 microconcentrators were obtained from Amicon; nitrocellulose discs were either Schleicher and Schuell BA85 or Millipore HATF filters; DEAE was from Toyosoda; horse heart cytochrome c (type 6) and UQ_0 were from Sigma; DNase was from Worthington; terbutryne was from Chem Service; and atrazine and o-phenanthroline (1,10-phenanthroline) were from Baker. All other chemicals were of reagent or HPLC grade.

Isolation of mutants

Mutants of the photosynthetic bacterium *Rb. sphaeroides* were isolated by their ability to grow photosynthetically under increased terbutryne (100 μmol) or atrazine (300 μmol) concentrations (Okamura 1984). *Rb. sphaeroides* 2.4.1 was grown anaerobically to early log-phase in Hutners medium. At this stage either 100 μmol terbutryne or 300 μmol atrazine was added (10 cm^3 tubes, tungsten lamp light intensity ~ 1 mW cm^{-2}, T = 30 °C). This temporarily stopped growth for a few weeks. The tubes were then grown to saturation and cells from each tube were spread on plates to isolate individual colonies. Single colonies were isolated and regrown to saturation in liquid culture with herbicide. Production of mutants resistant to o-phenanthroline using this method failed; growth did not resume after its addition, presumably because o-phenanthroline chelates the metals that are required for photosynthetic growth.

DNA cloning

DNA was isolated from anaerobically grown cells using a procedure similar to that described by Williams et al. (1983) with the following modifications: an extraction with phenol:chloroform (1:1) preceded the CsCl gradient. The CsCl was removed after the band extraction using Centricon 30 microconcentrators rather than dialysis. This was followed by a phenol:chloroform (1:1) extraction, a chloroform extraction, and ethanol precipitations in TE (10 mmol Tris-HCl pH 8, 1 mmol EDTA) with 2.5 mol ammonium acetate (75% ethanol) followed by a precipitation in TE with 0.3 mol sodium acetate (67% ethanol). The DNA at this stage had an optical absorbance ratio of $A_{260}/A_{280} \sim 2$.

All enzymatic reactions were performed as suggested by the manufacturers. Genomic DNA from the IM229 and SP223 mutants was digested with the restriction enzymes BamHI and HindIII and from the YG222 mutant with PstI. The digestions were stopped by a phenol:chloroform (1:1) extraction followed by ethanol precipitation. The fragments of the IM229 DNA were ligated into the pUC8 plasmid which was digested with BamHI and HindIII. The pUC119 plasmid was digested with BamHI and HindIII for the ligation of the SP223 DNA fragments and with PstI for the YG222 DNA fragments. All restriction enzyme digestions were checked on 0.8% (w/v) agarose gels using ethidium bromide to visualize the DNA bands. The digested vector DNA was dephosphorylated using alkaline phosphatase before starting the ligation. The ligation mix was adjusted to give a concentration of about $30\,\mu g\,cm^{-3}$ of vector DNA with a molar ratio ranging from 1:1 to 3:1 of chromosomal fragments to vector DNA.

The ligation mixes were either transformed directly into $CaCl_2$ treated *E. coli* (Maniatis et al. 1982) or were cleaned with a phenol:chloroform (1:1) extraction followed by an ethanol precipitation and transformed into cells treated by the Hanahan procedure (Hanahan 1985) to improve the transformation efficiencies. The *E. coli* hosts used were JM103 (Messing 1983), JM105 or JM109 (Yanisch-Perron et al. 1985). The transformed cells were grown in either LB, 2YT (Miller 1972) or SOB broth (Hanahan 1985) for 45–60 minutes and then plated on nitrocellulose discs on top of either LB, 2YT or SOB plates with $50\,\mu g\,cm^{-3}$ ampicillin. Cells were grown until the colony size was ~ 1 mm; this took 14–20 hours. At least two copies of the library were made and screened as described by Hanahan and Meselson (1980). The oligonucleotide, complementary to a region of the gene coding for L (*pufL*), Ala[141] through Phe[146] (5'-GCCTGGGGCTATGCCTTC-3'), was labelled using T4 polynucleotide kinase and $[\gamma^{-32}P]ATP$, which had a typical specific activity of $2–3 \times 10^5\,cpm\,pmole^{-1}$. The incubation time for hybridization was 12–36 hours at 42 °C. After hybridization the filters were rinsed in a solution of 0.9 mol NaCl with 0.09 mol sodium citrate three times at 23 °C for 30 minutes each and once at 42 °C for three minutes. The filters were air dried and exposed to Kodak XAR-5 film with an intensifying screen at − 80 °C.

Colonies which lined up with the positive signals on the film were isolated and rescreened. Plasmid DNA from these colonies was prepared using either the boiling procedure of Holmes and Quigley (1981) as modified by Crouse et al. (1983) or the alkaline-SDS procedure of Birnboim and Doly (1979) to confirm that the clone had the appropriate restriction enzyme recognition sites. Large scale plasmid preparation was performed as described by Maniatis et al. (1982).

Subfragments of the cloned BamHI-HindIII fragment from the IM229 mutant were cloned into the M13 phage to isolate single stranded template for sequencing. Eight subclones were needed to sequence the entire structural coding genes for the L and M subunits (*pufL* and *pufM* respectively). The following subclones were used: a PstI-PvuII fragment ($\sim$ 1 kbp) cloned into M13mp18, a PvuII-SalI fragment ($\sim$ 450 bp) cloned into M13mp18 and M13mp19, two SalI-XhoI fragments of opposite orientation ($\sim$ 400 bp) cloned into M13mp18, an XhoI-NruI fragment ($\sim$ 425 bp) cloned into M13mp18 and M13mp19 and an NruI fragment ($\sim$ 1.5 kbp) cloned into M13mp18. These clones were identified using the complementation test (Messing 1983) with subclones that had been made for the sequencing of *pufL* and *pufM* from *Rb. sphaeroides* 2.4.1 (Williams et al. 1983; Williams et al. 1984). The use of pUC119 for the subsequent cloning of the *puf* operon from SP223 and YG222 eliminated the need for subcloning into M13 phage.

DNA sequencing

The IM229 mutant genes were sequenced by the dideoxy method of Sanger (1977) utilizing the M13 recombinant phage with its universal primer. The SP223 and YG222 mutants were sequenced by the dideoxy method of Sanger (1977) using template DNA prepared by superinfection of the clone with M13K07 and the following oligonucleotide primers:

1. 5'-CCGACTGCAAGCGGAGAG-3' (complementary to a region 18 to 7 bases upstream of *pufL*),
2. 5'-TACCCGCCGGCCCTTGAA-3' (from Tyr^{67} to Glu^{72} of L),
3. 5'-GCCTGGGGCTATGCCTTC-3' (from Ala^{141} to Phe^{146} of L),
4. 5'-TTCACGAACGCGCTGGCTCTGGC-3' (from Phe^{181} to Ala^{188} of L),
5. 5'-TGGGTCGACTGGTGGCAA-3' (from Trp^{259} to Gln^{264} of L),
6. 5'-CTCTTCTCGGGCCTGATG-3' (from Leu^{60} to Met^{65} of M),
7. 5'-GCGCTGGGCATGGGCAAG-3' (from Ala^{139} to Lys^{144} of M),

8. 5′-TCGCGATGCACGGTGCGACCA-3′ (from Phe[216] to Thr[222] of M),
9. 5′-GCCGGTAATGATCATGCA-3′ (for sequencing the other strand near all of the mutation sites; from Gly[252] to Cys[247] of L).

The DNA was labelled in the sequencing reactions with ^{35}S using [α-^{35}S]dCTP in the presence of the large fragment of E. coli DNA polymerase I (Klenow subfragment). The DNA sequence analysis followed the procedure described by Williams et al. (1983).

Reaction center purification

RCs were isolated from each of the mutant strains using the following procedures: Bacteria were lysed using a French press and chromatophores were collected by centrifugation. RCs were solubilized from these chromatophores using LDAO by the method of Cogdell et al. (1975) and were further purified by an $(NH_4)_2SO_4$ precipitation followed by either DEAE chromatography (Feher and Okamura 1978) for SP223 and YG222 RCs or cytochrome c sepharose affinity chromatography (Brudvig et al. 1983) for IM229 RCs. The RCs had an optical absorbance ratio $A_{280}/A_{802} \simeq 1.3$ for SP223 and YG222 and $\simeq 1.7$ for IM229. In the final step the RCs were concentrated and dialyzed against TLE (10 mmol Tris-HCl pH 8, 0.025% LDAO, 1 mmol EDTA). The purified RCs from these mutants contained only one quinone (Q_A) as determined from either the amount of cytochrome c (cyt c^{2+}) oxidized by the RCs under saturating continuous illumination (10 μmol cyt c^{2+}, 10 mmol Pipes pH 6.8, 0.025% LDAO, 1–2 μmol RCs, T = 20 °C, I = 1 W cm^{-2} of white light) or from the percentage of fast and slow recovery of the stabilized donor, (the recovery time from Q_B is $\simeq 1$ s; the recovery from Q_A is $\simeq 100$ ms) (see for example Okamura et al. 1982). One quinone RCs of Rb. sphaeroides 2.4.1 and R26 were prepared as described by Okamura et al. (1975) and assayed as described above.

UQ_0 binding

To determine the occupancy of the Q_B site, we prepared one quinone RCs and measured the rate of cyt c^{2+} photooxidation under continuous illumination (see for example Okamura et al. 1982; I = 1 W cm^{-2}, white light) in the presence of varying concentrations of exogenous UQ_0. (Conditions: 10 μmol cyt c^{2+}, 10 mmol Pipes pH 6.8, 0.025% LDAO, 1 μmol RCs, T = 20 °C). Light absorbed by the RC causes an electron to flow from the donor (D) through an intermediate acceptor (Bphe) and primary quinone acceptor (Q_A) to the secondary quinone acceptor (Q_B). In the presence of cyt c^{2+} the oxidized donor is reduced allowing more electrons to flow from the donor to the quinons (see Fig. 1). Thus, the oxidation of cyt c^{2+}, monitored at 550 nm with a modified Cary 14 (McElroy et al. 1974, Kleinfeld et al. 1984),

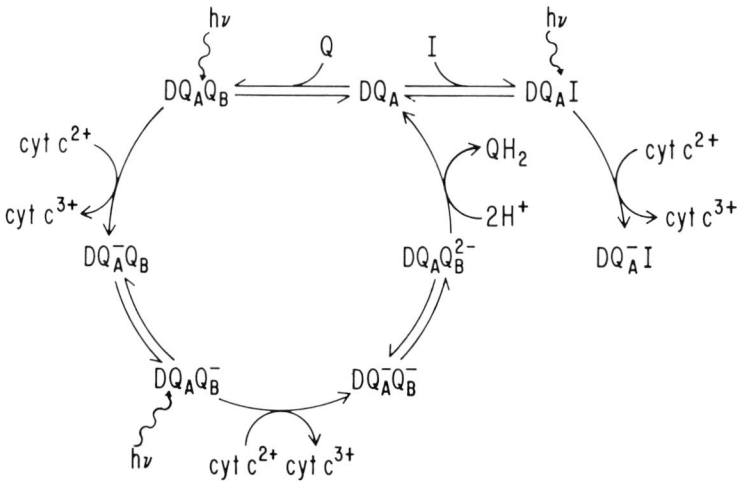

Fig. 1. Electron transfer scheme of the RC showing competitive inhibition between the secondary quinone (Q) and herbicide (I). When Q is bound, the RC can undergo a light induced catalytic cycle oxidizing cytochrome (cyt c^{2+} → cyt c^{3+}) and reducing exogenous quinone. Each Q_B accepts two electrons and two protons; when fully reduced to QH_2 it exchanges with oxidized Q from the quinone pool. For each electron cycled through the RC, one cyt c^{2+} is oxidized to cyt c^{3+}. When herbicide is bound to the RC, at the Q_B site, electron transfer from Q_A^- is inhibited and cyt c^{2+} photooxidation stops after one electron is transferred. Thus, the photooxidation rate provides a convenient assay for the fraction of RCs that have a bound Q_B. Cytochrome c photooxidation (cyt c^{2+} → cyt c^{3+}) was monitored optically at 550 nm as a function of time (see Fig. 3).

provides a measure of the electron turnover rate. RCs were cross illuminated with a projection lamp through 1 inch of water and a Corning 2-64 filter. The cytochrome turnover rate (in cyt $RC^{-1}s^{-1}$) was calculated at various quinone concentrations (Fig. 3). The data were fitted with a Michaelis-Menten equation (see for example Rawn 1983):

$$V = \frac{[UQ_0]}{K_Q + [UQ_0]} \times V_m, \qquad (1)$$

where V is the cytochrome photooxidation rate at the UQ_0 concentration ($[UQ_0]$), V_m is the maximum cytochrome photooxidation rate and K_Q is the concentration of UQ_0 needed to reach one half of V_m; it was used as a measure of quinone binding. Since we are using a *kinetic* assay to determine an *equilibrium* constant, K_Q represents an approximation of the actual binding constant for UQ_0. However if equilibrium is reached on the time scale of the electron transfer, K_Q represents the real binding constant.

Inhibitor binding
The inhibition of terbutryne or o-phenanthroline was studied using the same cyt c^{2+} photooxidation assay as described in the previous section (Fig. 1). The turnover rate was measured for different inhibitor concentrations in the presence of 100 μmol UQ_0 (the quinone concentration that approximately equals the value of K_Q for RCs of the wild type strains 2.4.1 or R26). The activity was defined as the observed photooxidation rate with inhibitor present (V) divided by the photooxidation rate without inhibitor (V_0). The % Activity versus herbicide concentration was fitted with the relation:

$$\% \text{ Activity } = \frac{V}{V_0} \times 100 = \left(1 - \frac{1}{1 + K_{inh}/[I]}\right) \times 100, \qquad (2)$$

where [I] is the concentration of inhibitor and K_{inh} is the inhibition constant found from this fit. A binding constant for the inhibitor (K_I) can be determined if direct competition between the inhibitor and Q_B is assumed (Wraight 1981, Brown et al. 1984, Diner et al. 1984) and equilibrium is reached on the time scale of the electron transfer to Q_B. The expression for K_I, which includes a correction for the number of active RCs at any time (which depends on the concentration of UQ_0, $[UQ_0]$, and the binding constant for the quinone, K_Q) is given by (see for example Rawn 1983):

$$K_I = \frac{K_{inh}}{1 + [UQ_0]/K_Q}. \qquad (3)$$

Results

Cloning and sequencing
Three herbicide resistant mutants of *Rb. sphaeroides* 2.4.1, that were phenotypically distinct from each other as judged by electron transfer and herbicide binding characteristics (Okamura 1984), were chosen for DNA sequence analysis. Since preliminary experiments have implicated either the L or M subunit as the site of the mutation(s) (Okamura et al. 1985), the structural genes coding for the L and M subunit proteins (*pufL* and *pufM*) were cloned from the three mutants.

Restriction enzyme digests of genomic DNA were cloned into either pUC8 (for IM229) or pUC119 (for SP223 and YG222). The use of pUC119 eliminated the need to subclone smaller fragments into the M13 phage for DNA sequencing. The libraries were screened with a synthetic oligonucleotide probe complementary to a region of *pufL*, which codes for Ala[141] through Phe[146] (5′-GCCTGGGGCTATGCCTTC-3′). Colonies that hybridized with the probe were obtained with a frequency of 3/7000, 7/13 000 and 4/5000 from the IM229, SP223 and YG222 libraries respectively. A 6.3 kb BamHI-HindIII genomic DNA fragment (from IM229 and SP223) or a 4.5 kb PstI fragment (from YG222) was isolated; these fragments include *pufL* and *pufM*.

The complete sequence of the noncoding strands of *pufL* and *pufM* from each mutant was determined using either the M13 phage system (for IM229) or internal oligonucleotide primers and pUC119 (for SP223 and YG222). The sequence of the mutant coding strand around the altered region as well as that of the parent 2.4.1. genes Williams et al. 1984 region were used to verify the mutation(s). The nucleotide changes and the deduced amino acid changes are shown in Fig. 2. All three mutants showed changes in the primary structure of the L subunit between the D and E helices (Allen et al. 1987b); no changes were found in the M subunit.

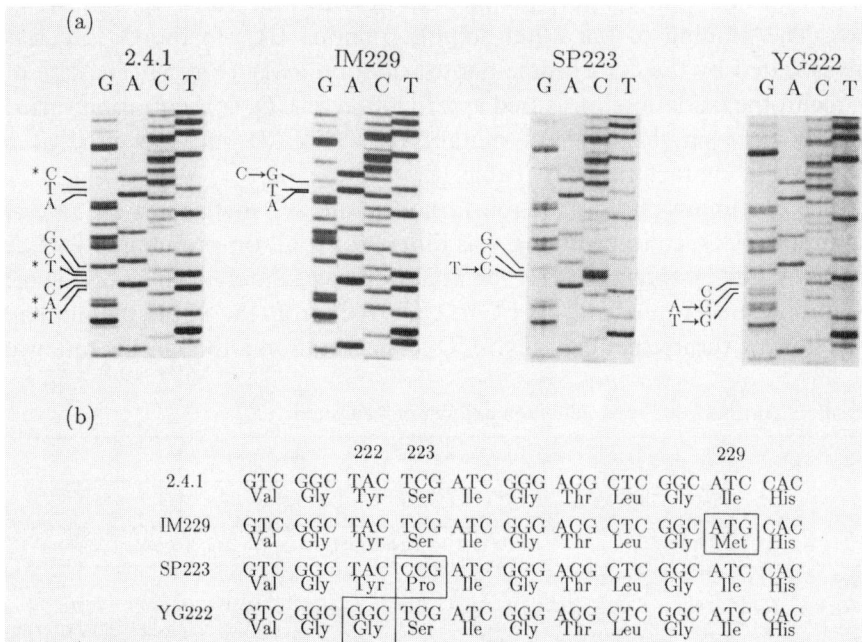

Fig. 2. Regions of the sequencing gels containing the mutations (a) and partial sequences of the mutated region (b) of *pufL* from *Rb. sphaeroides*. The changed nucleotides are indicated next to the gel (a): a C → G transversion was seen in *pufL* from the IM229 mutant, a T → C transition was seen for the SP223 mutant and two changes (a T → G transversion and a A → G transition) were seen in *pufL* from the YG222 mutant. (The gel is read upwards). The resulting amino acid replacements are boxed in (b). All three mutants show changes in the region between Tyr[222] and Ile[229] i.e: Ile[229] → Met (IM229), Ser[223] → Pro (SP223), and Tyr[222] → Gly (YG222).

A single nucleotide change was seen in *pufL* (Fig. 2) from the IM229 mutant (ATC → ATG), resulting in the replacement at position 229 of the L subunit of an isoleucine (Ile) (Williams et al. 1984) with the longer methionine residue (Met). This mutation was independently reported by Gilbert et al. (1985) and Schenck et al. (1986) for herbicide resistant mutants of *Rb. sphaeroides*. The cloned DNA from the SP223 mutant had two nucleotide changes, one at position 38 of the M subunit (CTG → CTC) resulted in no amino acid substitution and the other at position 223 of the L subunit (Williams et al. 1984) (TCG → CCG; Fig. 2) resulted in the replacement of serine (Ser) with proline (Pro). Two nucleotide changes were found in *pufL* from YG222 (TAC → GGC). Both were located in the same codon and resulted in the replacement at position 222 of the L subunit of tyrosine (Tyr) (Williams et al. 1984) with glycine (Gly).

UQ₀ binding studies

RCs with one quinone were prepared as described in Materials and Methods. The binding of the water soluble quinone UQ_0 to the Q_B site was investigated by the cytochrome photooxidation assay (Fig. 1). The rate of cytochrome oxidation, measured as a function of UQ_0 concentration varied greatly between the different mutant RCs (Fig. 3) and was used as a phenotypic characterization (Table 1).

The maximum photooxidation turnover rate, V_m (defined as the rate at saturating UQ_0 concentration), was different for all three mutant RCs (see Fig. 4; Table 1), showing that each mutation has a different effect on the net electron transfer rate from cyt c^{2+} to UQ_0. RCs from the SP223 mutant had the slowest turnover rate at high UQ_0 concentration although the rate was

Table 1. Turnover rates and inhibition and binding constants.

Strain[a]	Mutation	Turnover rate V_m^b (cyt/RC/sec)	Inhibition constants		Binding constants		
			K_{inh}^c		K_Q^d (mM UQ_0)	K_I^e	
			Ter (μM)	O-phen (μM)		Ter (μM)	O-phen (μM)
Wild type	–	260	0.1	24	0.09	0.05	11
IM229	Ile[229] → Met	350	12	270	1.9	11	260
SP223	Ser[223] → Pro	> 14	> 300	150	> 2.0	> 290	~ 140
YG222	Tyr[222] → Gly	60	> 300	620[f]	0.7	> 260	540

[a] Wild type refes to either 2.4.1 or R26, both of which give the same values. SP223 and YG222 were previously called 5a and 11a (Okamura et al. 1984) and S223P and Y222G (Paddock et al. 1987).

[b] Maximum cytochrome turnover rate ($\pm 25\%$).

[c] Inhibition constants ($\pm 30\%$). Assay uses < 0.1 μmol RC, 0.025% LDAO, 20 mmol cyt c^{2+}, 10 mmol Pipes pH 6.8, and 0.1–1.0 mmol EDTA under continuous illumination ($I = 1$ W cm^{-2}).

[d] Concentration of UQ_0 needed to reach half maximum of cytochrome turnover rate ($\pm 30\%$).

[e] Estimated inhibitor binding constant using equation 2 ($\pm 30\%$).

[f] The previously reported value of 300 μmol (Paddock et al. 1987) may have been due to a contamination with Wild type RCs.

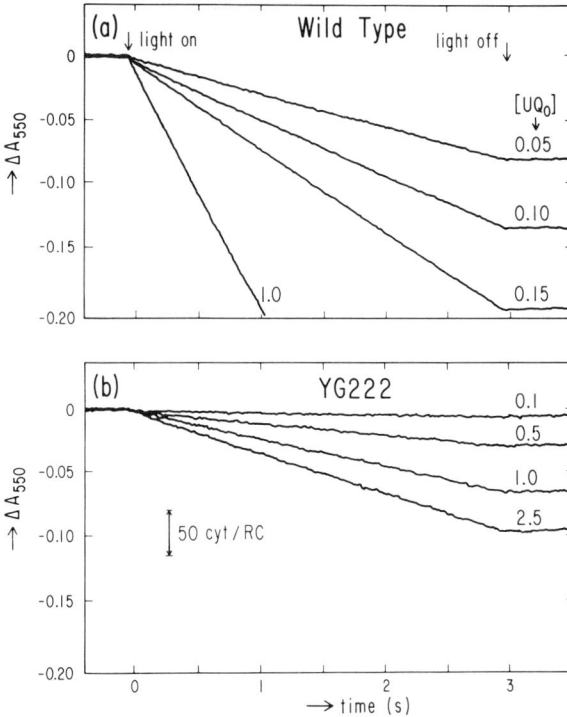

Fig. 3. Cytochrome photooxidation as a function of time for different quinone concentrations, $[UQ_0]$. The change in optical absorbtion from the oxidation of cyt $c^{2+} \rightarrow$ cyt c^{3+} is shown for RCs from the wild type strains (a), R26 and 2.4.1 (they give the same result), for different concentrations of UQ_0 in [mmol] (20 μmol cyt c, 10 mmol Pipes pH 6.8, 0.025% LDAO, 35 nmol RCs, T = 20 °C, I = 1 W cm^{-2}) and for RCs from the mutant YG222 (b) with UQ_0 concentrations in [mmol] as shown (conditions as above). The number of cyt c^{2+} oxidized per reaction center was determined from the relation (see for example Okamura et al. 1982):

$$\text{cyt c oxidized/RC} = 14.4 \, (\Delta A_{550}/A_{802}).$$

The rate of cyt c^{2+} photooxidation is calculated from the number of cyt c^{2+} oxidized per unit time.

never saturated up to the limit of solubility of UQ_0 ($V_m > 14$ cyt RC^{-1}s^{-1}). RCs from the YG222 mutant had a V_m about four times slower (60 cyt RC^{-1}s^{-1}) than the rate obtained for RCs isolated from the wild type strains 2.4.1 or R26 ($V_m = 260$ cyt RC^{-1}s^{-1}). The binding curve for RCs isolated from the IM229 mutant was fit with a V_m value slightly higher (350 cyt RC^{-1}s^{-1}) than V_m obtained from 2.4.1 or R26 (Table 1).

An approximate binding constant (K_Q) for UQ_0 to the RCs was determined from the fit of the cytochrome photooxidation rate (Fig. 4) with equation 1; K_Q is the concentration of UQ_0 needed to reach a cytochrome turnover rate of one half V_m. All three mutants exhibited reduced binding (see Table 1). The least affected mutation (YG222) still had an eight times

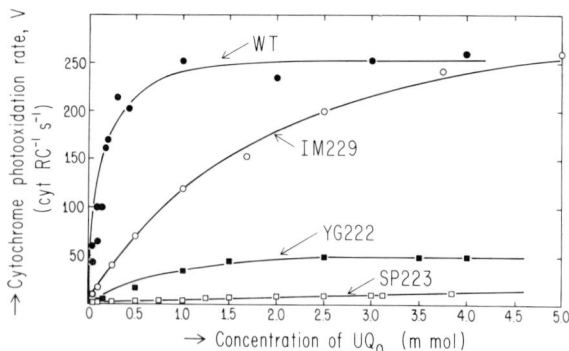

Fig. 4. Cytochrome photooxidation rate as a function of UQ_0 concentration for the wild type, WT (R26 and 2.4.1, which give the same result), and the mutant RCs (IM229, SP223, YG222). Data were obtained as described in Fig. 3. The binding constant, K_Q, was determined from the fit of the data to equation (1) and is the concentration of UQ_0 needed to reach 50% of the maximum rate, V_m (see Table 1). All of the mutant RCs show a reduced affinity for UQ_0 (higher K_Q than for the wild type strains). Note that the rate for neither the SP223 mutant nor the IM229 mutant is saturated at 4 mmol UQ_0.

lower binding constant ($K_Q = 0.7$ mmol) than RCs from the wild type strains 2.4.1 or R26 ($K_Q = 0.09$ mmol).

Herbicide binding studies

The reduction of the electron transfer rate, as measured by the cyt c^{2+} photooxidation assay, upon addition of inhibitors is shown for the wild type and the three mutants in Fig. 5. The solid lines represent theoretical fits to equation (2) with K_{inh} being equal to the herbicide concentration at 50% Activity. Values of the inhibition constants, K_{inh}, and the binding constants K_I (equation 3), are summarized in Table 1.

RCs from the SP223 and YG222 mutants were most resistant to ter-butryne. In both cases the turnover rate could not be reduced to 50% of V_0 within the limit of terbutryne solubility ($K_{inh} > 300 \mu$mol; $K_I > 260 \mu$mol). RCs from the IM229 mutant were not as resistant to terbutryne as the other two mutants but its K_I value ($= 11 \mu$mol) was still more than $100 \times$ larger than that of the wild type ($K_I = 0.05 \mu$mol). These results confirm that the mutated RCs confer terbutryne resistance to the bacterium. For o-phenanthroline the relative ordering of the mutants according to their resistance is different from that for terbutryne. RCs from the YG222 mutant showed the greatest resistance with a $K_I = 540 \mu$mol. RCs from both the SP223 and IM229 mutants are not as resistant to o-phenanthroline as they are to terbutryne (compare the relative values for K_I in Table 1), but they still show a significantly decreased binding of o-phenanthroline relative to UQ_0.

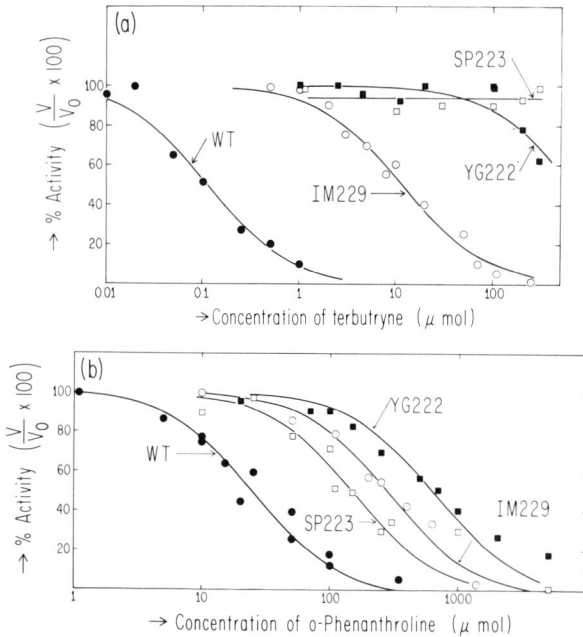

Fig. 5. Cytochrome turnover rate (% Activity = $(V/V_0) \times 100$) as a function of inhibitor concentration, terbutryne (a) and o-phenanthroline (b), for RCs from the three mutants (IM229, SP223, and YG222) and the wild type (WT) strains (R26 and 2.4.1). (Same conditions as described in Fig. 3). The relative resistance levels to herbicide inhibition are evident from these plots. All of the mutant RCs show an appreciable increase in resistance to inhibition by terbutryne and some resistance to inhibition by o-phenanthroline. Solid lines represent fits of the data to equation (2). Values of the inhibition constants, K_{inh}, determined from these plots and inhibitor binding constants, K_I (equation 3), are summarized in Table 1. Note the different scales in (a) and (b).

Summary and discussion

We have determined the changes in amino acid sequence of three herbicide resistant mutants of *Rb. sphaeroides* 2.4.1. The mutants were selected for increased resistance to the herbicide terbutryne. All mutants showed also increased resistance to o-phenanthroline. The mutated residues, Tyr^{222}, Ser^{223}, Ile^{229}, are located on the L subunit in the segment of amino acids between and including part of the D and E helices, which form the quinone binding pocket for Q_B as determined from the X-ray structure analysis of *Rb. sphaeroides* (Allen et al. 1987a, 1987b). An analogous pocket seen in the X-ray crystal structure of *R. viridis* (Michel et al. 1986a) (which lacks Q_B) serves as the binding site for the herbicide terbutryne and the inhibitor o-phenanthroline.

The reaction centers of the mutants were characterized with respect to electron-transfer rate, resistance to inhibition by terbutryne and o-phenanthroline and binding of quinone and inhibitors. A cytochrome photooxidation assay (Fig. 1) was used to determine the electron transfer rate (Fig. 3). Assuming that the mode of action of terbutryne and o-phenanthroline is to compete for the quinone binding site (Tischer and Strotman 1977, Pfister and Arntzen 1979, Velthuys 1981, Wraight 1981, Vermaas et al. 1983, Brown et al. 1984, Diner et al. 1984, Vermaas et al. 1984, Kyle 1985), we obtained from the reduction in electron transfer rate approximate values for the binding constants, K_Q, of quinone (equation 1) and inhibition constants, K_{inh}, for terbutryne and o-phenanthroline (equation 2). The approximate binding constants, K_I, of the inhibitors were obtained using equation (3) and the K_Q values. The characteristics of the mutants are summarized in Table 1.

All three mutants showed reduced binding of terbutryne and, except for IM229, a decrease in the maximum electron-transfer rate. The competitive inhibition assays were performed with the water soluble quinone UQ_0 that lack the isoprenoid chain. More detailed studies with the native quinone UQ_{10} will be published elsewhere.

The structure of the Q_B binding site, based on the X-ray diffraction analysis of the RCs from Rb. sphaeroides R26, is shown schematically in Fig. 6 (Allen et al., manuscript in preparation) together with the suggested positions of Q_B (a), terbutryne (b) and o-phenanthroline (c). At one end of the pocket (proximal to the Fe^{2+}) one of the keto groups of Q_B is H-bonded to His[190], the residue that form a ligand to Fe^{2+}. At the other end of the pocket (distal to the Fe^{2+}) the second keto group of Q_B interacts with Ser[223] that has been mutated to Pro in the SP223 mutant. Another mutated residue, Ile[229], is in close contact with the quinone ring at the proximal side of the pocket. Other residues near Q_B include Leu[193], which makes contact with the quinone ring, Phe[216], which contacts the isoprenoid side chain, Glu[212], which is near the methoxy groups at the bottom of the binding pocket, and the carbonyl group of Ile[224], which contacts the quinone ring (not shown in Fig. 6 for simplicity of the figure). The remainder of the pocket is formed by the backbone atoms that connect these residues. The third mutated residue, Tyr[222], is located outside of the Q_B binding pocket and does not make direct contact with Q_B. However it is close to residues Ser[223] and Ile[224] which are involved in the binding of terbutryne (Michel et al. 1986a). The hydroxyl group of Tyr[222] makes a H-bond with the backbone of Asn[44] in the M subunit.

In R. viridis o-phennthroline binds close to the Fe^{2+} (localised mainly at the proximal side of the pocket), interacting exclusively with the L subunit

(Fig. 6c). Its two nitrogen atoms share an H-bond with the imidazole nitrogens of His[190]. It is also in close contact with Ile[229] and Leu[193] (Michel et al. 1986a). Terbutryne binds further from the Fe^{2+} interacting mainly with residues at the distal side of the pocket making H-bonds with Ser[223] and the peptide nitrogen of Ile[224] (Fig. 6b). It makes close contacts with Val[220], Ile[229], Phe[216] and Glu[212] of the L subunit (Michel et al. 1986a).

The change of the quinone and herbicide binding in the RCs of the mutants can be qualitatively understood in terms of the structure of native RCs. The largest decrease in binding of UQ_0 was observed in the SP223 mutant (Ser[223] → Pro). This is probably due to the loss of interaction between the quinone and the serine hydroxyl group (see Fig. 6a) as well as possible changes in conformation of the residues near the distal end of the pocket. The changes in UQ_0 binding was smaller in the IM229 (Ile[229] → Met) and smallest in the YG222 mutant (Tyr[222] → Gly) reflecting a decreased alteration of the quinone binding pocket for these mutations. The IM229 is likely to move Q_B farther away from His[190], decreasing the binding energy for quinone (smaller K_Q, see Table 1). The increase in distance between Q_B and His[190], a ligand to the Fe^{2+}, is consistent with the observation that RCs from this mutant have an altered EPR spectrum due to $Q_B^- Fe^{2+}$ and a decreased stabilization of the $Q_A Q_B^-$ state with respect to the $Q_A^- Q_B$ (Paddock et al., unpublished). The YG222 mutation may effect the quinone binding by altering the interactions with the nearby residue Ser[223].

The inhibition constants, K_{inh}, and inhibitor binding constants, K_I, for terbutryne binding of the SP223 and YG222 mutant RCs are large compared to those of the IM229 mutant RCs. This follows from the position of the terbutryne binding site near the distal end of the quinone binding pocket as has also been found in the structure of *R. viridis* (Michel et al. 1986a).

The K_{inh} and K_I values for o-phenanthroline binding are similar for all mutant RCs and are increased less compared to the wild type than those for terbutryne binding; this is not surprising since these mutants were selected for increased resistance to terbutryne. The changes in binding are probably due to alterations in the contacts between o-phenanthroline and the residues lining the binding pocket. It should be noted that the pattern of inhibition when UQ_0 is used as an acceptor is different compared to that observed when UQ_{10} is used (Paddock et al., manuscript in preparation). This indicates the importance of the isoprenoid tail in the quinone binding as has been shown earlier by McComb and Wraight (1983) and Warncke et al. (1987).

At present it is not possible to quantitatively account for the changes in binding affinities of the mutant RCs. This is because the overall three-dimensional effect of the mutations cannot be predicted. Even small changes in

(a)

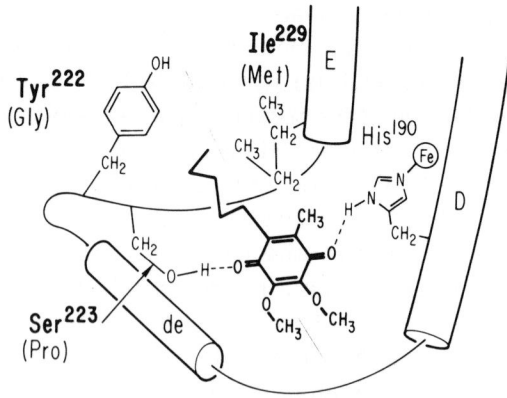

(b)

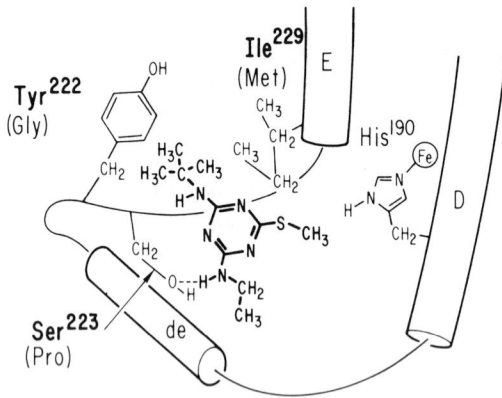

(c)

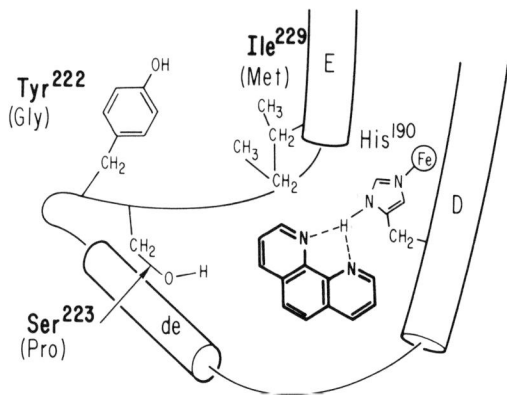

structure can have dramatic effects on binding affinities as seen in the SP229 mutant (Ile^{229} → Met) where an Ile residue is replaced by a slightly longer Met residue. In addition, in some instances even a qualitative understanding of the effects of the mutations is difficult at present. For instance it is difficult to explain why either the YG222 mutation (Tyr^{222} → Gly) or the SP223 mutation (Ser^{223} → Pro) has an effect on o-phenanthroline binding since o-phenanthroline is believed not to make direct contact with either Tyr^{222} or Ser^{223} (Michel et al. 1986a). Presumably these effects are due to small changes in the three-dimensional structure that propagate across distances of several angstroms. Detailed analysis of the three-dimensional structure of the mutant RCs may be required to explain the changes in binding affinities.

All three mutations described in this work involve residues that are conserved between *Rb. sphaeroides* (Williams et al. 1983, 1984, 1986), *Rb. capsulatus* (Youvan et al. 1984) and *R. viridis* (Michel et al. 1986b). This suggests the importance of these residues for optimal functioning of the RC and is consistent with the observed reduced binding constant for quinones when these residues are mutated. It is possible that more conservative mutations may produce herbicide resistance without decreased photosynthetic efficiency.

A topic of recent interest is the relationship between the RC from photosystem II (PSII) from higher organisms and bacteria (see for example Barber 1987, Evans 1987, Trebst 1987). Recent evidence indicates the proteins subunits D1 and D2 of PSII are analogous to the L and M subunits in bacterial RC. This analogy is partially based on the amino acid sequence homology between the L, M and D_1, D_2 subunits (Williams et al. 1984, Youvan et al. 1984, Michel et al. 1986b, Williams et al. 1986, Trebst 1986). Strong support for this proposal has come from the isolation of an active

Fig. 6. Schematic representation of the quinone (a) and inhibitor (b, c) binding sites. These sites are based on the X-ray diffraction analysis of the RCs from *Rb. sphaeroides* (Allen et al. 1987a) and *R. viridis* (Michel et al. 1986a). The sites are made up of a pocket that is lined with residues from part of the D and E helices of the L subunit and the loop joining the helices (in particular, residues His^{190}, Ile^{229}, Leu^{193}, Glu^{212}, Ser^{223}, Phe^{216}, and the backbone of Ile^{224}). Only residues which are in hydrogen bonding distance to the molecules, Ser^{223} and His^{190}, and the sites of the mutations, Ile^{229}, Ser^{223}, Tyr^{222} (in bold print with the mutation listed beneath in parentheses), are shown for simplicity. The suggested position of UQ_0, terbutryne and o-phenanthroline are shown in a, b and c, respectively. The putative hydrogen bonds are indicated by dashed lines. Those for UQ_0 are based on preliminary structural information (Allen et al., manuscript in preparation). Terbutryne binds more on the distal side (away from the Fe^{2+}) side of the quinone pocket, whereas o-phenanthroline binds more on the proximal side (closer to the Fe^{2+}). The view of the pocket in this sketch is normal to the two fold axis from the donor to the quinones rotated such that the Q_B is in the plane of the page.

PSII RC which contains only D1 and D2 polypeptides and 2 subunits of cyt b559 (Namba and Satoh 1987, Okamura et al. 1987). Mutants resistant to triazine herbicides (like terbutryne or atrazine) have been isolated from both bacteria and PSII containing organisms. The bacterial mutants contain an altered L subunit (Gilbert et al. 1985, Bylina and Youvan 1987; Paddock et al. 1987, Schenck et al. 1986, Sinning and Michel 1987), whereas the PSII mutants contain an altered D1 subunit (Hirschberg and McIntosh 1983, Erickson et al. 1984, Goloubinoff et al. 1984, Hirschberg et al. 1984, Erickson and Rochaix 1985, Erickson et al. 1985, Golden and Haselkorn 1985, Johanningmeier et al. 1987; for a review see Trebst 1987). The mutation sites of D1 are located in a region that is homologous to the location of the mutation sites of the bacterial L subunit, indicating that the structure of the Q_B and herbicide binding sites are similar in the PSII and bacterial RCs.

A detailed comparison of the mutated residues and the concomittant changes in function also show great similarities between PSII and bacterial RCs. A mutation commonly found in PSII RCs involves a change of Ser^{264} in D1 (Hirschberg and McIntosh 1983, Erickson et al. 1984, Goloubinoff et al. 1984, Hirschberg et al. 1984, Golden and Haselkorn 1985). This residue is homologous to the Ser^{223} of the L subunit in bacterial RCs (Williams et al. 1986). The mutation of Ser^{264} (to either Ala or Gly) in D1 result in a reduced binding constant of herbicides as well as altered electron transfer properties (Hirschberg and McIntosh 1983, Erickson et al. 1984, Hirschberg et al. 1984, Robinson et al. 1987). This is similar to the effects that we observed in the SP223 ($Ser^{223} \rightarrow$ Pro) mutant (see Table 1). Another mutation site, Leu^{275} of the D1 protein (Erickson and Rochaix 1985) is one helical turn away from Leu^{271}, the D1 equivalent (Williams et al. 1986) to Ile^{229} (the IM229 mutation site). These mutation sites may be homologous. No mutation of Tyr^{262}, corresponding to Tyr^{222} of L (Williams et al. 1986), has been found in the D1 protein although other mutations in the D1 protein occur in the vicinity of this residue. Some of these mutations (e.g., Ala^{251}, Val^{219}) involve residues that are not in direct contact with the quinone binding pocket similar to the bacterial YG222 mutant. The homology between the mutated residues in the L subunit and the D1 subunit provide further evidence of the structural and functional similarities between PSII and bacterial RCs (Brown et al. 1984, Williams et al. 1984, Michel et al. 1986, Trebst 1986). Thus, insights gained about the structure and function of bacterial RCs should in general be applicable to higher photosynthetic organisms.

Acknowledgements

We thank JoAnn Williams for assistance with DNA cloning and sequencing during the early stages of this project and many helpful discussions, JP Allen and DC Rees for unpublished structural information on the RC and Jeff Vieira for kindly supplying the pUC119 plasmid and the M13K07 helper phage. This work was supported by a grant from the US Department of agriculture (grant 82-CRCR-1-1043).

References

Allen JP, Feher G, Yeates TO, Komiya H and Rees DC (1987a) Structure of the reaction center from *Rhodobacter sphaeroides* R26: The cofactors. Proc Natl Acad Sci USA 84: 5730–5734

Allen JP, Feher G, Yeates TO, Komiya H and Rees DC (1987b) Structure of the reaction center from *Rhodobacter sphaeroides* R26: The protein subunits. Proc Natl Acad Sci USA 84: 6162–6166

Barber J (1987) Rethinking the structure of the photosystem two reaction center. Trends Biochem Sci 12: 123–124

Birnboim HC and Doly J (1979) A rapid alkaline extraction procedure for screening recombinant plasmid DNA. Nucl Acid Res 7: 1513–1523

Brown AE, Gilbert CW, Guy R and Arntzen CJ (1984) Triazine herbicide resistance in the photosynthetic bacterium *Rhodopseudomonas sphaeroides*. Proc Natl Acad Sci USA 81: 6310–6314

Brudvig GW, Worland ST and Sauer K (1983) Procedure for rapid isolation of photosynthetic reaction centers using cytochrome c affinity chromatography. Proc Natl Acad Sci USA 80: 683–686

Bylina EJ and Youvan DC (1987) Genetic engineering of herbicide resistance: saturation mutagenesis of isoleucine 229 of the reaction center L subunit. Z Naturforsch 42c: 769–774

Cogdell RJ, Monger TG and Parson WW (1975) Carotenoid triplet states in reaction centers from *Rhodopseudomonas sphaeroides* and *Rhodospirillum rubrum*. Biochim Biophys Acta 408: 189–199

Crouse GF, Fritschauf A and Lehrach H (1983) An integrated and simplified approach to cloning into plasmids and single-stranded phages. Meth Enz 101: 78–89

DeVitry C and Diner BA (1984) Photoaffinity labeling of the azidoatrazine receptor site in reaction centers of *Rhodopseudomonas sphaeroides*. FEBS Lett 167: 327–331

Diner BA, Schenck CC and deVitry C (1984) Effect of inhibitors, redox state and isoprenoid chain length on the affinity of ubiquinone for the secondary acceptor binding site in the reaction centers of photosynthetic bacteria. Biochim Biophys Acta 766: 9–20

Erickson JM, Rahire M, Bennoun P, Delepelaire P, Diner BA and Rochaix JD (1984) Herbicide resistance in *Chlamydomonas reinhardtii* results from a mutation in the chloroplast gene for the 32-kilodalton protein of photosystem II. Proc Natl Acad Sci USA 81: 3617–3621

Erickson JM and Rochaix JD (1985) In: Galan GA (ed.) Abstracts. First International Congress of Plant Molecular Biology, p 54/OR-25-02. Athens, USA: The University of Georgia Center for Education for the International Society for Plant Molecular Biology

Erickson JM, Rahire M and Rochaix JD (1985) Herbicide resistance and cross-resistance; changes at three distinct sites in the herbicide-binding protein. Science 228: 204–207

Evans MCW (1987) Plant reaction centre defined. Nature 327: 284–285

Feher G and Okamura MY (1978) Chemical composition and properties of reaction centers. In: Clayton RK and Sistrom WR (eds) The Photosynthetic Bacteria pp. 349–386. New York: Plenum Press

Gilbert CW, Williams JGK, Williams KAL and Arntzen CJ (1985) Herbicide action in photosynthetic bacteria. In: Steinbeck KE, Bonitz S, Arntzen CJ and Bogorad L (eds) Molecular Biology of the Photosynthetic Apparatus, pp 67–71. Cold Spring Harbor: Cold Spring Harbor Laboratory

Golden SS and Haselkorn R (1985) Mutation to herbicide resistance maps within the *psbA* gene of *Anacystis nidulans* R2. Science 229: 1104–1107

Goloubinoff P, Edelman M and Hallick RB (1984) Chloroplast-coded atrazine resistance in *Solanum nigrum*: *psbA* loci from susceptible and resistant biotypes are isogenic except for a single codon change. Nucl Acid Res 12: 9489–9496

Hanahan D (1985) Techniques for transformation of *E. coli*. In: Glover DM (ed) DNA Cloning, Vol 1, pp. 109–135. Washington DC: IRL Press

Hanahan D and Meselson M (1980) Plasmid screening at high colony density. Gene 10, 63–67

Hirschberg J and McIntosh L (1983) Molecular basis of herbicide resistance in *Amaranthus hybridus*. Science 222, 1346–1349

Hirschberg J, Bleecker A, Kyle DJ, McIntosh L and Arntzen CJ (1984) The molecular basis of triazine-herbicide resistance in higher-plant chloroplasts. Z. Naturforsch. 39c, 412–420

Holmes DS and Quigley M (1981) A rapid boiling method for the preparation of bacterial plasmids. Anal Biochem 114, 193–197.

Johanningmeier U, Bodner U and Wildner GF (1987) A new mutation in the gene coding for the herbicide-binding protein in *Chlamydomonas*. FEBS Lett 211, 221–224

Kleinfeld D, Okamura MY and Feher G (1984) Electron transfer in reaction centers of *Rhodopseudomonas sphaeroides*. Determination of the charge recombination pathway of $D^+Q_AQ_B^-$ and free energy and kinetic relations between $Q_A^-Q_B$ and $Q_AQ_B^-$. Biochem Biophys Acta 766, 126–140

Kyle DJ (1985) The 32000 dalton Q_B protein of photosystem II. Photochem Photobiol 41, 107–116

Maniatis T, Fritsch EF and Sambrook J (1982) Molecular Cloning – A Laboratory Manual. Cold Spring Harbor: Cold Spring Harbor Press

McComb JC and Wraight CA (1983) Activity of analogues as primary and secondary quinones in photosynthetic reaction centers (Abstract). Biophys J 41, 39a

McElroy JD, Mauzerall DC and Feher G (1974) Characterization of primary reactants in bacterial photosynthesis. II. Kinetic studies of the light-induced signal (g = 2.0026) and the optical absorbance changes at cryogenic temperatures. Biochem Biophys Acta 333, 261–278

Messing J (1983) New M13 vectors for cloning. Meth Enz 101, 20–78

Messing J and Vieira J (1982) A new pair of M13 vectors for selecting either DNA strand of double-digested restriction fragments. Gene 19, 269–276

Michel H, Epp O and Deisenhofer J (1986a) Pigment-protein interactions in the photosynthetic reaction centre from *Rhodopseudomonas viridis*. EMBO J 5, 2445–2451

Michel H, Weyer KA, Gruenberg H, Dugner I, Oesterhelt D and Lottspeich F (1986b) The 'light' and 'medium' subunits of the photosynthetic reaction centre from *Rhodopseudomonas viridis*: isolation of the genes, nucleotide and amino acid sequence. EMBO J 5, 1149–1158

Michel-Beyerle ME (ed.) (1985) Antennas and Reaction Centers of Photosynthetic Bacteria: Structure, Interactions, and Dynamics. New York: Springer-Verlag

Miller JH (1972) Experiments in Molecular Genetics. Cold Spring Harbor: Cold Spring Harbor Press

Nanba O and Satoh K (1987) Isolation of a photosystem II reaction center consisting of D-1 and D-2 polypetpides and cytochrome b-559. Proc Natl Acad Sci USA 84, 109–112

Okamura MY (1984) On the herbicide site in bacterial reaction centers. In: Thornber JP, Staehelin LA and Hallick RB (eds) Biosynthesis of the Photosynthetic Apparatus: Molecular Biology, Development and Regulation, pp 381–390. New York: Alan R Liss, Inc

Okamura MY, Isaacson RA and Feher G (1975) The primary acceptor in bacterial photosynthesis: the obligatory role of ubiquinone in photoactive reaction centers of *Rhodopseudomonas sphaeroides*. Proc Natl Acad Sci USA 72, 3491–3495

Okamura MY, Debus RJ, Kleinfeld D and Feher G (1982) Quinone binding sites in reaction centers from photosynthetic bacteria. In: Trumpower BC (ed) Functions of Quinones in Energy Conserving Systems, pp 299–317. New York: Academic Press

Okamura MY, Abresch EC and Debus RJ (1985) Reaction centers from triazine resistant strains of *Rhodopseudomonas sphaeroides*: localization of the mutation site by protein hybridization experiments. Biochim Biophys Acta 810, 110–113

Okamura MY, Satoh K, Isaacson RA and Feher G (1987) Evidence of the primary charge separation in the $D_1 D_2$ complex of photosystem II from spinach: EPR of the triplet state. In: Biggins J (ed) Progress in Photosynthesis Research, Vol 1, pp I.4.379–I.4.381. Dordrecht: Martinus Nijhoff Publishers

Ort DR and Melandri BA (1982) Mechanism of ATP synthesis. In: Govindjee (ed.) Photosynthesis: Energy Conversion by Plant and Bacteria, Vol 1, pp 537–587. New York: Academic Press

Ort DR (1986) Energy transduction in oxygenic photosynthesis: an overview of structure and mechanism. In: Staehelin LA and Arntzen CJ (eds) Encyclopedia of Plant Physiology, Vol 19: Photosynthesis III: Photosynthetic Membranes and Light Harvesting Systems, pp 143–196. New York: Springer-Verlag

Paddock ML, Williams JC, Rongey SH, Abresch EC, Feher G and Okamura MY (1987) Characterization of three herbicide resistant mutants of *Rhodopseudomonas sphaeroides* 2.4.1: structure-function relationship. In: Biggins J (ed.) Progress in Photosynthesis Research, Vol 3 pp III.11.775–III.11.778. Dordrecht: Martinus Nijhoff Publishers

Pfister K and Arntzen CJ (1979) The mode of action of photosystem II-specific inhibitors in herbicide-resistant weed biotypes. Z Naturforsch 34c, 996–1009

Rawn JD (1983) Biochemistry. New York: Harper and Row Publishers

Robinson H, Golden S, Brusslan J and Haselkorn R (1987) Functioning of photosystem II in mutant strains of the cyanobacterium *Anacystis nidulans* R2. In: Biggins J (ed.) Progress in Photosynthesis Research, Vol 4, pp IV.12.825–IV.12.828. Dordrecht: Martinus Nijhoff Publishers

Sanger F, Nicklen S and Coulson AR (1977) DNA sequencing with chain-terminating inhibitors. Proc Natl Acad Sci USA 74, 5463–5467

Schenck CC, Sistrom WR and Capaldi RA (1986) structure-function studies in reaction centers: characterization of an herbicide resistance mutation in *Rhodopseudomonas sphaeroides* (Abstract). Biophys J 49, 486a

Sinning I and Michel H (1987) Sequence analysis of mutants from *Rhodopseudomonas viridis* resistant to the herbicide terbutryn. Z Naturforsch 42c, 751–754

Stein RR, Castellvi AL, Bogacz JP and Wraight CA (1984) Herbicide-quinone competition in the acceptor complex of photosynthetic reaction centers from *Rhodopseudomonas sphaeroides*: a bacterial model for PS-II-herbicide activity in plants. J Cell Biochem 24, 243–259

Tischer W and Strotmann H (1977) Relationship between inhibitor binding by chloroplasts and inhibition of photosynthetic electron transport. Biochim Biophys Acta 460, 113–125

Trebst A (1986) The topology of the plastoquinone and herbicide binding peptides of photosystem II in the thylakoid membrane. Z Naturforsch 41c, 240–245

Trebst A (1987) The three-dimensional structure of the herbicide binding niche on the reaction center polypeptides from photosystem II. Z Naturforsch 42c, 742–750

Velthuys BR (1981) Electron-dependent competition between platoquinone and inhibitors for binding to photosystem II. FEBS Lett 126, 277–281

Vermass WFJ, Arntzen CJ, Gu L-Q and Yu C-A (1983) Interactions of herbicide and azidoquinones at a photosystem II binding site in the thylakoid membrane. Biochim Biophys Acta 723, 266–275

Vermaas WFJ, Renger G and Arntzen CJ (1984) Herbicide/quinone binding interactions in photosystem II. Z Naturforsch 39c, 368–373

Warncke K, Gunner MR, Braun BS, Yu C-A and Dutton PL (1987) Effects of hydrocarbon tail structure on the affinity of substituted quinones for the Q_A and Q_B sites in reaction center protein of *Rhodopseudomonas sphaeroides* R26 (Abstract). Biophys J 51, 124a.

Williams JC, Steiner LA, Ogden RC, Simon MI and Feher G (1983) Primary structure of the M subunit of the reaction center from *Rhodopseudomonas sphaeroides*. Proc Natl Acad Sci USA 80, 6505–6509

Williams JC, Steiner LA, Feher G and Simon MI (1984) Primary structure of the L subunit of the reaction center from *Rhodopseudomonas sphaeroides*. Proc Natl Acad Sci USA 81, 7303–7307

Williams JC, Steiner LA and Feher G (1986) Primary structure of the reaction center from *Rhodopseudomonas sphaeroides*. Proteins 1, 312–325

Wraight CA (1981) Oxidation-reduction physical chemistry of the acceptor quinone complex in bacterial photosynthetic reaction centers: evidence for a new model of herbicide activity. Israel J Chem 21, 348–354

Yanisch-Perron C, Vieira J and Messing J (1985) Improved M13 phage cloning vectors and host strains: nucleotide sequences of the M13mp18 and pUC19 vectors. Gene 33, 103–119.

Youvan DC, Bylina EJ, Alberti M, Begusch H, and Hearst JE (1984) Nucleotide and deduced polypeptide sequences of the photosynthetic reaction center, B870 antenna, and flanking polypeptides from *Rhodopseudomonas capsulata*. Cell 37, 949–957

Govindjee et al. (eds), Molecular Biology of Photosynthesis: 343–352
© 1988 Kluwer Academic Publishers

Minireview

Molecular genetics of herbicide resistance in cyanobacteria

JUDY BRUSSLAN & ROBERT HASELKORN

Department of Molecular Genetics and Cell Biology, University of Chicago, Chicago, IL 60637, USA

Received 29 October 1987; accepted 30 December 1987

Key words: cyanobaceria, herbicides, photosystem II, cross-resistance, dominance

Abstract. Cyanobacteria offer an excellent model system for studies of herbicide resistance in higher plants. Mutants resistant to classical and non-classical herbicides have been isolated, and in some cases the amino acid alteration(s) are known. Mutations in plants, algae, photosyntehtic bacteria, and cyanobacteria are compared. Data concerning the question of dominance or recessiveness of herbicide resistance in cyanobacteria are also discussed.

Abbreviations: DCPIP – dichlorophenolindophenol, LD_{50} – lethal dose for 50% killing

Herbicides have been developed during the last century to inhibit processes and pathways which are unique to plants: photosynthesis, chlorophyll biosynthesis, and essential amino acid biosynthesis (Brian 1976). Although not designed for this purpose, these herbicides are also lethal to cyanobacteria which share many of the unique pathways of plants. Indeed, these pathways probably originated in ancient cyanobacteria which were the ancestors of today's chloroplast.

The herbicides which inhibit photosynthesis have been studied extensively. These include the classical herbicides: triazines, ureas, triazinones, and uracils; and the non-classical herbicides which include the phenol-type (Trebst 1987). The classical herbicides block electron transfer between the plastoquinones Q_A and Q_B by competing with Q_B for the Q_B-binding site. Binding competition was demonstrated directly between atrazine and 6-azido-Q_0C_{10}, a synthetic quinone which mimics Q_B (Vermaas et al. 1983). Q_B, the second stable electron acceptor of photosystem II, receives two electrons from Q_A before it leaves the Q_B-binding site as Q_BH_2. When Q_B is only partially reduced, as Q_B^-, its affinity for the Q_B-binding site is high and herbicides do not compete effectively. A binary oscillation of herbicide binding affinity is observed, dependent on the number of single flashes given to intact chloroplasts. After an odd number of flashes, most of the plasto-

Table 1. Cross resistance in herbicide-resistant mutants (level of resistance relative to wild type).

Herbicide	Mutation									
	F211 → S	V219 → 1	A251 → V	F255 → Y	F255 → Y	G256 → E	S264 → A	S264 → A	S264 → G	L275 → F
Diuron	10[f]	17[a]	5[b]	0.5[a]	2[c]	10[d]	100[a]	150[c]	2[e]	5[d]
Atrazine		2	25	15	47	80	10	18	1000	1
Bromacil		1		1		70		300	20	5
Metribuzin			1000					5000	300	
Ioxynil					0.8			0.5	0.6	

[a] Erickson et al. 1985; Chlamydomonas rheinhardtii.
[b] Johanningmeier et al. 1987; Chlamydomonas rheinhardtii.
[c] Hirschberg, personal communication; Synechococcus 7942.
[d] Rochaix and Erickson 1987; Chlamydomonas rheinhardtii.
[e] Arntzen et al. 1982; Amaranthus hybridus.
[f] Buzby et al. 1987; Synechococcus 7002.

References are provided for herbicide-resistant strains in which resistance relative to wild type has been measured. In Synechococcus 7002 there is also a V219 → I change (Gingrich 1987) and in Synechococcus 7942 there is a F255 → L, S264 → A → A double mutation (J Hirschberg, personal communication). Mutants of Synechocystis 6714 have also been isolated (C Astier, personal communication) for which relative resistance values are provided in the text.

quinone is in the semiquinone form, tightly bound to the Q_B-binding site, and the number of herbicide binding sites is decreased (Laasch et al. 1984).

The Q_B-binding site is believed to be located on the D1 protein. The D1 protein forms part of the core of photosystem II which is involved in charge separation. This was demonstrated by isolation of a particle containing the D1 protein, the D2 protein, cytb559, 5 chlorophyll a, 2 pheophytin a, and one β-carotene: this particle is capable of pheophytin photoreduction (Nanba and Satoh 1987). It has been postulated, based on primary structure as well as functional analogy, that the D1 and D2 proteins are homologous to the L and M proteins, respectively, of the *Rhodopseudomonas viridis* reaction center (Trebst 1986). X-ray diffraction data from crystals of reaction centers from photosynthetic bacteria indicate that L and M bind 4 bacteriochlorophylls, 2 bacteriopheophytins, 2 quinones, and one non-heme iron (Deisenhofer et al. 1985), Michel et al. 1986a, Chang et al. 1986, Allen et al. 1987). L and M both cross the photosynthetic membrane five times, and the Q_B-binding site is located between the membrane-spanning helices IV and V of the protein. Homologous helices are probably located on the D1 protein (see below).

The use of triazine herbicides has inevitably led to the selection of herbicide-resistant weeds (Gressel et al. 1982). Weeds resistant to atrazine are also cross resistant to the triazinone herbicide metribuzin and the uracil herbicide bromacil. The D1 protein was implicated as the target of these herbicides due to the labelling of D1 by azido derivatives of atrazine and triazinone (Trebst 1986). Also, herbicide resistance is inherited maternally (Galloway and Mets 1983) which is consistent with D1 being encoded by a plastid gene. Indeed, the cloning and sequencing of a *psbA* gene from the chloroplast DNA of a herbicide-resistant mutant of *Amaranthus hybridus* revealed a single basepair change which produced a D1 protein mutated to gly at ser264 (Hirschberg and McIntosh 1983).

The photosynthetic reaction centers of cyanobacteria are very similar to those of higher plants (Bryant 1986). In fact, the *psbAl* gene product, D1, from *Synechococcus* 7942 has 85% similarity to the *psbA* gene product of *Pisum sativum* (Golden and Haselkorn 1985, Oishi et al. 1984). Photosynthetic herbicide-resistant mutants of cyanobacteria have been isolated (Golden and Sherman 1984, Astier et al. 1986, Buzby et al. 1987, Gingrich et al. 1988, Hirschberg et al. 1987a), and many of these mutants contain the same amino acid changes in the D1 protein that were seen in herbicide-resistant algae and higher plants (see Table 1). In some of these herbicide-resistant mutants, electron transfer from Q_A to Q_B is slowed (Astier et al. 1986, Robinson et al. 1987), as has been observed in herbicide-resistant mutants

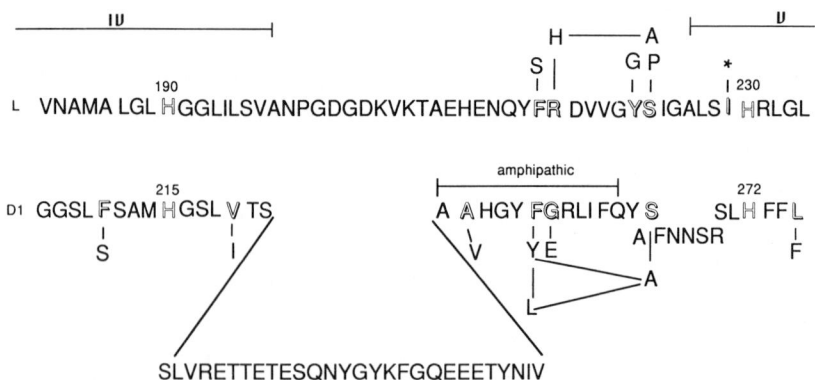

Fig. 1. Comparison of the amino acid sequence of the L subunit of R. viridis (Michel et al. 1986b) and the D1 protein encoded by psbAl of Synechococcus 7942 (Golden and Haselkorn 1985). The histidines which bind the non-heme iron are outlined, and the sequences are aligned as in Trebst 1987. Helices IV and V, as well as a putative amphipathic helix in the D1 protein are indicated. Amino acids in which herbicide resistant mutations have been localized are shadowed, and the amino acid substitution is noted above or below L or D1, respectively. Bars connecting 2 amino acid substitutions indicate double mutations. D1-S264 → A is found as a single or as a double mutation, but D1-F255 → L is found only as a double mutation (J Hirschberg, personal communication). L-S223 → P is found as a single mutation (Paddock et al. 1987), but L-S223 → A has been isolated only as a double mutation (Sinning and Michel 1987). The asterisk above 1229 of the L subunit indicates the site of saturation mutagenesis in R. capsulatus (Bylina and Youvan 1987).

of higher plants and algae (Arntzen et al. 1982, Erickson et al. 1985). For these reasons, cyanobacteria offer a legitimate model for studies of resistance to photosynthetic herbicides.

Cyanobacteria offer many advantages over higher plants and algae due to their fast growth, simple genome organization (Doolittle 1979), ability to be transformed with DNA, and their efficient homologous recombination system (Tandeau de Marsac 1987). A large number of cells can be mutagenized and herbicide-resistant mutants can be selected. If a resistant strain contains an alteration in the DNA then total chromosomal DNA from this strain can be used to transform a herbicide-sensitive strain to herbicide-resistance. In this way, isogenic strains differing solely in the nucleotide(s) responsible for the herbicide resistance phenotype can be constructed. If one is studying a herbicide which affects D1, then the psbA gene(s) from the resistant transformant can be cloned and sequenced in order to determine which amino acid(s) have been changed. The psbA gene from the herbicide-resistant strain can then be used to transform sensitive cells to herbicide resistance, unequivocally demonstrating that alteration of D1 alone results in herbicide resistance (Golden and Haselkorn 1985). Cyanobacterial thylakoids can be isolated and the Hill reaction, measuring electron transfer to

DCPIP, can be used to determine the LD_{50} of many types of herbicides (Astier et al. 1986, Hirschberg et al. 1987a).

Herbicide-resistant mutants of plants, algae, and cyanobacteria; their relative resistance to herbicides and changes in amino acids are listed in Table 1 and Fig. 1. It appears that mutations to herbicide resistance are repeatedly found at certain amino acids (D1-V219, D1-F255, and D1-S264). Two of these amino acids, D1-F255 and D1-S264, have terbutryn-resistant homologues in the L subunit of *R. sphaeroides* and/or *R. viridis* (Paddock et al. 1987, Sinning and Michel 1987) at positions L-F215 and L-S223. In these cases the amino acid changes are different: L-F216 → S instead of D1-F255 → Y and L-S223 → P instead of D1-S264 → A or G. A L-S223 → A has been isolated, but is accompanied by a change at L-R217 → H. In *R. sphaeroides* there is also a change at L-Y222 → G and L-1229 → M.

A decrease in the rate of electron transport from Q_A to Q_B has been osberved in D1-S264 mutants of *Amaranthus hybridus* (Arntzen et al. 1982), *Chlamydomonas rheinhardtii* (Erickson et al. 1985), *Synechococcus* 7942 (Robinson et al. 1987), but not in the grass *Phalaris paradoxa* (Hirschberg et al. 1987b). Electron transfer is also altered in D1-G256 → E, but not in D1-V219 → 1, D1-A251 → V, D1-F255 → Y or D1-L275 → F of *Chlamydomonas rheinhardtii* (Erickson et al. 1985, Rochaix and Erickson 1987). In *R. sphaeroides*, mutants L-Y222 → G and L-S223 → P have altered electron transport, but L-1229 → M has normal Q_A to Q_B electron transfer (Paddock et al. 1987).

The different patterns of cross resistance indicate that D1-S264 is important in the binding of many herbicides. It interacts with diuron and atrazine as well as the triazinones and other uracils. D1-S264 → G is more resistant to atrazine than to diuron, while D1-S264 → A is more resistant to diuron. The different consequences of mutation of D1-S264 → G or → A are not evident in the model shown in Fig. 1. An alternative structure for D1 in which S264 is located within the amphipathic helix, so that mutation to G or A might have distinguishable effects consistent with the data in Table 1, has been proposed by Kleier et al. (1987). D1-F255 interacts with atrazine, and D1-V219 interacts with diuron. The D1-A251 is involved primarily in triazinone and atrazine binding while the D1-L275 mutation is involved in diuron and bromacil binding, although alteration of this amino acid only changes resistance five-fold. The change at D1-F211 affects binding to atrazine. In all of the mutants in which it has been tested, sensitivity to the non-classical phenolic herbicide, ioxynil, is enhanced.

Mutants with different patterns of cross resistance have been isolated in *Synechocystis* 6714 (C. Astier, personal communication). DCMU II_B (Astier et al. 1986) is 300 × resistant to diuron, as compared to wild type, but only

5 × resistant to atrazine. Az-IV is 120 × resistant to diuron and 30 × resistant to atrazine. These patterns resemble D1-S264 → A mutants except that electron transport is not altered. Two mutants of *Synechocystis* 6714, resistant to ioxynil, but sensitive to diuron and atrazine, have also been isolated. Iox-I and Iox-II are both 5 × resistant to ioxynil, but only Iox-I is cross resistant to bromoxynil, another phenolic herbicide. Electron transport is normal in these strains, also. At present, *psbA* and *psbD*, which encode D1 and D2 respectively, are being cloned from the four mutant strains as well as from wild type. The amino acid alterations responsible for these new patterns of resistance should be known soon.

Double mutants have been isolated in *Synechoccocus* 7942 (J. Hirschberg, personal communication). Analysis of each mutation alone or in combination indicates that the effect of some herbicide resistant mutations are additive. For example, compared to wild type, D1-S264 → A is 18 × more resistant to atrazine while D1-F255 → Y is 47 × more resistant to atrazine. The double mutant is 350 × more resistant. For other herbicides this pattern is not observed. For example, resistance to monuron is increased 570 × in the D1-264 mutant and 1.5 × in the D1-255 mutant, yet only 250 × in the double mutant.

The positions of these substitutions on D1 encoded by *psbAl* from *Synechococcus* 7942 and on the L subunit of *R. viridis* are indicated in Fig. 1. One striking difference between L and D1 is in the region between helices IV and V, the Q_B-binding pocket. This region is highly conserved among plants, algae, and cyanobacteria, and thus must serve a specific function. The D1-F211 mutation is one turn of helix IV (four amino acids) from D1-H215 while the D1-L275 mutation is one turn of helix V (three amino acids) from D1-H272. These amino acids must face into the Q_B-binding pocket.

Bylina and Youvan, using *R. capsulatus*, have performed saturation mutagenesis on 1229 of the L subunit, the amino acid residue next to L-H230, which is bound to the non-heme iron. It was found that the amino acid residues L and M at position 229 confer atrazine resistance, while V and A substitutions are herbicide-sensitive. Only these four substitutions are fully competent photosynthetically. These results suggest that hydrophobic residues of moderate size function best at this position. Homologous mutations in algae and cyanobacteria have not yet been isolated, but they could be constructed using site-specific mutagenesis. D1-L271, adjacent to D1-H272, which binds the non-heme iron, would be an excellent candidate for mutagenesis. One of the three glutamic acids at positions D1-242-244 could be involved in binding the non-heme iron (Trebst and Draber 1986), and thus these amino acids ought to be examined as well.

One major question that can be answered by cyanobacterial studies is whether resistance is dominant or recessive to herbicide sensitivity. In plants, algae, and cyanobacteria grown in the light, the D1 protein turns over (Edelman et al. 1984, P. Goloubinoff, personal communication). This turnover is inhibited by diuron in wild type strains, but is unaffected by diuron in resistant strains. It has been believed that a chloroplast or cell containing both diuron-resistant and diuron-sensitive D1 proteins would eventually contain 100% non-functional diuron-sensitive D1 proteins in its reaction centers when the cell is treated with diuron. This is because diuron-sensitive D1 proteins cannot turn over in the presence of diuron. For this reason, diuron sensitivity is expected to be dominant to diuron resistance.

Two different approaches to the question of dominance, both using *Synechococcus* 7942, have been pursued. *Synechococcus* 7942 contains three functional copies of the *psbA* gene (Golden et al. 1986). Three strains, each containing two genes with a wild type D1-S264 and one gene with a D1-S264 → A mutation, in each of the three *psbA* gene positions, were constructed (Brusslan and Haselkorn 1987). The phenotype of these three strains is resistance to diuron, independent of light intensity. Thus the location of the mutant gene in the gene family does not affect cell phenotype, even if the resistance allele is in *psbAIII*, the least abundantly transcribed gene of the family. Fluorescence titration, measuring variable fluorescence one μsec after a saturating flash as a function of diuron concentration, indicates that there are two populations of functional reaction centers in a *psbAIII* diuron-resistant strain of *Synechococcus* 7942. At a low concentration of diuron, 6 nM, approximately one-third of the reaction centers fluoresce with fast-rise kinetics, demonstrating that Q_A to Q_B transfer has been blocked. (This is the diuron concentration required to block transfer in sensitive wild type reaction centers). The other two-thirds of the reaction centers do not fluoresce until a higher concentration of diuron, 180 nM, is reached (Robinson et al. 1987). If there are two populations of D1 proteins in the cell, and the phenotype of the cell is herbicide resistance, then it follows that resistance to diuron is dominant to sensitivity. The implications of this result for the interpretation of the turnover data remain to be determined.

On the other hand, a strain which contains a duplication of the *psbAI* gene has been constructed (Pecker et al. 1987). One allele of the *psbAI* gene is herbicide-resistant while the other allele is herbicide-sensitive. The phenotype of this strain is herbicide sensitivity, independent of light intensity; yet, in vitro Hill reactions measured in this strain indicate an intermediate value of herbicide resistance, and DNA isolated from this strain can be used to transform wild type cells to diuron resistance. The duplication strain has four *psbA* genes, of which only one is a diuron-resistant allele, yet the herbicide-resistant strains isolated in this lab contain one diuron-resistant allele (*psbAI*) and two sensitive alleles (*psbAII* and *psbAIII*) (Hirschberg et al. 1987a). This discrepancy may be caused by a position effect resulting from duplication or by the presence of four *psbA* genes, of which three are herbicide-sensitive. It will be interesting to examine the fluorescence properties of the *psbAI* duplication strain in a manner similar to that done on the *psbAIII* diuron resistant strain mentioned above. The phenotype of a *psbAII* or a *psbAIII* duplication may provide further insight into these conflicting results.

Cyanobacteria can also be used to study other herbicides. Gabaculin prevents cholorphyll biosynthesis by inhibiting glutamyl semialdehyde transaminase, an enzyme involved in an early step of chlorophyll biosynthesis (Kannangara and Schouboe 1985). A mutant of *Synechococcus* 7942 has been isolated which is resistant to a concentration of gabaculin that is 10 fold higher than is required to kill wild type cells (David Paterson, personal communication). The mutation is due to an alteration of the DNA as demonstrated by transformation of a gabaculin-sensitive strain to gabaculin resistance by chromosomal DNA isolated from the resistant strain. An analysis of the gene conferring herbicide resistance should indicate whether the transaminase itself is mutated, whether there is a mutation which results in increased expression of the transaminase as occurs in *Chlamydomonas rheinhardtii* (Kahn and Kannangara 1987), or whether a different gene confers gabaculin resistance.

References

Allen JP, Feher G, Yeates O, Komiya H and Rees DC (1987) Structure of the photosynthetic reaction center from *Rhodobacter sphaeroides* R-26: the cofactors. Proc Natl Acad Sci USA 84: 5730–5734

Arntzen CJ, Pfister K and Steinback KE (1982) The mechanism of chloroplast triazine resistance: alterations in the site of herbicide action. In: LeBaron HM and Gressel J (eds) Herbicide Resistance in Plants, pp 185–214. New York: John Wiley and Sons

Astier C, Meyer I, Vernotte C and Etienne AL (1986) Photosystem II electron transfer in highly herbicide resistant mutants of *Synechocystis* 6714. FEBS Lett 207: 234–238

Brian RC (1976) The history and classification of herbicides. In: LJ Audus (ed) Herbicides: Physiology, Biochemistry, Ecology, pp 1–54. London: Academic Press

Brusslan J and Haselkorn R (1987) Herbicide resistance in the *psbA* multigene family of *Synechococcus* PCC 7942. In: Von Wettstein D and Chua NH (eds) Plant Molecular Biology. New York: Plenum Publ pp. 367–375

Bryant D (1986) The cyanobacterial photosynthetic apparatus: comparisons to those of higher plants and photosynthetic bacteria. In: Platt T and Li WKW (eds) Photosynthetic Picoplankton. Canad Bull Fisheries Aquatic Sci 214: 423–500

Buzby JS, Mumma RO, Bryant DA, Gingrich J, Hamilton RH, Porter RD, Mullin CD and Stevens SE Jr (1987) Genes with mutations causing herbicide resistance from the cyanobacterium PCC 7002. In: Biggins J (ed) Progress in Photosynthesis Research, Vol IV, pp 757–760. Dordrecht: Martinus Nijhoff Publishers

Bylina EJ and Youvan DC (1987) Genetic engineering of herbicide resistance: saturation mutagenesis of isoleucine 229 of the reaction center L subunit. Z Naturforsch 42c: 751–754

Chang CH, Tiede D, Tang J, Smith U, Norris JR and Schiffer M (1986) Structure of *Rhodopseudomonas sphaeroides* R-26 reaction center. FEBS Lett 205: 82–86

Deisenhofer J, Epp O, Miki K, Huber R and Michel H (1985) Structure of the protein subunits in the photosynthetic reaction centre of *Rhodopseudomonas viridis* at 3A resolution. Nature 318: 619–624

Doolittle WF (1979) The cyanobacterial genome, its expression, and the control of that expression. Adv Microbiol Physiol 20: 1–102

Edelman M, Mattoo AK and Marder JB (1984) Three hats of the rapidly-metabolized 32 kilodalton protein of the thylakoids. In: Ellis RJ (ed) Chloroplast Biogenesis, pp 283–302. Cambridge: Cambridge University Press

Erickson JM, Rahire M, Rochaix J-D and Mets L (1985) Herbicide resistance and cross-resistance: changes at three distinct sites in the herbicide-binding protein. Science 228: 204–207

Galloway RE and Mets LJ (1983) Atrazine, bromacil, and diuron resistance in *Chlamydomonas*. Plant Physiol 74: 469–474

Gingrich J, Buzby JS, Stirewalt VL, and Bryant DA (1988) Genetic analysis of two new mutations resulting in herbicide resistance in the cyanobacterium *Synechococcus* PCC 7002. This volume.

Golden SS and Sherman LA (1984) Biochemical and biophysical characterization of herbicide-resistant mutants of the unicellular cyanobacterium, *Anacystis nidulans* R2. Biochim Biophys Acta 764: 239–246

Golden SS and Haselkorn R (1985) Mutation to herbicide resistance maps within the *psbA* gene of *Anacystis nidulans* R2. Science 229: 1104–1107

Golden SS, Brusslan J and Haselkorn R (1986) Expression of a family of *psbA* genes encoding a photosystem II polypeptide in the cyanobacterium *Anacystis nidulans* R2. EMBO J 5: 2789–2798

Gressel J, Ammon HU, Fogelfors H, Gasquez J, Kay QON and Kees H (1982) Discovery and distribution of herbicide-resistant weeds outside North America. In: LeBaron HM and Gressel J (eds) Herbicide Resistance in Plants, pp 33–55. New York: John Wiley and Sons

Hirschberg J and McIntosh L (1983) Molecular basis of herbicide resistance in *Amaranthus hybridus*. Science 222: 1346–1348

Hirschberg J, Ohad N, Pecker I and Rahat A (1987a) Isolation and characterization of herbicide resistant mutants in the cyanobacterium *Synechococcus* R2. Z Naturforsch 42c: 758–761

Hirschberg J, Yehuda AB, Pecker I and Ohad N (1987b) Mutations resistant to photosystem II herbicides. In: Von Wettstein D and Chua NH (eds) Plant Molecular Biology, pp 357–366. New York: Plenum Publ

Johanningmeier U, Bodner U and Wildner GF (1987) A new mutation in the gene coding for the herbicide-binding protein in *Chlamydomonas*. FEBS Lett 211: 221–224

Kahn A and Kannangara C Gamini (1987) Gabaculine-resistant mutants of *Chlamydomonas rheinhardtii* with elevated glutamate 1-semialdehyde aminotransferase activity. Carlsberg Res Commun 52: 73–81

Kannangara C Gamini and Schouboe A (1985) Biosynthesis of Δ-aminolevulinate in greening barley leaves. VII Glutamate 1-semialdehyde accumulation in gabaculine treated leaves. Carlsberg Res Commun 50: 179–191

Kleier DA, Andrea A, Hegedus JKJ, Gardner GN, and Cohen B (1987) The topology of the

32 KD herbicide binding protein of photosystem II in the thylakoid membrane. Z. Naturforsch. 42c: 733–738

Laasch H, Schreiber U and Urbach W (1984) Binding of radioactively labelled DCMU in dependence of the redox state of the photosystem II acceptor side. In: Sybesma C (ed.) Adv Photosyn Res, Vol IV, pp 25–28. The Hague: Martinus Nijhoff Publ

Michel H, Epp O and Deisenhofer J (1986a) Pigment-protein interactions in the photosynthetic reaction centre from *Rhodopseudomonas viridis*. EMBO J 5: 2445–2451

Michel H, Weyer KA, Greenberg H, Dunger I, Oesterhelt D and Lottspeich F (1986b) The 'light' and 'medium' subunits of the photosynthetic reaction centre from *Rhodopseudomonas viridis*: isolation of the genes, nucleotide and amino acid sequence. EMBO J 5: 1149–1158

Nanba O and Satoh K (1987) Isolation of a photosystem II reaction center consisting of D-1 and D-2 polypeptides and cytochrome b-559. Proc Natl Acad Sci USA 84: 109–112

Oishi KK, Shapiro DR and Tewari KK (1984) Sequence organization of a pea chloroplast DNA gene coding for a 34,500-dalton protein. Mol Cell Biol 4: 2556–2563

Paddock ML, Williams JC, Rongey SH, Abresch EC, Feher G and Okamura MY (1987) Characterization of three herbicide-resistant mutants of *Rhodopseudomonas sphaeroides* 2.4.1: structure-function relationship. In: Biggins J (ed.) Prog Photosyn Res, Vol III, pp 775–778. Dordrecht: Martinus Nijhoff Publ

Pecker I, Ohad N and Hirschberg J (1987) The chloroplast-encoded type of herbicide resistance is a recessive trait in cyanobacteria. In: Biggins (ed.) Prog Photosyn Res, Vol III, pp 811–814. Dordrecht: Martinus Nijhoff Publ

Robinson H, Golden SS, Brusslan J and Haselkorn R (1987) Functioning of photosystem II in mutant strains of the cyanobacterium *Anacystis nidulans* R2. In: Biggins J (ed.) Prog Photosyn Res, Vol IV, pp 825–828. Dordrecht: Martinus Nijhoff Publ

Rochaix J-D and Erickson JM (1988) Function and assembly of photosystem II: genetic and molecular analysis. Trends in Biochem 13: 56–59

Sinning I and Michel H (1987) Sequence analysis of mutants from *Rhodopseudomonas viridis* resistant to the herbicide terbutryn. Z Naturforsch 42c: 751–754

Tandeau de Marsac N and Houmard J (1987) Advances in cyanobacterial molecular genetics In: Fay P and VanBaalen C (eds) The Cyanobacteria: A Comprehensive Review. Elsevier Scientific Publ, in press

Trebst A and Draber W (1986) Inhibitors of photosystem II and the topology of the herbicide and Q_B binding polypeptide in the thylakoid membrane. Photosyn Res 10: 381–392

Trebst A (1987) The three-dimensional structure of the herbicide binding niche on the reaction center polypeptides of photosystem II. Z Naturforsch 42c: 742–750

Vermaas WFJ, Arntzen CJ, Gu L-Q and Yu C-A (1983) Interactions of herbicides and azidoquinones at a photosystem II binding site in the thylakoid membrane. Biochim Biophys Acta 723: 266–275

Govindjee et al. (eds), Molecular Biology of Photosynthesis: 353–369
© 1988 Kluwer Academic Publishers

Regular paper

Genetic analysis of two new mutations resulting in herbicide resistance in the cyanobacterium *Synechococcus* sp. PCC 7002

JEFFREY C. GINGRICH,[1] JEFFREY S. BUZBY,[2] VERONICA L. STIREWALT & DONALD A. BRYANT[3]
Department of Molecular and Cell Biology, Penn State University, University Park, PA 16802 USA; [1] Present address: Chemical Biodynamics Division, Lawrence Berkeley Laboratory, University of California, Berkeley, CA 94720 USA; [2] Present address: Department of Biology, University of California, Los Angeles, CA 90024 USA; [3] Author for correspondence

Received 6 October 1987; accepted 3 December 1987

Key words: cyanobacteria, photosynthesis, herbicide resistance, D1 protein, *psbA* gene, *Synechococcus* sp. PCC 7002

Abstract. Two herbicide-resistant strains of the cyanobacterium *Synechococcus* sp. PCC 7002 are compared to the wild-type with respect to the DNA changes which result in herbicide resistance. The mutations have previously been mapped to a region of the cyanobacterial genome which encodes one of three copies of *psbA*, the gene which encodes the 32 kDa Q_b-binding protein also known as D1 (Buzby et al. 1987). The DNA sequence of the wild-type gene was first determined and used as a comparison to that of the mutant alleles. A point mutation at codon 211 in the *psbA1* coding locus (T*T*C to T*C*C) results in an amino acid change from phenylalanine to serine in the D1 protein. This mutation confers resistance to atrazine and diuron at seven times and at two times the minimal inhibitory concentration (MIC) for the wild-type, respectively. A mutation at codon 211 resulting in herbicide resistance has not previously been described in the literature. A second point mutation at codon 219 in the *psbA1* coding locus (G*T*A to *A*TA) results in an amino acid change from valine to isoleucine in the D1 protein. This mutation confers resistance to diuron and atrazine at ten times and at two times the MIC for the wild-type, respectively. An identical codon change conferring similar herbicide resistance patterns has previously been described in *Chlamydomonas reinhardtii*. The atrazine-resistance phenotype in *Synechococcus* sp. PCC 7002 was shown to be dominant by plasmid segregation analysis.

Abbreviations: At^R – atrazine resistance, Du^R – diuron resistance, Km^R – kanamycin resistance, Ap^R – ampicillin resistance, MIC – minimum inhibitory concentration

Introduction

Members of the triazine class of herbicides (e.g., atrazine and 3-(3,4-di-chlorophenyl)-1,1-dimethylurea (DCMU or diuron) have been shown to disrupt transfer of electrons within the Photosystem II reaction center by competing for the second quinone (Q_b) binding site within the reaction center (Kyle 1985). Photoaffinity labeling of photosynthetic membranes with azido-atrazine has shown that the herbicide-binding domain is on a 32 kDa protein known as D1 (Pfister et al. 1981). In eukaryotes, the gene for D1 (*psbA*) is chloroplast-encoded and its sequence has been determined for a number of plants (Zurawski et al. 1982, Link and Langridge 1984, Oishi et al. 1984, Spielman and Stutz 1984, Sugita and Sugiura 1984) and algae (Erickson et al. 1984, Karabin et al. 1984). In cyanobacteria D1 is encoded by a small multigene family. Derived amino acid sequences for D1 in cyanobacteria are approximately 90% homologous to the proteins encoded by chloroplasts (Curtis and Haselkorn 1984, Mulligan et al. 1984, Golden et al. 1986).

Resistance to diuron and the triazine class of herbicides has been shown to occur as the result of single amino acid substitutions in D1. In higher plants herbicide-resistance has been shown to occur as the result of a change at residue 264 from a serine to a glycine (Hirschberg and McIntosh 1983, Goloubinoff et al. 1984). In *Chlamydomonas reinhardtii* and in *Synechococcus* sp. PCC 7942, a serine to alanine change in residue 264 results in herbicide resistance (Erickson et al. 1985, golden and Haselkorn 1985). In *C. reinhardtii* changes in residues 219, 251, 255, 256, and 275 have also been shown to result in herbicide resistance (Erickson et al. 1985, Johanningmeier et al. 1987, Rochaix and Erickson 1987). Models for the herbicide-binding domain in D1 are becoming more refined as additional mutations resulting in herbicide resistance are characterized.

In order to analyze additional genetic lesions which result in herbicide resistance and to define further the region of the D1 protein involved in herbicide binding, we have chosen to study herbicide resistance in a transformable cyanobacterial strain, *Synechococcus* sp. PCC 7002. Cyanobacteria offer an ideal model system in which to study the process of oxygenic photosynthesis. These prokaryotes contain both of the photosystems characteristic of higher plants (Bryant 1987). The genes studied to date which encode proteins of the photosynthetic apparatus are generally 80–90% identical to their higher plant counterparts (Bryant 1987). Furthermore, *Synechococcus* sp. PCC 7002 is a transformable photoheterotroph (Stevens and Porter 1980, Lambert and Stevens, 1986); hence mutations can be constructed and analyzed by complementation — methods not yet available in eukaryotic systems. *Synechococcus* sp. PCC 7002 provides two additional advantages as compared to plants: the generation time for this species is about four hours, and large numbers of cells can easily be manipulated or screened by standard bacteriological methods.

In a previous paper we described the isolation and characterization of spontaneous mutants in *Synechococcus* sp. PCC 7002 which are resistant to a number of different herbicides (Buzby et al. 1987) Two of these mutants, one resistant to atrazine and one resistant to diuron, were further characterized. The genetic loci conferring herbicide resistance were cloned from these mutants by complementation of the wild-type strain using the biphasic shuttle vector plasmid pAQE19 (Buzby et al. 1985). The DNA fragments which conferred herbicide-resistance to the wild-type, a *Hin*dIII fragment (Du^R-1) and an *Eco*RI fragment (At^R-1), both mapped to a region of DNA which encodes one of three copies of the gene for the D1 protein (*psbA1*) in *Synechococcus* sp. PCC 7002.

In this paper we describe the sequence of the wild-type *psbA1* gene and the nucleotide changes found in the mutant genes. Neither of these mutations has been previously reported in cyanobacteria, and one of them has not been reported from any source previously. These mutations are compared to previously described mutations and are discussed relative to their location within the predicted secondary structure of the D1 protein.

Materials and methods

Materials. Restriction endonucleases, T4 DNA ligase, DNA polymerase I, and the Klenow fragment of DNA polymerase I were purchased from Bethesda Research Laboratories (Gaithersburg, MD). (α-^{32}P)dATP and (α-^{35}S)dATP were purchased from New England Nuclear, Corp. (Boston, MA). Nitrocellulose was purchased from Schleicher and Schuell, Inc. (Keene, NH). Phosphoramidites, columns, and other reagents for DNA synthesis were purchased from Fisher Scientific Co. (Pittsburgh, PA).

Strains and Culture Conditions. Wild-type *Synechococcus* sp. PCC 7002 (*Agmenellum quadruplicatum* strain PR-6; Rippka et al. 1979) was grown at 37 °C in liquid culture medium A with 1 mg ml^{-1} NaNO$_3$ (Stevens and Porter 1980). The cultures were continuously illuminated and supplemented with 1% CO$_2$ in air. Recombinant plasmids containing the herbicide-resistance genes in the biphasic cloning vector pAQE19 (Buzby et al. 1985) were isolated as previously described (Buzby et al. 1987) and propagated in *E. coli* strain RDP145 (Buzby et al. 1983) by virtue of their ability to confer resistance to kanamycin.

Cloning and Sequencing Methods. *Synechococcus* sp. PCC 7002 DNA was isolated from exponential phase cultures in liquid medium as described (de Lorimier et al. 1984). A phage lambda-Charon 4A library of *Eco*RI fragments of *Synechococcus* sp. PCC 7002 DNA was prepared by the method described (Murphy et al. 1987). Recombinant lambda phage containing the wild-type *psbA* genes were isolated from this library by plaque hybridization (Maniatis et al. 1982) using a nick-translated *psbA* gene probe from *Amaranthus hybridus* (kindly provided by Dr. J.G.K. Williams, E.I. du Pont de Nemours and Co., Wilmington, DE). All hybridizations included at least a 6-hour prehybridization in 6X SET (1X SET = 0.15 M NaCl, 0.03 M Tris-HCl, 1 mM EDTA, pH 8.0)-5X Denhardt's solution (Maniatis et al. 1982)-0.5% SDS. Hybridizations were performed in the same solution at 62° overnight, and the filters were washed four times for thirty minutes at room temperature with 6X SET. Hybridization-positive lambda clones were purified, and DNA was prepared from these clones as described (Maniatis et al. 1982, Silhavy et al. 1984). The cyanobacterial DNA inserts were further mapped by restriction enzyme digestion and Southern blot hybridization. Appropriate DNA fragments from the clones were further subcloned into the plasmid vector pUC18 (Yanisch-Perron et al. 1985).

DNA sequencing of the pUC18 subclones was performed by the dideoxy, plasmid-sequencing method on base-denatured templates as described by Cantrell and Bryant (1987). All template DNAs were purified by centrifugation on two CsCl-ethidium bromide equilibrium gradients (Maniatis et al. 1982). Oligonucleotide sequencing primers (15 nucleotides in length) were synthesized on an Applied Biosystems Model 380A DNA synthesizer.

Results and discussion

In a previous communication we described the isolation and characterization of a number of spontaneous mutant strains of *Synechococcus* sp. PCC 7002 which are resistant to different herbicides (Buzby et al. 1987). These herbicides include the PSII-targeted herbicides atrazine (Pfister et al. 1981), diuron (Mattoo et al. 1981), and bromoxynil (Moreland 1980); diclofop-methyl, an herbicide that affects fatty acid biosynthesis (Hoppe 1985); and acifluorfen, a herbicide that affects ATP synthase and/or carotenoid biosynthesis (Boeger 1984). Total DNA from two of the herbicide-resistant strains, one selectively resistant to atrazine and one selectively resistant to diuron, was isolated and digested with restriction endonucleases *Eco*RI and *Hind*III, respectively, and inserted into the biphasic cloning vector pAQE19 (Buzby et al. 1985). This vector plasmid is capable of replication in either *Synechococcus* sp. PCC 7002 or *E. coli* and carries both K_m^R and A_p^R selectable markers. Wild-type *Synechococcus* sp. PCC 7002 was transformed with recombinant plasmids carrying DNA from the mutant strains. Double selection on kanamycin (which selects for the plasmid vector) and the

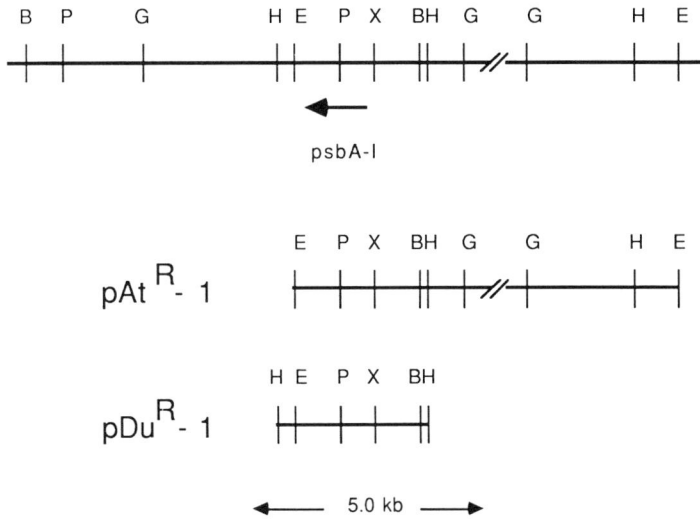

Fig. 1. Partial restriction enzyme digest maps of the region of *Synechococcus* sp. PCC 7002 DNA encoding *psbA1*(top) and the inserts in plasmids pAt^R-1 (middle) and pDu^R-1 (bottom). The location and orientation of *psbA1* is indicated. pAt^R-1 confers atrazine resistance to the wild-type and pDu^R-1 confers diuron resistance to the wild-type. Ddiagonal lines between the *Bg*III sites at the right side of the *psbA1* indicate a 10 kb break in the map shown. B = *Bam*HI; E = *Eco*RI; G = *Bg*III; H = *Hind*III; P = *Pst*I; X = *Xba*I.

herbicide (which selects for the herbicide-resistance gene) allowed the recovery of two transformed strains which carry plasmids capable of conferring resistance to the herbicides atrazine and diuron.

The inserts of the plasmids carry cyanobacterial DNA fragments corresponding to a region of *Synechococcus* sp. PCC 7002 DNA which encodes one of three copies (Gingrich and Bryant, unpublished results) of the gene for the D1 protein. Figure 1 shows the restriction endonuclease map of the region surrounding this *psbA* allele, denoted *psbA1*, and the map of the cyanobacterial DNA inserts contained in the atrazine and diuron resistance-conferring plasmids. The recombinant plasmid which carries DNA capable of transforming the wild-type to atrazine-resistance, from here on referred to as pAt^R-1, contains a 17.5 kb *Eco*RI fragment of cyanobacterial DNA. The *psbA1* gene is encoded near one end of this *Eco*RI fragment. Plasmid pAt^R-1 confers resistance to atrazine at a level of $10\,\mu g\,ml^{-1}$ and resistance to diuron at $0.5\,\mu g\,ml^{-1}$. These concentrations are seven times and two times the minimum inhibitory concentration (MIC) for the wild-type strain. The second plasmid, capable of transforming wild-type to diuron-resistance (from here on referred to as pDu^R-1), contains a 2.8 kb *Hind*III fragment of cyanobacterial DNA. The *psbA1* gene is near the center of this fragment (see Figure 1). Plasmid pDu^R-1 confers resistance to diuron at a level eight times

```
XbaI                                                              -250
TCTAGACGAG CTTGGGCAGT CGCTAGGGTC AACCAACGAT CCTGACCTGG ATCAACACGA AAATTGCAAT
                                 -200
ATGAACAGCG ATTGAAACAT TCGTAAGTTG GCACCAAGGT GAAAGCTGGA CTGTAGGTCA CAACAGCGTT
            -150                                                  -100
TGCCTGGTTA CTGCCGCCGC CTACCGGGCG ATCAACCCTT GACAGATCCA GCATGGCCCC TCCGCAAATC
                                        -50
AAAACATTTC CTTAAAACAC TTTACAAAAA GCAATAGTTT CATTAATATA GCTACTAAGG TTAAGCACCT
                                                                    30
TGATAATTTC ATAGCAATCA TAACGAT ATG ACT ACT ACA CTA CAA CAG CGC GGA AGC GCT TCC
                            Met Thr Thr Thr Leu Gln Gln Arg Gly Ser Ala Ser
                            60                                      90
TTG TGG GAG AAG TTC TGT CAG TGG ATC ACA AGC ACC GAG AAC CGC ATC TAT GTC GGT TGG
Leu Trp Glu Lys Phe Cys Gln Trp Ile Thr Ser Thr Glu Asn Arg Ile Tyr Val Gly Trp
                    120                                     150
TTC GGC GTC CTG ATG ATT CCT ACC CTT CTC ACT GCT ACT ACC TGC TTC ATC ATT GCG TTC
Phe Gly Val Leu Met Ile Pro Thr Leu Leu Thr Ala Thr Thr Cys Phe Ile Ile Ala Phe
                    180                                     210
ATC GCA GCT CCT CCC GTT GAC ATC GAC GGT ATC CGT GAG CCC GTC GCA GGT TCT CTT CTC
Ile Ala Ala Pro Pro Val Asp Ile Asp Gly Ile Arg Glu Pro Val Ala Gly Ser Leu Leu
                    240                                     270
TAC GGT AAC AAC ATC GTC TCT GGC GCA GTT GTA CCT TCT TCT AAC GCA ATT GGT CTC CAC
Tyr Gly Asn Asn Ile Val Ser Gly Ala Val Val Pro Ser Ser Asn Ala Ile Gly Leu His
                    300                                     330
TTC TAC CCC ATC TGG GAA GCT GCT TCC TTA GAT GAG TGG TTG TAC AAC GGT GGC CCT TAC
Phe Tyr Pro Ile Trp Glu Ala Ala Ser Leu Asp Glu Trp Leu Tyr Asn Gly Gly Pro Tyr
                    360                                     390
CAG TTG GTA ATT TTC CAC TTC CTC ATT GGT GTT TTC TGC TAC ATG GGT CGT GAG TGG GAA
Gln Leu Val Ile Phe His Phe Leu Ile Gly Val Phe Cys Tyr Met Gly Arg Glu Trp Glu
                    420                                     450
CTT TCT TAC CGC CTC GGT ATG CGT CCC TGG ATC TGT GTT GCG TTC TCT GCT CCC GTA GCA
Leu Ser Tyr Arg Leu Gly Met Arg Pro Trp Ile Cys Val Ala Phe Ser Ala Pro Val Ala
                    480                                     510
GCA GCA ACT GCA GTA TTC CTC ATC TAC CCC ATC GGT CAA GGT TCC TTC TCT GAT GGT ATG
Ala Ala Thr Ala Val Phe Leu Ile Tyr Pro Ile Gly Gln Gly Ser Phe Ser Asp Gly Met
                    540                                     570
CCT TTG GGT ATT TCT GGT ACG TTC AAC TTC ATG ATC GTA TTC CAG GCA GAG CAC AAC ATC
Pro Leu Gly Ile Ser Gly Thr Phe Asn Phe Met Ile Val Phe Gln Ala Glu His Asn Ile
                    600                                     630
CTG ATG CAC CCC TTC CAC ATG CTT GGT GTG GCT GGT GTA TTC GGT GGT TCT TTG TTC TCC
Leu Met His Pro Phe His Met Leu Gly Val Ala Gly Val Phe Gly Gly Ser Leu Phe Ser
                    660                                     690
GCA ATG CAC GGT TCT CTC GTA ACC TCT TCT TTG GTA CGT GAG ACC ACT GAA ACC GAA TCT
Ala Met His Gly Ser Leu Val Thr Ser Ser Leu Val Arg Glu Thr Thr Glu Thr Glu Ser
                    720                                     750
CAG AAC TAC GGT TAC AAG TTC GGT CAA GAG GAA GAA ACT TAC AAC ATC GTT GCA GCC CAC
Gln Asn Tyr Gly Tyr Lys Phe Gly Gln Glu Glu Glu Thr Tyr Asn Ile Val Ala Ala His
                    780                                     810
GGC TAC TTC GGT CGT TTG ATC TTC CAA TAT GCA TCT TTC AAC AAC AGC CGT TCC TTG CAC
Gly Tyr Phe Gly Arg Leu Ile Phe Gln Tyr Ala Ser Phe Asn Asn Ser Arg Ser Leu His
                    840                                     870
TTC TTC TTG GGT GCA TGG CCT GTA GTC GGT ATC TGG TTC ACT GCT CTT GGT GTA TCT ACC
Phe Phe Leu Gly Ala Trp Pro Val Val Gly Ile Trp Phe Thr Ala Leu Gly Val Ser Thr
                    900                                     930
ATG GCA TTC AAC CTG AAC GGT TTC AAC TTC AAC CAG TCC ATC CTT GAC TCT CAA GGT CGT
Met Ala Phe Asn Leu Asn Gly Phe Asn Phe Asn Gln Ser Ile Leu Asp Ser Gln Gly Arg
                    960                                     990
GTA ATC AAC ACC TGG GCG GAC ATT CTG AAC CGT GCG AAC CTC GGT TTT GAA GTA ATG CAC
Val Ile Asn Thr Trp Ala Asp Ile Leu Asn Arg Ala Asn Leu Gly Phe Glu Val Met His
GAG CGT AAC GCT CAC AAC TTC CCC TTA GAC TTA GCA GCT GGC GAG CAA GCA CCT GTG GCT
Glu Arg Asn Ala His Asn Phe Pro Leu Asp Leu Ala Ala Gly Glu Gln Ala Pro Val Ala
                    1080                                    1110
CTG CAA GCA CCT GCA ATC AAC GGT TAA TAACTGT TTGAGTCTTC GGGCTTAGTT GAGAAAAGAT
Leu Gln Ala Pro Ala Ile Asn Gly
                                            1160
GATTGCTTTG ACAGTAAAAC TGAATAAGTA ATGATTGGCG AAGGCCCCTA CCAGAGATGG TGGGGGTTTT
        1210                                                1260
TGCTTGCTTT TTTTGGGTC AATTTCGGAT TGTTGGCATT TTCAGCTAAG GAAAATTGGG ATTTTAGCTA
                                            1310
GATGCTGCTG AGCAAATATT TTCATCATTC ACGTTTGTAA TGGCTGAGGC GTCAAACTCG
        EcoR1
CCACCACTGAATTC
```

and atrazine to a level 1.7 times the MIC for wild-type ($2.0\,\mu g\,ml^{-1}$ and $2.5\,\mu g\,ml^{-1}$, respectively).

In order to determine the mutations present in pAt^R-1 and pDu^R-1, we have cloned and sequenced the wild-type copy of the *psbA1* gene (the same allele which confers herbicide resistance in the mutants). The wild-type copy of the gene was obtained from a library of *Eco*RI fragments of *Synechococcus* sp. PCC 7002 DNA inserted in the bacteriophage lambda vector Charon 4A. Approximately five thousand phage were screened by plaque hybridization. Seventeen phage were isolated which hybridized to a *psbA*-specific probe derived from the *A. hybridus* gene. Of these seventeen, three recombinant phages contained a 17.5 kb DNA insert with a restriction digestion map which was indistinguishable from that of the DNA insert harbored in pAt^R-1. The remaining phages contained inserts which carry one of the other two *psbA* alleles. From one of the lambda clones containing *psbA1*, a 1.64 kb *Eco*RI-*Xba*I fragment which hybridized to the *psbA* gene probe was subcloned into the plasmid cloning vector pUC18. This fragment was sequenced on both strands by the Sanger, dideoxy method without further subcloning. Synthetic oligonucleotide primers were used to prime synthesis following the initial determination of the sequence using the M13 forward and reverse sequencing primers. Primers were synthesized corresponding to the sequence determined at a spacing of approximately 200 nucleotides along both strands (see Fig. 5). An open reading frame homologous to other *psbA* genes was found to lie entirely within the pUC18 clone. the sequence of the entire *Eco*RI-*Xba*I fragment which was subcloned into pUC18 is shown in Fig. 2.

The *psbA1* coding sequence starts 308 nucleotides from the *Xba*I restriction site and ends 254 nucleotides before the *Eco*RI site. The coding sequence is 360 codons in length, identical to other cyanobacterial *psbA* genes which have been sequenced (Curtis and Haselkorn 1984, Mulligan et al. 1984, Golden et al. 1986). As indicated in Fig. 2, sequences which resemble the consensus *E. coli* promoter sequences can be found in the sequence upstream from the translational start codon for the *psbA1* gene. The position of the sequences indicated relative to mRNA initiation sites determined by S1 nuclease mapping for other cyanobacterial *psbA* genes is

---

Fig. 2. DNA sequence of the region of DNA encoding *psbA1* from *Synechococcus* sp. PCC 7002. The DNA sequence shown is that of an *Xba*I–*Eco*RI fragment cloned in pUC18. The sequence translated is that of the *psbA1* gene product and is based upon homology to other *psbA* gene products. Putative transcription initiation sequences analogous to *E. coli* promoter sequences are underlined (approx. -50 and -70). Inverted repeat sequences downstream from the gene are underlined (nucleotides 1158 to 1193) which may act to terminate transcription or stabilize the mRNA.

similar — generally 50 to 90 nucleotides from the translation start codon (Curtis and Haselkorn 1984, Mulligan et al. 1984, Golden et al. 1986). All proposed transcription initiation signals appear to be similar to the consensus *E. coli* promoter sequences (Curtis and Haselkorn 1984, Mulligan et al. 1984, Golden et al. 1986). The putative *Synechococcus* sp. PCC 7002 *psbA1* "-35" promoter sequences (TTTACA) is identical to two of the three -35 promoter sequences in *Synechococcus* sp. PCC 7942, and the third sequence differs by only one nucleotide (TTCACA; Golden et al. 1986). The putative "-10" sequence from *Synechococcus* sp. PCC 7002 (TAATAT) is less related to the "-10" sequences from *Synechococcus* sp. PCC 7942 (*psbA1* = TCTCCT; *psbA2* = TAGATT; *psbA3* = TATTAT; Golden et al. 1986). Inverted repeat sequences which may act as transcription termination signals or which may play a role in mRNA stabilization are found at the 3' end of the gene. Similar inverted repeat sequences are found in the 3' flanking sequences of all other *psbA* genes which have been sequenced (Curtis and Haselkorn 1984, Mulligan et al. 1984, Golden et al. 1986). The *psbA* genes in *Synechococcus* sp. PCC 7002 are not cotranscribed with other genes, since Northern blot analysis indicates that all *psbA* transcripts in this cyanobacterium have a length of approximiately 1.2 kb (Gasparich and Bryant, unpublished results).

The predicted amino acid sequence for the D1 protein encoded by the *psbA1* gene is shown in Fig. 3. The sequence is compared to D1 proteins from other cyanobacteria (*Synechococcus* sp. PCC 7942 (Golden et al. 1986) and *Anabaena* sp. PCC 7120 (Curtis and Haselkorn 1984)); to those from higher plants (spinach and tobacco (Zurawski et al. 1982)); and to those from eukaryotic algae (*Chlamydomonas reinhardtii* (Erickson et al. 1984) and *Euglena gracilis* (Karabin et al. 1984)). The predicted amino acid sequences of these widely divergent species are from 86% to 93% identical. The *Synechococcus* sp. PCC 7002 *psbA1* gene product is most similar to other cyanobacterial D1 proteins (88 to 93% identity), and is less homologous to eukaryotic D1 proteins (86–88% identity), as derived from the gene sequences. The region of least homology among D1 proteins is found in the 30 amino acids at the amino termini. Since the amino-terminal sequences are so divergent in D1 proteins, there has been speculation that the translation of the protein starts at amino acid residue Met-37. Recent evidence does indicate, however, that the amino termini contain these 30 amino acids in the mature polypeptide (Yoram et al. 1987). The cyanobacterial *psbA* gene products are also highly homologous in their carboxyl termini. However, this region is quite different from that found in the eukaryotic D1 proteins, where a portion of the sequence is either deleted or consists of a different sequence, or both (see Fig. 3). Apart from the two terminal regions of the

```
                  10         20         30         40         50         60
Syn 7002-I    MTTTLQQRGS ASLWEKFCQW ITSTENRIYV GWFGVLMIPT LLTATTCFII AFIAAPPVDI
Syn 7942-1    SI REQRR DNV DR  E    V    D                  I    V
Syn 7942-2,3  A  R E      QQ   E    V    D    L             I    V
Anab 625          S    NV  R    T                     A    V          V
Anab 620      A    K    NV  Q   E         N  L I       A
Spin/Nic      AI ER E   E    GR   N           L I           SV
C.reinhardii  AI ER EN S    AR   E            L I     I    C     SV
E.gracilis    ISPV KKYAR P    YR   A VA KK    L              A  V

                  70         80         90        100        110        120
Syn 7002-I    DGIREPVAGS LLYGNNIVSG AVVPSSNAIG LHFYPIWEAA SLDEWLYNGG PYQLVIFHFL
Syn 7942-1              M    I                                              V
Syn 7942-2,3            M    I                                              V
Anab 625                I    I
Anab 620                I    I
Spin/Nic           S        I    II T A                V              E IVL
C.reinhardii       S        IT    I T                                  IVC
E.gracilis         S    F   IT    T                         T          IVC   F

                 130        140        150        160        170        180
Syn 7002-I    IGVFCYMGRE WELSYRLGMR PWICVAFSAP VAAATAVFLI YPIGQGSFSD GMPLGISGTF
Syn 7942-1    L IS    Q                   Y      LS  F
Syn 7942-2,3                              Y
Anab 625          CA  L  Q                Y      L S
Anab 620      T    L                      L         T
Spin/Nic      L  A          F          A  Y
C.reinhardii  L  Y          F          A  Y      S    V
E.gracilis      ICS         F          A  Y      S   IV   L

                 190        200        210        220        230        240
Syn 7002-I    NFMIVFQAEH NILMHPFHML GVAGVFGGSL FSAMHGSLVT SSLVRETTET ESQNYGYKFG
Syn 7942-1       F
Syn 7942-2,3     F
Anab 625                                               I
Anab 620                                               N
Spin/Nic                                       I    N    A E   R
C.reinhardii                                   I    N    A E   R
E.gracilis                                     L    N    I V

                 250        260        270        280        290        300
Syn 7002-I    QEEETYNIVA AHGYFGRLIF QYASFNNSRS LHFFLGAWPV VGIWFTALGV STMAFNLNGF
Syn 7942-1                                                      SM I
Syn 7942-2,3                                        A           S  I
Anab 625                                  Q        A    I
Anab 620                            H     Q        A    I
Spin/Nic                                           A              I
C.reinhardii                                       A    I    L
E.gracilis       I     A                           AV

                 310        320        330        340        350        360
Syn 7002-1    NFNQSILDSQ GRVINTWADI LNRANLGFEV MHERNAHNFP LDLAAGEQAP VALQAPAING
Syn 7942-1         V          K     V      M                    AT    T  S H
Syn 7942-2,3       V                V      M                    AT    T
Anab 625                            I      M                    V     T
Anab 620                            I      M                    V     IS
Spin/Nic      VV                    I      M              I --- ---- ST
C.reinhardii  VV           L        I      M              STNSSS NN--------
E.gracilis    VI                           M             ------ ----------
```

Fig. 3. Amino acid sequences of the D1 protein as derived from nucleotide sequences. Syn 7002 = *Synechococcus* sp. PCC 7002. Syn 7942-1 and 7942-2,3 = *Synechococcus* sp. PCC 7942 (previously designated *Anacystis nidulans* R2; Golden et al. 1986). Genes 2 and 3 predict identical amino acid sequences for the Di protein. Anab 625 and 620 = *Anabaena* sp. PCC 7120. 625 and 620 are two of four *psbA* genes found in *Anabaena* sp. PCC 7120 (Curtis and Haselkorn 1984). Spin/Nic = *Spinacea oleracea* (Spin) and *Nicotiana debneyi* (Nic) which are identical in amino acid sequence (Zurawski et al. 1982). *C. reinhardii* = *Chlamydomonas reinhardtii* (Erickson et al. 1984). *E. gracilis* = *Euglena gracilis* (Karabin et al. 1984).

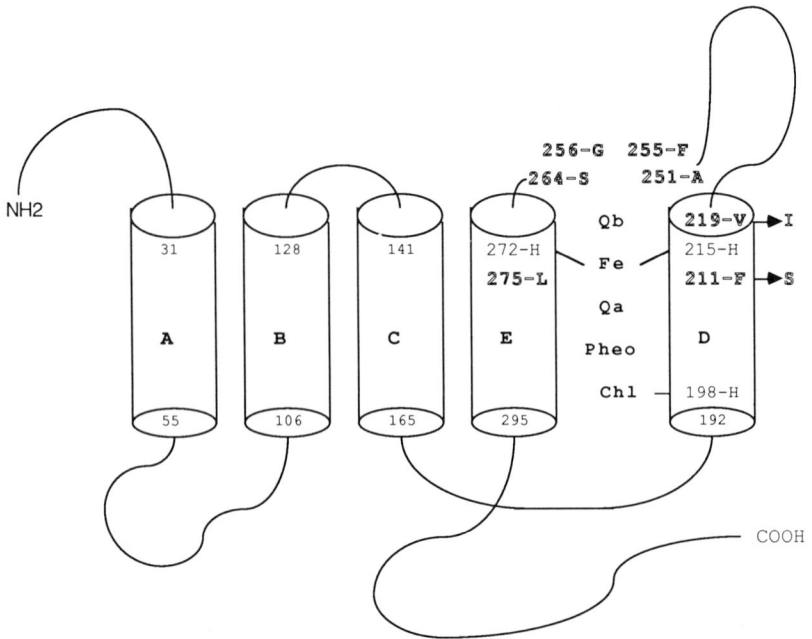

Fig. 4. Structural model for the D1 protein. This model is based upon models previously published (Sayre et al. 1986, Trebst and Draber 1986, Kleier et al. 1987, Trebst 1987). the protein consists of five trans-membrane helices (A through E). Residues which delinete the helices are indicated (e.g., 31 and 55). Residues proposed to be involved in binding the special chlorophyll p680 (H-198) and the non-heme iron (H-215 and H-272) are also indicated. Residues changed resulting in herbicide resistance are in outline.

protein, the D1 proteins are remarkably conserved, including essentially invariant regions from residues 85 to 115, 160 to 230, and 240 to 275. All of these regions are proposed to be exposed on the membrane surface (see Fig. 4) and are likely to include important functional domains of the protein.

All proposed chlorophyll and non-heme iron binding residues in the protein (e.g., His 198, 215, and 272, see Fig. 4) are also conserved amino acids in the *Synechococcus* sp. PCC 7002 *psbA1* gene product. In addition, the number and relative position of charged residues at the carboxyl terminus of the protein, which may be involved in manganese binding, are present in the *Synechococcus* sp. PCC 7002 *psbA1* gene product.

In the D1 protein, a number of different amino acid substitutions from residues 219 to 275 have previously been shown to confer herbicide resistance to a number of cyanobacterial, algal, and higher plant species (Hirschberg and McIntosh 1983, Goloubinoff et al. 1984, Golden and Haselkorn 1985, Johanningmeier et al. 1987, Rochaix and Erickson 1987). With this in

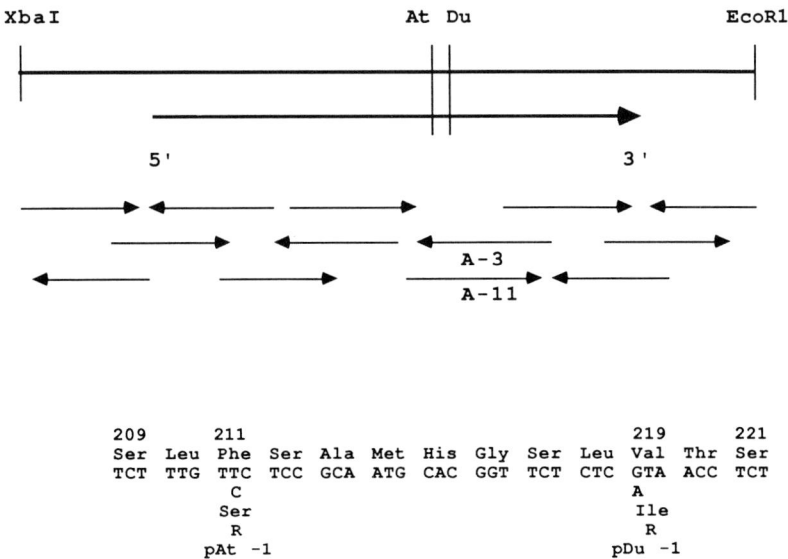

Fig. 5. Sequencing strategy for the *psbA1* gene from *Synechococcus* sp. PCC 7002. The top of the figure shows the location of the wild-type gene with respect to the *Eco*RI–*Xba*I fragment which was sequenced. At and Du indicate the location of the mutations which confer resistance to atrazine (At) and diuron (Du). The arrows below the gene indicate the sequence obtained and the strand from which the sequence was obtained using oligonucleotide primers. A-3 and A-11 indicate the sequence from these primers used to sequence the mutant alleles contained on pAt^R-1 and pDu^R-1. The DNA sequence below indicates the wild-type sequence from codons 209 to 221 and the changes present on pAt^R-1 and pDu^R-1 which result in herbicide resistance.

mind, we sequenced this region of the mutant alleles carried on pAt^R-1 and pDu^R-1. For this purpose we chose synthetic oligonucleotide primers (A-1 and A-11 in Fig. 5) which hybridize to the wild-type *psbA1* gene near this region. These oligonucleotides were then used to prime synthesis on pAt^R-1 and pDu^R-1 to determine their DNA sequence in this region. The sequencing strategy and results are shown in Figs 5 and 6. The DNA sequence determined on pAt^R-1 shows a single nucleotide change with respect to the wild-type in codon 211. This nucleotide change T*T*C to T*C*C, results in a phenylalanine to serine change in the herbicide-resistant protein. In pDu^R-1, a single base change in codon 219, *G*TA to *A*TA, results in a valine to isoleucine amino acid change in the herbicide-resistant protein. A summary of these and other published mutations which result in herbicide resistance is shown in Table 1.

Changes in the *psbA* gene product which result in herbicide resistance have been found to occur in two regions of the D1 protein. One of these

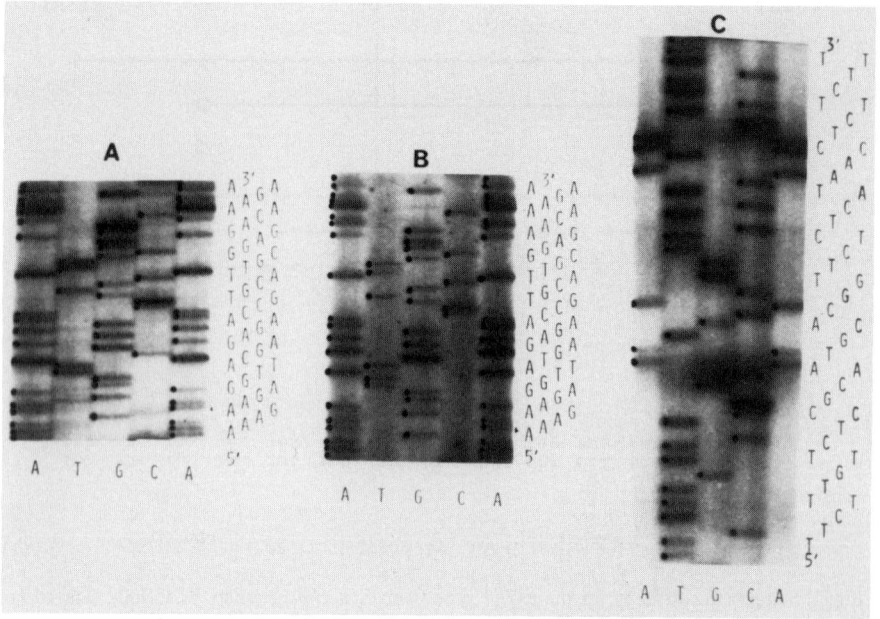

Fig. 6. Autoradiogram of DNA sequencing gel demonstrating the nucleotide sequence changes in the *psbAl* alleles borne by plasmids pAt^R-1 and pDu^R-1. The region shown corresponds to nucleotides 624 to 668 in Fig. 2. A. Nucleotide sequence obtained on pAt^R-1 template using oligonucleotide primer A-3 (see Fig. 3). B. Nucleotide sequence obtained on pDu^R-1 template using oligonucleotide primer A-3. C. Nucleotide sequence obtained on pDu^R-1 template using oligonucleotide primer A-11 (see Fig. 5). The sequence shown in panel C is the complement of that shown in panel B.

regions is from residues 251 to 275 while the other is in two residues, 211 and 219. Models for the structure of the D1 protein in the membrane, including the location of the mutations which result in herbicide resistance within this structure, are shown in Figs 4 and 7. These models are based upon hydropathy profiles of the protein (Trebst and Draber 1986, Kleier et al. 1987), antibody binding studies (Sayre et al. 1986), and analogy to the *Rhodopseudomonas viridis* reaction center structure which has been determined to a 3 Å resolution (Deisenhofer et al. 1985, Michel et al. 1986). The mutations described here, residues 211 and 219, are predicted to be either within (211) or at one end (219) of a membrane-spanning alpha helix, helix D in Figs 4 and 7. Both of these mutations occur near a histidine residue (215) which is proposed to coordinate a non-heme iron atom. The non-heme iron atom in turn helps coordinate the binding of a mobile plastoquinone electron carrier, Q_b. Herbicides such as atrazine and diuron act at this site of photosynthetic electron transport by preventing transfer of electrons to Q_b. The changes at residues 211 and 219 may affect binding of the herbicide directly. Alter-

Table 1. Summary of mutations causing herbicide resistance

Organism	Herbicide		Amino acid change
	AT	DU	
Syn. sp. PCC 7002	7X	2X	211 Phe to Ser
	2X	10X	219 Val to Ile
C. reinhardii	2X	17X	219 Val to Ile
Syn. sp. PCC 7942	10X	100X	264 Ser to Ala
C. reinhardii	100X	10X	264 Ser to Ala
A. hybridus	1000X	1X	264 Ser to Gly
S. nigrum	1000X	1X	264 Ser to Gly
C. reinhardii	25X	5X	251 Ala to Val
C. reinhardii	15X	0.6X	255 Phe to Tyr
C. reinhardii	84X	10X	256 Gly to Asp
C. reinhardii	1X	5X	275 Leu to Phe

Summary showing residue changes in the *psbA* gene product (Q_b-binding protein) which lead to resistance to the herbicides atrazine (AT) and diuron (DU). The resistance factors are relative to the minimal inhibitory concentration for the wild-type organism. References: Hirschberg and McIntosh (1983), Goloubinoff et al. (1984), Erickson et al. (1985), Golden and Haselkorn (1985), Buzby et al. (1987), Johanningmeier et al. (1987), Rochaix and Erickson (1987).

natively, the mutations may change the conformation of the protein in the vicinity of histidine-215 which in turn would affect the binding of the herbicides.

A mutation at residue 219 which confers DCMU-resistance in the eukaryotic alga *Chlamydomonas reinhardtii* has previously been reported (Erick-

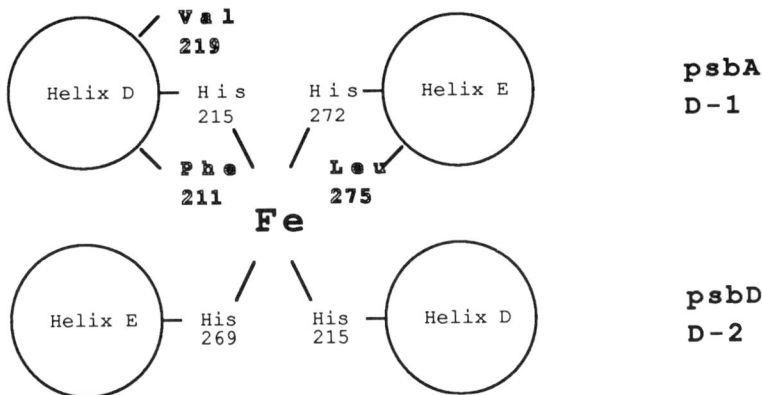

Fig. 7. Proposed model for the histidine residues involved in coordination of the non-heme iron atom (based on Trebst and Draber 1986, Trebst 1987). View is looking down at the plane of the membrane at trasmembrane alpha helices D and E of the D1 and D2 proteins. Histidine residues are indicated. In outline are the residues changed which result in herbicide resistance. Their organization relative to the histidine residues is based on a 100° shift in orientation of residues along an alpha helix.

son et al. 1985; see Table 1). This mutation results in the same amino acid substitution described here for *Synechococcus* sp. PCC 7002. In addition, both of these mutations result in a similar phenotype: resistance to diuron at 10–17 times that of the wild-type with a cross-tolerance to atrazine at twice the wild-type level (see Table 1).

Based on the model shown in Fig. 7, the change at residue 211 is at a similar position on helix D as the change in *C. reinhardtii* at residue 275 in helix E (Rochaix and Erickson 1987). In *Synechococcus* sp. PCC 7002 the change at residue 211 confers resistance to atrazine at seven times and diuron at two times the minimal inhibitory concentration for the wild-type. The mutation at residue 275 in *C. reinhardtii* confers resistance to diuron at five times the level of the wild-type with no cross-tolerance to atrazine (see Table 1). This difference may be reflective of changes in different regions of the binding site for the different herbicies, since both atrazine and diuron are proposed to have overlapping binding sites on the D1 protein (Trebst and Draber 1986).

To date, a serine to alanine or glycine change in residue 264 has been found to produce the most profound effect with respect to herbicide tolerance (Table 1). In plants, this change appears to cause specific resistance to atrazine, but in cyanobacteria and eukaryotic algae the change also results in significant cross-tolerance to diuron. The cause of these differences is unclear at present. Mutations in residues 251, 255, or 256 in *C. reinhardtii* show less dramatic resistance to herbicides than that at residue 264 (see Table 1). Each of these mutations is in a region of the D1 protein which is proposed to be exposed on the stromal side of the membrane between the membrane spanning helices D and E of the protein (see Fig. 4).

In higher plants resistance to atrazine is maternally inherited (Darr et al. 1981). This result is consistent with the occurrence of the *psbA* gene, encoding the $32\,kDa\,Q_b$ binding protein, in the chloroplast genome (Kyle 1985). Attempts to define whether the herbicide resistance phenotype is dominant or recessive in cyanobacteria have generated conflicting results. Pecker et al. (1987) have reported that atrazine resistance is phenotypically recessive in *Synechococcus* sp. PCC 7942 (Strain R2). However, Brusslan et al. (1987) concluded that diuron resistance was dominant in the same strain.

Dominance of the atrazine-resistance phenotype was examined in *Synechococcu* sp. PCC 7002 by allowing $At^R\,Km^R$ transformants, in which these markers were borne on plasmid pAt^R-1, to segregate under nonselective growth conditions. The resulting colonies were tested for resistance to both kanamycin and atrazine by replica plating. Of those colonies which had lost resistance to kanamycin (88/200), 85% (75/88) had also become atrazine-sensitive. This result indicates that these transformants were phenotypically heterozygous for the *psbA* gene and that the atrazine-resistant *psbA1* allele is, therefore, dominant. The transformants (13/88) which remained atrazine-resistant following segregation were probably the result of gene conversion events between the plasmid and chromosomal *psbA1* alleles. Gene conversion events of this type have previously been observed to occur in *Synechococcus* sp. PCC 7002. Porter et al. (1986) reported that transformants heterozygous for an Arg⁻ pheotype were converted back to wild-type phenotype in more than 85% of the segregants under similar conditions.

We are now in the process of isolating additional herbicide-resistant strains of *Synechococcus* sp. PCC 7002 which have distinctive resistance and cross-tolerance patterns to several photosystem II-targeted herbicides. We hope that characterization of such mutants will allow us to understand better the herbicide-binding domains in the photosystem II reaction center.

Acknowledgements

The authors wish to thank Prof. David M. Prescott, Dept. of MCD Biology, University of Colorado, Boulder, in whose laboratory some of this work was carried out. This work was supported in part by an NIH postdoctoral fellowship (GM 09402-03) awarded to J.C.G. and NSF grant DMB-8504294 awarded to D.A.B.

References

Boeger P (1984) Multiple modes of action of diphenyl ethers. Z Naturforsch 39c: 468–475

Brusslan JA, Golden SS and Haselkorn R (1987) Diuron resistance in the *psbA* multigene family of *Anacystis nidulans* R2. In: Biggins J (ed.) Progress in Photosynthesis Research, vol. IV, pp 821–824. Dordrecht: Martinus Nijhoff/Dr W Junk Publishers

Bryant DA (1987) The cyanobacterial photosynthetic apparatus: Comparison to those of higher plants and photosynthetic bacteria. In: Platt T and Li WKW (eds) Photosynthetic Picoplankton, pp 423–500. Ottawa, Canadian Bulletin of Fisheries and Aquatic Sciences Vol. 214, Dept Fisheries and Oceans

Buzby JS, Porter RD and Stevens SE Jr (1983) Plasmid transformation in *Agmenellum quadruplicatum* PR-6: Construction of biphasic plasmids and characterization of their transformation properties. J Bacteriol 154: 1446–1450

Buzby JS, Porter RD and Stevens SE Jr (1985) Expression of the *Escherichia coli lacZ* gene on a plasmid vector in a cyanobacterium. Science 230: 805—807

Buzby JS, Mumma RO, Bryant DA, Gingrich J, Hamilton RH, Porter RD, Mullion CA, and Stevens SE Jr (1987) Genes with mutations causing herbicide resistance from the cyanobacterium *Synechococcus* sp. PCC 7002. In: Biggins J (ed.) Progress in Photosynthesis Research, Vol IV, pp 757–760. Dordrecht: Martinus Nijhoff/Dr W Junk Publishers

Cantrell A and Bryant DA (1987) Molecular cloning and nucleotide sequence of the *psaA* and *psaB* genes of the cyanobacterium *Synechococcus* sp. PCC 7002. Plant Mol Biol 9: 453–468

368

Curtis SE and Haselkorn R (1984) Isolation, sequence and expression of two members of the 32 kd thylakoid membrane protein gene family from the cyanobacterium *Anabaena* 7120. Plant Mol Biol 3: 249–258

Darr S, Machado VS and Arntzen CJ (1981) Uniparental inheritance of a chloroplast photosystem II polypeptide controlling herbicide binding. Biochim Biophys Acta 634: 219–228

Deisenhofer J, Epp O, Miki K, Huber R and Michel H (1985) Structure of the protein subunits in the photosynthetic reaction center of *Rhodopseudomonas viridis* at 3 Å resolution. Nature 318: 618–624

De Lorimier R, Bryant DA, Porter RD, Liu W-Y, Jay E and Stevens SE Jr (1984) Genes for the α and β subunits of phycocyanin. Proc Nat Acad Sci USA 81: 7946–7950

Erickson JM, Rahire M and Rochaix J-D *Chlamydomonas reinhardii* gene for the 32,000 mol. wt. protein of photosystem II contains four large introns and is located entirely within the chloroplast inverted repeat. EMBO J 3: 2753–2762

Erickson JM, Rahire M, Rochaix J-D and Mets L (1985) Herbicide resistance and cross-resistance: Changes at three distinct sites in the herbicide-binding protein. Science 228: 204–207

Golden SS and Haselkorn R (1985) Mutation to herbicide resistance maps within the psbA gene of *Anacystis nidulans* R2. Science 229: 1104–1107

Golden SS, Brusslan J and Haselkorn R (1986) Expression of a family of psbA genes encoding a photosystem II polypeptide in the cyanobacterium *Anacystis nidulans* R2. EMBO J 5: 2789–2798

Goloubinoff P, Edelman M and Hallick RB (1984) Chloroplast-encoded atrazine resistance in *Solanum nigrum*: psbA loci from susceptible and resistance biotypes are isogenic except for a single codon change. Nucl Acids Res 12: 9489–9496

Hirschberg J and McIntosh L (1983) Molecular basis of herbicide resistance in *Amaranthus hybridus*. Science 222: 1346–1349

Hoppe HH (1985) Differential effect of diclofop-methyl on fatty acid biosynthesis in leaves of sensitive and tolerant plant species. Pestic Biochem Physiol 23: 297–308

Johanningmeier J, Bodner U and Wildner GF (1987) A new mutation in the gene for the herbicide-binding protein in *Chlamydomonas*. FEBS Lett 211: 221–224

Karabin GD, Farley M and Hallick RB (1984) Chloroplast gene for M_r 32,000 polypeptide of photosystem II in *Euglena gracilis* is interrupted by four introns with conserved boundary sequences. Nucl Acids Res 12: 5801–5812

Kleier DA, Andrea TA, Hegedus KJ, Gardner GM and Cohen B (1987) The topology of the 32 kDa herbicide binding protein of photosystem II in the thylakoid membrane. Z Naturforsch 42c: 733–738

Kyle DJ (1985) The 32,000 dalton Q_b protein of photosystem II. Photochem Photobiol 41: 107–116

Lambert DH and Stevens SE Jr (1986) Photoheterotrophic growht of *Agmenellum quadruplicatum* PR-6. J Bacteriol 165: 654–656

Link G and Langridge U (1984) Structure of the chloroplast gene for the precursor of the M_r 32,000 photosystem II protein from mustard (*Sinapis alba* L.). Nucl Acids Res 12: 945–958

Maniatis T, Fritsch EF and Sambrook J (1982) Molecular Cloning: A Laboratory Manual. Cold Spring Harbor: Cold Spring Harbor Laboratory Publications

Mattoo AK, Pick U, Hoffman H and Edelman M (1981) The rapidly metabolized 32,000-dalton polypeptide of the chloroplast is the "proteinaceous shield" regulating photosystem-II electron transport and mediating diuron herbicide sensitivity. Proc Nat Acad Sci USA 78: 1572–1576

Michel H, Epp O and Deisenhofer J (1986) Pigment-protein interactions in the photosynthetic reaction centre from *Rhodopseudomonas viridis*. EMBO J 5: 2445–2451

Moreland DE (1980) Mechanisms of action of herbicides. Ann Rev Plant Physiol 31: 597–638

Mulligan B, Schultes N, Chen L and Bogorad L (1984) Nucleotide sequence of a multiple-copy gene for the B protein of photosystem II of a cyanobacterium. Proc Nat Acad Sci USA 81: 2693–2697

Murphy RC, Bryant DA, Porter RD and Tandeau de Marsac N (1987) Molecular cloning and characterization of the *recA* gene from the cyanobacterium *Synechococcus* sp. PCC 7002. J Bacteriol 169: 2739–2747

Oishi KK, Shapiro DR and Tewari KK (1984) Sequence organization of a pea chloroplast DNA gene coding for a 34,500-dalton protein. Mol Cell Biol 4: 2556–2563

Pecker I, Ohad N, and Hirschberg J (1987) The chloroplast-encoded type of herbicide resistance is a recessive trait in cyanobacteria. In: Biggins J (ed.) Progress in Photosynthesis Research, Vol. III, pp 811–814. Dordrecht: Martinus Nijhoff/Dr W Junk Publishers

Pfister K, Steinback KE, Gardner G and Arntzen CJ (1981) Photoaffinity labelling of an herbicide receptor protein in chloroplast membranes. Proc Natl Acad Sci USA 78: 981–985

Porter RD, Buzby JS, Pilon A, Fields PI, Dubbs JM and Stevens SE Jr (1986) Genes from the cyanobacterium *Agmenellum quadruplicatum* isolated by complementation: Characterization and production of merodiploids. Gene 41: 249–260

Rippka R, Deruelles J, Waterbury JB, Herdman M and Stanier RY (1979) Generic assignments, strain histories, and properties of pure cultures of cyanobacteria. J Gen Microbiol 111: 1–61

Rochaix J-D and Erickson JM (1987) Genetic and molecular analysis of photosystem II. Trends Biochem Sci. (in press)

Sayre RT, Andersson B and Bogorad L (1986) The topology of a membrane protein: The orientation of the 32 kg Q_b-binding chloroplast thylakoid protein. Cell 47: 601–608

Silhavy TJ, Berman ML and Enquist LW (1984) Experiments with Gene Fusions. Cold Spring Harbor: Cold Spring Harbor Laboratory Publications

Spielman A and Stutz E (1984) Nucleotide sequence of soybean chloroplast DNA regions which contain *psbA* and *trnH* genes and cover the ends of the large single copy region and one end of the inverted repeats. Nucl Acids Res 11: 7157–7167

Stevens SE Jr and Porter RD (1980) Transformation in *Agmenellum quadruplicatum*. Proc Natl Acad Sci USA 77: 6052–6056

Sugita M and Sugiura M (1984) Nucleotide sequence and transcription of the gene for the 32,000 dalton thylakoid membrane protein from *Nicotiana tabacum*. Mol Gen Genet 195: 308–313

Trebst, A (1987) The three-dimensional structure of the herbicide binding niche on the reaction center polypeptide of photosystem II. Z Naturforsch 42c: 742–750

Trebst, A and Draber, W (1986) Inhibitors of photosystem II and the topology of the herbicide and Q_b binding polypeptide in the thylakoid membrane. Photosyn Res 10: 381–392

Yanisch-Perron C, Viera J and Messing J (1985) Improved M13 phage cloning vectors and host strains: Nucleotide sequences of the M13mp18 and pUC19 vectors. Gene 33: 103–119

Yoram E, Goloubinoff P and Edelman M (1987) The amino terminal region delimited by Met_1 and Met_{37} is an integral part of the 32 kDa herbicide binding protein. Plant Mol Biol 8: 337–343

Zurawski G, Bohnert HJ, Whitfeld PR and Bottomley W (1982) Nucleotide sequence of the gene for the M_r 32,000 thylakoid membrane protein from *Spinacea oleracea* and *Nicotiana debneyi* predicts a totally conserved translation product of M_r 38,950. Proc Natl Acad Sci USA 79: 7699–7703

Govindjee et al. (eds), Molecular Biology of Photosynthesis: 371–387
© 1988 Kluwer Academic Publishers

Regular paper

Nucleotide sequence of the genes encoding cytochrome b-559 from the cyanelle genome of *Cyanophora paradoxa*

AMANDA CANTRELL & DONALD A. BRYANT[1]
Department of Molecular and Cell Biology, S-101 Frear Building, Penn State University, University Park, PA 16802 USA; [1]to whom correspondence should be sent

Received 10 October 1987; accepted 22 December 1987

Key words: Cytochrome b-559, Photosystem II, *Cyanophora paradoxa*, nucleotide sequence, chloroplasts, cyanelles

Abstract. *Cyanophora paradoxa* is a flagellated protozoan which possesses unusual, chloroplast-like organelles referred to as cyanelles. The *psbE* and *psbF* genes, which encode the two apoprotein subunits of cytochrome b-559, have been cloned from the cyanelle genome of *C. paradoxa*. The complete nucleotide sequences of these genes and their flanking sequences were determined by the chain-termination, dideoxy method. The *psbE* gene is composed of 75 codons and predicts a polypeptide of 8462 Da that is seven to nine residues smaller than most other *psbE* gene products. The *psbF* gene consists of 43 codons and predicts a polypeptide of 4761 Da. Two open reading frames, whose sequences are highly conserved among cyanobacteria and numerous higher plants, were located in the nucleotide sequence downstream from the *psbF* gene. The first open reading frame, denoted *psbI*, is composed of 39 codons, while the second open reading frame, denoted *psbJ*, is composed of 41 codons. The predicted amino acid sequences of the *psbI* and *psbJ* gene products predict proteins of 5473 and 3973 Da respectively. These proteins are probably integral membrane proteins anchored in the membrane by a single, transmembrane alpha helix. The *psbEFIJ* genes are probably co-transcribed and constitute an operon as found for other organisms. Each of the four genes is preceded by a polypurine sequence which resembles the consensus ribsosome binding sequences for *Escherichia coli*.

Introduction

Perhaps the most elusive and enigmatic component of the thylakoid membranes of oxygenic photosynthetic organisms is cytochrome b-559 (for reviews see Cramer and Whitmarsh 1977, Butler 1978, Cramer and Crofts 1982, Bendall 1982, Satoh 1985 and Bryant 1987). As determined by immunoelectron microscopy, this heme-containing membrane protein is almost exclusively located in the grana stacks of the chrloroplasts in higher plants (Rao et al. 1986). This result is consistent with those obtained by fractionation methods in which the cytochrome co-purifies with fractions highly enriched in Photosystem II (Berthold et al. 1981, Murata et al. 1984). More recent studies have clearly established that cytochrome b-559 is an intrinsic component of the most highly purified Photosystem II reaction center preparations from cyanobacteria (Satoh et al. 1985, Yamagishi and Satoh 1985, Ohno et al. 1986), *Chlamydomonas reinhardtii* (De Vitry et al. 1987), and higher plants (Nanba and Satoh 1987, Barber et al. 1987, Ghanotakis et al. 1987, Satoh et al. 1987, Yamada et al. 1987).

Cytochrome b-559 can exist in both high-potential ($+350\,\text{mV}$) and low-potential ($+110\,\text{mV}$) forms (e.g., see Butler 1978, or Hervas et al. 1985). The precise stoichiometry of cytochrome b-559 relative to the P680 reaction center of Photosystem II is still not known with certainty. In less highly purified, oxygen-evolving preparations, most workers have found approximately equal amounts of the high-potential and low-potential forms of the cytochrome and have reported approximately two hemes per P680 center (e.g., see Murata et al. 1984, Yamamoto et al. 1984, and Murata and Miyao 1985). However, several other groups have found only one heme per P680 center (Ghanotakis et al. 1984; see discussion in Ghanotakis and Yocum 1985). Uniform ^{14}C-labelling of the polypeptides of the PS II core complex of *Dunaliella salina* has recently provided strong evidence that there is a single cytochrome b-559 molecule per P680 reaction center core complex (Gounaris et al. 1987).

Although many theories have been put forth, the actual physiological role of cytochrome b-559 remains unknown. Possible functions include the following: (1) a redox function in the water oxidation process; (2) a function in electron and/or proton transfer in the major transport pathways; (3) a function in cyclic electron flow around the Photosystem II reaction center; (4) an electron transport function along a side pathway from the plastoquinone pool; (5) a function as a proton acceptor from the water oxidation complex; and (6) a structural role in Photosystem II (Cramer and Whitmarsh 1977, Butler 1978, Cramer and Crofts 1982, Bendall 1982, Falkowski et al. 1986, Thompson et al. 1986). Whatever the true function(s) of the cytochrome in PS II, it is clear from gene inactivation studies in the cyanobacterium *Synechococcus* sp. PCC 6803 that the cytochrome does play an essential structural and/or functional role in the PS II reaction center (Pakrasi et al. 1987). It is also clear that the high-potential form of the cytochrome is not required for the water oxidation process (Briantais et al. 1985, Ghanotakis et al. 1987).

The apoproteins of cytochrome b-559 and the genes that encode them, designated *psbE* and *psbF*, have recently been extensively characterized. Metz et al. (1983) and Widger et al (1984) identified a 9–10 kDa polypeptide associated with maize and spinach cytochrome b-559. Zielinski and Price (1980) had reported that cytochrome b-559 was a product of chloroplast protein synthesis. This was confirmed by Westhoff et al. (1985), who showed by cell-free, coupled transcription-translation and RNA-programmed, hybrid-selection translation that the *psbE* gene is located in the large single-copy region of the spinach chloroplast genome between the *petA* and *psbB* genes. The nucleotide sequence of this region revealed the presence of not one but two open reading frames (Herrmann et al. 1984). The first of these predicted an apoprotein of 83 residues; the predicted amino acid sequence matched the amino-terminal amino acid sequence of the spinach 9 kDa apoprotein determined by Widger et al. (1984). This open reading frame contained a single histidine residue, however, and it had been shown by electron paramagnetic resonance, optical, and resonance-Raman spectroscopic measurements that the heme prosthetic group of cytochrome b-559 has bis(histidine) axial ligation (Babcock et al. 1985). The second open reading frame, found immediately 3' to the *psbE* gene, predicted a polypeptide of 39 residues that also contained a single histidine residue (Herrmann et al. 1984). Hence, it was proposed that the cytochrome b-559 promoter is an ($\alpha\beta$) heterodimer in which the α and β subunits are cross-linked by the histidines which are axially liganded to the heme group (Herrmann et al. 1984). Support for this model came from a re-investigation of the polypeptide composition of the purified spinach cytochrome b-559 (Widger et al. 1985). These workers reported that a 4.3 kDa polypeptide, which existed in a 1:1 molar stoichiometry with the 9 kDa apoprotein, could be isolated from their cytochrome preparations. The amino-terminal amino acid sequence and amino acid composition of this polypeptide confirmed that the open reading frame downstream from *psbE*, viz. *psbF*, encodes a second apoprotein subunit of cytochrome b-559.

As part of studies on the origins of chloroplasts, we have been examining the organization of the cyanelle genome of *C. paradoxa*. In previous studies (Lambert et al. 1985) we had localized the *psbE* and *psbF* genes to an 18.0 Kb region of the large, single-copy region of the cyanelle genome. In this report we present the complete nucleotide sequence of the genes encoding cytochrome b-559. During the course of this work we recognized that two highly conserved open reading frames occur downstream from the *psbF* genes of all higher plant chloroplast genomes examined to date. The possibility that these open reading frames encode components of the Photosystem II reaction center is discussed.

Materials and methods

Materials

Restriction endonucleases and other DNA modifying enzymes were obtained from Bethesda Research Laboratories (Gaithersburg, MD), New England Bio-Labs (Beverley, MA), and International Biotechnologies, Inc. (New Haven, CT), and were used according to the specifications of the supplier. Deoxynucleotides and dideoxynucleotides were obtained from Pharmacia Molecular Biologicals (Piscataway, NJ). (α-^{32}P)dATP (> 600 Ci mmol^{-1}) and (γ-^{32}P)ATP (1000–3000 Ci mmol^{-1}) were supplied by NEN Research Products Boston, MA). Reagents for oligonucleotide biosynthesis were supplied by Applied Biosystems, Inc. (Foster City, CA), Beckman Instruments, Inc. (Palo Alto, CA), or Fisher Scientific Co. (Pittsburg, PA). Nitrocellulose for hybridization experiments was obtained from Schleicher & Schuell (Keene, NH). Acrylamide, bis-acrylamide, and other reagents for DNA sequencing were obtained from Bio-Rad (Richmond, CA). Antibiotics were purchased from Sigma Chemical Co. (St. Louis, MO). Technical-grade cesium chloride was used for all DNA preparations; all other chemicals were reagent-grade or higher.

DNA isolation and subcloning

The construction of a partial library of the cyanelle DNA of *C. paradoxa* has been described (Lambert et al. 1985). Synthetic oligonucleotide probes, whose sequences were based upon the nucleotide sequence of the spinach *psbE* and *psbF* genes (81-mers; kindly provided by Dr. H. Pakrasi, E.I. DuPont de Nemours, Co., Wilmington, DE) were used as hybridization probes to localize the *psbE* and *psbF* genes to an 18.0 Kb *Pst*I fragment (plasmid pCpcPst18.0) of the cyanelle genome. The *psbE* gene probe corresponds to nucleotides 54 to 135 and the *psbF* gene probe corresponds nucleotides 279 to 359 of the spinach gene sequences determined by Herrmann et al. (1984; see Fig. 2). Additional hybridization experiments showed that both probes also hybridized to a 750 bp *Eco*RI fragment derived from plasmid pCpcPst18.0. The 750 bp *Eco*RI fragment was purified after agarose gel electrophoresis of a preparative-scale *Eco*RI digestion of this plasmid as described (Cantrell and Bryant 1987b). The purified *Eco*RI fragment was ligated into the *Eco*RI site of plasmid pUC9 (Viera and Messing 1982) to create plasmid pCpcPst18.1; the plasmid was transformed into *E. coli* strain RDP145 (Buzby et al. 1983). Small scale-plasmid extractions for plasmid screening from overnight cultures of *E. coli* were performed as described (Holmes and Quigley 1981). Large-scale plasmid extractions and purifications were performed as described by Seifert and Porter (1984). Other DNA manipulations were performed as described by Maniatis et al. (1982).

Hybridization analyses

DNA restriction fragments were transferred to nitrocellulose filters as described (Southern 1975, Maniatis et al. 1982). Oligonucleotide probes were 5′ end-labeled with T4 polynucleotide kinase as described (Maniatis et al. 1982). Hybridization conditions were as described by Bryant and Tandeau de Marsac (1987); hybridizations with the synthetic 81-mers were performed at 50 °C. Blots were fluorographed at − 70 °C using Kodak X-Omat AR X-ray film and DuPont Lightning-Plus intensifying screens.

DNA sequence analyses

DNA sequence analyses were performed by the chain termination method of Sanger et al. (1977) on base-denatured plasmid templates (Hattori and Sakaki 1986) using the strategy described by Bryant et al. (1985). In this strategy initial sequences are obtained using the M13 forward and reverse primers; additional sequencing primers (pentadecanucleotides) are synthesized based upon this sequence information, and this process is then repeated until the sequence of the insert is complete. Oligonucleotides were synthesized on an Applied Biosystems Model 380A DNA synthesizer. Other conditions for DNA sequencing were as described (Cantrell and Bryant 1987b). Computer analysis of the DNA sequence information obtained was carried out using the programs of Conrad and Mount (1982), the Pustell programs (International Biotechnologies, Inc., New Haven, CT), and programs written by Dr R.D. Porter (Penn State University, personal communication).

Results and discussion

The *psbE* and *psbF* genes, encoding the cytochrome b-559 apoproteins, were localized on the cyanelle genome of *C. paradoxa* using synthetic 81-mer oligonucleotide probes whose sequences were based upon the spinach chloroplast *psbE* and *psbF* gene sequences determined by Herrmann et al. (1984). The probes, which included the coding sequences for the hydrophobic, putative membrane-spanning domain of each gene product, each hybridized to an 18.0 Kb *Pst*I fragment of the cyanelle genome which had been cloned into the *E. coli* plasmid vector pBR322 (Lambert et al. 1985). The *psbE* and *psbF* genes were further localized by restriction endonuclease digestion and Southern blot hybridization analyses of the pBR322 clone (plasmid pCpcPst18.0). The synthetic oligonucleotide probes were found to hybridize exclusively to a 750 bp *Eco*RI fragment which was isolated and cloned into the *Eco*RI site of the plasmid pUC9 to yield plasmid pCpcPst18.1. DNA sequence analysis of the insert in plasmid pCpcPst18.1, using the M13 forward and reverse sequencing primers with double-stranded, base-denatured templates, revealed that this subclone contained the entire sequence for the *psbE* gene but contained only the 5′ portion of the *psbF* gene (see Fig. 1). The remainder of the psbF gene and regions downstream from this gene were sequenced by using the parental plasmid, pCpcPst18.0, as a template and by using synthetic oligonucleotide sequencing primers based upon the previously determined sequences from the two clones. Figure 1 summarizes the sequencing strategy employed and indicates the relative orientations of the four open reading frames detected at this locus in the *C. paradoxa* cyanelle genome. As shown in Fig. 1, the nucleotide sequence for this region was determined on both strands except for a portion of the *psbJ*(ORF40) gene.

The nucleotide sequence of the region encoding the *C. paradoxa* cyanellar *psbE* and *psbF* genes, along with their deduced amino acid sequences, are shown in Fig. 2. The *psbE* and *psbF* genes occur in a cluster which contains four open reading frames. The *psbE* gene is composed of 75 codons and is located at the 5′ end of this cluster; it extends from nucleotides 1–225 in Fig. 2. The *psbF* gene is located 39 bp downstream from the *psbE* gene and is composed of 43 codons (nucleotides 265–393). Two additional open reading frames occur in the sequences downstream from the *psbF* gene. The first of these, denoted *psbI* (ORF38), occurs 9 bp downstream from the stop codon

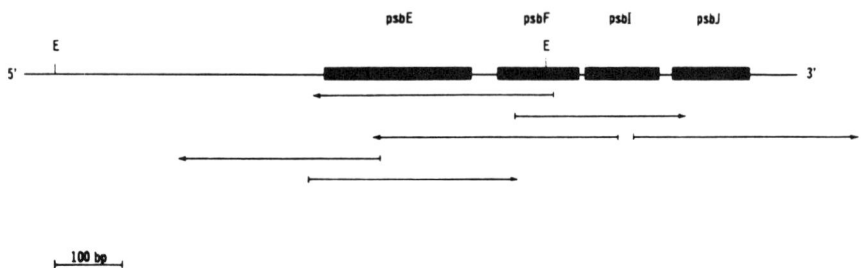

Fig. 1. Restriction map and sequencing strategy for the *psbEFIJ* coding region of the *C. paradoxa* cyanelle genome. A portion of the 18.0 Kb *Pst*I insert of pCpcPst18.0 that contains the four open reading frames of this cluster is shown. The 750 bp *Eco*RI insert of pCpcPst18.1 is demarcated by the two *Eco*RI restriction endonuclease sites indicated (E). The thick bars represent the locations of the locations of the *psbE*, *psbF*, *psbI*, and *psbJ* genes. The arrows below indicate the sequences obtained and strand from which they were obtained. The vertical hashmarks at the bases of the arrows indicate primer positions.

C. PARADOXA PSBE, PSBF, ORF38, AND ORF40.

```
    -240       -230       -220       -210       -200       -190       -180       -170       -160       -150
5'TCTTTAAGTAGTCAGATATCTGGAAATACTTGGAAAAAAGAAACAAGTACAGACAATATTCCTTCAAATTTCTATAAAATTAGTTTTATAGATAAAGAAGTA

    -140       -130       -120       -110       -100       -90        -80        -70        -60        -50
GGTTTATACTAGGAAATCAGGGAACACTTTTACGATATGTATCATTATAATAGAATAATTAATATCTTATAATTTGAATAAAATCTTTTATTATACATAAAA

  -40        -30        -20        -10        1          10         20         30         40
GCATCATATATATATATATATATAATTTTTAGGAGTGATTTAT ATG TCT GGA GGA ACT ACT GGC GAA CGC CCA TTT TCT GAC ATT GTT
                                             M   S   G   G   T   T   G   E   R   P   F   S   D   I   V

        50         60         70         80         90         100        110        120
ACT AGT ATT CGT TAT TGG GTT ATT CAT ACT GTA ACT ATT CCA TTC TTT ATT GTT GCA GGT TGG CTT TTT GTA AGT ACT
 T   S   I   R   Y   W   V   I   H   T   V   T   I   P   F   F   I   V   A   G   W   L   F   V   S   T

        130        140        150        160        170        180        190        200
GGT TTA GCT TAT GAT GTA TTT GGT ACT CCA AGA CCA GAT GAA TAC TTC ACT GAA GAA CGT CAA GAA GTA CCA ATT ATT
 G   L   A   Y   D   V   F   G   T   P   R   P   D   E   Y   F   T   E   E   R   Q   E   V   P   I   I

        210        220        230        240        250        260        270        280
AAT CAA CGT TTT TCA ACT AAT TAA TCGTTATTAATTTATAGTTTCTAAAAAGAGGAAAAGGAA ATG AAT AAT CCT AAT CAA CCG GTT
 N   Q   R   F   S   T   N   END                                         M   N   N   P   N   Q   P   V

 290        300        310        320        330        340        350        360
TCT TAT CCA ATT TTT ACA GTT AGA TGG TTA GCA ATT CAT GCT ATT GGA ATT CCA GCT GTA TTT TTT ATT GGA TCT ATT
 S   Y   P   I   F   T   V   R   W   L   A   I   H   A   I   G   I   P   A   V   F   F   I   G   S   I

     370        380        390        400        410        420        430        440
ACT GCA ATG CAA TTT ATT CAA CGA TAG GAGATATAT ATG GTT AGC CAA AAT CCT AAT AGA CAA AAA GTT GAA TTA AAT
 T   A   M   Q   F   I   Q   R   END           M   V   S   Q   N   P   N   R   Q   K   V   E   L   N

     450        460        470        480        490        500        510        520
CGT ACT TCC CTA TTC TGG GGA TTA CTT TTA ATT TTC GTA TTA GCA ATT TTA TTC TCT AGC TAC ATT TTT AAC TAA ATT
 R   T   S   L   F   W   G   L   L   L   I   F   V   L   A   I   L   F   S   S   Y   I   F   N   END

     530        540        550        560        570        580        590        600
TCTTCAATAAGGAGTCTTT ATG GCA AAT ACT GGT GGA CGC ATT CCT TTA TGG CTT GTT GCT ACA GTT GCA GGT TTA GCA GCT
                    M   A   N   T   G   G   R   I   P   L   W   L   V   A   T   V   A   G   L   A   A

  610        620        630        640        650        660        670        680
ATT GGT GTA CTA GGA ATC TTT TTC TAT GGT GGT TAC TCT GGT TTA GGT TCC TCT ATT TAA TAAAATCAATTTAAGATTTTTTG
 I   G   V   L   G   I   F   F   Y   G   G   Y   S   G   L   G   S   S   I   END

 690        700        710        720        730        740        750        760        770        780        790
TAATAAATTAATTAAAAAAAAATAGAGAGTTAAATTTTTTTAACTCTCTATTCTTTAGACATTTATAACTAATAAAAGAAATTAAGCAATTTCATCTGTTTCA
      ──>    <──
         800        810        820        830
ATATAAACCAAATAATAATGCCATAACATTGCTGGTAAGACA
```

Fig. 2. Nucleotide sequence and the deduced amino acid sequences for the *psbE, psbF, psbI,* and *psbJ* genes as discussed in the text. Polypurine sequences which resemble putative ribosome-binding sequences are underlined. The sequence underlined by arrows downstream from *psbJ* and centered at nucleotide 694 could form a stable hairpin structure. Such hairpin structures could play a role in mRNA stabilization or in the termination of transcription by RNA polymerase.

of the *psbF* gene and is composed of 39 codons (nucleotides 403–519 in Fig. 2.). The second open reading frame, denoted *psbJ*(ORF40), is found 22 bp downstream from the stop codon of *psbI*(ORF38) and is composed of 41 codons (nucleotides 542–664 in Fig. 2). Arguments for designating these latter two open reading frames as *psbI* and *psbJ*, indicating a function or structural role in Photosystem II will be presented below. Each of the four open reading frames is preceded by a polypurine sequence which closely resembles the sequence suggested to form the ribosome binding site as defined by Shine and Dalgarno (1974). This sequence, A-G-G-A-G/A is found 5–8 bp upstream from the translation start codon of each open reading frame; quite similar sequences have been observed to precede ten other genes encoded by the cyanellar DNA of *C. paradoxa* (Bryant et al. 1985, Lemaux and Grossman 1985; Annarella, Stirewalt and Bryant, unpublished observations). An inverted repeat occurs between nucleotides 681 and 708; this sequence could form a hairpin structure that might play a role in the termination of transcription by RNA polymerase or might play a role in the stabilization of the mRNA. A number of sequences which resemble the *E. coli* consensus promoter sequences (Harley and Reynolds 1987) can be found in the sequence upstream from the *psbE* gene.

The deduced amino acid sequences for the four open reading frames are shown in Figs. 2, 3, and 4. Figure 3 shows an alignment of the deduced amino acid sequences of the *psbE* and *psbF* gene products from the cyanelle genomes of *C. paradoxa* with those of the chloroplast genomes of spinach (Herrmann et al. 1984); pea (Willey and Gray, personal communication); liverwort (Ohyama et al. 1986); tobacco (Carrillo et al. 1986, Shinozaki et

```
                      1      10        20        30        40        50        60        70
A. C. PARADOXA:       MSGGTTGERPFSDIVTSIRYWVIHTVTIPFFIVAGWLFVSTGLAYDVFGTPRPDEYFTEERQEVPIINQRFSIN*
   SYNECHOCYSTIS:      -                SI' MLFI              A              QT   L  LQE YDI QEIQEFNG
   SPINACH:            -S   S A  I      SI  SLFI         S   N   S  GI L TG  DSLEQLDEFSF
   PEA:                -S   S A  I    I SI  SLFI         S   N   T  GI L TG  DSLEQLDEFSF
   TOBACCO:            -S   S A  I      SI  SLFI         S   N   S  GI L TG  DPLEQIDLFSF
   LIVERWORT:          -N     A  I      SI  SLFI         S   N   N     L TG  NSLEQIDEFIF
   PRIMROSE:           -S G S    I      SI  SLFI         S   N   S  GI L TG  DSLEQIDEFSF
   WHEAT:              -S   S A  I      SI  SLFI         S   N   S  GI L TD  DSLEQLDEFSF
   E. GRACILIS:        A -S      I      S   SLF G  I   IV I      S   T  QA L SD  NALEQMDQFIF

                      1      10        20        30        40
B. C. PARADOXA:       MNNPNQPVSYPIFTVRWLAIHAIGIPAVFFIGSITAMQFIQR*
   SYNECHOCYSTIS: MATQ      T         V TLAV S   V A A
   SPINACH:            T---IDRT         GLAV T S L    S
   PEA:                T---IDRT       V GLAV T S L    S
   TOBACCO:            T---IDRT         GLAV T   L    S              PSBF
   LIVERWORT:          T---IDRT       V GLAV T   L A S
   PRIMROSE:           T---IDRT       V GLAV T S L    S  G
   WHEAT:              T---IDRT         GLAV T   L    S
   E. GRACILIS:        TT- KDTR     F   V LA T    L    S
```

Fig. 3. Comparisons of the deduced amino acid sequences for the *psbE* and *psbF* gene products. Only the residues which differ from those found at the equivalent position in the *C. paradoxa* cyanelle products are shown. Dashes denote deletions introduced to maximize the homology. The brackets above the sequences indicate hydrophobic stretches of amino acids which could form trans-membrane alpha helices. A. Comparison of the deduced amino acid sequences for the *psbE* gene products. B. Comparison of the deduced amino acid sequences for the *psbF* gene products. Data for *C. paradoxa* (this work); *Synechocystis* sp. PCC 6803 (H. Pakrasi, personal communication); spinach (Herrmann et al. 1984); pea (Willey and Gray, personal communication); tobacco (Carrillo et al. 1986, Shinozaki et al. 1986); liverwort (Ohyama et al. 1986); primrose (Carrillo et al. 1986); wheat (Hird et al. 1986); and *Euglena gracilis* (Cushman et al. 1987).

```
                          ├───────┼───────┤
             1      10       20      30     40
A.  C. PARADOXA:  M-VSQNPNRQKVELNRTSLFWGLLLIFVLAILFSSYIFN:

    SYNECHOCYSTIS:  DRNS    P         YL    VA G       F

    PEA:            TQS  E N          Y          V   N F

    TOBACCO:        TQS  E N          Y          V   N F    PSBI
                                                            ───
    LIVERWORT:      TQP  K S          Y          V   N F

    E. GRACILIS:    AQT S KKT         Y          V
```

```
                          ├───────┼───────────┤
             1      10       20      30     40
B.  C. PARADOXA:  MANT--GGRIPLWLVATVAGLAAIGVLGIFFYGGYSGLGSSI:

    SYNECHOCYSTIS:  F-A  E     V GV   IG      L    A A      M

    PEA:                T       IIG   IVV  LI L    S        V

    TOBACCO:            T       IIG   ILV  LI      S        L   PSBJ
                                                                ───
    LIVERWORT:      D   T V     IG    ILV  LV      S        L

    E. GRACILIS:    S SENT        L II     AL AL   S        L
```

Fig. 4. Comparisons of the deduced amino acid sequence for the *psbI* and *psbJ* gene products. Only the residues which differ from those found at the equivalent position in the *C. paradoxa* cyanelle products are shown. Dashes denote deletions introduced to maximize the homology. The brackets above the sequences indicate hydrophobic stretches of amino acids which could form trans-membrane alpha helices. A. Comparison of the deduced amino acid sequences for the *psbI* gene products. B. Comparison of the deduced amino acid sequences for the *psbJ* gene products. Data for *C. paradoxa* (this work); *Synechocystis* sp. PCC 6803 (Pakrasi, personal communication); pea (Willey and Gray, personal communication); tobacco (Shinozaki et al. 1986); liverwort (Ohyama et al. 1986); *Euglena gracilis* (Cushman et al. 1987).

al. 1986); primrose (Carrillo et al. 1986); *Euglena gracilis* (Cushman et al. 1987); and with that of the cyanobacterium *Synechocystis* sp. PCC 6803 (Pakrasi, personal communication). Identity values for the various pair-wise combinations of the *psbE* and *psbF* gene products are shown in Table 1. The identity values were calculated as the percentage of identical amino acids which occur within the region of overlap between the compared sequences. The *psbE* gene product of *C. paradoxa* is 67–78% identical to the *psbE* gene products of higher plants, *E. gracilis*, and the cyanobacterium, but is seven to nine residues shorter. The *psbF* gene product of *C. paradoxa* is 57–71% identical to the *psbF* gene products of the same organisms. Homology values, which include allowances for substitutions by biochemically similar amino acids range from 83–98% for the same pair-wise comparisons.

A comparison of the predicted polypeptide sequences reveals a high degree of conservation in the regions previously assigned functional and structural importance in the cytochrome b-559 apoproteins (Herrmann et al. 1984, Cramer et al. 1985). The hydrophobic, putative membrane-spanning α-helical regions of the proteins, which contain the histidine residues suggested to axially ligand the cytochrome b-559 heme prosthetic group, are

Table 1. Amino acid sequence identity values for the products of the *psbE* and *psbF* genes of *C. paradoxa*, *Synechocystis* sp. PCC 6803 and several other eucaryotes. The values above and to the right of the main diagonal indicate percentage identity values for the *psbE* gene products. The values below and to the left to the main diagonal indicate percentage identity values for the *psbF* gene products.

					psbE				
	C. paradoxa	*Synechocystis*	Spinach	Pea	Tobacco	Liverwort	Primrose	Wheat	*E. gracilis*
C. paradoxa		78	74	73	74	77	74	74	67
Synechocystis	71		65	68	67	68	67	71	63
Spinach	69	55		98	99	90	98	99	71
Pea	60	67	97		96	88	95	96	70
Tobacco	64	71	97	95		88	98	99	71
Liverwort	60	79	92	95	95		87	88	74
Primrose	57	64	97	97	92	92		96	70
Wheat	64	71	98	95	100	95	92		71
E. gracilis	64	66	74	76	79	76	74	76	
					psbF				

conserved in both *psbE* and *psbF* polypeptide families. In the *C. paradoxa* sequences these regions occur from positions 20 to 46 in the *psbE* gene product and from positions 17 to 41 in the *psbF* gene product. The proposed heme-liganding histidine residues occur five amino acids into these hydrophobic regions, to the N-terminal end, for all of the polypeptides compared. These histidine residues occur at positions 24 and 21 in the *C. paradoxa psbE* and *psbF* gene products, respectively. Although some sequence variation in the hydrophobic regions is osberved, the substitutions which are found are biochemically similar in all cases and strictly maintain the hydrophobic nature of the sequences. The amino- and carboxyl-terminal sequences of the polypeptides are quite hydrophilic in nature and are not as highly conserved. The carboxyl terminus of the *psbE* gene product is rich in aspartic acid and glutamic acid residues and probably carries a net negative charge. The periplasmic surfaces of the *Rhodopseudomonas viridis* and *Rhodobacter sphaeroides* reaction centers carry a net negative charge, while their cytoplasmic surfaces are neutral to slightly positively charged (Michel et al. 1986, Yeates et al. 1987). Structural models for the *psbA* and *psbD* gene products suggest that the lumenal surface of the PS II reaction center would also carry a net negative charge and that the cytoplasmic surface would be neutral to slightly negative (Trebst 1986). Hence, the distribution of charges on the *psbE* gene product might suggest that the carboxyl terminus of the protein extends into the lumenal space with the amino-terminal portion of the polypeptide extending into the cytoplasm. This arrangement would place the heme group closer to the cytoplasmic surface of the thylakoid membrane than to the lumenal surface (if there is only one heme per P680/PS II reaction center).

In a preliminary report describing these results, arguments were presented for the occurrence of two additional genes in the *psbE-psbF* locus (Cantrell and Bryant 1987a). The 38 and 40 codon open reading frames discovered immediately downstream from *psbF* in the cyanelle genome of *C. paradoxa* were tentatively designated *psbH* and *psbI*, respectively. However, the locus designation *psbH* was simultaneously assigned to the 9 kDa phosphoprotein of PS II by Gray and coworkers (1987) and Herrmann and coworkers (Westhoff et al. 1986). For this reason and for those discussed below, we now propose to call these two open reading frames *psbI* and *psbJ*, respectively; other workers have referred to these open reading frames as ORF38(39) and ORF40(42).

The two highly conserved open reading frames, *psbI* and *psbJ*, could encode polypeptides of 38 or 39 residues and 40 to 42 residues, respectively; these genes have now been located and completely sequenced for several organisms as shown in Fig. 4. Although these open reading frames were not originally reported to occur in the sequence downstream from the *psbF* gene of the cyanobacterium *Synechocystis* sp. PCC 6803, a re-investigation of this sequence data reveals the *psbI* and *psbJ* genes to be present (Pakrasi, personal communication). Identity values for the various pair-wise combinations of the *psbI* and *psbJ* gene products are shown in Table 2. As shown in Fig. 4 and Table 2, the predicted products of the *psbI* and *psbJ* genes are very highly conserved among both eucaryotes and procaryotes. The high degree of conservation of these putative polypeptides across such broad phylogenetic distances strongly indicates their structural and/or functional importance.

In *E. gracilis*, the sequences encoding *psbI* (ORF38) and *psbJ* (ORF42) are cotranscribed with the *psbE* and *psbF* genes. Similarly, transcript analyses performed with spinach, tobacco, and *Oenothera* sp. indicate that these sequences are cotranscribed with *psbE* and *psbF* (assuming that these sequences actually occur in the same relative positions in spinach and *Oenothera* sp.; Carrillo et al. 1986). Although transcript analyses have not been performed with RNA from the cyanelles of *C. paradoxa*, two observations suggest that the *psbI* and *psbJ* sequences are actually expressed. Firstly, there are no palindromic sequences resembling transcription termination or mRNA stabilization sequences in the region downstream from the *psbF* gene until the region downstream from the *psbJ* gene. Secondly, both the *psbI* and

Table 2. Amino acid sequence identity values for the products of the *psbI* and *psbJ* genes of *C. paradoxa*, *Synechocystis* sp. PCC 6803, and several other eucaryotes. The values above and to the right of the main diagonal indicate the percentage identity values for the *psbI* gene products. The values below and to the left of the main diagonal indicate the percentage identity values for the *psbJ* gene sequences.

	psbI					
	C. paradoxa	*Synechocystis*	Pea	Tobacco	Liverwort	*E. gracilis*
C. paradoxa		72	76	76	76	76
Synechocystis	68		72	72	69	64
Pea	70	68		100	92	76
Tobacco	70	63	93		92	84
Liverwort	73	68	83	90		84
E. gracilis	67	57	62	62	60	
	psbJ					

psbJ initiator methionine codons are preceded by consensus Shine-Dalgarno-type polypurine sequences at an appropriate distance upstream from the translation start codon (see Fig. 2). Cotranscription of the *psbI* and *psbJ* genes with the *psbE* and *psbF* genes, which are known to encode components of the Photosystem II reaction center core, could certainly suggest that these polypeptides are functionally associated with PS II.

The predicted polypeptide products of the *psbI* and *psbJ* genes are notably devoid of 'functional amino acids' (e.g., histidines, cysteines, etc.) but are rich in hydrophobic amino acids. The predicted sequence for each gene contains an hydrophobic stretch of amino acids greater than 22 amino acids in length which could form a membrane-spanning alpha helix. This observation suggests that these polypeptides are components of the thylakoid membrane. Cushman et al. (1987) have reached a similar conclusion from their analyses of the predicted gene products from *E. gracilis*. A computer search has failed to identify polypeptides sharing any striking homologies to the predicted polypeptides of either the *psbI* and *psbJ* genes. However, there are several small polypeptides with molecular masses less than 10 kDa which are found in association with Photosystem II in spinach (Ljungberg et al. 1986). Although the function and site of synthesis of these polypeptides are unknown at present, it is possible two of these polypeptides are the products of the *psbI* and *psbJ* genes. This possibility is being explored at present by the isolation and amino acid sequencing of these small polypeptides (Andersson, personal communication). De Vitry et al. (1987) have reported the presence of at least two polypeptides smaller than 9 kDa in very highly purified preparations of Photosystem II reaction centers from *C. reinhardtii*. These proteins are intrinsic membrane proteins synthesized in the chloroplast and are largely missing in mutants deficient in PS II centers; one or both of these polypeptides may be phosphorylated. Although one of these polypeptides is presumably the *psbF* gene product, the other(s) may be products of *psbI* and/or *psbJ*. Strong suggestive evidence linking the *psbI* and *psbJ* genes to the Photosystem II reaction center has recently been obtained by Pakrasi and coworkers for *Synechocystis* sp. PCC 6803 (Pakrasi, personal communication). Mutants constructed in the *psbI-psbJ* region do not produce functional Photosystem II reaction centers and are nonphotosynthetic.

Acknowledgements

The authors would like to thank Dr H. Pakrasi for the synthetic oligonucleotides used to map and clone the *psbE* and *psbF* genes from *Cyanophora paradoxa* and for the communication of unpublished results. We also thank Drs David Willey, John Gray, John Cushman, and Carl Price for providing unpublished results. This work was supported by National Science Foundation grant DMB-8504294 and Project 2612 of the Agriculture Experiment Station of Penn State University.

References

Babcock GT, Widger WR, Cramer WA, Oertling WA and Metz JG (1985) Axial ligands of chloroplast cytochrome b-559: identification and requirement for a heme-cross-linked polypeptide structure. Biochem 24: 3638–3645

Barber J, Chapman DJ and Telfer A (1987) Characterization of a PS II reaction centre isolated from the chloroplasts of *Pisum sativum*. FEBS Lett 220: 67–73

Bendall DS (1982) Photosynthetic cytochromes of oxygenic organisms. Biochim Biophys Acta 683: 119–151

Berthold DA, Babcock GT and Yocum CF (1981) A highly resolved, oxygen-evolving, photosystem II preparation from spinach thylakoid membranes. FEBS Lett 134: 231–234

Briantais J-M, Vernotte C, Miyao M, Murata N and Picaud M (1985) Relationship between O_2 evolution capacity and cytochrome b-559 high-potential form in Photosystem II particles. Biochim Biophys Acta 808: 348–351

Bryant DA (1987) The cyanobacterial photosynthetic apparatus: comparison to those of higher plants and photosynthetic bacteria. In: Platt T and Li WKW (eds) Photosynthetic Picoplankton, pp 423–500. Ottawa: Canad Bull Fisheries Aquatic Sciences, Vol 214, Dept. Fisheries and Oceans, Canada

Bryant DA, de Lorimier R, Lambert DH, Dubbs JM, Stirewalt VL, Stevens SE Jr, Tam J and Jay E (1985) Molecular cloning and nucleotide sequence of the α and β subunits of

384

allophycocyanin from the cyanelle genome of *Cyanophora paradoxa*. Proc Natl Acad Sci USA 82: 3242–3246

Bryant DA and Tandeau de Marsac N (1987) Isolation of genes encoding components of the photosynthetic apparatus. In: Packer L and Glazer AN (eds) Methods in Enzymology, Vol. XXX, in press. New York: Academic Press

Butler WL (1978) On the role of cytochrome b-559 in oxygen evolution in photosynthesis. FEBS Lett 95: 19–25

Buzby JS, Porter RD and Stevens SE Jr (1983) Plasmid transformation in *Agmenellum quadruplicatum* PR-6: Construction of biphasic plasmids and characterization of their transformation properties. J Bacteriol 154: 1446–1450

Cantrell A and Bryant DA (1987a) Molecular cloning and nucleotide sequences of the genes encoding cytochrome b-559 from the cyanelle genome of *Cyanophora paradoxa*. In: Biggins J (ed.) Progress in Photosynthesis Research, Vol. IV, pp 659–662. Dordrecht: Martinus Nijhoff/Dr W Junk Publishers

Cantrell A and Bryant DA (1987b) Molecular cloning and nucleotide sequence of the *psaA* and *psaB* genes of the cyanobacterium *Synechococcus* sp. PCC 7002. Plant Mol Biol 9: 453–468

Carrillo N, Seyer P, Tyagi A and Herrmann RG (1986) Cytochrome b-559 genes from *Oenothera hookeri* and *Nicotiana tabacum* show a remarkably high degree of conservation as compared to spinach. The enigma of cytochrome b-559: highly conserved genes and proteins but no known function. Curr Genet 10: 619–624

Conrad B and Mount DW (1982) Microcomputer programs for DNA sequence analysis. Nucl Acids Res 10: 31–38

Cramer WA and Crofts AR (1982) Electron and proton transport. In: Govindjee (ed.) Photosynthesis: Energy Conversion by Plants and Bacteria, pp 387–467. New York: Academic Press

Cramer WA and Whitmarsh J (1977) Photosynthetic cytochromes. Ann Rev Plant Physiol 28: 133–172

Cramer WA, Widger WR, Herrmann RG and Trebst A (1985) Topography and function of thylakoid membrane proteins. TIBS 10: 125–129

Cushman JC, Christopher DA, Little MC, Hallick RB and Price CA (1987) The organization and expression of the *psbE*, *psbF*, ORF38, and ORF42 loci on the *Euglena gracilis* chloroplast genome. Curr Genet (in press)

De Vitry C, Diner BA and Lemoine Y (1987) Chemical composition of photosystem II reaction centers (PS II): Phosphorylation of PS II polypeptides. In: Biggins J (ed.) Progress in Photosynthesis Research, Vol. II, pp 105–108. Dordrecht: Martinus Nijhoff/Dr W Junk Publishers

Falkowski PG, Fujita Y, Ley A and Mauzerall D (1986) Evidence for cyclic electron flow around photosystem II in *Chlorella pyrenoidosa*. Plant Physiol 81: 310–312

Ghanotakis DF, Babcock GT and Yocum CF (1984) Structural and catalytic properties of the oxygen-evolving complex. Correlation of polypeptide and manganese release with the behavior of Z^+ in chloroplasts and a highly resolved preparation of the PS II complex. Biochim Biophys Acta 765: 388–398

Ghanotakis DF, Demetriou DM and Yocum CF (1987) Isolation and characterization of an oxygen-evolving Photosystem II reaction center core preparation and a 28 kDa Chl a-binding protein. Bichim Biophys Acta 891: 15–21

Ghanotakis DF and Yocum CF (1985) Polypeptides of photosystem II and their role in oxygen evolution. Photosynth Res 7: 97–114

Gounaris K, Pick U and Barber J (1987) Stoichiometry and turnover of photosystem II polypeptides. FEBS Lett 211: 94–98

Gray JC, Blyden ER, Eccles CJ, Dunn PPJ, Hird SM, Hoglund A-S, Kaethner TM, Smith AG, Willey DL and Dyer TA (1987) Chloroplast genes for photosynthetic membrane components. In: Biggins J (ed.) Progress in Photosynthesis Research, Vol. IV, pp 617–624. Dordrecht: Martinus Nijhoff/Dr W Junk Publishers

Harley CB and Reynolds RP (1987) Analysis of *E. coli* promoter sequences. Nucl Acids Res 15: 2343–2358

Hattori M and Sakaki Y (1986) Dideoxysequencing method using denatured plasmid templates. Analyt Biochem 152: 232–238

Herrmann RG, Alt J, Schiller B, Widger WR and Cramer WA (1984) Nucleotide sequence of the gene for apocytochrome b-559 on the spinach plastid chromosome: implications for the structure of the membrane protein. FEBS Lett 176: 239–244

Hervas M, Ortega JM, de la Rosa MA, de la Rosa FF and Losada M (1985) Location and function of cytochrome b-559 in the chloroplast noncyclic electron transport chain. Physiol Veg 23: 593–604

Hird SM, Willey DL, Dyer TA and Gray JC (1986) Location and nucleotide sequence of the gene from cytochrome b-559 in wheat chloroplast DNA. Mol Gen Genet 203: 95–100

Holmes DS and Quigley M (1981) A rapid boiling method for the preparation of bacterial plasmids. Analyt Biochem 114: 193–197

Lambert DH, Bryant DA, Stirewalt VL, Dubbs JM, Stevens SE Jr and Porter RD (1985) Gene map for the *Cyanophora paradoxa* cyanelle genome. J Bacteriol 164: 659–664

Lemaux PG and Grossman AR (1985) Major light-harvesting polypeptides encoded in polycistronic transcripts in a eukaryotic alga. EMBO J 4: 1911–1919

Ljungberg J, Henrysson T, Rochester CP, Akerlund H-E and Andersson B (1986) The presence of low-molecular-weight polypeptides in spinach photosystem II core preparations. Isolation of a 5 kDa hydrophilic polypeptide. Biochim Biophys Acta 767: 145–152

Maniatis T, Fritsch EF and Sambrook J (1982) Molecular Cloning: A Laboratory Manual. 545 pp, Cold Spring Harbor: Cold Spring Harbor Laboratory

Metz JG, Ulmer G, Bricker TM and Miles D (1983) Purification of cytochrome b-559 from oxygen-evolving photosystem II preparations of spinach and maize. Biochim Biophys Acta 725: 203–209

Michel H, Weyer KA, Gruenberg H, Dunger I, Oesterhelt D and Lottspeich F (1986) The 'light' and 'medium' subunits of the photosynthetic reaction centre from *Rhodopseudomonas viridis*: isolation of the genes, nucleotide and amino acid sequence. EMBO J 5: 1149–1158

Murata N and Miyao M (1985) Extrinsic membrane proteins in the photosynthetic oxygen-evolving complex. TIBS 10: 122–124

Murata N, Miyao M, Omata T, Matsunami H and Kuwabara T (1984) Stoichiometry of the components in the photosynthetic oxygen evolution system of photosystem II particles prepared with Triton X-100 from spinach chloroplasts. Biochim Biophys Acta 765: 363–369

Nanba O and Satoh K (1987) Isolation of a photosystem II reaction center consisting of D-1 and D-2 polypeptides and cytochrome b-559. Proc Natl Acad Sci USA 84: 109–112

Ohno T, Satoh K and Kastoh S (1986) Chemical composition of purified oxygen-evolving complexes from the thermophilic cyanobacterium *Synechococcus* sp. Biochim Biophys Acta 852: 1–8.

Ohyama K, Fuguzawa H, Kohchi T, Shirai H, Sano T, Sano S, Umesono K, Shigi Y, Takeuchi M, Chang Z, Aoto S-I, Inokuchi H and Ozeki H (1986) Chloroplast gene organization deduced from the complete sequence of liverwort *Marchantia polymorpha* chloroplast DNA. Nature 322: 572–574

Pakrasi HB, Williams JGK and Arntzen CJ (1987) Genetically engineered cytochrome b-559 mutants of the cyanobacterium *Synechocystis* 6803. In: Biggins J (ed.) Progress in Photosynthesis Research, Vol IV, pp 813–816. Dordrecht: Martinus Nijhoff/Dr W Junk Publishers

386

Rao LVM, Usharani P, Butler WL and Tokuyasu KT (1986) Localization of cytochrome b-559 in the chloroplast thylakoid membranes of spinach. Plant Physiol 80: 138–141

Sanger F, Nicklen S and Coulson AR (1977) DNA sequencing with chain-terminating inhibitors. Proc Natl Acad Sci USA 74: 5463–5467

Satoh, K (1985) Protein-pigments and photosystem II reaction center. Photochem Photobiol 42 845–853

Satoh K, Fujii Y, Aoshima T and Tado T (1987) Immunological identification of the polypeptide bands in the SDS-polyacrylamide gel electrophoresis of photosystem II preparations. FEBS Lett 216: 7–10

Satoh K, Ohno T and Katoh S (1985) An oxygen-evolving complex with a simple subunit structure – 'a water-plastoquinone oxidoreductase' – from the thermophilic cyanobacterium *Synechococcus* sp. FEBS Lett 180: 326–330

Seifert HS and Porter RD (1984) Enhanced recombination between *λplac5* and miniF*lac*: the *tra* regulon is required for recombination enhancement. Mol Gen Genet 193: 269–274

Shine J and Dalgarno L (1974) The 3′-terminal sequence of *Escherichia coli* 16S ribosomal RNA: complementarity to nonsense triplets and ribosome binding sites. Proc Natl Acad Sci USA 71: 1342–1346

Shinozaki K, Ohme M, Tanaka M, Wakasugi T, Hayashida N, Matsubayashi T, Zaita N, Chungsongse J, Obokata J, Yamaguchi-Shinozaki K, Ohto C, Torazawa K, Meng BY, Sugita M, Deno H, Kamogashira T, Yamada K, Kusuda J, Takaiwa F, Katoh A, Todoh N, Shimada H and Sugiura M (1986) The complete nucleotide sequence of the tobacco chloroplast genome: its gene organization and expression. EMBO J 5: 2043–2049

Southern E (1975) Detection of specific sequences among DNA fragments separated by gel electrophoresis. J Mol Biol 98: 503–517

Thompson LK, Sturtevant JM and Brudwig GW (1986) Differential scanning calorimetric studies of Photosystem II: evidence for a structural role for cytochrome b-559 in the oxygen-evolving complex. Biochem 25: 6161–6169

Trebst A (1986) The topology of the plastoquinone and herbicide binding peptides of Photosystem II in the thylakoid membrane. Z Naturforsch 41c: 240—245

Viera J and Messing J (1982) The pUC plasmids, an M13 mp7-derived system for insertion mutagenesis and sequencing with synthetic universal primers. Gene 19: 259–268

Westhoff P, Alt J, Widger WR, Cramer WA and Herrmann RG (1985) Localization of the gene for apocytochrome b-559 on the plastid chromosome of spinach. Plant Mol Biol 4: 103–110

Westhoff P, Farchaus JW and Herrmann RG (1986) The gene for the M_r 10,000 phosphoprotein associated with photosystem II is part of the *psbB* operon of the spinach plastid chromosome. Curr Genet 11: 165–169

Widger WR, Cramer WA, Hermodson M and Herrmann RG (1985) Evidence for a hetero-oligomeric structure of the chloroplast cytochrome b-559. FEBS Lett 191: 186–190

Widger WR, Cramer WA, Hermodson M, Meyer D and Gullifor M (1984) Purification and partial amino acid sequence of the chloroplast cytochrome b-559. J Biol Chem 259: 3870–3876

Yamada Y, Tang X-S, Itoh S and Satoh K (1987) Purification and properties of an oxygen-evolving Photosystem II reaction-center complex from spinach. Biochim Biophys Acta 891: 129–137

Yamagishi A and Katoh S (1985) Further characterization of the two Photosystem II reaction center complex preparations from the thermophilic cyanobacterium *Synechcoccus* sp. Biochim Biophys Acta 807: 74–80

Yamamoto Y, Tabata K, Isogai Y, Nishimura M, Okayama S, Matsuura K and Itoh S (1984) Quantitative analysis of membrane components in a highly active O_2-evolving photosystem II preparation from spinach chloroplasts. Biochim Biophys Acta 767: 493–500

Yeates TO, Komiya H, Rees DC, Allen JP and Feher G (1987) Structure of the reaction center from *Rhodobacter sphaeroides* R-26: Membrane-protein interactions. Proc Natl Acad Sci USA 84: 6438–6442

Zielinski RE and Price CA (1980) Synthesis of thylakoid membrane proteins by chloroplasts isolated from spinach. Cytochrome b-559 and P700-chlorophyll a protein. J Cell Biol 85: 435–445

Govindjee et al. (eds), Molecular Biology of Photosynthesis: 389–405
© 1988 Kluwer Academic Publishers

Regular paper

Protein composition of the photosystem II core complex in genetically engineered mutants of the cyanobacterium *Synechocystis* sp. PCC 6803

W.F.J. VERMAAS[1,2], M. IKEUCHI[1] and Y. INOUE[1]

[1]*Solar Energy Research Group, RIKEN, Hirosawa 2-1, Wako-shi, Saitama 351-01, Japan;*
[2]*Arizona State University, Department of Botany, Tempe AZ 85287-1601, USA (address for correspondence)*

Received 10 August 1987; accepted 17 December 1987

Key words: directed mutagenesis, photosynthesis, Photosystem II proteins, protein complex assembly, protein function

Abstract. The presence of four photosystem II proteins, CP47, CP43, D1 and D2, was monitored in mutants of *Synechocystis* sp. PCC 6803 that have modified or inactivated genes for CP47, CP43, or D2. It was observed that: (1) thylakoids from mutants without a functional gene encoding CP47 are also depleted in D1 and D2; (2) inactivation of the gene for CP43 leads to decreased but significant levels of CP47, D1 and D2; (3) deletion of part of both genes encoding D2, together with deletion of part of the CP43-encoding gene causes a complete loss of CP47 and D1; (4) thylakoids from a site-directed mutant in which the His-214 residue of D2 has been replaced by asparagine do not contain detectable photosystem II core proteins. However, in another site-directed mutant, in which His-197 has been replaced by tyrosine, some CP47 as well as breakdown products of CP43, but no D1 and D2, can be detected. These data could indicate a central function of CP47 and D2 in stable assembly of the photosystem II complex. CP43, however, is somewhat less critical for formation of the core complex, although CP43 is required for a physiologically functional photosystem II unit. A possible model for the assembly of the photosystem II core complex is proposed.

Introduction

Light-induced plastoquinone reduction by water, a key process in photosynthesis, depends on functional integrity of the photosystem II (PS II) complex. This complex consists of a "core" of at least five integral membrane proteins: the chlorophyll-binding proteins CP47 and CP43, the 32–34 kDa proteins D1 and D2, and cytochrome b_{559} (Satoh 1985; Arntzen and Pakrasi 1986). In addition, one to several extrinsic proteins also participate in PS II function. The precise role of the individual proteins of the complex is still the subject of speculation. Moreover, it is as yet largely unknown how the assembly of the PS II core complex depends on the presence of each individual polypeptide of that complex.

A rather fruitful approach to study the assembly processes in PS II has been to isolate and characterize mutants with altered PS II properties. Although characterization of mutants that were obtained by random mutagenesis or spontaneous mutations have yielded some important insight into the requirements for PS II core complex function or assembly (Miles et al. 1979, Bishop 1982, Metz et al. 1986, Jensen et al. 1986), it is often difficult to distinguish whether an observed change in one of the proteins should be attributed to modification of its structural gene, or to the pleiotropic effects of a mutation in another gene.

Recent advances in molecular genetics have allowed the construction of cyanobacterial mutants in which the gene encoding a component of the PS II complex has been specifically inactivated or modified (Williams 1987, Golden et al. 1987, Sherman et al. 1984). Using the photoheterotrophic cyanobacterium *Synechocystis* sp. PCC 6803 (hereafter *Synechocystis* 6803), which can be propagated on glucose-containing medium in the light in the absence of PS II activity, mutants with inactivated genes for CP47, cyt b_{559}, D2 and/or CP43 have been developed by specific interruption of these genes (Williams 1987, Vermaas et al. 1987a, Pakrasi et al. 1987, J.G.K. Williams, W.F.J. Vermaas and C.J. Arntzen, unpublished results). Moreover, site-directed mutations have been introduced to replace crucial histidine residues of the D2 protein (Vermaas et al. 1987b).

In many *Synechocystis* 6803 mutants with modified PS II, the overall PS II function (light-driven electron transport from water to plastoquinone) was inactivated completely by the gene modifications. In various mutants, CP47 and/or CP43 were still present (Vermaas et al. 1986, Vermaas et al. 1987c, Pakrasi et al. 1987), albeit in apparently decreased quantities in some of these mutants. However, since there are many proteins in the thylakoids from *Synechocystis* 6803, and since we have not yet succeeded in obtaining a pure PS II preparation from this organism, we could not observe whether the two 32–34 kDa proteins were still present in these mutants. At the time of those experiments no suitable antibodies against D1 or D2 were available to us. An unambiguous interpretation of the impaired PS II activity of the different mutants was not possible because of the uncertainty concerning the presence of the D1 and D2 proteins.

To address this problem, we have raised polyclonal antibodies against the spinach D1 and D2 proteins, and these antibodies show good cross-reactivity with *Synechocystis* 6803 D1 and D2. The results of immunoreactions of antibodies against D1, D2, CP43 and CP47 with thylakoid proteins from the various directed PS II mutants are reported in this paper. The data suggest that both CP47 and D2 appear to be required for stable integration and assembly of the PS II core complex.

Materials and methods

Synechocystis sp. PCC (Pasteur Culture Collection) 6803 was cultivated in BG-11 medium as described previously (Vermaas et al. 1987a). Mutant cells of this organism were grown in BG-11 supplemented with 5 mM glucose and the appropriate antibiotic(s) against which a mutant was resistant. The final concentration of antibiotic was 20, 30 and 7 μg/ml for kanamycin, spectinomycin and chloramphenicol, respectively. For obligate photoheterotrophic growth, a PS II electron transport inhibitor (25 μM atrazine) was added.

For preparation of thylakoids, cells were harvested and resuspended in 25 mM HEPES (pH = 7.5), 0.3 M sorbitol, 25 mM NaCl, 1 mM phenylmethylsulfonylfluoride, 1 mM ε-aminocaproic acid and 1 mM benzamidine (resuspension medium) to a concentration of 1 mg chlorophyll per ml. The cells were passed twice through a French Press at 1500 kg/cm². Unbroken cells were removed by centrifugation at 5000 × g for 10 min. The thylakoid fraction was collected by centrifugation at 37,000 rpm in a RP50-2 rotor (Hitachi) for 1 hour, and resuspended in the resuspension medium.

Sodium dodecyl sulfate-polyacrylamide gel electrophoresis (SDS-PAGE) of thylakoid proteins was performed essentially according to the method described by Chua (1980); however, both stacking and resolving gel contained 5.5 M urea, and 12.5% acrylamide was used for the resolving gel. Thylakoids were solubilized in a buffer containing 60 mM Tris/HCl (pH 8.0), 2% lithium dodecyl sulfate, 60 mM dithiothreitol and 20% (w/v) sucrose. After running the gels, thylakoid proteins were blotted (30 V; 16 hrs) onto nitrocellulose (BioRad) in 25 mM Tris, 192 mM glycine, 20% (v/v) methanol and 0.01% (w/v) sodium dodecyl sulfate. Coomassie Brilliant Blue staining of the blotted gel and Amido Black staining of the blot showed that this procedure yielded an effective protein transfer to the nitrocellulose.

To raise antibodies, CP43 and CP47 were prepared by isoelectric focusing of PS II membrane proteins in the presence of β-octylglucopyranoside. Spinach PS II membrane particles (Berthold et al. 1981) were treated with 0.8 M Tris/HCl (pH 8.5) and solubilized with 60 mM β-octylglucopyranoside in 10% (w/v) sucrose, 20 mM Tricine/NaOH (pH 8.5), 10 mM NaCl and 0.1% (w/v) Ampholine (pH 3–10, LKB). The sample was layered in the middle of a linear gradient of 0.05–0.2% (w/v) Ampholine (pH 4-6.5) and

0-20% (w/v) sucrose containing 60 mM β-octylglucopyranoside and electrofocused at 300 V for 12 hrs at 0° C. A green band at pI 5.4 was collected and centrifuged at 224,000 × g for 1 hr. Both CP43 and CP47 remained in the supernatant. This supernatant was diluted with 20 mM MES/NaOH (pH 6.5) and 10 mM NaCl to make a final concentration of 35 mM β-octylglucopyranoside, layered on a 30% (w/v) sucrose cushion containing 35 mM β-octylglucopyranoside, 20 mM MES/NaOH (pH 6.5) and 10 mM NaCl, and centrifuged at 224,000 × g for 5 hrs. CP47 and CP43 were recovered in the green precipitate and in the green band in the supernatant, respectively. D1 and D2 were prepared from PS II reaction center complexes (Ikeuchi et al., 1985) by gel electrophoresis. Protein bands were stained with 1% (w/v) Coomassie Brilliant Blue R-250 in water for 10 min, D1 and D2 bands were cut out, and then again subjected to gel electrophoresis. D1 and D2 were eluted from the gel strips by electrophoresis (Mendel-Hartvig 1982), dialyzed against water, and lyophilized. 50-100 μg of each protein was injected into a rabbit according to Chua et al. (1982).

The nitrocellulose thylakoid protein blots were probed with one of these polyclonal antibodies, which was detected using an assay kit (BioRad, Richmond CA) employing horse radish peroxidase (for D1 and D2 antibodies) or alkaline phosphatase (for CP43 and CP47 antibodies) conjugated to antibodies (from goat) against the rabbit antibody F_c portion, following the procedures recommended by the manufacturer of the assay kit.

Results

Mutant characteristics

Some properties of the mutants that have been used in this study have been summarized in Table 1. The genetic characteristics of some mutants (*psbB*-I1, *psbDC*-D1, *psbD*-S197Y and *psbD*-S214N) have been published (Vermaas et al. 1987a,b). The name of each mutant is composed of the targeted gene(s) for mutagenesis, followed by an abbreviation of the type of mutation made (Insertion, Deletion, Chimaeric construction, or Site-directed mutation), and a number. For site-directed mutations, the number represents the amino acid residue that has been mutated, and is followed by the one-letter code of the residue into which it has been changed. The nature of the genetic changes in the others has been summarized below.

The first five mutants listed in Table 1 were generated by transformation of wild-type *Synechocystis* 6803 with a construction containing the spinach *psbB* gene (coding for CP47) flanked on both sides by the flanking regions

Table 1. Site(s) of genetic modification in the *Synechocystis* 6803 mutants.

mutant	gel lane (Fig. 2–5)	modified gene	nature of modification	PS II ET	CP47	CP43	D2	D1
psbB-C1	2	⎫	part of gene replaced by *psbB*	+	++	++	++	++
psbB-C2	3	⎪	from spinach. It is not yet known	+	++	++	++	++
psbB-C3	4	⎬ *psbB*	how far the spinach sequence ex-	–	–	++	+	+
psbB-C4	5	⎪	tends in each case. *neo* insertion	–	–	++	+	+
psbB-C5	6	⎭	just beyond the 3′ end of *psbB*	+	++	++	++	++
psbB-I1	7	*psbB*	*neo* insertion in *psbB* 182 bp before 3′ end	–	–	+	–	–
psbC-I1	8	*psbC*	*neo* insertion in *psbC* 154 bp before the 3′ end	–	+	–	+	+
psbC-I2	9	beyond *psbC*	*neo* insertion 172 bp beyond the 3′ end of *psbC*	+	++	++	++	++
psbDC-D1	10	*psbD1* *psbD2* *psbC*	part of *psbD1* and *psbC* deleted and replaced by DNA fragment leading to Cm^R. Part of *psbD2* deleted and replaced by DNA fragment leading to Sm^R	–	–	–		–
psbD-S197Y	11	*psbD1* *psbD2*	single-base change in *psbD1*; leads to replacement of His-197 by Tyr in D2; part of *psbD2* replaced by Sm^R. *neo* inserted 172 bp beyond *psbC* 3′	–	+	*	–	–
psbD-S214N	12	*psbD1* *psbD2*	as *psbD*-S197Y, but containing a single-base change leading to replacement of His-214 by Asn in D2, and containing the wild-type His-197 in D2	–	–	–	–	–

CP47 and CP43 are encoded by *psbB* and *psbC*, respectively. The two copies of *psbD*, *psbD1* and *psbD2*, both code for the D2 protein. PS II ET: electron transport from water to plastoquinone. The absence of electron transport is indicated by "–", its presence by "+". Apparently normal levels of any of the four proteins is indicated by "++", decreased levels by "+", and apparent absence of any of the proteins from the thylakoid by "–". *: breakdown product detectable. Cm^R: chloramphenicol resistance; Sm^R: spectinomycin resistance; *neo*: DNA fragment containing the neomycin phosphotransferase gene leading to kanamycin resistance.

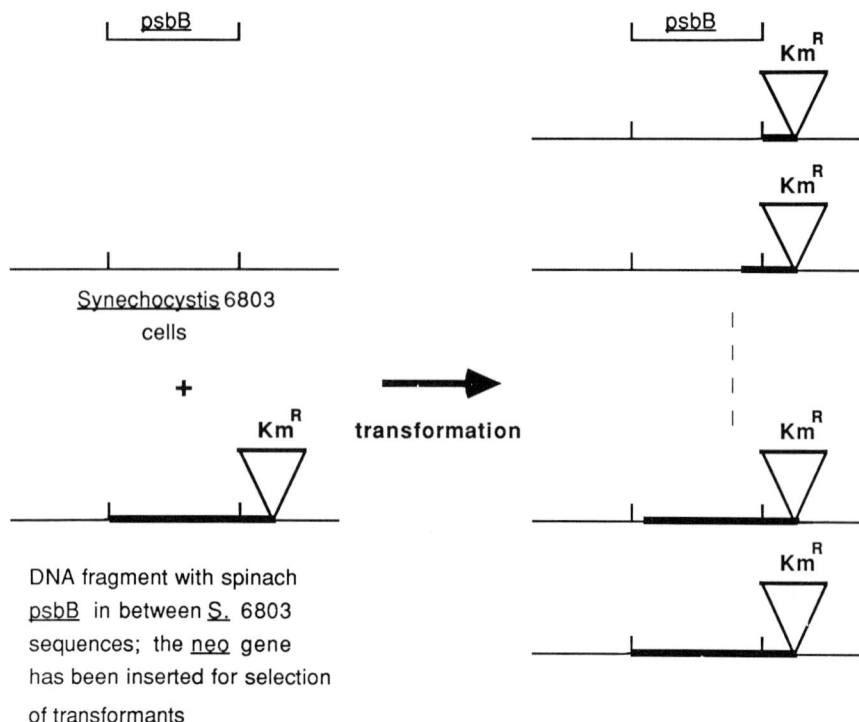

Fig. 1. Scheme for construction of cyanobacterial transformants that contain a chimeric spinach/*Synechocystis* 6803 *psbB* gene. Wild-type *Synechocystis* 6803 cells were transformed with a DNA fragment that contained: (1) a 0.2 kb *HpaI/NcoI* fragment upstream of *psbB* obtained from *Synechocystis* 6803 (——); (2) an *NcoI/DraI* fragment from spinach containing spinach *psbB* and a short (0.1 kb) downstream region (━━); (3) a 0.4 kb *DraI/KpnI* *Synechocystis* 6803 fragment from downstream of *psbB* (——); and (4) a DNA fragment containing a neomycin phosphotransferase gene (——) leading to kanamycin resistance (Km^R) between (2) and (3). The *NcoI* site in (1) had been induced by site-directed mutagenesis of a single base just outside the *psbB* coding region in order to allow an in-phase linkage of (1) and (2) at the *psbB* start site. For a precise location of the restriction sites, see the sequences as published by Vermaas et al. (1987a) and Morris and Herrmann (1984). After transformation, kanamycin-resistant transformants were selected. To obtain such transformants, cross over had to take place (1) in the homologous region downstream of *psbB* and the DNA fragment containing the neomycin phosphotransferase gene, and (2) in a region of homology between spinach and cyanobacterial *psbB*, or in the homologous region upstream of *psbB*. Various possible transformants have been indicated in the Figure. In principle a large number of different transformants is expected to be created, depending on where the cross over in *psbB* occurred.

of *psbB* from *Synechocystis* 6803 (see Fig. 1). For convenient selection of transformants, a DNA fragment containing the neomycin phosphotrans-ferase gene (the *neo* gene) had been inserted at the spinach/*Synechocystis* 6803 junction downstream of *psbB*. Integration of spinach *psbB* into the

1 2 3 4 5 6 7 8 9 10 11 12 M_r(kDa)

-94
-67

CP47→

-43

-30

-20

-14

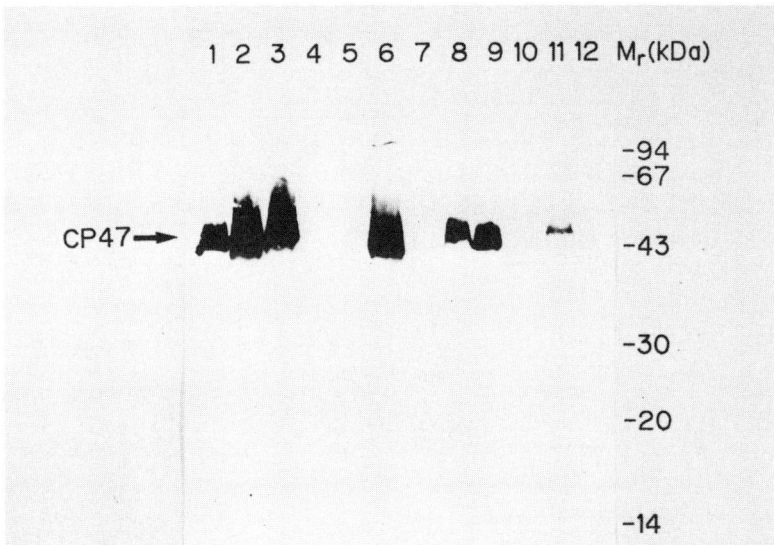

Fig. 2. Western blot of thylakoid proteins from wild type and mutant *Synechocystis* 6803 probed with rabbit polyclonal antibodies raised against the spinach CP47 protein. Per lane, 3 μg chlorophyll was loaded. The location of the molecular mass markers is indicated. Lane 1: wild type; lane 2-6: spinach/cyanobacterial *psbB* hybrids *psbB*-C1, *psbB*-C2, *psbB*-C3, *psbB*-C4, and *psbB*-C5, respectively; lane 7: *psbB*-I1 (interrupted *psbB*); lane 8: *psbC*-I1 (interrupted *psbC*); lane 9: *psbC*-I2 (interruption 172 bp beyond the end of *psbC*); lane 10: *psbDC*-D1 (deletion of a large part of both copies of *psbD* and of *psbC*); lane 11: *psbD*-S197Y (site-directed mutation of His-197 of D2 into Tyr); lane 12: *psbD*-S214N (site-directed mutation of His-214 of D2 into Asn). See Table 1 for more details on the mutants.

cyanobacterial chromosome leading to kanamycin-resistant transformants can occur either by cross over in both the completely homologous 5′ and 3′ flanking regions of the *Synechocystis* 6803 *psbB* gene, or both somewhere within *psbB* in a region showing high homology between the spinach and *Synechocystis* 6803 sequences and in the 3′ flanking region of the gene. Restriction mapping of transformant DNA showed that the various kanamycin-resistant transformants differed in the length of the spinach DNA that had been integrated into the cyanobacterial chromosome and that had replaced the corresponding part of the *Synechocystis* 6803 gene: in *psbB*-C2 and *psbB*-C5 only a limited portion in the 3′ region of the gene is of spinach origin, whereas in *psbB*-C1, *psbB*-C3 and *psbB*-C4 most of the *psbB* gene is from spinach (W.F.J. Vermaas and C. Bunch, unpublished results).

The mutant *psbC*-I1 has been interrupted by a DNA fragment containing the *neo* gene at the *Sma*I site 154 bp before the 3′ end of the gene. This mutant is devoid of oxygen-evolving activity, but shows primary charge separation (W. Vermaas, H. Koike, P. Mathis and Y. Inoue, unpublished

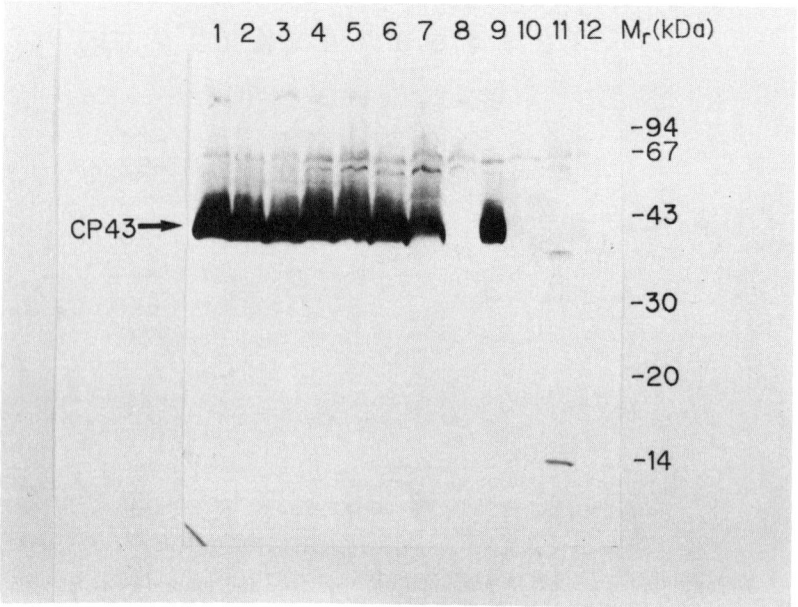

Fig. 3. Western blot of thylakoid proteins from wild type and mutant *Synechocystis* 6803 probed with rabbit polyclonal antibodies raised aginst the spinach CP43 protein. The lanes were loaded as in Fig. 2.

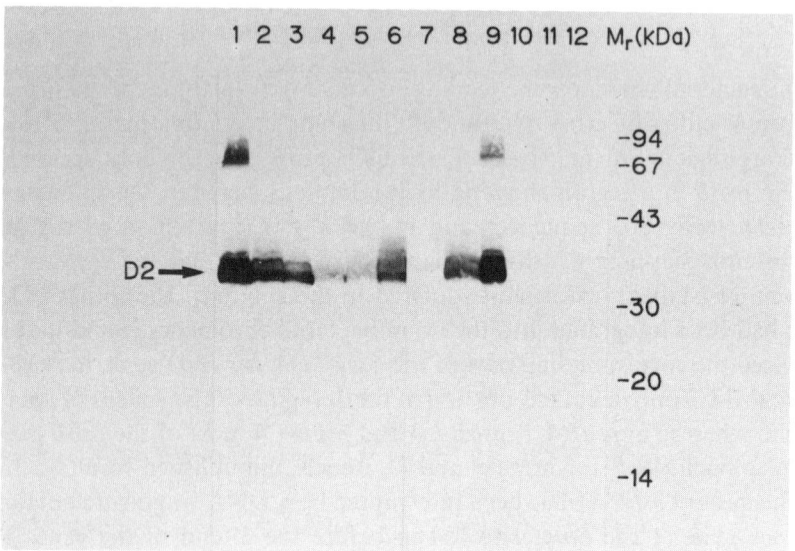

Fig. 4. Western blot of thylakoid proteins from wild type and mutant *Synechocystis* 6803 probed with rabbit polyclonal antibodies raised against the spinach D2 protein. The lanes were loaded as in Fig. 2.

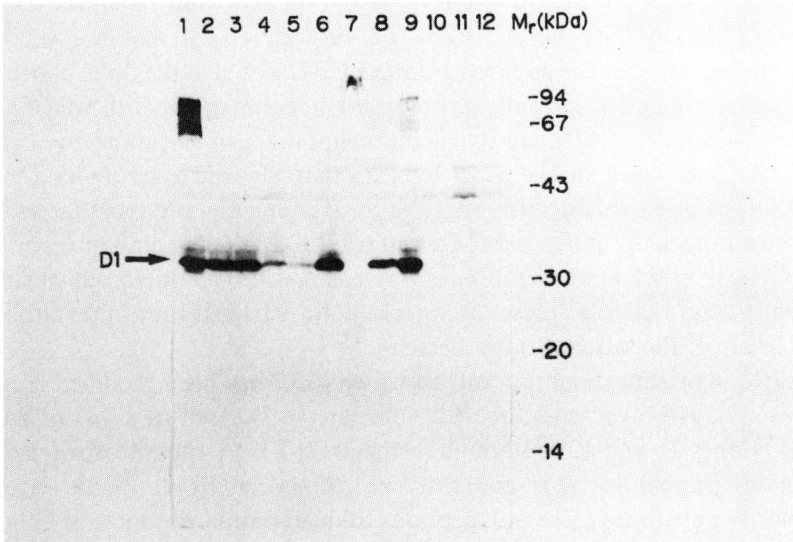

Fig. 5. Western blot of thylakoid proteins from wild type and mutant *Synechocystis* 6803 probed with rabbit polyclonal antibodies raised against the spinach D1 protein. The lanes were loaded as in Fig. 2.

data). To ascertain that the physiological effects observed were related primarily to the lack of an intact CP43 protein in the mutant, and not to synthesis inhibition of other proteins possibly encoded downstream of *psbC* under the same promoter, a DNA fragment with the *neo* gene was inserted at the *Xmn*I site 172 bp downstream of *psbC* rather than in *psbC* itself (mutant *psbC*-I2). In this case, no inhibition of oxygen-evolving activity was observed, suggesting that the inhibition of water-splitting in *psbC*-I1 indeed is due to the inhibition of synthesis of intact CP43.

Protein composition of the mutants

The thylakoid fraction from *Synechocystis* 6803 contains a large number of different proteins, making the evaluation of stained protein gels ambiguous. For this reason, blots of the protein gels were probed with antibodies specific for one of the PS II core complex components (CP47, CP43, D1 or D2). We have not yet raised antibodies against cyt b_{559}. The results are shown in Figs. 2–5.

CP47 appears to be present in the three photosynthetically competent mutants containing a hybrid *psbB* (Fig. 2; lanes 2,3 and 6) as well as in the mutants *psbC*-I1 and *psbC*-I2 (lanes 8 and 9), and some is present in the site-directed D2 mutant *psbD*-S197Y (lane 11) as was noticed before (Vermaas et al. 1987c). No CP47 was detected in the membrane fraction from the *psbD/psbC* double-deletion strain (*psbDC*-D1) and from *psbD*-S214N (the asn-214 D2 mutant) (lanes 10 and 12, respectively). It has not yet been established whether the apparently larger CP47 signal in the three photosynthetically competent hybrid *psbB* mutants in comparison with wild type is due to a larger CP47 quantity in the membrane, or is caused by a more diffuse CP47 band in the *psbB* hybrids than in wild type under the gel conditions used. It is striking that the *psbB* chimaeric mutants that are not photosynthetically active, *psbB*-C3 and *psbB*-C4, lack detectable amounts of CP47 in the thylakoid membrane, whereas it will be pointed out in subsequent paragraphs that these mutants contain CP43, D1 and D2, although the levels of the latter two are decreased.

CP43 is present in all mutants in which *psbB* has been modified (Fig. 3; lanes 2-7). However, site-directed mutations in D2 lead to a loss of native CP43 (lanes 11 and 12), although in the *psbD*-S197Y mutant smaller polypeptides present in trace amounts are recognized by antibodies against CP43. We attribute these polypeptides to degradation products of CP43 in the thylakoid, only accumulating in detectable quantities in the *psbD*-S197Y mutant. As might be expected, interruption of the *psbC* gene or deletion of this gene leads to the absence of proteins cross-reacting with the CP43 antibody in the thylakoid membrane (lanes 8 and 10). Insertion of the *neo* gene downstream of the *psbD1/C* operon does not lead to loss of CP43 (lane 9) or of any of the other PS II core polypeptides.

In the mutants studied, the D2 and D1 proteins appear to be "coupled" in the sense that all mutants either contain or do not contain both D1 and D2. The D1 and D2 proteins, or polypeptides that specifically cross-react with the D1 or D2 antibodies, could not be detected in the site-directed *psbD* mutants (Figs. 4 and 5; lanes 11 and 12), the *psbD/C* deletion mutant (*psbDC*-D1; lane 10), and the mutant in which *psbB* had been interrupted (*psbB*-I1; lane 7). Decreased levels of both D1 and D2 appeared to be present in the photosynthetically inactive *psbB* hybrid mutants (lanes 4 and 5). In the other mutants used in this study only relatively small deviations from wild-type D1 and D2 levels in the thylakoid could be detected. However, in the wild type and (to a lesser extent) in the *psbC*-I2 mutant (lanes 1 and 9, respectively) the D1 and D2 antibodies reproducibly recognized proteins in the higher molecular weight region (possibly oligomers of D1 and/or D2), whereas this is not the case in any of the other mutants. It is not yet clear what the reason for this different behaviour may be.

In both wild type and D2-containing mutants, typically two major bands (with apparent molecular weights of 34 and 36 kDa) that cross-react with the D2 antibody can be discerned, although the photographic reproductions have caused the two bands to merge in most lanes in Fig. 4. Similar D2 doublets observed in eukaryotic photosynthetic systems have been attributed to phosphorylation of D2: the upper band represents phosphorylated D2, the lower band non-phosphorylated D2 (Delepelaire, 1984). This suggests that *Synechocystis* 6803 D2 may also be phosphorylated.

Discussion

The data presented above show that modification of one of the genes encoding a component of the PS II core complex in most cases has drastic effects on the presence and/or stoichiometry of other PS II core components. From the results shown in Fig. 2-5, the following suggestions can be made regarding the roles of the individual core proteins in stable assembly and/or synthesis of other core proteins in PS II:
1. CP47 may be needed for stable integration of D1 and D2 and for PS II activity.
2. Integration of CP43 is relatively independent of the presence of other core proteins: after interruption of *psbC* the other polypeptides are stably integrated, whereas after interruption of the *psbB* gene CP43 is present.
3. The D2 residues proposed to bind PS II cofactors appear to be of direct importance to the stability of the PS II complex. In other words, the PS II cofactors (or prosthetic groups) may play a crucial role in the assembly and stabilization of the PS II complex.
4. In each of the mutants the relative amounts of D1 and D2 in the membrane seem to be comparable.

Before elaborating on these principles, it should be noted that we cannot yet distinguish whether the effects of gene modification on the presence of other proteins in the complex result from effects at the transcriptional, translational or post-translational level, since we have no data on protein synthesis rates nor on the occurrence of mRNA coding for the PS II core components in the mutants. In the discussion mainly translational and/or post-translational control mechanisms will be considered, since these appear to occur most frequently (Jensen et al. 1986, Bennoun et al. 1986, Erickson et al. 1986). However, it should be kept in mind that other control mechanisms cannot yet be ruled out.

CP47
Until recently, CP47 was assumed to bind P680 (Nakatani et al. 1984, Yamagishi and Katoh 1985, Satoh 1986). Now, however, it seems more likely that it primarily acts as an antenna protein close to the reaction center

(see, for example, Vermaas et al. (1987d) for a review). CP47 appears to be more intricately involved with the PS II reaction center complex than CP43 is, because it is relatively easy to prepare active reaction center particles containing CP47 but without CP43 (Yamagishi and Katoh 1984), whereas the photochemically active reaction center preparation that lack CP47 is also free of CP43 (Nanba and Satoh 1987). However, the data shown here indicate that CP47 is not "just" an antenna protein. The mutants psbB-C3 and psbB-C4, both without PS II activity, contain the putative reaction center proteins D1 and D2 (although in reduced amounts as compared to wild type) as well as CP43, but do not accumulate CP47 in their thylakoids. This implies that the presence of CP47 in the membrane is required, either for structural or for functional reasons, to obtain an active PS II complex. The reduced D1 and D2 contents in mutants psbB-C3 and psbB-C4, in the absence of significant steady-state concentrations of CP47 in the membrane, may seem inconsistent with the results obtained with psbB-I1, in which no intact CP47 can be synthesized due to insertion of a DNA fragment encoding neomycin phosphotransferase within the coding sequence of the psbB gene (Vermaas et al. 1987a) and in which no membrane-bound D1 or D2 is accumulated. A possible explanation for this apparent discrepancy is that CP47, which might be synthesized in psbB-C3 and psbB-C4, but not in psbB-I1, is required for stable insertion of D1 and D2, but that subsequent turn over of CP47 in the PS II complex does not destabilize D1 and D2 integrated in this complex. Regardless of the validity of this hypothesis, the function of CP47 clearly is not limited to light harvesting for PS II.

It is equally obvious that CP47 is not the only factor that could regulate assembly of other polypeptides: the site-directed D2 mutants are good examples of organisms in which a specific gene modification leads to destabilization and/or inhibition of formation of other PS II components. Similarly, psbA (encoding D1) and psbD mutants from Chlamydomonas reinhardii lack other PS II core components (Bennoun et al. 1986, Erickson et al. 1986). However, it is difficult to compare these data with published information on other C. reinhardii mutants that were defective in synthesis and/or assembly of PS II components (Jensen et al. 1986) as well as in certain maize mutants (Leto et al. 1985), since the mutation sites in those mutants have not yet been established. Moreover, since some of these C. reinhardii mutations, as well as the maize mutation, are nuclear encoded, these mutations did not occur in any of the PS II core genes.

CP43

PS II retains its primary functions after extraction of CP43 (Yamagishi and Katoh 1984, Boska et al. 1986), which suggests that CP43 is not essential for

PS II function and primarily serves as an antenna without direct effects on the PS II reaction center. On one hand, this interpretation appears to be an oversimplification: after interruption of the *psbC* gene overall PS II activity is absent, although PS II electron transport with diphenylcarbazide as electron donor occurs (W. Vermaas, H. Koike and Y. Inoue, in preparation). On the other hand, the absence of dramatic effects of inactivation of *psbC* (and, thus, inhibition of CP43 formation) on the composition of the remainder of the PS II complex indeed suggests that CP43 is not extensively involved in assembly of the PS II core complex or in regulation of the synthesis of its components. However, it should be noted that in the CP43-less mutant the quantity of PS II core complexes in the thylakoid appears to be significantly reduced in comparison with wild type, indicating that CP43 may have a stabilizing or regulatory effect on the PS II complex. Interestingly, CP43 not only appears to be rather unimportant for assembly of the PS II complex, but other components of the complex in some cases also seem to be unimportant for incorporation of CP43 into the membrane: in the CP47 mutant *psbB*-I1, CP43 is still present in the membrane, whereas CP47, D1 and D2 are not. However, this phenomenon is not observed in the site-directed D2 mutants, where an intact CP43 is absent from the membrane, together with D1 and D2, whereas in *psbD*-S214N also CP47 is absent. A model that could explain this seeming paradox is presented at the end of the Discussion.

D2

Studies on *C. reinhardii* mutants have shown that deletion or modification of *psbA* or *psbD*, encoding D1 and D2, respectively, in most cases leads to a rapid turnover of the other PS II core proteins (Bennoun et al., 1986; Erickson et al., 1986) without a significant effect on transcription and translation. The site-directed mutations in the *psbD1* gene (and deletion of a large part of the other *psbD* gene, *psbD2*) described here may resemble the effects of modified *psbD* or *psbA* genes in *C. reinhardii*. In many respects the site-directed D2 mutations also resemble the double-deletion mutant (*psbDC*-D1), in which both *psbD* genes as well as the *psbC* gene linked to one of the *psbD* genes (*psbD1*) have been inactivated. This implies that the single amino acid change in D2 is sufficient to completely destabilize the D2 protein, under the reasonable assumption that the single base change will not have effects on translation and turnover of the mRNA or on the transcription of the gene.

The obvious question to raise at this point is why the mutation of the His-197 or His-214 residue has such dramatic effects on the stability of D2, and, hence, the other PS II core complex proteins. As was discussed in

Vermaas et al. (1987c), these histidine residues are predicted to be involved in the binding of P680, and of Q_A and Fe^{2+}, respectively, on the basis of homology with the reaction center proteins from *Rhodopseudomonas viridis* for which the crystal structure has been resolved (Deisenhofer et al. 1984 and 1985, Michel et al. 1986). This leads to the suggestion that binding of these prosthetic groups to D2 is a crucial factor in stabilization of D2, and thus PS II, in the membrane. A cofactor-binding requirement for protein stabilization is not uncommon. A similar phenomenon has been observed in antenna complexes (Bennett 1981, Youvan and Daldal 1986, Klug et al. 1986). It is not yet clear why the two site-directed mutants differ in the amount of CP43 breakdown products and CP47 in the membrane. Whatever is the mechanism, it is obvious that stable membrane-integrated D2 is crucial for assembly of other components of the PS II core complex. This is as would be expected from the proposed function of the D2 polypeptide as one of the reaction center proteins (Nanba and Satoh 1987, Vermaas et al. 1987c).

A possible scenario of complex assembly

The data discussed above indicate how much the PS II core proteins appear to depend on each other for assembly into a stable PS II complex. At this moment it is still too early to formulate a comprehensive model of the sequence of events of complex assembly in wild type and mutants, since we lack information regarding mRNA levels and translation, and protein synthesis rates. However, based on the data outlined in this paper we will describe a possible scenario for complex assembly by which the observations can be explained.

We propose that D1 and D2 first are incorporated into the membrane as an unstable pre-complex, which is stabilized by CP47 incorporation. CP43 may become part of the pre-complex even in the absence of CP47. If CP47 is absent, D1 and D2 turn over very rapidly whereas CP43 turns over much less rapidly (mutant *psbB*-I1). In the presence of CP47, none of the components turn over very rapidly, and all PS II core components accumulate in the thylakoid (wild type and mutants *psbC*-I2, *psbB*-C1, *psbB*-C2 and *psbB*-C5). Greatly reduced stability of CP47 in the thylakoid (mutants *psbB*-C3 and *psbB*-C4) leads to intermediate levels of D1 and D2. To explain this, we propose that, after formation of the membrane-integrated D1/D2/CP47 complex, turnover of CP47 does not lead to a greatly increased turnover rate of D1 and D2. This hypothesis implies that the reason for decreased D1 and D2 levels in the mutants *psbB*-C3 and *psbB*-C4 is the low concentration of integrated (or integratable) CP47 rather than increased turnover rates of integrated D1 and D2. Obviously, experiments can be designed to test these hypotheses.

Absence of CP43 has a relatively small effect on the stability of the other core components in the membrane (mutant *psbC*-I1). If synthesis of the D2 protein cannot occur, the D1/D2 complex cannot form, and therefore CP47 is not incorporated into the membrane (mutant *psbDC*-D1). If D2 cannot bind its proper prosthetic groups, the D1/D2 complex will turn over very rapidly, without allowing significant incorporation of CP47 and/or CP43. Assuming that there are differences in D1/D2 stability depending on which prosthetic groups can still be bound by D2, the differences in CP43-break-down-product and CP47 content between the two site-directed D2 mutants may thus be explained by a slightly higher stability of the D1/D2 complex of the *psbD*-S197Y mutant than of the *psbD*-S214N mutant (but both stabilities being too low to detect any D1 or D2 by Western blotting).

This scenario described above can explain the results presented in this paper as well as the results from *C. reinhardii psbA* and *psbD* mutants (Bennoun et al. 1986, Erickson et al. 1986). However, other scenarios (perhaps including "docking proteins") could be imagined that would explain the data equally well. Therefore, this model has been proposed as a basis for further experimentation and testing. It should be stressed that this model is only meant to describe *de novo* synthesis of PS II complexes, and does not address the problems raised by unequal rates of turnover of the various components of the PS II complex. Moreover, it should be kept in mind that cyt b_{559} has not been evaluated in this study, but should be taken into account since it is known to be an integral part of the PS II core complex (Satoh 1985, Arntzen and Pakrasi 1986, Nanba and Satoh 1987).

In conclusion, the *Synechocystis* 6803 PS II mutants provide valuable experimental material for studies of PS II function and assembly. However, to obtain more data for formulation and testing of detailed hypotheses on the function of each of the proteins in the assembly process, it will be necessary to generate more mutants as well as to characterize protein synthesis rates and mRNA levels of the existing mutants.

Acknowledgements

This study was supported in part by a grant on Solar Energy Conversion by means of Photosynthesis awarded to the Institute of Physical and Chemical Research (RIKEN) by the Science and Technology Agency of Japan (M.I. and Y.I.), and by a grant from the University Research Fund at Arizona State University (W.V.). At RIKEN, Wim Vermaas was supported by the US-Japan Program of Cooperation on Photoconversion and Photosynthesis. W. V. thanks Candace Bunch for her technical assistance.

References

Arntzen C J and Pakrasi H B (1986) Photosystem II reaction center: polypeptide subunits and functional cofactors. In: Staehelin L A and Arntzen C J (eds) Encyclopedia of Plant Physiology, New Series, Vol. 19, pp 457–467. Berlin: Springer

Bennett J (1981) Biosynthesis of light-harvesting chlorophyll a/b proteins. Eur J Biochem 118: 61–70

Bennoun P, Spierer-Herz M, Erickson J, Girard-Bascou J, Pierre Y, Delosme M and Rochaix J-D (1986) Characterization of photosystem II mutants of *Chlamydomonas reinhardii* lacking the *psbA* gene. Plant Mol Biol 6: 151–160

Berthold D A, Babcock G T and Yocum C F (1981) A highly resolved, oxygen-evolving Photosystem II preparation from spinach thylakoid membranes. FEBS Lett 134: 231–234

Bishop N I (1982) Isolation of mutants of *Scenedesmus obliquus* defective in photosynthesis. In: Edelman M, Hallick R B and Chua N-H (eds) Methods in Chloroplast Molecular Biology, pp 51–63. Amsterdam: Elsevier

Boska M, Yamagishi A and Sauer K (1986) EPR signal II in cyanobacterial Photosystem II reaction-center complexes with and without the 40 kDa chlorophyll-binding subunit. Biochim Biophys Acta 850: 226–233

Chua N-H (1980) Electrophoretic analysis of chloroplast proteins. Meth in Enzymol 69: 434–446

Chua N-H, Bartlett S G and Weiss M (1982) Preparation and characterization of antibodies to chloroplast proteins. In: Edelman M, Hallick R B and Chua N-H (eds) Methods in Chloroplast Molecular Biology, pp 1063–1080. Amsterdam: Elsevier

Deisenhofer J, Epp O, Miki K, Huber R and Michel H (1984) X-ray structure analysis of a membrane-protein complex: electron density map at 3 Å resolution and a model of the chromophores of the photosynthetic reaction center from *Rhodopseudomonas viridis*. J Mol Biol 180: 385–398

Deisenhofer J, Epp O, Miki K, Huber R and Michel H (1985) Structure of the protein subunits in the photosynthetic reaction centre of *Rhodopseudomonas viridis* at 3 Å resolution. Nature 318: 618–624

Delepelaire P (1984) Partial characterization of biosynthesis and integration of the Photosystem II reaction centers in the thylakoid membrane of *Chlamydomonas reinhardii*. EMBO J 3: 701–706

Erickson J M, Rahire M, Malnoe P, Girard-Bascou J, Pierre Y, Bennoun P and Rochaix J-D (1986) Lack of the D2 protein in a *Chlamydomonas reinhardtii psbD* mutant affects photosystem II stability and D1 expression. EMBO J 5: 1745–1754

Golden S S, Brusslan J and Haselkorn R (1987) Genetic engineering of the cyanobacterial chromosome. Meth in Enzymol 153: 215–231

Ikeuchi M, Yuasa M and Inoue Y (1985) Simple and discrete isolation of an O_2-evolving PS II reaction center complex retaining Mn and the extrinsic 33 kDa protein. FEBS Lett 185: 316–322

Jensen K H, Herrin D L, Plumley F G and Schmidt G W (1986) Biogenesis of Photosystem II complexes: transcriptional, translational, and posttranslational regulation. J Cell Biol 103: 1315–1325

Klug G, Liebetanz R and Drews G (1986) The influence of bacteriochlorphyll synthesis on formation of pigment-binding proteins and assembly of pigment protein complexes in *Rhodopseudomonas capsulata*. Arch Microbiol 146: 284–291

Leto K J, Bell E and McIntosh L (1985) Nuclear mutation leads to an accelerated turnover of chloroplast-encoded 48 kd and 34.5 kd polypeptides in thylakoids lacking photosystem II. EMBO J 4: 1645–1653

Mendel-Hartvig I (1982) A simple and rapid method for the isolation of peptides from sodium dodecyl sulfate-containing polyacrylamide gels. Anal Biochem 121: 215–217

Metz J G, Pakrasi H B, Seibert M and Arntzen C J (1986) Evidence for a dual function of the herbicide-binding D1 protein in photosystem II. FEBS Lett 205: 269–274

Michel H, Epp O and Deisenhofer J (1986) Pigment-protein interactions in the photosynthetic reaction centre from *Rhodopseudomonas viridis*. EMBO J 5: 2445–2451

Miles D, Markwell J P and Thornber J P (1979) Effect of nuclear mutation in maize on photosynthetic activity and content of chlorophyll-protein complexes. Plant Physiol 64: 690–694

Morris J and Herrmann R G (1984) Nucleotide sequence of the gene for the P680 chlorophyll *a* apoprotein of the Photosystem II reaction center from spinach. Nucl Acids Res 12: 2837–2850

Nakatani H Y, Ke B, Dolan E and Arntzen C J (1984) Identity of the photosystem II reaction center polypeptide. Biochim Biophys Acta 765: 347–352

Nanba O and Satoh K (1987) Isolation of a photosystem II reaction center consisting of D-1 and D-2 polypeptides and cytochrome *b*-559. Proc Natl Acad Sci USA 84: 109–112

Pakrasi H B, Williams J G K and Arntzen C J (1987) Genetically engineered cytochrome b559 mutants of the cyanobacterium, *Synechocystis* 6803. In: Biggins J (ed.) Progress in Photosynthesis Research, Vol. 4, pp. 813–816. Dordrecht: Martinus Nijhoff Publishers

Satoh K (1985) Protein-pigments and Photosystem II reaction center. Photochem Photobiol 42: 845–853

Satoh K (1986) Photosystem II particles largely depleted in the two intrinsic polypeptides in the 30 kDa region from *Synechococcus* sp.; identification of a subunit which carries the photosystem II reaction center. FEBS Lett 204: 357–362

Sherman L A, Golden S S and Vann C (1984) Transformation and cloning vectors of the cyanobacterium Anacystis nidulans. In: Thornber J P, Staehelin L A, Hallick R B (eds) Biosynthesis of the Photosynthetic Apparatus pp 357–379. New York: A R Liss, Inc

Vermaas W F J, Williams J G K, Rutherford A W, Mathis P and Arntzen C J (1986) Genetically engineered mutant of the cyanobacterium *Synechocystis* 6803 lacks the photosystem II chlorophyll-binding protein CP-47. Proc Natl Acad Sci USA 83: 9474–9477

Vermaas W F J, Williams J G K and Arntzen C J (1987a) Sequencing and modification of *psb*B, the gene encoding the CP-47 protein of Photosystem II, in the cyanobacterium *Synechocystis* 6803. Plant Mol Biol 8: 317–326

Vermaas W F J, Williams J G K, Chisholm D A and Arntzen C J (1987b) Site-directed mutagenesis in the photosystem II gene *psb*D, encoding the D2 protein. In: Biggins J (ed.) Progress in Photosynthesis Research, Vol. 4, pp 805–808. Dordrecht: Martinus Nijhoff Publishers

Vermaas W F J, Williams J G K and Arntzen C J (1987c) Site-directed mutations of two histidine residues in the D2 protein inactivate and destabilize Photosystem II in the cyanobacterium *Synechocystis* 6803. Z. Naturforsch 42c: 762–768

Vermaas W F J, Pakrasi H B and Arntzen C J (1987d) Photosystem II and inhibition by herbicides. In: Newman D and Wilson K (eds) Models in Plant Physiology and Biochemistry, Vol. I, pp 9–12. Boca Raton: CRC Press

Williams J G K (1988) Construction of specific mutations in the Photosystem II photosynthetic reaction center by genetic engineering methods in the cyanobacterium *Synechocystis* 6803. Meth in Enzymol (in press)

Yamagishi A and Katoh S (1984) A photoactive Photosystem II reaction center complex lacking a chlorophyll-binding 40 kDa subunit from the thermophilic cyanobacterium *Synechococcus* sp. Biochim Biophys Acta 765: 118–124

Yamagishi A and Katoh S (1985) Further characterization of the two Photosystem II reaction center complex preparations from the thermophilic cyanobacterium *Synechococcus* sp. Biochim Biophys Acta 807: 74–80

Youvan D C and Daldal F (eds) (1986) Microbial Energy Transduction: Genetics, Structure and Function of Membrane Proteins. Cold Spring Harbor: Cold Spring Harbor Laboratory

Govindjee et al. (eds), Molecular Biology of Photosynthesis: 407–421
© 1988 Kluwer Academic Publishers

Regular paper

The Q_B site modulates the conformation of the photosystem II reaction center polypeptides

ACHIM TREBST, BRIGITTE DEPKA, BERND KRAFT & UDO JOHANNINGMEIER
Department of Biology, Ruhr-University of Bochum, P.O. Box 10 21 48, D-4630 Bochum 1, FRG

Received 20 October 1987; accepted 30 January 1988

Key words: chloroplast, herbicide, plastoquinone, Q_B site, trypsin, thylakoid

Abstract. The sensitivity of the D-1 and D-2 polypeptide subunits of photosystem II towards trypsin treatment of the thylakoid membrane has been probed with specific antibodies. As long known, electron flow from water to ferricyanide becomes inhibitor insensitive after this trypsin treatment. We show that under these conditions the D-2 polypeptide is cut by trypsin at arg 234. Also the D-1 polypeptide is cut, probably at arg 238. When short time trypsination of the membrane is done in the presence of inhibitors, electron flow also becomes inhibitor insensitive and the D-2 polypeptide is still cut, but the D-1 polypeptide is cut only under certain conditions. A protection of the D-1 polypeptide is possible with inhibitors of photosystem II of the DCMU/triazine-type and with an artificial acceptor quinone, but not with inhibitors of the phenol-type. In hexane extracted membranes plastoquinone has been removed from the Q_B site. Both the D-1 and D-2 polypeptides are more trypsin sensitive in such preparations. The D-1, but not the D-2 polypeptide is protected when plastoquinone has been readded to the membrane before the trypsin digestion.

The results show that plastoquinone, artificial quinones and inhibitors of photosystem II at the Q_B site, but also carotene to a lesser extent, have an effect on the conformation of both the D-1 and D-2 polypeptide. it is postulated that the amino acid sequence around arginine 238 of the D-1 polypeptide is part of the Q_B binding niche. Furthermore this sequence is modified or its conformation is changed if the Q_B site is occupied by either plastoquinone or a DCMU-type inhibitor because under these conditions arginine 238 is less accessible to the trypsin. If the Q_B site, however, is empty, the amino acid sequence with arg 238 is very trypsin sensitive. This property of modulation or the conformation of the amino acid sequence of the D-1 polypeptide by the state of the Q_B site is likely to be relevant also for the events in the rapid turnover of the D-1 polypeptide.

Abbreviations: BNT – 2-bromo-4-nitro-thymol, DCMU – dichlorophenyldimethylurea, PMSF – phenylmethylsulfonylfluoride, SDS — sodium dodecylsulfate

Introduction

The trypsin sensitivity of photosynthetic electron flow via Q_B in photosystem II is long known (Regitz and Ohad 1976). After trypsin treatment of the thylakoid membrane oxygen is still evolved and ferricyanide reduced, but the reaction is no longer inhibitor sensitive (Regitz and Ohad 1976, Renger 1976, Renger et al. 1976, Renger 1977, Trebst 1979, Böger and Kunert 1979, Mattoo et al. 1981, Steinback et al. 1981, 1982, Trebst and Depka 1985). Renger (1976) explained this by suggesting that trypsin induces an accessibility of Q_A to the hydrophilic ferricyanide bypassing the inhibitor sensitive Q_B site. It has been assumed (Renger 1977, Trebst 1979, Böger and Kunert 1979, Mattoo et al. 1981, Steinback et al. 1981, 1982, Trebst and Depka 1985) that these effects are due to a trypsin attack on the herbicide binding protein D-1 because of the known trypsin sensitivity of this polypeptide (Mattoo et al. 1981, Steinback et al. 1982). The trypsin effect has been used to establish that the D-1 polypeptide is the product of the *psbA* gene (Mattoo et al. 1981).

Recent developments on the architecture of photosystem II suggested that the D-1 protein is not only the Q_B and herbicide binding protein, but is also part of the reaction center of photosystem II, together with the Q_A binding polypeptide D-2 (Deisenhofer et al. 1984, Trebst 1986, Nanba and Satoh 1987). We have recently reported on the high trypsin sensitivity of the D-2 polypeptide in the membrane probed with site directed antibodies that showed that the polypeptide is cut at arg 234 (Geiger et al. 1987). We wish to extend and document further on the effect of trypsin on the D-2 polypeptide. The results indicate that the main cause for the DCMU insensitivity of the electron flow in trypsin treated thylakoid membranes is a trypsin cut of the D-2 subunit rather than of the D-1 subunit. In trypsination of the membrane in the presence of excess plastoquinone or of inhibitors of the DCMU-type, the D-1 polypeptide remains stable, whereas the D-2 polypeptide is still cut. On the other hand, in the presence of the phenol-type inhibitors, both the D-1 and D-2 polypeptide are trypsin sensitive. The effects of plastoquinone and of inhibitors on the D-1 subunit stability are interpreted to show a modulation of photosystem II conformation by the Q_B binding site.

Materials and methods

Spinach thylakoids were prepared according to standard methods (Robinson and Yocum 1980) in 0.4 M NaCl and 20 mM tris buffer pH 8.0 and the chloroplasts broken in 20 mM tris buffer pH 8.0 + 0.15 M NaCl. Their photosynthetic activity (DCMU sensitive) was 190 μmoles O_2/mg chlorophyll/h with methylviologen as acceptor.

For trypsination according to (Völker et al. 1985, Trebst and Depka 1985) spinach thylakoid membranes with 300 μg chlorophyll in 3 ml ± inhibitor were incubated with 100 μg trypsin (Boehringer) in 300 mM MES buffer pH 6.5 and 6 mM $MgCl_2$ at room temperature for the time indicated. The reaction was stopped with 1.5 mg aprotinin (Trasylol^R Bayer) and 2 mM PMSF (the use of Trasylol^R is essential, as PMSF alone will not stop trypsin action completely and degradation continues on the gel). The photosynthetic activity of the membrane after two minutes trypsination (now DCMU insensitive) was 150 μmoles O_2/mg chlorophyll/h with ferricyanide as acceptor.

For hexane extraction according to Hirayama and Kabata (1977) thylakoid membranes with 30–50 mg chlorophyll in 50 ml 5 mM hepes buffer pH 6.5 were quickly frozen at $-180°$, freeze dried and then extracted with 250 ml hexane (pro analysis) for 10 minutes at room temperature under nitrogen. The extract obtained was used for the reconstitution experiments and was chromatographed for the preparation of pure plastoquinones A, B and C, tocopherol and menaquinone. The particles were washed with 100 ml hexane. For readdition of the extract or of the quinones and other components in Figs. 6 and 7 (the compounds dissolved in 3 ml hexane) were added to 1–2 mg dry extracted thylakoid membranes, stirred for 5 minutes and then the hexane evaporated to dryness of the membrane. The dried powder was resuspended in 10 mM tricine buffer pH 8.0, 0.4 M sucrose and 10 mM NaCl tested for function and then used in the trypsination experiments.

For the Western blots (Towbin et al. 1979) spinach thylakoid membranes were dissolved in 5% SDS, 15% glycerin, 50 mM tris pH 6.8, 2% mercaptoethanol at room temperature and the polypeptides separated by polyacrylamide gel (15%) electrophoresis at room temperature for 48 h. The gels were blotted for 4 h and 0.4 A at 10 °C on nitrocellulose in 25 mM tris, 192 mM glycine and 20% methanol. After saturation with 3% gelatine in tris buffer pH 7.5 the first antibody was allowed to react over night at room temperature in 1% gelatine. After washing in tris and 0.05% tween 20 the second antibody (horseradish peroxidase conjugated) was allowed to react in 1% gelatine for 1 h, and developed with HPR colour development and 0.005% H_2O_2. The antibody preparations against the expressed fusion protein of a construct with a segment of the psbA gene, corresponding to amino acids 167–353 were described by Johanningmeier (1987) and the D-2 antibodies against a synthetic oligopeptide, corresponding to amino acids 235–241 by Geiger et al. (1987). For an approximate estimation of the antibody stain the photographs of the gels of the Western blots were scanned in an ISCO gel scanner at 580 nm and the integral of the peaks calculated in percent of the control.

Results

It is well established that after trypsin treatment of the thylakoid membrane photosynthetic electron flow through photosystem II from water to ferricyanide is not inactivated, but still proceeds with an appreciable rate, although lower than the control. The ferricyanide Hill reaction obtained after trypsin treatment of the membrane is, however, insensitive to photosystem II inhibitors, like DCMU and others (Regitz and Ohad 1976, Renger 1976, Renger et al. 1976, Renger 1977, Trebst 1979, Böger and Kunert 1979, Mattoo et al. 1981, Steinback et al. 1981, 1982, Trebst and Depka 1985). This is shown again in Figs. 1 and 2 and in Table 1 to indicate the principal properties of the thylakoid system studied in this paper. The DCMU and bromonitrothymol (BNT) insensitivity of electron flow after trypsin treatment of the membrane is clearly seen in Fig. 1. If the inhibitor sensitivity is measured after trypsin treatment of the membrane in the presence of the inhibitors, then the loss of inhibition is not complete (Fig. 2), as already indicated for DCMU in the results by Renger et al. (1977) and Mattoo et al. (1981), but is more appreciable in our experiments. Maximal

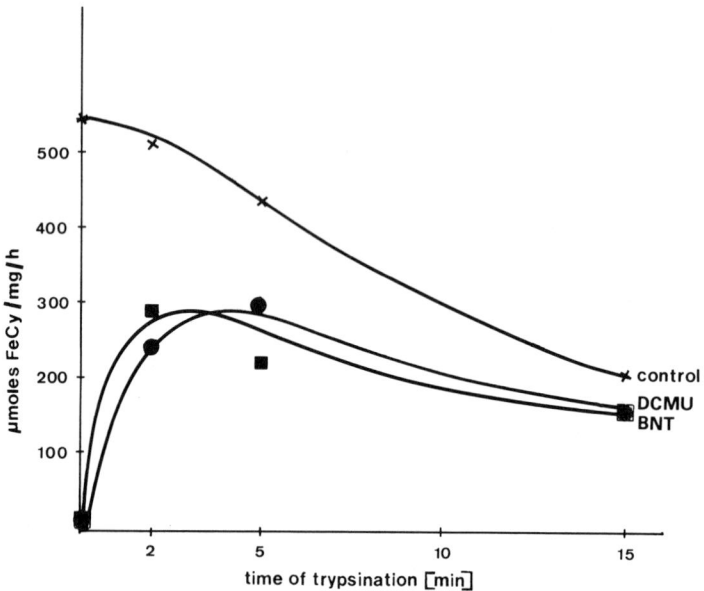

Fig. 1. Inhibitor sensitivity of photosynthetic electron flow after trypsination of the thylakoid membrane. Ferricyanide reduction with isolated spinach thylakoid preparations before and after trypsination was followed spectrophotometrically as described in Methods. Ten μM DCMU or BNT respectively were added when indicated.

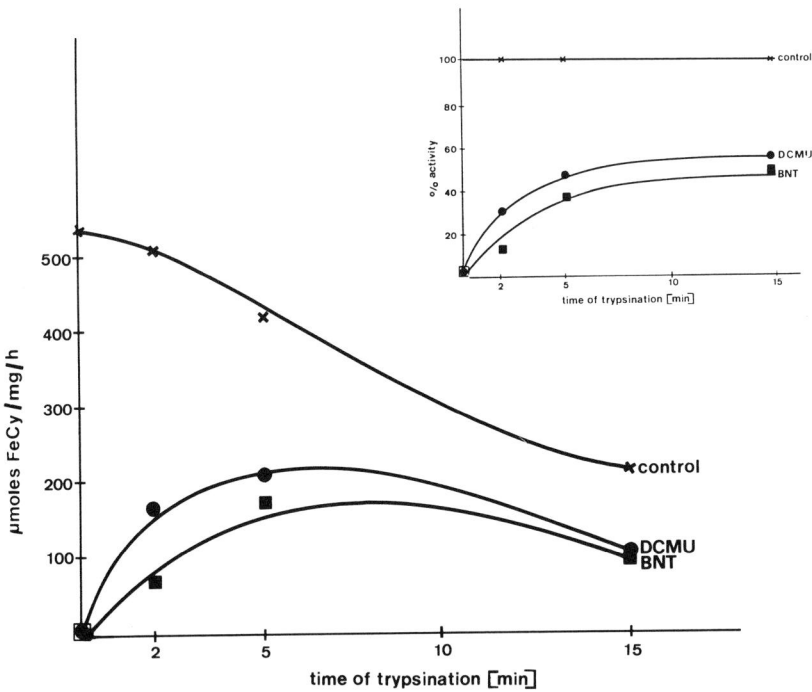

Fig. 2. Inhibitor insensitivity of photosynthetic electron flow after trypsination of the membrane in the presence of 10 mM inhibitor. Conditons as in Fig. 1, but note that the inhibitors were already present during trypsination and also added during activity measurement.

inhibition is only 50% even at very high concentrations of either DCMU or BNT (Fig. 2).

Western blots of the thylakoid polypeptides separated on SDS gels before and after trypsin treatment probed with D-1 and D-2 antibodies show that both the D-1 and D-2 polypeptides – in addition to the light harvesting complex and subunits of the coupling system as seen in the comassie blue stain – have been cut by trypsin (Fig. 3). After 5 minutes trypsination of the membrane (under our conditions of Völker et al. (1985) and Trebst and Depka (1985)) both polypeptides have disappeared. A stable degradation product of the D-2 polypeptide of about 10 kDa is spotted with the site directed D-2 antibody (against amino acids 235–241) as already reported (Geiger et al. 1987). In the D-1 experiments sometimes an unstable band at 30.5 kDa is observed with the antibody against a fusion protein with amino acids 167 to 353 in E. coli (Johanningmeier 1987). Another degradation product of the D-1 protein appears at about 8 kDa in the experiments when BNT had been added (see Fig. 3) that disappears after longer trypsination times. If the trypsination of the thylakoid membrane is done in the presence

Table 1. Content of the D-1 and D-2 polypeptide subunits in spinach thylakoids after trypsination in the absence or presence of photosystem II inhibitors.

μM	Inhibitor	D-1 Time of trypsination				D-2 Time of trypsination				Photosynthetic activity (μmoles FeCy/mg chlorophyll.h)	
		0′	2′	5′	15′	0′	2′	5′	15′	Before	After 5′ trypsination
	none	100	70	11	0	100	40	12	5	540	432
10	DCMU		100	100	41		85	42	21	0	312
20	atrazine		100	74	26		73	48	17	0	300
20	metribuzin		64	36	29		55	50	18	0	313
10	thiazclyliden-ketonitrile[a]		100	74	0		46	37	5	0	230
300	phenanthroline		45	45			32	17	0	0	153
20	hydroxypyridine[b]		46	21	0		32	0	0	0	272
10	BNT		28	0	0		38	11	0	0	250
100	methylenedioxy-dimethylbenzo-quinone		41								

[a] 2[4-(2,4-dichlorophenyl)-2,3-dihydro-thiazol-2-yliden]-3-oxo-4-phenyl-butyronitrile.
[b] 2-fluoro-tribromo-4-hydroxypyridine.

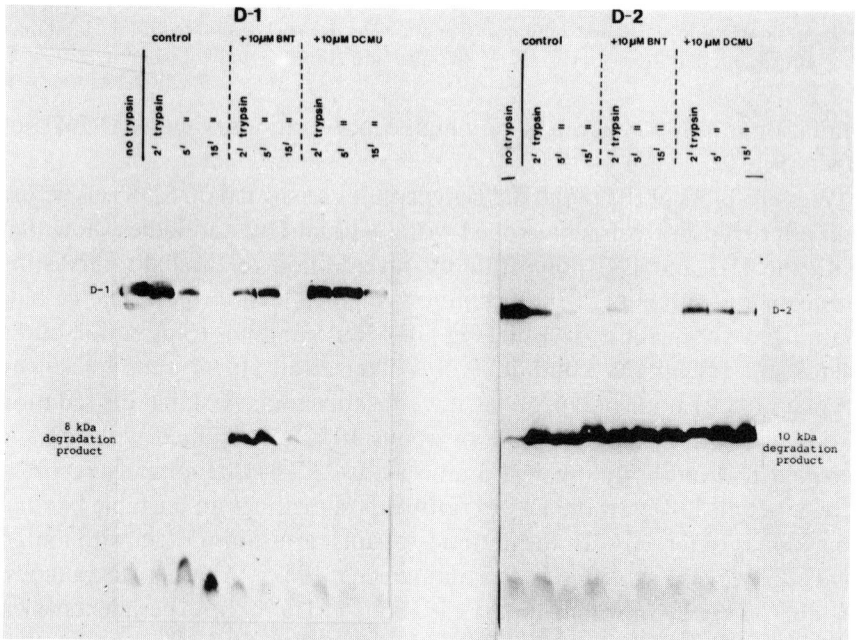

Fig. 3. Western blot with D-1 and D-2 antibodies of thylakoid membrane polypeptides separated by PAGE electrophoresis after trypsination of the membrane in the absence or presence of 10 mM inhibitor.

of DCMU, however, the result is quite different. The D-1 polypeptide has not disappeared in short time trypsination experiments and is protected even at 15 minutes trypsination. A similar observation has been made by Mattoo et al. (1981) following radioactive labeling of the D-1 polypeptide. The protection by DCMU is concentration dependent (Fig. 4). The degradation of the D-2 polypeptide by trypsin in the presence of DCMU is also less severe than the control, but still appreciable. Other inhibitors of photosystem II, however, BNT (Trebst and Draber 1979), and a bromohydroxypyridine have no protective, possibly even an enhancing effect on the trypsin sensitivity of either the D-1 or D-2 polypeptide (Fig. 3 and Table 1).

The extent of D-1/D-2 polypeptide disappearance (Fig. 5) was measured from the stain of the Western blots like in Fig. 3. It shows clearly the difference in protection of the D-1 vs the D-2 polypeptide by the presence of DCMU during the trypsin treatment for 5 minutes. Different inhibitors of photosystem II fall into two categories in protection of the D-1 protein from trypsin attack. Whereas the 'classical' inhibitors of the DCMU/triazine-type protect the D-1 protein, the phenol-type inhibitors (Trebst and Draber 1979, Trebst 1987) like phenanthroline, hydroxypyridine derivatives (Trebst et al. 1985) or thiazolyliden-ketonitrile (Bühmann et al. 1987) do not (Table 1). The effect of protection is time dependent and not absolute. DCMU and BNT are representing the extremes in strong stabilization vs no protection. There is also a stabilizing effect on the D-2 polypeptide.

The inhibitors of photosystem II are assumed to displace plastoquinone from the Q_B site in the membrane. Also artificial electron acceptors for photosystem II substitute for plastoquinone at the Q_B site. Indeed, as the data in Table 1 show, the artificial acceptor methylenedioxydimethyl-p-

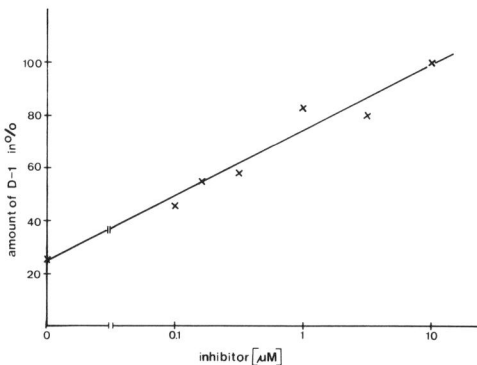

Fig. 4. Dependence of the protection of the D-1 polypeptide from trypsination on the concentration of DCMU present. Indicated is the amount in percent of control of the D-1 polypeptide that is still present in the membrane after trypsination.

414

benzoquinone shows a small, but definite protection of the D-1 polypeptide when present during the trypsin treatment of the membrane. The effect of plastoquinone itself on the trypsin sensitivity of the D-1 protein has been probed in hexane extracted chloroplasts. It has been often studied that thylakoids extracted with hexane under mild conditions have lost the plasto-quinone pool and Q_B, but retain Q_A (Cox and Bendall 1974, Knaff et al. 1977, Hirayama and Kabata 1977). Electron flow in hexane extracted thylakoid membranes can be restored by addition of pure plastoquinone or the extract (which contains carotenes, tocopherols, vitamin K_1 and other lipophilic compounds in addition to the three plastoquinone A, B and C). In hexane extracted thylakoid membranes the sensitivity of the D-1 and D-2 polypeptide towards trypsin treatment seems increased (Figs. 6a and 6b). Readdition of excess plastoquinone to the membrane before the trypsin treatment protects markedly the D-1 polypeptide (Fig. 6a) and the D-2

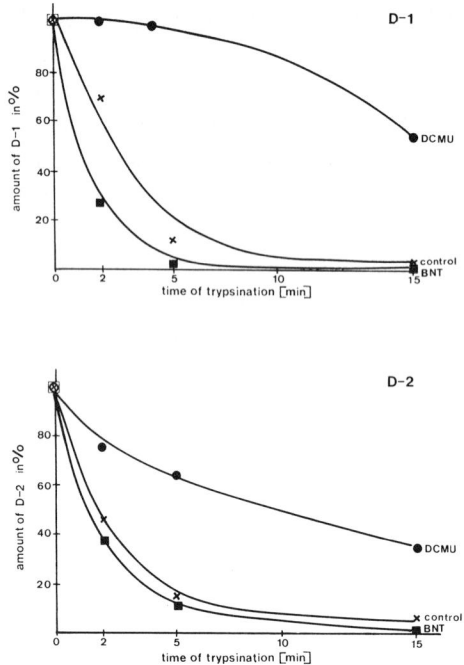

Fig. 5. Protection of the D-1 and D-2 polypeptide from trypsination in the presence of DCMU, but not of BNT (10 mM each). The experiments are similar, but not identical to those in Table 1. Indicated is the amount of polypeptide in percent of the control that is still present after the trypsination, calculated from the stains in Western blots like in Fig. 3.

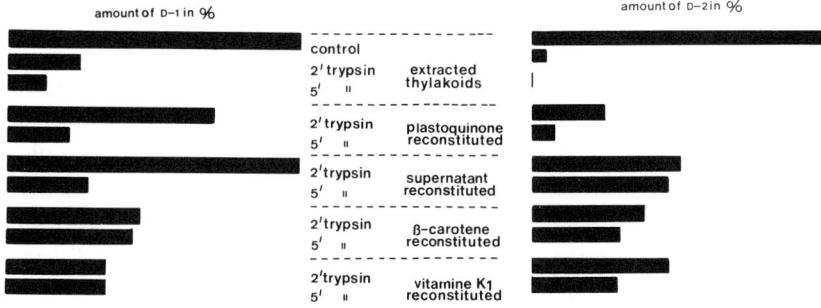

Fig. 6. Trypsin sensitivity of the D-1 (left) and of the D-2 (right) polypeptide, probed with an antibody, in hexane extracted thylakoid membranes (1 mg chlorophyll and its protection by the presence of 0.6 μmoles plastoquinone, 0.23 μmoles β-carotene, 0.17 μmoles menaquinone or supernatant containing 0.4 μmoles plastoquinone, 0.2 μmoles β-carotene and 0.12 μmoles menaquinone.

polypeptide (Fig. 6b). The D-1 polypeptide is protected – time dependent – by plastoquinone and the supernatant, but also to a varying extent by the β-carotene and menaquinone (vitamin K_1) fraction (separated from the hexane supernatant by thin layer chromatography). The supernatant is the most effective in protecting the D-1 polypeptide against trypsin. Increasing the plastoquinone concentratoin added back to the membrane does not lead to further protection (therefore data not shown). Also not shown is an experiment in which various amounts of plastoquinone + carotene or menaquinone were added, as this also does not yield additive effects. As is it assumed that the plastosemiquinone is the most stable derivative of plastoquinone in the binding niche and therefore protects in the controls, addition of ferricyanide or dithionite in order to remove the semiquinone or of plastohydroquinone were compared with plastoquinone in the protection either in thylakoids or hexane treated thylakoids. The data are not presented, as the results are not clear cut. In short time trypsinations ferricyanide, dithionite and also plastohydroquinone had a protective effect, whereas in 15 minutes trypsination there was increased sensitivity (removal of the quinone from the site?). As known thylakoid membranes contain both plastoquinones A, B and C as well as tocoquinone and tocopherol. These quinones equally protect the D-1 polypeptide from trypsination in hexane extracted chloroplasts. For matters of control the artificial electron donor for PS II diphenylcarbazide or the artificial acceptor for photosystem I methylviologen were tested; as expected they do not protect.

Discussion

The long-known (Regitz and Ohad 1976, Renger 1976, Renger et al. 1976 Renger 1977, Trebst 1979, Böger and Kunert 1979, Mattoo et al. 1981, Steinback et al. 1981, 1982, Trebst and Depka 1985) DCMU insensitivity of oxygen evolution and ferricyanide reduction in trypsin treated thylakoid membranes is due to an acquired accessibility of Q_A to ferricyanide so that electron flow would bypass the inhibitor sensitive reaction from Q_A to Q_B (Renger 1976). There is also a loss of inhibitor binding affinity to the Q_B protein D-1 (Trebst 1979, Steinback et al. 1981). This protein is long known to be trypsin sensitive (Mattoo et al. 1981, Steinback et al. 1982). Therefore it has been assumed so far that trypsination of the membrane effects the D-1 polypeptide, although this does not easily explain the exposure of the Q_A site. The results in this paper with immuno-blotting show that indeed the effect of trypsin on photosystem II is not necessarily on the D-1, but can be primarily on the D-2 polypeptide. There are conditions where the D-1 polypeptide is not cut, as DCMU-type inhibitors protect the D-1 polypeptide during trypsin treatment. The D-2 polypeptide is cut (when in the membrane, as in the experiments here) specifically at arg 234, giving rise to a degradation product of 10 kDa, as already described (Geiger et al. 1987). This effect of trypsin on the D-2 polypeptide supports the explanation of Renger (1976) for the events in trypsin treatment, as the recent notion assigns the D-2 polypeptide the role of the Q_A binding protein (Deisenhofer et al. 1984, Trebst 1986, Nanba and Satoh 1987). The trypsin sensitivity of the D-2 polypeptide explains that inhibitor insensitivity is always obtained, whether the D-1 polypeptide is cut or not.

The D-2 polypeptide takes also part in the reaction center chlorophyll binding of photosystem II. Nevertheless electron flow from water to Q_A, remains undisturbed even after longer trypsin treatment, i.e. these reactions remain almost intact, even after all the D-2 polypeptide has been cut at arg 234 on the matrix site. The same argument is true for the D-1 polypeptide. It is also participating in reaction center activity (Deisenhofer et al. 1984, Trebst 1986, Nanba and Satoh 1987) – actually providing the active arm in the symmetric arrangement of the D-1/D-2 reaction center core and probably also the immediate electron donor tyrosine for the reaction center (Debus et al. 1987). It follows from the results that the reaction center and its function on the donor side of PS II remains intact, even after the amino acid sequences of their binding polypeptides have been cut on the matrix exposed site. The results furthermore confirm the prediction from a folding model (Trebst 1986, 1987) that the arginines in D-1 and D-2 discussed here are accessible from the matrix side. The D-1 polypeptide is cut by trypsin very probably at arg 238 according to the studies of Marder et al. (1984), probably also at arg 16 or 23 (see Marder et al. 1984), as there is in some experiments a degradation product at 30.5 kDa. An interesting intermittent band at 8 kDa seems to be stabilized by BNT. It is so far unclear what sequence it has and why it is stabilized.

All inhibitors of PS II, of the DCMU/triazine-type as well as of the phenol-type do not inhibit any more electron flow from water to ferricyanide in trypsin treated membranes (Regitz and Ohad 1976, Renger 1976, Renger et al. 1976, Renger 1977, Trebst 1979, Böger and Kunert 1979, Mattoo et al. 1981, Steinback et al. 1981, 1982, Trebst and Depka 1985) and they have lost their binding site(s) (Trebst 1979, Steinback et al. 1981). However, there is a marked qualitative difference in their influence on the trypsin sensitivity of polypeptides. Several inhibitors of the DCMU-type protect the D-1 polypeptide appreciably – as already known (Renger 1977, Mattoo et al. 1981) for DCMU itself – whereas those of the phenol-type do not. Actually bromonitrothymol and bromohydroxypyridines seems to increase trypsin sensitivity of the reaction center polypeptides. The protecting effect is time and concentration dependent and varies quantitatively among the different inhibitors of this class, but it seems clear to us that the phenol-type inhibitors (as defined by Trebst and Draber (1986), Trebst 1987) have appreciably less protective capacity – DCMU and BNT representing the extremes. There is also a protecting effect of the inhibitors of the DCMU-type inhibitors on the D-2 protein. Although gradual, the different behaviour between DCMU and the phenol-type inhibitors is another example of the difference of the two inhibitor families that have been grouped recently as a serine and as a histidine family (Trebst 1987). Another difference can now be specified: whereas compounds of the DCMU-type protect the amino acid sequence around arginine 238 of the D-1 polypeptide, the phenol-type compounds do not. As this is time dependent, there may be a relation to the different residence time of the two types of inhibitors on the membrane (Vermaas et al. 1984). Also carotene stabilizes somewhat the D-1 polypeptide when readded to carotene free hexane extracted thylakoids indicating a general stabilizing effect. Menaquinone also protects because it is known that naphthoquinones bind to the Q_B site and many of them are therefore inhibitors at the Q_B site (Oettmeier et al. 1986).

Interesting is the influence of plastoquinone on the trypsin sensitivity of the D-1 polypeptide. Both artificial quinone electron acceptors for photosystem II as well as the plastoquinones (A, B and C) itself (tested in hexane extracted thylakoids) have a protective effect on the D-1 polypeptide towards trypsin treatment. The protection towards the trypsin cut, assumed to be at arginine 238 of the D-1 polypeptide (Marder et al. 1984), by both

DCMU and its analogues as well as by plastoquinone and other quinone acceptors of PS II allows to generalize. The state of the Q_B binding site has an effect on the conformation and chemistry of the sequence around that arginine 238. If the site is occupied by plastoquinone and artificial acceptor quinones or an inhibitor (of the DCMU-type, but not of the phenol-type), certain amino acid residues in the amino acid sequence in the binding niche are either directly modified by interactions with the quinones or inhibitors preventing chemically a trypsin cut or a conformational change is induced hiding the sequence from the hydrophilic trypsin. It follows that the amino acid sequence around arg 238 may be indeed part of the Q_B and DCMU binding niche. The participation of the sequence around arg 238 in DCMU binding in the Q_B site domain is likely the explanation why the equivalent L subunit of the reaction center of purple bacteria is not sensitive to DCMU because this sequence of the D-1 polypeptide is much shorter in the L subunit. It indicates a major folding difference around the Q_B site of PS II compared with the bacterial system. The folding model for the reaction center polypeptides (Deisenhofer et al. 1984, Trebst 1986) allows that this lengthy amino acid sequence after helix IV with the arg 238 addressed here could fold easily atop the Q_B between helix IV and the parallel helix and even sideways to them. The equivalent amino acid sequence between helix IV and the parallel helix of also the D-2 polypeptide folds back, in this case above the Q_A site. Likely the D-1 and D-2 polypeptides touch each other at these equivalent sequences. This explains the effect of Q_B and inhibitors bound to D-1 on also the D-2 polypeptide in the protection of arginine 234 of the D-2 protein.

The arginine 238 in the D-1 polypeptide cut by trypsin is in the sequence recently identified by Greenberg et al. (1987) to be part of the cleavage site for the endogenous protease responsible for the degradation in the rapid turnover of the D-1 protein. The properties of the trypsin sensitivity of the D-1 polypeptide are therefore relevant for the mechanism of substrate preparation for the endogenous protease involved in rapid turnover. The state of the Q_B site would modulate the protease sensitivity of the cleavage site of the polypeptide: The condition where D-1 is 'rapidly turning over' is like that for trypsin sensitivity, i.e. an empty Q_B site. The site occupied by either plastoquinone or DCMU would protect the protein from turnover because the cleavage site is modified by the compounds in the site. Indeed it is well documented that DCMU prevents rapid turnover (Mattoo et al. 1984). This provides a simple mechanistic explanation when the protease attacks the D-1 polypeptide or not: Q_B modulates the rapid turnover because the cleavage site is part of the Q_B binding site and the site is chemically modified when the quinone inhibitor interacts with the amino acid residues in this site, or a conformational change prevents access of the protease to the site. DCMU protects the D-1 polypeptide from rapid turnover for an extended time, whereas it does not in the trypsin cut. The reason for the latter is that the trypsin cut is also and preferentially at the D-2 polypeptide and therefore has an additional disturbing effect on the conformation of photosystem II and the stability of the D-1 polypeptide cannot be sustained.

As already discussed above, the loss of inhibition of electron flow to ferricyanide by DCMU and other inhibitors is primarily due to the trypsin cut on the D-2 protein. It is not proportional to the D-1 cut, as assumed earlier (Regitz and Ohad 1976, Renger 1976, Renger et al. 1976, Renger 1977, Trebst 1979, Böger and Kunert 1979, Mattoo et al. 1981, Steinback et al. 1981, 1982) particularly obvious when there is little loss of D-1 protein during trypsination in the presence of DCMU or plastoquinone. But the loss of inhibitor sensitivity is also not directly proportional to the cut of the D-2 polypeptide. This might be another indication for the existence of the so far missing polypeptide in photosystem II, equivalent to the H subunit in PS II could account for the loss of inhibitor binding. Furthermore the loss of the rate of electron flow by trypsin treatment occurs even earlier than the loss of DCMU sensitivity of electron flow and is clearly earlier than the cut in the D-2 polypeptide. Therefore still another component of photosystem II has to be trypsin sensitive to account for the fast loss of electron flow activity. For example Cramer et al. (1981) reported that cytochrome b_{559} changes its midpoint potential after trypsin treatment of the membrane even before a DCMU insensitivity is obtained. But there are, of course, more trypsin sensitive polypeptides in photosystem II that may be responsible (see for example Trebst and Depka 1985).

The results extend our knowledge on the mode of action of herbicides and on details of the topology of their binding niche on the D-1 polypeptide. In this they point also to the importance for the efficiency of a herbicide in sustaining the inhibition of electron flow by the herbicide. Those inhibitors become herbicides in vivo that block also the repair of the D-1 polypeptide in preventing the rapid turnover. This has been argued for already by Mattoo et al. (1986) and is given further support here by showing that certain inhibitors of photosystem II may not block the rapid turnover, therefore do not sustain inhibition of photosynthesis and therefore are not herbicides.

Acknowledgement

The support by the Deutsche Forschungsgemeinschaft is gratefully acknowledged.

References

Böger P and Kunert K-J (1979) Differential effects of herbicides upon trypsin-treated chloroplasts. Z Naturforsch 34c: 1015–1025

Bühmann U, Herrmann EC, Kötter C, Trebst A, Depka B and Wietoska H (1987) Inhibition and photoaffinity labeling of photosystem II by thiazolyiden-ketonitriles. Z Naturforsch 42c: 704–712

Cox RP and Randell DS (1974) The functions of plastoquinone and β-carotene in photosystem II of chloroplasts. Biochim Biophys ACta 347: 49–59

Cramer WA, Whitmarsh J and Widger W (1981) On the properties and function of cytochrome b-559 and f in chloroplast electron transport. In: Akoyunoglou G (ed) Photosynthesis II. Electron Transport and Photophosphorylation, pp 509–522. Philadelphia: Balaban Int Science Services

Debus RJ, Barry A, Babcock GT and McIntosh L (1987) Site directed mutagenesis identifies a tyrosine radical involved in the photosynthetic oxygen evolving system. Proc Natl Acad Sci USA 85: 427–430

Deisenhofer J, Epp O, Miki K, Huber R and Michel H (1984) X-ray structure analysis of a membrane protein complex. Electron density map at 3 A resolution and a model of the chromatophores of the photosynthetic reaction center from *Rhodopseudomonas viridis*, J Mol Biol 180: 385–398

Geiger R, Berzborn RJ, Depka B, Oettmeier W and Trebst A (1987) Site directed antisera to the D-2 polypeptide subunit of photosystem II. Z Naturforsch 42c; 491–498

Greenberg BM, Gaba V and Mattoo AK (1987) Identification of a primary in vivo degradation product of the rapidly-turning-over 32 kd protein of photosystem II, EMBO J 6: 2865–2869

Hirayama O and Kabata K (1977) Lipid extraction and reconstitution of lyophilized chloroplasts, Agric Biol Chem 41: 2423–2426

Johanningmeier U (1987) Expression of the psbA gene in E. coli, Z Naturforsch 42c: 755–757

Knaff DB, Malkin R, Myron JC and Stoller M (1977) The role of plastoquinone and β-carotene in the primary reaction of plant photosystem II, Biochim Biophys Acta 459: 402–411

Marder JB, Goloubinoff P and Edelman M (1984) Molecular architecture of rapidly metabolized 32-kilodalton protein of photosystem II. J Biol Chem 259: 3900–3908

Mattoo AK, Hoffmann-Falk H, Marder JB and Edelman M (1984) Regulation of protein metabolism: Coupling of photosynthetic electron transport to in vivo degradation of the rapidly metabolized 32-kilodalton protein of the chloroplast membranes, Proc Natl Acad Sci USA 81: 1380–1384

Mattoo AK, Marder JB, Gaba V and Edelman M (1986) Control of 32 kDa thylakoid protein degradation as a consequence of herbicide binding to its receptor. In: Regulation of Chloroplast Differentiation, pp 607–613. New York: Alan R Liss

Mattoo AK, Pick U, Hoffmann-Falk H and Edelman M (1981) The rapidly metabolized 32,000-dalton polypeptide of the chloroplast is the 'proteinaceous shield' regulating photosystem II electron transport and mediating diuron herbicide sensitivity, Proc Natl Acad Sci 78: 1572–1576

Nanba O and Satoh K (1987) Isolation of a photosystem II reaction center consisting of D-1 and D-2 polypeptides and cytochrome b-559, Proc Natl Acad Sci USA 84: 109–112

Oettmeier W, Dierig C and Masson K (1986) QSAR of 1,4-naphthoquinones as inhibitors of photosystem II electron transport, Quant Struct-Act Relat 5: 50–54

Regitz G and Ohad I (1976) Trypsin-sensitive photosynthetic activities in chloroplast membrans from *Chlamydomonas reinhardi*, y-1. J Biol Chem 251: 247–252

Renger G (1976) Studies on the structural and functional organization of system II of photosynthesis. The use of trypsin as a structurally selective inhibitor at the outer surface of the thylakoid membrane, Biochim Biophys Acta 440: 287–300

Renger G (1977) The use of trypsin as a structurally selective modifier of the thylakoid membrane. In: Packer L, Papageorgiou GC and Trebst A (eds) Bioenergetics of Membranes, pp 339–350. Amsterdam Oxford New York: Elsevier/North-Holland Biomedical Press

Renger G, Erixon K, Döring G and Wolff C (1976) Studies, on the nature of the inhibitory effect of trypsin on the photosynthetic electron transport of system II in spinach chloroplasts, Biochim Biophys Acta 440: 278–286

Robinson HH and Yocum CF (1980) Cyclic photophosphorylation reactions catalyzed by ferredoxin, methyl viologen and anthraquinone sulfonate, Biochim Biophys Acta 590: 97–106

Steinback KE, Pfister K and Arntzen CJ (1981) Trypsin mediated removal of herbicide binding sites within the photosystem II complex. Z Naturforsch 36c: 98–108

Steinback KE, Pfister K and Arntzen CJ (1982) Identification of the receptor site for triazine herbicides in chloroplast thylakoid membranes. In: Moreland DE, St John JB and Hess FD (eds) Biochemical Responses Induced by Herbicides. ACS Symposium Series 181, pp 37–55. Washington: American Chemical Society

Towbin H, Staehelin T and Gordon J (1979) Electrophoretic transfer of proteins from polyacrylamide gels to nitrocellulose sheets: Procedure and some applications, Proc Natl Acad Sci USA 76: 4350–4354

Trebst A (1979) Inhibition of photosynthetic electron flow by phenol and diphenylether herbicides in control and trypsin-treated chloroplasts, Z Naturforsch 34c: 986–991

Trebst A (1986) The topology of the plastoquinone and herbicide binding peptides of photosystem II in the thylakoid membrane, Z Naturforsch 41c: 240–245

Trebst A (1987) The three-dimensional structure of the herbicide binding niche on the reaction center polypeptides of photosystem II. Z Naturforsch 42c: 742–750

Trebst A and Depka B (1985) The architecture of photosystem II in plant photosynthesis. Which peptide subunits carry the reaction center of PS II? In: Michel-Beyerle ME (ed) Springer Series in Chemical Physics 42. Antennas and Reaction Centers of Photosynthetic Bacteria. Structure, Interactions and Dynamics. pp 216–224. Berlin Heidelberg New York Tokyo: Springer Verlag

Trebst A, Depka B, Ridley SM and Hawkins AF (1985) Inhibition of photosynthetic electron transport by halogenated 4-hydroxy-pyridines, Z Naturforsch 40c: 391–399

Trebst A and Draber W (1979) Structure activity correlations of recent herbicides in photosynthetic reactions, In: Geissbühler H (ed) Advances in Pesticide Science, Part 2, pp 223–234. Oxford New York: Pergamon Press

Trebst A and Draber W (1986) Photosynthesis Res. 10: 381–392

Vermaas M, Renger G and Dohnt G (1984) Exchange kinetics of herbicides and quinones at their common binding site near Q in photosystem II, In: Sybesma C (ed) Advances in Photosynthesis Research, Vol IV, pp 1.13–1.16. The Hague/ Boston/Lancaster: Martinus Nijhoff/Dr W Junk Publishers

Völker M, Ono T, Inoue Y and Renger G (1985) Effect of trypsin on PS-II particles. Correlation between Hill-activity, Mn-abundance and peptide pattern, Biochim Biophys Acta 806: 25–34

Govindjee et al. (eds), Molecular Biology of Photosynthesis: 423–439
© 1988 Kluwer Academic Publishers

Regular paper

Photoregulation of *psbA* transcript levels in mustard cotyledons

J.E. HUGHES[1] & G. LINK[2]

Pflanzliche Zellphysiologie, Ruhr-Universität Bochum, P/F 102148, D 4630 Bochum 1, FRG;
[1]*present address: Department of Botany, University of Georgia, Athens, GA 30602, USA;*
[2]*author for reprint requests*

Received 1 August 1987; accepted 10 December, 1987

Key words: chloroplast gene expression, photoreceptors, seedling development, *Sinapis alba*, thylakoid membranes

Abstract. We have investigated the photoreceptors potentially involved in the light regulation of the transcript levels of the *psbA* gene coding for D1, the 32 kD Q_B-binding protein of PSII. In cotyledons of 4 day old mustard seedlings, increasing fluence rates of continuous white light from ca. 0.1 to 250 μmol m^{-2} s^{-1} (400–700 nm) lead to a five-fold increase in transcript level from ca. 0.7 to 2.8 mg/g total RNA. The blue (< 500 nm) component of this light did not contribute substantially to this effect, thus ruling out cryptochrome as the receptor responsible. Although phytochrome involvement was apparent from red/far-red reversibility, even multiple red pulses failed to elicit a comparable increase in transcript level to that seen under continuous white light. Although DCMU successfully inhibited delayed fluorescence quenching, it had no effect on transcript levels, thus ruling out photoregulation via electron transport and later components of the photosynthetic system. By contrast, Norflurazon, which leads to photobleaching of chlorophyll and hence disruption of thylakoid membrane assembly, completely abolished the light effect on *psbA* transcript level. We infer that photoregulation of the *psbA* transcript is principally related to thylakoid development, which is in turn critically dependent on photoconversion of protochlorophyllide to chlorophyll, but also associated with other processes such as phytochrome-regulated LHCP availability. Photocontrol of *psbA* expression is discussed in relation to that of the nuclear *cab* and *rbcS* genes.

Abbreviations: D1 – 32 kD Q_B-binding protein, DCMU – 3-(3,4'-dichlorophenyl)-1,1-dimethylurea, FR – far-red light, HIR – high irradiance response, LHCP – light-harvesting chlorophyll a/b binding protein, NFZ – Norflurazon, 4-chloro-5-(methylamino)-2-(α,α,α-trifluoro-*m*-tolyl)-3(2*H*)-pyridinone, Pfr – far-red absorbing form of phytochrome, PSII – photosystem II, R – red light, SDS – sodium lauryl sulphate, W – continuous white light

Introduction

The concerted regulation of synthesis of chloroplast proteins involved in the two photosystems is essential for the efficient maintenance of photosynthesis. Little is known of the control mechanism, however. Light regulation is of special interest because of its relevance to the photon-capturing process itself and, additionally, light has great utility as a physiological tool to probe metabolic control. Here we report studies related to the photocontrol of a PSII component in mustard (*Sinapis alba* L.) seedlings.

The D1 protein of PSII is the chloroplast-encoded product of the *psbA* gene (Zurawski et al. 1982; for review see Whitfeld and Bottomley 1983). The primary polypeptide is processed (Grebanier et al. 1978) to a thylakoid-bound, ca. 32 kD product which binds Q_B, a plastoquinone electron acceptor on the reducing side of PSII. Its functioning is specifically inhibited by triazine (Pfister et al. 1981) and diphenyl urea (Mattoo et al. 1981) herbicides. From its high turnover rate in bright light, it was thought that D1 might act as an oxidation shield for the P680 PSII reaction center (Mattoo et al. 1984), but by analogy to purple photosynthetic bacteria, Trebst (1986) has proposed that D1 and its sister protein D2 might be central components of the reaction centre complex. Recent experimental evidence of Nanba and Satoh (1987) supports this view.

Light stimulation of *psbA* transcript levels was first demonstrated in maize (Bedbrook et al. 1978). Such an effect would be in harmony with the photolability of the protein (Mattoo et al. 1984). Attempts to define the photoreceptor(s) involved have, however, yielded conflicting results related perhaps to major differences between species or, at least in part, to the different experimental protocols employed (for discussion see Tobin and Silverthorne 1985, Harpster and Apel 1985).

Gross distinctions between the photoreceptor system(s) which might be involved can be made on the basis of established physiology. Photosynthetic regulation of expression via photoconversion of protochlorohyllide to chlorophyll and/or via some aspect of photon capture, water-splitting, electron transport, or CO_2 fixation is conceivable. Amongst the techniques which could be applied to investigate this is the application of herbicides like DCMU (which blocks electron transfer via Q_B) and Norflurazon (which leads to chlorophyll photobleaching). Similarly, phytochrome is unique in usually exhibiting effects following a pulse of red light but not if the pulse is immediately followed by a far-red pulse (the control being a far-red pulse alone, *not darkness*). Phytochrome effects generally saturate at much lower light levels than does photosynthesis but significant photoconversion of protochlorophyllide might result from red pulses – no far-red reversibility would be seen, however. By contrast to the phytochrome and photosynthesis systems, the photoreceptor cryptochrome (see Senger and Schmidt 1986) is sensitive to only the UV/A-blue region of the spectrum, so its action can be distinguished on this basis.

We have recently reported kinetic studies on mustard seedlings (Hughes et al. 1987), showing that light begins to stimulate *psbA* transcript accumulation only about 48 h after sowing. We have now extended these findings to a quantitative comparison of the effects of different photoreceptors. In the photoregulation of *psbA* we now show that, although phytochrome does play a role, the principle light effect seems related to the assembly of the thylakoid membrane as affected by protochlorophyllide photoconversion to chlorophyll.

Materials and methods

Plant material
Mustard (*Sinapis alba* L., var "Albatros") seeds of good quality and uniformity were carefully selected and sown in batches of fifty, uniformly spaced on moist filter paper in air-tight transparent plastic boxes ca. 6 cm square. These were placed in darkness for 12 h before being subjected to the appropriate irradiation protocol; control plants were held in darkness. Temperatures within the boxes were measured using a shielded thermometer and maintained at 25 °C. At higher fluence rates this was achieved by forced cooling beneath a sheet of Plexiglas. Representative healthy seedlings were selected 96 h after sowing and the cotyledons removed and stored in liquid nitrogen.

Nucleic acid extraction
Total cellular nucleic acids were extracted from a known number of powdered cotyledons in an aqueous/organic system (40 vol aqueous phase – 50 mM Tris/Cl pH 7.6, 2.5 mM $MgCl_2$, 1% NaCl with 2% Na-triisopropyl-naphthalene sulphonate: organic phase – 24 vol Tris-buffer saturated phenol pH 8, 24 vol chloroform, 1 vol isoamyl alcohol). This was repeated and followed by extraction with chloroform/isoamyl alcohol alone. Nucleic acids were precipitated at 1M NH_4 acetate, 55% isopropanol. Following centrifugation, the pellet was resuspended in 5 mM Tris/Cl pH 8.0, 0.25 mM EDTA (low EDTA buffer), briefly centrifuged and the supernatant treated with DNase I (Worthington) as described (Hughes et al. 1987). Total RNA concentration was determined at 260 nm (1 OD $= 40 \mu g\,cm^{-3}$), and the extracts stored in small aliquots at -80 °C.

Electrophoresis, blotting and hybridisation
Electrophoresis in formaldehyde -1.2% Agarose gels and Northern blotting of RNA followed Maniatis et al. (1982). Manifold (Schleicher &

Schuell) dot-blots were prepared from dilution series of 1 μg lyophilised aliquots of RNA in 10 × SSC, 17% formaldehyde.

The 1 kb *psbA* clone pSA452a, a derivative of pSA452 has been described (Hughes et al. 1987). The *psbA* sequence was inserted into the transcription vector pSPT18 (Pharmacia) giving pSPT452a. Purified plasmid DNA was extracted using the alkaline-lysis procedure, restricted 3′ of the insert and transcribed using SP6 (sense strand) or T7 (antisense strand) polymerase to generate mRNA or cRNA as appropriate. Labelled cRNA for use as a probe was synthesised at 30 °C for 2 h using T7 polymerase buffer (BRL), 400 μM each of ATP, CTP and GTP, 20 μM unlabelled UTP, 50 μCi α-^{32}P-labelled UTP (800 Ci/mmol; Amersham), 5 mM DTT, 30 U T7 RNA polymerase (BRL), 10 U RNase inhibitor (Sigma) with 1 μg DNA template per 20 μl reaction mixture. Unlabelled mRNA for calibration purposes was synthesised similarly in SP6 polymerase buffer (BRL), 400 μM each of the four nucleotides, 1 mM DTT, 30 U SP6 RNA polymerase (BRL), 10 U RNase inhibitor (Sigma) with 1 μg DNA template per 50 μg DNA template per 50 μl reaction mixture. Template was removed by adding 1 μg DNase I (Worthington) for ten minutes before extracting with phenol/chloroform and isopropanol precipitation. The pellets were disolved in 300 μl low EDTA buffer. Incorporation into cRNA was measured by scintillation counting following TCA precipitation; mRNA concentration was determined by OD at 260 nm. Transcript length was determined on formaldehyde/agarose gels.

(Pre-)Hybridisation and washing were carried out at 65 °C according to Zinn et al. (1983). Prehybridisation (2–6 h): 5 × SSC, 8 × Denhardt's solution, 50 mM Na/H PO$_4$ pH 6.5, 1% SDS, 250 μg cm^{-3} denatured Herring sperm DNA, 500 μg cm^{-3} yeast RNA, 50% formamide. Hybridisation (ca. 12 h): as for pre-hybridisation, plus 100-fold excess probe and dextran sulphate to 10%. Wash protocol: 3 × 15 min in 2 × SSC, 0.1% SDS, then once in 0.2 × SSC, 0.1% SDS for 20 min. Blots were air dried and fluorographed at −80 °C using Kodak X-omat AR film. Dot-blot hybridisation signals from dilution series were quantified by scintillation counting of dots stamped out of the nitrocellulose. Standard errors were calculated from 2–3 independent RNA extracts each blotted at least twice.

Light regimes and herbicide treatment
A. Continuous white light (W) at 250 μmol m^{-2}s^{-1}, ca. 15 klux, was provided by a bank of 20 metal halide – mercury discharge lamps ('Powerstar', 400 W HQIL; Osram) above an air-cavity window and 4 mm clear Plexiglas sheet (Röhm; Darmstadt, FRG). Fluence rates were reduced to 100 and 25 μmol m^{-2}s^{-1} by interposing single layers on thin drawing paper

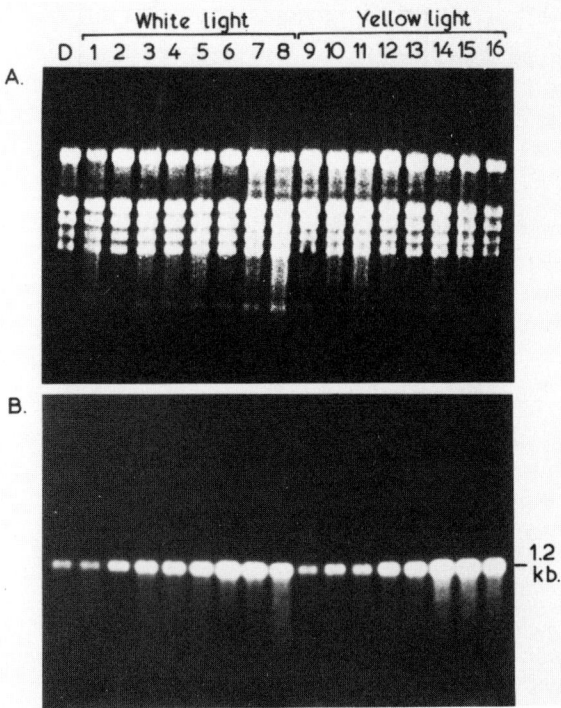

Fig. 1. Fluence rate effects on *psbA* message. Ethidium bromide stained gel (A) and Northern flurograph (B) of cotyledon RNA from 4-day-old seedlings grown in darkness (D) or continuous white (1–8) or yellow (9–16) light of different fluence rates between 0.1 and 250 μmol m^{-2} s^{-1}.

and thick blotting paper, respectively. Lower fluence rates were provided in adjacent compartments of a long box in which the light level was successively attenuated. Light entering this horizontally from an angled mirror was partly scattered downwards into the first compartment by a sheet of thin drawing paper at 45° and partly diffusely transmitted further along the box to a second angled sheet, again deflecting part of the light downwards and transmitting a portion, etc. Light levels were measured using a Gamma Scientific spectroradiometer (San Diego, USA); spectral effects were less than 15%, 400–800 nm.

B. Using similar levels of white light, yellow 303 Plexiglas was substituted for clear Plexiglas between the light source and the plants for irradiation with continuous yellow light.

C. Red light (R) pulses of 4 min duration 0.4 μmol m^{-2} s^{-1} 600–700 nm were provided by a bank of six red fluorescent discharge tubes (Philips TL 20W/15) above a red 501 Plexiglas window to a light-tight chamber. Far-red light (FR) pulses of 4 min duration, 4 μmol m^{-2} s^{-1} 720–900 nm were

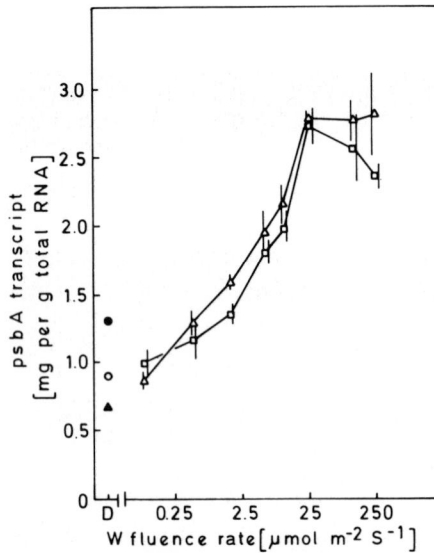

Fig. 2. Fluence rate dependence of *psbA* from dot-blot hybridisation. D = Darkness; (▲) continuous darkness; (△) continuous white light; (□) continuous yellow light. For comparison with Fig. 4: (○) single red pulse; (●) multiple red pulses.

provided by a bank of ten incandescent filament tubes (60 W, Sürola; Südlicht) filtered through layers of red 501 and blue 627 Plexiglas and 'Deep Blue' Cinemoid above a connecting chamber. Light pulse treatments were given either once 12 h after sowing or every 12 h after sowing. In reversion experiments, tranfer from R to FR was achieved within 5 s. Analytical darkness was preserved outside irradiation periods.

D. DCMU and NFZ were dissolved in water from $10\,mg\,cm^{-3}$ ethanol solution to concentrations of 0.25 and $5\,\mu M$, respectively. Mustard seed was imbibed and seedlings grown on these solutions in darkness or at $250\,\mu mol\,m^{-2}s^{-1}$ as above, but on Vermiculite rather than filter paper. Fluorescence kinetics were measured following a 633 nm He/Ne laser pulse using a photomultiplier and storage oscilloscope.

Results

A. Continuous white light

Seedlings were grown under a wide range of W fluence rates and harvested after 4 days. Total RNA levels were not significantly affected by fluence rate (ca. 40 μg per cotyledon pair) in agreement with Hughes et al. (1987). Figure 1 shows a typical ethidium bromide-stained gel and corresponding Northern blot fluorographs of 10 μg aliquots of the extracted RNA. The *psbA* transcript signal clearly increases with fluence rate. This increase was quantified by dot-blotting of extracted total seedling RNA alongside measured amounts of SP6-synthesized *psbA* transcript series, hybridised at identical stringency.

The log [W fluence rate]/response curve of the *psbA* transcript (Fig. 2) increased linearly from the lower limit of ca. 0.1 to at least $25 \, \mu mol \, m^{-2} s^{-1}$. Transcript concentration tended to level off above this, although greater heterogeneity within the data rules out a definitive conclusion that the response saturates here.

B. Continuous yellow light

Seedlings were grown under similar conditions to those above, except that the blue (< 500 nm) component was removed by a yellow filter. There was no significant difference between the total RNA levels under yellow and white light. Northern data obtained with RNA from seedlings grown under yellow light are shown in Fig. 1 (right); again, a substantial increase in *psbA* signal with fluence rate is apparent. Dot-blots showed an essentially similar log [W fluence rate]/response curve to that for W (Fig. 2).

C. Light pulse effects

Gels and northern fluorographs of RNA from 4-day-old seedlings subjected to light pulse treatments at different times are shown in Fig. 3, quantitative dot-blot data is summarised in Fig. 4. A single FR pulse 12 h after sowing was ineffective relative to dark controls. A single R pulse led to a ca. 40% increase in hybridisable mRNA, whereas a subsequent single FR pulse reduced the signal to control levels. Multiple FR pulses (every 12 h) induced significant increases in *psbA* transcript (ca. 70% greater than dark controls). Multiple R pulses induced a somewhat greater increase (ca. twice the dark control level), and this was FR-reversible to approximately the level of FR alone.

D. Herbicide effects

Figure 5 shows gels and Northern fluorographs of RNA from 4-day-old seedlings grown on vermiculite saturated with either pure water or dilute solutions of DCMU or NFZ under $250 \, \mu mol \, m^{-2} s^{-1}$ and in darkness. Quantitative dot-blots indicated that both herbicides had small (ca. 15%) inhibitory effects on *psbA* transcript levels in darkness. In light, however,

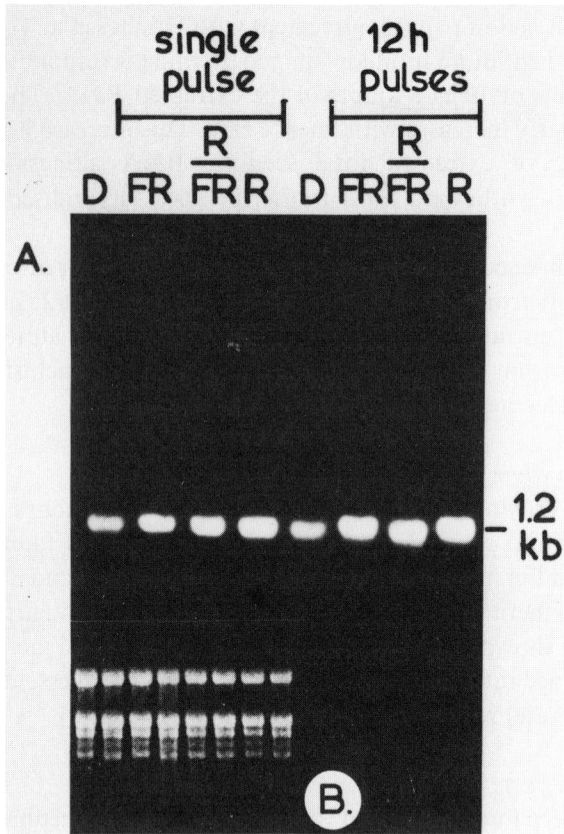

Fig. 3. Red/far-red light pulse effects on *psbA* message in cotyledon RNA from 4 day old seedlings. A. Northern fluorograph. B. (Inset) ethidium bromide stained gel. D, dark controls; FR, 4 min far-red pulse; R, 4 min red pulse; R/FR, 4 min red pulse, < 5 s dark, 4 min far-red pulse. Pulses given either once, 12 h after sowing or every 12 h after sowing.

whereas DCMU and water controls showed a ca. 7-fold increase, NFZ resulted in a transcript level similar to that in darkness.

Discussion

Substantial changes in transcript levels with increasing fluence rates of white light possibly showing threshold and saturation fluence regions indicate photoregulation. Hughes et al. (1987) investigated the kinetics of chloroplast mRNA levels in mustard seedlings. Relative to total cotyledon RNA, the kinetic of *psbA* transcript level in light follows that in darkness upto ca. 48 h after sowing; thereafter, the level in light continues to increase strongly,

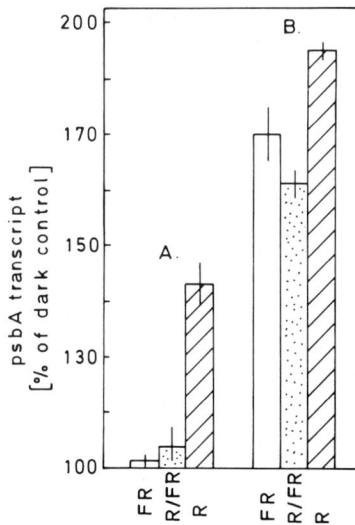

Fig. 4. R/FR light pulse dependence of *psbA* message from dot-blot hybridisation (relative to dark controls). A. Single pulse(s) given 12 h after sowing. B. Pulses given every 12 h after sowing.

while in darkness it flattens off and even declines slightly after ca. 72h; at 96 h the transcript level at $250 \, \mu\text{mol} \, \text{m}^{-2} \, \text{s}^{-1}$ exceeds that in darkness about 5-fold. The total extractable cotyledon RNA increases substantially during this period, but is scarcely affected by light (Hughes et al. 1987). Similarly, chloroplast DNA per cotyledon (copy number) increases but is not photoregulated (Dietrich et al. 1987). Many chloroplast transcripts are *differentially* photoregulated, while tending to increase in abundance relative to total RNA (Hughes et al. 1987, Dietrich et al. 1987, Link 1984). Such differential photocontrol of chloroplast mRNA abundance has also been reported in spinach (Deng and Gruissem 1987) and barley (Mullet and Klein 1987, Klein and Mullet 1987).

It is apparent from Figs. 1 and 2 that whereas W fluence rates between 0.1 and at least $25 \, \mu\text{mol} \, \text{m}^{-2} \, \text{s}^{-1}$ lead to an increase in transcript level, the role of the blue light component appears to be of no special importance. The yellow filter reduces the fluence rate $< 500 \, \text{nm}$ at least 100-fold, yet the log [W fluence rate]/response curve is not significantly shifted to the right. Thus a significant role for cryptochrome is not evident in our system. This is in contrast to previous findings for the *psbA* transcript in *Spirodela* (Gressel 1978) and in cell cultures of *Nicotiana* (Richter 1984, Richter and Wessel 1985). In both cases the effect of W was explained on the basis of cryptoch-

432

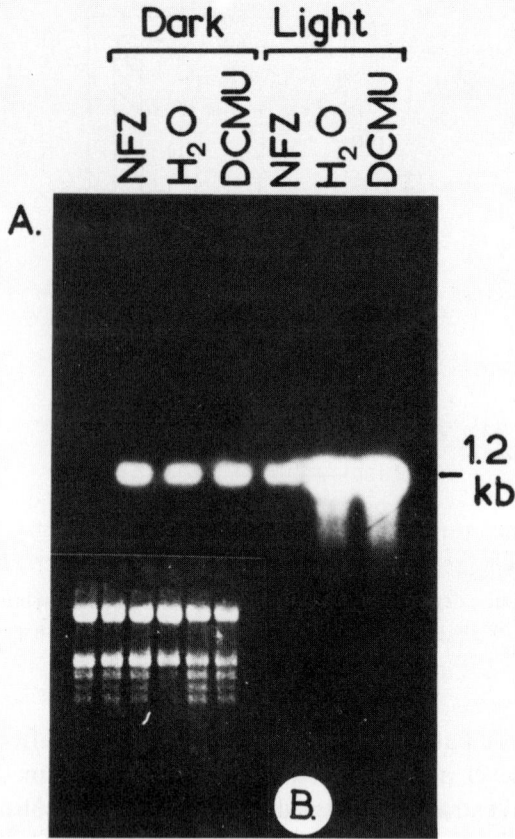

Fig. 5. Effects of herbicides (NFZ 5 μM; DCMU 0.25 μM) on *psbA* message in cotyledon RNA from 4-day-old seedlings grown in darkness and white light (250 μmol m^{-2} s^{-1}). A. Northern fluorograph. B. (Inset) ethidium bromide stained gel.

rome control via the blue spectral component. Amongst the manifold differences between the three systems investigated, it might be significant that in both earlier cases the plant material was dark-grown, whereas in the present study seedlings were irradiated from 12 h after sowing, as the radicle emerged and before etiolation began. In etiolated tissue, the physiologically-active form of phytochrome, Pfr, is unstable ($t_{1/2}$ ca. 1 h). Thus, under continuous irradiation establishing a high proportion of Pfr, the whole phytochrome pool rapidly becomes depleted; maximal responses in etiolated tissue are thought to result from certain light treatments establishing a low but persistent level of Pfr. This so-called high irradiance response (HIR) has been investigated in neither the *Nicotiana* cell culture nor the *Spirodela* system. In mustard, where the HIR has been extensively characterised, there is indeed evidence for an HIR of *psbA* transcript levels (Link 1984). On the other hand, in non-etiolated material grown under white or red light, the HIR is abolished (Beggs et al. 1980), Pfr probably being quite stable ($t_{1/2}$ ca. 1 day; see Holmes and Jabben 1982). Blue light might affect *psbA* transcript levels via the HIR in dark-adapted systems. Indeed, red light apparently antagonises the effect of blue (Richter and Wessel 1985). We emphasise, however, that the origin of the observed differences might lie elsewhere.

On the basis of the substantial response to continuous ruby-red light (known to induce HIR's in this system) and of partial R/FR reversibility, it was concluded that phytochrome is involved in *psbA* expression in mustard cotyledons (Link 1982). This was supported by work on pea (Thompson et al. 1983) and maize (Zhu et al. 1985). Our present results (Figs. 3 and 4) extend these findings. FR reversibility of a single R pulse indicates phytochrome action. It should be noted, however, that lack of reversibility does not necessarily rule out phytochrome as the photoreceptor for a response. Mandoli and Briggs (1981) found that extremely low levels of Pfr might also be physiologically potent. Even long-wavelength FR can saturate this very low fluence response, so R/FR reversibility is, of course, absent. Such responses have been demonstrated for transcript levels of a nuclear gene (Kaufman et al. 1985). The origin of very low fluence phytochrome responses has been critically discussed by Brockmann et al. (1987).

Twelve hours after sowing is unlikely to be the optimal timing for an R pulse to be effective but was chosen to avoid complications resulting from etiolation later. R establishes a similar level of Pfr to W but, while Pfr might then be quite stable in our system, the full potential of the phytochrome effect might only be seen with multiple R pulses maintaining the Pfr level. However, demonstrating reversibility with the necessary multiple FR pulses would become complicated by HIR phenomena. Indeed, Fig. 4 shows that the effect of multiple FR pulses either alone or following R pulses was significant relative to dark controls, as observed previously (Link 1982). In preliminary experiments, multiple pulses of FR containing a ruby-red component led to significantly greater effects, in accordance with HIR involvement.

Much more importantly, however, multiple R pulses failed to elicit an increase in transcript level comparable to that in continuous white light (Fig. 2). High levels of Pfr maintained by such pulses were substantially less effective than moderate W fluence rates. Moreover, the fluence rates over which the response to W is approximately linear are probably several orders greater than would be necessary to establish phytochrome photoequilibrium within hours of the start of irradiation. Hence it is unlikely that phytoch-

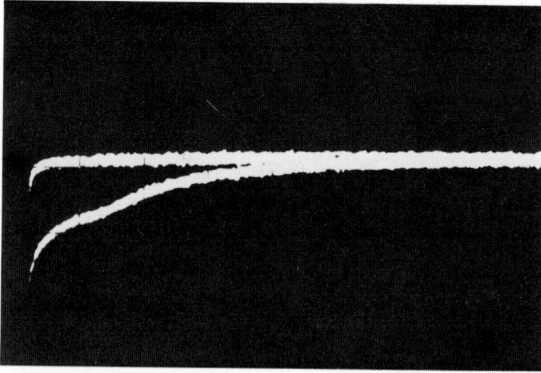

Fig. 6. Induced fluorescence kinetics of cotyledons from 4-day-old seedlings grown at $250\,\mu\mathrm{mol\,m^{-2}\,s^{-1}}$ with (upper curve) and without (lower curve) $0.25\,\mu$M DCMU.

rome is the dominant photoreceptor under normal irradiation conditions. A similar situation might exist in pea for *psbA*, *rbcL* and particularly *rbcS* transcripts (Thompson et al. 1983). Here subsequent W had a substantial effect in addition to that of multiple R pulses. This was not the case for *cab* transcripts (see below).

What is the evidence that the photosynthetic system is involved in regulating *psbA* transcript level? Turnover of D1, the *psbA* product, is thought to be related to photosynthetic electron transfer (Trebst 1986), so the photosynthetic system itself is an attractive potential photoreceptor in regulating *psbA* expression. In the present study, low concentrations of specific inhibitors of the photosynthetic system were used to investigate this possibility (Fig. 5).

In principle, photocontrol via photosynthesis could result from transduction of a signal at almost any point in the system from the antenna chlorophyll to the Calvin cycle. The herbicide DCMU specifically inhibits the binding of Q_B to the D1 protein and was used to test for *psbA* transcript regulation by process at the reducing side of PSII, including D1. The DCMU effect was minimal. The question immediately arises, whether the herbicide reached its target effectively. Figure 6 shows induced fluoresence kinetics of cotyledons from seedlings grown in W in the presence and absence of DCMU. It is clear that DCMU inhibits the quenching of Q_B; thus, photocontrol of *psbA* transcript at or downstream of D1 can probably be ruled out.

The herbicide Norflurazon specifically inhibits phytoene conversion to carotenoids and xanthophylls (Bartels and McCullough 1972). In their absence, chlorophyll is readily photo-oxidised following triplet formation

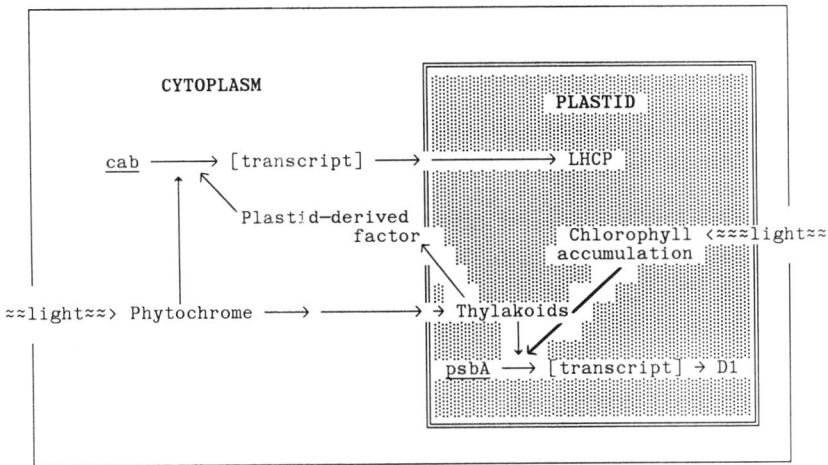

Fig. 7. Hypothetical schemes of light regulation of *cab* and *psba* expression (see text).

and widespread destruction of the photosynthetic apparatus ensues. In the present study, Norflurazon completely inhibited light-induced accumulation of the *psbA* transcript.

This effect is strikingly similar to those reported for the expression of the nuclear *cab* genes encoding the light-harvesting chlorophyll a/b binding protein(s) (LHCP) in maize, mustard, and barley (Mayfield and Taylor 1984, Oelmüller and Mohr 1986, Batschauer et al. 1986). Genetic lesions at certain points in the chlorophyll biosynthesis pathway had similar effects to Norflurazon, while at others none was apparent. While the LHCP itself requires chlorophyll and other thylakoid components in order to accumulate, phytochrome seems to be the predominant photoreceptor controlling *cab* transcript levels (Apel 1979, Tobin 1981, Thompson et al. 1983). Phytochrome action is unlikely to be affected by Norflurazon or the genetic lesions. These factors are also unlikely to influence LHCP transcripts directly – dark controls show that photobleaching is essential for Norflurazon to affect the mRNA level. An attractive solution to this paradox is that a subtle phytochrome-controlled effect on organelle development acting indirectly via a 'plastid-derived factor' (Mayfield and Taylor 1984) is necessary for LHCP message to accumulate (see Fig. 7). Recent inhibitor studies (Oelmüller et al. 1986) are consistent with this view, although the interpretation of chloramphenicol inhibition of plastid protein synthesis should be cautious; both plastids and mitochondria are affected and artefacts arising from the widespread metabolic consequences are difficult to rule out (see Ellis 1981). The putative plastid factor alone might elicit the photocontrol of LHCP message, although work with isolated nuclei and exogenous phytochrome (Mösinger et al. 1987) argues against this. In *Chlamydomonas* at least, this plastid factor might be related to the level of the protochlorophyllide precursor Mg-protoporphyrin methylester (Johanningmeier and Howell 1984).

Whether the control system of *psbA* expression is indeed related to that of *cab* remains to be seen. As we have shown here, phytochrome is not the predominant photoreceptor involved in regulating *psbA* transcript level. It is reasonable to suggest on the basis of our data that photoconversion of protochlorophyllide to chlorophyll is the most significant light-mediated event. Light effects on *psbA* transcript levels first appear 48 h after sowing (Hughes et al. 1987), when chlorophyll begins to accumulate (Schmidt et al. 1987). Hence, the more extensive development of the thylakoid system associated with chlorophyll binding to LHCP might be an important requirement for further photocontrol. In this context, the data (Thompson et al. 1983, Jenkins et al. 1983) from pea for *rcbS* (the nuclear-encoded gene(s) for the small subunit of ribulose bisphosphate carboxylase) implies that substantial thylakoid development is necessary to elicit the transcript levels seen under white light. It is clear that any analysis of light-regulated gene expression should consider the developmental state of the tissue investigated and be wary of assigning control to a single chosen photoreceptor.

Our data provides no clue as to the mechanism of light regulation of *psbA* transcript level. In principle, transcription, RNA processing and stability could each be affected independently. There is evidence from other chloroplast systems that the latter predominates (Altman et al. 1984, Deng and Gruissem 1987, Mullet and Klein 1987). Run-on transcription and pulse labelling experiments with lysed plastids from mustard cotyledons should prove valuable in addressing this aspect.

Acknowledgements

We thank E. Klippel for expert technical assistance, G. Wildner for carrying out the delayed fluorescence measurements, H. Neuhaus for constructing pSPT452a, and U. Johanningmeier for helpful discussion. J.E. Hughes is the holder of a Senior Fellowship of the European Community Biotechnology Directorate. This work was supported by the Deutsche Forschungsgemeinschaft (Li 261/7-1) and the Fonds der Chemischen Industries, FRG.

References

Altman A, Cohen BN, Weissbach H and Brot N (1984) Transcriptional activity of isolated maize chloroplasts. Arch Biochem Biophys 235: 26–33

Apel K (1979) Phytochrome-induced appearance of mRNA activity for the apoprotein of the light-harvesting chlorophyll a/b protein of Barley (*Hordeum vulgare*). Eur J Biochem 97: 183–188

Bartels PG and McCullough C (1972) A new inhibitor of carotenoid synthesis in higher plants: 4-chloro-5-(dimethylamino)-2-α,α,α, (trifluoro-m-tolyl)-3(2H)-pyridazinone (Sandoz 6706). Biochem Biophys Res Com 48: 16–22

Batschauer A, Mösinger E, Kreuz K, Dörr I and Apel K (1986) The implication of a plastid-derived factor in the transcriptional control of nuclear genes encoding the light-harvesting chlorophyll a/b protein. Eur J Biochem 154: 625–634

Bedbrook JR, Link G, Coen DM, Bogorad L and Rich A (1978) Maize plastid gene expressed during photoregulated development. Proc Natl Acad Sci USA 75: 3060–3064

Beggs CJ, Holmes MG, Jabben M and Schäfer E (1980) Action spectra for the inhibition of hypocotyl growth by continuous irradiation in light and dark-grown *Sinapis alba* seedlings. Plant Physiol 66: 615–618

Brockmann J, Rieble S, Kazarinova-Fukshansky N, Seyfried M and Schäfer E (1987) Phytochrome behaves as a dimer in vivo: opinion. Plant Cell Environ 10: 105–112

Deng X-W and Gruissem W (1987) Control of plastid gene expression during development; the limited role of transcriptional regulation. Cell 49: 379–387

Dietrich G, Detschey S, Neuhaus H and Link G (1987) Temporal and light control of plastid transcript levels for proteins involved in photosynthesis during mustard (*Sinapis alba* L.) seedling development. Planta 172: 393–399

Ellis JR (1981) Chloroplast proteins: Synthesis, transport, and assembly. Ann Rev Plant Physiol 32: 111–137

Grebanier AE, Coen DM, Rich A and Bogorad L (1978) Membrane proteins synthesized but not processed by isolation maize chloroplasts. J Cell Biol 78: 734–746

Gressel J (1978) Light requirements for the enhanced synthesis of a plastid mRNA during *Spirodela* greening. Photochem Photobiol 27: 167–169

Harpster M and Apel K (1985) The light-dependent regulation of gene expression during plastid development in higher plants. Physiol Plant 64: 147–152

Holmes MG and Jabben M (1982) Phytochrome in light-grown plants. In: Shropshire W Jr and Mohr H (eds) Encyclopedia of Plant Physiology, N.S., 16B: Photomorphogenesis, pp 704–722. Berlin/Heidelberg/New York: Springer-Verlag

Hughes JE, Neuhaus H and Link G (1987) Transcript levels of two adjacent chloroplast genes during mustard (*Sinapis alba* L.) seedling development are under differential temporal and light control. Plant Mol Biol 9: 355–363

Jenkins GI, Hartley MR and Bennett J (1983) Photoregulation of chloroplast development: transcriptional, translational and post-translational controls? Phil Trans R Soc Lond (B) 303: 419–431

Johanningmeier U and Howell SH (1984) Regulation of light-harvesting chlorophyll-binding protein mRNA accumulation in *Chlamydomonas reinhardii*. J Biol Chem 259: 13541–13549

Kaufman LS, Briggs WR and Thompson WF (1985) Phytochrome control of specific mRNA levels in developing pea buds. Plant Physiol 78: 388–393

Klein RR and Mullet JE (1987) Control of gene expression during higher plant chloroplast biogenesis. Protein synthesis and transcript levels of *psbA*, *psaA–psaB*, and *rbcL* in dark-grown and illuminated barley seedlings. J Biol Chem 262: 4241–4348

Link G (1982) Phytochrome control of plastid mRNA in mustard (*Sinapis alba* L.). Planta 154: 81–86

Link G (1984) Hybridisation study of developmental plastid gene expression in mustard (*Sinapis alba*) with cloned probes from most plastid DNA regions. Plant Mol Biol 3: 81–86

Mandoli DF and Briggs WR (1981) Phytochrome control of two low-irradiance responses in etiolated oat seedlings. Plant Physiol 67: 733–739

Maniatis T, Fritsch EF and Sambrook J (eds) (1982) Molecular Cloning, A Laboratory Manual. New York: Cold Spring Harbor Laboratory

Mattoo AK, Pick U, Hoffmann-Falk H and Edelman M (1981) The rapidly metabolized 32,000-dalton polypeptide of the chloroplast is the "proteinaceous shield" regulating photosystem II electron transport and mediating diuron herbicide sensitivity. Proc Natl Acad Sci USA 78: 1572–1576

Mayfield SP and Taylor WC (1984) Carotenoid-deficient maize seedlings fail to accumulate light-harvesting chlorophyll a/b binding protein (LHCP) mRNA. Eur J Biochem 144: 79–84

Mösinger E, Batschauer A, Vierstra R, Apel K and Schäfer E (1987) Comparison of the effects of endogenous native phytochrome and in-vivo irradiation on in-vitro transcription in isolated nuclei from barley (*Hordeum vulgare*). Planta 170: 505–514

Mullet JE and Klein RR (1987) Transcription and RNA stability are important determinants of higher plant chloroplast RNA levels. EMBO J 6: 1571–1579

Nanba O and Satoh K (1987) Isolation of a photosystem II reaction center consisting of D1 and D2 polypeptides and cytochrome b559. Proc Natl Acad Sci USA 84: 109–112

Oelmueller R, Levitan I, Bergfeld R, Rajasekhar VK and Mohr H (1986) Expression of nuclear genes as affected by treatments acting on the plastids. Planta 168: 482–492

Oelmueller R and Mohr H (1986) Photooxidative destruction of chloroplasts and its consequences for expression of nuclear genes. Planta 167: 106–113

Pfister K, Steinback KE, Gardner G and Arntzen CJ (1981) Photoaffinity labeling of an herbicide receptor protein in chloroplast membranes. Proc Natl Acad Sci USA 78: 981–985

Richter G (1984) Blue light control of the level of two plastid mRNAs in cultured plant cells. Plant Mol Biol 3: 271–276

Richter G and Wessel K (1985) Red light inhibits blue light-induced chloroplast development in cultured plants cells at the mRNA level. Plant Mol Biol 5: 175–182

Schmidt, Drumm-Herrel H, Oelmüller R and Mohr H (1987) Time course of competence in phytochrome-controlled appearance of nuclear-encoded plastidic proteins and messenger RNAs. Planta 170: 1–8

Senger H and Schmidt W (1986) Cryptochrome and UV receptors. In: Kendrick R and Kronenberg GHM (eds) Photomorphogenesis in Plants, pp 137–158. Dordrecht/Boston/Lancaster: Martinus Nijhoff Publishers

Thompson WF, Everett M, Polans NO, Jorgensen RA and Palmer JD (1983) Phytochrome control of RNA levels in developing pea and mung-bean leaves. Planta 158: 487–500

Tobin EM (1981) Phytochrome-mediated regulation of messenger RNAs for the small subunit of ribulose 1,5-bisphosphate carboxylase and the light-harvesting chlorophyll a/b binding protein in *Lemna gibba*. Plant Mol Biol 1: 35–51

Tobin EM and Silverthorne J (1985) Light regulation of gene expression in higher plants. Ann Rev Plant Physiol 36: 569–593

Trebst A (1986) The topology of the plastoquinone and herbicide binding peptides of photosystem II in the thylakoid membrane. Z Naturforsch 41: 240–245

Whitfeld PR and Bottomley W (1983) Organisation and structure of chloroplast genes. Ann Rev Plant Physiol 34: 279–310

Zhu YS, Kung SD and Bogorad L (1985) Phytochrome control of levels of mRNA complementary to plastid and nuclear genes of maize. Plant Physiol 79: 371–376

Zinn K, DiMaio D and Maniatis T (1983) Identification of two distinct regulatory regions adjacent to the human beta-interferon gene. Cell 34: 865–879

Zurawski G, Bohnert HJ, Whitfield PR and Bottomley W (1982) Nucleotide sequence of the gene for the Mr 32,000 thylakoid membrane protein from *Spinacia oleracea* and *Nicotiana debneyi* predicts a totally conserved primary translation product of Mr 38,950. Proc Natl Acad Sci USA 79: 7699–7703

Govindjee et al. (eds), Molecular Biology of Photosynthesis: 441–484
© 1988 Kluwer Academic Publishers

Minireview/hypothesis

The molecular mechanism of the Bicarbonate effect at the Plastoquinone reductase site of Photosynthesis

DANNY J. BLUBAUGH[1] & GOVINDJEE[2]

Departments of Physiology & Biophysics and Plant Biology, University of Illinois, 289 Morrill Hall, 505, South Goodwin Avenue, Urbana, IL 61801, USA

Received 10 October 1987; accepted 30 March 1988

Key words: Photosystem II, quinone reductase, bicarbonate effect, protonation, (spinach)

Abstract. It has been known for some time that bicarbonate reverses the inhibition, by formate under HCO_3^--depletion conditions, of electron transport in thylakoid membranes. It has been shown that the major effect is on the electron acceptor side of photosystem II, at the site of plastoquinone reduction. After presenting a historical introduction, and a minireview of the bicarbonate effect, we present a hypothesis on how HCO_3^- functions in vivo as (a) a proton donor to the plastoquinone reductase site in the D1-D2 protein; and (b) a ligand to Fe^{2+} in the Q_A-Fe-Q_B complex that keeps the D1-D2 proteins in their proper functional conformation. They key points of the hypothesis are: (1) HCO_3^- forms a salt bridge between Fe^{2+} and the D2 protein. The carboxyl group of HCO_3^- is a bidentate ligand to Fe^{2+}, while the hydroxyl group H-bonds to a protein residue. (2) A second HCO_3^- is involved in protonating a histidine near the Q_B site to stabilize the negative charge on Q_B. HCO_3^- provides a rapidly available source of H^+ for this purpose. (3) After donation of a H^+, CO_3^{2-} is replaced by another HCO_3^-. The high pKa of CO_3^{2-} ensures rapid reprotonation from the bulk phase. (4) An intramembrane pool of HCO_3^- is in equilibrium with a large number of low affinity sites. This pool is a H^+ buffering domain functionally connecting the external bulk phase with the quinones. The low affinity sites buffer the intrathylakoid $[HCO_3^-]$ against fluctuations in the intracellular CO_2. (5) Low pH and high ionic strength are suggested to disrupt the HCO_3^- salt bridge between Fe^{2+} and D_2. The resulting conformational change exposes the intramembrane HCO_3^- pool and low affinity sites to the bulk phase.

Two contrasting hypotheses for the action of formate are: (a) it functions to remove bicarbonate, and the low electron transport left in such samples is due to the left-over (or endogenous) bicarbonate in the system; or (b) bicarbonate is less of an inhibitor and so appears to relieve the inhibition by formate. Hypothesis (a) implies that HCO_3^- is an essential requirement for electron transport through the plastoquinones (bound plastoquinones Q_A and Q_B and the plastoquinone pool) of photosystem II. Hypothesis (b) implies that HCO_3^- does not play any significant role in vivo. Our conclusion is that hypothesis (a) is correct and HCO_3^- is an essential requirement for electron transport on the electron acceptor side of PS II. This is based on several observations: (i) since HCO_3^-, not CO_2, is the active species involved (Blubaugh and Govindjee 1986), the calculated concentration of this species (220 μM at pH 8, pH of the stroma) is much higher than the calculated dissociation constant (Kd) of 35–60 μM; thus, the likelihood of bound HCO_3^- in ambient air is high; (ii) studies on HCO_3^- effect in thylakoid samples with different chlorophyll concentrations suggest that the "left-over" (or "endogenous") electron flow in bicarbonate-depleted chloroplasts is due to "left-over" (or endogenous) HCO_3^- remaining bound to the system (Blubaugh 1987).

442

Abbreviations: DCMU – 3-(3,4-dichlorophenyl) – 1, 1-dimethylurea (common name: diuron), PSII – photosystem II, Q_A – first plastoquinone electron acceptor of PSII, Q_B – second plastoquinone acceptor of PS II.

Introduction

Photosynthesis involves the oxidation of H_2O and the reduction of CO_2. The energy required for this reaction is supplied by the light that is absorbed by the photosynthesis pigments and transferred with great efficiency to the reaction center chlorophylls (see e.g. Pearlstein 1982, van Grondelle and Amesz 1986). In the early days of photosynthesis research, it was not known whether the O_2 that is evolved came from H_2O or from CO_2. From the overall chemical equation for photosynthesis

$$H_2O + CO_2 + \text{light} \rightarrow O_2 + 1/n \, (CH_2O)_n \qquad (1)$$

it was supposed by most early workers that CO_2 was the source of evolved O_2, with the chlorophyll somehow catalyzing the transfer of carbon from CO_2 to H_2O. An example of this was the Willstätter-Stoll hypothesis, which had hydrated CO_2 reacting with chlorophyll to yield formaldehyde and O_2, with the formaldehyde then undergoing enzymatic condensation in the conversion to carbohydrates (for an excellent description of this hypothesis and its subsequent loss of favor, see van Niel (1949)). It is now firmly established that the evolved O_2 comes from H_2O (for a historical precedence, see Wurmser (1987)). However, before introducing the role of bicarbonate in photosystem II (PS II), we review the major lines of evidence for H_2O as the substrate for O_2 evolution, since some controversy over this question has continued to the present day, fueled by the requirement for bicarbonate in PS II.

The first major line of evidence for H_2O as the source of photosynthetic O_2 came from comparative studies between the green and purple sulfur-reducing bacteria and green plants (van Niel 1931, also see van Niel 1941, Wraight 1982, Wurmser 1987). In these bacteria, the overall reaction for photosynthesis is

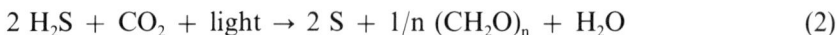

$$2 \, H_2S + CO_2 + \text{light} \rightarrow 2 \, S + 1/n \, (CH_2O)_n + H_2O \qquad (2)$$

Eqn. 1 from green plants can be rewritten as

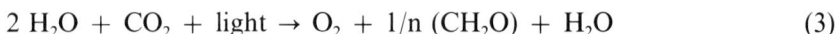

$$2 \, H_2O + CO_2 + \text{light} \rightarrow O_2 + 1/n \, (CH_2O) + H_2O \qquad (3)$$

The similarity between Eqns. 2 and 3 is obvious; this led van Niel to propose that photosynthesis is a light-catalyzed oxidation-reduction reaction of the form

$$2 H_2A + CO_2 + light \rightarrow 2 A \text{ or } A_2 + 1/n (CH_2O)_n + H_2O \qquad (4)$$

In the green and purple photosynthetic bacteria A would be sulphur, whereas in higher plants, algae and cyanobacteria it would be oxygen. Thus, the implication was clear that the source of photosynthetically derived O_2 is H_2O and not CO_2.

This generalized equation for photosynthesis gained strong support with the demonstration that a variety of compounds can be photosynthetically oxidized by the purple sulfur bacteria: elemental sulfur, sulfite and thiosulfate can all be oxidized to sulfate (van Niel 1931); elemental selenium is oxidized to selenate (Saposhnikov 1937, cited by van Niel 1941); various organic substances can be used as hydrogen donors (c.f. Foster 1940); and some species can use molecular hydrogen (c.f. French 1937). Only when H_2O is the hydrogen donor is O_2 evolved. Thus, plant photosynthesis as a special case of the generalized equation became widely accepted after the pioneering work of van Niel (also see Wurmser 1987).

Nevertheless, it could be argued that what occurs in photosynthetic bacteria is not necessarily what occurs in higher plants, algae and cyanobacteria. Metzner (1975), for example, argues that since plants (including cyanobacteria), unlike the photosynthetic bacteria, require two photosystems to transfer electrons from the primary donor to the terminal acceptor, that it is not valid to compare them. Furthermore, the bacterial photosystem requires electron sources that are relatively easy to oxidize, compared to the O_2-generating PS II. On the other hand, Gaffron (1940) succeeded in adapting cultures of the green alga *Scenedesmus* to the utilization of H_2, so that they photosynthesized without evolution of O_2. The H_2/CO_2 quotient was the same as in purple photosynthetic bacteria. Thus, at least in these species, oxygenic photosynthesis seems to fit the generalized equation of van Niel. Nevertheless, the hypothesis that O_2 comes from H_2O needed to be tested. The test became possible with the availability of oxygen isotopes.

The isotopic composition of the evolved O_2 from photosynthesizing *Chlorella* cells that were supplied with H_2O or HCO_3^- enriched in ^{18}O matched that of the water, not of the HCO_3^-; thus, it was concluded that the O_2 came from the H_2O (Ruben et al. 1941). Such experiments are complicated by the fact that CO_2 and H_2O are always in rapid equilibrium with H_2CO_3, according to the reaction

$$CO_2 + H_2O \leftrightarrow H_2CO_3 \tag{5}$$

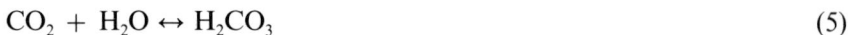

While CO_2 dissolved in H_2O will reach chemical equilibrium very rapidly ($t_{1/2} \approx 19$ s; Gibbons and Edsall 1963), isotopic equilibrium between all the oxygens takes about 1000 times longer (Mills and Urey 1940). This is because for isotopic equilibrium every molecule must come to equilibrium with every other molecule; therefore, the hydration reaction must occur many times. Chemical equilibrium, on the other hand, is dependent only on the ratio of forward to reverse rate constants; as soon as $[H_2CO_3]$ reaches its equilibrium value, the forward and reverse rates are equal, and chemical equilibrium is achieved. The time required to reach isotopic equilibrium is very sensitive to the pH, since the greater the ratio of $[HCO_3^-]/[CO_2]$ (i.e. the higher the pH), the fewer the number of hydration reactions in a given time. Because of the very slow oxygen exchange at the temperature (25 °C) and pH (~ 10) of the ^{18}O experiments with *Chlorella*, the original investigators felt justified in ignoring this complication, as the oxygen exchange was much slower than the rate of photosynthesis. However, Kamen and Barker (1945) pointed out that it was an unproven assumption that the isotope exchange is no more rapid inside the cells, or inside the chloroplasts, than in the outside medium. Assuming an internal pH of 6 or less, these authors calculated that the randomization of ^{18}O would be rapid enough to invalidate the conclusions.

Because of its heavier mass, CO_2 is slightly more enriched in ^{18}O than is H_2O at isotopic equilibrium (Webster et al. 1935). Dole and Jenks (1944) were able to sensitively measure this difference, and they showed that in photosynthesizing cells in which isotopic equilibrium was maintained with carbonic anhydrase, the evolved O_2 had nearly the same enrichment as the H_2O, but was clearly less enriched than the CO_2, thus confirming the original conclusions. Since this experiment was done at isotopic equilibrium, it was not subject to the same criticism as the first experiments.

This line of evidence still had difficulties. It has long been presumed that the source of the earth's atmospheric O_2 is photosynthesis, yet the ^{18}O content of atmospheric O_2 is considerably greater than that of natural water (Dole 1935). Green and Voskuyl (1936) pointed out that the ^{18}O content of atmospheric O_2 is what would be predicted if photosynthetic O_2 were derived from the water and CO_2 together. Yosida et al. (1942) claimed to have measured an ^{18}O content of photosynthetic O_2 evolved from aqueous plants that indicated one-third of the O_2 came from CO_2. They were able to account for this observation using a modified Willstätter-Stoll hypothesis for O_2 evolution, in which carbonic acid (i.e. hydrated CO_2) yields formaldehyde and O_2 through a peroxide intermediate. The experiment of Dole and Jenks

(1944), described above, contradicts these results and shows clearly that, at least in *Chlorella*, the O_2 evolved is much closer in isotopic composition to the H_2O than it is to the CO_2. Still, they did show a slight increase in ^{18}O abundance compared to the H_2O; this they noted is precisely what would be predicted if there were an efficient oxygen exchange between O_2 and H_2O, although from the earlier measurements of Ruben et al. (1941) it was concluded that such an exchange did not occur. Given the complications of oxygen exchange reactions, the greater ^{18}O content of atmospheric O_2 compared to natural H_2O, and the small number of species tested, these ^{18}O labeling experiments have not escaped controversy. We refer the reader to Rabinowitch (1945, 1951, 1956) for further references and details. In our opinion the original experiments of Ruben et al. (1941) and of Dole and Jenks (1944) were carefully done. However, we note that Warburg (1964) and Metzner (1975) have challenged their validity. Thus, further research is still necessary.

Stemler and Radmer (1975) measured the isotopic O_2 released from HCO_3^--depleted thylakoids after the addition of $HC^{18}O_3^-$ and found that only $^{16}O_2$ was evolved. This experiment was superior to the original experiments of Ruben et al. (1941) in that the appearance of $C^{18}O_2$ and $C^{16}O_2$ were monitered along with the evolution of $^{16}O_2$, and it was shown that the isotopic exchange reactions were too slow for HCO_3^- or CO_2 to be the source of photosynthetic O_2. Also, since the thylakoids had been depleted of native $HC^{16}O_3^-$, the $^{16}O_2$ could only have come from $H_2^{16}O$. However, since the mass spectrometer used in this experiment was not sensitive enough to moniter single turnovers of the reaction centers, these authors could not rule out the remote possibility that the first few molecules of evolved O_2 might have been derived from the reactivating HCO_3^-. This qestion was apparently settled by Radmer and Ollinger (1980) who showed that even after an actinic flash, the isotopic composition of the evolved O_2 matches that of the H_2O and not of the HCO_3^-. It would seem, then, that there is little reason to doubt that H_2O is the substrate for O_2 evolution. However, the validity of even these experiments has been questioned by Stemler (1982), who holds open the possibility that PS II may be able to catalyze the hydration of CO_2 (and therefore the isotopic exchange reactions) at the O_2 evolving site.

The third major line of evidence that H_2O is the substrate for photosynthetic O_2 evolution was the classical observation by Hill (1937, 1939) that broken chloroplasts could be made to evolve O_2 in the light by the addition of ferric oxalate. This was the first successful observation of photosynthetic O_2 evolution in a system other than whole cells or an intact leaf. It was not necessary for Hill to supply the chloroplasts with CO_2; in fact, he made his observation with the chloroplasts evacuated, so any CO_2 remaining was very low in concentration. At first it was thought that some O_2 had to be present, but Hill and Scarisbrick (1940a) later found that this was simply because the small amount of ferric oxalate added was quickly reduced by organic acids present, and the O_2 was necessary to reoxidize the ferrous oxalate to ferric oxalate. If no exhaustion of the ferric salt was allowed to occur, the evolution of O_2 could continue for several hours. From the behavior of the reaction with various inhibitors, Hill and Scarisbrick (1940b) concluded that the O_2 was evolved in the "light" reaction, and that CO_2 was reduced in dark reactions that were disrupted during chloroplast isolation. The ferric salts stimulated O_2 evolution by acting as an oxidant (electron acceptor) for the "light" reactions. This physical separation of the reactions leading to O_2 evolution from the reactions catalyzing CO_2 fixation was strong evidence for the non-involvement of CO_2 in O_2 evolution. Furthermore, the observed reduction of the added electron acceptor (now referred to as a "Hill oxidant"; electron transport in isolated thylakoids was called the "Hill reaction", we presume, by French et al., (1946)) confirmed that the "light" reactions of photosynthesis were oxidation-reduction reactions, with H_2O being the terminal reductant. Thus, the evolution of O_2 could be seen as a simple oxidation:

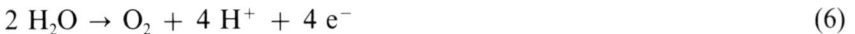

$$2\,H_2O \rightarrow O_2 + 4\,H^+ + 4\,e^- \tag{6}$$

To sum up the discussion so far, three major lines of evidence have contributed to the current picture of H_2O as the source of photosynthetic O_2 evolution: (i) by comparison with photosynthetic bacteria, which use light energy to reduce CO_2 and oxidize a hydrogen donor, plant photosynthesis is presumed to operate by a similar mechanism, with H_2O as the hydrogen donor; (ii) ^{18}O labeling experiments have shown that the O_2 that is evolved is labeled to the same extent, or to nearly the same extent, as the H_2O, but not as the CO_2; and (iii) the "light" reactions of photosynthesis can be separated from the dark reactions; O_2 is evolved in the "light" reactions, while CO_2 is reduced in the dark reactions. The first two lines of evidence have difficulties inherent in the method that make them less than conclusive: (i) photosynthetic bacteria have major fundamental differences from photosynthetic plants, such as in the gross architecture, in the number of photosystems, etc., and it has been argued that a direct comparison is not necessarily valid; and (ii) the complications of natural oxygen exchange between CO_2 and H_2O have cast doubt on the validity of the ^{18}O experiments in the minds of some researchers.

Given these objections to the first two lines of evidence, the third line of evidence, the independence of the Hill reaction on CO_2, became a cornerstone argument for H_2O as the source of photosynthetic O_2. It was therefore of great importance when Warburg and Krippahl (1960) showed conclusively that the Hill reaction was reversibly inhibited by removal of CO_2 from the membranes. Although the Hill reaction did not require that CO_2 be supplied to the membranes, Hill did not actually remove CO_2 that might have already been bound to the membranes. When this CO_2 is removed, the Hill reaction ceases. The implication is that if O_2 evolution cannot occur in the absence of CO_2, then CO_2 could remain a candidate for the source of evolved O_2. Even if H_2O is also shown to be involved, this does not preclude hydrated CO_2 (i.e. H_2CO_3 or HCO_3^-) as the source, as originally proposed by Willstätter and Stoll. This hypothesis, and variations of it, have been proposed throughout the nearly three decades since Warburg and Krippahl's discovery (e.g. see Warburg 1964, Stemler 1982).

Despite the above objections, the scientific community has not seriously questioned H_2O as the source of evolved O_2 since the time of Hill and Bendall's very successful "Z-scheme" (Hill and Bendall 1960; for reviews, see Govindjee 1975, 1982). The accumulated evidence for the light reactions of photosynthesis being a series of oxidation-reductions, with H_2O as the ultimate source of electrons, is overwhelming, and the picture is not likely to be overturned easily. It is possible, of course, to incorporate CO_2 into the mechanism of O_2 evolution without a major overhaul of our current model – the mechanism of O_2 evolution is still one of the least understood aspects of photosynthesis (see reviews by Renger and Govindjee 1985, Govindjee et al. 1985, Babcock 1987, Renger 1987) – but models such as that championed by Warburg (1964), in which CO_2 has more than a regulatory or catalytic role, are probably out of vogue forever. One effect of the controversy over the past 25 years seems to have been not so much to cast doubt on the role of H_2O, as to have raised suspicion or caused indifference to the role of HCO_3^-. This is unfortunate, because although a major effect of HCO_3^- on O_2 evolution can be ruled out, it very clearly does play a major role in the reduction of plastoquinone (see reviews by Govindjee and van Rensen 1978, Vermaas and Govindjee 1981a, 1982a, van Rensen and Snel 1985, Govindjee and Eaton-Rye 1986) and is, therefore, deserving of more attention than it has received. This paper is one attempt to redress this imbalance. We first examine below what is known about the role of HCO_3^- in PSII.

The requirement for HCO_3^- in PS II

The discovery
Credit for the discovery of the HCO_3^- requirement in the light reactions of

photosynthesis is generally given to Warburg and Krippahl (1958, 1960). However, such a requirement was suggested several years earlier by Boyle (1948), who quite by accident noticed that O_2 production was halted in the H_2O to p-benzoquinone Hill reaction in ground spinach leaves when KOH was placed in the center well of a manometer vessel to take up CO_2. This observation was shown later to be an artifact (Abeles et al. 1961), due to distillation of quinone from the main compartment of the reaction vessel into the KOH solution. Under such conditions, O_2 uptake occurs in the center well at a rate sufficient to reabsorb all of the O_2 produced by the Hill reaction. Thus, no net O_2 evolution would have been observed by Boyle, regardless of whether CO_2 were required or not. The conclusion that CO_2 was required was, therefore, unwarranted. Under the same conditions as Boyle, Abeles et al. (1961) found no effect on the Hill reaction by having CO_2 present or absent in the vessel. On the other hand, the same authors were able to confirm the observation by Warburg and Krippahl, noting, under their conditions, a consistent difference in O_2 evolution when atmospheric CO_2 was present or absent. In these experiments the quinone concentration was lower, and there was no KOH in the center well. This, along with differences in pH, probably accounts for the failure to observe a CO_2 dependence under Boyle's conditions.

Other researchers may have noticed the requirement for HCO_3^- earlier, without fully recognizing it as such. For instance, Franck (1945) reported that broken chloroplasts evolved more O_2 when they were supplied with CO_2 than when they were flushed with N_2. Admission of CO_2 to CO_2-free suspensions always caused a sudden increase in the rate of O_2 production. Franck took this observation to justify that he was looking at real photosynthesis, with CO_2 as oxidant. However, from his protocol it is clear that he was using broken chloroplasts, which would have been missing the stromal enzymes necessary for CO_2 reduction. He did not add a Hill oxidant to his preparations; only small amounts of O_2 were evolved, but the rate was significantly higher when CO_2 was present. Later, under the same experimental conditions, Brown and Franck (1948) found that when $^{14}CO_2$ was used, there was no accumulation of the radiolabel in the chloroplasts. Therefore, they concluded that the stimulation by CO_2 was not due to CO_2 fixation, but to some other, possibly catalytic, role for CO_2.

This observation did not attract much attention until Warburg and Krippahl (1958, 1960) rediscovered the requirement for CO_2. They, too, showed that the Hill reaction was inhibited by CO_2 removal and strongly stimulated by addition of CO_2 at low partial pressure, and that there was no net reduction of CO_2 occurring simultaneously. These authors suggested that the CO_2 requirement reflects a catalytic function for CO_2 in the mechanism of O_2 evolution. Earlier, Burk and Warburg (1950) had postulated an elaborate scheme for photosynthesis which contradicted much of the collective wisdom of the photosynthesis community, and this new finding of a CO_2 requirement was quickly pounced upon as evidence for the scheme (Warburg et al. 1959). In Warburg's scheme, which did not separate the photochemical process from CO_2 metabolism, a photochemical reaction consumes one molecule of CO_2 and yields one molecule of O_2 per quantum, then a thermochemical back reaction consumes two-thirds of the released O_2 and releases two-thirds of the consumed CO_2. To explain the Hill reaction, in which no net reduction of CO_2 occurs (Brown and Franck 1948, Warburg and Krippahl 1958), Warburg and Krippahl postulated that unlike whole cells, isolated chloroplasts cannot retain the reduced CO_2, which is reoxidized by the Hill reagent. In Warburg and Krippahl's scheme the precursor of O_2 is a phosphorylated peroxide of carbonic acid, produced by the action of illuminated chlorophyll on the Hill oxidant, CO_2, H_2O, and phosphate.

Perhaps because of Warburg's insistence that H_2O is not the source of photosynthetic O_2, this observation of a CO_2 requirement, unlike the earlier observation of Franck (1945), attracted much notice, and was rapidly confirmed by several researchers (Abeles et al. 1961, Stern and Vennesland 1962, Izawa 1962, Good 1963). The CO_2 effect was shown to be a general phenomenon, observable with a wide variety of Hill reagents and with a wide variety of species (Stern and Vennesland 1962). Several observations, especially by Norman Good, at this time argued against the scheme of Warburg and Krippahl: (i) the stimulatory effect of CO_2 on the Hill reaction was much reduced in weak light, compared to strong light, suggesting that CO_2 was not involved in a photochemical reaction, but in a non-photochemical step (Izawa 1962, Good 1963); (ii) the correlation of the CO_2 dependence with the presence of small anions suggested that HCO_3^-, not CO_2, was the important substance (Good, 1963); (iii) whereas Warburg's scheme has photophosphorylation intimately connected with CO_2 metabolism, uncouplers of phosphorylation do not relieve the impairment caused by CO_2 depletion, indicating a site of action remote from phosphorylation (Good 1963); and (iv) whereas one would expect a greater CO_2 dependence with weaker Hill oxidants if the oxidant is involved in CO_2 metabolism, no such trend was observed (Good 1963). Nevertheless, Warburg continued to present his scheme as though it were established fact (c.f. Warburg 1964).

The site of action
Because of the non-independence of the HCO_3^- effect on light intensity, it was concluded that HCO_3^- acts at a non-photochemical step of the Hill

reaction (Izawa 1962, Good 1963). The first attempt to locate this site of action was by Punnett and Iyer (1964), who looked at the effect of CO_2 on photophosphorylation. They observed that by adding relatively high concentrations of HCO_3^- to non-HCO_3^--depleted chloroplasts, they could accelerate the Hill reaction, as well as enhance the rate of phosphorylation. The ATP:2e$^-$ ratio was also increased, particularly when the pH was above 7. Thus, one of the effects of added CO_2 appeared to be to improve the coupling between electron transport and phosphorylation. However, as pointed out by Batra and Jagendorf (1965), the stimulation of the Hill reaction by high [HCO_3^-] in the absence of either uncouplers or ADP and phosphate seems to argue, if anything, for a looser coupling. The apparent contradiction of these two observations they found difficult to rationalize. Punnett and Iyer suggested that CO_2 may increase the efficiency of formation of a high energy intermediate resulting from electron transport (now understood to be a pH difference across the membrane), but Batra and Jagendorf found that added CO_2 actually decreases the yield of the high energy state of the chloroplasts, which suggested to them that the high energy state may be in competition with the formation of ATP (it is now understood that ATP synthesis occurs at the expense of the transmembrane pH difference).

Batra and Jagendorf extended the observations of Punnett and Iyer and showed that the effect observed by them is actually a different effect than the HCO_3^- dependence observed by Franck (1945) and by Warburg and Krippahl (1958, 1960): (i) the *Punnett and Iyer effect* requires a relatively high [HCO_3^-] added to non-HCO_3^--depleted chloroplasts, whereas the *Franck/Warburg effect* requires much lower concentrations of HCO_3^- added to HCO_3^--depleted chloroplasts; (ii) uncouplers eliminate the stimulation of the Hill reaction by HCO_3^- in non-depleted chloroplasts (Batra and Jagendorf 1965), whereas uncouplers have no effect on the HCO_3^- dependence of depleted chloroplasts (Stern and Vennesland 1962, Good 1963, Khanna et al. 1977); (iii) added HCO_3^- stimulates phosphorylation under conditions of cyclic electron flow around PS I, supported by pyocyanine, with or without CMU (p-chlorophenyl-1,1-dimethyl urea) to block electron donation by PS II, whereas the removal of CO_2 by depletion has no effect on pyocyanine supported phosphorylation (Batra and Jagendorf 1965); and (iv) the *Franck/Warburg effect* appears to represent a requirement for HCO_3^-, in that the rate of electron transport is depressed by removal of CO_2 and is restored by adding back the HCO_3^-, whereas the *Punnett and Iyer effect* is a true stimulation, in that removal of CO_2 does not inhibit phosphorylation and cyclic electron transport (Batra and Jagendorf 1965). To this list can be added several other observations: the pH optimum for the *Franck/Warbuck effect* is around pH 6.5 (Khanna et al. 1977, Vermaas and van Rensen 1981),

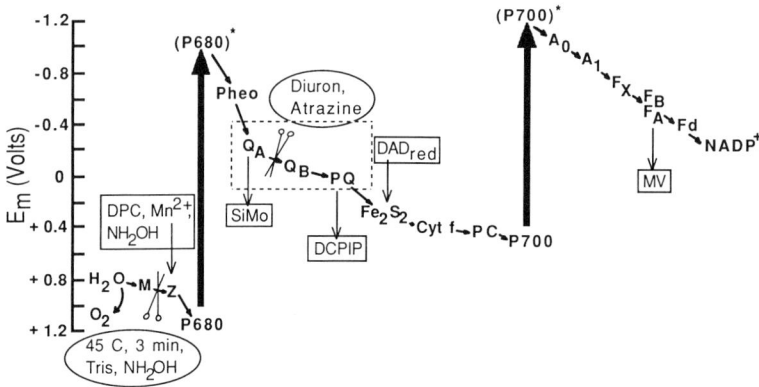

Fig. 1. Sites of electron donors, acceptors and inhibitors in the noncyclic electron transfer pathway of plant photosynthesis. The dashed box encloses the steps that are inhibited by removal of HCO_3^-. The solid boxes enclose exogenous redox compounds which act as artificial electron donors or acceptors at the step indicated by the arrows. DPC is diphenylcarbazide; SiMo is silicomolybdate; DCPIP is 2,6-dichlorophenolindophenol; DAD_{red} is reduced diaminodurene; MV is methylviologen. The arrows point in the direction of electron flow. The ellipses enclose treatments or compounds which interrupt electron flow at the step indicated by the scissors. The photosynthetic components are placed vertically according to their approximate midpoint redox potential (E_m). The exogenous compounds enclosed by solid boxes or ellipses are placed for diagramatic convenience, not according to their redox potentials. Electron flow is initiated with a photon or an exciton at the reaction center chlorophyll *a* molecules of photosystem II (P_{680}) or photosystem I (P_{700}). The asterisks indicate excited states. M is the charge accumulator and the oxygen evolving complex; Z, thought to be tyrosine-160 of the D_1 reaction center polypeptide, is the electon donor to P_{680}; Pheo represents pheophytin; Q_A is a plastoquinone (PQ) molecule permanently bound on the D_2 polypeptide; Q_B is a PQ molecule transiently bound on the D_1 polypeptide -- after reduction to plastoquinol, Q_B (H_2) exchanges with another PQ molecule; PQ represents a pool of non-bound plastoquinone; Fe_2S_2 represents the Rieske iron-sulphur center; Cyt f stands for cytochrome f; PC is a plastocyanin; A_o is suggested to be a chlorophyll molecule; A_1 is possibly a quinone, phylloquinone; F_A, F_B and F_X are thought to be Fe-S centers; Fd is soluble ferredoxin; $NADP^+$ is nicotinamide adenine dinucleotide phosphate (also see Govindjee and Eaton-Rye 1986).

whereas Punnett and Iyer observed the maximal effect between pH 7.0 and 7.8; the *Franck/Warburg effect* requires a dark incubation for HCO_3^- to bind (Stemler and Govindjee 1973, Vermaas and van Rensen 1981), whereas the binding of HCO_3^- to a low affinity site occurs preferentially in the light (Blubaugh and Govindjee 1984); and herbicide binding appears to overlay HCO_3^- added to depleted thylakoids, but not HCO_3^- binding to a low affinity site in non-depleted thylakoids (Blubaugh and Govindjee 1984).

The first attempt to locate the site of impairment in HCO_3^- depleted chloroplasts (*Franck/Warburg effect*) was the study by Stemler and Govindjee (1973), which seemed to show that HCO_3^- depletion had no effect on the rate of electron transport from the artificial PS II electron donor diphenylcarbazide (DPC) to the electron acceptor 2,6-dichlorophenolindophenol

(DCPIP) (Fig. 1). DPC is believed to donate electrons to the primary electron donor to the reaction center of PS II, Z. Therefore it was concluded then that the effect of HCO_3^- depletion on the H_2O to DCPIP reaction was due to a HCO_3^- site prior to Z; that is, at the O_2 evolving locus itself. However, this conclusion was soon abandoned as the result was reinterpreted to be due to the rate-limiting donation of electrons by DPC, which obscures the HCO_3^- effect (Wydrzynski and Govindjee 1975). This scenario was repeated a decade later when Fischer and Metzner (1981) concluded that HCO_3^- was required at the O_2 evolving site, in part because they could not observe a HCO_3^- effect in thylakoids using artificial electron donors to PS II (hydroxylamine, Mn^{2+}, a tetramethylbenzidine and tetraphenylboron). Eaton-Rye and Govindjee (1984) showed that for at least two of these (hydroxylamine and benzidine), the electron transport rates supported by these donors is no greater in non-depleted controls than the rates typically obtained by HCO_3^- depletion; thus, this approach cannot be used to assign a location for the HCO_3^- impairment.

Another problem with DPC as an electron donor is that it has an apparent effect on the membrane structure; in thylakoids that have begun to break down, DPC appears to stimulate energy trapping by PS II, perhaps by linking physically separated components (Harnischfeger 1974). This increase in the quantum yield can mask an impairment of electron transport. However, the HCO_3^- effect can be seen with DPC and other artificial electron donors to PS II, if one looks at chlorophyll a fluorescence instead of electron transport (for recent reviews on chlorophyll a fluorescence, see Govindjee et al. 1986). Wydrzynski and Govindjee (1975) showed that HCO_3^- depletion accelerates the rise of the chlorophyll a fluorescence transient in a manner similar to the herbicide diuron (3-(3,4-dichlorophenyl)-1, 1-dimethylurea, DCMU), which is known to "block" electron transport after the first stable plastoquinone acceptor, Q_A (Fig. 1). In contrast, treatments which are known to impair the O_2 evolving mechanism, such as mild heat treatment, Tris treatment, etc., were shown to eliminate the variable chlorophyll a fluorescence. These effects are predictable from the understanding that chlorophyll a fluorescence is a moniter of $[Q_A^-]$ (Duysens and Sweers 1963, Murata et al. 1966). Since HCO_3^- depletion produces a transient similar to treatment with diuron, Wydrzynski and Govindjee were the first to conclude that HCO_3^- depletion causes a block on the acceptor side of PSII, after Q_A. In support of this argument, they showed that DPC, as well as other artificial PS II donors, restore the variable fluorescence to heat-treated and Tris-treated chlororplasts, but the effects of HCO_3^- depletion and restoration remain, even with these donor systems. Similarly, Eaton-Rye and Govindjee (1984) showed that when hydroxylamine is used to simultaneously inhibit O_2 evolution and to donate electrons to PSII, the decay of chlorophyll a fluorescence after a flash, which moniters the reoxidation of Q_A^-, is reversibly inhibited by HCO_3^- depletion. Thus, they reaffirmed the location of the HCO_3^- requirement to be on the acceptor side of PS II, after Q_A.

Initially, Stemler and Govindjee (1974) had interpreted the effect of HCO_3^- depletion on the chlorophyll a fluorescence transient as supportive of an impairment on the O_2 evolving side of PS II. This is because, although they observed an acceleration of the rise from the initial level F_0 to the intermediate hump I (F_I), they had observed a slower rise from F_I to the maximum fluorescence level F_{max} (F_P). They had argued that a block after Q_A would have caused a higher fluorescence at all times. However, this transient can now be better understood as a partial block after Q_A, due to a partial HCO_3^- depletion in their experiments. The accelerated rise from F_0 to F_I is due to the faster accumulation of Q_A^-, while the slower rise from F_I to F_{max} represents the filling of the plastoquinone (PQ) pool, which is slowed by the impairment; only when the PQ pool is reduced can $[Q_A^-]$ accumulate to its maximum level (Vermaas and Govindjee 1981a). A thorough HCO_3^- depletion causes a complete, or nearly complete, block between Q_B and the PQ pool, causing a fluorescence transient which is indeed higher at all times, up to F_{max} (Vermaas and Govindjee 1982b).

It is now well established that the major site of impairment caused by HCO_3^- depletion is on the electron acceptor side of PS II (for a review of this side, see Vermaas and Govindjee 1981b). By the use of artificial electron donors and acceptors, Khanna et al. (1977) suggested that the site was between Q_A and PQ. The PS I electron transport, as measured by O_2 uptake during electron transport from reduced diaminodurene (DAD_{red}) to methylviologen (MV) (DAD_{red} donates electrons after the PQ pool; MV accepts electrons from the terminal side of PS I and passes them to O_2; see Fig. 1), did not show any bicarbonate effect. Since the rates of electron flow were very high indeed, it is firmly established that HCO_3^- is not involved in these reactions. The PS II electron transport prior to Q_A, as measured by O_2 evolution during electron transport from H_2O to silicomolybdate (SiMo), also remained uninhibited by HCO_3^- depletion. However, the PS II reduction of oxidized DAD, which efficiently accepts electrons from the PQ pool, did show a strong HCO_3^- dependence. Although the SiMo result cannot be taken to prove the absence of the HCO_3^- effect in this reaction, again, due to the low rates of electron flow, all the results (see later) taken together suggest that the site of inhibition is after Q_A, but before the PQ pool. Graan (1986) has challenged the generally accepted premise that SiMo accepts electrons from Q_A (e.g. Giaquinta and Dilley 1975, Zilinskas and Govindjee

1975). He argues that all available evidence concerning SiMo involvement with PS II is also consistent with SiMo simply replacing diuron from the Q_B binding site (see also Böger 1982). The apparent ability to replace diuron is dependent on the redox state of SiMo; the reduced form apparently binds not all all or much less tightly than the oxidized form (Graan 1986). Thus, SiMo may be functioning like benzoquinone and other electron acceptors which replace PQ, except that the binding affinity of the oxidized form is high enough to outcompete diuron. Therefore, many reported observations throughout the literature, including the absence of a HCO_3^- effect in the H_2O-to-SiMo reaction, may have to be re-evaluated if Graan is correct. However, if the reduction of SiMo is rate limiting, then an impairment of electron transport after Q_A by HCO_3^- depletion would not be expected to be seen.

Regardless of what the final outcome concerning SiMo will be, there remains ample evidence for the involvement of HCO_3^- in electron transport between Q_A and the PQ pool. Jursinic et al. (1976) were the first to demonstrate that HCO_3^- depletion slowed the oxidation of Q_A^-, and, consequently reduction of the next electron acceptor Q_B, as monitered by the decay of chlorophyll a fluorescence yield after an actinic flash, from a half-time of about 0.5 ms to approximately 2.6 ms. When the decay was determined as a function of flash number (Govindjee et al. 1976), it was discovered that the oxidation of Q_A^- was even slower after the third and subsequent flashes, with a half-time of about 150 ms. Since Q_B acts as a "two electron gate" (see e.g. review by Vermaas and Govindjee 1981b), this suggests that two electrons can still flow through Q_A to reduce Q_B to Q_B^{2-}, and that the reoxidation of Q_B^{2-} then becomes rate limiting. Thus, HCO_3^- depletion appears to slow down electron flow from Q_A to Q_B and to block the exchange of Q_B^{2-} with the PQ pool.

There also appeared to be a 30-50% inhibition of charge separation in these and other repetitive flash experiments (Stemler et al. 1974, Jursinic et al. 1976, Siggel et al. 1977), which prompted the suggestion that HCO_3^- depletion also inactivates a portion of the PS II reaction centers (Jursinic et al. 1976). However, an alternative explanation was offered by Jursinic and Stemler (1982), who found that a very slow component of the fluorescence decay, with a half-time of 1-2 s, was increased two-to-three-fold in HCO_3^- depleted samples. They suggested that in a significant portion of the reaction centers of HCO_3^- depleted chloroplasts, Q_A^- was not reoxidized in the dark time between flashes, thus keeping the reaction centers in a photosynthetically closed state. Since the increase of this very slow component occurred even after the first flash, they concluded that it was a component of the Q_A^- to Q_B electron transfer, and they suggested that HCO_3^- depletion may alter the redox potential (Em, 7) of Q_A with respect to Q_B, or reduce a local field that stabilizes Q_B^-. Vermaas and Govindjee (1982b) did not find any effect of HCO_3^- on the redox potential of Q_A/Q_A^-. It has been suggested that HCO_3^- depletion does destabilize Q_B^- by preventing the protonation of a nearby protein group (Eaton-Rye 1987). It is also possible that this slow component represents some inactive PS II centers (e.g. Graan and Ort 1986, Garab et al. 1987), and that HCO_3^- depletion somehow raises the number of such centers, perhaps by inhibiting the binding of PQ (e.g. Eaton-Rye 1987, see also Blubaugh 1987). It is possible that these slow centers are inactive they don't have HCO_3^- bound to them, an idea that is also shared by J. Whitmarsh, (personal communication). In normal active centers, PQ binding and release must occur with a half-time less than 1 ms (Crofts et al. 1984), in order to account for the observed reduction time of the PQ pool (Stiehl and Witt 1969).

Robinson et al. (1984) confirmed the slower chlorophyll fluorescence decay of HCO_3^- depleted thylakoids, but obtained much faster rates, overall, than were reported by Govindjee et al. (1976). Presumably, this was due to a slower flash frequency (1 Hz, instead of 33 Hz) that permitted most of the very slow component to decay between flashes. After one or two flashes, Q_A^- decays with a half-time of 1.2 ms in HCO_3^- depleted thylakoids, compared to 0.23 ms in the control samples. After 3 flashes the half-time is increased to 10 ms. Eaton-Rye (1987) has extended these observations to show that at pH 7.5, the half-time of Q_A^- decay in HCO_3^- depleted thylakoids continues to increase after each flash up to 5–7 flashes, as opposed to only 3 flashes at pH 6.5. Also, after one or two flashes, the half-time is greater at the acidic pH, whereas after 4 or more flashes the half-time is greater at the alkaline pH. This behavior was explained by two separate pH-dependent processes. From a kinetic analysis of the pH dependence of the decay rate after 1–2 flashes, it was suggested that the binding of PQ is inhibited by HCO_3^- depletion, with the greatest effect occurring at acidic pH. A simultaneous inhibition of the protonation of Q_B^- was presumed to occur, which becomes more severe at alkaline pH because of the further reduced availability of H^+.

The inhibition of Q_A^- reoxidation by HCO_3^- depletion has also been shown by following the decay of the absorbance change at 320 nm, which is due to absorption by the semiquinones Q_A^- and Q_B^-, with comparable results to those obtained by the fluorescence decay experiments described above (Siggel et al. 1977, Farineau and Mathis 1983).

The site of HCO_3^- action has also been located at the quinone reactions by the interaction between HCO_3^- binding and herbicide binding. Khanna et al. (1981) showed that HCO_3^- depletion decreased the binding affinity of atrazine. Similarly, a variety of atrazine-type herbicides have been shown to inhibit HCO_3^- binding (van Rensen and Vermaas 1981, Vermaas et al. 1982,

Snel and van Rensen 1983). Most of these herbicides appear not to be competitive with HCO_3^-, but bind closely enough to be affected by it. Since these herbicides are believed to inhibit PS II by replacing PQ from the Q_B site (e.g. Oettmeier and Soll 1983), the binding of HCO_3^- at or near Q_B is presumed.

While the effect of HCO_3^- depletion on the acceptor side of PS II has been firmly established, an effect on the donor side has been a source of controversy. Numerous observations have been reported to support the idea of a major effect on the O_2 evolving complex (for a review, see Stemler 1982). However, most of these observations have been explained without the need to invoke a significant site on the donor side of PS II (e.g. Vermaas and Govindjee 1981a, 1981b, 1982a). A small effect is, however, well known in which HCO_3^- can replace Cl^- to some extent on the electron donor side of PS II (see e.g. Critchley et al. 1982, Jursinic and Stemler 1988). On the other hand, some of the effects on the S-state transitions of the O_2 evolving complex (Stemler et al. 1974, Stemler 1980) may be due to interactions of the S-states with the Q_B site (see e.g. Diner 1977, Govindjee et al. 1984). Some of the observations have been shown to be artifactual, such as the apparent insensitivity of electron transport supported by artificial PS II donors (discussed above); and an effect on the kinetics of O_2 evolution after a flash could not be confirmed (Vermaas and Gonvindjee 1982a). One observation, that the rate of $H^{14}CO_3^-$ binding appears to be dependent on the pH of the lumen, rather than the external pH (Stemler 1980), has an alternative interpretation (see *The Model*). No firm evidence has, to date, been shown for a significant involvement of HCO_3^- on the donor side of PS II. In fact, numerous studies suggest a non-involvement of HCO_3^- on the donor side (Stemler and Radmer 1975, Jursinic et al. 1976, Khanna et al. 1977, van Rensen and Vermaas 1981, Khanna et al. 1981); this, of course, does not contradict the inefficient and small effect of HCO_3^-, like other monovalent anions, in replacing Cl^- on the donor side of PS II.

The active species

As was discussed in previous sections, the requirement for CO_2 in the Hill reaction was originally thought to indicate an involvement of CO_2 in the O_2 evolving mechanism (Warburg et al. 1959, see also Stemler and Govindjee 1973). Warburg and coworkers (1959) developed an elaborate scheme to show how a phosphorylated peroxide of carbonic acid (hydrated CO_2) could be the precursor to O_2 evolution. This scheme assumed that CO_2 or H_2CO_3 was the species required.

Good (1963) was the first to examine the effect of various anions on the bicarbonate (HCO_3^-) dependence, and found that small monovalent anions increased the dependence of the chloroplasts on HCO_3^-. Particularly effective were formate (HCO_2^-) and acetate ($CH_3CO_2^-$), which suggested to Good that the HCO_3^- ion is the important substance, not CO_2. Stemler and Govindjee (1973) took advantage of this suggestion to obtain the first method for reproducibly obtaining what was then considered to be a large (4–5 fold) HCO_3^- dependence of ferricyanide or 2,6-dichlorophenol indophenol (DCPIP) reduction by isolated chloroplasts. Their treatment consisted of low pH to favor the conversion of HCO_3^- to CO_2 and high salt (250 mM NaCl, 40 mM Na acetate) to compete with HCO_3^- for its binding site. Their maximum restored rates, however, were still largely inhibited with respect to non-treated chloroplasts, an effect which today can be understood as a consequence of irreversible damage by low pH (see e.g. Vermaas and Govindjee 1982b).

A subsaturating [HCO_3^-] showed a larger stimulation of the Hill reaction at pH6.8 than it did at pH 5.8, supporting the suggestion that HCO_3^- is the active species (Stemler and Govindjee 1973). This experiment was not conclusive, however, since it did not rule out the possibility of a pH dependence on the binding affinity of the active site. The authors favored the conclusion that HCO_3^- is the active species, however, by another argument: They showed that HCO_3^- stimulation of the Hill reaction only occurred when the chloroplasts were incubated with HCO_3^- in the dark. No stimulation occurred while the chloroplasts were illuminated, though a subsequent dark period would restore the activity. It was suggested that HCO_3^- is released in the light at a rate that corresponds with its binding. Since CO_2 is uncharged and non-polar, it would not be expected to bind other than by covalent attachment, while the suggested exchange is more consistent with an ionic binding. Thus, HCO_3^- as the active species was thought to be more likely. However, CO_2 can form relatively unstable carbamate complexes with protein amino groups, which decompose readily. Such a carbamate formation, for instance, has been demonstrated for the regulation of ribulose-1,5-bisphosphate carboxylase by CO_2 (Lorimer et al. 1976).

The pH profile of the HCO_3^- dependence shows an optimum around pH 6.5 (Khanna et al. 1977, Vermaas and van Rensen 1981). While the measurements of Stemler and Govindjee (1973) were confirmed by the pH profile, the drop-off of the dependence as the pH is increased above 6.5 argued against HCO_3^- as the sole active species, since [HCO_3^-] would be expected to increase with increasing pH up to around pH 8.4. Again, this argument ignores any possible pH effects on the binding environment. However, because of the close proximity of the pH optimum to the pKa of HCO_3^-/CO_2

(pKa = 6.4), it was suggested that both CO_2 and HCO_3^- are required (Vermaas and van Rensen 1981).

Evidence that CO_2 is involved in the "bicarbonate effect" was obtained by Sarojini and Govindjee (1981a, 1981b) by measuring the lag time between the addition of CO_2 or HCO_3^- and the onset of O_2 evolution. At a low assay temperature (5 °C) to slow the interconversion of CO_2 and HCO_3^-, and an alkaline pH of 7.3 to give a large ratio of $[HCO_3^-]$ to $[CO_2]$, the lag time was considerably shorter when CO_2 was added, compared to HCO_3^-. When cabonic anhydrase was present to accelerate the interconversion of CO_2 and HCO_3^-, the lag times became nearly identical for either species added. These results were interpreted to mean that either CO_2 was the species that was bound, or that CO_2 was required for diffusion to the active site.

Stemler (1980) concluded that CO_2, not HCO_3^-, is the binding species because in non-depleted chloroplasts inhibited by 100 mM HCO_2^-, the addition of 50 mM HCO_3^- caused a further inhibition of O_2 evolution at pH 8.0, whereas HCO_3^- partially restored the activity at pH 7.3. Because of the much larger $[CO_2]$ at the lower pH, it was concluded that CO_2 is the binding species and that HCO_3^-, like HCO_2^-, is inhibitory. However, these measurements, like the earlier ones at pH 5.8 and 6.8 (Stemler and Govindjee 1973), are not sufficient to determine the active species, since they ignore possible pH effects on the binding environment.

Stemler (1980) observed that the rate of ^{14}C labeling of HCO_3^- depleted chloroplasts by $H^{14}CO_3^-$ decreases with increasing pH over the pH range of 6.0 to 7.8, provided that at least a 5 min incubation period is given at the pH before the $H^{14}CO_3^-$ is added. When the incubation was omitted, the rate of ^{14}C labeling was pH independent for at least two minutes. These observations led Stemler to conclude that not only is CO_2 the binding species, but that the binding ocurs on the inside surface of the thylakoid membrane (i.e. at the O_2 evolving locus), as an incubation is necessary to allow the internal pH to equilibrate after a pH jump. However, the rate of ^{14}C labeling under these experimental conditions is greater at pH 6.0 than it is at pH 6.8, whereas the equilibrium amount of ^{14}C bound (Stemler 1980), as well as the Hill activity (Khanna et al. 1977, Vermaas and van Rensen 1981) is greater at the higher pH. Another explanation for the higher rate of binding at low pH is offered in the model).

That HCO_3^- is indeed the active species was demonstrated conclusively by taking advantage of the pH dependence of the equilibrium ratio of $[CO_2]$ to $[HCO_3^-]$ to effectively hold one concentration constant while varying the other (Blubaugh and Govindjee 1986). The restoration of the Hill activity to HCO_3^- depleted thylakoids was shown to be dependent only on $[HCO_3^-]$ and was independent of $[CO_2]$, $[H_2CO_3^{2-}]$, or $[CO_3^{2-}]$, over the pH range studied (6.3–6.9), which spanned both sides of the pH optimum. Although these results indicate that HCO_3^- is the binding species, they leave open as a possible role for CO_2 diffusion to the binding site (Sarojini and Govindjee 1981a, b), since they were performed under equilibrium conditions.

The number of binding sites

Stemler (1977) measured the binding of $H^{14}CO_3^-$ to isolated thylakoids and determined that there were two pools of HCO_3^-: a high affinity pool at a concentration of approximately one HCO_3^- per 300–400 chlorophyll molecules and a low affinity pool at a concentration as large or larger than that of the bulk chlorophyll. Depletion of the high affinity pool was correlated with the loss of Hill activity, whereas the role of the low affinity pool, presumed to be largely empty under physiological conditions, remained undetermined.

The low affinity pool. Blubaugh and Govindjee (1984) demonstrated an acceleration of the chlorophyll *a* fluorescence rise in the presence of diuron when excess HCO_3^- was added to non-depleted thylakoids, which they attributed to the binding of HCO_3^- or CO_2 to a very low affinity site separate from the site responsible for activation of the Hill activity. In addition to a large difference in affinity, the authors also observed (1) that the low affinity site appeared to bind HCO_3^- or CO_2 preferentially in the light, whereas the high affinity site binds HCO_3^- preferentially in the dark (see also Stemler and Govindjee 1973, Vermaas and van Rensen 1981), and (2) the high affinity site appeared to be overlaid by diuron (see also Stemler 1977, Vermaas and van renson 1981), which was not the case for the low affinity site.

Other effects of excess HCO_3^- added to non-depleted thylakoids have been reported: (1) it inhibits the photosystem II reduction of silicomolybdate (Barr and Crane 1976); even though (2) it accelerates whole-chain electron transport (Punnet and Iyer 1964, Batra and Jagendorf 1965, Barr and Crane 1976); and (3) it stimulates photophosphorylation (Punnet and Iyer 1964, Batra and Jagendorf 1965). Several of these effects require a significantly larger concentration of HCO_3^- than is required to stimulate the Hill reaction in HCO_3^- depleted thylakoids, suggesting involvement of the low affinity site.

Excess HCO_3^- (or CO_2) has been proposed to effect a conformational change in the coupling factor CF_1 (Cohen and MacPeek 1980). Little is known about this site, and there is the possibility that this effect is non-specific, in that carboxylic acids in general appear to stimulate phosphorylation (Nelson et al. 1972). However, at least one effect of HCO_3^- on CF_1, the increased inhibition of phosphorylation by N-ethylmaleimide in the presence of HCO_3^- and the decreased ability of adenylates to protect against this inhibition, is specific for HCO_3^- or CO_2 (Cohen and MacPeek 1980).

This effect on CF_1, however, cannot fully explain the stimulation of electron transport by excess HCO_3^-, since this stimulation occurs whether or not ADP and P_i are present (Punnett and Iyer 1964). It is also difficult to rationalize how a conformational change of CF_1 would accelerate the chlorophyll a fluorescence rise in the presence of diuron (Blubaugh and Govindjee 1984). Therefore, another low affinity site probably exists. Although HCO_3^-, not CO_2, is the species that restores electron transport to HCO_3^- depleted thylakoids (Blubaugh and Govindjee 1986), the species responsible for these other effects is still an open question.

The high affinity pool. Eaton-Rye (1987) has proposed two HCO_3^- sites in order to explain apparent effects on both PQ binding and protonation of Q_B^-. One site is suggested to be on Fe^{2+} in the Q_A-Fe-Q_B complex of PS II, as proposed by Michel and Deisenhofer (1988), and is responsible for maintaining the proper conformation for efficient PQ binding. The second site is suggested to be more difficult to deplete of HCO_3^- and is responsible for delivering a H^+ to a protein group near Q_B^-.

We have obtained evidence for two HCO_3^- sites, both necessary for Hill activity, from a kinetic study of the dependence on HCO_3^- (Blubaugh 1987): The amount of Hill activity restored to HCO_3^- depleted thylakoids by a half-saturating $[HCO_3^-]$ was non-linear with respect to [Chl], an indication that some endogenous HCO_3^- was present in the depleted samples. Since the basal activity in the absence of any added HCO_3^- was less than 7% of the fully restored rate, it was suggested that the small residual activity commonly observed after HCO_3^- depletion is due to endogenous HCO_3^- not removed during the depletion procedure, and that HCO_3^- is an essential requirement for the Hill activity. When corrected for the endogenous $[HCO_3^-]$, the activity vs. $[HCO_3^-]$ data for these same membranes no longer suggested Michaelis-Menten kinetics, but instead suggested two HCO_3^- sites with cooperative binding ($n > 1.4$).

The large low affinity pool, discussed earlier, does not appear to be correlated with restoration of the Hill activity, as the activity is largely restored even when the low activity pool is mainly empty (Stemler 1977). Therefore, the two cooperative sites suggested to control the Hill activity must be separate from this low affinity pool.

The role of HCO_3^-

Q_B is known to be a transiently bound PQ molecule which, after becoming doubly reduced to plastoquinol (PQH_2), exchanges with the PQ pool, as was first proposed independently by Velthuys (1981) for green plants and by Wraight (1981) for purple bacteria (ubiquinone replaces PQ in the bacteria). The steps at which protonation of Q_B occurs, however, is not fully elucidated. Diner (1977) proposed, based on thermodynamic considerations of the equilibrium constant for dismutation of duroquinol, compared to the equilibrium constants for electron transfer through the quinones of PS II, that the protonation of Q_B should occur at the level of Q_B^{2-}. Similarly, Pulles et al. (1976) found from the difference absorption spectrum that Q_B^- is unprotonated, and Fowler (1977) observed a binary oscillation in H^+ uptake corresponding to the production of Q_B^{2-}. However, the oscillations observed by Fowler were very small, and were not seen at all by others (Hope and Moreland 1979, Förster et al. 1981, see, however, van Rensen 1988). To account for an unprotonated Q_B^- and a lack of binary oscillation in H^+ uptake, Förster et al. (1981), proposed the protonation of a protein group to stabilize Q_B^-, as was proposed earlier for photosynthetic bacteria (Wraight 1979). Crofts et al. (1984) similarly proposed a scheme in which a protein group near Q_A and Q_B needs to be protonated before Q_B^- can accept a second electron from Q_A^-. They showed that the pKa for this group appears to shift from about 6.4 to approximately 7.9 when Q_B is reduced to Q_B^-, and the oxidation of Q_A^- by Q_B^- is slowed down when this group is unprotonated. This is analogous to what occurs in photosynthetic bacteria, although the pKa's are different (Wraight 1979).

It is tempting to assign HCO_3^- a role in this protonation (*e.g.* Stemler 1977), since the absence of HCO_3^- slows down the reduction of Q_B (Jursinic et al. 1976, Siggel et al. 1977), and the pKa of CO_2/HCO_3^- is about 6.4. The pKa of HCO_3^-/CO_3^{2-} is 10.2 in aqueous solution, though a bound HCO_3^-/CO_3^{2-} would be expected to have a lower pKa, due to stabilization of the negative charge upon binding to a positively charged group, and could be as low as 7.9. Such a speculation has been made (Vermaas 1984, van Rensen 1988). However, it is not likely that HCO_3^- is the group that is undergoing the pKa shift observed by Crofts et al. (1984). For one thing, at pH's below 6.4 the putative protein group is already protonated before Q_B^- formation, and Q_A^- to Q_B^- electron ransfer is not impaired (Crofts et al. 1984). If HCO_3^- is protonated it decomposes to form CO_2, which would leave the HCO_3^- site empty, and electron transfer from Q_A^- to Q_B^- would be impaired. It is possible, however, that HCO_3^- is responsible for providing a ready H^+ to this group when its pKa shifts to 7.9.

Eaton-Rye (1987) measured the kinetics of Q_A^- oxidation after one to ten actinic flashes, as a function of pH, in HCO_3^- depleted thylakoids, and observed that at pH 7.5 two turnovers of the Q_B "two-electron gate" (i.e. four flashes) were necessary before the maximum inhibition of Q_A^- to Q_B electron flow could be seen on subsequent flashes. By contrast, at pH 6.5 the inhibition was maximum after only one turnover of Q_B (i.e. after two flashes). The maximum inhibition at pH 7.5 was also greater than that at pH 6.5, but this difference was diminished as the flash frequency was increased. To explain these observations, it was suggested that the protonation of the putative protein group near Q_B was the rate-limiting step at alkaline pH when HCO_3^- was not present. This explains the greater inhibition at pH 7.5: at pH 6.5, sufficient H^+s were supposedly available from sources other than HCO_3^- to ameliorate the inhibition, but with increasing flash frequency the time for such a H^+ to arrive became limiting. The greater inhibition at pH 7.5 would also prevent the advancement of the two electron gate during a flash in some centers, to account for the observation that two turnovers of Q_B are necessary to see the full inhibition. (Although HCO_3^- is presumed to supply a H^+ to this protein group, removal of HCO_3^- would not be expected to deprotonate the group.) These centers, then would not be inhibited until the second turnover.

An analysis of the fast phase of Q_A^- oxidation after the first two actinic flashes revealed another pH-dependent impairment, in which Q_A^- to Q_B^- electron flow was slower, not at higher but at lower pH (Eaton-Rye 1987). As the fast phase is presumably due to centers with Q_B already bound prior to the flash (Crofts et al. 1984), this observation was taken to indicate either an impairment in the binding of PQ to the Q_B site (Govindjee and Eaton-Rye 1986, Eaton Rye 1987) or some other unspecified conformational change that is more pronounced at lower pH. Thus in addition to its involvement in the protonation reactions, a second, possibly structural, role for HCO_3^- is implicated (see also Vermaas and Rutherford 1984).

That HCO_3^- depletion may impair PQ binding at the Q_B site was also suggested by the effect of HCO_2^- incubation on the binding of a photoreactive PQ analog, 6-azido-5-decyl-2,3-dimethoxy-p-benzoquinone (6-azido-Q_0C_{10}; Blubaugh 1987). This analog can functionally replace PQ, but becomes covalently attached to the Q_B site during ultraviolet (UV) irradiation (Vermaas et al. 1983). After this covalent attachment, the rise of the Chl a fluorescence transient is accelerated (Vermaas et al. 1983, Blubaugh 1987). However, when HCO_2^- was present during the 6-azido-Q_0C_{10}/UV treatment, followed by HCO_3^- addition, the Chl a fluorescence rise was not accelerated (Blubaugh 1987). This suggests that HCO_2^- incubation, which is known to remove HCO_3^- from its binding sites, may impair the binding of 6-azido-Q_0C_{10} at the Q_B site.

Physiological significance

Reproducibly large effects of HCO_3^- depletion were not observed in thylakoids until low pH and high salt concentrations were used during the depletion procedure (Stemler and Govindjee 1973). Since 1973, high concentrations of formate, (HCO_2^-) have been routinely used to aid in depleting samples of HCO_3^-. With the measurement by several laboratories of a dissociation constant for the HCO_3^- * PS II complex of $80 \mu M$ (Stemler and Murphy 1983, Snel and van Rensen 1984, Jursinic and Stemler 1986), it has been suggested that the stimulatory effect of HCO_3^- is no more than a simple reversal of an inhibitory effect of HCO_2^- (Stemler and Murphy 1983). However, the argument was based on the observation that the intracellular $[CO_2]$ is around $5 \mu M$ in photosynthesizing chloroplasts (Hesketh et al. 1982). It is now established that HCO_3^-, not CO_2, is the activating species (Blubaugh and Govindjee 1986). At a pH of 8 (the approximate stromal pH during photosynthesis), the $[HCO_3^-]$ in equilibrium with $5 \mu M$ CO_2 would be $220 \mu M$. Also, the Kd value of $80 \mu M$ is overestimated by at least a factor of two, since its determination did not take into account the fact that nearly half of the added HCO_3^- was converted into CO_2. Thus, the calculated Kd should be around $40 \mu M$. Therefore, there is no reason to presume that the activating sites are empty under physiological conditions. On the other hand, while a HCO_3^- dependence in the absence of HCO_2^- has been observed (Good 1963, Robinson et al. 1984, Eaton-Rye et al. 1986, see also chapter 2 in Blubaugh 1987), the effect is never as dramatic as when inhibitory anions are present. This does not, however, necessarily imply that HCO_3^- is not required – it could simply reflect the difficulty of removing HCO_3^- without the assistance of a competitor. It has recently been suggested, from a kinetic study of the HCO_3^- effect, that HCO_3^- is an essential activator and that the small amount of activity remaining after HCO_3^- depletion is due to endogenous HCO_3^- that was not removed (Blubaugh 1987).

The observation of a HCO_3^- effect on PS II *in vivo* is difficult to distinguish, due to the obvious requirement for CO_2 in the Calvin cycle. Nevertheless, Ireland et al. (1987) have attempted to do this in leaves by examining the effect on chlorophyll *a* fluorescence of a small decrease in the intracellular CO_2 concentration from an already low level. Although CO_2 fixation decreased slightly from an already low level, they observed a much larger increase in $[Q_A^-]$, and concluded that HCO_3^- is involved in the quinone reactions *in vivo*. (Also see Garab et al. 1983.)

In the model which follows, HCO_3^- is assumed to be a physiological requirement; CO_2 is a diffusing species.

The model

In this working model (see Fig. 2), one HCO_3^- (Site A, see Table 1) is a bidentate ligand to the Fe^{2+}, as proposed by Michel and Deisenhofer (1988), and forms a salt bridge necessary for the functional configuration of the reaction center. Perhaps, it is H-bonded to a residue on the D_2 protein, as a replacement for glutamate-232 of the M subunit of the bacterial reaction center (Michel and Deisenhofer 1988). Disruption of this salt bridge is suggested to alter the distance between Q_A and Q_B, resulting in a slower electron transfer, to make Q_A^- more accessible to direct oxidation, and to alter the binding affinity of plastoquinone to the Q_B site.

We propose that a second high-affinity HCO_3^- (Site B, see Table 1) is bound to an arginine in the D_1 protein, as suggested by Shipman (1981), and is involved in protonating a histidine at the Q_B site. The histidine would be the group whose pKa has been observed to shift from 6.4 to 7.9 upon

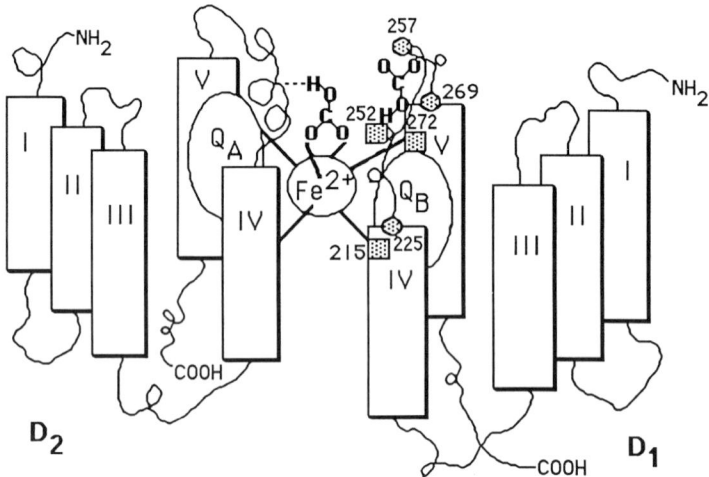

Fig. 2. A model of the Q_A-Fe-Q_B complex of the D_1-D_2 reaction center proteins, showing the HCO_3^- ligand to Fe^{2+} (site A, see text) and the possible arginine/histidine pairs on D_1 that comprise the second high-affinity HCO_3^- site (site B, see text). At site A, the HCO_3^- is shown H-bonded to a residue on the D_2 protein, as a structural requirement for the native configuration of the reaction center. In photosynthetic bacterial reaction centers, which do not appear to exhibit a HCO_3^- dependence (RJ Shopes, D Blubaugh, CA Wraight and Govindjee, unpublished), such a bridge is supplied by glutamate-232 of the M subunit (Michel and Deisenhofer 1986). At site B, the HCO_3^- is bound to an arginine and undergoes acid/base chemistry with a nearby histidine. The possible arginines are indicated by the stippled circles (arginines 225, 257, 269), the possible histidines by stippled squares (histidines 215, 252, 272); arginine 257/hisitidine 252 are the most likely pair (see text). The locations of these residues are patterned after the folding sequence of Trebst (1987). The rectangles represent the membrane spanning regions of the proteins.

formation of Q_B^- (Crofts et al. 1984). This pKa shift, induced by the negative charge on Q_B^-, encourages protonation of the histidine by the HCO_3^-, whose own pKa is lowered by the electron withdrawing effects of the arginine (Fig. 3). The possible arginines are indicated by the stippled circles in Fig. 2 (Arg 225, 257, 269); the possible histidines are indicated by the stippled squares (His 215, 252, 272). The locations of these residues are patterned after the folding sequence suggested by Trebst (1987). There are thus three arginine/histidine pairs that could be involved. In two of these (Arg 257/His 252 and Arg 269/His 272), the arginine and histidine would be separated by a single helical turn, while in the third (Arg 225/His 215) they would be separated by two helical turns. Since histidines 215 and 272 are already liganded to Fe^{2+}, we consider that histidine-252 is most likely to undergo the acid/base chemistry proposed here. Therefore, we favor **arginine-257** and **histidine-252** as the catalytic site (Site B) for this second HCO_3^-. We further suggest that the HCO_3^- ions at Site A and Site B bind cooperatively.

The CO_3^{2-} resulting after deprotonation of the HCO_3^- at Site B is suggested to bind less tightly because of the greater delocalization of the charge over the molecule (Fig. 3, see also Blubaugh and Govindjee 1986). Another HCO_3^- from an intramembrane pool displaces CO_3^{2-}, ensuring irreversibility of the protonation reaction. Interaction of the new HCO_3^- with the N of histidine may favor H^+ transfer from histidine to Q_B^-. CO_3^{2-}, with a pKa of 10.2, would readily pick up a H^+.

A large number of low-affinity sites (Site C, see Table 1; see also Stemler 1977, Blubaugh and Govindjee 1984) may act as a buffer of the intrathylakoid $[HCO_3^-]$, to keep the arginine loaded during rapid turnover of the reaction center. Consumption of H^+ during illumination would drive the equilibrium toward HCO_3^-, leading to a net influx of CO_2 (and an effective influx of H^+) and loading of the low-affinity sites (Fig. 4). Efflux of HCO_3^- (as CO_2) in the dark would be limited by the availability of H^+. Low pH and high ionic strength, both necessary for effective HCO_3^- depletion, are suggested to disrupt the $Fe^{2+} \cdot D_2$ salt bridge (Site A) and to expose the low affinity sites to the bulk phase (Fig. 5).

Basic assumptions of the model and their justifiction
Assumption 1: HCO_3^- is the active species. It was demonstrated previously (Blubaugh and Govindjee 1986) that restoration of the Hill activity to HCO_3^- depleted thylakoids depends only on the equilibrium $[HCO_3^-]$; it is independent of the equilibrium $[CO_2]$, $[H_2CO_3]$ or $[CO_3^{2-}]$.

Assumption 2: There are two cooperative sites of HCO_3^- binding. Although it has been previously reported that the restoration of the Hill activity to

Table 1. Proposed sites of HCO_3^--binding in photosystem II, their functions and supporting evidence. For details, see text.

Site	Location	Function	Supportive arguments
A	Ligand to Fe^{2+}, H-bonded to residue on D_2 protein	• Structural; salt bridge holds reaction center in proper conformation • Disruption of salt bridge slows electron transfer from Q_A^- to Q_B, impairs PQ binding, permits direct oxidation of Q_A^-, allows exchange of HCO_3^- pool (sites C) with bulk phase • Cooperative binding with site B	• EPR signal at g = 1.82 increases dramatically upon HCO_3^- depletion (Vermaas & Rutherford, 1984) • Formate prevents oxidation of Fe^{2+} by exogenous quinones (Zimmermann & Rutherford, 1986) • In bacterial reaction centers, glutamate-232 of M subunit is bidentate ligand to Fe^{2+}; the region containing this residue is deleted in D_2 protein of plants (Michel & Deisenhofer, 1988) • Two possible sites of cooperative binding; endogenous HCO_3^- possibly exchanges with bulk phase at low pH, high ionic strength (Blubaugh 1987) • In presence of formate, azido analog of PQ becomes less attached to Q_B site; may also oxidize Q_A^- directly (Blubaugh 1987)
B	D_1 protein, at one of following arginine/histidine pairs: Arg-225/His-215 Arg-257/His 252* Arg-269/His-272 *The pair favored in this model	• Protonation of Q_B^- • HCO_3^- provides rapid source of H^+ to histidine to stabilize Q_B^- • Released CO_3^{2-} is recipient of H^+ • Cooperative binding with site A	• Arginine-HCO_3^- complex would have proper dipole moment for herbicide binding (Shipman, 1981) • The herbicide diuron overlays HCO_3^- (Stemler, 1977, Blubaugh and Govindjee, 1984) • Bromoacetate modifies a residue (histidine, cysteine or tyrosine) near HCO_3^- binding site (Stemler, 1985)

| C | Large number of low affinity sites in thylakoid membrane, in equilibrium with site B | • Not saturated, ordinarily
• Provides pool of HCO_3^- for rapid replacement of CO_3^{2-} at site B, after protonation of histidine
• Buffers intrathylakoid $[HCO_3^-]$ | • Large pool of low affinity $H^{14}CO_3^-$ binding sites (Stemler, 1977)
• High concentrations of HCO_3^- stimulate electron transport (Punnett & Iyer, 1964; Barr & Crane, 1976) and accelerate Chl a fluorescence rise in presence of diuron (Blubaugh & Govindjee, 1984); these effects are due to a separate site of HCO_3^- action (Batra & Jagendorf, 1965; Blubaugh & Govindjee, 1984) |

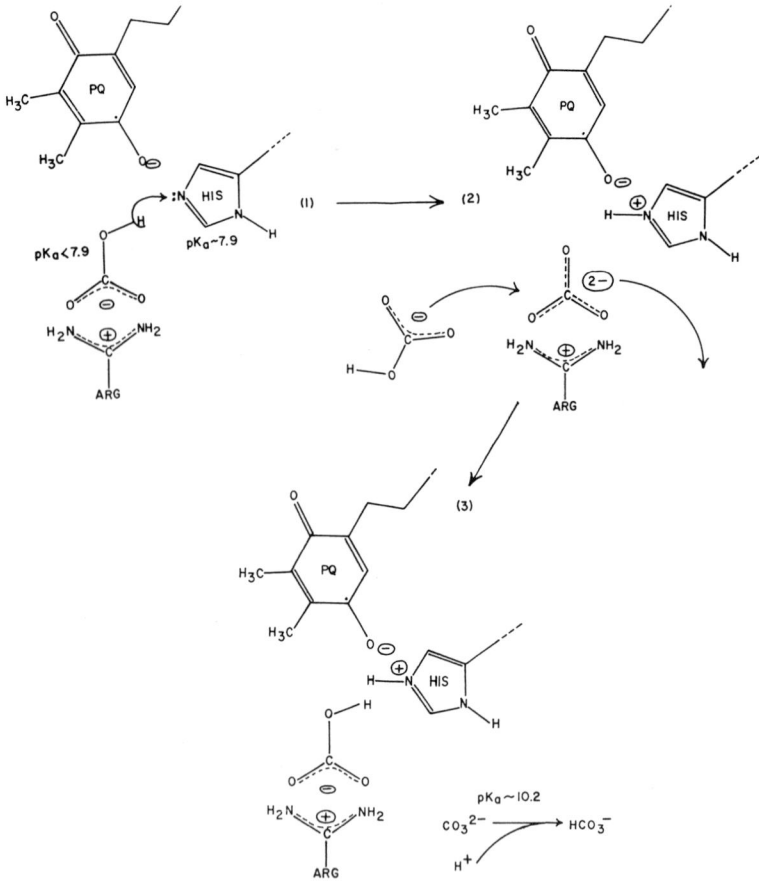

Fig. 3. A working model for the involvement of HCO_3^- in the protonation of Q_B^-. Step 1: (start at top left) When Q_B (PQ) is reduced to Q_B^-, the pKa of the histidine (HIS) shifts from 6.4 to 7.9 (Crofts et al. 1984), in response to the repulsive interaction between the negative charge on Q_B^- and the lone pair of electrons on histidine. The pKa of the bound HCO_3^- is presumed to be lowered from 10.2 to about 7.9, due to the electron withdrawing effect of the arginine (ARG). Therefore, there would be an equilibrium sharing of a H^+ between CO_3^{2-} and histidine. Step 2: (See top right) Replacement of CO_3^{2-} by another HCO_3^-, would drive the equilibrium toward histidine and ensure irreversibility of the protonation reaction. Step 3: (see middle and bottom) The pKa of the free CO_3^{2-} is 10.2, so reprotonation would be very rapid. Steric effects between the OH of HCO_3^- and the H on histidine would favor the transfer of H^+ from histidine to Q_B^- or Q_B^{2-},

HCO_3^- depleted thylakoids exhibits Michaelis-Menten kinetics (e.g. Snel and van Rensen 1983), we have recently observed that even in well-depleted thylakoids exhibiting a large HCO_3^- dependence (basal activity $< 7\%$ of the fully restored rate), a significant amount of endogenous HCO_3 remains (Blubaugh 1987). When the kinetic data was corrected for the estimated

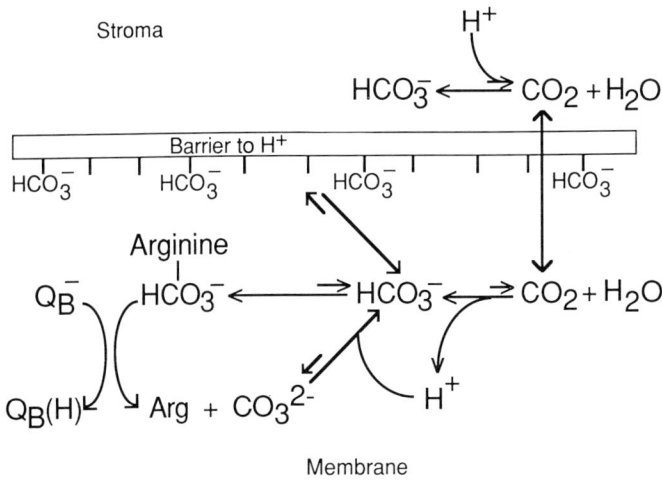

Fig. 4. A working model for the involvement of a sequestered intrathylakoid pool of HCO_3^- in the protonation of Q_B^-. The membrane provides an effective barrier to H^+ and HCO_3^- diffusion. CO_2 diffuses (see Sarojini and Govindjee, 1981a,b) into the sequestered space, driven by the consumption of H^+ as Q_B is turned over. Free HCO_3^- is in equilibrium with the high-affinity binding site on the arginine near Q_B (site B, see text), with the intracellular $[CO_2]$, with free CO_3^{2-}, and with a large number of postulated low affinity sites (sites C, see text), represented here as small tic marks. The longer arrows indicate the direction in which equilibrium would lie at a pH of 8. In the stroma, H^+ is taken up in the conversion of HCO_3^- to CO_2, and is effectively transported by the diffusion of CO_2. The low affinity sites buffer the intrathylakoid $[HCO_3^-]$. Although their affinity is low, there is a large number of them, so many HCO_3^- molecules are bound. In the dark, efflux of the HCO_3^- pool would be limited by the availability of H^+ for conversion of HCO_3^- to CO_2.

endogenous $[HCO_3^-]$, a Hill coefficient of $N_{app} > 1.4$ was obtained, indicating at least two sites with cooperative binding.

Assumption 3: A bidentate ligand to Fe^{2+} exists (Site A). HCO_3^- depletion has been observed to induce a large increase in the EPR signal at $g = 1.82$ (Vermaas and Rutherford 1984), attributed to the $Q_A^- * Fe^{2+}$ complex (Rutherford and Zimmermann 1984). In bacterial reaction centers, the Fe^{2+} forms two ligands to the carboxyl group of a glutamate residue on the M subunit and one ligand to each of four histidines, two of which are in the L subunit, and two of which are in the M subunit (Michel et al. 1986). In PS II reaction centers, the Q_B protein, D_1, has a high degree of similarity with the L subunit, while the D_2 protein (to which D and Q_A are believed to bind) is highly similar to the M subunit (e.g. Trebst and Draber 1986, Trebst 1987). However, one major difference is that D_2 is lacking an extra loop which, in the M subunit, carries the glutamate ligand to the Fe^{2+}. Michel and Deisenhofer (1988) have suggested that in PS II, HCO_3^- takes the place of the

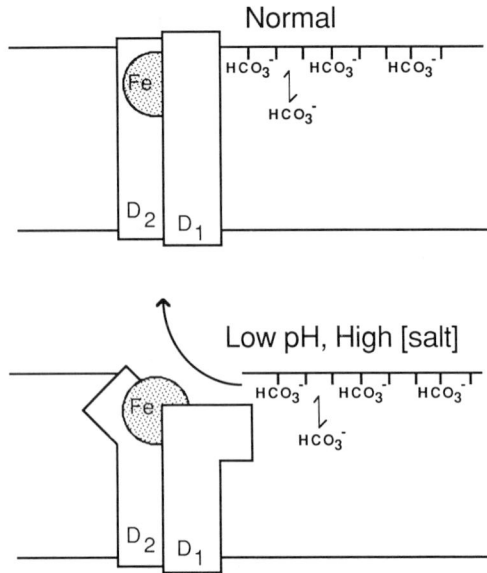

Fig. 5. A model of the D_1-D_2 reaction center proteins and intrathylakoid HCO_3^- pool. At low pH and high ionic strength, a conformational change exposes the low affinity HCO_3^- sites (sites C, see text) to the bulk phase. The conformational change may be due to disruption of a Fe^{2+}-HCO_3^--D_2 salt bridge (Fig. 1).

glutamate ligand. This would explain why a HCO_3^- dependence has never been observed in the photosynthetic bacteria (R.J. Shopes, D. Blubaugh, C.A.Wraight and Govindjee, unpublished). To complete the analogy, the HCO_3^- in PS II would presumably form a salt bridge between the Fe^{2+} and some residue in the D_2 protein.

Assumption 4: A binding site at an arginine near Q_B exists (Site B). One HCO_3^- site appears to be overlaid by the herbicide diuron. Diuron prevents the effect of HCO_3^- on the variable Chl a fluorescence in HCO_3^- depleted thylakoids if it is added first, but it does not reverse the effect of HCO_3^- if it is added second (Blubaugh and Govindjee 1984). Diuron in believed to replace PQ from the Q_B binding site (e.g. Oettmeier and Soll 1983). One thing that PS II herbicides appear to have in common is a flat polar component with a dipole moment of 3–5 Debyes (Shipman 1981). Thus, Shipman (1981) proposed that the herbicides bind electrostatically to a strongly polar binding site within a hydrophobic surface on the Q_B protein. HCO_3^-, bound to an arginine, could provide an appropriate electric field for such an interaction, and was considered by Shipman to be a likely part of the herbicide binding environment. The observation that diuron appears to overlay the HCO_3^- in PS II (Stemler 1977, Blubaugh and Govindjee 1984, Eaton-Rye and Govindjee 1988b), supports this notion.

Assumption 5: Acid/base chemistry occurs with a histidine near Site B . Because of the very close proximity of this HCO_3^- to Q_B, and because a protein group near Q_B is protonated upon reduction of Q_B to Q_B^- (Crofts et al. 1984), it is likely that the hydroxyl group of this HCO_3^- is involved in an acid/base reaction with the protein group. Eaton-Rye (1987) has also suggested that HCO_3^- is required at this site for the protonation of a site to stabilize Q_B^-. In our model, a histidine is shown as the protein group to be protonated for three reasons (1) the observed pKa is 6.4 (Crofts et al. 1984); (2) there is a histidine cluster around the Q_B site (Michel et al. 1986, Trebst and Draber 1986); and (3) bromoacetate, a protein modifier (histidine, lysine, tyrosine, and cysteine residues), which competes with HCO_3^- for its binding site, becomes covalently attached to a residue near the site (Stemler 1985). The pKa of this group shifts to about 7.9 upon formation of Q_B^- (Crofts et al. 1984). The negative charge of Q_B^- in close proximity to the histidine would cause such a shift in the pKa by stabilizing the positive charge of the protonated histidine. The pKa of the hydroxyl group of HCO_3^- is 10.2 in aqueous solution, but because of the electron withdrawing effect of the arginine, its pKa would be shifted lower. Thus, upon reduction of Q_B to Q_B^-, a H^+ transfer could occur from HCO_3^- to the histidine.

Assumption 6: CO_3^{2-} dissociates from Site B to be replaced by a HCO_3^- from an intrathylakoid pool. Blubaugh and Govindjee (1986) suggested that CO_3^{2-} binds less tightly than HCO_3^-. This was inferred from the difference in the extent of inhibition by nitrite (NO_2^-) versus nitrate (NO_3^-) (e.g. Stemler and Murphy 1985), which closely resemble HCO_3^- and CO_3^{2-}, respectively, in their electronic structures. Thus, it is presumed that the CO_3^{2-} resulting after H^+ transfer is displaced by another HCO_3^-. Although CO_3^{2-} has a greater negative charge for interaction with the arginine at Site B, a significant portion of the charge is distributed away from the arginine, toward, in our model, the negative charge on Q_B^-, which we would expect to be destabilizing. In the case of HCO_3^-, however, all of the charge would be directed toward the arginine. Although it is possible that the CO_3^{2-} could remain in place and simply pick up a H^+ from another source, there would be a distinct advantage to the system if CO_3^{2-} were to be replaced as proposed. Once released from its site, the pKa of CO_3^{2-} would return to 10.2, and reprotonation would be very rapid. Furthermore, the rapid replacement of CO_3^{2-} with HCO_3^- would ensure that the H^+ remains on the histidine; the equilibrium sharing of a H^+ between HCO_3^- and histidine would be shifted far in favor of the histidine. Such an exchange will occur most readily if a pool of HCO_3^- is available. The uptake of a H^+ from the bulk phase could, of course, occur with any of the CO_3^{2-} molecules in the pool and would be faster than if the CO_3^{2-} had to diffuse to the outer surface.

Assumption 7: A large number of low affinity sites exist (Sites C). Stemler (1977) demonstrated the existence of a large low affinity pool of HCO_3^-. This HCO_3^- could be the intrathylakoid pool proposed here. However, the number of HCO_3^- "sites" was found to at least approach, and perhaps exceed, the number of chlorophyll molecules (Stemler 1977). Similarly, Blubaugh (1987), from kinetic evidence for endogenous HCO_3^- remaining in HCO_3^- depleted thylakoids, estimated that perhaps as many as several thousand HCO_3^- molecules per PS II remained non-exhangeable with the bulk phase at moderate pH. From consideration of the possible size of the intrathylakoid $[HCO_3^-]$, which must remain in equilibrium with the intracellular $[CO_2]$, a large number of low affinity binding sites is proposed. By this means, a large number of intrathylakoid HCO_3^- molecules could exist at a moderate free $[HCO_3^-]$. The advantage to the system would be the buffering of the free $[HCO_3^-]$ against a drop in the intracellular $[CO_2]$ during photosynthesis. Although we show the low affinity sites binding HCO_3^-, CO_2 could also be bound as carbamate complexes with protein amino groups.

Assumption 8: A barrier to H^+ and HCO_3^- diffusion exists. The proteinaceous environment around Q_B has long been known to be a barrier to H^+ diffusion (e.g. Renger 1976). CO_2, rather than HCO_3^- appears to be the species which diffuses to the active sites (Sarojini and Govindjee 1981a, 1981b).

Discussion of the model
This model proposes two high affinity sites of HCO_3^- binding (Sites A and B) and numerous low affinity sites (Sites C). The HCO_3^- at Site A forms a salt bridge between the Fe^{2+} and a residue on the D_2 reaction center protein (Fig. 2), and is probably an essential requirement. The HCO_3^- at Site B is bound to an arginine (most likely arginine-257) of the D_1 reaction center protein and protonates a histidine (most likely histidine-252) in D_1, as part of a mechanism for protonation of Q_B^- (Fig. 3).

Such a specific proposal allows the application of molecular genetics tools (such as site-directed mutagenesis) to be used to test our model. After losing a H^+ to the histidine, the CO_3^{2-} at Site B exchanges with a HCO_3^- from an intrathylakoid pool. The low affinity Sites C buffer the intrathylakoid $[HCO_3^-]$ to ensure a rapid exchange with CO_3^{2-} at Site B, and perhaps to protect against fluctuations in the intracellular $[CO_2]$ (Fig. 4).

Neither CO_2, H_2CO_3 nor CO_3^{2-} have any apparent effect on the restoration of the Hill activity by HCO_3^- (Blubaugh and Govindjee 1986). Thus, the active species is presumed to be HCO_3^- at both sites A and B. Since the inhibitory formate (HCO_2^-) is identical to HCO_3^- in all respects except for a missing hydroxyl group, it can be presumed that the hydroxyl group is important for functioning at both sites. At Site A the hydroxyl group is suggested to be involved in an H-bond with a residue on the D_2 protein, to provide the proper conformation to the PS II reaction center. This salt bridge would be adversely affected by protonation during a low pH treatment, or by shielding of its charge by high salt concentrations. Blubaugh (1987) suggested that some ionizable group, when protonated or charge shielded, causes the release of otherwise-held HCO_3^-. It is possible that low pH or high ionic strength, both of which facilitate HCO_3^- depletion, disrupts the Fe^{2+}-HCO_3^--D_2 salt bridge, inducing a conformational change which permits the exchange of the intramembrane pool of HCO_3^- with the bulk phase (Fig. 5). Thus, low pH and high ionic strength alone are likely to induce a partial HCO_3^- depletion by permitting the HCO_3^- in the vicinity of Q_B to exchange with a large volume.

Although HCO_3^- was recently shown to be the binding species (Blubaugh and Govindjee 1986), Stemler (1980), earlier, concluded that CO_2 is the active species and that it binds on the lumen side of the membrane, because the rate of $H^{14}CO_3^-$ binding was pH independent for the first few minutes after a pH jump, but then showed a marked pH dependence after a 5 min incubation at the new pH, with the binding being greater at lower pH values. All that is necessary to reconcile this experiment with the model presented here is to postulate that the salt bridge is disrupted by low pH on a time scale of minutes or that the diffusion of HCO_3^- out of the sequestered space is slow. Immediately after a jump to low pH, the HCO_3^- binding sites would still be sequestered. After an incubation at low pH, however, the sites would become exchangeable with the bulk phase, and the rate of $H^{14}CO_3^-$ binding would be accelerated.

The pH optimum of the HCO_3^- effect is about 6.5 (Khanna et al. 1977, Vermaas and van Rensen 1981). The ascending arm is probably due to the fact that HCO_3^- is the active species (Blubaugh and Govindjee 1986), and the pKa of CO_2/HCO_3^- is about 6.35. The descending arm may reflect the pKa of the group to which HCO_3^- is H-bonded in the salt bridge. This pKa would be about 6.8, suggesting a histidine residue. At pH values significantly below this, both the removal and the binding of HCO_3^- would be facilitated (see Stemler 1977). When the conformational change occurs, the affinity for HCO_3^- at Site B would also be lowered because of cooperativity between Sites A and B (Blubaugh 1987).

At Site B, the hydroxyl group on HCO_3^- is suggested to undergo acid/base chemistry with a histidine. Stemler (1985) showed that bromoacetate, a protein modifier, as well as a structural analog of HCO_3^-, binds covalently in PS II, and that prior to covalent attachment it inhibits $H^{14}CO_3^-$ binding competitively, whereas after covalent attachment $H^{14}CO_3^-$ binding is inhibited noncompetitively. The model presented here is consistent with this observation: Bromoacetate undergoes a nucleophile substitution reaction with histidine, lysine, tyrosine and cysteine residues (Korman and Clarke 1955). Ordinarily, the reaction is very inefficient, but it can be accelerated considerably if the amino acid is "activated"; for instance, the half-time of the reaction is about 12 h with free histidine (Korman and Clarke 1955), as opposed to less than 1 h with a histidine in ribonuclease (Barnard and Stein 1959). The reaction proceeds significantly within several hours in PS II, and is faster in the light (Stemler 1985). Structurally, the C-Br bond of bromoacetate corresponds to the O-H bond of HCO_3^-. The orientation of this bond towards a histidine, as well as the pKa shift of the histidine upon Q_B^- formation, would, by increasing the nucleophilicity of the histidine, increase the reaction rate. After covalent attachment (Br is replaced by the histidyl residue), the carboxyl end would be pulled away from the arginine, but would remain close enough to continue to noncompetitively inhibit HCO_3^- binding at Site B.

Site B is very close to the site of Q_B binding, such that diuron binding to the Q_B site can also overlay the HCO_3^- site (Stemler 1977, Blubaugh and Govindjee 1984). Stemler and Murphy (1983) reported that diuron binding eliminates half of the HCO_3^- binding sites, consistent with this model (Site B eliminated, but not Site A). It has been similarly shown previously that HCO_3^- depletion lowers the binding activity of diuron-type herbicides (Khanna et al. 1981) and vice versa (van Rensen and Vermaas 1981, Vermaas et al. 1982, Snel and van Rensen 1983). This interaction is only partially competitive, and was explained by Vermaas and van Rensen (1981) as due to an overlapping of sites.

Since the two high affinity HCO_3^- sites (Sites A and B) are probably cooperative, anything that affects the binding at one site would be expected to affect the binding at the other through allosteric interactions. Thus, diuron would be expected to alter the binding affinities of HCO_3^- at both sites. Stemler and Murphy (1984) reported that the binding of atrazine to a high affinity site (i.e. the Q_B site) seems to remove some HCO_3^- "neither competitively nor noncompetitively but somehow directly". Similarly, the binding of HCO_3^- at site A would be expected to affect diuron binding, even though the sites are spatially separated. The apparent ability of silicomolybdate (SiMo) to reduce the binding affinity of diuron (Böger 1982, Graan

1986) may be due to SiMo removing HCO_3^- from the Fe^{2+} (Site A), rather than due to competition between SiMo and diuron for the same site, as was suggested by Böger (1982). That SiMo does remove bound HCO_3^- was shown previously by Stemler (1977), who observed that when the diuron was added before the SiMo, about half as much HCO_3^- was removed. This latter observation is predicted by this model; by overlaying site B, diuron would prevent the dissociation of half of the high-affinity HCO_3^-.

HCO_3^- may be required at Site A to help hold the reaction center proteins D_1 and D_2 together, via the Fe^{2+}. Removal of this HCO_3^- would be expected to induce a significant conformational change in PS II, which could disrupt electron flow through the quinones. Allostery is most often associated with polymeric enzymes. The involvement of HCO_3^- in a salt bridge to maintain the active configuration of D_1 and D_2 is, therefore, consistent with the suggested cooperativity (Blubaugh 1987) between the two HCO_3^- sites (Sites A and B). To fully explain the cooperativity, HCO_3^- binding at the arginine near Q_B (Site B) should also induce a conformational change that brings D_1 and D_2 closer to their native structure, thereby favoring the binding of HCO_3^- at Fe^{2+} (Site A).

Recently, it has been shown that some exogenous quinones, when reduced to their semiquinone form, can extract their second needed electron from the Fe^{2+}, and that formate blocks this oxidation of the Fe^{2+} (Zimmermann and Rutherford 1986). It is plausible that by removing the HCO_3^- ligand to the Fe^{2+} (Site A), formate induces a conformational change that increases the distance between the Fe^{2+} and the Q_B site, thus making electron transfer less likely. This same conformational change may then allow the exogenous quinones to accept electrons directly from Q_A^- (see Blubaugh 1987). SiMo, which might act as a metal chelator, may have a similar effect; by removing HCO_3^- from Site A, it would be able to both expose Q_A^- and act as the electron acceptor. The existence of such a conformational change upon addition of SiMo was suggested previously (Zilinskas 1975). Excess bicarbonate inhibits the photosynthetic reduction of SiMo (Barr and Crane 1976), perhaps by simple competition for the liganding site on the Fe^{2+}. Such a conformational change might also slow electron transfer from $Pheo^-$ to Q_A, as was recently observed for disulfiram (tetraethylthiuramdisulfide), which, as a metal chelator, was suspected of binding to the Fe^{2+} (Blubaugh and Govindjee 1988).

Q_{400}, first identified as a high-potential component of PS II by Ikegami and Katoh (1973), and which is now known to be the Fe^{2+} (Petrouleas and Diner 1986), is oxidized when chloroplasts are incubated with ferricyanide (FeCy) in the presence of diuron (Ikegami and Katoh 1973, Wraight 1985). Like the deceleration of the chlorophyll a fluorescence rise in the presence of diuron in HCO_3^- depleted thylakoids (Vermaas and Govindjee 1982b, Blubaugh and Govindjee 1984), the oxidation of Q_{400} by FeCy in the presence of diuron is also dependent on the order of addition (Ikegami and Katoh 1973, Bowes et al. 1979). Although the oxidation of Q_{400} can be partially observed when diuron is added first, under conditions involving the unstacking and then restacking of the grana (Wraight 1985), observation of the full effect requires that FeCy be added before the diuron. This suggests that FeCy, in addition to acting as a PS I electron acceptor, may be binding to the HCO_3^- site near Q_B (Site B). It may oxidize the Fe^{2+} from this position or, alternatively, removal of HCO_3^- from this site may permit Q_B^- to oxidize the Fe^{2+}. The ability of some exogenous quinones to oxidize Fe^{2+} from the Q_B site when they become reduced to the semiquinone (Zimmermann and Rutherford 1986) may likewise be due to an alteration in the binding affinity of HCO_3^- at Site B by these exogenous quinones. Although FeCy may bind to the HCO_3^- site near Q_B (Site B), it could be exerting an effect on the Fe^{2+} allosterically through the HCO_3^- site on the Fe^{2+} (Site A). FeCy is also able to oxidize Q_A^- directly in the presence of diuron at high salt concentrations (Itoh 1978) or at low pH (Itoh and Nishimura 1977), two treatments which, as discussed earlier, are likely to remove the liganding HCO_3^- from the Fe^{2+}. This supports the proposition above that a conformational change induced by the removal of HCO_3^- from one site permits the direct oxidation of Q_A^- by exogenous acceptors at the Q_B site. High salt concentrations actually decrease the oxidation of Q_A^- by FeCy when the pH is already low (Itoh 1978). This is understandable with the above model: when HCO_3^- is already gone from site A, high salt will have no additional effect, other than perhaps a competition with the FeCy for site B.

The affinity of the Q_B site for PQ may be lowered by the removal of HCO_3^- (Blubaugh 1987, see, also, Eaton-Rye 1987). This could be due to effects of HCO_3^- at either site A or B. Eaton-Rye (1987) has speculated that HCO_3^- binding at the Fe^{2+} (site A) affects PQ binding, while another HCO_3^- near Q_B (Site B) is involved in the protonation of Q_B^-. If HCO_3^- binding tightens PQ binding, it follows necessarily that PQ binding also tightens HCO_3^- binding. This may be sufficient to explain the preferential binding of HCO_3^- in the dark as opposed to the light (Stemler and Govindjee 1973, Stemler 1979), particularly when formate is present (Vermaas and van Rensen 1981). In HCO_3^- depleted thylakoids, Q_B will be largely reduced in the light (PQ^{2-}) and slow to exchange with the PQ pool. The absence of oxidized PQ in the Q_B site would make the affinity for HCO_3^- at site A less than it would be in the dark. While formate has a carboxyl group that could be a ligand to Fe^{2+} (Site A), it would not be able to H-bond with D_2, so it would be ineffective at providing the salt bridge that HCO_3^- can provide. Thus, formate may compete with HCO_3^- for Site A without having the same cooperativity with PQ binding (formate would also not have the same cooperativity with another formate at site B; this might explain why a relatively small $[HCO_3^-]$ (e.g. 10 mM) can outcompete a much larger concentration of formate (e.g. 100 mM). Therefore, in the presence of formate the difference in HCO_3^- binding affinity in the light versus dark would be even more significant, since the HCO_3^- would have the added hindrance of a competitor whose binding is uninfluenced by the redox state of Q_B.

Acknowledgement

We thank the National Science Foundation (PCM 83-06061) for financial support. D.B. was supported by a service award from the National Institute of Health (PHS 5-T32GM7283). During the preparation of this paper, Govindjee was supported by an Interdisciplinary Grant from the McKnight Foundation. We are obliged to Dr. Paul Jursinic for a critical reading of this manuscript.

References

Abeles FB, Brown AH, Mayne BC (1961) Stimulation of the Hill reaction by carbon dioxide. Plant Physiol 36: 202–207.

Babcock, GT (1987) The photosynthetic oxygen evolving process. In: Amesz J (ed.) Photosynthesis, Chapter 6, pp 125–158. Elsevier Science Publishers B.V. (Biomed Div).

Barnard EA and Stein WD (1959) The histidine residue in the active centre of ribonuclease: I. A specific reaction with bromoacetic acid. J Mol Biol 1: 339–349.

Barr R, and Crane FL (1976) Control of photosynthesis by CO_2: evidence for a bicarbonate-inhibited redox feedback in photosystem II. Proc Indiana Acad Sci 85: 120–128.

Batra PP and Jagendorf AT (1965) Bicarbonate effects on the Hill reaction and photophosphorylation. Plant Physiol 40: 1074–1079.

Blubaugh DJ (1987) The mechanism of bicarbonate activation of plastoquinone reduction in photosystem II of photosynthesis. Ph. D. Thesis, University of Illinois at Urbana-Champaign.

Blubaugh DJ and Govindjee (1984) Comparison of bicarbonate effects on the variable chlorophyll a fluorescence of CO_2-depleted and non-CO_2-deleted thylakoids in the presence of Diuron. Z Naturforsch 39c: 378–381.

Blubaugh DJ and Govindjee (1986) Bicarbonate, not CO_2, is the species required for the stimulation of photosystem II electron transport. Biochim Biophys Acta 848 147–152.

Blubaugh DJ and Govindjee (1988b) Sites of inhibition by disulfiram in thylakoid membranes. Plant Physiol, in press.

Böger P (1982) Replacement of photosynthetic electron transport inhibitors by silicomolybdate. Physiol plant 54: 221–224.

Bowes, JM, Crofts AR and Itoh S (1979) A high potential acceptor for photosystem II. Biochim Biophys Acta 547: 320–335.

Boyle FP (1948) Some factors involved in oxygen evolution from triturated spinach leaves. Science 108: 359–360.

Brown AH and Franck J (1948) On the participation of carbon dioxide in the photosynthetic activity of illuminated chloroplast suspensions. Arch Biochem 16: 55–60.

Burk D and Warburg O (1950) 1-Quanten-Mechanismus und Energie-Kreisprozess bei der Photosynthese. Die Naturwissenschaften 37: 560–569.

Cohen WS and MacPeek WA (1980) A proposed mechanism for the stimulatory effect of bicarbonate ions on ATP synthesis in isolated chloroplasts. Plant Physiol 66: 242–245.

Critchley C, Baianu IC, Govindjee and Gutowsky HS (1982) The role of chloride in O_2 evolution by thylakoids from salt-tolerant higher plants. Biochim Biophys Acta 682: 436–445.

Crofts AR, Robinson HH and Snozzi M (1984) Reactions of quinones at catalytic sites; a diffusional role in H-transfer. In: C Sybesma (ed.) Advances in Photosynthesis Research, pp461–468 The Hague: Martinus Nijhoff.

Diner BA (1977) Dependence of the deactivation reaction of photosystem II on the redox state of plastoquinone pool A, varied under anaerobic conditions: equilibria on the acceptor side of photosystem II. Biochim Biophys Acta 460: 247–258.

Dole M (1935) The relative atomic weight of oxygen in water and in air. J Am Chem Soc 57:2731–2731.

Dole M and Jenks G (1944) Isotopic composition of photosynthetic oxygen. Science 100: 409–409.

Duysens LNM and Sweers HE (1963) Mechanism of two photochemical reactions in algae as studied by means of fluorescence. In: Jap Soc of Plant Physiol (eds.)*Studies on Microalgae and Photosynthetic Bacteria*, pp 353–372. Tokyo: The University of Tokyo Press.

Eaton-Rye JJ (1987) Bicarbonate reversible anionic inhibition of the quinone reductase in photosystem II. Ph.D. Thesis, University of Illinois at Urbana.

Eaton-Rye JJ and Govindjee (1984) A study of the specific effect of bicarbonate on photosynthetic electron transport in the presence of methyl viologen. Photobiochem Photobiophys 8: 279–288.

Eaton-Rye JJ, Blubaugh DJ and Govindjee (1986) Action of bicarbonate on photosynthetic electron transport in the presence or absence of inhibitory anions. In: GC Papageorgiou, J Barber and S Papa (eds) *Ion Interactions in Energy Transfer Biomembranes* pp 263–278. New York: Plenum Publishing Corporation.

Farineau J and Mathis P (1983) Effect of bicarbonate on electron transfer between plastoquinones in photosystem II. In: Y Inoue, AR Crofts, Govindjee, N Murata, G Renger and K Satoh (eds) *The Oxygen-Evolving System of Plant Photosynthesis,* pp 317–325. New York: Academic Press.

Fischer K and Metzner H (1981) Bicarbonate effects on photosynthetic electron transport: I Concentration dependence and influence on manganese reincorporation. Photobiochem Photobiophys 2: 133–140.

Förster V, Hong Y-Q and Junge W (1981) Electron transfer and proton pumping under excitation of dark-adapted chloroplasts with flashes of light. Biochim Biophys Acta 638: 141–152.

Foster JW (1940) The role of organic substrates in photosynthesis of purple bacteria. J Gen Physiol 24: 123–134.

Fowler CF (1977) Proton translocation in chloroplasts and its relationship to electron transport between the photosystems. Biochim Biophys Acta 459: 351–363.

Franck J (1945) Photosynthetic activity of isolated chloroplasts. Rev Mod Phys 17: 112–119.

French CS (1937) The quantum yield of hydrogen and carbon dioxide assimilation in purple bacteria. J Gen Physiol 20: 711–735.

French CS, Holt AS, Powell and Anson HL (1946) The evolution of oxygen from illuminated suspensions of frozen, dried and homogenized chloroplasts. Science 103: 505–506.

Gaffron H (1940) Carbon dioxide reduction with molecular hydrogen in green algae. Am J Bot 27: 273–283.

Garab G, Sanchez Burgos AA, Zimányi L and Faludi-Dániel A (1983) Effect of CO_2 on the energization of thylakoids in leaves of higher plants. FEBS Lett 154: 323–327.

Garab G, Chylla RG and Whitmarsh J (1987) Photosystem II: evidence for active and inactive complexes. In: M Gibbs (ed) Hungarian-USA Binational Symposium on Photosynthesis, pp 37–47.

Giaquinta RT and Dilley RA (1975) A partial reaction in photosystem II: reduction of silicomolybdate prior to the site of dichlorophenyl-dimethylurea inhibition. Biochim Biophys Acta 387: 288–305.

Gibbons BH and Edsall JT (1963) Rate of hydration of carbon dioxide and dehydration of carbonic acid at 25 °C. J Biol Chem 238: 3502–3507.

Good NE (1963) Carbon dioxide and the Hill reaction. Plant Physiol 38: 298–304.

Govindjee (ed.) Bioenergetics of Photosynthesis. New York: Academic Press.

Govindjee (ed.) (1982) Photosynthesis, Vol 1. Energy Conversion by Plants and Bacteria, New York: Academic Press.

Govindjee and Eaton-Rye JJ (1986) Electron transfer through photosystem II acceptors: interaction with anions. Photosynth Res 10: 365–379.

Govindjee and van Rensen JJS (1978) Bicarbonate effects on the electron flow in isolated broken chloroplasts. Biochim Biophys Acta 505: 183–213.

Govindjee, Pulles R, Govindjee R, van Gorkom and Duysens LNM (1976) Inhibition of the reoxidation of the secondary electron acceptor of photosystem II by bicarbonate depletion. Biochim Biophys Acta 449: 602–605.

Govindjee, Nakatani HY, Rutherford AW and Inoue (1984) Evidence from thermolumine-scence for bicarbonate action on the recombination reactions involving the secondary quinone electron acceptor of photosystem II. Biochim Biophys Acta 766: 416–423.

Govindjee, Kambara T and Coleman W (1985) The electron donor side of photosystem II: the oxygen evolving complex. Photochem Photobiol 42: 187–210.

Govindjee, Amesz J and Fork DC (eds) (1986) Light Emission by Plants and Bacteria, Orlando FL: Academic Press.

Graan T (1986) The interaction of silicomolybdate with the photosystem II herbicide-binding site. FEBS Lett 206: 9–14.

Graan T and Ort DR (1986) Detection of oxygen evolving photosystem II centers inactive in plastoquinone reduction. Biochim Biophys Acta 852: 320–330.

Greene CH and Voskuyl RJ (1936) An explanation of the relatively large concentration of ^{18}O in the atmosphere. J Am Chem Soc 58: 693–694.

Harnischfeger G (1974) Studies on the effect of diphenylcarbazide in isolated chloroplasts from spinach. Z. Naturforsch 29c: 705–709.

Hesketh JD, Woolley JT and Peters DB (1982) Predicting Photosynthesis. In: Govindjee (ed) Photosynthesis, Vol. II, Development, Carbon Metabolism, and Plant Productivity, pp 387–418. New York: Academic Press.

Hill R (1937) Oxygen evolved by isolated chloroplasts. Nature (London) 139: 881–882.

Hill R (1939) Oxygen produced by isolated chloroplasts. Proc Roy Soc London B127: 192–210.

Hill R and Bendell F (1960) Function of the two cytochrome components in chloroplasts: a working hypothesis. Nature (London) 186: 136–137.

Hill R and Scarisbrick R (1940a) Production of oxygen by illuminated chloroplasts. Nature (London) 146: 61–62.

Hill R and Scarisbrick R (1940b) The reduction of ferric oxalate by isolated chloroplasts. Proc Roy Soc London B129: 238–255.

Hope AB and Moreland A (1979) Proton translocation in isolated spinach chloroplasts after single-turnover actinic flashes. Aust J Plant Physiol 6: 289–304.

Ikegami I and Katoh S (1973) Studies on chlorophyll fluorescence in chloroplasts: II. Effect of ferricyanide on the induction of fluorescence in the presence of 3-(3,4-dichlorophenyl)-1,1-dimethylurea. Plant Cell Physiol 829–836.

Ireland CR, Baker NR and Long SP (1987) Evidence for a physiological role of CO_2 in the regulation of photosynthetic electron transport in intact leaves. Biochim Biophys Acta 893: 434-443.

Itoh S (1978) Membrane surface potential and the reactivity of the system II primary electron acceptor to charged electron carriers in the medium. Biochim Biophys Acta 504: 324–340.

Itoh S and Nishimura M (1977) pH dependent changes in the reactivity of the primary electron acceptor of system II in spinach chloroplasts to external oxidant and reductant. Biochim Biophys Acta 460: 381–392.

Izawa S (1962) Stimulatory effect of carbon dioxide upon the Hill reaction as observed with the addition of carbonic anhydrase to reaction mixture. Plant Cell Physiol 3: 221–227.

Jursinic P and Stemler A (1982) A seconds range component of the reoxidation of the primary photosytem II acceptor, Q: effects of bicarbonate depletion in chloroplasts. Biochim Biophys Acta 681: 419–428.

Jursinic P and Stemler A (1986) Correlation between the binding of formate and decreased rates of charge transfer through the photosystem II quinones. Photochem Photobiol 43: 205–212.

Jursinic P and Stemler A (1988) Multiple anion effects on photosystem II in chloroplast membranes. Photosynth Res 15: 41–56.

Jursinic P, Warden J and Govindjee (1976) A major site of bicarbonate effect in system II reaction: evidence from ESR signal II_{vf}, fast fluorescence yield changes and delayed light emission. Biochim Biophys Acta 440: 322–330.

Kamen MD and Barker HA (1945) Inadequacies in present knowledge of the relation between photosynthesis and ^{18}O content of atmoshperic oxygen. Proc Natl Acad Sci USA 31: 8–15.

Khanna R, Govindjee and Wydrzynski (1977) Site of bicarbonate effect on Hill reaction: evidence from the use of artificial electron acceptors and donors. Biochim Biophys Acta 462: 208–214.

Khanna R, Pfister K, Keresztes A, van Rensen JJS and Govindjee (1981) Evidence for a close spatial location of the binding sites for CO_2 and for photosystem II inhibitors. Biochim Biophys Acta 634: 105–116.

Korman S and Clarke HT (1955) Carboxymethylamino acids and peptides. J Biol Chem 221: 113–131.

Lorimer GH, Badger MR and Andrews TJ (1976) The activation of ribulose-1,5,-bisphos-phate carboxylase by carbon dioxide and magnesium irons: equilibria, kinetics, a suggested mechanism, and physiological implications. Biochem 15: 529–536.

Metzner H (1975) Water decomposition in photosynthesis? A critical reconsideration. J Theor Biol 51: 201–231.

Michel H and Deisenhofer J (1988) Relevance of the photosynthetic reaction center from purple bacteria to the structure of photosystem II. Biochemistry 27: 1–7

Michel H, Epp O and Deisenhofer J (1986) Pigment-protein interactions in the photosynthetic reaction centre from *Rhodopseudomonas viridis*. EMBO J 5: 2445–2451.

Mills A and Urey HC (1940) The kinetics of isotopic exchange between carbon dioxide, bicarbonate ion, carbonate ion and water. J Am Chem Soc 62: 1019–1026.

Murata N, Nishimura M and Takamiya A (1966) Fluorescence of chlorophyll in photosynthetic systems: II. Induction of fluorescence in isolated spinach chloroplasts. Biochim Biophys Acta 120: 23–33.

Nelson N, Nelson H and Racker E (1972) Partial resolution of the enzymes catalyzing photophosphorylation: XI. Magnesium-adenosine triphosphatase properties of heat-activated coupling Factor 1 from chloroplasts. J. Biol Chem 247: 6505–6510.

Oettmeier W and Soll HJ (1983) Competition between plastoquinone and 3-(3,4-dichlorophenyl)-1,1-dimethylurea at the acceptor side of photosystem II. Biochim Biophys Acta 724: 287–290.

Pearlstein RM (1982) Chlorophyll singlet excitons. In: Govindjee (ed.) *Photosynthesis, Vol. 1. Energy Conversion by Plants and Bacteria*, pp 293–330. New York: Academic Press.

Petroules V and Diner B (1986) Identification of Q_{400}, a high potential electron acceptor of photosystem II, the iron of the quinone-iron acceptor complex. Biochim Biophys Acta 849: 264–275.

Pulles MPJ, van Gorkom HJ and Willemsen JG (1976) Absorbance changes due to the charge-accumulating species in system 2 of photosynthesis. Biochim Biophys Acta 449: 536–540.

Punnett T and Iyer RV (1964) The enhancement of photophosphorylation and the Hill reaction by carbon dioxide. J Biol Chem 239: 2335–2339.

Rabinowitch E (1945, 1951, 1956) *Photosynthesis and Related Processes*, Vol. I, pp 54-56, 61-67; Vol II (parts 1 and 2), pp 1529–1530, 1915–1918 (part 2) New York: Interscience.

Radmer R and Ollinger O (1980) Isotopic composition of photosynthetic O_2 flash yields in the presence of $H_2^{18}O$ and $HC^{18}O_3^-$. FEBS Lett 110: 57–61.

Renger G (1976) Studies on the structural and functional organization of system II of photosynthesis: the use of trypsin as a structurally selective inhibitor at the outer surface of the thylakoid membrane. Biochim Biophys Acta 440: 203–224.

Renger G (1987) Mechanistic aspects of photosynthetic water cleavage. Photosynthetica 21: 203–224.

Renger G and Govindjee (1985) The mechanism of photosynthetic water oxidation. Photosynth Res 6: 33–55.

Robinson HH, Eaton-Rye JJ, van Rensen JJS and Govindjee (1984) The effects of bicarbonate depletion and formate incubation on the kinetics of oxidation-reduction reactions of the photosystem II quinone acceptor complex. Z Naturforsch 39c: 382–385.

Ruben S, Randall M, Kamen MD and Hyde JL (1941) Heavy oxygen (^{18}O) as a tracer in the study of photosynthesis. J Am Chem Soc 63: 877–878.

Rutherford AW and Zimmermann JL (1984) A new EPR signal attributed to the primary plastoquinone acceptor in photosystem II. Biochim Biophys Acta 767: 168–175.

Sarojini G and Govindjee (1981a) On the active speices in bicarbonate stimulation of Hill reaction in thylakoid membranes. Biochim Biophys Acta 634: 340–313.

Sarojini G and Govindjee (1981b) Is CO_2 an active species in stimulating the Hill reaction in thylakoid membranes? In: G. Akoyunoglou (ed.) *Photosynthesis, Vol. 2, Electron Transport and Photophosphorylation*, pp. 143–149. Philadelphia: Balaban International Science Services.

Shipman LL (1981) Theoretical study of the binding site and mode of action for photosystem II herbicides. J. Theor Biol 90: 123–148.

Siggel U, Khanna R, Renger G and Govindjee (1977) Investigation of the absorption changes of the plastoquinone system in broken chloroplasts: the effect of bicarbonate-depletion. Biochim Biophys Acta 462: 196–207.

Snel JFH and van Rensen JJS (1983) Kinetics of the reactivation of the Hill reaction in CO_2-depleted chloroplasts by addition of bicarbonate in the absence and in the presence of herbicides. Physiol Plant 57: 422–427.

Snel JFH and van Rensen JJS (1984) Reevaluation of the role of bicarbonate and formate in the regulation of photosynthetic electron flow in broken chloroplasts. Plant Physiol 75: 146–150.

482

Stemler A (1977) The binding of bicarbonate ions to washed chloroplast grana. Biochim Biophys Acta 460: 511–522.

Stemler A (1979) A dynamic interaction between the bicarbonate ligand and photosystem II reaction center complexes in chloroplasts. Biochim Biophys Acta 545: 36–45.

Stemler A (1980) Forms of dissolved carbon dioxide required for photosystem II activity in chloroplast membranes. Plant Physiol 65: 1160–1165.

Stemler A (1982) The functional role of bicarbonate in photosynthetic light reaction II. In Govindjee (ed.) *Photosynthesis, Vol. II, Development, Carbon Metabolism, and Plant Productivity*, pp 513–558. New York: Academic Press.

Stemler A (1985) Carbonic anhydrase: Molecular insights applied to photosystem II research in thylakoid membranes. In: WJ Lucas and JA Berry (eds.) *Inorganic Carbon Uptake by Aquatic Photosynthetic Organisms*. American Society of Plant Physiologists, pp 377–387.

Stemler A and Govindjee (1973) Bicarbonate ion as a critical factor in photosynthetic oxygen evolution. Plant Physiol 52: 119–123.

Stemler A and Govindjee (1974) Effects of bicarbonate ion on chlorophyll *a* fluorescence transients and delayed light emission from maize chloroplasts. Photochem Photobiol 19: 227–232.

Stemler A and Murphy J (1983) Determination of the binding constant of $H^{14}CO_3^-$ to the photosystem II complex in maize chloroplasts: effects of inhibitors and light. Photochem Photobiol 38: 701–707.

Stemler A and Murphy J (1984) Inhibition of HCO_3^- binding to photosystem II by atrazine at a low-affinity herbicide binding site. Plant Physiol 76: 179–182.

Stemler A and Radmer R (1975) Source of photosynthetic oxygen in bicarbonate-stimulated Hill reaction. Science 190: 457–458.

Stemler A, Babcock GT and Govindjee (1974) The effect of bicarbonate on photosynthetic oxygen evolution in flashing light in chloroplast fragments. Proc Nat Acad Sci USA 71: 4679–4683.

Stern BK and Vennesland B (1962) The effect of carbon dioxide on the Hill reaction. J Biol Chem 237: 596–602.

Stiehl HH and Witt HT (1969) Quantitative treatment of the function of plastoquinone in photosynthesis. Z Naturforsch 24b: 1588–1598.

Trebst A (1987) The three-dimensional structure of the herbicide binding niche on the reaction center polypeptides of photosystem II. Z Naturforsch 42c: 742–750.

Trebst A and Draber W (1986) Inhibitors of photosystem II and the topology of the herbicide and Q_B binding polypeptide in the thylakoid membrane. Photosynth Res 10: 381–392.

van Grondelle R and Amesz J (1986) Excitation energy transfer in photosynthetic systems. In: Govindjee, J Amesz and DC Fork (eds.) *Light Emission by Plants and bacteria*, pp 191–224. Orlando: Academic Press.

van Niel CB (1931) On the morphology and physiology of the purple and green sulphur bacteria. Arch Microbiol 3: 1–102.

van Niel CB (1941) The bacterial photosynthesis and their importance for the general problem of photosynthesis. Adv Enzymol 1: 263–328.

van Niel CB (1949) The comparative biochemistry of photosynthesis. In J Franck, WE Loomis (eds.) *Photosynthesis in Plants* pp 437–495. Ames, Iowa: Iowa State College Press.

van Rensen JJS (1988) Involvement of bicarbonate in the protonation of the secondary quinone electron acceptor of photosystem II via the non-haem iron of the quinone-iron acceptor complex. FEBS Lett 226: 347–351.

van Rensen JJS and Snel JFH (1985) Regulation of photosynthetic electron transport by bicarbonate, formate, and herbicides in isolated broken and intact chloroplasts. Photosynth Res 6: 231–246.

van Rensen JJS and Vermaas WFJ (1981) Action of bicarbonate and photosystem 2 inhibiting herbicides on electron transport in pea grana and in thylakoids of a blue-green alga. Physiol Plant 51: 106–110.

Velthuys BR (1981) Electron-dependent competition between plastoquinone and inhibitors for binding to photosystem II. FEBS Lett 126: 277–281.

Vermaas WFJ (1984) The interaction of quinones, herbicides and bicarbonate with their binding environment at the acceptor side of photosystem II in photosynthesis. Ph.D. Thesis, Agricultural University Wageningen, The Netherlands.

Vermaas WFJ and Govindjee (1981a) Unique role(s) of carbon dioxide and bicarbonate in the photosynthetic electron transport system. Proc Indian Nat Sci Acad B47: 581–605.

Vermaas WFJ and Govindjee (1981b) The acceptor side of photosystem II in photosynthesis. Photochem Photobiol 34: 775–793.

Vermaas WFJ and Govindjee (1982a) Bicarbonate or CO_2 as a requirement for efficient electron transport on the acceptor side of photosystem II. In: Govindjee (ed.) *Photosynthesis, Vol. II, Development, Carbon Metabolism, and Plant Productivity*, pp 541–558. New York: Academic Press.

Vermaas WFJ and Govindjee (1982b) Bicarbonate effects on chlorophyll *a* fluorescence transients in the presence and absence of diuron. Biochim Biophys Acta 680: 202–209.

Vermaas WFJ and Rutherford AW (1984) EPR measurements on the effects of bicarbonate and triazine resistance on the acceptor side of photosystem II. FEBS Lett 175: 243–248.

Vermaas WFJ and van Rensen JJS (1981) Mechanism of bicarbonate action on photosynthetic electron transport in broken chloroplasts. Biochim Biophys Acta 636: 168–174.

Vermaas WFJ, van Rensen JJS and Govindjee (1982) The interaction between bicarbonate and the herbicide ioxynil in the thylakoid membrane and the effects of amino acid modification on bicarbonate action. Biochim Biophys Acta 681: 242–247.

Vermaas WFJ, Arntzen CJ, Gu L-Q and Yu C-A (1983) Interactions of herbicides and azidoquinones at a photosystem II binding site in the thylakoid membrane. Biochim Biophys Acta 723: 266–275.

Warburg O (1964) Prefatory chapter. Ann Rev Biochem 33: 1–18.

Warburg O and Krippahl G (1958) Hill-Reaktionen. Z Naturforsch 13b: 509–514.

Warburg O and Krippahl G (1960) Notwendigkeit der Kohlensaure für die chinon und ferricyanid-Reactionen in grünen Grana. Z Naturforsch 15b: 367–369.

Warburg O, Krippahl G, Gewitz HS and Volker W (1959) Uber den chemischen Mechanismus der Photosynthese. Z Naturforsch 14b: 712–724.

Webster LA, Wahl MH and Urey HC (1935) The fractionation of the oxygen isotopes in an exchange reaction. J. Chem Phys 3: 129–129.

Wraight CA (1979) Electron acceptors of bacterial photosynthetic reaction centers: II. H^+ binding coupled to secondary electron transfer in the quinone acceptor complex. Biochim Biophys Acta 548: 309–327.

Wraight CA (1981) Oxidation-reduction physical chemistry of the acceptor quinone complex in bacterial photosynthetic reaction centers: evidence for a new model of herbicide activity. Isr J Chem 21: 348–354.

Wraight CA (1982) Current attitudes in photosynthesis research. In: Govindjee (ed.) *Photosynthesis*, Volume I, pp 17–61. New York: Academic Press.

Wraight CA (1985) Modulation of herbicide-binding by the redox state of Q_{400}, an endogenous component of photosystem II. Biochim Biophys Acta 809: 320–330.

Wurmser R (1987) Letter to the editor. Photosynth Res 13: 91–93.

Wydrzynski T and Govindjee (1975) A new site of bicarbonate effect in photosystem II of photosynthesis: evidence from chlorophyll fluorescence transients in spinach chloroplasts. Biochim Biophys Acta 387: 403–408.

484

Yoshida T, Morita N, Tamiya H, Nakayama H and Huzisige H (1924) Über den Gehalt des Assimilationssauerstoffs an schwerem Isotop. Ein Beitrag zur kenntnis des Mechanismus der Photosynthese. Acta Phytochim 13: 11–18.

Zilinskas BA (1975) Photosystem II reactions in thylakoid membranes. Ph.D. Thesis, University of Illinois at Urbana-Champaign.

Zilinskas B and Govindjee (1975) Silicomolybdate and silicotungstate mediated dichlorophenyldimethylurea-insensitive photosystem II reaction: electron flow, chlorophyll a fluorescence and delayed light emission changes. Biochim Biophys Acta 387: 306–319.

Zimmermann J-L and Rutherford AW (1986) Photoreductant-induced oxidation of Fe^{2+} in the electron-acceptor complex of photosystem II. Biochim Biophys Acta 851: 416–423.

Govindjee et al. (eds), Molecular Biology of Photosynthesis: 485–496
© 1988 Kluwer Academic Publishers

Minireview

Photosystem I complex

PATRICIA REILLY & NATHAN NELSON
Roche Institute of Molecular Biology, Roche Research Center, Nutley, New Jersey 07110, USA

Received 2 October 1987; accepted 30 March 1988

Key words: biogenesis, photosystem I, reaction center, structure

Abstract. Photosystem I is an integral component of the thylakoid membrane which catalyzes the photoreduction of ferredoxin using plastocyanin or cytochrome c as electron donor. In higher plants, the photosystem I complex is composed of eight protein subunits, chlorophyll a, carotenoids, phylloquinone and bound iron sulfur clusters. The molecular biology and biochemistry of the complex are discussed in relation to the structure and function of the individual components. The mechanisms involved in the assembly of the components into a functional complex are also discussed.

Introduction

The second photosystem in the photosynthetic electron transport chain was named photosystem I for historical reasons (see Thornber, 1986 for a historical account of the study of chlorophyll-protein complexes). The first reaction center to be isolated and characterized in detail was photosystem I (see Bengis and Nelson 1975, 1977, Mullet et al. 1980) due to a simpler composition and greater stability than photosystem II. Later on, photosystem II was isolated by detergent treatment at low pH and has now been studied in detail (see Berthold et al. 1981, Ghanotakis and Yocum 1985). A major advance in the study of photosystem II has resulted from its extensive amino acid similarity to the bacterial reaction center which has been crystallized and its structure determined (Deisenhofer et al. 1984, 1985, Allen et al. 1987, Yeates et al. 1987, also see Trebst 1986). Nanba and Satoh (1987) have isolated a photosystem II reaction center consisting of D-1 and D-2 polypeptides and cytochrome b-559. Thus, in recent years, our understanding of the function of photosystem II has progressed rapidly. Similar progress has not occurred in the study of photosystem I, which has no (significant) similarity to the bacterial reaction center. Although there have been reports of the crystallization of photosystem I (Ford et al. 1987, Witt et al. 1987), understanding the structure and organization of this complex depends on obtain-

Table 1. Polypeptides of photosystem I.

Subunits	Coded in	Properties/Function
Ia; Ib	Chloroplast	Contains reaction center chlorophyll *a* P700; primary acceptor A_0 ($\equiv$ chlorophyll) (?); acceptor A_1 ($\equiv$ phylloquinone) (?); and the center X (F_X). Amino acid sequence is known; 83 kDa polypeptides.
II	Nucleus	There is some hint that it may contain Fe-S center F_A or F_B (bound ferredoxin), but most evidence suggests that subunit VII may be the site for these centers.
III	Nucleus	This may function in the reduction of $P700^+$ by plastocyanin; amino acid sequence is known; it is a basic protein
IV	Nucleus	Unknown
V	Chloroplast (?)	Unknown
VI	Nucleus	Equivalent to subunit III of *Chlamydomonas*; unknown.
VII	Chloroplast	This may contain Fe-S center F_A or F_B; this subunit is equivalent to subunit IV of *Chlamydomonas*; its molecular mass is 8.5 kDa.

ing better crystals which can be used in diffraction studies. Therefore, for the time being, the structure and function of photosystem I must be discussed based on biochemical and photochemical studies. For earlier reviews on photosystem I, see Malkin (1982, 1987) and on photochemistry and electron carriers in photosystem I, see Rutherford and Heathcote (1985). The light harvesting complexes (LHCPI$_a$ and LHCPII) are not discussed in this overview. I present here my personal view on composition and function, molecular biology and structure and biogenesis of photosytem I.

Composition and function

The photosystem I complex can be defined as the minimal structural unit which catalyzes the photoreduction of ferredoxin with plastocyanin or cytochrome c as the electron donor. Such a protein complex was first isolated from Swiss chard chloroplasts and subsequently from a wide variety of higher plants (Nelson 1987). Photosystem I complex, isolated from higher plants, contains eight different polypeptides, designated as subunits Ia, Ib through VII in order of decreasing molecular weight, which range from 83 kDa to about 8 kDa (see Table 1). A wider interpretation of photosystem I complex includes all of those subunits plus four extra polypeptides, some of which are photosystem I specific light harvesting chlorophyll-proteins (Mullet et al. 1980). All of the subunits are found as one copy per protein complex with a molecular mass of about 350 kDa (Nelson 1987). The stoichiometry of units was challenged by Lundell et al. (1985) who determined 4 copies of subunits Ia plus Ib per one subunit II. Each protein complex contains one P-700, about forty chlorophyll a molecules and a few β-carotene molecules, one of which is in close proximity to the P-700 and is highly oriented with respect to this pigment (Junge et al. 1977). The photosystem I complex also contains three iron sulfur clusters which function as secondary electron acceptors (Malkin 1982). Recently phylloquinone was proposed as an electron acceptor in photosystem I (Brettel et al. 1986, Mansfield and Evans 1986, Malkin 1986, Petersen et al. 1987). However, direct involvement of phylloquinone in the reducing side of photosystem I was challenged by Ziegler et al. (1987). Very recently direct biochemical evidence for the role of the quinone was provided by Biggins and Mathis (1988).

It is considered likely that the sequence of electron carriers in photosystem I is:

P700 $\rightarrow$ A_0 (chlorophyll) $\rightarrow$ A_1 (phylloquinone) $\rightarrow$

X (or F_X, iron sulfur center) $\rightarrow$

Iron sulfur center A or B (F_A or F_B, bound ferredoxin) $\rightarrow$

Ferredoxin.

Subunit I consists of two homologous proteins designated Ia and Ib, and functions as the primary photochemical reaction center. Subunit Ia plus Ib, when isolated, contain all of the pigments in the complex and are capable of catalyzing the primary photochemistry of photosystem I (Bengis and Nelson 1975). This subunit should contain the primary electron acceptor A_0 (chlorophyll) (see e.g., Fenton et al. 1979, Wasielewski et al. 1987) and the acceptor A_1 (phylloquinone). Subunit I should also contain the second secondary electron acceptor, because the isolated subunit is capable of stable P-700 photooxidation at room temperature. The non-heme iron cluster, X (or F_X) is the most likely candidate for this function (Golbeck and Cornelius 1986). The amino acid sequence of the subunit I proteins are known, subunit Ia contains 4 cysteine residues and subunit Ib contains 2. The limited number of cysteines indicates that only one iron sulfur cluster can be bound to this subunit; therefore, the other iron sulfur centers should be located on other subunits. The distribution of the cysteines in the linear sequence suggests that X may be shared between subunit Ia and Ib.

The function of the other six subunits is not clear. Subunit III may function in the reduction of P700$^+$ by plastocyanin (Bengis and Nelson 1977). Depletion of this subunit from the photosystem I complex results in inhibition of reduction by plastocyanin, which could be partially overcome by the addition of high concentrations of magnesium (Nechushtai and Nelson 1981a). Recent sequence analysis of subunit III indicates that it is a basic protein (R.G. Herrmann, personal communication), which may explain the ability of magnesium to substitute in this reaction. Subunit II appears to be the most important polypeptide out of these secondary subunits. There is strong immunological cross-reactivity among subunit II from cyanobacteria, Prochloron, through green algae and up to higher plants (Nechushtai et al. 1983, Schuster et al. 1985). This immunological cross-reactivity indicates amino acid sequence similarity and conserved function throughout evolution. Biochemical studies provided some evidences that subunit II may contain one of the bound iron sulfur protein (ferredoxin) cluster A or B (Nechushtai and Nelson 1981b, Bonnerjea et al. 1985, P. Reilley and coworkers, unpublished). This suggestion is supported by recent cross-linking experiments showing interaction between ferredoxin and subunit II of photosystem I complex (Zanetti and Merati 1987). However, without the amino acid sequence it is too early to reach a final conclusion on the location of these iron sulfur clusters. A recent study suggests that subunit VII may be the binding site for A (F_A) and B (F_B) clusters, based on similarity of cysteine positions in the chloroplast encoded gene with other ferredoxins (Hoj et al. 1987). The biochemical evidence to support subunit VII in this function is based on the distribution of sulfur in SDS-polyacrylamide gels, which has also been used to suggest that subunits III and IV are bound ferredoxins based on the distribution of iron in gels (Hoj et al. 1986). Therefore, it is not certain whether subunit VII contains both clusters A (F_A) and B (F_B) of the bound ferredoxin, or subunit II bears one of them. The function of the remaining subunits is not known, although some of them may function in the assembly of the complex.

A photosystem I complex containing five different polypeptides was isolated from cyanobacteria and green algae (Nechushtai and Nelson 1981, Nechushtai et al. 1983, Takahashi et al. 1982). Although these preparations are not capable of NADP-photoreduction, they resemble the preparations from higher plants in several respects. First, subunits Ia and Ib have high sequence similarity to those of higher plants and subunits I and II are immunologically cross-reactive. Second, a complex containing P-700 reaction center can be prepared in a state which is photochemically identical to that of higher plants (Bengis and Nelson 1975, Nechushtai and Nelson 1981b). It is interesting to note that even the purification procedure developed for higher plants is applicable to isolation of photosystem I from cyanobacteria and green algae; however, during the purification of the algal complex, one of the subunits may be lost resulting in the lack of NADP-photoreduction activity.

Molecular biology and structure

Studies of photosystem I at the molecular genetic level have not progressed as rapidly as photosystem II, where most of the genes have been cloned and sequenced. The best characterized subunit in photosystem I are subunits Ia and Ib. The genes for these two polypeptides are encoded in the plastid genome. The location of these genes has been mapped and they have been sequenced in a number of higher plant species (Fish et al. 1985a, Kirsh et al. 1986), *Chlamydomonas* (Kuck et al. 1987) and a cyanobacterium (Bryant et al. 1987). Subunits II, III and possibly IV and VI are nuclear encoded in higher plants, with the rest coded for in the chloroplast. Although the entire chloroplast genome has been sequenced (Shinozaki et al. 1986, Ohyama et al. 1986), only the gene coding for subunit VII was definitely identified on the chloroplast chromosome (Hoj et al. 1987).

In higher plants the genes for the subunit I apoproteins are located together on the chloroplast genome, with a small 25 bp spacer between them. The genes are cotranscribed as a single mRNA (Fish et al. 1985a), which in spinach also contains the transcript for the ribosomal protein S14 (Kirsh et al. 1986). The polycistronic nature of the mRNA found in higher plants may provide a mechanism for coordinate expression of these proteins which are found in a 1:1 ratio in the photosystem I complex. There is a single transcriptional start site upstream from the psaA gene, but there are potential ribosome binding sites before each initiator AUG. A similar gene organization for the two subunit I genes was observed in the cyanobacterium, *Synechococcus* 7002 (Bryant et al. 1987). However, the genes are separated by a longer spacer at 173 bp, with a 39 codon open reading frame in this region.

The genes psaA and psaB, for Ia and Ib, have recently been mapped and sequenced in the green algae *Chlamydomonas* (Kuck et al. 1987). This organism exhibits striking differences in the organization of these genes. Each gene maps to different locations on the chloroplast chromosome and the psaA gene is split into three exons which are far apart on the chromosome. The transcription and translation of these genes appears to require mRNA splicing and is quite unique (Kuck et al. 1987).

Fig. 1. Photosystem I, purified from *Synechocystis* 6803, in 0.1 M Tris HCl, pH 8.0, 0.1% dodecyl-β-D-maltoside at a chlorophyll concentration of 1 mg/ml, was crystallized by vapor diffusion against 0.2 M ammonium acetate, 0.1 M Tris, 2.0 M Na, K phosphate. The crystal shown is 40 by 45 μ.

The subunit I polypeptides have been highly conserved throughout evolution. Compared to maize, spinach is 95% homologous, *Chlamydomonas* 83% and *Synechococcus* 78%. This high degree of conservation reflects its important role in the binding of pigments and its function in the primary photochemistry of photosytem I. Subunits Ia and Ib are also homologous to each other. Optimal alignment of the maize polypeptides gives identical residues at 45% of the positions, with an additional 9% of the amino acids being conservative changes (Fish et al. 1985b). The high degree of homology between these two polypeptides suggest that they may have arisen through gene duplication.

Knowledge of the gene and amino acid sequence can lead to predictions about the structure of the protein and possible functional domains. In the case of photosytem I, many of these predictions remain to be supported by biochemical studies. Each of the subunit I proteins probably have 11 transmembrane α-helical domains (Fish et al. 1985b) as predicted from hydropathy plots. A large percentage of the histidine residues are predicted to be within the transmembrane domains, and probably are the coordinating ligands to the large number of chlorophyll *a* molecules bound to these polypeptides. There are also two cysteine residues in both Ia and Ib, which are located in a large completely conserved domain. These residues may be the ligands to the iron-sulfur cluster, X (F_X), which is thought to be located on this subunit.

The fine structure of photosystem I reaction center can be resolved only when good crystals become available. From the large amount of chlorophyll present in each complex, thus far isolated, one can predict that the crystal must be optically thick. Such a crystal is depicted in Fig. 1. In this experiment, highly purified photosystem I complex from *Synechocystis* 6803 was crystallized in the presence of 0.1% dodecyl maltoside. Crystallization resulted in small cubic crystals with the expected optical thickness. The challenge now is to come up with well diffracting crystals and to solve the three dimensional structure of the reaction center.

Biogenesis

As with all other protein complexes located in the chloroplast membrane, the subunits of photosytem I are synthesized on both cytoplasmic and chloroplast ribosomes (Herrmann et al. 1985, Nelson 1987). The subunits that are synthesized on cytoplasmic ribosomes are transported across the chloroplast membranes and coordinately assembled with their counterparts, which were synthesized in the chloroplast stroma. Thus, the synthesis and assembly of photosytem I is quite complex and involves many regulated steps.

The first studies to indicate the site of synthesis of the various subunits involved protein synthesis inhibitors which have differential effects on cytoplasmic and chloroplast ribosomes (Nechushtai et al. 1981). In higher plants, subunits Ia, Ib, and VII are synthesized in the chloroplast and the rest of the subunits are synthesized in the cytoplasm. Photosystem I, isolated from *Chlamydomonas*, contains only five different subunits. Subunits Ia, Ib, and IV are chloropolast products, while subunit II is synthesized in the cytoplasm (Nechushtai and Nelson 1981b). In this study, which used ^{35}S-sulfate labeling, the site of synthesis of subunit III, which is equivalent to subunit VI in higher plants, was not determined because it lacks cysteine and methionine. However, recent studies of the genes for these subunits have confirmed the site of synthesis of these polypeptides and have shown that subunit III is nuclear encoded (Tanaka and Tsuji 1985).

The chloroplast gene products are inserted into the membrane cotranslationally and usually no post-translational removal of a signal peptide is required (Herrmann et al. 1985). On the other hand, all of the nuclear gene products are synthesized as larger precursors and transported into the chloroplast via an energy-dependent vectorial processing mechanism (Chua and Schmidt 1979, Grossman et al. 1980). During this process the precursors are transported across the membrane into the stroma in an ATP-dependent manner, the amino-terminal signal sequence is removed by a specific protease and the protein is inserted into the membrane in another ATP-dependent step (Cline 1986, Chitnis et al. 1987). The various polypeptides are then assembled into the functional protein complex concomitant with the insertion of the pigment and other functional groups.

The synthesis and assembly of the various subunits of photosystem I are regulated by several factors, the most important of which is light. Dark grown plants have no photosynthetic activity; however, following illumination, chlorophyll *a* is synthesized and partial reactions catalyzed by photosystem I are observed in greening leaves (Bradbeer 1981). Phytochrome, which responds to red-far red light, initiates the transcription of nuclear genes encoding photosystem I subunits (Apel 1979, Gallagher and Ellis 1982). Light also induces the transcription of chloroplast genes for photosystem I by an unknown mechanism. The quality of light appears to play a role in the regulation of chloroplast gene expression and the assembly of complexes in the membrane (Glick et al. 1986).

Under most conditions, chlorophyll-protein complexes cannot be detected in chloroplast membranes prior to the appearance of chlorophyll (Bradbeer 1981, Thornber 1986). In chlorophyll-deficient mutants, the apo-proteins of photosystem I and II are synthesized at normal rates but are rapidly degraded indicating a requirement for chlorophyll for stable assembly of these complexes (Bellemare et al. 1982). Subunit I of photosystem I was detected in etiolated leaves using antibodies as a probe (Nechushtai and Nelson 1985). However, Klein and Mullet (1986) did not detect this subunit in etiolated plants. The quantity of this subunit was variable, depending on both the quality of the darkness and the supply of cytokinins from the roots and therefore under safe darkness subunit I may be missing. Mutants lacking one of the low molecular subunits do not form the stable photosystem I complex (Herrmann et al. 1985). Therefore, a partially assembled complex may be stable in etioplasts and not in developed chloroplasts reflecting differences in the mechanism of assembly of protein complexes during greening and mature chloroplasts.

The mechanism of assembly of photosystem I appears to be different than that of other chloroplast membrane complexes. The proton-ATPase complex and cytochrome b_6-f are assembled via a concerted mechanism (Nelson and Riezman 1984, Nelson 1987), while photosystem II is assembled in a two stage mechanism (Liveanu et al. 1986). Photosystem I is assembled via a step by step mechanism. Upon illumination of etiolated plants, subunit II was the first to be detected using subunit-specific antibodies (Nechushtai and Nelson 1985). At later times during illumination, accumulation of subunit III was observed, followed by the other subunits. The levels of subunit I present in the dark grown plastids do not increase significantly during the first two hours of illumination. It was concluded that subunit II may serve as the template for assembly of the complex, and the regulation of assembly, therefore, resides in the nucleus.

Understanding the structure, function and biogenesis of photosystem I requires much more in-depth study. Three dimensional crystals are required to elucidate the structural relationships among the pigment and protein components. Solving the structure will then provide insight into how the components function to catalyze the photoreaction. For advanced studies of the biogenesis of photosystem I reaction center, a convenient and efficient transformaton of chloroplasts containing cells is badly needed. Recent studies in our laboratory suggest that both goals are not too far to reach.

References

Allen JP, Feher G, Yeates TO, Komiya H and Rees DC (1987) Structure of the reaction center from *Rhodobacter sphaeroides* R-26: the protein subunits. Proc Natl Acad Sci USA 84: 6162–6166

Apel K (1979) Phytochrome-induced appearance of mRNA activity for the apoprotein of the light-harvesting chlorophyll a/b protein of barley (*Hordeum vulgare*). Eur J Biochem 97: 183–188

Bengis C and Nelson N (1975) Purification and properties of photosystem I reaction center from chloroplasts. J Biol Chem 250: 2783–2788

Bengis C and Nelson N (1977) Subunit structure of chloroplast photosystem I reaction center. J Biol Chem 252: 4564–4569

Berthold DA, Babcock GT and Yocum CF (1981) A highly resolved, oxygen-evolving photosystem II preparation from spinach thylakoid membranes. FEBS Lett 134: 231–234

Biggins J and Mathis P (1988) Functional role of vitamin K_1 in photosystem I of the cyanobacterium Synechocystis 6803. Biochemistry 27: 1494–1500

Bonnerjea J, Ortiz W and Malkin R (1985) Identification of a 19-kDa polypeptide as an Fe-S center apoprotein in the photosystem I primary electron acceptor complex. Arch Biochem Biophys 240: 15–20

Bradbeer JW (1981) Development of photosynthetic function during chloroplast biogenesis. In: Hatch MD and Boardman NK (eds) The Biochemistry of Plants, pp 423–472, New York: Academic Press

Brettel K, Setif P and Mathis P (1986) Flash-induced absorption changes in photosystem I at low temperature: evidence that the electron acceptor A, is vitamin K_1. FEBS Lett 203: 220–224

Bryant DA, DeLorimer R, Guglielmi G, Stirewalt VL, Cantrell A and Stevens SE (1987) The cyanobacterial photosynthetic apparatus: a structural and functional analysis employing molecular genetics. In: Biggens J (ed.) Progress in Photosysthesis Research, Volume IV, pp 749–755, Dordrecht: Martinus Nijhoff Publishers

Chitins PR, Nechushtai R and Thornber JP (1987) Insertion of the precursor of the light-harvesting chlorophyll a/b-protein into the thylakoids requires the presence of a developmentally regulated stromal factor. Plant Mol Biol 10: 3–11

Chua N-H and Schmidt GW (1979) Transport of proteins into mitochondria and chloroplasts. J Cell Biol 81: 461–483

Cline K (1986) Import of proteins into chloroplasts. Membrane integration of thylakoid precursor protein reconstituted in chloroplast lysates. J Biol Chem 261: 14804–14810

Deisenhofer J, Epp O, Miki K, Huber R and Michel H (1984) X-ray structure analysis of a membrane protein complex. J Mol Biol 180: 385–398

Deisenhofer J, Epp O, Miki K, Huber R and Michel (1985) Structure of the protein subunits in the photosynthetic reaction center of *Rhodopseudomonas viridis* at 3 Å resolution. Nature 318: 618–624

Fenton JM, Pellin MJ, Govindjee and Kaufmann KJ (1979) Primary photochemistry of the reaction center of photosystem I. FEBS Lett 100: 181–190

Fish LE, Kuck U and Bogarad L (1985a) Two partialy homologous adjacent light-inducible maize chloroplast genes encoding polypeptides of the P700 chlorophyll a protein complex of photosystem I. J Biol Chem 260: 1413–1421

Fish LE, Kuck U and Bogarad L (1985b) Analysis of the two partially homologous P700 chlorophyll a proteins of maize photosystem I: predictions based on the primary sequences and features shared by other chlorophyll proteins. In: Steinback KE, Bonitz S, Arntzen CJ and Bogarad L (eds) Molecular Biology of the Photosynthetic Apparatus, pp 111–120, New York: Cold Spring Harbor Laboratory

Ford RC, Picot D and Garavito RM (1987) Crystallization of the photosystem I reaction center. EMBO J 6: 1581–1586

Gallagher TF and Ellis RJ (1982) Light-stimulated transcription of genes for two chloroplast polypeptides in isolated pea leaf nuclei. EMBO J 1: 1493–1498

Ghanotakis DF and Yocum CF (1985) Polypeptides of photosystem II and their role in oxygen evolution. Photosyn Res 7: 97–114

Glick RE, McCauley SW, Gruissem W and Melis A (1986) Light quality regulates expression of chloroplast genes and assembly of photosynthetic membrane complexes. Proc Natl Acad Sci USA 83: 4287–4291

Goldbeck JH and Cornelius JM (1986) Photosystem I charge separation in the absence of centers A and B. Biochim Biophys Acta 849: 16–24

Grossman AR, Bartlett SG and Chua N-H (1980) Energy-dependent uptake of cytoplasmically-synthesized polypeptides by chloroplasts. Nature (Lond.) 285: 625–628

Herrmann RG, Westhoff P, Alt J, Tittgen J and Nelson N (1985) Thylakoid membrane proteins and their genes. In: van Vloten-Doting L, Groot GSP and Hall TC (eds) Molecular Form and Function of the Plant Genome, pp 233–255, New York: Planum Publishing Corp

Hoj PB and Moller BL (1986) The 110-kDa reaction center protein of photosystem I P700-chlorophyll a-protein 1, is an iron-sulfur protein. J Biol Chem 261: 14292–14300

Hoj PB, Svendsen I, Scheller HV and Moller BL (1987) Identification of a chloroplast encoded 9 kDa polypeptide as a 2 (4FE-4S) protein carrying center A and B of photosystem I. J Biol Chem 262: 12676–12684

Junge W, Schaffernicht H and Nelson N (1977) On the mutual orientation of pigments in photosystem I particles of green plants. Biochim Biophys Acta 462: 73–85

Kirsch W, Seyer P and Herrmann RG (1986) Nucleotide sequence of the clustered genes for two P700 chlorophyll a apoproteins of the photosystem I reaction center and the ribosomal protein S14 of the spinach plastid chromosome. Curr Genet 10: 843–855

Klein RR and Mullet JE (1986) Regulation of chloroplast-encoded chlorophyll-binding protein translation during higher plant chloroplast biogenesis. J Biol Chem 261: 11138–11145

Kuck U, Choquet Y, Schneider M, Dron M and Bennoun P (1987) Structural and transcriptional analysis of two homologous genes for the P700 chlorophyll a-apoproteins in *Chlamydomonas reinhardii*: evidence for in vivo transplicing. EMBO J 6: 2185–2195

Liveanu V, Yocum CF and Nelson N (1986) Polypeptides of the oxygen-evolving photosystem II complex. J Biol Chem 261: 5296–5300

Lundell DJ, Glazer AN, Melis A and Malkin R (1985) Characterization of a cyanobacterial photosystem I complex. J Biol Chem 260: 646–654

Malkin R (1982) Photosystem I. Ann Rev Plant Physiol 33: 455–479

Malkin R (1986) On the function of two vitamin K_1 molecules in the PSI electron acceptor complex. FEBS Lett 208: 343–346

Malkin R (1987) Photosystem I. Topics in Photosynthesis Research 8: 495–525

Mansfield RW and Evans MLW (1986) UV optical different spectrum associated with the reduction of electron acceptor A_1 in photosystem I of higher plants. FEBS Lett 203: 225–229

Michel H (1982) 3-Dimensional crystals of a membrane protein complex. The photosynthetic reaction center from *Rhodopseudomonas viridis*. J Mol Biol 158: 567–572

Mullet JE, Burke JJ and Arntzen CJ (1980) Chlorophyll proteins of photosystem I. Plant Physiol 65: 814–822

Mullet JE, Grossman AR and Chua N-H (1982) Synthesis and assembly of the polypeptide subunits of photosystem I. Cold Spring Harbor Symp 46: 979–984

Nanba O and Satoh K (1987) Isolation of a photosystem II reaction center consisting of D-1 and D-2 polypeptides and cytochrome b-559. Proc Natl Acad Sci USA 84: 109–112

Nechushtai R and Nelson N (1981a) Photosystem I reaction centers from chlamydomonas and higher plant chloroplasts. J Bioenerg Biomembranes 13: 295–306

Nechushtai R and Nelson N (1981b) Purification properties and biogenesis of *Chlamydomonas reinhardii* photosystem I reaction center. J Biol Chem 256: 11624–11628

Nechushtai R and Nelson N (1985) Biogenesis of photosystem I reaction center during greening. Plant Mol Biol 4: 377–384

Nechushtai R, Nelson N, Mattoo A and Edelman M (1981) Site of synthesis of subunits to photosystem I reaction center and the proton-ATPase in spirodela. FEBS Lett 125: 115–119

Nechushtai R, Muster P, Binder A, Liveanu V and Nelson N (1983) Photosystem I reaction center from the thermophilic cyanobacterium *Mastigocladus laminosus*. Proc Natl Acad Sci USA 80: 1179–1183

Nelson N (1987) Structure and function of protein complexes in the photosynthetic membrane. In: Amesz J (ed.) New Comprehensive Biochemistry, Volume 15, pp 213–231, Amsterdam: Elsevier Science Publishers

Nelson N and Riezman H (1984) Biogenesis of energy-transducing systems. In: Ernster L (ed.) Bioenergetics, pp 351–377, Amsterdam: Elsevier Science Publishers

Ohyama K, Fukuzawa H, Kohchi T, Shirai H, Tohru S, Sano S, Umesono K, Shiki Y, Takeuchi M, Chang Z, Aota S-I, Inokuchi H and Ozeki H (1986) Chloroplast gene organization deduced from complete sequence of liverwort *Marchantia polymorpha* chloroplast DNA. Nature 322: 572–574

Petersen J, Stehlik D, Gast P and Thurnauer M (1987) Comparison of the electron spin polarized spectrum found in plant photosystem I and in iron-depleted bacterial reaction centers with time-resolved K-band EPR, evidence that the photosystem I acceptor A_1 is quinone. Photosynth Res 14: 15–30

Rutherford AW and Heathcote P (1985) Primary photochemistry in photosystem I. Photosynth Res 6: 295–316

Schuster G, Nechushtai R, Nelson N and Ohad I (1985) Purification and composition of photosystem I reaction center of Prochloron sp., an oxygen-evolving prokaryote containing chlorophyll b. FEBS Lett 191: 29–33

Shinozaki K, Ohme M, Tanaka M, Wakasugi T, Hayashida N, Matsubayashi T, Zaita N, Chunwongse J, Obokata J, Yamaguchi-Shinozaki K, Ohto C, Torozawa K, Meng BY, Sujita M, Deno H, Kamogashira T, Yamada K, Kusuda J, Takaiwa F, Kato A, Tohdoh N, Shimada H and Sugiura M (1986) The complete sequence of the tobacco chloroplast genome: its gene organization and expression. EMBO J 5: 2043–2049

Takahashi Y, Koike H and Katoh S (1982) Multiple forms of chlorophyll-protein complexes from a thermophilic cyanobacterium Synechococcus sp. Arch Biochem Biophys 219: 209–218

Tanaka A and Tsuji H (1985) Appearance of chlorophyll-protein complexes in greening barley seedlings. Plant Cell Physiol 26: 893–902

Thornber JP (1986) Biochemical characterization and structure of pigment-proteins of photosynthetic organisms. In: Staehelin LA and Arntzen CJ (eds) Encl. Plant Physiol. New Series. Volume 19. Photosynthesis III: Photosynthetic Membranes, pp 98–142, Berlin: Springer-Verlag

Trebst A (1986) The topology of the plastoquinone and herbicide binding peptides of photosystem II in the thylakoid membrane. Z Naturforsch 41c: 240–245

Wasielewski MR, Fenton JM and Govindjee (1987) The rate of formation of $P700^+$-A_0^- in photosystem I particles from spinach as measured by picasecond transient absorption spectroscopy. Photosynth Res 12: 181–190

Witt I, Witt HT, Gerken S, Saenger W, Dekker JP and Rogner M (1987) Crystallization of reaction center I of photosynthesis, low-concentration crystallization of photoactive protein complexes from the cyanobacterium Synechoccus sp. FEBS Lett 221: 260–264

Yeates TO, Komiya H, Rees DC, Allen JP and Feher G (1987) Structure of the reaction center from *Rhodobacter sphaeroides* R-26: membrane-protein interactions. Proc Natl Acad Sci USA 84: 6438–6442

Zanetti G and Merati G (1987) Interaction between photosystem I and ferredoxin: Identification by chemical cross-linking of the polypeptide which binds ferredoxin. Eur J Biochem 169: 143–146

Ziegler K, Lockau W and Nitschke W (1987) Bound electron acceptors of photosystem I: evidence against the identity of redox center A with phylloquinone. FEBS Lett 217: 16–20

Govindjee et al. (eds), Molecular Biology of Photosynthesis: 497–516
© 1988 Kluwer Academic Publishers

Minireview

Synthesis and assembly of the cytochrome *b-f* complex in higher plants

DAVID L. WILLEY & JOHN C. GRAY

Botany School, University of Cambridge, Downing Street, Cambridge CB2 3EA, UK

Received 11 October 1987; accepted 5 January 1988

Key words: Chloroplast genes, cytochrome *b*-563, cytochrome *f*, Rieske FeS protein

Abstract. The cytochrome *b-f* complex is composed of four polypeptide subunits, three of which, cytochrome *f*, cytochrome *b*-563 and subunit IV, are encoded in chloroplast DNA and synthesised within the chloroplast, and the fourth, the Rieske FeS protein, is encoded in nuclear DNA and synthesised in the cytoplasm. The assembly of the cytochrome *b-f* complex therefore requires the interaction of subunits encoded by different genomes. A key role for the nuclear-encoded Rieske FeS protein in the assembly of the complex is suggested by a study of cytochrome *b-f* complex mutants. The assembly of individual subunits of the complex may be regulated by the availability of prosthetic groups. The genes for the chloroplast-encoded subunits and cDNA clones for the Rieske FeS protein have been isolated and characterised. Cytochrome *f* and the Rieske FeS protein are synthesised initially with *N*-terminal presequences required for their correct assembly within the chloroplast. The deduced amino acid sequences of the four subunits have been used to suggest models for the arrangement of the polypeptides in the thylakoid membrane.

I Introduction

The cytochrome *b-f* complex is the simplest and probably the best characterised of the multisubunit complexes that catalyse the light reactions of photosynthesis in the chloroplast thylakoid membrane. The cytochrome *b-f* complex mediates the transfer of electrons from plastoquinol to plastocyanin, and is involved in non-cyclic electron flow from photosystem II to photosystem I (Wood and Bendall 1976, Hurt and Hauska 1981), as well as in cyclic electron flow around photosystem I (Lam and Malkin 1982). The complex contains cytochrome *f* (33–38 kDa), cytochrome *b*-563 (19.5–23.5 kDa), the Rieske FeS protein (18–20 kDa) and a polypeptide of 15–17 kDa (subunit IV) with no identified redox centre (Hurt and Hauska 1981, 1982, Clark and Hind 1983, Phillips and Gray 1983). In addition the presence of small polypeptides ($\sim$ 5 kDa) in the complex has been suggested (Hurt and Hauska 1982) but has not yet been substantiated. An association of ferredoxin-NADP$^+$ reductase (FNR) with the complex has been reported (Clark and Hind 1983, Clark et al. 1984, Phillips and Gray 1984c) and has been suggested to be functionally significant for the role of the complex in cyclic electron flow (Shahak et al. 1981). However FNR is not an intrinsic component of the cytochrome b-f complex and the association of FNR with the purified complex may be fortuitous (C.J. Eccles and J.C. Gray, unpublished).

II Sites of synthesis of components of the cytochrome b-f complex

Several experimental approaches have been used to determine the subcellular sites of synthesis of the components of the cytochrome b-f complex. The aim of these experiments was to determine whether the polypeptides were synthesised on chloroplast 70S ribosomes or on cytoplasmic 80S ribosomes. The subcellular site of synthesis of cytochrome f and cytochrome b-563 in greening bean leaves has been investigated using selective inhibitors of protein synthesis (Gregory and Bradbeer 1973). The appearance of cytochrome f and cytochrome b-563 was shown to be inhibited by treatment with D-*threo* chloramphenicol, suggesting that the two cytochromes are synthesised on chloroplast 70S ribosomes. Cytochromes f and b-563 have also been shown to be absent in heat-treated rye plants which are deficient in chloroplast 70S ribosomes (Feierabend 1979). However, these experiments examined spectrally-detected cytochromes and it is not clear if the absence of cytochromes is due to inhibition of protein or haem synthesis, or lack of components of an assembly system.

The synthesis of the protein components of the cytochrome b-f complex has been investigated using light-driven protein synthesis by isolated pea chloroplasts incubated in the presence of radiolabelled amino acids. The cytochrome f polypeptide was shown to be a product of protein synthesis by isolated pea chloroplasts using immunoprecipitation and peptide mapping of the labelled cytochrome f (Doherty and Gray 1979). Subsequently it was shown that cytochrome f, cytochrome b-563 and subunit IV polypeptides were synthesised and assembled into the cytochrome b-f complex by isolated pea chloroplasts (Phillips and Gray 1984a). These experiments, and the inhibitor studies, have suggested that three components of the complex, cytochrome f, cytochrome b-563 and subunit IV are synthesised on chloroplast 70S ribosomes. This has been confirmed in spinach by translation of chloroplast RNA and immunoprecipitation of polypeptides with antibodies to cytochrome f, cytochrome b-563 and subunit IV (Alt et al. 1983).

Studies with spinach indicate that the Rieske FeS protein is synthesised from poly(A) RNA as a 26 kDa precursor, 7 kDa larger than the mature protein (Alt et al. 1983), suggesting that the protein is synthesised on cytoplasmic ribosomes. Tittgen et al. (1986) have shown that the mRNA for the Rieske FeS protein may be hybrid-selected with a cDNA clone isolated from a λgtll library and translated in a rabbit reticulocyte lysate to give the 26 kDa precursor. This precursor was imported into chloroplasts and processed to a 19 kDa polypeptide (Tittgen et al. 1986).

III Genes for components of the cytochrome *b-f* complex

The genes encoding the subunits of the cytochrome *b-f* complex are distributed between the nuclear and chloroplast genomes. The chloroplast genes *pet*A, *pet*B and *pet*D for cytochrome *f*, cytochrome *b*-563 and the subunit IV polypeptide respectively have been isolated and well characterised. The nuclear gene encoding the Rieske FeS protein has not yet been isolated, although cDNA clones from spinach (Tittgen et al. 1986, Steppuhn et al. 1987) and pea (A.H. Salter, B.J. Newman and J.C. Gray, unpublished) have been isolated and characterised.

Gene for cytochrome f
The location of the structural gene for cytochrome *f* in chloroplast DNA was suggested by two plastome mutants of *Oenothera hookeri* which were deficient in spectrally-detected cytochrome *f* (Hallier and Heber 1977) and by the demonstration that the cytochrome *f* polypeptide shows a maternal mode of inheritance in interspecific F_1 hybrids of *Nicotiana* (Gray 1980). The location of the structural gene for cytochrome *f* was first demonstrated in pea by coupled transcription-translation of pea chloroplast DNA and of cloned restriction fragments of pea chloroplast DNA in a cell-free system from *Escherichia coli* (Willey et al. 1983). Pea chloroplast DNA and a 17.3 kbp *Pst*I restriction fragment were shown to direct the synthesis of a 39 kDa polypeptide which could be immunoprecipitated by antibodies to cytochrome *f*. The 39 kDa polypeptide was shown to be similar to authentic pea cytochrome *f* (37.3 kDa) by peptide mapping of the products of partial proteolytic digestion. This suggested that cytochrome *f* was synthesised as a higher molecular weight form in the *E. coli* cell-free system. The gene for cytochrome *f* was subsequently located in chloroplast DNA from spinach by hybrid select translation and cell-free coupled transcription-translation (Alt et al. 1983) and in wheat chloroplast DNA by coupled transcription-translation (Willey et al. 1984b). The gene has also been localised in chloroplast DNA from *Vicia faba* (Ko et al. 1984), *Nicotiana tabacum* (Lin and Kung

1984, Shinozaki et al. 1986), barley (Oliver 1984), tomato (Phillips 1985), *Glycine max* (Singh et al. 1985), *Vigna radiata, Pelargonium hortorum, Coriandrum sativum* (Palmer 1985), *Capsicum annuum* (Gounaris et al. 1986), *Oenothera hookeri* (Tyagi and Herrmann 1986), *Ginkgo biloba* (Palmer and Stein 1986), rice (Wu et al. 1986), lettuce and *Barnadesia* (Asteraceae) (Jansen and Palmer 1987) by Southern hybridisation with heterologous probes from the pea or spinach genes. In the *E. coli* cell-free coupled transcription-translation system, pea, spinach and wheat cytochrome *f* were all synthesised as higher molecular weight precursors larger than the mature polypeptide (Alt et al. 1983, Willey et al. 1983, 1984b). The precursors were estimated to be 1.7 kDa larger than mature pea cytochrome *f* and 4 kDa larger in the case of spinach and wheat cytochrome *f*. The nature of these presequences has been established by nucleotide sequence analysis of the cytochrome *f* genes.

Nucleotide sequence analysis of the gene for pea cytochrome *f* revealed an open reading frame of 320 amino acids. This open reading frame has been identified as the gene for cytochrome *f* by comparison of the deduced amino acid sequence with the partial *N*-terminal amino acid sequence determined for pea cytochrome *f* (Willey et al. 1984a) and with sequences for spinach cytochrome *f* (Ho and Krogmann 1980; D. Krogmann, personal communication). Comparison of the deduced amino acid sequence with the determined *N*-terminal sequence of pea cytochrome *f* (Willey et al. 1984a) indicated that the cytochrome *f* gene encoded an *N*-terminal 35 amino acid presequence and a mature polypeptide of 285 amino acids. The cytochrome *f* genes from spinach (Alt and Herrmann 1984), wheat (Willey et al. 1984b), *Oenothera hookeri* (Tyagi and Herrmann 1986), rice (Wu et al. 1986), tobacco (Shinozaki et al. 1986) and *Vicia faba* (Ko and Straus 1987) have also been sequenced and show that the gene encodes a mature polypeptide of 285 amino acids with an *N*-terminal presequence of 35 amino acids in spinach, wheat, tobacco, *Vicia faba* and rice and 33 amino acids in *Oenothera hookeri*.

The nucleotide sequence of the cytochrome *f* gene predicts a protein with a putative haem-binding site, Cys-Ala-Asn-Cys-His, located near the *N*-terminus at amino acid residues 21–25. A hydropathy plot (Kyte and Doolittle 1982) of the deduced amino acid sequences shows two highly hydrophobic regions, one preceding the mature *N*-terminus and the other located near the *C*-terminus. The *N*-terminal hydrophobic sequence, not present in the mature protein, and covering residues -19 to -8 from the mature *N*-terminus, may be acting as a signal sequence to direct cytochrome *f* to the thylakoid membrane (see Section V). The *C*-terminal hydrophobic sequence (amino acids 251–270) is involved in the anchoring of cytochrome f in the thylakoid membrane by a putative membrane-spanning α-helix (see Section IV).

In pea chloroplasts the gene for cytochrome f is cotranscribed with an upstream open reading frame as a 3.6 kb RNA (D.L. Willey and J.C. Gray, unpublished). This open reading frame of 231 amino acids, located immediately 5′ to the gene for cytochrome f encodes a putative membrane protein (Willey et al. 1984a). An homologous open reading frame located 5′ to the gene for cytochrome f is found in chloroplast DNA from tobacco (Shinozaki et al. 1986) and wheat (D.L. Willey and J.C. Gray, unpublished). Maps of this region of chloroplast DNA are shown in Fig. 1. A hydropathy plot of the deduced amino acid sequence of the pea open reading frame revealed a hydrophobic region at the N-terminus covering amino acid residues 6–27, and an extended hydrophobic region covering amino acid residues 110–231 suggesting that the open reading frame may encode a

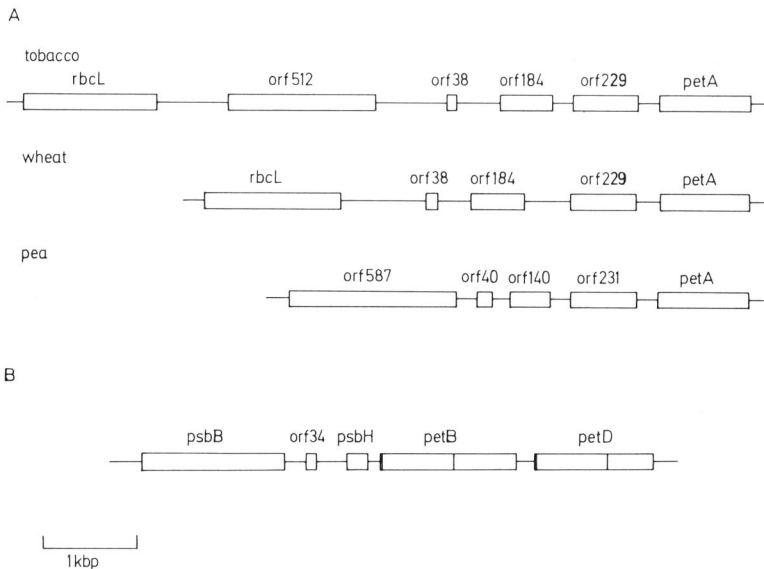

Fig. 1. Arrangement of the genes for components of the cytochrome b-f complex in chloroplast DNA. A. The organisation of the gene for cytochrome f (*pet*A) and the upstream open reading frames. The organisation in tobacco is taken from Shinozaki et al. (1986); the organisation in wheat and pea is from work carried out in the laboratories of T.A. Dyer and J.C. Gray. Pea orf 231 is homologous to orf 229 in wheat and tobacco; pea orf 140 shows very limited homology to wheat and tobacco orf 184 which are very similar; pea orf 40 is homologous to wheat and tobacco orf 38; pea orf 587 is homologous to tobacco orf 512, a similar open reading frame is not present in this region of wheat chloroplast DNA. B. The organisation of the genes for cytochrome b-563 (*pet*B) and subunit IV (*pet*D), together with the genes for components of photosystem II (*psb*B and *psb*H). This organisation, including a small open reading frame between *psb*B and *psb*H, is found in tobacco, wheat and maize.

membrane protein. The *N*-terminal hydrophobic region has a putative signal peptidase cleavage site (von Heijne 1983) and may encode a signal sequence required for insertion of the polypeptide into the thylakoid membrane. A comparison of the deduced amino acid sequences of the homologous polypeptides from pea, wheat and tobacco reveals that there are 2 histidine residues located in putative membrane-spanning regions which are conserved in all three sequences and which may be involved in binding a haem group. This suggests that the open reading frame may encode a *b*-type cytochrome, possibly the low potential form of cytochrome *b*-559 (Anderson and Boardman 1973).

Genes for cytochrome b-563 and subunit IV

The genes, *pet*B and *pet*D, for the cytochrome *b*-563 and subunit IV polypeptides of the complex have been localised in chloroplast DNA from spinach (Alt et al. 1983), pea (Phillips and Gray 1984b, Courtice et al. 1985, Berends et al. 1986), barley (Oliver 1984), wheat (Courtice et al. 1985), tomato (Phillips 1985), *Vigna radiata, Pelargonium hortorum, Coriandrum sativum* (Palmer 1985), *Capsicum annuum* (Gounaris et al. 1986), tobacco (Shinozaki et al. 1986), *Ginkgo biloba* (Palmer and Stein 1986), *Vicia faba* (Michalowski et al. 1987), lettuce, *Barnadesia* (Jansen and Palmer 1987) and maize (Rock et al. 1987). The genes were first located in spinach chloroplast DNA by hybrid select translation and coupled transcription-translation of cloned chloroplast DNA restriction fragments (Alt et al. 1983). The genes were located close together on the spinach chloroplast genome, approximately 15 kbp from the gene for cytochrome *f* (Alt et al. 1983) suggesting that the genes for cytochrome *f*, cytochrome *b*-563 and the subunit IV polypeptide are not all cotranscribed and must be under separate transcriptional control. In pea the gene for the subunit IV polypeptide was located by coupled transcription-translation of cloned restriction fragments of pea chloroplast DNA (Phillips and Gray 1984b). The gene is located approximately 36 kbp away from the gene for cytochrome *f* and is transcribed from the opposite strand of chloroplast DNA. The gene for pea cytochrome *b*-563 is located 196 bp 5' to the gene for the subunit IV polypeptide (C.J. Eccles and J.C. Gray, unpublished). Maps of this region of chloroplast DNA are shown in Fig. 1.

The *pet*B and *pet*D genes have been sequenced in spinach (Heinemeyer et al. 1984), pea (Phillips and Gray 1984b, C.J. Eccles and J.C. Gray, unpublished), tobacco (Shinozaki et al. 1986), maize (Rock et al. 1987) and wheat (S.M. Hird and J.C. Gray, unpublished). The presence of introns in both the *pet*B and *pet*D genes was first suggested for the lower plant *Marchantia* *polymorpha* by a computer search for consensus intron boundary sequences (Ohyama et al. 1986). Similar sequence comparisons suggested the presence of introns in both genes in maize (Rock et al. 1987), tobacco (Shinozaki et al. 1986), wheat (S.M. Hird and J.C. Gray, unpublished) and pea (C.J. Eccles and J.C. Gray, unpublished). In *Marchantia* the presence and location of the introns has been demonstrated by RNA sequencing (Fukuzawa et al. 1987). Sl nuclease analysis and electron microscopy of DNA-RNA hybrids has confirmed the location of the introns in the pea *pet*B and *pet*D genes (C.J. Eccles and J.C. Gray, unpublished). In all species the 5′ exons of the genes are very small; 6 bp in the *pet*B gene and 8 bp in the *pet*D gene. The introns in both genes are approximately 750 bp and resemble yeast mitochondrial group II introns (Michel and Dujon 1983). When account is made of the single intron within each gene the nucleotide sequence analysis indicates polypeptides of 215 amino acid residues for cytochrome *b*-563 and 160 amino acid residues for the subunit IV polypeptide. Rock et al. (1987) have suggested that alternative forms of these polypeptides may be translated from unspliced transcripts of the genes. In maize, a product of 232 amino acid residues may be produced from an unspliced transcript of the *pet*B gene. However the inframe methionine codon and coding sequence in the intron is not conserved in the pea or wheat *pet*B genes (C.J. Eccles, S.M. Hird and J.C. Gray, unpublished). Translation of the unspliced *pet*D transcript in maize would produce a polypeptide of 174 amino acid residues (Rock et al. 1987). However because this *N*-terminal sequence is not as highly conserved as that encoded by the spliced transcript and because the sequence is not homologous to the *Nostoc pet*D gene product, Rock et al. (1987) suggest that this protein may not be expressed.

The cytochrome *b*-563 polypeptide is homologous to the *N*-terminal sequence of mitochondrial cytochrome *b* whereas the subunit IV polypeptide is homologous to the *C*-terminal region of mitochondrial cytochrome *b* (Phillips and Gray 1984b, Widger et al. 1984). A comparison of the deduced amino acid sequence of spinach cytochrome *b*-563 with other cytochrome *b* sequences has revealed 4 conserved histidine residues which are suggested to be involved in haem-binding (Widger et al. 1984). The arrangement of these sequences in the thylakoid membrane is discussed in Section IV.

Gene for Rieske FeS protein

The gene for the Rieske FeS protein is located in nuclear DNA (Tittgen et al. 1986). cDNA clones encoding the Rieske FeS protein have been isolated from spinach and pea cDNA libraries in the expression vector λgtll (Tittgen et al. 1986; A.H. Salter, B.J. Newman and J.C. Gray, unpublished). The amino acid sequences deduced from the nucleotide sequences of the spinach and pea cDNA clones have been identified as the Rieske FeS protein by comparison with partial amino acid sequence of spinach Rieske FeS protein (Pfefferkorn and Meyer 1986). The deduced amino acid sequence of the spinach (Steppuhn et al. 1987) and pea (A.H. Salter and J.C. Gray, unpublished) proteins shows the presence of two highly conserved sequences, Cys-Thr-His-Leu-Gly-Cys and Cys-Pro-Cys-His, located near the C-terminus of the polypeptide which may be involved in coordinating the iron sulphur centre. A hydropathy plot (Kyte and Doolittle 1982) of the deduced amino acid sequence of the mature spinach and pea proteins shows the presence of a single extended hydrophobic region located near the N-terminus covering amino acid residues 17–55 which is suggested to anchor the protein in the thylakoid membrane. The arrangement of the protein in the membrane is discussed in Section IV.

Nucleotide sequence analysis of the spinach cDNA clone indicates that the Rieske FeS protein is synthesised initially with an N-terminal extension of 68 amino acid residues (Steppuhn et al. 1987). The presequence shows some homology with presequences of other nuclear-encoded chloroplast proteins (Karlin-Neumann and Tobin 1986) with a net positive charge and a large number of amino acids with hydroxyl side chains. This presequence presumably functions as a cleavable transit sequence to direct the polypeptide to the chloroplast and across the envelope membranes (Tittgen et al. 1986).

IV Membrane topology of components of the cytochrome b-f complex

The determination of the nucleotide sequences of the genes and the identification of intron sequences has allowed the prediction of the primary structures of the four polypeptides of the complex and of their organisation in the thylakoid membrane. Models for the membrane topology of these polypeptides are shown in Fig. 2.

Cytochrome f was suggested to be held in the membrane by a single hydrophobic membrane-spanning α-helix located near the C-terminus of the polypeptide (Willey et al. 1984a). The N-terminal haem-containing region was predicted to form a large hydrophilic globular domain in the thylakoid lumen, where it could interact with plastocyanin. A short C-terminal sequence of 15 amino acid residues was predicted to be exposed to the stroma. This membrane topology has been largely confirmed by partial proteolytic digestion of cytochrome f in various different thylakoid membrane preparations from pea (Willey et al. 1984a).

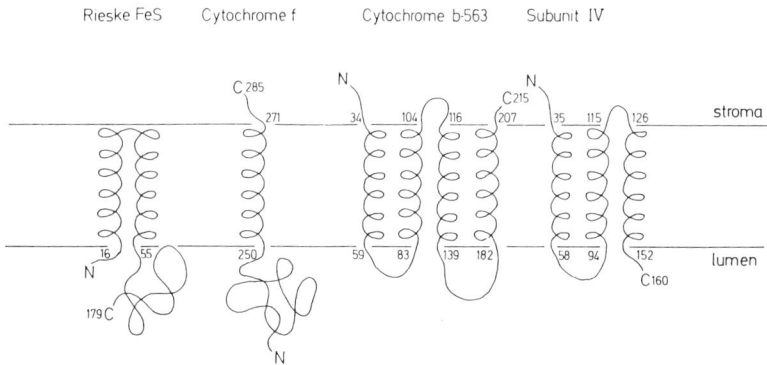

Fig. 2. Predicted topology of the polypeptides of the cytochrome *b-f* complex in the chloroplast thylakoid membrane. Membrane-spanning regions are shown as α-helices; the arrangement of extra-membranous regions is purely schematic. The *N*- and *C*-termini of the polypeptides are shown, and the polypeptides are numbered from the *N*-terminal residue.

The model for the arrangement of cytochrome *b*-563 shown in Fig. 2 is different from that originally proposed by Widger et al. (1984). In the original model for the predicted arrangement of the cytochrome *b*-563 polypeptide in the thylakoid membrane, five transmembrane α-helical sequences were proposed (Widger et al. 1984). However, closer inspection of the deduced amino acid sequences of the polypeptide from spinach, tobacco, pea and wheat reveals that the predicted membrane span IV contains an aspartate, a glutamate and two or more proline residues making it very unlikely that this sequence forms a membrane-spanning α-helix. This suggests that the cytochrome *b*-563 polypeptide folds with only four trans-membrane sequences, positioning the *N*- and *C*-termini of the polypeptide on the same side of the membrane (Fig. 2). This folding pattern may also be applied to mitochondrial cytochrome *b*, apparently resolving a paradox concerning the location of amino acid residues altered in diuron-resistant yeast mutants (di Rago et al. 1986). Diuron is one of a number of compounds which inhibit electron transfer through the cytochrome $b-c_1$ complex by occupying ubiquinone-binding sites. These inhibitors may be grouped into two classes: those, such as diuron and antimycin A, which block the ubiquinone reduction site on the matrix side of the complex and those, such as mucidin and myxathiazol, which block the ubiquinol oxidation site on the side of the complex facing the intermembrane space. Using a model for the folding of mitochondrial cytochrome *b* based on the five membrane span model of cytochrome *b*-563, the altered amino acid residues in the diuron-resistant mutants were placed on either side of the membrane (di Rago et al. 1986), and this seems unlikely in view of the location of the diuron-binding site. However the proposed model with four membrane spans in this region of the polypeptide would place all the altered amino acid residues on the same side of the membrane. This suggests that in the wild type, cytochrome b residues Ile 17, Asn 31 and Phe 225 are all involved in forming the ubiquinone-binding site on the matrix side of the mitochondrial membrane. The amino acid residue in the chloroplast complex corresponding to Phe 225 is located in subunit IV (residue Tyr 27), and suggests that the N-terminus of the subunit IV polypeptide is responsible for forming part of a quinone-binding site. By analogy with the mitochondrial cytochrome b, this would be the plastoquinone-binding site on the stromal side of the thylakoid membrane. This would place the N- and C-termini of cytochrome b-563 and the N-terminus of subunit IV on the stromal side of the membrane (Fig. 2). Models for the arrangement of subunit IV in the membrane with three transmembrane spans have been proposed (Phillips and Gray 1984b, Widger et al. 1984) and these would therefore place the C-terminus of the polypeptide in the lumen of the thylakoid membrane (Fig. 2). This arrangement is supported by the presence of a highly conserved sequence in subunit IV and mitochondrial cytochrome b (Phillips and Gray 1984b). In yeast, this sequence is encoded in the fourth exon of the split gene for cytochrome b (Nobrega and Tzagaloff 1980), which has been shown to be the site of a mutation, $muc3$, which results in mucidin resistance (Subik and Takacsova 1978). Mucidin, and related inhibitors such as myxathiazol, inhibit electron transfer by interfering with a binding site for ubiquinol and preventing the reduction of the Rieske FeS centre and cytochrome b. These inhibitors act on the outside of the mitochondrial inner membrane and the mucidin-resistant mutant is therefore important for indicating residues involved in forming the quinol-binding site on this side of the membrane. The corresponding plastoquinol-binding site in the chloroplast cytochrome complex would be located on the lumenal side of the thylakoid membrane and the conserved sequence in subunit IV would therefore be expected to be located on this side of the membrane (Fig. 2). This suggests that subunit IV is responsible for forming part of the plastoquinol-binding site allowing electron transfer to the Rieske FeS centre and cytochrome b-563. Oettmeier et al. (1982) have shown, using an azido-plastoquinone photoaffinity label, that plastoquinone binds predominantly to cytochrome b-563 and the Rieske FeS protein, but that at higher concentrations of the photoaffinity label subunit IV becomes labelled.

The orientation of the cytochrome b-563 and subunit IV polypeptides in the thylakoid membrane has been investigated by proteolytic digestion of inside-out and right-side-out vesicles of pea chloroplast membranes (Mansfield and Anderson 1985). These experiments indicated that the C-termini of both polypeptides were accessible to carboxypeptidase digestion in right-side-out vesicles. This is not in accord with the model presented in Fig. 2. However, the interpretation of their data rests crucially on the specificity of the antisera used to detect the cytochrome b-563 and subunit IV, and on the absence of chymotrypsin contamination of the carboxypeptidase, and these points have not been satisfactorily established.

A further consequence of the four membrane span model for the topology of cytochrome b-563 is a reassignment of the histidine residues proposed as ligands for the two haem groups (Widger et al. 1984). In the original model the two haem groups were proposed to be held by four histidine residues located in the two antiparallel transmembrane spans II and V, with one haem held by His 100 and His 187 and the other held by His 86 and His 202. However in the four membrane span model, the original span V has the opposite direction across the membrane resulting in the haem on the stromal side of the membrane held by His 100 and His 202 and the haem on the lumenal side of the membrane being held by His 86 and His 187.

The arrangement of the Rieske FeS protein in the thylakoid membrane is not clear because of the presence of a single extended hydrophobic region near the N-terminus. Two alternative models for the arrangement of the protein in the thylakoid membrane can be proposed. One model has a short 16 amino acid N-terminal sequence exposed to the stroma and the C-terminal iron-sulphur binding-site located in the intrathylakoid space where it may interact with the haem group of cytochrome f (Steppuhn et al. 1987). However, an alternative arrangement with the N- and C-termini located on the same side of the thylakoid membrane is possible, by comparison with the mitochondrial Rieske FeS protein which has both the N- and C-termini located on the outside of the inner mitochondrial membrane (Schagger et al. 1987). The length of the N-terminal hydrophobic sequence would allow for two membrane-spanning α-helices and the presence of a potential β-turn sequence (Pro-Tyr-Gly-Ser) within the hydrophobic sequence suggests that this sequence may fold to form a hairpin loop to hold the polypeptide in the membrane with the N- and C-termini both exposed in the lumen (Fig. 2). The arrangement of the Rieske FeS protein in the thylakoid membrane has been tested experimentally using protease digestion of right-side-out and inside-out thylakoid vesicles. The products of proteolytic digestion were then identified with antibodies against the Rieske FeS protein (Mansfield and Anderson 1985). From their experimental data Mansfield and Anderson (1985) suggest that the Rieske FeS protein is a transmembrane polypeptide with the C-terminus exposed to the stroma and the N-terminus in the lumen. This membrane topology is not in agreement with the model based on the deduced amino acid sequence (Fig. 2) although their experimental evidence is equivocal.

V Membrane insertion and assembly of the cytochrome *b-f* complex

The mechanisms of insertion of the polypeptides of the cytochrome *b-f* complex into the thylakoid membrane have received little attention. Of the chloroplast-encoded polypeptides of the complex, only cytochrome *f* has been shown to be synthesised in a precursor form which may be related to membrane insertion. As discussed in section III, cytochrome *f* is synthesised with a 33–35 amino acid residue *N*-terminal extension which shows some similarity to signal sequences of bacterial and animal proteins (Willey et al. 1984b). The *N*-terminal extension of the pea protein has been shown to function as a signal sequence in *E. coli* where it directed the products of gene-fusions with *lac*Z to the inner membrane (Rothstein et al. 1985). The demonstration that cytochrome *f* is synthesised by thylakoid-bound ribosomes (Gray et al. 1984) suggests that the signal sequence is functional in chloroplasts. The signal sequence may cause the binding of ribosomes to the thylakoid membrane and then facilitate the translocation of the large hydrophilic portion of the polypeptide (amino acids 1–250) across the thylakoid membrane. The *C*-terminal hydrophobic sequence (amino acids 251–270) would then act as a stop-transfer signal to hold the cytochrome *f* polypeptide in the thylakoid membrane.

The *N*-terminal sequences of the mature cytochrome *b*-563 and subunit IV polypeptides have not yet been published so it is not known whether they are synthesised initially as higher molecular weight precursors. However translation of spinach chloroplast RNA in a rabbit reticulocyte lysate produced cytochrome *b*-563 and subunit IV polypeptides of the mature size (Alt et al. 1983) suggesting that little or no post-translational processing takes place. The deduced amino acid sequences of the *pet*B and *pet*D gene products do not reveal any *N*-terminal extensions analogous to the cytochrome *f* presequence which suggests a different mechanism for the insertion of these two polypeptides into the thylakoid membrane. The polypeptides may contain internal hydrophobic signal sequences and stop-transfer sequences which are used for the correct insertion and arrangement of the polypeptides in the membrane. Experiments using DNA probes from the *pet*B and *pet*D genes against RNA isolated from soluble or membrane-bound polysomes have shown that the RNA for the *pet*B and *pet*D genes is associated with the membrane-bound polysomes (C.J. Eccles, R. Wilson and J.C. Gray, unpublished) indicating that there are probably internal non-cleavable signal sequences in the polypeptides for directing ribosomes to the thylakoid membrane.

The mechanism of insertion of the Rieske FeS protein into the thylakoid membrane is unknown and the formulation of possible mechanisms is complicated by the lack of information on the organisation of the polypeptide in the thylakoid membrane (see Section IV). The transit sequence of the Rieske FeS protein, which presumably contains information for transfer across the chloroplast envelope, does not show the two domain structure of transit sequences of lumenal proteins, such as plastocyanin and the polypeptides of the oxygen evolving complex (Smeekens et al. 1985, Tyagi et al. 1987, Jansen et al. 1987). The transit sequences of these proteins contain an *N*-terminal region responsible for translocation across the chloroplast envelope and a *C*-terminal region which probably constitutes a signal sequence for transfer across the thylakoid membrane (Smeekens et al. 1985). In the case of the Rieske FeS protein the translocation of much of the mature protein across the thylakoid membrane may utilise an uncleaved internal signal sequence.

The assembly of the components into a functional cytochrome *b-f* complex is poorly understood. Phillips and Gray (1984a) showed that newly-synthesised polypeptides of cytochrome *f*, cytochrome *b*-563 and subunit IV were assembled into the complex in isolated pea chloroplasts. However assembly of the cytochrome *b*-563 polypeptide into the complex was found to require the addition of 5 mM Mg-ATP to the isolated chloroplast system, suggesting that assembly of cytochrome *b*-563 requires an additional energy source not available in the illuminated isolated chloroplasts. The step in the assembly requiring this additional ATP is not known. In the absence of added ATP, the newly synthesised cytochrome *f* and subunit IV polypeptides were assembled into the complex, perhaps suggesting that subunit exchange with pre-existing complexes was being observed. However, assembly de novo of complete complexes in the isolated chloroplasts cannot be ruled out because these chloroplasts may contain pre-existing pools of assembly-competent cytochrome *b*-563 and the cytoplasmically-synthesised Rieske FeS protein. The presence of unassembled pools of subunits in chloroplasts remains to be established.

A key role for the Rieske FeS protein in the assembly process is suggested by a study of mutants lacking components of the cytochrome *b-f* complex. A mutant of *Lemna perpusilla* has been described which was originally shown to lack the high potential Rieske FeS protein (Malkin and Posner 1978) and has subsequently been shown to lack all the cytochrome *b-f* polypeptides (Lam and Malkin 1985), although it is not known if the mutation is in a nuclear or cytoplasmic gene. Nuclear mutants of maize which lack spectrally-detected cytochrome *f* or cytochrome *b*-563 (Miles 1982) and nuclear mutants of *Chlamydomonas* which lack cytochrome *f* and cytochrome *b*-563 (M. Sanguansermsri and D.S. Bendall, unpublished) have been characterised and suggest that the assembly of the cytochrome complex may be mediated by a nuclear gene product. However, a number of mutants deficient in only one or more of the chloroplast-encoded components have also been described. A collection of *Oenothera* plastome mutants (Stubbe and Herrmann 1982) have been characterised using Western blot analysis to detect cytochrome *b-f* polypeptides (Herrmann et al. 1985). The plastome mutants Iμ and IIλ were found to lack specifically cytochrome *f*, mutants Iϕ and IIμ lack both cytochrome *f* and cytochrome *b*-563 and the mutant IIθ lacks cytochrome *f*, cytochrome *b*-563 and the subunit IV polypeptide (Herrmann et al. 1985). All these *Oenothera* plastome mutants have detectable Rieske FeS polypeptide. No mutants have been described which lack only the nuclear-encoded Rieske FeS protein. This suggests that the Rieske FeS protein is necessary for the further assembly of the complex, and in its absence the other components are not assembled, and may be degraded. In the case of the cytochrome *b-f* complex, the Rieske FeS protein may be the 'key assembly protein of nuclear origin' suggested by Miles (1982).

VI Regulation of the synthesis of cytochrome *b-f* complex components

The synthesis of the components of the complex and their assembly into a functional cytochrome *b-f* complex are likely to be closely linked. The formation of a functional complex requires not only the polypeptides but also the prosthetic groups for the cytochromes and the FeS centre. The appearance of redox centres of the cytochrome complex during chloroplast development has been studied in barley (Henningsen and Boardman 1973, Plesnicar and Bendall 1973, Baltimore and Malkin 1977), french bean (Gregory and Bradbeer 1973) and pea (A. Doherty and J.C. Gray, unpublished) and shows that there are major differences between synthesis in cereals and in legumes. In barley, cytochrome *f*, cytochrome *b*-563 and the Rieske FeS centre are present in comparable amounts in both etiolated and greened leaves and there is little change upon illumination (Henningsen and Boardman 1973, Plesnicar and Bendall 1973, Baltimore and Malkin 1977). In french bean and pea, however, cytochrome *f* and cytochrome *b*-563 are present in low amounts in etiolated leaves and the amounts increase upon illumination (Gregory and Bradbeer 1973; A. Doherty and J.C. Gray, unpublished). More recent studies using specific antibodies against the four subunits of the complex have confirmed that the corresponding proteins show similar developmental patterns. In wheat, cytochrome *f*, cytochrome *b*-563 and the Rieske FeS protein are detectable in dark-grown plants and there is little change upon illumination (Takabe et al. 1986). In pea, cytochrome *f*, cytochrome *b*-563, the Rieske FeS protein and subunit IV are all present in very low amounts in dark-grown plants and the amount of each subunit increases substantially upon illumination (Takabe et al. 1986; B.J. Newman and J.C. Gray, unpublished).

Examination of the amounts of transcripts from each of the genes during greening of dark-grown pea plants indicates a marked difference in the accumulation of transcripts from the Rieske FeS protein gene compared to the transcripts from the three chloroplast genes (P. Dupree and J.C. Gray, unpublished). Transcripts of the Rieske FeS protein gene are present in very low amounts in dark-grown plants and rapidly accumulate on greening, whereas the transcripts of the chloroplast genes are present in easily detectable amounts in dark-grown plants and accumulate to a much smaller extent on greening (P. Dupree and J.C. Gray, unpublished). The presence of substantial amounts of transcripts of the chloroplast genes but only very low amounts of the proteins suggests that the accumulation of the chloroplast-encoded polypeptides is regulated at a post-transcriptional stage in pea seedlings. Preliminary experiments, using gabaculine to inhibit the synthesis of 5-aminolaevulinate, an intermediate in the synthesis of haem, in wheat seedlings indicate a link between the accumulation of the cytochrome f polypeptide and haem availability (R. Wilson and J.C. Gray, unpublished). This appears to be similar to the translational control of the synthesis of chloroplast-encoded chlorophyll-proteins of photosystems I and II (Klein and Mullet 1986, 1987), which appears to be mediated by the light-dependent synthesis of chlorophyll. As haem is synthesised by the same pathway in chloroplasts (Bhaya and Castelfranco 1986) and is possibly repressed at the same step in the dark, it seems likely that chlorophyll-proteins and cytochromes may share features of translational control mechanisms.

If, as suggested in the previous section, the presence of the Rieske FeS protein is necessary for assembly of the chloroplast-encoded polypeptides, then the expression of the gene for the Rieske FeS protein may be the main point of regulation of the synthesis and assembly of the complex. The availability of cDNA clones for the Rieske FeS protein will allow an examination of the regulation of the expression of this nuclear gene, and the ability to produce transgenic plants containing additional or altered genes for the Rieske FeS protein will allow hypotheses for its role in the assembly of the complex to be tested.

VII Conclusions

The synthesis and assembly of the cytochrome b-f complex is clearly a complicated process with interactions between products encoded by nuclear

512

and chloroplast genomes. The existence of nuclear mutants which lack chloroplast-encoded components of the complex suggests that the nucleus has a dominant role in the synthesis and assembly process. A thorough characterisation of these nuclear mutants may provide important insights into the mechanisms of synthesis and assembly of the complex. Now that specific gene probes and antisera are available for each component of the complex a thorough study of the appearance of mRNA and protein during chloroplast development can be undertaken. This should provide useful information on the regulation of synthesis and whether the control of synthesis occurs at the transcriptional or post-transcriptional level. The assembly of the components of the complex also requires further investigation using chloroplast systems in vitro. Chloroplast systems may be useful for studies on the N-terminal processing of the precursors to cytochrome f and the Rieske FeS protein, membrane insertion of the polypeptides, the addition of redox centres to cytochrome f, cytochrome b-563 and the Rieske FeS protein and the energetics of assembly of cytochrome b-563 into the complex.

Acknowledgements

We would like to thank Paul Dupree, Chris Eccles, Sean Hird, Barbara Newman, Hugh Salter and Rebecca Wilson for access to their unpublished results, and Hugh Salter for help with computing. This work was supported by various SERC grants.

References

Alt J, Westhoff P, Sears BB, Nelson N, Hurt E, Hauska G and Herrmann RG (1983) Genes and transcripts for the polypeptides of the cytochrome $b6/f$ complex from spinach thylakoid membranes. EMBO J 2: 979–986

Alt J and Herrmann RG (1984) Nucleotide sequence of the gene for pre-apocytochrome f in the spinach plastid chromosome. Curr Genet 8: 551–557

Anderson JM and Boardman NK (1973) Localization of low potential cytochrome b-559 in photosystem I. FEBS Lett 32: 157–160

Baltimore BG and Malkin R (1977) Appearance of membrane-bound iron-sulfur centers and the photosystem I reaction center during greening of barley. Plant Physiol 60: 76–80

Berends T, Kubicek Q and Mullet JE (1986) Localization of the genes coding for the 51 kDa PSII chlorophyll apoprotein, apocytochrome $b6$, the 65–70 kDa PSI chlorophyll apoproteins and the 44 kDa PSII chlorophyll apoprotein in pea chloroplast DNA. Plant Mol Biol 6: 125–134

Bhaya D and Castelfranco PA (1986) Synthesis of a putative c-type cytochrome by intact, isolated pea chloroplasts. Plant Physiol 81: 960–964

Clark RD and Hind G (1983) Isolation of a five-polypeptide cytochrome b-f complex from spinach chloroplasts. J Biol Chem 258: 10348–10354

Clark RD, Hawkesford MJ, Coughlan SJ, Bennett J and Hind G (1984) Association of ferredoxin-NADP$^+$ oxidoreductase with the chloroplast cytochrome b-f complex. FEBS Lett 174: 137–142

Courtice GRM, Bowman CM, Dyer TA and Gray JC (1985) Localisation of genes for components of photosystem II in chloroplast DNA from pea and wheat. Curr Genet 10: 329–333

Di Rago J-P, Perea X and Colson A-M (1986) DNA sequence analysis of diuron-resistant mutations in the mitochondrial cytochrome b gene of *Saccharomyces cerevisiae*. FEBS Lett 208: 208–210

Doherty A and Gray JC (1979) Synthesis of cytochrome f by isolated pea chloroplasts. Eur J Biochem 98: 87–92

Feierabend J (1979) Role of cytoplasmic protein synthesis and its coordination with the plastidic protein synthesis in the biogenesis of chloroplasts. Ber Deutsch Bot Ges 92: 553–574

Fukuzawa H, Yoshida T, Kohchi T, Okumura T, Sawano Y and Ohyama K (1987) Splicing of group II introns in mRNAs coding for cytochrome b_6 and subunit IV in the liverwort *Marchantia polymorpha* chloroplast genome: exon specifying a region coding for two genes with the spacer region. FEBS Lett 220: 61–66

Gounaris I, Michalowski CB, Bohnert HJ and Price CA (1986) Restriction and gene maps of plastid DNA from *Capsicum annuum* (comparison of chloroplast and chromoplast DNA). Curr Genet 11: 7–16

Gray JC (1980) Maternal inheritance of cytochrome f in interspecific *Nicotiana* hybrids. Eur J Biochem 112: 39–46

Gray JC, Phillips AL and Smith AG (1984) Protein synthesis by chloroplasts. In: Ellis RJ (ed) Chloroplast Biogenesis, pp 137–163. Cambridge: Cambridge University Press

Gregory P and Bradbeer JW (1973) Plastid development in primary leaves of *Phaseolus vulgaris:* the light-induced development of the chloroplast cytochromes. Planta 109: 317–326

Hallier UW and Heber UW (1977) Cytochrome f deficient plastome mutants of *Oenothera*. In: Miyachi S, Katoh S, Fujita Y and Shibata K (eds) Photosynthetic Organelles, special issue of Plant and Cell Physiology, pp 257–273. Japan: Jap Soc Plant Physiol

Heinemeyer W, Alt J and Herrmann RG (1984) Nucleotide sequence of the clustered genes for apocytochrome $b6$ and subunit 4 of the cytochrome b/f complex in the spinach plastid chromosome. Curr Genet 8: 543–549

Henningsen KW and Boardman NK (1973) Development of photochemical activity and the appearance of the high potential form of cytochrome b-559 in greening barley seedlings. Plant Physiol 51: 1117–1126

Herrmann RG, Westhoff P, Alt J, Tittgen J and Nelson N (1985) Thylakoid membrane proteins and their genes. In: Vloten-Doting LV, Groot GSP and Hall TC (eds) Molecular Form and Function of the Plant Genome, pp 233–256, NATO ASI series. New York: Plenum Publishing Corporation

Ho KK and Krogmann DW (1980) Cytochrome f from spinach and cyanobacteria. J Biol Chem 255: 3855–3861

Hurt E and Hauska G (1981) A cytochrome f/b_6 complex of five polypeptides with plasto-quinol-plastocyanin oxidoreductase activity from spinach chloroplasts. Eur J Biochem 117: 591–599

Hurt E and Hauska G (1982) Identification of the polypeptides in the cytochrome b_6/f complex from spinach chloroplasts with redox-center-carrying subunits. J Bioenerg Biomembr 14: 405–424

Jansen T, Rother C, Steppuhn J, Reinke H, Beyreuther K, Jansson C, Andersson B and Herrmann RG (1987) Nucleotide sequence of cDNA clones encoding the complete '23 kDa' and '16 kDa' precursor proteins associated with the photosynthetic oxygen-evolving complex from spinach. FEBS Lett 216: 234–240

Jansen RK and Palmer JD (1987) Chloroplast DNA from lettuce and *Barnadesia* (Asteraceae): structure, gene localization, and characterization of a large inversion. Curr Genet 11: 553–564

Karlin-Neumann GA and Tobin EM (1986) Transit peptides of nuclear-encoded chloroplast proteins share a common amino acid framework. EMBO J 5: 9–13

Klein RR and Mullet JE (1986) Regulation of chloroplast-encoded chlorophyll-binding protein translation during plant chloroplast biogenesis. J Biol Chem 261: 11138–11145

Klein RR and Mullet JE (1987) Control of gene expression during higher plant chloroplast biogenesis: protein synthesis and transcript levels of *psb*A, *psa*A-*psa*B, and *rbc*L in dark-grown and illuminated barley seedlings. J Biol Chem 262: 4341–4348

Ko K, Straus NA and Williams JP (1984) The localization and orientation of specific genes in the chloroplast chromosome of *Vicia faba*. Curr Genet 8: 359–367

Ko K and Straus NA (1987) Sequence of the apocytochrome f gene encoded by the *Vicia faba* chloroplast genome. Nucl Acids Res 15: 2391

Kyte J and Doolittle RF (1982) A simple method for displaying the hydropathic character of a protein. J Mol Biol 157: 105–132

Lam E and Malkin R (1982) Ferredoxin-mediated reduction of cytochrome b-563 in a chloroplast b-563/f complex. FEBS Lett 141: 98–101

Lam E and Malkin R (1985) Characterization of a photosynthetic mutant of *Lemna* lacking the cytochrome b_6/f complex. Biochim Biophys Acta 810: 106–109

Lin CM and Kung SD (1984) *Nicotiana* chloroplast genome 8. Localization of genes for subunits of ATP synthase, the cytochrome b-f complex and the 32 kD protein. Theor Applt Genet 68: 213–218

Malkin R and Posner HB (1978) On the site of function of the Rieske iron-sulfur center in the chloroplast electron transport chain. Biochim Biophys Acta 501: 552–554

Mansfield RW and Anderson JM (1985) Transverse organization of components within the chloroplast cytochrome b-563/f complex. Biochim Biophys Acta 809: 435–444

Michalowski C, Breunig KD and Bohnert HJ (1987) Points of rearrangements between plastid chromosomes: location of protein coding regions on broad bean chloroplast DNA. Curr Genet 11: 265–274

Michel F and Dujon B (1983) Conservation of RNA secondary structures in two intron families including mitochondrial-, chloroplast- and nuclear-encoded members. EMBO J 2: 33–38

Miles D (1982) The use of mutations to probe photosynthesis in higher plants. In: Edelman M, Hallick RB and Chua N-H (eds) Methods in Chloroplast Molecular Biology, pp 75–107. Amsterdam: Elsevier Biomedical Press

Nobrega FG and Tzagaloff A (1980) Assembly of the mitochondrial membrane. DNA sequence and organisation of the cytochrome b gene in *Saccharomyces cerevisiae* D273-10B. J Biol Chem 255: 9828–9837

Oettmeier W, Masson K, Soll H-J, Hurt E and Hauska G (1982) Photoaffinity labelling of plastoquinone binding sites in chloroplast cytochrome b_6/f complex. FEBS Lett 144: 313–317

Ohyama K, Fukuzawa H, Kohchi T, Shirai H, Sano T, Sano S, Umesono K, Shiki Y, Takeuchi M, Chang Z, Aota S-i, Inokuchi H and Ozeki H (1986) Chloroplast gene organization deduced from complete sequence of liverwort *Marchantia polymorpha* chloroplast DNA. Nature 322: 572–574

Oliver RP (1984) Location of the genes for cytochrome f, subunit IV of the b_6/f complex, the α-subunit of CF$_1$ ATP-synthase and subunit III of the CF$_0$ ATP-synthase on the barley chloroplast genome. Carlsberg Res Commun 49: 555–557

Palmer JD (1985) Comparative organization of chloroplast genomes. Ann Rev Genet 19: 325–354

Palmer JD and Stein DB (1986) Conservation of chloroplast genome structure among vascular plants. Curr Genet 10: 823–833

Pfefferkorn B and Meyer HE (1986) N-terminal amino acid sequence of the Rieske iron-sulfur protein from the cytochrome b_6/f-complex of spinach thylakoids. FEBS Lett 206: 233–237

Phillips AL (1985) Localisation of genes for chloroplast components in tomato plastid DNA. Curr Genet 10: 153–161

Phillips AL and Gray JC (1983) Isolation and characterization of a cytochrome b-f complex from pea chloroplasts. Eur J Biochem 137: 553–560

Phillips AL and Gray JC (1984a) Synthesis of components of the cytochrome b-f complex by isolated pea chloroplasts. Eur J Biochem 138: 591–595

Phillips AL and Gray JC (1984b) Location and nucleotide sequence of the gene for the 15.2 kDa polypeptide of the cytochrome b-f complex from pea chloroplasts. Mol Gen Genet 194: 477–484

Phillips AL and Gray JC (1984c) Synthesis of components of the cytochrome b-f complex. In: Sybesma C (ed) Advances in Photosynthesis Research, Vol IV, pp 571–574. The Hague: Martinus Nijhoff/Dr W Junk Publishers

Plesnicar M and Bendall DS (1973) The photochemical activities and electron carriers of developing barley leaves. Biochem J 136: 803–812

Rock CD, Barkan A and Taylor WC (1987) The maize plastid psbB-psbF-petB-petD gene cluster: spliced and unspliced petB and petD RNAs encode alternative products. Curr Genet 12: 69–77

Rothstein SJ, Gatenby AA, Willey DL and Gray JC (1985) Binding of pea cytochrome f to the inner membrane of Escherichia coli requires the bacterial secA gene product. Proc Natl Acad Sci USA 82: 7955–7959

Schagger H, Borchart U, Machleidt W, Link TA and Von Jagow G (1987) Isolation and amino acid sequence of the 'Rieske' iron sulfur protein of beef heart ubiquinol:cytochrome c reductase. FEBS Lett 219: 161–168

Shahak Y, Crowther D and Hind G (1981) The involvement of ferredoxin-NADP$^+$ reductase in cyclic electron transport in chloroplasts. Biochim Biophys Acta 636: 234–243

Shinozaki K, Ohme M, Tanaka M, Wakasugi T, Hayashida N, Matsubayashi T, Zaita N, Chunwongse J, Obokata J, Yamaguchi-Shinozaki K, Ohto C, Torazawa K, Meng BY, Sugita M, Deno H, Kamogashira T, Yamada K, Kusuda J, Takaiwa F, Kato A, Tohdoh N, Shimada H and Sugiura M (1986) The complete nucleotide sequence of the tobacco chloroplast genome: its gene organisation and expression. EMBO J 5: 2043–2049

Singh GP, Wallen DG and Pillay DTN (1985) Positioning of protein-coding genes on the soybean chloroplast genome. Plant Mol Biol 4: 87–93

Smeekens S, de Groot M, van Binsbergen J and Weisbeek P (1985) Sequence of the precursor of the chloroplast thylakoid lumen protein plastocyanin. Nature 317: 456–458

Steppuhn J, Rother C, Hermans J, Jansen T, Salnikow J, Hauska G and Herrmann RG (1987) The complete amino-acid sequence of the Rieske FeS-precursor protein from spinach chloroplasts deduced from cDNA analysis. Mol Gen Genet 210: 171–177

Stubbe W and Herrmann RG (1982) Selection and maintenance of plastome mutants and interspecific genome/plastome hybrids from Oenothera. In: Edelman M, Hallick RB and Chua N-H (eds) Methods in Chloroplast Molecular Biology, pp 149–165. Amsterdam: Elsevier Biomedical Press

Subik J and Takacsova G (1978) Genetic determination of ubiquinol-cytochrome c reductase. Mitochondrial locus muc3 specifying resistance of Saccharomyces cerevisiae to mucidin.

516

Mol Gen Genet 161: 99–108

Takabe T, Takabe T and Akazawa T (1986) Biosynthesis of P700-chlorophyll *a* protein complex, plastocyanin, and cytochrome b_6/f complex. Plant Physiol 81: 60–66

Tittgen J, Hermans J, Steppuhn J, Jansen T, Jansson C, Andersson B, Nechushtai R, Nelson N and Herrmann RG (1986) Isolation of cDNA clones for fourteen nuclear-encoded thylakoid membrane proteins. Mol Gen Genet 204: 258–265

Tyagi AK and Herrmann RG (1986) Location and nucleotide sequence of the pre-apocytochrome *f* gene on the *Oenothera hookeri* plastid chromosome (Euoenothera plastome I). Curr Genet 10: 481–486

Tyagi A, Hermans J, Steppuhn J, Jansson C, Vater F and Herrmann RG (1987) Nucleotide sequence of cDNA clones encoding the complete '33 kDa' precursor protein associated with the photosynthetic oxygen-evolving complex from spinach. Mol Gen Genet 207: 288–293

Von Heijne G (1983) Patterns of amino acids near signal-sequence cleavage sites. Eur J Biochem 133: 17–21

Widger WR, Cramer WA, Herrmann RG and Trebst A (1984) Sequence homology and structural similarity between cytochrome *b* of mitochondrial complex III and the chloroplast b_6/f complex: position of the cytochrome *b* hemes in the membrane. Proc Natl Acad Sci USA 81: 674–678

Willey DL, Huttly AK, Phillips AL and Gray JC (1983) Localization of the gene for cytochrome *f* in pea chloroplast DNA. Mol Gen Genet 189: 85–89

Willey DL, Auffret AD and Gray JC (1984a) Structure and topology of cytochrome *f* in pea chloroplast membranes. Cell 36: 555–562

Willey DL, Howe CJ, Auffret AD, Bowman CM, Dyer TA and Gray JC (1984b) Location and nucleotide sequence of the gene for cytochrome *f* in wheat chloroplast DNA. Mol Gen Genet 194: 416–422

Wood PM and Bendall DS (1976) The reduction of plastocyanin by plastoquinol-1 in the presence of chloroplasts: a dark electron transfer reaction involving components between the two photosystems. Eur J Biochem 61: 337–344

Wu N-h, Cote JC and Wu R (1986) Nucleotide sequence of the rice cytochrome *f* gene and the presence of sequence variation near this gene. Gene 50: 271–278

Govindjee et al. (eds), Molecular Biology of Photosynthesis: 517–542
© 1988 Kluwer Academic Publishers

Regular paper

Genes encoding ferredoxins from *Anabaena* sp. PCC 7937 and *Synechococcus* sp. PCC 7942: structure and regulation

JAN VAN DER PLAS, ROLF DE GROOT, MARTIN WOORTMAN, FONS CREMERS, MIES BORRIAS, GERARD VAN ARKEL & PETER WEISBEEK
Department of Molecular Cell Biology and Institute of Molecular Biology, University of Utrecht, Padualaan 8, 3584CH Utrecht, The Netherlands

Received 22 September 1987; accepted 15 February 1988

Key words: *Anabaena* sp. PCC 7937, cyanobacteria, ferredoxin, gene-expression, iron-regulation, recombination, *Synechococcus* sp. PCC 7942

Abstract. The gene encoding ferredoxin I (*petF1*) from the filamentous cyanobacterium *Anabaena* sp. PCC 7937 (*Anabaena variabilis* ATCC 29413) was cloned by low stringency hybridization with the ferredoxin cDNA from the higher plant *Silene pratensis*. The *petF1* gene from the unicellular cyanobacterium *Synechococcus* sp. PCC 7942 (*Anacystis nidulans* R2) was cloned by low stringency hybridization with the *petF1* gene from *Anabaena* sp. PCC 7937. One copy of the *petF* genes was detected in both organisms, and a single transcript of about 630 b was found for *Synechococcus* sp. PCC 7942. Both the *Synechococcus* sp. PCC 7942 and the *Anabaena* sp. PCC 7937 *petF1* genes contain a 297 bp open reading frame coding for a small acidic protein, consisting of 98 amino-acid residues, with a molecular mass of about 10.5 kDa.

The ferredoxin content of *Synechococcus* sp. PCC 7942 is strongly reduced under iron-limited growth conditions. The slight decrease in the amount of ferredoxin transcript found under iron limitation does not account for the more severe reduction in ferredoxin protein observed. The main regulation of the ferredoxin content probably is effected at the level of translation and/or degradation. Although ferredoxin expression can be strongly reduced by iron stress, the ferredoxin function seems to be indispensable, as *Synechococcus* sp. PCC 7942 appeared refractory to yield mutants lacking the *petF1* gene.

Abbreviations: b – bases, bp – basepair(s), kb – 10^3 basepairs, SSC – standard saline citrate

Introduction

Reduced ferredoxin has a central function in many light-dependent processes in both cyanobacteria and plants. Ferredoxin is the last component of their photosynthetic electron transport chains. It is not only involved in $NADP^+$ photoreduction and cyclic photophosphorylation, but also in such diverse processes as nitrogen-fixation, nitrate- and nitrite-reduction, gluta-mate synthesis, and fatty acid metabolism (Hall and Rao 1977, Ho and Krogmann 1982, Smeekens et al. 1985). In addition, ferredoxin is involved in the regulation of many other metabolic activities via the thioredoxin system (Buchanan 1984).

Plant-type ferredoxins are small non-haem iron-sulfur proteins of about 11 kDa, containing one [2Fe–2S] cluster bound by four cysteinyl-sulfur bonds (Hall and Rao 1977). The amino-acid sequences of ferredoxins from many cyanobacteria and plants have been determined and used for the construction of phylogenetic trees (Matsubara and Hase 1983). Eighteen amino-acids were found to be invariant and centered mainly around the four cysteine residues that participate in the [2Fe–2S] cluster binding. Sub-stitutions near these sites are usually equivalent amino-acids, while most non-conservative changes are encountered in other parts of the protein.

Two types of ferredoxin have been found in several plants and in some cyanobacteria (Matsubara and Hase 1983). The differences in primary structure require that there are two distinct structural genes involved. It is most plausible that the individual ferredoxins contribute to different functions within the organism. The biological significance of these differen-ces, however, has not been established yet, with one possible exception. A second ferredoxin has been isolated from heterocysts of *Anabaena* sp. PCC 7937, which seems better suited to act as electron donor to nitrogenase when compared to the ferredoxin isolated from vegetative cells of the same organism (Schrautemeier and Böhme 1985, Böhme and Schrautemeier 1987).

Under conditions of iron limitation, algae and cyanobacteria have been reported to replace ferredoxin by flavodoxin, a flavoprotein (Bothe 1977, Hutber et al. 1977). Cyanobacterial cells from natural blooms frequently do not contain detectable quantities of ferredoxin (Ho et al. 1979). The regulat-ory processes which cause the shift from ferredoxin synthesis to flavodoxin production are unknown.

We pursued the isolation of the genes for ferredoxin in order to start the investigation both of the dynamic process of ferredoxin/flavodoxin ex-change and of the relation between structure and function in ferredoxin. In this paper we report on the isolation and the DNA sequence of a *petF1* gene from the filamentous cyanobacterium *Anabaena* sp. PCC 7937 and of the correspondig gene from *Synechococcus* sp. PCC 7942. The transcription and the regulation of the *Synechococcus* sp. PCC 7942 *petF1* gene were analysed and mutagenesis experiments with recombinational deletion vectors were performed. A preliminary report of the coding sequences of the *petF1* genes has appeared (Van der Plas et al. 1986a,b).

Materials and methods

Materials

Nitrocellulose filter (PH79) and DEAE membrane (NA45) were manufactured by Schleicher & Schuell (Dassel, FRG). Oligonucleotides for probe labeling (random hexamers) were from Pharmacia (Uppsala, Sweden). $\alpha^{32}P$-dCTP (specific activity 1.1×10^{11} Bq/mol) and $\alpha^{35}S$-dATP (specific activity 1.85×10^{10} Bq/mol) were purchased from Amersham (Little Chalfont, UK).

Organisms and growth conditions

The cyanobacterial strains *Anabaena* sp. PCC 7937 (*Anabaena variabilis* ATCC 29413) (Duyvesteyn et al. 1983), *Synechococcus* sp. PCC 7942 (in this study the small-plasmid-cured derivative *Anacystis nidulans* R2-SPc has been used; Kuhlemeier et al. 1983), *Synechococcus* sp. PCC 6301, *Synechocystis* sp. PCC 6803, and *Calothrix* sp. PCC 7601 were grown in BG11 medium (i.e. with $25 \mu mol/l$ Fe^{3+}) (Rippka et al. 1979). Iron-limited growth conditions were obtained by replacement of ferric ammonium citrate in BG11 medium with equal molar amounts of ammonium citrate. The concentration of iron originating from impurities in the other components of the BG11 medium was less than $0.1 \mu mol/l$. Iron-starved cells were grown for at least 100 generations in BG11 medium from which the Fe^{3+} component was omitted.

Escherichia coli PC 2495, a *recA⁻*, *hsdS⁻* derivative of JM 101 *supE, thi⁻*, Δ(*lac-proAB*), [F', *traD36, proAB, lacIᵠZ* M15] (Vieira and Messing 1982) constructed by E. Kampert of this department, and *E. coli* HB 101, F⁻, *hsdS20, recA13, ara-14, proA2, lacY1, galK2, rpsL20, xyl-5, mtl-1, supE44*, λ⁻ (Maniatis et al. 1982) were grown as described (Maniatis et al 1982).

Western blotting

Collected cyanobacterial cells were suspended to a density of 0.16 g wet weight per cm^3 in 25 mmol/l Tris-HCl, 10 mmol/l EDTA, pH 8.0, and disrupted by sonication for *Synechococcus* sp. PCC 7942, by a combined lysozyme and osmotic shock treatment for *Anabaena* sp. PCC 7937, *Synechococcus* sp. PCC 6301 and *Calothrix* sp. PCC 7601, or by French Press treatment for *Synechocystis* sp. PCC 6803. After removal of the celldebris by centrifugation (10 min, 10.000 × g, 4 °C) an ammonium sulphate fractionation was performed on the supernatant. The 60–100% saturation precipitate was dissolved in 10 mmol/l potassium phosphate buffer pH 7.6 and dialyzed against the same buffer. Protein concentration of the extracts was determined as described (Bradford 1976).

These partially purified protein extracts were used for SDS-PAGE (Laemmli 1970) and electro-blotted onto nitrocellulose filters (Towbin et al. 1979). Alternatively, *Synechococcus* sp. PCC 7942 cells were directly boiled in sample buffer and subjected to SDS-PAGE. The blots were probed with polyclonal rabbit antibodies raised against spinach ferredoxin. Immuno-reactions were visualized by incubation with anti-rabbit IgG antibody conjugated to horse radish peroxidase, followed by enzymatic color development through incubation with peroxide and 3,3′,5,5′-tetramethylbenzidine. Lanes containing molecular mass markers were cut from the filters beforehand and stained with amido-black.

Electron microscopic investigation
Synechococcus sp. PCC 7942 mid-exponential phase cells were fixed with 2% paraformaldehyde and 0.5% glutaraldehyde in phosphate buffered saline (PBS). Preparation of ultra-thin cryo sections, immunolabeling and staining of the cryo sections was done as described previously (Van Bergen en Henegouwen and Leunissen 1986). The sections were examined on a Philips EM301 electron miscroscope.

Southern analysis and colony hybridization
Large scale preparations of total DNA from *Synechococcus* sp. PCC 7942 and *Anabaena* sp. PCC 7937 were made according to Mazur et al. (1980) and Curtis and Haselkorn (1983). Small scale isolations were performed as described by Dzelzkalns et al. (1984) with the addition of a lithium chloride precipitation step (Cathala et al. 1983) after the RNase treatment. Restricted DNAs were separated in agarose gels with Tris-borate-EDTA as electrophoresis buffer (Maniatis et al. 1982). Blotting of the separated DNA from the agarose gels to nitrocellulose filters was performed bidirectionally (Meinkoth and Wahl 1984). After baking in vacuo at 80 °C for 30 min the blots were washed at room temperature in 4 × SSC (1 × SSC is 150 mmol/l sodium chloride plus 15 mmol/l trisodium citrate) for 30 min and prehybridized at 50 °C for at least 30 min in a mix of 6.6 × SSC, 10 × Denhardt's solution, 0.1% SDS, 0.05% PPi and 0.1 mg/cm^3 denatured herring sperm DNA (0.25 cm^3 hybridization mix/cm^2 nitrocellulose) (Maniatis et al. 1982, Meinkoth and Wahl 1984). Hybridization was performed by adding denatured probe DNA, labeled to high specific activity ($\geqslant 1.8 \times 10^{16}$ Bq/g DNA) with ^{32}P by nick translation (Maniatis et al. 1982) or by random priming with oligodeoxyribonucleotides (Feinberg and Vogelstein 1983), to the prehybridization medium and continuing the incubation at 50 °C for 16 h. The hybridized blots were washed for 10–30 min at 50 °C, initially three times in 5 × SSC + 0.1% SDS and once in 5 × SSC. Next, the washed

filters were blotted dry, wrapped in plastic foil while still moist and auto-radiographed at $-20\,°C$, using X-ray film and intensifying screens (Eastman Kodak, Rochester, NY, USA). If necessary the stringency of hybridization was increased by repeating the washing steps at a higher temperature.

Ordered replicas of a library of 504 cosmid clones of *Synechococcus* sp. PCC 7942 DNA in *E. coli* HB 101, stored in microtiterplates ($-70\,°C$, 15% glycerol), were made on nitrocellulose filters with a stainless steel replicating device and incubated overnight at $37\,°C$. After colony-filter processing (Maniatis et al. 1982), hybridization was performed exactly as described above for the Southern blots.

DNA cloning

Plasmid DNAs were prepared with the alkaline lysis method (Maniatis et al. 1982) for large scale isolations. 'Minipreps' of plasmid DNA were done by a modified "boiling" procedure (Maniatis et al. 1982): after boiling and spinning the lysates, the supernatant was first extracted with phenol/chloroform, before the DNA was precipitated with 0.6 volumes isopropanol in the presence of 2.5 mol/l ammonium acetate. For cloning or for probe preparation DNA fragments were isolated from agarose gels with DEAE-membrane (Lizardi et al. 1984). Dephosphorylation of linearized vector DNAs with calf intestinal phosphatase (Boehringer, Mannheim, FRG) was carried out 30 min at $37\,°C$. The phosphatase was inactivated by adding 20 mmol/l EGTA and heating the sample 15 min at $68\,°C$, followed by phenol /chloroform extraction. DNA ligations were done according to Dugaiczyk et al. (1975). A cosmid library consisting of 504 cosmid clones, was constructed by ligating the 35–55 kb fraction of partially *Sau*3A-cleaved *Synechococcus* sp. PCC 7942 DNA, fractionated on a 10–30% sucrose gradient, to *Bam*HI-linearized and phosphatase-treated DNA of cosmid vector pJB8, followed by in vitro packaging into λ phage particles and transduction to *E. coli* HB 101 (Ish-Horowicz and Burke 1981, Maniatis et al. 1982). Transformation of *E. coli* was performed by the calcium chloride method as described (Maniatis et al. 1982). Transformation of *Synechococcus* sp. PCC 7942 was carried out as described (Van den Hondel et al. 1980), except that after plating the cells, the plates were first incubated in the dark for 16 h.

DNA sequence analysis

DNA sequences were determined by the dideoxy chain termination method of Sanger et al. (1980) with ^{35}S-dATP as radiolabel and with standard primers. Most regions were sequenced several times and in both orientations. Templates for sequencing were single-strand DNAs of in vivo packaged pEMBL8, -9, -18 and -19 clones (Dente et al. 1983). Analysis of the sequences produced was performed with aid of the computer program MICROGENIE (Beckman Instruments, Palo Alto, CA, USA) and of the sequence analysis package of Stephens (1985).

Transcription analysis
RNA was isolated from mid-exponential phase cultures of *Synechococcus* sp. PCC 7942 according to Dzelzkalns (1984) and was analysed on agarose gels run in 25 mmol/l potassium phosphate buffer, after denaturation with glyoxal-DMSO according to Thomas (1983). The RNA was blotted to nitrocellulose (Meinkoth and Wahl 1984) and hybridized with a homologous probe at 65 °C in aqueous medium as described for the Southern blots.

Optical density measurements
The integrated optical density of bands identified on autoradiographs of Northern blots and on prints from Western blots was determined with the IBAS automatic image analysis system (Kontron/Zeiss, FRG). The integrated optical densities were used as parameter for specific mRNA and protein content, respectively.

Results

Homology at the protein level
In Western blots, partially purified ferredoxin protein extracts from *Anabaena* sp. PCC 7937, *Calothrix* sp. PCC 7601, *Synechococcus* sp. PCC 6301, *Synechococcus* sp. PCC 7942 and *Synechocystis* sp. PCC 6803 all gave a single, strongly cross-reacting band with approximately the same electrophoretic mobility as that for spinach ferredoxin when probed with antibodies raised against spinach ferredoxin (results not shown). This result agrees well with the functional exchangeability of ferredoxins in in vitro tests (Hall and Rao 1977, Ho and Krogmann 1982) and with the similarity in ferredoxin amino-acid sequences (Matsubara and Hase 1983). The observed conservation of the plant-type ferredoxin proteins was exploited for the isolation of cyanobacterial genes encoding ferredoxins.

Homology at the DNA level
Total *Anabaena* sp. PCC 7937 DNA was digested with the restriction enzymes *Cla*I, *Eco*RI, *Hind*III and *Xba*I, and *Synechococcus* sp. PCC 7942 DNA was digested with *Bam*HI, *Eco*RI, *Hind*III and *Pst*I. The restricted

DNAs were separated on agarose gels and transferred to nitrocellulose filters by Southern blotting. The blots were hybridized with a DNA probe derived from the *petF1* gene of *Silene pratensis* (Smeekens et al. 1985). This probe consisted of a 236 bp *Hin*fI fragment of the coding region for the mature protein and was labeled with ^{32}P by nick translation. The digests of *Anabaena* sp. PCC 7937 DNA gave one major hybridization signal at 55 °C, namely a 5.3 kb *Cla*I, a 20.0 kb *Eco*RI, a 2.7 kb *Hin*dIII, and a 4.9 kb *Xba*I fragment (Fig. 1A).

The blots with the *Synechococcus* sp. PCC 7942 chromosomal digests did not give unique signals. Therefore it was decided to isolate the *petF*-specific DNA from *Anabaena* sp. PCC 7937 first and to use this for the detection of the corresponding gene from *Synechococcus* sp. PCC 7942.

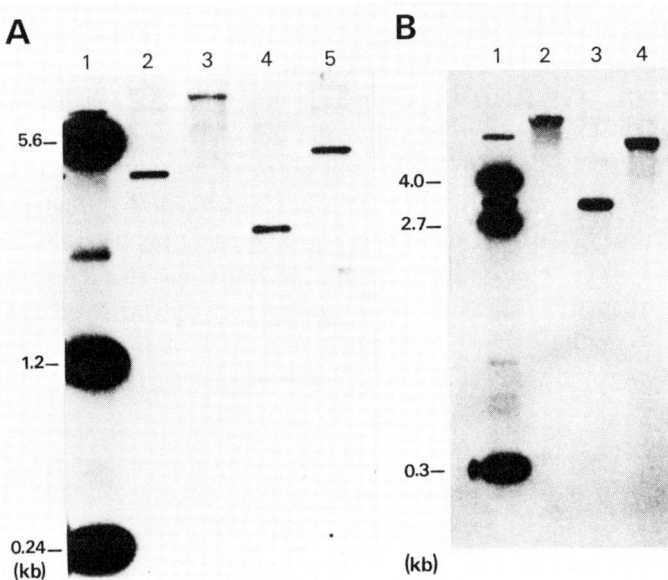

Fig. 1. Chromosomal DNA of *Anabaena* sp. PCC 7937 probed with *S. pratensis petF1* DNA and chromosomal DNA of *A. nidulans* probed with the *petF1* DNA from *Anabaena* sp. PCC 7937. A. Total *Anabaena* sp. PCC 7937 DNA (10 µg) was digested with *Cla*I (lane 2), *Eco*RI (land 3), *Hin*dIII (lane 4) and *Xba*I (lane 5) and run on a 0.8% agarose gel. Lane 1 contained a positive control mixture of three fragments (5.6, 1.2 and 0.24 kb respectively) from the ferredoxin cDNA clone pFD1 of *S. pratensis* (Smeekens et al. 1985). In the amount applied of each of these fragments ca 1 ng *petF1* probe DNA was present. The DNA was probed with the nick-translated 240 bp *Hin*fI fragment from pFD1. B. Total *Synechococcus* sp. PCC 7942 DNA (10 µg) was digested with *Bam*HI (lane 2), *Eco*RI (lane 3) and *Hin*dIII (lane 4), and run on an 0.8% agarose gel. Lane 1 contained a positive control mixture of three fragments (4.0, 2.7 and 0.3 kb respectively) from pVA1 (see Fig. 2). In the amount applied of each of these fragments ca. 1 ng of the *petF1* probe DNA was present. The DNA was probed with the 300 bp *Dra* I fragment from the *Anabaena* sp. PCC 7937 *petF1* gene, labeled by random oligo-priming.

The petF1 *gene from* Anabaena *sp. PCC 7937*

DNA fragments in the range of 4.5–5.5 kb were isolated from an *Xba*I digest of chromosomal *Anabaena* sp. PCC 7937 DNA by preparative agarose gel electrophoresis. These fragments were ligated to *Xba*I-linearized and phosphatase-treated pUC18 (Vieira and Messing 1982) and transformed to *E. coli* PC 2495. The colony library obtained was grown in 64 mixes of 12 clones. Plasmid DNAs of each of these mixed cultures were isolated, digested with *Xba*I, separated on agarose gels and transferred to nitrocellulose filters by Southern blotting. The resulting blots were hybridized with the *S. pratensis petF1* probe at 55 °C. Out of the 64 mixed plasmid preparations 10 were found to contain a 4.9 kb *Xba*I fragment which hybridized to the probe. Of the corresponding mixed cultures one was plated for single colonies. The separate plasmid DNAs subsequently obtained were screened by Southern hybridization analysis. This resulted in the detection of the plasmid pVA1, which contained the 4.9 kb *Xba*I fragment.

This 4.9 kb *Xba*I DNA fragment was analyzed for restriction enzyme cleavage sites. The few enzymes that did cut were *Acc*I, *Bal*I, *Cla*I, *Dra*I, *Eco*RV, *Hinc*II, and *Hind*III. Physical maps of the fragment constructed with these enzymes are shown in Fig. 2. A 1.7 kb *Eco*RV—*Hind*III subfragment, hybridizing with the plant *petF1* gene probe, was used to construct a detailed map for the enzymes *Dra*I and *Hinc*II (Fig. 2B). The hybridizing sequence within this fragment was found to be a 300 bp *Dra*I fragment.

The 1.7 kb *Eco*RV—*Hind*III fragment and some of its subfragments were cloned into pEMBL plasmid vectors and sequenced. Translation of the

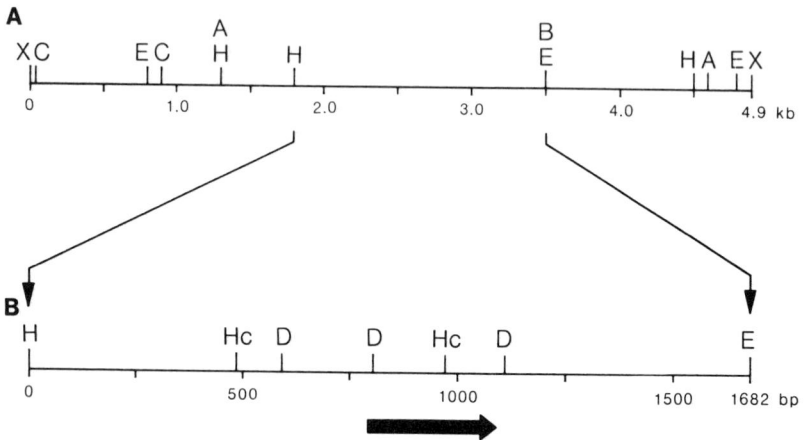

Fig. 2. Restriction maps of the 4.9 kb *Xba*I insert of pVA1 (A) and the sequenced *Eco*RV-*Hind*III region (B). The arrow in fig. 2B indicates the *petF1* coding region from *Anabaena* sp. PCC 7937. A = *Acc*I, B = *Bal*I, C = *Cla*I, D = *Dra*I, E = *Eco*RV, H = *Hind*III, Hc = *Hinc*II, X = *Xba*I.

A

```
    -781                            -751                            -721
A AGC TTG ATG TAA GCC ACC TAA AAG TTG AAT GTG TTC GAC GTG GAG TAA TTT TTG CTG TGC TTT GGC TAA GGC AAT CAG TTC TTC TGC
    -691                            -661                            -631
TTC AGT TAA ATC TAC AGA TAA AGG ATA TTC TAC AAT TAC GTG TTT GCC GGC AGT TAG AGC AGC ACG GCG ATC GCA CCA TGA TCA CGA TTG
    -601                            -571                            -541
ATG GTA CAA ATA ACC ACT AAA TCT ATA TCT CGT TCT ACT AAC TGT TGC CAG CCA GTT ATC GCT TCA GCC TGG TAA TCT TGA GCC AAA
    -511                            -481                            -451
GCC TGG GTT CGC TCC AGT GTA CTA CCC GCT ATG GCG ACT AGG TGC GAT CGC TTA TCC TCC AAA AAG GCT TCT GCC CGC AGT TTT GCT GCA
    -421                            -391                            -361
TAC CCT GTC CAA CTA TGC CAA TAC GTA TTG TTG CTT GTG CCA.AAG CTA CCT CCG ACT CCT ATA TAA TCC TCA CCA GTC CTG CTA CCA AGC
    -331                            -301                            -271
ACA GTA ATA AAT TTT TAT TTT TCA AAA AAA GCT AGG TTA ACG CAG TAC CAG CCT AAC TGA AGG TTA TCC CTG TTC CAG CAAC TTT TTT ATC
    -241                            -211                            -181
ATA TTC TTG ATT CAA GCC ATA AAT TTT TCT AAT AAC TGT GGG TTT AAC GTC TTT AAA GTA TAA AAA AAA TTT ATT GTT ATC TGC TAA CTA
    -151                            -121                            -91
AAT TTC ATT ACT AAT CGT GGT GGA TTT CAA GTT ATA TAC TTT GAT TTT TCC CGT AGT ATC AGA ATT GAA CTA AAT TTA ATT CTG TGG CTA
    -61                            -31                            -1
ATC CCC TGA GAA TAG CCG CTA AGT TCT GCT TTA GCA TAA CTT ATA CTG CCG ATT ACA AAA GAG AGG ATT ACG GAA ATG GCA ACT TTT AAA
                                                                    M   A   T   F   K
    30                            60                            90
GTT ACA TTG ATC AAC GAA GCA GAA GGA ACC AGC AAC ACA ATT GAC GTT CCT GAT GAT GAG TAT ATT TTA GAC GCT GCC GAA GAA CAG GGT
V   T   L   I   N   E   A   E   G   T   S   N   T   I   D   V   P   D   D   E   Y   I   L   D   A   A   E   E   Q   G
    120                            150                            180
TAT GAC CTA CCC TTT TCC TGT CGT GCA GGT GCT TGC TCC ACC TGC GCC GGT AAA CTA GTA TCC GGT ACT GTT GAC CAG TCT GAC CAA TCA
Y   D   L   P   F   S   C   R   A   G   A   C   S   T   C   A   G   K   L   V   S   G   T   V   D   Q   S   D   Q   S
    210                            240                            270
TTC TTG GAT GAC GAT CAA ATC GAA GCT GGA TAT GTA TTG ACC TGT GTT GCT TAT CCA ACC TCT GAT GTA ACC ATC.CAA ACC CAC AAA GAA
F   L   D   D   D   Q   I   E   A   G   Y   V   L   T   C   V   A   Y   P   T   S   D   V   T   I   Q   T   H   K   E
    300                            330                            360
GAA GAC CTC TAC TAA GAG TCA AGT TAG CTT TAA AAA TAG AGG CTA GAG GCT AGC AGC TAC AGA CTA GTC CCT AGC CTT TTT ATT TTT TTT
E   D   L   Y
    390                            420                            450
CAT TAG GGA ACA GTC TAT TGG GAC TGG GTG TAA GGG TGT AAG GGT ATA GGG GTG TAA TAG TTT CAA ACA TTT ATA CTC TTT TCA ACC CTT
    480                            510                            540
GAT ATT TCG TTT TCA TCC GTA AGT CCC ACA AAA AAT AAA TTA TCC AAA ATT GAT GGT TTG GTA GGG TGC GTC AAT AGA AAT CAT TTC TGA
    570                            600                            630
GTG TAT TTA GGC TCT ATC GCA CTG ACG GAC ACT ACA TTT TGG ATA TTT TTT GAT CTG AAA GTC CCT TAA AAT TTG ATT TCT TCA TTC AGG
    660                            690                            720
CTG TGA AGT TGT GTA CCA GGA TAA TAA AAT TGG AGA ATT TTG CTG TTA GAC CAA CCT AGT TTA GCT AAA GTT TGA GCG CCA GTT TGA CTT
    750                            780                            810
AAG CCC ACT CCA TGT CCT AAA CCA CCA CCA ATA AAA GCG TAT CCC ACA ACT CTT CTT TGC CTT TGT TGA GAG GTA CTA TGT AAA AAA GCG
    840                            870
TGC TTC TTG GCG CAG CAA AGG CAC TAC GCA CTT CAT CTT TGT GTA GGG TAA AAA TGC CGA TAT C
```

B

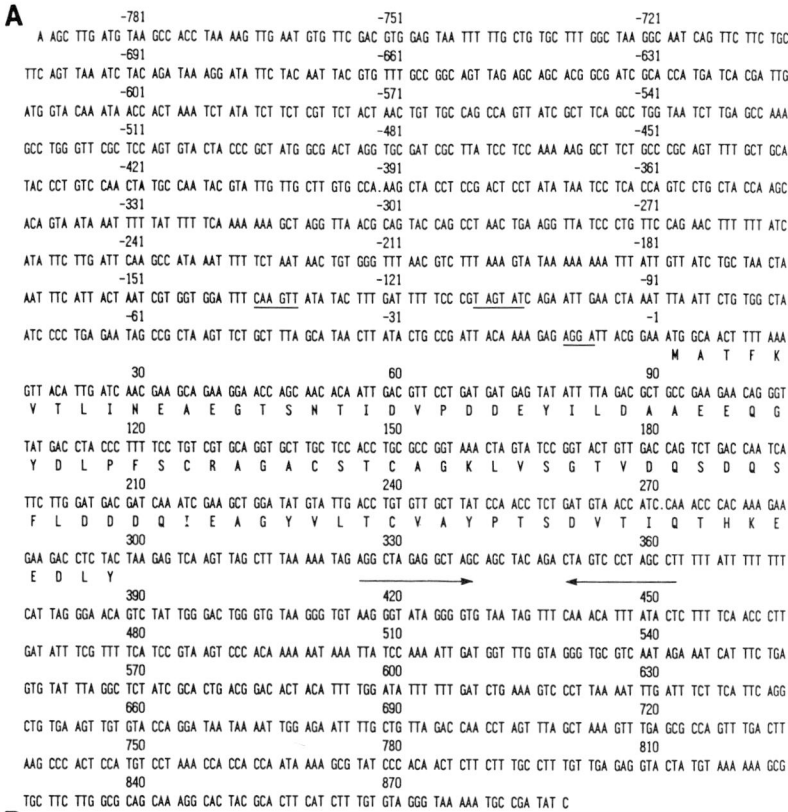

Fig. 3. Nucleotide sequence of the *Anabaena* sp. PCC 7937 *petF1* gene. A. The sequence corresponds to the 1.7 kb *Hind*III-*Eco*RV fragment from pVA1. The putative ribosome binding site (AGGA) and putative promoter sequences are underlined (see also Fig. 12). The sequence underlined by arrows, downstream of the coding region and centered at nucleotide 344, indicates a potential stable hairpin structure. B. Scan for stopcodons in the six possible reading frames of the 1682 bp sequence. Numbering of the bp is according to that in A. The arrow in frame 2 indicates the ferredoxin coding sequence.

1.7 kb sequence (Fig. 3A) in all six possible reading frames resulted in the identification of the coding sequence for a plant-type ferredoxin in the second reading frame (Fig. 3B). This open reading frame of 297 bp is preceded by a putative ribosome binding site AGGA (Tomioka and Sugiura 1983) 8 nucleotides upstream of the initiator codon ATG. Downstream of the gene is a palindromic sequence with the potential to form a structure in the mRNA that is similar to a ϱ-independent transcription termination signal of E. coli (Rosenberg and Court 1979), or that is involved in mRNA stabilization (Csiszàr et al. 1987). The calculated free energy of formation of this stem-loop structure is -6.9×10^4 J/mol (Tinoco et al. 1973).

The petF1 gene from Synechococcus sp. PCC 7942

The petF1 gene from Anabaena sp. PCC 7937 was used to isolate the corresponding gene of Synechococcus sp. PCC 7942. Southern blots of total Synechococcus sp. PCC 7942 DNA digested with BamHI, EcoRI and HindIII were hybridized with a ^{32}P-labeled 300 bp DraI fragment containing almost the entire coding sequence of the Anabaena sp. PCC 7937 petF1 gene (Fig. 2B), in order to investigate the DNA sequence similarity of the petF1 genes. At 60 °C one major hybridizing fragment was found with the BamHI, EcoRI and HindIII digests, with sizes of 11, 3.2 and 6.5 kb respectively (Fig. 1B). Subsequently, nitrocellulose replica filters of a Synechococcus sp. PCC 7942 cosmid library were screened with the same probe. One colony gave a strong and reproducible hybridization reaction. It contained a cosmid, designated pA612, of about 50 kb in size. Restriction and hybridization

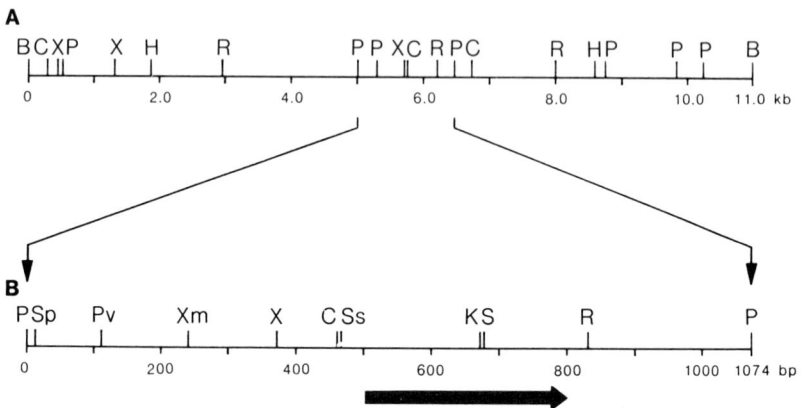

Fig. 4. Restriction maps of the 11 kb BamHI fragment (A) and of the 1.1 PstI fragment from cosmid pA612 (B). The arrow in Fig. 4b indicates the ferredoxin coding region from Synechococcus sp. PCC 7942. B = BamHI, C = ClaI, H = HindIII, K = KpnI, P = PstI, Pv = PvuII, R = EcoRI, S = SaII, Sp = SphI, Ss = SstI, X = XbaI, Xm = XmaIII.

A

```
              -481                          -451                          -421
CT GCA GCC GCT GCA TGC GCA GGG CGT TCG ATT TGA CGG CTT TTT GTG CAA CAT CTT GGC GCA CAT CAT CCA AGC TCT AAC ACC GAC GCT
              -391                          -361                          -331
GTC TGA GTT GGC TAG TCC TGG CAG CTG GGC AAT CTT TAG CGG CTT GCT AAC CAG TCA AGC CGA CAC TGT CAG CGT CAC TTT GGA AGA GTA
              -301                          -271                          -241
CGG TTG GGT GAT CCG CGA TCG CGC CAG TCA GGG AGA TTG GTG TCG TTT GGT CGC GGA TTT TCG GCC GGA ACG ATA AAT CTC ACT AAT GCT
              -211                          -181                          -151
TAG CTT AGA GGG CTT ACT GGG AGC GGG CCG AGT TTG AGC CGT GAT TAC CCC TAC GAA CTT TCC GGC CAC GCT CCA TTG CTT AGA CAT AAA
              -121                          -91                           -61
ATT CCC TTA TGT CTA GAC TGG CGA TTG ATA GCA TTT CTC GCG CAG TTC GCC CTT TGG CAA CCC ATA GTA TCA ATG GGA AAG GTA CGG
              -31                           -1                            30
GCA GGC TGT CAA TCG ATG AGC TCT GCC ACC CCA AAA GCG ATA GAG GAC ACG CTC ATG GCA ACC TAC AAG GTT ACG CTC GTC AAT GCT GCC
                                                                           M   A   T   Y   K   V   T   L   V   N   A   A
              60                            90                            120
GAA GGC TTG AAC ACC ACG ATC GAC GTG GCT GAC GAT ACC TAC ATC TTG GAC GCC GCT GAA GAG CAA GGC ATT GAC CTG CCT TAC TCC TGC
 E   G   L   N   T   T   I   D   V   A   D   D   T   Y   I   L   D   A   A   E   E   Q   G   I   D   L   P   Y   S   C
              150                           180                           210
CGT GCT GGT GCT TGC TCG ACC TGT GCT GGC AAA GTC GTC TCT GGT ACC GTC GAC CAA TCG GAT CAA TCC TTC TTG GAT GAC GAC CAA ATT
 R   A   G   A   C   S   T   C   A   G   K   V   V   S   G   T   V   D   Q   S   D   Q   S   F   L   D   D   D   Q   I
              240                           270                           300
GCA GCA GGC TTT GTC CTG ACC TGC GTC GCC TAT CCG ACC TCC GAT GTG ACG ATC GAA ACC CAC AAA GAA GAA GCC CTC TAC TAA GTC TTG
 A   A   G   F   V   L   T   C   V   A   Y   P   T   S   D   V   T   I   E   T   H   K   E   E   A   L   Y
              330                           360                           390
CTT CGA CTG CTT CAA TCC TTA GAA TTC AAA TCA AAT TGC GGC TTC CAA ATT GGG AGT CGC TTT TTT ATC GCC GTA GTC AGC AGC GAT CGC

              420                           450                           480
CTC ACT ACC GTC CAG AAA CAG CCT GCG ATC GCC CTG TAA CGC CCC TCC CCT AGC AAT CCT GAC TCG GTA GAG TTC AGG CCA AGG ACT

              510                           540                           570
CAA GGT TGA GGC GAT CGT TCA GCA ACG TCT TGC TCT AGG CGT CCG GCG TGA CTG TTG TGG GGA AGA GGC GAT CGC CTT CCT GCA G
```

B

```
Fr 1
Fr 2
Fr 3
Fr 4
Fr 5
Fr 6

   -503        -288        -73         141         356         571
```

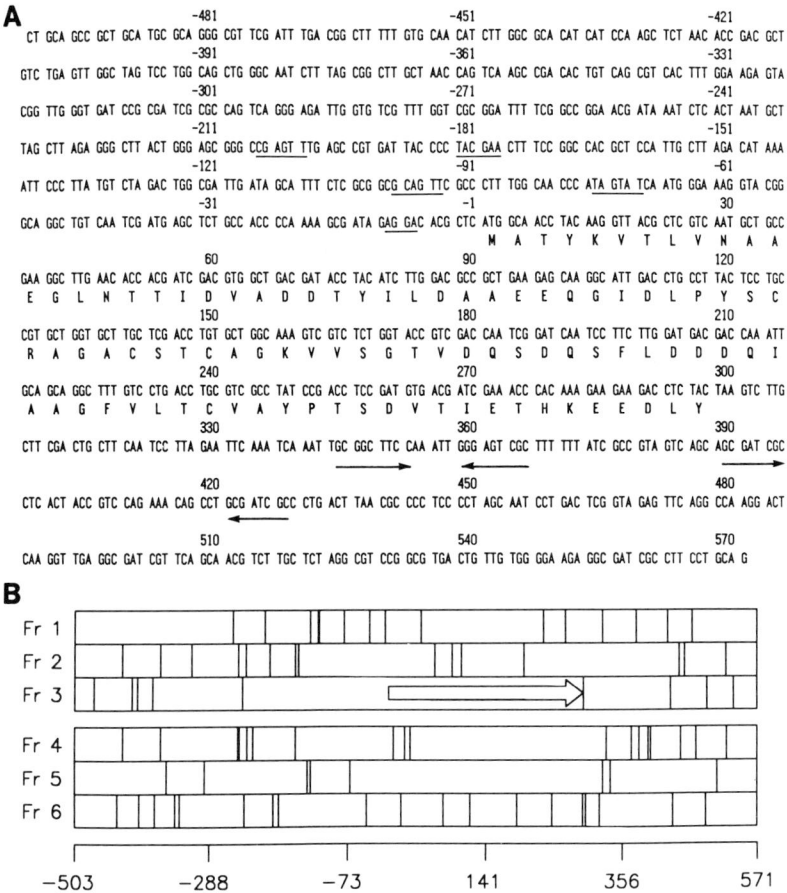

Fig. 5. Nucleotide sequence of the *Synechococcus* sp. PCC 7942 *petFl* gene. A. The sequence of the 1.1 kb *Pst*I fragment from pA612 is given. The ribosome binding site (AGGA) and putative promoter sequences for 'P1' and 'P2' (see also Fig. 12) are underlined. The sequences underlined by arrows, downstream of the coding region and centered at nucleotides 355 and 409, respectively, indicate potential stable hairpin structures. B. Scan for stopcodons in the six possible reading frames of the 1074 bp sequence. Numbering of the bp is according to that in A. The *petFl* coding sequence is indicated by the arrow in reading frame 3.

analysis of pA612 showed hybridizing *Bam*HI, *Eco*RI and *Hin*dIII fragments (data not shown) of the same size as those reacting in Southern blots of chromosomal DNA of *Synechococcus* sp. PCC 7942 (Fig. 1B). The 11 kb *Bam*HI fragment was subcloned in pUC18 and subsequently mapped (Fig. 4A). The 1.1 kb *Pst*I fragment that contains the hybridizing sequence was further subcloned in pEMBL vectors for mapping (Fig. 4B) and sequencing (Fig. 5A).

Translation of the 1.1 kb sequence (Fig. 5A) in all six possible reading

528

frames resulted in the identification of the coding sequence for a plant-type ferredoxin in the third reading frame (Fig. 5B). This open reading frame of 297 bp is preceded by a putative ribosome binding site AGGA (Tomioka and Sugiura 1983) 7 nucleotides upstream of the initiator codon ATG. The non-coding region following the gene contains two sequences (see Fig. 5A) able to form base-paired structures in mRNA with calculated free energies

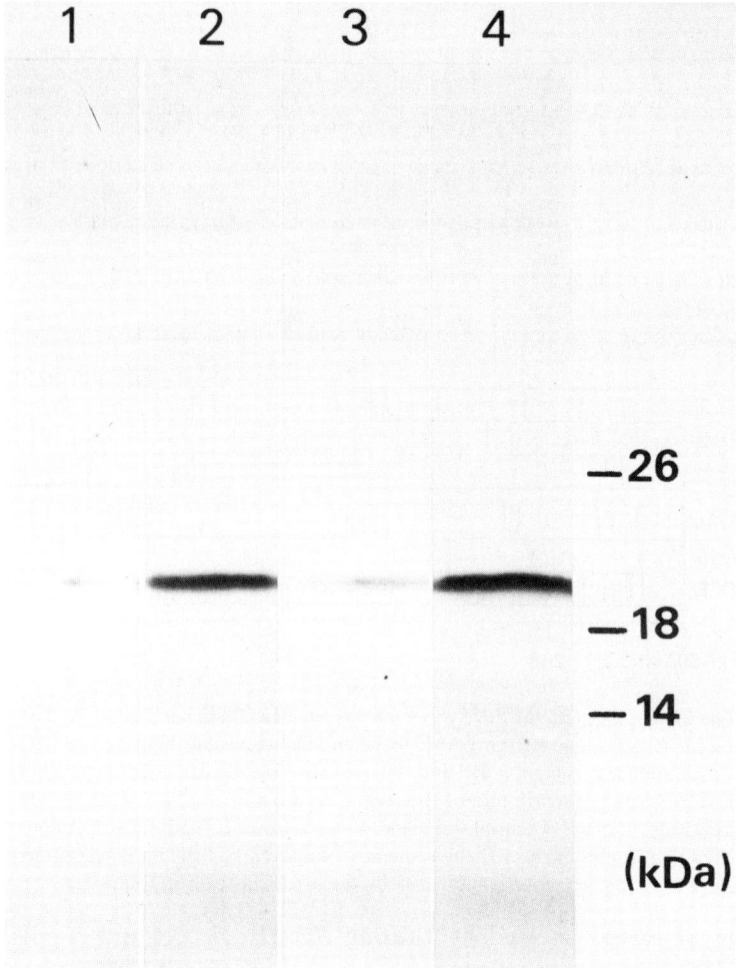

Fig. 6. Protein analysis of *Synechococcus* sp. PCC 7942 grown with or without Fe^{3+}. Cells were harvested by centrifugation, boiled in sample buffer and used for SDS-PAGE. After electrophoretic transfer to nitrocellulose, the proteins were incubated with anti-ferredoxin antibodies. Antibody reaction was visualized via horse radish peroxidase coupled second antibodies. Lanes 1 and 3: cells grown in BG11 minus Fe^{3+}. Lanes 2 and 4: cells grown under standard conditions with complete BG11 medium. In lanes 1 and 2: 5.10^7 cells and in lanes 3 and 4 10^8 cells were used, respectively.

of formation (Tinoco et al. 1973) of -4.4×10^4 and $-6.9 \times {}^4$ J/mol, respectively.

Regulation of gene expression

Iron limitation is known to induce in cyanobacteria a shift from the synthesis of ferredoxin to the production of flavodoxin (Bothe 1977, Hutber et al. 1977). To determine the level at which this regulation occurs, both the ferredoxin protein content and the level of mRNA encoding ferredoxin were studied.

The relative amounts of ferredoxin protein present in *Synechococcus* sp. PCC 7942 cells grown either under iron limitation or under standard conditions were measured immunologically. In Western blot analysis (Fig. 6) the samples from cells grown in the presence of iron reproducibly gave signals approximately 30 times stronger than the samples from iron-limited cultures, as determined by optical density measurement. As an alternative approach, electron microscopic analysis of cells from the two types of cultures described above was performed. Cryo sections were incubated with ferredoxin antibodies, after which antibody-antigen complexes were made visible under the electron microscope with protein A-gold (Van Bergen en Henegouwen and Leunissen 1986). Gold-labeling is found almost exclusively in the cytoplasm (Fig. 7). From this observation in can be concluded that ferredoxin is localized throughout the cytoplasm, rather than associated with thylakoid or cytoplasmic membranes. The gold-labeling of cells from iron-limited cultures when compared to that in cells from complete medium, is reduced to the same degree as the corresponding signal in the Western blots (Fig. 6).

The specific *petF1* mRNA levels were determined by Northern analysis. Total RNA was extracted from *Synechococcus* sp. PCC 7942 cells grown with and without Fe^{3+} added to the culture medium. Northern blots of total RNA hybridized with the 360 bp *SstI-EcoRI petF1* gene fragment from *Synechococcus* sp. PCC 7942 as probe, showed a single signal, corresponding to a transcript of about 630 b (Fig. 8). The hybridization signal found with total RNA from iron-limited cells was 2–3 times weaker than obtained with the same amount of total RNA extracted from cells grown in standard medium.

In conclusion, the slight decrease in the amount of *petF1* transcript found under iron limitation does not account for the significantly stronger reduction in ferredoxin protein content observed.

Attempted deletion of the petF1 gene

The *petF1* gene from *Synechococcus* sp. PCC 7942 was used in mutagenesis

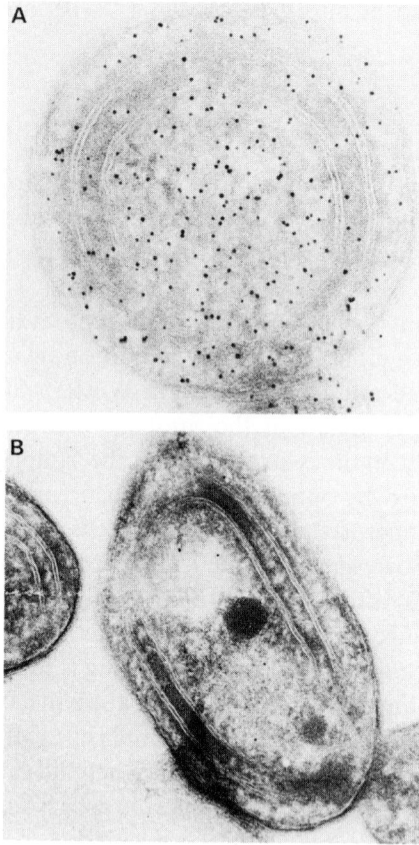

Fig. 7. Electron microscopic analysis of *Synechococcus* sp. PCC 7942 cells grown with or without Fe^{3+}. Cryo sections of cells, grown in A: BG11 medium and B: BG11 medium without iron, were incubated with anti-ferredoxin antibodies, followed by incubation with protein A-gold for electron microscopic investigation.

experiments, in order to investigate whether this organism can grow in the absence of ferredoxin and, if so, to create concomitantly a host-strain for studies on ferredoxin function. For the construction of a *petF1* deletion mutant by exploiting homologous recombination, a plasmid was made (Fig. 9A) in which the *petF1* coding sequence was replaced by a fragment from pUC4-K bearing the *neo* gene (Vieira and Messing 1982). The vector does not replicate in *Synechococcus* sp. PCC 7942 and stable kanamycin-resistant cells can only be formed when the vector or part of it integrates into the chromosome.

Before transformation, the plasmid pFDdel II (Fig. 9A) was linearized in the vector sequences with *Sca*I, in order to reduce the frequency of chro-

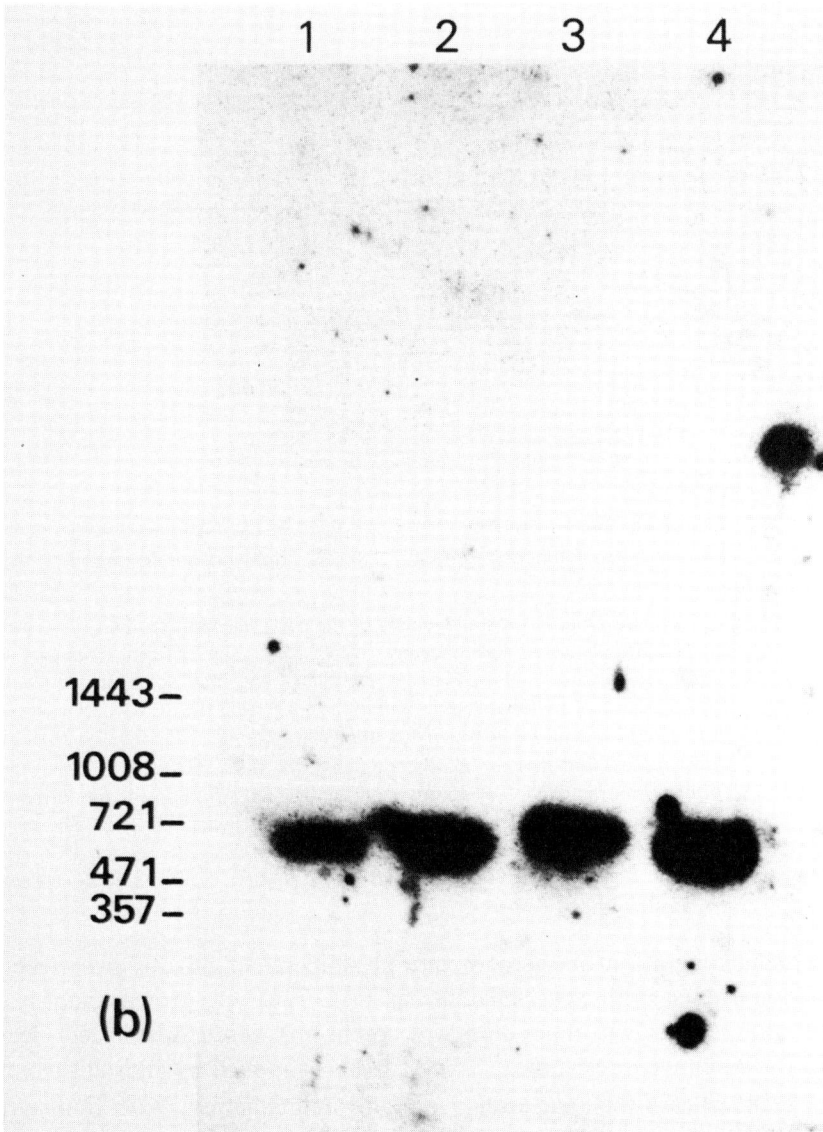

Fig. 8. Transcription analysis of *Synechococcus* sp. PCC 7942 cells grown with or without Fe^{3+}. Total RNA was separated on an 1.1% agarose gel, blotted and hybridized with an *Synechococcus* sp. PCC 7942 *petF1* DNA probe, consisting of the radioactively labeled 360 bp *Sst*I-*Eco*RI fragment. pEMBL8-*Taq*I DNA fragments were denatured and used as size markers. Lanes 1 and 3: RNA from cells grown inBG11 medium minus FE^{3+}. Lanes 2 and 4: RNA from cells grown in complete BG11 medium. In lanes 1 and 2 5 μg and in lanes 3 and 4 10 μg total RNA was loaded onto the gel

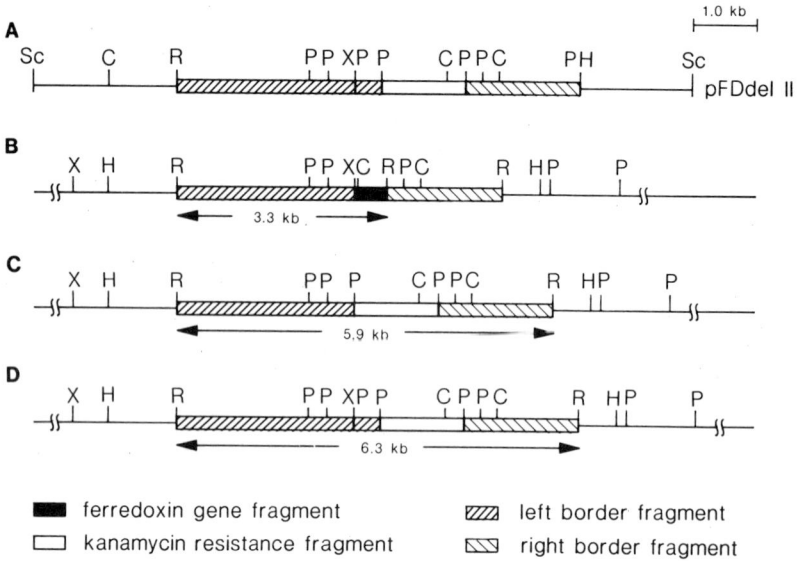

Fig. 9. Plasmid vector pFDdel II for mutagenesis of the *Synechococcus* sp. PCC 7942 *petFl* gene. Fragments flanking the *petFl* gene were ligated to the *neo* gene fragment from pUC4-K and inserted into the polylinker of pEMBL19. Details of the final construct pFDdel II (10.3 kb) are shown in Fig. 9A as linearized by *Sca*I in the *bla* gene of the pEMBL19 part. Note that the left border has a small duplication (it contains two copies of the 0.37 kb *Pst*I-*Xba*I fragment in direct repeat) and that the *Eco*RI site delimiting the *petFl* gene fragment to be deleted in vivo, has been removed. For comparison the original situation for the ferredoxin gene in the chromosome of *Synechococcus* sp. PCC 7942 is drawn in fig. 9B. The results expected from recombination between pFDdel II and the *Synechococcus* sp. PCC 7942 genome are drawn in Figs. 9C and 9D. C = *Cla*I, H = *Hin*dIII, Sc = *Sca*I, R = *Eco*RI, P = *Pst*I, X = *Xba*I.

mosomal integration of the complete plasmid by single-crossover recombination events. Both *Synechococcus* sp. PCC 7942 cells grown under iron-limitation and grown under standard conditions were transformed. Selection for kanamycin resistance ($7.5 \, \mu g/cm^3$) was applied by adding the antibiotic underneath the agar after overnight preincubation. With both types of cells rather large numbers of very tiny, slowly growing colonies (transformation frequency: 10^{-3}–$10^{-4}/\mu g$ DNA), as well as a few larger colonies (transformation frequency: 10^{-6}–$10^{-7}/\mu g$ DNA) were detected. The cells with small-colony phenotype could not be subcultured further under kanamycin pressure together with iron limitation.

Chromosomal DNA was extracted from wildtype *Synechococcus* sp. PCC 7942 and from a limited number of the larger type transformants. The chromosomal DNAs were digested with *Eco*RI, separated on an agarose gel and blotted to two nitrocellulose filters (Fig. 10). One filter was probed with

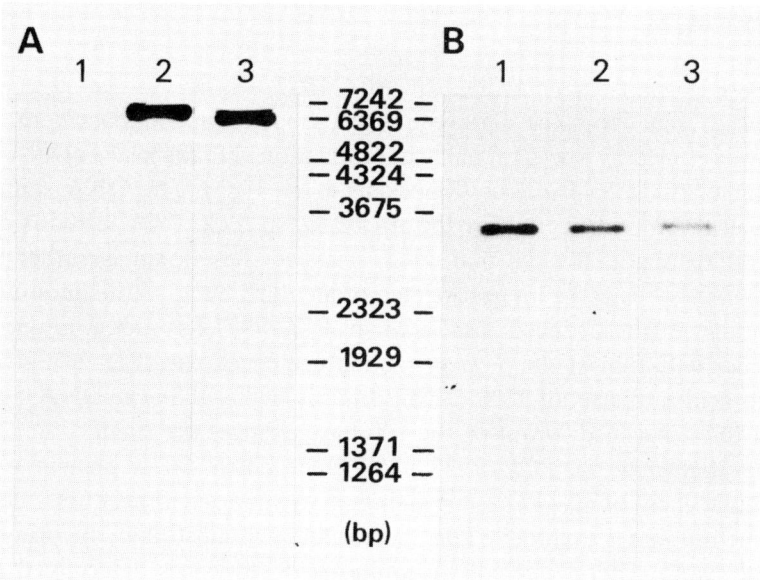

Fig. 10. Southern hybridization analysis of kanamycin-resistant *Synechococcus* sp. PCC 7942 recombinants produced by transformation with pFDdelII. Southern blots of chromosomal DNA (10 μg) from wildtype *Synechococcus* sp. PCC 7942 (lane 1) and from two different types of kanamycin-resistant recombinants (lanes 2 and 3) are shown. The blot in panel A was probed with a radioactively labeled *neo* gene fragment from PUC4-K and the blot in panel B was probed with the radioactively labeled 360 bp *Sst*I-*Eco*RI fragment from the *Synechococcus* sp. PCC 7942 *petF1* gene. the size markers indicated are from a *Bst*EII digest of phage lambda DNA.

a radioactively labeled *neo* gene fragment. The other filter was probed with the radioactively labeled 360 bp *Sst*I-*Eco*RI fragment, which is completely contained within the 460 bp *Xba*I-*Eco*-RI fragment to be deleted from the chromosome. The DNAs from all the transformants hybridized with the *neo* gene probe on one of the two expected fragments of 6.3 kb and 5.9 kb. Since the left border of pFDdel II has a small duplication of th 0.37 kb *Pst*I-*Xba*I fragment, recombination can take place in or adjacent to this duplication (Figs. 9C and 9D). The *neo* gene therefore must have been integrated correctly. Nevertheless, all strains still contained the 3.3 kb wildtype *petF1* gene fragment as in the wildtype control (Fig. 10B). So the transformants, which were colony-purified before further analysis, were heterozygous with respect to the *petF1* region. It is likely that segregation of the *petF1* gene causes loss of viability.

534

Discussion

The petF1 *gene from* Anabaena *sp. PCC 7937*

The *petF1* gene from the filamentous cyanobacterium *Anabaena* sp. PCC 7937 was cloned by low-stringency hybridization with the *petF1* cDNA from the higher plant *S. pratensis* (Smeekens et al. 1985). The DNA similarity between the coding region for the *Anabaena* sp. PCC 7937 ferredoxin and the coding region for the mature *S. pratensis* ferredoxin is rather homogeneously spread (the longest stretch of identical nucleotides amounts 11). Alignment of the deduced *Anabaena* sp. PCC 7937 ferredoxin amino-acid sequence with that of the *S. pratensis* mature ferredoxin (Fig. 11), gives an amino-acid similarity of 71%. This is slightly higher than the DNA similarity of 65%, because 50 out of the 100 base changes are silent.

The *Anabaena* sp. PCC 7937 *petF1* gene codes for a protein with a molecular mass of 10.6 kDa, consisting of 98 amino-acid residues when the first methionine is excluded. This methionine will most likely be processed, for all cyanobacterial ferredoxin amino-acid sequences determined begin with alanine (Matsubara and Hase 1983). The observed molecular mass of ferredoxin (Fig. 6) is higher than the calculated one because of the anoma-

```
                              1    10    20    30    40    50      60    70    80    90   98                    A   B
Anabaena sp. PCC7937      AT FKVTLINEAEGTSNTIDVPDDEYILDAAEEQGYDLPFSCRAGACSTCAGKLVSGTV DQSDQSFLDDDQIEAGYVLTCVAYPTSDVTIQTHKEEDLY   14   0
Silene pratensis          .. Y.....TKES.-TV.F.C...V.V..E...E.I...Y.....S..S...V.A.S. .................W....A...SA....E.....E.TA   27  29

Synechococcus sp.PCC7942  AT YKVTLVNAAEGLNTTIDVADDTYILDAAEEQGIDLPYSCRAGACSTCAGKVVSGTV DQSDQSFLDDDQIAAGFVLTCVAYPTSDVTIETHKEEDLY    0  14
Chlorogloeopsis fritschii .. ......I.D....Q..E.D..........A.L..........IK.... ..............E..Y...........C.......E.             13  16
Anabaena sp. PCC7937      .. F...I.E...TSN...P..E..........Y...F.............L.... .............E..Y............Q.........           14   0
Mastigocladus laminosus   .. ......I.E....K..E.P..Q........A...............LI... .............E..Y...........CV......E..           14  14
Aphanizomenon flos-aquae  .. ......I-D...TT....CP..........A.L.............L.T..I ...........VE..Y...................             15  14
Nostoc strain MAC I       ..V......-DQ..TE....P..E....I..D..L.....................I..... .............EK.Y...........LK........   17  18
Anabaena sp. PCC7120      .. F...I.E...TKHE.E.P..E..........Y...F.............L.... .............E..Y...........V.Q........           17   5
Nostoc muscorum           .. F...I.E...TKHE.E.P..E........E.Y...F.............L.... .............E..Y...........V.Q........           18   5
Spirulina platensis       .. ......I.E...I.E...CD..........A.L.............TIT...I ...........E..Y...........C..K..Q..G..           18  18
Spirulina maxima          .. ......ISE...I.E...CD..........A.L.............IT..SI ...........E..Y...........C..Q..Q..G..           19  18
Anabaena variabilis       .. F...I.E...TKHE.E.P..E........E.Y...F.............L.... .............E..Y...........CV.Q.......           19   6
Aphanothece halophitica   .S ......I.EEM...E..E.P..E....V...E.................IKE.EI ...........E..Y...........A..C..I..Q..E..    23  22
Synechocystis sp. PCC6714 .S .T.K.II-PD.-ENS.ECS..........A.L.............ITA.S. ............E..Y...........C..........           24  24
Synechococcus sp.         .. ......R-PD.SE....PE.E....V....L....F.............LLE.E. ............EK.........R..CK.L.NQ..E..       25  24
Nostoc strain MAC II      .. ....R.F.....DE..E.P..E.......A.L....F...S.S..S.N.ILKK. .....N..........NCE....R.DAIA                 28  29
Aphanothece sacrum I      .S .....KT-PD.-DNV.T.P..E....V...E.L...............L..PA PDE........Q..YI.........G.CV........A..        29  27
Aphanothece sacrum II     .. ......I.EE...I.AILE...QT...G..A.L...S....S.......L...AAPN.D..A.......L...W.M......G.C..M..Q.SEVL      33  36
Synechococcus sp. PCC6301 .. .Q.EVIYOGQSQTF.A.-S.QS-V..S.QAA.V...A..LT.V.T...ARIL..E. ..P.AMGVGPEPAKQ.YT.L......R..LK......DE..ALQFGQPG  62  62
```

Fig. 11. Comparison of ferredoxin amino-acid sequences. The ferredoxin amino-acid sequences for *Synechococcus* sp. PCC 7942 and *Anabaena* sp. PCC 7937 are shown. The sequence of *Anabaena* sp. PCC 7937 is aligned to that of *S. pratensis* (Smeekens et al. 1985) and the differences are indicated. the sequence of *Synechococcus* sp. PCC 7942 is compared with the sequences known from other cyanobacteria (Alam et al. 1986, Chan et al. 1983, Cozens and Walker 1987, Matsubara and Hase 1983). For each ferredoxin the number of amino-acid residues that differ from *Synechococcus* sp. PCC 7942 ferredoxin (column A) and from *Anabaena* sp. PCC 7937 ferredoxin (column B) are given. At the bottom the 24 residues shared by all cyanobacterial ferredoxins are indicated. Cysteine residues involved in [2Fe-2S]-cluster formation are nos 41, 46, 49 and 79.

lous electrophoretic mobility of ferredoxin in this gel system. This effect might be produced by dimerization of ferredoxin (Böhme and Schrautemeier 1987), or by the highly negative charge of the protein, which prevents proper binding of SDS (Smeekens et al. 1985), or by both. The protein deduced from the DNA sequence contains 4 cysteinyl residues and their positions correspond to those in the other ferredoxin sequences (Fig. 11). Besides the 4 cysteinyl residues, which are required for the chelation of the two iron atoms of the [2Fe–2S] cluster, 20 amino-acid residues are conserved in all the known ferredoxins of cyanobacterial origin. The *Anabaena* sp. PCC 7937 ferredoxin, like all other known plant-type ferredoxins, has an acidic character, as predicted by the high content of aspartic acid and glutamic acid residues (20x) relative to that of arginine and lysine residues (4x). The amino-acid sequence of the ferredoxin from *Anabaena* sp. PCC 7937 is most similar to the three almost identical amino-acids sequences of the ferredoxins from *Nostoc muscorum* (Matsubara and Hase 1983), from an unspecified *Anabaena variabilis* strain (Chan et al. 1983) and of the ferredoxin encoded by the *petF1* gene of *Anabaena* sp. PCC 7120 (Alam et al. 1986) (Fig. 11).

In many plants and some cyanobacteria two types of ferredoxin are found, differing in amino-acid sequence and redox potential (Matsubara and Hase 1983). By convention the type I designation is given to the major ferredoxin component in the cell. When the amino-acid sequence of the *Anabaena* sp. PCC 7937 ferredoxin was compared with those of the type I and II ferredoxins from *Aphanothece sacrum* and *Nostoc* sp. strain MAC (Matsubara and Hase 1983), a higher degree of similarity to the type I sequences was found. Therefore the *Anabaena* sp. PCC 7937 ferredoxin can be assigned to the type I ferredoxins.

The *Anabaena* sp. PCC 7937 *petF1* gene probably encodes the ferredoxin found in vegetative cells of this strain, because the deduced amino-acid composition agrees almost completely with that determined for the purified protein (Böhme and Schrautemeier 1987). This protein is also found in heterocysts of this strain as the minor ferredoxin species. The amino-acid composition of the major ferredoxin from heterocysts, thought to be specialized in electron transfer to nitrogenase (Schrautemeier and Böhme 1985), appears to be quite different (Böhme and Schrautemeier 1987).

In the *Anabaena* sp. PCC 7937 genome *petF1* is probably present as a single-copy gene, as only one major hybridization signal was found, both with the plant probe and with the *petF1* gene itself as probe (results not shown). The genome of *Anabaena* sp. PCC 7937, however, is expected to possess at least two different *petF* genes, as suggested by the presence of two clearly different ferredoxin proteins in the heterocysts (Schrautemeier and

Böhme 1985, Böhme and Schrautemeier 1987). The second ferredoxin most likely is so divergent that a corresponding second hybridization signal can not readily be detected under the conditions applied. The *petF1* gene probably is not part of an operon, as directly adjacent to the *petF1* reading frame no other open reading frames were found (Fig. 3).

Alignment of the DNA sequence of the *petF1* gene and some flanking sequences (790 bp in total) from *Anabaena* sp. PCC 7937 with that from *Anabaena* sp. PCC 7120 (Alam et al. 1986) resulted in an overall similarity of 92% (93% for the coding region alone). Hence even the non-coding flanking sequences are strongly conserved. Further sequence comparisons might reveal whether the high similarity found is incidental, or whether it is indicative of close strain relatedness despite distinct differences in restriction endonuclease content (Duyvesteyn et al. 1983) and in heterotrophic growth capacity (Wolk and Schafer 1976, Rippka et al. 1979).

The petF1 *gene from* Synechococcus *sp. PCC 7942*

The *petF1* gene from the unicellular cyanobacterium *Synechococcus* sp. PCC 7942 was cloned by low-stringency hybridization with the *petF1* DNA from *Anabaena* sp. PCC 7937. The DNA similarity between the coding regions of the *petF1* genes of the two organisms is 76%. The 3′ part is somewhat more conserved, with a continuous stretch of 29 identical nucleotides (including the TAA stopcodon). Outside the coding regions sequence similarity is insignificant. Alignment of the deduced *Synechococcus* sp. PCC 7942 ferredoxin amino-acid sequence with that of *Anabaena* sp. PCC 7937 ferredoxin (Fig. 11) reveals 14 differences, 5 of which are exchanges for equivalent amino-acids. The amino-acid similarity of 86% is much higher than the DNA homology (76%), because 49 out of 72 base changes are silent.

The *Synechococcus* sp. PCC 7942 *petF1* gene codes for a small acidic protein with a molecular mass of 10.4 kDa, consisting of 98 amino-acid residues, the first methionine being excluded. It also contains 4 cysteinyl residues, in positions corresponding to those in the other ferredoxin sequences shown (Fig. 11). Its acidic character is inferred from the high content of aspartic acid and glutamic acid residues (18x) relative to arginine and lysine residues (4x). The codon usage resembles that seen in genes already sequenced from this organism (Tandeau de Marsac and Houmard 1987).

The *Synechococcus* sp. PCC 7942 ferredoxin must be assigned to the type I ferredoxins for the same reason as discussed for the *Anabaena* sp. PCC 7937 ferredoxin. Remarkably the *Synechococcus* sp. PCC 7942 ferredoxin shows much greater similarity with the type I ferredoxins of the filamentous cyanobacteria than with the type I ferredoxins of the unicellular strains *Aphanothece sacrum, Aphanothece halophitica, Synechocystis* sp. PCC 6714 and *Synechococcus* sp. (Fig. 11). This reflects the observed heterogeneity in the group of unicellular cyanobacteria (Rippka et al. 1979).

For two reasons *Synechococcus* sp. PCC 7942 is now expected to possess a second *petF* gene. The detection of a second, minor, ferredoxin component for *Anacystis nidulans* has been reported (Sakihama and Shin 1987), and a coding sequence for another [2Fe–2S] ferredoxin has been identified in *Synechococcus* sp. PCC 6301 (Cozens and Walker 1987), a strain very closely related to *Synechococcus* sp. PCC 7942. The latter gene, of which the expression is unknown, is located near the *atpC* gene and is clearly different from *petF1* described by Reith et al. (1986) and by us (Van der Plas et al. 1986b; this paper) (Fig. 11). The similarity in amino-acid sequence between the gene products of this second *petF* gene and the *petF1* gene amounts to only 42% (the DNA similarity is 57% at the most). The deduced amino-acid sequence also differs considerably from that of the type II ferredoxins (Fig. 11). In spite of these indications for a second *petF* gene, only one major hybridization signal was found with the *Anabaena* sp. PCC 7937 *petF1* probe. Using the *Synechococcus* sp. PCC 7942 *petF1* gene itself as probe for hybridization at low stringency (45 °C) with total *Synechococcus* sp. PCC 7942 DNA digested with *Eco*RI and *Hin*dIII, additional signals of 4.9 and 2.5 kb, respectively, were observed (results not shown).

Directly adjacent to the *petF1* reading frame no other open reading frames are found (Fig. 5). Together with the detection of a single messenger for ferredoxin with a length 630 b, this leads to the conclusion that the *petF1* gene is not part of an operon. This length of the *petF1* specific messenger is found reproducibly for *Synechococcus* sp. PCC 7942 under our conditions. In Fig. 8 pEMBL8-*Taq*I fragments were used as size markers, but the same transcript length was determined in Northern blots with the size markers phage ØX174 RF-DNA fragments, rRNAs, or RNA transcripts made in vitro with SP6 polymerase (data not given). This mRNA size differs significantly from the 430 b determined for *Synechococcus* sp. PCC 7942 by Reith et al. (1986). As the DNA sequence they reported is almost identical to ours (we find one extra C at position 396 and one extra G at position 553), the discrepancy in size can not be explained just by strain differences. Little is known about the transcription start and termintion processes in cyanobacteria. A mRNA of 430 b would fit a transcription start at position − 64, as determined by S1-mapping (Reith et al. 1986), and a transcription stop directly after the first putative terminator structure downstream from the stopcodon. Directly in front of the start site the sequence TAGTAT is found, resembling the *E. coli* consensus − 10 region (Rosenberg and Court 1979), but the − 35 region shows no familiar consensus sequence (fig. 12. 'P2'). This presumable *Synechococcus* sp. PCC 7942 promoter has considerable homology with the putative promoter region of *Anabaena* sp. PCC 7120 *petF1* (Alam et al. 1986) (Fig. 12). The corresponding sequence is also conserved in *Anabaena* sp. PCC 7937 (Figs 3 and 12). However, at position −178 of the *Synechococcus* sp. PCC 7942 *petF1* sequence another possible promoter region was found (Fig. 5A and Fig. 12. 'P1'), which again resembles the *Anabaena* sequences. The −35 region of P1 matches better than that of P2, while the −10 region of P2 fits better than that of P1. Assuming that transcription starts from P1 at about −170, transcription termination after the second, more stable, palindromic structure would result in a messenger of 600 b, in agreement with our transcript length of about 630 b. It can not be completely ruled out, however, that the difference in size observed for the *petF1* transcripts corresponds only to different transcription stops.

The main regulation of the cell's ferredoxin content by the availability of Fe^{3+} most probably occurs at the level of translation and/or degradation. The reduction by a factor of 2–3 in the amount of *petF1* transcript found during growth under iron limitation (Fig. 8) is not sufficient to explain the 30-fold decrease in protein content (Figs 6 and 7). The small decrease in transcript level, as related to the amount of total RNA (mainly rRNA), might be caused by a small general reduction in transcript levels. As the cells remain viable and keep growing at an almost unchanged rate under conditions of iron limitation, the cells must be able to compensate for the drastically lowered ferredoxin levels. Flavodoxin is reported to take over the function of ferredoxin under such circumstances (Bothe 1977, Hutber et al. 1977, Ho and Krogmann 1982). However, it has not been established whether flavodoxin can fully replace ferredoxin in vivo. The presence of ferredoxin, even though at a very reduced level, might be essential for certain specific functions.

Indispensability of the petF1 *gene*
The data from the deletion mutagenesis experiments suggest that a cell

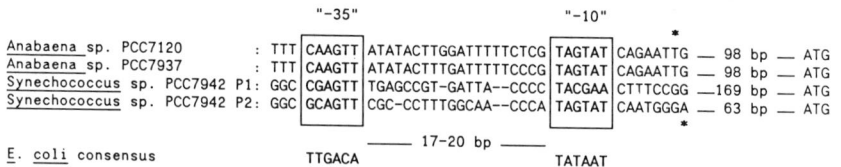

```
                                   "-35"                              "-10"
                                                                       *
Anabaena sp. PCC7120      : TTT CAAGTT ATATACTTGGATTTTTCTCG TAGTAT CAGAATTG — 98 bp — ATG
Anabaena sp. PCC7937      : TTT CAAGTT ATATACTTTGATTTTTCCCG TAGTAT CAGAATTG — 98 bp — ATG
Synechococcus sp. PCC7942 P1: GGC CGAGTT TGAGCCGT-GATTA--CCCC TACGAA CTTTCCGG —169 bp — ATG
Synechococcus sp. PCC7942 P2: GGC GCAGTT CGC-CCTTTGGCAA--CCCA TAGTAT CAATGGGA — 63 bp — ATG
                                                                       *
                                   ——————— 17-20 bp ———————
E. coli consensus                  TTGACA                      TATAAT
```

Fig. 12. Comparison of putative *petF1* promoter sequences. Putative *petF1* promoter regions for *Anabaena* sp. PCC 7120 (Alam et al. 1986), *Anabaena* sp. PCC 7937 (this study, Fig. 3A), and *Synechococcus* sp. PCC 7942 (P1: this study, Fig. 5A; P2: Reith et al. 1986) are aligned for optimal homology. the presumed '−10' and '−35' regions are boxed. The 5′ ends determined for the *petF1* transcripts of *Anabaena* sp. PCC 7120 (Alam et al. 1986) and *Synechococcus* sp. PCC 7942 (Reith et al. 1986) are marked with an asterisk.

lacking an intact *petF1* gene is not viable. Although the *neo* gene was inserted properly in the chromosomal DNA, thereby replacing the *petF1* gene, the kanamycin-resistant transformants still contained an intact *petF1* gene. These results are interpreted as an indication that the *petF1* ferredoxin is essential to the cell, even under conditions of iron-limitation, when the synthesis of flavodoxin is induced. The precise genetic make-up of these transformants is not yet understood. The intact and the impaired gene copy probably reside on different copies of the chromosome, because *Synechococcus* sp. PCC 7942 is thought to contain up to 16 genome equivalents per growing cell (Mann and Carr 1974). In experiments to eliminate the *petF1* gene, those cells seem to be selected in which a partial heterozyzgous situation ensures the cell of both the necessary *petF1* and *neo* gene products.

The mutagenesis technique by homologous recombination has proven to be effective for inactivation of *Synechococcus* sp. PCC 7942 genes: for the *met1* gene (Kuhlemeier et al. 1985; Van der Plas et al., to be published) and for up to two of the three expressed copies of the *psbA* genes (Golden et al. 1986). It seems that the desired mutants can only be produced if the mutagenized function is dispensable, i.e. as long as it can be compensated e.g. by the uptake of methionine added to the growth medium or by the expression of a remaining intact copy in the case of *psbA*. Attempts to inactivate the third *psbA* gene copy (Golden et al. 1986) or a single copy 'recA'-like gene (Borrias et al., to be published) failed and gave results similar to those for the *petF1* gene. Further evidence regarding the dispensability of ferredoxin may come from conditional ferredoxin mutants, in which expression of the *petF1* gene is directed by a controlable promoter.

Acknowledgements

The authors thank Ans van Pelt, Gigi Jonk and Renske Oosterhoff-Teertstra for their help in the sequence determination, Sjef Smeekens for providing the *S. pratensis petF1* cDNA clone, Jenneke Huizingh and Geert de Vrieze for their technical assistance, Evelien Kampert for the construction of *E. coli* PC 2495, Maarten Terlou from the Department for Image Processing and Design for the optical density measurements, Dick Smit and Ronald Leito for help in preparing the figures, Mark Tuyl, René Pieterse and Bart Klein for their contribution to the blotting and mapping experiments, and Johan Hageman for his advice and support in handling computers and software.

540

References

Alam J, Whitaker RA, Krogmann DW and Curtis SE (1986) Isolation and sequence of the gene for ferredoxin I from the cyanobacterium *Anabaena* sp. Strain PCC 7120. J Bacteriol 168: 1265–1271

Böhme H and Schrautemeier B (1987) Comparative characterization of ferredoxins from heterocysts and vegetative cells of *Anabaena variabilis*. Biochim Biophys Acta 891: 1–7

Bothe H (1977) Flavodoxin. In: Trebst A and Avron M (eds) Photosynthesis (Encyclopedia of Plant Physiology 5), pp 216–221. Berlin: Springer

Bradford MM (1976) A rapid and sensitive method for the quantitation of microgram quantities of protein utilizing the principle of protein-dye binding. Anal Biochem 72: 248–254

Buchanan BB (1984) The ferredoxin/thioredoxin system: a key element in the regulatory function of light in photosynthesis. BioScience 34: 378–383

Cathala G, Savouret J, Mendez B, West BL, Karin M, Martial JA and Baxter JD (1983) A method for the isolation of intact, translationally active ribonucleic acid. DNA 2: 329–335

Chan TM, Hermodson MA, Ulrich EL and Markley JL (1983) Nuclear magnetic resonance studies of two-iron-two-sulfur ferredoxins. 2. Determination of the sequence of the *Anabaena variabilis* ferredoxin II, assignment of aromatic resonances in proton spectra and effects of chemical modifications. Biochemistry 22: 5988–5995

Cozens AL and Walker JE (1987) The organization and sequence of the genes for ATP synthase subunits in the Cyanobacterium Synechococcus 6301; support for an endosymbiotic origin of chloroplasts. J Mol Biol 194: 359–383

Csiszàr K, Houmard J, Damerval T and Tandeau de Marsac N (1987) Transcriptional analysis of the cyanobacterial *gvpABC* operon in differentiated cells: occurrence of an antisense RNA complementary to three overlapping transripts. Gene 60: 29–37

Curtis SE and Haselkorn R (1983) Isolation and sequence of the gene for the large subunit of ribulose-1,5-bisphosphate carboxylase from the cyanobacterium *Anabaena* 7120. Proc Natl Acad Sci USA 80: 1835–1839

Dente L, Cesareni G and Cortese R (1983) pEMBL: a new family of single stranded plasmids. Nucleic Acids Res 11: 1645–1655

Dugaiczyk A, Boyer HW and Goodman HM (1975) Ligation of *Eco*RI endonuclease-generated DNA fragments into linear and circular structures. J Mol Biol 96: 171–184

Duyvesteyn MGC, Korsuize J, De Waard A, Vonshak A and Wolk CP (1983) Sequence-specific endonucleases in strains of *Anabaena* and *Nostoc*. Arch Microbiol 134: 276–281

Dzelzkalns VA, Owens GC and Bogorad L (1984) Chloroplast promoter driven expression of the chloramphenicol acetyl transferase gene in a cyanobacterium. Nucleic Acids Res 12: 8917–8925

Feinberg AP and Vogelstein B (1983) A technique for radiolabelling DNA restriction endonuclease fragments to high specific activity. Anal Biochem 132: 6–13

Golden SS, Brusslan J and Haselkorn R (1986) Expression of a family of *psbA* genes encoding a photosystem II polypeptide in the cyanobacterium *Anacystis nidulans* R2. EMBO J 5: 2789–2798

Hall DO and Rao KK (1977) Ferredoxin. In: Trebst A and Avron M (eds) Photosynthesis (Encyclopedia of Plant Physiology 5), pp 206–216. Berlin: Springer

Ho KK, Ulrich EL, Krogmann DW and Gomez-Lojero C (1979) Isolation of photosynthetic catalysts from cyanobacteria. Biochim Biophys Acta 545: 236–248

Ho KK and Krogmann DW (1982) Photosynthesis. In: Carr NG and Whitton BA (eds) The Biology of Cyanobacteria, pp 191–214. Oxford: Blackwell

Hutber GN, Hutson KG and Rogers LJ (1977) Effect of iron deficiency on levels of two ferredoxins and flavodoxin in a cyanobacterium. FEMS Microbiol Lett 1: 193–196

Ish-Horowicz D and Burke JF (1981) Rapid and efficient cosmid cloning. Nucleic Acids Res 9: 2989–2998

Kuhlemeier CJ, Thomas AAM, Van der Ende A, Van Leen RW, Borrias WE, Van den Hondel CAMJJ and Van Arkel GA (1983) A host-vector system for gene cloning in the cyanobacterium *Anacystis nidulans* R2. Plasmid 10: 156–163

Kuhlemeier CJ, Hardon EM, Van Arkel GA and Van de Vate C (1985) Self-cloning in the cyanobacterium *Anacystis nidulans* R2: fate of a cloned gene after reintroduction. Plasmid 14: 200–208

Laemmli UK (1970) Cleavage of structural proteins during the assembly of the head of bacteriophage T4. Nature 277: 680–685

Lizardi PM, Binder R and Short SA (1984) Preparative isolation of DNA and biologically active mRNA from diethylaminoethyl membrane. Gene Anal Techn 1: 33–39

Maniatis T, Fritsch EF and Sambrook J (1982) Molecular Cloning: A Laboratory Manual. Cold Spring Harbor, NY: Cold Spring Harbor Laboratory

Mann N and Carr NG (1974) Control of macromolecular composition and cell division in the blue-green alga *Anacystis nidulans*. J Gen Microbiol 83: 399–405

Matsubara H and Hase T (1983) Phylogenetic consideration of ferredoxin sequences in plants, particularly algae. In: Jensen U and Fairbrothers DE (eds) Proteins and nucleic acids in plant systematics, pp 168–181. Berlin: Springer

Mazur BJ, Rice D and Haselkorn R (1980) Identification of blue-green algal nitrogen fixation genes by using heterologous DNA hybridization probes. Proc Natl Acad Sci USA 77: 186–190

Meinkoth J and Wahl G (1984) Hybridisation of nucleic acids immobilized on solid supports. Anal Biochem 138: 267–284

Reith ME, Laudenbach DE and Straus NA (1986) Isolation and nucleotide sequence analysis of the ferredoxin I gene from the cyanobacterium *Anacystis nidulans* R2. J Bacteriol 168: 1319–1324

Rippka R, Deruelles J, Waterbury JB, Herdman M and Stanier RY (1979) Generic assignments, strain histories and properties of pure cultures of cyanobacteria. J Gen Microbiol 111: 1–61

Rosenberg M and Court D (1979) Regulatory sequences involved in the promotion and termination of RNA transcription. Ann Rev Genet 13: 319–353

Sakihama N and Shin M (1987) Evidence from high-pressure liquid chromatography for the existence of two ferredoxins in plants. Arch Biochem Biophys 256: 430–434

Sanger F, Coulson AR, Barrell BG, Smith AJH and Roe BA (1980). Cloning in single-stranded bacteriophage as an aid to rapid DNA sequencing. J Mol Biol 143: 161–178

Schrautemeier B and Böhme H (1985) A distinct ferredoxin for nitrogen fixation isolated from heterocysts of the cyanobacterium *Anabaena variabilis*. FEBS Lett 184: 304–308

Smeekens S, Van Binsbergen J and Weisbeek P (1985) The plant ferredoxin precursor: nucleotide sequence of a full length cDNA clone. Nucleic Acids Res 113: 3179–3194

Stephens RM (1985) A sequencers sequence analysis package for the IBM PC. Gene Anal Techn 2: 67–75

Tandeau de Marsac N and Houmard J (1987) Advances in cyanobacterial molecular genetics. In: Fay P and Van Baalen C (eds) The Cyanobacteria, pp 251–302. Amsterdam: Elsevier

Thomas PS (1983) Hybridization of denatured RNA transferred or dotted to nitrocellulose paper. Methods Enzymol 100: 255–266

Tinoco J, Borer PN, Dengler B, Levine MD, Uhlenbeck OC, Crothers DM and Cralla J (1973) Improved estimation of secondary structure in ribonucleic acids. Nature (London) New Biol 246: 40–41

Tomioka N and Sugiura M (1983) The complete nucleotide sequence of a 16S ribosomal RNA gene from a blue-green alga, *Anacystis nidulans*. Mol Gen Genet 191: 46–50

Towbin H, Staehelin T and Gordon J (1979) Electrophoretic transfer of proteins from polyacrylamide gels to nitrocellulose sheets: procedure and some applications. Proc Natl Acad Sci USA 76: 4350–4354

Van Bergen en Henegouwen PMP and Leunissen JLM (1986) Controlled growth of colloidal gold particles and implications for labelling efficiency. Histochem 85: 81–87

Van den Hondel CAMJJ, Verbeek S, Van der Ende A, Weisbeek P, Borrias WE and Van Arkel GA (1980) Introduction of transposon Tn901 into a plasmid of *Anacystis nidulans*: preparation for cloning in cyanobacteria. Proc Natl Acad Sci USA 77: 1570–1574

Van der Plas J, De Groot RP, Weisbeek PJ and Van Arkel GA (1986a) Coding sequence of a ferredoxin gene from *Anabaena variabilis* ATCC 29413. Nucleic Acids Res 14: 7803

Van der Plas J, De Groot RP, Woortman MR, Weisbeek PJ and Van Arkel GA (1986b) Coding sequence of a ferredoxin gene from *Anacystis nidulans* R2 (*Synechococcus* PCC 7942). Nucleic Acids Res 14: 7804

Vieira J and Messing J (1982) The pUC plasmids, an M13mp7-derived system for insertion mutagenesis and sequencing with synthetic universal primers. Gene 19: 259–268

Wolk CP and Schafer PW (1976) Heterotrophic micro- and macrocultures of a nitrogen-fixing cyanobacterium. Arch Microbiol 110: 144–147

Govindjee et al. (eds), Molecular Biology of Photosynthesis: 543–564
© 1988 Kluwer Academic Publishers

Minireview

Structure, organization and expression of cyanobacterial ATP synthase genes

STEPHANIE E. CURTIS

Department of Genetics, Box 7614, North Carolina State University, Raleigh, NC 27695-7614, USA

Received 12 October 1987; accepted 25 January 1988

Key words: amino acid sequence, *Anabaena*, ATP synthase, cyanobacteria, DNA sequence, *Synechococcus*

Abstract. The genes encoding the nine polypeptides of the ATP synthase from *Synechococcus* sp. PCC 6301, a unicellular cyanobacterium, and *Anabaena* sp. PCC 7120, a filamentous cyanobacterium, have recently been isolated and their sequences determined. These represent the first such sequences available from procaryotic organisms that perform oxygenic photosynthesis. Similar to the organization in chloroplasts, the ATP synthase genes of both cyanobacteria are arranged in two gene clusters which are not closely linked in the chromosome. Three of the genes located in one cluster in cyanobacteria, however, are localized in the nuclear rather than the chloroplast genomes of plants. The cyanobacterial ATP synthase genes are ordered in the same manner as those in the single gene cluster of *Escherichia coli*. Cyanobacteria contain an additional gene denoted *atpG* which appears to be a duplicated and diverged form of the *atpF* gene. The larger cyanobacterial cluster, *atp* 1, is comprised of eight ATP synthase subunit genes arranged in the order *atpI–atpH–atpG–atpF–atpD–atpA–atpC*. An overlap between the *atpF* and *atpD* gene coding regions observed in *Anabaena* sp. PCC 7120 is absent in both *Synechococcus* sp. PCC 6301 and *E. coli*. The second cluster of genes, *atp* 2, contains the remaining two ATP synthase genes in the order *atpB–atpE*. Unlike the situation in many chloroplast genomes, this gene pair does not overlap in either cyanobacterial species. In *Anabaena* sp. PCC 7120, *atp* 1 and *atp* 2 each comprise an operon and the transcription initiation sites for each gene cluster have been identified. The cyanobacterial ATP synthase subunits are much more closely related in sequence to the equivalent polypeptides from chloroplasts than they are to those of *E. coli*. The similarity in chloroplast and cyanobacterial ATP synthase subunit sequences and gene oreganization argue strongly for an endosymbiotic origin for plant chloroplasts.

Introduction

The ATP synthase is a ubiquitous membrane enzyme complex present in both procaryotic and eucaryotic cells that functions to couple ATP synthesis to proton translocation across a membrane. The enzyme complex is localized in bacterial cytoplasmic membranes, in the thylakoid membranes of chloroplasts and in the inner membranes of mitochondria (reviewed in Futai and Kanazawa 1983, Hatefi 1985, Merchant and Selman 1985). The multisubunit nature and widespread occurrence of the ATP synthase render it an excellent subject for evolutionary studies. Examination of the genes encoding the ATP synthase subunits present an opportunity to study the evolution of a set of related genes across a broad evolutionary scale.

The endosymbiont theory suggests that the cyanobacteria are related to the evolutionary progenitors of plant chloroplasts (Margulis 1981). Recently, sequences of the genes encoding all of the ATP synthase subunit genes from two cyanobacterial species have been determined and provide a basis for examining the relatedness of ATP synthase genes from cyanobacteria and chloroplasts. This review presents a discussion of the structure and organization of the cyanobacterial ATP synthase genes and a comparison of these genes with their homologs in *E. coli* and plant chloroplasts. For a more comprehensive discussion which includes ATP synthase genes in other eubacteria and mitochondria, readers are referred to a review by Walker and Tybulewicz (1985).

ATP synthase structure and function

The ATP synthases of all species are similarly structured, consisting of two major components denoted F_1 and F_0 (Fig. 1; McCarty and Hammes 1987). The F_0 component forms a transmembrane proton channel while the F_1 is membrane extrinsic and functions to couple ATP synthesis to proton trans-

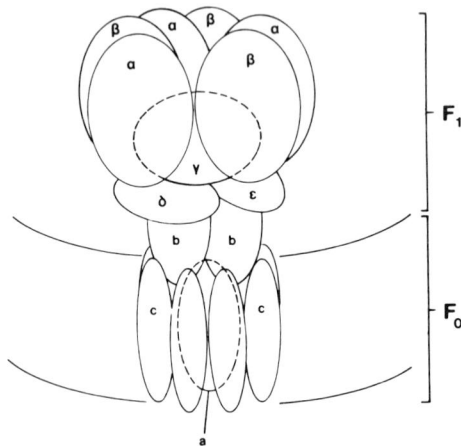

Fig. 1. Model of *E. coli* ATP synthase based on the model presented by Kanazawa and Futai 1983. The chloroplast and cyanobacterial ATP synthases have a similar structure but may have one *b* and one *b'* subunit rather than two identical *b* subunits as shown.

location across the membrane. The enzyme is also capable of the reverse reaction, i.e. ATP hydrolysis.

Five polypeptides denoted by the Greek letters α,β,γ, δ, and ε form the F_1 portion of the ATP synthase in all species examined. The F_1 subunits of E. coli are assembled with a stoichiometry of $\alpha_3\beta_3\gamma_1\delta_1\varepsilon_1$ (Foster and Fillingame 1982). The composition of the chloroplast F_1 (CF_1) appears to be similar to that of E. coli, although the stoichiometry of the δ polyptide has not been unequivocally determined (Merchant and Selman 1985).

The F_1 subunits have been assigned functions based on biochemical and genetic studies in E. coli and chloroplasts (for reviews, see Futai and Kanazawa 1983 and Merchant and Selman 1985). Of the five F_1 subunits, β and α are the most highly conserved and both contain nucleotide binding sites. The β (and possibly the α) subunits contain the active site(s) of the complex. The δ subunit is required for functional binding of the F_1 to the F_0. The roles of the γ and ε subunits are less clearly defined, but in chloroplasts both are believed to be involved in the regulation of ATPase activity (McCarty and Hammes 1987).

Three components (denoted a, b and c in a stoichiometry of $a_1b_2c_{10}$) form the F_0 portion of the E. coli ATP synthase (Foster and Fillingame 1982). The chloroplast F_0 (CF_0) is more complex and has been more difficult to characterize with regard to subunit composition and stoichiometry than its E. coli counterpart. Analysis of ATP synthase subunits by gel electrophoresis in the presence of urea (Pick and Racker 1979, Westhoff et al. 1985), amino acid sequencing of CF_0 subunits (Bernzborn et al. 1987) and gene sequence analysis (Cozens et al. 1987, Hennig and Herrmann 1986) have established that CF_0 is composed of four subunits, three of which are equivalent to those of E. coli F_0. The chloroplast subunits, designated I–IV, and their E. coli subunit equivalents are listed in Table 1. For simplicity, the E. coli subunit nomenclature is used in this article. The fourth CF_0 subunit, which lacks an equivalent in E. coli, is deisgnated b′ (Walker and Tybulewicz 1985).

Information on the functions of the F_0 subunits is derived from studies in E. coli (Futai and Kanazawa 1983). The c subunit, which is extremely hydrophobic and highly conserved among different organisms, is believed to form the transmembrane proton channel of the F_0. The a and b subunits appear to be involved in the binding of F_0 to F_1.

Although not as extensively studied, cyanobacterial ATP synthases are clearly similar in structure and function to those of chloroplasts and E. coli (Lubberding et al. 1983, Hicks and Yocum 1986a, b). The five Spirulina platensis coupling factor subunits have been shown to be similar in molecular mass to those of chloroplast CF_1 (Hicks and Yocum 1986a) and such information has been confirmed by DNA sequence analysis of cyanobac-

Table 1. Nomenclature for chloroplast ATP synthase subunits and genes and *E. coli* subunit equivalents

E. coli subunit designations	Chloroplast subunit designations	Gene locus designations for chloroplast and cyanobacterial subunits[a]
α	α	*atpA*
β	β	*atpB*
γ	γ	*atpC*
δ	δ	*atpD*
ε	ε	*atpE*
a	IV	*atpI*
b	I	*atpF*
NE[b]	II[c]	*atpG*
c	III	*atpH*

[a]Hallick and Bottomley 1983.
[b]NE = no subunit equivalent.
[c]Subunit denoted *b′* in cyanobacteria (Cozens and Walker 1987).

terial F_1 genes. DNA sequence analysis of cyanobacterial ATP synthase subunit genes (see later sections) has also established that cyanobacterial F_0 is most likely composed of four subunits equivalent to those of the chloroplast enzyme.

Isolation of ATP synthase genes from cyanobacteria

The DNA sequences encoding the ATP synthase subunit genes from two cyanobacterial species, *Synechococcus* sp. PCC 6301 (Cozens and Walker 1987) and *Anabaena* sp. PCC 7120 (Curtis 1987, McCarn et al. 1988) have recently been isolated and characterized. *Synechococcus* sp. PCC 6301 (formerly denoted *Anacystis nidulans*) is a unicellular cyanobacterium, while *Anabaena* sp. PCC 7120 is filamentous and capable of aerobic nitrogen fixation. As is characteristic of cyanobacteria, both organisms perform oxygenic photosynthesis virtually identical to that of plant chloroplasts. Because of the strong conservation of particular ATP synthase subunits, it has been possible to employ heterologous DNA probes from both purple bacteria (Cozens and Walker 1987) and chloroplasts (Cozens and Walker 1987, Curtis 1987 and McCarn et al. 1988) as a means of identifying and isolating equivalent genes in cyanobacteria. In one case the reverse was also performed – a cyanobacterial gene was used to identify the corresponding pea chloroplast gene (Cozens et al. 1986).

The DNA sequences of all of the ATP synthase genes from both *Anabaena* sp. PCC 7120 (Curtis 1987, McCarn et al. 1988) and *Synechococcus* sp. PCC

6301 (Cozens and Walker 1987) have been determined. As was noted above, little information on cyanobacterial ATP synthases is available, including amino acid composition or sequence. Accordingly, gene identifications are based on several criteria: 1. similarity of the primary sequences of the gene translation products with those from other organisms; 2. similarity of the predicted secondary strucures of gene translation products with those from other organisms; and 3. gene locations relative to the arrangement in other organisms.

The ATP synthase genes of *Anabaena* sp. PCC 7120 are single copy in the genome. This suggests that ATP synthase complexes assembled from the same subunits are utilized for ATP synthesis during both photosynthesis and respiration. Photosynthetic electron transport in cyanobacteria is believed to have carriers in common with respiratory electron transport, and there is evidence that photosynthetic membranes may contain sites for respiratory activity (Wolk 1980). It is thus plausible that the same ATP synthase complexes generate ATP during both respiration and photosynthesis in cyanobacteria, similar to the situation in the purple nonsulfur bacteria (Baccarini-Melandri and Melandri 1978).

Organization of ATP synthase genes

The arrangement of ATP synthase genes in *Anabaena* sp. PCC 7120 and *Synechococcus* sp. PCC 6301 is identical and has features in common with the organization of these genes in both *E. coli* and chloroplasts. The ATP synthase genes of *E. coli* (denoted *unc* or *atp*) are clustered on the bacterial chromosome (Futai and Kanazawa 1983; Fig. 2). Preceding the eight ATP synthase subunit genes and in the same operon is an additional gene denoted *unc*I, which encodes a hydrophobic protein of unknown function. Cyanobacteria have genes equivalent to all of those in the *unc* operon (including *unc*I) but differ from *E. coli* by having an additional subunit gene denoted *atp*G (Walker and Tybulewicz 1985). The cyanobacerial ATP synthase genes are found in the same order as those in *E. coli* but are split into two loci which are not closely linked in the chromosome. A large cluster of genes denoted *atp* 1 (Cozens and Walker 1987), contains eight genes in the order *atpI–atpH–atpG–atpF–atpD–atpA–atpC* and the second cluster, denoted *atp* 2, consists of the remaining two genes in the order *atpB–atpE* (Fig. 2).

As is common for multi-subunit complexes of organelles, the ATP synthase subunits of chloroplasts are encoded in both the chloroplast and nuclear genomes of plants. The ATP synthase genes of most but not all (e.g. *Chlamydomonas reinhardii*; Woessner et al. 1987) chloroplast species studied

I	a	c	b	δ		α		γ		β		ε	

E. coli

I	a	c	b'	b	δ		α		γ		β		ε	

cyanobacteria

a		c		b		α		β		ε	

chloroplasts

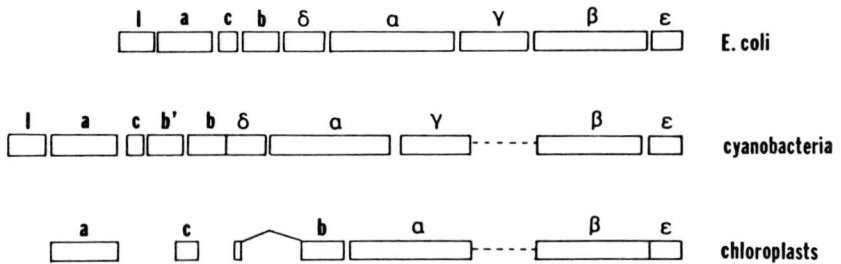

Fig. 2. Arrangements of genes encoding ATP synthase subunits in *E. coli*, chloroplasts and cyanobacteria. Polypeptide products (*E. coli* nomenclature) are given above the genes (boxes). The two gene clusters in cyanobacteria and chloroplasts are many kilobase pairs apart in the genome. The *atpF* gene of chloroplasts contains an intron (denoted by a split gene) and cyanobacteria have a duplicated *b* gene (*atpF*) denoted *b'* (*atpG*) (Walker and Tybulewicz 1985). Not all chloroplast genomes have the organization shown.

are found in two clusters in the chloroplast genome (Palmer 1985). Analogous to the arrangement in cyanobacteria, *atpB* and *atpE* are closely linked and map some distance from the remaining ATP synthase genes (Fig. 2). The second array of genes is ordered in the same manner as those of the *atp* 1 locus of cyanobacteria; however, the genes encoding the *b'* (*atpG*), δ (*atpD*) and γ (*atpC*) subunits (and perhaps *uncI*?) are absent and localized in the nuclear genome. Thus the organization of cyanobacterial ATP synthase genes is truly intermediate to that observed in *E. coli* and most chloroplast species: relative to *E. coli*, the genes are found in two clusters, one of which contains an additional gene; relative to chloroplasts, the large cyanobacterial gene cluster contains additional ATP synthase genes which have a nuclear location in plants.

Additional information on the evolution of photosynthetic organelles is derived from studies of the ATP synthase genes of *Cyanophora paradoxa*. *C. paradoxa* is a eucaryotic protozoan with photosynthetic organelles, termed cyanelles, that retain a peptidoglycan layer and strong ultrastructural resemblances to cyanobacteria but have a genome size similar to that of chloroplasts (Wasmann et al. 1987). The ATPase genes of *C. paradoxa* map to two regions of the cyanelle genome (Bohnert et al. 1985, Ko et al. 1985, Lambert et al. 1985). Recent DNA sequence analysis of the large cluster has shown genes to be present in the order *atpH–atpG–atpF–atpD–atpA*; there is no evidence that the *atpI* and *atpC* genes reside on the cyanelle genome and these are presumably nuclear in location (Bryant 1986; D. Bryant, personal communication). The *C. paradoxa* arrangement is thus intermediate between that of cyanobacteria and green plant chloroplasts and provides additional clues to photosynthetic organelle lineages.

Comparison of two cyanobacterial ATP synthase gene clusters

Although very similar, loci encoding the ATP synthases of *Synechococcus* sp. PCC 6301 and *Anabaena* sp. PCC 7120, display distinctive DNA sequence features. One striking difference between the two cyanobacteria occurs in the coding sequences of the *atpF* and *atpD* genes: in *Anabaena* sp. PCC 7120 these genes overlap while in *Synechococcus* sp. PCC 6301 they are separated by a spacer of twelve base pairs. The nature of the *Anabaena* sp. PCC 7120 *atpF/atpD* gene overlap is similar to that observed for the *atpB* and *atpE* genes of some chloroplast species (Krebbers et al. 1982, Zurawski et al. 1982, Howe et al. 1985) in which the last two codons of the upstream gene overlap with the first two codons of the downstream gene (Fig. 3). Overlap between the *atpB* and *atpE* is not found in either cyanobacterial species. Interestingly, although the *atpF/atpD* gene overlap is absent in *Synechococcus* sp. PCC 6301, it is also observed in the cyanelle genome of *C. paradoxa* (D. Bryant, personal communication).

A distinctive feature of the chloroplast *atpF* genes thus far studied (Bird et al. 1985, Hennig and Herrmann 1986, Ohyama et al. 1986, Shinozaki et al. 1986) is the presence of an intron in the coding region. The *atpF* genes of cyanobacteria and *C. paradoxa* cyanelles lack an intron (D. Bryant, personal communication). Although frequently observed in chloroplast genes, there is no evidence as yet that any cyanobacterial genes are interrupted by introns.

In general, the ATP synthase genes of *Synechococcus* sp. PCC 6301 are much more closely spaced on the chromosome than are those of *Anabaena* sp. PCC 7120. The *Synechococcus* sp. PCC 6301 spacer lengths, on the order of ten to fifty base pairs, are more similar to those observed between *E. coli* ATP synthase genes while those of *Anabaena* sp. PCC 7120, most of which are several hundred base pairs in length, are similar to spacer lengths

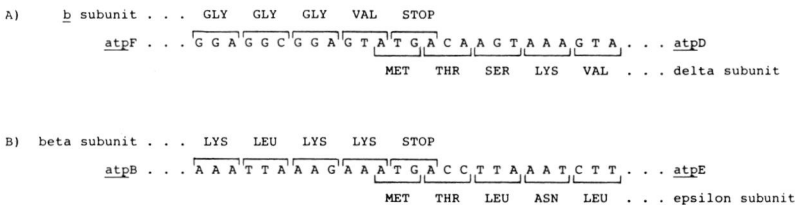

```
A)      b subunit . . .  GLY   GLY   GLY   VAL   STOP
        atpF . . . G G A  G G C  G G A  G T A  T G A  C A A G T A A A G T A . . . atpD
                                         MET   THR   SER   LYS   VAL  . . . delta subunit


B)  beta subunit . . .  LYS   LEU   LYS   LYS   STOP
        atpB . . . A A A  T T A  A A G  A A A  T G A  C C T T A A A T C T T . . . atpE
                                         MET   THR   LEU   ASN   LEU  . . . epsilon subunit
```

Fig. 3. Nucleotide and amino acid sequences showing the overlap between A) the *atpF* and *atpD* genes of the cyanobacterium *Anabaena* sp. PCC 7120 (McCarn et al.1988) and B) the *atpB* and *atpE* genes of spinach chloroplasts (Zurawski et al. 1982). The *atpB* and *atpE* gene overlap is not observed in all chloroplast genomes and the *atpD/atpF* gene overlap is not observed in the cyanobacterium *Synechococcus* sp. PCC 6301.

observed between some chloroplast ATP synthase genes. Many of the *Anabaena* sp. PCC 7120 intergenic spacer regions contain numerous clusters of short direct repeats. Whether these repeated elements have functional significance or merely represent duplication events that lengthen the spacer regions relative to those in *Synechococcus* sp. PCC 6301 is not clear.

The *Synechococcus* sp. PCC 6301 and *Anabaena* sp. PCC 7120 ATP synthase gene clusters are followed by DNA sequences which can potentially form stable hairpin structures (i.e. inverted repeats). These sequences share features with *E. coli* transcription termination signals and may serve an analogous function. Alternatively, these sequences could enhance the stability of the mRNA. In both *Anabaena* sp. PCC 7120 and *Synechococcus* sp. PCC 6301 an additional inverted repeat is present between the *atpB* and *atpE*. Interestingly, the inverted repeat between the *Anabaena* sp. PCC 7120 *atpB* and *atpE* consists of two sets of direct repeats which are complementary to each other; the same inverted repeat sequence is also found between the *Anabaena* sp. PCC 7120 *atpA* and *atpC* genes in the same relative position in the spacer.

Transcription of cyanobacterial ATP synthase genes

The ATP synthase gene clusters of *Anabaena* sp. PCC 7120 have each been shown to be transcribed into a single polycistronic messenger RNA, as is the gene cluster of *E. coli* (Futai and Kanazawa 1983). Comparable data for *Synechococcus* sp. PCC 6301 is not available although presumably *atp* 1 and *atp* 2 also constitute operons in this organism. There is no evidence in *Anabaena* sp. PCC 7120 or *E. coli* that the polycistronic mRNAs are processed to generate suboperon transcripts that contain coding sequences for one or more genes. This implies that the stoichiometry of subunits in the assembled ATP synthase complex is achieved by a mechanism(s) other than regulation at the level of gene transcription. There is evidence in *E. coli* that translational control is responsible for the subunit stoichiometry. Certain genes in the operon, whose products are present in lower amounts in the ATP synthase complex, are preceded by sequences capable of forming stable hairpin structures (Brusilow et al. 1982). If the hairpin structures were formed, the ribosome binding site and translational start for the downstream gene would be folded up in mRNA secondary structure, thereby inhibiting efficient translation. There is additional evidence that the *atpH* gene of *E. coli*, whose product is present in high copy number in the complex, is preceded by a translational enhancer that mediates very efficient translation of this subunit (McCarthy et al. 1985). Although the *atpE* genes of *Anabaena* sp. PCC 7120 and *Synechococcus* sp. PCC 6301 and the *atpC* gene of *Anabaena* sp. PCC 7120 are preceded by inverted repeats, these are not located in the same relative position as those observed in *E. coli*, i.e. overlapping translational signals for the downstream gene. Thus the significance of these sequences, and whether the hairpin structures actually exist in the mRNA in vivo, is open to question. Since there is no evidence for suboperon-length transcripts, it is unlikely that these sequences represent sites of mRNA processing or transcription termination. Sequences similar to the translational enhancer region identified for the *E. coli atpH* gene (McCarthy et al. 1985) are not observed in either *Synechococcus* sp. PCC 6301 or *Anabaena* sp. PCC 7120.

The 5' endpoints of the *Anabaena* sp. PCC 7120 *atp* 1 and *atp* 2 transcripts have been identified. The *atp* 1 transcript has a single endpoint which maps 140 base pairs upstream from the first gene of the operon (McCarn et al. 1988). Two RNA endpoints are identified at 65 and 200 base pairs upstream from *atp* 2 (Curtis, 1987). In *E. coli*, conserved sequences which constitute the promoter are observed at ten ('–10') and thirty-five ('–35') base pairs upstream from the transcription initiation sites (Reznikoff et al. 1985). A sequence similar to the '–10' consensus sequence of *E. coli* promoters is found upstream from each of the *Anabaena* sp. PCC 7120 transcript endpoints, but an equivalent *E. coli* '–35' consensus sequence is absent (Fig. 4).

The transcription initiation sites for the *Synechococcus* sp. PCC 6301 ATP synthase gene clusters have not been identified, but sequences upstream from *atp* 1 and *atp* 2 that resemble *E. coli* '–10' consensus sequences have

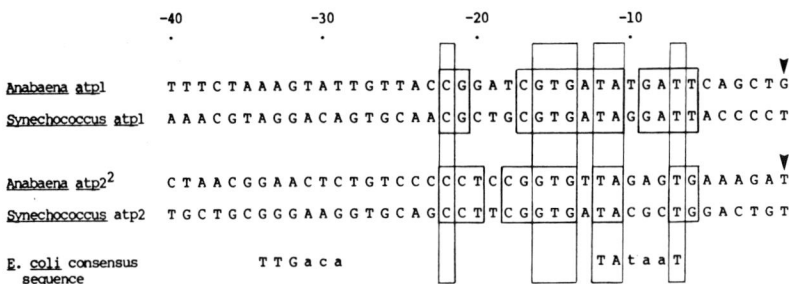

Fig. 4. Comparison of sequences upstream from *Synechococcus* sp. PCC 6301 (Cozens and Walker 1987) and *Anabaena* sp. PCC 7120 (Curtis 1987, McCarn et al. 1988) ATP synthase gene operons (atp1 and atp2). Arrowheads indicate the transcription initiation sites determined for the *Anabaena* sp. PCC 7120 operon (Curtis 1987, McCarn et al. 1988). Numbers indicate the number of nucleotides upstream from the transcription start sites in *Anabaena* sp. PCC 7120. Conserved sequences around '–10' are boxed. The lower line shows the consensus sequences for E. coli promoters (Reznikoff 1985).

been noted (Cozens and Walker 1987). Comparison of these regions with those upstream from the transcription initiation sites of *Anabaena* sp. PCC 7120 *atp* 1 and *atp* 2 indicates a block of homology around '–10' that may be important for transcription (Fig. 4). In similar comparisons with other *Anabaena* sp. PCC 7120 photosynthesis genes (Curtis and Haselkorn 1984, Nierzwicki-Bauer et al. 1984, Alam et al. 1986), no sequence at the '–35' position or elsewhere is conserved. Some cyanobacterial genes (e.g. *psbA* genes in *Synechococcus* sp. PCC 7942; Golden et al. 1987) have been shown to have typical *E. coli* promotor sequences upstream from the transcription initiation sites. It is clear, however, that at least in *Anabaena* sp. PCC 7120, the sequences which constitute promoters remain to be determined.

Amino acid sequences of ATP synthase subunits

Characterization of the cyanobacterial ATP synthase genes has provided information on the subunit composition of cyanobacterial ATP synthases as well as the primary amino acid sequences of the various polypeptide constituents. Comparison of the protein sequences from the cyanobacterial subunits with one another and with those from other organisms yields information on regions of the proteins which are conserved and presumably important for subunit structure and function.

The individual subunits from *Anabaena* sp. PCC 7120 and *Synechococcus* sp. PCC 6301, as would be expected, are more similar to each other than they are to the corresponding subunits from either *E. coli* or chloroplasts. It is apparent from cyanobacterial subunit comparisons, however that the subunits have undergone variable rates of evolution and that several of the polypeptides (e.g. δ, b, b') have diverged to a considerable degree (Table 2). The degree and variability of subunit divergence can be seen in a qualitative manner in Figs. 5 and 6 in which the available protein sequences from cyanobacteria, *E. coli* and chloroplasts are compared in a manner which emphasizes differences in amino acid sequence.

Among the F_1 subunits, α and β are very highly conserved, probably

Table 2. Percent identity between ATP synthase subunits from *Anabaena* sp. PCC 7120 and *Synechococcus* sp. PCC 6301.[a]

α	β	γ	δ	ε	a	b	b'	c	*uncI* product[b]
78	87	80	49	70	80	41	58	96	52

[a]Percent identical amino acid residues after alignment for maximum similarity.
[b]Translation product of cyanobacterial genes equivalent to *E. coli uncI* gene.

A) ALPHA SUBUNIT

```
                                                                           50
wheat chloroplast  - M A T L R V D E I H K I V R E L I E Q Y N R K V G I E N I G R V V Q V G D G I A R I I G L G E I M
Anabaena 7120      M S I S I   P       S S   I Q Q Q           D Q E   K V A   V   T   L                     Y       E K A
Synechococcus 6301   V S I   P       S S   I   Q Q       S Q D   K V     V   T   L             Y       Q Q V
E. coli            - Q   N S T       S E L I K Q R   A   F   V V S E A H   E   T I   S   S       V I   H     A D C


                                                                          100
wheat chloroplast  S G E L V E F A E G T R G I A L N L E S K N V G I V L M G D G L M I Q E G S P V K A T G R I P Q I P
Anabaena 7120      A     L     E D     V       Q         E D       A           R E         T   T           A       G
Synechococcus 6301               E D     T             E D       A           R N         T         K   A
E. coli            Q     M I S L P G N R Y A           R D S     A   V       P Y A D L A     M K         C       L E V


                                                                          150
wheat chloroplast  V S F A Y L G R V V N A L A K P I D G K G E I I A S E S R L I E S P A P S I I S R R S V Y E P L Q T
Anabaena 7120      G     L I       D     G R A         D K                   G A         H       M
Synechococcus 6301 G P   V         S P   G A L           A   T   N           G A         H       M
E. coli            G R G L         T   G A         P L D H D G F S A V   A I       G V   E     Q       D Q   V


                                                                          200
wheat chloroplast  G L I A I D S M I P I G R G Q R E L I I G D R Q T G K T A V A T D T I L N Q K G Q G V I C V Y V A I
Anabaena 7120        I T                                           T   I       I           E D   V
Synechococcus 6301   I T     V A                                   I   I                   E D
E. coli            Y K   V                                         L   I   A   I     R D S   I K   I


                                                                          250
wheat chloroplast  G Q R A S S V A Q V V T T F H E E G A M E Y T I V V A E M A D S P A T L Q Y L A P Y T G A A L A E Y
Anabaena 7120          K       T       N     Q L Q K       D     V       A G   S E           F             T I
Synechococcus 6301      K           N I I E V L R   R     L D     V       A N   S E       S     A         R M P V   L M G


                                                                          300
wheat chloroplast  F M Y R E R H T L I I Y D D L S K Q A Q A Y R Q M S L L L R R P P G R E A Y P G D V F Y L H S R L L
Anabaena 7120          K G K A     V                                       G                     I
Synechococcus 6301     K G K A     V           T                 V       I                   F
E. coli            R D   G E D A                             V         I                     F


                                                                          350
wheat chloroplast  E R A A K L N S L L - - - - - - G E - - - - - G S M T A L P I V E T Q S G D V S A Y I P T N V I S I
Anabaena 7120                S D E             K               I     A
Synechococcus 6301              S D A             G             V I     A               P V
E. coli                R V   A E Y V E A P T K     V K G K T     L       I       A         P V


                                                                          400
wheat chloroplast  T D G Q I P L S A D L F N A G I R P A I N V G I S V S R V G S A A Q I K A M K Q V A G K S K L E L A
Anabaena 7120                S                       V   P                     T     K       I
Synechococcus 6301            S             S L               V   P                 T   I K I     T L
E. coli                E T N                       V   P                 G         T Y I     M L S   G I R T A


                                                                          450
wheat chloroplast  Q F A E L Q A F A Q P A S A L D K T S Q N Q L A R G R R L R E L L K Q S Q A N P L P V E E Q I A T I
Anabaena 7120          D D                   D       A T   D           Q               N Q       S A     V I L
Synechococcus 6301     D     A               D       A T             Q         V T       K   Y A   M S   A Q   S L V L
E. coli            Y R     A       S         D       D A T R K     D H   Q   V T           K Y A M S A Q   S L V L


                                                                          500
wheat chloroplast  Y T G T R G Y L D S L E I E Q V N K P L D E L R K H L K D T K - P Q F Q E I I S S S K T F T E Q A E
Anabaena 7120          A   I N             D I P V D K   T T   T K D       D Y     S G V N   Y       D V Q   K K A L G D D E E K
Synechococcus 6301     A   V K   L I       E I P V N     T A   V S         S Y     T S   -   E   I   K V Q         Q L D D A
E. coli            F A A E             A D V   L S K I G S   E A A     L A Y V D R D H A     L M         N G - T G G Y N D E I


wheat chloroplast  I L L K E A I Q E Q L E R F S L Q - - - - - -
Anabaena 7120      A   -   A   L E D Y K K T   K A T A
Synechococcus 6301 A             L L K E A I A E V K K N I L A A V
E. coli            G K     G -   L - - - D S   K A T Q S W
```

B) BETA SUBUNIT

```
                                                                              50
spinach chloroplast  M R I N P T T S D P G V S T L E K K N L G R I A Q I I G P V L N V A F P P G K M P N I Y N A L I V K
Anabaena 7120        - - - - - - - - - - M   T A       T I Y T           V D K     N     L Q           T I
Synechococcus 6301   - - - - - - - - - - M   T T       T I Y R V     V D E     A     L Q           V I
E. coli              - - - - - - - - - - - - - - - - M A T K V V     A   V D E     Q D A V   R V D     E V Q

                                                                              100
spinach chloroplast  G R D T A G Q P M N V T C E V Q Q L L G N N R V R A V A M S A T D G L T R G M E V I D T G A P L S V
Anabaena 7120        T N E     Q L   L V         D Q I       S         V   L   V         I
Synechococcus 6301   K N E     D L S             D R K       S   T     V   L             I
E. coli              N G N - - - - - E R L V L       Q   G G I     T I   G S S     R   L D   K   L E H   I E V

                                                                              150
spinach chloroplast  P V G G P T L G R I F N V L G E P V D N L R P V D T R T T S P I H R S A P A F T Q L D T K L S I F E
Anabaena 7120        K A                           R G     N N Q E   L       P     K L   E E       P V
Synechococcus 6301   E A                           E G     N A A         D     K L   D   E       P K V
E. coli              K A           M               M K G E I G E E E R W A       A     S Y E E     S N S Q E L L

                                                                              200
spinach chloroplast  T G I K V V N L L A P Y R R G G K I G L F G G A G V G K T V L I M E L I N N I A K A H G G V S V F G
Anabaena 7120        D         T                               I M             T Q         A
Synechococcus 6301   I D     A       Q                                 Q         E
E. coli              I D   M C   F A K       V                       N M         R     I E   S   Y         A

                                                                              250
spinach chloroplast  G V G E R T R E G N K L Y M E M K E S G V I N E Q N I A E S K V A L V Y G Q M N E P P G A R M R V G
Anabaena 7120                   N     I D - - - - - - - - - - -   N
Synechococcus 6301                 Q   F     N   D K     N
E. coli              F   H     T   N V I D - - - - - -         S                 N   L       A

                                                                              300
spinach chloroplast  L T A L T M A E Y F R D V N E Q D V L L F I D N I F R F V Q A G S E V S A L L G R M P S A V G Y Q P
Anabaena 7120        S G                   K
Synechococcus 6301   S               H     K
E. coli              G             K     - E G R         V         Y   Y T L     T

                                                                              350
spinach chloroplast  T L S T E M G S L Q E R I T S T K E G S I T S I Q A V Y V P A D D L T D P A P A T T F A H L D A T T
Anabaena 7120        G   D V   Q             T                                         G
Synechococcus 6301   G   D V   Q             L
E. coli              A E     V           T         V                 S               V

                                                                              400
spinach chloroplast  V L S R G L A A K G I Y P A V D P L D S T S T M L Q P R I V G E E H Y E I A Q R V K E T L Q R Y K E
Anabaena 7120              S             G             N   D     N T   R A   Q S
Synechococcus 6301           S                           S         R T   R A   Q S
E. coli              Q I   S L               R Q   D   L V   Q     D T   R G   Q S I         Q

                                                                              450
spinach chloroplast  L Q D I I A I L G L D E L S E E D R L T V A R A R K I E R F L S Q P F F V A E V F T G S P G K Y V G
Anabaena 7120                           L           V               K                             K
Synechococcus 6301                         Q             K                             K
E. coli              K           M         K   V       Q                             S

spinach chloroplast  L A E T I R G F Q L I L S G E L D S L P E Q A F Y L V G N I D E A T A K A M N L E M E S K L K K
Anabaena 7120        E D   K   K           D             D   N   I     E K I K G - - - - - -
Synechococcus 6301   E D   S   N R   N     D             D Q   I E   G A K   K A - - - - - -
E. coli              K D         K G   M E     Y H       M   S   E     V E     K K - - - - - - - -
```

→

Fig. 5. Comparison of the derived amino acid sequences for the F$_1$ subunits of spinach or wheat chloroplasts, *E. coli* and two cyanobacterial species: A) α, B) β, C) γ, D) δ, and E) ε. Sequences are aligned to maximize homologies. At each position, only amino acids that differ from the top sequence are shown. References: wheat chloroplast α, Howe et al. 1985; spinach chloroplast β and ε, Zurawski et al. 1982; *Anabaena* sp. PCC 7120 β and ε, Curtis 1987; *Anabaena* sp. PCC 7120 α, γ and δ, McCarn et al. 1988; *Synechococcus* sp. PCC 6301 (all subunits), Cozens and Walker 1987; *E. coli* β, ε and γ, Saraste et al. 1981 and Kanazawa et al. 1982; *E. Coli* α, Gay and Walker 1981.

C) GAMMA SUBUNIT

```
                                                                              50
Anabaena 7120     M P N L K S I R D R I Q S V K N T K K I T E A M R L V A A A R V R R A Q E Q V I A T R P F A D R L A
Synechococcus 6301  A   A     E     S K   A       Q   Q         K       E M     S K M   K S   D R M A   S       Y   E T M R
E. coli             A G A   E     S K   A       Q   Q         K       E M     S K M   K S   D R M A   S       Y   E T M R
```

```
                                                                              100
Anabaena 7120     Q V L Y G L Q R - R L R F E D V D L P L L K K R E V K S V G L L V I S G D R G L C G G Y N T N V I R
Synechococcus 6301    A   Q Q -   Q N       Q             T A       V                     S
E. coli             K   I G H   A H G N - - - L E Y K H   Y   E D     D       R     Y     V T                 L   I   L F K
```

```
                                                                              150
Anabaena 7120     R A E N R A K E L K A E G L D Y T P V I V G R K A E Q Y F R R R E Q P I D A S Y T G L E Q I P T A D
Synechococcus 6301    Q       R     S Q       K           G           Q             T       S             Q
E. coli             K L L A E M K T W T D K   V Q C D L A M I   S   G V S F   P P N S V G G N V V A Q V T G M G D N P S
```

```
                                                                              200
Anabaena 7120     E - - A N K I A D E L L S L F L S E K V D R I E L V Y T R F V S L V S S R P V I Q T L L P L D T Q G
Synechococcus 6301             D                       G T   V           K L       A N     V                 P
E. coli             L S E L I G P V K V M   Q A Y D E G R L   K L Y I   S N K   I N T M   Q V   T I S Q               - - -
```

```
                                                                              250
Anabaena 7120     L E A A D D E I F R L T T R G G Q F Q V E R Q T V T S Q A R P L P R D M I F E Q D P V Q I L D S L L
Synechococcus 6301    A S S                       S T       E K L     E V A                             A         S A
E. coli             - A S D       D L - - - - - - - - - - - - - K H K   W - - - - - -     Y L Y   P     K A L     D T
```

```
                                                                              300
Anabaena 7120     P L Y L S N Q L L R A L Q E S A A S E L A A R M T A M S N A S E N A G E L I K S L S L S Y N K A R Q
Synechococcus 6301                       A                         N S       D     N A   V G Q   T   V
E. coli             R R   V E S   V Y Q G V V   N L       Q           V       K A   T D   G G S   I K E   Q   V
```

```
Anabaena 7120     A A I T Q E L L E V V G G A E A L T -
Synechococcus 6301                     A               N G
E. coli             S       T I S     A   V - -
```

D) DELTA SUBUNIT

```
                                                                              50
Anabaena 7120     M T S K V A N T E V A Q P Y A Q A L L S I A K S K S L T E E F G T D A R T L L N L L T E N Q Q L R N
Synechococcus 6301    T S Q L F D - - -       E     M A       R E Q G   E D R     E     A L F R S T   A A S A D       H
E. coli             - - - M S E F I T     R       K   A P D F   V E H Q S V   R W Q -   M L A F A A E V T K N E Q M A E
```

```
                                                                              100
Anabaena 7120     F I D N P F I A A E N K K A L I K Q I L - S E A S P Y L R N F L L L L V D K R R I F F L P E I L Q Q
Synechococcus 6301    L L E     T L F S S Q       V L N     V F G   S V H     L V L         N           R N       A     D G   A D R
E. coli             L L S G A - L A P E T L A E S F I A V C G E Q L D E N G Q   L I R V M A E N G     L N A       D V     E Q
```

```
                                                                              150
Anabaena 7120     Y L A L L R A L N Q T V L A E V T S A V A L T E D Q Q Q A V T E K V L A L T K A R Q V E L A T K V D
Synechococcus 6301    Q           K R N V     R D S           P       A V   V I           K Q     G   A G       I E S Q
E. coli             F I H     R A V S E A     A E V D     I       A A     S   Q   L A K I S A A M - E L R L S     K     K   N C K I
```

```
Anabaena 7120     S D L I G G V I I K V G S Q V I D S S I R G Q L R R L S L R L S N S
Synechococcus 6301    A       L                       L A                 K     I I   S A A -
E. coli             K S V M A           R A   D M   I   G V     R E       A D V   Q S -
```

E) EPSILON SUBUNIT

```
                                                                              50
spinach chloroplast  - - M T L N L C V L T P N R S I W N S E V K E I I L S T N S G Q I G V L P N H A P T A T A V D I G I
Anabaena 7120        T V R   I S     D K T V   D A       D V     P S T T       L I S G           L S       L   T V
Synechococcus 6301   - -   S   T V R     I A   D   T V   D A P A Q   V       P S T T       L I     G       L L S     L   T V
E. coli              M A     Y H   D     V S A E Q Q M F S G L   E K   Q V T G S E     E L   I Y   G             L L     I K P   M
```

```
                                                                              100
spinach chloroplast  L R I R - - L N D Q W L T L A L M G G F A R I G N N E I T I L V N D A E R G S D I D P Q E A Q Q T L
Anabaena 7120        V   T S K S Q N   Q A I     L       E V E E D   V           G G       D T   N L E     R T A Y
Synechococcus 6301   V   - - A D K E     A I   V L       E V E     V     V A         D K     L E     R A A F
E. coli              I     V - K Q H G H E E F I Y   S     I L E V Q P G N V     V     A D T   I     Q   L   E A R   M E A K
```

```
spinach chloroplast  E I A E A N L - R K A E G K R Q K - I E A N L A L R R A R T R V E A S N T I S S - -
Anabaena 7120        S Q   Q T K   N Q V P A   D     A Q Q     Q   F K       A   F Q   T G G L A - - -
Synechococcus 6301   S Q   D E R   K G V K   D D     G K F Q   T Q   Y       A   L Q   A G G L V S V -
E. coli              R K     E H I - S S S H   D V D Y - A Q     S A E L A K     I A Q L R V I E L T K K A M
```

reflecting constraints on amino acid sequence related to their role in catalysis and regulation. In both proteins there are many regions which are invariant in sequence among the organisms compared (Fig. 5A, B). Some of these highly conserved regions have homology with the nucleotide binding sites of ATP-requiring enzymes (Walker et al. 1982).

γ appears to be the most conserved of the remaining F_1 subunits although comparative data from the plant nuclear genes is not available. Among cyanobacteria and *E. coli*, portions of the protein near the amino and carboxy termini are the most similar (Fig. 5C). One of the interesting features of the cyanobacterial γ sequences is the presence of only one cysteine residue (position 91, Fig. 5C) which falls in a region that is conserved between cyanobacteria and *E. coli*. In chloroplasts an intramolecular disulfide bond formed between two of the four cysteins of the γ subunit is reduced upon light activation by the thioredoxin system (Moroney et al. 1984, Nalin and McCarty 1984). Studies of the ATPase from *S. platensis* indicate that the cyanobacterial enzyme is regulated similarly to the chloroplast enzyme (Hicks and Yocum 1986b). Since neither cyanobacterial gamma γ sequence contains more than one cysteine which could participate in intramolecular disulfide bonding, regulation of the cyanobacterial enzyme may be mediated by a mechanism different form that in chloroplasts.

Both the δ subunit, for which partial amino acid sequences are available

A) δ SUBUNIT

```
                                                        .                              50
spinach chloroplast  - - - - - - - - - - - - - - - M N V L S Y S I N P L K G L Y A I S G V E V G Q H F Y W Q I G G F Q I H D
Anabaena 7120               - - - - - M L N F L N F Y S V P L A E L       K  L         N L K L  G
Synechococcus 6301     M G S A T L P S D L M S M P T L L E  S S V L P L A E L            N Y R L  G
E. coli             M A S E N M T P Q D Y I G H H L N N L Q L D L R T F S L V D P Q N P P A T F W T I N  D S M F F S V

                                                        .                             100
spinach chloroplast  K A L I T S W V V I A I L L G S A A I A V R S P Q T I P T G G Q N F F E Y V L E F I R D V S K T Q I
Anabaena 7120        Q V F L      F   G V  V L A S V A  S S N V K R    S I  L L   A         L A  N
Synechococcus 6301   Q V F L      F   A  V V L S L L  N  N L R   S L   M    D      N L A R
E. coli             V L G L L F L V L F R S V A K K   T S G  P G K P Q T A I E L V I G  V N G S V K D M Y H G  S K L

                                                        .                             150
spinach chloroplast  G E - E Y R P W V P F I G T M F L F I F V S N W S G A L L P W K I I Q L P H G E L A A P T N D I N T
Anabaena 7120        K          V   L        V  F L  H    E    T      S
Synechococcus 6301   K          L        L       I    L K   S       S
E. coli              - P L A L T I F V W V F L M N L M D L L P I D L L P Y I A E H V L G L P A L R V V  S A   V   V

                                                        .                             200
spinach chloroplast  T V A L A L L A S V A Y F Y A G L T K K G L G Y F G K Y I - - Q P T - - - - - - P I L L P I N I L E
Anabaena 7120          T  L          F S          N  V - -    V - - - - - - S F M    F K  I
Synechococcus 6301     T  L          F S R        N  V - - H    - - - - - - V M     P K
E. coli             L S M    G V F I L I L F Y S I K M    I  G  T  E L T L  P F N H W A F I  V N   I L E G V S

                                                        .                             250
spinach chloroplast  D F T K P L S L S F R L F G N I L A D E L V V V V L V S L V P L V V P I P V M P L G L - F T S - - -
Anabaena 7120                                     A        L       P      L     A F P
Synechococcus 6301                                A        L       F    L A I
E. coli             L L S    V   G L        M Y  G    I F I L  A G  L  W W S Q W I L N V P W A I  H I L I I

                                                        .
spinach chloroplast  G I Q A L I F A T L A A A Y I G E S L E G H H - - - - - - - -
Anabaena 7120        A                   A      A M   D   H G E E H E E H H
Synechococcus 6301   A                   S      A V   E   G E E H A E - - -
E. coli              T L   F    M V  T I V  L S M A S   E   - - - - - - - - -
```

B) B SUBUNIT

```
                                                                 .           ▼        .              .                        50
spinach chloroplast  - - - - - - - M K N V T D S F V P L G H W P S A G S P G F N T D I L A T N L I N L S V V L G V L I F
Anabaena 7120        M G T F L L L   A E A S A V G G E   A E G G A E   G       L     N     D           A I I I T     P V
Synechococcus 6301   - - - - - - - - - - M S   W I L   A   A E T S   -     L   L   L F E         A I I I   L   V Y
E. coli              - - - - - - - - - - - - - - - - - - - - - - - - - V N L   A T     G Q A - - - - A I A P V L P V L

                                          *                                             .                           100
spinach chloroplast  F G K G V L S D L L D N R K Q R I L N T I R N S E E L R G K A I E Q L - - - - - - - - E K A R A - L
Anabaena 7120            K   G N T   K T   R E N   E T A   K   A   Q R A A D   A K           K E   Q Q K
Synechococcus 6301   A   R   P   G N   S N   R A A   E A E   E V     K L A S S A Q A             S Q   Q T Q
E. coli              C M K Y V W P P   M - - - A A   E K R Q K E I A D G L A S   E R A H K D L D L A K A S -   T D Q

                                                                                                       150
spinach chloroplast  K K V E M D A D Q P R V N G Y S E I E R E K M N L I N - - S T Y K T L E Q F E N Y K - N E T I Q F -
Anabaena 7120        E Q A Q A E   E R I K K S A Q D N A Q T A G Q A I   A - - Q A A V D I   R L Q E A G A A D L N A E L
Synechococcus 6301   E A   A E   A R L L   E A K A R A A A V R Q E I L D - - K A A A D V   R L K A T A - A Q D V S T -
E. coli              A K A E   Q V I I E Q A N K R - - - - R S Q I L D E A K A E A E Q   R T K I V A Q A Q A E I E A

                                                                                 .                                    200
spinach chloroplast  E Q Q K A I N Q V R Q R - - - - V F Q Q A L Q G A L G T L N S C L N N E L H L R T I N A N I G M P G
Anabaena 7120        D R A I   Q L R Q   V V - - - - A L         K V E S E   Q G G I S E D A Q K T L   D R S   A Q L
Synechococcus 6301       R V L D E L   R Y - - - - A V A         S R V E T Q   S Q Q   D E A A Q Q   L   D R S L A T L -
E. coli              R K R   R E E L   K Q V A I L A V A G   E K I I E R - - S V - - D E A A N S D I V D K L V A E L -

spinach chloroplast  A M N E I T D
Anabaena 7120        G G V - - - -
Synechococcus 6301   - ▼ - - - - -
E. coli              - - - - - - -
```

C) B' SUBUNIT

```
                                                                                                            50
Anabaena 7120        M T H W I T L L A V E K V A K E G G L F D L D A T L P L M A I Q F L L L L A L I L N A T L Y K P L G K
Synechococcus 6301       N A   M I       E A V Q E A               V   I   V   V F L       V F       F

                                                                                                            100
Anabaena 7120        A I D G R N E Y V R N N Q L E A Q E R L S K A E K L A E A Y E Q E L A G A R R Q A Q T I I A D A Q A
Synechococcus 6301   V L   D   D Q F     G G R Q D   K A     A E V K A   T A Q           A T   K   S   A L     E     T

                                                                                                            150
Anabaena 7120        E A Q K I A A E K V A A A Q K E A Q A Q R E Q A A G E I E Q Q K Q Q A L A S L E Q Q V D A L S R Q I
Synechococcus 6301       G R     Q Q L       R           Q Q     D     A V   Q     D               H

Anabaena 7120        L E K L L G A D L V K Q R
Synechococcus 6301       D       A R A - - - - -
```

D) C SUBUNIT

```
                                                                                                            50
spinach chloroplast  M N P L I A A A S V I A A G L A V G L A S I G P G V G Q G T A A G Q A V E G I A R Q P E A E G K I R
Anabaena 7120            D       V S       L       A           A       I       N
Synechococcus 6301       D S   T S       L       A           A       I       S
E. coli              E N   N M D L L Y M       A V M M         A   A A I   I   I L G   K F L       A           D L I P L L

spinach chloroplast  G T L L L L S L A F M E A L T I Y G L V V A L A L L F A N P F V
Anabaena 7120                                                  V               A
Synechococcus 6301                                              V               A
E. coli              T Q F F I V M G L V D   I P M I A V G L G   Y V M     V A - -
```

E) UNCI-EQUIVALENT PRODUCT

```
                                                                                                            50
Anabaena 7120        M Q E F Y Q L Y Q E L V L I T L V L T G V V F I S V W I P Y S L N I A L N Y L L G A C T G V V Y L R
Synechococcus 6301       A   Y   A   Q R Q   L Q V       I C   V   I G A     W A         T   A S         M G   L L

                                                                                                            100
Anabaena 7120        M L A K D V E R L G R E K Q S L S K T R L A L L M A L I L L A A R W N Q L Q I M P I F L G F L T Y K
Synechococcus 6301       G   A       I     E R R R Q F G   S           P V V     V         Q Y   E L       V

Anabaena 7120        A T L I I Y V V R V A F I S D S P K L R Q P
Synechococcus 6301       A       W   T L   A V I P T A E N S - - - -
```

Fig. 6. Comparison of the derived amino acid sequences of the F_0 subunits from spinach chloroplasts, *E. coli* and two cyanobacterial species: A) *a*, B) *b*, C) *b'*, D) *c* and E) translation product of *uncI* equivalent (function unknown, see text). The asterisk in B) indicates the amino acid whose codon is interrupted by an intron in the *b* genes of chloroplasts (Bird et al. 1985, Hennig and Herrmann 1986, Ohyama et al.1986, Shinozaki et al. 1986); the arrowhead in B) indicates the position equivalent to the site of processing of the wheat protein (Bird et al. 1985). Sequences are aligned to maximize homologies. At each position, only amino acids that differ from the top sequences are shown. References: spinach chloroplast, Hennig and Herrmann 1986; *E. coli*, Gay and Walker 1981; *Anabaena* sp. PCC 7120, McCarn et al. 1988; *Synechococcus* sp. PCC 6301, Cozens and Walker 1987.

from spinach and maize (Berzborn et al. 1986), and ε subunits are poorly conserved relative to other F_1 subunits (Fig. 5D, E). In both δ and ε there are short regions that are invariant in the organisms compared. These regions are concentrated toward the carboxy terminus of the δ protein; in ε they fall in the middle of the primary sequence and toward the amino terminus. It has been suggested that the conserved portions of these subunits represent regions that interact with the α and β subunits (Walker et al. 1982).

Little was known about the subunit composition of cyanobacterial F_0 prior to sequence analyses of cyanobacterial ATP synthase genes. Characterization of the ATP synthase genes revealed that four open reading frames with sequence or secondary structure similarities to F_0 subunits from other organisms are found in cyanobacteria. This implies that the cyanobacterial ATP synthases are probably very similar to those of chloroplasts, which also have four F_0 subunits.

The *c* subunit is the most invariant of the ATP synthase subunits (Fig. 6D). Among the cyanobacterial and chloroplast subunits there are very few positions at which the amino acid sequences differ, and most of these represent conservative replacements. The highly hydrophobic nature of the *c* subunit relates to its presumptive role in forming the proton channel within the membrane. The great degree of sequence conservation implies that the amino acid sequence of this subunit has been very strongly constrained during evolution. The *a* subunits, although not as conserved as *c* subunits, also display a great deal of sequence similarity (Fig. 6A). The greatest variation is observed in the amino terminal third of the polypeptide; the carboxy terminal two-thirds of the protein contain a number of regions that are very highly conserved among all of the organisms compared (Fig. 6A).

The *b* and *b'* polypeptides, which appear to result from gene duplication and divergence (Cozens and Walker 1987) are perhaps the most interesting of the F_0 components. Of the two subunits, *b* shares the greatest homology with the *b* subunit of *E. coli* and the equivalent subunit in chloroplast (subunit I). Although the degree of primary sequence conservation is not

the activation, whereas addition of DTT suppresses the inhibitory effect of DCMU. It was concluded that operation of PS I is sufficient for energization of the C_i-transporting system, while PS II provides reducing equivalents needed for activation of the system. Energization of the system by PS I suggests that the transport is driven by ATP produced by cyclic photophosphorylation.

Hyperpolarization of the cytoplasmic membrane upon addition of HCO_3^- indicates that the C_i transporting mechanism involves a primary electrogenic pump (Kaplan et al. 1982), though mechanistic details are lacking. It is not yet known, for instance, whether C_i uptake is due to a primary, active HCO_3^- pump or to a secondary active transporter powered by a primary pump for another ion. Attempts to detect C_i stimulated ATPase activity in the cytoplasmic membrane, which might account for the observed activity of C_i transport, have so far been unsuccessful (see Kaplan, 1985). It is worth noting, however, that the mechanistic possibilities of HCO_3^--H^+ symport (or HCO_3^--OH^- antiport) secondary to a H^+-extrusion pump have been discounted (Zenvirth et al. 1984). The paucity of data concerning Na^+ involvement in C_i uptake accommodates various mechanisms (Kaplan 1985, Reinhold et al. 1984): Na^+-HCO_3^- symport secondary to an active extrusion of Na^+ (though evidence for Na^+ flux during C_i uptake is lacking (Miller and Canvin 1985); a direct effect of Na^+ on the affinity of the transporter for HCO_3^-; regulation of intracellular pH during CO_2 fixation via a presumed Na^+-H^+ antiport system. It is clear from these various possibilities that the mechanism of C_i uptake is completely obscure, and alternate approaches will be required before gaining a useful understanding of this process.

Biochemical studies of the C_i transport mechanism

Identification of the protein(s) involved in C_i transport is important for elucidating the molecular mechanism of the transport. However, difficulties attend this identification: C_i transport is measurable only with intact cells or spheroplasts; there is no known biochemical activity that is specific to the transport system and measurable in a cell-free system; no specific inhibitor of the transport is known. Under these circumstances, analysis of the adaptation process of H-cells to low CO_2 conditions has been used as an alternative approach to ascertain what proteins might be involved in C_i transport. Since de novo protein synthesis is required for the adaptation (Marcus et al. 1982, Omata and Ogawa 1986), it was expected that analysis of protein synthesis during adaptation would reveal proteins (not necessarily transporter(s)) involved in the transport activity.

great even between cyanobacteria (Table 2; Fig. 6B), hydropathy profiles of the proteins from cyanobacteria, chloroplasts and *E. coli* are very similar (Cozens and Walker 1987). The chloroplast *b* (I) polypeptides have an amino terminal extension which is processed to form the mature subunit (Bird et al. 1985). Sequence comparisons (Fig. 6B) suggest that the cyanobacterial *b* proteins have an analogous extension.

In both *Synechococcus* sp. PCC 6301 and *Anabaena* sp. PCC 7120, an open reading frame with weak homology to *b* subunit sequences is observed between genes encoding the *c* and *b* subunits in the *atp* 1 locus. It has been proposed that the open reading frame encodes a second form of the *b* subunit denoted *b'* that is equivalent to CF_0 subunit II. This designation was based in part on the similarity of the hydropathy profiles of the *b'* and *b* sequences (Cozens and Walker 1987). Recent information on the amino sequence of CF_0 subunit II (Berzborn et al. 1987), which is nuclear-encoded in plants, indicates that this designation is correct. It has been suggested that the ATP synthases of cyanobacteria and chloroplasts may contain one *b* and one *b'* subunit rather than two identical *b* subunits as is observed in *E. coli* (Cozens and Walker 1987). Although the *b* and *b'* primary sequences are not highly conserved even between cyanobacteria (Table 2), the similarity in hydropathy profile patterns suggests that maintenance of subunit confirmation is important in the binding of F_0 and F_1.

The first gene in the *unc* operon of *E. coli* (*uncI*) encodes a hydrophobic protein of unknown function which does not appear to be required for ATP synthase function (Futai and Kanazawa 1983). The *atp* 1 operons of *Synechococcus* sp. PCC 6301 and *Anabaena* sp. PCC 7120 each have an *uncI*-homologous sequence in the analogous position upstream from the *atpI* gene. Similar to the situation for the *b* subunits, the primary sequence of the protein is not highly conserved, but the hydrophobic profiles of the proteins from cyanobacteria and *E. coli* are similar. Since an *uncI* equivalent is present in cyanobacteria, it will certainly not be surprising if plants are found to have an *uncI* homolog as well.

Genomic organization of cyanobacterial ATP synthase gene clusters

The ATP synthase gene clusters of both *Anabaena* sp. PCC 7120 and *Synechococcus* sp. PCC 6301 are at least ten kilobase pairs apart in the chromosome (Cozens and Walker 1987, McCarn et al. 1988). Considerable DNA sequence data is available for regions outside the *atp* 1 and *atp* 2 loci of both cyanobacteria. In *Synechococcus* sp. PCC 6301 there is an unidentified open reading frame (URF) approximately 200 base pairs upstream from *atp* 1 (Cozens and Walker 1987). An URF with sequence similarity is observed upstream from the *Anabaena* sp. PCC 7120 *atp* 1 but at a distance of approximately 700 base pairs (McCarn and Curtis, unpublished). Thus as was observed within the ATP synthase gene clusters, the spacer regions between genes in *Anabaena* sp. PCC 7120 appear to be much larger than in *Synechococcus* sp. PCC 6301.

There are no open reading frames of significant length on either DNA strand for 1.3 kbp downstream for *atp* 1 in *Anabaena* sp. PCC 7120 (McCarn and Curtis, unpublished). In contrast, a region of comparable length downstream from the *Synechococcus* sp. PCC 6301 *atp* 1 contains a ferredoxin gene followed by the beginning of an URF greater than 600 base pairs in length (Cozens and Walker 1987). The *Synechococcus* sp. PCC 6301 ferredoxin gene is distinct from one characterized from *Anabaena* sp. PCC 7120 (Alam et al. 1986). An *Anabaena* sp. PCC 7120 ferredoxin gene equivalent to that characterized in *Synechococcus* sp. PCC 6301 may map much further downstream from *atp* 1 or in another chromosomal location.

Additional differences between *Synechococcus* sp. PCC 6301 and Anabaena sp. PCC 7120 are observed around the *atp* 2 loci. A long and a short URF are found upstream from *Synechococcus* sp. PCC 6301 *atp* 2 on the opposite DNA strand (Cozens and Walker 1987). Neither of these is observed in *Anabaena* sp. PCC 7120, in which an URF of greater than 600 base pairs is found upstream from *atp* 2 on the same DNA strand (Curtis 1987). There is little DNA sequence information for the region downstream from *atp* 2 in *Anabaena* sp. PCC 7120. The comparable region in *Synechococcus* sp. PCC 6301 contains the end of an open reading frame on the opposite strand of DNA which has similarity to a family of cation translocating ATPases (Cozens and Walker 1987).

From the information available, it appears that the genomic organization of ATP synthase gene clusters in *Synechococcus* sp. PCC 6301 and *Anabaena* sp. PCC 7120 may be somewhat different. In both cyanobacteria a similar URF is observed upstream from *atp* 1, however other URFs observed in *Synechococcus* sp. PCC 6301 are not found near *atp* 1 and *atp* 2 in *Anabaena* sp. PCC 7120 in the same relative position as in *Synechococcus* sp. PCC 6301 but much further from the ATP synthase genes.

Evidence for the endosymbiont hypothesis from ATP synthase genes

The endosymbiont hypothesis suggests that the organelles of eucaryotes evolved from free living bacteria that came to reside within a host cell; the evolutionary progenitors of plant chloroplasts in particular are believed to be related to the cyanobacteria (Margulis 1981). Gray and Doolittle (1982) have outlined the types of evidence that can be used in support of an endosymbiotic origin for chloroplasts and mitochondria. They suggest that one form of proof can be derived from the demonstration that homologous proteins encoded in different genomes of the cell have originated from different bacterial lineages.

The information derived from characterization of ATP synthase genes from chloroplasts, mitochondria and cyanobacteria is supportive of an endosymbiotic origin for plant organelles. The chloroplast ATP synthase is clearly more similar to that of cyanobacteria with regard to subunit composition and subunit structure than it is to ATP synthases of plant mitochondria; the plant mitochondrial ATP synthase is more closely related to enzymes from animal mitochondria and the purple photosynthetic bacteria (Cozens and Walker 1987). An analysis of ATP synthase gene organization in chloroplasts and cyanobacteria also supports this view. Thus, the evidence that the ATP synthases of plant mitochondria and chloroplasts are probably derived from different bacterial lineages is strongly supportive of an endosymbiotic origin for these organelles.

Acknowledgement

Research on photosynthesis genes in the author's laboratory is supported by grant number DMB-8614434 from the National Science Foundation.

References

Alam J, Whitaker RA, Krogmann DW and Curtis SE (1986) Isolation and sequence of the gene for ferredoxin I from the cyanobacterium *Anabaena* sp. strain PCC 7120. J Bacteriol 168: 1265–1271

Baccarini-Melandri A and Melandri BA (1978) Coupling factors. In: Clayton RK and Sistrom WR (eds) The Photosynthetic Bacteria, p 651. New York: Plenum Press

Berzborn RJ, Finke W, Otto J, Volker M, Meyer HE, Nier W, Oworah-Nkruma R, Block R (1986) Partial protein sequence of the subunit delta from spinach and maize CF_1 and topographical studies on the binding region between CF_1 and CF_0. In: Biggins J (ed) Progress in Protosynthesis Research, Vol. III, pp 99–102. Dordrecht: Martinus Nijhoff Publishers

Berzborn RJ, Otto J, Finke W, Meyer HE and Block J (1987) Conclusions from N-terminal amino acid sequences of subunits delta from spinach and maize CF_1 and of subunits I and II from spinach CF_0. Biolog Chem Hoppe Seyler 36: 351–552

Bird CR, Koller B, Auffret AD, Huttley AK, Howe CJ, Dyer TA and Gray JC (1985) The wheat chloroplast gene for CF_0 subunit I of ATP synthase contains a large intron. EMBO J 4: 1381–1388

562

Bohnert HJ, Michalowshi S, Bevacqua S, Mucke H and Loffelhardt W (1985) Cyanelle DNA from *Cyanophora paradoxa*. Physical mapping and locating of protein coding regions. Mol Gen Genet 201: 565–574

Brusilow WSA, Klionsky DJ and Simoni RD (1982) Differential polypeptide synthesis of the proton-translocating ATPase of *Escherichia coli*. J Bacteriol 151: 1363–1371

Bryant DA (1986) The cyanobacterial photosynthetic apparatus: comparison of those of higher plants and photosynthetic bacteria. In: Platt T and Li WKW (eds) Photosynthetic Picoplankton, pp 423–500 Canad Bull Fisheries Aquatic Sci, Vol 214. Ottawa: Dept Fisheries Oceans

Cozens AL, Walker JE, Phillips AL, Huttly AK and Gray JC (1986) A sixth subunit of ATP synthase, an F_0 component, is encoded in the pea chloroplast genome. EMBO J 5:217–222

Cozens AL and Walker JE (1987) The organization and sequence of the genes for ATP synthase subunits in the cyanobacterium *Synechococcus* 6301. J Mol Biol 194: 359–383

Curtis SE and Haselkorn R (1984) Isolation, sequence and expression of two members of the 32 kd thylakoid membrane protein gene family from the cyanobacterium *Anabaena* 7120. Plant Mol Biol 3: 249–258

Curtis SE (1987) Genes encoding the β and ε subunits of the proton-translocating ATPase from *Anabaena* sp. strain PCC 7120. J Bacteriol 169: 80–86

Foster DL and Fillingame RH (1982) Stoichiometry of subunits in the H^+-ATPase of *Escherichia coli*. J Biol Chem 257: 2009–2015

Futai Y and Kanazawa H (1983) Structure and function of proton translocating ATPase (F_0F_1): biochemical and molecular biological approaches. Micorbiol Rev 47: 285–312

Gay NJ and Walker JE (1981) The atp operon: nucleotide sequence of the region encoding the α-subunit of *Escherichia coli* ATP-synthase. Nucl Acids Res 9: 2187–2194

Gay NJ and Walker JE (1981) The atp operon: nucleotide sequence of the promoter and the genes for the membrane proteins, and the δ subunit of *Escherichia coli* ATP-synthase. Nucl Acids Res 9: 3919–3926

Golden SG, Brusslan J and Haselkorn R (1986) Expression of a family of *psb*A genes encoding a photosystem II polypeptide in the cyanobacterium *Anacystis nidulans* R2. EMBO J 5: 2789–2798

Gray MW and Doolittle WF (1982) Has the endosymbiont hyposthesis been proven? Micro-biol Rev 46: 1–42

Hallick RB and Bottomley W (1983) Proposal for the naming of chloroplast genes. Plant Mol Biol Rep 1: 38–43

Hatefi Y (1985) The mitochondrial electron transport and oxidative phosphorylation system. Ann Rev Biochem 54: 1015–1069

Hennig J and Herrmann RG (1986) Chloroplast ATP synthase of spinach contains nine nonidentical subunit species, six of which are encoded by plastid chromosomes in two operons in a phylogenetically conserved arrangement. Mol Gen Genet 8: 543–549

Hicks DB and Yocum CF (1986) Properties of the cyanobacterial coupling factor ATPase from *Spirulina platensis*. I. Electrophoretic characterization and reconstitution of photophosphorylation. Arch Biochem Biophys 245: 220–229

Hicks DB and Yocum CF (1986) Properties of the cyanobacterial coupling factor ATPase from *Spirulina platensis*. II. Activity of the purified and membrane-bound enzyme. Arch Biochem Biophys 245: 230–237

Howe CJ, Fearnley IM, Walker JE, Dyer TA and Gray JC (1985) Nucleotide sequences of the genes for the α, β and ε subunits of wheat chloroplast ATP synthase. Plant Mol Biol 4: 333–345

Kanazawa H, Kayano T, Kiyasu T, and Futai M (1982) Nucleotide sequence of the genes for β and ε subunits of proton-translocating ATPase of *Escherichia coli*. Biochem Res Commun 105: 1257–1264

Ko K, Jaynes JM and Strauss N (1985). Homology between the cyanelle DNA of *Cyanophora paradoxa* and the chloroplast DNA of *Vicia faba*. Plant Science 42: 115–123

Krebbers ET, Larrinua IM, McIntosh L and Bogorad L (1982) The maize chloroplast genes for the β and ε subunits of the photosynthetic coupling factor CF1 are fused. Nucl Acids Res 10: 4985–5002

Lambert DH, Bryant DA, Stirewalt VL, Dubbs JM, Stevens JR SE and Porter RD (1985) Gene map for the *Cyanophora paradoxa* cyanelle genome. J Bacteriol 164: 659–664

Lubberding HJ, Zimmer G, Van Walraven HS, Schrickx J and Kraayenhof R (1983) Isolation, purification and characterization of the ATPase complex from the thermophilic cyanobacterium *Synechococcus* 6716. Eur J Biochem 137: 95–99

Margulis L (1981) Symbiosis in cell evolution, pp 37–62. San Francisco: WH Freeman & Co

McCarn DF, Whitaker RA, Alam J, Vrba JM and Curtis SE (1988) The genes encoding the α, γ, δ and four F_0 subunits of the ATP synthase constitute an operon in the cyanobacterium *Anabaena* sp. strain PCC 7120. J Bacteriol, in press

McCarthy JEG, Schairer HU and Sebald W (1985) Translational initiation frequency of *atp* genes from *Escherichia coli*: identification of an intercistronic sequence that enhances translation. EMBO J 4: 519–526

McCarty RE and Hammes GG (1987) Molecular architecture of chloroplast coupling factor 1. Trends Biochem Sci 12: 234–237

Merchant S and Selman BR (1985) Photosynthetic ATPases: purification, properties, subunit isolation and function. Photosyn Res 6: 3–31

Moroney JV, Fullmer CS and McCarty RE (1984) Characterization of the cysteinyl-containing peptides of the γ subunit of coupling factor 1. J Biol Chem 259: 7281–7285

Nalin CM and McCarty RE (1984) Role of a disulfide bond in the γ subunit in activation of the ATPase of chloroplast coupling factor 1. J Biol Chem 259: 7275–7280

Ohyama K, Fukuzawa H, Kohchi T, Shirai H, Sano T, Sano S, Umesono K, Shiki Y, Takeuchi M, Chang Z, Aota S, Inokuchi H and Ozeki H (1986) Chloroplast gene organization deduced from complete sequence of liverwort *Marchantia polymorpha* chloroplast DNA. Nature 322: 572–574

Palmer JD (1985) Comparative organization of chloroplast genomes. Ann Rev Genet 19: 325–354

Pick U and Racker E (1979) Purification and reconstitution of the N,N'-dicyclohexycarbodiimide-sensitive ATPase complex from spinach chloroplasts. J Biol Chem 254: 2793–2799

Reznikoff WS, Siegele DA, Cowing DW and Gross CA (1985) The regulation of transcription initiation in bacteria. Ann Rev Genet 9: 355–388

Saraste M, Gay NJ, Eberle A, Runswick MJ and Walker JE (1981) The *atp* operon: nucleotide sequence of the genes for the γ, β, and ε subunits of *Escherichia coli* ATP synthase. Nucl Acids Res 9: 5286–5296

Shinozaki K, Deno H, Wakasugi T and Suguira M (1986) Tobacco chloroplast gene coding for subunit I of proton-translocating ATPase: comparison with the wheat subunit I and *E. coli* subunit *b*. Curr Genet 10: 421–423

Walker JE, Eberle A, Gay NJ, Runswick MJ and Saraste M (1982) Conservation of structure in proton-translocation ATPases of *Escherichia coli* and mitochondria. Biochem Soc Trans 10: 203–206

Walker JE and Tybulewicz VLJ (1985) Comparative genetics and biochemistry of light-driven ATP synthases. In: Arntzen C, Bogoard L, Bonitz S and Steinbeck K (eds) Molecular biology of the Photosynthetic Apparatus, pp 141–153. Cold Spring Harbor, NY: Cold Spring Harbor Laboratory Press

Wassman CC, Loffelhardt W and Bohnert JH (1987) Cyanelles: Organization and molecular biology. In: Fay P and Van Baalen C (eds) The Cyanobacteria, pp 303–324. New York: Elsevier/North Holland Biomedical Press

Westhoff P, Alt J, Nelson N, Herrman RG (1985) Genes and transcripts for the ATP synthase CF_0 subunits I and II from spinach thylakoid membranes. Mol Gen Genet 199: 290–299

Woessner JP, Gillham NW and Boynton JE (1987) Chloroplast genes encoding subunits of the H^+-ATPase complex of *Chlamydomonas reinhardtii* are rearranged compared to higher plants: sequence of the *atpE* gene and location of the *atpF* and *atpI* genes. Plant Mol Biol 8: 151–158

Wolk CP (1980) Cyanobacteria (Blue-green algae) In: The biochemistry of Plants, Vol -1, pp 659–687. New York: Academic Press

Zurawski G, Bottomley W and Whitfeld PR (1982) Structures of the genes for the β and ε subunits of spinach chloroplast ATPase indicate a dicistronic mRNA and an overlapping translation stop/start signal. Proc Natl Acad Sci USA 79: 6260–6264

Govindjee et al. (eds), Molecular Biology of Photosynthesis: 565–582
© 1988 Kluwer Academic Publishers

Minireview

The chloroplast genes encoding subunits of the H^+-ATP synthase

GRAHAM S. HUDSON & JOHN G. MASON
CSIRO, Divison of Plant Industry, GPO Box 1600, Canberra, A.C.T. 2601, Australia

Received 31 December 1987; accepted 12 February 1988

Key words: ATP synthase, chloroplast, evolution, gene expression, operon

Abstract. Three CF_1 and three CF_0 subunits of the chloroplast H^+-ATP synthase are encoded on the chloroplast genome. The chloroplast *atp* genes are organized as two operons in plants but not in the green alga, *Chlamydomonas reinhardtii*. The *atp*BE or β operon shows a relatively simple organisation and transcription pattern, while the *atp*IHFA or α operon is transcribed into a large variety of mRNAs. The *atp* genes are related to those of cyanobacteria and, more distantly, to those of non-photosynthetic bacteria such as *E. coli*, suggesting a common origin of most F_1F_0 ATP synthase subunits. Both the chloroplast and cyanobacterial ATP synthases have four F_0 subunits, not three as in the *E. coli* complex. The proton pore of the CF_0 is proposed to be formed by the interaction of subunits III and IV.

1. Introduction

H^+-ATP synthase is essential for electron transport and photophosphorylation during photosynthesis. It has an F_1F_0 structure analogous to the proton-translocating ATP synthases of bacteria and mitochondria, that is, a hydrophobic sector (CF_0) within the thylakoid membrane and a hydrophilic moiety (CF_1) protruding into the stroma. The complex couples the phosphorylation of ADP with the transmembrane proton gradient generated during light-driven electron transport. The catalytic site for ATP synthesis is located on the CF_1 portion while the CF_0 subunits act as a proton channel for translocation of protons across the thylakoid membrane. A number of authors have reviewed the structure and function of the chloroplast ATP synthase complex and its subunits over recent years (Shavit 1980, Nelson 1981, Strotmann and Bickel-Sandkötter 1984, Merchant and Selman 1985).

Like other multimeric complexes of the chloroplast thylakoids, the ATP synthase is the product of the two genetic systems. Some subunits are encoded and synthesised in the chloroplast, while the remainder are encoded in the nucleus, synthesised in the cytosol and transported into the organelle (Herrmann et al. 1985). This review covers the organisation, expression and evolution of the chloroplast *atp* genes encoding six subunits of the ATP synthase.

2. *E. coli* ATP synthase

Before discussing the chloroplast ATP synthase and *atp* genes, it is necessary to first briefly examine the ATP synthase of *E. coli* because of the wealth of genetic, biochemical and molecular data on this complex (Futai and Kanazawa 1980, Cross 1981, Dunn and Heppel 1981, Fillingame 1981, Senior and Wise 1983, Hoppe and Sebald 1984). Although there are fundamental differences between chloroplasts and *E. coli* in the mechanism by which the enzymes are regulated, in particular with respect to light activation of CF_1CF_0, they appear to be very similar in structure and catalytic mechanism.

E. *coli* F_1F_0 is composed of eight different types of polypeptides, five in the F_1 factor (α_3, β_3, γ_1, δ_1 and ε_1) and three in F_0 (a_1, b_2 and c_{6-10}), with the stoichiometry believed to be as indicated (Foster and Fillingame 1982, Hoppe and Sebald 1984). The subunits are encoded by the *unc* or *atp* operon, and the DNA sequence of this region has been completely determined (Kanazawa and Futai 1982, Walker et al. 1984). The operon carries a single gene for each subunit, with the order of genes encoding subunits as follows: uncI, *a*, *c*, *b*, δ, α, γ, β, ε where the first gene (*uncI* or *atpI*) encodes a polypeptide of unknown function. Thus the genes for the F_0 and F_1 subunits are clustered respectively at the 5' and 3' ends of the operon.

The operon is transcribed off a single promotor upstream of *uncI* although a second weak promotor may exist within the *uncI* gene (Walker et al. 1984). Mechanisms involving differential translation initiation or elongation have been proposed to account for the different subunit stoichiometries required in the complex (Walker et al. 1984). Proof of translational control has been obtained by identification of a sequence involved in enhancement of translation initiation of the gene for subunit *c* (McCarthy et al. 1985). Conflicting data exist on the assembly pathway of the complex, in particular whether the F_0 and F_1 sectors can assemble independently or not (Cox et al. 1981, Aris et al. 1985).

The functions of the individual subunits, particuarly those of F_1, have been intensively studied (Futai and Kanazawa 1980, Dunn and Heppel 1981, Senior and Wise 1983, Walker et al. 1984). The α and β subunits carry the binding sites for adenine nucleotides, and with the γ subunit form the core of F_1. The δ and ε subunits participate in the binding of F_1 to F_0, and the ε subunit also acts as an inhibitor of F_1 catalytic activity (Klionsky et al. 1984). The three F_0 subunits are predicted to be transmembrane polypeptides (Hoppe and Sebald 1984, Kanazawa and Futai 1982, Walker et al. 1984). The c subunit is intimately associated with proton conductance and is the site of interaction with the inhibitor, N,N'-cyclohexylcarbodiimide (DCCD) (Sebald and Hoppe 1981, Hoppe and Sebald 1984). The b subunit carries an extrinsic portion which binds the F_1 (Perlin et al. 1983, Walker et al. 1984). The a subunit is thought to interact with the c subunit in the formation of the proton pore (Cain and Simoni 1986, Cox et al. 1986).

3. Subunit compositon of the chloroplast ATP synthase

The structure of CF_1CF_0 appears to be analogous to the *E. coli* enzyme except that it probably contains nine types of subunits, not eight (see Table 1). CF_1 can be readily detached from CF_0 by EDTA treatment of thylakoid membranes (Pick and Racker 1979, Nelson 1981). It contains five subunits called α, β, γ, δ and ε. The α and β subunits are nucleotide-binding polypep-

Table 1. Composition of the chloroplast H^+-ATP synthase from spinach.

Subunit	Apparent M.W.[a] (kd)	Probable no. per complex[b]	Gene name	Gene location	Predicted M.W.[c] (kd)	*E. coli* subunit homologue
CF_1						
α	59	3	*atp*A	chloroplast	56	α
β	52	3	*atp*B	chloroplast	54	β
γ	37	1	*atp*C	nucleus		γ
δ	20	1	*atp*D	nucleus		δ
ε	16	1	*atp*E	chloroplast	15	ε
CF_0						
I	18	1	*atp*F	chloroplast	21(19[d])	b
II	16	1	*atp*G	nucleus		b?
III	8	5–10	*atp*H	chloroplast	8	c
IV	19	1	*atp*I	chloroplast	27(25[d])	a

[a]From SDS gel electrophoresis data (Westhoff et al. 1985).
[b]Based on the composition of H^+-ATP synthase from *E. coli* (Sebald and Hoppe 1981, Foster and Fillingame 1982), *Vicia faba* chloroplasts and *Avena sativa* chloroplasts (Süss and Schmidt 1982).
[c]From nucleotide sequence data (Zurawski et al. 1982, Hudson et al. 1987).
[d]Processed molecular weight (Bird et al. 1985, Fromme et al. 1987).

tides (Nelson 1981, Merchant and Selman 1985), arranged around the γ subunit (Akey et al. 1983). The γ subunit is the site of light-dark regulation of ATP synthase activity through the reduction or formation of a disulphide bond (Arana and Vallejos 1982, Nalin and McCarty 1984). The δ and ε subunits are not required for the binding of CF_1 to thylakoid membranes (Patrie and McCarty 1984), but ε is essential for ATP synthesis and to prevent proton leakage through CF_0 (Richter et al. 1984). The ε subunit is also an inhibitor of ATP synthase (Nelson 1981, Finel et al. 1984, Richter et al. 1984). Gene mapping and DNA sequencing have shown that the genes for α, β and ε are located on the chloroplast genome and encode polypeptides homologous to the subunits of the same name in *E. coli* (see Sections 5, 6). Subunits γ and δ are encoded in the nucleus and synthesised in the cytosol as precursor proteins of higher molecular weight (Nelson et al. 1980, Watanabe and Price 1982, Tittgen et al. 1986). Isolation of their cDNAs has been reported (Tittgen et al. 1986; JG Mason, unpublished data). The γ subunit shows homology to the *E. coli* γ subunit and possesses an N-terminal transit peptide sequence for transport into the chloroplast (JG Mason, unpublished results).

By comparson, the number of types and functions of CF_0 subunits have been less well defined. Three polypeptides designated subunits I, II and III were originally observed on polyacrylamide gels (Pick and Racker 1979, Nelson et al. 1980, Westhoff et al. 1985) but a fourth component of 19 kd could sometimes be resolved (Pick and Racker 1979; Süss 1980, Westhoff et al. 1985); see Table 1. The smallest polypeptide, subunit III, is the DCCD-binding proteolipid equivalent to subunit *c* of *E. coli* F_0, and its amino acid sequence has been determined (Sebald and Wachter 1980). This subunit is associated with proton translocation (Sebald and Hoppe 1981). DNA sequencing studies have revealed the presence of genes for subunits I, III and an additional CF_0 subunit, IV, on the chloroplast genome (see Section 5). Amino acid sequencing of the isolated 19 kd polypeptide has correlated this component with the subunit IV predicted from the DNA sequence (Fromme et al. 1987). Support for the presence of four types of CF_0 subunits has come from the identification of four F_0 subunit genes for the ATP synthase of a cyanobacterium (Cozens et al. 1986, Cozens and Walker 1987). Subunits I and IV are analogous to subunits *b* and *a* of *E. coli* (see Section 7). Subunit II is synthesised in the cytosol as a precursor protein of 23 kd (Westhoff et al. 1985, Tittgen et al. 1986), and may also be analogous to subunit *b* of *E. coli* (Cozens and Walker 1987). A nuclear mutation which affects CF_0 assembly, and may be in the structural gene for subunit II, has been mapped to the short arm of chromosome I of maize (Echt et al. 1987).

Prior to the discovery of subunit IV, the composition of the CF_1CF_0 was estimated as $\alpha_3\beta_3\gamma_1\delta_1\varepsilon_1 I_1 II_1 III_5$ (Süss and Schmidt 1982). Given the presence of an additional polypeptide, subunit IV, and the data for the composition of *E. coli* F_1F_0, the probable composition is $\alpha_3\beta_3\gamma_1\delta_1\varepsilon_1 I_1 II_1 III_n IV_1$ where n is at least five.

4. Structure of the chloroplast genome and arrangement of *atp* genes

The chloroplast maintains its own genome and apparatus for DNA replication, transcription and translation. The genome from plant chloroplasts is as multicopy, closed circular DNA molecule of 120 to 220 kbp (Palmer 1985). A general feature of the genome, except in some legumes including pea, is the presence of two large inverted repeats which divide the DNA molecule asymmetrically into large and small single copy regions (Fig. 1). These repeats contain the rRNA operons as well as other genes. The complete DNA sequences of two chloroplast genomes, from tobacco and liverwort, have been determined and shown to contain about 120 genes for tRNAs, rRNAs and proteins involved mostly in either transcription, translation or photosynthesis (Ohyama et al. 1986, Shinozaki et al. 1986b).

The six *atp* genes of the subunits of the ATP synthase are found at two loci, *atp*BE and *atp*IHFA, separated by 40 kbp in the large single copy

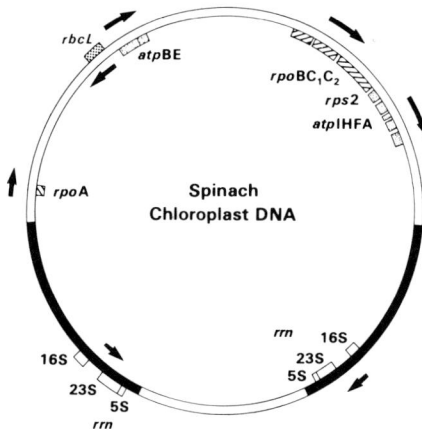

Fig. 1. Map of the spinach chloroplast genome showing the position of the operons for ATP synthase subunits (*atp*BE and *atp*IHFA). Also shown are the rRNA operons (*rrn*) situated within the large, inverted repeats (shaded), and the genes *rps*2, *rpo*A, *rpo*B, *rpo*C₁, *rpo*C₂ and *rbc*L (see text). The arrows indicate the direction of transcription of the genes.

region of the spinach and tobacco chloroplast genomes (Fig. 1). The gene names and their encoded subunits are listed in Table 1. Other plant chloroplasts show the same organization of genes, although the orientation and distances between the operons may vary due to rearrangements in the genomes (Palmer 1985). In the green alga, *Chlamydomonas reinhardtii*, the arrangement of the chloroplast *atp* genes is quite different (Woessner et al. 1987), in that the genes are separated and may be monocistronic. This indicates that the organisation of *atp* genes as two operons is not essential for the correct biogenesis of chloroplast ATP synthase.

5. The α operon

The *atp*H and *atp*A genes for subunits III and α were initially mapped to a locus distant from the *atp*BE operon (Westhoff et al. 1981, Howe et al. 1982, 1983, Deno et al. 1983). Nucleotide sequencing and RNA mapping subsequently showed that these two genes are associated with two others, *atp*I and *atp*F for subunits IV and I, to form an operon (spinach (Alt et al. 1983, Hennig and Herrmann 1986, Hudson et al. 1987); tobacco (Deno et al. 1983, 1984, Shinozaki et al. 1986a, b); pea (Cozens et al. 1986, Hudson et al. 1987); liverwort (Ohyama et al. 1986)). Furthermore, these *atp* genes are situated distal to the genes for three RNA polymerase subunits and one ribosomal protein, respectively the *rpo*B, $rpoC_1$, $rpoC_2$ and *rps*2 genes (Hudson et al. 1988); see Fig. 1). The relationship of these genes is not clear, but the three *rpo* genes and *rps*2-*atp*IHFA appear to form separate transcription units, with some leakage of *rpo* transcripts into *rps*2 (Hudson et al. 1987, 1988). The most notable features of the *atp*IHFA operon are the relatively large intergenic regions separating individual genes (ranging from 69 to 696 bp in spinach) and the presence of a single intron (766 bp in spinach) splitting *atp*F into two exons. The precise ends of this intron have been mapped by isolation of a cDNA (Hudson et al. 1987) and by S1 mapping in wheat (Bird et al. 1985). The intron resembles other group II introns found in chloroplast and fungal mitochondrial genes (Michel and Dujon 1983).

Our understanding of expression of these genes is relatively poor. RNA mapping has revealed a complex pattern of mono-, di- and polycistronic transcripts spanning the *rps*2 and *atp* genes in spinach and pea (Hudson et al. 1987); see Fig. 2B. The 5′ termini of the spinach mRNAs have been mapped to positions upstream of *rps*2, *atp*I and *atp*H. No obvious promoter-like sequences are associated with these termini apart from a sequence (TTTACG–18 bp–TATATT) associated with the –373 bp end near *atp*H. It

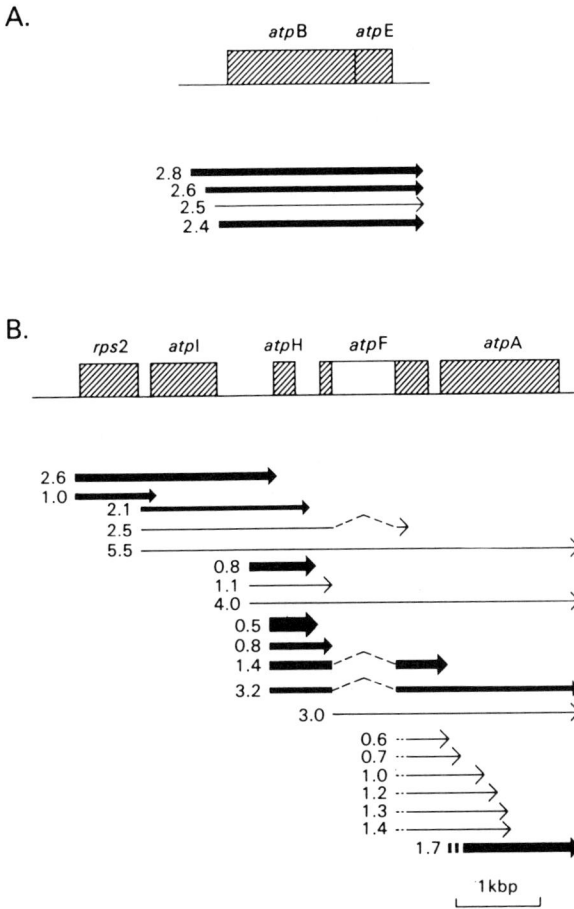

Fig. 2. Transcription of the spinach chloropast *atp* genes. A) *atp*BE (after Zurawski et al. 1982, Mullet et al. 1985), B) *atp*IHFA (after Hudson et al. 1987). The boxes filled with hatched lines indicate coding sequences, while the open region in *atp*F indicates the intron in this gene. The numbers give the approximate size of the transcripts in kilobases, and the thickness of the lines reflects the relative abundance of each of the transcript sizes. Note that the 3' ends of these transcripts have not been accurately mapped.

is therefore likely that many of these mRNAs have resulted from cleavage of larger primary transcripts. It is also notable that transcripts crossing *atp*F are a mixture of unspliced, spliced and truncated molecules. The significant levels of unspliced or short *atp*F and *atp*A RNAs may lead to the synthesis of truncated polypeptides. The most abundant transcripts are those of *atp*H, consistent with CF_0III being the most abundant subunit in the complex.

6. The β operon

The *atp*BE or β operon is located approximately 700 bp upstream from, but on the opposite strand to, the *rbc*L gene encoding the large subunit of ribulose bisphosphate carboxylase (Fig. 1). It encodes the β and ε subunits of CF_1, and in most higher plants the two coding regions overlap for four bases such that the UGA stop codon of *atp*B is preceded by an A to form the initiator codon of *atp*E (maize (Krebbers et al. 1982); spinach (Zurawski et al. 1982); tobacco (Shinozaki et al. 1983); barley (Zurawski and Clegg 1984); wheat (Howe et al. 1985)). A chloroplast mutation leading to fusion of the β and ε subunits has been proposed as a defect in translation of *atp*BE (Sears and Herrmann 1985). In pea (Zurawski et al. 1986) a duplication of a 5 bp sequence has resulted in the creation of a new stop codon 22 bp upstream from the *atp*E initiator codon so that the genes do not overlap. The genes also do not overlap in sweet potato (Kobayashi et al. 1987) and liverwort (Ohyama et al. 1986). These observations and the fact that one ε and three β subunits are required per CF_1 complex suggest that translation of *atp*E is not tightly coupled to translation of *atp*B.

RNA mapping has shown the maize β operon to be transcribed into a single, dicistronic mRNA (Krebbers et al. 1982). In wheat and tobacco, separate transcripts may exist for the two genes (Shinozaki et al. 1983, Howe et al. 1985). The spinach β operon is transcribed into four mRNAs which differ at their 5′ termini within the untranslated region proximal to *atp*B (Mullet et al. 1985); see Fig. 2A. By contrast, maize and pea show only single 5′ ends. The *atp*B promoters from maize, spinach and pea have been located by in vitro transcription using crude chloroplast lysates (Bradley and Gatenby 1985, Gruissem and Zurawski 1985, Orozco et al. 1985). The promoter in spinach lies upstream of the most distant 5′ end (454 bp from the *atp*B initiation codon), and is similar in sequence to the promoters mapped in other species. It features two separate sequence elements, TTGACA and TGTATA (TAGTAT in maize and other monocots, TGTAAA in pea; Zurawski and Clegg 1987), which are separated by 17 bp and are analogous to the '−35' and '−10' promoter elements of *E. coli* genes. Other chloroplast genes possess similar promoter elements (Gruissem and Zurawski 1985, Hanley-Bowdoin and Chua 1987). It is unclear if the multiple transcripts in spinach result from RNA processing or the presence of additional promoters (Mullet et al. 1985, Orozco et al. 1985), and if such a long untranslated 5′ region is crucial to the expression of *atp*BE.

In common with other chloroplast genes, transcriptional activity for the spinach *atp*BE operon increases then decays during chloroplast development in the light (Deng and Gruissem 1987). Synthesis of ATP synthase subunits also increases during chloroplast development (de Heij et al. 1984). Apart from these developmental effects, it seems likely that control over *atp*BE expression is at the post-transcriptional level in the mature chloroplast (Deng and Gruissem 1987). There is some evidence for rapid degradation of unassembled subunits (Biekmann and Feierabend 1985) providing a mechanism for balancing unequal nuclear and chloroplast syntheses of subunits. However, it is unclear if expression of the two chloroplast *atp* operons and the nuclear *atp* genes are specifically coordinated, or are simply tied to the general development of the chloroplast and the leaf cell.

7. Prediction of subunit structure and function from the DNA sequences

The CF_1 α and β subunits encoded by *atp*A and *atp*B are highly homologous to the α and β subunits of eubacterial and mitochondrial H^+-ATP synthases (Table 2). They are also weakly homologous to each other (18% identical, excluding gaps, for the spinach chloroplast subunits) as are the *E. coli* and mitochondrial subunits (Walker et al. 1984, 1985) so that their genes must have evolved from a single ancestral gene. Both chloroplast proteins carry binding sites for adenine nucleotides, and regions of the amino acid sequences involved in binding sites have been identified by homology with nucleotide binding proteins or ATP synthase subunits from other organisms (Howe et al. 1985). Specific amino acid residues in the bovine mitochondrial β subunit which are modified by affinity reagents are also conserved in the chloroplast β subunit (Howe et al. 1985, Walker et al. 1985). Other residues in β which may be the binding site for Mg^{2+} were also identified (Howe et al. 1985). The β subunit probably carries the catalytic site for ATP synthesis

Table 2. Comparison of amino acid sequences of F_1F_0 subunits relative to those of the spinach chloroplast complex.

Organism	% Amino acid sequence identity Subunit					
	α	β	ε	a(IV)	b(I)	c(III)
Tobacco chloroplast[a]	95	95	87	96	91	99
Pea chloroplast[b,c]	91	93	81	90	83	99
Liverwort chloroplast[d]	86	89	60	81	51	98
Cyanobacterium[e]	74	81	43	70	26	86
E. coli[f]	54	66	26	12	19	29
Bovine mitochondria[f,g,h,i]	60	68	23	10	9	33

[a]Shinozaki et al. 1986b; [b]Hudson et al. 1987; [c]Zurawski et al. 1986; [d]Ohyama et al. 1986; [e]Cozens and Walker 1987; [f]Walker et al. 1984; [g]Walker et al. 1985; [h]Walker et al. 1987; [i]Anderson et al. 1982.

(Nelson 1981, Merchant and Selman 1985) and its importance is reflected in its high level of conservation across all organisms (Table 2).

By comparison, the ε subunit is poorly conserved between chloroplasts and eubacteria, but is moderately well conserved within the chloroplasts (Table 2). Few amino acid residues have been assigned functional roles. An essential glycine residue in the *E. coli* ε subunit which affects ATP synthase activity (Cox et al. 1987) is conserved in the chloroplast polypeptide.

The three chloroplast-encoded CF_0 subunits are all predicted to be integral membrane proteins. Subunit I features a two-domain structure (Fig. 3) with a membrane anchorage segment and a C-terminal hydrophilic stalk of two opposing α-helices protruding into the stroma (Bird et al. 1985, Hudson et al. 1987). Although the amino acid sequence is poorly homologous to the subunit *β* of *E. coli* (Table 2), it is likely that the secondary structures of the two polypeptides are similar. The CF_1 is thought to be attached to the CF_0 via this stalk, as proposed for the *E. coli* F_1F_0 (Perlin et al. 1983, Walker et al. 1984). The polypeptide also has an N-terminal leader peptide of seventeen residues in spinach (six in pea) which is presumably removed post-translationally (Bird et al. 1985). The role of this leader peptide is probably to allow correct insertion of subunit I into the thylakoid membrane, and implies the presence of a protease for some polypeptides destined for the thylakoid membrane. Cytochrome *f* is also synthesised in the chloroplast with a leader peptide which is removed post-translationally (Alt and Herrmann 1984, Wiley et al. 1984), but most other thylakoid polypeptides expressed in the chloroplast apparently do not undergo proteolytic processing.

Subunit III possesses no leader peptide since the predicted amino acid sequence from spinach *atp*H (Alt et al. 1983, Hudson et al. 1987) matches exactly that determined for the protein purified by Sebald and Wachter (1980). The spinach subunit III and *E. coli* subunit *c* are 29% identical. The predicted secondary structure is two membrane spanning α-helices connected by reverse turn (Sebald and Hoppe 1981); see Fig. 3. The subunit features buried proline and glutamic acid residues in the first and second α-helices respectively. These residues are thought to be important in the packing of the proteolipid subunits and in the formation of the proton pore (Hudson et al. 1987). The *E. coli* subunit *c* has a buried aspartic acid which is the site of interaction with DCCD (Sebald and Hoppe 1981).

Subunit IV is a 27 kd hydrophobic protein whose amino acid shows limited homology to subunit *a* of *E. coli* F_0, mostly at the carboxy terminus. The first eighteen residues are absent in the purified protein (Fromme et al. 1987) and these probably act as a leader peptide for membrane insertion. The predicted secondary structure for subunit IV is five transmembrane α-helices, with the fourth having an amphipathic character (Hudson et al. 1987); see Fig. 3. The proton pore is proposed to be formed by interaction

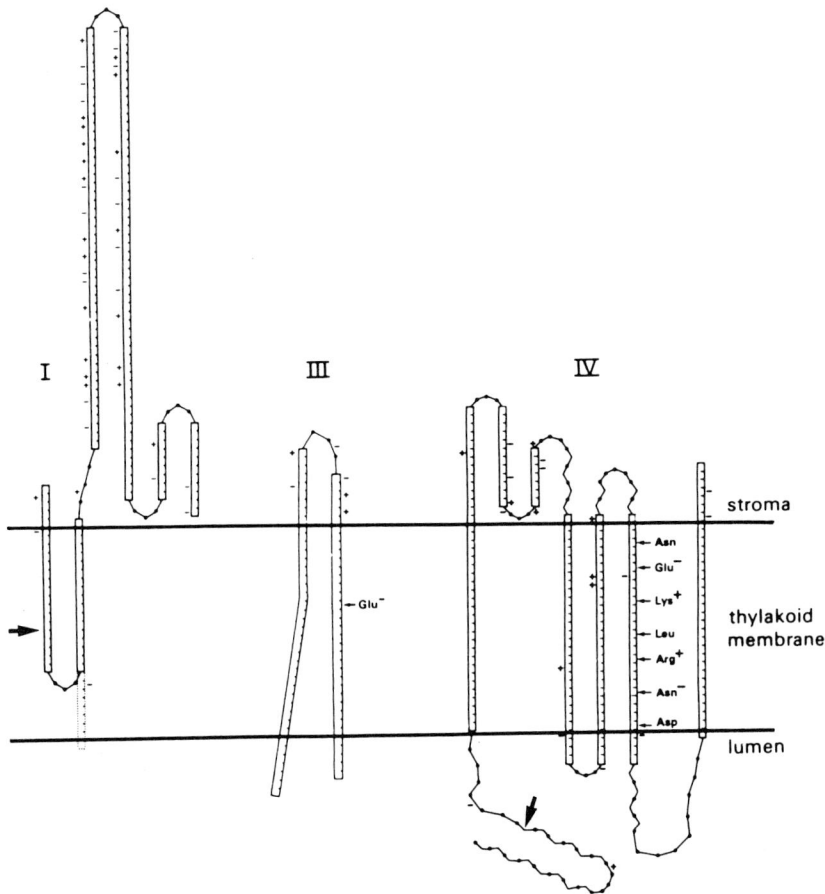

Fig. 3. Predicted secondary structures of chloroplast CF$_0$ subunits I, III and IV. The thick arrows show the predicted sites of proteolytic processing of subunits I and IV (Bird et al. 1985, Fromme et al. 1987). The amino acid residues shown in subunits III and IV may participate in the proton pore of CF$_0$ (Hudson et al. 1987).

of this fourth α-helix with the buried glutamic acid of a subunit III (Cox et al. 1986, Hudson et al. 1987). Based on a model of *E. coli* F$_0$ (Cox et al. 1986), the CF$_0$ can be envisaged as a central core of subunits I, II and IV surrounded by a layer of multiple subunits III, only one of which is interacting with subunit IV at a particular moment.

8. Evolution of the ATP synthase complex

Interspecies comparisons of the *atp* genes and their encoded products have revealed general patterns of both conservation and change at four levels.

First, comparison of chloroplast *atp* nucleotide sequences has shown that these genes are highly conserved (Zurawski and Clegg 1984, 1987, Hudson et al. 1987, Rodermel and Bogorad 1987). The majority of changes in the coding regions are synonymous substitutions at the third positions of codons, and the calculated rates of substitution show that chloroplast genes are evolving more slowly than mammalian nuclear and mitochondrial genes (Rodermel and Bogorad 1987, Zurawski and Clegg 1987). Noncoding sequences vary considerably due to the duplication or deletion of short runs of nucleotides. Thus the *atp* genes and other chloroplast genes in general are very stable, and are probably evolving more slowly than nuclear genes encoding chloroplast proteins.

Second, there is a conservation of *atp* gene organisation, reflecting the eubacterial origin of chloroplast (Gray and Doolittle 1982, Woese 1987) and the rarity of rearrangements in the chloroplast genome (Palmer 1985). The *E. coli unc* operon encodes, in order, the uncI polypeptide, a, c, b, δ, α, γ, β, ε (Kanazawa and Futai 1982, Walker et al. 1984). The cyanobacterium *Synechococcus* 6301 has two *atp* operons encoding the equivalent of uncI, a, c, b', b, δ, α, γ and β, ε respectively (Cozens and Walker 1987). The b' subunit is an additional b-like subunit, which may have arisen through duplication of the ancestral gene encoding subunit b. This organization of two operons closely resembles the organisation of the chloroplast α and β operons: a(IV), c(III), b(I), α and β, ε. The missing genes for subunits γ, δ and b' (equivalent to subunits γ, δ and II) may have been transferred to the nucleus following endosymbiosis. Further proof for the close ancestry of chloroplasts and cyanobacteria (Gray and Doolittle 1982, Woese 1987) is demonstrated by the close homology of the cyanobacterial and chloroplast polypeptides (Cozens et al. 1986, Cozens and Walker 1987), including the apparent possession of leader peptides by the cyanobacterial subunits b and b' (Cozens and Walker 1987). This organisation of *atp* genes is indicative of the cyanobacterial-chloroplast lineage, but is not universal since the *atp* genes in the chloroplasts of *Chlamydomonas reinhardtii* map to individual loci (Woessner et al. 1987).

Third, there is considerable variation in the conservation of the mapping structures of the subunits. Table 2 summarises the percentage of identical amino acid sequences for the six spinach chloroplast-encoded polypeptides with those from other species. The CF_1 subunits α and β are highly conserved across the whole spectrum of F_1F_0 complexes, indicating their importance in catalytic function. The CF_0 subunits III and IV are highly conserved within the chloroplast group but not between chloroplasts, mitochondria and eubacteria. Their conservation reflects their predicted importance in proton pore function, but the variation between the different groups may arise from fundamental differences in the environments of the thylakoid, inner mitochondrial and bacterial cytoplasmic membranes in which these F_0 polypeptides reside. Subunits I and ε are not well conserved across all species, suggesting lower constraints on their evolution.

Fourth, there is some variation in subunit complexity of the F_1F_0 enzymes. Anaerobic bacteria appear to possess a relatively simple F_1 structure (Clarke et al. 1979, Biketov et al. 1982, Mileykovskaya et al. 1987). The H^+-ATPase from *Lactobacillus casei* possesses two types of F_1 and three types of F_0 subunits, with the F_1 containing six copies of a 43 kd polypeptide believed to be the homologue of the α and β subunits (Biketov et al. 1982, Mileykovskaya et al. 1987). This simplicity may reflect the role of the complex as an ATP-driven proton pump without the associated ATP synthase activity. In the aerobe *E. coli* eight types of subunits are found, with two (α and β) resulting from gene duplication (Walker et al. 1984). In the cyanobacterium *Synechococcus* 6301 and chloroplasts, there are apparently nine types of subunits with the extra subunit having arisen by duplication of the gene for subunit *b* (Cozens and Walker 1987). In mitochondria, there appear to be thirteen subunits, with the additional subunits apparently bearing little resemblance to *E. coli* subunits (Walker et al. 1987). In this case, recruitment of new polypeptides to the F_1F_0 may have occurred but the advantage of this increase in complexity is not known.

9. Future research

The next few years should see the characterisation of the nuclear *atp* genes and studies on the import of the encoded subunits II, δ and γ into the chloroplast. Further work on the chloroplast genes will be necessary to unravel the complexities of expression, particularly for the α operon. Of major interest in the long term are the problems of 1. how correct and coordinated expression and assembly of the ATP synthase subunits occur, and 2. how these processes are linked to the development of the plastid and the plant.

Acknowledgements

We thank Allan Downie and Paul Whitfeld for comments, and Denese McCann for typing of the manuscript.

578

References

Akey CW, Crepean RH, Dunn SD, McCarty RE and Edelstein SJ (1983) Electron microscopy and single molecule averaging of subunit-deficient F_1—ATPases from *Escherichia coli* and spinach chloroplasts. EMBO J 2: 1409–1413

Alt J and Herrmann RG (1984) Nucleotide sequence of the gene for pre-apocytochrome *f* in the spinach plastid chromosome. Curr Genet 8: 551–557

Alt J, Winter P, Sebald W, Moser JG, Schedel R, Westhoff P and Herrmann RG (1983) Localization and nucleotide sequence of the gene for the ATP synthase proteolipid subunit on the spinach plastid chromosome. Curr Genet 7: 129–138

Anderson S, de Bruijn MHL, Coulson AR, Eperon IC, Sanger F and Young IG (1982) Complete sequence of bovine mitochondrial DNA. Conserved features of the mammalian mitochondrial genome. J Mol Biol 156: 683–717

Arana JL and Vallejos RH (1982) Involvement of sulfhydryl groups in the activation mechanism of the ATPase activity of chloroplast coupling factor 1. J Biol Chem 257: 1125–1127

Aris JP, Klionsky DJ and Simoni RD (1985) The F_0 subunits of the *Escherichia coli* F_1F_0-ATP synthase are sufficient to form a functional proton pore. J Biol Chem 260: 11207–11215

Biekmann S and Feierabend J (1985) Synthesis and degradation of unassembled polypeptides of the coupling factor of photophosphorylation CF_1 in 70S ribosome-deficient rye leaves. Eur J Biochem 152: 529–535

Biketov SF, Kasho VN, Kozlov IA, Mileykovskaya YI, Ostrovsky DN, Skulachev VP, Tikhonova GV and Tsuprun VL (1982) F_1-like ATPase from anaerobic bacterium *Lactobacillus casei* contains six similar subunits. Eur J Biochem 129: 241–250

Bird CR, Koller B, Auffret AD, Huttly AK, Howe CJ, Dyer TA and Gray JC (1985) The wheat chloroplast gene for CF_0 subunit I of ATP synthase contains a large intron. EMBO J 4: 1381–1388

Bradley D and Gatenby AA (1985) Mutational analysis of the maize chloroplast ATPase-β subunit gene promoter: the isolation of promoter mutants in *E. coli* and their characterization in a chloroplast *in vitro* transcription system. EMBO J 4: 3641–3648

Cain BD and Simoni RD (1986) Impaired proton conductivity resulting from mutations in the *a* subunit of F_1F_0 ATPase in *Escherichia coli*. J Biol Chem 261: 10043–10050

Clarke DJ, Fuller FM and Morris JG (1979) The proton-translocating adenosine triphosphatase of the obligately anaerobic bacterium *Clostridium pasteurianum*. 1. ATP phosphohydralase activity. Eur J Biochem 9: 597–612

Cox GB, Downie JA, Langman L, Senior AE, Ash G, Fayle DRH and Gibson F (1981) Assembly of the adenosine triphosphatase complex in *Escherichia coli*: assembly of F_0 is dependent on the formation of specific F_1 subunits. J Bact 148: 30–42

Cox GB, Fimmel AL, Gibson F and Hatch L (1986) The mechanism of ATP synthase: a reassessment of the functions of the *b*- and *a*-subunits. Biochim Biophys Acta 849: 62–69

Cox GB, Hatch L, Webb D, Fimmel AL, Lin Z-H, Senior AE and Gibson F (1987) Amino acid substitutions in the ε-subunit of the F_1F_0-ATPase of *Escherichia coli*. Biochim Biophys Acta 890: 195–204

Cozens AL and Walker JE (1987) The organization and sequence of the genes for ATP synthase subunits in the cyanobacterium *Synechococcus* 6301. Support for an endosymbiont origin of chloroplasts. J Mol Biol 194: 359–383

Cozens AL, Walker JE, Phillips AL, Huttly AK and Gray JC (1986) A sixth subunit of ATP synthase, an F_0 component, is encoded in the pea chloroplast genome. EMBO J 5: 217–222

Cross RL (1981) The mechanism and regulation of ATP synthesis by F_1-ATPases. Ann Rev Biochem 50: 681–714

De Heij HT, Jochemsen A-G, Willemsen PTJ and Groot GSP (1984) Protein synthesis during chloroplast development in *Spirodela oligorhiza*. Coordinated synthesis of chloroplast-encoded and nuclear-encoded subunits of ATPase and ribulose-1,5-bisphosphate carboxylase. Eur J Biochem 138: 161–168

Deng X-W and Gruissem W (1987) Control of plastid gene expression during development: the limited role of transcriptional regulation. Cell 49: 379–387

Deno H, Shinozaki K and Suguira M (1983) Nucleotide sequence of tobacco chloroplast gene for the α subunit of proton-translocating ATPase. Nucleic Acids Res 10: 7511–7520

Deno H, Shinozaki K and Sugiura M (1984) Nucleotide sequence and transcription of tobacco chloroplast genes for CF_0 subunits of proton-translocating ATPase. Gene 32: 195–201

Dunn SD and Heppel LA (1981) Properties and functions of the subunits of the *Escherichia coli* coupling factor ATPase. Arch Biochem Biophys 210: 421–436

Echt CS, Polacco ML and Neuffer MG (1987) A nuclear encoded chloroplast ATP synthase mutant of *Zea mays* L. Mol Gen Genet 208: 230–234

Fillingame RH (1981) Biochemistry and genetics of bacterial H^+-translocating ATPases. Curr Topics Bioenerg 11: 35–106

Finel M, Rubinstein M and Pick U (1984) Preparation of an ε-deficient chloroplast coupling factor 1 having a high ATPase activity. FEBS Lett 166: 85–89

Foster DL and Fillingame RH (1982) Stoichiometry of subunits in the H^+-ATPase complex of *Escherichia coli*. J Biol Chem 257: 2009–2015

Fromme P, Gräber P and Salnikow J (1987) Isolation and identification of a fourth subunit in the membrane part of the chloroplast ATP-synthase. FEBS Lett 218: 27–30

Futai M and Kanazawa H (1980) Role of subunits in proton-translocating ATPase (F_0-F_1). Curr Topics Bioenerg 10: 181–215

Gray MW and Doolittle WF (1982) Has the endosymbiont hypothesis been proven? Micro Rev 46: 1–42

Gruissem W and Zurawski G (1985) Analysis of promoter regions for the spinach chloroplast *rbc*L, *atp*B and *psb*A genes. EMBO J 4: 3375–3383

Hanley-Bowdoin L and Chua N-H (1987) Chloroplast promoters. Trends in Biochem Sci 12: 67–70

Hennig J and Herrmann RG (1986) Chloroplast ATP synthase of spinach contains nine nonidentical subunit species, six of which are encoded by plastid chromosomes in two operons in a phylogenetically conserved arrangement. Mol Gen Genet 203: 117–128

Herrmann RG, Westhoff P, Alt J, Tittgen J and Nelson N (1985) Thylakoid membranes and their genes. In: Van Vloten-Doting L, Grott GSP and Hall TC (eds) Molecular Form and Function of the Plant Genome. New York: Plenum Press

Hoppe J and Sebald W (1984) The proton conducting F_0-part of bacterial ATP synthases. Biochim Biophys Acta 768: 1–27

Howe CJ, Auffret AD, Doherty A, Bowman CM, Dyer TA and Gray JC (1982) Location and nucleotide sequence of the gene for the proton-translocating subunit of wheat chloroplast ATP synthase. Proc Natl Acad Sci USA 79: 6903–6907

Howe CJ, Bowman CM, Dyer TA and Gray JC (1983) The genes for the alpha and proton-translocating subunits of wheat chloroplast ATP synthase are close together on the same strand of chloroplast DNA. Mol Gen Genet 190: 51–55

Howe CJ, Fearnley IM, Walker JE, Dyer TA and Gray JC (1985) Nucleotide sequences of the genes for the alpha, beta and epsilon subunits of wheat chloroplast ATP synthase. Plant Mol Biol 4: 333–345

Hudson GS, Mason JG, Holton TA, Koller B, Cox G, Whitfeld PR and Bottomley W (1987) A gene cluster in the spinach and pea chloroplast genomes encoding one CF_1 and three CF_0 subunits of the H^+-ATP synthase complex and the ribosomal protein S2. J Mol Biol 196: 283–298

Hudson GS, Holton TA, Whitfeld PR and Bottomley W (1988) The spinach chloroplast *rpo*BC genes encode three subunits of the chloroplast RNA polymerase. J Mol Biol, in press

Kanazawa H and Futai M (1982) Structure and function of H⁺-ATPase: What we have learned from *E. coli* H⁺-ATPase. Ann NY Acad Sci 402: 45–64

Klionsky DJ, Brusilow WSA and Simoni RD (1984) In vivo evidence for the role of the ε subunit as an inhibitor of the proton-translocating ATPase of *Escherichia coli*. J Bact 160: 1055–1060

Kobayashi K, Nakamura K and Asahi T (1987) CF₁ ATPase β- and ε-subunit genes are separated in the sweet potato chloroplast genome. Nucleic Acids Res 15: 7177

Krebbers ET, Larrinua IM, McIntosh L and Bogorad L (1982) The maize chloroplast genes for the β and ε subunits of the photosynthetic coupling factor CF₁ are fused. Nucleic Acids Res 10: 4985–5002

McCarthy JEG, Schairer HU and Sebald W (1985) Translation initiation frequency of *atp* genes from *Escherichia coli*: identification of an intercistronic sequence that enhances translation. EMBO J. 4: 519–526

Merchant S and Selman BR (1985) Photosynthetic ATPases: purification, properties, subunit isolation and function. Photosynth Res 6: 3–31

Michel F and Dujon B (1983) Conservation of RNA secondary structure in two intron families including mitochondrial-, chloroplast- and nuclear-encoded members. EMBO J 2: 33–38

Mileykovskaya EI, Abuladze AN and Ostrovsky DN (1987) Subunit composition of the H⁺-ATPase complex from anaerobic bacterium *Lactobacillus casei*. Eur J Biochem 168: 703–708

Mullet JE, Orozco EM JR and Chua N-H (1985) Multiple transcripts for higher plant *rbc*L and *atp*B genes and localization of the transcription initiation site of the *rbc*/L gene. Plant Mol Biol 4: 39–54

Nalin CM and McCarty RE (1984) Role of a disulphide bond in the γ subunit in activation of the ATPase of chloroplast coupling factor 1. J Biol Chem 259: 7275–7280

Nelson N (1981) Proton-ATPase of chloroplasts. Curr Topics Bioenerg 11: 1–33

Nelson N, Nelson H and Schatz G (1980) Biosynthesis and assembly of the proton-translocating adenosine triphosphatase complex from chloroplasts. Proc Natl Acad Sci USA 77: 1361–1364

Ohyama K, Fukuzawa H, Kohchi T, Shirai H, Sano T, Sano S, Umesono K, Shiki Y, Takeuchi M, Chang Z, Aota S, Inokuchi H and Ozeki H (1986) Chloroplast gene organization deduced from complete sequence of liverwort *Marchantia polymorpha* chloroplast DNA. Nature 322: 572–574

Orozco EM Jr, Mullet JE and Chua N-H (1985) An in vitro system for accurate transcription initiation of chloroplast protein genes. Nucleic Acids Res 13: 1283–1302

Palmer JD (1985) Comparative organization of chloroplast genomes. Ann Rev Genet 19: 325–354

Patrie WJ and McCarty RE (1984) Specific binding of coupling factor 1 lacking the δ and ε subunits to thylakoids. J Biol Chem 259: 11121–11128

Perlin DS, Cox DN and Senior AE (1983) Integration of F₁ and the membrane sector of the proton-ATPase of *Escherichia coli*. J Biol Chem 258: 9793–9800

Pick U and Racker E (1979) Purification and reconstitution of the *N,N*′-dicyclohexylcarbodiimide-sensitive ATPase complex from spinach chloroplasts. J Biol Chem 254: 2793–2799

Richter ML, Patrie WJ and McCarty RE (1984) Preparation of the ε subunit and ε subunit-deficient chloroplast coupling factor 1 in reconstitutively active forms. J Biol Chem 259: 7371–7373

Rodermel SR and Bogorad L (1987) Molecular evolution and nucleotide sequences of the maize plastid genes for the α subunit of CF$_1$ (atpA) and the proteolipid subunit of CF$_0$ (atpH). Genetics 116: 127–139

Sears B and Herrmann RG (1985) Plastome mutation affecting the chloroplast ATP synthase involves a post-transcriptional defect. Curr Genet 9: 521–528

Sebald W and Hoppe J (1981) On the structure and genetics of the proteolipid subunit of the ATP synthase complex. Curr Topics Bioenerg 12: 1–60

Sebald W and Wachter E (1980) Amino acid sequence of the proteolipid subunit of the ATP synthase from spinach chloroplasts. FEBS Lett 122: 307–311

Senior AE and Wise JG (1983) The proton-ATPase of bacteria and mitochondria. J Membrane Biol 73: 105–124

Shavit N (1980) Energy transduction in chloroplasts: structure and function of the ATPase complex. Ann Rev Biochem 49: 111–138

Shinozaki K, Deno H, Kato A and Sugiura M (1983) Overlap and cotranscription of the genes for the beta and epsilon subunits of tobacco chloroplast ATPase. Gene 32: 195–201

Shinozaki K, Deno H, Wakasugi T and Sugiura M (1986a) Tobacco chloroplast gene coding for subunit I of proton-translocating ATPase: comparison with the wheat subunit I and E. coli subunit b. Curr Genet 10: 421–423

Shinozaki K, Ohme M, Tanaka M, Wakasugi T, Hayashida N, Matsubayashi T, Zaita N, Chunwongse J, Obokata J, Yamaguchi-Shinozaki K, Ohto C, Torazawa K, Meng BY, Sugita M, Deno H, Kamogashira T, Yamada K, Kusuda J, Takaiwa F, Kato A, Tohdoh N, Shimada H and Sugiura M (1986b) The complete nucleotide sequence of the tobacco chloroplast genome: its gene organization and expression. EMBO J 5: 2043–2049

Strotmann H and Bickel-Sandkötter S (1984) Structure, function, and regulation of chloroplast ATPase. Ann Rev Plant Physiol 35: 97–120

Süss K-H (1980) Identification of chloroplast thylakoid membrane polypeptides. ATPase complex (CF$_1$–CF$_0$) and light-harvesting chlorophyll a/b-protein (LHCP) complex. FEBS Lett 112: 255–259

Süss K-H and Schmidt O (1982) Evidence for an α$_3$, β$_3$, γ, δ, I, II, ε, III$_5$ subunit stoichiometry of chloroplast ATP synthetase complex (CF$_1$–CF$_0$). FEBS Lett. 144: 213–218

Tittgen J, Hermans J, Steppuhn J, Jansen T, Jansson C, Andersson B, Neuchushtai R, Nelson N and Herrmann RG (1986) Isolation of cDNA clones for fourteen nuclear-encoded thylakoid membrane proteins. Mol Gen Genet 204: 258–265

Walker JE, Saraste M and Gay NJ (1984) The unc operon. Nucleotide sequence, regulation and structure of ATP-synthase. Biochim Biophys Acta 768: 164–200

Walker JE, Fearnley IM, Gay NJ, Gibson BW, Northrop FD, Powell SJ, Runswick MJ, Saraste M and Tybulewicz VLJ (1985) Primary structure and subunit stoichiometry of F$_1$-ATPase from bovine mitochondria. J Mol Biol 184: 677–701

Walker JE, Runswick MJ and Poulter L (1987) ATP synthase from bovine mitochondria. The characterization and sequence analysis of two membrane-associated sub-units and of the corresponding cDNAs. J Mol Biol 197: 89–100

Watanabe A and Price CA (1982) Translation of mRNAs for subunits of chloroplast coupling factor 1 in spinach. Proc Natl Acad Sci USA 79: 6304–6308

Westhoff P, Nelson N, Bunemann H and Herrmann RG (1981) Localization of genes for coupling factor subunits on the spinach plastid chromosome. Curr Genet 4: 109–120

Westhoff P, Alt J, Nelson N and Herrmann RG (1985) Genes and transcripts for the ATP synthase CF$_0$ subunits I and II from spinach thylakoid membranes. Mol Gen Genet 199: 290–299

Wiley DL, Howe CJ, Auffret AD, Bowman CM, Dyer TA and Gray JC (1984) Location and nucleotide sequence of the gene for cytochrome f in wheat chloroplast DNA. Mol Gen Genet 194: 416–422

Woese CR (1987) Bacterial evolution. Micro Rev 51: 221–271

Woessner JP, Gillham NW and Boynton JE (1987) Chloroplast genes encoding subunits of the H$^+$-ATPase complex of *Chlamydomonas reinhardtii* are rearranged compared to higher plants: sequence of the *atp*E gene and location of the *atp*F and *atp*I genes. Plant Mol Biol 8: 151–158

Zurawski G and Clegg MT (1984) The barley chloroplast DNA *atp*BE, *trn*M2, and *trn*V1 loci. Nucleic Acids Res 12: 2549–2559

Zurawski G and Clegg MT (1987) Evolution of higher-plant chloroplast DNA-encoded genes: implications for structure-function and phylogenetic studies. Ann Rev Plant Physiol 38: 391–418

Zurawski G, Bottomley W and Whitfeld PR (1982) Structures of the genes for the β and ε subunits of spinach chloroplast ATPase indicate a dicistronic mRNA and an overlapping translation stop/start signal. Proc Natl Acad Sci USA 79: 6260–6264

Zurawski G, Bottomley W and Whitfeld PR (1986) Sequence of the genes for the β and ε subunits of ATP synthase from pea chloroplasts. Nucleic Acids Res 14: 3974

Govindjee et al. (eds), Molecular Biology of Photosynthesis: 583–591
© 1988 Kluwer Academic Publishers

Regular paper

Structural gene regions of *Rhodobacter sphaeroides* involved in CO₂ fixation

PAUL L. HALLENBECK & SAMUEL KAPLAN[1]

[1]*Department of Microbiology, University of Illinois, 407 South Goodwin Avenue, Urbana, IL 61801, USA; [1]author for correspondence*

Received 6 October 1987; accepted February 25 1988

Key words: CO₂ fixation, gene expression

Abstract. From studies conducted in both our laboratory and by Gibson, Tabita and colleagues, as well as drawing on the recent studies with *Alcaligenes eutrophus*, we describe two genetic regions which have been identified on the chromosome of *Rhodobacter sphaeroides* which code for a number of enzymes involved in CO₂ fixation. One region was found to contain the genes coding for fructose 1,6-bisphosphatase (*fbpA*), phosphoribulokinase (*prkA*), a 37 kDa polypeptide (*cfxA*), and form I ribulose 1,5-bisphosphate carboxylase/oxygenase (*rbcL, S*). These genes appear to be expressed in the same transcriptional direction and are tandomly arranged. A second, apparently unlinked region of the chromosome contains a duplicate (with respect to functionality of gene products) but not identical set of these same four genes. Although the gene order in both regions is apparently identical, there is approximately 4 kb of DNA separating the 3'- end of *prkB* and the beginning of *cfxB*. The specific genetic organizations and proposed roles of these two genetic regions are discussed.

Introduction

Rhodobacter sphaeroides is a Gram-negative nonsulfur purple facultative photoheterotrophic bacterium. This bacterium has the remarkable ability of being able to grow under a wide range of environmental conditions exhibiting chemoheterotrophic, photoheterotrophic, and photoautotrophic growth (Madigan and Gest 1979). Under the latter two growth conditions the synthesis of several of the Calvin cycle enzymes is derepressed in *R. sphaeroides* (Tabita 1981, Jouanneau and Tabita 1986, Hallenbeck and Kaplan unpublished observations). Further, the level of at least one of these enzymes can vary under photoheterotrophic growth conditions depending on the oxidation state of the carbon source and the incident light intensity to the culture (Zhu and Kaplan 1985, Tabita 1981). This change in enzyme level is at least partially under transcriptional control (Zhu and Kaplan 1985). The unique enzymes of the above pathway, namely, phosphoribulokinase (PRK) (ATP:D-ribulose 5-phosphate-1-phosphotransferase; EC 2.7.1.19) and ribulose 1,5-bisphosphate carboxylase-oxygenase (RuBPCase)(3-phospho-D-glycerate carboxylase [dimerizing]; EC 4.1.1.39) are physiologically dedicated to the Calvin cycle (Tabita 1981). PRK is the last step in the reductive pentose phosphate cycle in that it phosphorylates ribulose 5-phosphate (RuMP) to regenerate the primary CO_2 acceptor ribulose 1,5-bisphosphate(RuBP). RuBPCase then carboxylates the primary acceptor to yield 3-phosphoglycerate(3-PGA). *R. sphaeroides* has the additionally unusual feature of possessing two forms of RuBPCase. Form I represents the plant or autotrophic prototype whereas the form II enzyme is only found in relatively few bacteria, mainly the nonsulfur purple group. The distinguishing feature between the form I and form II enzymes is that the latter has no small subunit in association (Gibson and Tabita 1977).

Since PRK and RuBPCase are known to be derepressed when grown photoheterotrophically on reduced substrates such as butyrate (Tabita 1981), it might be reasonable to assume that the synthesis of alkaline fructose 1,6-bisphosphatase(FBPase)(D-fructose 1,6-bisphosphate 1-phosphohydrolase;EC 3.1.1.11) and glyceraldehyde 3-phosphate dehydrogenase (GAPDH) (D-glceraldehyde-3-phosphate:NAD$^+$ oxidoreductase; EC 1.2.1.12) as well as other physiologically related enzymes may also be subject to regulatory control. Indeed, RuBPCase, PRK, FBPase, and GAPDH have higher specific activities when cultured autotrophically in the related organism *Rhodospirillum rubrum* (Anderson and Fuller 1967). However, unlike PRK and RuBPCase, FBPase and GAPDH as well as other physiologically related enzymes are also expressed in chemoheterotrophically growing cells. FBPase is essential in *R. sphaeroides* cells grown either chemoheterotrophically or photoautotrophically, being that it catalyses the conversion, during gluconeogenesis, of fructose 1,6-bisphosphate(FBP) to fructose 6-phosphate(FMP) which amongst other things is required for cell wall biosynthesis. FMP is also a precursor for the regeneration of RuBP used in the Calvin cycle. GAPDH serves in the oxidation of glyceraldehyde 3-phosphate to 3-phosphoglyceroyl phosphate in glycolysis, as well as the reverse reaction in gluconeogenesis. In the reductive pentose phosphate cycle GAPDH catalyses the reduction of 3-phophoglyceroyl phosphate to glyceraldehyde 3-phosphate. Whether FBPase and GAPDH function as unique gene products, one for gluconeogenesis and one for the Calvin cycle is presently unknown.

R. sphaeroides must have the capacity to post-translationally regulate the above mentioned enzymes as well as regulate their synthesis, given the enormous complexity inherent in the ability of this organism to grow on a diversity of carbon sources and environmental conditions. How the synthesis and activity of RuBPCase, PRK, FBPase and GAPDH as well as

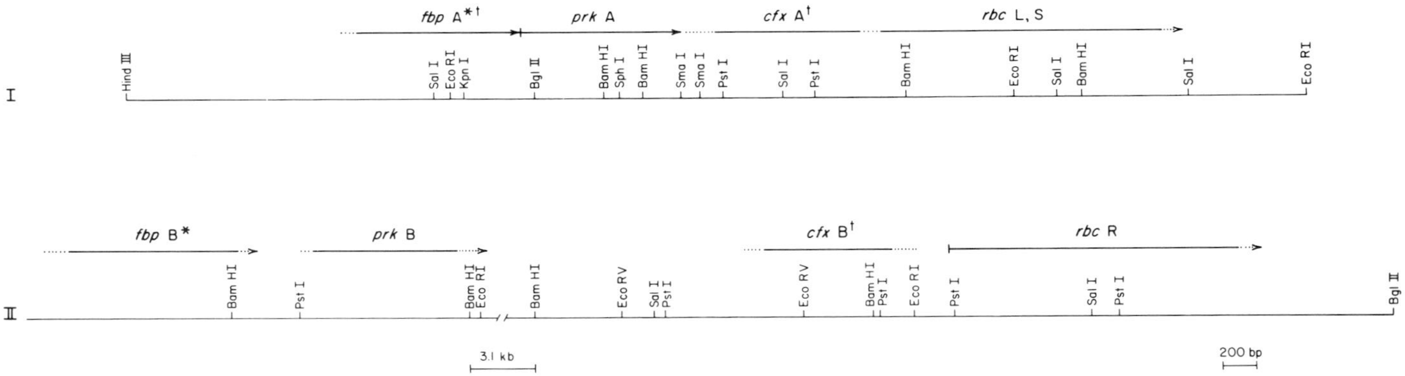

Fig. 1. Restriction maps of two genetic regions carrying carbon fixation genes in *R. sphaeroides*. The arrows indicate the direction of transcription and the approximate size and location of these genes. *Kindly provided by J.L. Gibson and F.R. Tabita, The University of Texas, Austin, TX, personal communication. † Unpublished results from the authors laboratory. The position of all of the other genes is from published information.

other key enzymes is regulated is an important topic for investigation. As a first step to understanding the complex regulation of Calvin cycle enzymes, the genes for some of the above mentioned enzymes have been identified and described in *R. sphaeroides*. The results presented below and in Fig. 1 represent a composite of the data derived from our laboratory and that of Tabita and coworkers.

Discussion

The gene coding for form II RuBPCase (*rbcR*), Fig. 1, region II, was originally isolated by Quivey and Tabita (Quivey and Tabita 1984) and Muller, Chory and Kaplan (Muller et al. 1985a, Muller et al. 1985b). Their approach involved the use of heterologous DNA hybridization using the *Rhodospirillum rubrum* form II gene (Somerville and Somerville 1984) as a probe in Southern hybridization analysis. Further studies demonstrated that *rbcR* could be expressed and its gene product characterized in *Escherichia coli* (Quivey and Tabita 1984, Tabita et al. 1986, Muller et al. 1985a, Muller et al. 1985b) as well as in an *R. sphaeroides* in vitro transcription-translation system (Chory et al. 1985). Subsequent hybridization experiments revealed that the region of DNA immediately upstream, which we designate *cfxB*, and including a proximal portion of *rbcR*, had significant homology to another region of DNA elsewhere on the *R. sphaeroides* chromosome, which we designated *cfxA* (Fig. 1) (Hallenbeck and Kaplan 1987, Muller et al. 1985a, Muller et al. 1985b). We identified the *cfxA* gene as coding for a polypeptide of approximately 37 kDa. Interestingly, in vitro transcription-translation analysis indicated that this polypeptide may undergo processing to a smaller gene product of 35.5 kDa (Hallenbeck and Kaplan 1987). Similar analysis of the *cfxB* gene revealed that it codes for a polypeptide of only 34 kDa which may also be processed (Hallenbeck and Kaplan unpublished observations). Comparison of the partial DNA sequence of *cfxB* and *cfxA* has shown a remarkable similarity between not only the DNA sequence of these two genes but also their deduced amino acid sequences. DNA sequence analysis of *cfxB* and *cfxA*, along with other recently obtained data (Hallenbeck and Kaplan, unpublished observations), suggest that *cfxB* is transcribed in the same direction as *rbcR* and similarly *cfxA* is transcribed in the same direction as the gene coding for Form I RuBPCase (Gibson and Tabita 1986) which is located immediately downstream of *cfxA*.

The location of the Form I RuBPCase gene (*rbcL,S*) (Gibson and Tabita 1987) with respect to *cfxA* was conclusively identified by hybridization to the

Anacystis nidulans rbcL,S genes (Shinozaki et al. 1983) and its position similarly inferred by Hallenbeck and Kaplan (1987). Subsequent DNA sequence determination confirmed that there is 45% homology between the deduced amino terminal 51 amino acids of the *R. sphaeroides rbcL* gene and the *A. nidulans rbcL* gene (Shinozaki et al. 1983; Hallenbeck and Kaplan unpublished observations). Gibson and Tabita have expressed the *R. sphaeroides rbcL,S* genes in *E. coli* (Gibson and Tabita 1986). Preliminary evidence suggests that the translational stop codons for both the open reading frames *cfxA* and *cfxB* are very near the start codons for *rbcL* and *rbcR* respectively (Hallenbeck and Kaplan, unpublished observations).

The gene for PRK (*prkA*) was recently cloned from *R. sphaeroides* (Hallenbeck and Kaplan 1987, Gibson and Tabita 1987). It was originally identified by expression and characterization of the gene product in *E. coli.* Amino acid sequence analysis of the amino terminal end of partially purified PRK isolated from *R. sphaeroides* as well as the PRK expressed from the cloned DNA in *E. coli* confirmed their identity. By comparing the amino acid sequence with the deduced amino acid sequence derived from DNA sequencing studies we were able to position the exact location of the 5' end of *prkA* and deletion analysis identified the approximate 3' end of *prkA* (Hallenbeck and Kaplan 1987). An internal portion of *prkA* from *R. sphaeroides* and a fragment of DNA carrying at least a portion of the PRK gene from *Alcaligenes eutrophus*, hybridized to a cosmid containing the *rbcR* gene (Hallenbeck and Kaplan 1987; Hallenbeck and Kaplan unpublished observations). Gibson and Tabita showed that this region of similarity encoded a second *prk* gene (*prkB*) (Gibson and Tabita 1987). It was also shown that *prkA*, when expressed in and purified from *E. coli* has an almost complete requirement for NADH as an activator of enzyme activity (Hallenbeck and Kaplan 1987) whereas *prkB* expressed and the gene product purified in an analogous manner, is still 75% active in the absence of NADH (Gibson and Tabita 1987).

Interestingly, there are approximately 4 kb of DNA which separate the 3' end of *prkB* and the 5' end of *cfxB* whereas the distance between the 3' end of *prkA* and the 5' end of *cfxA* is at most 150 bp (Fig. 1) (Hallenbeck and Kaplan unpublished observations). Within this 4 kb region lies the gene coding for GAPDH (Gibson and Tabita 1988). However, we have no evidence to suggest that *cfxB* codes for GAPDH. Although the gene organization in *A. entrophus* (Klintworth et al. 1985, Bowien et al. 1987) for both sets of *prk* and *rbc* genes, one on a 450 kb megaplasmid and one on the chromosome, are in the opposite orientation to that described above, there is 3.5 kb of DNA between the 3' end of *rbcS* and the 5' end of *prk*. Whether this region contains other genes involved in carbon dioxide fixation such as the gene for GAPDH as reported by Tabita (Gibson and Tabita 1988) is unknown.

There are two genes coding for FBPase, one upstream of *prkA* (*fbpA*) and the other upstream of *prkB* (*fbpB*) (Gibson and Tabita 1988). During the course of DNA sequencing of *prkA* we identified a portion of an open reading frame immediately to the left of *prkA*. When comparing the deduced amino acid sequence to the amino acid sequence of FBPases from pig kidney, spinach chloroplast, *Saccharomyces cerevisiae*, and *E. coli* considerable homology was observed within several conserved regions (Marcus et al. 1986). However, preliminary data indicates that the *R. sphaeroides* FBPase has the most divergent amino acid sequence of those FBPase amino acid sequences thus far known and compared. The location of the 5′ end of the gene for FBPase (*fbpA*) is approximately 1 kb upstream of the 5′ end of *prkA* and is transcribed in the same direction as the latter. Notably, the apparent stop codon of *fbpA* is located within the ribosome binding site of *prkA* (Hallenbeck and Kaplan unpublished results). Like *prkA*, *cfxA*, and *rbcL,S*, there is a counterpart to the *fbpA* gene, designated *fbpB* which is located upstream of and transcribed in the same direction as *prkB* (Gibson and Tabita 1988).

Why these genes are arranged as indicated in Fig. 1 is the subject of ongoing research. We have some evidence which can be interpreted to suggest that *prkA, cfxA*, (Hallenbeck and Kaplan 1987) and *rbcR* (Muller et al. 1985b) have their own promoters. However, other evidence indicates that proper expression and regulation of these gene products requires expression of a promoter(s) or regulatory region further upstream of these same structural genes. For example, a 'knockout' mutation in the structural gene for *prkA*, which can grow photoautotrophically only very slowly, cannot be complemented *in trans* with a plasmid containing the *prkA* structural gene and 480 bp of upstream DNA (Hallenbeck and Kaplan unpublished observations). This construct however does express *prkA* in an *R. sphaeroides in vitro* transcription-translation system (Hallenbeck and Kaplan 1987). Likewise, a plasmid containing a translational fusion of β-galactosidase to the amino terminal end of *prkA* and the same 480 bp of upstream DNA, expresses β-galactosidase in *R. sphaeroides* only when grown anaerobically (Hessler, Hallenbeck and Kaplan unpublished observations). Although this may imply normal expression of *prkA* by a promoter within the 480 bp, further experimentation is required to elucidate the precise mechanism(s) involved in regulating gene expression. It will certainly be of interest to ascertain the level of expression of *prkA* and *prkB*, under varying growth conditions. Presumably, total PRK activity with respect to whole cell protein, is at least controlled by oxygen, reducing power, and CO_2 concentration as is RuBPCase activity (Tabita 1981).

The regulation of expression of *rbcR* by light, oxygen, and reducing power can be transcriptionally controlled (Zhu et al. 1985). Similar regulation of expression of RuBPCase, and possibly PRK by reducing power in *A. eutrophus* has been proposed (Im et al. 1983). Evidence that the transcriptional control of *rbcR* expression requires DNA *in cis* much further upstream of the 5′ end of the structural gene is derived from *lac* fusion experiments. An in frame fusion of the DNA coding for the amino terminal end of Form II RuBPCase and 400 bp of upstream DNA to the DNA coding for β-galactosidase was constructed and placed *in trans*. When this construct was analyzed in *R. sphaeroides* grown aerobically, high level expression of β-galactosidase was observed (Hallenbeck, Sorenson, and Kaplan unpublished results). However, no *rbcR* mRNA can be detected in wild type *R. sphaeroides* cells grown aerobically (Zhu et al. 1985). Since *rbcR* specific transcripts of between 1.4–2.3 kb have been observed in *R. sphaeroides* grown photoheterotrophically (Zhu et al. 1985), we must conclude that either initiation or control of transcription originates from a region upstream of *cfxB*.

Like Form II RuBPCase, PRK B is the predominant form of the enzyme when cells are cultured photoheterotrophically on reduced substrates such as succinate (Hallenbeck and Kaplan unpublished observations). However, cells grown photoautotrophically contain more Form I RuBPCase than Form II RuBPCase, (Jouanneau and Tabita 1986). Thus it might be concluded that the genes in region I are expressed and the gene products utilized primarily for autotrophic growth whereas those genes in region II are expressed when cells are grown photoheterotrophically (Falcone et al. 1988). These enzymes, coded for by genes in region II, may function under the latter condition as a sink for excess reducing power since the majority of fixed carbon is from succinate and not CO_2 (Tabita 1981). DMSO, an alternate electron sink, greatly diminishes the expression of PRK and reduces the cellular level of RuBPCase (Hallenbeck, McEwan and Kaplan unpublished observations) when cells are growing photoheterotrophically.

R. sphaeroides strains have been constructed which contain a mutation in *prkA* and these grow very slowly photoautotrophically with respect to the wild type indicating that the analogous, apparently functional gene (*prkB*) is unable to compensate for the loss (Hallenbeck and Kaplan unpublished results). Although the function of *cfxA* and *cfxB* is unknown their locations suggest that one or both play a role in the regulation of carbon dioxide fixation. Evidence that the two regions, comprising the genes for carbon dioxide fixation in *A. eutrophus* have shared regulatory signals comes from the fact that a Tn5 induced mutation in the chromosome completely abolishes synthesis of PRK and RuBPCase from either set of genes (Bowien 1987). The fact that *prkA* and *prkB* are located on separate *DraI* fragments of between 630 to 700 kb makes the possibility of either gene being located on a plasmid remote unless there is an as yet unidentified plasmid in *R. sphaeroides* that is exceedingly large (Wu Yong Qiang and Kaplan unpublished observations). It should also be noted that the four enzymes discussed, namely FBPase, GAPDH, PRK, and RuBPCase as well as others not discussed here, are all subject to complex post-translational and/or allosteric control (Bassham et al. 1982, Bowien et al. 1987, Hallenbeck et al. 1987, Gibson et al. 1987, Jouanneau et al. 1987) in addition to the complex positional and transcriptional control discussed here. The elucidation of the regulation of carbon dioxide fixation genes coded for in these two genetic regions (which together with any other genes involved in CO_2 fixation we designate the *cfx* regulon) will certainly yield valuable information towards our understanding carbon metabolism in general in *R. sphaeroides* as well as other living systems.

Acknowledgements

We gratefully acknowledge J.L. Gibson, F.R. Tabita and coworkers for the use of their unpublished data.

References

Anderson L and Fuller RC (1967) Photosynthesis in *Rhodospirillum rubrum*. II. Photoheterotrophic carbon dioxide fixation. Plant Physiol 42: 491

Bassham JA and Buchanan BB (1982) Carbon dioxide fixation pathways in plants and bacteria, p. 141–189. In: Govindjee (ed.) Photosynthesis: Development, Carbon Metabolism, and Plant Productivity, Vol. II. New York: Academic Press, New York

Bowien B, Gusemann M, Klintworth R and Windhörth U (1987) Metabolic and molecular regulation of the CO_2-assimilating enzyme system in aerobic chemoautotrophs, p. 21–27. In: Van Verseveld HW and Duine JA (eds) Microbial Growth on C_1 Compounds. Dordrecht: Martinus Nijhoff Publishers

Chory J, Muller E and Kaplan S (1985) DNA-directed *in vitro* synthesis and assembly of the form II D-ribulose-1,5-bisphosphate carboxylase/oxygenase from *Rhodopseudomonas sphaeroides*. J Bacteriol 161: 307–313

Falcone DL, Quivey RG Jr and Tabita FR (1988) Transposon mutagenesis and physiological analysis of strains containing inactivated form I and form II ribulose bisphosphate carboxylase/oxygenase genes in *Rhodobacter sphaeroides*. J Bacteriol 170: 5–11

Gibson JL and Tabita FR (1977) Different molecular forms of D-ribulose 1,5-bisphosphate carboxylase from *Rhodopseudomonas sphaeroides*. J Biol Chem 252: 943–949

Gibson JL and Tabita FR (1986) Isolation of the *Rhodopseudodomas sphaeroides* form I ribulose 1,5-bisphosphate carboxylase/oxygenase large and small subunit genes and expression of active hexadecameric enzyme in *Escherichia coli*. Gene 44: 271–278

Gibson JL and Tabita FR (1987) Organization of phosphoribulokinase and ribulose bisphosphate carboxylase.oxygenase genes in *Rhodopseudomonas (Rhodobacter) sphaeroides*. J Bacteriol 169: 3685–3690

Gibson JL and Tabita FR (1988) Localization and mapping of CO_2 fixation genes within two gene clusters within *Rhodobacter sphaeroides*. J Bacteriol (in press)

Hallenbeck PL and Kaplan S (1987) Cloning of the gene for phosphoribulokinase activity from *Rhodobacter sphaeroides* and its expression in *Escherichia coli*. J Bacteriol 169: 3669–3678

Im DS and Friedrich CG (1983) Fluoride, hydrogen, and formate activate ribulose bisphosphate carboxylase formation in *Alcaligenes eutrophus*. J Bacteriol 154: 803–808

Jouanneau Y and Tabita FR (1986) Independent regulation of synthesis of form I and form II ribulose bisphosphate carboxylase-oxygenase in *Rhodopseudomonas sphaeroides*. J Bacteriol 165: 620–624

Jouanneau Y and Tabita FR (1987) *In vivo* Regulation of Form I Ribulose 1,5-Bisphosphate Carboxylase/Oxygenase from *Rhodopseudomonas sphaeroides*. Arch Biochem Biophys 254: 290–303

Klintworth R, Husemann M, Salnikov J and Bowien B (1985) Chromosomal and plasmid locations for phosphoribulokinase genes in *Alicaligenes eutrophus*. J Bacteriol 164: 954–956

Madigan MT and Gest H (1979) Growth of the photosynthetic bacterium *Rhodopseudomonas capsulata* chemoautotrophically in darkness with H_2 as the energy source. J Bacteriol 127: 530–542

Marcus F, Gontero B, Harrsch PB and Rittenhouse J (1986) Amino acid sequence homology among fructose-1,6-bisphosphatases. Biochem Biophys Res Commun 135: 374–381

Muller ED, Chory J and Kaplan S (1985a) Cloning and characterization of the gene product of the form II ribulose-1,5-bisphosphate carboxylase gene of *Rhodopseudomonas sphaeroides*. J Bacteriol 161: 469–472

Muller ED, Chory J and Kaplan S (1985b) Cloning and *in vitro* transcription-translation of the Form II ribulose-1,5-bisphosphate carboxylase gene of *Rhodopseudomonas sphaeroides*, pp. 319–324. In: Steinback KE, Banitz SB, Arntzen CJ and Bogorad LB (eds) Molecular biology of the photosynthetic apparatus. Cold Spring Harbor Laboratory

Quivey RJ Jr and Tabita FR (1984) Cloning and expression in *Escherichia coli* of the form II ribulose 1-,5-bisphosphate carboxylase/oxygenase gene from *Rhodopseudomonas sphaeroides*. Gene 31: 91–101

Shinozaki K, Yamada C, Takakata N and Sugiura M (1983) Molecular cloning and sequence analysis of the cyanobacterial gene for the large subunit of ribulose-1,5-bisphosphate carboxylase/oxygenase. Proc Natl Acad Sci USA 80: 4050–4054

Somerville CR and Somerville SC (1984) Cloning and expression of the *Rhodospirillum rubrum* ribulose-bisphosphate carboxylase gene. Mol Gen Genet 193: 214–219

Tabita FR (1981) Molecular regulation of carbon dioxide assimilation in autotrophic microorganisms, pp. 70–81. In: Dalton H (ed.) Microbial growth on C_1 compounds. London: Heydon and Sons

Tabita FR, Gibson JL, Mandy WJ and Quivey Jr RG (1986) Synthesis and assembly of a novel recombinant ribulose bisphosphate carboxylase/oxygenase. Bio/Technology 4: 138–141

Zhu YS and Kaplan S (1985) Effects of light, oxygen and substrates on steady-state levels of mRNA coding for ribulose 1,5-bisphosphate carboxylase and light-harvesting and reaction center polypeptides on *Rhodopseudomonas sphaeroides*. J Bacteriol 162: 925–932

Govindjee et al. (eds), Molecular Biology of Photosynthesis: 593–606
© 1988 Kluwer Academic Publishers

Minireview

Uptake and utilization of inorganic carbon by cyanobacteria

JOHN PIERCE[1] & TATSUO OMATA[2]
[1] *Central Research and Development Department, E.I. Du Pont de Nemours and Company, Experimental Station, Building 402, Room 2230, Wilmington, Delaware 19898, USA;* [2] *Solar Energy Research Group, The Institute of Physical and Chemical Research (RIKEN), Wako-shi, Saitama, 351-01, Japan*

Received 6 October 1987; accepted 9 December 1987

Key words: carbonic anhydrase, carboxysomes, CO_2 transport, cyanobacteria, inorganic carbon utilization, ribulose bisphosphate carboxylase

Abstract. In the cyanobacteria, mechanisms exist that allow photosynthetic CO_2 reduction to proceed efficiently even at very low levels of inorganic carbon. These inducible, active transport mechanisms enable the cyanobacteria to accumulate large internal concentrations of inorganic carbon that may be up to 1000-fold higher than the external concentration. As a result, the external concentration of inorganic carbon required to saturate cyanobacterial photosynthesis in vivo is orders of magnitude lower than that required to saturate the principal enzyme (ribulose bisphosphate carboxylase) involved in the fixation reactions. Since CO_2 is the substrate for carbon fixation, the cyanobacteria somehow perform the neat trick of concentrating this small, membrane permeable molecule at the site of CO_2 fixation. In this review, we will describe the biochemical and physiological experiments that have outlined the phenomenon of inorganic carbon accumulation, relate more recent genetic and molecular biological observations that attempt to define the constituents involved in this process, and discuss a speculative theory that suggests a unified view of inorganic carbon utilization by the cyanobacteria.

Abbreviations: C_i – Inorganic carbon, H-cells – Cells grown under high CO_2, L-cells – Cells grown under low CO_2, RuBP – Ribulose-1,5-bisphosphate, WT – Wild type

Introduction

Cyanobacteria generally utilize CO_2 as their primary carbon source. The supply of CO_2 often limits their growth, and this limitation can be broadly seen to derive from both physical and chemical factors. The low partial pressure of CO_2 in the atmosphere ($\sim 0.035\%$) and the slow diffusion of CO_2 through water conspire to lower intrinsic rates of delivery of CO_2 to the cyanobacterium. In addition, the primary agent of CO_2 fixation, ribulose bisphosphate (RuBP) carboxylase, suffers from a low intrinsic affinity for CO_2 and is also inhibited by the presence of O_2, an alternate substrate that is produced in large quantity by the primary light reactions of photosynthesis.

The detailed manner by which the cyanobacteria utilize inorganic carbon (C_i) is not known at present, but a variety of biochemical and physiological experiments have sketched the broad outlines of the problem. The techniques of molecular biology and genetic analysis are just now being used to address these questions. In this review, we will briefly outline the biochemical and physiological experiments on C_i transport, accumulation, and fixation that form the basis of our present knowledge of C_i utilization by the cyanobacteria. In addition, we will mention the more recent genetic experiments that have have provided interesting insights into the process. Finally, we will discuss some more speculative aspects of C_i utilization that may be of use in providing a logical framework for future research in the area. The interested reader is referred to a recent symposium proceedings for further information (Lucas and Berry 1985)

RuBP carboxylase and CO₂ fixation by cyanobacteria

The cyanobacteria fix CO_2 by the Calvin cycle of reductive photosynthesis, and are therefore susceptible to the usual inefficiencies associated with this type of metabolism. (Although there is evidence for appreciable levels of phospho-enol pyruvate carboxylase in the cyanobacteria (Codd and Stewart 1973, Colman et al. 1976), detailed, quantitative studies of the flux of carbon in cyanobacterial photosynthesis indicate that this C-4 type of carbon fixation does not predominate (Creach et al. 1981).) The inefficiency is due to the molecular properties of RuBP carboxylase, which catalyzes the condensation of RuBP with CO_2 in the commencing reaction of the Calvin cycle as well as the oxygenation of RuBP with O_2 in the commencing reaction of photorespiration.

Cyanobacterial RuBP carboxylases are large (530 kD), hexadecameric enzymes (L_8S_8) comprising 8 large catalytic subunits (51 kD) and 8 small subunits (15 kD) of unknown function (Takabe et al. 1976; Andrews et al. 1981). These two polypeptides are encoded by adjacent structural genes which are co-transcribed onto a single mRNA (Shinozaki and Sugiura 1985). It is presumably by this method of co-transcription that the 1:1 stoichiometry of large and small subunits is maintained by the organism. This simple arrangement has also made it possible to express both subunits of cyanobacterial RuBP carboxylases in *E. coli*, whereby large quantities of the enzyme have been made available for study (Gatenby et al. 1985, Tabita and Small 1985). The cyanobacterial RuBP carboxylases possess extraordinarily low affinities for CO_2 (200–300 μM) and extraordinarily high maximum turnover numbers (12 s^{-1}) compared to RuBP carboxylases from other sources (Badger 1980, Andrews and Abel 1981, TJ Andrews, personal communication). In addition, the relative specificity of these enzymes for CO_2 versus O_2 is low ($\sim$50) compared with that of higher plant enzymes (Jordan and Ogren 1981, 1983). As a result, the oxygenase activity of these enzymes is appreciable in solutions equilibrated in air.

A recent study showed that *Synechocystis* 6803 contained an amount of RuBP carboxylase that was $\sim$1.6–2 fold higher than that required to maintain maximum rates of light-driven CO_2 fixation (Pierce et al. 1988). This amount of extractable carboxylase activity (measured at saturating concentrations of substrates) was equivalent to the maximum rate of photosynthetic electron transport (measured in the presence of methylviologen). Since we are not certain of the activation level of this enzyme, nor of the exact concentrations of intracellular CO_2 and O_2 maintained during steady state metabolism, it is not clear whether RuBP carboxylase activity is limiting to overall photosynthesis. However, an enzyme with an intrinsic $K_m(CO_2) = 300\,\mu$M operating in air-saturated solutions ([O_2] $\sim 255\,\mu$M) would, due to O_2 inhibition, possess an effective $K_m(CO_2) \sim 600\,\mu$M. Under these conditions, the organism would be required to maintain a steady state internal [CO_2] of 600–1000 μM ($\sim$30–50 mM total internal C_i at an internal pH of 7.8) just to support the maximum observed rates of photosynthesis.

The measurement of extractable enzyme activities and quantitative extrapolation of these to in vivo activities is, of course, subject to numerous difficulties. However, regardless of whether the activity of RuBP carboxylase is cleanly rate limiting to photosynthesis, it has become clear that there is no overabundance of this activity. In studies with *Synechocystis* 6803, the natural RuBP carboxylase gene was replaced with that of the RuBP carboxylase from *Rhodospirillum rubrum* (Pierce et al. 1988). As a result of this mutagenesis, the mutant organism grew photosynthetically by virtue of the action of the foreign carboxylase. Maximum rates of photosynthetic electron transport were similar to WT levels, but the maximum rate of light-driven CO_2 fixation was $\sim$2-fold lower and the amount of extractable RuBP carboxylase activity was $\sim$3-fold lower in this mutant than in WT cells. In addition, the maximum growth rate of the mutant was $\sim$2-fold lower than that of the WT cells. It seems clear that photosynthesis in the mutant is limited by RuBP carboxylase. Further, since a three-fold diminution in carboxylase activity was associated with a two fold diminution in photosynthesis, it appears that levels of RuBP carboxylase activity in WT cells are not in large excess, but are rather fairly well matched to photosynthetic electron transport rates.

The low affinity of cyanobacterial RuBP carboxylases for CO_2 is exacerbated by the high levels of O_2 produced during photosynthetic electron transport. Even at air levels of CO_2 and O_2, these enzymes spend ~ 30–35% of their time functioning as RuBP oxygenases, and the phosphoglycolate so produced must be metabolized to avoid further metabolic complications (phosphoglycolate is a fairly potent inhibitor of triose phosphate isomerase). For instance, phosphoglycolate was shown to accumulate to high levels when *Anabaena variabilis*, grown under high CO_2, was placed in CO_2 free air for a short time (Marcus et al. 1983). However, in cells which had adapted to growth in air, only very low levels of phosphoglycolate were observed. Clearly, the cyanobacteria must be able to metabolize this compound, although the detailed manner in which they do so is unclear. Possible mechanisms include dephosphorylation to glycolate with subsequent excretion or respiratory metabolism (Codd and Sallal 1978, Bergman et al. 1984, 1985). Although the details of glycolate metabolism in the cyanobacteria are complex (indeed, multiple pathways may be operative (Codd and Stewart 1973), it is clear that the cell loses both energy and carbon as a result of having to dispatch the phosphoglycolate produced by RuBP oxygenase activity.

Uptake of C_i by cyanobacteria

To enhance their capacity to fix CO_2, the cyanobacteria have elaborated an active mechanism for accumulating intracellular C_i (Kaplan et al. 1980, Miller and Colman 1980, Badger and Andrews 1982). An internal C_i concentration of over 100 mM and an accumulation ratio ($[C_i]_{in}/[C_i]_{out}$ as high as 1000 have been reported (Badger and Andrews 1982, Miller and Colman 1980, Badger et al. 1985). With such elevated internal concentrations, the rate of RuBP carboxylation is increased, the rate of RuBP oxygenation is decreased, and the losses associated with phosphoglycolate production are minimized. The accumulation of intracellular C_i is made possible by an active transport system for C_i. C_i transport activity is higher in cells grown under air levels (0.035%) of CO_2 (L-cells) than in cells grown under elevated levels (1 to 5% in air) of CO_2 (H-cells) (Kaplan et al. 1980). Transfer of H-cells to low CO_2 conditions increases their C_i-transporting activity and photosynthetic affinity for C_i, and de novo protein synthesis is required for this adaptation (Marcus et al. 1982, Omata and Ogawa 1986).

It has been shown that both CO_2 and HCO_3^- are actively removed from the medium by cyanobacterial cells and used for photosynthesis (Badger and Andrews 1982, Volokita et al. 1984, Miller and Canvin 1985). It is not known whether CO_2 and HCO_3^- uptake are due to common or different transporters. Recent observations have shown that H-cells utilize mainly CO_2, while L-cells utilize both CO_2 and HCO_3^- (Abe et al. 1987, Miller and Canvin 1987). Although CO_2 is taken up faster than HCO_3^- at lower concentrations of C_i in the medium (Badger and Andrews 1982, Volokita et al. 1984), HCO_3^- appears to be the C_i species which arrives at the inner side of the cytoplasmic (plasma) membrane, regardless of the C_i species supplied (Volokita et al., 1984, Badger et al. 1985, Ogawa and Kaplan 1987). That is, CO_2 transport appears to occur with a concomitant hydration. This important result suggests that a carbonic anhydrase-like moiety may be present in the vicinity of the transporter and may participate in CO_2 uptake.

Na^+ plays an important role in C_i transport, though the Na^+ requirement for C_i uptake appears to vary depending on the growth conditions experienced by the cells. In L-cells of *Anabaena variabilis* (Kaplan et al. 1984, Reinhold et al. 1984) and *Anacystis nidulans* (Miller and Canvin 1985), the affinity for HCO_3^- transport increases markedly as Na^+ concentrations are increased through the millimolar range. In *A. nidulans*, however, Na^+-independent HCO_3^- uptake has also been demonstrated in cells from standing cultures (Espie and Canvin 1987). CO_2 uptake by *A. nidulans* is independent of Na^+ in L-cells (Miller and Canvin 1985), but variations in $[Na^+]$ though the μM range do affect CO_2 uptake in H-cells (Miller and Canvin 1987). In contrast to the case in *A. nidulans*, CO_2 uptake by L-cells of *A. variabilis* is stimulated by low levels of Na^+ (Abe et al. 1987). It has yet to be determined whether these different modes of C_i uptake are due to different C_i transporters, or perhaps to modification of common transporter(s).

Mechanism of C_i transport

C_i uptake is light-dependent, though there have been contradictory observations as to the requirement for PS I and PS II reactions. It is known that C_i uptake is inhibited by DCMU (Miller and Colman 1980, Badger and Andrews 1982, Ogawa et al. 1985). However, in the presence of iodoacetamide, which inhibits CO_2 fixation, C_i uptake is insensitive to DCMU and shows action spectra typical of PS I reactions (Ogawa and Ogren 1985, Ogawa et al. 1985). Recently, it was shown that light is required not only for energization but also for a time-dependent activation of the C_i-transporting system. This activation, which occurs over a 5–10 min period when dark adapted *A. nidulans* cells are exposed to light, requires only a very low activity of PS II and is similar to the induction of photosynthesis that is observed during dark–light transitions (Kaplan et al. 1987). DCMU inhibits

A large amount of a 42-kD protein is present in the cytoplasmic membrane of L-cells of *A. nidulans* (Omata and Ogawa 1985). This protein was shown to be the only major protein that is actively synthesized during adaptation from high CO_2 to low CO_2 conditions (Omata and Ogawa 1986). Studies with high CO_2-requiring mutants of *A. nidulans* R2 also indicated that the synthesis of the protein is one of the characterizing features of the adaptation (Ogawa et al. 1987, Omata et al. 1987). These observations suggested that the 42-kD protein might be an essential component for the adaptation. Recently, the structural gene for this protein was cloned and used to construct a defined mutant of *A. nidulans* R2 which is totally deficient in the protein (Omata et al. in preparation). Unexpectedly, the resulting mutant was still capable of adapting to low CO_2 conditions with an increase in C_i-transporting activity. The results indicate that neither adaptation nor C_i-transport require this protein, and the physiological role of the 42-kD protein remains obscure. Proteins that *are* essential for adaptation and transport may prove to be minor components which went un-undetected in the previous studies.

High CO_2-requiring mutants of cyanobacteria

Another approach for studying C_i transport and accumulation processes is to obtain mutants which require high CO_2 for growth. Among the high CO_2-requiring mutants isolated from *A. nidulans* R2 and *Synechocystis* 6803 (Carlson and Pierce, unpublished observations) after chemical mutagenesis, the mutants from *A. nidulans* R2 have been biochemically and physiologically characterized (Marcus et al. 1986, Ogawa et al. 1987, Omata et al. 1987). The mutant E1 (Marcus et al. 1986) has C_i-transporting activity, wild type RuBP carboxylase activity, and is capable of adaptive transformation to low CO_2 conditions (Omata et al. 1987). Yet, it is incapable of photosynthesis under low CO_2. Although it can produce and maintain elevated levels of intracellular C_i, the mutant appears to be incapable of utilizing this pool. It has been suggested that the mutant lacks intracellular carbonic anhydrase activity which converts the intracellular pool of C_i (mostly HCO_3^-) into CO_2. However, this hypothesis is difficult to prove since even wild type levels of carbonic anhydrase are so low that accurate measurements of its activity are quite difficult. Though its genotype is unknown, the E1 mutant has proven to be useful for studies of the C_i-transporting mechanism, since C_i uptake can be measured in the effective absence of CO_2 fixation (Ogawa and Kaplan 1987).

Another mutant, RK1, was also shown to be defective in its ability to utilize the intracellular C_i pool (Ogawa et al. 1987). In contrast to the E1 mutant, RK1 showed no sign of adaptation nor increased amounts of the 42-kD protein upon exposure to low CO_2 conditions. The mutant may be deficient in a component which transmits the low CO_2 signal for inducing the functional and compositional changes that are observed in the wild type strain during adaptation to low CO_2. Interestingly, revertants of RK1 capable of growth in air regained the ability to adapt to low CO_2 conditions and also synthesized increased amounts of the 42-kD protein under inducing conditions (Omata and Pierce, unpublished observations). This result suggests that, although the 42-kD protein is clearly not required for growth in air (see above), the 42-kD gene may be somehow linked with the gene responsible for the RK1 phenotype.

At present, no mutants with specific defects in C_i transport have been obtained. Cloning of these transport-specific genes and the genes responsible for the mutations in E1 and RK1 would provide valuable information on the mechanisms of C_i utilization by cyanobacteria. Since *Anacystis nidulans* R2 and related strains are transformable by foreign DNA, it should be quite straightforward to clone the genes responsible for a given phenotype by complementation with plasmid libraries containing wild type DNA. We may reasonably expect reports along these lines in the near future.

Carbonic anhydrase and the CO_2 diffusion problem

As previously discussed, it appears that the major species of C_i reaching the cytoplasm of cyanobacteria is in the form of HCO_3^-, and it is this form that accumulates to high internal concentrations. Since CO_2 is the species utilized by RuBP carboxylase, the HCO_3^- must be dehydrated prior to use, and a possible role for carbonic anhydrase in CO_2 utilization is easily envisioned. However, (and despite clear evidence for the involvement of carbonic anhydrase in CO_2 utilization by micro-algae such as *Chlamydomonas reinhardtii* (for a review, see Aizawa and Miyachi 1986), experimental evidence for carbonic anhydrase in cyanobacteria remained elusive until fairly recently. In a careful study using sophisticated techniques, Badger et al. (1985) showed that there was very low, but detectable carbonic anhydrase activity in a *Synechococcus* species. Importantly, these workers also provided theoretical arguments indicating that, even though carbonic anhydrase activity was low, it was nevertheless required by this organism to maintain known rates of CO_2 fixation. Rather higher levels of carbonic anhydrase activity have been found in *Anabaena variabilis*, although the distribution of the enzyme (soluble or insoluble activity) varied among different strains of

A. variabilis (Yagawa et al. 1984). Recently a particulate carbonic anhydrase activity has also been recovered from *Chlorogloeopsis fritschii* cells (Lanaras et al. 1985).

The requirement of RuBP carboxylase for CO_2 and the presence of carbonic anhydrase activity in cyanobacteria pose interesting problems in understanding CO_2 utilization in these organisms. If C_i accumulation is to be effective in supplying useful concentrations of CO_2, then the internal CO_2 concentration must be large compared to external concentrations. To avoid a large penalty due to diffusion of this CO_2 out of the cell, a permeability barrier must exist. In fact, by assuming that a cyanobacterial cell contains a single impermeability barrier (the plasmalemma), a very low CO_2 permeability (10^{-5} cm s^{-1}) can be measured (Zenvirth and Kaplan 1984, Badger et al. 1985). This value is extraordinarily low for a polar lipid bilayer, and prompts a further concern emphasized by Badger et al. If one makes the reasonable assumption that the permeability of O_2 is of the same order as that for CO_2, then the O_2 released by photosynthesis would build up to concentrations of a few mM (~ 2–6 atmospheres) when photosynthesis is saturated by C_i. The dilemma, of course, is that such high levels of O_2 would be likely to cause oxidative damage and would be expected to severely inhibit CO_2 fixation by RuBP carboxylase, the enhancement of which appears to be a major purpose for C_i accumulation in the first place.

A unified view of C_i utilization by cyanobacteria

Recently, an interesting hypothesis has been advanced that obviates the apparent dilemma of a toxic buildup of oxygen without requiring a lipid bilayer that is selectively impermeable to CO_2. This hypothesis (Reinhold et al. 1987) ascribes a major, functional role to carboxysomes, the subcellular inclusion bodies in which cyanobacterial RuBP carboxylase is normally sequestered. These polyhedral bodies, which are found in a wide range of autotrophic organisms (Codd and Marsden 1984), contain a very limited distribution of polypeptides, of which up to 50% appears to be RuBP carboxylase. The hypothesis of Reinhold et al. suggests that both RuBP carboxylase and carbonic anhydrase are localized in cyanobacterial carboxysomes, and that this localization can serve to allow profitable use of C_i that is transported into the cytoplasm. Briefly, the HCO_3^- that builds up to high levels in the cytoplasm may diffuse into the carboxysome, where it is dehydrated to CO_2 by carbonic anhydrase and subsequently utilized by RuBP carboxylase. In this manner, CO_2 is produced and utilized in the same subcellular location away from the site of photosynthetic O_2 generation. The small amount of CO_2 produced by non-enzymatic dehydration in the cytoplasm and the large amount of photosynthetically produced O_2 would be free to diffuse across a permeable plasmalemma. It is this slow production of cytoplasmic CO_2 which would give rise to the apparently low rate of CO_2 diffusion out of the cell. In this manner, the problems of high internal O_2 tensions and low permeability of the plasmalemma to CO_2 outlined above would be avoided.

To be sure, there are difficulties associated with this model for CO_2 utilization by cyanobacteria: 1) the model requires a carboxysome membrane that is relatively impermeable to CO_2, or an internal carboxysome structure which maintains a higher CO_2 concentration in the vicinity of RuBP carboxylase than at the carboxysome membrane (Reinhold et al. 1987); 2) the model requires a carboxysomal location for carbonic anhydrase, though as mentioned above, carbonic anhydrase has been found in both soluble and particulate fractions.[1] However, regardless of the detailed mechanism of CO_2 utilization, other evidence and considerations tend to support a functional role for carboxysomes. To date, all cyanobacteria which have been shown to actively accumulate C_i also contain carboxysomes. Furthermore, since a large fraction of the total cyanobacterial RuBP carboxylase resides in the carboxysome (Coleman et al. 1982), and since there is no overabundance of RuBP carboxylase activity (see above), it would appear that carboxysomal CO_2 fixation is required. Carboxysomes isolated by differential centrifugation are indeed capable of RuBP dependent CO_2 fixation (Codd and Stewart 1976), and therefore appear to be permeable to RuBP. This is an important finding (consistent with evidence for a cytoplasmic location for ribulose 5-phosphate kinase (Hawthornthwaite et al. 1985) and would indicate that most of the reactions of the Calvin cycle could take place cytoplasmically. Permeability to such a highly charged molecule as RuBP may indicate either a specific transporter or a very unusual carboxysome membrane. In addition, a mutant of *Synechocystis* 6803 has been obtained which requires high CO_2 levels for growth, is

[1] These findings should not be taken as strictly disallowing the model. Although the model's absolute requirement for a carboxysomal location of carbonic anhydrase should provide a rigorous test of its veracity, the experimental difficulties in performing the test are great. As mentioned above, only a very low (and easily overlooked) level of carbonic anhydrase activity is required to support cyanobacterial photosynthesis. Difficulties in experimental interpretation would be compounded if the carbonic anhydrase-like activity responsible for the concomitant production of HCO_3^- during CO_2 transport (see above) were, under certain circumstances, to masquerade as a particulate carbonic anhydrase. Carboxysomes are easily broken during cell disruption, and carboxysomal carbonic anhydrase would appear as a soluble activity. Finally, it is conceivable that carbonic anhydrase exists in more than one location.

extremely sensitive to changes in the CO_2/O_2 ratio supplied during growth, and is unable to grow at all in air. This mutant also lacked microscopically observable carboxysomes (Pierce et al. 1988). All of these phenotypes were due to a single genetic lesion, the replacement of the WT carboxylase gene with the corresponding gene from *Rhodospirillum rubrum*. Altogether, the mutant's extreme sensitivity to O_2 was much higher than could be expected from the differences in gaseous substrate specificity between the *R. rubrum* and WT cyanobacterial enzymes, and it is interesting to speculate (in the manner of Reinhold et al.) that this sensitivity is due to the absence of functional carboxysomes. For instance, it may be that the *R. rubrum* enzyme, as a consequence of its dimeric, quaternary structure (c.f., the hexadecameric cyanobacterial enzyme), was not sequestered in the carboxysome, thereby rendering these subcellular structures un-observable. If carboxysomes do indeed play a role in effective CO_2 utilization, their functional absence in the mutant might explain its anomalous O_2 sensitivity.

The model of Reinhold et al. is important in providing a unified view of C_i utilization in cyanobacteria and in suggesting useful experiments concerning the role of carboxysomes in C_i utilization. In particular, it states that effective accumulation of C_i requires not only an active transport mechanism, but also a diffusional barrier provided by the carboxysome. Experiments designed to validate the model should be most informative, even if they result in the model's dismissal.

Concluding remarks

In this review, we have sketched the broad outlines of C_i utilization by the cyanobacteria. It is our opinion that, due essentially to their procaryotic nature, the mechanisms for C_i utilization in the cyanobacteria will prove to be quite different from that in higher plants and eucaryotic algae. It will also be evident from our liberal use of speculation, that many aspects of C_i utilization by the cyanobacteria remain obscure. It would appear however, that the irreducible requirements for effective assimilation of CO_2 include an energy dependent transport process, a barrier to outward diffusion of internal CO_2, and the ability to catalyze the dehydration of HCO_3^-. These aspects of C_i utilization are beginning to be addressed by genetic and molecular biological studies. These approaches complement the earlier physiological studies and should provide for a molecular understanding of the temporal and spatial nature of C_i utilization by the cyanobacteria.

604

References

Abe T, Tsuzuki M and Miyachi S (1987) Transport and fixation of inorganic carbon during photosynthesis in cells of *Anabaena* grown under ordinary air III. Some characteristics of the HCO_3^- transport system in cells grown under ordinary air. Plant Cell Physiol 28: 867–874

Aizawa K and Miyachi S (1986) Carbonic anhydrase and CO_2 concentrating mechanisms in microalgae and cyanobacteria. FEMS Microbiol Rev 39: 215–233

Andrews TJ and Abel KM (1981) Kinetics and subunit interactions of ribulose bisphosphate carboxylase-oxygenase from the cyanobacterium, *Synechococcus* sp. J Biol Chem 256: 8445–8451

Andrews TJ, Abel KM, Menzel E and Badger MR (1981) Molecular weight and quaternary structure of ribulose bisphosphate carboxylase from the cyanobacterium, *Synehococcus* sp. Arch Microbiol 130: 344–348

Badger MR (1980) Kinetic properties of RuBP carboxylase from *Anabaena variabilis*. Arch Biochem Biophys 201: 247–254

Badger MR and Andrews TJ (1982) Photosynthesis and inorganic carbon usage by the marine cyanobacterium, *Synechococcus* sp. Plant Physiol 70: 517–523

Badger MR, Bassett M and Comins HN (1985) A model for HCO_3^- accumulation and photosynthesis in the cyanobacterium *Synechoccus* sp. Plant Physiol 77: 465–471

Bergman B, Codd GA and Hällbom L (1984) Glycollate excretion by N_2-fixing cyanobacteria treated with photorespiratory inhibitors. Z Pflanzenphysiol 113: 451–460

Bergman B, Codd GA, Hällbom L and Codd GA (1985) Effects of amino-oxyacetate and aminoacetonitrile on glycolate and ammonia release by the cyanobacterium *Anabaena cylindrica*. Plant Physiol 77: 536–539

Codd GA and Marsden WJN (1984) The carboxysomes (polyhedral bodies) of autotrophic prokaryotes. Biol Rev 59: 389–422

Codd GA and Sallal A-KJ (1978) Glycolate oxidation by thylakoids of the cyanobacteria *Anabaena cylindrica*, *Nostoc muscorum*, and *Chlorogloea fritschii*. Planta 139: 177–181

Codd GA and Stewart WDP (1973) Pathways of glycollate metabolism in the blue—green alga *Anabaena cylindrica*. Arch Microbiol 124: 149–154

Codd GA and Stewart WDP (1976) Polyhedral bodies and ribulose 1,5-diphosphate carboxylase of the blue-green alga *Anabaena cylindrica*. Planta 130: 323–326

Coleman JR, Seeman JR and Berry JA (1982) RuBP carboxylase in carboxysomes of blue-green algae. Carnegie Inst of Wash Ybk 81: 83–87

Colman B, Cheng K-H and Ingle RK (1976) The relative activities of PEP carboxylase and RuDP carboxylase in the blue-green algae. Plant Science Letters 6: 123–127

Creach E, Codd GA and Stewart WDP (1981) Primary products of photosynthesis and studies of carboxylating enzymes in the filamentous cyanobacterium *Anabaena cylindrica*. In: G. Akoyunoglou (ed.) Photosynthesis IV. Regulation of Carbon Metabolism, pp 49–56. Balaban International Science Services, Philadelphia, USA

Espie GS and Canvin DT (1987) Evidence for Na^+-independent HCO_3^- uptake by the cyanobacteriuim *Synechococcus leopoliensis*. Plant Physiol 84: 125–130

Gatenby AA, van der Vies S and Bradley D (1985) Assembly in *E. coli* of a functional multisubunit ribulose bisphosphate carboxylase from a blue green alga. Nature 314: 617–620

Hawthornthwaite AM, Lanaras T and Codd GA (1985) Imuno-electronmicroscopic localization of Calvin cycle enzymes in *Chlorogloeopsis fritschii*. J Gen Microbiol 131: 2497–2500

Jordan DB and Ogren WL (1981) Species variation in the specificity of ribulose bisphosphate carboxylase/oxygenase. Nature 291: 513–515

Jordan DB and Ogren WL (1983) Species variation in the kinetic properties of ribulose 1,5-bisphosphate carboxylase/oxygenase. Arch Biochem Biophys 227: 425–433

Kaplan A (1985) Adaptation to low CO_2 levels: Induction and the mechanism for inorganic carbon uptake. In: Lucas WJ and Berry JA (eds) Inorganic Carbon Uptake by Aquatic Photosynthetic Organisms, pp 325–328. Waverly Press/American Society of Plant Physiologists

Kaplan A, Badger MR and Berry JA (1980) Photosynthesis and the intracellular inorganic carbon pool in the bluegreen alga Anabaena variabilis: Response to external CO_2 concentration. Planta 149: 219–226

Kaplan A, Volokita M, Zenvirth D and Reinhold L (1984) An essential role for sodium in the bicarbonate transporting system of the cyanobacterium Anabaena variabilis. FEBS Lett 176: 166–168

Kaplan A, Zenvirth D, Marcus Y, Omata T and Ogawa T (1987) Energization and activation of inorganic carbon uptake by light in cyanobacteria. Plant Physiol 84: 210–213

Kaplan A, Zenvirth D, Reinhold L and Berry JA (1982) Involvement of a primary electrogenic pump in the mechanism for HCO_3^- uptake by the cyanobacterium Anabaena variabilis. Plant Physiol 69: 978–982

Lanaras T, Hawthornthwaite AM and Codd GA (1985) Localization of carbonic anhydrase in the cyanobacterium Chlorogloeopsis fritschii. FEMS Microb Lett 26: 285–288

Lucas WJ and Berry JA (1985) Inorganic carbon uptake by aquatic photosynthetic organisms. Waverly Press, Baltimore, pp 1–480

Marcus Y, Harel E and Kaplan A (1983) Adaptation of the cyanobacterium Anabaena variabilis to low CO_2 concentration in their environment. Plant Physiol 71: 208–210

Marcus Y, Schwarz R, Friedberg D and Kaplan A (1986) High CO_2 requiring mutant of Anacystis nidulans R2. Plant Physiol 82: 610–612

Marcus Y, Zenvirth D, Harel E and Kaplan A (1982) Induction of HCO_3^- transporting capability and high photosynthetic affinity to inorganic carbon by low concentration of CO_2 in Anabaena variabilis. Plant Physiol 69: 1008–1012

Miller AG and Canvin DT (1985) Distinction between HCO_3^- and CO_2-dependent photosynthesis in the cyanobacterium Synechococcus leopoliensis based on the selective response of HCO_3^- transport to Na^+. FEBS Lett 187: 29–32

Miller AG and Canvin DT (1987) Na^+-stimulation of photosynthesis in the cyanobacterium Synechococcus UTEX 625 grown on high levels of inorganic carbon. Plant Physiol 84: 118–124

Miller AG and Colman B (1980) Active transport and accumulation of bicarbonate by a unicellular cyanobacterium. J Bacteriol 143: 1253–1259

Ogawa T and Kaplan A (1987) The stoichiometry between CO_2 and H^+ fluxes involved in the transport of inorganic carbon in cyanobacteria. Plant Physiol 83: 888–891

Ogawa T and Ogren WL (1985) Action spectra for accumulation of inorganic carbon in the cyanobacterium Anabaena variabilis. Photochem Photobiol 41: 583–587

Ogawa T, Kaneda T and Omata T (1987) A mutant of Synechococcus PCC 7942 incapable of adapting to low CO_2 concentration. Plant Physiol 84: 711–715

Ogawa T, Miyano A and Inoue Y (1985) Photosystem-I-driven inorganic carbon transport in the cyanobacterium, Anacystis nidulans. Biochem Biophys Acta 808: 77–84

Omata T and Ogawa T (1985) Changes in the polypeptide composition of the cytoplasmic membrane in the cyanobacterium Anacystis nidulans during adaptation to low CO_2 conditions. Plant Cell Physiol 26: 1075–1081

Omata T and Ogawa T (1986) Biosynthesis of a 42-kD polypeptide in the cytoplasmic membrane of the cyanobacterium Anacystis nidulans strain R2 during adaptation to low CO_2 concentration. Plant Physiol 80: 525–530

Omata T, Ogawa T, Marcus Y, Friedberg D and Kaplan A (1987) Adaptation to low CO_2 levels in a mutant of *Anacystis nidulans* R2 which requires high CO_2 for growth. Plant Physiol 83: 892–894

Pierce J, Carlson TJ and Williams JGK (1988) Anomalous oxygen sensitivity in a cyanobacterial mutant requiring the expression of ribulose bisphosphate carboxylase from a photosynthetic anaerobe. Submitted to Proc Nat Acad Sci

Reinhold L, Volokita M, Zenvirth D and Kaplan A (1984) Is HCO_3^- transport in *Anabaena* a Na^+ symport? Plant Physiol 76: 1090–1092

Reinhold L, Zviman M and Kaplan A (1987) Inorganic carbon fluxes and photosynthesis in cyanobacteria — a quantitative model. In: Biggins J (ed.) Progess in Photosynthesis Research, Vol 4, pp 289–296. Dordrecht: Martinus Nijhoff

Shinozaki K and Sugiura M (1985) Genes for the large and small subunit of ribulose bisphosphate carboxylase/oxygenase constitute a single operon in a cyanobacterium *Anacystis nidulans*. 6301. Mol Gen Genet 200: 27–32

Spalding MH, Spreitzer RJ and Ogren WL (1983) Reduced inorganic carbon transport in a CO_2-requiring mutant of *Chlamydomonas reinhardtii*. Planta 159: 261–266

Tabita FR and Small CL (1985) Expression and assembly of active cyanobacterial ribulose-1,5-bisphosphate carboxylase/oxygenase in *Escherichia coli* containing stoichiometric amounts of large and small subunits. Proc Nat Acad Sci USA 82: 6100–6103

Takabe T, Nishimura M and Akazawa T (1976) Presence of two subunit types in ribulose 1,5-bisphosphate carboxylase from blue-green algae. Bioch Biophys Res Commun 68: 537–544

Volokita M, Zenvirth D, Kaplan A and Reinhold L (1984) Nature of inorganic carbon species actively taken up by the cyanobacterium *Anabaena variabilis*. Plant Physiol 76: 599–602

Yagawa Y, Shiraiwa Y and Miyachi S (1984) Carbonic anhydrase from the blue-green alga (cyanobacterium) *Anabaena variabilis*. Plant and Cell Physiol 25: 775–783

Zenvirth D and Kaplan A (1981) Uptake and efflux of inorganic carbon in *Dunaliella salina*. Planta 152: 8–12

Zenvirth D, Volokita M and Kaplan A (1984) Evidence against H^+-HCO_3^- symport as the mechanism for HCO_3^- transport in the cyanobacterium *Anabaena variabilis*. J Membrane Biol 79: 271–274

Govindjee et al. (eds), Molecular Biology of Photosynthesis: 607–619
© 1988 Kluwer Academic Publishers

Minireview

Synthesis and assembly of bacterial and higher plant Rubisco subunits in *Escherichia coli*

ANTHONY A. GATENBY
Central Research and Development Department, Experimental Station, E.I. du Pont de Nemours & Co., Wilmington, DE 19898 U.S.A.

Received 2 September 1987; accepted 17 December 1987

Key words: Carbon fixation, enzyme assembly, gene expression, recombinant DNA

Abstract. The synthesis in *Escherichia coli* of both the large and small subunits of cereal ribulose bisphosphate carboxylase/oxygenase has been obtained using expression plasmids and bacteriophages. The level and order of synthesis of the large and small subunits were regulated using different promoters, resulting in different subunit pool sizes and ratios that could be controlled in attempts to optimize the conditions for assembly. Neither assembly nor enzyme activity were observed for the higher plant enzyme. In contrast, cyanobacterial large and small subunits can assemble to give an active holoenzyme in *Escherichia coli*. By the use of deletion plasmids, followed by infection with appropriate phages, it can be demonstrated that the small subunit is essential for catalysis. However, the small subunit is not required for the assembly of a large subunit octomer core in the case of the *Synechococcus* enzyme; self-assembly of the octomer will occur in an *rbcS* deletion strain. The cyanobacterial small subunits can be replaced by wheat small subunits to give an active enzyme in *Escherichia coli*. The hybrid cyanobacterial large/wheat small subunit enzyme has only about 10% of the level of activity of the wild-type enzyme, reflecting the incomplete saturation of the small subunit binding sites on the large subunit octomer, and possibly a mismatch in the subunit interactions of those small subunits that do bind, giving rise to a lower rate of turnover at the active sites.

Abbreviations: IPTG–isopropyl-β-D-thiogalactopyranoside, L–large subunit, Rubisco–ribulose bisphosphate carboxylase/oxygenase, S–small subunit

Introduction

One objective of the synthesis and assembly of Rubisco in *Escherichia coli* is to obtain an understanding of the assembly process itself. Once assembly can be demonstrated, then experiments can be designed to understand how the subunits interact with each other, and the importance of particular structures in catalysis, and its possible modification. Following the initial report on the successful expression of a chloroplast *rbcL* gene in *E. coli* (Gatenby et al. 1981) there have been rapid developments which have expanded the range of *rbcL* and *rbcS* genes from both plant and prokaryotic sources that can be expressed in *E coli*, resulting in the synthesis of Rubisco large (L) and small (S) subunits. This has enabled experimental systems to be developed that result in the assembly of catalytically active dimeric (Somerville and Somerville 1984), hexameric (Quivey and Tabita 1984) and hexadecameric (reviewed by Bradley et al. 1986) enzymes. These systems can now be used to introduce site specific mutations to investigate the active site, to examine the assembly pathway, and the formation of hybrid enzymes. The isolation from its normal cell of a pathway leading from gene expression to subunit assembly, and its reconstitution in *E. coli* enables a precise approach to finding what the requirements are for successful assembly by mutating, using different subunit genes and otherwise modifying the system. Another feature, particularly important where subunits are encoded by a multigene family, is that assembly from a cloned gene will provide an enzyme with identical, well characterized subunits of a particular type, rather than the plant derived enzyme which may be a heterogeneous mixture of subunits giving rise to an average value when used in enzyme assays. This review examines some of the strategies that have been adopted to asesemble active Rubisco in *E. coli*.

Expression of higher-plant Rubisco genes in *E. coli*

The maize and wheat chloroplast *rbcL* genes were the first plant genes to be expresed in *E. coli* (Gatenby et al. 1981). The original experiments were designed to detect the synthesis of the L subunit when the appropriate chloroplast DNA fragment was cloned into the plasmid pBR322. The L subunit synthesized had a similar M_r to that observed in chloroplasts, suggesting that translation was initiating correctly at the chloroplast ribosome binding site. Furthermore, the levels of expression observed were similar for both orientations of the *rbcL* gene in the plasmid, indicating that transcription was probably initiating within chloroplast DNA, and was not the result of readthrough transcription from plasmid promoters. This was later substantiated by the observation that plasmid encoded lambda P_L promoters could not direct transcription through the maize *rbcL* gene unless a phage encoded transcription antitermination protein was present (Gatenby and Cuellar 1985), and that chloroplast promoters could be recognized by the bacterial RNA polymerase (Bradley and Gatenby 1985). It was clear, therefore, that the close similarities between the chloroplast and bacterial transcription and translation signals enabled the plant genes to be readily expressed in *E. coli*. The expression of chloroplast *rbcL* genes in *E. coli* has also been observed for tobacco (Fluhr et al. 1983, Zhu et al. 1984), petunia (Bovenberg et al. 1984) and *Chlamydomonas* (Zhu et al. 1984).

The levels of synthesis of the L subunit, using chloroplast promoters, is low. It was estimated that only about 100–200 monomers were synthesized with the plasmids pZmB1A and pZmB1B using chloroplast promoters (Ellis and Gatenby 1984). The level of synthesis could be substantially improved by increasing the rate of transcription of the gene. This was initially achieved by cloning the maize *rbcL* gene into bacteriophage lambda, and then relying on the early P_L promoter, or the late P'_R promoter to transcribe the gene during lytic infection (Gatenby et al. 1981). Although *rbcL* expression levels achieved 0.5–1.0% of total protein synthesis following phage infection, the protein did not accumulate to high levels due to the transitory nature of expression in the lytic cycle. By using a plasmid vector containing the lambda P_L promoter, a transcription antiterminator, and the maize *rbcL* gene, it was possible to induce synthesis of L subunits by thermal denaturation of the *c*I repressor. Under these conditions synthesis of the L subunit continued for more than 120 min and the percentage of L in the cell increased after induction to reach a maximum of 2% of total *E. coli* protein (Gatenby and Castleton 1982). This represents approximately 60,000 subunit monomers per bacterial cell. High level expression of the maize *rbcL* gene has also been achieved by Somerville et al. (1983, 1986), using a different strategy. In their experiment a *lac* promoter was used and a hybrid ribosome binding site was formed using the sites of *lacZ* and *rbcL* fused together.

The levels of expression obtained with plasmid pPBI3 (Gatenby and Castleton 1982) enabled studies to be carried out on the properties of the L subunit synthesized in the absence of the S subunit. An advantage with this approach is that the use of denaturing conditions which are used to separate the L and S subunits in the holoenzyme, can be avoided. Pulse-chase experiments showed that the L subunit was stable in *E. coli* for at least 4 h (Gatenby 1984). The protein was also insoluble, whether expressed at high or low levels. Attempts to detect enzyme activity, or to form a stable enzyme.metal.CO_2.[^{14}C] carboxyarabinitol bisphosphate quaternary complex were unsuccessful. The presence of the substrate ribulose bisphosphate, or the positive effector fructose 1,6-bisphosphate, did not improve solubility, although it was found that pH 11–12 solubilized L slowly suggesting that aggregate structures may have formed. It would appear that maize L adopts an inactive, insoluble conformation after, or during, synthesis. Similar results have been reported by others for maize L (Somerville et al. 1983, 1986).

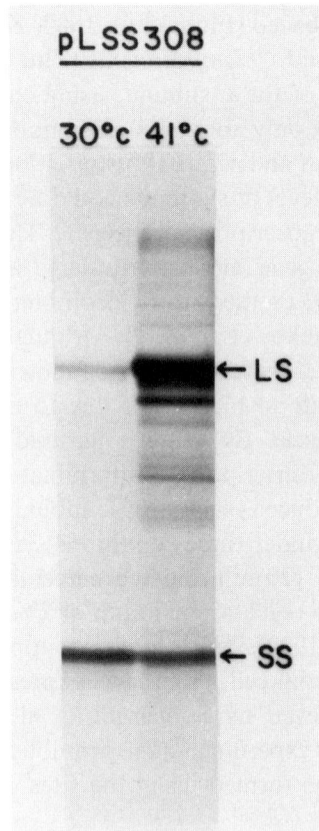

Fig. 1. Fluorograph showing the synthesis of Rubisco subunits in *E. coli* following induction of plasmid pLSS308. Cells were labelled with [^{35}S] methionine in the presence of IPTG to induce wheat small subunit (SS) synthesis, and either at 30 °C or 41 °C to obtain repression or induction respectively of maize large subunit (LS) synthesis. The ^{35}S-labelled polypeptides were incubated with anti-wheat RuBPCase serum and protein A Sepharose before electrophoresis on polyacrylamide gels containing SDS. Reprinted from Gatenby et al. (1987).

In an attempt to overcome the problem of L insolubility, and to assemble a plant holoenzyme, experiments were designed to co-express the L and S subunit genes in *E. coli* (Bradley et al. 1986, Gatenby et al. 1987). These experiments required that the nuclear encoded S subunit gene be cloned into an *E. coli* vector such that the S polypeptide is synthesized without the amino-terminal transit peptide, which is presumably only necessary for transport of the protein into chloroplasts (Broglie et al. 1983). These constructions have been made by subcloning a region of the wheat *rbcS* gene, which encodes the mature S subunit, from a cDNA clone into plasmid and M13 vectors (Bradley et al. 1986, van der Vies et al. 1986). The resulting expression clones (p565, pLSS308, and M13-72) produce the S subunit as a *lacZ::rbcS* fusion protein in *E. coli*. The amino-terminals of the S fusion protein contain 11 (p565 and pLSS308) or 10 (M13-72) amino acids from β-galactosidase. The first four amino acids of the mature S polypeptide have also been deleted. Transcription and translation of the *rbcS* gene fusions are initiated from the *lacZ* promoter and ribosome-binding sites respectively, resulting in synthesis of an S polypeptide of 15.6 kDa, which migrates slightly slower than authentic wheat isolated from chloroplasts (14.7 kDa) as expected from the fusion at its N-terminus.

The *rbcS* expression clones were used to study plant holoenzyme formation in two ways, either using heterologous subunit mixtures, or homologous subunits. In plasmid pLSS308 the wheat *rbcS* gene was placed under the control of P*lac*. and the maize *rbcL* genes were placed under the control of P_L. The use of different promoters allows the independent expression of the *rbcS* and *rbcL* genes. The expression of *rbcL* from pLSS308 relied on translational intiation at the chloroplast ribosome binding site (Gatenby et al. 1981), and temperature induced transcription obtained by shifting cultures to 41–42 °C in the presence of the anti-terminating protein pN (Gatenby and Cuellar 1985). Fig. 1 shows the result of a typical induction experiment in which S synthesis is constitutive, but the synthesis of L has been increased at 41 °C. The advantage of this approach is that by altering the pool sizes of the two subunits it might be possible to arrive at an optimal ratio of subunit types for assembly to occur. When the S subunit is synthesized in the absence of the L subunit it is found in the soluble fraction, but the polypeptide is unstable and has a half-life of less than 15 min (Gatenby et al. 1987). Its size on sucrose gradients indicates a monomeric or dimeric form. When L subunit synthesis is induced in cells containing the S subunit both subunits are found predominantly in the insoluble fraction and are fully stable for more than 120 min, suggesting that aggregation of the subunits may occur. The two subunits do not assemble to form an active holoenzyme in vivo even when nascent L subunits are synthesized in a pool of mature S subunits.

To test if the synthesis of homologous subunits in *E. coli* would provide a more favorable opportunity for assembly, the wheat *rbcS* and *rbcL* genes were expressed together (Bradley et al. 1986). A bacteriophage M13 clone (M68) expressing the wheat L subunit under the control of P*lac* was constructed. Cells were allowed to first synthesize the S subunit by means of a plasmid-borne *rbcS* gene (p565), and were then infected with the bacteriophage M68 (*rbcL*). Phage M68 was also used to infect plasmid-free *E coli* cells to examine the properties of the wheat L subunit in the absence of S. In contrast to the maize L subunit insolubility in *E. coli*, about 60% of

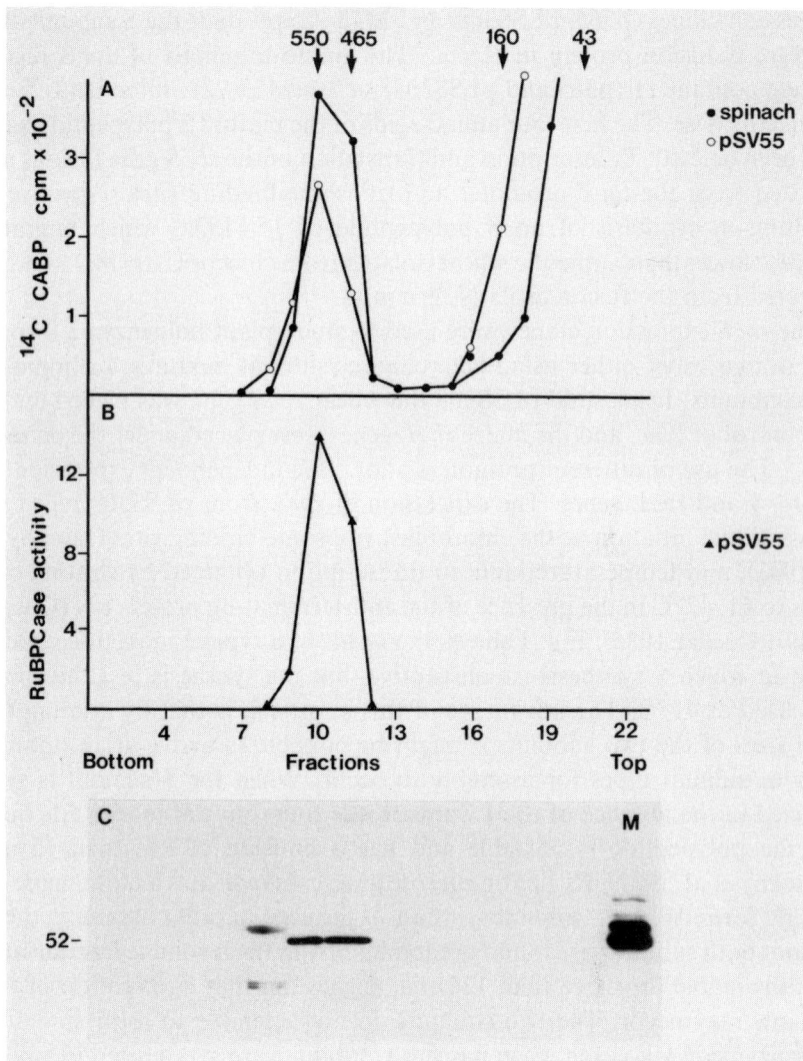

Fig. 2. Velocity sedimentation in sucrose gradients of cyanobacterial Rubisco synthesized and assembled in *E. coli*. Soluble protein samples were prepared from *E. coli* cells that had synthesized the *Synechococcus* L and S subunits following induction of plasmid pSV55. In panel A the bacterial extract, or purified spinach Rubisco, was incubated with the transition state analogue [^{14}C] carboxyarabinitol bisphosphate before centrifugation. In panel B the distribution of Rubisco activity was measured in a similar sucrose gradient using the induced bacterial extract. In panel C an immunoblot analysis of the fractions from panel B was carried out using anti-Rubisco serum. The M track is a maize leaf track used as a marker. The M_r markers used to calibrate the sucrose gradients are shown by arrows. Reprinted from Gatenby et al. (1985).

the wheat L subunit was found in the soluble fraction (Bradley et al. 1986). The S subunit did not appear to influence the amount of L that was found in the soluble fraction. It was also observed that about 80% of the wheat S subunit was in the soluble fraction from *E. coli* cells. Rubisco activity was undetectable. To examine subunit assembly, the soluble proteins from cells expressing both L and S were subjected to velocity sedimentation on sucrose gradients. Most of the S subunit remained at the top of the gradient. The wheat L subunit, although distributed throughout the gradient, sedimented with a peak abundance indicating a molecular mass of 750 kDa. This probably represents a large protein aggregate. Therefore, even with a homologous combination of subunits, it was not possible to detect assembly or activity of the higher plant enzyme.

Synthesis of bacterial Rubisco genes in *E. coli*

An attractive feature of synthesizing bacterial Rubisco in *E. coli* is that the prokaryotic structure of the genes can facilitate their expression. A range of different forms of active Rubisco has been synthesized in *E. coli* using the isolated genes from several different species. These include the homodimeric L subunit enzyme from *Rhodospirillum rubrum* (Somerville and Somerville 1984, Larimer et al. 1986), the hexadecameric L subunit enzyme from *Rhodopseudomonas sphaeroides* (Quivey and Tabita 1984) and the hexadecameric enzyme containing both L and S subunits from cyanobacteria (Christeller et al. 1985, Gatenby et al. 1985, Gurevitz et al. 1985, Tabita and Small 1985), *Chromatium* (Viale et al. 1985) and *Rhodopseudomonas sphaeroides* (Gibson and Tabita 1986).

The prokaryotic nature of cyanobacteria also simplifies the expression of their genes in *E. coli*, because the absence of an organelle, and therefore an S subunit transit peptide, enables the direct synthesis of mature S. In addition, the *rbcL* and *rbcS* genes are physically linked on a contiguous fragment of DNA. For expression of the *Synechococcus* Rubisco genes they were initially transcribed from a bacteriophage lambda P_L promoter following temperature induction (Gatenby et al. 1985). Synthesis of the L and S subunits could be detected, together with enzyme activity and [^{14}C] carboxyarabinitol bisphosphate binding. In addition, sucrose gradient centrifugation resolved a peak of enzyme activity that was similar in size to the spinach holoenzyme (Fig. 2). The large size indicated that the L_8 structure was formed, but an $L_8 S_n$ complex where n is less than 8 would be difficult to resolve from $L_8 S_8$ because of the small size of S. Although S must be present to obtain an active enzyme (Andrews and Abel 1981), it was suspected that the S binding sites on the L octomer were not fully saturated because of two observations. First, the specific activity in cell extracts is lower than expected based on the known amount of L_8 present (as determined by [^{14}C] carboxyarabinitol bisphosphate binding). Second, the amount of ^{35}S present in the purified L and S polypeptides indicated that $L_8 S_n$ structures were formed where n = 2 or 3.

More recently it has been shown that transfer of the appropriate *Synechococcus* DNA fragment from the P_L expression plasmid to a *Plac* expression plasmid increases the amount of Rubisco synthesized and enables either the assembly of $L_8 S_8$ structures, or structures almost completely saturated with S subunits (Bradley et al. 1986, van der Vies et al. 1986). Christeller et al. (1985) demonstrated that cloning the *Anacystis nidulans rbcL* and *rbcS* genes separately resulted in the absence of enzyme activity, indicating that both L and S subunits were required for activity. We have obtained similar results using plasmid or phage encoded *Synechococcus rbcL* and *rbcS* genes (van der Vies et al. 1985). Plasmid pDB50 contains both *rbcL* and *rbcS* and directs the synthesis of active Rubisco. If *rbcS* is deleted, the resulting plasmid (pDB53) synthesized only the L subunit which is inactive. Since this deletion does not result in any addition, or noticeable reduction, in the amount of soluble L subunit synthesized, it is concluded that the product of *rbcS* is essential for activity. If this is correct, infection *E. coli* cells containing the S subunit plasmid delection (pDB53) with M13 phages encoding the *rbcS* gene should restore Rubisco activity by complementation. This is exactly what happens. Following phage infection, Rubisco activity can be detected. These deletion and complementation experiments have also enabled us to obtain information about the assembly process of Rubisco in *E. coli*. It appears that the major soluble protein structures formed in cells synthesizing the L subunit alone (pDB53) are oligomeric forms, principally an L octomer and an L dimer (van der Vies et al. 1986). Therefore, the formation of a *Synechococcus* L_8 structure does not require S subunits to be present. The L_2 structure may represent an intermediate which then self-assembles to the L_8 form, or may be the result of disassembly of the L octomer. When both cyanobacterial L and S subunits are synthesized in the same cell the L dimer is not present, and the main L peak shifts further down the gradient indicating the binding of S subunits. The presence of S subunits appears to reduce the pool of unassembled L_2 and promote formation of L_8.

Assembly of a hybrid Rubisco in *E. coli*

Since it was possible to demonstrate complementation of Rubisco activity

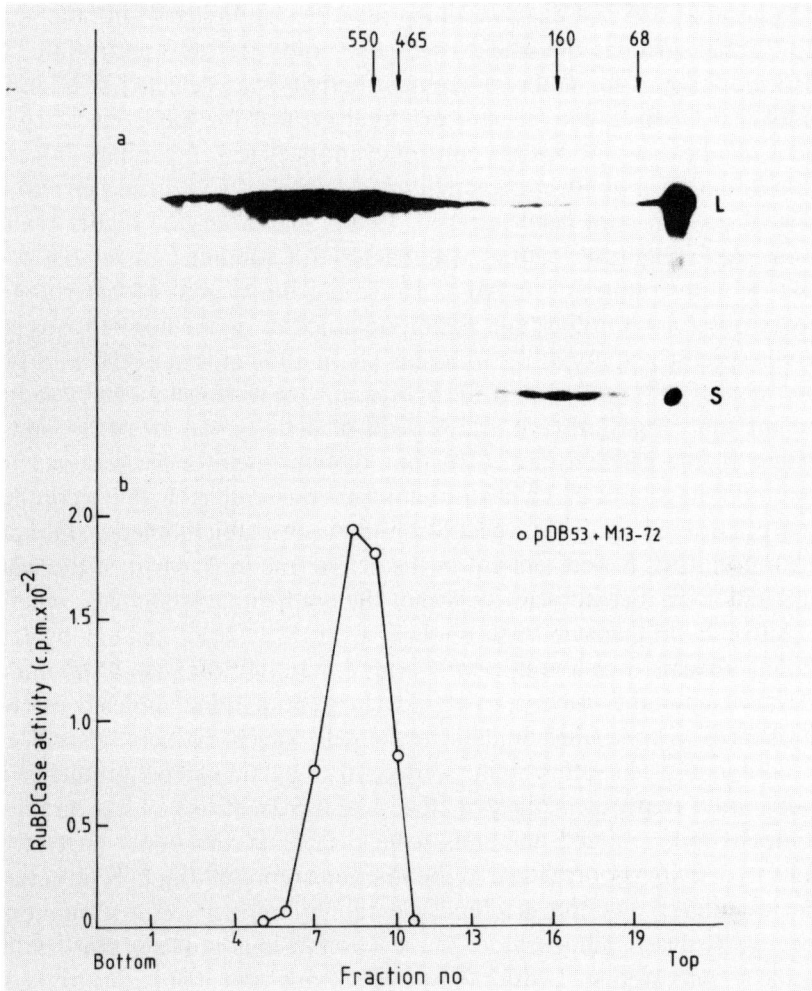

Fig. 3. Analysis of heterologous Rubisco assembly in *E. coli*. A soluble protein extract of *E. coli* cells containing plasmid pDB53 (encoding the cyanobacterial *rbcL* gene), and infected with the phage ML3-72 (encoding the wheat *rbcS* gene) was fractionated on a sucrose gradient. Panel a shows an immunoblot of the fractions using anti-Rubisco serum, and panel b shows the determination of Rubisco activity. The gradient was calibrated with M_r markers shown by arrows. Wheat leaf protein was used as a marker in the right lane of panel a. Reprinted from van der Vies et al. (1986).

in an S subunit plasmid deletion strain by infection with M13 *rbcS*[+] phages encoding a homologous S subunit (van der Vies et al. 1986), it was decided to attempt complementation using the heterologous wheat *rbcS* gene. The plasmid pDB53 encodes the *Synechococcus rbcL* gene transcribed by P*lac*

(Gatenby et al. 1985). Cells containing pDB53 were infected with the phage M13-72 that encodes the wheat *rbcS* gene, and after an appropriate time cells were assayed for Rubisco activity. Activity was detected at a level of 3 nmol of CO_2 fixed per min per mg total protein for the hybrid enzyme. Under identical conditions the control cyanobacterial *rbcS* phage infection resulted in a level of Rubisco activity of 32 nmol of CO_2 fixed per min per mg total protein (van der Vies et al. 1986). Therefore, the higher plant S subunit can substitute for the cyanobacterial S subunit in assembly of an active Rubisco molecule in vivo. However, the degree of heterologous activity observed is not as high as that obtained using the homologous gene and the specific activity of the hybrid is about 10% of that of the wild-type.

Extracts of cells synthesizing the hybrid enzyme were examined by sucrose gradient centrifugation (Fig. 3). A peak of activity was observed and its position in the gradient indicated that the enzyme is a *Synechococcus* L subunit octomer, and that wheat subunits have assembled with it to produce activity. The binding of the wheat S subunit is non-stoichiometric since only trace amounts of S were present in the active fractions. Most of the wheat S subunits were located near the top of the gradient. It is not clear whether the active hybrid enzyme has an L_8S_8 or $L_8S_{<8}$ structure, but wheat S subunits appear to be less efficiently bound than homologous S subunits to the cyanobacterial L_8 subunit core, and a large pool of incompletely assembled wheat S subunits accumulate in the cell. When extracts from *E. coli* containing only wheat S subunits were centrifuged on sucrose gradients, the S subunits were present only in the two top fractions of the gradient. Although both L and S subunits were present in the upper part of the gradient, activity is only found in the fractions representing L_8S_x structures, indicating that the L_8 core is a fundamental requirement in the formation of an active Rubisco holoenzyme. These observations indicate that the low specific activity obtained with the hybrid enzyme is probably due to incomplete saturation of the binding sites on the L subunit octomers. Those that do bind possibly have a mismatch in their subunit interactions leading to a lower rate of turnover at the active site.

Experiments to assemble a plant/cyanobacterial hybrid enzyme in vivo were an intermediate step that revealed that there is no intrinsic reason why the wheat S subunit in *E. coli* could not bind to an L_8 structure to form an active site. The fact that cyanobacterial L_8 can function as an acceptor for wheat S, but plant L cannot, suggests that the principle obstacle for higher plant assembly may be the formation of a suitable L_8 core for S subunit binding.

The complexity of Rubisco assembly

In the prokaryotic cyanobacteria the *rbcL* and *rbcS* genes are closely linked together on a contiguous piece of DNA and are co-transcribed (Shinozaki and Sugiura 1985). The gene products are present together from their time of synthesis in the same cell compartment and do not need to be transported through membranes. This arrangement may have favored a facile self-assembly mechanism which enables the holoenzyme to form. This is supported by the ease with which cyanobacterial Rubisco can self-assemble into an active enzyme when the genes are expressed in *E. coli* (van der Vies et al. 1986). It is also known that a cyanobacterial Rubisco can be dissociated in vitro to form an L_8 core devoid of S, and that S can be added back to give a functional enzyme, using either homologous (Andrews and Ballment 1983) or heterologous S subunits (Andrews and Lorimer 1985). From these data and the work of van der Vies et al. (1986) in which self-assembly of an L_8 core occurs in *E. coli*, it would appear that formation of an L_8 core may be the critical first step in assembly. It is to this L_8 core that S can be attached to obtain high levels of activity. An alternative model for assembly has been suggested by Gurevitz et al. (1985) in which assembly proceeds using L + S heterodimers.

In plants the assembly of Rubisco appears more complex. The *rbcL* and *rbcS* genes have different coding sites and the *rbcS* gene product has to be transported through the chloroplast membrane before Rubisco assembly can occur. In addition, in vitro dissociation studies of plant Rubisco reveal that it is not possible to obtain an L_8 core that remains soluble in aqueous solutions (Voordouw et al. 1984), and maize L synthesized in *E. coli* is insoluble (Gatenby 1984). It has also not been possible to obtain assembly of the plant L and S subunits expressed in *E. coli* (Bradley et al. 1986, Gatenby et al. 1987). The hypothesis has been invoked that the chloroplast synthesized L subunit requires a large subunit binding protein to function as a molecular chaperone to assist in holoenzyme formation (Ellis 1987). Formal proof of the role of this binding protein in assembly has not yet been presented, but its suggested function becomes increasingly attractive as a mechanism for mediating Rubisco assembly.

Acknowledgements

I thank Doug Bradley, Judy Castleton, Rick Cueller, Steve Rothstein, Mike Saul and Saskia van der Vies for contributing to the research described in this review, which was carried out primarily at the Plant Breeding Institute,

618

Cambridge, England, and also at the University of Wisconsin, Madison, Wisconsin, U.S.A.

References

Andrews TJ and Abel KM (1981) Kinetics and subunit interaction of ribulose bisphosphate carboxylase/oxygenase from the cyanobacterium *Synechococcus* sp. J Biol Chem 256; 8445–8451

Andrews TJ and Ballment B (1983) The function of the small subunits of ribulose bisphosphate carboxylase-oxygenase. J Biol Chem 258: 7514–7518

Andrews TJ and Lorimer GH (1985) Catalytic properties of a hybrid between cyanobacterial large subunits and higher plant small subunits of ribulose bisphosphate carboxylase-oxygenase. J Biol Chem 260: 4632–463

Bovenberg WA, Koes RE, Kool AJ and Nijkamp HJJ (1984) Physical mapping, nucleotide sequencing and expression in *E. coli* minicells of the gene for the large subunit of ribulose bisphosphate carboxylase from *Petunia hybrida*. Curr Genet 8: 231–241

Bradley D and Gatenby AA (1985) Mutational analysis of the maize chloroplast ATPase-β subunit gene promoter: the isolation of promoter mutants in *E. coli* and their characterization in a chloroplast in vitro transcription system. EMBO J 4: 3641–3648

Bradley D, van der Vies SM and Gatenby AA (1986) Expression of cyanobacterial and higher-plant ribulose 1,5-bisphosphate carboxylase genes in *Escherichia coli*. Phil Trans R Soc London B 313: 447–458

Broglie R, Coruzzi G, Lamppa G, Keith B and Chua N-H (1983) Structural analysis of nuclear genes coding the precursor to the small subunit of wheat ribulose-1,5-bisphosphate carboxylase. Biotechnology 1: 55–61

Christeller JT, Terzaghi BE, Hill DF and Laing WA (1985) Activity expressed from cloned *Anacystis nidulans* large and small subunit ribulose bisphosphate carboxylase genes. Plant Mol Biol 5: 257–263

Ellis RJ (1987) Proteins as molecular chaperones. Nature 328: 378–379

Ellis RJ and Gatenby AA (1984) Ribulose bisphosphate carboxylase: properties and synthesis. *In*: PJ Lea, GR Stewart (eds) The genetic manipulation of plants and its application to agriculture. Oxford Univ Press, Oxford, pp 41–60

Fluhr R, Fromm H and Edelman M (1983) Clone bank of *Nicotiania tabacum* chloroplast DNA: mapping of the alpha, beta and epsilon subunits of the ATPase coupling factor, the large subunit of ribulose bisphospate carboxylase, and the 32-kDal membrane protein. Gene 25: 271–280

Gatenby AA (1984) The properties of the large subunit of maize ribulose bisphosphate carboxylase/oxygenase synthesized in *Escherichia coli*. Eur J Biochem 144: 361–366

Gatenby AA and Castleton JA (1982) Amplification of maize ribulose bisphosphate carboxylase large subunit synthesis in *E. coli* by transcriptional fusion with the lambda *N* operon. Mol Gen Genet 185: 424–429

Gatenby AA, Castleton JA and Saul MW (1981) Expression in *E. coli* of maize and wheat chloroplast genes for large subunit of ribulose bisphosphate carboxylase. Nature 291: 117–121

Gatenby AA and Cuellar RE (1985) Antitermination is required for readthrough transcription of the maize *rbcL* gene by a bacteriophage promoter in *Escherichia coli*. Eur J Biochem 153: 355–359

Gatenby AA, van der Vies SM and Bradley D (1985) Assembly in *E. coli* of a functional multi-subunit ribulose bisphosphate carboxylase form a blue-green alga. Nature 314: 617–620

Gatenby AA, van der Vies SM and Rothstein SJ (1987) Co-expression of both the maize large and wheat small subunit genes of ribulose-bisphosphate carboxylase in *Escherichia coli*. Eur J Biochem 168: 227–231

Gibson JL and Tabita FR (1986) Isolation of the *Rhodopseudomonas sphaeroides* form 1 ribulose 1,5-bisphosphate carboxylase/oxygenase large and small subunit genes and expression of the active hexadecameric enzyme in *Escherichia coli*. Gene 44: 271–278

Gurevitz M, Somerville CR and McIntosh L (1985) Pathway of assembly of ribulosebisphosphate carboxylase/oxygenase from *Anabaena* 7120 expressed in *Escherichia coli*. Proc Natl Acad Sci USA 82: 6546–6550

Larimer FW, Machanoff R and Hartman FC (1986) A reconstruction of the gene for ribulose bisphosphate carboxylase from *Rhodospirillum rubrum* that expressed the authentic enzyme in *Escherichia coli*. Gene 41: 113–120

Shinozaki K and Sugiura M (1985) Genes for the large and small subunit of ribulose bisphosphate carboxylase/oxygenase constitute a single operon in a cyanobacterium *Anacystis nidulans* 6301. Mol Gen Genet 200: 27–32

Somerville C, Fitchen J, Somerville S, McIntosh L and Nargang F (1983) Enhancement of net photosynthesis by genetic manipulation of photorespiration and RuBP carboxylase/oxygenase. In: K Downey, RW Voellmy, F Ahmed, J Schultz (eds) Advances in Gene Technology: Molecular Genetics of Plants and Animals. Academic Press, New York, pp 295–309

Somerville CR, McIntosh L, Fitchen J and Gurevitz M (1986) The cloning and expression in *Escherichia coli* of RuBP carboxylase/oxygenase large subunit genes. Meth Enzymol 118: 491–433

Somerville CR and Somerville SC (1984) Cloning and expression of the *Rhodospirillum rubrum* ribulose bisphosphate carboxylase gene in *E. coli*. Molec Gen Genet 193: 214–219

Tabita FR and Small CL (1985) Expression and assembly of active cyanobacterial ribulose-1,5-bisphosphate carboxylase/oxygenase in *Escherichia coli* containing stoichiometric amounts of large and small subunits. Proc Natl Acad Sci USA 82: 6100–6103

Quivey RG and Tabita FR (1984) Cloning and expression in *Escherichia coli* of the form II ribulose 1,5-bisphosphate carboxylase/oxygenase gene from *Rhodopseudomonas sphaeroides*. Gene 31: 91–101

Van der Vies SM, Bradley D and Gatenby AA (1986) Assembly of cyanobacterial and higher plant ribulose bisphosphate carboxylase subunits into functional homologous and heterologous enzyme molecules in *Escherichia coli*. EMBO J 5: 2439–2444

Viale AM, Kobayahsi H, Takabe T and Akazawa T (1985) Expression of genes for subunits of plant-type Rubisco from *Chromatium* and production of the enzymically active molecule in *Escherichia coli*. FEBS Lett 192: 283–288

Voordouw G, van der Vies SM and Bouwmeister PP (1984) Dissociation of ribulose bisphosphate carboxylase/oxygenase from spinach by urea. Eur J Biochem 141: 313–318

Zhu YS, Lovett PS, Williams DM and Kung SD (1984) *Nicotania* chloroplast genome 7. Expression in *E. coli* and *B. subtilis* of tobacco and chlamydomonas chloroplast DNA sequences coding for the large subunit of RuBP carboxylase. Theor Appl Genet 67: 333–336

Govindjee et al. (eds), Molecular Biology of Photosynthesis: 621–644
© 1988 Kluwer Academic Publishers

Minireview

Organization and expression of the genes encoding ribulose-1,5-bisphosphate carboxylase in higher plants

THIANDA MANZARA & WILHELM GRUISSEM

Department of Botany, University of California, Berkeley, CA 94720, USA

Received 13 October 1987; accepted 3 December 1987)

Key words: higher plants, rbcL, rbcS, ribulose-1,5-bisphosphate carboxylase, Rubisco, SSU

Abstract. The chloroplast gene encoding the large subunit (rbcL) and the nuclear multigene family encoding the small subunit (rbcS) of ribulose-1,5-bisphosphate carboxylase have been extensively studied in several higher plants. Comparisons among the characterized genes at the structural and organizational levels, as well as at the level of gene expression and regulation, show evolutionarily conserved and divergent features. Individual members of the rbcS gene families of tomato, petunia, and pea are differentially expressed, and show developmental, organ-specific, and light regulation depending upon the species examined. Upstream nucleotide sequences from rbcS genes of several dicotyledonous plants were aligned in order to identify those sequences which may be important in regulating rbcS expression.

Introduction

The enzyme ribulose-1,5-bisphosphate carboxylase (Rubisco) catalyzes the first step in photosynthetic CO_2 fixation in higher plants. Rubisco is a chloroplast enzyme consisting of 16 subunits (Jensen and Bahr 1977), eight large subunits of molecular weight approximately 55,000, and eight small subunits of molecular weight approximately 14,000 (Blair and Ellis 1973). The large subunit of the enzyme contains the catalytic site (Lorimer and Miziorko 1980, Lorimer 1981), is encoded by the chloroplast genome (Chan and Wildman 1972, Coen et al. 1977), and is synthesized on chloroplast ribosomes (Blair and Ellis 1973). The small subunit is encoded by the nuclear genome as a small multigene family of 5 to 15 members in most higher plants (Berry-Lowe et al. 1982, Broglie et al. 1983, Coruzzi et al. 1983, Dunsmuir et al. 1983, Wimpee et al. 1983). The small subunit protein is synthesized on cytoplasmic ribosomes as a larger precursor polypeptide which is cleaved to the mature form during import into the chloroplasts, where the holoenzyme is then assembled (Kawashimia and Wildman 1972, Highfield and Ellis 1978, reviewed by Ellis 1981).

In the past several years workers in a number of laboratories have been engaged in studying the molecular basis of Rubisco gene expression. Because Rubisco is a highly abundant enzyme in plants, its biochemical properties were characterized early on, and the protein was purified and partially sequenced. This facilitated the identification of the genes for the large subunit (*rbc*L) and the small subunit (*rbc*S), which were among the first chloroplast and nuclear genes, respectively, to be cloned from higher plants. Since then, *rbc*L and *rbc*S have been sequenced, and their expression has been extensively characterized in several higher plants. The expression of *rbc*S and *rbc*L provides an important model system to study the developmental and light regulation of genes for photosynthetic proteins. For example, our understanding of *rbc*L expression will be applicable to the broader question of how chloroplast gene expression is regulated as a whole. Likewise, understanding the regulation of expression of the *rbc*S gene family will provide important information for the more general question of how nuclear gene expression is regulated, and in particular, how expression of nearly identical members of a multigene family can be regulated differentially. Finally, learning how approximately equimolar amounts of small and large subunit polypeptides are synthesized, given the fact that the gene dosage of *rbc*S and *rbc*L in a plant cell are vastly different, will help us understand how gene expression is coordinated between the nuclear and chloroplast compartments.

In this review we will compare the structure and organization of the *rbc*L and *rbc*S genes from a number of higher plants. In addition, we will review what is known about the in vivo expression of *rbc*S and *rbc*L at the mRNA level, and will examine how expression of these genes is regulated. We will emphasize the work on tomato that has recently been completed in our laboratory, especially with regard to the *rbc*S gene family. We will compare and contrast the in vivo expression data for the *rbc*S genes of tomato, petunia, and pea, which represent the higher plants for which detailed information is available on the expression of the individual *rbc*S genes. In an effort to identify upstream sequences which may influence *rbc*S gene expression, we will compare the 5′ flanking regions of all *rbc*S genes which have been sequenced from dicotyledonous plants, and will discuss experiments which have addressed the question of the functional significance of *rbc*S conserved upstream sequences.

Structure and organization of the *rbc*L and *rbc*S genes

Evidence that the *rbc*L gene was chloroplast encoded and that *rbc*S was nuclear encoded initially came from the results of genetic crosses (Chan and Wildman 1972, Kawashima and Wildman 1972), and from studies in which the large subunit polypeptide was synthesized in vitro using isolated chloro-plasts (Blair and Ellis 1973) or purified chloroplast RNA (Hartley et al. 1975, Sagher et al. 1976). The first *rbc*L gene isolated from a higher plant was cloned from maize chloroplast DNA by Coen et al. (1977), who used a rabbit-reticulocyte in vitro transcription-translation system to screen cloned maize chloroplast DNA fragments for production of a large subunit-sized polypeptide. The identity of the polypeptide product produced by the cloned DNA was then verified serologically and by proteolytic cleavage. Since that time, *rbc*L has been mapped on the chloroplast genome in a number of higher plants (compiled by Crouse et al. 1985). In chloroplast genomes which have the typical inverted repeat structure, the *rbc*L gene is located in the large single copy region next to, and of opposite polarity from the polycistronic genes encoding the beta- and epsilon-subunits of the chloro-plast ATPase (*atp*B/E). The *rbc*L gene has been sequenced in spinach (Zurawski et al. 1981), petunia (Aldrich et al. 1986a), alfalfa (Aldrich et al. 1986b), maize (McIntosh et al. 1980), pea (Zurawski et al. 1986), tobacco (Shinozaki and Sugiura 1982), barley (Zurawski et al. 1984), *Nicotiana otophora* and *N. acuminata* (Lin et al. 1986), and has been partially sequen-ced in tomato (Manzara, Chonoles-Imlay and Gruissem, unpublished). In all higher plants studied to date, the *rbc*L coding sequence is continuous, approximately 1.4 kb in length, and encodes a polypeptide of approximately 475 amino acids. The coding sequences are highly conserved between *rbc*L genes of different species, with the majority of nucleotide changes occurring at the 3′ ends of the genes. In addition, many of the base changes are silent, resulting in even higher homology between the amino acid sequences. For example, the homology between the *rbc*L nucleotide sequences of maize versus spinach is 84%, while the homology between the corresponding amino acid sequences is 90% (Zurawski et al. 1981). In closely related species, such as petunia and tobacco, the amino acid sequences are 97% conserved, and the nucleotide sequences of the 5′ and 3′ flanking regions are also highly homologous (Aldrich et al. 1986a). In less related species, the nucleotide sequence conservation between the 5′ and 3′ flanking regions is considerably lower, although functional domains are conserved (Zurawski and Clegg 1987). Evolutionary relationships based on *rbc*L sequence data from the 5′ flanking and coding regions of a number of higher plants have been determined by Ritland and Clegg (1987).

In all higher plants studied to date, the small subunit is encoded as a multigene family in the nuclear genome. For example, the *rbc*S multigene families contain at least ten members in soybean (Berry-Lowe et al. 1982), five members in pea (Coruzzi et al. 1983), eight members in petunia (Dun-smuir et al. 1983), thirteen members in duckweed (Wimpee et al. 1983), and

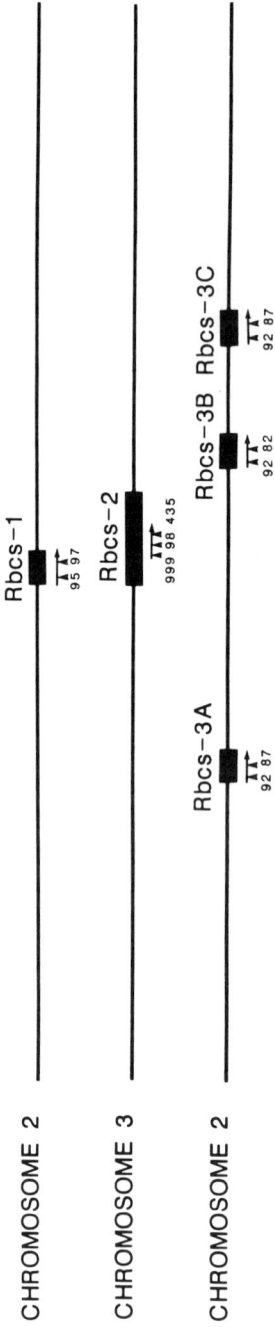

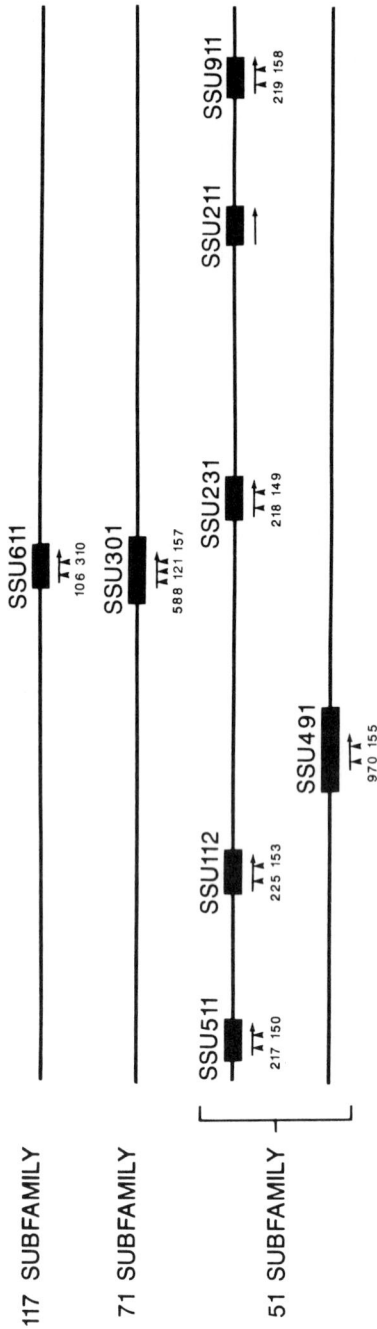

Fig. 1. Organization of the *rbc*S genes of tomato and petunia. The solid boxes indicate the location and size of each gene, including the intron regions. The horizontal arrows below each gene indicate the direction of transcription. The triangles represent introns, with the size of each intron indicated. The organization of the tomato genes was determined by Vallejos et al. (1986) and Sugita et al. (1987). The organization of the petunia genes was determined by Dean et al. (1985a, 1987a).

five members in tomato (Sugita et al. 1987). The first *rbc*S gene was identified from a pea cDNA library by Bedbrook et al. (1980), who selected clones containing *rbc*S by hybridization to polyA mRNA fractions differentially enriched for *rbc*S mRNA, either by size fractionation or by preparation of mRNA from light versus dark grown pea leaves. The identity of the putative *rbc*S cDNA clones was then verified by DNA sequence analysis which confirmed that the deduced amino acid sequences were identical to the known sequence of the purified small subunit polypeptide. Since then, additional *rbc*S cDNA clones have been isolated and sequenced from a number of higher plants including petunia (Dunsmuir et al. 1983), duckweed (Stiekema et al. 1983), wheat (Smith et al. 1983), *Nicotiana sylvestris* (Pinck et al. 1984), tomato (McKnight et al. 1986, Pickersky et al. 1986), white campion (Smeekens et al. 1986), sunflower (Waksman and Freyssinet 1987), and *Flaveria trinervia* (Adams et al. 1987). Genomic clones have been isolated and sequenced from tobacco (Mazur and Chui 1985, Poulsen et al. 1986), soybean (Berry-Lowe et al. 1982, Grandbastien et al. 1986), maize (Lebrun et al. 1987), wheat (Broglie et al. 1983), duckweed (Wimpee 1984), petunia (Tumer et al. 1986, Dean et al. 1985b, 1987a), pea (Coruzzi et al. 1984, Timko et al. 1985, Fluhr et al. 1986a), and tomato (Pichersky et al. 1986, Sugita et al. 1987).

The most completely characterized *rbc*S gene families to date are petunia, pea and tomato where most of the genes have been sequenced and the organization and/or the chromosomal localization of the genes has been determined. In tomato, the five *rbc*S genes are encoded on three separate genetic loci termed *Rbcs-1*, located on chromosome 2, *Rbcs-2* (chromosome 3), and *Rbcs-3* (chromosome 2) (Vallejos et al. 1986). As illustrated in Fig. 1, *Rbcs-1* and *Rbcs-2* each encode a single gene, while *Rbcs-3* contains three tandemly arranged genes spanning a 10 kb region, termed *Rbcs-3A*, *Rbcs-3B*, and *Rbcs-3C* (Sugita et al. 1987). The four genes contained at *Rbcs-1* and *Rbcs-3* are similar to one another in structure, each containing two introns of approximately 100 bp, the first occurring between codons two and three, and the second between codons 41 and 42 of the mature polypeptide. In contrast, the *Rbcs-2* gene contains three introns, two of which are located at sites identical to the introns of the other four genes, and the third of which occurs within codon 59. The first and third introns of *Rbcs-2* are composed of 999 bp and 435 bp, respectively, but do not appear to encode open reading frames. Petunia, like tomato, is a member of the family *Solanaceae*, and the organization of the eight petunia *rbc*S genes is very similar to the organization of the tomato genes (see Fig. 1). In both species, the *rbc*S genes are encoded at three genetic loci, with a single two-intron gene at one locus, a single three-intron gene at a second locus, and a tandem array of two-intron genes at the third locus (Dean et al. 1985b, Sugita et al. 1987). In contrast, the five member pea *rbc*L gene family maps to a single genetic locus on chromosome five (Polans et al. 1985), and all five genes contain two introns (Coruzzi et al. 1984, Timko et al. 1985, Fluhr et al. 1986a). *Rbc*S genes containing one intron have been reported in wheat (Broglie et al. 1983), in maize (Lebrun et al. 1987), and in six members of the duckweed family (Wimpee 1984, Tobin et al. 1985). Thus far, three intron *rbc*S genes have been sequenced from tomato, petunia, and tobacco, and in all cases, the intron positions are identical with respect to the coding sequence of the mature polypetpide. This is also true of all two intron *rbc*S genes (sequenced from soybean, pea, petunia and tomato), where the intron positions corres- pond to the positions of introns 1 and 2 of the three intron genes. In contrast, the position of the intron in single intron genes varies; the introns of the wheat and maize gene correspond in position to intron 1 and the introns of the duckweed genes correspond in position to intron 2 of the dicot genes.

Although the structures of the *rbc*S genes may vary between species and even within species, the nucleotide and amino acid sequences of the coding regions are conserved among gene families, and are highly conserved within a given *rbc*S gene family. When sequence comparisons are made between the members of the tomato *rbc*S gene family, several general patterns emerge which are also applicable to the petunia, duckweed, and pea *rbc*S gene families. First, the nucleotide sequences are conserved, ranging from 86% to 100% among the tomato *rbc*S genes. Second, many of the nucleotide sub- stitutions are silent, resulting in even higher amino acid sequence conserva- tism (91% to 100% in tomato). Third, the majority of the amino acid substitutions occur in the transit peptide sequences, such that in tomato, the five genes encode four different precursor polypeptides, which are cleaved to produce only three different mature polypeptides. Fourth, the sequence conservation is greater between genes at the same locus than it is between genes at different loci. For example in pea, where the *rbc*S genes are encoded at a single locus, the maximum variation between amino acid sequences is only 1% (i.e. two of 180 amino acids) (Fluhr et al. 1986a). Fifth, the tomato *rbc*S genes encode polypeptides of 180 or 181 amino acids, with 57 or 58 amino acids contained in the transit peptide sequence and 123 amino acids contained in the mature polypeptide. This is also true of the petunia, pea and tobacco *rbc*S genes, while the *rbc*S genes of monocot species appear to be more variable. Whereas in all six of the duckweed genes sequenced to date the transit peptide is 57 amino acids and the mature polypeptide is 120 amino acids, the corresponding sequences are 47 and 122 amino acids in the maize gene, and 46 and 128 amino acids in the wheat gene.

Table 1. The distribution of total *rbc*S transcripts in organs of tomato, petunia, and pea. The transcript level in each organ is expressed as a percentage of the level in leaves, which is designated as 100% in all three plants.

Tissue	Tomato	Petunia	Pea
Leaves	100	100	100
Stem	3.2	2.3	2.7
Root	ND	0.2	0.7
Petals	–	3.0	8.5
Fruit			
immature green	6.5	–	–
mature green	1.6	–	–
red	ND	–	–
Etiolated seedlings	4.6	–	–
Sepals	–	11.2	–
Stigmas/anthers	–	0.25	–
Pericarp	–	–	50
Seeds	–	–	8.0

ND = not detected

Comparison of the organization and structures of *rbc*S genes suggests that specific evolutionary changes have occurred in the *rbc*S gene families (reviewed by Pichersky et al., in preparation). For example, because the intron positions are conserved in the two and three intron *rbc*S genes, and because the intron positions differ in the one intron *rbc*S genes, it is likely that the ancestral *rbc*S gene contained two introns, and that both duplication and deletion of introns has occurred during evolutionary time. Alternatively, it is possible that the ancestral *rbc*S gene contained three introns, and that the third intron was preferentially lost. Duplication and/or deletion of both *rbc*S genes and *rbc*S gene loci is similarly indicated by the fact that while both the tomato and the petunia *rbc*S genes are encoded by three genetic loci, the pea genes are encoded by only one genetic locus, and by the fact that while the tomato *Rbcs-3* locus contains three genes, the orthologous locus in petunia contains six genes (see Fig. 1). We have already mentioned that the nucleotide sequence conservation is greater between genes at a single locus than between genes at different loci. At the same time, the amino acid sequence variation observed within members of a given *rbc*S gene family is generally less than the variation observed between orthologous genes of diverse species, suggesting concerted evolution of the *rbc*S gene families. In tomato, the high sequence conservation and the tandem arrangement of the genes in the *Rbcs-3* locus may indicate gene conversion as the mechanism responsible for maintaining sequence homogeneity within this locus, while selection appears to be the mechanism for maintaining homogeneity between genetic

Table 2. Expression patterns of individual rbcS genes in organs of tomato, petunia and pea. For each organ, the transcript level of the most highly expressed gene is designated as 1, and the levels of the remaining genes are expressed as a proportion of the level of the most highly expressed gene.

Tomato

Gene	Immature leaf	Mature leaf	Stem	Fruit	Etiolated seedlings	Root
Rbcs-1	0.44	0.28	0.45	0.60	1	ND
Rbcs-2	0.78	0.42	0.46	1	0.08	ND
Rbcs-3A	0.40	0.31	0.15	ND	0.24	ND
Rbcs-3B	1	0.83	1	ND	ND	ND
Rbcs-3C	0.50	1	0.20	ND	ND	ND

Petunia

Gene	Leaf	Sepal	Stigma/ anther	Petal	Stem
SSU301	1	1	1	1	1
SSU611	0.50	0.63	0.15	0.27	0.22
SSU491	0.14	0.08	0.15	0.02	0.03
SSU112⎱ SSU911⎰	0.11	0.08	0.15	0.01	0.03
SSU231	0.32	–	–	–	–
SSU211	0.04	–	–	–	–
SSU511	ND	–	–	–	–

Pea

Gene	Leaf	Stem	Root	Petal	Pericarp	Seed
rbcS-3A	1	1	ND	0.90	0.98	1
rbcS-3C	0.85	0.68	ND	1	1	0.34
rbcS-E9⎫ rbcs-8.0⎬ rbcs-3.6⎭	0.65	0.62	ND	0.06	0.64	0.06

ND = not detected

loci (Pichersky et al. 1986, Sugita et al. 1987). The selective pressure to maintain the sequence of the mature *rbc*S polypeptide appears to be higher than the selective pressure to maintain the sequence of the transit peptides, as was revealed by the sequence comparisons discussed earlier.

In vivo expression patterns of *rbc*S and *rbc*L

The pattern of in vivo *rbc*S gene expression has been studied in pea (Coruzzi et al. 1984, Fluhr et al. 1986a), petunia (Dean et al. 1985b, 1987b), and tomato (Sugita and Gruissem 1987) both at the level of total *rbc*S expression and at the level of individual members of the *rbc*S gene family. When total *rbc*S mRNA levels are compared between different organs of these three plants, differential accumulation of *rbc*S transcripts is observed (Table 1). Although different organs were assayed for *rbc*S expression in the three plants, there are some obvious similarities in the expression patterns. In all three plants, *rbc*S expression is highest in leaves, is substantially lower in all other tissues, and is virtually non-existent in roots. In order to address the question of whether these unique *rbc*S expression patterns are due to differential expression of some or all members of the *rbc*S gene families, differences in DNA sequence and/or transcript lengths have been exploited in order to measure the relative expression levels of individual *rbc*S genes. Table 2 summarizes the data for tomato, petunia, and pea, with the level of expression of each gene indicated as a proportion of the most highly expressed gene in a given tissue. From the data, the following conclusions can be drawn: a.) In petunia one gene, SSU301, is the most highly expressed in all tissues, and although the ranking of the genes is not substantially different in the different tissues, the proportion of the total expression contributed by each gene with respect to SSU301 changes from tissue to tissue. b.) In pea, two genes, *rbc*S-3A and *rbc*S-3C, are the most highly expressed in all tissues, but again, the relative proportions of the other genes change from tissue to tissue. c.) The data for tomato are quite different in that the most highly expressed gene varies from tissue to tissue. Also, in etiolated seedlings and in fruit, *Rbcs-3B* and *-3C* are not expressed, but are the most highly expressed genes in mature leaves. d.) The variation from the most highly expressed gene to the least expressed gene in mature leaf is 25-fold or more for petunia, approximately 10-fold for pea, and only 4-fold for tomato. Thus, it appears that differential expression of the *rbc*S genes both among and within plant organs is typical, but that the pattern of differential expression can vary dramatically between plant species.

The data presented in Tables 1 and 2 for the tomato *rbc*S genes imply that in addition to differential and organ-specific expression, the *rbc*S genes show distinct developmental regulation. During tomato leaf development, the level of the expression of the *rbc*S genes changes such that although *Rbcs-2* and *-3B* are the most highly expressed genes in immature leaves, the most highly expressed genes in mature leave are *Rbcs-3B* and *-3C*. Developmental regulation of both total *rbc*S and of *rbc*L gene expression has also been observed during tomato fruit development and ripening. Following pollination, the levels of *rbc*S and *rbc*L mRNA in tomato fruit pericarp increase for a period of 15 days, and then decrease, such that in red, ripe fruit, *rbc*S is undetectable and only low levels of *rbc*L transcript are found (Piechulla et al. 1986). This decrease in expression appears to be developmentally regulated and is not simply a consequence of overall senescence, since the expression of ripening-specific genes is activated during this time, and constitutive expression of other genes is maintained. Another example of developmental regulation of rbcS and rbcL gene expression is evident in etiolated seedlings of amaranth and maize grown in the dark, in which the steady-state levels of rbcS and rbcL mRNA increase independently of light during development of the seedlings (Nelson et al. 1984, Berry et al. 1985). In tomato, a subset of the rbcS multigene family consisting of Rbcs-1, -2 and -3A, is responsible for expression in etiolated seedlings (see Table 2), and in particular, the expression of Rbcs-1 may be developmentally regulated (see below).

Although rbcS and rbcL transcripts can be detected in dark grown seedlings of some higher plants, dramatic increases in the steady-state mRNA levels in response to light have been reported (reviewed by Tobin and Silverthorne 1985). Both rbcL and total rbcS transcripts from tomato (Sugita and Gruissem 1987), maize (Nelson et al. 1984, Sheen and Bogorad 1985, 1986), amaranth (Berry et al. 1985), mustard (Oelmüller et al. 1986) and pea (Sasaki et al. 1981, 1984, Thompson et al. 1983) have been shown to increase upon exposure of etiolated seedlings to light, although the increase in rbcS has been shown to be greater than the increase in rbcL. In tomato, this increase represents an increase in the steady-state mRNA levels of all five rbcS genes, but the two genes which are not expressed in etiolated seedlings, and are most highly expressed in mature leaf, Rbcs-3B and -3C, showed a significant lag in transcript accumulation. Similarly, in pea, transcripts from two of the genes, rbcS-3A and rbcS-3C, increased rapidly during greening, while transcripts from two other genes, rbcS-E9 and rbcS-8.0, showed a lag in induction (Fluhr et al. 1986a). While the lag in tomato gene induction in cotyledons was from three to nine hours, the lag for the pea gene induction in leaves was greater than 24 hours. The fact that this lag in induction was not observed when mature pea leaves were transferred to the dark, and then reexposed to light was interpreted as evidence for developmental regulation of the two pea genes (reviewed by Kuhlemeier et al. 1987a). In maize, light may also act to modify expression of rbcS and rbcL gene expression in a cell-type specific manner. Although rbcS and rbcL transcripts are found in both bundle sheath and mesophyll cells of etiolated maize leaves, these mRNAs decrease in mesophyll cells and persist in bundle sheath cells when the leaves are illuminated (Sheen and Bogorad 1985, 1986).

While light stimulates expression of rbcS and rbcL, lack of light reduces expression levels. In the case of rbcS, differential control of the individual genes in the dark is evident. For example, when mature greenhouse grown tomato plants were transferred to total darkness for 24 hours, it was found that rbcL transcript levels did not drop substantially (less than twofold),

Dark Light

0 3 6 12 24 3 6 24

1

2

3A

3B

3C

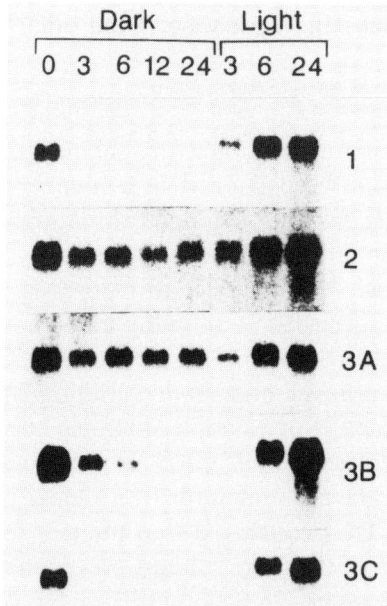

Fig. 2. Regulation of *rbc*S transcripts in light/dark-shifted tomato plants. Six-week old, greenhouse grown tomato plants were placed in absolute darkness for 24 hours, after which they were placed in the greenhouse for an additional 24 hours. Leaves were harvested 0, 3, 6, 12 and 24 hours after plants were transferred to the dark, and 3, 6, and 24 hours after plants were transferred back to the greenhouse. Total RNAs (10 µg) were hybridized with gene specific oligonucleotide probes complementary to the 3′ untranslated region of each mRNA.

while the level of total *rbc*S transcripts showed a marked decrease. When the levels of the individual *rbc*S transcripts were studied under these conditions (see Fig. 2), it was found that after 12 hours in the dark, the transcripts of *Rbcs-1*, *-3B*, and *-3C* had decreased to undetectable levels, whereas *Rbcs-2*, and *-3A* only decreased to 33% and 40%, respectively, of the levels in the light (Sugita and Gruissem 1987). This is particularly interesting in terms of the differential expression in the dark of *Rbcs-1* in cotyledons and in leaves. Although *Rbcs-1* is expressed in the dark in etiolated seedlings, it is not expressed in the dark in mature leaves, implying that the regulation of *Rbcs-1* expression in the dark is under developmental and/or organ-specific control. While *Rbcs-1* expression is independent of light in cotyledon tissue, it is strictly light-regulated in mature leaf tissue. Based on expression patterns in the dark, the five tomato *rbc*S genes can be divided into three groups: *Rbcs-1* which is the only gene expressed in the dark in cotyledons but not in leaves, *Rbcs-2* and *-3A* which are expressed in the dark in cotyledons and in mature leaves, and *Rbcs-3B* and *-3C* are not expressed in the dark in either tissue. It will be of interest to determine what factors are responsible for the differences in expression in these three groups of genes, and whether similar groups of genes can be identified from other *rbc*S gene families.

Regulation of *rbc*L gene expression

The changes in *rbc*L gene expression with light induction and during chloroplast development have been studied in a number of higher plants and transcriptional, post-transcriptional, and translational control of *rbc*L gene expression have been proposed (Berry et al. 1985, Inamine et al. 1985, Rodermel and Bogorad 1985, Berry et al. 1986, Sasaki et al. 1987, reviewed by Tobin and Silverthorne 1985). In many cases, however, transcriptional control of *rbc*L gene expression has not been adequately distinguished from differential mRNA turnover and from variation in DNA levels during chloroplast development. For example, Inamine et al. (1985) have shown that the threefold increase in *rbc*L mRNA levels observed during greening of pea seedlings could be accounted for by a similar increase in chloroplast DNA copy number. They concluded that the ten- to twenty-fold increase observed in the levels of the large subunit polypeptide was due to translational control of expression. On the other hand, Sasaki et al. (1987) studied changes in the chloroplast DNA copy number and in *rbc*L mRNA levels in greening pea seedlings subjected to different light intensities, and have found that increases in *rbc*L mRNA levels were not necessarily correlated with increase in chloroplast DNA copy number. They have suggested that *rbc*L gene expression is regulated at the transcriptional and post-transcriptional levels, although the issue of mRNA stability was not addressed. Recently, Deng and Gruissem (1987) have used a run-on transcription system derived from spinach plastids to provide evidence that chloroplast gene expression is regulated primarily at the post-transcriptional level. They have found that although the general transcriptional activity of the chloroplast genome increases in response to illumination, the transcriptional activities of most chloroplast genes, including *rbc*L, do not change relative to one another. Furthermore, they found that the changes in the transcriptional activity of the chloroplast genome were independent of chloroplast DNA copy number, and that the relative transcriptional activities of certain genes, including *rbc*L, correlated with the promoter strengths as determined by Gruissem and Zurawski (1985a). Thus, it appears that transcriptional control of chloroplast gene expression occurs at a global level, with relative transcriptional efficiencies being dictated by the strength of individual promoter regions, and that the expression of individual genes is controlled primarily at the post-transcriptional level.

The spinach and maize *rbc*L promoter regions have been functionally defined using chloroplast in vitro transcription extracts derived from these plants (Gruissem and Zurawski 1985b, Hanley-Bowdoin and Chua 1987).

Furthermore, the promoter region has been verified for higher plant chloroplast *rbc*L genes by Zurawski and Clegg (1987) using a method termed evolutionary filtering, in which upstream regions were compared for sequences conserved among all *rbc*L genes. The *rbc*L promoter region contains two critical sequence elements termed cpt1 and cpt2 which are analogous to the prokaryotic '-35' and '-10' consensus sequences, respectively, and which have been shown to be of functional significance in chloroplast gene expression (Gruissem and Zurawski 1985b). Although multiple transcription start sites have been reported for several *rbc*L genes (Palmer et al. 1982, Crossland et al. 1984, Poulsen 1984, Mullet et al. 1985), in vitro capping experiments have shown that there is only one primary transcript, and that shorter transcripts are most likely processing products (Crossland et al. 1984, Mullet et al. 1985). In most cases, the *rbc*L promoter region as defined above corresponds to the start sites of these primary transcripts (Zurawski and Clegg 1987).

Regulation of *rbc*S gene expression

From the analysis of *rbc*S multigene families in higher plants, several general patterns of *rbc*S expression emerge: a) Light affects the expression of the *rbc*S genes of most, but not all of the higher plants studied. b) The effect of light is mediated by phytochrome and one or more blue-light receptors, whose influence may vary with the developmental stage of the leaf. c) Light regulation of *rbc*S gene expression occurs primarily at the level of transcription, but differential mRNA stabilities have also been reported (Flores and Tobin 1986, reviewed by Tobin and Silverthorne 1985, and by Kuhlemeier et al. 1987a). In addition to light regulation, the in vivo expression data indicate that *rbc*S gene expression is organ-specific and developmentally regulated. Because regulation of *rbc*S expression appears to be primarily at the level of transcription, most workers have focused on analyzing the upstream regions of the *rbc*S genes to find specific DNA sequences responsible for directing the distinct patterns of expression (Dean et al. 1985b, Fluhr et al. 1986a). We have identified several upstream sequences from the tomato *rbc*S genes which correlate with some of the unique expression patterns that we have observed. It was of interest, therefore, to determine whether these sequences were present in other *rbc*S genes. An alignment of the tomato *rbc*S gene sequences along with the *rbc*S sequences presently available for other dicotyledonous plants is shown in Fig. 3. We have found that the two tomato genes which are constitutively expressed in the dark, *Rbcs-2* and *-3A*, share at least three common sequences: a CT rich sequence, designated 1 in Fig. 3, the sequence 5'-GGATGAGATAAGAYTA, and the

```
                1                        2                  3              4        5
Le Rbcs-1  -                                    GATCATTTTTTTGAAACAGAATAAAAAAAAATAGAGACATTTTTTTT  TTAAATAGAGGGCGTAAAATATT
Le Rbcs-2  - TACATTCATATCCTCTTCCTACCCCCATCTTGGATGAGATAAGATTAACGAGGTCTTA  CACGTGTCAC CTCTA
Le Rbcs-3A - AACCATTTTCACTCATTCCTTACCCCTTTTAGGATGAGATAAGACTATTCTCATTCTGA CACGTGGCAC CCTTTC
Le Rbcs-3B - CTCTATGGAGTTTTTTCATCAATTTTTTTTTCTTTTTTAAACTGTATTTTTAAAAAAATATTGAATAAAACATGTCCTATTCA
Le Rbcs-3C - AACTCGTCAGAAAGAAAAAGCAAAAGCAACAAAAAAATTGCAAGTATTTTTTAAAAAAGAAAAAAAAACATATCTTGTTTTG
Ph SSU491  - CGTCTTTACAAAAAGAATTAACTTGTTCTATTAGTTACAAATAATAGCGGCAAATATTTTATTCAAAAATACATGCCCCATTGA
Ph SSU112  - TTGTTTAAAAGTGATTAAAATTGTCCTTCTAAAATTAAGAAGAATAGCGGCAAATATTTTATTTAAT                A
Ph SSU911  - TATCATAAAAAATGAAAACTTGTCCGTCGATC              TAATTTTAGACATAATAAAAACATATTCCATTCA
Ph SSU511  -              GAAGTAAAG TTAATGAGCTTCCGC        CACGTGGCAC
Ph SSU611  - ACCATTCATATCCTCTTCCTACCCCCATT TCGGATGAGATAAGATTATTGAGCTACTGA  CACGTGTCAC C     TCCATTCTGGTTAGTTAGTGAAAAAAATG
Ph SSU301  - CACATTCATATCCTCACTTCCTAC TCCTATATCGGATGAGATAAGATTACTAAGTGCTTC  CACGTGGCAC C    TCCATTGTGGTGACAT AATGAAGA
Np rbcS-8B - CACATTCATATCCTCTTCCTACCCC ATCTAGGATGAGATAAGATTACT AGGT  CTTA  CACGTGGCAC C    TCCATTGTGGTGACTA AATGAAGAAT
Nt         - CACATTCATATCCTCTTCCTACCCC ATCTAGGATGAGATAAGATTACT AGGT  CTTA  CACGTGGCAC C    TCCATTGTGGTGACTA AATGAAGAGT
Ps rbcS-8.0-                                                                               TATCCTA GTGTGGTAAT
Ps rbcS-E9 - TGATCATGTTTAATCCTTATACTGTTGTTGTAGTTTTTTCAGTTAGCTTAGTGGGCATCTTA  CACGTGGCAT TATTATCCTA TTGGTGGCTAAT
Ps rbcS-3.6-                                                                               TATCCTA TTGGTGGCAAAT
Ps rbcS-3A - ACCACACATCACACATTTACACTCTTCACA TGAAAAGATAAGATCAGTGAGGTAATATC CACACATGCAC TG   TCCTA TTGGTGGCTAAT
Ps rbcS-3C - ACTAGCACTCACACATTTATGTTCTTCACACATGAAAAGATAAGGTCA   TCATC  CACGTGGCAC TC   TCCATA GTGATGCATAAT
Gm SRS1    -                                     CTCC GAGAGGAATGAATAATGGAAACTC CATGTGTCAC CTCC TCATA TGGTATCCAAC
Gm SRS4    - ACCACCATCACACATTTTACGTTCTTCCAAGGAAGA GATAAGA TAATGAAGCCTCCTC  CACGTGTCAC TTCC ACAT  GGTACCTAAC
```

```
                8
Le Rbcs-3A - TTGTGGCTTAATTA ATATATCTAATTATTATTATAGCTCACCCACCCTCCACGCCCAAATTAATGTCATTAAGATGGAGTTATAATTCTACTTAATAGATTC
```

```
               2              6   7                 8                    9             10
Le Rbcs-1  - TTTTTAAAAAAGGCACTTAGCTCCAATTTCTTACCTTTC  ATGTGGCCATTAAA CTTTGTAATATA        TCAA      CAACCA
Le Rbcs-2  -                                        TTGTGGTGACTTAA                                 AAAAA
Le Rbcs-3A - GATAAA                 ATT CT ACTTTTTGAA ATGTGAACAAGGC                                  CA
Le Rbcs-3B - GATAAG                                  GAGTGTGTGAATTTC AGAGGCTATTAAT TTTGAA ATG TCAA      GAGCCACA
Le Rbcs-3C - GATAAG GAC GAG   TGAGGGGTTAAAATT       CAGTGGCCATTGAT       TTTGTA ATG CCAA      GAACCACA
Ph SSU491  - GATAAG GACTGAATGTAGTGTCAGGGGTTAAAATTC     ATGTGGCCACAAT       TTAGTA ATG TCAA      GAACCACA
Ph SSU112  - GATAAG GAATGAGTGTTATTTCAGGGGTTAAAATTC     ATGTGGCCACTAAT       TTTGTA ATG TCAA      GAACCACA
Ph SSU911  - GATAAG GACTGAGTGTAGTGCGAGGGGTT AAATTC   ATGTGGGGCTTCAA    T   TTAGTA AT  TCAA GAACCAGT GAACCACA
Ph SSU511  - ACTATTTTTATG ACTAAATATCTTTTTCCTAACCTTC   ATGTGGCCATAAC       TTTGTA AAA TCAA      GAAC ACA
Ph SSU611  - GGCTCATAGCT AAAAAATATCA TTTTC TCTTTC     ATGTGGCCATTAAC       TTTGTA AAA TCAA      GAACCACA
Ph SSU301  - GGGTCTTAGCTCCAAAAATA CATT TCCAA CCTTTCAT GTGTGGATATTAAA        TTTGTA ATA TCAA      GAACCACA
Np rbcS-8B - TGGCTTAGCACC AAAAATAA TTT TCCAA CCTTTCAT GTGTGGATATTAAG A      TTGTATAATGTATCAA           GAACCACA
Nt         - TGGCTTAGCTC AAAATATAA TTT TCCAA CCTTTCAT GTGTGGATATTAAG T      TTTGTGTAGTGAATCAA          GAACCACA
Ps rbcS-8.0- GATAAG GTTAG CACACAAAACTT TTCAA TCTT   GTGTGGTTAATATG ACTGCAAAGT AGTTTT
Ps rbcS-E9 - GATAAG GTTAA CACACAAAACTT TTCAA TCTT     GTGTGGTTAATATG ACTGCAAAGT AGT TT
Ps rbcS-3.6- GATAAG GTTAGGACACAA CTT TTCAA TCTT      GTGTGGTTAATATG ACTGCAAAGT AGT TT
Ps rbcS-3A - GATAAG GCTAG CACACAAAA T TTCATC ATCTT     GTGTGGTTAATATG ACTGCAAACTTACT TT
Ps rbcS-3C - TGTATG GCTAGTATACAATTTTCTT TTCAA ACTC    GTGTGGCCATTGATG GTTCCATTGATG
Gm SRS1    - AATAAG GCTACCATTCGAAAATTTTCCTC GACTC     GTGTGGCTATATG CTGTAATGTCA
Gm SRS4    - GATAAG GCTACCATTN AAAATTTTCCTC  ACTC     GTGTGGCCAATATG CTGTAATGTCA
```

```
               11    12     13      14      15              16          17      18
Le Rbcs-1  -         AA  TCCAAT GGTCGCCTTCATCTA AGATGAGGCTTCTTTTGTTTC TATCCGTTAGATTTTAAAAAACGTCT AAAAACCTTATCA
Le Rbcs-2  -         AT  TCCAAT CTTTCATATGTAGATATT AAGTAA TTGTATAATGTTATCAAGAACCACATAACATATCA AAAAACCTTATCA TTTCA
Le Rbcs-3A -         TGA TCCAAT GGTTAC            AAATG GGTTGGTTAA TTTGTGTCCGTTAGATGGGAAAGTTAAGT GAAACCTTATCA
Le Rbcs-3B -         TAA TCCAAT GGTTATGGTTGCTCTT  AGATGAGGTTATTGCTTTTAGGT           GAAACCTTATCA TCA
Le Rbcs-3C -         AAA TCCAAT GGTTACCATTCCTGTA AGATGAGGTTTGCTAACTCTTTTTGTCCGTTAGATAG GAAGCCTTATCA
Ph SSU491  -         TTA TCCAAA TGGTTACAATTCCTCGCTTTTAAGATGAGGTTTCCTCCA CTTGTGTCCGTCAGATGGGAAAGAGAT GAAACTTTATCA
Ph SSU112  -         TAA TCCAAT GGTTACAATTCCTCTAATT AGTAGGTTTCCTTG TTTGTGTCCGTTAGATG AGAAAAGGATG T GAAACTTTATCA
Ph SSU911  - ACACCACA TAA TCCAAT TGTTACCCTCCTCTA AGATGAGGTTTGTTTA CTTGTATCCGTTAGATGTAGAGGTT ATG T GAAAGCTTAACA
Ph SSU511  -         TAA TCCAAT GGTTACAATTCATCTA AGATGAGGCTTGTTTA          GTTA TGT  GAAACCTTTTCA
Ph SSU611  -         TAA TGCAAA GTTTACGTTATCATTT AG  GAA                 GTTA TGT   GAAACCTTTTCA
Ph SSU301  -         TAA TCCAAT GGTTAGCTTTATTCCA AGA GAGGTTAGTTGA TTTTTGTCCGTTGA TATGTGAAATATGT TA AAAAACCTTATCA
Np rbcS-8B -         TAA TCCAAT GGTTAGCTTTATTCCA AGATGAGGGGTTGTTGA TTTTTGTCCGTTGA TATGCGAAATATGT T AAAAACCTTATCA
Nt         -         TGA TCCAAA GGTTAGCTTTATTCCA AGATGAGGGGTTGTTGA TTTTTGTCCGTCGA TATAGGAAAATTG T AAAACCTTATCA
Ps rbcS-8.0- ATCATTT CAC AA  TCCAAC AAATGTGTA CTAGGCAGCAATCAATTACCACAATTTTAAGACAACAA      TATTGGAAATAGA AAAATCAATACA
Ps rbcS-E9 - ATCATTTTCAC AA  TCCAAC AAAT GGTTCTAGGCAGTGGCTACCACAATATTAAGACATAA            TATTGGAAATAGA AAAATCAATACA
Ps rbcS-3.6- ATCATTT CAC AA  TCTAAC AAGATTGCTACTAGGCAGTGGCTAATTACCACAATATTAAGACCATAA          TATTGGAAATAGA TAAATAAAAACA
Ps rbcS-3A - ATCATTTTCAC TA  TCTAAC AAGATTGGTTACTAGGCAGTGGCTAAGTACCACAATATTAAGACCATAA          TATTGGAAATAGA TAAATAAAAACA
Ps rbcS-3C - ATCATTTTTCAC AA  TCCAAA ACCATTGGTTA        TACTT                 GATCAGAAACC   ATTTCTGAAACA
Gm SRS1    - ATCACTTATTC GAA TCCAAC GGTTGTAACTTTTCGGCCAACCAATCCTCTCTCCATTTCACACA      ATTTGGATTAGTACTACACAAATCGACACTA
Gm SRS4    - ATCACTTATTC AA  TCCAAN GGTTGTAACTTCTCAGCAACCAATCATCCCTCCATTTCACACC      ATCCGGATTAGTACTACACAAATC ACACTA
```

```
               19               20
Le Rbcs-1  - TTTATATAAA GGGACGATACCACTGTGTAATAA GCATCTT TTTAAAAAAAAATATAGTTTCTTTGAAATTAAAAAAAAAAAAAAACATTTTTATAGCTAAGTAAGTAAACGCA
Le Rbcs-2  - TTTATATAAA AGGATAGTGGACATCAAAAGGTT CATATT GAACCAAAAAAAGAGAGAAGAAGCAAT
Le Rbcs-3A - TTTATATATA GAGGGAGAGACTAGAAAG       CAATA  ACCCTCCTTGAGTTCAAGATAAGCACTTGGTTTTCAGCA
Le Rbcs-3B - CTATATATA  CAAGGGGGATACTAGAGACCAATT ATTGTC AACA
Le Rbcs-3C - CTATATATA  CAAGCGTCCTAATAACCTCCTTA GTAACC AATTATTTCACCA
Ph SSU491  - CTATAAATA  GAAGTGGGACTAGGAAGCAAGGA CCATTG ACCCTCCTTGAGAACAAAGCTCAAGGGGAACCAAGGATTTATTTTCAGAA
Ph SSU112  - TTATATAAA  GAGGGGGAAGACTACGTACCGCA ATAACT TTTCTAAGGATATTTCAGCA
Ph SSU911  - CTATAAATA  GATATTGCAAATGTCAAGGGAAG CAATAG CAATTATATTTAGCA
Ph SSU511  - TTATATATA  GAAGGGGGCACTATACAT       CAATA  ACCCTCTTGAAGCAAAGGTGGGAAAGGGAAACACAAAAATATAAGCTAACGATTCTTTAGCA
Ph SSU611  - TTATATAAA  GGACAGGAGTCATAGTGCAATGA CCATCA TAAAGCA
Ph SSU301  - TTATATAAA  GGGTGGTGGTGGGCAATACAAAG TCAGTG TGAAGTGTTTAAAGGAAAAAGCTTTGGAAGAAGCAAAATCTTCTAACT
Np rbcS-8B - TTATATATA  GAGTGGTGGTGGGCAACGATGCATAG ACCATC TTGGAAGTTTAAAAGGAAAAGCGAAAAGGGAAAAAGGAAGAAGAAATCTTTCGTCTTAAGTGTAATTAACA
Nt         - TTATATATA  GGGTGGTGGTGGGCAACTATGCAATG ACCATA TTGGAAGTT    AAAAGGAAAAGAGAGAAAAGAAGAAATCTTTCTGTCT AAGTGTAATTAACA
Ps rbcS-8.0- TTATATATA  GCAAGTTTTAGTACAAGCTTTGC AATTCA ACCACAAGAACTAACAAAGTCAGAAAAA
Ps rbcS-E9 - TTATATATA  GCAAGTTTTAGTACAAGCTTTGC AATTCA ACCACAAGAACTAAGAAAGTCAGAAAAA
Ps rbcS-3.6- TTATATATA  GCAAGTTTTAGTACAAGCTTTGC AATTCA TACA GAA  GTGAGAAAA
Ps rbcS-3A - TTATATATA  GCAAGTTTTAGCAGAAGCTTTGC AATTCA TACA GAA  GTGAGAAAA
Ps rbcS-3C - TTATATATA  CCTATAGAACGTTTGCAACCTAA TTATAA CAACA
Gm SRS1    - TTATATATA  GCAAGTTTGAGCAGAAGCTTGGA TATCTG GCAGCAGAAAAACAAGTAGTTGAGAACTAAGAAGAAGAA
Gm SRS4    - TTATATATA  GTAAGTTTGAGCAGAAGCTTGGA TATCTG GCAGCAGAAGAACAAGTAGTTGAGAACTAAGAAGGAGAAGCAA
```

sequence 5'-CACGTGG/TCAC, designated 2 and 3, respectively, in Fig. 3. These sequences are also conserved in the petunia genes SSU301 and SSU611 (Dean et al. 1985b), and in the tobacco genes. In the soybean and pea genes, sequence 3 is conserved, while sequences 1 and 2 are not conserved or are only partly conserved. In addition, a partial sequence 2, 5'-GATAAG, is found 3' to sequence 3 in several of the *rbc*S genes. Three tomato genes, *Rbcs-1*, *-3B* and *-3C*, do not contain sequences 1, 2 and 3. These represent a subset of the *rbc*S gene family whose expression rapidly decreases to undetectable levels when greenhouse-grown plants are placed in the dark (see Fig. 2). The three genes contain a sequence designated 5 in Fig. 3, which is also conserved among three of the petunia genes, SSU491, SSU112, and SSU911. Although the correlation between expression pattern in the dark and the presence of these sequences in the tomato *rbc*S genes is intriguing, it is unknown whether such correlations exist for the other *rbc*S gene families, since, to our knowledge, experiments similar to that shown in Fig. 2 have not been executed for petunia, tobacco, pea, or soybean. Sequences 1, 2 and 3, however, are included in a region from the *N. plumbaginifolia rbc*S-*8B* genes which has been shown to function as an enhancer in transgenic tobacco plants (Poulsen and Chua 1987). The functional significance of these sequences in tomato is currently under investigation in our laboratory.

For the additional conserved sequences indicated in Fig. 3 no correlative relationship with the available differential expression data is indicated. These conserved sequences can be divided into three categories. The first category is comprised by those sequences which are conserved among all *rbc*S genes, and are likely to play a fundamental role in *rbc*S gene expression (sequences 8, 12 and 19). These include the eukaryotic 'TATA' box (19), a sequence similar to the eukaryotic 'CAAT' box (12), and the sequence 5'-GTGTGGTTAA/CTATG, termed 'box II' (8) (Fluhr et al. 1986b), that has been shown to function as a silencer in the dark for the pea *rbc*S genes (see below). A 'box II'-like sequence has been found in all *rbc*S genes sequenced to date, all of which have been shown to display reduced expression levels in the dark. It will be of interest to determine whether this sequence is also found in *rbc*S genes which are less responsive to light, such

---

Fig. 3. Alignment of the upstream regions of the *rbc*S genes from tomato, petunia, tobacco, pea and soybean. Conserved sequences are underlined or boxed and numbered. Numbers corresponding to underlined sequences are underlined. Sequence 20 represents the region encompasssing the transcription start sites of several *rbc*S genes (transcription start sites have been mapped for most, but not all, of the genes). The last nucleotide of each sequence represents the nucleotide immediately preceeding the ATG. Le, *Lycopersicon esculentum* (tomato); Ph, *petunia hybrida* (petunia); Np, *Nicotiana plumbaginifolia* (tobacco); Nt *Nicotiana tabacum* (tobacco); Ps, *Pisum sativum* (pea); Gm, *Glycine max* (soybean).

as barley, castor bean and cucumber (reviewed by Tobin and Silverthorne 1975). The second category of conserved sequences includes those which are conserved among all members of a specific *rbc*S gene family or among all *rbc*S genes of closely relates species (i.e. among members of the family Solanaceae, or members of the family Leguminosae), and may be important in a specific aspect of gene expression (sequences 7, 11 and 18). These include the 'box I' sequence 5'-TTCAA (6), which may be a component of a positive regulatory factor in the pea *rbc*S genes (Green et al. 1987), the 'box III' sequence 5'-ATCATTTTCAC, which appears to be analogous to 'box II' in pea (Green et al. 1987), and the consensus sequence 5'-GAAACCTTATCA found in tomato, petunia (Dean et al. 1985b) and tobacco. The significance of this last sequence is not clear at this time, but it is conceivable that it may function in some aspect of *rbc*S gene expression that is unique to plants which have photosynthetically active cotyledons as opposed to cotyledons which serve primarily as storage organs. The third category of sequences includes those which are conserved among some members of closely related and/or diverse gene families, and may be responsible for some as yet unrecognized aspect of differential *rbc*S gene expression (sequences, 4, 6, 9, 10, 13, 14, 15, 16, and 17). These conserved sequences may or may not have any functional significance, however, it is possible that they are related to some aspect of differential gene expression which has not yet been recognized. Alternatively, these sequences may not be involved in regulating gene expression, but may simply reflect differences in the lineages of the genes.

In order to test the functional significance of conserved upstream sequences from the *rbc*S genes, experiments have been done in which a chimeric gene, usually consisting of an 'intact' or deleted *rbc*S upstream region transcriptionally fused to a bacterial reporter gene, has been introduced into plant cells, and the expression has been studied in the resulting transgenic plant. The majority of these experiments have used pea upstream regions expressed in transgenic tobacco plants, and have demonstrated that the upstream regions of the *rbc*S genes are sufficient to direct light-regulated and organ-specific expression in the heterologous transgenic plants (reviewed by Kuhlemeier et al. 1987a). Deletion mutagenesis of several upstream regions has defined enhancer-like elements which are capable of directing light-regulated and organ-specific expression in either orientation with respect to the promoter element. In *Nicotiana plumbaginifolia*, Poulsen et al. (1986) found that an upstream sequence comprising bases $-1,038$ to $+32$ was sufficient to confer light-regulated and organ-specific expression on the bacterial chloramphenicol acetyltransferase (CAT) gene expressed in the homologous plant system. Deletion mutagenesis of this region defined an enhancer element contained within bases -312 to -192 (Poulsen and Chua

1987). Fluhr et al. (1986b) found that a region from − 330 to − 50 of the pea *rbcS-3A* gene and a region from − 317 to − 82 of the pea *rbcS-E9* gene would direct light-regulated and tissue-specific expression of the CAT gene in transgenic petunia when fused to the cauliflower mosaic virus 358 (CaMV 35S) promoter region. Within this region of the pea *rbcS-3A* gene, a light regulatory element of 58 base pairs was identified which silenced expression in the dark of the CAT gene under the control of the normally constitutive CaMV 35S enhancer element (Kuhlemeier et al. 1987b). The 58 bp element appears to be repeated a number of times in the *rbc*S upstream regions and encompasses the three sequence boxes I, II, and III, (see above), which are conserved among the pea *rbc*S genes. When DNA fragments containing boxes I, II, and II were used in gel retardation and DNase I footprinting experiments using pea nuclear extracts, a nuclear factor termed GT-1 was identified which was shown to interact with box II and box III sequences, and which was present in nuclear extracts from both light and dark grown peas. If GT-1 is involved in light regulation of *rbc*S gene expression, it presumably interacts with another factor which alters its activity in the light or in the dark. Undoubtedly additional DNA binding factors will be identified in the near future which interact with other *rbc*S upstream sequence elements. Interestingly, it appears that different regulatory elements may function in monocots and dicots, as expression of a wheat *rbc*S gene was not detected in transgenic tobacco plants (Keith and Chua 1986). Furthermore, when the wheat gene was expressed in transgenic tobacco under the control of the CaMV 35S promoter, inefficient splicing and inaccurate polyadenylation of the wheat *rbc*S mRNA was observed, and the effect on splicing could not be accounted for by nucleotide sequence differences between the introns of the two plants (Keith and Chua 1986).

Although several workers have shown that the upstream regions of the *rbc*S genes are sufficient for light regulated and organ-specific expression, Dean et al. (in preparation) have shown that sequences 3′ to the ATG of the petunia *rbc*S genes influence the level of *rbc*S gene expression. This could be due to the presence of enhancer elements or due to post-transcriptional regulation or *rbc*S gene expression, but clearly all aspects of *rbc*S gene expression are not directed solely by the 5′ upstream regions.

Summary

Although numerous questions remain to be answered, we are well on our way to elucidating some of the mechanisms governing the regulation of *rbc*S and *rbc*L gene expression. In vitro transcription experiments have indicated that light regulation of *rbc*S gene expression occurs primarily at the level of transcription, suggesting that sequences of the 5′ upstream regions are important in this regard, although other regions may also be involved. The nucleotide sequences of many *rbc*S genes are now available, making it possible to compare the upstream regions, and to tentatively identify sequences which may be important in directing *rbc*S gene expression. Detailed information is available on the in vivo expression of the rbcS gene families from petunia, pea and tomato, allowing classification of the genes based on light regulation and tissue-specific expression, and allowing correlative associations to be made between conserved sequences and expression pattern. Plant transformation has become routine for many species of dicotyldonous plants, allowing the functional importance of specific nucleotide sequences to be ascertained in vivo. In vitro techniques such as DNA retardation and footprinting can then be used to identify sequences which interact with DNA binding proteins, or alternatively, to identify sequences to be mutagenized and evaluated in vivo. The next important steps will be to isolate specific DNA binding proteins, to determine the hierarchy protein-protein interactions, and to determine how DNA binding proteins exert their action. Similarly for *rbc*L, in vitro run-on experiments have indicated that regulation of chloroplast gene expression occurs primarily at the post-transcriptional level, focusing attention on stability and processing of chloroplast RNA and on translational regulation of gene expression. Promoter strength has been implicated as the factor dictating relative transcriptional efficiencies of chloroplast genes, and factors influencing promoter strength have been evaluated in vitro. Because chloroplast transformation has not yet been accomplished for higher plants, it has not been possible to evaluate in vivo the significance of specific DNA sequences in chloroplast gene expression.

A number of questions still remain before Rubisco gene expression is completely understood. First, it is still not clear how the overall expression of the chloroplast genome is regulated. Specifically, it is unknown whether specific nuclear factors are involved, or whether more global forms of regulation are employed. Secondly, although communication between the chloroplast and nucleus is thought to occur, it is not known whether communication is interactive or uni-directional. Thirdly, it must be determined whether common regulatory molecules are responsible for controlling gene expression in both the nuclear and chloroplast compartments. It is clear that Rubisco gene expression will continue to serve as an important model system for dissecting the coordinate regulation of chloroplast and nuclear gene expression in higher plants.

Acknowledgements

We thank Dr Jonathon Narita and Leslie Wanner for reading the manuscript, and Dr Caroline Dean for providing unpublished results. This work has been supported by a grant from NIH (GM33813) to W.G.

References

Adams CA, Babcock M, Leung F and Sun SM (1987) Sequence of a ribulose-1,5-bisphosphate carboxylase/oxygenase cDNA from the C_4 dicot *Flaveria trinervia*. Nucleic Acids Res 15: 1987

Aldrich J, Cherney B, Merlin E and Palmer J (1986a) Sequence of the *rbc*L gene for the large subunit of ribulose bisphosphate carboxylase-oxygenase from petunia. Nucleic Acids Res 14: 9534

Aldrich J, Cherney B, Merlin E and Palmer J (1986b) Sequence of the *rbc*L gene for the large subunit of ribulose bisphosphate carboxylase-oxygenase from alfalfa. Nucleic Acids Res 14: 9535

Bedbrook JR, Smith SM and Ellis RJ (1980) Molecular cloning and sequencing of cDNA encoding the precursor to the small subunit of chloroplast ribulose-1,5-bisphosphate carboxylase. Nature 287: 692–697

Berry JO, Nikolau BJ, Carr JP and Klessig DF (1985) Transcriptional and post-transcriptional regulation of ribulose-1,5-bisphosphate carboxylase gene expression in light- and dark-grown amaranth cotyledons. Molec and Cell Biol 5: 2238–2246

Berry JO, Nikolau BJ, Carr JP and Klessig DF (1986) Translational regulation of light-induced ribulose-1,5-bisphosphate carboxylase gene expression in amaranth. Molec and Cell Biol 6: 2347–2353

Berry-Lowe SL, McKnight TD, Shah DM and Meagher RB (1982) The nucleotide sequence, expression and evolution of one member of a multigene family encoding the small subunit of ribulose-1,5-bisphosphate carboxylase in soybean. J Molec Appl Genet 1: 483–498

Blair GE and Ellis RJ (1973) Protein synthesis in chloroplasts I. Light-driven synthesis of large subunit of Fraction I protein by isolated pea chloroplasts. Biochim Biophys Acta 319: 223–234

Broglie R, Coruzzi, G, Lamppa G, Keith B and Chua NH (1983) Structural analysis of nuclear genes coding for the precursor to the small subunit of wheat ribulose-1,5-bisphosphate carboxylase. Biotechnology 1: 55–61

Chan PH and Wildman SG (1972) Chloroplast DNA codes for the primary structure of the large subunit of fraction I protein. Biochim Biophys Acta 277: 677–680

Coen DM, Bedbrook JR, Bogorad L and Rich A (1977) Maize chloroplast DNA fragment encoding the large subunit of ribulosebisphosphate carboxylase. Proc Natl Acad Sci USA 74: 5487–5491

Coruzzi G, Broglie R, Cashmore A and Chua NH (1983) Nucleotide sequences of two pea cDNA clones encoding the small subunit of ribulose-1,5-bisphosphate carboxylase and the major chlorophyll a/b-binding thylakoid polypeptide. J Biol Chem 258: 1399–1402

Coruzzi G, Broglie R, Edwards C and Chua NH (1984) Tissue-specific and light-regulated expression of a pea nuclear gene encoding the small subunit of ribulose-1,5-bisphosphate carboxylase. EMBO J 3: 1671–1679

Crossland LD, Rodermel SR and Bogorad L (1984) Single gene for the large subunit of ribulosebisphosphate carboxylase in maize yields two differentially regulated mRNAs. Proc Natl Acad Sci USA 81: 4060–4064

Crouse EJ, Schmitt JM and Bohnert HJ (1985) Chloroplast and cyanobacterial genomes, genes and RNAs: A compilation. Plant Molec Biol Reporter 3: 42–89

Dean C, van den Elzen P, Tamaki S, Dunsmuir P and Bedbrook J (1985a) Linkage and homology analysis divides the eight genes for small subunit of petunia ribulose-1,5-bisphosphate carboxylase into three gene families. Proc Natl Acad Sci USA 82: 4964–4968

Dean C, van den Elzen P, Tamaki S, Dunsmuir P and Bedbrook J (1985b) Differential expression of the eight genes of the petunia ribulose bisphosphate carboxylase small subunit multi-gene family. EMBO J 4: 3055–3061

Dean C, van den Elzen P, Tamaki S, Black M, Dunsmuir P and Bedbrook J (1987a) Molecular characterization of the rbcS multi-gene family of Petunia (Mitchell). Mol Gen Genet 206: 465–474

Dean C, Favreau M, Dunsmuir P and Bedbrook J (1987b) Confirmation of the relative expression levels of the Petunia (Mitchell) rbcS genes. Nucleic Acids Res 15: 4655–4668

Deng XW and Gruissem W (1987) Control of plastid gene expression during development: The limited role of transcriptional regulation. Cell 49: 379—387

Dunsmuir P, Smith S and Bedbrook J (1983) A number of different nuclear genes for the small subunit of RuBPCase are transcribed in petunia. Nucleic Acids Res 11: 4177–4183

Ellis RJ (1981) Chloroplast proteins: Synthesis, transport, and assembly. Ann Rev Plant Physiol 32: 111–137

Flores S and Tobin EM (1986) Benzladenine modulation of the expression of two genes for nuclear-encoded chloroplast proteins in Lemna gibba: Apparent post-transcriptional regulation. Planta 168: 340–349

Fluhr R, Moses P, Morelli G, Coruzzi G and Chua NH (1986a) Expression dynamics of the pea rbcS multigene family and organ distribution of the transcripts. EMBO J 5: 2063–2071

Fluhr R, Kuhlemeier C, Nagy F and Chua NH (1986b) Organ-specific and light-induced expression of plant genes. Science 232: 1106–1112

Fluhr R and Chua NH (1986) Developmental regulation of two genes encoding ribulose-bisphosphate carboxylase small subunit in pea and transgenic petunia plants: Phytochrome response and blue-light induction. Proc Natl Acad Sci USA 83: 2358–2362

Grandbastien MA, Berry-Lowe S, Shirley BW and Meagher RB (1986) Two soybean ribulose-1,5-bisphosphate carboxylase small subunit genes share extensive homology even in distant flanking sequences. Plant Molec Biol 7: 451–465

Green PJ, Kay SA and Chua NH (1987) Sequence-specific interactions of a pea nuclear factor with light-responsive elements upstream of the rbcS-3A gene. EMBO J 6: 2543–2549

Gruissem W and Zurawski G (1985a) Analysis of promoter regions for the spinach chloroplast rbcL, atpB, and psbA genes. EMBO J 4: 3375–3383

Gruissem W and Zurawski G (1985b) Identification and mutational analysis of the promoter for a spinach chloroplast transfer RNA gene. EMBO J 4: 1637–1644

Hanley-Bowdoin L and Chua NH (1987) Chloroplast promoters. TIBS 12: 67–70

Hartley MR, Wheeler A and Ellis RJ (1975) Protein synthesis in chloroplasts. J Mol Biol 91: 67–77

Highfield PE and Ellis RJ (1978) Synthesis and transport of the small subunit of chloroplast ribulose bisphosphate carboxylase. Nature 271: 420–424

Inamine G, Nash B, Weissbach H and Brot N (1985) Light regulation of the synthesis of the large subunit of ribulose-1,5-bisphosphate carboxylase in peas: evidence for translational control. Proc Natl Acad Sci USA 82: 5690–5694

Jensen RG and Bahr JT (1977) Ribulose-1,5-bisphosphate carboxylase-oxygenase. Annu Rev Plant Physiol 28: 379–400

Kawashima N and Wildman SG (1972) Studies on fraction I protein IV. Mode of inheritance of primary structure in relation to whether chloroplast or nuclear DNA contains the code for a chloroplast protein. Biochim Biophys Acta 262: 42–49

Keith B and Chua NH (1986) Monocot and dicot pre-mRNAs are processed with different efficiencies in transgenic tobacco. EMBO J 5: 2419–2425

Kuhlemeier C, Green PJ, Chua NH (1987a) Regulation of gene expression in higher plants. Annu Rev Plant Physiol 38: 221–257

Kuhlemeier C, Fluhr R, Green PJ and Chua NH (1987b) Sequences in the pea *rbcS-3A* gene have homology to constitutive mammalian enhancers but function as negative regulatory elements. Genes and Development 1: 247–255

Lebrun M, Waksman G and Freyssinet G (1987) Nucleotide sequence of a gene encoding corn ribulose-1,5-bisphosphate carboxylase/oxygenase small subunit (*rbcS*). Nucleic Acids Res 15: 4360

Lin CM, Liu ZQ and Kung SD (1986) *Nicotiana* chloroplast genome: X. Correlation between the DNA sequences and the isoelectric focusing patterns of the LS of Rubisco. Plant Molec Biol 6: 81–87

Lorimer GH and Mizioko HM (1980) Carbamate formation on the epsilon-amino group of a lysyl residue as the basis for the activation of ribulosebisphosphate carboxylase by CO_2 and Mg^{2+}. Biochemistry 19: 5321–5328

Lorimer GH (1981) Ribulosebisphosphate carboxylase: amino acid sequence of a peptide bearing the activator carbon dioxide. Biochemistry 20: 1236–1240

Mazur BJ and Chui CF (1985) Sequence of a genomic clone for the small subunit of ribulose bis-phosphate carboxylase-oxygenase from tobacco. Nucleic Acids Res 13: 2373–2386

McIntosh L, Poulsen C and Bogorad L (1980) Chloroplast gene sequence for the large subunit of ribulose bisphosphate carboxylase of maize. Nature 288: 556–560

McKnight TD, Alexander DC, Babcock MS and Simpson RB (1986) Nucleotide sequence and molecular evolution of two tomato genes encoding the small subunit of ribulose-1,5-bisphosphate carboxylase. Gene 48: 23–32

Mullet JE, Orozco EM and Chua NH (1985) Multiple transcripts for higher plant *rbc*L and *atp*B genes and localization of the transcription initiation site of the *rbc*L gene. Plant Molec Biol 4: 39–54

Nelson T, Harpster MH, Mayfield SP and Taylor WC (1984) Light-regulated gene expression during maize leaf development. J Cell Biol 98: 558–564

Oelmüller R, Dietrich G, Link G and Mohr H (1986) Regulatory factors involved in gene expression (subunits of ribulose-1,5-bisphosphate carboxylase) in mustard (*Sinapis alba* L.) cotyledons. Plant 169: 260–266

Palmer JD, Edwards H, Jorgensen RA and Thompson WF (1982) Novel evolutionary variation in transcription and location of two chloroplast genes. Nucleic Acids Res 10: 6819–6832

Pichersky E, Bernatzky R, Tanksley SD and Cashmore AR (1986) Evidence for selection as a mechanism in the concerted evolution of *Lycopersicon esculentum* (tomato) genes encoding the small subunit of ribulose-1,5-bisphosphate carboxylase/oxygenase. Proc Natl Acad Sci USA 83: 3880–3884

Pichersky E, Cashmore AR, Tanksley S, Bernatsky R, Gruissem W, Sugita M, Manzara T (in preparation) The CAB and RBCS genes: Evolution of two gene families encoding photosynthetic proteins in higher plants

Piechulla B, Pichersky E, Cashmore AR and Gruissem W (1986) Expression of nuclear and plastid genes for photosynthesis-specific proteins during tomato fruit development and ripening. Plant Molec Biol 7: 367–376

Pinck M, Guilley E, Durr A, Hoff M, Pinck L and Fleck J (1984) Complete sequence of one of the mRNAs coding for the small subunit of ribulose bisphosphate carboxylase of *Nicotiana sylvestris*. Biochimie 66: 539–545

Polans NO, Weeden NF and Thompson WF (1985) Inheritance, organization, and mapping of *rbc*S and cab multigene families in pea. Proc Natl Acad Sci USA 82: 5083–5087

Poulsen C (1984) Two mRNA species differing by 258 nucleotides at 5′ end are formed from the barley chloroplast *rbc*L gene. Carlsberg Res Commun 49: 89–104

Poulsen C, Fluhr R, Kauffman JM, Boutry M and Chua NH (1986) Characterization of an *rbc*S gene from *Nicotiana plumbaginifolia* and expression on an *rbc*S-CAT chimeric gene in homologous and heterologous nuclear background. Mol Gen Genet 205: 193–200

Poulsen C and Chua NH (1987) Dissection of upstream sequences involved in regulated expression of the *Nicotiana plumbaginifolia* rbcS-8B gene in homologous nuclear background. Abstract from NATO Plant Molecular Biology Meeting, June 10–19

Ritland K and Clegg MT (1987) Evolutionary analysis of plant DNA sequences. Amer Nat 130: S74–S100

Rodermel SR and Bogorad L (1985) Maize plastid photogenes: mapping and photoregulation of transcript levels during light-induced development. J Cell Biol 100: 463–476

Sagher D, Grosfeld H and Edelman M (1976) Large subunit ribulosebisphosphate carboxylase messenger RNA from *Euglena* chloroplasts. Proc Natl Acad Sci USA 73: 722–726

Sasaki Y, Ishiye M, Sakihama T and Kamikubo T (1981) Light-induced increase of mRNA activity coding for the small subunit of ribulose-1,5-bisphosphate carboxylase. J Biol Chem 256: 2315–2320

Sasaki Y, Tomoda Y and Kamikubo T (1984) Light regulates the gene expression of ribulosebisphosphate carboxylase at the levels of transcription and gene dosage in greening pea leaves. FEBS Lett 173: 31–35

Sasaki Y, Nakamura Y and Matsuno R (1987) Regulation of gene expression of ribulose bisphosphate carboxylase in greening pea leaves. Plant Molec Biol 8: 375–382

Sheen JY and Bogorad L (1985) Differential expression of the ribulose bisphosphate carboxylase large subunit gene in bundle sheath and mesophyll cells of developing maize leaves is influenced by light. Plant Physiol 79: 1072–1076

Sheen JY and Bogorad L (1986) Expression of the ribulose-1,5-bisphosphate carboxylase large subunit gene and three small subunit genes in two cell types of maize leaves. EMBO J 5: 3417–3422

Shinozaki K and Suguira M (1982) The nucleotide sequence of the tobacco chloroplast gene for the large subunit of ribulose-1,5-bisphosphate carboxylase/oxygenase. Gene 20: 91–102

Smeekens S, van Oostgen J, de Groot M, Weisbeek P (1986) Silene cDNA clones for a divergent chlorophyll-a/b-binding protein and a small subunit of ribulosebisphosphate carboxylase. Plant Molec Biol 7: 433–440

Smith SM, Bedbrook J and Speirs J (1983) Characterisation of three cDNA clones encoding different mRNAs for the precursor to the small subunit of wheat ribulosebisphosphate carboxylase. Nucleic Acids Res 11: 8719–8734

Stiekema WJ, Wimpee CF and Tobin EM (1983) Nucleotide sequence encoding the precursor of the small subunit of ribulose-1,5-bisphosphate carboxylase from *Lemna gibba* L.G-3. Nucleic Acids Res 11: 8051–8061

Sugita M and Gruissem W (1987) Developmental, organ-specific and light-dependent expression of the tomato ribulose-1,5-bisphosphate carboxylase small subunit gene family. Proc Natl Acad Sci USA 84: 7104–7108

Sugita M, Manzara T, Pichersky E, Cashmore A, and Gruissem W (1987) Genomic organization, sequence analysis and expression of all five genes encoding the small subunit of ribulose-1,5-bisphosphate carboxylase/oxygenase from tomato. Mol Gen Genet 209: 247–256

Thompson WF, Everett M, Polans NO, Jorgensen RA and Palmer JD (1983) Phytochrome control of RNA levels in developing pea and mung-bean leaves. Planta 158: 487–500

Timko MP, Kausch AP, Hand JM, Cashmore AR, Herrera-Estrella L, Van den Broeck G and Van Montagu M (1985) Structure and expression of nuclear genes encoding polypeptides of the photosynthetic apparatus. In: Steinbeck KE, Bonitz S, Arntzen CJ and Bogorad L (eds) Molecular biology of the photosynthetic apparatus, pp 381–396. NY: Cold Spring Harbor Laboratory Press

Tobin EM and Silverthorne J (1985) Light regulation of gene expression in higher plants. Ann Rev Plant Physiol 36: 569–593

Tobin EM, Wimpee CF, Karlin-Neumann GA, Silverthorne J and Kohorn BD (1985) Phytochrome regulation of nuclear gene expression. In: Steinbeck KE, Bonitz S, Arntzen CJ and Bogorad L (eds) Molecular biology of the photosynthetic apparatus, pp 373–380. NY: Cold Spring Harbor Laboratory Press.

Tumer NE, Clark WG, Tabor CJ, Hironaka CM, Fraley RT and Shah D (1986) The genes encoding the small subunit of ribulose-1,5-bisphosphate carboxylase are expressed differentially in petunia leaves. Nucleic Acids Res 14: 3325—3342

Vallejos CE, Tanksley SD, Bernatzky R (1986) Localization in the tomato genome of DNA restriction fragments containing sequences homologous to the rRNA (45S), the major chlorophyll a/b binding polypeptide and the ribulose bisphosphate carboxylase genes. Genetics 112: 93–105

Waksman G and Freyssinet G (1987) Nucleotide sequence of a cDNA encoding the ribulose-1,5-bisphosphate carboxylase/oxygenase from sunflower (Helianthus annuus). Nucleic Acids Res 15: 1328

Wimpee CF, Stiekma WJ and Tobin EM (1983) Sequence heterogeneity in the RuBP carboxylase small subunit gene family of Lemna gibba. In: Goldberg RB (ed.) Plant Molecular Biology, UCLA Symposium on Molecular and Cellular Biology, New Series Vol. 1-2, pp 391–401. NY: Alan R. Liss, Inc

Wimpee CF (1984) Organization and expression of light-regulated genes in Lemna gibba L. G-3. Ph.D. thesis, University of California, Los Angeles

Zurawski G, Perrot B, Bottomley W and Whitfeld PR (1981) The structure of the gene for the large subunit of ribulose-1,5-bisphosphate carboxylase from spinach chloroplast DNA. Nucleic Acids Res 9: 3251–3270

Zurawski G, Clegg MT and Brown AHD (1984) The nature of the nucleotide sequence divergence between barley and maize chloroplast DNA. Genetics 106: 735–749

Zurawski G, Bottomley W and Whitfeld PR (1986) Sequence of the gene for the large subunit of ribulose-1,5-bisphosphate carboxylase from pea chloroplasts. Nucleic Acids Res 14: 3975

Zurawski G and Clegg MT (1987) Evolution of higher plant chloroplast DNA-encoded genes: Implications for structure-function and phylogenetic studies. Ann Rev Plant Physiol 38: 391–418

Govindjee et al. (eds), Molecular Biology of Photosynthesis: 645–660
© 1988 Kluwer Academic Publishers

Minireview

Cloning, expression and directed mutagenesis of the genes for ribulose bisphosphate carboxylase/oxygenase

BRUCE A. MCFADDEN & CHRISTOPHER L. SMALL

Biochemistry/Biophysics Program, Washington State University, Pullman, WA 99164–4660, USA

Received 24 August 1987; accepted 21 December 1987

Key words: large subunit gene, small subunit gene, ribulose bisphosphate carboxylase/oxygenase, cloning, expression, in vitro directed mutagenesis

Abstract. The dominant natural form of ribulose-1,5-bisphosphate carboxylase/oxygenase (RuBisCO) is composed of large (L) 55-kDa and small (S) 15-kDa subunits. This enzyme (as the L_8S_8 form) is widely distributed among oxygenic photosynthetic species and among chemosynthetic bacteria. Another form lacking small subunits is found as an L_2 dimer in *Rhodospirillum rubrum* or an L oligomer of uncertain aggregation state from *Rhodopseudomonas sphaeroides*. The present article reviews two basically different approaches in cloning the *R. rubrum* gene for RuBisCO. One results in high level expression of this gene product fused with a limited aminoterminal stretch of β-galactosidase and the other results in expression of wild-type enzyme in *Escherichia coli*. Also reviewed are a number of reports of cloning and assembly of the L_8S_8 enzyme in using *E. coli* L and S subunit genes from *Anacystis nidulans*, *Anabaena* 7120, *Chromatium vinosum* and *Rps. sphaeroides*.

In vitro oligonucleotide-directed mutagenesis has been applied to the gene for RuBisCO from *R. rubrum*. In terms of contributing new information to our understanding of the catalytic mechanism for RuBisCO, the most significant replacement has been of *lys* 166 by a number of neutral amino acids or by *arg* or *his*. Results establish that *lys* 166 is a catalytically essential residue and illustrate the power of directed mutagenesis in understanding structure-function correlates for RuBisCO.

Oligonucleotide-directed mutagenesis has also been applied to the first and second conserved regions of the S subunit gene for RuBisCO from *A. nidulans*. In the latter region, corresponding amino acid changes of *trp* 55 and *trp* 58 to *phe*, singly or together, had little or no effect upon enzyme activity. In contrast, mutagenesis in the first conserved region leading to the following pairs of substitutions: *arg*10 *arg* 11 to *gly* 10 *gly*11; *thr*14 *phe* 15 *ser* 16 to *ala* 14 *phe* 15 *ala* 16; *ser* 16 *tyr* 17 to *ala* 16 *asp* 17; or *pro* 19 *pro* 20 to *ala* 19 *ala* 20, are all deleterious.

Advances are anticpated in the introduction and expression of interesting modifications of S (and L) subunit genes in plants. A new method of introducing and expressing foreign genes in isolated etiochloroplasts is identified.

Abbreviations: RuBisCO – ribulose bisphosphate carboxylase/oxygenase, 2-CABP – 2-carboxyarabinitol-1,5-bisphosphate, 4-CABP – 4-carboxyarabinitol-1,5-bisphosphate

Introduction

The enzyme ribulose bisphosphate carboxylase/oxygenase (RuBisCO) is normally composed of large (L) 55-kDa and small (S) 15-kDa subunits. The dominant quarternary structure of this enzyme is an $L_8 S_8$ form which can be isolated from many autotrophic bacteria, higher plants, and green and nongreen algae (McFadden 1980, McFadden et al. 1986). In green plants and algae, the precursor of small subunits is encoded by the nucleus and synthesized on cytoplasmic ribosomes with a long (ca. 5-kDa) N-terminal leader sequence that is removed after transport into chloroplasts (Musgrove and Ellis 1986). In chloroplasts, mature S subunits assemble with L subunits encoded by chloroplast DNA to form RuBisCO in a process which may require a L subunit binding protein (Cannon et al. 1987, Musgrove and Ellis 1986). In a nongreen chromophytic alga such as *Olisthodiscus luteus*, RuBisCO is synthesized in the chloroplasts (Reith and Cattolico 1985) and DNA from these organelles contains genes for both L and S subunits (Reith and Cattolico 1986).

In some prokaryotes such as cyanobacteria, the $L_8 S_8$ enzyme is encoded by adjacent L and S subunit genes which are co-transcribed (Shinozaki and Sugiura 1983, Nierzwicki-Bauer et al 1984). Presumably the S subunit gene is absent in *Rhodospirillum rubrum* because dimeric RuBisCO from this source lacks small subunits (McFadden et al. 1986). Also of interest among the prokaryota is the occurrence in *Rhodopseudomonas sphaeroides* of two forms of RuBisCO: I, $L_8 S_8$ and II, an L oligomer of uncertain aggregation state (Gibson and Tabita 1977).

Expression of RuBisCO genes has been achieved in myriad studies by various laboratories and a review, even of the current literature, is beyond the scope of the present article. Therefore, this contribution will focus upon recent studies of the cloning and expression of L and S subunit genes. Largely, these investigtions have sought to elucidate structural gene and flanking sequences and to obtain high-level expression of RuBisCO. Because the L and S subunit genes in green plants are not in tandem but in different compartments, they have not yet been successfully cloned and expressed together in a recipient organism in terms of the synthesis of active RuBisCO. The present article will emphasize expression of active RuBisCO via the cloning of L and S subunit genes from several procaryotic sources in *Escherichia coli*. Also considered will be very recent studies of in vitro site-directed mutagenesis of the L and S subunit genes. Large subunits harbor catalytic activity (or potential; see McFadden et al. 1986) and have been the focus of initial studies of directed mutagenesis.

Experimental approaches

Cloning studies
In general, L and S subunit genes have been cloned downstream from a *lac* promoter in a recombinant plasmid of high copy number. Expression has been obtained in a suitable *E. coli* host after transformation by the plasmid. More detail will be given in the *Results* section.

Coupled in vitro transcription-translation
The development of a coupled transcription-translation system using extracts of *Rps. sphaeroides* has been described (Chory and Kaplan 1982). Briefly the in vitro protein synthesizing system is active from either chemoheterotrophically- or photoheterotrophically-grown cells in the exponential phase. Characteristics have been compared with an analogous system from *E. coli* (Chory and Kaplan 1982). The use of coupled transcription-translation to obtain differential expression of the form II RuBisCO gene from *Rps. sphaeroides* in a homologous system will be described under *Results*.

Directed in vitro mutagenesis of the L subunit gene
Directed mutagenesis has been accomplished with the L subunit gene from *R. rubrum*.

In one approach, the L subunit gene was cut within a recombinant plasmid, PRR 2119, at a *Stu*1 restriction site 30-bases upstream from the intended mutation site (which was to confer an *asp* 188 → *glu* 188 change). The resultant blunt ended DNA was then digested with a 3′ → 5′ exonuclease (T4 DNA polymerase) to expose the target site on the 5′ → 3′ strand. A 20-base synthetic deoxyoligonucleotide, complementary to the region except for a single mismatch, was then annealed to the linear template (Gutteridge et al. 1984). Finally the 3′ → 5′ strand was repaired and the circular plasmid was reformed by ligation. The mutated plasmid was then expressed in *E. coli*. In other approaches (Estell et al. 1985, Terzaghi et al. 1986), the entire RuBisCO gene from *R. rubrum* or a region of it (to be subjected to mutagenesis) was transferred from plasmid PRR 2119 (Somerville and Somerville 1984) into the replicative form of a bacteriophage M13-derived vector. These vetors are of known base sequence, have multiple cloning sites within the β-galactosidase gene, and produce progeny that contain (+) strands of single stranded (ss) DNA (Messing 1983). Plaques with an insert at a cloning site are readily recognized by an assay that detects inactivation of the β-galactosidase gene. Messing (1983) has discussed the utility of the M13 phages in sequence studies of inserts using the Sanger dideoxy method. The M13 phages are especially useful because ssDNA, which contains either the sense of antisense strand of an inserted gene (depending upon the insert orientation), can be easily obtained from a pure recombinant plaque. Thus one strand can be used for sequence studies or can be annealed with a suitable complementary synthetic deoxyoligonucleotide (termed a mutagenic primer) in studies of mutagenesis. Zoller and Smith (1983, 1984) have discussed the details of directed mutagenesis using M13 bacteriophage. Basically, after phosphorylation at the 5' end, the synthetic oligonucleotide is annealed to a ssDNA preparation, in addition to a universal sequencing primer, and complementary intervening sequences are synthesized using DNA polymerase I (large fragment). After ligation, the heteroduplex (or nearly complete heteroduplex) is used to transfect *E. coli*. Mutated recombinant phage particles are identified by plaque lift hybridization using ^{32}P-labeled mutagenic primer as a probe. The basis of this method of detection of mutant DNA is that the mutagenic oligonucleotide will form a more stable duplex with a mutant clone having a perfect match than with a wild-type clone bearing a mismatch. Hybridization is conducted at increasing stringency and reveals putative mutant DNA (Zoller and Smith 1984). After plaque purification of suspected mutants, ssDNA is isolated and sequenced as necessary to confirm the base substitution(s). The mutated gene is then transferred from the RF form of the recombinant M13 bacteriophage into an appropriate cloning site in a related pUC plasmid (Zoller and Smith 1984). The mutated recombinant pUC derivative is then expressed. Using this general approach, RuBisCO from *R. rubrum* has been altered by a single amino acid substitution in several positions. Results will be considered in the following section.

Directed in vitro mutagenesis of the S subunit gene
Directed mutagenesis of the S subunit gene from *Anacystis nidulans* has been obtained using the following approach.

Prior to mutagenesis of this subunit gene, a *Pst*1 fragment of chromosomal DNA from *Anacystis nidulans* carrying the L and S subunit genes for RuBisCO (Shinozaki et al. 1983) was inserted into pUC7 to yield pUC7LS100 (Voordouw et al. 1987). The latter plasmid was then used to construct pUC19LS101 carrying an insert containing the L and S subunit genes with an internal *Eco*R1 site corresponding to the carboxyterminal region of the L subunit and *Sac*1 and *Xba*1 sites flanking the insert. In parallel, an M13 derivative (M13K18S100) was constructed which contained the S subunit gene for RuBisCO from *A. nidulans* flanked by *Eco*R1 and *Sac*1 sites. Single stranded DNA from this bacteriophage was used as a template for synthetic oligonucleotides containing a mismatched pair of bases in 17- and 18-mers (Voordouw et al. 1987). Intended substitutions were (after expression) of *trp* 55 or *trp* 58 by *phe* in the S subunit. Mutant phages were prepared and identified as described earlier in this section. Expression of RuBisCO with the altered S subunit gene was obtained after transfer of the *Eco*R1, *Sac*1 fragment from M13K18S100 to *Eco*R1, *Sac*1-cut pUC19LS101 (Voordouw et al. 1987).

Results

Cloning studies

RuBisCO was first expressed in *E. coli* in 1984 by placing the L subunit gene from *R. rubrum* under *lac* control in a recombinant plasmid pRR2119 and transforming *E. coli* JM103. RuBisCO accounted for ca. 12% of the protein in crude heat-treated extracts of IPTG-induced cultures of the recipient (Somerville and Somerville 1984). Recently Larimer et al. (1986) determined that the resultant enzyme contains a 24-residue N-terminal sequence of β-galactosidase fused to the aminoterminus of RuBisCO from *R. rubrum* confirming previous speculation (Nargang et al. 1984). These investigators constructed a new clone of the *R. rubrum* RuBisCO gene in the plasmid pFL34 and obtained expression of the wild-type enzyme in *E. coli*. This construction entailed removing all *lac* Z-coding sequences and a portion of the 5'-noncoding leader of the RuBisCO gene (Larimer et al. 1986). In the process, expression of RuBisCO was considerably reduced in comparison with that from pRR2119 but purification of the enzyme by immunoaffinity chromatography promises to be relatively rapid (Niyogi et al. 1986). Thus laboratories now may choose to work either on the fusion protein or wild-type RuBisCO from *R. rubrum* both of which seem to be closely similar in catalytic and physical properties.

High-level expression of the interesting *Rps. sphaeroides* form II enzyme, which also lacks small subunits, was first obtained in *E. coli* in 1984. In that case, a 3-kb *Eco*R1 fragment was inserted into pUC8 and was shown to be under *lac* control in the hybrid plasmid pRQ52 (Quivey and Tabita 1984). Of significance is the recent evidence that expression of pRQ52 in *E. coli* results in enzyme that is identical at the aminoterminus to form II RuBisCO from *Rps. sphaeroides* (Tabita et al 1986). Thus construction of the hybrid plasmid fortuitously avoided subsequent expression of the gene product as a fusion protein with β-galactosidase.

In parallel research, considerable attention was focused upon cloning and expressing genes for the L_8S_8 form of the enzyme in *E. coli*. Success was first achieved in 1985 for L and S subunit genes from the cyanobacteria *Anacystis nidulans* (Christeller et al. 1985, Gatenby et al. 1985, Tabita and Small 1985) and *Anabaena* 7120 (Gurevitz et al. 1985) and from *Chromatium vinosum* (Viale et al. 1985). In the former studies, a *Pst*1 fragment, originally isolated from *A. nidulans* DNA, was inserted downstream from: a) the P_L promoter of λ bacteriophage conferring thermal induction (plasmid pSV55, see Gatenby et al. 1983); or b) the *lac* promoter in pUC9 (plasmid pCS75, see Tabita and Small 1985); or c) the *lac* promoter in the phage M13mp10 (Christeller et al. 1985). In the latter two cases, induction with IPTG was established. The 2.2-kb *Pst*1 fragment contained 93 bp between L and S subunit genes (Shinozaki and Sugiura 1983). Gurevitz et al. (1985) inserted the RuBisCO genes from *Anabaena* 7120 into pUC19 to create a hybrid plasmid pANX105. Expression of L subunits in *E. coli* vastly exceeded that of S subunits because most transcripts terminated in the 545-bp region between L and S subunit genes (Nierzwicki-Bauer et al. 1984). Nevertheless ca. 0.6% of the soluble protein in *E. coli* was fully functional L_8S_8 RuBisCO containing stoichiometric amounts of both subunits (Gurevitz et al. 1985). A much higher yield of the L_8S_8 enzyme from *Chromatium vinosum* has been achieved after inserting a 3-kbp DNA fragment in a *Sma*1-digested plasmid pKK223-3 to yield a hyrid plasmid pCKS1. This latter plasmid could be expressed in *E. coli* to give yields of RuBisCO as high as 15% of the soluble protein (Viale et al. 1985). Recently, a 4-kb *Sma*1 fragment from *Rps. sphaeroides* DNA has been inserted into pUC8. IPTG-dependent expression of the resultant hybrid plasmid (pJG29), in terms of form I RuBisCO was obtained in *E. coli* (Gibson and Tabita 1986).

In several of these studies of the L_8S_8 enzyme it was established that RuBisCO could not be detected after expression of the L subunit gene alone. Also, active RuBisCO could not be formed after cloning L and S subunit genes independently followed by mixing of the *E. coli* extracts (Christeller et al. 1985). It has been established, however, that expression of the *A. nidulans* L-S subunit gene tandem under control of the P_L promoter results in RuBisCO with substoichiometric amounts of S subunit. Moreover, enzyme activity could be increased by enhancing the expression of the S subunit gene (Gatenby et al. 1985). In some studies, expression occurs of both wild-type cyanobacterial RuBisCO and a fusion protein with a segment of β-galactosidase or another protein encoded by the plasmid. This critically depends upon the plasmid (or phage) construction. For example, expression in *E. coli* of the plasmid pSynRES1 (or its equivalent, pCS75; see Tabita and Small 1985) arising from insertion of the *A. nidulans Pst*1 fragment into pUC9 results in both wild-type RuBisCO and a fusion protein in a ratio of ca. 2:1. The latter contains 10 amino acid residues from the N-terminus of β-galactosidase and 24 from the 5′-flanking region of the L subunit gene fused to the N-terminus of the normal L subunit (Gutteridge et al. 1986).

Another example is a fusion protein between 181 N-terminal amino acids of β-lactamase and the L subunit which has been detected in addition to normal L (and S) subunits after thermal induction of the plasmid pSV55 (Gatenby et al. 1985). Obviously in some plasmid constructions, a segment of the plasmid β-galactosidase (or β-lactamase) gene is in frame with the L subunit gene and translation results in fusion. Such constructions should be avoided unless thorough characterization of the fusion proteins is anticipated.

Very recently, the L–S gene tandem has been excised from chromosomal DNA of *Alcaligenes eutrophus* by *Sau*3A digestion and fragments inserted at the *Bam*H1 site of plasmid pRK310. One plasmid, pAE312 (2.3-kb insert), yielded the highest RuBisCO activity when expressed in *Pseudomonas aeruginosa* 2036. However, after sodium dodecylsulfate-polyacrylamide gel electrophoresis, western blotting revealed one species of L subunit like that of the enzyme from *A. eutrophus* and one larger L-subunit species in an almost equivalent amount. Although S subunit was detectable (Andersen and Wilke-Douglas 1987), it may have been substoichiometric. The possibility that the larger L-subunit form was a fusion protein was not further investigated.

Coupled in vitro transcription–translation
In 1985 Kaplan and co-workers described the in vitro synthesis and assembly of form II RuBisCO from *Rps. sphaeroides* in extracts of the same organism. Excellent expression was obtained using a plasmid, pL110, with a RuBisCO gene insert in a pBR322 derivative carrying a *trp* promoter or using a plasmid, pL14, which has the RuBisCO gene in the opposite orientation (Chory et al. 1985). Of considerable interest was the finding that the L_4 form of the enzyme had the highest specific activity. Also of interest was the finding that a 9-11 fold enhancement of enzyme activity occurred upon the addition of plasmid pL14 to the coupled *transcription–translation* system derived from cells which had been grown chemoheterotrophically either aerobically or anaerobically; in contrast the analogous system derived from cells which had been grown photoheterotrophically was stimulated only 50% in the presence of pL14.

Although these latter results were complicated by the presence of both form I and form II enzyme in extracts after photoheterotrophic growth (Chory et al. 1985), the data collectively demonstrate the potential of using coupled transcription and translation in studying the in vitro expression of RuBisCO genes. Especially attractive is the use of a homologous system such that endogenous factors which may be required for synthesis and/or assembly can be identified and studied. An analogous system developed for the cyanobacteria, especially *A. nidulans*, could be advantageous in studies of RuBisCO synthesis and assembly.

Directed in vitro mutagenesis of the L subunit gene

The first ammino acid replacement by directed mutagenesis was of *asp* 188 (incorrectly specified as *asp* 198) → *glu* 188 in the fusion protein between β-galactosidase and RuBisCO from *R. rubrum* (Gutteridge et al. 1984). The substitution was designed to test the participation of the carboxylate of *asp* 188 in chelation of Mn^{2+}. The mutant enzyme had slightly altered kinetic properties. The electron spin resonance spectra of the complex between active enzyme (in the Mn^{2+} form) and either 2-carboxyarabinitol-1,5-bisphosphate (2-CABP), a 'transition-state analog', or 4-CABP were subtly different for the mutant vs. wild-type enzyme. The observed slightly altered environment of Mn^{2+} was indeed consistent with coordination between the Mn^{2+} and *asp* 188 but did not prove this interaction.

A much more striking result was obtained when *lys* 191 was replaced with *glu* in the same enzyme (Estelle et al. 1985). The resultant enzyme bound 2-CABP but was completely inactive. In contrast to that for the 'wild-type' enzyme, 2-CABP binding was not enhanced in the presence of Mg^{2+} and CO_2. These observations were consistent with previous research which had suggested that *lys* 191 was essential to the first step in activation in that this residue bound activator CO_2 to form the carbamate.

In another study of the same enzyme, *met* 330 was changed to *leu*. The resultant enzyme bound 2-CABP reversibly whereas the 'wild-type' enzyme bound this compound irreversibly (Terzaghi et al. 1986). Unfortunately, the substitution was not very diagnostic in terms of enzyme function.

Recognizing that directed mutagenesis is best conducted on the wild-type RuBisCO gene, Hartman and colleagues have recently constructed a plasmid, pFL34, in which there is no segment of the β-galactosidase gene immediately upstream from the RuBisCO gene of *R. rubrum* (Larimer et al. 1986). Directed mutagenesis of this wild-type gene yielding the replacement of *his* 291 by *ala*, resulted in an enzyme which had a somewhat lower k_{cat} and a higher K_m^{RuBP}. The data clearly established that *his* 291 is not a catalytically essential amino acid (Niyogi et al. 1986) in spite of the implied placement of this residue in the active-site domain (Igarashi and McFadden 1985). In parallel research on this gene for RuBisCO of *R. rubrum*, *lys* 166 has been very recently replaced with *gly*, *ala*, *ser*, *gln*, *arg*, *cys* or *his*. All seven of the mutant proteins were severely deficient in carboxylase activity (Hartman et al. 1987a). Although low, the carboxylase activity displayed by some of the mutant proteins established that *lys* 166 is not required for substrate binding. Some of the mutant proteins bound 2-CABP in the presence of CO_2 and

Mg^{2+} suggesting that they underwent the carbamylation reaction required for activation of wild-type protein. The authors concluded that since *lys* 166 was required neither for activation nor for substrate binding, it must be essential to catalysis.

Using the plasmid pFLL34, known to express wild-type *R. rubrum* RuBisCO, Hartman et al. (1987b) have also replaced *glu* 48 with *gln* via site-directed mutagenesis. This substitution did not prevent subunit association, carbamylation by CO_2 or 2-CABP binding; however, bound 2-CABP exchanged with the free ligand under conditions in which no exchange occurred using the wild-type quaternary complex. The altered protein was, however, virtually inactive establishing that *glu* 48 is essential for catalysis. Most recently, this research group has substituted an *asp* for active-site *lys* 166 to yield completely inactive enzymic monomers (Lee et al. 1987). Parallel chemical cross-linking experiments have established an intersubunit contact between *cys* 58 and *lys* 166 (Lee et al. 1987). Together, the two mutagenesis results (Hartman et al. 1987b, Lee et al. 1987) and chemical cross-linking (Lee et al. 1987) suggest that the catalytic site of RuBisCO from *R. rubrum* is positioned at an interface between subunits. A functional catalytic site requires, directly or indirectly, regions of both subunits at this interface.

Directed in vitro mutagenesis of the S subunit gene

There are three conserved regions of the S subunit including residues 10–21 (Shinozaki and Sugiura 1983, Torres-Ruiz and McFadden 1987), 54–63, and 88–104 (Shinozaki and Sugiura 1983) in RuBisCO from *A. nidulans* when these regions are aligned with counterparts from higher plants or *C. vinosum*. Although the same three regions are somewhat conserved in the S subunit sequene deduced for the *Alcaligenes eutrophus* enzyme, there is more divergence (Andersen and Caton 1987). Of speial note in the first conserved region is the complete conservation in all aligned S subunit sequences of residues corresponding to *thr* 14, *ser* 16, *leu* 18 and *pro* 19 in *A. nidulans* S subunits.

Stretch 12–21 in the S subunit from *A. nidulans* is hydrophobic according to Shinozaki and Sugiura (1983) although it is preceded by a pair of basic residues. The middle conserved region is α-helical (Shinozaki and Sugiura 1983). Recently, Voordouw et al. (1987) have replaced *trp* 55 and *trp* 58 (incorrectly designated 54 and 57) by *phe*. The second mutant RuBisCO species (expressed in *E. coli*) was an L_8S_8 enzyme with the same molecular masses for L and S subunits as in wild-type enzyme. Both mutant proteins had the same $K_m^{CO_2}$ values as that for wild-type enzyme. However, the V_{max} values for both mutants were reduced by ca. 60% (Voordouw et al. 1987).

Directed in vitro mutagenesis of the S subunit gene (in progress)
We have recently undertaken directed mutagenesis in the first conserved region of the S subunit gene in RuBisCO from *A. nidulans*. Specifically, the 930-base pair *Eco*R1 fragment that contains the S subunit gene in pCS75 has been subcloned into M13mp18 and four unique double amino acid substitutions have been engineered by oligonucleotide-directed mutagenesis (Zoller and Smith 1983, 1984). Putative mutants, identified by stringent hybridization selection, have been purified and sequenced by the dideoxynucleotide method of Sanger (described in Messing 1983). Sequence information has confirmed that the desired substitutions were introduced and that bordering sequences were not altered.

The four altered S subunit genes have been liberated from the mp 18 vector by *EcoR1* digestion and individually ligated downstream from the L subunit gene to restore the original construction of pCS75. Final reconstructions were subjected to restriction analysis. Digestion with *EcoR1*, *Pst*1, *Bam*H1, *Sph*1, and combinations of these endonucleases verified that the mutant plasmids were identical to pCS75 on the basis of identical banding patterns on agarose gels. The 2.1-kbp fragment containing the L and S subunit genes was then purified from pCS75 and the four mutant plasmids. These fragments were individually ligated into pUC19 to yield one plasmid that contains unaltered L and S subunit genes, pCS88, and four mutant

Table 1. The consequences of substitution of amino acid pairs in the S subunit of RuBisCO from *A. nidulans*.

Plasmid	Mutation in the S subunit gene		Specific activity in crude extracts[a]
pCS88	None		0.0690 (100)
pCS75ΔSS	No S subunit gene		0.0000 (0)
pCS88A	19 20 pro pro	→	19 20 ala ala	0.0490 (71)
pCS88C	16 17 ser tyr	→	16 17 ala asp	0.0048 (7)
pCS88D	14 15 16 thr phe ser	→	14 15 16 ala phe ala	0.0042 (6)
pCS88E	10 11 arg arg	→	10 11 gly gly	0.0220 (32)

[a] Specific activity is defined as: μMoles RuBP-dependent CO_2 fixation/min/mg protein measured in *E. coli* extracts at 30 °C (Tabita and Small 1985). Crude extracts represent the soluble fraction from a bacterial culture which was harvested by centrifugation, sonically disrupted and centrifuged at 100,000 g for 1 hour.

plasmids, pCS88A, pCS88C, pCS88D and pCS88E. When expressed, these pUC19 derivatives do not produce a fusion protein with β-galactosidase.

Crude extracts of *E. coli* cultures which had been transformed with pCS88, pCS75ΔSS (pCS75 less the 930-base pair *Eco*R1 fragment containing the S subunit gene), or the four S subunit mutant plasmids were examined and results are summarized in Table 1. As evident, all paired substitutions in the first conserved region of the S subunit gene from *A. nidulans* are deleterious leading to 39–94% reduction in V_{max}. Particularly deleterious are the substitutions:

$$\frac{14 \quad 16}{ala \;\; ala} \text{ for } \frac{14 \quad 16}{thr \;\; ser} \text{ and } \frac{16 \quad 17}{ala \;\; asp} \text{ for } \frac{16 \quad 17}{ser \;\; tyr}.$$

These striking results suggest that this first conserved region is critically required in RuBisCO from *A. nidulans* for one or more of the following: assembly, activation or catalysis. Characterization of the resultant altered enzymes and an examination of the consequences of derivative single substitutions in this region of the S subunit are in progress.

Discussion

In the past three years, efforts to clone and express the RuBisCO genes have been very encouraging. Large amounts of RuBisCO have been obtained and L and S subunit genes have been inserted in a variety of plasmids suitable for oligonucleotide-directed mutagenesis of the inserts. In cloning the RuBisCO gene for *R. rubrum*, two different approaches have been used. One has the advantage of high-level expression in *E. coli* but results in a fusion protein between a 24-residue N-terminal sequence form β-galactosidase and the aminoterminus of RuBisCO (Somerville and Somerville 1984, Larimer et al. 1986). The other approach results in much lower expression of the L_2 enzyme in *E. coli* but the gene product is wild-type RuBisCO (Larimer et al. 1986). Amino-terminal sequencing has established that one other cloned gene, that for form II RuBisCO from *Rps. sphaeroides*, an enzyme which also lacks small subunits, results in expression of wild-type enzyme (Tabita et al. 1986). In cloning and expression of the cyanobacterial L and S subunit genes in terms of the L_8S_8 enzyme, the yield after expression in *E. coli* is higher when the length of the spacer sequence between these genes is reduced (Nierzwicki-Bauer et al. 1984, Tabita and Small 1985). Expression, however, may be considerably elevated in the native cyanobacterial environment stressing the need for development of homologous cloning systems. Deletion of the resident L and/or S subunit genes would be very helpful in such systems.

Directed mutagenesis studies have been focused upon the gene specifying the L$_2$ enzyme from *R. rubrum*. Most definitive in our understanding of the catalytic mechanism has been the replacement of *his* 291 and *lys* 166. Although both of these residues are probably in the active-site domain, the results suggest that *his* 291 is non-essential whereas *lys* 166 is essential to catalytic activity (Hartman et al. 1987a, Niyogi et al. 1986). In fact, Hartman and co-workers have suggested that *lys* 166 may function to deprotonate ribulose bisphosphate in the catalytic mechanism (Hartman et al. 1987a).

Only one report of directed mutagenesis of the S subunit gene for RuBisCO has previously appeared (Voordouw et al. 1987). Two mutations in the middle conserved region of this gene are not particularly deleterious. The gene for the S subunit from this source is an excellent target for mutagenesis because of the ease of expression of the fully active L$_8$S$_8$ enzyme in *E. coli* (Tabita and Small 1985). Our data obtained with four pairs of amino acid substitutions in the first conserved region of the S subunit suggest that this region is vital to the final expression of enzyme activity (Table 1). However, studies of derivative single amino acid substitutions remain to be completed.

It seems likely that a large variety of known point mutations in L and S subunit genes will soon become available. Especially interesting modifications of each RuBisCO gene will lead to isolable gene products that contribute to enhanced carboxylase activity or an increase in the ratio of carboxylase:oxygenase. Either type of change may increase plant productivity if the genes can be reinserted into the nuclear or chloroplast genome. To date, genetic engineering of the chloroplast has proven to be a formidable problem. However, our recent introduction and expression of bacterial genes for β-lactamase, RuBisCO S subunit and chloramphenicol acetyltransferase in isolated cucumber etioplasts may open the way to new approaches in plant genetic engineering (Daniell and McFadden 1987). If stable gene expression can be achieved after reintroduction of transformed etioplasts into protoplasts from a homologous plant, the potential will be opened to regenerate altered plants. The effect of RuBisCO upon photosynthesis could be examined in such plants. Thus the introduction and expression of interesting modifications of the gene for the S subunit of RuBisCO may become a reality. Prospects to introduce and express altered genes for the chloroplast-encoded L subunit of RuBisCO are more daunting. This is because the number of genomes per individual chloroplast varies between 180 and 10 in mature and young spinach leaves, respectively (Scott and Possingham 1983). Each genome probably contains one L subunit gene

(Coen et al. 1977). Thus interesting modifications of L subunit genes would be highly diluted in a given transformed chloroplast even if multiple copies of altered L subunit genes could be incorporated. The magnitude of the problem can be further appreciated on the basis of data that suggests that there are 1500–9500 chloroplasts genomes/cell (Lamppa and Bendich 1979, Scott and Possingham 1980, 1983∫ and one L subunit gene (for RuBisCO) per genome (Coen et al. 1977) in comparison to 6–12 copies of the nuclear-encoded S subunit gene per cell (Corruze et al. 1984). Nevertheless it may be possible to introduce functional modifications of the L subunit gene via chloroplast engineering if the expression of resident genes can be reduced while enhancing that of the introduced foreign L subunit gene. We anticipate that this new approach to the remodeling of plant plastid genomes will be a fertile area of research in coming years.

Acknowledgements

Research in our laboratory has been supported in part by NIH grant GM-19,972. We thank S.N. Mogel and R. Haining for excellent technical assistance in our studies of directed mutagenesis of the S subunit gene from *A. nidulans*.

References

Andersen K and Wilke-Douglas M (1987) Genetic and physical mapping and expression in *Pseudomonas aeruginosa* of the chromosomally encoded ribulose bisphosphate carboxyase genes of *Alcaligenes eutrophus*. J Bacteriol 169: 1997–2004

Andersen K and Caton J (1987) Sequence analysis of the *Alcaligenes eutrophus* chromosomally encoded ribulose bisphosphate carboxylase large and small subunit genes and their gene products. J Bacteriol 169: 4547–4558

Cannon S, Wang P and Roy H (1987) Inhibition of ribulose bisphosphate carboxylase assembly by antibody to a binding protein. In: Biggins J (ed.) Progress in Photosynthesis Research, Vol III, pp 5.423–5.426. Boston: Martinus Nijhoff

Chory J and Kaplan S (1982) The in vitro transcription–translation of DNA and RNA templates by extracts of *Rhodopseudomonas sphaeroides*. J Biol Chem 257: 15110–15121

Chory J, Muller ED and Kaplan S (1985) DNA-directed in vitro synthesis and assembly of the form II D-ribulose-1,5-bisphosphate carboxylase/oxygenase from *Rhodopseudomonas sphaeroides*. J Bacteriol 161:307–313

Christeller JT, Terzaghi BE, Hill DF and Laing WA (1985) Activity expressed from cloned *Anacystis nidulans* large and small subunit ribulose bisphosphate carboxylase genes. Plant Mol Biol 5: 257–263

Coen DM, Belbrook JR, Bogorad L and Rich A (1977) Maize chloroplast DNA fragment encoding the large subunit of ribulosebisphosphate carboxylase. Proc Natl Acad Sci USA 74: 5487–5491

658

Coruzzi G, Broglie R, Edwards C and Chua N-H (1984) tissue-specific and light-regulated expression of a pea nuclear gene encoding the small subunit of ribulose-1,5-bisphosphate carboxylase. EMBO J 3: 1671–1679

Daniell H and McFadden BA (1987) Uptake and expression of bacterial and cyanobacterial genes by isolated cucumber etioplasts. Proc Natl Acad Sci USA 84: 6349–6353

Estelle M, Hanks J, McIntosh L and Somerville C (1985) Site-specific mutagenesis of ribulose-1,5-bisphosphate carboxylase/oxygenase. Evidence that carbamate formation at *lys* 191 is required for catalytic activity. J Biol Chem 260: 9523–9526

Gatenby AA, Van der Vies SM and Bradley D (1985) Assembly in *E. coli* of a functional multi-subunit ribulose bisphosphate carboxylase from a blue-green alga. Nature 314: 617–620

Gibson JL and Tabita FR (1977) Different molecular forms of D-ribulose-1,5-bisphosphate carboxylase from *Rhodopseudomonas sphaeroides*. J Biol Chem 252: 943–949

Gibson JL and Tabita FR (1986) Isolation of the *Rhodopseudomonas sphaeroides* form I ribulose 1,5-bisphosphate carboxylase large and small ambient genes and expression of the active hexadecameric enzyme in *Escherichia coli*. Gene 44: 271–278

Gurevitz M, Somerville CR and McIntosh L (1985) Pathway of assembly of ribulosebisphosphate carboxylase/oxygenase from *Anabaena* 7120 expressed in *Escherichia coli*. Proc Natl Acad Sci USA 82: 6546–6550

Gutteridge S, Sigal I, Thomas B, Arentzen R, Cordova A and Lorimer G (1984) A site-specific mutatin within the active site of ribulose-1,5-bisphosphate carboxylase of *Rhodospirillum rubrum*. EMBO J 3: 2737–2743

Gutteridge, S, Phillips AL, Kettleborough CA, Parry MA and Keys AJ (1986) Expression of bacterial Rubisco genes in *Escherichia coli*. Phil Trans R Soc Lond B 313: 433–445

Hartman FC, Soper TS, Niyogi SK, Mural RJ, Foote RS, Mitra S, Lee EH, Machanoff R and Larimer FW (1987a) Function of Lys-166 of *Rhodospirillum rubrum* ribulosebisphosphate carboxylase/oxygenase as examined by site-directed mutagenesis. J Biol Chem 262: 3496–3501

Hartman FC, Larimer FW, Mural RJ, Machanoff R and Soper TS (1987b) Essentiality of glu-48 of ribulose bisphosphate carboxylase/oxygenase as demonstrated by site-directed mutagenesis. Biochem Biophys Res Commun 145: 1158–1163

Igarishi I, McFadden BA and El-Gul T (1985) Active-site histidine in spinach ribulose bisphosphate carboxylase/oxygenase modified by diethyl-pyrocarbonate. Biochemistry 24: 3957–3962

Lamppa GK and Bendich AJ (1979) Changes in chloroplast DNA levels during development of pea (*Pisum Sativum*). Plant Physiol 64: 126–130

Larimer FW, Machanoff R and Hartman FC (1986) A reconstruction of the gene for ribulose bisphosphate carboxylase from *Rhosospirillum rubrum*. Gene 41: 113–120

Lee EH, Soper TS, Mural RJ, Stringer CD and Hartman FC (1987) An intersubunit interaction at the active site of ribulose bisphosphate carboxylase/oxygenase as revealed by cross-linking and site-directed mutagenesis. Biochemistry 26: 4599–4604

McFadden BA (1980) A perspective of ribulose bisphosphate carboxylase/oxygenase, the key catalyst in photosynthesis and photorespiration. Act Chem Res 13: 394–399

McFadden BA, Torres-Ruiz J, Daniell H and Sarojini G (1986) Interaction, functional relations and evolution of large and small subunit in RuBisCO from prokaryota and eukaryota. Phil Trans R Soc Lond B 313: 347–358

Messing J (1983) New M13 vectors for cloning. In: Wu R, Grossman L and Moldave K (eds) Methods in enzymology, Vol 101, pp 20–78. New York: Academic Press

Musgrove JE and Ellis RJ (1986) The Rubisco large subunit binding protein. Phil Trans R Soc Lond B 313: 419–428

Nargang F, McIntosh L and Somerville C (1984) Nucleotide sequence of the ribulose bis-phosphate carboxylase gene from *Rhodospirillum rubrum*. Mol Gen Genet 193: 220–224

Nierzwicki-Bauer SA, Curtis SE and Haselkorn R (1984) Co-transcription of genes encoding the small and large subunits of ribulose-1,5-bisphosphate carboxylase in the cyanobacterium *Anabaena* 7120. Proc Natl Acad Sci USA 81: 5961–5965

Niyogi SK, Foote RS, Mural RJ, Larimer FW, Mitra S, Soper TS, Machanoff R and Hartman FC (1986) Nonessentiality of histidine 291 of *Rhodospirillum rubrum* ribulose-bisphosphate carboxylase/oxygenase as determined by site-directed mutagenesis. J Biol Chem 261: 10087–10092

Quivey RG and Tabita FR (1984) Cloning and expression in *Escherichia coli* of the form II ribulose 1,5-bisphosphate carboxylase/oxygenase gene from *Rhodopseudomonas sphaeroides*. Gene 31: 91–101

Reith M and Cattolico RA (1985) In vivo chloroplast protein synthesis by the chromophytic alga *Olisthodiscus luteus*. Biochemistry 24: 2556–2561

Reith M and Cattolico RA (1986) Inverted repeat of *Olisthodiscus luteus* chloroplast DNA contains genes for both subunits of ribulose-1,5-bisphosphate carboxylase and the 32,000-dalton Q_B protein: phylogenetic implications. Proc Natl Acad Sci USA 83: 8599–8603

Scott NS and Possingham JV (1980) Chloroplast DNA in expanding spinach leaves. J Exp Botany 31: 1081–1092

Scott NS and Possingham JV (1983) Changes in chloroplast DNA levels during growth of spinach leaves. J Exp Botany 34: 1756–1767

Shinozaki K, Yamada C, Takahata N and Sugiura M (1983) Molecular cloning and sequence analysis of the cyanobacterial gene for the large subunit of ribulose-1,5-bisphosphate carboxylase/oxygenase. Proc Natl Acad Sci USA 80: 4050–4054

Shinozaki and Sugiura M (1983) The gene for the small subunit of ribulose-1,5-bisphosphate carboxylase/oxygenase is locted close to the gene for the large subunit in the cyanobacterium *Anacystis nidulans* 6301. Nucleic Acids Res 11: 6957–6964

Shinozaki K and Sugiura M (1985) Genes for the large and small subunits of ribulose-1,5-bisphosphate carboxylase/oxygenase constitute a single operon in a cyanobacterium *Anacystis nidulans* 6301. Mol Gen Genet 200: 27–32

Somerville CR and Somerville SC (1984) Cloning and expression of the *Rhodospirillum rubrum* ribulose bisphosphate carboxylase gene in *E. coli*. Mol Gen Genet 193: 214–219

Tabita FR and Small CL (1985) Expression and assemblyof active cyanobacterial ribulose-1,5-bisphosphate carboxylase/oxygenase in *Escherichia coli* containing stoichiometric amounts of large and small subunits. Proc Natl Acad Sci USA 82: 6100–6103

Tabita FR, Gibson JL, Mandy WJ and Quivey RG Jr (1986) Synthesis and assembly of a novel recombinant ribulose bisphosphate arboxylase/oxygenase. Bio/Technology 4: 138–141

Terzaghi BE, Laign WA, Christeller JT, Petersen GB and Hill DF (1986) Ribulose 1,5-bisphosphate carboxylase. Effect on the catalytic properties of changing methionine-330 to leucine in *Rhodospirillum rubrum* enzyme. Biochem J 235: 839–846

Torres-Ruiz JA and McFadden BA (1987) The nature of L_8 and L_8S_8 forms of ribulose bisphosphate carboxylase/oxygenase from *chromatium vinosum*. Arch Biochem biophys 254: 63–68

Viale AM, Kobayashi H, Takabe T, Aakazawa T (1985) Expression of genes for subunits of plant-type RuBisCO from *Chromatium* and production of the enzymically active molecule in *Escherichia coli*. FEBS Lett 192: 283–288

Voordouw G, De Vries PA, Van Den Berg WAM and De Clerk EPJ (1987) Site-directed mutagenesis of the small subunit of ribulose-1,5-bisphosphate carboxylase/oxygenase from *Anacystis nidulans*. Eur J Biochem 163: 591–598

Zoller MJ and Smith M (1983) Oligonucleotide-directed mutagenesis of DNA fragments cloned into M13 vectors. In: Wu R, Grossman L and Moldave K (eds) Methods in Enzymology, Vol 100, pp 468–500. New York: Academic Press

Zoller MJ and Smith M (1984) Laboratory methods. Oligonucleotide-directed mutagenesis: a simplemethod using two oligonucleotide primers and a single stranded DNA template. DNA 3: 479–488

Govindjee et al. (eds), Molecular Biology of Photosynthesis: 661–675
© 1988 Kluwer Academic Publishers

Minireview

The Rubisco subunit binding protein

R. JOHN ELLIS & SASKIA M. VAN DER VIES
Department of Biological Sciences, University of Warwick, Coventry CV4 7AL, UK

Received 21 September 1987; accepted 30 September 1987

Key words: binding protein, chloroplast protein synthesis, large subunit, molecular chaperone, ribulose bisphosphate carboxylase-oxygenase, small subunit

Abstract. Chloroplasts contain an abundant soluble protein that binds non-covalently newly synthesized large and small subunits of the enzyme ribulose bisphosphate carboxylase-oxygenase. This binding protein has been purified from *Pisum sativum* and *Hordeum vulgare* in the form of a dodecamer consisting of equal amounts of two types of subunit. These subunits are synthesized as higher molecular mass precursors by cytoplasmic ribosomes before import into the chloroplast. Antibodies raised against the purified binding protein from *Pisum sativum* detect polypeptides not only in extracts of plastids from several plant species but also in cell extracts of several bacterial species. The oligomeric binding protein dissociates reversibly into monomeric subunits in the presence of 1–5 mmol/liter MgATP. For one type of subunit the cDNA sequence has been isolated and determined and reveals homology with certain bacterial proteins.

These observations are discussed in relation to the idea that the binding protein is an example of a general class of proteins termed "molecular chaperones" which are required for the correct assembly of certain oligomeric proteins such as the carboxylase from their subunits.

Abbreviations: BP – Binding protein, Rubisco – Ribulose bisphosphate carboxylase-oxygenase

I. The importance of the assembly of ribulose bisphosphate carboxylase-oxygenase

A. The problem

The enzyme ribulose bisphosphate carboxylase-oxygenase (Rubisco) is a major target for genetic engineering because of its pivotal role in determining the relative rates of photosynthesis and photorespiration. Molecular biologists would dearly love to be able to alter the properties of this enzyme to see whether the rate of net photosynthesis by crop plants can be increased, with consequent improvements in economic yield (Ellis and Gatenby 1984, Andrews and Lorimer 1988). Among the many technical problems to be solved before this question can be answered is the devising of a system in which mutated genes for Rubisco can be expressed to produce an altered active enzyme. The conventional test-bed for producing mutated proteins is the bacterium *Escherichia coli*. Once a given protein has been characterised sufficiently to allow the cloning of its encoding nucleotide sequence as complementary DNA (cDNA), it is a relatively straightforward matter to transfer this DNA to *E. coli* as part of an infecting bacteriophage or plasmid (Old and Primrose 1985). The transformed bacterial cells will then in many cases synthesize the foreign protein in a biologically active form. Once the synthesis of active protein has been achieved, specific mutations can be introduced at known positions in the polypeptide chain and the effects on the properties of the protein assessed. What is the problem in applying this approach to Rubisco?

In all eukaryotes the Rubisco protein is an oligomer consisting of eight large subunits bound to eight small subunits. The synthesis of this oligomer requires a complex series of events involving both the nuclear-cytoplasmic and chloroplast genetic systems (Ellis and Gatenby 1984, Ellis and Gray 1986, Gutteridge and Gatenby 1987). The large subunit carries the catalytic site for both carboxylase and oxygenase activities, and is synthesized inside the chloroplast from a gene present in the chloroplast genome. The small

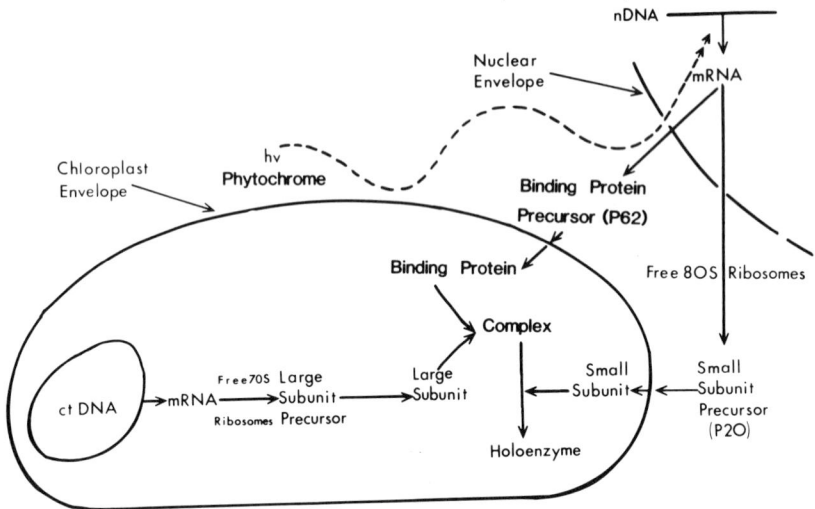

Fig. 1. Proposed role of the binding protein in the assembly of Rubisco in higher plants. Rubisco large subunits are synthesized from chloroplast genes (ct DNA) whereas Rubisco small subunits and binding protein subunits are synthesized from nuclear genes (nDNA). The symbols P20 and P62 refer to the approximate molecular masses $\times\ 10^{-3}$ of the precursor polypeptides. The wavy line indicates the role of light in the stimultion of small subunit gene transcription. Reprinted from Ellis (1987) with kind permission.

subunit is required for enzymic activity for some undefined reason, and is synthesized by cytoplasmic ribosomes as a higher molecular mass precursor from a family of genes present in the nuclear genome. This precursor enters the chloroplast by an ATP-dependent protein transport mechanism; in the stromal compartment the aminoterminal extension is cleaved by a specific protease, and the mature small subunits assemble with the large subunits to produce the active holoenzyme.

Encoding cDNA sequences for both the large and small Rubisco subunits have been prepared from a number of higher plant species. When these cDNA molecules are introduced into *E. coli* in a form suitable for their expression, the bacterial cells synthesize both types of Rubisco subunit. However these subunits do not associate with one another to form the oligomer and no enzymic activity is detectable (Bradley et al. 1986, Gatenby et al. 1987). This failure of assembly is currently blocking all attempts to pursue directed mutagenesis studies on Rubisco from higher plants; it is only with Rubisco from certain prokaryotic organisms that success has been achieved. The article by A.A. Gatenby in this volume discusses the use of *E. coli.* to express Rubisco cDNA sequences from both eukaryotes and prokaryotes and should be consulted for further information.

B. A possible solution

Studies in the authors' laboratory led to the suggestion that another chloroplast protein may be required for the assembly of Rubisco from its subunits in higher plants (Barraclough and Ellis 1980). This suggestion was based on the observation that when isolated intact chloroplasts are incubated under conditions that permit protein synthesis, the Rubisco large subunits that are made associate non-covalently with another pre-existing protein present in the stromal compartment. This latter protein we have termed the "Rubisco large subunit binding protein"; however, recent studies in both our laboratory and independently that of A.A. Gatenby have shown that this protein also binds Rubisco small subunits (see Section IIIE). Accordingly we now propose to rename this protein the Rubisco subunit binding protein (BP). This protein is nuclear-encoded and its subunits are imported across the chloroplast envelope after synthesis by cytoplasmic ribosomes. Figure 1 represents our current model for the involvement of BP in Rubisco synthesis in higher plants.

The proposal that another protein may be required for the assembly of Rubisco in higher plant chloroplasts may seem to violate one of the current principles of molecular biology, namely that all the information required to specify the structure and function of a protein is contained within its amino acid sequence. However, a number of examples are now known where the folding of certain polypeptide chains and their assembly into oligomeric structures requires the presence of other proteins that are not components of the final structure (Pelham 1986). This latter class of proteins has been termed "molecular chaperones" since their role appears to be to prevent the formation of "incorrect" structures that may be produced by the transient appearance of hydrophobic or charged domains during the synthesis of the polypeptide (Ellis 1987).

It should be noted in this connection that large subunits isolated in vitro from purified higher plant Rubisco by detergents or extremes of pH undergo self-aggregation and are insoluble in aqueous media, whereas those obtained from prokaryotic Rubisco remain soluble and capable of recombining with small subunits to form active enzyme (Andrews and Lorimer 1985). It may be that this difference between prokaryotes and eukaryotes in the behaviour of the Rubisco large subunits requires the involvement of a molecular chaperone to mediate Rubisco assembly in eukaryotes. The synthesis of the two types of Rubisco subunit in different intracellular compartments in higher plants may exacerbate this problem, since the assembly of large subunits with small subunits depends upon an adequate supply of small subunits entering the chloroplast. Alternatively, the successful synthesis of active Rubisco from prokaryotic genes in *E. coli* may reflect the presence of molecular chaperones in this bacterium that can function in the assembly of prokaryotic Rubisco. There is some evidence for the occurrence of proteins related to BP in several prokaryotes including *E. coli* (see Section III).

It must be emphasized that although the evidence for the involvement of BP in Rubisco assembly is increasing, it is still circumstantial; the definitive demonstration that BP is obligatorily required for Rubisco assembly still eludes us. However, it is important to pursue this hypothesis, not only for its scientific interest, but also in case it provides the solution to the current block that prevents attempts to produce mutated forms of higher plant Rubisco. The emphasis of our research is directed at the characterization of the binding protein so that cloned cDNA sequences can be expressed in the same *E. coli* cell as the Rubisco large and small subunit sequences. We regard this approach as a good test of the idea that the assembly of active Rubisco from higher plants requires the presence of the binding protein. The remainder of this article reviews our research on the Rubisco subunit binding protein.

II. The discovery of the Rubisco subunit binding protein

When intact chloroplasts isolated from young seedlings of *Pisum sativum* are incubated in a suitable osmoticum with K^+ ions, ^{35}S-methionine and light as energy source, they synthesize a number of labelled polypeptides, the most abundant being the Rubisco large subunit (Blair and Ellis 1973). Analysis of the physical state of these newly synthesized large subunits by electrophoresis on non-denaturing polyacrylamide gels reveals that only a minority of these subunits comigrate with the pre-existing Rubisco holoenzyme, suggesting that this minority is assembling into the oligomeric form of Rubisco. Time course studies show that the synthesis of these labelled large subunits in isolated chloroplasts proceeds at a faster rate than their assembly into Rubisco; indeed synthesis largely ceases before assembly begins. In addition, the proportion of newly synthesized large subunits that comigrates with the holoenzyme depends upon the composition of the medium used to incubate the chloroplasts. Assembly of labelled large subunits into Rubisco is best seen in a medium containing sorbitol as the osmoticum; if the sorbitol is replaced by KCl, assembly is largely prevented whereas synthesis is unaltered (Barraclough and Ellis 1980).

Unassembled large subunits remain in the supernatant fraction when the chloroplasts are osmotically lysed and the thylakoids removed by centrifugation. It is important to realize that these newly synthesized large subunits can be detected only by their radioactivity, since the amount made by isolated chloroplasts is far too small to be visualized by protein stains. However, analysis of the supernatant fraction by electrophoresis on nondenaturing polyacrylamide gels reveals that these large subunits comigrate with an abundant staining protein which behaves on these gels as if it had a relative molecular mass of about 700,000 (Ellis 1977). This comigration is not fortuitous since it is exact and observed over a range of gel concentrations.

The initial interpretation of these observations was that the staining protein is an aggregated form of unassembled large subunits made in the plant before the chloroplasts were isolated, so that comigration of the newly synthesized labelled large subunits with this protein is to be expected (Ellis 1977). However, Roger Barraclough showed that this interpretation is incorrect; when the stained and labelled protein band is excised from the non-denaturing gel and analyzed by electrophoresis on gels containing the detergent SDS, the staining protein migrates with a relative molecular mass (60,000) larger than that of the labelled large subunits (52,000). The staining 60 kDa protein is not itself labelled when isolated chloroplasts are incubated with radioactive amino acids. It was concluded that newly synthesized Rubisco large subunits made by isolated chloroplasts are bound noncovalently to another chloroplast protein that occurs in extracts as an oligomer of 60 kDa subunits. Quantitative time course studies showed that the proportion of labelled large subunits bound to this protein declines with

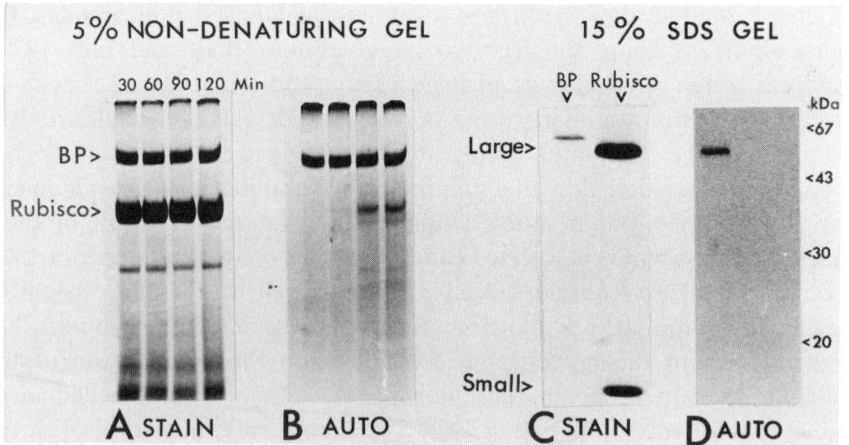

Fig. 2. Binding of the newly synthesized Rubisco large subunits to another stromal protein. Isolated intact chloroplasts of *Pisum sativum* were illuminated at 20 °C in a medium containing sorbitol as osmoticum and ^{35}S-methionine. Samples were removed at intervals, the chloroplasts lysed in hypotonic buffer, and the stromal fractions electrophoresed on a non-denaturing polyacrylamide gel. The gel was stained in Coomassie Blue (A) and an autoradiograph made (B). The staining bands marked BP and Rubisco were excised from the 30 min track and analysed separately on an SDS polyacrylamide gel. The SDS gel was stained (C) and autoradiographed (D). Note that (1) labelled large subunits comigrate exactly with the staining band of binding protein (cf. A and B); (2) these large subunits can be visualized by their radioactivity but not by staining (cf. C and D); (3) the binding protein is visible as a staining band on the SDS gel but is not labelled (cf. C and D). Symbols: BP, binding protein; Rubisco, holoenzyme of ribulose bisphosphate carboxylase-oxygenase; large and small, large and small subunits of Rubisco. This experiment was performed by R.A. Johnson.

time in isolated chloroplasts as the proportion comigrating with the Rubisco holoenzyme increases. The hypothesis was advanced that this combination of large subunits with the binding protein may be required for the assembly of the Rubisco holoenzyme (Baraclough and Ellis 1980). A recent repetition of the basic observations that led to the discovery of the binding protein is shown in Fig. 2.

III. Properties of the Rubisco subunit binding protein

A. Purification

The Rubisco subunit binding protein has been purified from *Pisum sativum* (Hemmingsen and Ellis 1986) and *Hordeum vulgare* (Musgrove, Johnson and Ellis 1987). Analysis of the products of partial digestion by protease reveals no similarity between the products obtained from Rubisco large subunits and those from the binding protein from *P. sativum* (Musgrove and Ellis 1986). Polyclonal antibodies raised against the *Pisum* binding protein do not cross-react with the Rubisco large subunit from *Pisum*, nor do anti-Rubisco antibodies cross-react with the binding protein. These observations confirm that the binding protein is not an aggregated form of Rubisco large subunits but is a distinct protein. There is a slight cross-reaction of anti-binding protein antibodies with the Rubisco small subunit; if the binding protein and small subunits bind to large subunits at some common sites, they may possess domains that are co-recognized by a subset of antibodies raised against the binding protein.

The apparent relative molecular mass of the purified *Pisum* binding protein is 720,000 as determined by Sephacryl columns. Analysis on SDS-containing polyacrylamide gels reveals the presence of two types of subunit, which migrate so closely together that they can be resolved only at high ratios of acrylamide to bis-acrylamide (Hemmingsen and Ellis 1986). The apparent relative molecular masses of these subunits are about 61,000 (the alpha subunit) and 60,000 (the beta subunit). Densitometric scanning of silver-stained gels shows that these two types of subunit are present in equal amounts (Musgrove and Ellis 1986). From these findings it was concluded that the binding protein has the likely composition $\alpha_6\beta_6$.

The binding protein from *H. vulgare* and *Triticum aestivum* has slightly larger subunits than that from *P. sativum*. The two subunits from *P. sativum* are immunologically distinct, show different partial protease digestion patterns and have different aminoterminal sequences (Musgrove, Johnson and Ellis 1987).

B. Distribution

Antibodies against the *P. sativum* binding protein detect 60 kDa polypeptides in extracts of leaves from spinach, wheat, barley and tobacco, as well as in extracts of leucoplasts from the endosperm of *Ricinus communis* seeds which contains Rubisco. Analyses of extracts from leaves of the C4 plant *Zea mays* show that the binding protein occurs in the bundle sheath cells but not in the mesophyll cells (Ellis, van der Vies and Hemmingsen 1987). There is thus a correlation between the occurrence of the binding protein and the occurrence of Rubisco in higher plants, which is consistent with our hypothesis that the binding protein is involved in Rubisco assembly.

Polypeptides immunologically related to the *Pisum* binding protein have also been detected in extracts in a range of bacterial species, including some which are not autotrophic and do not contain Rubisco (Hemmingsen, Dennis and Ellis 1987). The significance of this observation is discussed in section V. Extracts of the cyanobacterium *Anacystis nidulans* failed to show any reaction with the *Pisum* binding protein antibodies.

C. Synthesis

The subunits of the binding protein are not labelled when isolated intact chloroplasts are incubated with radioactive amino acids. When polysomes from *Pisum* leaves are run-off in a wheat-germ translation system in the presence of ^{35}S-methionine and chloramphenicol, an ^{35}S-labelled protein can be precipitated by antibodies to the binding protein. Chloramphenicol is added to prevent translation by chloroplast polysomes which may be present in the extract. The immunoprecipitated protein has a slightly higher molecular mass than the binding protein subunits when analyzed on SDS-containing polyacrylamide gels (Hemmingsen and Ellis 1986). We have been unable to resolve this higher molecular mass product into two bands, suggesting that the precursors to the alpha and beta subunits have identical mobilities. The observation that cytoplasmic polysomes synthesize a higher molecular mass product is consistent with the behaviour expected for chloroplast proteins which are encoded in nuclear genes and synthesized in the cytoplasm in precursor form.

Because the accumulation of Rubisco by leaves of *P. sativum* is greatly stimulated by light acting at the level of transcription (Gallagher and Ellis 1982), studies have been performed to ascertain the effect of light on the accumulation of the Rubisco subunit binding protein. Rocket immuno-electrophoresis reveals that etiolated *Pisum* plants contain low but detectable amounts of both Rubisco and binding protein. When the etiolated plants are exposed to continuous white light the amount of each protein per shoot increases, but the increase in Rubisco is greater than the increase in binding protein. Comparison of etiolated plants with plants grown for the same time of eight days under a photoperiod of 12 h shows that the content of Rubisco per shoot is increased 30-fold by growth in the light, but that the increase in the content of the binding protein is 7-fold (Lennox and Ellis 1986). Thus the expression of the nuclear genes for the Rubisco subunit binding protein in *P. sativum* is not subject to the same high degree of photoregulation as the genes for Rubisco.

D. Reversible dissociation by MgATP

The addition of MgATP to stromal extracts of *Pisum* chloroplasts causes dissociation of the binding protein dodecamer (Bloom, Milos and Roy 1983). We have confirmed this observation and shown that this dissociation produces the monomeric form of each subunit but is not accompanied by the stable phosphorylation or adenylation of either subunit (Hemmingsen and Ellis 1986, Hemmingsen, Dennis and Ellis 1987, Musgrove, Johnson and Ellis 1987). The dissociation by MgATP is highly specific; CTP, UTP, GTP, ADP, AMP, 3′-5′ cyclic AMP, NADPH and pyrophosphate at 1–5 mmol/liter with Mg^{2+} present at the same concentration, do not cause dissociation, nor does the β-γ methylene analogue of ATP. Mg^{2+} ions are required for dissociation by ATP but can be replaced by Ca^{2+} ions. If the concentration of ATP is lowered by allowing protein synthesis to occur in the stromal extracts, the dissociation is reversed and the dodecameric form of the binding protein increases with time. This reversal of dissociation is not seen if the extract is dialysed, which suggests that dialysable factors may be required for reversal. Immunoblotting techniques show that some of the monomeric subunits are present in the absence of added MgATP, while some of the dodecameric form is detectable even in the presence of 10 mmol/ liter MgATP. We propose that the binding protein undergoes a reversible dissociation between the dodecameric and monomeric forms, MgATP causing this equilibrium to shift towards the monomeric subunits. This equilibrium can be written as:-

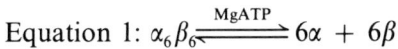

Equation 1: $\alpha_6 \beta_6 \underset{\longleftarrow}{\overset{MgATP}{\rightleftharpoons}} 6\alpha + 6\beta$

The position of this equilibrium in vivo is unknown. The concentration of ATP within the chloroplast in vivo is estimated to be in the range 1–5 mmol/ liter, which would be expected to favour dissociation. However, the concentration of binding protein inside the chloroplast we estimate at about 10 mg/cm^3, which is 20- to 100-fold higher than that at which it occurs in stromal extracts. Equation 1 suggests that increasing the concentration of binding protein strongly favours association of the subunits. Thus results obtained with stromal extracts cannot be extrapolated directly to the in vivo situation, and more of the dodecameric form may occur inside the chloroplast in vivo than is observed in stromal extracts to which MgATP has been added.

The fate of the attached newly-synthesized Rubisco large subunits on treatment of the binding protein with MgATP is unclear. The large subunits move near the top of sucrose density gradients after MgATP treatment, suggesting that they occur as dimers, but on non-denaturing gradient polyacrylamide gels these large subunits migrate as a continuous smear extending over an apparent molecular mass range of 2 to 4 $\times$ 10^5. We do not understand this behaviour and we are not able to determine whether these "re-

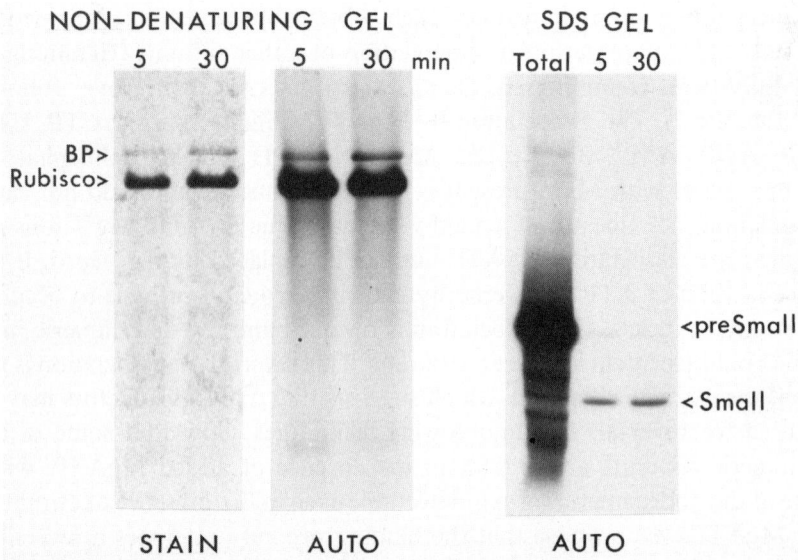

Fig. 3. Binding of mature small subunits to the Rubisco subunit binding protein. Labelled small subunit precursor from *Pisum sativum* was made in a wheat-germ translation from RNA transcribed from an SP6 plasmid containing the precursor cDNA sequence (Krieg and Melton 1984). We are grateful to N-H. Chua for supplying the cDNA for the pea small subunit precursor. The labelled precursor was incubated with intact isolated *Pisum* chloroplasts in the presence of excess unlabelled methionine and light as energy source. Samples were removed at 5 and 30 min and stromal extracts analyzed by electrophoresis on non-denaturing and SDS polyacrylamide gels. The gels were stained with Coomassie Blue and subjected to autoradiography. Symbols: Total, complete translation products from SP6 plasmid; preSmall, precursor to Rubisco small subunit; Small, mature small subunit. Other symbols as in Fig. 2.

leased" large subunits are attached to subunits of the binding protein. However another laboratory has reported that newly synthesized large subunits leave association with the binding protein and comigrate with the Rubisco holoenzyme when stromal extracts are treated with MgATP (Milos and Roy 1984, Milos, Bloom and Roy 1985). This transfer of labelled large subunits to the holoenzyme is inhibited by the addition of antisera to the binding protein (Cannon, Wang and Roy 1986). These observations are consistent with the reported requirement for ATP of the assembly of large subunits into Rubisco holoenzyme in intact isolated chloroplasts (Bloom, Milos and Roy 1983) and with our hypothesis that the binding protein is involved in the assembly of Rubisco. One limitation of these type of experiments is that it is not clear whether the comigration of newly-synthesized large subunits with pre-existing Rubisco holoenzyme represents true assembly or whether the comigration results from an exchange with preformed large subunits present in the holoenzyme, or even from binding to the complete holoenzyme without exchange. Possible ways of distinguishing between assembly and exchange in future experiments are discussed in Section V.

E. Binding of Rubisco small subunits

If the binding protein functions by holding Rubisco large subunits in the correct conformation to interact with small subunits, it might be expected that a transient association of small subunits with the binding protein occurs as part of the assembly process. We have detected such an association by supplying labelled small subunit precursors to intact isolated chloroplasts and analyzing the stromal fraction by non-denaturing polyacrylamide gel electrophoresis. The majority of the imported labelled precursor rapidly enters the Rubisco holoenzyme as mature small subunits, but a small proportion comigrates with the binding protein (Fig. 3). Identical observations have been made independently in another laboratory (Lubben et al. 1987). Its for this reason that we have renamed the protein "the Rubisco subunit binding protein". Whether the small subunits are attached to the binding protein subunits directly or to the bound large subunits is unknown.

IV. Cloning

A cDNA sequence encoding the alpha subunit of the binding protein from *Ricinus communis* has been isolated from a lambda gtll cDNA expression library (Hemmingsen, Dennis and Ellis 1987). The recombinant plasmid (pG3BP2) contains 90% of the mature polypeptide nucleotide sequence and its aminoterminal region shows homology at the amino acid level with that of the alpha subunit from *P. sativum*. There is no apparent sequence homology with either the large or small subunits of Rubisco. However, there is good homology (50%) at the amino acid level with two bacterial proteins, one from *E. coli* and the other from *Mycobacterium leprae*. The similarity to the mycobacterial protein is especially remarkable; the sequences are in register from the mature aminoterminus of each polypeptide and only minor compensations are needed to keep them in register throughout the entire sequence. Unfortunately the functions of neither bacterial protein are known.

We have screened a lambda gtll cDNA expression library of *T. aestivum*, kindly supplied by C. Raines and T.A. Dyer, with antibodies raised against the binding protein from *P. sativum*. A cDNA fragment of 1.85 kbp shows strong homology with the alpha subunit cDNA sequence from *R. communis*, as judged by hybridization (Ellis, van der Vies and Hemmingsen 1987). This fragment is currently being sequenced. We are also attempting to isolate a full-length cDNA for the beta subunit of the binding protein from *T. aestivum* so that both subunits can be produced in the same *E. coli* cells that are synthesizing Rubisco subunits from wheat.

V. The future

A. Testing the hypothesis

It is our view that the most important aspect of research on the binding protein to be tackled in future experiments is the testing of the hypothesis that this protein is obligatorily required for Rubisco assembly. There are three possible ways by which this testing might be achieved:-

1. The development of a soluble extract derived from chloroplasts which will both synthesize the Rubisco large subunit and assemble it into the holoenzyme. We have recently developed such a system and are planning to determine whether removal of the binding protein from this extract by immunological methods will prevent assembly but not synthesis. If this result is observed, then the effect of adding back purified binding protein to the extract before synthesis of the large subunit can be tried to see if this addition restores assembly. The limitation of this type of experiment is that the extract contains large quantities of preformed Rubisco so that it is not possible to be certain that the comigration of newly-synthesized large subunits with the holoenzyme reflects assembly or exchange or sticking.

2. The synthesis in vitro of Rubisco large and small subunits from cloned cDNA sequences located in a suitable plasmid such as SP6. If it is found that the appearance of correctly-sized oligomer containing equal amounts of large and small subunits is dependent upon the addition of purified binding protein, the hypothesis would be strongly supported. The advantage of this approach is that there is no pool of preformed Rubisco to complicate the interpretation of the experiment.

3. The expression of cDNA sequences for both subunits of the binding protein in the same *E. coli* cells that are expressing sequences for both subunits of Rubisco. Success in this type of experiment would not only provide strong support for the hypothesis but would also permit the use of in vitro mutagenic techniques to monitor the effect of specific amino acid changes on the kinetic properties of the enzyme from higher plants. The advantage of this approach over the previous one is that sufficient Rubisco molecules can be assembled to allow detection by their enzymic activity.

All three approaches are currently being investigated at Warwick.

B. *The wider picture*

Is the suggested requirement of another protein for Rubisco assembly peculiar to this enzyme, perhaps reflecting its extreme abundance and the properties of its large subunit, or is it one example of a general class of assembly proteins? What is the significance of the immunological and sequence similarities between the binding protein and certain proteins in bacteria that do not contain Rubisco? We suggest that we may have stumbled on a representative of a class of proteins that Pelham (1986) has proposed is involved in the assembly and disassembly of oligomeric proteins, and which have been termed "molecular chaperones" (Ellis 1987). Such molecular chaperones would be expected to occur in all cells which synthesize oligomeric proteins whose subunits have the potential to interact incorrectly. The Rubisco subunit binding protein may have evolved from this general class of proteins to mediate Rubisco assembly specifically. Such an origin would account for the similarities between the binding protein from higher plants and certain bacterial proteins.

If this general argument is correct, there may exist other molecular chaperones involved transiently in the assembly of oligomeric proteins, especially where the constituent subunits originate in different subcellular compartments where problems of folding, stability and transport may be more acute. We would like to suggest that researchers studying the synthesis of chloroplast and mitochondrial proteins consider the use of non-denaturing gradient polyacrylamide gels to detect the possible involvement of binding proteins in the assembly of oligomeric proteins since this technique permits the detection of non-covalently linked protein complexes with high resolution.

Note added in proof:
The protein from *Escherichia coli* to which the Rubisco subunit binding protein is 50% homologous at the amino acid level has been identified as the product of the *gro* EL gene. The *gro* EL protein is a molecular chaperone required for the correct assembly into an oligomeric structure of one of the proteins of the head of phage lambda.

Acknowledgements

We should like to thank T.J. Andrews, A.A. Gatenby and H. Roy for access to unpublished data and papers in press, and the Science and Engineering Research Council (UK) for financial support.

674

References

Andrews TJ and Lorimer GH (1985) Catalytic properties of a hybrid between cyanobacterial large subunits and higher plant small subunits of ribulose bisphosphate carboxylase-oxygenase. J Biol Chem 260: 4632–4636

Andrews TJ and Lorimer (1988) Rubisco: structure, mechanism and prospects for improvement. In: Hatch MD (ed.) The biochemistry of plants. Academic Press (forthcoming)

Barraclough R and Ellis RJ (1980) Protein synthesis in chloroplasts IX. Assembly of newly-synthesized large subunits into ribulose bisphosphate carboxylase in isolated intact pea chloroplasts. Biochim biophys Acta 608: 19–31

Blair GE and Ellis RJ (1973) Protein synthesis in chloroplasts. I. Light-driven synthesis of the large subunit of Fraction I protein by isolated pea chloroplasts. Biochem biophys Acta 319: 223–234

Bloom MV, Milos P and Roy H (1983) Light-dependent assembly of ribulose-1,5-bisphosphate carboxylase. Proc Nat Acad Sci USA 80: 1013–1017

Bradley D, van der Vies SM and Gatenby AA (1986) Expression of cyanobacterial and higher plant ribulose 1,5-bisphosphate carboxylase genes in *Escherichia coli*. Phil Trans R Soc Lond B 313: 447–458

Cannon S, Wang P and Roy H (1986) Inhibition of ribulose bisphosphate carboxylase assembly by antibody to a binding protein. J. Cell Biol 1903: 1327–1335

Ellis RJ (1977) Protein synthesis by isolated chloroplasts. Biochim Biophys Acta 463: 185–215

Ellis RJ (1987) Proteins as molecular chaperones. Nature 328: 378–379

Ellis RJ and Gatenby AA (1984) Ribulose bisphosphate carboxylase-oxygenase: properties and synthesis. In: Lea PJ and Steward GR (eds). The genetic manipulation of plants and its application to agriculture, pp 41–60, Ann Proc Phytochem Soc Eur Vol 23. Oxford: Clarendon Press

Ellis RJ and Gray JC (1986) Ribulose bisphosphate carboxylase-oxygenase, Proc of a Royal Society Discussion Meeting. Phil Trans R Soc Lond B313: 305–469. London: The Royal Society

Ellis RJ, van der Vies, SM and Hemmingsen SM (1987) The Rubisco large subunit binding protein — a molecular chaperone? In: von Wettstein D and Chuva N-H (eds) Proceedings of the NATO Advanced Study Institute Plant Molecular Biology 87. Carlsberg Laboratory, Copenhagen June 10–19, 1987, pp 33–40. New York: Plenum Publishing Co.

Gallagher TF and Ellis RJ (1982) Light-stimulated transcription of genes for two chloroplast polypeptides in isolated pea leaf nuclei. EMBO J 1: 1493–1498

Gatenby AA, van der Vies SM and Rothstein S (1987) Co-expression of both the maize large and wheat small subunit genes of ribulose bisphosphate carboxylase in *Escherichia coli*. Eur J Biochem 168: 227–231

Gutteridge S and Gatenby AA (1987) The molecular analysis of the assembly, structure and funtion of Rubisco. In: Miflin BJ (ed.) Oxford Surveys of Plant Molecular and Cell Biology. Oxford University Press (forthcoming).

Hemmingsen SM and Ellis RJ (1986) Purification and properties of ribulose bisophosphate carboxylase large subunit binding protein. Plant Physiol 80: 269–276

Hemmingsen SM, Dennis DT and Ellis RJ (1987) The Rubisco large subunit binding protein. In: Bohnert HJ and Jensen R (eds) Proceedings of the Conference Rubisco 87. Tucson Arizona (forthcoming)

Krieg PA and Melton DA (1984) Functional messenger RNAs are produced by SP6 in vitro transcription of cloned cDNAs. Nucl Acid Res 12: 7057–7070

Lennox CR and Ellis RJ (1986) The carboxylase large subunit binding protein: photoregulation and reversible dissociation. Biochem Soc Trans 14: 9–11

Lubben TH, Gatenby AA, Ahlquist P and Keegstra K (1987) Imported large subunits of ribulose-1,5-bisphosphate carboxylase/oxygenase, but not imported coupling factor beta subunits, are assembled into holoenzyme in isolated chloroplasts. EMBO J (submitted)

Milos P and Roy H (1984) ATP-released large subunits participate in the assembly of RuBP carboxylase. J Cell Biochem 24: 153–162

Milos P, Bloom MV and Roy H (1985) Methods for studying the assembly of ribulose bisphosphate carboxylase. Pl Mol Biol Reporter 3: 33–42

Musgrove JE and Ellis RJ (1986) The Rubisco large subunit binding protein. Phil Trans R Soc Lond B 313: 419–428

Musgrove JE, Johnson RA and Ellis RJ (1987) Dissociation of the ribulose bisphosphate carboxylase large subunit binding protein into dissimilar subunits. Eur J Biochem 163: 529–534

Old RW and Primrose SB (1985) Principles of Gene Manipulation, 3rd ed., Oxford: Blackwell Scientific Publications

Pelham HRB (1986) Speculations on the functions of the major heat-shock and glucose-regulated proteins. Cell 46: 959–961

Govindjee et al. (eds), Molecular Biology of Photosynthesis: 677–698

Minireview

The value of mutants unable to carry out photorespiration

RAY D. BLACKWELL,[1] ALAN J.S. MURRAY,[1] PETER J. LEA,[1]
ALAN C. KENDALL,[2] NIGEL P. HALL,[2] JANICE C. TURNER[2] &
ROGER M. WALLSGROVE[2]
[1]*Department of Biological Sciences, University of Lancaster, Lancaster LA1 4YQ, UK (for correspondence and reprints); [2]Biochemistry Department, Rothamsted Experimental Station, Harpenden, Herts AL5 2JQ, UK*

Received 6 October 1987; accepted 28 October 1987

Key words: *Hordeum*, mutants, nitrogen metabolism, photosynthesis, photorespiration, *Pisum*

Abstract. Manipulation of the CO_2 concentration of the atmosphere allows the selection of photorespiratory mutants from populations of seeds treated with powerful mutagens such as sodium azide. So far, barley lines deficient in activity of phosphoglycollate phosphatase, catalase, the glycine to serine conversion, glutamine synthetase, glutamate synthase, 2-oxo-glutarate uptake and serine: glyoxylate aminotransferase have been isolated. In addition one line of pea lacking glutamate synthase activity and one barley line containing reduced levels of Rubisco are available. The characteristics of these mutations are described and compared with similar mutants isolated from populations of *Arabidopsis*. As yet, no mutant lacking glutamine synthetase activity has been isolated from *Arabidopsis* and possible reasons for this difference between barley and *Arabidopsis* are discussed. The value of these mutant plants in the elucidation of the mechanism of photorespiration and its relationships with CO_2 fixation and amino acid metabolism are highlighted.

Abbreviations: GS_1 – cytoplasmic glutamine synthetase, GS_2 – chloroplastic glutamine synthetase, PFR – Photon fluence rate, Rubisco – Ribulose-1,5-bisphosphate carboxylase/oxygenase, RuBP – Ribulose-1,5-bisphosphate, SGAT – serine:glyoxylate aminotransferase

Introduction

The chloroplast enzyme ribulose-1,5-bisphosphate carboxylase/oxygenase (Rubisco) catalyses both the carboxylation and oxygenation of RuBP (Bowes et al. 1971, see Ogren et al. 1986 for a review). Carboxylation of RuBP produces two molecules of phosphoglycerate which are used in the photosynthetic reduction cycle. Oxygenation produces one molecule of phosphoglycerate, plus one molecule of phosphoglycollate which is used in the interlocking photorespiratory carbon and nitrogen cycles (Fig. 1). In most temperate C_3 plants at ambient levels of CO_2 (0.03%), one molecule of

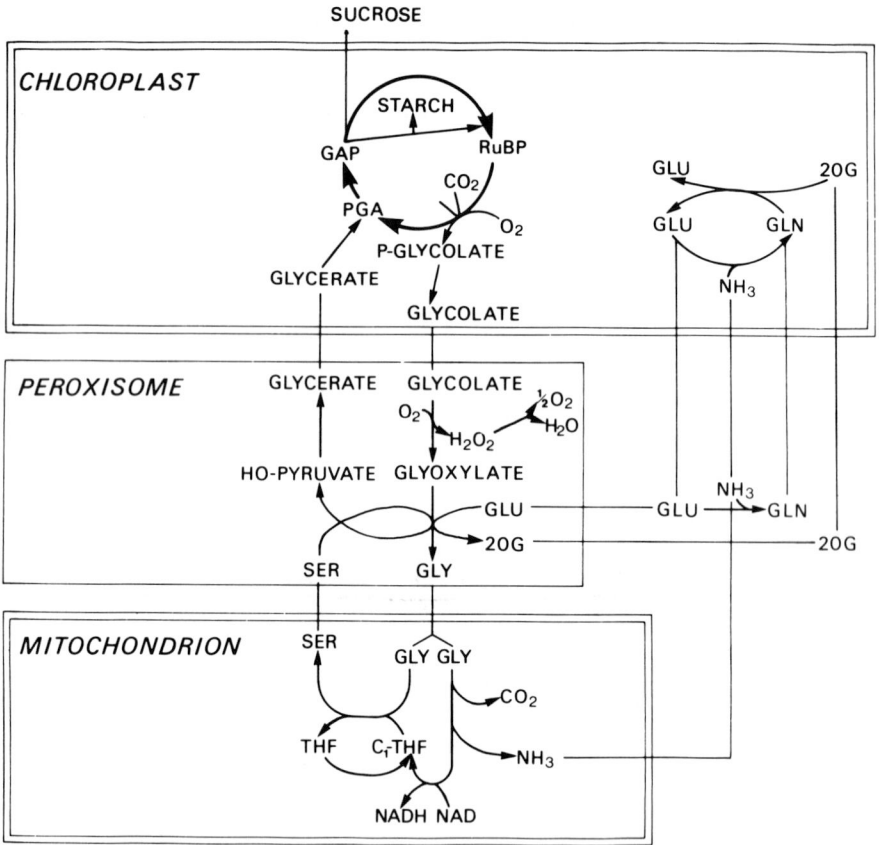

Fig. 1. Photorespiratory carbon and nitrogen cycles in C_3 plants.

RuBP is oxygenated for every 2.1–2.6 molecules carboxylated (Keys 1986). Phosphoglycollate is subsequently metabolized by a sequence of reactions in the chloroplasts, peroxisomes and mitochondria, which permits the recovery of some of the carbon by the photosynthetic reduction cycle. Glycine is decarboxylated and deaminated in the mitochondria, the ammonia being rapidly reassimilated via the glutamate synthase cycle (Keys et al. 1978). Most of the CO_2 released is lost to the atmosphere. Values of this photorespiratory loss have been calculated to be as high as 75% of the net photosynthetic rate (Zelitch 1979), but more recent determinations employing photorespiratory mutants (Somerville and Somerville 1983; see later discussion) and ^{18}O and ^{14}C (Gerbaud and André 1987) have put the value nearer to 40%. Keys (1986) has calculated that, under photorespiratory conditions, an additional 30 molecules of ATP and 18 molecules of NADPH are

Table 1. Mutant lines of barley lacking enzymes of the photorespiratory carbon and nitrogen cycle.

Enzyme deficiency	Mutant line
Rubisco protein	RPr 84/29
Phosphoglycollate phosphatase	RPr 84/38, 84/86, 84/90 LaPr 85/36
Catalase	RPr 79/1, 79/4, 83/2, 83/13, 83/21, 84/70, 84/87
Serine:glyoxylate aminotransferase	LaPr 85/84
Glycine:serine interconversion	RPr 83/202; LaPr 85/55, 87/30
Chloroplast glutamine synthetase	RPr 83/20, 83/31, 83/32, 84/12, 84/34 84/50, 84/83, 84/84 LaPr 85/14, 85/80, 85/67, 86/38, 86/44, 86/52
Ferredoxin dependent glutamate synthase	RPr 82/1, 82/9, 83/28, 84/13, 84/42, 84/64, 84/82 LaPr 85/56, 85/73, 86/14, 86/22, 86/70, 86/77; LaPr 86/33*
Chloroplast 2-oxoglutarate transport	RPr 79/2, 83/19, 83/109

* A mutant of pea.
Photorespiratory mutants isolated at Rothamsted are designated RPr and those at Lancaster LaPr.

consumed for each molecule of sucrose synthesized. It is likely that such severe losses of carbon and energy have a major effect on the yields of C_3 crop plants.

If plants are grown under conditions in which the level of CO_2 is so high that RuBP oxygenation is prevented, then there is no flux of carbon through the photorespiratory cycle. Somerville and Ogren utilized this technique to isolate conditional lethal mutants of *Arabidopsis* which grew normally in CO_2-enriched air, but showed symptoms of severe stress after 3–4 days illumination in normal air. By feeding the mutants plants $^{14}CO_2$ and carrying out specific enzyme determinations, they were able to identify seven different lesions in the photorespiratory cycle, two of which are involved in the conversion of glycine to serine (Somerville 1984, 1986). Whilst these mutants were not of direct value in the generation of plants with reduced rates of photorespiration, they unequivocally confirmed the mechanism of the photorespiratory carbon and nitrogen cycle.

A similar programme of mutant selection, supported by the AFRC, using the important agronomic crop barley was initiated at Rothamsted and is now also being carried out at the Univerity of Lancaster. To date, seven different classes of photorespiratory mutants of barley have been isolated, plus one of pea (Table 1). In this review we will discuss the progress made with these mutants at both Rothamsted and Lancaster.

Rubisco-deficient mutant

In addition to mutants with defects in photorespiratory carbon and nitrogen metabolism, the screening method used allows the selection of mutants which require additional CO_2 for growth because of other defects in metabolism. A mutant of this type in *Arabidopsis* lacked activation in vivo of Rubisco, even though the purified enzyme from this mutant was indistinguishable from the wild type (Somerville et al. 1982). The existence of this mutant allowed the identification of a hitherto unknown enzyme which has been recently characterized as Rubisco activase. This enzyme is now known to exist in 15 diverse species of plants including spinach, pea, maize, tomato and *Chlamydomonas* as well as *Arabidopsis* (Salvucci et al. 1987).

A different type of mutant requiring CO_2 has been selected in barley and partially characterized (Hall et al. 1986). In air this mutant, RPr 84/29, did not die; it merely grew more slowly than the wild type.

Rubisco normally accounts for over 50% of the total protein content of barley leaves. In RPr 84/29, however, only 10% of the extracted protein was found to be Rubisco. Not only was the total amount of this protein reduced in the mutant, but the specific activity of Rubisco was also reduced to 25% of the wild type. Rubisco protein from both the wild type and RPr 84/29 was partially purified using sucrose gradients (Gutteridge et al. 1982). Comparison of the kinetic constants of the purified enzyme from the wild type and the mutant, however, revealed no differences (unpublished results).

Table 2. Specific activity of Rubisco in 26 day old barley plants. Leaf extracts were prepared at 4 °C and Rubisco assayed by ^{14}C incorporation into PGA at 13 °C as described by Boyle and Keys (1982).

Leaf number	Maris Mink	RPr 84/29
4	396.4	154.5
3	402.7	199.3
2	432.5	233.7
1	439.2	235.2

All values are in $nmol\,min^{-1}\,mg^{-1}$ protein.

The difference in specific activity of Rubisco in leaves of mutant and wild type was more marked in recently emerged leaves (Table 2; leaf 1 being the oldest, leaf 4 the youngest) indicating a lower rate of synthesis of Rubisco in the mutant. In this respect RPr 84/29 was similar to developmental mutants of maize with reduced Rubisco levels (Harpster et al. 1984). In vivo labelling studies of RPr 84/29 support the idea of impaired Rubisco synthesis. Leaf segments were fed with ^{35}S-methionine, extracted and subjected to SDS gel electrophoresis. In RPr 84/29 the ratio of incorporation of label into the large sub-unit compared to the small sub-unit of Rubisco was 1.9 while in the wild type it was 3.7. The soluble free amino acid content of leaves of RPr 84/29 was 3-fold higher than that of corresponding leaves of the wild type. Within this pool of amino acids the levels of asparagine and arginine were 6- and 11-fold higher respectively. We have not yet resolved whether the mutation in RPr 84/29 affects synthesis of all chloroplast proteins or only the large sub-unit of the Rubisco protein.

Phosphoglycollate phosphatase

The isolation of an *Arabidopsis* mutant deficient in phosphoglycollate phosphatase (Somerville and Ogren 1979) proved that phosphoglycollate was the first product of the photorespiratory cycle and was the source of photorespiratory glycollate and hence CO_2 production. The subsequent isolation and genetic analysis of similar mutants in barley provided evidence of the existence of isozymes of phosphoglycollate phosphatase. Polyacrylamide gel electrophoresis of leaf extracts from wild-type barley, followed by activity staining, revealed three bands of phosphoglycollate phosphatase activity. All three bands were absent from similarly treated extracts of the mutant, RPr 84/90 (Hall et al. 1987a). A previous report had suggested that two isoenzymes of phosphoglycollate phosphatase were present in *Phaseolus vulgaris* leaves (Verin-Vergeau 1979). The possibility that different amounts of citrate and isocitrate associate with the enzyme, as in tobacco (Christeller and Tolbert 1978), giving the protein various negative charges (and mobilities on electrophoresis), needs to be considered.

Heterozygous plants, containing only one mutant gene (*PcoA/pcoA*) grew normally in air. The gas exchange characteristics of such plants were indistinguishable from the wild type, while the rate of CO_2 fixation of the homozygous mutant was severely inhibited in air (Hall et al. 1987a). Heterozygous plants contained approximately 50% of the wild type activity of phosphoglycollate phosphatase (Fig. 2). This pattern of enzyme activity indicated that the amount of active enzyme produced in heterozygous plants

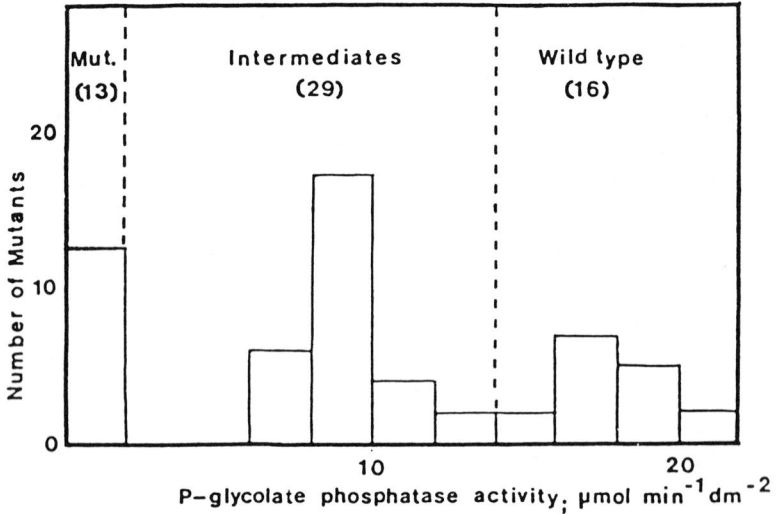

Fig. 2. Trimodal distribution of phosphoglycollate phosphatase activity in the F_2 progeny from a cross of wild type and a phosphoglycollate phosphatase-deficient mutant line.

was determined by the number of functional alleles, and not by metabolic considerations (Turner et al. 1986).

Catalase

During photorespiration, oxidation of glycollate in the peroxisome produces hydrogen peroxide, which is degraded in the same organelle by catalase. The first photorespiratory mutant characterized in barley (Kendall et al. 1983) lacked catalase activity. Neither of the two catalase isoenzymes present in extracts of wild-type leaves were detected in this mutant, suggesting that both isoenzymes were under the control of the same gene. Other workers had proposed that the active catalase enzyme was composed of four subunits of equal molecular weight (Scheiffer et al. 1976, Lamb et al. 1978, Scandalios 1979). Therefore, either the two bands seen after starch gel electrophoresis were an artefact of the separation procedure or post-translational modification of the protein had to be considered (Kendall et al. 1983). A further eight catalase-deficient mutants have now been isolated and the six so far tested were all allelic to the original mutant, RPr 79/4 (Turner et al. 1986).

The catalase-deficient mutants, in contrast to all other photorespiratory mutants so far isolated, showed no short-term inhibition of photosynthesis on transfer from non-photorespiratory to photorespiratory conditions (Kendall et al. 1983). Illuminated spinach chloroplasts can reduce hydrogen peroxide using glutathione and ascorbate as intermediate electron carriers (Anderson et al. 1983, Foyer et al. 1983). Membrane disruption was evident in these mutants in air (Parker and Lea 1983). Presumably, this followed the accumulation of high levels of peroxide in the leaves. Exposure to air caused a rapid 5–10-fold increase in the total glutathione content of the leaf (Smith et al. 1984), suggesting that glutathione synthesis protected the leaf from peroxide toxicity and that cellular damage (Parker and Lea 1983) only occurred when glutathione production declined. Studies of the intracellular distribution of these elevated glutathione levels in the catalase-deficient leaves, showed that almost all the increase in the chloroplast fraction was due to oxidized glutathione, with a relatively constant pool of the reduced form (Smith et al. 1985). We suggested that oxidation of reduced glutathione by peroxide in the mutant relieved feedback inhibition of the biosynthetic enzymes, a proposal consistent with that put forward by Rennenberg (1982).

Metabolism of ^{14}C-glycollate in the catalase-deficient mutant in air, was similar to that observed in the wild-type (Kendall et al. 1983). This contrasted with the proposal of some workers (Grodzinski and Butt 1976, Walton and Butt 1981, Grodzinski and Woodrow 1981) that photorespiratory CO_2 could be produced by the non-enzymic oxidation of glyoxylate by hydrogen peroxide. The catalase-deficient mutant confirms that the major role of catalase in C_3 plant leaves is to detoxify the hydrogen peroxide liberated in the conversion of glycollate to glyoxylate (Tolbert 1979).

Serine:glyoxylate aminotransferase

During photorespiration, serine is produced in the mitochondria and is subsequently used to transaminate glyoxylate in a peroxisomal reaction catalysed by the enzyme serine:glyoxylate aminotransferase (SGAT) (Keys et al. 1978). Extensive characterisations of SGAT deficient mutants from two species, *Arabidopsis* (Somerville and Ogren 1980a) and barley (Murray et al. 1987) have been published. The characteristics of this mutation in both species are very similar.

Although the SGAT-deficient barley mutant was recovered as a strong growing plant with a large leaf area, the rate of CO_2 fixation in air fell to 20–30% of the wild type rate. Under non-photorespiratory conditions (1% O_2, 0.035% CO_2) the rates for the mutant and the wild type were the same. As pointed out by Somerville and Ogren for the *Arabidopsis* mutant, the poor photosynthetic performance of these plants is consistent with the idea that any carbon entering photorespiration must be returned to the Calvin cycle to maintain the rate of CO_2 fixation. As this review emphasises, investigations with all other photorespiratory mutants have further strengthened this opinion (see conclusions).

Somerville and Ogren (1980a) measured CO_2-efflux into CO_2-free 50% O_2 by *Arabidopsis* lacking SGAT. They found that CO_2 fixation was maintained at 60% of the wild type rate for at least 50 minutes after the onset of photorespiration, and suggested that the chloroplast was mobilizing storage carbohydrates at a rate high enough to maintain near normal levels of photorespiration, while only achieving about one-fifth of the normal rate of CO_2 fixation. Furthermore, the exhaustion of stored carbohydrate could be responsible for the senescence of these mutants under photorespiratory conditions.

Under certain atmospheric conditions it has been argued by Sivak and Walker (1986) and Sharkey et al. (1986) that CO_2 fixation can become independent of the ambient O_2 level. This situation occurs under high PFR in atmospheres containing elevated levels of CO_2 ($> 0.07\%$). The rate of CO_2 fixation is then limited by the transport of phosphate back into the chloroplast after sucrose synthesis in the cytoplasm. One essential component of this hypothesis is that photorespiration still proceeds. Before the existence of photorespiratory mutants only very indirect evidence for the continued operation of photorespiration under these conditions was available. Murray et al. (1987) were able to demonstrate that, as expected, under high light and CO_2, increasing the oxygen concentration flowing over the leaf did not decrease the CO_2 fixation rate of wild type barley. Under the same conditions, the photosynthetic rate of the SGAT mutant was sensitive to oxygen concentration. Since CO_2 fixation for this mutant in non-photorespiratory conditions is similar to the wild type, the observed decrease in CO_2 fixation demonstrates that photorespiration is still occurring, albeit at a lower rate. In the mutant the limitation following transfer to 21% O_2 is probably the lack of carbon recycling to the Calvin cycle from photorespiration.

Although only one peroxisomal form of the serine:glyoxylate aminotransferase enzyme exists in spinach leaves (Rehfeld and Tolbert 1972), analysis of wild type barley extracts after ion exchange chromatography, using DEAE-Sephacel, revealed two major and one minor peaks of SGAT activity (Murray et al. 1987). The mutant plant was shown to lack all forms of activity demonstrating convincingly that extreme caution is required when inferring isoenzyme patterns from the consideration of ion exchange chromatography alone. This cautionary note is strengthened by the results of a backcross to wild type and subsequent analysis of the F_2 generation

(Murray et al. 1987), which showed that a deficiency in SGAT was inherited as a single nuclear recessive mutation, thus completely ruling out any possibility that three separate mutations occurred in this plant all acting to remove serine:glyoxylate activity.

Serine:glyoxylate aminotransferase transaminates only half the glyoxylate flowing through the photorespiratory pathway. Another enzyme, glutamate:glyoxylate aminotransferase, catalyses the production of the rest of the glycine. As glutamate:glyoxylate aminotransferase is present in the mutant in normal amounts, it seems likely that half of the glyoxylate produced would still be metabolised to glycine in SGAT deficient mutants. Consequently they would still produce ammonia and CO_2 at 50% of the wild type rate in the subsequent mitochondrial conversion of glycine to serine. Methionine sulphoximine has been used to reveal the production of photorespiratory ammonia by inhibiting glutamine synthetase activity (Walker et al. 1984). Using this technique, Murray et al. (1987) demonstrated the expected low rate of ammonia release by SGAT deficient barley mutants, relative to the wild type. However, CO_2 release from a leaf fed [1-^{14}C]glyoxylate was higher for the SGAT-deficient mutant than for the wild type. This was assumed to result from the direct oxidation of accumulated glyoxylate to produce formate and CO_2. This decarboxylation is thought to occur only when insufficient aminodonors are available to transaminate the glyoxylate (Walton and Butt 1980, Somerville and Ogren 1981b).

Earlier work suggested asparagine as an aminodonor to the photorespiratory cycle (Ta et al. 1984), probably entering by the action of asparagine:glyoxylate aminotransferase. Asparagine, however does not accumulate in this mutant in air suggesting it is not an important photorespiratory aminodonor in barley. Recent experiments with ^{15}N labelled amino acids suggested that, in pea, only 11% of the nitrogen metabolized from asparagine was directed into the photorespiratory pathway, compared with 60% from metabolised serine (Ta and Joy 1986). Asparagine:glyoxylate aminotransferase activity is also absent from the SGAT deficient mutants, confirming that these activities are catalysed by the same enzyme, along with serine:pyruvate aminotransferase activity (Rehfeld and Tolbert 1972).

Glycine-serine conversion

During photorespiration a complex sequence of mitochondrial reactions is involved in the metabolism of glycine. The first reaction, the glycine decarboxylase (cleavage) system involves four separate proteins (Walker and

Oliver 1986) and catalyses the oxidation of glycine yielding CO_2, NH_3 and 5.10-methylene tetrahydrofolate (Kisaki and Tolbert 1970). The mitochondrial isozyme of serine transhydroxymethylase then transfers the C_1 group of 5.10-methylene tetrahydrofolate to a second molecule of glycine to produce serine which is used as an aminodonor in the peroxisome. The complex that carries out both reactions has recently been isolated in an active form from spinach mitochondria (Neuberger et al. 1986). Experiments with *Arabidopsis* mutants deficient in glycine decarboxylase and serine transhydroxymethylase activity demonstrated clearly that the only significant source of photorespiratory CO_2 from leaves is the decarboxylation of glycine (Somerville and Ogren 1981a, b). No convincing evidence exists to support the alternative view (Grodzinski and Butt 1976, Grodzinski and Woodrow 1981) that, in vivo, a significant amount of CO_2 is evolved due to non-enzymatic decarboxylation of glyoxylate. However, this reaction can be demonstrated in isolated organelles or reconstituted in vitro systems (Walton and Butt 1981).

Three barley mutants, RPr 83/202, LaPr 85/55 and 87/30, unable to convert glycine to serine have been isolated (Lea et al. 1984, Blackwell et al. 1987a). On exposure to air, the rate of CO_2-fixation by these mutants was rapidly inhibited. After ten minutes exposure to $^{14}CO_2$ in air, 65% of the radioactivity in extracts prepared from leaves of RPr 83/202 was detected as glycine and only 6% as sucrose. In contrast, extracts from the wild type contained 13% of recovered counts in glycine and 44% in sucrose. In the mutant leaves, metabolism of ^{14}C-glycine was severely reduced, with 90% of the total radioactivity recovered as glycine, 1.5% in sucrose and none in serine. Wild type leaves accumulated only 50% of the ^{14}C in glycine, but 30% in sucrose and 8% in serine. Radioactivity in the evolved CO_2, when the leaves were fed ^{14}C-glycine, was 10-fold higher in the wild type than in the mutant.

Treatment with mutagens often results in plant which are slow growing and require careful back-crossing to improve. Full characterization of RPr 83/202 was therefore not attempted, since another mutation in the plant (presumably unrelated to that preventing the conversion of glycine to serine) affected normal development. No seed was obtained and so no genetic studies were possible. This mutant was isolated as a slow-growing plant with grass-like appearance and a poor root system. Ear development never proceeded past the stage of maximum primordia and consequently propagation has only been possible by removing and rooting tillers. So far, attempts to induce ear development (by manipulation of growth temperature, daylength and light intensity and by the application of various growth regulators) have failed.

A full biochemical and genetic characterization of LaPr 85/55 and 87/30 is under way. These lines are strong growing and seed set is normal at elevated levels of CO_2. LaPr 85/55 and 87/30 contain wild type levels of serine transhydroxymethylase activity indicating that these mutations affect the glycine decarboxylase complex. Consistent with this idea is the low level (2% of wild type) of ammonia production by leaves of these mutants in the presence of methionine sulphoximine in air and the inability of these plants to metabolise ^{14}C-glycine (Blackwell et al. 1987a).

Mutations of the photorespiratory nitrogen cycle

Glutamate synthase and/or glutamine synthetase

A continued flux of carbon around the photorespiratory cycle is dependent on a sufficient supply of aminodonors for the transamination of glyoxylate in the peroxisomes. Within the cycle, the production of the necessary amino acids to act as aminodonors is interdependent and linked to the reincorporation of ammonia derived from photorespiration (Keys et al. 1978).

Since 1980, mutants deficient in glutamate synthase have been available in *Arabidopsis thaliana* (Somerville and Ogren 1980b), but more recently, glutamate synthase-deficient plants have been isolated from barley (Kendall et al. 1986, Blackwell et al. 1987a) and pea (Blackwell et al. 1987a). These photorespiratory mutants, which are not viable in air, coupled with the subsequent isolation of similar plants deficient in chloroplastic glutamine synthetase activity (Wallsgrove et al. 1987, Blackwell et al. 1987b), have demonstrated beyond doubt the importance of these two enzymes in the reassimilation of ammonia derived from photorespiration. Since both types of mutant contained normal levels of leaf glutamate dehydrogenase activity (Kendall et al. 1986, Wallsgrove et al. 1987), the involvement of glutamate dehydrogenase in the reincorporation of ammonia from photorespiration must be minimal.

In leaves of wild type plants ammonia does not accumulate, but if either glutamine synthetase or glutamate synthase activity is missing the plants accumulate ammonia when exposed to air (Kendall et al. 1986, Blackwell et al. 1987b; Fig. 3). For the continued operation of the photorespiratory cycle, it is important that ammonia released in the mitochondria is reassimilated inside the chloroplasts. In higher plants, glutamine synthetase exists in different forms, separable by ion exchange chromatography (McNally et al. 1983). In barley, the leaves contain two forms of glutamine synthetase. The cytoplasmic isozyme (GS_1) accounts for approximately 10% of the total

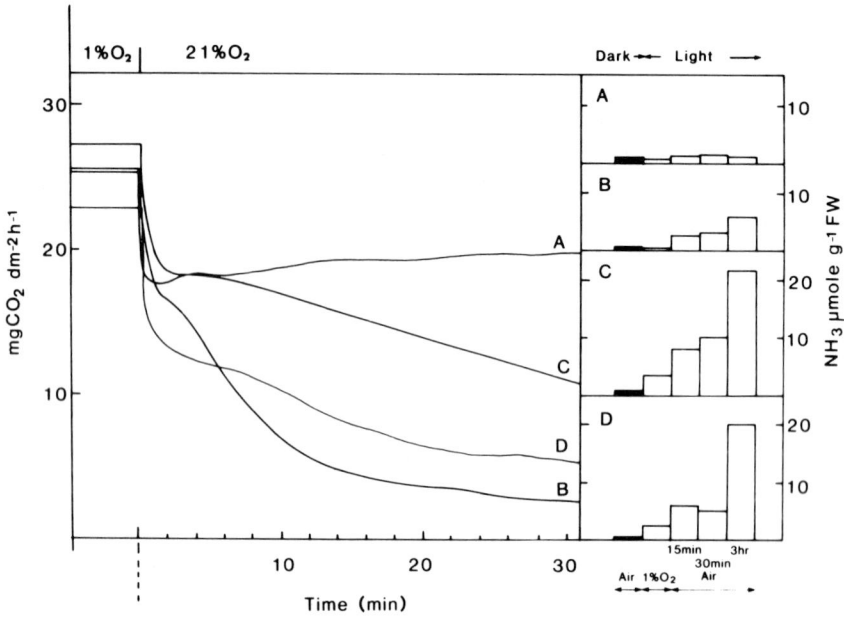

Fig. 3. Rates of CO_2 fixation and ammonia accumulation in detached leaves of (A) wild type, (B) glutamate synthase-deficient, (C) glutamine synthetase-deficient, (D) glutamine syn- and glutamate synthase-deficient. Leaf sections were illuminated ($1000\,\mu mol\,m^{-2}\,s^{-1}$) for 60 min in 1% O_2, 0.035% CO_2, and then the gas stream was switched to 21% O_2, 0.035% CO_2. At intervals leaf sections were removed and frozen in liquid N_2 prior to ammonia determination (from Kendall et al. 1987).

activity, with the major chloroplastic form (GS_2) providing the remainder (Mann et al. 1980, Blackwell et al. 1987b, Wallsgrove et al. 1987). So far, all glutamine synthetase-deficient plants are deficient in GS_2 activity. Most contain little cross-reacting material to antibody to chloroplastic glutamine synthetase (Wallsgrove et al. 1987). The sole exception, RPr 84/12, had apparently wild-type levels of cross-reacting material.

To date, 14 lines deficient in glutamine synthetase activity and 13 lines deficient in glutamate synthase activity have been isolated. Six of the RPr lines deficient in glutamine synthetase activity have been shown to be allelic as have three of the RPr glutamate synthase-deficient lines. All the lines of each mutant share common physiological and biochemical characteristics. However, the two types of mutant (glutamine synthetase or glutamate synthase) exhibit widely different characteristics. Initial results suggested that the glutamate synthase-deficient pea resembles the glutamate synthase-deficient barley (Blackwell et al. 1987a). Recently, barley mutants lacking either glutamine synthetase or glutamate synthase have been crossed and plants lacking both of these enzyme activities have been isolated and characterized from the subsequent generation (Kendall et al. 1986, Murray et al. 1988).

Plants exhibiting any of these mutations fixed CO_2 at the wild type rate under non-photorespiratory conditions (1% O_2, 0.035% CO_2). After transfer to air, the rate at which the wild type fixed CO_2 dropped within 30 sec by 30% and remained constant for at least a further two hours. Mutants lacking either glutamine synthetase or glutamate synthase, however, exhibited widely differing patterns during the initial 30 min in air, although both were unable to maintain the wild type rate for any length of time (Fig. 3). The gas exchange characteristics of plants lacking both enzyme activities more nearly resembled the parent lacking only glutamate synthase than the parent lacking glutamine synthetase.

The block in the photorespiratory pathway was emphasized in glutamate synthase-deficient mutants when these plants were fed $^{14}CO_2$ for 10 min in air. CO_2 fixation had been fully induced under non-photorespiratory conditions for 60 min. Not only was a smaller amount of ^{14}C incorporated into the mutant, but the distribution of radioactivity amongst the various carbon containing fractions was dramatically different from the wild type. The proportions of ^{14}C detected in the photorespiratory intermediates including glycollate, glycine, serine and glycerate were severely reduced. After seven days in air the glutamate synthase-deficient plants showed a large increase in incorporation into malate compared with the wild type. In addition these plants accumulated malate whereas glutamine synthetase-deficient plants did not. Thus, there seems very little correlation between ammonia accumulation and enhanced malate synthesis, as was originally suggested (Kendall et al. 1986). Somerville and Ogren (1980b) also detected an accumulation of ^{14}C in the organic acid fraction of an extract from leaves of an *Arabidopsis* mutant lacking glutamate synthase activity. However, when the *Arabidopsis* mutant was labelled with $^{14}CO_2$ at the onset of illumination, the labelling pattern of ^{14}C containing compounds was essentially the same as detected in the wild type.

A possible explanation for the inability of these mutants to maintain a wild type rate of CO_2 fixation for any appreciable time in air is that leaves of both lines accumulate ammonia (Kendall et al. 1986, Blackwell et al. 1987a, Wallsgrove et al. 1987) to levels much higher than those known to uncouple electron transport in vitro (Krogmann et al. 1959). However, several lines of evidence from these mutants suggested that the large accumulation of ammonia was not directly responsible for the decline in the rate of CO_2 fixation. First, both the extent to which the fixation rate was depressed and the level to which ammonia accumulated within leaves of the glutamine synthetase-deficient mutant were altered by changing the photon

Table 3. Content of photorespiratory amino acids in barley leaves three hours after transfer to air from 0.7% CO_2. Amino acids were measured as OPA derivatives by HPLC (Murray et al. 1987).

Amino acid	Wild type	Glutamine synthetase-deficient	Glutamate synthase-deficient	Glutamine × glutamate synthetase synthase
Glycine	648	n.d	n.d	n.d
Serine	2289	n.d	n.d	n.d
Glutamate	6354	350	492	441
Glutamine	4143	n.d	10564	1378
Alanine	4165	233	n.d	n.d

All values are in nmoles g^{-1} FW; n.d. = not detected.

fluence rate (PFR). Up to a PFR of $350 \, \mu mol \, m^{-2} s^{-1}$ there was a linear increase in both the rate of CO_2 fixation and ammonia accumulation, but at a photon fluence rate above $400 \, \mu mol \, m^{-2} s^{-1}$ the CO_2 fixation rate was progressively inhibited while the content of ammonia remained virtually unchanged (Blackwell et al. 1987b). Secondly, the glutamine synthetase-deficient lines accumulate ammonia as much as 10-fold more rapidly than the glutamate synthase-deficient plants, yet the latter had a much more rapid drop in CO_2 fixation (Kendall et al. 1986, Wallsgrove et al. 1987, Blackwell et al. 1987b). Thirdly, feeding glutamine to the cut end of the glutamine synthetase-deficient leaf overcame the effects of the mutation on the rate of CO_2 fixation, and also stimulated the level of ammonia accumulation (Blackwell et al. 1987b). Finally, concentrations of ammonia lower than those detected in these mutants in air can totally uncouple photophosphorylation in the chloroplast in vitro (Krogmann et al. 1959). However, during an extensive investigation of the chlorophyll a fluorescence characteristics of these plants Sivak et al. (1988) failed to find any evidence of uncoupling of electron transport in vivo. It is, therefore, necessary to assume that initially ammonia is sequestered harmlessly in another compartment of the cell, thus lowering its effective concentration within the chloroplast. Recently, Grumbles (1987) has also concluded that ammonia accumulation, by an *Arabidopsis* mutant lacking ferredoxin-dependent glutamate synthase activity, was not a major component in initiating the decline in CO_2 fixation in these plants.

By analogy with earlier work using glutamine synthetase inhibitors such as methionine sulphoximine which also lead to the accumulation of ammonia and inhibition of CO_2 fixation (see Lea and Ridley 1988, for review), Wallsgrove et al. (1987) speculated that a more likely explanation involved the probable lack of aminodonors in the leaves. Subsequently, a detailed analysis of the amino acid pool sizes of these mutants (Table 3) grown under identical conditions and compared with levels of amino acids in the wild type did indeed demonstrate a lack of aminodonors (Blackwell et al. 1987a). After transfer to air the build-up of glycine and serine detected in the wild type was absent from the mutants. The glutamate synthase-deficient plant accumulated higher than normal levels of glutamine while virtually the only amino acid detected in the glutamine synthetase-deficient plant was glutamate. A large accumulation of glutamine was also detected by Kendall et al. (1986) in leaves of a plant lacking glutamate synthase activity following seven days in air. The plants containing both mutations were, therefore, in a particularly difficult position. Unable to synthesize glutamine these plants could not accumulate the high levels of glutamine detected in the glutamate synthase-deficient plant. Their inability to synthesize glutamate at the wild type rate was reflected in a reduced amount of glutamate on transfer to air.

Chloroplast oxoglutarate uptake

The synthesis of glutamate by both ferredoxin dependent and NAD(P)H dependent glutamate synthase, is carried out in the chloroplast (Wallsgrove et al. 1979, Matoh and Takahashi 1981). The 2-oxoglutarate required for this synthesis must be transported across the chloroplast envelope into the stroma. Mutants of both *Arabidopsis* and barley have been isolated that are deficient in chloroplast 2-oxoglutarate uptake (*dct* mutant; Somerville and Ogren 1983, Wallsgrove et al. 1986). In both species, these *dct* mutants resemble glutamate synthase-deficient lines but with some differences. On transfer to air, glutamine levels increased while glutamate levels fell in leaves and the rate of CO_2 fixation declined, although noticeably less rapidly than for glutamate synthase mutants. As with both GS_2-deficient and glutamate synthase-deficient plants, detached leaves of *dct* mutants fixed CO_2 at the same rate as wild type in 1% O_2. Labelling studies with either $^{14}CO_2$ or ^{14}C 2-oxoglutarate, confirmed that these mutants had greatly reduced ability to synthesise glutamate, and accumulated 2-oxoglutarate (Somerville and Ogren 1983, Wallsgrove et al. 1986). The *dct* mutation segregated as a single recessive nuclear gene in both species.

The normal growth and development of these mutants in non-photorespiratory conditions demonstrated that there is a particular 2-oxoglutarate carrier that is not required except in photorespiratory nitrogen cycling. Other studies have suggested the presence of two dicarboxylate carriers in chloroplasts, with different but overlapping specificities (Dry and Wiskich 1983, Woo et al. 1987). For *Arabidopsis*, a chloroplast membrane polypeptide has been identified that is missing in the *dct* mutant (Somerville and Somerville 1985).

Glutamine uptake as a possible limitation to photorespiration
Leaves of glutamine synthetase-deficient plants still contain wild type levels of the cytoplasmic form of GS activity (Wallsgrove et al. 1987, Blackwell et al. 1987b), and thus retain a limited ability to synthesize small amounts of glutamine and hence glutamate. The limitation of photorespiration could, therefore, involve the rate of glutamine uptake into the chloroplast (Barber and Thurman 1978). The double mutant (glutamine synthetase × g-lutamate synthase) dies more rapidly in air than either parent, and again the limitation of a low rate of glutamine uptake by the chloroplast may exacerbate the problem of low rates of glutamate synthesis. Similar arguments may apply to the chloroplast oxoglutarate uptake mutants (Wallsgrove et al. 1986) which, as discussed earlier, may be considered as functionally equivalent to 'leaky' glutamate synthase-deficient plants, even though they do not accumulate ammonia in air.

There is, as yet, no clear evidence as to the source of the high levels of ammonia or glutamine detected in these mutants. When glutamate supply is limiting, all available nitrogen is probably diverted into glutamate via transamination reactions. As stressed earlier, this is reflected in the low levels of amino acids in leaves of these plants in air (Blackwell et al. 1987b). In the alga *Chlamydomonas* it was found that protein catabolism was a significant source of photorespiratory nitrogen when glutamine synthetase was inhibited using methionine sulphoximine (Cullimore and Sims 1980). No similar mechanism has yet been described in higher plants.

Conclusions

Seven different lesions in the photorespiratory pathway have so far been identified in barley, and one in pea. Two of these lesions, in catalase and glutamine synthetase, have not previously been detected in *Arabidopsis* (Somerville 1984). No mutants have been found with lesions in glycolate oxidase, glutamate:glyoxylate aminotransferase, hydroxypyruvate reductase or glycerate kinase, in any of the species. In addition, only one lesion has been identified in a membrane transport system (chloroplast 2-oxoglutarate uptake). Somerville (1984) suggested that certain mutants (e.g. glycolate oxidase-deficient) might not lead to obvious lesions that would be picked out in the screen, but merely slow down the rate of growth of the plant. From the characteristics of our current mutants we would, however, argue that any lesion in photorespiratory carbon or nitrogen cycling would produce visible symptoms and ultimately kill the affected plant. It is, therefore, important to continue the search for the mutations that are currently unavailable. Why then have some of these lesions not been found? It may be that for some or all of the enzymes involved, duplicate or multiple genes exist, and no single mutation can lead to loss of all activity. Alternatively, these enzyme activities may have other functions, not related to photorespiration, so that a mutation would be lethal under all conditions.

Modification of the screening system has been proposed to pick up other mutations, both in *Arabidopsis* (Somerville 1986) and barley (Hall et al. 1987b). In addition, there are still many barley mutants, only viable in high CO_2, that have not been characterized. They appear to have wild-type activities of all the enzymes of the pathway, and similar $^{14}CO_2$ labelling patterns to that of the wild-type plant. Identifying the lesions in these plants may provide further insights into the photorespiration pathway.

The different classes of mutation appear with very different frequencies. Ferredoxin-dependent glutamate synthase-deficient mutants appear to be relatively common in all three species tested so far. In barley, glutamine synthetase mutants are perhaps the most frequent, but such mutants have not been found in *Arabidopsis*, probably because this is one of the species containing only a single GS isoform in the leaf (McNally et al. 1983). Loss of chloroplast GS would thus produce a glutamine auxotroph, non-viable even in high CO_2. Genetic analysis of glutamate synthase and GS mutants suggests both enzymes are coded for by single genes, and it is somewhat surprising that mutations in these genes occur much more frequently than mutations affecting the conversion of glycine to serine, where a minimum of five polypeptides and at least five genes are involved.

With the exception of the catalase mutants, lesions in the photorespiration pathway lead to decreasing rates of CO_2 fixation in air. In leaves of mutant plants lacking phosphoglycollate phosphatase, serine:glyoxylate aminotransferase and the glycine:serine interconversion, there will be a direct block in the cycle which will prevent glycerate returning to the chloroplast. In mutants lacking glutamine synthetase, glutamate synthase and the 2-oxoglutarate transporter, there will be a lack of aminodonors for the conversion of glyoxylate to glycine, which again prevents the synthesis of phosphoglycerate in the chloroplast. Blocking the pathway will deplete the Calvin cycle intermediates, and hence decrease RuBP synthesis and its subsequent carboxylation (see Walker et al. 1986, for discussion of Calvin cycle regulation). This hypothesis can be tested by a careful analysis of changes in the pools of Calvin cycle intermediates on transfer of the mutant plants to air.

It has been shown that growing plants in low levels of CO_2 leads to chlorosis and premature senescence (Widholm and Ogren 1969). Similar symptoms are seen with the mutants, and in both cases low rates of CO_2

fixation are implicated. In the absence of CO_2, energy cannot be dissipated by the synthesis of ATP and NADPH and, although some electrons may be directed to oxygen by the Mehler reaction, activated oxygen species can form resulting in photoinhibition (Powles 1984). Dodge (1988) has discussed the role of such activated oxygen species in plant death. Both ammonia and hydroxylamine, an analogue of ammonia, can exacerbate photoinhibition (Krause and Laasch 1987, Cleland et al. 1987). Possibly, ammonia accumulation by glutamine synthetase and glutamate synthase-deficient plants may promote the senescence of these plants, thus explaining the ease with which such mutants are detected.

Photorespiration mutants have confirmed the view that it will only be possible to decrease photorespiration by genetic manipulation of Rubisco, or introduction of CO_2-concentrating mechanisms into C_3 plants. Blocking the flux of carbon or nitrogen through the pathway is an effective way of killing plants, not increasing their photosynthetic efficiency. A major role for these mutant lines will be as recipients for cloned genes. Possibilities exist for the shortcircuiting of the photorespiratory pathway or for the insertion of CO_2 concentrating mechanisms. Unfortunately reliable transformation techniques are presently not available for barley and are only just beginning for pea.

Acknowledgement

We are indebted to the AFRC for continued support of this work.

References

Anderson JW, Foyer CH and Walker DA (1983) Light-dependent reduction of hydrogen peroxide in intact spinach chloroplasts. Biochem Biophys Acta 724: 69–74

Barber DJ and Thurman DA (1978) Transport of glutamine into isolated pea chloroplasts. Plant Cell Environ 1: 297–303

Blackwell RD, Murray AJS and Lea PJ (1987a) The isolation and characterisation of photorespiratory mutants of barley and pea. In: Biggins J (ed.) Progress in Photosynthesis Research vol 3, pp 625–628. Dordrecht: Martinus Nijhoff

Blackwell RD, Murray AJS and Lea PJ (1987b) Inhibition of photosynthesis in barley with decreased levels of glutamine synthetase activity. J Exp Bot 38: 1799–1809

Bowes G, Ogren WL and Hageman RH (1971) Phosphoglycollate production catalysed by ribulose diphosphate carboxylase. Biochem Biophys Res Commun 45: 716–722

Boyle FA and Keys AJ (1982) Regulation of RuBP carboxylase activity associated with photoinhibition of wheat. Photosynth Res 3: 105–111

Christeller JT and Tolbert NE (1978) Phosphoglycollate phosphatase purification and properties. J Biol Chem 253: 1780–1785

Cleland RE, Critchley C and Melis A (1987) Alteration of electron flow around P_{680}: The effect of photoinhibition. In: Biggins J (ed.) Progress in Photosynthesis Research vol 4, pp 27–30. Dordrecht: Martinus Nijhoff

Cullimore JV and Sims AP (1980) An association between photorespiration and protein catabolism: studies with *Chlamydomonas*. Planta 150: 392–396

Dodge AD (1988) Herbicides that interact with photosystem 1. Free radical formation and damage mechanisms. In: Dodge AD (ed.) Herbicides and Plant Metabolism. Cambridge: Cambridge University Press (in press)

Dry IB and Wiskich JT (1983) Characterisation of dicarboxylate stimulation of ammonia, glutamine and 2-oxoglutarate-dependent O_2-evolution in isolated pea chloroplasts. Plant Physiol 72: 291–296

Foyer CH, Rowell J and Walker DA (1983) Measurement of the ascorbate content of spinach leaf protoplasts during illumination. Planta 157: 239–244

Gerbaud A and André M (1987) An evaluation of the recycling in measurements of photorespiration. Plant Physiol 83: 933–937

Grodzinski B and Butt VS (1976) Hydrogen peroxide production and the release of carbon dioxide during glycollate oxidation in leaf peroxisomes. Planta 133: 261–266

Grodzinski B and Woodrow L (1981) Serine synthesis and CO_2 release from intermediates of the glycolate pathway. In: Akoyunoglou G (ed.) Photosynthesis vol 4, pp 551–559. Philadelphia: Balaban International Science Services

Grumbles RM (1987) The effects of glutamate synthase deficiency and ammonia on *Arabidopsis* metabolism. J Plant Physiol 130: 363–371

Gutteridge S, Millard BN and Parry MAJ (1982) The reaction between active and inactive forms of wheat ribulose bisphosphate carboxylase and effectors. Eur J Biochem 126: 597–602

Hall NP, Kendall AC, Lea PJ, Turner JC and Wallsgrove RM (1987a) Characteristics of a photorespiratory mutant of barley (*Hordeum vulgare* L.) deficient in phosphoglycollate phosphatase. Photosynth Res 11: 89–96

Hall NP, Kendall AC, Turner JC, Wallsgrove RM and Keys AJ (1987b) Selection screen for novel photorespiratory mutants of barley. Plant Physiol 83: 11S

Hall NP, Kendall AC, Turner JC, Wallsgrove RM and Keys AJ (1986) A barley mutant deficient in RuBP carboxylase. Plant Physiol 80: 281S

Harpster MH, Mayfield SP and Taylor WC (1984) Effects of pigment-deficient mutants on the accumulation of photosynthetic proteins in maize. Plant Mol Biol 3: 59–71

Kendall AC, Keys AJ, Turner JC, Lea PJ and Miflin BJ (1983) The isolation and characterisation of a catalase-deficient mutant of barley (*Hordeum vulgare*, L.) Planta 159: 505–511

Kendall AC, Wallsgrove RM, Hall NP, Turner JC and Lea PJ (1986) Carbon and nitrogen metabolism in barley (*Hordeum vulgare*) mutants lacking ferredoxin-dependent glutamate synthase. Planta 168: 316–323

Kendall AC, Bright SWJ, Hall NP, Keys AJ, Lea PJ, Turner JC and Wallsgrove RM (1987) Barley photorespiration mutants. In: Biggins J (ed.) Progress in Photosynthesis Research vol 3, pp 629–632. Dordrecht: Martinus Nijhoff

Keys AJ, Bird IF, Cornelius MJ, Lea PJ, Wallsgrove RM and Miflin BJ (1978) The photorespiratory nitrogen cycle. Nature 275: 741–743

Keys AJ (1986) Rubisco, its role in photorespiration. Phil Trans Royal Soc Lond B 313: 325–336

Kisaki T and Tolbert NE (1970) Glycine as a substrate for photorespiration. Plant Cell Physiol 11: 247–258

Krause GH and Laasch H (1987) Photoinhibition of photosynthesis. Studies on mechanisms of damage and protection in chloroplasts. In: Biggins J (ed.) Progress in Photosynthesis Research vol 4, pp 19–24. Dordrecht: Martinus Nijhoff

696

Krogmann DW, Jagendorf AT and Avron M (1959) Uncouplers of spinach chloroplast photosynthetic phosphorylation. Plant Physiol 34: 272–277

Lamb JE, Riezman H, Becker WM and Leaver CM (1978) Regulation of glyoxysomal enzymes during germination of cucumber. Plant Physiol 62: 754–760

Lea PJ, Kendall AC, Keys AJ and Turner JC (1984) The isolation of a photorespiratory mutant of barley unable to convert glycine to serine. Plant Physiol 75: 881S

Lea PJ and Ridley SM (1988) Glutamine synthetase and its inhibition. In: Dodge AD (ed.) Herbicides and Plant Metabolism. Cambridge: Cambridge University Press (in press)

Mann AF, Fentem PA and Stewart GR (1980) Tissue localisation of barley (*Hordeum vulgare* L.) glutamine synthetase isoenzymes. FEBS Lett 110: 265–267

Matoh T and Takahashi E (1981) Glutamate synthase in greening pea shoots. Plant Cell Physiol 22: 727–731

McNally SF, Hirel B, Gadal P, Mann AF and Stewart GR (1983) Glutamine synthetases of higher plants. Evidence for a specific isoform content related to their possible physiological role and their compartmentation within the leaf. Plant Physiol 72: 22–25

Murray AJS, Blackwell RD, Joy KW and Lea PJ (1987) Photorespiratory N donors, amino-transferase specificity and photosynthesis in a mutant of barley deficient in serine:glyoxylate aminotransferase activity. Planta 172: 106–113

Murray AJS, Blackwell RD, Lea PJ and Joy KW (1988) Photorespiratory aminodonors, sucrose synthesis and the induction of CO_2 fixation in barley deficient in glutamine synmetase and/or glutamate synthase. J Exp Bot (in press)

Neuberger M, Bourguignon J and Douce R (1986) Isolation of a large complex from the matrix of pea leaf mitochondria involved in rapid transformation of glycine into serine. FEBS Lett 207: 18–22

Ogren WL, Salvucci ME and Portis AR Jr (1986) The regulation of Rubisco activity. Phil Trans Royal Soc Lond B 313: 337–346

Parker ML and Lea PJ (1983) Ultrastructure of the mesophyll cells of leaves of a catalase-deficient mutant of barley (*Hordeum vulgare*, L.). Planta 159: 512–517

Powles SB (1984) Photoinhibition of photosynthesis induced by visible light. Ann Rev Plant Physiol 35: 15–44

Rehfeld DW and Tolbert NE (1970) Aminotransferases in peroxisomes from spinach leaves. J Biol Chem 247: 4803–4811

Rennenberg H (1982) Glutathione metabolism and possible biological roles in higher plants. Phytochemistry 21: 277–2781

Salvucci ME, Werneke JM, Ogren WL and Portis Jr AR (1987) Purification and species distribution of Rubisco activase. Plant Physiol 84: 930–936

Scandalios JG (1979) Control of gene expression and enzyme differentiation. In: Scandalios JG (ed.) Physiological Genetics, pp 63–78. Academic Press

Scheiffer S, Tüfel W and Kindl H (1976) Plant microbody proteins, I. Purification and characterisation of catalase from leaves of *Lens culinaris*. Hoppe-Seyler. Z Physiol Chem Biol 357: 163–175

Sharkey TD, Stitt M, Heinke D, Gerhardt R, Raschke K and Heldt HW (1986) Limitation of photosynthesis by carbon metabolism. II O_2-insensitive CO_2 uptake results from limitation of triose phosphate utilisation. Plant Physiol 81: 1123–1129

Sivak MN, Lea PJ, Blackwell RD, Murray AJS, Hall NP, Kendall AC, Turner JC and Wallsgrove RM (1988) Some effects of oxygen on photosynthesis by photorespiratory mutants of barley (*Hordeum vulgare* L.) I. Response to changes in oxygen concentration. J Exp Bot (in press)

Sivak MN and Walker DA (1986) Photosynthesis in vivo can be limited by phosphate supply. New Phytol 102: 499–512

Smith IK, Kendall AC, Keys AJ, Turner JC and Lea PJ (1984) Increased levels of glutathione in a catalase deficient mutant of barley. Plant Sci Lett 37: 29–33

Smith IK, Kendall AC, Keys AJ, Turner JC and Lea PJ (1985) The regulation of the biosynthesis of glutathione in leaves of barley (*Hordeum vulgare* L.). Plant Science 41: 11–17

Somerville CR (1984) The analysis of photosynthetic carbon dioxide fixation and photorespiration by mutant selection. In: Miflin BJ (ed.) Oxford surveys of Plant Molecular and Cell Biology, pp 103–131. Oxford: Oxford University Press

Somerville CR (1986) Physiological and molecular genetics of higher plants. In: 21st Ann Rep MSU-DOE Plant Research Laboratory, pp 92–95. Michigan State University

Somerville CR and Ogren WL (1979) A phosphoglycollate phosphatase deficient mutant of *Arabidopsis*. Nature 280: 833–836

Somerville CR and Ogren WL (1980a) Photorespiration mutants of *Arabidopsis thalania* deficient in serine-glyoxylate aminotransferase activity. Proc Natl Acad Sci USA 77: 2684–2687

Somerville CR and Ogren WL (1980b) Inhibition of photosynthesis in *Arabidopsis* mutants lacking leaf glutamate synthase activity. Nature 286: 257–259

Somerville CR and Ogren WL (1981a) Photorespiration-deficient mutants of *Arabidopsis thalania* lacking mitochondrial serine transhydroxymethylase activity. Plant Physiol 67: 666–671

Somerville CR and Ogren WL (1981b) Mutants of the cruciferous plant *Arabidopsis thalania* lacking glycine decarboxylase activity. Biochem J 202: 373–380

Somerville SC and Ogren WL (1983) An *Arabidopsis thaliana* mutant defective in chloroplast dicarboxylate transport. Proc Natl Acad Sci USA 80: 1290–1294

Somerville CR, Portis AR and Ogren WL (1982) A mutant of *Arabidopsis thalania* which lacks activation of RuBP carboxylase in vivo. Plant Physiol 70: 381–387

Somerville SC and Somerville CR (1983) Effect of oxygen and carbon dioxide on photorespiratory flux determined from glycine accumulation in a mutant of *Arabidopsis*. J Exp Bot 34: 415–421

Somerville SC and Somerville CR (1985) A mutant of *Arabidopsis* deficient in chloroplast dicarboxylate transport is missing an envelope protein. Plant Sci Lett 37: 317–320

Ta TC, Joy KW and Ireland RJ (1984) Amino acid metabolism in pea leaves: Utilization of nitrogen from amide and amino groups of [^{15}N]-Asparagine. Plant Physiol 74: 822–826

Ta TC and Joy KW (1986) Metabolism of some amino acids in relation to the photorespiratory nitrogen cycle of pea leaves. Planta 169: 117–122

Tolbert NE (1979) Glycollate metabolism by higher plants and algae. In: Gibbs M, Latzko E, eds. Encyclopedia of Plant Physiology vol 6: Photosynthesis II, pp 338–353. Berlin: Springer

Turner JC, Hall NP, Kendall AC, Wallsgrove RM and Bright SWJ (1986) Genetic analysis of photorespiratory mutants of barley In: Meeting of photosynthesis AFRC, p 41

Verin-Vergeau C, Baldy P and Cavalié G (1979) La phosphoglycollate phosphatase des feuilles de haricot: deux formes isofunctionnelles. Phytochemistry 18: 1279–1282

Walker DA, Leegood RC and Sivak MN (1986) Ribulose bisphosphate carboxylase-oxygenase: its role in photosynthesis. Phil Trans Royal Soc Lond B 313: 306–324

Walker JL and Oliver DJ (1986) Glycine decarboxylase multienzyme complex: Purification and partial characterisation from pea leaf mitochondria. J Biol Chem 261: 2214–2221

Walker KA, Keys AJ and Givan CV (1984) Effect of L-methionine sulphoximine on the products of photosynthesis in wheat (*Triticum aestivum*) leaves. J Exp Bot 35: 1800–1810

Wallsgrove RM, Kendall AC, Hall NP, Turner JC and Lea PJ (1986) Carbon and nitrogen metabolism in a barley (*Hordeum vulgare*) mutant with impaired chloroplast dicarboxylate transport. Planta 168: 324–329

698

Wallsgrove RM, Keys AJ, Bird IF, Cornelius MJ, Lea PJ and Miflin BJ (1980) The location of glutamine synthetase in leaf cells and its role in the reassimilation of ammonia released in photorespiration. J Exp Bot 31: 1005–1017

Wallsgrove RM, Lea PJ and Miflin BJ (1979) Distribution of the enzymes of nitrogen assimilation within the pea leaf cell. Plant Physiol 63: 232–236

Wallsgrove RM, Lea PJ and Miflin BJ (1982) The development of NAD(P)H-dependent and ferredoxin dependent glutamate synthase in greening pea and barley leaves. Planta 154: 473–476

Wallsgrove RM, Turner JC, Hall NP, Kendall AC and Bright SWJ (1987) Barley mutants lacking chloroplast glutamine synthetase-biochemical and genetical analysis. Plant Physiol 83: 155–158

Walton NJ and Butt VS (1981) Metabolism and decarboxylation of glycollate and serine in leaf peroxisomes. Planta 153: 225–231

Widholm JM and Ogren WL (1969) Photorespiratory-induced senescence of plants under conditions of low carbon dioxide. Proc Natl Acad Sci USA 63: 668–675

Woo KC, Flügge UI and Heldt HW (1987) A two-translocator model for the transport of 2-oxoglutarate and glutamate in chloroplasts during ammonia assimilation in the light. Plant Physiol 84: 624–632

Zelitch I (1979) Photorespiration studies with whole tissue. In: Gibbs M, Latzko E (eds). Encyclopaedia of Plant Physiology vol 6, pp 353–367. Berlin: Springer-Verlag

Govindjee et al. (eds), Molecular Biology of Photosynthesis: 699–711
© 1988 Kluwer Academic Publishers

Regular paper

Gene expression during CAM induction under salt stress in *Mesembryanthemum*: cDNA library and increased levels of mRNA for phosphoenolpyruvate carboxylase and pyruvate orthosphosphate dikinase

JÜRGEN M. SCHMITT,[1] CHRISTINE MICHALOWSKI[2] & HANS J. BOHNERT[2]

[1]*Botanisches Institut der Universität Würzburg, 8700 Würzburg, FRG;* [2]*Biochemistry Department, University of Arizona, Tucson AZ 85721, USA*

Received 9 September 1987; accepted 17 December 1987

Key words: Crassulacean acid metabolism, gene expression, *Mesembryanthemum crystallinum*, mRNA levels, soil salinity

Abstract. *Mesembryanthemum crystallinum* responds to high salinity in the soil by shifting the mode of carbon assimilation from the C3 mode to Crassulacean acid metabolism (CAM). Several enzymes of carbon metabolism have increased apparent activities in the CAM mode, including phosphoenolpyruvate carboxylase (PEPcase) and pyruvate orthophosphate dikinase (PPDK). We have identified cDNA clones for PEPcase and PPDK by immunological screening of a cDNA library constructed in the protein expression vector *lambda* gt11. The clones were characterized by immunoblotting and RNA blotting techniques. RNA blotting showed that during CAM induction the steady-state level of mRNAs for both PEP case and PPDK increased.

Abbreviations: IPTG–isopropyl thiogalactoside, PEP–phosphoenolpyruvate, PEPcase–phosphoenolpyruvate carboxylase, PPDK–pyruvate orthophosphate dikinase, Xgal–5–bromo–4–chloro–3–indolyl-*beta*-D-galactopyranoside

Introduction

Crassulacean acid metabolism (CAM) can be regarded as an adaptation to arid or seasonally arid environment. CAM plants open their stomates at night to fix carbon dioxide when evaporative water loss is minimal. The first step in the CAM pathway, the formation of oxaloacetate by carboxylation of PEP, is catalysed by PEPcase. PPDK resynthesizes PEP during the day for gluconeogenesis (see Ting 1985, for a recent review).

In a few species, CAM is facultative (for review see Kluge and Ting 1978). The shift from C3 to CAM can be developmentally programmed or triggered by environmental stimuli. In *Mesembryanthemum crystallinum*, irrigation with saline solutions induces CAM (Winter and von Willert 1972). CAM induction is characterized by increased apparent activities of several enzymes involved in the metabolism of carbon compounds, including PEPcase and PPDK (Holtum and Winter 1982). For example, the activity of PEPcase increases about 40 fold. The increase in PEPcase activity is linearly correlated with enzyme mass as determined immunochemically (Höfner et al., 1987). De novo synthesis of enzyme protein and an increase in the level of translatable mRNA during the induction period has been demonstrated (Höfner et al., 1987, Ostrem et al., 1987a).

We are interested in the regulatory pathway by which this increase of mRNA is induced. It is not known, how the synthesis of enzyme protein is regulated. In order to characterize the induction process molecularly, by measuring the amounts of mRNA and the time course of the increase, gene isolation was necessary. We constructed a cDNA library in the expression vector *lambda* gt11. We describe here the isolation and identification of cDNA clones specific for PEPcase and PPDK from a *lambda* gt11 cDNA library. The clones were initially obtained from the expression vector library by screening with antibodies raised against purified enzymes and the nature of the cloned DNA was verified. The clones served as molecular probes for DNA/RNA hybridization experiments. Some of the results have previously been presented at a conference (Ostrem et al., 1987b).

Materials and methods

Plant material. Mesembryanthemum crystallinum L. was grown from seed as described by Ostrem et al. (1987a).

RNA isolation. Isolation of RNA from leaves and preparation of polyadenylated RNA was performed as described by Ostrem et al. (1987a).

cDNA cloning. Eight-week old plants were used to extract leaf RNA for cDNA cloning. Leaves from 4 individual plants, previously stressed for 8 days by irrigation with 0.4 M NaCl, were harvested at 4 time points (1 h and 6 h after the beginning of the light period, 1 h and 6 h after the beginning of the dark period) and pooled for the extraction of RNA. cDNA was synthesized from 1 μg of polyadenylated RNA according to Gubler and Hoffman (1983), using chemicals and enzymes purchased from Amersham. Methylation of internal Eco RI sites and addition of Eco RI decameric linkers was done essentially as described by Huynh et al. (1985), using

Table 1. List of the clones used.

Clone	Vector	Insert Size kbp	Fusion Protein Size kDa	Identification	Reference
McCAM5	*lambda* gt11	0.5	124	anti-PEPcase	this work
McCAM7	*lambda* gt11	0.7	132	anti-PEPcase	this work
pMcCAM7	pUC18	0.7	n.a.	McCAM7	this work
McCAM8	*lambda* gt11	1.5	159	anti-PEPcase	this work
McCAM9	*lambda* gt11	1.4	121	anti-PEPcase	this work
pMcPEP1	pUC18	1.0	n.a.	McCAM7	this work
McCAM401	*lambda* gt11	0.6	135	anti-PPDK	this work
pMcCAM401	pUC19	0.6	n.a.	McCAM401	this work
McCAM402	*lambda* gt11	0.3	126	anti-PPDK	this work
McCAM403	*lambda* gt11	1.5	165	anti-PPDK	this work
pMccab6	pUC18	0.8	n.a.	pFab31	Smith and Bedbrook (unpublished)
pMcss5	pUC18	0.8	n.a.	pPSR6	Cashmore (1983)

n.a.: not applicable.

enzymes and linkers from New England Biolabs. The cDNAs were size fractionated using a Biogel A50m column (0.2 × 12 cm; Bio Rad). The size class containing double stranded cDNA ranging in size from approximately 0.5–5 kb was ligated into *lambda* gt11 and packaged, using dephosphory-lated arms and packaging extracts purchased from Promega Biotec. Ap-proximately 400 000 plaques were obtained, of which approximately 85% contained inserts as judged by the IPTG/Xgal color assay (Huynh et al. 1985).

An additional library was constructed in a plasmid vector, starting from 1 μg of polyadenylated RNA. In this case, the cDNAs were not size-frac-tionated. They were blunt-end ligated into dephosphorylated SmaI or Hin-cII ends of pUC18 at a molar ratio of ends of approximately 3:1 (vector:in-sert). After transformation *E. coli* 71-18 or MC1061 cells (Messing et al. 1977, Casadaban and Cohen 1980) were streaked on LB plates containing 100 μg/ml ampicillin. The colonies were replicated onto nitrocellulose filters and screened by hybridization using probes specific for the chlorophyll a/b binding protein, the small subunit of ribulose bisphosphate carboxylase and PEPcase (see Table 1 for sources of probes).

Immunological identification of cDNA clones. Approximately 20,000–30,000 plaques were plated per 85 mm dish to screen for clones specific for PEPcase and PPDK as described by Huynh et al. (1985). Serum raised against PEPcase from *Pennisetum americanum* (designated "D5", Ostrem et al.

1987a) was used at a dilution of 1:1500. Serum raised against PPDK from maize leaves was kindly provided by Dr K. Aoyagi, Berkeley (Aoyagi and Bassham 1984) and used at a dilution of 1:1000. Filters were incubated in the antisera overnight. Immunoreactive plaques were identified as described under *Immunoblotting*. Washing and blocking of filters was as described by Ostrem et al. (1987a). Recombinant phage were cloned by 4 to 6 sequential replatings and used to generate recombinant lysogens in *E. coli* Y1089 as described by Huynh et al. (1985).

Expression of fusion protein. Recombinant lysogens were induced to express fusion protein by a temperature shift to 42 °C and addition of IPTG to 2 millimol/l to the culture as described by Huynh et al. (1985). Cells were harvested by low-speed centrifugation, resuspended in 1/50 volume of 50 mM Tris-Cl, pH 7.2 and freeze-thawed to achieve lysis. To the lysates were added 1 millimol/l magnesium chloride and 10 μg/ml each DNase and RNase. After incubation for 10 min. at ice-bath temperature, aliquots were analyzed by SDS gel electrophoresis and immunoblotting.

Isolation of lambda DNA. *Lambda* DNA was isolated as described by Carlock (1986).

Immunoblotting. Proteins were transferred to nitrocellulose membranes (Schleicher and Schuell, BA83) after SDS gel electrophoresis (Laemmli 1970) according to Towbin et al. (1979). Blocking and washing of the filters was as described by Ostrem et al. (1987a). Incubation in the first antiserum was done overnight. Bands tagged with antibody were detected using goat anti-rabbit IgG coupled to horseradish peroxidase (Bio-Rad) and staining with 4-chloronaphtol and hydrogen peroxide as suggested by Bio-Rad.

Northern hybridizations. RNA was fractionated on 1% agarose gels containing methylmercuric hydroxide (Chandler et al. 1979), transferred to Nytran membranes (Schleicher and Schuell) according to Bittner et al. (1980) and used for hybridization. Filters were prehybridized and hybridized in 0.25% non-fat dry milk (Johnson et al. 1984), 6 × SSC, 50% formamide at 42 °C. Stringency washes were in 0.1 × SSC, 0.1% SDS at 55–60 °C. Probes were labelled by nick-translation.

Other techniques. Other procedures were performed using standard techniques as described in Maniatis et al. (1982).

Results

Identification of cDNA clones
Lambda gt11 is a cloning vector which allows the insertion of DNA at a restriction endonuclease recognition site which is located at the 3′ end but still within the coding region of the lacZ gene specifying the enzyme *beta* galactosidase. The gene is under the control of the lac promoter which can be induced by adding the synthetic substrate IPTG. Any inserted DNA fragment will be transcribed as a fusion mRNA and, depending on the insertion of this DNA, will be expressed as a fusion protein (Huynh et al. 1985). Phage-mediated lysis of cells releases these proteins which may be blotted to nitrocellulose filters. Immunological methods may then be used to screen for phage which over-expressed a fusion protein which is recognized by the specific antibody. As a cDNA may be inserted in two orientations and as cloning in the correct reading frame is statistically achieved in one third of all clones, one out of six clones containing a particular sequence may produce a positive antibody reaction.

Polyclonal rabbit antisera often contain a significant proportion of antibodies reacting with *E. coli* proteins. The recommended dilution of antisera for screening of cDNA libraries is 1:100 (Huynh et al. 1985). We have used immunoblots of crude extracts from *E. coli* and *M. crystallinum* to titrate the antisera intended for screening. With increasing dilution of the sera, the signals obtained with bacterial protein decreased more rapidly than the specific signals of PEPcase and PPDK. As a compromise between sensitivity and specificity, dilutions of 1:1500 for anti-PEPcase serum and 1:1000 for anti-PPDK serum were chosen. These antiserum dilutions produced a faint, nonspecifically colored image of the plaque pattern but strongly colored spots of positive plaques. Sera were used up to five times, which resulted in an improved signal/background ratio as non-specific antibodies were gradually removed.

We screened approximately 60 000 plaques using anti-PEPcase serum and identified 13 immunoreactive plaques. Approximately 20 000 plaques were screened using anti-PPDK serum, yielding 25 positive plaques. Of these, four PEPcase- and three PPDK-specific plaques were purified to homogeneity by several cycles of replating (Table 1).

Clones specific for the chlorophyll a/b binding protein (pMccab6), for the small subunit of RuBPcase (pMcss5) and a clone with a larger insert specific for PEPcase (pMcPEP1) were identified by hybridization of homologous and heterologous gene probes, respectively, to colonies of the plasmid cDNA library (Table 1).

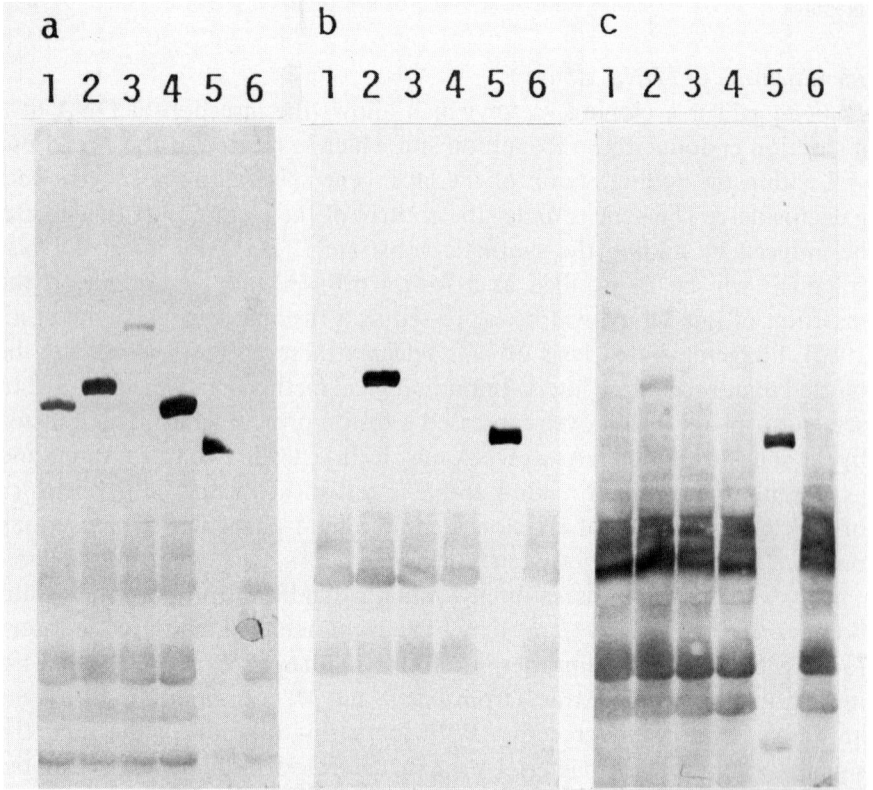

Fig. 1. Immunoblot analysis of fusion proteins expressed in *lambda* gt11 lysogens using PEPcase antisera derived from different rabbits. a: Anti-*Pennisetum americanum* PEPcase serum "D5" (Ostrem et al. 1987a), which had previously been used for the screening of the library, diluted 1:250. b: Anti-maize serum "K5" (Höfner et al. 1987), diluted 1:250. c: Anti-maize serum "P9", diluted 1:250. Samples were separated on a large gel (Schmitt 1979). Lanes 1: *lambda* McCAM5. Lanes 2: *lambda* McCAM7. Lanes 3: *lambda* McCAM8. Lanes 4: *lambda* McCAM9. Lanes 5: Crude extract from leaves performing CAM, approximately 45 μg of total protein. Lanes 6: *lambda* gt11. *E. coli* lysate containing approximately 250 μg of protein was loaded per lane. The apparent molecular weights of the fusion proteins are given in Table 1.

Verification of antibody-identified cDNA clones

The antisera used to identify clones coding for PEPcase and PPDK were monospecific as judged by immunoblot analysis of crude extracts from leaves and immunoprecipitation of radioactively labelled polypeptides (Aoyagi and Bassham 1984, Ostrem et al. 1987a). We wanted to confirm the preliminary identification of the clones by immunoblotting. Crude extracts from recombinant lysogens previously induced to express fusion proteins were separated on SDS gels, transferred to nitrocellulose and probed with the antisera used to screen the libraries. Fusion proteins migrating slower than *beta* galactosidase were recognized in all clones tested, while *beta* galactosidase expressed from the parental vector *lambda* gt11 was non-reactive (Figs. 1a, 2). With an independent serum against PPDK (Baer and Schrader 1985; kindly provided by G. Baer), the same PPDK fusion proteins were recognized (data not shown). In contrast, PEPcase antisera from two different rabbits reacted with only one out of the four PEPcase fusion proteins tested, pMcCAM7 (Fig. 1b, c). When antibodies adsorbed to the fusion

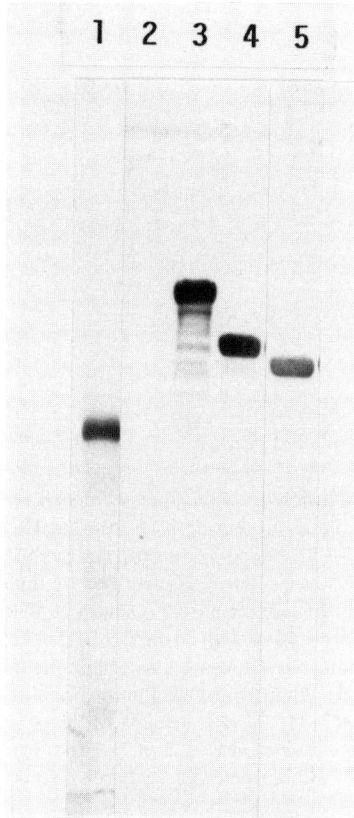

Fig. 2. Immunoblot analysis of fusion proteins expressed in *lambda* gt11 lysogens using anti-PPDK serum (1:2000 dilution). Total protein derived from IPTG-induced *E. coli* lysogens harbouring the clones indicated was loaded and separated on a small gel (Biometra, Göttingen, FRG). Lane 1: Crude extract from leaves performing CAM, approximately 7.5 μg of total protein. Lane 2: *lambda* gt11. Lane 3: *lambda* McCAM403. Lane 4: *lambda* McCAM401. Lane 5: *lambda* McCAM402. The apparent molecular weights of the fusion proteins are given in Table 1.

706

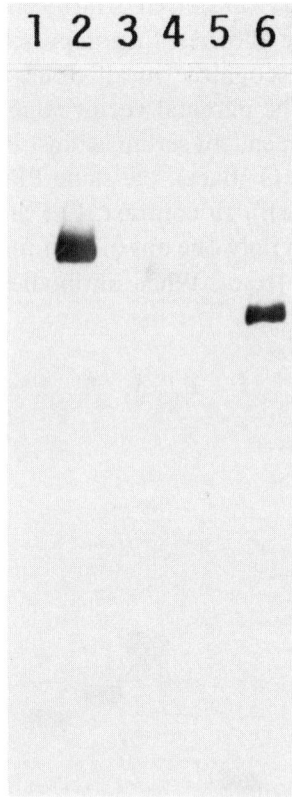

Fig. 3. Immunoblot analysis of fusion proteins expressed in *lambda* gt11 lysogens using affinity purified PEPcase antiserum. Total protein derived from an IPTG-induced *E. coli* lysogen harbouring *lambda* McCAM7 was loaded onto a preparative SDS gel. Proteins were transferred to nitrocellulose, the fusion protein was identified by immunodetection and cut out. Anti-*Pennisetum americanum* PEPcase serum D5 (Ostrem et al. 1987a) was adsorbed to the filter-bound protein. Filters were washed with 20 millimol/1 Tris, 150 millimol/1 NaCl, pH 7.2. Bound immunoglobulins were released using 100 millimol/1 glycine-HCl, pH 2.5. After neutralization with Tris base, the affinity purified immunoglobulins were diluted 1:10 and used as probes. Lane 1: *lambda* McCAM5. Lane 2: *lambda* MCcAM7. Lane 3: *lambda* McCAM8. Lane 4: *lambda* McCAM9. Lane 5: *lambda* gt11. Lane 6: Crude extract from leaves performing CAM, approximately 7.5 μg of total protein. *E. coli* lysate containing approximately 25 μg of protein was loaded per lane.

protein from *lambda* McCAM7, previously blotted to nitrocellulose, were eluted by acid treatment and used to probe an immunoblot, only authentic PEPcase and *lambda* McCAM7 fusion proteins were recognized by the affinity purified immunoglobulins (Fig. 3). This result shows that the fusion protein derived from *lambda* McCAM7 shares no epitopes with the other putative PEPcase fusion proteins, but is immunologically homologous with

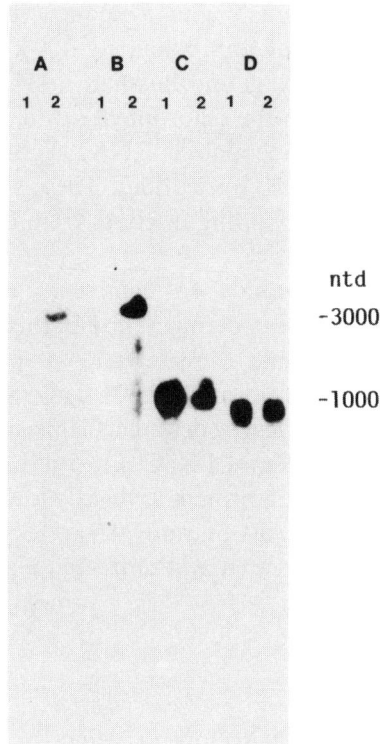

Fig. 4. Northern blot analysis of total RNA from unstressed plants (1) and plants stressed for 5 days by irrigation with 0.4 M NaCl (2). A: 40 μg of RNA per lane was probed with PPDK-specific clone pMcCAM401. B: 20 μg of RNA per lane was probed with PEPcase clone pMcPEP1. C: 10 μg of RNA per lane was probed with clone pMccab6 specific for the chlorophyll a/b binding protein. D: 5 μg of RNA per lane was probed with clone pMcss5 specific for the small subunit of RuBPcase.

authentic PEPcase, confirming the correct identification of *lambda* McCAM7. In some experiments, however, antibodies immunopurified from *lambda* McCAM7 protein showed a very weak reaction with protein from *lambda* McCAM9.

The insert of *lambda* McCAM8 whose fusion protein did not react with immunopurified antibodies from *lambda* McCAM7 fusion protein was cloned into pUC18, labelled by nick translation and used to probe Northern blots of total leaf RNA. It reacted with RNA approximately 2 kb in size, too small to accommodate the coding region of PEPcase (data not shown). In contrast, pMcPEP1 and pMcCAM401 recognized discrete RNAs of approximately 3900 and 3300 nucleotides, respectively (Fig. 4). These sizes are sufficient to accommodate the coding regions of PEPcase and PPDK in addition to some untranslated 3' and 5' sequences. Both proteins have apparent molecular masses of approximately 100 kDa (Sims and Hague 1981, Höfner et al. 1987, Aoyagi and Bassham 1984, Harpster and Taylor 1986). The clones specific for the chlorophyll a/b binding protein and the small subunit of RuBPcase recognized RNAs slightly larger and slightly smaller, respectively, than 1000 nucleotides. These sizes are similar to published values for the corresponding mRNAs from other species.

Concentrations of mRNAs during CAM induction
Fig. 4 shows a Northern blot experiment comparing RNAs isolated from untreated and salt-treated plants. Messenger RNA specific for PEPcase and PPDK is near the limit of detection in RNA derived from plants irrigated with nutrient medium, but clearly distinguishable in RNA isolated from plants irrigated with 400 millimol/l NaCl for 5 days. In contrast, mRNA for the chlorophyll a/b binding protein is slightly lower in concentration in plants induced to perform CAM and mRNA for SSU is present in about equal concentration in salt treated and untreated plants.

Discussion

The identification of the cDNA clones specific for PEPcase and PPDK, respectively, is based on the recognition of their fusion proteins by at least two independent antisera. In addition, immunoglobulins affinity purified from *lambda* McCAM7 fusion protein recognized a protein migrating at the position of authentic PEPcase in *Mesembryanthemum* (Fig. 3). Reliance on only one antiserum may lead to incorrect identification as demonstrated by the fact that the DNA of the putative PEPcase clone *lambda* McCAM8 hybridized with a discrete RNA species approximately 2000 nucleotides in length, i.e. too small to code for PEPcase. The nature of the protein for which this cDNA codes is not known. It is possible, that this protein contains a structural domain which is shared by PEPcase.

Clones with coding sequences for PEPcase have been isolated from *E. coli*, Anabaena, Anacystis and maize (Fujita et al. 1984, Katagiri et al. 1985, Kodaki et al. 1985, Harpster and Taylor 1986, Harrington et al. 1986; Hudspeth et al. 1986, Izui et al. 1986). Clones with coding sequences for PPDK have been isolated from maize (Hague et al., 1983, Hudspeth et al. 1986).

PEPcase activity is regulated differently in C4 and CAM plants. In C4 plants, the enzyme is active during the light period. In CAM plants, PEPcase is active at night, but inactive during the day. Futile carbon dioxide cycling is thus avoided (for review see Ting 1985).

The clones specifying PEPcase and PPDK described in this paper represent the first clones from CAM plants. It will be interesting to see, whether the different regulation of PEPcase activity in CAM and C4 plants can be traced back to the primary structure of the enzymes.

Holtum and Winter (1982) failed to detect PPDK in C3 tissue of *Mesembryanthemum*, but were able to measure enzyme activity in extracts from CAM leaves. Our data show (Fig. 4) that the level of PPDK mRNA is considerably lower in leaves from control plants as compared to leaves from plants irrigated with 0.4 M NaCl. Likewise, PEPcase activity is drastically increased in CAM leaves as compared to C3 leaves (Holtum and Winter 1982, Höfner et al. 1987). The increase in protein concentration is again reflected in the mRNA levels of leaves from salt-treated plants. Ostrem et al. (1987a) have shown that the level of translatable mRNA for PEPcase increases during CAM induction. Taken together, these data suggest that the induction of CAM-specific enzymes by high soil salinity may be a response at the level of transcription of genes. Two possibilities remain, however, providing alternative mechanisms by which the increase in mRNA might be explained. It might either be due to altered processing of large pre-mRNA molecules (Kamalay and Goldberg 1984) which would have gone undetected under our experimental conditions. Alternatively, decreased turnover of continuously synthesized mRNAs might be responsible for the net accumulation.

Brulfert et al. (1985) using a heterologous probe from maize (Harpster and Taylor 1986), have shown that induction of CAM brought about by a change in photoperiod in Kalanchoe is accompanied by increased levels of PEPcase mRNA.

It is unclear, whether PEPcase enzymes functioning in the C3 and CAM modes, respectively, are identical in amino acid sequence and whether they are derived from different genes. Experiments to identify cDNA clones from all members of the presumed gene family for PEPcase are in progress to clarify this point by restriction site mapping and sequence analysis.

The cDNA clones for PEPcase and PPDK have been identified as *lambda* gt11 clones by virtue of their expression in *E. coli*. The availability of clones directing the efficient synthesis of fusion proteins of these two enzymes will greatly facilitate the isolation of preparative amounts of protein which can be used for immunization to generate new antiserum. Antibodies against the *beta* galactosidase part of the chimaeric proteins can be removed from the antiserum by affinity chromatography or absorption with authentic *beta* galactosidase protein, rendering the antiserum monospecific.

710

Acknowledgements

This work was supported by grants from Deutsche Forschungsgemeinschaft (Schm 490/3), USDA (CRGP-8500395 and CRSR-2-2748) and NSF (PCM-8316188). Travel was funded by NATO (Collaborative Research Grant RG84/0230). We are grateful to Kazuko Aoyagi and Gianni Baer for providing antisera against PPDK and to Mechtild Piepenbrock for skilled technical assistance.

References

Aoyagi K and Bassham JA (1984) Pyruvate orthophosphate dikinase of C_3 seeds and leaves as compared to the enzyme of maize. Plant Physiol 75: 387–392

Baer GR and Schrader LE (1985) Stabilization of pyruvate, Pi dikinase regulatory protein in maize leaf extracts. Plant Physiol 77: 608–611

Bittner M, Kupferer P and Morris CF (1980) Electrophoretic transfer of proteins and nucleic acids from slab gels to diazobenzyloxymethyl cellulose or nitrocellulose sheets. Anal Biochem 102: 459–471

Brulfert J, Vidal J, Keryer E, Thomas M, Gadal P and Queiroz O (1985) Phytochrome control of phosphoenolpyruvate carboxylase synthesis and specific RNA level during photoperiodic induction in a CAM plant during greening in a C4 plant. Physiol Veg 23: 921–928

Carlock LR (1986) Analyzing lambda libraries. Focus 8: 6–8

Casadaban MJ and Cohen SN (1980) Analysis of gene control signals by DNA fusion and cloning in Escherichia coli. J Mol Biol 138: 179–207

Cashmore AR (1983) Nuclear genes encoding the small subunit of ribulose-1,5-bisphosphate carboxylase. In: Kosuge T, Meredith CP and Hollaender A (eds) Genetic engineering of plants – an agricultural perspective, pp 29–38, New York, Plenum

Chandler PM, Rimkus D and Davidson N (1979) Gel electrophoretic fractionation of RNAs by partial denaturation with methylmercuric hydroxide. Anal Biochem 99: 200–206

Fujita N, Miwa T, Ishijima S, Izui K and Katsuki H (1984) The primary structure of phosphoenolpyruvate carboxylase of Escherichia coli. Nucleotide sequence of the ppc gene and deduced amino acid sequence. J Biochem 95: 909–916

Gubler U and Hoffman BJ (1983) A simple and very efficient method for generating cDNA libraries. Gene 25: 263–269

Hague DR, Uhler M and Collins PD (1983) Cloning of cDNA for pyruvate, Pi dikinase from maize leaves. Nucl Acids Res 11: 4853–4865

Harpster MH and Taylor WC (1986) Maize phosphoenolpyruvate carboxylase cloning and characterization of mRNAs encoding isozymic forms. J Biol Chem 261: 6132–6136

Harrington TR, Glick BR and Lem NW (1986) Molecular cloning of the phosphoenolpyruvate carboxylase gene of Anabaena variabilis. Gene 45: 113–116

Höfner R, Vazquez-Moreno L, Winter K, Bohnert HJ and Schmitt JM (1987) Induction of Crassulacean acid metabolism in Mesembryanthemum crystallinum by high salinity: Mass increase and de novo synthesis of PEP-Carboxylase. Plant Physiol 83: 915–919

Holtum JAM and Winter K (1982) Activity of enzymes of carbon metabolism during the induction of Crassulacean acid metabolism in Mesembryanthemum crystallinum L. Planta 155: 8–16

Hudspeth RL, Glackin CA, Bonner J and Grula JW (1986) Genomic and cDNA clones for maize phosphoenolpyruvate carboxylase and pyruvate, orthophosphate dikinase: Expression of different gene-family members in leaves and roots. Proc Natl Acad Sci USA 83: 2884–2888

Huynh TV, Young RA and Davis RW (1985) Constructing and screening cDNA libraries in lambda gt10 and lambda gt11. In: Glover DM (ed) DNA cloning Volume I, a practical approach, pp 49–78, Oxford: IRL Press

Izui K, Ishijima S, Yamaguchi Y, Katagiri F, Murata T, Shigesada K, Sugiyama T and Katsuki H (1986) Cloning and sequence analysis of cDNA encoding active phosphoenolpyruvate carboxylase of the C4-pathway from maize. Nucl Acids Res 14: 1615–1628

Johnson DA, Gautsch JW, Sportsman JR and Elder JH (1984) Improved technique utilizing nonfat dry milk for analysis of proteins and nucleic acids transferred to nitrocellulose. Gene Anal Techn 1: 3–8

Kamalay JC and Goldberg RB (1984) Organ-specific nuclear RNAs in tobacco. Proc Natl Acad Sci USA 81: 2801–2805

Katagiri F, Kodaki T, Fujita N, Izui K and Katsuki H (1985) Nucleotide sequence of the phosphoenolpyruvate carboxylase gene of the cyanobacterium Anacystis nidulans. Gene 38: 265–269

Kluge M and Ting IP (1978) Crassulacean acid metabolism. Analysis of an ecological adaptation. Berlin, Springer.

Kodaki T, Katagiri F, Asano M, Izui K and Katsuki H (1985) Cloning of phosphoenolpyruvate carboxylase gene from a cyanobacterium, Anacystis nidulans, in Escherichia coli. J Biochem (Tokyo) 97: 533–539

Laemmli UK (1970) Cleavage of structural proteins during the assembly of the head of bacteriophage T4. Nature 227: 680–685

Maniatis T, Fritsch EF and Sambrook J (1982) Molecular cloning a laboratory manual. Cold Spring Harbor Laboratory

Messing J, Gronenborn B, Müller-Hill B and Hofschneider PH (1977) Filamentous coliphage M13 as a cloning vehicle: Insertion of a HindII fragment of the lac regulatory region in M13 replicative form in vitro. Proc Natl Acad Sci USA 74: 3642–3646

Ostrem JA, Olson SW, Schmitt JM and Bohnert HJ (1987a) Salt stress increases the level of translatable mRNA for phosphoenolpyruvate carboxylase in Mesembryanthemum crystallinum. Plant Physiol 84: 1270–1275

Ostrem JA, Michalowski CB, Schmitt JM, Olson SW and Bohnert HJ (1987b) Salt stress induces transcription of phospoenolpyruvate carboxylase (PEPC) in Mesembryanthemum crystallinum. J Cell Biochem 11: 41

Sims TL and Hague DR (1981) Light-stimulated increase of translatable mRNA for phosphoenolpyruvate carboxylase in leaves of maize. J Biol Chem 256: 8252–8255

Schmitt JM (1979) Purification of hordein polypeptides by column chromatography using volatile solvents. Carlsberg Res Commun 44: 431–438

Ting IP (1985) Crassulacean acid metabolism. Ann Rev Plant Physiol 36: 595–622

Towbin H, Staehelin T and Gordon J (1979) Electrophoretic transfer of proteins from polyacrylamide gels to nitrocellulose sheets: procedure and some applications. Proc Natl Acad Sci USA 76: 4350–4354

Winter K and von Willert DJ (1972) NaCl induzierter Crassulaceen-Säurestoffwechsel bei Mesembryanthemum crystallinum. Z Pflanzenphys 67: 166–170

Govindjee et al. (eds), Molecular Biology of Photosynthesis: 713–734
© 1988 Kluwer Academic Publishers

Minireview

Transport of proteins into chloroplasts

THOMAS H. LUBBEN[2], STEVEN M. THEG & KENNETH
KEEGSTRA[1]
Botany Department, University of Wisconsin, Madison, Wisconsin 53706, USA;
[1] *corresponding author*
[2] *present address: Central Research and Development, E402, E.I., DuPont de Nemours &
Co., Wilmington, DE 19898*

Received 13 October 1987; accepted 16 December 1987

Abstract. The import of cytoplasmically synthesized proteins into chloroplasts involves an interaction between at least two components; the precursor protein, and the import apparatus in the chloroplast envelope membrane. This review summarizes the information available about each of these components. Precursor proteins consist of an amino terminal transit peptide attached to a passenger protein. Transit peptides from various precursors are diverse with respect to length and amino acid sequence; analysis of their sequences has not revealed insight into their mode of action. A variety of foreign passenger proteins can be imported into chloroplasts when a transit peptide is present at the amino terminus. However, foreign passenger proteins are not imported as efficiently as natural passenger proteins, and some chimeric precursor proteins are not imported into chloroplasts at all. Therefore, the passenger protein, as well as the transit peptide, influences the import process. Import begins by binding of the precursor to the chloroplast surface. It has been suggested that this binding is mediated by a receptor, but evidence to support this hypothesis remains incomplete and a receptor protein has not yet been characterized. Protein translocation requires energy derived from ATP hydrolysis, although there are conflicting reports as to where hydrolysis occurs and it is unclear how this energy is utilized. The mechanism(s) whereby proteins are translocated across either the two envelope membranes or the thylakoid membrane is not known.

Abbreviations: EPSP – 5-enolpyruvyulshikimate-3-phosphate, LHCP – Chlorophyll a/b binding protein of the light-harvesting complex, NPT-II – Neomycin phosphotransferase II, PC – Plastocyanin, Pr – Precursor, Rubisco – Ribulose-1,5,-bisphosphate carboxylase /oxygenase, SS – Small subunit of Rubisco

Introduction and overview

Most chloroplast proteins are encoded in the nucleus and synthesized in the cytoplasm. The process responsible for delivery of these cytoplasmically synthesized proteins into chloroplasts has received considerable attention in the past decade. Although several proteins have been examined with respect to protein targeting, the one most extensively studied is the small subunit (SS) of ribulose-1,5-bisphosphate carboxylase/oxygenase (Rubisco). It was the first chloroplast protein for which a larger precursor protein (prSS) was identified (Dobberstein et al. 1977; Cashmore et al. 1978), and for which posttranslational import into chloroplasts was demonstrated (Chua and Schmidt 1978; Highfield and Ellis 1978). The extra amino acids on prSS and other precursors are present at the amino terminus (see Schmidt and Miskind 1986, for review). Chua and Schmidt (1979) proposed that the amino terminal extensions of chloroplast precursors be called transit peptides to distinguish them from signal peptides present on the proteins transported into the endoplasmic reticulum. Work in the past few years has demonstrated that a transit peptide is necessary for transport of precursors into chloroplasts (Mishkind et al. 1985, Anderson and Smith 1986). Moreover, it has been demonstrated in several cases that a transit peptide is sufficient to cause the transport of foreign passenger proteins into chloroplasts (Cashmore et al. 1985, Schreier and Schell 1986, della-Cioppa et al. 1987b).

A key development that allowed significant advances in understanding of protein transport into chloroplasts was the demonstration that import could be reconstituted in vitro (Highfield and Ellis 1978, Chua and Schmidt 1978). In these early studies a mixture of radioactive precursor proteins, obtained by translation of polyadenylated mRNA, was incubated with isolated intact chloroplasts. The precursor to small subunit was import into chloroplasts, processed to its proper size and assembled with large subunit to give the holoenzyme (Chua and Schmidt 1978, Highfield and Ellis 1978). This in vitro assay has subsequently been used to examine the import of numerous precursor porteins (Grossman et al. 1982) including the light-harvesting chlorophyll a/b protein (LHCP) (Schmidt et al. 1981), plastocyanin (PC) (Smeekens et al. 1986), Fd (Smeekens et al 1986), and 5-enolpyruvylshikimate-3-phosphate (EPSP) synthase (della-Cioppa et al 1986). The left panel of figure 1 shows an example of the results of such an assay, using proteins made by in vitro translation of polyadenylated mRNA. The precursor and imported forms of SS and LHCP are clearly visible and can be easily identified. Other precursors and improted molecules are present, but are not readily identifiable in such a mixture of labelled proteins.

The utility of the in vitro import assay was further enhanced by the development of in vitro transcription systems (Krieg and Melton 1984). These systems allow the synthesis of precursor proteins from cloned genes via sequential in vitro transcription and translation reactions (see Anderson and Smith 1986, for example). One advantage of this technique is that it allows experiments to be performed with radiochemically pure precursors, rather than with the mixture of precursors that results from the translation of total polyadenylated mRNA (Fig. 1, lane 1). This not only eliminates the possibility of competition between precursors, but also simplifies data analysis by eliminating the need for immunoprecipitation of imported proteins. An additional and probably more significant advantage is that it allows the

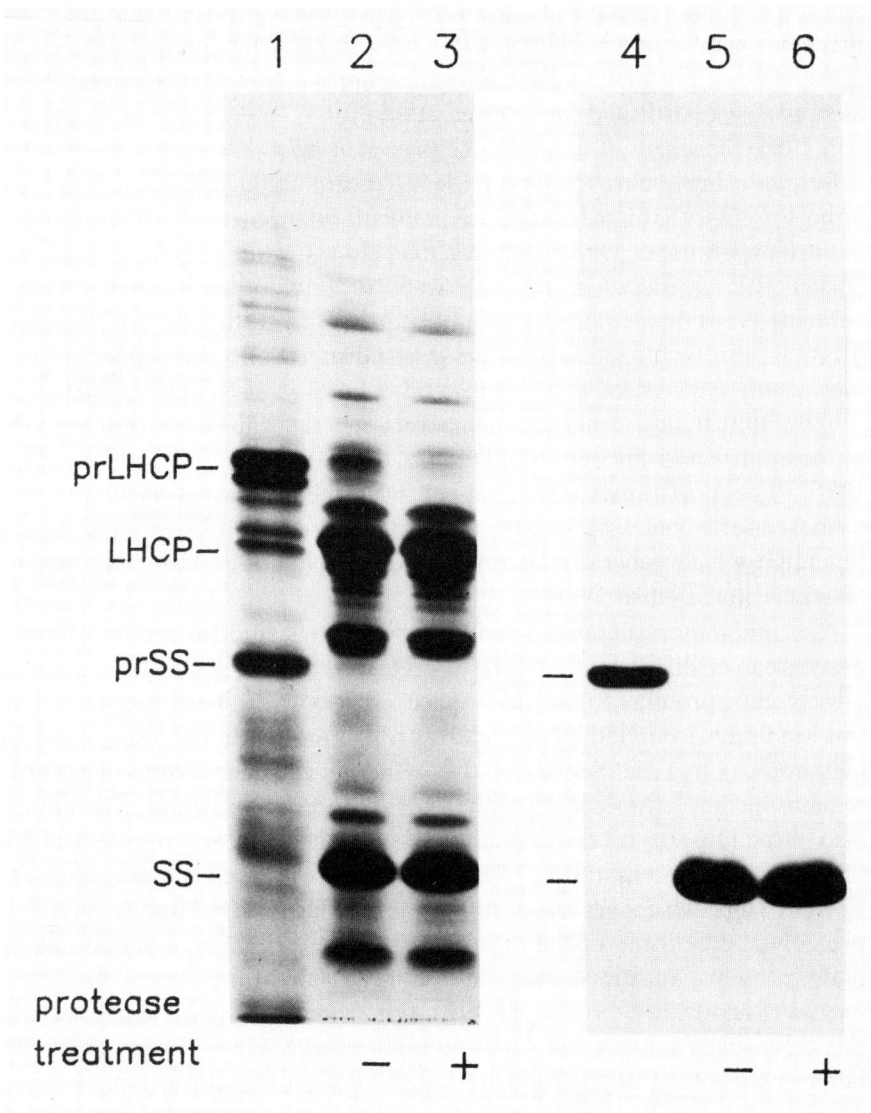

Fig. 1. Import of proteins into chloroplasts. Pea chloroplasts were incubated in vivo with radiolabelled precursor proteins made by in vitro translation of either total polyadenylated mRNA, isolated from peas (left panel), or mRNA coding for prSS, produced by in vitro transcription with SP6 RNA polymerase (right panel). After incubation, chloroplasts were either treated with protease to remove external proteins (lanes 3 and 6) or not protease-treated (lanes 2 and 5). Chloroplast were re-isolated and analyzed by sodium dodecyl sulfate gel electrophoresis and autofluorography (Cline et al. 1985). The fluorogram is shown. Lanes 1 and 4 show aliquots of the in vitro translations.

relatively rapid production of altered precursor proteins via modification of precursor genes by recombinant DNA techniques. The abilitiy of the modified precursor proteins to be imported and properly localized can be examined using reconstituted import assays (see for example Lubben and Keegstra 1986, Smeekens et al. 1986; Reiss et al. 1987).

Studies using in vitro assays have led to several important insights into the import process that would have been difficult or impossible to obained from in vitro experiments. One example is the demonstration that in vitro import occurs after synthesis of the precursor protein is complete (Chua and Schmidt 1978). Indeed, it still has not been confirmed that in vivo import of precursors into chloroplasts occurs posttranslationally. Transport of cytoplasmically synthesized proteins into mitochondria also occurs posttranslationally and many similarities exist between the import of proteins into mitochondria and chloroplasts (Douglas et al. 1986, Schmidt and Mishkind 1986). The posttranslational import of proteins into chloroplasts and mitochrondria contrasts with the transport of proteins into the endoplasmic reticulum which generally occurs as the polypeptide is being synthesized (Wickner and Lodish 1985).

A second important insight derived from in vitro studies is that the import process can be divided into several discrete steps. The first step is the binding of precursor proteins to the chloroplast surface. It has been suggested that this binding is mediated by a receptor protein present in the outer envelope membrane (Chua and Schmidt 1979), although evidence for this suggestion remains incomplete (see discussion in the second section below). The second step in the import process is translocation of the precursor protein accross the two envelop membranes. This step requires energy in the form of ATP. A third step, which occurs during or immediately after translocation, is proteolytic processing of the precursor to remove the transit peptide. With many proteins, additional steps are required for them to reach their final location and achieve their active form. For example, SS must assemble with the chloroplast-synthesized large subunit of rubisco to form the active holoenzyme. Other imported proteins must have prosthetic groups or metal ions added, while still others must be inserted into or across another membrane before reaching their final location. In this brief review we will focus primarily on the binding and translocation events. Although processing, intraorganellar targeting, and assembly are extremely interesting (and some of these steps are being studied in our laboratory), they are covered by other authors in this volume.

Each of the first two steps in the import process involves the interaction of two separate components; the precursor protein and the import apparatus present in the chloroplast envelope membranes. Because of the availability of cloned genes for precursor proteins, some progress has been made in defining the structural features that are important for their import. Some of the major observations and conclusions regarding the essential features of precursor structure are presented in the first section. The second section describes what is known about the chloroplast components which make up the transport machinery, and the mechanisms responsible for translocation of proteins across the two envelope membranes.

Structure of precursor proteins

Transit peptides

The term 'transit peptide' was originally proposed for the amino terminal extension present on chloroplast precursor proteins because it 'is likely to be involved in the post-translational transport mechanism' (Chua and Schmidt 1979). Work from several laboratories has demonstrated that this prediction is correct. The necessity of a transit peptide for precursor import has been demonstrated in two ways. First, when part or all of the transit peptide was removed from prSS by the action of a partially purified stromal protease, the resulting protein could not be imported into chloroplasts (Mishkind et al. 1985). Secondly, mature SS, produced by transcription and translation of a cloned SS gene which lacked a transit peptide, was not imported into chroroplasts (Anderson and Smith 1986).

Efforts to define critical regions of transit peptides have employed several different strategies. One has been to compare their amino acid sequences in an effort to detect conserved regions. Analyses based on a limited number of precursors led to proposals of a 'common framework' for transit peptides (Mishkind et al. 1985, Karlin-Neumann and Tobin 1986). These hypotheses identify three regions of similarity; one near the amino terminus, a second in the middle of the transit peptide near a putative intermediate processing site, and a third at the carboxyl terminus (Fig. 2). The amino terminal and central regions have been proposed to be involved in the binding or translocation steps (Mishkind et al. 1985, Karlin-Neumann and Tobin 1986), whereas the region near the carboxyl terminus has been proposed to be required for proteolytic removal of the transit peptide (Mishkind et al. 1985, Karlin-Neumann and Tobin 1986). The central region may also be involved in an intermediate processing step (Mishkind et al. 1985, Robinson and Ellis 1985). To be valid as a general working model such as hypothesis must accommodate all transit peptides. As the sequences of additional transit peptides have been determine, some have been found that do not fit the proposed framework (Fig. 2). For example, the transit peptide for the

Primary Sequences of Chloroplast Transit Peptides

Stromal Proteins

Soybean prSS   MASSMISSPA VTTVNRAGAG MVAPFTGLKS MAGFPTRKTN NDITSIASNG GRVQC

Wheat prSS   MAPAVMASSA TTVAPFQGLK STAGLPISCR SGSTGLSSVS NGGRIRC

S. *pratensis* prFd   MASTLSTLSV SASLLPKQQP MVASSLPTNM GQALFGLKAG SRGRVTAM

Spinach prACP   MASLSATTTV RVQPSSSSLH KLSQGNGRCS SIVCLDWGKS SFPTLRTSRR RSFISA

Lumenal Proteins

Spinach 16 kDa   MAQAMASMAG LRGASQAVLE GSLQISGSNR LSGPTTSRVA VPKMGLNIRA QQVSAEAETS RRAMLGFVAA GLASGSFVKA VLA

Spinach 33 kDa   MAASLQASTT FLQPTKVASR NTLQCRSTQN VCKAFGVESA SSGGRLSLSL QSDLKELANK CVDATKLAGL ALATSALIAS GANA

Spinach 23 kDa   MASTACFLHH HAAISSPAAG RGSAAQRYQA VSIKPNQIVC KAQKQDDNEA NVLNSGVSRR LALTVLIGAA AVGSKVSPAD A

S. *pratensis* prPC   MATVTSSAAV AIPSFAGLKA SSTTRAATVK VAVATPRMSI KASLKDVGVV VAATAAAGIL AGNAMA

Thylakoid Membrane

pea prLHCP   MAASSSSSMA LSS PTLAGK QLKLNPSSQE LGAARFT

Shared Framework

                  MA*SM*SS--------P*F*G*K*----P-----------S----GGRV

Fig. 2. Amino acid sequences of transit peptides. The amino acid sequences of vaarious transit peptides of nuclear-encoded proteins from the chloroplast stroma, thylakoid membrane and thylakoid lumen are shown (soybean prSS, Berry-Lowe et al. (1982); wheat prSS, Broglie et al. (1983); S. pratensis prFd, Smeekens et al. (1985b); spinach acyl carrier protein precursor (prACP), Scherer and Knauf (1987); spinach 16 kDa) protein and 23 kDa protein, Jansen et al. (1987); spinach 33 kDa protein, Tyagi et al. (1987); S. pratensis prPC, Smeekens et al. (1985a); pea prLHCP, Cashmore (1984). Regions which were included in the amino acid framework of Mishkind et al. (1985) and Karlin-Neumann and Tobin 91986) are underlined; a consensus sequence of this proposed framework is shown at the bottom of the figure.

precursor to acyl carrier protein possesses only the amino terminal region (Scherer and Knauf 1987). The transit peptides for the 33 kDa (Tyagi et al. 1987) and the 23 kDa and 16 kDa polypeptides (Jansen et al. 1987) of the water splitting complex do not contain similarities in the central and carboxyl-terminal portions. However, it should be borne in mind that precursors that do not fit the proposed common framework may be transported via a different import pathway involving different receptors and import apparatus. As described in more detail below, Reiss et al. (1987) deleted both the central conserved domain and the amino-terminal end of the prSS transit peptide without severely impairing import. This provided direct experimental evidence that the current hypotheses are inadequate.

The disparity in size and primary structure which exists among transit peptides may indicate that it is not the amino acid sequence, but rather secondary or tertiary structural features which are important in their function. In the case of mitochondrial precursors, where more transit sequences are known, it has also been pointed out that common primary structural features are not obvious (von Heijne 1986, Roise et al. 1986). For these precursors, it has been suggested that the critical secondary structural feature is an amphipathic helix (von Heijne 1986, Roise et al. 1986). However, folding predictions indicate that transit peptides of chloroplast precursors do not readily fold into amphipathic helices. Thus, it seems unlikely that this hypothesis will apply to chloroplast precursors.

An experimental approach to defining critical regions of transit peptides is to modify, substitute or delete various amino acids of the transit peptide, and then assay the modified precursor for import activity. In one such study prSS was translated from isolated mRNA using amino acid analogues which substituted for either arg, pro, lys, thr or leu (Robinson and Ellis 1984). When arg, pro or leu were replaced, both import and processing of prSS were greatly reduced. Replacement of lys reduced import somewhat. Replacement of lys greatly reduced the import of a different precursor, prLHCP. One problem with this type of study is that amino acid substitutions occur throughout the precursor, and are not confined to the transit peptide. Therefore, it is possible that the observed changes were due in part to substitutions in the mature peptide.

Deletions have been made from various regions of the transit peptide of prSS, and their influence upon import has been examined. Reiss et al. (1987) prepared a series of mutant precursors which had sections deleted from the three regions which have been proposed to be important for import (Mishkind et al. 1985, Karlin-Neumann and Tobin 1986). All mutants showed reduced import compared with prSS, but deletions in different regions affected import quite differently. Deletion of five residues from the amino terminus resulted in little inhibition of import, although deletion of the first 24 amino acids nearly abolished import. The first five amino acids comprise a major portion of the amino-terminal region of the proposed common framework for transit peptides (Mishkind et al. 1985, Karlin-Neumann and Tobin 1986). A ten amino acid deletion from the central region also only slightly affected import. These observations provide evidence that at least a portion of the amino terminal region, and the entire central region, are not essential for import. Deletions made from the carboxyl-terminal region severely inhibited import and resulted in abnormal processing.

Deletions have also been made in the transit peptide of prFd. In these studies, different amounts of the carboxyl-terminal region of the transit peptide, or the amino-terminal region of the mature peptide, were deleted, and their effects on import, processing and binding were measured (Smeekens 1986). A deletion of two amino acids from the transit peptide had no deleterious effect on import. However, larger deletions in this region from either the transit peptide or mature peptide reduced or abolished import and binding. Thus, with prFd as well as prSS, the region near the transit peptide/mature peptide junction appears to be essential for binding and/or import.

Although the investigations with altered or mutated transit peptides have provided some valuable insights into transit peptide function, these types of studies have several inherent technical and conceptual shortcomings. One problem is that the substitutions or deletions in the transit peptide may change the conformation of the entire precursor in such a way that the transit peptide is sterically hindered from properly interacting with the putative receptor or other component of the import apparatus. The solubility or stability of the precursor may also be changed. A second problem is that extent of import, rather than initial rates, are often measured in such studies. Often the import of these altered proteins is assayed at a single time point, typically 30 min or 1 hour. Time course studies have shown that the rate of import may cease to be linear after 5—10 minutes with faster-importing molecules such as prSS (Lubben and Keegstra 1986). Other molecules, particularly slower-importing chimeric precursors, may be taken up at initial rates for up to 30 min or possibly longer (Lubben, unpublished data). A third problem is that the in vitro import assay shares the same advantages and disadvantages common to all in vitro biochemical systems, namely, that it does not exactly duplicate in vivo conditions. For example, during our studies with prSS, we have noticed that import ceases before all the precursors have been imported into chloroplasts. We have been observed that for different preparations of precursor, the percentage which can be imported is different (Lubben and Keegstra 1986). The factors that prevent complete import in the in vitro system are presently unknown. One possibility is that the unimported precursors are denatured with respect to import, and that for different precursor preparations, different amounts are in a conformation which is import-incompetent. Therefore, when reduced rates of import are observed, these rates may reflect the reduced effective concentration of *active* precursor rather the activity of that (native) precursor in the import process. Thus, it is important to determine both the rate and extent of import of both control and modified precursors.

A more serious problem with studies using mutated transit peptides is a conceptual one that is best depicted by a comparison of import studies to enzyme studies. It is not likely that much useful information would come from studies in which different portions of an enzyme were deleted, and the enzyme activity assayed at a single time point. Certainly, if changes of the magnitude of those used to study transit peptide function were made to the amino acid sequence of an enzyme, many incorrect conclusions would be drawn concerning which regions were involved in catalytic activity. In order for useful information on vital regions to result from substitution or deletion studies of a protein, these changes should be made with a knowledge, or at least a hypothesis, of how the conformation of the protein and its active site or region(s) will be affected. Unfortunately, such knowledge is not currently available for precursor proteins.

Passenger proteins

The term passenger protein is used here to refer to any protein or peptide that is carboxyl-terminal to a transit peptide and can be imported into chloroplasts. We use it to include natural passenger proteins such as SS, as well as foreign proteins that are not normally imported into chloroplasts in nature.

It is now well established that, at least in some cases, the addition of a transit peptide is sufficient to mediate import of a foreign passenger protein into chloroplasts. Evidence to support this somewhat surprising conclusion comes from both in vitro and in vivo and studies using several different transit peptides and several different foreign passenger proteins (Table 1). Thus, it appears that transit peptides from a variety of nuclear-encoded chloroplast proteins can be used to target foreign proteins to chloroplasts. This has important implications for those wishing to alter the metabolism of chloroplasts by genetic engineering. The utility of this strategy has been extended by the observation that transit peptides can also be used to import some proteins normally synthesized within chloroplasts. The beta subunit of chloroplast coupling factor ATPase and the large subunit of rubisco from *Synechococcus* (which shows strong sequence homology with higher plant large subunit), have been imported into chloroplasts (Lubben et al. 1988). This approach may offer an alternative to transformation of the chloroplast genome when changes in chloroplast-encoded properties are desired.

We have also found that certain chimeric precursors are not imported into chloroplasts, even though they contain an amino terminal transit peptide. One of the more interesting examples from our laboratory is a series of chimeric proteins containing different amount of prSS fused to the coat protein of brome mosaic virus (Lubben, Gatenby, Ahlquist and Keegstra, manuscript in preparation). The longest chimeric precursor protein, which contained the coat protein attached to nearly the entire prSS molecule (172 of 178 amino acids), was not imported into chloroplasts. Surprisingly, two other chimeras, in which the coat protein was attached to either a transit peptide alone or to a transit peptide plus 13 amino acids of mature SS, were imported, although poorly. Because both proteins (SS and coat protein)

Table 1. Chimeric precursors imported into chloroplasts.

Transit peptide from	Passenger protein	Reference
prSS	NPT-II	van de Broek et al. (1985) Schreier et al. (1985) Kuntz et al. (1986) Wasmann et al. (1986)
	soybean heatshock protein	Lubben and Keegastra (1986)
	Synechococcus Rubisco large subunit	Lubben et al. (1988)
	Maize coupling factor beta subunit	Lubben et al. (1988)
	Brome mosaic virus coat protein	Lubben, Gatenby, Ahlquist and Keegstra (manuscript in preparation)
prFd	plastocyanin	Smeekens et al. (1986)
	mitochondrial superoxide dismutase	Smeekens et al. (1987)
prPC	ferredoxin	Smeekens et al. (1986)
	mitochondrial superoxide dismutase	Smeekens et al. (1987)
prEPSP synthase	bacterial EPSP synthase	della-Cioppa et al. (1987a)

which make up the non-imported chimera can be imported in other constructions, they cannot merely contain amino acid sequences which prevent import. From these observations we conclude that the determinants of importability are not simple functions of the primary structure that are read in a linear fashion. Rather, the three dimensional configuration of the entire precursor molecule is probably important in determining importability. A mechanistic explanation for these observations could be that the precursors fold in such a way that the transit peptide is sterically hindered from properly interacting with the putative receptor or other components of the important apparatus. It is also possible that interactions between distal segments of the protein may inhibit unfolding, thereby limiting import. Regardless of the mechanisms responsible, a consequence of these observations is that the importability of a precursor cannot be predicted by its amino acid sequence.

Other studies have investigated the influence of a 23 amino acid portion of the mature peptide, in addition to the transit peptide, to direct the import of NPT-II in vitro (Wasmann et al. 1986) and in vivo (Kuntz et al. 1986). Each group examined two chimeric proteins which differed from each other only in that one contained a 23 amino acid portion of the SS mature peptide and the other contained none of the mature peptide. The protein containing the 23 amino acid mature peptide portion was imported more efficiently in vivo (Wasmann et al. 1986), whereas the other was imported more efficiently in vivo (Kuntz et al. 1986). It was suggested that precursor degradation in the cytoplasm was responsible for the differences between the in vitro and in in vivo results (Kuntz et al. 1986). Thus, in addition to effects on protein import, the passenger protein may also affect the precursor molecule with respect to other intracellular processes, such as proteolytic degradation.

These results illustrate that in vitro and vivo assays of import provide different and complementary information regarding protein targetting. In vivo assays with transgenic plants provide an integrative picture of protein targeting taking into account several factors. These include protein synthesis rates, resistance of precursors to proteolysis in the cytoplasm before import, efficiency of precursor import, and stability of the foreign protein within chloroplasts. On the other hand, in vitro assays provide information regarding the import process directly.

Studies with prSS and prFd have shown that natural passenger proteins influence the import process. As mentioned above, Smeekens (1986) found that import was reduced by a deletion of four or more amino acids from the amino terminus of the mature peptide of prFd. In our laboratory, we have found that a 47 amino acid deletion from the carboxyl-terminus of prSS caused a slight reduction in the rate of import (unpublished data).

Localization of proteins within chloroplasts

So far, we have discussed proteins which are imported from the cytoplasm to the stroma. However, cytoplasmically synthesized proteins are transported into other aqueous compartments, such as the thylakoid lumen and the envelope intermembrane space, as well as into the three chloroplast membranes. The determinants responsible for proper localization of these proteins are poorly understood. Smeekens et al. (1986) have demonstrated that transport into the thylakoid lumen occurs in two steps. They suggested that the transit peptide of prPC contains two domains; the function of one is to transport the precursor across the envelope membranes, and the function of the second is to transport an intermediate across the thylakoid membrane. Cline (1986) has demonstrated that insertion of LHCP into thylakoid membranes can be studied separately from import, and that this step probably occurs following import of the precursor into the stroma. Although more work needs to be done with these and other proteins, it appears that localization of proteins within chloroplasts can be separated from the transport of precursors across the envelope membranes.

The import process

Binding of precursors to chloroplasts

As mentioned in the introduction, the first step in protein import is the binding of precursors to the surface of chloroplasts. Cline et al. (1985) demonstrated that precursor binding can be assayed separately from translocation across the envelope membranes. They showed that binding of either prSS or prLHCP without transport could be achieved by incubating the chloroplasts with the precursors in the absence of ATP (see below). Chloroplasts treated in this manner could be washed to remove unbound precursor, and then assayed for transport of the bound species upon addition of exogenous ATP.

The initial characterization of the binding of prSS to chloroplasts was extended by Friedman and Keegstra (manuscript in preparation). They found that binding was both saturable and specific. Under the conditions used for the assays, binding approached saturation at a precursor concentration of 5–10 nM, with between one and three thousand precursor molecules bound to each chloroplast. The binding was specific in that prSS was bound, but SS was not. These results are consistent with the idea that prSS binding is mediated by specific receptor proteins located in the outer envelope membrane.

Membrane receptors for protein import

A number of attempts have been made to identify the receptor(s) involved in protein import. Pfisterer et al. (1982) examined the binding of a mixture of precursor proteins to isolated chloroplast envelope membranes, and more recently, Bitsch and Kloppstech (1986) provided that precursors could be bound to solubilized envelope proteins reconstituted into lipid vesicles. These latter authors also attempted to cross-link prSS to proteins on the surface of the chloroplast envelopes (Kloppstech and Bitsch 1986), however no clear receptor protein emerged from that study. Cornwell and Keegstra (1987) also used cross-linkers in an effort to identify the receptor for prSS. Utilizing a heterobifunctional, photoactivatable cross-linking reagent, they found a specific interaction of prSS with a 66 kD protein on the surface of chloroplasts. This protein was postulated to be the receptor, or part of the receptor complex, that mediates the binding of prSS to the chloroplast surface. A great deal of additional work is needed to confirm this hypothesis.

Energy requirements

Energy requirements for import into the stroma. The energy requirement for the import of nuclear-encoded proteins into chloroplasts was first addressed by Grossman et al. (1980). They demonstrated that chloroplast protein import was stimulated in the light over that observed in the dark, and that uncouplers were effective inhibitors of that stimulation. DCMU, which inhibits linear, but not cyclic, photophosphorylation, did not eliminate protein import. Exogenously added ATP was able to support protein import into dark-incubated chloroplasts, either in the absence or presence of uncouplers. These experiments led the authors to propose that protein import into isolated chloroplasts requires ATP as the sole energy source. As mentioned above, this latter approach, wherein exogenously added ATP was used to over come the uncoupler-induced inhibition of import, was later employed by Cline et al. (1985) to separate the binding and translocation steps.

Until this past year, studies on the energy requirements for the protein import into organelles have focused primarily upon mitochondria. Although initial studies indicated that ATP was required for protein import into mitochondria (Nelson and Schatz 1979), subsequent studies indicated that instead, the electrical component of the proton electrochemical potential gradient was required (Schleyer et al. 1982, Gasser et al. 1982). The latter authors showed that the ability of mitochondria to import proteins was correlated with the presence of a transmembrane electric field, and not with the high intramitochondrial levels of ATP. Pfanner and Neupert (1985) later demonstrated that protein import could be driven by a diffusion potential in the absence of either electron transport or added ATP. More recently, the energy requirements of protein import into mitochondria were reinvestigated, and a requirement for ATP was observed (Pfanner and Neupert 1986, Eilers et al. 1987). Thus the emerging consensu is that both ATP and a transmembrane electric field are necessary to drive the import of proteins into mitochondria. A requirement for ATP hydrolysis may be common to all protein translocation systems; such a requirement has also been reported for the transport of proteins across the endoplasmic reticulum membrane (Rothblatt and Meyer 1986, Hansen et al. 1986, Waters and Blobel 1986, Schlenstedt and Zimmermann 1987), movement of secretory proteins through the Golgi apparatus (Balch and Keller 1986, Balch et al. 1986), and for protein secretion from E. coli (Chen and Tai 1986, Yamane *et al.* 1987). In the last case a protonmotive force also appears to be involved.

Investigations into the energy requirements for protein import into chloroplasts have resulted in four reports on the subject within the last 12 months. In each instance prSS was the precursor investigated. The major difference between the studies concerned the types of plastids chosen for the import experiments: Flügge and Hinz (1986) used chloroplasts isolated from spinach, Schindler et al. (1987) and Pain and Blobel (1987) investigated chloroplasts isolated from peas, and Boyle et al. (1986) used leucoplasts preparted from castor bean endosperm. Since they all agreed on many of their respective conclusions, it will be most convenient to refer to the results of certain experiments collectively. However, the question of the location of the site of ATP utilization was answered differently by the different research groups, and in this instance, the choice of experimental material may be responsible for some of the reported discrepancies.

The first question addressed by all four groups was whether hydrolysis of ATP was required for protein import. This was found to be the case. Other nucleotides (GTP, CTP, UTP and AMP) were found to be ineffective sustitutes for ATP in supporting protein uptake by plastids in the dark (Flügge and Hinz 1986, Schindler et al. 1987, Boyle, Hemmingsen and Dennis, personal communication), as were a number of non-hydrolyzable ATP analogs (Flügge and Hinz 1986, Schindler et al. 1987, Pain and Blobel 1987). Interestingly, both CTP and GTP inhibited light-dependent import in an unknown manner which could be overcome by the addition of ATP (Pain and Blobel 1987).

One question that was reconsidered in the recent studies was whether, as with mitochondria, a protonmotive force is required in addition to ATP to support protein import. These authors reasoned that perhaps exogenously added ATP was hydrolyzed by ATPases present in either the thylakoid or envelope membranes, resulting in the generation of a transmembrane electric field which could in turn drive import. However, it was found that the ATP-dependent import of prSS into dark-incubated chloroplasts was unaffected by agents which collapse either or both the pH and electrical components of the protonmotive force (Flügge and Hinz 1986, Schindler et al. 1987, Pain and Blobel 1987). This again confirmed that ATP was the only energy source necessary for import into chloroplasts. In this respect at least, there is a significant difference between the energy requirements for protein import into chloroplasts and mitochondria.

In contrast to the agreement beteen the various research groups concerning the two questions described above, different conclusions have been reached concerning the location of ATP utilization during protein import. One strategy for examining this problem was to generate ATP inside plastids by the addition of glycolytic or Calvin-Bensen cycle intermediates. Three sets of experiments involved chloroplasts, and fourth involved leucoplasts

(Boyle et al. 1986). In chloroplasts, import of prSS was supported by internally-generated ATP (Flügge and Hinz 1986, Schindler et al. 1987, Pain and Blobel 1987). These results were not unexpected since light can stimulate import and ATP produced by light is also generated in the stroma. In leucoplasts internally generated ATP was unable to drive prSS import (Boyle, Hemmingsen and Dennis, personal communication). However, the evidence that ATP was in fact produced under the experimental conditions was indirect and came from the observation that ATP-dependent fatty acid synthesis was stimulated in a manner similar to that observed when exogenous ATP was added.

Although internally generated ATP supported protein import in chloroplasts, it was unknown whether this ATP was actually utilized internally or after it had been transported to the external medium. To address this question, ATP was generated in the stroma while external ATP was consumed by exogenous membrane-impermeable ATP traps. Presumably, ATP-mediated protein import would occur under these conditions only if it was driven by ATP inside the chloroplasts. However, the results of the experiments seem to depend upon the choice of enzyme used to consume ATP. In Pain and Blobel's experiments (1987), in which either glucose/hexokinase or apyrase was used, import of prSS into pea chloroplasts occurred normally, which led the authors to conclude that ATP was hydrolyzed in the stroma. When Schindler et al. (1987) performed this same experiment with glucose/hexokinase they obtained the opposite result, namely, that protein import was inhibited. They concluded that ATP was required externally for protein import. Flügge and Hinz (1986) performed the most careful study of this question, measuring the actual ATP levels in the chloroplast under conditions of the import assays. Furthermore, they reduced possible complications which could result from rapid transport of ATP across the envelope by using spinach chloroplasts, whcih display lower adenylate translocator activity than do pea chloroplasts. These authors found that when either glucose/hexokinase or fructose-6-phosphate/fructose-6-phosphate kinase were used as external ATP consuming systems, the import of prSS was not inhibited. However, when alkaline phosphatase was used as the external ATP trap, the import of prSS was inhibited and correlated with the external, rather than the internal, ATP concentration. They concluded that ATP outside the stroma appeared to be required for import, although an alternate explanation could involve the participation of an externally accessible phosphorylated intermediate, if this intermediate was dephosphorylated by the alkaline phosphatase. Therefore, although internally generated ATP will transport proteins into chloroplasts, the precise location of ATP hydrolysis is not yet clear.

Energy requirement for targeting to thylakoids. In the four studies described above, prSS was chosen as the experimental precursor. It is noteworthy that mature SS resides in the stroma and never has to cross the energy-transducing thylakoid membranes. The possibility remains that a requirement for a protonmotive force is required for the import of proteins from the stroma into the lumen, but not for translocation from the external medium into the stroma. This question is currently under investigation in our laboratory. Preliminary evidence suggests that, as is the case for prSS import, a protonmotive force is not required to transport proteins into the lumen (Theg, Olsen and Keegstra, unpublished data).

Mechanism of protein translocation
The actual mechanism by which proteins are translocated across the various membranes in chloroplasts remains obscure. At least two fundamental and related questions exist which must be resolved before a clear mechanism can emerge: 1) Are translocator proteins present in the membranes that mediate the movement of proteins across them? 2) What is the conformation of the precursor during binding and translocation?

Translocator proteins have been proposed to be involved in translocation of proteins across the endoplasmic reticulum membrane (Blobel and Dobberstein 1975, Blobel 1980, Singer et al. 1987). The latter authors have proposed a general mechanism whereby these putative translocator proteins mediate the transport of proteins across all cellular membranes. However, the existence of translocators remains hypothetical, and in the years since their existence was first postulated (Blobel and Dobberstein 1975), they have not been identified in the endoplasmic reticulum nor in any other protein translocation system. Because of this lack of biochemical evidence, particularly for the well-studied endoplasmic reticulum system, it seems unwise to postulate the existence of universal translocator proteins. In addition, the significant differences that exist between different protein transport systems argue against the existence of a universal mechanism. In any case, there is currently no evidence for translocator proteins that mediate chloroplast protein import.

The other fundamental question concerning the mechanism of protein translocation concerns the conformation of the precursor during import. Eilers and Schatz (1986) addressed this question for mitochondria and have proposed that protein unfolding must preceed translocation. This hypothesis is based largely on the observation that the import of a chimeric construction using the transit peptide from subunit IV of cytochrome oxidase fused to dihydrofolate reductase was inhibited by the folate antagonist, methotrexate. Methotrexate also imparted some protection against dihydrofolate reductase degredation by thermolysin. Additional evidence for the requirement of unfolding prior to translocation was recently provided by Verner and Schatz (1987), who showed that a prematurely terminated chimeric precursor, which presumably was incompletely folded, was imported into mitochondria without the usual ATP requirement. While these results are consistent with the notion the protein unfolding occurs during import into mitochondria, they in fact demonstrate only that conformational changes of these precursors accompany the import process.

In chloroplasts, recent evidence suggests that if proteins are unfolded during import, they are not exposed to a hydrophobic environment during translocation. Lubben et al. (1987) fused two different hydrophobic stretches of amino acids that function as stop-transfer sequences (Blobel 1980) to the carboxyl terminus of prSS. Neither chimera was halted during translocation by an interaction of the hydrophobic stop transfer region with the membranes, which indicates the regions were probably never exposed to the lipid bilayer. These results argue against the threading of an unfolded protein directly through the lipid bilayer or other hydrophobic environment. They also suggest that if translocator proteins are a universal feature of protein transport, then a putative chloroplast translocator protein must be sufficiently different from the endoplasmic reticulum translocator protein such that it could not recognize the stop-transfer domains of either chimera.

Future questions

The last few years have seen significant advances in our understanding of protein import into chloroplasts. One important consequence of this process is the ability to direct foreign proteins into chloroplasts. Despite these significant advances, the foregoing discussions point out that many important questions regarding both precursor structure and the mechanism(s) involved in transport of proteins into chloroplasts remain unanswered. Some are common to all protein translocation systems, while others are unique to chloroplast protein import. One important question regarding precursor structure is whether the ability to mediate transport is contained entirely within the amino acid sequences of transit peptides (Karlin-Neumann and Tobin 1986), or whether higher order structures are necessary. If, as appears likely, particular secondary or tertiary structures are needed, the task of identifying these structural features will be difficult. Another problem relevant to both precursor structure and mechanism of import concerns the conformational changes precursors undergo during transport. Are proteins completely unfolded, and threaded through the envelop membranes, or are proteins translocated in some folded conformation? If proteins are wholly or partially unfolded, do enzymes (for example, ATP-dependent unfoldases) participate in the unfolding process? Is the energy required for protein translocation used to unfold the protein or is it used in some other steps? Another series of unanswered questions concern the envelope membrane proteins involved in the transport process. Are there receptors which bind precursor proteins prior to translocation? Are there translocator proteins, which transport precursor proteins, either as folded or unfolded polypeptides?

These, and many other, questions must be answered in order to gain a deeper insight into protein transport. In contemplating the problem of how large proteins are transported across two lipid bilayer membranes, one is faced with realization that many observations concerning protein transport do not fit well with our current models regarding general membrane structure and dynamics. In order to understand protein transport into chloroplasts, it may be necessary to revise some features of these models.

Acknowledgements

The authors thank K. Cline for the experiment shown in Fig. 1 and Willow Ealy for assistance in preparing this manuscript. Work from the authors' laboratory was supported by grants from the NSF, the Division of Biological Energy Research at DOE and the Competitive Research Grants Office at USDA.

References

Anderson S and Smith SM (1986) Synthesis of the small subunit of ribulosebisphosphate carboxylase from genes cloned into plasmids containing the SP6 promoter. Biochem J 240: 709–715

Balch WE, Elliott MM and Keller DS (1986) ATP-coupled transport of vesicular stomatitis virus G protein between the endopoasmic reticulum and the golgi. J Biol Chem 261: 14681–14689

Balch WE and Keller DS (1986) ATP-coupled transport of vesicular stomatitis virus G progein: Functional boundaries of secretory compartments. J Bio Chem 261: 14690–14696

Berry-Lowe SL, McKnight TD, Shah DM and Meagher RB (1982) The nucleotide sequence, expression, and evolution of one member of a multigene family encoding the small subunit of ribulose-1,5-bisphosphate carboxylase in soybean. J Mol Appl Genet 1: 483–498

Bitsch A and Kloppstech K (1986) Transport of proteins into chloroplasts. Reconstitution of the binding capacity for nuclear-coded precursor proteins after solubilization of envelopes with detergents. Eur J Cell Biol 40: 160–166

Blobel G (1980) Intracellular protein topogenesis. Proc Natl Acad Sci USA 77: 1496–1500

Blobel G and Dobberstein B (1975) Transfer of proteins across membranes. I. Presence of proteolytically processed nascent immunoglobulin light chains on membrane-bound ribosomes of murine myeloma. J Cell Biol 67: 835–851

Boyle SA, Hemmingsen SM and Dennis DT (1986) Uptake and processing of the precursor to the small subunit of ribulose 1,5-bisphosphate carboxylase by leucoplasts from the endosperm of developing caster oil seeds. Plant Physiol 81: 817–822

Broglie R, Coruzzi G, Lamppa G, Keith B and Chua N–H (1983) Structural analysis of nuclear genes coding for the precursor to the small subunit of wheat ribulose-1,5-bisphosphate carboxylase. Bio/Tech 1: 55–61

Cashmore AR (1984) Structure and expression of a pea nuclear gene encodeing a chlorophyll a/b-binding polypeptide. Proc Natl Acad Sci USA 81: 2960–2964

Cashmore AR, Broadhurst MK and Gray RE (1978) Cell-free synthesis of leaf protein: identification of an apparent precursor of the small subunit of ribulose-1,5-bisphosphate carboxylase. Proc Natl Acad Sci USA 75: 655–659

Cashmore A, Szabo L, Timko M, Kausch A, Van den Broeck G, Schreier P, Bohnert H, Herrera-Estrella L, Van Montagu M and Schell J (1985) Import of polypeptides into chloroplasts. Bio/Tech 3: 803–808

Chen L and Tai PC (1986) Effects of nucleotides on ATP-dependent protein translocation into Escherichia coli membrane vesicles. J Bact 168: 828–832

Chua N–H and Schmidt GW (1978) Post-translational transport into intact chloroplasts of a precursor to the small subunit of ribulose-1, 5-bisphosphate carboxylase. Proc Natl Acad Sci USA 75: 6110–6114

Chua N–H and Schmidt GW (1979) Transport of proteins into mitochondria and chloroplasts. J Cell Biol 81: 461–483

Cline K (1986) Import of proteins into chloroplasts: Membrane integration of a thylakoid precursor protein in chloroplast lysates. J Bio Chem 261: 14804–14810

Cline K, Werner-Washburne M, Lubben TH and Keegstra K (1985) Precursors to two nuclear-encoded chloroplast proteins bind to the outer envelope membrane before being imported into chloroplasts. J Biol Chem 260: 3691–3696

Cornwell KL and Keegstra K (1987) Evidence that a chloroplast surface protein is associated with a specific binding site for the precursor to the small subunit of ribulose-1,5-bisphosphate carboxylase. Plant Physiol 85: 780–785

della-Cioppa G, Bauer SC, Klein BK, Shah DM, Fraley RT and Kishore GM (1986) Translocation of the precursor of 5-enolpyruvylshikimate-3-phosphate synthase into chloroplasts of higher plants in vitro. Proc Natl Acad Sci USA 83: 6873–6877

della-Cioppa G, Bauer SC, Taylor ML, Rochester DE, Klein BK, Shah DM, Fraley RT and Kishore GM (1987a) Targeting a herbicide-resistant enzyme from Escherichia coli to chloroplasts of higher plants. Bio/Tech 5: 579–584

della-Cioppa G, Kishore GM, Beachy RN and Fraley RT 91987b) Protein trafficking in Plant Cells. Pl Physiol 84: 965–968

Dobberstein B, Blobel G and Chua N–H (1977) In vitro synthesis and processing of a putative precursor for the small subunit of ribulose-1,5-bisphosphate carboxylase. Proc Natl Acad Sci USA 74: 1082–1085

Douglas MG, McCammon MT and Vassarotti A (1986) Targeting proteins into mitochondria. Micro Rev 50: 166–178

Eilers M and Schatz G (1986) Binding of a specific ligand inhibits import of a purified precursor protein into mitochondria. Nature 322: 228–232

Eilers M, Oppliger W and Schatz G (1987) Both ATP and an energized inner membrane are required to import a purified precursor protein into mitochondria. EMBO J 6: 1073–1077

Flügge UI and Hinz G (1986) Energy dependence of protein translocation into chloroplasts. Eur J Biochem 160: 563–570

Gasser SM, Daum G and Schatz G (1982) Import of proteins into mitochondria: Energy dependent uptake of precursors by isolated mitochondria. J Biol Chem 257: 13034–13041

Grossman A, Bartlett S and Chua N–H (1980) Energy-dependent uptake of cytoplasmically synthesized polypeptides by chloroplasts. Nature 285: 625–628

Grossman AR, Bartlett SG, Schmidt GW, Mullet JE and Chua N–H (1982) Optimal conditions for post-translational uptake of proteins by isolated chloroplasts: In vitro synthesis and transport of plastocyanin, ferredoxin-NADP oxidoreductase, and fructose-1,6-bisphosphatase. J Biol Chem 257: 1558–1563

Hansen W, Garcia PD and Walter P (1986) in vitro protein translocation across the yeast endoplasmic reticulum: ATP-dependent posttranslational translocation of the prepro-α-factor. Cell 45: 397–406

Highfield PE and Ellis RJ (1978) Synthesis and transport of the small subunit of chloroplast ribulose bisphosphate carboxylase. Nature 271: 420–424

Jansen T, Rother C, Steppuhn J, Reinke H, Beyreuther K, Jansson C, Andersson B and Herrmann RG (1987) Nucleotide sequence of cDNA clones encoding the complete '23 kDa' and '16 kDa' precursor proteins associated with the photosynthetic oxygen-evolving complex from spinach. FEBS Lett 216: 234–240

Karlin-Neumann GA and Tobin EM (1986) Transit peptides of nuclear-encoded chloroplast proteins share a common amino acid framework. EMBO J 5: 9–13

Kloppstech K and Bitsch A (1986) Crosslinking of envelope proteins presumably involved in the binding of nuclear coded chloroplast precursor proteins. In: Regulation of Chloroplast Differentiation, pp 235–240. Alan R. Liss, Inc.

Krieg PA and Melton DA (1984) Functional messenger RNAs are produced by SP6 in vitro transcription of cloned cDNAs. Nucl Acids Res 12: 7057–7071

Kuntz M, Simmons A, Schell J and Schreier PH (1986) Targeting of protein to chloroplasts in transgenic tobacco by fusion to mutated transit peptide. Mol Gen Genet 205: 454–460

Lubben TH and Keegstra K (1986) Efficient in vitro import of a cytosolic heat shock protein into pea chloroplasts. Proc Natl Acad Sci USA 83: 5502–5506

Lubben T, Bansberg J and Keegstra K (1987) Stop-transfer regions do not halt translocation of proteins into chloroplasts. Science 238: 1112–1114

Lubben T, Gatenby A, Ahlquist P and Keegstra K (1988) Imported large subunits of ribulose-1,5-bisphosphate carboxylase/oxygenase, but not imported coupling factor beta subunti, are assembled into holoenzyme in isolated chloroplasts. EMBO J (in press)

Mishkind ML, Wessler SR and Schmidt GW (1985) Functional determinants in transit sequences: import and partial maturation by vascular plant chloroplasts of the ribulose-1,5-bisphosphate carboxylase small subunit of Chlamydomonas. J Cell Biol 100: 226–234

Nelson N and Schatz G (1979) Energy-dependent processing of cytoplasmically made precursors to mitochondrial proteins. Biochemistry 76: 4365–4369

Pain D and Blobel G (1987) Protein import into chloroplasts requires a chloroplast ATPase. Proc Natl Acad Sci USA 84: 3288–3292

Pfanner N and Neupert W (1985) Transport of proteins into mitochondria: a potassium difussion potential is able to drive the import of ADP/ATP carrier. EMBO J 4: 2819–2825

Pfanner N and Neupert W (1986) Transport of F_1-ATPase subunit β into mitochondria depends on both a membrane potential and nucleoside triphosphates. FEBS Lett 209: 152–156

Pfisterer J, Lachmann P and Kloppstech K (1982) Transport of proteins into chloroplasts. Binding of nuclear-coded chloroplast proteins to the chloroplast envelope. Eur J Biochem 120: 143–148

Reiss B, Wasmann CC and Bohnert HJ (1987) Regions in the transit peptide of SSU essential for transport into chloroplasts. Mol Gen Genet 209: 116–121

Robinson C and Ellis RJ (1984) Transport of proteins into chloroplasts: The effect of incorporation of amino acid analogues on the import and processing of chloroplast polypeptides. Eur J Biochem 142: 343–346

Robinson C and Ellis RJ (1985) Transport of proteins into chloroplasts: The precursor of small subunit of ribulose bisphosphate carboxylase is processed to the mature size in two steps. Eur J Biochem 152: 67–73

Roise D, Horvath SJ, Tomich JM, Richards JH and Schatz G (1986) A chemically synthesized pre-sequence of an imported mitochondrial protein can form an amphiphilic helix and perturb natural and artificial phospholipid bilayers. EMBO J 5: 1327–1334

Rothblatt JA and Meyer DI (1986) Secretion in yeast: translocation and glycosylation of prepro-α-factor *in vitro* can occur via an ATP-dependent post-translational mechanism. EMBO J 5: 1031–1036

Scherer DE and Knauf VC (1987) Isolation of a cDNA clone for the acyl carrier protein-I of spinach. Plant Mol Bio 9: 127–134

Schindler C, Hracky R and Soll J (1987) Protein transport in chloroplasts: ATP is prerequisit. Z Naturforsch 42c: 103–108

Schlenstedt G and Zimmermann R (1987) Import of frog prepropeptide GLa into microsomes requires ATP but does not involve docking protein or ribosomes. EMBO J 6: 699–703

Schleyer M, Schmidt B and Neupert W (1982) Requirement of a membrane potential for the posttranslational transfer of proteins into mitochondria. Eur J Biochem 125: 109–116

Schmidt GW, Bartlett SG, Grossman AR, Cashmore AR and Chua N–H (1981) Biosynthetic pathways of two polypeptide subunits of the light-harvesting chloropyll a/b protein complex. J Cell Biol 91: 468–478

Schmidt GW and Mishkind ML (1986) The transport of proteins into chloroplasts. Annu Rev Biochem 55: 879–912

Schreier PH and Schell J (1986) Use of chimaeric genes harbouring small subunit transit peptide sequences to study transport in chloroplasts. Phil Trans R Soc Lond B313: 429–432

Schreier PH, Seftor EA, Schell J and Bohnert HJ (1985) The use of nuclear-encoded sequences to direct the light-regulated synthesis and transport of a foreign protein into plant chloroplasts. EMBO J 4: 25–32

Singer SJ, Maher PA and Yaffe MP (1987) On the translocation of proteins across membranes. Proc Natl Acad Sci USA 84: 1015–1019

Smeekens S (1986) Transport of nuclear-encoded chloroplast proteins. Ph.D. Thesis, Rijksuniversiteit te Utrecht, Utrecht, The Netherlands.

Smeekens S, Bauerle C, Hageman J, Keegstra K and Weisbeek P (1986) The Role of the transit peptide in the routing of precursors toward different chlorplast compartments. Cell 46: 365–375.

Smeekens S, de Groot M, van Binsbergen J and Weisbeek P (1985a) Sequence of the precursor of the chloroplast thylakoid lumen protein plastocyanin. Nature 317: 456–458.

Smeekens S, van Binsbergen J and Weisbeek P (1985b) The plant ferredoxin precursor; nucleotide sequence of a full length cDNA clone. Nucl Acids Res 13: 3179–3194

Smeekens S, van Steeg H, Bauerle C, Bettenbroek H, Keegstra K and Weisbeek P (1987) Import into chloroplasts of a yeast mitochondrial protein directed by ferredoxin and plastocyanin transit peptides. Plant Mol Bio 9:377–388

Tyagi A, Hermans J, Steppuhn J, Jannson C, Vater F and Herrmann RG (1987) Nucleotide sequence of cDNA clones encoding the complete "3 kDa" precursor protein associated with the photosynthetic oxygen-evolving complex from spinach. Mol Gen Genet 207: 288–293

van den Broeck G, Timko MP, Kausch AP, Cashmore AR, Van Montagu M and Herrera-

Estrella L (1985) Targeting of a foreign protein to chloroplasts by fusion to the transit peptide from the small subunit of ribulose 1,5-bisphosphate carboxylase. Nature 313: 358–363.

Verner K and Schatz G (1987) Import of an incompletely folded precursor protein into isolated mitochondria requires an energized inner membrane, but no added ATP. EMBO J 6: 2449–2456.

von Heijne G (1986) Mitochondrial targeting sequences may form amphiphilic helices. EMBO J 5: 1335–1342

Wasmann CC, Reiss B, Bartlett SG and Bohnert HJ (1986) The importance of the transit peptide and the transported protein for protein import into chloroplasts. Mol Gen Genet 205: 446–453

Waters MG and Blobel G (1986) Secretory protein translocation in a yeast cell-free system can occur posttranslationally and requires ATP hydrolysis. J cell Biol 102: 1543–1550.

Wickner WT and Lodish HF (1985) Multiple mechanisms of protein insertion into and across membranes. Science 230: 400–407.

Yamane K, Ichihara S and Mizushima S (1987) In vitro translocation of protein across Escherichia coli membrane vesicles requires both the proton motive force and ATP. J Biol Chem 262: 2358–2362.

Govindjee et al. (eds), Molecular Biology of Photosynthesis: 735–744
© 1988 Kluwer Academic Publishers

Minireview

Protein transport towards the thylakoid lumen: post-translational translocation in tandem

SJEF SMEEKENS & PETER WEISBEEK
Department of Molecular Cellbiology and Institute of Molecular Biology, University of Utrecht, Padualaan 8, 3584 CH Utrecht, The Netherlands

Received 3 September 1987; accepted 4 December 1987

Key words: chloroplast, protein transport, thylakoid, lumen, transit peptide, processing, plastocyanin

Abstract. Many proteins found in the chloroplast are synthesized in the cytoplasm as precursor molecules containing transit peptides. Proteins targeted to the stroma must pass through the two envelope membranes to reach their destination. Proteins located in the chloroplast lumen also have to be transferred across the thylakoid membrane. That is, lumen proteins must cross three biological membranes in order to reach their final location. Recent evidence shows that the routing of plastocyanin towards the lumen involves two post-translational transport processes mediated by two different regions of the transit peptide and two different processing proteases. It is postulated that the genetic information for the plastocyanin precursor, which already contained a signal peptide, was transferred from the endosymbiont to the nucleus. Then a chloroplast-specific targeting-peptide was added.

Introduction

Proteins that are functional in other cellular compartments than where they are synthesized must contain information for compartmentalization. In addition, mechanisms are necessary that respond specifically to this information by transporting such proteins toward their proper compartment. In chloroplasts at least six different compartments can be distinguished (Fig. 1): the outer and inner membrane with the intermembrane space, the chloroplast stroma, and the continuous thylakoid membrane system which encloses the thylakoid lumen. Nuclear-encoded chloroplast-specific proteins are translated on cytoplasmic ribosomes as larger precursor molecules, which contain an amino-terminal extension, the transit peptide (Ellis 1981, Schmidt and Mishkind 1986). Following import, the transit peptide is cleaved off by a specific stromal protease. The available evidence indicates that protein import into chloroplasts is a completely post-translational process.

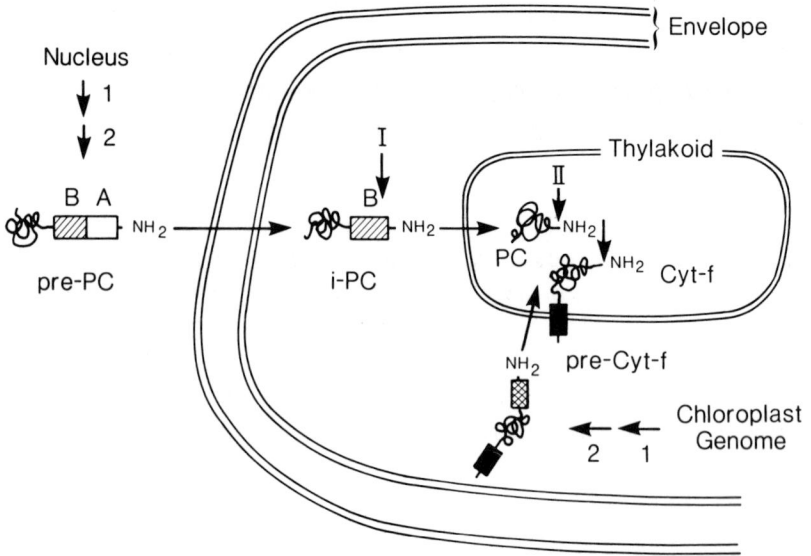

Fig. 1. Routing of the thylakoid lumen-specific plastocyanin (PC) and cytochrome f (cyt-f) precursor proteins, encoded by the nuclear- and plastid genomes, respectively. Pre-PC, precursor; i-PC, intermediate; A, chloroplast import domain; B, thylakoid transfer domain. roman numerals I and II indicate stroma- and thylakoid-associated processing activities, respectively. Transcription (1) and translation (2) are indicated by arrows. the cross-hatched and black boxes in the cytochrome f precursor indicate the signal sequence and the stop transfer sequence, respectively.

Transit peptides are 35 to 80 amino acid residues in length and consist mainly of hydrophobic or hydroxy amino acid residues interspersed with basic amino acids (arginine or lysine) (Schmidt and Mishkind 1986). Little homology is observed between different transit peptides (Karlin-Neumann and Tobin 1986). That the transit peptide contains information for chloroplast import is illustrated by experiments in which plant cells were transformed with a hybrid gene consisting of the transit peptide of the small subunit of ribulose-1,5-bisphosphate carboxylase/oxygenase linked to the coding sequence of the bacterial neomycin phosphotransferase (NPT II), a protein normally not found in chloroplasts. In transformed plants the NPT activity could be detected inside the chloroplasts (Schreier et al. 1985, Van den Broeck et al. 1985). No chloroplast-associated NPT activity was detected when the transit sequence was deleted. Chloroplast import of another foreign protein, the yeast mitochondrial enzyme superoxide dismutase, mediated by transit peptides for ferredoxin and plastocyanin has also been demonstrated (Smeekens et al. 1987).

Many imported proteins need only to reach the chloroplast stroma. These proteins need to be translocated only across the envelope. However, several other imported proteins, e.g. some components of the photosystems, the electron transport chain or the ATP-synthase complex must be incorporated into or transported across the thylakoid membrane. Therefore, in addition to chloroplast targeting information such proteins must contain information for intra-organellar routing. The additional information could reside either in the transit peptide or in the mature protein or in both parts of the protein. The availability of cDNA clones coding for precursors to proteins located in different chloroplast compartments (Smeekens et al. 1985a, b) and of in vitro expression and chloroplast import systems (Cline et al. 1985) has made a detailed study of the different aspects of chloroplast intra-organellar routing mechanisms possible. Import and intra-organell transport has been studied with the lumen protein plastocyanin. (Smeekens et al. 1986, Hageman et al. 1986)

Pre-plastocyanin is transported to the lumen in two steps

The import of pre-plastocyanin and its subsequent routing toward the lumen could be achieved using in vitro synthesized radio-labeled precursor proteins and isolated intact chloroplasts. The plastocyanin protein was directed by this system toward the thylakoids where it is present in the protease-resistant fraction. This routing involves two distinct steps, both of which occur post-translationally. At first, pre-plastocyanin binds to the chloroplast envelope and is translocated across the envelope into the stroma where the amino-terminal part of the transit peptide is cleaved off (Fig. 1). Next, this stromal intermediate recognizes the thylakoid membrane through the remaining part of the transit peptide and is then translocated across the thylakoid membrane into the lumen where it is processed to the mature size.

Based on these observations a 'two domain transit peptide model' has been proposed (Smeekens et al. 1986). The plastocyanin transit peptide is composed of a chloroplast import domain, which mediates transport to the stroma, and a thylakoid transfer domain, involved in transport across the thylakoid membrane (Figs. 1 and 2). Deletions of varying sizes in the thylakoid transfer domain of the transit peptide interfere only with the thylakoid transport step and not with import into the chloroplast.

Successive removal of these domains requires the presence of the proper processing enzymes at the right place. A stromal transit peptidase processes pre-plastocyanin to the intermediate size. Processing to the mature size

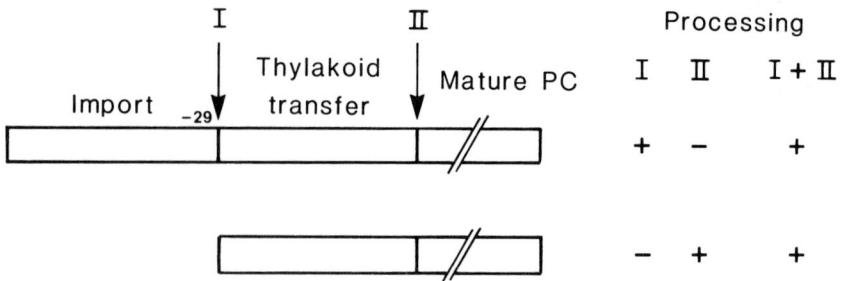

Fig. 2. In vitro processing activities of the stroma-located (I) and thylakoid-associated (II) processing enzymes when incubated with the plastocyanin precursor protein (upper part) and the intermediate protein (lower part). +, processing; −, no processing. The arrows indicate the two processing sites in the plastocyanin precursor. The import domain and the thylakoid transfer domain of this precursor are indicated. Note that the thylakoid-located enzyme is not active on the complete precursor.

requires an additional protease, which subsequently was found to be present in the thylakoids from which it can be isolated by Triton X-100 solubilization (Hageman et al. 1986). This protease clearly differs from the stromal transit peptidase, since it can only process the plastocyanin intermediate to its mature size while it does not recognize the intact precursor (Fig. 2).

These results make it highly unlikely that contact sites between the thylakoids and the envelope are necessary for transport of proteins into the thylakoids. A similar conclusion was reached for the routing of the chlorophyll a/b-binding protein. It can be inserted in vitro into isolated thylakoid membranes and this appears to be dependent only on factors present in the stroma (Cline 1986).

Other thylakoid lumen proteins also have a two-domain transit peptide

Since the identification of the two functional domains in the transit peptide of the white campion plastocyanin precursor, the amino acid sequences for transit peptides of two other plastocyanin precursors (Fig. 3A) (Vorst et al. 1988, Rother et al. 1986) and of several other lumenal proteins (Fig. 3B) have become available. The latter proteins include the three components of the PS II-associated water-splitting complex (Mayfield et al. 1987, Tyagi et al. 1987, Jansen et al. 1987) and of the inducible cytochrome c-552 of *Chlamydomonas* (Merchant and Bogorad 1987). In all these transit peptides two domains that correspond to the proposed plastocyanin chloroplast import domain and thylakoid transfer domain can be distinguished. The sequence homologies (Fig. 3) indicate a targeting mechanism for these

A

CHLOROPLAST IMPORT   THYLAKOID TARGETING

```
pc    A.th     MAAITSAT VTIPSFTGLKLAVSSKPKTLSTISRSSSATRAPPKLALKSSLKDFGVIAVATAASIVLAGNAMA
pc    s.pr     MATVTSSAAVAIPSFAGLK A SSTTR AATVKVAVAT PRMSIKASLKDVGVVVAATAAAGILAGNAMA
pc    S.oc     MATVASSAAVAVPSFTGLK A SGSIK  PTTAKIIPTTAVPRLSVKASLKNVGAAVVATAAAGLLAGNAMA

pc    A.var                                        MKLIAASLRRLSLAVLTVLLVVSSFAVFTPSAAA

cyt f P.sat                            MQTRNAFSWIKKEITRSISVLLMIYIITRAPISNA
            -70    -60    -50    -40    -30    -20    -10    -1
```

B

S.oc WATERSPLITTING ENZYMES

```
33 kd         MAASLQASTTFLQPTKVASRNTLQLRSTQNVCKAFGVESASSG GRLSLSLQSDLKELANKCVDATKLAGLALATSALIASGANA
23 kd MASTACFLHHHAAISSPAAGRGSAAQRYQAVSIKPNQIVCKAQKQDDNEANVLNS  GVSRRLALTVLIGAAAVGSKVSPADA
16 kd MAQAMASMAGLRGASQAVLEGSLQISGSNRLSGPTTSRVAVPKMGLNIRAQQVSAEAETSRRAMLGFVAAGLASGSFVKAVLA
      -80    -70    -60    -50    -40    -30    -20    -10    -1
```

C. reinhardtii

```
OEE 2         MATALCNKAFAAAPVARPASRRSAVVV RASGSDVSRRAALAGFAGAAALVSSPANA
cyt-c552      MLQLANRSVRAKAARASQSARSVSCAAA KRGADVAPLTSALAVTASILLTGAASASA
              -50    -40    -30    -20    -10    -1
```

Fig. 3. Comparison of thylakoid targeting sequences. A. The transit peptides of three different plant plastocyanin transit sequences and the signal sequence of a cyanobacterial plastocyanin and the chloroplast-encoded cytochrome f. B. The nuclear-encoded water-splitting enzymes (the 33, 23 and 16 kd proteins and the OEE2 protein) and cytochrome c-552. Strongly charged residues are indicated above the sequence. The thick line indicates the region involved in import; the thin line marks the thylakoid transfer domain. The arrows point to the final processing site in the precursor protein. A.th, *Arabidopsis thaliana*; S.pr, *Silene pratensis*; S.oc, *Spinacia oleracea*; P.sat, *Pisum sativum*; C. reinhardtii, *Chlamydomonas reinhardtii*.

proteins that is very similar to the proposed two-step model for plastocyanin. It appears likely that all imported proteins which are directed to the thylakoid lumen use a common routing mechanism.

Protein translocation across the thylakoid membrane

The transport of the plastocyanin intermediate from the stroma into the lumen is functionally equivalent to transport of thylakoid-specific proteins that are encoded by chloroplast DNA and therefore synthesized inside the chloroplast. The best studied example of such a chloroplast-encoded protein is the precursor for cytochrome f (Fig. 1). This protein is synthesized in the chloroplast stroma on thylakoid-bound ribosomes (Willey et al. 1984) as a precursor with a cleavable signal peptide (Fig. 3A) and, possibly co-translationally, inserted into the thylakoid membrane. The larger part of this protein (containing the heme-binding domain) is transported across the thylakoid membrane into the lumen whereas the carboxy-terminus is anchored to the thylakoid membrane with a stop/transfer sequence. When this precursor is synthesized in E. coli it behaves like a membrane protein and its transport depends on E. coli secA gene products (Rothstein et al. 1985).

From this result the interesting question arises whether the imported plastocyanin makes use of the same mechanism for thylakoid translocation as the chloroplast encoded cytochrome f or whether there are two or more independent pathways to cross the thylakoids. If the same mechanism is used then the thylakoid transfer domain of an imported molecule must be recognized in the same way as the signal sequence of a thylakoid protein that is encoded in plastid DNA. It also means that the thylakoid transfer domain of an imported protein uses the plastid (co-translational?) transport machinery for its post-translational transport over the thylakoid membrane. Such a post-translational membrane transfer involving a vectorial type of signal sequences has been described for the bacterial export system (Oliver 1985) and, more recently, also in the yeast and animal co-translational transport systems (Perara et al. 1986, Mueckler and Lodish 1986, Rothblatt and Meyer 1986). A prediction of such a model is that the processing enzyme located in thylakoids that can process plastocyanin to its final size, is a signal peptidase that can also process the cytochrome f precursor.

The primary structure of the known thylakoid transfer domains support the concept that a common mechanism for transport of imported and chloroplast-encoded proteins across the thylakoids exists. The sequence is similar to a signal sequence (Von Heijne 1982) with a positively charged amino terminus followed by a hydrophobic core region and a processing site typical for signal sequences at the carboxy terminus (Fig. 3).

This concept may be tested, e.g. by placing a chloroplast import domain in front of the cytochrome f precursor and probing its correct assembly into the thylakoid membrane following import into the chloroplast. If correct, pre-cytochrome f, when introduced in the stroma in this way, should properly integrate into the membrane and should be processed correctly. In addition, the thylakoid-located protease that processes the plastocyanin intermediate to its mature size should also process the cytochrome f precursor to its mature size. Such experiments are under way in our laboratory and their results will add to our insight into thylakoid-specific protein transport.

Evolutionary economy of transport pathways

The use of such a plastid-specific targeting mechanism for the routing of imported lumen proteins also appears logical in an evolutionary context. Thylakoid transfer could be accomplished without the need to evolve a new mechanism when most of the genetic material from the prokaryotic progenitor of the extant chloroplast was transferred to the nucleus. Following this view the complete genetic information for such proteins, including the information for thylakoid targeting, was transferred to the nucleus. The only change required was the addition of a chloroplast import domain. Recent experiments indicate that the efficiency of generating such an organelle-specific targeting sequence may be surprisingly high. It was found that random DNA sequences, when fused to the DNA sequence coding for a passenger protein, can provide targeting signals that promote transport of a passenger to specific locations with a frequency of up to 25% (Baker and Schatz 1987, Kaiser et al. 1986).

Support for this idea comes from the sequence of the gene for plastocyanin of a cyanobacterium, a possible ancestor of the chloroplast that has recently been isolated and characterized in our laboratory (J. van der Plas, personal communication). This protein is made as a precursor that contains a 34 amino acid residues long cleavable leader sequence that resembles a signal peptide and that might be involved in its transport across the cyanobacterial thylakoid membrane. The general characteristics of this signal sequence is very similar to what we find in the thylakoid transfer domain of its eukaryotic counterpart (Fig. 3A) although the primary amino acid sequence is rather different. It consists of a positively charged polar head, a long hydrophobic region and a typical sequence motive near the procesing

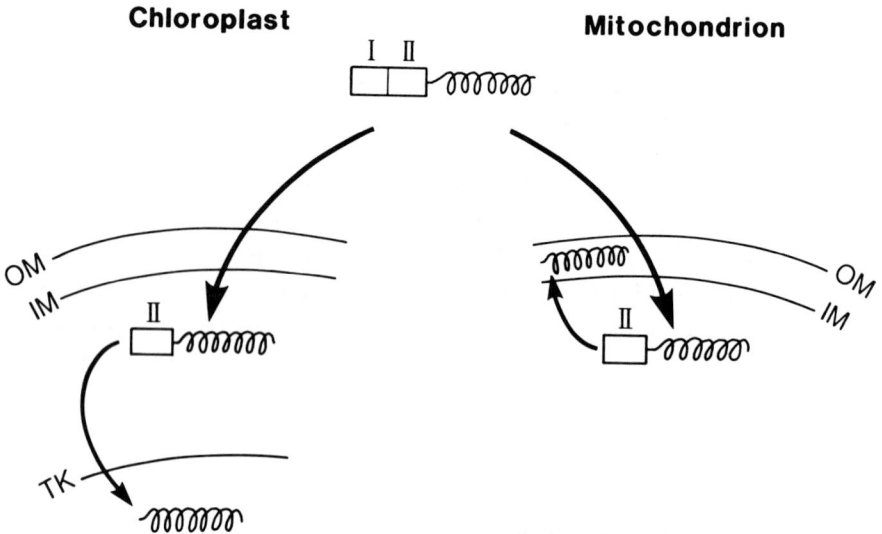

Fig. 4. Comparison of the transport pathways of chloroplast proteins targeted to the thylakoid lumen and mitochondrial proteins directed towards the intermembrane space. Both types of precursor contain a two domain routing peptide (domains I and II) that is cleaved off in two separate steps by different peptidases located in different compartments. OM, outer membrane, IM, inner membrane, TK, thylakoid membrane.

site. The hydrophobic regions of the eukaryotic thylakoid transfer domains are longer than the corresponding regions in the prokaryotic export signals. These findings stress that the specific translocation of imported proteins to thylakoids in chloroplasts and of proteins synthesized in the prokaryotic cyanobacteria appear to share a common mechanism.

A similar suggestion for the coupling of an ancient transport system to a newly evolved import mechanism has also been made for transport of nuclear-encoded mitochondrial proteins that are found in the intermembrane space or on the outer surface of the inner membrane (Hartl et al. 1986). The Fe/S protein of ubiquinol-cytochrome c reductase is transported to the outer surface of the inner membrane in two steps, reminiscent of the plastocyanin transport to the thylakoids (Fig. 4). First, the Fe/S protein is imported into the mitochondrial matrix where part of its presequence is cleaved off by the specific matrix processing enzyme. Next, this intermediate is transported back across the inner membrane to the intermembrane space where it is cleaved to its mature size by a different protease. This presequence has the same functionally significant two-domain structure as thylakoid-specific transit peptides such as plastocyanin. A two-domain presequence and a two-step processing mechanism have also been found in other mitochondrial intermembrane space proteins (Kaput et al. 1982, Sadler et al. 1984, Van Loon et al. 1986).

Conclusion

The two-step transport mechanism by which plastocyanin is translocated from the cytoplasm into the thylakoid lumen may very well reflect the basic pattern by which the symbiont/organelle responded to the transfer of genetic information to the nucleus. The information for transfer across the organellar envelope was conferred to the proteins which were now synthesized in the cytoplasma by the addition of an N-terminal extension or eventually by alterations in the mature portion of the protein. Once imported the proteins are linked to the same targeting pathway they used to follow when they were synthesized inside the organelle.

Although this appears an attractive hypothesis explaining the evolution of such protein targeting, we certainly need more experimental evidence about the actual mechanisms before definite conclusions can be draw. One prediction of the model is that proteins targeted for the inner membrane of the chloroplasts will also have a two-domain transit peptide and that the second domain will be a signal peptide that is specific for the inner membrane.

References

Baker A and Schatz G (1987) Sequences from a prokaryotic genome or the mouse dihydrofolate reductase gene can restore the import of a truncated precursor protein into yeast mitochondria. Proc Natl Acad Sci USA 84: 3117–3121

Cline K, Werner-Washburne M, Lubben T and Keegstra K (1985) Precursors to two nuclear-encoded chloroplast proteins bind to the outer envelope membrane before being imported into chloroplasts. J Biol Chem 260: 3691–3696

Cline K (1986) Import of proteins in chloroplasts. Membrane integration of a thylakoid precursor protein reconstituted in chloroplast lysates. J Biol Chem 261: 14804–14810

Ellis JR (1981) Chloroplast proteins: synthesis, transport and assembly. Ann Rev Plant Physiol 32: 111–137

Hageman J, Robinson C, Smeekens S and Weisbeek P (1986) A thylakoid-located processing protease is required for complete maturation of the lumen protein plastocyanin. Nature 324: 567–569

Hartl F-U, Schmidt B, Wachter E, Weiss H and Neupert W (1987) Transport into mitochondria and intramitochondrial sorting of the Fe/S protein of ubiquinol-cytochrome c reductase. Cell: 939–950

Jansen T, Rother C, Steppuhn J, Reinke H, Beyreuther K, Jansson C, Andersson B and Herrmann R (1987) Nucleotide sequence of cDNA clones encoding the complete 23 kda and 16 kda precursor proteins associated with the photosynthetic oxygen-evolving complex from spinach. FEBS Lett 216: 234–240

Kaput J, Goltz S and Blobel G (1982) Nucleotide sequence of the yeast nuclear gene for cytochrome c peroxidase precursor. J Biol Chem 257: 15054–15058

Kaiser C, Preuss D, Grisafi P and Botstein D. (1986) Science 235: 312–317

Karlin-Neumann GA and Tobin EM (1986) Transit peptides of nuclear encoded chloroplast proteins share a common amino acid framework. EMBO J 5: 9–13

744

Mayfield S, Rahire M, Frank G, Zuber H and Rochaix J-D (1987) Cloning of the OEE2 gene from Chlamydomonas. Proc Natl Acad Sci USA 84: 749–753

Merchant S and Bogorad L (1987) The Cu(II)-repressible plastidic cytochrome c. Cloning and sequence of a cDNA for Pre-apoprotein. J Biol chem 262: 9062-9067

Mueckler M and Lodish HF (1986) The human glucose transporter can insert posttranslationally into microsomes. Cell 44: 629–637

Oliver D (1985) Protein secretion in Escherichia coli. Ann Rev Microbiol 39: 615–648

Perara E, Rothman RE and Lingappa VR (1986) Uncoupling of translocation from translation: implications for transport of proteins across membranes. Science 232: 348–352

Rothblatt JA and Meyer DI (1986) Secretion in yeast: translocation and glycolysation of prepro-alpha-factor in vitro can occur via an ATP-dependent post-translational mechanism. EMBO J 5: 1031–1036

Rother C, Jansen T, Tyagi A, Tittgen J and Herrmann R (1986) Plastocyanin is encoded by an uninterrupted nuclear gene in spinach. Curr Genet 11: 171–176

Rothstein SJ, Gatenby AA, Willey DL and Gray JC (1985) Binding of pea cytochrome f to the inner membrane of Escherichia coli requires the bacterial secA gene product. Proc Natl Acad Sciences 82: 7955–7959

Sadler I, Suda K, Schatz G, Kaudewitz F and Haid A (1984) Sequencing of the nuclear gene for the cytochrome cl precursor reveals an unusually complex amino-terminal presequence. EMBO J 3: 2137–2143

Schmidt GW and Mishkind M (1986) The transport of proteins into chloroplasts. Ann Rev Biochem 55: 879–912

Schreier PH, Seftor EA, Schell J and Bohnert HJ (1985) The use of nuclear encoded sequences to direct the light regulated synthesis and transport of a foreign protein into plant chloroplasts. EMBO J 4: 25–32

Smeekens S, van Steeg H, Bauerle C, Bettenbroek H, Keegstra K and Weisbeek P (1987) Import into chloroplasts of a yeast mitochondrial protein directed by ferredoxin and plastocyanin transit peptides. Plant Mol Biol 9: 377–388

Smeekens S, Van Binsbergen J and Weisbeek P (1985a) The plant ferredoxin precursor: nucleotide sequence of a full length cDNA clone. Nucl Acids Res 13: 3179–3194

Smeekens S, De Groot M, Van Binsbergen J and Weisbeek P (1985b) Sequence of the precursor of the chloroplast lumen protein plastocyanin. Nature 317: 456–458

Smeekens S, Bauerle C, Hageman J, Keegstra K and Weisbeek P (1986) The role of the transit peptide in the routing of precursors towards different chloroplast compartments. Cell 46: 365–375

Tyagi A, Hermans J, Steppuhn J, Jansson C, Vater F and Herrmann R (1987) Nucleotide sequence of cDNA clones encoding the complete 33 kda precursor protein associated with the photosynthetic oxygen-evolving complex from spinach. Mol Gen Genet 207: 288–293

Van den Broeck G, Timko MP, Kausch AP, Cashmore AR, Montagu M van and Herrera-Estrella L (1985) Targeting of foreign protein to chloroplasts by fusion to the transit peptide from the small subunit of ribulose-1,5-biphosphate carboxylase. Nature 313: 358–363

Van Loon A, Brandli A and Schatz G (1986) The presequence of two imported mitochondrial proteins contain information for intracellular and intramitochondrial sorting. Cell 44: 8012–812

Von Heyne G (1982) Signal sequences are not uniformly hydrophobic. J Mol Biol 159: 537–541

Vorst O, Oosterhoff-Teertstra R, Van Kan P, Smeekens S and Weisbeek P (1988) Plastocyanin of Arabidopsis thaliana; isolation and characterization of the gene and chloroplast import of the precursor protein. Gene (in press)

Willey DL, Auffret AD and Gray JC (1984) Structure and topology of cytochrome f in pea chloroplast membranes. Cell 36: 555–562.

Govindjee et al. (eds), Molecular Biology of Photosynthesis: 745–776
© 1988 Kluwer Academic Publishers

Minireview

Recent developments in chloroplast protein transport

MICHAEL L. MISHKIND & SCOTT E. SCIOLI
Department of Biochemistry and Microbiology, Lipman Hall, Cook College,
Rutgers University, New Brunswick, New Jersey 08903, USA

Received 14 December 1987; accepted 3 January 1988

Key words: mitochondria, organelle biogenesis, transit peptide, chloroplast, protein transport

Abstract. Most proteins located in chloroplasts are encoded by nuclear genes, synthesized in the cytoplasm, and transported into the organelle. The study of protein uptake by chloroplasts has greatly expanded over the past few years. The increased activity in this field is due, in part, to the application of recombinant DNA methodology to the analysis of protein translocation. Added interest has also been gained by the realization that the transport mechanisms that mediate protein uptake by chloroplasts, mitochondria and the endoplasmic reticulum display certain characteristics in common. These include amino terminal sequences that target proteins to particular organelles, a transport process that is mechanistically independent from the events of translation, and an ATP-requiring transport step that is thought to involve partial unfolding of the protein to be translocated. In this review we examine recent studies on the binding of precursors to the chloroplast surface, the energy-dependent uptake of proteins into the stroma, and the targeting of proteins to the thylakoid lumen. These aspects of protein transport into chloroplasts are discussed in the context of recent studies on protein uptake by mitochondria.

Abbreviations: CAT – chloramphenicol acetyl transferase, CCCP – carbonylcyanide *m*-chlorophenylhydrazone, DHFR – dihydrofolate reductase, EPSP – 5-*enol*-pyruvylshikimate-3-phosphate, ER – endoplasmic reticulum, LHCP – light harvesting chlorophyll *a/b* apoprotein, NPT – neomycin phosphotransferase, oATP – adenosine-2′,3′-dialdehyde-5′-triphosphate, P_i-inorganic phosphate Rubisco – ribulose-1,5-bisphosphate carboxylase/oxygenase, SSU – small subunit of ribulose-1,5-bisphosphate carboxylase/oxygenase, SRP – signal recognition particle

Introduction

Chloroplasts synthesize a small subset of the proteins required for photosynthesis and other processes that occur within the organelle. The vast majority of polypeptides in the organelle are encoded by nuclear genes, synthesized in the cytoplasm, and transported across the two lipid bilayers of the chloroplast envelope (Ellis 1981). The mechanisms that govern protein transport and targeting are thus of exceptional importance in the biogenesis of chloroplasts.

Analysis of protein uptake by chloroplasts began with the development of cell-free systems in which transport events could be reconstituted (Highfield and Ellis 1978, Chua and Schmidt 1978). These systems became practical, in large part, as a result of the availability of methods for isolating physiologically intact chloroplasts and for the in vitro translation of mRNA (for review see Mishkind et al. 1987). Transport of proteins across the envelope (Highfield and Ellis 1978) followed by their assembly into functional complexes (Chua and Schmidt 1978) was found to occur in mixtures that contained isolated chloroplasts and the in vitro translation products of cytoplasmic mRNAs that encode chloroplast proteins. In the decade since these reports appeared, several steps in the transport process have been elucidated (for review see Cashmore et al. 1985, Schmidt and Mishkind 1986, della-Cioppa et al. 1987). Cytoplasmic ribosomes discharge chloroplast proteins as soluble precursors with extensions of amino acids, termed transit or leader peptides, at their amino termini. The precursors bind and cross the limiting membranes of the organelle in the absence of concomitant protein synthesis. Transport across the envelope into the stroma requires the hydrolysis of ATP. During transport, or shortly thereafter, soluble endoproteases cleave the transit peptides from the precursors. Certain imported polypeptides are routed from the stroma to assembly sites in the thylakoid membrane or lumen.

Eukaryotic cells contain several membrane-bound compartments in addition to chloroplasts whose complement of polypeptides is completely or largely the product of cytoplasmic protein synthesis. These include mitochondria (Douglas et al. 1986, Schatz 1987), the endoplasmic reticulum (ER) (Walter and Lingappa 1986), and peroxisomes (Lazarow and Fujiki 1985). Mitochondria, in a manner similar to chloroplasts, synthesize a small number of polypeptides within the organelle. Until recently, distinct mechanisms were thought to mediate protein transport across different membranes. Transport of proteins into chloroplasts and mitochondria was found to occur after translation is completed; hence, protein translocation into these organelles was considered to occur by a posttranslational mechanism. In contrast, transport of polypeptides across the ER membrane was observed to be linked to the process of translation; thus a cotranslational mechanism was suggested in which the events of protein synthesis directly mediate the transfer of polypeptides across the membrane. Studies that have appeared over the past few years, however, suggest that similar mechanisms operate during critical steps in protein transport across a wide variety of membranes. Common aspects include amino terminal sequences that target proteins to particular organelles, a transport process that is mechanistically independent from the events of protein synthesis, and an ATP-requiring transport step that is thought to involve partial unfolding of the protein to be translocated.

Given this new perspective, we have chosen to review recent reports on protein transport into chloroplasts in the context of important findings from other protein transport systems. The similarity in transport mechanisms suggests that results obtained from diverse systems will be directly relevant to an understanding of protein transport into chloroplasts. The studies discussed here add numerous instructive details to the steps in the transport process. Insights contributed by these findings will be instrumental in determining the still elusive mechanism by which proteins traverse lipid bilayers.

Background

With the advent of in vitro translation systems, it was established that most secreted proteins are synthesized as precursors with leader peptides at their amino termini. The role of these peptide extensions in targeting the precursors to the ER led to their characterization as 'signal sequences' (Blobel and Dobberstein 1975a). Translocation of the precursors across the ER membrane was found to require ongoing protein synthesis (Blobel and Dobberstein 1975b). The observation that proteolytic removal of the amino terminal leader sequence (Blobel and Dobberstein 1975a) occurs prior to polypeptide chain termination verified the cotranslational nature of the process.

Further characterization of the system led to the discovery that a ribonucleoprotein complex termed signal recognition particle (SRP) arrests the translation of the precursors of secreted proteins by binding to the signal sequence shortly after it emerges from the ribosome (Walter et al. 1984). The subsequent binding of the arrested translation complex to a receptor (termed SRP receptor or docking protein) on the cytoplasmic surface of the ER was found to restore translational elongation and initiate transfer of the nascent polypeptide across the ER. Translational arrest was presumed to occur before elongating polypeptide chains assume three-dimensional conformations that could interfere with protein transfer across the hydrophobic lipid bilayer. Since transport and protein synthesis appeared to be obligatorily linked, the signal hypothesis proposed that transport proceeds by the vectorial insertion of a nascent polypeptide through the ER membrane. Translational elongation was proposed to supply the energy that directs a polypeptide across the membrane (Blobel and Dobberstein 1975a).

In dramatic contrast to protein transfer into the ER, studies performed in the late 1970s revealed that transport of proteins into chloroplasts and mitochondria occurs in the absence of ongoing protein synthesis (for review see Cashmore et al. 1985, Schmidt and Mishkind 1986). Ribosomes were removed from cell-free translation products containing organellar precursor proteins. These precursors retained capacity for transport into the isolated organelles. Since the translation apparatus was not required, protein synthesis and translocation were clearly mechanistically separate for the cytoplasmically synthesized chloroplast and mitochondrial proteins. The energetic and cell biological problems of protein transport appeared so different from those encountered in the ER that two independent transport mechanisms were postulated – posttranslational for chloroplasts and mitochondria, cotranslational for the ER.

Recent studies demonstrate that despite the difference in translational requirements for protein transfer, similar mechanisms appear to mediate protein transfer into the ER, chloroplasts and mitochondria (Schatz 1986, Singer et al. 1987). It was analysis of protein transport in prokaryotes, however, that gave initial indications that co- and posttranslational transport might occur by similar mechanisms. Kinetic analysis of secretion in *Escherichia coli* demonstrated that although transport might occur cotranslationally, the actual transfer of polypeptide domains across the cytoplasmic membrane is not directly linked to the events of translation. For example, although transport of the maltose binding protein occurs while the polypeptide is a nascent chain associated with ribosomes, transfer across the membrane can begin after as much as 80% of the polypeptide is synthesized (Randall 1983). Furthermore, initiation of transfer is not linked to a particular stage in the synthesis of the protein; it occurs co- or posttranslationally and seems to depend on the availability of transfer sites on the inner surface of the cytoplasmic membrane (Josefsson and Randall 1981). Similarly, the secretion of β-lactamase in *E. coli* is not obligatorily linked to the events of translation. The protein is secreted posttranslationally from cells grown in poor medium (Chen et al. 1985) or irradiated with UV light (Koshland and Botstein 1982) but is cotranslationally secreted from cells grown in rich medium (Chen et al. 1985).

Evidence has also accumulated which demonstrates that events in the transport of proteins into the ER are not as closely associated with translational elongation as was initially considered. Although ribosomes might be required to maintain a secreted protein in a conformation compatible with transport, translational elongation does not provide the energy that directs the protein across the ER (Walter and Lingappa 1986). Most convincing are reports that the precursor to the yeast mating factor can be transported into ER-derived vesicles posttranslationally (Hansen et al. 1986, Rothblatt and Meyer 1986, Waters and Blobel 1986). It thus appears that the transfer of proteins across ER membranes, whether in nascent form or fully synthesized, is not linked directly to the events of translation. The tendency of certain proteins such as invertase or IgG kappa light chains to require cotranslational conditions for transport is more likely a consequence of the characteristics of the protein than an indication of the mechanics of protein transport (Rothblatt and Meyer 1986).

The overriding question in the study of protein transport concerns the mechanism by which hydrophilic polypeptides cross the hydrophobic environment of lipid bilayers. As discussed above, initial formulations of the signal hypothesis dealt with this problem by suggesting cotranslational insertion of short stretches of polypeptide through the lipid bilayer. Thus, transport would occur before polypeptides fold into bulky three-dimensional structures. Given the posttranslational nature of transport, however, at least some protein folding likely occurs prior to the initiation of transfer across the lipid bilayer.

Proteins refractory to posttranslational transport may attain tightly folded conformations during or shortly after synthesis. For example, Maher and Singer (1986) have suggested that polypeptides stabilized by intra-chain disulfide bonds require that transport be cotranslational. Transfer of polypeptide domains across the membrane could then occur before disulfide bonds form. In contrast, translocation of proteins that lack intra-chain disulfide bonds need not be closely linked to translational elongation. Evidence in support of this hypothesis included the observation that posttranslational transport of preprolactin into ER-derived vesicles occurs only when disulfide bonds are maintained in a reduced form by dithiothreitol (Maher and Singer 1986).

The potential that certain proteins are intractable to transport has been proposed as an explanation for the maintenance of genomes in chloroplasts and mitochondria. Thus, genes that remain in the organelle would be predicted to encode proteins that could not be efficiently imported from the cytoplasm. A related hypothesis has been proposed by von Heijne (1986a) in which a gene is confined to an organelle genome if the gene product carries a determinant that would be recognized by the endoplasmic reticulum as a signal sequence. Polypeptides such as these would be targeted inefficiently to chloroplasts or mitochondria. An alternate hypothesis suggests that certain genes may remain in the organelle because their products require cotranslational assembly steps.

Recent studies invalidate suggestions such as these. Two chloroplast-encoded polypeptides, when linked to the ribulose-1,5-bisphosphate carboxylase (Rubisco) small subunit (SSU), transit peptides, are imported in vivo into chloroplasts or mitochondria. The Rubisco large subunit (Hurt et al. 1986) and the quinone-binding thylakoid membrane protein Q_B (Cheung et al. 1988), follow the posttranslational transport pathway into mitochondria or chloroplasts when synthesized in the cytoplasm as precursors with amino terminal transit peptides. Furthermore, the Q_B polypeptide functionally assembles into the photosystem II reaction center. Thus, this protein does not require a cotranslational assembly step for insertion into the thylakoid membrane.

The capability of chloroplasts to import polypeptides that have extensively folded prior to the initiation of transport was dramatically demonstrated in the analysis of the import of the precursor of 5-*enol*-pyruvylshikimate-3-phosphate (EPSP) synthase into isolated chloroplasts (della-Cioppa et al. 1986). The precursor enzyme was obtained by in vitro transcription of a full-length cDNA followed by cell-free translation of the transcript. Isolated chloroplasts efficiently imported the precursor. Remarkably, the in vitro translation products contained activity that catalyzed the conversion of phosphoenolpyruvate and shikimate-3-phosphate to EPSP. The precursor thus folded in such a way as to form the active site of the enzyme but also to maintain competence for transport. At least partial unfolding of this enzymatically active 55 kD precursor is likely to be required for its transport across the chloroplast envelope.

A requirement for partial unfolding was demonstrated by Eilers and Schatz (1986) in their analysis of the transport of a chimeric protein composed of a mitochondrial-targeting leader sequence fused to mouse dihydrofolate reductase (DHFR). The folate antagonist methotrexate blocked import of the fusion protein into isolated mitochondria. The authors contend that a tight complex formed between the drug and the DHFR moiety of the precursor that could not be unfolded by the transport machinery of the mitochondria.

Outlines of a common pathway of protein transport are beginning to emerge. We will review recent studies on the transport of proteins from the cytoplasm into chloroplasts in the context of this broader perspective.

Binding of precursor proteins to the chloroplast envelope

After release from ribosomes, chloroplast precursor proteins bind to the cytoplasmic surface of the chloroplast outer membrane. Earlier studies (for review see Cashmore et al. 1985, Schmidt and Mishkind 1986) demonstrated that binding does not require energy and is mediated by protein-containing components present at the surface of the outer envelope membrane. The failure of the mature form of an imported polypeptide to bind isolated chloroplasts or to be transported (della-Cioppa et al. 1986) strongly suggested that the transit peptide mediates binding of the precursor to the envelope.

Detailed quantitative analysis of the binding of a precursor to an organelle surface has been achieved only in mitochondria. These experiments were facilitated by the availability of large amounts of the outer membrane protein porin (Kleene et al. 1987, Pfaller and Neupert 1987). The analysis exploited the property that porin is not proteolytically processed during its assembly in the outer membrane. Hence, rather than generating precursors by in vitro translation, mitochondria were used as a source of the polypeptide in a form that retained many of its transport characteristics. At low temperatures, porin bound to protease-sensitive sites in the mitochondrial outer membrane. The authors point out that since the affinity of binding ($K_a = 1$–$5 \times 10^8 \, \text{M}^{-1}$) was similar to that of hormones for receptors in the plasma membrane, the sites may represent receptors for import. Binding to the high affinity sites is followed by assembly of porin as an integral protein in the mitochondrial outer membrane.

The availability of large amounts of precursor polypeptides has also permitted competition studies designed to determine whether different imported polypeptides bind to the same class of receptors. Pfaller and Neupert (1987) found that porin effectively competes with the precursors of the ADP/ATP carrier for binding to the outer membrane and for import. These data suggest that although the ADP/ATP carrier is transported to the inner mitochondrial membrane, it may share initial binding sites with porin. In contrast, apocytochrome c, prior to its transport to the outer surface of the inner membrane, binds sites distinct from those occupied by the ADP/ATP carrier (Zimmermann et al. 1981).

As a step towards the biochemical characterization of receptors, selective binding of precursors to vesicles reconstituted from chloroplast and mitochondrial membrane preparations has been accomplished. Binding is rapid and of high affinity, specific for sites on the outer membrane, and specific for precursor forms of organellar proteins. In chloroplasts, Pfisterer et al. (1982) found that envelope vesicles bound translation products from mRNA of greening plants. Light-harvesting chlorophyll a/b apoproteins (LHCPs) and the SSU precursor were enriched in the vesicles. Binding of chloroplast precursors did not require ATP and was sensitive to protease digestion. Bound, radiolabeled precursors were competitively removed by excess unlabeled protein. Bitsch and Kloppstech (1986) reconstituted the binding of nuclear-encoded chloroplast precursors from solubilized envelopes. Similarly, Riezman et al. (1983) were able to reconstitute cytochrome b_2 precursor binding activity from solubilized mitochondrial outer membrane vesicles. The identification of a potential receptor polypeptide in chloroplast envelopes has recently been reported (Pain et al. 1988, see 'Future Prospects', for a discussion).

One approach for identifying receptor proteins has been to screen antibodies raised to outer membrane preparations for inhibitory activity towards precursor binding. Ohba and Schatz (1987a) reported that antibodies directed to a 45 kD outer membrane protein inhibited import of precursors by isolated mitochondria. Mild trypsinization destroys the import capacity of isolated mitochondria, presumably by disruption of a proteinaceous import receptor. Mitoplasts, mitochondria from which the outer membrane is removed, were generated from trypsinized mitochondria. While the mitoplasts displayed restored capacity for protein uptake, import became insensitive to the antibody to the 45 kD outer membrane protein (Ohba and Schatz 1987b). This polypeptide thus may participate in binding precursors at an early stage in the import process.

Covalent crosslinking reagents have also been employed to detect receptors for protein import. Preliminary chemical crosslinking studies with chloroplast envelopes indicate the presence of receptor proteins that bind precursor polypeptides (Cornwell and Keegstra 1987). In mitochondria, Gillespie (1987) crosslinked a 30 kD outer membrane protein to a synthetic transit peptide. The binding of the transit peptide was saturable and reversible. Scatchard analysis predicted that the 30 kD protein constitutes 4–10% of the protein in the outer mitochondrial membrane. Mild trypsin treatment that eliminated import by mitochondria also prevented crosslinking of the 30 kD protein to the transit peptide. A relationship of the 30 kD polypeptide to the 45 kD outer membrane protein described above, however, has not been reported.

Proteinaceous receptors appear to be a general component of protein transport systems. A signal sequence receptor has been detected in the ER that binds the signal sequence after its release from the signal recognition particle (Wiedmann et al. 1987). This 35 kD integral membrane glycoprotein may play a role analogous to the chloroplast and mitochondrial import receptors discussed above.

Energy requirements for transport

The role of ATP as an energy source for protein transport into organelles (for review, see Hurt 1987) was established initially in chloroplasts (Grossman et al. 1980, for review see Cashmore et al. 1985, Schmidt and Mishkind 1986). Flugge and Hinz (1986) have extended earlier studies with a quantitative analysis of the ATP requirements for protein import. When chloroplasts are incubated in the dark, protein import capacity steeply declines and the concentration of ATP falls to about one-third of that present under illuminated conditions. The residual pool of ATP in dark-adapted chloroplasts probably remains tightly bound to thylakoid ATP synthase complexes and thus not available to support transport (Flugge and Hinz 1986). Illumination or addition of exogenous ATP to darkened plastids reestablishes both stromal ATP levels and transport capacity. Transport is restored as well when ATP is supplied in the form of metabolites (e.g. dihydroxyacetone phosphate/oxaloacetate/Pi) that enter the stroma and promote ATP synthesis in darkness (Flugge and Hinz 1986, Pain and Blobel 1987). The uptake of the Rubisco SSU precursor by isolated spinach chloroplasts displayed an apparent K_a for MgATP of about 0.9 mM. Maximal import occurred at about 5 mM MgATP (Flugge and Hinz 1986). The nucleotide requirement is specific for ATP; other nucleotides including dATP and ADP do not support protein uptake while GTP, CTP and certain non-hydrolyzable analogues are somewhat inhibitory (Flugge and Hinz 1986, Pain and Blobel 1987).

Studies of protein transport into non-photosynthetic plastids such as etioplasts (Schindler and Soll 1986) and leucoplasts (Boyle et al. 1986, Strzalka et al. 1987) substantiate the ATP requirement for transport. These plastid types require exogenous ATP to support protein uptake both in the light and the dark. The ATP levels that optimally support protein uptake into isolated leucoplasts (0.5 to 1.0 mM) are in the same range as those found in the cytoplasm of plant cells (Flugge and Hinz 1986). Thus, non-photosynthetic cells probably maintain ATP at levels adequate to support protein import into plastids.

While it is not understood how the hydrolysis of ATP is coupled to transport, the participation of an electrochemical gradient can be ruled out. The protonophore carbonylcyanide m-chlorophenylhydrazone (CCCP) and the ionophores nigericin, monensin, and A23187 effectively block protein import (Cline et al. 1985a, Flugge and Hinz 1986, Pain and Blobel 1987). These reagents deplete the organnele of ATP. Exogenously supplied ATP, in the presence of CCCP or the ionophores, restores protein import to uninhibited levels. Thus, a pH gradient is not required for protein uptake. Moreover, the ionophores nigericin, monensin and A23187 selectively exchange protons for, respectively, K^+, Na^+, and divalent cations. Gradients of these cations therefore do not support protein import.

An important step towards understanding the mechanism of protein import will be to localize the site where transport-linked ATP hydrolysis occurs. Towards this end, recent studies have sought to distinguish whether the ATP that supports protein import is required inside or outside the organelle. Flugge and Hinz (1986) manipulated internal and external ATP concentrations in isolated chloroplasts. As discussed above, incubation in darkness depletes internal ATP levels. External application of ATP replenishes internal pools; when ATP is supplied with glycerate, however, the stromal ATP concentration remains at the level found in darkness. Glycerate presumably enters the stroma where its conversion to 3-phosphoglycerate depletes internal ATP pools. Despite the low level of ATP present in the stroma of glycerate-treated plastids, protein import is similar to that observed in untreated controls and illuminated plastids. A role for ATP at the external surface of chloroplasts is reinforced by studies that utilized an inhibitor of the envelope ATP transporter (Flugge and Hinz 1986). The ATP derivative adenosine-2′,3′-dialdehyde-5′-triphosphate (oATP) substantially inhibits ATP exchange between chloroplasts and the external medium but does not affect stromal ATP levels. This ATP derivative inhibits light-driven protein import but has little effect on import into darkened chloroplasts supplied with external ATP. Although more detailed analysis will be required to verify the conclusion, these data suggest that during in vitro transport photosynthetically generated ATP must be exported from the chloroplast for use in protein import. In agreement with the site of ATP utilization, Flugge and Hinz (1986) demonstrated that divalent cations are required for protein import (Cline et al. 1985a) at a location outside the chloroplast.

Experiments performed with agents that deplete external pools of ATP provide evidence in apparent contradiction with the results described above. Neither glucose/hexokinase, fructose/fructose 6-phosphate kinase, nor apyrase effectively inhibit protein import when ATP is supplied from within the chloroplast. Protein import continues under these conditions if ATP synthesis is driven in the light by photosynthesis or in the dark by dihydroxyacetone phosphate/oxaloacetate/P_i (Flugge and Hinz 1986, Pain and Blobel 1987). Since the outer envelope membrane does not pose a diffusion barrier to small molecules such as nucleoside triphosphates, these consuming systems should deplete ATP in the inter-membrane space as well as in the external medium (Pain and Blobel 1987). These data can be reconciled if transport utilizes ATP bound at sites in the inter-membrane space where the non-penetrating enzymes of the ATP-consuming systems do not gain access.

Another non-penetrating enzyme, alkaline phosphatase, markedly inhibits protein import (Flugge and Hinz 1986). Unlike the other ATP-consuming systems, however, this enzyme dephosphorylates proteins as well as nucleoside triphosphates. The inhibitory activity of alkaline phosphatase thus may be due to its reaction with a phosphorylated protein component of the transport apparatus present in solution or at the outer aspect of the outer membrane.

In mitochondria, both nucleoside triphosphates and an energized inner membrane are required for protein import (Pfanner and Neupert 1986, for review see Hurt 1987). The ATP is used primarily at the outer aspect of the inner membrane; non-hydrolyzable analogs could not be substituted. Although several nucleotides support protein uptake by mitochondria (Pfanner et al. 1987, Eilers et al. 1987, Chen and Douglas 1987), Chen and Douglas (1987) argue that the mitochondrial transport apparatus is specific for ATP. The broader specificity is likely the consequence of a nucleoside diphosphokinase present in the mitochondrial outer membrane and inter-membrane space that phosphorylates ADP at the expense of nucleoside triphosphates (Chen and Douglas 1987).

ATP appears to be required both to initiate transport into mitochondria and facilitate movement of the polypeptide across the lipid bilayer (Chen and Douglas 1987, Pfanner et al. 1987). For example, transport of the β-subunit of the F_1-ATPase into isolated mitochondria is blocked at an intermediate step in assays conducted at 4 °C (Schleyer and Neupert 1985). The bulk of the polypeptide remains exposed at the exterior surface of the mitochondrion, whereas its amino terminus extends through a contact site between the outer and inner membranes into the matrix. The formation of this transfer intermediate requires both ATP and a membrane potential. In contrast, subsequent transfer of the polypeptide to the matrix requires ATP hydrolysis but occurs in the absence of a membrane potential (Chen and Douglas 1987, Pfanner et al. 1987). Pfanner et al. (1987) found that higher levels of ATP are necessary to translocate the polypeptide across the inner membrane than to mediate the formation of the transfer intermediate. An ATP requirement has also been noted for the incorporation of the ATP/ADP carrier and porin into the outer mitochondrial membrane (Pfanner et al. 1987). Subsequent transfer of the ATP/ADP carrier to its site of function in the inner membrane requires a membrane potential but not ATP (Pfanner and Neupert 1987).

Unlike protein uptake by chloroplasts and the ER (Walter and Lingappa 1986) and protein secretion by E. coli (Chen and Tai 1987), protein transport into mitochondria requires a membrane potential in addition to ATP hydrolysis. The membrane potential-requiring step correlates with the association of transported proteins with the inner membrane. The necessity for an electrochemical gradient may be a consequence of the unique properties of this highly specialized energy-transducing membrane system.

Hydrolysis of ATP has emerged as a critical step in protein transport in chloroplasts and mitochondria, as well as in E. coli (Chen and Tai 1987) and the ER (Walter and Lingappa 1986). A role for ATP in the unfolding of precursors at an early stage of the import process is suggested by recent mitochondrial studies. Partially unfolded precursors were found not to require ATP for import into isolated mitochondria (Pfaller and Neupert 1987, Verner and Schatz 1987). Presumably, tightly folded precursors must be converted to less compact forms in order for transport to initiate. The transport apparatus may therefore include ATP-dependent unfolding enzymes, associated with the organelle surface or in a soluble form in the cytoplasm. Pfanner et al. (1987) provided preliminary evidence for soluble ATP-dependent unfolding activity in reticulocyte lysates. Rothman and Kornberg (1986) have postulated that ATP-driven unfolding enzymes may serve not only in protein transport but also in other cellular processes that require protein unfolding such as ATP-dependent protein degradation.

Other potential roles for ATP include as a substrate for a membrane-bound protein-translocating enzyme or as a phosphoryl donor for phosphorylation of a translocation component (Flugge and Hinz 1986, Pain and Blobel 1987, Hurt 1987). Although precursors themselves may be phosphorylated during transport, preliminary experiments failed to detect phosphorylation of the Rubisco SSU precursor during import into chloroplasts (Flugge and Hinz 1986).

Properties of transit peptides

Transit peptides and signal peptides
The amino terminal extensions of chloroplast precursor proteins were termed transit sequences to distinguish these hydrophilic peptides that mediate posttranslational transport from the hydrophobic amino terminal signal sequences that mediate cotranslational transfer of proteins into the ER (Chua and Schmidt 1979). Despite the recently identified similarities in transport mechanisms, the terms transit and signal sequence (or peptide) retain their utility. The dramatic differences in structure of the leader sequences of chloroplast and mitochondrial precursors on the one hand, and secreted proteins on the other, clearly separate two classes of targeting peptides (Douglas 1987). Leader peptides of chloroplast and mitochondrial precursor proteins are rich in basic, hydroxylated, and hydrophobic amino acids (Roise et al. 1986, Allison and Schatz 1986, Schmidt and Mishkind 1986, Karlin-Neumann and Tobin 1986). Acidic amino acids are typically absent, but one or more glycines or prolines are often found. Extended hydrophobic domains are not observed; instead, positively charged residues are interspersed throughout the peptide separated by hydrophobic or hyd-

roxylated residues. Proteins that are transported to the thylakoid lumen, however, contain a region towards the carboxy end of their transit peptides that is similar to the hydrophobic domain found in signal sequences (see below).

In contrast, signal peptides are characterized by a dramatically different pattern of amino acids (von Heijne 1985). A short amino terminal domain contains several positively charged amino acid residues, and at least one alpha-helix destabilizing residue. A hydrophobic domain, 14–21 residues long (predominantly alanine, valine, and leucine), follows. The central region of the hydrophobic domain usually contains one glycine or proline residue, both of which are known to be involved in beta-turn structure of proteins. At the cleavage site of eukaryotic signal peptides small or neutral amino acids (glycine, alanine, serine, threonine) are found frequently (95 and 68% at positions -1 and -3 with respect to the processing site, respectively). Signal peptides are predicted to have a beta-sheet structure (43%) in aqueous solution but alpha-helical structure (46%) in non-polar solvents. A beta-turn potentially precedes the cleavage site (Duffaud et al. 1986).

Physical properties of transit peptides
The physical properties of chloroplast transit peptides have not been reported. Given the similarities, however, recent studies of mitochondrial leader peptides should be at least partially applicable to chloroplasts. Although no significant sequence homology can be found among the leader peptides of mitochondrial precursors, most have the potential to form amphiphilic helices at membrane-water interfaces (Tamm 1986, von Heijne, 1986b). The presence of negatively charged phospholipids facilitates the uptake of the transit peptide into lipid monolayers. The helix incorporates with its long axis parallel to the monolayer (Tamm 1986). In studies designed to test the role of these helices in transport, synthetic peptides corresponding to all or part of the leader sequence of the precursor of subunit IV of yeast cytochrome oxidase were assayed for their effects on phospholipid bilayers (Roise et al. 1986). The peptides were soluble in water and surface active; they inserted into phospholipid monolayers and caused phospholipid vesicles to become leaky. Gillespie et al. (1985) observed that a synthetic leader peptide that corresponded to the 27 amino terminal residues of pre-ornithine carbamyl transferase (pOCT) could collapse the electrochemical gradient across the mitochondrial inner membrane.

The targeting and transport capacity of artificial leader sequences has been analyzed. Allison and Schatz (1986) designed several peptides that were similar to mitochondrial leader sequences in charge and hydrophobicity but substantially different in amino acid composition. Each of these peptides, when fused to the amino terminus of subunit IV of yeast cytochrome oxidase, directed the polypeptide into mitochondria (Allison and Schatz, 1986). These experiments conclusively demonstrate that higher order structures of a leader peptide, rather than its primary sequence, are critical in membrane transport.

Transit peptides of stromal proteins

Inspection of the amino acid composition and sequence of transit peptides from various species reveals certain features held in common (Mishkind et al. 1985, Schmidt and Mishkind 1986, Karlin-Neumann and Tobin 1986). Transit peptides of SSUs from vascular plants and the alga *Chlamydomonas reinhardtii* contain segments that display similar primary structures. A section especially rich in serine directly follows the amino terminal met-ala dipeptide. The central region of SSU transit sequences contains the tripeptide gly–leu–lys (with the substitution of ala for gly in *C. reinhardtii*). Invariably, pro and val precede the tripeptide at positions -2 and -4, respectively. A third conserved domain is found near the carboxyl terminus of SSU (as well as ferredoxin and LHCP) transit sequences from vascular plants. Located two residues upstream from the processing site where the mature polypeptide is generated, this domain contains a tripeptide in which each amino acid in the sequence carries a characteristic sidechain: small uncharged (gly or ala) – basic (arg or lys) – hydrophobic (val, ile, or phe). The absence of this tripeptide at the carboxyl terminus of the transit sequence of *C. reinhardtii* pSSU is consistent with differences in transit peptidases in vascular plants and the alga (for review see Schmidt and Mishkind 1986).

The functional significance of the conserved domains in the small subunit transit peptides is not known. Although similar structures are found in pre-ferredoxin, a polypeptide targeted to the stroma, the leader peptides of LHCP precursors contain the tripeptide associated with the processing site but lack the central domain conserved among the SSU precursors (Schmidt and Mishkind 1986, Karlin-Neumann and Tobin 1986, Kohorn and Tobin 1986). The LHCP transit peptides, ranging in length from 34–37 residues, are somewhat smaller than their counterparts on the SSU (44 to 57 residues). Although the LHCP precursors assemble as integral membrane proteins in the thylakoids, their leader sequences resemble those of proteins targeted to the stroma: large hydrophobic domains are absent, basic and hydroxylated amino acids predominate.

Targeting information that directs proteins to non-photosynthetic plastids is only beginning to be examined. One example is the waxy protein

(UDP-glucose starch glycosyl transferase), a polypeptide bound to starch grains in the stromal compartment of amyloplasts. Its leader sequence of 72 amino acids is larger than the transit peptides of other proteins targeted to the stroma but displays the typical abundance of basic and hydroxylated residues (Klosgen et al. 1986). A tripeptide (gly–arg–phe) located six residues upstream from the processing site is similar to tripeptides located near the processing sites of other precursors.

Transit peptides of thylakoid lumen proteins
Several precursors traverse the chloroplast envelope and the thylakoid membrane enroute to sites of action within the thylakoid lumen. The multiple transport events experienced by these precursors are reflected in leader peptides (58–87 residues) somewhat larger than those that target proteins to the stroma. Furthermore, these sequences can be divided into two distinct structural domains. Leader peptides from the precursors of plastocyanin (Smeekens et al. 1985, Rother et al. 1986), apocytochrome *c*–552 (Merchant and Bogorad 1987), and the extrinsic polypeptides of the oxygen-evolving complex (Jansen et al. 1987, Tyagi et al. 1987, Mayfield et al. 1987) display this bipartite organization. Amino terminal domains resemble those of other chloroplast precursors: basic and hydroxylated residues are prevalent. Unique to this class of precursors, however, is a region rich in hyrophobic and uncharged residues found between the mature polypeptide and the amino terminal domain of the leader sequence. Translocation into the stroma appears to be mediated by the amino terminal domain whereas the hydrophobic region probably functions as a thylakoid-targeting segment that directs the polypeptide to the thylakoid lumen. The putative intermediate processing site at the junction of the hydrophilic and hydrophobic domains may form amphipathic beta-sheets (Tyagi et al. 1987). This region should contain a determinant recognized by the stromal processing enzyme that cleaves polypeptides after transfer across the envelope.

Recent studies confirm the dual function of the plastocyanin transit peptide. The experimental evidence has been recently reviewed (Weisbeek et al. 1986, Weisbeek and Smeekens 1988). By recombinant DNA methods a chimeric protein was generated that contained the transit peptide of plastocyanin fused to the amino terminus of mature ferredoxin (Smeekens et al. 1986). The plastocyanin transit peptide mediated the transport of the fusion protein into isolated chloroplasts. Some of the imported fusion proteins were recovered as fully processed polypeptides from the thylakoid lumen. Thus, targeting information in the plastocyanin transit peptide directed the stromal polypeptide ferredoxin to the thylakoid lumen. Fusion proteins were also recovered from the stromal compartment but these polypeptides had been processed to molecular weights intermediate in size between precursor and mature forms of the fusion protein (Smeekens et al. 1986).

These data were clarified by Hageman et al. (1986) who established that a water-soluble protease, recovered from thylakoids, processes the plastocyanin intermediate to its mature molecular weight. The stromal protease that removes the entire transit peptide from SSU precursors processes pre-plastocyanin to the intermediate molecular weight form. The intermediate, and not the full-length precursor, is the substrate for the thylakoid-associated enzyme. Thus, plastocyanin appears to be processed in two steps; the first is associated with transfer of the precursor from the cytoplasm to the stroma and the second occurs during or after transfer into the thylakoid lumen. A similar two-step processing pathway for the nuclear-encoded polypeptides of the water oxidation complex was inferred by Chia and Arntzen (1986) upon analysis of a tobacco mutant that lacks photosystem II.

Weisbeek and Smeekens (1988) note that the hydrophobic domain of the plastocyanin transit peptide is similar to the hydrophobic leader sequence present on the chloroplast-synthesized thylakoid membrane protein cytochrome f. This leader sequence is thought to mediate the transfer of the polypeptide into the thylakoid membrane. The authors propose that the two-domain leader peptides on plastocyanin and other imported thylakoid lumen polypeptides arose as a consequence of the endosymbiotic origin of chloroplasts. The stromal-targeting domain might have been added upon transfer of a gene for plastocyanin or other lumen polypeptides from the endosymbiont to the nucleus. The likelihood of such a transfer is suggested by the abundance of sequences in cytoplasmic proteins that can function as mitochondrial transport signals (Baker and Schatz 1987, Hurt and Schatz 1987, Vassarotti et al. 1987b) if located at the amino terminus of a passenger protein. A probable consequence of these events, as suggested by Weisbeek and Smeekens (1988), is that plastocyanin and cytochrome f may utilize the same transport pathway for entry into the thylakoid membrane.

A similar evolutionary scenario was proposed by Hartl et al. (1986) to explain the leader peptide and transport pathway of the mitochondrial Fe/S protein of ubiquinol-cytochrome c reductase. These authors hypothesized that the first step in targeting of certain polypeptides involves delivery of the protein to the compartment analogous to the one in which it was synthesized in the ancestor of the putative endosymbiont. Thereafter, the targeting machinery characteristic of the endosymbiont would direct the polypeptide to its site action. In the case of the Fe/S protein, targeting involves import to the matrix from the cytoplasm followed by export of a partially processed form of the protein to the other surface of the inner membrane.

An alternate targeting procedure involves 'stop transfer' sequences (Blobel 1980, Colman and Robinson 1986, Hurt and van Loon 1986). For example, cytochrome c_1 is transported from the cytoplasm to the inner mitochondrial membrane. The leader peptide contains a matrix-targeting domain at its amino terminus that initiates transfer of the polypeptide across the inner membrane. A stop transfer domain, composed of 19 uncharged amino acids followed by a negatively charged region, appears to block further transfer of the protein through the inner membrane (van Loon and Schatz 1987). A potential role for stop transfer sequences in chloroplast protein import has been brought into question. Lubben et al. (1987) found that heterologous stop transfer sequences from ER-targeted proteins, when incorporated into the precursor of the SSU, fail to prevent translocation of the polypeptide to the stroma. In contrast, a stop transfer sequence that lacked effect in chloroplasts was found to direct a mitochondrial precursor to the inner membrane rather than the matrix (Nguyen and Shore 1987). Definitive evaluation of a role for these sequences in chloroplast protein import, however, will require analysis of the translocation pathways of cytoplasmically synthesized envelope proteins.

Functional analysis of transit peptides
Functional analysis by in vitro mutagenesis is beginning to reveal whether the conserved sequence characteristics of transit peptides are critical for transport. The initial studies in which SSU transit peptides were shown to direct heterologous proteins into chloroplasts utilized two distinct gene constructs. In one, the coding region for the transit peptide was fused directly to the gene for the passenger protein neomycin phosphotransferase II (NPT II) (Van den Broeck et al. 1985). The other construct contained information for the transit sequence as well as the first 23 amino acids from the amino terminus of the mature SSU linked to the NPT II gene (Schreier et al. 1985). The chimeric genes were introduced into plants by *Agrobacterium tumefaciens*-mediated transformation. That the products of both of these gene fusions were transported in vivo into chloroplasts proved the sufficiency of the transit sequence in supporting import as well as the dispensable nature of specific determinants in the mature polypeptide.

Despite the impressive results obtained with transformed plants, the dissection of transit peptides into functionally important domains has relied primarily on in vitro systems. The speed and convenience of in vitro analysis permits large numbers of constructs to be analyzed. Moreover, in vivo data obtained from transformed plants can be difficult to interpret. For example, failure to recover a protein from chloroplasts may be due to several factors that are difficult to distinguish. Mutations may affect the interaction of a protein with the transport apparatus. Alternatively, amino acid changes may render the polypeptide susceptible to degradation by cytoplasmic proteases, thereby reducing the amount of the polypeptide that enters the chloroplast (see Kuntz et al. 1986, for discussion).

Recent studies performed with isolated chloroplasts have utilized efficient cell-free transcription/translation systems (e.g. Melton et al. 1984). Chimeric or mutated genes constructed by recombinant DNA methods or cDNAs that encode chloroplast protein precursors are incorporated into plasmids that have been engineered for transcription of the inserted DNA. Transcription initiates at bacteriophage promoters (such as SP6 or T7) that are specific for particular RNA polymerases. The transcripts obtained in vitro are translated in wheat germ or reticulocyte cell-free systems to generate precursor proteins. These polypeptides are then incubated with chloroplasts to assay transport.

Cell-free transport experiments performed with chimeric protein constructs have shown that although transit peptides alone mediate the import of a protein into chloroplasts, characteristics of the protein linked to the transit peptide can drastically alter the rate at which import occurs. Wasmann et al. (1986) assayed the transport efficiency of the transit peptide-NPT II fusion proteins that had been examined in transformed plants described above. The construct that contained the transit peptide and 23 amino acids of the mature SSU polypeptide entered isolated chloroplasts at least 10-fold more rapidly than the polypeptide in which the transit peptide was linked directly to NPT II. Differences in transport efficiency were also reported by Lubben and Keegstra (1986), who linked a 17.5 kD cytoplasmic heat shock protein to the Rubisco SSU transit peptide. The chimeric protein entered chloroplasts at 40% of the efficiency of the small subunit precursor.

The dramatic variation in transport efficiency observed when different proteins are linked to SSU transit peptides has led to the suggestion that although the transit peptide carries functional determinants sufficient for transport, the secondary structure of the precursor is also of critical importance (Lubben and Keegstra 1986, Wasman et al. 1986). Whereas nuclear-encoded chloroplast protein precursors likely attain conformations that foster import, heterologous polypeptides may influence folding of precursors such that transport is inhibited. Certain three-dimensional structures may interfere with movement of the precursor across lipid bilayers or hinder binding of precursors to sites on the exernal surface of the organelle. In addition, determinants present in the mature polypeptide may be required for optimal transport rates.

The influence of passenger proteins on transport efficiencies into yeast mitochondria has been examined. The 34 N-terminal amino acids (the 26 residue leader sequence plus 8 amino acids from the mature polypeptide) from the precursor to mitochondrial manganese-superoxide dismutase were fused to invertase and DHFR. The DHFR fusion protein efficiently entered mitochondria but most of the invertase fusion protein accumulated in the cytoplasm (Van Steeg et al. 1986). Invertase appears to be exceptionally intractable to posttranslational transport. As discussed above, invertase requires ongoing translation for transport into the ER lumen (Rothblatt and Meyer 1986). Invertase may assume a rigid conformation incompatible with transport across membranes.

Recent mutagenesis studies on the SSU transit peptide clearly demonstrate the influence of the passenger protein on transport dynamics (Reiss et al. 1987). Transit peptides with defined deletions were linked either to the SSU or to NPT II. Deletions in the amino terminal domain (in which the 5 amino terminal residues were retained) or in the conserved central region of the transit peptide completely blocked or drastically reduced the level of transport in the NPT II constructs. In contrast, transport activity was relatively unaffected when identical deletions were introduced into the authenthic SSU precursor. In particular, removal of the central domain greatly inhibited the transport capacity of the NPT II construct (Wasmann et al. 1986) but had little effect when the SSU was the passenger protein (Reiss et al. 1987). Only when the amino terminal 24 residues were removed was transport of the small subunit greatly reduced.

Cytoplasmically synthesized proteins destined for the mitochondrial matrix traverse the outer and inner mitochondrial membranes en route to their site of action. The transport pathway of these polypeptides thus is similar to that of the SSU. Functional determinants in the leader sequences of several matrix proteins have been examined by in vitro mutagenesis. Unexpectedly, the location of regions critical for transport differs among precursors. Transport of pre-ornithine transcarbamylase into isolated rat liver mitochondria is blocked upon removal of 15 residues from the central region of its 32 amino acid leader peptide. Substitution of a glycine for an arginine residue located in this central region also blocked transport (Horwich et al. 1986). Whether the central region alone can support transport was not determined. In contrast, the region critical for transport of the precursor of yeast cytochrome oxidase subunit IV is located at the amino terminus of its leader sequence (Hurt et al. 1985).

Since the secondary structure of precursors probably influences transport dynamics, the effect of mutations introduced into transit peptides must be analyzed in the context of the conformation of the precursor as a whole as well as that of the transit peptide. Interpretation of in vitro mutagenesis experiments is further complicated by the possibility that components of the transport apparatus have overlapping or even redundant functions. Thus, deletions or substitutions introduced within functionally important determinants may not perturb transport. Redundancies have been demonstrated for the precursor of the mitochondrial F_1-ATPase beta-subunit (Bedwell et al. 1987). Any one of three domains, two in the leader sequence and one near the amino terminus of the mature polypeptide, support import of the polypeptide into mitochondria. The domains contain basic amino acid residues and appear to act cooperatively.

In vitro mutagenesis has been useful in identifying regions in transit peptides that are required for accurate proteolytic maturation of precursors. The pattern that has emerged from analysis of both chloroplast and mitochondrial precursors is that domains required for transport are distinct from those that specify proteolytic processing. Residues critical for processing can be deleted or mutated without loss of transport function. Furthermore, determinants required for accurate processing are distal from the amino acids at the processing site itself. In mitochondrial precursor proteins, critical determinants may be located in the leader sequence (ornithine transcarbamylase, Nguyen et al. 1987) or in the mature portion of the polypeptide (F_1-ATPase beta-subunit, Vassarotti et al. 1987b). Precursors which lack these domains are transported into mitochondria but accumulate as unprocessed or incorrectly processed polypeptides. The unprocessed F_1-ATPase beta-subunit assembles into a functional ATPase. In contrast, removal of the four amino terminal residues from the leader sequence of yeast cytochrome oxidase subunit IV results in loss of transport capacity and processing determinants (Hurt et al. 1987).

Analysis of processing determinants among chloroplast proteins has been limited to the SSU precursor. Small deletions (Kuntz et al. 1986) or insertions (Reiss et al. 1987) introduced at the processing site where the mature SSU arises have little effect on transport or processing. In contrast, a deletion mutant that lacks the 17 residues at the carboxyl terminus of the transit peptide retains transport capacity but is processed to a series of four discrete polypeptides intermediate in molecular weight between the precursor and mature forms of the small subunit. Two of these intermediates are at least partially sensitive to externally added protease and thus may represent species stalled at intermediate transport steps.

In an earlier study in which a carboxymethylated SSU precursor was imported into isolated pea chloroplasts, Robinson and Ellis (1984b, 1985) recovered a series of SSU processing intermediates similar to those described in the deletion study (Reiss et al. 1987). Presumably, a modified cysteine residue at the carboxyl terminus of the transit peptide occluded the processing site. A processing intermediate was also observed when unmodified pea SSU precursor was incubated with partially purified pea transit peptidase; this form appeared only transiently during processing and was converted to

the molecular weight of the mature SSU (Robinson and Ellis 1984b). These results suggest that the SSU precursor experiences two proteolytic processing events, the first occurring during or shortly after transport of the precursor through the envelope.

Two-step maturation of the SSU precursor is also suggested by in vitro transport experiments in which chloroplast protein precursors of algal origin were transported into vascular plant chloroplasts (Mishkind et al. 1985). The SSU precursor from the unicellular green alga *C. reinhardtii* is imported into pea chloroplasts and processed to form with a molecular weight intermediate between precursor and mature forms of the polypeptide. Cleavage occurs within the central region of transit peptide that is conserved between the algal and vascular plant SSU precursors. This domain thus appears to contain determinants recognized by the proteolytic maturation apparatus in vascular plants. A functional role for this domain is also suggested by the observation that import into isolated chloroplasts is dramatically inhibited when the pea SSU precursor is synthesized with beta-hydroxyleucine in place of leucine (Robinson and Ellis 1985, Bartlett et al. 1986). The only leucine residue in the transit peptide occurs within the conserved central region.

The functional significance for two proteolytic maturation sites in the SSU precursor is not known. Unlike thylakoid lumen polypeptides which are processed upon entry into the stroma and cleaved a second time when transferred across the thylakoid membrane, the SSU does not require a second transport event after its delivery to the stroma. Multiple processing sites, however, have also been observed within the leader sequences of the mitochondrial matrix polypeptide ornithine transcarbamylase (Sztul et al. 1987). The processing events are catalyzed by matrix-localized proteolytic activities. The intermediate is reported to span the inner and outer mitochondrial membranes. Observations consistent with membrane-spanning intermediates of SSU-NPT II fusion proteins in transformed plants have also been reported (Kuntz et al. 1986). Anderson and Smith (1986) suggest that intermediate processing sites might correspond to determinants recognized by proteases that degrade the transit peptide rather than regions associated with transport.

Future prospects

Targeting in plant cells
Transit peptides confer considerable selectivity in the delivery of proteins to chloroplasts and mitochondria. Tobacco plants were generated which ex-

pressed heterologous polypeptides targeted to the cytoplasm, chloroplasts or mitochondria (Boutry et al. 1987). The polypeptides were chloramphenicol acetyl transferase (CAT) or chimeras composed of CAT linked either to a chloroplast or plant mitochondrial transit peptide (from the SSU and ATPase β-subunit precursors, respectively). Analysis of the subcellular location of CAT in the transformed plants demonstrated that the heterologous polypeptides accumulated in the compartments appropriate to the transit peptide to which they were attached. Targeting to mitochondria was examined in most detail; recovery of CAT activity from various subcellular fractions closely followed that of the mitochondrial matrix marker enzyme, malate dehydrogenase. Although the delivery of proteins to chloroplasts and mitochondria in plants thus appears quite selective, these experiments were limited to the assay of steady-state enzyme levels. Consequently, the uptake and rapid degradation of a mistargeted polypeptide would not have been detected.

The striking structural similarity among transit peptides has prompted studies designed to test whether a chloroplast leader sequence can direct a polypeptide to mitochondria in yeast. In one report, the yeast mitochondrial enzyme manganese superoxide dismutase, when linked to transit peptides from either plastocyanin or ferredoxin, was not observed to be transported into isolated yeast mitochondria (Smeekens et al. 1987). In contrast, the 31 amino terminal residues of the transit peptide from the *C. reinhardtii* SSU precursor mediated the import of proteins into yeast mitochondria both in vitro and in vivo (Hurt et al. 1986). Although proteins linked to authentic mitochondrial-targeting sequences displayed considerably higher transport rates, these data raise questions about the specificity of targeting in plant cells. Do the leader peptides of the precursors to chloroplast and mitochondrial proteins in plants display distinct binding affinities for receptors on the cytoplasmic surface of the organelles? Alternatively, do components present in the plant cell cytoplasm selectively deliver precursor proteins to chloroplasts and mitochondria? Does mistargeting occur in plant cells? In vivo assays utilizing either transformed plants or transient expression systems will be useful for determining transit peptide domains that function in selective targeting.

Suborganellar transport pathways
Considerable information is currently available on import pathways of mitochondrial proteins. For example, a growing body of evidence implicates contact sites between the inner and outer membranes in transfer of proteins from the cytoplasm to the matrix (e.g. Schwaiger et al. 1987). Ultrastructural analysis of chloroplasts reveals similar dense 'contact' zones between the inner and outer envelope membranes (Carde et al. 1982, Cline et al. 1985b).

A recent study that employed anti-idiotypic antibodies suggests these zones contain binding sites for precursor proteins. Anti-idiotypic antibody methods, although in use for some time in the study of hormone receptors, represent a new and potentially powerful approach in the analysis of receptors for imported chloroplast proteins.

In this study, Pain et al. (1988) tested the possibility that a polyclonal antiserum directed towards a transit peptide might contain antibodies that bind the transit peptide in a manner similar to that of a receptor located on the cytoplasmic surface of the chloroplast envelope. Moreover, the antigen-binding site (or idiotypic region) of these antibodies could potentially elicit an antiserum that would recognize such a receptor. Pain et al (1988) raised antibodies to a synthetic peptide that corresponded to the 30 carboxy-terminal residues of the transit peptide from the pea SSU precursor. The anti-idiotypic antiserum elicited by the antibodies to the transit peptide bound a 30 kD envelope protein that displays properties consistent with a role as a receptor for SSU precursors. For example, monovalent Fab fragments produced from the antiserum inhibited the uptake of the SSU precursor into isolated chloroplasts. Ultrastructural immunocytochemistry localized the 30 kD polypeptide to contact zones between the inner and outer envelope membranes, a structure that has been suggested as a transport site (Dobberstein et al. 1977, Chua and Schmidt 1979).

Detailed characterization of the antibodies and the potential receptor molecule should provide insights into early transport steps. A critical question concerns the location of sites within the transit peptides of the SSU and other imported proteins that are recognized by the receptor. Also of interest will be the evaluation of whether the receptor assembles with other polypeptides to form a transport apparatus (perhaps, in the suggestion of Pain et al. (1988), as a pore complex located at membrane contact sites). Another important problem concerns the apparent cross-reactivity of the anti-idiotypic antibodies with Rubisco large subunits. Pain et al. (1988) suggest that if this binding proves specific, receptor sites on the large subunits for transit peptides might function in holoenzyme assembly. There will be interest as well to determine the relationship of the envelope polypeptide recognized by the anti-idiotypic antibodies to previously characterized envelope membrane proteins, in particular the abundant 30 kD phosphate translocator located in the inner membrane (Werner-Washburne et al. 1983, Block et al. 1983).

An understanding of the pathway followed by proteins as they traverse the envelope will require the development of experimental systems in which transport is blocked at intermediate steps. Certain mutations introduced into the transit peptide (e.g. Kuntz et al. 1986) may lead to the accumulation of transport intermediates.

The study of protein transport in chloroplasts has focused on the transfer of proteins across the envelope. Despite its central importance in the development of the photosynthetic apparatus, the mechanism of protein transport into and across the thylakoid membrane has received comparatively little attention. Evidence suggests that certain proteins are transferred to the thylakoids by a mechanism similar to that which mediates protein export in prokaryotes. For example, the precursor of cytochrome f, when expressed in *E. coli*, utilizes the bacterial transport apparatus for export across the cytoplasmic membrane (Rothstein et al. 1985).

Detailed characterization of protein uptake by thylakoids requires a system in which transport can be reconstituted with purified components. Cline (1986) and Chitnis et al. (1987) have described such a system for the uptake and assembly into isolated thylakoids of precursors to the LHCP. They found that integration of LHCP and other precursor proteins into the membrane requires a soluble stromal component and the magnesium salt of ATP. These data raise the intriguing possibility that transfer of polypeptides to thylakoid membranes includes an ATP-dependent unfolding step similar to that suggested for transport across other membrane systems. Characterization of the stromal components that promote protein incorporation into thylakoids should provide important insights into the transfer mechanism.

In contrast to other thylakoid membrane precursors, polypeptide domains that target LHCPs to the thylakoids have not been resolved. Unlike the thylakoid lumen polypeptides described above, transit peptides of the LHCP precursors lack hydrophobic targeting signals and appear to contain only information for the transfer of the polypeptide from the cytoplasm to the stroma. Rather, information for membrane integration may reside in the mature polypeptide. Cline (1986) suggests that a soluble form of LHCP apoprotein traverses the stoma enroute to the thylakoid membrane. The observation that LHCP precursors can be processed in vitro by soluble, presumably stomal, proteases (Marks et al. 1985, Lamppa and Abad 1987) is consistent with this hypothesis. Proteolytic maturation, however, is not a prerequisite for assembly into functional light-harvesting complexes (Cline 1986, Chitnis et al. 1986).

Other processes, such as the assembly of co-factors with apoproteins, may also be critical in transport pathways. By analogy with the import of cytochrome c into mitochondria, Merchant and Bogorad (1987) speculate that the incorporation of cytochrome into the thylakoid membrane may be linked to the covalent attachment of heme to the apoproteins. In mitochondria, conformational changes accompany the synthesis of holocytochrome c have been proposed to energize the translocation of this polypeptide across the outer membrane (for review, see Douglas et al. 1986). Similarly, the assembly of LHCPs in the thylakoid membrane may be facilitated by the association of the apoproteins with chlorophyll (Kohorn et al. 1986).

Components of the transport apparatus
Biochemical characterization of the transport apparatus in chloroplasts is just beginning. Potentially critical components include cytoplasmic factors that promote early steps in the transport process, receptors that bind precursors at the organelle surface, pore complexes that facilitate passage of polypeptides across the envelope, and proteases that cleave leader peptides from imported proteins.

The participation of the ribonucleoprotein complex, termed signal recognition particle, in the transport of proteins across the ER (Walter et al. 1984) suggests that cytosolic factors may facilitate chloroplast and mitochondrial protein import. Although not ruled out, a role for RNA is discounted by the observation that ribonuclease treatment does not perturb protein uptake into isolated chloroplasts (Bartlett et al. 1986). Soluble protein factors, however, have been implicated in the transfer of proteins into mitochondria (Argan et al. 1983, Miura et al. 1983, Ohta and Schatz 1984). Functions for transport factors are yet to be determined; prominent candidates include cytoplasmic (Pfanner et al. 1987) or membrane-bound (Eilers and Schatz 1988) protein unfolding enzymes.

Several distinct proteolytic maturation steps occur during protein import into chloroplasts. None of the proteases that catalyze these steps, however, have been purified to homogeneity. A partially purified preparation of the protease (Robinson and Ellis 1984a) that processes the precursor of the SSU to its mature form has been characterized (Robinson and Ellis 1984a, 1984b, for review see Ellis and Robinson 1985, Schmidt and Mishkind 1986). Detailed analysis of the substrate specificity, kinetics, and other properties of this enzyme, however, will require preparations of great purity. Determination of substrate specificity will also be of interest for the protease that processes the plastocyanin intermediate upon its transport into the thylakoid lumen. Hageman et al. (1986) suggest that this enzyme is analogous to prokaryotic signal peptidases and may process chloroplast-synthesized as well as imported polypeptides that function in the thylakoid lumen. The multiple forms of LHCPs have been suggested to arise, at least in part, as a consequence of proteolytic processing (Kohorn et al. 1986). These processing events, however, have not been analyzed in organelle-free extracts.

770

Acknowledgements

New Jersey Agricultural Experiment Station, Publication No. D 01506 01-88. Preparation of the review was supported by a grant DCB-8616291 from the National Science Foundation. The authors thank T. Chase, J. Jenkins, J. Reed-Scioli and B. Zilinskas for critical reading of the manuscript.

References

Allison DS and Schatz G (1986) Artificial mitochondrial presequences. Proc Natl Acad Sci USA 83: 9011–9015

Anderson S and Smith SM (1986) Synthesis of the small subunit of ribulose-bisphosphate carboxylase from genes cloned into plasmids containing the SP6 promoter. Biochem J 240: 709–715

Argan C, Lusty CJ and Shore GC (1983) Membrane and cytosolic components affecting transport of the precursor for ornithine carbamyltransferase into mitochondria. J Biol Chem 258: 667–670

Baker A and Schatz G (1987) Sequences from a prokaryotic genome or the mouse dihydrofolate reductase gene can restore the import of a truncated precursor protein into yeast mitochondria. Proc Natl Acad Sci USA 84: 3117–3121

Bartlett SG, Landry SJ and Pomarico SM (1986) Transport of proteins into chloroplasts. Curr Top Plant Biochem Physiol Vol 5, University of Missouri, Columbia, pp 105–115

Bedwell DM, Klionsky DJ and Emr SD (1987) The yeast F1-ATPase precursor B subunit contains functionally redundant mitochondrial import information. Mol Cell Biol 7: 4038–4047

Bitsch A and Kloppstech K (1986) Transport of proteins into chloroplasts. Reconstitution of the binding capacity for nuclear-coded precursor proteins after solubilization of envelopes with detergents. Eur J Cell Biol 40: 160–166

Blobel G (1980) Intracellular protein topogenesis. Proc Natl Acad Sci USA 77: 1496–1500

Blobel G and Dobberstein B (1975a) Transfer of proteins across membranes. I. Presence of proteolytically processed and unprocessed nascent immunoglobulin light chains on membrane-bound ribosomes of murine myeloma. J Cell Biol 67: 835–851

Blobel G and Dobberstein B (1975b) Transfer of proteins across membranes. II. Reconstitution of functional rough microsomes from heterologous components. J Cell Biol 67: 852–862

Block MA, Dorne A-J, Joyard J and Douce R (1983) Preparation and characterization of membrane fractions enriched in outer and inner envelope membranes from spinach chloroplasts. I. Electrophoretic and immunochemical analysis. J Biol Chem 258: 13273–13280

Boutry M, Nagy F, Poulsen C, Aoyagi K and Chua N-H (1987) Targeting of bacterial chloramphenicol acetyltransferase to mitochondria in transgenic plants. Nature 328: 340–342

Boyle SA, Hemmingsen SM and Dennis DT (1986) Uptake and processing of the precursor to the small subunit of ribulose 1,5-bisphosphate carboxylase by leucoplasts from the endosperm of developing castor oil seeds. Plant Physiol 81: 817–822

Carde J-P, Joyard J and Douce R (1982) Electron microscopic studies of envelope membranes from spinach plastids. Biol Cell 44: 315–324

Cashmore A, Szabo L, Timko M, Kausch A, Van den Broeck G, Schreier P, Bohnert H, Herrera-Estrella L, Van Montagu M and Schell J (1985) Import of polypeptides into chloroplasts. Biotechnology 3: 803–808

Chen W-J and Douglas MG (1987) Phosphodiester bond cleavage outside mitochondria is required for the completion of protein import into the mitochondrial matrix. Cell 49: 651–658

Chen L and Tai PC (1987) Evidence for the involvement of ATP in co-translational protein translocation. Nature 328: 164–166

Chen L, Rhoads D and Tai PC (1985) Alkaline phosphatase and OmpA protein can be translocated posttranslationally into membrane vesicles of *Escherichia coli*. J Bacteriol 161: 973–980

Cheung AV, Bogorad L, Van Montagu M and Schell J (1988) Relocating a gene for herbicide tolerance: A chloroplast gene is converted into a nuclear gene. Proc Natl Acad Sci 85: 391–395

Chia CP and Arntzen CJ (1986) Evidence for two-step processing of nuclear-encoded chloroplast proteins during membrane assembly. J Cell Biol 103: 725–731

Chitnis PR, Harel E, Kohorn BD, Tobin EM and Thornber JP (1986) Assembly of the precursor and processed light-harvesting chlorophyll *a/b*-protein of *Lemna* into the light-harvesting complex II of barley etiochloroplasts. J Cell Biol 102: 982–988

Chitnis PR, Nechushtai R and Thornber JP (1987) Insertion of the precursor of the light-harvestig chlorophyll *a/b*-protein into the thylakoids requires the presence of a developmentally regulated stromal factor. Plant Mol biol 10: 3–11

Chua N-H and Schmidt GW (1978) Post-translational transport into intact chloroplasts of a precursor to the small subunit of ribulose-1,5-bisphosphate carboxylase. Proc Natl Acad Sci USA 75: 6110–6114

Chua N-H and Schmidt GW (1979) Transport of proteins into mitochondria and chloroplasts. J Cell Biol 81: 461–483

Cline K (1986) Import of proteins into chloroplasts: membrane integration of a thylakoid precursor protein reconstituted in chloroplast lysates. J Biol Chem 261: 14804–14810

Cline K, Werner-Washburne M, Lubben TH and Keegstra K (1985a) Precursors to two nuclear-encoded chloroplast proteins bind to the outer envelope membrane before being imported into chloroplasts. J Biol Chem 260: 3691–3696

Cline K, Keegstra K and Staehelin LA (1985b) Freeze-fracture electron microscopic analysis of ultrarapidly frozen envelope membranes on intact chloroplasts and after purification. Protoplasma 125: 111–123

Colman A and Robinson C (1986) Protein import into organelles: hierarchical targeting signals. Cell 46: 321–322

Cornwell KLK and Keegstra K (1987) Evidence that a chloroplast surface protein is associated with a specific binding site for the precursor to the small subunit of ribulose-1,5-bisphosphate carboxylase. Plant Physiol 85: 780–785

della-Cioppa G, Bauer SC, Klein BK, Shah DM, Fraley RT and Kishore GM (1986) Translocation of the precursor of 5-enolpyruvylshikimate-3-phosphate (EDSP) synthase into chloroplasts of higher plants in vitro. Proc Natl Acad Sci USA 83: 6873–6877

della-Cioppa G, Kishore GM, Beachy RN and Fraley RT (1987) Protein trafficking in plant cells. Plant Physiol 84: 965–968

Dobberstein B, Blobel G and Chua N-H (1977) In vitro synthesis and processing of a putative precursor for the small subunit of ribulose-1,5-bisphosphate carboxylase of *Chlamydomonas reinhardtii*. Proc Natl Acad Sci USA 74; 1082–1085

Douglas MG (1987) Hydrophobic and hydrophilic signals in protein sorting. Protein Engineering 1: 80–81

Douglas MG, McCammon M and Vassarotti A (1986) Targeting proteins into mitochondria. Microbiol Rev 50: 166–178

Duffaud GD, Lehnhardt SK, March PE and Inouye M (1986) Structure and function of the signal peptide. In Current Topics in Membranes and Transport, Vol 24, Academic Press, London, pp 65–104

Eilers M and Schatz G (1986) Binding of a specific ligand inhibits import of a purified precursor protein into mitochondria. Nature 322: 228–232

Eilers M, Oppliger W and Schatz G (1987) Both ATP and an energized inner membrane are required to import a purified precursor protein into mitochondria. EMBO J 6: 1073–1077

Eilers M and Schatz G (1988) Protein unfolding and the energetics of protein translocation across biological membranes. Cell 52: 481–483

Ellis RJ (1981) Chloroplast proteins: synthesis, transport and assembly. Ann Rev Plant Physiol 32: 111–137

Ellis RJ and Robinson C (1985) Post-translational transport and processing of cytoplasmically-synthesized precursors of organellar proteins. In the Enzymology of Post-translational Modification of Proteins, Vol 2, Academic Press, London, pp 25–39

Flugge UI and Hinz G (1986) Energy dependence of protein translocation into chloroplasts. Eur J Biochem 160: 563–570

Gillespie LL (1987) Identification of an outer mitochondrial membrane protein that interacts with a synthetic signal peptide. J Biol Chem 262: 7939–7942

Gillespie LL, Argan C, Taneja AT, Hodges RS, Freeman KB and Shore GC (1985) A synthetic signal peptide blocks import of precursor proteins destined for the mitochondrial inner membrane or matrix. J Biol Chem 260: 16045–16048

Grossman A, Bartlett SG and Chua N-H (1980) Energy-dependent uptake of cytoplasmically-synthesized polypeptides by chloroplasts. Nature 285: 625–628

Hageman J, Robinson C, Smeekens S and Weisbeek P (1986) A thylakoid processing protease is required for complete maturation of the lumen protein plastocyanin. Nature 324: 567–569

Hansen W, Garcia PD and Walter P (1986) In vitro protein translocation across the yeast endoplasmic reticulum: ATP-dependent post-translation translocation of the prepro-α-factor. Cell 45: 397–406

Hartl F-U, Schmidt B, Wachter E, Weiss H and Neupert W (1986) Transport into mitochondria and intramitochondrial sorting of the Fe/S protein of ubiquinol-cytochrome c reductase. Cell 47: 939–951

Highfield PE and Ellis RJ (1978) Synthesis and transport of the small subunit of chloroplast ribulose bisphosphate carboxylase. Nature 271: 420–424

Horwich AL, Kalousek F, Fenton WA, Pollock RA and Rosenberg LI (1986) Targeting of pre-ornithine transcarbamylase to mitochondria: Definition of critical regions and residues in the leader peptide. Cell 44: 451–459

Hurt EC (1987) Unravelling the role of ATP in post-translational protein translocation. Trends Biochem Sci 12: 369–370

Hurt EC and van Loon APGM (1986) How proteins find mitochondria and intramitochondrial compartments. Trends in Biochem Sci 11: 204–207

Hurt EC, Pesold-Hurt B, Suda K, Oppliger W and Schatz G (1985) The first twelve amino acids (less than half of the pre-sequence) of an imported mitochondrial protein can direct mouse cytosolic dihydrofolate reductase into the yeast mitochondrial matrix. EMBO J 4: 2061–2068

Hurt EC, Soltanifar N, Goldschmidt-Clermont M, Rochaix J-D and Schatz G (1986) The cleavable pre-sequence of an imported chloroplast protein directs attached polypeptides into yeast mitochondria. EMBO J 5: 1343–1350

Hurt EC and Schatz G (1987) A cytosolic protein contains a cryptic mitochondrial targeting signal. Nature 325: 499–503

Hurt EC, Allison DS, Muller U and Schatz G (1987) Amino-terminal deletions in the presequence of an imported mitochondrial protein block, the targeting function and proteolytic cleavage of the presequence at the carboxy terminus. J Biol Chem 262: 1420–1424

Jansen T, Rother C, Steppuhn J, Reinke H, Beyreuther K, Jansson C, Andersson B and Herrmann RG (1987) Nucleotide sequence of cDNA clones encoding the complete '23 kDa' and '16 kDa' precursor proteins associated with the photosynthetic oxygen-evolving complex from spinach. FEBS Lett 216: 234–240

Josefsson, L-G and Randall LL (1981) Differential exported protein in *E. coli* show differences in the temporal mode of processing in vivo. Cell 25: 151–157

Karlin-Neumann G and Tobin EM (1986) Transit peptides of nuclear-encoded chloroplast proteins share a common amino acid framework. EMBO J 5: 1343–1350

Kleene R, Pfanner N, Pfaller R, Link TA, Sebald W, Neupert W and Tropschug M (1987) Mitochondrial porin of *Neurospora crassa*: cDNA cloning, in vitro expression and import into mitochondria. EMBO J 6: 2627–2633

Klosgen RB, Gierl A, Schwarz-Sommer Z and Saedler H (1986) Molecular analysis of the waxy locus of *Zea mays*. Mol Gen Genet 203: 237–244

Kohorn BD and Tobin EM (1986) Chloroplast import of light-harvesting chlorophyll *a/b*-proteins with different amino termini and transit peptides. Plant Physiol 82: 1172–1174

Kohorn BD, Harel E, Chitnis PR, Thornber JP and Tobin EM (1986) Functional and mutational analysis of the light-harvesting chlorophyll *a/b* protein of thylakoid membranes. J Cell Biol 102: 972–981

Koshland D and Botstein D (1982) Evidence for posttranslational translocation of β-lactamase across the bacterial inner membrane. Cell 30: 893–902

Kuntz M, Simons A, Schell J and Schreier PH (1986) Targeting of protein to chloroplasts in transgenic tobacco by fusion to mutated transit peptide. Mol Gen Genet 205: 454–460

Lamppa GK and Abad MS (1987) Processing of a wheat light-harvesting chlorophyll *a/b* protein precursor by a soluble enzyme from higher plant chloroplasts. J Cell Biol 105: 2641–2648

Lazarow PB and Fujiki Y (1985) Biogenesis of peroxisomes. Ann Rev Cell Biol 1: 489–530

Lubben TH and Keegstra K (1986) Efficient in vitro import of a cytosolic heat shock protein into pea chloroplasts. Proc Natl Acad Sci USA 83: 5502–5506

Lubben TH, Bansberg J and Keegstra K (1987) Stop transfer regions do not halt translocation of proteins into chloroplasts. Science 238: 112–114

Maher PA and Singer SJ (1986) Disulfide bonds and the translocation of proteins across membranes. Proc Natl Acad Sci USA 83: 9001–9005

Marks DB, Keller BJ and Hoober JK (1985) In vitro processing of precursors of thylakoid membrane proteins of *Chlamydomonas reinhardtii* y-1. Plant Physiol 79: 108–113

Mayfield SP, Rahire M, Frank G, Zuber H and Rochaix J-D (1987) Expression of the nuclear gene encoding oxygen-evolving enhancer protein 2 is required for high levels of photosynthetic oxygen evolution in *Chlamydomonas reinhardtii*. Proc Natl Acad Sci USQA 84: 749–753

Melton DA, Krieg PA, Rebagliatai MR, Maniatis T, Zinn K and Green MR (1984) Efficient in vitro synthesis of biologically active RNA and RNA hybridization probes from plasmids containing a bacteriophage SP6 promoter. Nucl Acids Res 12: 7035–7056

Merchant S and Bogorad L (1987) The Cu(II)-repressible plastidic cytochrome *c*. Cloning and sequence of a complementary DNA for the pre-apoprotein. J biol Chem 262: 9062

Mishkind ML, Wessler SR and Schmidt GW (1985) Functional determinants in transit sequences: Import and partial maturation by vascular plant chloroplasts of the ribulose-1,5-bisphosphate carboxylase small subunit of *Chlamydomonas*. J Cell Biol 100: 226–234

Mishkind ML, Greer KS and Schmidt GW (1987) Cell-free reconstitution of protein transport into chloroplasts. Met Enzymol 148: 274–294

Miura S, Mori M and Tatibana M (1983) Transport of ornithine carbamyltransferase precursor into mitochondria. Stimulation by potassium ion, magnesium ion, and a reticulocyte cytosolic protein(s). J Biol Chem 258: 6671–6674

Nguyen M and Shore GC (1987) Import of hybrid vesicular stomatitis G protein to the mitochondrial inner membrane. J Biol Chem 262: 3929–3931

Nguyen M, Argan C, Sheffield W, Bell A, Shields D and Shore G (1987) A signal sequence domain essential for processing, but not import, of mitochondrial pre-ornithine carbamyl transferase. J Cell Biol 104: 1193–1198

Ohba M and Schatz G (1987a) Protein import into yeast mitochondria is inhibited by antibodies raised against 45-kD proteins of the outer membrane. EMBO J 6: 2109–2115

Ohba M and Schatz G (1987b) Disruption of the outer membrane restores protein import to trypsin-treated yeast mitochondria. EMBO J 6: 2117–2122

Ohta S and Schatz B (1984) A purified precursor polypeptide requires a cytosolic protein fraction for import into mitochondria. EMBO J 3: 651–657

Pain D and Blobel G (1987) Protein import into chloroplasts requires a chloroplast ATPase. Proc Natl Acad Sci USA 84: 3288–3292

Pain D, Kanwar YS and Blobel G (1988) Identification of a receptor for protein import into chloroplasts and its localization to envelope contact zones. Nature 3313: 232–237

Pfaller R and Neupert W (1987) High affinity binding sites involved in the import of porin into mitochondria. EMBO J 6: 2635–2642

Pfanner N and Neupert W (1986) Transport of F_1-ATPase subunit beta into mitochondria depends on both a membrane potential and nucleoside triphosphates. FEBS Lett 209: 152–156

Pfanner N and Neupert W (1987) Distinct steps in the import of ADP/ATP carrier in mitochondria. J Biol Chem 262: 7528–7536

Pfanner N, Tropschug M and Neupert W (1987) Mitochondrial protein import: nucleoside triphosphates are involved in conferring import-competence to precursors. Cell 49: 815–823

Pfisterer J, Lachmann P and Kloppstech K (1982) Transport of proteins into chloroplasts. Binding of nuclear-coded chloroplast proteins to the chloroplast envelope. Eur J Biochem 126: 143–148

Randall LL (1983) Translocation of domains of nascent periplasmic proteins across the cytoplasmic membrane is independent of elongation. Cell 33: 231–240

Reiss B, Wasmann CC and Bohnert H (1987) Regions in the transit peptide of SSU essential for transport into chloroplasts. Mol Gen Genet 209: 116–121

Riezman H, Hay R, Witte C, Nelson N and Schatz G (1983) Yeast mitochondrial outer membrane specifically binds cytoplasmically-synthesized precursors of mitochondrial proteins. EMBO J 2: 1113–1118

Robinson C and Ellis RJ (1984a) Transport of proteins into chloroplasts. Partial purification of a chloroplast protease involved in the processing of imported precursor polypeptides. Eur J Biochem 142: 337–342

Robinson C and Ellis RJ (1984b) Transport of protein into chloroplasts. The precursor of small subunit of ribulose bisphosphate carboxylase is processed to the mature size in two steps. Eur J Biochem 142: 343–346

Robinson C and Ellis RJ (1985) Transport of proteins into chloroplasts. The effect of incorporation of amino acid analogues on the import and processing of chloroplast polypeptides. Eur J Biochem 152: 67–73

Roise D, Horvath SJ, Tomich JM, Richards JH and Schatz G (1986) A chemically synthesized pre-sequence of an imported mitochondrial protein can form an amphiphilic helix and perturb natural and artificial phospholipid bilayers. EMBO J 5: 1327–1334

Rothblatt JA and Meyer DI (1986) Secretion in yeast: translocation and glycosylation of prepro-α-factor in vitro can occur via an ATP-dependent post-translational mechanism. EMBO J 5: 1031–1036

Rother C, Jansen T, Tyagi A, Tittgen J and Herrmann RG (1986) Plastocyanin is encoded by an uninterrupted nuclear gene in spinach. Curr Genet 11: 171–176

Rothman JE and Kornberg RD (1986) An unfolding story of protein translocation. Nature 322: 209–210

Rothstein SJ, Gatenby AA, Willey DL and Gray JC (1985) Binding of pea cytochrome *f* to the inner membrane of *Escherichia coli* requires the bacterial secA gene product. Proc Natl Acad Sci USA 82: 7955–7959

Schatz G (1986) Protein translocation. A common mechanism for different membrane systems? Nature 321: 108–109

Schatz G (1987) Signals guiding proteins to their correct locations in mitochondria. Eur J Biochem 165: 1–6

Schindler C and Soll J (1986) Protein transport in intact, purified pea etioplasts. Arch Biochem Biophys 247: 211–220

Schleyer M and Neupert W (1985) Transport of proteins into mitochondria: translocational intermediates spanning contact sites between outer and inner membranes. Cell 43: 339–350

Schmidt GW and Mishkind ML (1986) The transport of proteins into chloroplasts. Ann Rev Biochem 55: 879–912

Schreier PH, Seftor EA, Schell J and Bohnert HJ (1985) The use of nuclear-encoded sequence to direct the light-regulated synthesis and transport of a foreign protein into plant chloroplasts. EMBO J 4: 25–32

Schwaiger M, Herzog V and Neupert W (1987) Characterization of translocation contact sites involved in the import of mitochondrial proteins. J Cell Biol 105: 235–246

Singer SJ, Maher PA and Yaffe MP (1987) On the translocation of proteins across membranes. Proc Natl Acad Sci USA 84: 1015–1019

Smeekens S, de Groot M, van Binsbergen J and Weisbeek P (1985) Sequence of the precursor of the chloroplast thylakoid lumen protein plastocyanin. Nature 317: 456–458

Smeekens S, Bauerle H, Hageman J, Keegstra K and Weisbeek P (1986) The role of the transit peptide in the routing of precursors toward different chloroplast compartments. Cell 46: 365–375

Smeekens S, van Steeg H, Bauerle C, Bettenbroek H, Keegstra K and Weisbeek P (1987) Import into chloroplasts of a yeast mitochondrial protein directed by ferredoxin and plastocyanin transit peptides. Plant Mol Biol 9: 377–388

Strzalka K, Ngernprasirtsiri J, Watanabe A and Akazawa T (1988) Sycamore amyloplasts can import and process precursors of nuclear encoded chloroplast proteins. Biochem Biophys Res Commun, 149: 799–806

Sztul ES, Hendrick JP, Kraus JP, Wall D, Kalousek F and Rosenberg LE (1987) Import of pre-ornithine transcarbamylase into mitochondria: Two step processing of the leader peptide. J Cell Biol 105: 2631–2639

Tamm LK (1986) Incorporation of a synthetic mitochondrial signal peptide into charged and uncharged phospholipid monolayers. Biochemistry 25: 7470

Tyagi A, Hermans J, Steppuhn J, Hansson C, Vater F and Herrmann RG (1987) Nucleotide sequence of cDNA clones encoding the complete '33 kDa' precursor protein associated with the photosynthetic oxygen-evolving complex from spinach. Mol Gen Genet 207: 288–293

Van den Broeck G, Timko MP, Kausch AP, Cashmore AR, Van Montagu M and Herrera-Estrella L (1985) Targeting of a foreign protein to chloroplasts by fusion to the transit peptide from the small subunit of ribulose 1,5-bisphosphate carboxylase. Nature 313: 358–363

van Loon APGM and Schatz G (1987) Transport of proteins to the mitochondrial intermem-

776

brane space: the 'sorting' domain of the cytochrome *c1* presequence is a stop-transfer sequence specific for the mitochondrial inner membrane. EMBO J 6: 2441–2448

Van Steeg H, Oudshoorn P, Van Hell B, Polman JEM and Grivell LA (1986) Targeting efficiency of a mitochondrial pre-sequence is dependent on the passenger protein. EMBO J 5: 3643–3650

Vassarotti A, Chen W-J, Smagula C and Douglas MG (1987a) Sequences distal to the mitochondrial targeting sequences are necessary for the maturation of the F_1-ATPase beta-subunit precursor in mitochondria. J Biol Chem 262: 411–418

Vassarotti A, Storoud R and Douglas M (1987b) Independent mutations at the amino terminus of a protein act as surrogate signals for mitochondrial import. EMBO J 6: 705–711

Verner K and Schatz G (1987) Import of an incompletely folded precursor protein into isolated mitochondria requires an energized inner membrane, but no added ATP. EMBO J 6: 2449–2456

von Heijne G (1985) Structural and thermodynamic aspects of the transfer of proteins into and across membranes. In Current Topics in Membranes and Transport, Vol 24, Academic Press, New York, pp 151–179

von Heijne G (1986a) Why mitochondria need a genome. FEBS Lett 198: 1–4

von Heijne G (1986b) Mitochondrial targeting sequences may form amphiphilic helices. EMBO J 5: 1335–1342

Walter P and Lingappa VR (1986) Mechanisms of protein translocation across the endoplasmic reticulum membrane. Ann Rev Cell Biol 2: 499–516

Walter P, Gilmore R and Blobel G (1984) Protein translocation across the endoplasmic reticulum. Cell 38: 5–8

Wasmann CC, Reiss B, Bartlett SG and Bohnert HJ (1986) The importance of the transit peptide and the transported protein for protein import into chloroplasts. Mol Gen Genet 205: 446–463

Waters MG and Blobel G (1986) Secretory protein translocation in a yeast cell-free system can occur posttranslationally and requires ATP hydrolysis. J Cell Biol 102: 1543–1550

Weisbeek P and Smeekens S (1988) Protein transport toward the thylakoid lumen: Post-translational translocation in tandem. Photosyn Res 16: 177–186

Weisbeek P, Hageman J, Cremers F, Keegstra K, Bauerle C and Smeekens S (1986) Nuclear-encoded chloroplast proteins: genes, transport and localization. Curr Top Plant Biochem Physiol, Vol 5, University of Missouri, Columbia, pp 88–104

Werner-Washburne M, Cline K and Keegstra K (1983) Analysis of pea chloroplast inner and outer envelope membrane proteins by two-dimensional gel electrophoresis and their comparison with stromal proteins. Plant Physiol 73: 569–575 .

Wiedmann M, Kurzchalia TV, Hartmann E and Rapoport TA (1987) A signal sequence receptor in the endoplasmic reticulum membrane. Nature 328: 831–833

Zimmermann R, Hennig B and Neupert W (1981) Different transport pathways of individual precursor proteins in mitochondria. Eur J Biochem 116: 455–460

Govindjee et al. (eds), Molecular Biology of Photosynthesis: 777–800
© 1988 Kluwer Academic Publishers

Minireview

Protein synthesis by isolated chloroplasts

A. GNANAM, C.C. SUBBAIAH & R. MANNAR MANNAN
*Department of Plant Sciences, School of Biological Sciences, Madurai Kamaraj University,
Madurai 625021, India*

Received 6 October 1987; accepted 10 March 1988

Key words: chloroplast biogenesis, gene expression, ribosomes, transcription, translation

Abstract. Isolated chloroplasts show substantial rates of protein synthesis when illuminated. This 'in organello' protein synthesis system has been advantageously utilised to elucidate the coding capacity of chloroplast and the regulation of chloroplast genes. The system is also being used recently to transcribe and translate homologous and heterologous templates. In this mini-review, we attempt to critically evaluate the available literature and present the current and the prospective lines of research.

Introduction

It is now common knowledge that chloroplasts, the organelles that carry out photosynthesis, are genetically autonomous, though to a limited extent. They possess their own DNA, the master molecule of genetic information, and the machinery necessary for its replication, transcription, as well as the translation of the transcribed messages into functional proteins. At the time when the presence of DNA in chloroplasts was being ascertained, a distinct type of ribosomes was also discovered to be present in the plastid (Lyttleton 1962) in the form of polysomes indicating that they were functional (Clark et al. 1964, Chen and Wildman 1966). With the demonstration of an independent protein synthetic machinery, and later the presence of genetic material itself in the chloroplasts, attention was turned to an investigation of the products of this organellar genome, i.e., which of the proteins are coded for and synthesized within the chloroplasts.

Until the development of the modern tools of recombinant DNA technology, this problem had been approached in three ways:
(i) identification of the proteins synthesized by isolated organelles;
(ii) dissecting the in vivo protein synthetic processes with specific translational inhibitors; and
(iii) analysis of cytoplasmic (extra-nuclear) gene mutations – mainly in algae like *Chlamydomonas* – which are amenable to analysis by the methods of classical genetics, the only technology available at that time.
Initial attempts using isolated chloroplasts to analyze the products of en-

dogenous templates were unrewarding, particularly because of the very low rates of incorporation of radioactive precursors; and also were fraught with problems of bacterial contamination. So the other two approaches, namely the use of compartment-specific protein synthetic inhibitors and analysis of photosynthetically deficient non-mendelian mutants, were favored for a long time. However, studies using inhibitors have not always been foolproof since these antibiotics are also known to affect many metabolic processes other than protein synthesis. Also, since photosynthetic mutants are generally pleiotropic, it was not easy to point out precisely the direct effect of the mutation in terms of the product of the defective gene, even if the locus could be assigned on the plastid genome map. Hence, unequivocal proof that a particular protein is synthesized in the chloroplast and is thus coded for by the plastid genome, should be possible by simply demonstrating its synthesis in the isolated organelle. Although the first credible report that isolated spinach chloroplasts could incorporate labeled amino acids into TCA precipitable proteins appeared as early as 1965 (Spencer 1965), the rates then obtained were too low to identify any authentic product (Kirk 1970).

The ability to obtain chloroplasts that were highly active in protein synthesis required a prior, slow development of procedures to isolate intact functional plastids, and Blair and Ellis (1973) were the first to exploit such methods. These authors showed that isolated pea chloroplasts, capable of high rates of CO_2 fixation, could also synthesize discrete protein products using light as an energy source. The large subunit of RubPcase was identified as a definitive component of these products for the first time in their studies. Since then, the process of chloroplast protein synthesis in terms of its mechanism as well as the products of this system have attracted great interest from many workers. So far, at least nineteen different plant species, including four algae – *Euglena, Chlamydomonas, Acetabularia* and *Olisthodiscus* besides a woody perennial namely cashew (*Anacardium occidentale* L.) have been shown to yield chloroplasts capable of substantial rates of protein synthesis (Table 1) although there is an inherent variation among different plant species in the rates of protein synthesis in organello. Several reviews have appeared updating the progress in this area from time to time (Ellis 1976, 1977, 1981; Ellis and Barraclough 1978, Ellis and Hartley 1982, Ellis et al. 1978, Margulies 1986). In the present article, we discuss recent aspects of this topic, besides compiling any new information in the conventional aspects. The areas focused on here are:

i) identification of products of chloroplast protein synthesis;

ii) attempts to use the system to understand the regulation of chloroplast gene expression during the plastid development;

iii) division of labor between the soluble and thylakoid bound ribosomes in the plastid;

iv) assembly and turnover of newly synthesized proteins in isolated chloroplasts; and

v) attempts to use chloroplasts or their lysate as an in vitro translation or coupled transcription-translation assay system for analyzing exogenous RNA or DNA templates.

Optimization of the system

There have been several modifications to the initially-developed technique (Blair and Ellis 1973, also see Ellis 1977), to suit a particular plant material as well as to maximize the efficiency of the system in terms of the rate, duration and close approximation to the in vivo translation process. Though crude chloroplast preparations were enough to carry out protein synthesis in plants like pea and spinach (Blair and Ellis 1973, Hartley et al. 1975), purification of intact chloroplasts over percoll gradients was necessary to obtain substantial rates of protein synthesis in species such as cashew

Table 1. Plant species yielding chloroplasts active in in organello protein synthesis

Plant species	Reference
Acetabularia cliftoni	Green (1980, 1982)
Anacardium occidentale	Subbaiah and Gnanam (1984, unpublished)
Chlamydomonas sp.	Leu et al. (1984a)
Cucumis sativus	Walden and Leaver (1982)
	Uma Bai et al. (1984)
	Daniell et al. (1986)
Euglena gracilis	Vasconcelos (1976)
	Ortiz et al. (1980)
Hordium vulgare	Klein and Mullet (1986)
Nicotiana tabacum	Archer et al. (1987)
Olisthodiscus luteus	Reith and Cattolico (1985)
Petunia hybrida	Colijn et al. (1982)
Phaseolus vulgaris	Drumm and Margulies (1970)
Pisum sativum	Blair and Ellis (1973)
Ricinus communis	Uma Bai et al. (1984)
Sorghum vulgare	Geetha and Gnaman (1980a, 1980b)
	Mannan et al. (1987)
Spinacea oleracea	Bottomley et al. (1976)
Triticum aestivum	Obokata (1984)
Vicia faba	Hachtel (1982)
Vigna sinensis	Krishnan et al. (1987)
Zea mays	Grebanier et al. (1979)

(Subbaiah and Gnanam 1984) and *Euglena* (Vasconcelos 1976). Even in species like spinach and sorghum, which readily yield chloroplasts active in protein synthesis, even from crude preparations (Hartley et al. 1975, Geetha and Gnanam 1980a), a purification step over percoll solutions further improved the incorporation of labeled amino acids (Morgenthaler and Mendiola-Morgenthaler 1976). In *Euglena*, chloroplasts obtained by gentle lysis of sphaeroplasts and purified over percoll gradients were superior (at least by 100-fold) in protein synthetic acitivity to the organelles prepared by direct homogenization of the whole cells (Ortiz et al. 1980). Work done on pea chloroplasts in the laboratory of Jagendorf requires a detailed examination (Fish and Jagendorf 1982, Fish et al. 1983, Nivison and Jagendorf 1984). Fish and Jagendorf (1982) achieved very high rates of protein synthesis (almost 200 nmoles of [^{3}H-leucine incorporation/mg chlorophyll) by systematically working out each of the various components and conditions – starting from handling of the plants to chloroplast preparation. Use of destarched young plants, rapid chilling of leaves from pre-illuminated plants after harvest, homogenization by polytron – using high tissue to buffer ratio – purification of chloroplasts over linear percoll gradients, a reaction medium containing sorbitol (350 mM), K^+ (30 mM), $MgCl_2$ (1 mM), $MnCl_2$ (1 mM) and EDTA (2 mM) all routinely resulted in at least 60–100 nmoles of [^{3}H]-leucine incorporation per mg chlorophyll (Fish and Jagendorf 1982). They could also prolong the active translation from the usual 20 min cited in previous reports to almost an hour by further modifications, namely: addition of other amino acids with sufficient free Mg^{2+}; use of lower light intensities; and addition of Pi and ATP (Nivison and Jagendorf 1984). Recent work in our laboratory has shown that addition of such polyamines as putrescine or spermidine further improves the rate of amino acid incorporation by isolated sorghum chloroplasts (Subbaiah et al. 1987).

While the above work concerns the quantitative improvement in the translational activity of isolated chloroplasts, work by Mullet et al. (1986) was aimed at simulating the in vivo protein synthetic pattern of these organelles., when isolated. Using almost similar procedures to those developed by Jagendorf and his colleagues, the above authors also obtained high rates of protein synthesis as well as a large number of labeled products resolved by SDS-PAGE and fluorography. However these authors could not confirm the assumption that all the products revealed in the fluorograph represented separate polypeptides. They tested this possibility by a pulse-chase experiment and found that many of the lower molecular mass polypeptides were, in reality, incomplete products paused at discrete points of mRNA when chloroplasts were labeled. These could be converted to full-length authentic polypeptides by a subsequent chase with cold methionine.

By introducing this chase step, these authors found that the in vitro translation profile of plastids closely resembled the in vivo labelled pattern.

Characteristics of protein synthesis by isolated chloroplasts

Initial characterization of the system was carried out by Ellis and his group using pea and spinach chloroplasts. We refer the reader to earlier reviews for details (Ellis 1977, 1981), while presenting here only the highlights and recent developments in this area.

The special feature of protein synthesis by isolated chloroplasts is that it is light-driven, requiring no additional energy source, unlike other well-known in vitro translation systems. High light intensities were used in early studies; however, Nivison and Jagendorf (1984) showed low light intensities (45 μmol/m^2.s) to be more effective, even if the initial rates of incorporation were low. Gomez-Silva and Schiff (1985) studied the dose response using light varying wavelengths and showed that in *Euglena* the chloroplast protein synthesis was saturated at 5 W/m^2 whereas photosynthetic CO$_2$ fixation required 15–30 W/m^2 for the optimal rate. Protein synthesis was more responsive to blue light but photosynthesis responded better to white or red light.

The chloroplast system is also functional in the dark if ATP is supplied, although the rates of such ATP driven protein synthesis were only 50% or less than those of the light driven synthesis (Bottomley et al. 1974, Siddell and Ellis 1975). However, working with sorghum, Geetha and Gnanam (1980a) showed that these rates could be as high as 85% of the light driven incorporation (0.5 to 1.0 nmol of phenylalanine or lysine/mg of chlorophyll/h). Fish et al. (1983) worked out the details and established the need for Mg^{2+} addition (since ATP would chelate endogenous Mg^{2+}) for rapid ATP-driven protein synthesis in the dark. By supplying equimolar amounts of Mg^{2+} and ATP these authors obtained even greater rates of [^{3}H] leucine incorporation in the dark than those driven by light. Light-dependent incorporation itself could be enhanced by 45% at lower intensities (45 μmol/m^2.s) and 18% at high intensities (900 μmol/m^2.s) by added ATP in the presence of Mg^{2+}.

Protein synthesis in isolated chloroplasts of a chromophytic alga *Olisthodiscus luteus* is an exception in that it was light-independent and also non-responsive to ATP supplied externally (Reith and Cattolico 1985). The authors assumed that a storage product in the chloroplast (most likely mannitol) would supply the necessary ATP. Though similar light-independent protein synthesis was earlier reported in the isolated chloroplasts of

Acetabularia (Goffeau and Brachet 1965), some work (Green 1980) showed that chloroplast protein synthesis was strictly light dependent in this species. Reith and Cattolico, in their paper (1985), also referred to an unpublished report of diatom plastids showing light-independent protein synthesis. Except in such isolated cases, light has been shown to be an absolute requirement for chloroplast protein synthesis in the absence of any added energy source.

The major light-dependent reaction of chloroplasts, namely CO_2 assimilation, requires both ATP and the reducing agent NADPH. However, since protein synthesis needs only ATP, the translational activity of isolated chloroplasts is expected to depend solely on their photochemical capacity to form ATP and should be independent of their potential to generate reducing action. In fact, early work by Ramirez et al. (1968) suggested that cyclic photophosphorylation (which generates only ATP but no NADPH and is insensitive to 3-(3,4-dichlorophenyl)-1,1-dimethylurea, DCMU) was the main source of energy for light-driven protein synthesis of spinach chloroplasts. However, the later reports showed that protein synthesis by isolated chloroplasts was fully or partially sensitive to DCMU (Blair and Ellis 1973, Colijn et al. 1982, Gomez-Silva and Schiff 1985) indicating the contribution of non-cyclic photophosphorylation to this process, although there is some evidence that DCMU acts indirectly on cyclic photophosphorylation (Arnon and Chain 1975, Hosler and Yocum 1987). We also observed a total inhibition of protein synthesis by DCMU in sorghum chloroplasts (Geetha and Gnanam 1980a) but only a partial effect in castor chloroplasts (Uma Bai et al. 1984). Interestingly, Fish and Jagendorf (1982) reported no effect of DCMU on chloroplast protein synthesis in pea (compare Blair and Ellis 1973) or even a slight stimulation (15%) at a very low concentration of the inhibitor. These contradictory effects might be due to multiple sites of action of DCMU in photosynthetic electron transport. In fact this is reflected by the dual effect of this herbicide on electron transport itself, both as an inhibitor (Izawa 1968) and as a stimulator (Ramanujam et al. 1981). We attempted to resolve this problem (i.e., which of the two – cyclic or non-cyclic phosphorylation – contributes to in vitro protein synthesis) by a different approach. We followed the sequence of events during greening of etioplasts isolated from cucumber cotyledons and showed that light-dependent protein synthesis commenced concomitantly with the development of cyclic phosphorylation, well before the appearance of non-cyclic phosphorylation (Uma Bai et al. 1984). However, this did not prove that cyclic phosphorylation is the one which contributes preferentially to protein synthesis when both cyclic and non-cyclic paths are operative in the fully developed chloroplast.

There is some evidence that photosynthetic CO_2 fixation and amino acid incorporation into proteins are competing reactions (for ATP) in isolated chloroplasts when these are illuminated; although in some species chloroplasts active in protein synthesis show poor CO_2 fixation, e.g. *Euglena* (Ortiz et al. 1980, Reith and Cattolico 1985). Substrates of Calvin cycle such as 3-PGA, ribose-5-P, $NaHCO_3$ and α-ketoglutarate plus glutamate showed substantial inhibition of protein synthesis in pea chloroplasts (Fish and Jagendorf 1982). Though bicarbonate was also inhibitory to spinach chloroplasts, 3-PGA and oxaloacetate stimulated protein synthesis (Gnanam et al. 1981). We explained the promotor effect of organic acids through faster regeneration of $NADP^+$ needed for the maintenance of electron transport and attendant phosphorylation. DL-glyceraldehyde, an inhibitor of the Calvin cycle, did not promote but inhibited protein synthesis (Fish and Jagendorf 1982). Though the site of action was not defined, Ca^{++} was also shown to inhibit amino acid incorporation by isolated chloroplasts (Bouthyette and Jagendorf 1981).

Isolated proplastids and etioplasts were shown to be proficient in protein synthesis; but with an absolute dependence on added ATP as energy source, since they lacked machinery for light-harvesting and attendant photochemistry (Siddell and Ellis 1975, Dockerty and Merrett 1979, Obokata 1984, Cushman and Price 1986, Krishnan et al. 1987). However, rates of amino acid incorporation by immature plastids were extremely low (Drumm and Margulies 1980, Siddell and Ellis 1975, Dockerty and Merret 1979, Miller and Price 1982, Miller et al. 1983) and in some cases not even detectable, e.g. cucumber etioplasts (Walden and Leaver 1981). Cushman and Price (1986), by taking advantage of the optimization experiments carried out on pea chloroplasts by Jagendorf et al. (see above), have obtained greatly increased rates of protein synthesis from proplastids of *Euglena*, even though the rates were still below those observed with chloroplasts. In this organism, more differentiated plastids (etioplasts?) showed detectable (though low) rates of light-dependent protein synthesis which greatly increased if plastids were isolated from cells exposed to light for even 1 h (Miller and Price 1982, Miller et al. 1983). On the contrary, plastids capable of light-driven protein synthesis could be obtained from etiolated wheat seedlings only after 3 h of greening (Obokata 1984). Work in our laboratory showed that etioplasts from hormone pretreated cucumber cotyledons were effecient either in ATP- or light-driven protein synthesis (Uma Bai et al. 1984); and the light-dependent protein synthesis of these plastids was linear for 8 h (Daniell et al. 1986).

Products of chloroplast protein synthesis

Though chloroplasts are made up of approximately 400 proteins – including both structural and functional ones, any contribution to these by the genome itself is very small. Theoretically a DNA molecule, the size of chloroplast genome, can code for only 20–30% of the total polypeptides found in this organelle, the rest of them being contributed by the nuclear genome. In fact, complete nucleotide sequence data of two chloroplast genomes, one from a lower plant, *Merchantia polymorpha*, and another from tobacco (Ohyama et al. 1986, Shinozaki et al. 1986) could also predict that the plastid genome can make only about 120–130 products, including stable RNAs. An updated list of polypeptides identified and later confirmed to be the products of authentic chloroplast genes is given in Table 2. So far, 15 membrane polypeptides, 2 soluble products, 18 ribosomal proteins and 3 envelope polypeptides are identified from the products of in organello protein synthesis.

We have found that isolated chloroplasts or etioplasts of *Vigna sinensis* would synthesize a set of high molecular weight proteins following heat shock. These chloroplast-produced heat shock proteins (hsps) were not synthesized in vivo if the heat shock was given to the whole leaf (Krishnan et al. 1987). However, in our recent experiments, we have observed the in vivo synthesis of these polypeptides if the leaves were exposed to a gradual temperature rise (Krishnasamy et al. 1988). We are currently looking at the origin of the message for these chloroplast-made hsps as well as their functional significance. Kloppstech et al. (1986) have also reported that an hsp in *Acetabularia* is coded for outside the nuclear genome. From their experiments using specific protein synthetic inhibitors, the authors presumed that the above polypeptide was synthesized within the chloroplast. This possibility can be tested by analyzing the products of protein synthesis of heat shocked chloroplasts in vitro in this species, as was done in *Vigna* by Krishnan et al. (1987).

Division of labor between the soluble and thylakoid-bound ribosomes in the plastid

Among the products of chloroplast protein synthesis, there appeared to be a certain specificity regarding their site of synthesis within the organelle. Such a suggestion emerged early in the 1970s from a previous observation from many laboratories that a substantial fraction of chloroplast ribosomes occurred bound to chloroplast membranes, at least a part of which were in

Table 2. Identified products of in vitro chloroplast protein synthesis.

Polypeptide	Plant species	References
Soluble products		
LSU of RuBisCO	Pea	Blair and Ellis (1973)
	Spinach	Bottomley et al. (1974)
	Euglena	Vasconcelos (1976)
	Acetabularia	Green (1980)
	Chlamydomonas	Leu et al. (1984)
Protein synthetic elongation factors		
T and G	Spinach	Ciferri et al. (1979)
Ef–Tu	*Euglena*	Spermulli (1972)
		Miller et al. (1983)
Thylakoid polypeptides		
α, β and ε subunits of CF_1	Pea	Mendiola-Morganthaler et al. (1976)
	Spinach	Nelson et al. (1980)
Subunit III of CF_0	Pea	Doherty and Gray (1980)
Cytochrome f	Pea	Doherty and Gray (1979)
Cytochrom b_{557}	Spinach	Zielenski and Price (1980)
P_{700} chlorophyll a apoproteins of PSI	Sorghum	Geetha and Gnanam (1980b) Mannan et al. (1987)
Chlorophyll a apo-proteins of PS II	*Acetabularia*	Green (1982)
	Vicia faba	Hachtel (1987)
D_1 protein	Pea	Blair and Ellis (1973)
	Spinach	Bottomley et al. (1974)
	Euglena	Vasconcelos (1976)
	Maize	Grebanier et al. (1978) Steinback et al. (1981)
	Chlamydomonas	Leu et al. (1984a)
D_2 protein	*Chlamydomonas*	Herrin et al. (1981)*
Envelope membrane polypeptide		
	Pea	Joy and Ellis (1975)
	Spinach	Morganthaler and Mendiola-Morgenthaler (1976)
Ribosomal proteins		
	Chlamydomonas	Schmidt et al. (1983)
	Spinach	Dorne et al. (1983)
	Vicia faba	Hachtel (1985)

* Thylakoid ribosomes.

the form of polysomes (Chen and Wildman 1970, Chua et al. 1973). Further-more, the dependence of their release from the membranes on puromycin, an elongation inhibitor, indicated the involvement of these bound polysomes in the synthesis of membrane proteins which in turn helped the ribosomes to bind the membranes (Margulies and Michaels 1974). Margulies et al. (1975) tested this possibility and demonstrated in *Chlamydomonas* that these polysomes were active in protein synthesis in vitro when supplemented with cell extract and the newly synthesized proteins were vectorially discharged into the membrane. The authors also proposed that these membrane bound polysomes could be important in the synthesis of thylakoid polypeptides. Jagendorf and co-workers (Alscher et al. 1978, Alscher-Herman et al. 1979, Yamamoto et al. 1981, Fish and Jagendorf 1982, Hurewitz and Jagendorf 1987) made detailed studies on the binding of ribosomes to thylakoid membrane and concluded that

i)   the binding was both ionic and by insertion of the nascent polypeptide chains;
ii) high salt and puromycin were necessary to release them;
iii) anoxia and transfer of plants or organelles to darkness would also release the bound polysomes;
iv) binding of ribosomes onto the membrane in light, required a product of 70S ribosomal protein synthesis as well as non-cyclic electron flow;
v)   the major effect of light on ribosome binding could be due to higher stromal pH; and
vi) ribosomes would get redistributed between stroma and thylakoids.

They also showed that the isolated membrane-bound polysomes would incorporate amino acids into proteins if supplemented with the S-30 fraction from *E. coli* and majority of the products were associated with the membranes (Alscher et al. 1978). Using this soluble factor-supplemented system, Michaels and his colleagues tested the products of these bound polysomes in *Chlamydomonas* and showed that these ribosomes synthesized α and β subunits of coupling factor, and D_1 and D_2 polypeptides of the PS II reaction center (Herrin and Michaels 1985a, 1985b, Herrin et al. 1981). Other workers also demonstrated that the thylakoid bound polysomes synthesized only membrane polypeptides such as the reaction center polypeptides of PS I (Minami and Watanabe 1984, Margulies et al. 1987) and PS II including *cyt* b_{559} (Minami et al. 1986), coupling factor subunits α and β and the 32 kDa herbicide binding protein (Minami and Watanabe 1984) while soluble proteins like the large subunit of RuBisCO were manufactured only by stromal ribosomes (Minami and Watanabe 1984, Leu et al. 1984b). However there was no compartmentalization of messengers for soluble and membrane polypeptides and both the populations of polysomes had similar mRNA members (Minami and Watanabe 1984, Leu et al. 1984b). Hence it was suggested that the thylakoid membranes might play a role in the expression of chloroplast mRNAs leading to a division of labor between the stromal and membrane bound polysomes (Leu et al. 1984b). However,

Bhaya and Jagendorf (1984, 1985a, 1985b), working with their optimized translation systems of chloroplast sub-fractions, did not find any such specificity in the synthesis of large subunit (LSU) of RuBisCO and α and β subunits of coupling factor CF_1, both being synthesized by stromal and thylakoid bound polysomes in pea. But they did observe such a specificity in the synthesis of the highly hydrophobic protein – subunit III of CF_0 – which was primarily the product of thylakoid ribosomes (Bhaya and Jagendorf 1984b). Hattori and Margulies (1986) also found that LSU of RuBisCO was an important product of thylakoid bound ribosomes in spinach and suggested that such a thylakoid association of LSU synthesis might be significant in view of its hydrophobic nature. The reverse of this argument may hold true with respect to the synthesis of CF_1 subunits on stromal ribosomes, since this multisubunit complex (CF_1) is hydrophilic and only lossely attached to the membrane. These reports caution us against generalizing the proposal of division of labor between the two populations of chloroplast polysomes, and further studies in this direction are warranted.

Gene expression during chloroplast biogenesis and by inducing agents

Many workers have utilized greening etiolated tissues or dark grown algal cells to study gene expression during chloroplast biogenesis, since the process of de-etiolation, though much telescoped in time, is almost similar to the normal chloroplast development (Silverthrone and Ellis 1980). Ellis and his group carried out pioneering studies in this area, utilizing the in organello protein synthesis system they had developed to monitor differential expression of plastid genome during light-induced development of etiolated pea shoots (Siddell and Ellis 1975) and during the normal development of spinach leaves (Silverthrone and Ellis 1980). They observed that, in chloroplasts isolated from different stages of greening, synthesis of LSU declined faster than that of peak D. These observations made in organello reflected the changes in vivo in the level of these two proteins and their transcripts during greening (Silverthrone and Ellis 1980). Walden and Leaver (1981) found that only quantitative changes occur in the rate of incorporation by developing plastids of cucumber cotyledons, without any evidence of differential translation of specific mRNAs. However, both qualitative as well as quantitative changes were observed in the activity of protein synthesis by plastids isolated at different periods of illumination of dark-grown *Euglena* cells (Miller and Price 1982, Miller et al. 1983, Cushman and Price 1986). Notable among them were
i)   a gradual increase in the synthesis of LSU and 32 kDa polypeptides;

ii) decrease in a 44 kDa polypeptide, possibly elongation factor Ef-Tu; and

iii) gradual disappearance of a set of high molecular weight polypeptides synthesized by the proplastids (Miller and Price 1982, Miller et al. 1983).

Later, using optimized conditions for in organello protein synthesis, Cushman and Price (1986) observed that undifferentiated proplastids manufactured a large number of polypeptides that were absent in the products of more mature proplastids or chloroplasts. These authors were not certain whether these additional products made by proplastids represented authentic individual polypeptides or differences in the post-translational modifications. Significant changes both in the soluble and membrane associated products of in organello synthesis were also reported during plastid biogenesis of light-grown wheat seedlings, although the exact identity of these products was not given except for LSU of RuBisCO (Obokata 1984). Klein and Mullet (1986, 1987) followed the sequence of events during chloroplast development in barley by assaying the products of chloroplast protein synthesis as well as by estimating the in vivo transcript levels for individual polypeptides. They showed that barley etioplasts, like their counterparts in pea and spinach, could synthesize LSU and α and β subunits of ATPase but not apoproteins of PS I and PS II or the 32 kDa herbicide binding protein, despite the presence of significant transcript levels for these polypeptides even before greening. These polypeptides appeared rapidly once the seedlings were illuminated, thus indicating that regulation occurred at the translation level. However, during later stages of leaf development, synthesis of Chl a-apoproteins and LSU by chloroplasts declined rapidly, but the 32 kDa polypeptide continued to be synthesized – as was reported in pea and spinach by Ellis and his associates. A general observation which also summarizes the changes occurring in the in organello protein synthesis during greening is a decrease in the ratio of incorporation into the soluble fraction as compared to that in the membrane fraction (Obokata 1984, Klein and Mullet 1986, 1987). Such a rapid increase in the synthesis of membrane polypeptides, with a concomitant decrease in soluble products during illumination of etiolated tissue (Klein and Mullet 1987) is consistent with the proposed in vivo phenomenon of light-induced redistribution of plastid ribosomes from stroma onto the membranes, and so also with the proposal that polysomes bound to the membrane primarily synthesize thylakoid polypeptides (Alscher-Herman et al. 1979).

Garcia et al. (1983) studied the role of chloroplast protein synthesis in the induction of leaf senescence in barley, since it had been shown earlier that chloramphenicol retarded leaf senescence (Sabater and Rodriguez 1978, Yu and Kao 1981). The authors monitored the changes in the pattern of polypeptides synthesized by isolated chloroplasts during senescence of detached leaf segments. Plastid protein synthetic activity greatly increased 10–20 h after incubation of leaf segments in the dark, though there was a marked loss of rRNA. The rise in the rate of incorporation during dark incubation was more pronounced if kinetin was added to the medium. At the same time, incubation in light decreased the protein synthetic activity of isolated chloroplasts and kinetin could restore it only to a small extent. There were also qualitative differences in the newly synthesized polypeptides. Dark incubation, which promoted senescence, induced the synthesis of a 66 kDa polypeptide. A polypeptide of almost similar molecular weight (62 kDa) was found to increase during the senescence of soybean cotyledons (Bricker and Newman 1980). By contrast, treatments that retarded senescence, such as kinetin application, enhanced the synthesis of an entirely different set of proteins, one of them being the LSU of RuBisCO (Garcia et al. 1983). However it is not known whether this differential expression of plastid genes represents a cause or an effect of leaf senescence.

The in organello protein synthesis was also used to study gene expression in chloroplasts induced by other environmental agents like temperature, or genotoxic agents such as ethidium bromide or herbicides (Krishnan et al. 1987, Subbaiah et al. 1987). Archer et al. (1987) have used the system to diagnose the molecular basis of a chloroplast coded virescent mutant in tobacco. From their experiments on in organello protein synthesis, they concluded that the inability of plastids from the mutant to make a 37.5 kDa polypeptide was probably the causal factor of this mutation and thus this polypeptide might be important for the assembly of pigment-protein complexes in the developing world.

Processing, assembly and turnover of in organello synthesized products

The majority of polypeptides manufactured in the chloroplast are synthesized in their authentic size except for the 32 kDa herbicide-binding protein. The primary amino acid sequence derived from the nucleotide sequence of the gene coding for this protein gave a sum of 353 amino acids with an expected molecular mass of 38 950 (Zurawski et al. 1982). Cohen et al. (1984) attempted to explain this discrepancy between the predicted and the authentic sizes of the polypeptide, when they located a second initiation codon, downstream to the first one and, in fact, it was at this second methionine that the translation was initiated in their in vitro experiments yielding a polypeptide of 34 000 molecular mass. Cohen et al. (1984) invoked a product-precursor relationship to explain the difference in the sizes of in vitro (heterologous), and in organello synthesized products. In fact such a precursor was identified among the products of protein synthesis by isolated maize chloroplasts as early as in 1978 (Grebanier et al. 1978) and unfortunately these chloroplasts were unable to process this polypeptide. This possibility was later tested by pulse chase experiments and it was shown that isolated chloroplasts would synthesize and post-translationally process the 34.6 kDa precursor to the authentic 32 kDa polypeptide (see *e.g.*, Minami and Watanabe 1985). Marder et al. (1984), who carried out detailed studies on this protein, showed that the processing occurred at the C-terminus of the precursor.

There were reports that other polypeptides such as LSU of RuBisCO (Langridge 1981), *Cyt* b_{559} (Zielinski and Price 1980) and the β-subunit of CF$_1$ (Watanabe and Price 1982) might also be synthesized as higher molecular weight precursors, but this could not be confirmed by later workers.

Many of the chloroplast proteins are multi-subunit complexes most of which contain polypeptides synthesized both in the cytoplasm as well as the chloroplast. Many studies have shown that polypeptides newly synthesized in isolated plastids would assemble into authentic complexes. Products such as the subunits of CF$_0$ and CF$_1$, LSU of RuBisCO and reaction center polypeptides could be extracted as part of their respective complexes by standard methods from the organelles previously labelled in vitro (Mendiola-Morgenthaler et al. 1976, Ellis 1977, Barraclough and Ellis 1980, Nelson et al. 1980, Zieliniski and Price 1980, Green 1980, Hacthel 1982, 1985, 1987, Bloom et al. 1983, Mannan et al. 1987). Nelson et al. (1980), who demonstrated the in organello assembly of new synthesized CF$_1$ subunits, suggested that isolated chloroplasts would contain pools of cytoplasmically synthesized subunits of this complex. Alternatively it could be that the newly synthesized polypeptides were exchanged with the pre-existing subunits in the complex, the only possibility that would explain the assembly of α and β subunits of CF$_1$ synthesized by ribosomes bound to washed thylakoids into authentic CF$_1$ complex in spinach and *Chlamydomonas* (Minami and Watanabe 1984, Herrin and Michaels 1985a). It was, however, exceptional that, in isolated maize chloroplasts, the newly formed CF$_1$ subunits were not correctly assembled (Grebanier et al. 1978). As mentioned earlier, these chloroplasts also failed to process the precursor of the 32 kDa polypeptide, for reasons unknown. Interestingly, the ribosomal proteins, synthesized by isolated spinach chloroplasts, assembled themselves into incomplete ribosomal particles which were stable (Dorne et al. 1984). The authors suggested that the formation of these particles could be an intermediate step in the assembly of complete ribosomes, a phenomenon similar to the stepwise assembly of ribosomes in *E. coli*.

The assembly of LSU into holoenzyme was worked out in detail by Ellis and his associates and has been shown to require binding another stromal protein before integrating with cytoplasmically formed small subunit SSU (for details see Ellis and Van der Vies 1988). Bloom et al. (1983) showed that assembly of LSU in organello was a light-dependent process mediated by ATP. The requirement for light was not verified for the assembly of other proteins.

Proteins that are newly formed by isolated chloroplasts are not entirely stable. A part of the fully mature polypeptides may be degraded without assembly due to a lack of cytoplasmically synthesized complementary polypeptides. There may also be incomplete or defective polypeptides that have to be removed by proteolysis. Such an instability of in organello synthesized proteins was first reported to occur in pea chloroplasts during ATP-driven amino acid incorporation (Fish et al. 1983). This was followed by detailed studies (Liu and Jagendorf 1984, Malek et al. 1984) which confirmed that about 20–35% of the labeled protein was degraded during a subsequent incubation period of 20–30 min after a pulse feeding of label for 10–30 min. This proteolysis was strictly ATP dependent and was either totally absent or very little in evidence in the dark. Addition of Mg^{2+}-ATP stimulated the degradation 2–3 fold in darkness or light. The protein loss was the greatest (34–40%) when ATP was added together with intense light. In *Euglena* proplastids or chloroplasts, the proteolytic loss was as much as 50–60% (Cushman and Price 1986). Hydrolysis was not specific for a particular protein, and thylakoid as well as soluble proteins were degraded at comparable rates (Liu and Jagendorf 1984, Malek et al. 1984, Cushman and Price 1986). Liu and Jagendorf (1985) have examined the role of this ATP-dependent proteolysis. They suggested that this system, mainly confined to the thylakoids, might hydrolyze mature but non-integrated polypeptides. This was indicated by the increased rates of ATP-dependent in organello degradation of in vivo labeled chloroplast polypeptides in the presence of cycloheximide. They could also localize another proteolytic system in the stroma, which was ATP-independent but required Mg^{2+} and attacked prematurely terminated polypeptides induced by puromycin, canavanine or kanamycin.

The turnover of the 32 kDa atrazine binding protein has attracted considerable attention. Ellis and his group were the first to highlight the special features of this polypeptide (Siddell and Ellis 1975, Ellis 1981) which is, so far, the best studied chloroplast protein (Kyle 1985). This was the major product of chloroplast protein synthesis in terms of incorporation in vitro but this protein would never accumulate like the LSU of RuBisCO (Ellis 1977). Thus it has been found to have the fastest turnover of all the chloroplast proteins. This is also reflected by in vivo labeling studies in which very short pulse labeling (5–30 min) in the light led to the rather exclusive visualization of this protein on the autotradiogram (Hoffman-Falk et al. 1983) whereas long labeling (> 12 h) generated an autoradiograph pattern similar to the Coomassie Blue pattern. The turnover is dependent on light and is proportional to the light intensity. Electron transport inhibitors such as DCMU would inhibit the degradation. The actual mechanism of hydrolysis is still being worked out, however.

There is also evidence that cytoplasmic protein synthesis regulates the translational activity of chloroplasts. Reardon and Price (1983) demonstrated that in organello protein synthesis in plastids isolated from *Euglena* was substantially reduced (40–90%) if the cells were pretreated for 2–4 h with cycloheximide (CHI), an inhibitor of cytoplasmic protein synthesis. The synthesis of soluble products was more sensitive to the treatment than that of membrane polypeptides, though there were no qualitative changes in the polypeptide patterns of the soluble fraction. However, qualitative changes were seen in thylakoid associated products – especially those that required a processing step – such as the precursor of Q_B or herbicide binding protein. In chloroplasts isolated from CHI treated cells, the precursor for this protein accumulated and so also did a few novel polypeptides larger than 65 kDa. These changes could be partly reversed if the chloroplasts from treated cells were incubated with cytoplasm from control cells. These data suggested that products of cytoplasmic protein synthesis were involved in the processing of chloroplast translation products. The severe decrease in the synthesis of soluble products in plastids from CHI treated cells could be due to the depletion of their cytoplasmically synthesized partners for assembly (Spreitzer et al. 1985). Alternatively, it could be due to rapid degradation of newly synthesized proteins, known to be triggered during such conditions (Liu and Jagendorf 1985).

Isolated chloroplasts as a potential in vitro transcription and/or translation system

Until the early 1970s, protein synthesis by isolated chloroplasts was used only to identify plastome (plastid genome) coded products. Later, attempts were made to exploit the system for translating exogenous messengers of both homologous and heterologous origin. More recent work has shown that the usefulness of the system can be further extended to the transcription of exogenous DNA templates as well as the coupling of this process to translation, thus qualifying it as a complete assay system for gene expression.

We have used pre-illuminated chloroplasts (in order to exhaust the endogenous templates) to translate exogenously provided mRNAs (Indira and Gnanam 1976, Geetha and Gnanam 1980a). We have been able to show that pre-incubation completely eliminates the background incorporation due to native transcripts and renders the organelles an efficient in vitro translation assay system for transcripts of different origin – both prokaryotic and eukaryotic. We effectively substituted this system for other well-known in vitro protein synthesis systems such as wheat germ or rabbit reticulocyte lysate in our studies to establish the origin of messages for different chloroplast proteins such as PS I apoprotein, castorbean chloroplast glycosidase and pyruvate dikinase (Geetha and Gnanam 1980a, 1980b, Geetha et al. 1981, Mariappan 1984, Valliammai 1985, Uma Bai 1985, Ananda Krishnan and Gnanam 1987). Cammerino et al. (1982), working basically on the same idea, used lysed chloroplasts after treatment with micrococcal nuclease to remove pre-existing mRNAs, since the lysates could not stand the pre-incubation step we used in our experiments. These lysates faithfully translated both exogenously supplied homologous and heterologous mRNAs. The lysed chloroplast system is more precise than the whole chloroplast preparations in terms of availability of exogenously added RNA and is also more open to manipulation to keep the requirements of cofactors etc., optimal. In spite of such a clear demonstration of the efficacy of isolated chloroplasts or their lysates for translating exogenous RNA templates and also the obvious advantages of the technique over other in vitro systems (as outlined in earlier publications, see Geetha and Gnanam 1980a) such as the ease and inexpensive nature of the preparation of material, others have continued using other in vitro translation systems such as wheat germ or *E. coli* S-30 fraction, even to test RNA from chloroplasts.

Bard et al. (1985) demonstrated that lysates of chloroplasts can be used as a DNA-dependent coupled transcription-translation system as efficient as a similar system established in *E. coli* (Bottomley and Whitfield 1979). The new system, according to the authors' claim, responds well to total ct-DNA or its cloned fragments, plasmid vectors and also *E. coli* chromosomal DNA as well. However confirmatory reports on both the in vitro systems (Cammerino et al. 1982, Bard et al. 1985) are lacking, both from other laboratories and from their own.

An immediate application of such a system, if it were well standardized, could be the identification of many open reading frames revealed by the sequence data of chloroplast genomes in *Marchantia* and tobacco (Ohyama et al. 1985, Shinozaki et al. 1985). Another important task is to study the expression of chloroplast genes, not just the developmental regulation in leaves but also in all the other parts of the plant. It is now known that plastids in other tissues also contain identical genomes but develop differentially to perform divergent functions depending on the tissue. The in vitro systems can be used to study the processes and factors involved in such tissue-specific regulation of gene expression.

Acknowledgements

The authors thank Ms. Mary Andrews for her dedicated typing of the manuscript. Part of the work reviewed here is supported by a grant no. 21(7)/84-STP-II from the Department of Science and Technology.

References

Alscher R, Patterson R and Jagendorf AT (1978) Activity of thylakoid-bound ribosomes in pea chloroplasts. Plant Physiol 62: 88–93

Alscher-Herman R, Jagendorf AT and Grumet R (1979) Ribosome-thylakoid association in peas. Influence of anoxia. Plant Physiol 64: 232–235

Ananda Krishnan V and Gnanam A (1988) Properties and regulation of Mg^{2+} dependent chloroplast inorganic pyrophosphatase from Sorghum vulgare leaves. Arch Biochem Biophys 260: 277–284

Archer EK, Hakanssan G and Bonnet HT (1987) The phenotype of a virescent chloroplast mutant in tobacco is associated with the absence of a 37.5 kD thylakoid polypeptide. Plant Physiol 83: 926–932

Arnon DI and Chain RK (1975) Regulation of ferredoxin-catalysed photosynthetic phosphorylation. Proc Natl Acad Sci USA 72: 4961–4965

Bard J, Bourque DP, Hildebrand M and Zaitlin D (1985) In vitro expression of chloroplast genes in lysates of higher plant chloroplasts. Proc Natl Acad Sci USA 82: 3983–3987

Barraclough R and Ellis RJ (1980) Protein synthesis in chloroplasts. IX. Assembly of newly synthesized large subunits into ribulose bisphosphate carboxylase in isolated intact pea chloroplasts. Biochim Biophys Acta 608; 19–31

Bhaya D and Jagendorf AT (1984a) Optimal conditions for translation by thylakoid-bound polysomes from pea chloroplasts. Plant Physiol 75: 832–838

Bhaya D and Jagendorf AT (1984b) Syunthesis or subunit III of CF_0 by thylakoid bound polysomes from pea chloroplasts. Plant Mol Biol 3(5): 277–280

Bhaya D and Jagendorf AT (1985) Synthesis of the xx and subunits or coupling factor 1 by polysomes from pea chloroplasts. Arch Biochem Biophys 237: 217–223

Blair GE and Ellis RJ (1973) Protein synthesis in chloroplasts. 1. Light-driven synthesis of the large subunit of fraction 1 protein by isolated pea chloroplasts. Biochim Biophys Acta 319: 223–234

Bloom MV, Miles P and Roy H (1983) Light dependent assembly of ribulose-1,5-biphosphate carboxylase. Proc Natl Acad Sci USA 80: 1013–1017

Bottomley W and Whitfield P (1979) Cell-free transcription and translation of total spinach chloroplast DNA. Eur J Biochem 93: 31–39

Bottomely W, Spencer D and Whitfeld PR (1974) Protein synthesis in isolated chloroplasts. Comparison of light driven and ATP-driven synthesis. Arch Biochem Biophys 164: 106–117

Bouthyette P-Y and Jagendorf AT (1981) Calcium inhibition of amino acid incorporation by pea chloroplasts and the question of loss of activity with age. In: Akoyunoglou G (ed.) Photosynthesis V Chloroplast Development, pp 599–609. Philadelphia: Balaban Intern Science Services

Bricker TM and Newman DW (1980) Quantitative changes in the chloroplast thylakoid polypeptide complement during senescence. Z Pflanzenphysiol 98: 339–346

Camerino G, Sai A and Ciferri O (1982) A chloroplast system capable of translating heterologous mRNAs. FEBS Lett 150(1): 94–98

Chen JL and Wildman SG (1966) Functional chloroplast polyribosomes from tobacco leaves. Science 155: 1271–1273

Chen JL and Wildman SG (1970) 'Free' and membrane bound ribosomes, and nature of products formed by isolated tobacco chloroplasts incubated for protein synthesis. Biochim Biophys Acta 209: 207–219

Chua N-H, Blobel G, Siekevitz P and Palade GE (1973) Attachment of chloroplast polysomes to thylakoid membranes in *Chlamydomonas reinhardtii*. Proc Natl Acad Sci USA 70: 1554–1558

Ciferri O, DiPasquale G and Tiboni O (1979) Chloroplast elongation factors are synthesized in the chloroplasts. Eur J Biochem 102: 331–335

Clark MF, Mathews REF and Ralph RK (1964) Ribosomes and polyribosomes in *Brassica pekinsis*. Biochim Biophys Acta 91: 289–304

Cohen BN, Bloom MV, Coleman T and Weissbach H (1984) Analysis of an in vitro chloroplasts gene expression using a simplified dipeptide synthesis system. In: Cardinale GJ (ed.) Annual Report 1983, Roche Institute of Molecular Biology, pp 77–78

Colijn CM, Kool AJ and Nijkamp HJJ (1982) Protein synthesis in *Petunia hybrida* chloroplasts isolated from leaves and cell cultures. Planta 155: 37–44

Cushman JC and Price CA (1986) Synthesis and turnover of proteins in proplastids and chloroplasts of *Euglena gracilis*. Plant Physiol 82(4): 972–977

Daniell H, Krishnan M, Uma Bai V and Gnanam A (1986) An efficient and prolonged in vitro translational system from isolated cucumber *Cucumis sativus* etioplasts. Biochem Biophys Res Commun 135(1): 248–255

Dockerty A and Merrett MJ (1979) Isolation and enzymic characterization of *Euglena* chloroplasts. Plant Physiol 63: 468–473

Doherty A and Gray JC (1979) Synthesis of cytochrome *f* by isolated pea chloroplasts. Eur J Biochem 98: 87–92

Doherty A and Gray JC (1980) Synthesis of a dicyclohexylcarbodimide binding proteolipid by isolated pea chloroplasts. Eur J Biochem 108: 131–136

Dorne A-M, Lescure A-M and Mache R (1984) Site of synthesis of spinach chloroplast ribosomal proteins and formation of incomplete ribosomal particles in isolated chloroplasts. Plant Mol Biol 3: 83–90

Drumm HE and Margulies MM (1970) In vitro protein synthesis by plastids of *Phaseolus vulgaris*. IV. Amino acid incorporation by etioplasts and effect of illumination of leaves on incorporation by plastids. Plant Physiol 45: 435–442

Ellis RJ (1976) The intact chloroplast. In: Barber J (ed.) Protein and Nucleic Acid Synthesis by Chloroplasts, pp 335–364. The Netherlands: Elsevier/North-Holland Biomedical Press

Ellis RJ (1977) Protein synthesis by isolated chloroplasts. Biochim Biophys Acta 463: 185–215

Ellis RJ (1981) Chloroplast proteins synthesis transport and assembly. Ann Rev Plant Physiol 32: 111–138

Ellis RJ and Barraclough R (1978) Synthesis and transport of chloroplast proteins inside and outside the cell. In: Akoyunoglou G, Argyroudi-Akoyunoglou JH (eds) Chloroplast Development, pp 185–194. Amsterdam: Elsever/North-Holland Biomedical Press

Ellis RJ and Hartley MR (1982) Preparation of higher plant chloroplasts active in protein and RNA synthesis. In: Edelman M, Hallick RB and Chua N-H (eds) Methods in Chloroplast Molecular Biology, pp 170–188. Amsterdam: Elsevier Biomedical Press

Ellis RJ, Highfeld PE and Silverthrone J (1978) The synthesis of chloroplast proteins by subcellular systems. In: Hall DO, Coombs J and Goodwin TW (eds) Photosynthesis II – Proceedings of the IV International Congress on Photosynthesis, pp 497–506. London: The Biochemical Society

Ellis RJ and Van der Vies SM (1988) The Rubisco subunit binding protein. Photosynth Res 16: 000–000

Fish LE and Jagendorf AT (1982) High rates of protein synthesis by isolated chloroplasts. Plant Physiol 70(4): 1107–1114

Fish LE, Deshaies R and Jagendorf AT (1983) A Mg^{2+} requirement for rapid ATP-driven protein synthesis by intact pea chloroplasts. Plant Sci Lett 31: 139–146

Garcia S, Martin M and Sabater B (1983) Protein synthesis by chloroplasts during the senescence of barley leaves. Physiol Plant 57: 260–266

Geetha V and Gnanam A (1980a) An in vitro protein synthesizing system with isolated chloroplasts of *Sorghum vulgare*: An alternate assay system for exogenous template RNA. J Biol Chem 255(2): 492–497

Geetha V and Gnanam A (1980b) Identification of P700 chlorophyll *a* protein complex as a product of chloroplast protein synthesis. FEBS Lett 111(2): 272–276

Geetha V, Mohamed AH and Gnanam A (1980) Cell-free synthesis of active ribulose-1,5-bisphosphate carboxylase in the mesophyll chloroplasts of *Sorghum vulgare*. Biochim Biophys Acta 606(1): 83–94

Gnanam A, Mariappan T and Manjula Devi JD (1981) In vitro translation of exogenous messengers with isolated chloroplasts of spinach. In: Akoyunoglou G (ed.) Photosynthesis V. Chloroplast Development, pp 841–846, Philadelphia: Balban International Science Services

Goffeau A and Brachet J (1965) DNA-dependent incorporation of amino acids into the proteins of chloroplasts isolated from anucleate *Acetabularia* fragments. Biochim Biophys Acta 95: 302–313

Gomez Silva B and Schiff JA (1985) The light requirement for protein synthesis and carbon dioxide fixation in highly purified intact *Euglena* chloroplasts. Plant Science 39: 111–119

Grebanier AE, Coen DM, Rich A and Bogorad L (1978) Membrane proteins synthesized but not processed by isolated maize chloroplasts. J Cell Biol 78: 734–746

Green BR (1980) Protein synthesis by isolated *Acetabularia* chloroplasts. In vitro synthesis of the apoprotein of the P700 chlorophyll *a* protein complex (CPI). Biochim Biophys Acta 609: 107–120

Green BR (1982) Protein synthesis by isolated *Acetabularia* chloroplasts synthesis of the two minor chlorophyll *a* complexes in vitro. Eur J Biochem 128: 543–546

Hachtel W (1982) Biosynthesis and assembly of thylakoid membrane proteins in isolated chloroplasts from *Vicia faba*. The P700 chlorophyll *a* protein. Z Pflanzenphysiol 107(5): 383–394

Hachtel W (1985) Biosynthesis and assembly of chloroplast ribosomal proteins in the isolated chloroplasts from *Vicia faba*. Biochem Physiol Pflanz 180(2): 115–124

Hachtel W (1987) Synthesis and assembly of thylakoid membrane proteins in isolated pea chloroplasts. The chlorophyll *a* proteins of the photosystem II reaction center. Plant Sci 48(1): 43–48

Hartley MR, Wheeler AM and Ellis RJ (1975) Protein synthesis in chloroplasts. V. Translation of messenger RNA for the large subunit of Fraction I Protein in a heterologous cell-free system. J Mol Biol 91: 67–77

Hattori T and Margulies MM (1986) Synthesis of large subunit of ribulose bisphosphate carboxylase by thylakoid-bound polyribosomes from spinach, *Spinacia oleracea* cultivar savoy-supreme chloroplasts. Arch Biochem Biophys 244(2): 630–640

Herrin D and Michaels A (1985a) In vitro synthesis and assembly of the peripheral subunits of coupling factor CF_1 (α and β) by thylakoid bound ribosomes. Arch Biochem Biophys 237: 224–236

Herrin D and Michaels A (1985b) The chloroplast 32 kD protein is synthesized on thylakoid bound ribosomes in *Chlamydomonas reinhardii*. FEBS Lett 184(1): 90–95

Herrin D, Michaels A and Hickey E (1981) Synthesis of a chloroplast membrane polypeptide on thylakoid bound ribosomes during the cell cycle of *Chlamydomonas reinhardii* 137[+]. Biochim Biophys Acta 655: 136–145

Hoffman-Falk H, Mattoo AK, Marder JB, Edelman M and Ellis RJ (1983) General occurrence and structural similarity of the rapidly synthesized 32 000 dalton protein of the chloroplast membrane. J Biol Chem 257: 4583–4587

Hosler JP and Yocum CF (1987) Regulation of cyclic phosphorylation during ferredoxin mediated electron transport. Effect of DCMU and the $NADPH/NADP^+$ ratio. Plant Physiol 83: 965–969

Hurewitz J and Jagendorf AT (1987) Further characterization of ribosome binding to thylakoid membranes. Plant Physiol 84: 31–34

Indira GM and Gnanam A (1976) Induction of chloroplast development–translation studies with temporally distinguished messengers. In: Bogorad L and Weil JH (eds) Acids Nucleiques et Synthese des Proteines Chez les Vegetaux, pp 617–621. Strasbourg: Colloq Int Cent Natl Rech Sci

Izawa S (1968) Effect of Hill reaction inhibitors on photosystem I. In: Shibata K, Takamiya A, Jagendorf AT and Fuller RC (eds) Comparative Biochemistry and Biophysics of Photosynthesis, pp 140–147. Pennsylvania: University Press, State College

Joy KW and Ellis RJ (1975) Protein synthesis in chloroplasts IV. Polypeptides of the chloroplast envelope. Biochim Biophys Acta 378: 143–151

Kirk JTO (1970) Biochemical aspects of chloroplast development. Ann Rev Plant Physiol 21: 11–42

Klein RR and Mullet JE (1986) Regulation of chloroplast encoded chlorophyll-binding protein translation during higher plant chloroplast biogenesis. J Biol Chem 261(24): 11138–11145

Klein RR and Mullet JE (1987) Control of gene expression during higher plant chloroplast biogenesis. Protein synthesis and transcript levels of psb A, psa A – psa B and rbcL in dark grown and illuminated barley seedlings. J Biol Chem 262(9): 4341–4348

Kloppstech K, Ohad I and Schweizer HG (1986) Evidence for an extranuclear coding site for a heat shock protein in *Acetabularia*. Eur J Cell Biol 42: 239–245

Krishnan M, Krishnasamy S, Mannar Mannan R and Gnanam A (1987) In vitro synthesis of heat shock proteins by chloroplasts. In: Biggins J (ed.) Progress in Photosynthesis Research, Vol IV, pp 593–596. Dordrecht: Martinus Nijhoff

Krishnasamy S, Mannan RM, Krishnan M and Gnanam A (1988) Heat shock response of chloroplast genome in *Vigna sinensis*. J Biol Chem 263: 5104–5109

Kyle DJ (1985) The 32 000 dalton Q_B protein of photosystem II. Photochem Photobiol 41: 107–116

Langridge P (1981) Synthesis of the large subunit of spinach ribulose bisphosphate carboxylase may involve a precursor polypeptide. FEBS Lett 123: 85–89

Leu S, Mendiola-Morgenthaler L and Boschetti A (1984a) Protein synthesis by isolated chloroplasts (ed.) Advances in Photosynthetic Research IV, pp 541–544. The Hague: Martinus-Nijhoff/Dr W Junk Publishers

Leu S, Bolli R, Mendiola-Morgenthaler L and Boschetti A (1984b) In vitro translation of different messenger RNA containing fractions of *Chlamydomonas reinhardii* chloroplasts. Planta 160(3): 204–211

Liu X-Q and Jagendorf AT (1984) ATP-dependent proteolysis in pea chloroplasts. FEBS Lett 166: 248–252

Liu X-Q and Jagendorf AT (1985) Roles for ATP-dependent and ATP-independent proteases of pea chloroplasts in regulation of the plastid translation products. Physiol Veg 23: 749–755

Lyttleton JW (1962) Isolation of ribosomes from spinach chloroplasts. Exp Cell Res 26: 312–317

Malek L, Bogorad L, Ayers AR and Goldberg AL (1984) Newly synthesized proteins are degraded by an ATP-stimulated proteolytic process in isolated pea chloroplasts. FEBS Lett 166: 253–257

Mannan RM, Krishnan M and Gnanam A (1987) Synthesis of polypeptides associated with PS I by the isolated *Sorghum vulgare* chloroplasts. In: Biggins J (ed.) Progress in Photosynthesis Research, Vol. IV, pp 589–592. Dordrecht: Martinus Nijhoff Publishers

Marder JB, Goloubinoff P and Edelman M (1984) Molecular architecture of the rapidly metabolized 32 Kilodalton protein of photosystem II. J Biol Chem 259: 3900–3908

Margulies MM (1986) Compartmentation of protein synthesis within the chloroplast. In: Akoyunoglou G (ed.) Regulation of Chloroplast Differentiation, pp 171–180. New York: Alan R Liss

Margulies MM and Michaels A (1974) Ribosomes bound to chloroplast membranes in *Chlamydomonas reinhardii*. J Cell Biol 60: 65–77

Margulies MM, Tiffany HL and Michaels A (1975) Vectorial discharge of nascent polypeptides attached to chloroplast thylakoid membranes. Biochem Biophys Res Commun 64: 735–739

Margulies MM, Tiffany HL and Hattori T (1987) Photosystem I reaction center polypeptides of spinach are synthesized on thylakoid bound ribosomes. Arch Biochem Biophys 254(2): 454–461

Mariappan T (1984) Protein synthesis in the chloroplasts. Ph.D. Thesis, Madurai Kamaraj University

Mendiola-Morgenthaler LR, Morgenthaler JJ and Price CA (1976) Synthesis of coupling factor CF_1 protein by isolated spinach chloroplasts. FEBS Lett 62: 96–100

Miller ME and Price CA (1982) Protein synthesis by developing plastids from *Euglena gracilis*. FEBS Lett 147: 156–160

Miller ME, Jurgenson JE, Reardon EM and Price CA (1983) Plastid translation in organello and in vitro during light induced development in *Euglena*. J Biol Chem 258: 14478–14484

Minami EI and Watanabe A (1984) Thylakoid membranes the translation site of chloroplast DNA-regulated thylakoid polypeptides. Arch Biochem Biophys 235(2): 562–570

Minami EI and Watanabe A (1985) Detection of a precursor polypeptide of the rapidly synthesized 32 000 dalton thylakoid protein in spinach chloroplasts. Plant Cell Physiol 26(5): 839–846

Minami EI, Shinohara K, Kuwabara T and Watanabe A (1986) In vitro synthesis and assembly of photosystem II proteins of spinach chloroplasts. Arch Biochem Biophys 244(2): 517–527

Morgenthaler JJ and Mediola-Morgenthaler L (1976) Synthesis of soluble thylakoid and envelope membrane proteins by spinach chloroplasts purified from gradients. Arch Biochem Biophys 172: 51–58

Mullet JE, Klein RR and Grossman AR (1986) Optimization of protein synthesis in isolated higher plant chloroplasts. Identification of paused translation intermediates. Eur J Biochem 155(2): 331–338

Nelson N, Nelson H and Schatz G (1980) Biosynthesis and assembly of the proton-translocating adenosine triphosphatase complex from chloroplasts. Proc Natl Acad Sci USA 77: 1361–1364

Nivison HT and Jagendorf AT (1984) Factors permitting prolonged translation by isolated pea *Pisum sativum* cultivar progress – No. 9 chloroplasts. Plant Physiol 75(4): 1001–1008

Obokata J (1984) Protein synthesis in isolated plastids during chloroplast development in wheat. Plant Cell Physiol 25(5): 821–830

Ohyama K, Fukuzawa H, Kockli T, Shirai H, Sano T, Sano S, Umesono K, Shiki Y, Takeuchi M, Chang Z, Aota S-I, Inokuchi H and Ozeki H (1986) Chloroplast gene organization deduced from complete sequence of liverwort *Marchantia polymorplea* chloroplast DNA. Nature 322: 572–574

Ortiz W, Reardon EM and Price CA (1980) Preparation of chloroplasts from *Euglena* highly active in protein synthesis. Plant Physiol 66: 291–294

Ramanujam P, Gnanam A and Bose S (1981) Stimulation of photosystem I electron transport by high concentration of 3-(3-4,dichlorophenyl)-1,1-dimethyl urea in uncoupled chloroplasts. Plant Physiol 68: 1485–1487

Ramirez JM, DelCampa FF and Arnon DI (1968) Photosynthetic phosphorylation as energy source for protein synthesis and carbon dioxide assimilation by choroplasts. Proc Natl Acad Sci USA 59: 606–611

Reardon EM and Price CA (1983) Cytoplasmic regulation of chloroplast translation in *Euglena gracilis*. Arch Biochem Biophys 226(2): 433–440

Reith ME and Cattolico RA (1985) In vitro chloroplast protein synthesis by the chromophytic alga *Olisthodiscus luteus*. Biochemistry 24: 2550–2556

Sabater B and Rodriguez MT (1978) Control of chlorophyll degradation in detached leaves of barley and oat through kinetin on chlorophyllase levels. Physiol Plant 43: 274–276

Schmidt RJ, Richardson CB, Gillham NW and Boynton JE (1983) Site of synthesis of chloroplast ribosomal proteins in *Chlamydomonas*. J Cell Biol 96: 1451–1463

Shinozaki K, Ohme M, Tanaka M, Wakasugi T, Hayashida N, Matsubayashi T, Zaita N, Chungwongse J, Obokata J, Yamaguchi-Shinozaki K, Ohto C, Torazawa K, Meng BY, Sugita M, Deno H, Kamogashira T, Yamada K, Kusuda J, Takaiwa F, Kato A, Tohdoh N, Shimada H and Sugiura M (1986) The complete nucleotide sequence of the tobacco chloroplast genome: its gene organization and expression. EMBO J 5: 2043–2049

Siddell SG and Ellis RJ (1975) Protein synthesis in chloroplasts. VI. Characteristics and products of protein synthesis in etiolated and developing chloroplasts from pea leaves. Biochem J 146: 675–685

Silverthrone J and Ellis RJ (1980) Protein synthesis in chloroplasts. 8. Differential synthesis of chloroplast proteins during spinach Spinacia-oleracea leaf development. Biochim Biophys Acta 607(2): 319–330

Spencer D (1965) Protein synthesis by isolated spinach chloroplasts. Arch Biochem Biophys 111: 381–390

Spermulli LL (1982) Chloroplast elongation factor Tu evidence that it is the product of a chloroplast gene in *Euglena gracilis*. Arch Biochim Biophys 214: 734–741

Spreitzer RJ, Goldschmidt-Clermont M, Ralive M and Rochaix JD (1985) Nonsense mutations in *Chlamydomonas* chloroplasts gene for the large subunit of ribulose-1,5-bisophosphate carboxylase/oxygenase. Proc Natl Acad Sci USA 82: 5460–5464

Steinback KE, McIntosh L, Bogorad L and Arntzen CJ (1981) Identification of the triazine receptor protein as a chloroplast gene product. Proc Natl Acad Sci USA 78: 7463–7467

Subbaiah CC and Gnanam A (1987) In organello protein synthesis by isolated cashew chloroplasts. In: Mehta SL (ed.) Proc Nat Symp Photosyn and Nitrogen Metabolism (in press)

Subbaiah CC, Kulandaivelu G and Gnanam A (1987) Effect of ethidium bromide and UV-C radiation on chloroplast development. In: Rajamanickam C (ed.) Proc Second Internat Symp on Biomembranes, pp 347–358. New Delhi: Today and Tomorrow's Publishers

Subbaiah CC, Mannan RM and Gnanam A (1987) Effect of polyamines on in organello protein synthesis by *Sorghum* chloroplasts. Biochem J (submitted)

Uma Bai V (1985) Development of chloroplast structure and function with reference to membrane proteins. Ph.D. Thesis, Madurai Kamaraj University

Uma Bai V, Daniell H and Gnanam A (1984) Role of photoelectron transport in the protein synthesis by isolated plastids. In: Sybesma C (ed.) Advances in Photosynthetic Research, Vol IV, pp 681–684. The Hague: Martinus-Nijhoff/Dr W Junk Publishers

Valliammai V (1985) Studies on Chloroplast Development of *Chlorella protothecoides*. Ph.D. Thesis, Madurai Kamaraj University

Vasconcelos AC (1976) Synthesis of proteins by isolated *Euglena gracilis* chloroplasts. Plant Physiol 58: 719–721

Walden R and Leaver CJ (1981) Synthesis of chloroplast proteins during germination and early development of cucumber, *Cucumis sativus* cultivar long green ridge. Plant Physiol 67(6): 1090–1096

Watanabe A and Price CA (1982) Translation of mRNAs for subunits of chloroplast coupling factor I in spinach. Proc Natl Acad Sci USA 79: 6304–6308

Yamamoto T, Burke J, Autz G and Jagendorf AT (1981) Bound ribosomes of pea chloroplast thylakoid membranes: Location and release in vitro by high salt, puromycin and RNase. Plant Physiol 67: 940–949

Yu SM and Kao CH (1981) Retardation of leaf senescence by inhibitors of RNA and protein synthesis. Physiol Plant 52: 207–210

Zielinski RE and Price CA (1980) Synthesis of thylakoid membrane proteins by chloroplasts isolated from spinach Spinacia oleracea cytochrome B-559 and P700 chlorophyll *a* protein. J Cell Biol 85(2): 435–445

Zurawski G, Bohnert HJ, Whitfeld PR and Bottomley W (1982) Nucleotide sequence of the gene for the Mr 32 000 thylakoid membrane protein from *Spinacea oleracea* and *Nicotiana debnevi* predicts a totally conserved translational product of Mr 38 950. Proc Natl Acad Sci USA 79: 7699–7703

Subject Index